First International Congress on
Adhesion Science and Technology - invited papers

Kashmiri L. Mittal

First International Congress on Adhesion Science and Technology

-

Invited Papers

Festschrift in Honor of Dr. K.L. Mittal
on the occasion of his 50th Birthday

Editors:
W.J. van Ooij
H.R. Anderson, Jr.

///VSP///
Utrecht, The Netherlands, 1998

VSP BV
P.O. Box 346
3700 AH Zeist
The Netherlands

Tel: +31 30 692 5790
Fax: +31 30 693 2081
E-mail: vsppub@compuserve.com
Home Page: http://www.vsppub.com

First published in 1998

ISBN 90-6764-291-6

Printed in The Netherlands by Ridderprint bv, Ridderkerk.

Contents

Mittal Festschrift, pp. xi–xiii
W. J. Van Ooij and H. R. Anderson, Jr. (Eds)

Preface

This Festschrift documents the Proceedings of the *First International Congress on Adhesion Science and Technology*, held in honor of Dr. Kash Mittal on the occasion of his 50th Birthday, in Amsterdam, October 16–20, 1995.

When the idea of organizing such an event to honor Dr. Mittal was conceived, we knew there would be a tremendous response from the international adhesion community, as he had made unparalleled contributions and had a sterling record of accomplishments. However, the actual response far surpassed our initial expectations, as can be gathered from the scientific program and the worldwide participation. When the announcement was sent out, the worldwide adhesion community expressed that it was a very fitting tribute to Kash to organize a conference in his honor. Immediately, many of the internationally renowned adhesionists expressed their keen interest in joining us on this occasion, and those who were unable to personally congratulate Kash expressed their deep regrets. All this interest signifies the respect and admiration the international adhesion community has for Kash. Frankly, we are not aware of any precedent where someone was recognized by organizing in his honor an international congress at the tender age of 50!

During the Opening Ceremony the following general items were covered. One of us (WJvO) welcomed the delegates and formally opened the Congress. Dr. J. J. M. Mulderink, Chairman of the Royal Dutch Chemical Society, talked about "Adhesion and the Dutch". Then Mr. I. P. Khosla, His Excellency The Ambassador of India, commented on "Science and technology in the economic reforms". This talk was followed by "Personal and professional vignettes of the first half-century life of Dr. Kash Mittal" by one of us (HRA). Then Dr. Kash Mittal himself reflected on his "Quarter Century romance with adhesion science" and the ceremony was concluded with an interesting presentation on "Reincarnation in adhesion" by Drs. J. R. F. Th. Groesbeek.

The final technical program comprised a total of 235 papers, the authors of which hailed from 38 countries, representing industrial and government research laboratories and academic groves. Prof. P. G. de Gennes (Nobel Laureate in Physics) and Prof. G. A. Somorjai delivered the plenary lectures. The program was graced with 63 invited papers and 170 contributed research papers, which were presented in oral/poster sessions. Many different ramifications of adhesion science and technology were discussed and thus the conference veritably reflected the whole gamut of adhesion, as Kash had made contributions in the various aspects of adhesion.

In addition to the technical presentations, Dr. Bandyopadhyay and Dr. Dwight recited poems to express their admiration for Kash. At the banquet, Drs. Groesbeek announced the initiation of two awards to honor Dr. Mittal in perpetuity, named: "The Mittal Award" and "The Mittal Lecture". Dr. Mittal was also honored by Prof. P. G. de Gennes who presented him with a plaque on behalf of the worldwide adhesion community. The texts of all these are presented in this Festschrift.

When we were ruminating over how to document the Proceedings, it immediately became apparent that to chronicle the full written accounts of all 235 papers would be almost impossible. Therefore, we decided to include in this Festschrift only the invited presentations. However, we thought, for posterity and to capture the full spirit of the Congress, it behooved us to list the titles, names and affiliations of authors of the other papers. Concomitantly, this Festschrift should serve as a full acta of this momentous occasion.

Now turning to this Festschrift, it opens with a Foreword by Prof. P. G. de Gennes, followed by the write-up by one of us (HRA), entitled "Kashmiri Lal Mittal: Industrial surface chemist/educator", in which he expatiates upon the educational background, the many and various contributions and accomplishments of Kash, augmented by his own impression (in the form of personal vignettes) of Kash. This essay brings out his zeal, energy, voracious reading habit, contagious enthusiasm and Kash as a role model to emulate. The next write-up, entitled "Adhesion and the Dutch", by Dr. J. J. M. Mulderink, comments on the role played by the field of adhesion in many human endeavors — in almost every walk of life.

The next segment of this Festschrift comprises the full accounts of the 49 invited papers, starting with "The molecular science of adhesion" by Dr. G. A. Somorjai, which is both very informative and illuminating and offers new insights and paves the way to fathom the adhesion phenomenon at a molecular/atomic level. This particular section is divided into seven parts: Part 1: Fundamental aspects of adhesion and general topics; Part 2: Contact angle, wettability and surface energetics; Part 3: Surface modification: Relevance to adhesion; Part 4: Adhesives and adhesive joints; Part 5: Adhesion aspects of polymeric coatings, and polymer-polymer interphase; Part 6: Metal-polymer and metal-ceramic adhesion; and Part 7: General papers. The topics covered include many different aspects of adhesion science and technology, and both fundamental and applied issues are addressed.

The section documenting the full papers is followed by the listing of titles, authors and affiliations of the other 185 papers which were included in the technical program of the Conference. This particular segment is divided into 12 sections: Section 1: Fundamental aspects of adhesion and general topics; Section 2: Contact angle, surface energetics and adhesion; Section 3: Surface modification: Relevance to adhesion; Section 4: Silanes and other adhesion promoters; Section 5: Adhesive joints: Characterization, testing and durability; Section 6: Adhesion aspects of polymeric coatings; Section 7: Adhesion aspects of thin films; Section 8: Adhesion aspects of metallized plastics; Section 9: Bioadhesion; Section 10: Particle adhesion; Section 11: Interfaces in composites; and Section 12: Adhesive materials: Characteristics and performance.

All in all, this Festschrift contains bountiful information and provides a panoramic view of the world of adhesion science and technology. We sincerely hope this Festschrift will serve as a Baedeker to those who want to make their maiden voyage in the wonderful discipline of adhesion and as a commentary on the contemporary R&D

activity for veteran researchers and technologists. It should also provide a fountainhead of new ideas, and as deeper insights in the phenomenon of adhesion are gained, new vistas will emerge.

Acknowledgements

Now comes the pleasant task of thanking those individuals who helped in more ways than one. An event of this magnitude and quality requires unflinching dedication of many individuals. Firstly, we would like to express our heart-felt gratitude to Drs. J. R. F. Th. Groesbeek (of VSP) for sponsoring this event. He and his team spent untold hours taking care of the multitude of aspects of local organization, and we must say they were most gracious hosts. Next, we wish to express our appreciation to the following industrial sponsors for their generous support: Akzo Nobel (The Netherlands), DSM Research (The Netherlands), Dow Corning Corporation (USA), H. F. Fuller Company (USA), 3M Company (USA), and Procter & Gamble (USA). This event was held under the auspices of the following societies: Adhesion Society of Japan, French Chemical Society, Israeli Polymers and Plastics Society, Italian Chemical Society, Royal Dutch Chemical Society and Swedish Chemical Society. We are thankful to them for their interest and support. Our appreciation goes to the members of the International Advisory Committee for their suggestions. Our thanks are extended to the unsung heroes (reviewers) for their time and efforts in offering many valuable comments, as peer review is a *sine qua non* to maintain the highest standard of a publication. Our gratitude and appreciation is also extended to the authors for their contribution, cooperation and enthusiasm, without which this Festschrift would not have been possible. Last, but not least, we are deeply thankful to the staff of VSP for embodying this Festschrift.

It has been a real pleasure and privilege for us to organize this Congress in honor of Kash and to make this Festschrift available to the adhesion community. It reflects a very fitting tribute to Kash, who has done so much and has been an effective ambassador in spreading the 'gospel' of adhesion science around the world. We are sure all of you will join us in wishing Kash another 50 years of unabated professional success and good health and happiness.

W.J. van Ooij
University of Cincinnati, OH

H.R. Anderson, Jr.
University of California at Davis

Mittal Festschrift, p. xv
W. J. Van Ooij and H. R. Anderson, Jr. (Eds)

Foreword

Adhesion is both important and difficult. The role of chemical creativity has been huge — from the early days with formol-phenol condensates, to the great times of acrylics, epoxy resins, and so forth.

In our days, we can make joints which resist extreme conditions; and also joints which form very fast. However, in spite of these major successes, we are still far from a full understanding of the adhesion processes. One great pioneer — Alan Gent — has clarified the role of the adhesion energy in a sequence of outstanding experiments. And a number of enthusiasts, like Dr. K. L. Mittal, have created an active, interactive community of adhesion scientists — through conferences and proceedings.

The present volume is dedicated to Mittal on his 50th birthday. It contains a wealth of information on science and on technology. We all wish him a great success.

Prof. P. G. de Gennes
College de France

Professor P. G. de Gennes presenting the plaque to Dr. K. L. Mittal

1st International
Congress on

ADHESION SCIENCE & TECHNOLOGY

IN HONOR OF DR. KASH MITTAL ON HIS 50TH BIRTHDAY

On the occasion of the 1st International Congress on Adhesion Science & Technology, held in honor of Dr. Kash Mittal on his 50th birthday, I, on behalf of the global adhesion scientific community, would like to recognize him for his commitment and dedication to the advancement of adhesion science and technology and for his exceptional achievements. All in the adhesion world would like to thank Dr. Mittal for his outstanding leadership and distinguished professional service.

Professor P. G. de Gennes

Text of the plaque presented to Dr. K. L. Mittal

Drs. J. R. F. Th. Groesbeek and Dr. K. L. Mittal with all Kash's 50 books on display

Mittal Festschrift, pp. xix–xx
W. J. Van Ooij and H. R. Anderson, Jr. (Eds)

Fun at the Adhesion Conference in Amsterdam, 1995

I am Dr. Bando
Teach polymers in the school
It's really very easy
As can see any fool.

Take 2 ethylene molecules
Make them a dimer
Repeat it ten thousand times
And you got a polymer.

Linear molecules, cross-links
Anything it can be
Thermosets or thermoplastics
(With) some crystallinity.

When I come to this conference
On science of adhesion
I find it very educative
From the session one.

"Atomic force the way to go"
Says Berkeley's Somorjai
"Don't discount acid-bases"
Pleads Dwight, who was at Lehigh.

Mittal spoke of 300 ways
To determine adhesion
Kendall says oh, no, no
"Fracture mechanics is the only one".

Compatibility apparently
Plays a major role
(And) rate of peeling of course controls
Formation of fibrils and holes.

Weak, or strong bonds?
Would it depend on interface and its extent?
The Yankees would say “interphase”
Of course they have to be different.

Now I know what adhesion means
Through electrostatics and all
Reptation of the long molecules
Or a rise in voltage and fall.

Surface engineering through plasma treatment
Sounds a good idea
Who better to talk about it
Than the guys from Australia.

I came here to show my results
On micro-macro correlation
Richard Wool says on Friday morning
There’ll be hardly anyone.

Nothing I can do about it
(But) I have learnt a great deal
And I must thank the program chairman
For his enormous zeal.

Also thank the beautiful girls
Who are always there
To make the conference a very good success
No efforts did they spare.

And finally

Who could ever really forget
(Drs.) Groesbeek with his smile
He seemed to be omnipresent
“Re-incarnation” is his style.

Dr. S. Bandyopadhyay, Australia

Mittal Festschrift, p. xxi
W. J. Van Ooij and H. R. Anderson, Jr. (Eds)

Ode to Kash

We have come from all around the world
To the beautiful city of Amsterdam
To show our honour to Dr. Kash Mittal
A truly great man, a truly great man.

What Kash's done for adhesion
Is beyond all imagination
It is his contribution
That's brought her to where she now stands.

When you and I write a book or two
We think what a job is done
After Kash has edited 50 through
He is still on the run.

Oh, dear Kash, we love you
We salute thee
There wouldn't be many more like you
In Adhesion History.

Oh, dear Kash
We love you
We love you
We love you.

Dr. S. Bandyopadhyay, Australia

Mittal Festschrift, p. xxiii
W. J. Van Ooij and H. R. Anderson, Jr. (Eds)

Happy Birthday 1995!

ICAST '95

There once was a Doctor named Mittal
At fifty and fit as a fiddle
He went to A'dam for
The last word on Adhesion
And found it is still a riddle.

With best regards,

Dave Dwight
17 September 1995

Mittal Festschrift, p. xxv
W. J. Van Ooij and H. R. Anderson, Jr. (Eds)

Special announcement

Due to the tremendous success of ICAST '95 and the very appropriate dedication of this congress to Dr. Kash Mittal, a new lecture has been founded:

The Kash Mittal Lecture

This lecture will be held every three years on the occasion of each future ICAST congress. A committee will be nominated to select the person who will be invited to deliver the next Kash Mittal Lecture.

Furthermore, VSP is pleased to announce a new award, which will be presented to the author of the best article published in the *Journal of Adhesion Science and Technology* in a given year:

The Kash Mittal Award

A committee will be formed to evaluate all papers published in the journal.

Mittal Festschrift, pp. xxvii–lxiii
W. J. Van Ooij and H. R. Anderson, Jr. (Eds)

Kashmiri Lal Mittal: Industrial surface chemist/educator

HERBERT R. ANDERSON, JR.*

Department of Chemical Engineering and Materials Science, University of California at Davis, Davis, CA 95616, USA

October 16–20, 1995 was a noteworthy period in time for the scientific community. During those few days, the first International Congress on Adhesion Science and Technology (ICAST '95) was held in Amsterdam, The Netherlands. Not only did the congress provide a forum for the presentation of adhesion-related research results from highly regarded scientists throughout the world, but also such invited scientists used this auspicious occasion to dedicate their presentations to honor Dr. Kashmiri Lal Mittal on the occasion of his 50th birthday. Such an honor was bestowed on Dr. Mittal for his many adhesion science-related contributions to the published literature, and as an educator of technologists through short courses to provide them with knowledge about the various facets of the interdisciplinary field of adhesion science and technology. Dr. Mittal's contributions to the published literature about various types of interfaces, e.g. solid-solid, during the last 25 years have led to his name being recognized by scientists throughout the world, and coupled with his receiving a number of awards presented by various scientific organizations, accentuate the high regard with which he is viewed by his peers. This Festschrift is further testimony to that stellar record as the ICAST '95 invited presenters have made special efforts to transform their earlier presentations into written accounts. Such invited papers have been combined in book form to memorialize this historic occasion and provide a tangible item to express their collective gratitude to Dr. Mittal for his efforts, and, concomitantly, it should serve the bountiful needs of the scientific community dependent on such instructive and reliable information about research results at the forefront of adhesion science and technology now and in the foreseeable future. This Festschrift also documents the titles, but not the full text, of other papers presented at ICAST '95.

It is an honor for me and it is with great pleasure that I take this opportunity to honor a friend and a collaborator by first summarizing the educational background and then the extensive scientific activities of Dr. Mittal up to the present time.

Dr. Kashmiri Lal Mittal was born on October 15, 1945 in a very small hamlet known as Kilrodh in the state of Haryana in India. Kilrodh did not have a grade

*IBM retiree.

school. As a result, he received his elementary education at a nearby larger village called Dadanpur, which required him to walk a mile each way. Since there was no electricity in his home village, Kash and his school mates did their homework with the aid of light from a cotton seed oil lamp. Otherwise, it was necessary for them to study at times by moonlight early in the morning.

After finishing his elementary education, he moved to Jhajjar Higher Secondary School, which was about 7 miles from his home. It is interesting to note that on each Monday, Kash together with his friends and cousins would normally walk to and live at the school throughout the week. They spent their weekends at home. Then during 1961–1964 he attended Nehru College Jhajjar, which was affiliated with Panjab University. It is noteworthy that the foundation of that college was laid by the late Prime Minister Jawahar Lal Nehru.

Thus, the above information highlights that Kash came from very humble beginnings. Similarly, it is of interest to note that Kash's father had his own business but he had no real formal education. In spite of such beginnings, Kash has amassed a record of scientific achievements, which, as enumerated below and because of their import, have impressed and have led to a professional stature regarded highly by the scientific community. Kash's success, to a great extent, is due to his motivation and enthusiasm, which, as will be seen from the text below, have catalyzed his use of his formal education in chemistry to become a scientist achieving world-wide fame and recognition early in his professional career.

Finally, on a personal note, Kash developed a cataract in his right eye in 1958 when he was just 13 years of age. The medical community could not provide a cogent explanation about the cause of such a critical affliction to his vision. Later, the same eye required several surgical operations in 1991, because it had developed severe glaucoma. Unfortunately, Kash has virtually lost all vision in his right eye. It is important to highlight Kash's attitude about his vision impairment, which has handicapped him since 1958. Kash says, 'If I could see with both eyes, I could read double and could contribute doubly'.

It is appropriate now to turn to a description of the salient elements of his professional career. His scientific activities have had a principal chemical focus on the nature of interfaces, e.g. the interdisciplinary field of adhesion science and technology. Such an emphasis has led to a bountiful and enviable scientific record for him. Dr. Mittal obtained his initial collegiate educational background by obtaining a Bachelor of Science (B.Sc.) degree in 1964 at Panjab University in Chandigarh, India. He extended his education at the Indian Institute of Technology, New Delhi, where he was awarded a Master of Science (M.Sc.) degree in chemistry in 1966. He then emigrated to the United States of America to attend the University of Southern California (USC) in Los Angeles, California, to study for a Doctor of Philosophy (Ph.D.) in chemistry, which was granted in 1970. It should be mentioned that he received outstanding grades in his course work at these universities. His M.Sc. and Ph.D. research projects focused on the chemical aspects of surfaces and interfaces associated with the ion exchange capacity of some clays and factors influencing the ultracentrifugal stability of oil-in-water emulsions, respectively. His research endeavors at USC were complemented by stints as a Teaching Assistant there and as a Lecturer in the Department of Chemistry at California State College at Dominguez Hills, CA.

His interests in surface and interface phenomena were sustained subsequently by accepting postdoctoral fellowships in the USA to carry out research projects at two other universities and one industrial corporation. Firstly, at Pennsylvania State University he carried out research concerned with the nature and behavior of pollutant ions in acid mine water using a light scattering technique. He then accepted a position at the University of Pennsylvania to study, among other facets of research, the corrosion and passivation of metals using ellipsometry. Finally, he broadened his research experiences during this formative stage of his career by accepting a 2-year postdoctoral position with IBM at its research facilities in San Jose, CA. At IBM, he was assigned to projects to carry out research concerned with the surface chemical facets of adhesion science. He used various characterization techniques to determine the effectiveness of proprietary processing techniques to promote the adhesion between two dissimilar solid materials and, concomitantly, ensure the durability of such bonds when exposed to various ambient conditions of temperature and humidity. It should be noted that these foci were relevant to the functional performance and environmental stability of magnetic tape, organic photoconducting media, and metal/polymer film laminates, among other multicomponent material systems associated with the then current and future IBM products.

Subsequently, Dr. Mittal began his industrial career in 1974, by accepting a position with IBM at its facilities in Poughkeepsie, NY. For 20 years, his experimental project assignments with various research, development, and manufacturing teams at three IBM locations utilized his substantial and ever-expanding knowledge of the chemical nature of phenomena at various types of interfaces. His responsibility centered on addressing adhesion-related technical roadblocks encountered with the functional performance and reliability and stability of advanced multilayer, multimaterial thin film structures associated with electronic subsystems, which were under development and/or being manufactured by IBM. Ultimately, such subsystems would become part of advanced computer systems. Such research foci resulted in Dr. Mittal and his collaborators concentrating on the determination of the chemistry of solid surfaces before and after such surfaces combined to form numerous interfaces of the aforementioned structures. It should be noted that Dr. Mittal's technical leadership led to a methodology for thin film structures, which, in many aspects, was pioneering at the time. It occurred when several integral parts of the methodology were juxtaposed in time. These were as follows: (1) the concept of acid-base interactions, which was being introduced by the late Dr. Frederick M. Fowkes of Lehigh University, Bethlehem, PA, to explain the intermolecular forces operating at an interface; (2) the feasibility of using dry processing with plasmas to clean or modify the chemistry of surfaces had been shown; (3) various analytical instruments, which hitherto were unknown, had become available to elicit the chemistry of surfaces; (4) with the introduction of thin films of high-temperature organic polymers as dielectric materials in multilayer electronic structures, mechanical stress considerations in contiguous layers of materials coupled with the strength of interfacial bonds and their durability in chemically aggressive environmental conditions became paramount, among other considerations.

Such a methodology was so versatile that it could be used to address the aforementioned technical roadblocks. The methodology made it possible to form an interface or, where necessary, an interphase between two materials upon joining them. Characteristically, Dr. Mittal and his collaborators employed a variety of instrumental techniques to determine the chemical nature and the energetics of solid surfaces prior to forming a joint with two such characterized materials. One such instrument used to determine the presence or absence of organic contamination on the surface of a metal or a ceramic involved the use of a contact angle goniometer. On polymer surfaces, following the teachings of W. A. Zisman, a series of liquids were used to determine the critical surface tension for wetting of that surface. Such information guided Dr. Mittal in formulating polymer solutions for coating overlayers on decontaminated solid surfaces. Such selected polymer overlayers had intrinsic physical and chemical properties which made them suitable as electrical insulators, photoresist materials, and/or protective coatings for metals in multilayer structures exposed to ambient (polluted) air. By placing emphasis on cleaning processes and characterization techniques to assess the state of cleanliness of the surfaces, Dr. Mittal, through internal and external short courses on contamination, described a methodology to prevent the premature mechanical failure of a joint due to the presence of a mechanically weak layer of organic contamination at its interface, which, typically, is readily adsorbed from the air. In order to gain insight about the chemical nature of modified and unmodified surfaces, other more sophisticated techniques using instruments for chemical analysis of surfaces, which at the time were at the forefront of instrumental analysis, namely X-ray photoelectron spectroscopy (XPS) and inelastic electron tunneling spectroscopy (IETS), were employed. In the latter case, it should be noted that such an instrument was not available commercially at the time. Due to the urgent needs of the project, Dr. Mittal became part of a technical team which, guided by his technical leadership, planned and carried out experiments after the IETS instrument was designed, built, and calibrated, yielding information about the chemistry of the surface of aluminum in a timely way.

Equally important was his attention to the benefits derived from using various specialized processes and materials to alter the chemical nature of the surfaces of materials, which, upon subsequent joining, were used to form multilayer, multimaterial structures. By tailoring the chemical nature of surfaces prior to joining them to simulate the building of multilayer, multimaterial structures used in computer subsystems, Dr. Mittal and his collaborators found that the tailored interfaces/interphases in such structures were resistant to failure through delamination, even though it had been calculated that substantial mechanical stresses were concentrated at the interfaces/interphases. Further, such tailored sites were found to resist corrosion and ultimately electrical failure of thin metal circuitry in the aforementioned structures while being exposed to a wide variety of ambient environmental conditions of temperature, humidity, and acidic gases. As an example, one of his first contributions to the field of adhesion science and technology more than two decades ago was to use and to show the virtues of using selected gaseous chemical compositions, which, when excited to create plasma conditions, were able to remove the ever-present and mechanically weak organic contamination on the surface layers of a variety of materials prior to joining such layers to form interfaces/interphases between them. In the case of organic polymer layers, Dr. Mittal and his collaborators employed plasma processing to modify the chemistry, namely acidity-basicity, of the original surface of the polymer layer. In so doing, he sought to maximize the acid-base interactions between two contiguous layers involving various combinations of materials relevant to multilayer structures. Another type of example was his use of specially chosen silane-based coupling agents to modify plasma-cleaned inorganic surfaces. Thereby he sought to have surface chemistries of materials such that their acid-base interactions would be at a maximum value when such materials were being joined and, concomitantly, yield a water-resistant interface/interphase.

In addition to his active involvement in research and development activities, which took place during the first decade of his industrial career, it is noteworthy that Dr. Mittal devoted endless hours to reading the published literature so that, with time, he would have a comprehensive knowledge about the field of adhesion science and technology, which, concurrently, provided him with an awareness of the more significant prior contributions of his scientific predecessors, who, initially, were considered by him to be his mentors. As a result of this dedication, Dr. Mittal's set of qualifications have, over the years, changed from that of a student to one where he is considered to be both an outstanding educator of adhesion science and a consultant capable of providing technical insight about and plans leading to solutions of adhesion-related issues encountered in a number of different industrial technologies. Such skill transformations have been guided, in part, by his zeal to make significant and noteworthy contributions to various applied fields of science dependent, in part, on his educational background in surface chemistry and consummate interest in material systems involving various types of interfaces, e.g. solid-solid, liquid-solid, etc.

When he was an IBM post-doctoral associate in the early 1970s, he was involved for the first time with the interdisciplinary field of adhesion science and technology and became aware about the dearth of published literature concerned with it. Further, he noted that this interdisciplinary field of science was not featured at any university as a comprehensive curriculum at the time. Thus, he concluded that the scientific literature and the education of material scientists about the various facets of adhesion science and technology were areas of opportunity, a niche, for him to aggrandize his reputation by making scientific and technological contributions to the published literature about phenomena at interfaces/interphases relevant to many types of industrial products.

Apart from his scientific interests, one of Dr. Mittal's hobbies is the study of the English language each day. He has devoted an immeasurable number of hours to this hobby in order to enhance his reading comprehension, writing skills, verbal virtuosity, and editing acumen. It is important to highlight that these linguistic strivings have led to his having an excellent command of the English language, his second language. Thus, over the last two decades, his entrepreneurial instincts have led him down several pathways, which, among others, included contributing to the educational needs of the technical community both inside and outside IBM. As an example, after his assessment that IBM's technologies needed emphasis in educational areas, Dr. Mittal's entrepreneurial instincts led him to justify and organize periodic internal technical meetings which emphasized adhesion science and technology. Thus, IBM's technical community was able to discuss proprietary research results at such meetings, which was attended by both peers and management. As a result of such pioneering meetings, Dr. Mittal was very instrumental in arousing an awareness within IBM that its technologies were critically dependent on adhesion science and technology. Thereafter, more emphasis was given to the degree and scope of adhesion-related experimental activities relevant to the functional needs of subsystems in key IBM applications. Concomitantly, Dr. Mittal paved the way leading to the education of IBM's technologists about the various facets of adhesion science and technology via short courses and specialized symposia featuring presentations by peers from various IBM locations. Such meetings became opportunities for him to educate other attendees about various facets of adhesion science and technology. In fact, Dr. Mittal's sustained efforts as an educator within IBM led to his being selected as an instructor in its Education Division in 1984. For a decade thereafter, he was able to devote full time to the education of IBM technologists in materials and processes for microelectronics, e.g. the chemistry of surface sciences. Thereby, Dr. Mittal was a principal force in equipping IBM technologists with knowledge critical to the development of new products, which had an ever-increasing dependence on multilayer, multimaterial thin film structures, which, in turn, required the formation of numerous mechanical stress-bearing solid-solid interfaces. It is important to mention that Dr. Mittal and his collaborators were challenged by the aforementioned technologies, which, in many respects, were at the forefront of knowledge about materials science and especially adhesion science and technology. Many of their publications, inventions, and innovations advanced the state of knowledge about solid-solid interfaces, the materials associated with such interfaces, their characterization, and the processes used to provide long-term chemical and mechanical stability to thin film structures, on which the products of the electronics industry have depended as it has sought to meet the needs of the ever-burgeoning information processing 'superhighway'.

Most of the contributions noted above were made for and directed towards the project needs of IBM technologies, and the edification of its directly-affected technical community by the development and presentation of specialized adhesion-related courses/curricula. However, it should be mentioned that Dr. Mittal was highly involved in numerous professional activities while satisfying his responsibilities at IBM from 1972 to 1994. His initiative in and dedication to identifying and justifying specialized international symposia have resulted in his having an incomparable record as Chairman and/or Organizer of such symposia (45 since 1974 and still counting; see Appendix I), which have featured research results at the forefront of the science and technology of interfacial phenomena associated with advanced materials systems. In some cases, such materials systems were prepared by novel processing techniques and characterized by instruments capable of providing information for the first time about the chemical and mechanical nature of their surfaces and/or interfaces. Scrutiny of the titles of the aforementioned symposia shows that the materials systems and their characterization were colloids, emulsions, coatings, surfactants, polymer surfaces, particles, among other things. These symposia were sponsored by such scientific organizations as the American Chemical Society, the Society of Plastics Engineers, the Electrochemical Society, the Fine Particle Society, Skill Dynamics (an IBM company), and the American Society for Testing and Materials, among others, which collectively show the diverse scientific communities interested in phenomena associated with interfaces/surfaces, the foci of Dr. Mittal's professional efforts. Further, Dr. Mittal conceived of using these symposia to honor selected scientists for their earlier outstanding work through invited presentations by their peers in their field(s) of endeavor. To that end, Dr. Mittal's initiative and dedication made it possible for him and his peers to honor those who had made significant contributions to advance the scientific knowledge about various types of surfaces and interfaces. These scientists were R. D. Vold and M. J. Vold for their work on interfacial and colloidal phenomena, F. M. Fowkes for highlighting the importance of acid-base interactions in making solid-solid interfacial bonds, E. P. Plueddemann for his creative synthesis of silane coupling agents (adhesion promoters) to provide durable bonds between numerous combinations of solid materials, and R. J. Good for his innovative work with colloids and surface chemistry. Still further, these symposia took place in 13 states and the District of Columbia in the United States, and seven other countries. As a result of these symposia, the scientific communities throughout many parts of the world have had opportunities to contribute to and/or be edified by the forefront knowledge of interfacial science and technology. Dr. Mittal carried this ever-expanding forefront knowledge to the benefit of an even-wider audience by arranging for and editing topical books (57 and still counting; see Appendix II), derived from the compilation of peer-reviewed publications submitted to him by the various symposia presenters. Suffice it to say that this gesture required Herculean efforts on his part when one considers all of his other undertakings.

He has been a leader in the education of others about the science and technology associated with various types of interfaces throughout his professional career of more than 25 years. These include short courses and, at times, courses a semester in length on such topics as surface and colloid chemistry, adhesion science and technology, surface contamination and cleaning, and adhesion of polymeric and metallic thin films. Such highly acclaimed and industrially relevant courses were given annually or semi-annually under the auspices of such organizations as the Center for Professional Advancement, the American Vacuum Society, and the State University of New York at New Paltz, to name a few. It should be noted that such courses had not been offered at universities during most of the past 25 years. Throughout this period, these courses (see Appendix III) were given in numerous cities in the USA and in a number of cities in other countries. As a result, it can be concluded that Dr. Mittal's educational thrusts as a course organizer and/or instructor have filled an important void in the spectrum of available materials science courses offered by educational organizations throughout the world. The popularity of these courses could be ascribed to their technological foci, which have grown in importance with time due to the ever-expanding dependence of commercial products on adhesion science and technology issues associated with structures derived from novel multiphase, composite material systems with their numerous interfaces, which in part control the structures' functional performances and their long-term stabilities in a variety of environmental conditions. It should be emphasized that the educational courses listed in Appendix III were the products of

Dr. Mittal's early insights about the educational needs of personnel in major industries of commerce, his initiative and motivation leading to courses about adhesion science and technology, and his leadership in justifying and conducting courses. The favorable regard with which such courses have been viewed by the multitude of attendees speaks to the priority needs of the technical community throughout the world during the last 25 years. It is noteworthy that the courses on adhesion science and technology given by Dr. Mittal have been taken by over 4000 technologists, which is an impressive record. Based on a multitude of ever-present new invitations to present the aforementioned courses and other educational endeavors, it is clear that Dr. Mittal's scientific reputation has blossomed during the last 25 years to one of sufficient renown to prompt the scientific community to hold a special symposium in his honor — ICAST '95. It should be highlighted that Dr. Mittal has devoted endless extracurricular hours to the development and updating of the aforementioned curricula and associated courses while he has attended to IBM's project demands during working hours throughout this period. He and selected fellow instructors used such courses to enhance the knowledge of non-IBM professionals about the interdisciplinary field of adhesion science and technology during this period of time. Needless to say, this Festschrift is another and more enduring way for the technical community to express its gratitude to Dr. Kashmiri L. Mittal.

Commentaries presented above were put in capsule form to highlight some of the innovative technical details associated with his project and educational activities, which have focused on phenomena associated with materials systems with various types of interfaces/interphases. However, Dr. Mittal is a 'multidimensional' scientist who, although engaged in industrial research and development work, has adopted an academic 'publish or perish' approach to informing the scientific community about his work and work carried out in collaboration with others. As a result, there are 66 publications in a variety of scientific journals (see Appendix IV). A perusal of Appendix IV reveals topics about various types of interfaces, e.g. solid-solid, liquid-liquid, etc., associated with a number of different materials systems. It should also be noted that such materials systems were relevant to a number of industries, e.g. cosmetics, polymers, and electronics, among others. Further, the publications resulted not only from his project activities while employed by IBM, but also from symposia and workshops, among others, which had been led/moderated by him and sponsored by such organizations as the National Science Foundation, the U.S. Department of Energy, the American Chemical Society, the Electrochemical Society, and the American Society for Testing Materials, among others. The reader is encouraged to read the published articles and scan the list of voluntary services shown in Appendix V to gain an appreciation of the scope and magnitude of Dr. Mittal's services to professional societies and governmental organizations. Those of us who have been associated with Dr. Mittal know that some of the aforementioned articles describe work that has 'pushed back the frontier' of adhesion science and technology, while contributing to IBM's introduction of new and more functional products based on structures composed of thin layers of materials. Some of the more recent computer systems have utilized structures having thin layers of high-temperature polymers, namely polyimides, which, because of their electrical properties, had the potential to enhance the computer's performance. Initially, such superior electrical insulators gave rise to several technical roadblocks when attempts were made to have such high-temperature polymeric films as part of multimaterial, multilayered structures for electronic applications. Such problems were solved by focusing on the chemical and mechanical nature of the interfaces/interphases, the processes used to create contamination-free interfaces/interphases, and the ability of the materials forming the interfaces/interphases to withstand mechanical stresses and resist environmentally-induced hydrolysis so that the multilayer structures would not fail because of the cracking of thin polymer films contiguous with or by delamination occurring at one or more of the interfaces/interphases. Finally, inspection of the entries in Appendix V shows that Dr. Mittal's professional stature, namely peer regard for his knowledge of interfacial phenomena, has grown during the last 25 years such that he is considered to be one of the leading experts in the interdisciplinary field of adhesion science and technology.

Another important aspect of Dr. Kashmiri L. Mittal's professionalism has been his dedication to presenting the results of his project activities at selected conferences and/or providing tutorials to professional audiences seeking to learn about various aspects of phenomena associated with interfaces. As a result of this dedication and the growing awareness of his reputation by the professional community during the last 25 years, it should be highlighted that he has given 169 lectures and seminars, and still counting, 31% of which were at the invitation of various scientific organizations. A perusal of Appendix VI shows that such invitations were made by diverse professional societies interested in various adhesion-related issues. Further, the listing reveals that many of the invitations were extended by professional organizations in many countries outside the USA. This information reflects the relevance of adhesion science and technology to the current technological needs of industry and the high esteem with which Dr. Mittal's expertise is regarded by a world-wide audience.

Dr. Mittal's thrusts to advance the diversity and relevance of contemporary adhesion research to materials-dependent technologies provided him the impetus to start a new, much needed journal. Concomitantly, he and Dr. W. J. van Ooij of the Materials Science Department at the University of Cincinnati, Cincinnati, OH founded and are the current Co-Editors of the *Journal of Adhesion Science and Technology*. This journal, which was founded to record and disseminate research results in a timely way throughout the world, has been published since its beginning in 1987 by VSP, which has its main office in Utrecht, The Netherlands. Since I have had the opportunity to be the Co-Editor of this Festschrift, I have had an appreciation of the tremendous workload that has been carried by Dr. Mittal as he has undertaken the varied tasks involved with the periodic publication of the aforementioned journal during the last decade. In addition, it should be noted in Appendix VII that he is, or has been, a member of nine Editorial Boards of other journals. Also, he has written 20 book reviews and has critiqued about 143 journal papers, book chapters, research proposals, and counting (see Appendix VIII). Once again, these additional and varied professional activities of Dr. Mittal provide further testimony to the prodigious level of effort that he has expended to contribute to and be at the forefront of the scientific literature, which has led to respect for his professional expertise concerning phenomena associated with interfaces by his peers. Such dedication to one's profession is exceedingly rare.

One other facet of Dr. Kashmiri L. Mittal's professionalism is his membership in and technical support of 20 scientific and other societies. Scrutiny of the list shown in Appendix IX reveals both honor societies and those professional societies that feature scientific interests in phenomena at interfaces formed between flat surfaces and/or with particles.

In the light of some of the activities and accomplishments cited above, Dr. Mittal has been the recipient of a number of scholarships, awards, and other honors, which are shown chronologically in Appendix X. While he has distinguished himself as a student and subsequently as an industrial surface chemist and educator, it is important to highlight initially the several awards that have resulted from his contributions during a quarter of a century to the field of adhesion science and technology, because it would be in keeping with ICAST '95, an international symposium held to honor him on his 50th birthday. Dr. Mittal's employment with IBM made it possible for him to make a number of technical contributions about phenomena at interfaces, which have been presented above and/or are listed *in toto* in the attached appendices. His creative efforts led to novel findings at times. Some of these had such merit that IBM sought legal protection for such findings through the filing of patent applications to the United States Patent Office, or through publications in *IBM's Journal of Research and Development*. Such work qualified him for IBM's First Invention Achievement Award in 1982. His interests in education motivated him to apprise the technical community about the import and nature of organic and/or particulate contamination on surfaces, which resulted in his receiving a Certificate of Appreciation from the Institute of Environmental Sciences in 1979. Similarly, the IBM Corporate Contamination Control Committee cited his service with a Certificate of Appreciation in 1985, to which they added a Recognition Plaque for Continued Leadership and Significant Technical Professional Contribution to the Field of Contamination Control Technology in 1989. His missionary efforts highlighting the role of interfaces in polymer-dependent technologies resulted in the Mid-Hudson Section of the Society of Plastics Engineers presenting him with an Outstanding Contribution Award in 1984. The American Society for Testing and Materials expressed its gratitude to Dr. Mittal for enhancing the knowledge base of its membership by awarding him the Charles B. Dudley Award and the Adhesives Award of Committee D-14 in 1990 and 1997, respectively. In recognition of the impact of his numerous and highly regarded scientific treatises and publications about various facets of the field of adhesion science and technology, he was honored by the Adhesion Society by being made a Robert L. Patrick Fellow of the society in 1994. Similarly, the *Adhesives Age* magazine honored him with their Adhesives Age Award and the International Microelectronics and Packaging Society (IMAPS) selected him for the John A. Wagnon Technical Achievement Award, both of which occurred in 1997. His outstanding contributions to dielectric science and technology resulted in his selection for the Thomas D. Callinan Award by the Dielectric Science and Technology Division of the Electrochemical Society in 1995. The International Surface and Colloid Science Community presented him with a Recognition Plaque for his Continued Leadership and Distinguished Professional Service in 1986. He received the VHP Award of Excellence for

Outstanding Professional Achievements from the Vishwa Hindu Parishad of America in 1987.

His record of achievements about and service to various professional scientific societies concerned with phenomena at interfaces has caught the attention of scientists throughout the world. As a result, his name is listed with other men and women of scientific distinction in a number of well-known and highly regarded publications, e.g. Fourth Edition of *Who's Who in Technology Today* (Appendix X shows other listings). While it is anticipated that Dr. Kashmiri L. Mittal will continue to make technical and educational contributions while being of service to scientific societies and industrial and governmental organizations in the future, it seems appropriate at this juncture in his career to finish with comments about the significance of two events, which were held in 1995 on the occasion of his 50th birthday, to honor him for his numerous achievements in and contributions to the interdisciplinary field of adhesion science and technology. As indicated above, one was ICAST '95. The other was the dedication of a special issue of *Particulate Science and Technology: An International Journal*. It is with pride that I indicate once again that this Festschrift provides further testimony about peer respect for Dr. Kashmiri L. Mittal's stellar record of achievements. The ICAST '95 invited presenters have made special efforts to transform their presentations made earlier into written papers. Such papers have been combined to form a book in order to memorialize the aforementioned historic occasion and to provide a tangible item to express their collective gratitude to Dr. Kashmiri L. Mittal for his efforts noted above, and, concomitantly, to serve the bountiful needs of the scientific community dependent on such instructive and reliable information about research results at the forefront of adhesion science and technology now and in the foreseeable future.

In conclusion, the above text has concentrated on the highlights of his achievements and contributions in the field of adhesion science and technology. However, it should be emphasized that Dr. Kashmiri L. Mittal has impacted other scientific areas significantly, namely surfactant chemistry, surface contamination and cleaning, and polyimides, by editing a number of books in those areas. Technologists working in these areas recognize his name immediately, and, such professionals have great esteem for his scientific reputation.

ADDENDUM

It is with pride that I have provided an abridged set of highlights of Dr. Kashmiri (Kash) L. Mittal's professional exploits, many of which are known throughout the scientific world through his personal contacts, publications, and educational courses. However, I would like to add an addendum, which describes my long-standing relationship with him. I met him first during a series of interviews that I had with candidates for a 2-year post-doctoral position with my research group at IBM San Jose, CA. While his educational background and experiences were of limited value initially, his credentials seemed adequate to enable him to become acquainted with and contribute subsequently to our plans to carry out projects. Such projects were directed towards gaining more information about the nature of materials at interfaces formed by several combinations of materials critical to the performance of information processing equipment at the time. Thus, I take great pride in having launched Kash's efforts leading to numerous and valuable contributions to the field of adhesion science and technology thereafter. About a year later, my relationship with Kash changed from that of a manager, and at times a mentor, to that of a collaborator. IBM granted me a 1-year sabbatical with an opportunity to pursue activities of my own choosing, which required some 'soul searching'. One main effort was directed towards enhancing my knowledge of and contributing to the information about the intermolecular forces prevalent at an interface by arranging a research program with the aforementioned Professor Frederick Fowkes. At that time, he had just begun his thrust at Lehigh University to determine the role of acid-base reactions in forming an interface. One other main effort of mine was directed towards assisting IBM's Education Department at San Jose, CA in the enhancement of its curriculum, which led to more relevant courses about polymer and adhesion science. Kash and I were instrumental in the formulation of these courses. A team of IBM employees served as instructors for the courses. It should be noted that such courses were of sufficient merit to qualify the graduates for credit(s) towards a degree with San Jose State University.

Upon completion of the 2-year post-doctoral position, Kash accepted a permanent position with an IBM organization located in Poughkeepsie, NY. While we were separated geographically at the time, we maintained contact as I published three papers on the characterization of acid-base reactions by means of electrical measurements and XPS. In the mid-1970s, I accepted a position at an IBM facility located in East Fishkill, NY. Once again, I was located a relatively short distance (15 miles) from Kash. A few months later, I was asked to set up a new polymer science group to carry out long-range research/development experimental programs. We were asked to address some of the more important polymer-dependent issues affecting IBM's semiconductor and packaging technologies. Needless to say, adhesion was one such issue.

Arrangements were made for Kash to join this new group so that he could apply his technical insights to the aforementioned technologies. I was Kash's manager but, more importantly, I had a chance to collaborate with him on project activities and be one of the instructors recruited by Kash to provide a short course on adhesion science and technology, which depended in part on the presentation of my research results on acid-base reactions. Once again, I had an opportunity to be witness to some of Kash's early efforts to fill a void in the professional world about adhesion science and technology through short courses. He has amassed a stellar record in this arena.

In the late 1970s, I had the good fortune to be elected Chairman of the Gordon Research Conference on the Science of Adhesion. Kash's expansive knowledge of scientists carrying out research at the forefront of adhesion science and technology aided me immensely in making contact with and inviting such scientists to be part of a technical program which was in the spirit of the Gordon Conferences. It gave me an opportunity to make Kash one of the Session Chairmen. Judging from the very favorable response of the attendees and their personal comments to me at the time, it was judged to be one of the best such conferences held describing research at the forefront of adhesion science and technology. Needless to say, I was indebted to Kash for his guidance and the scientists who provided the presentations. Once again, my collaboration with Kash has given rise to a positive result.

My experiences with the management of the Gordon Conferences during the above period of time made me realize that they considered a conference about adhesion science and technology to be of such low interest and relevance that consideration was being given not to offer it in the future. Such a revelation prompted me and six other scientists to co-found the Adhesion Society and I was elected to be its second President. Kash was one of the Charter members of this new society and again I was aided by him in the formulation of technical programs during my 2-year tenure. Subsequently, I had the honor of recommending him for, and developing information which was used to make him, the Robert L. Patrick Fellow of the Adhesion Society.

These are some of the more important highlights which relate to my meeting Kash about a quarter of a century ago and provide more detail about our collaboration and friendship. At this juncture, I gain a great deal of satisfaction as a 'professional father' knowing that it was I who pointed him in a scientific direction early on and, with his intrinsic motivation, he has since created an enviable record of accomplishments.

APPENDIX I

Symposia organization

Chairman and Organizer of the International Symposium on Interfacial and Colloidal Phenomena honoring Professors R. D. Vold and M. J. Vold held at the American Chemical Society Meeting, Los Angeles, CA, April 2–5, 1974

Chairman and Organizer of the International Symposium on Micellization, Solubilization, and Microemulsions held at the 7th Northeast Regional Meeting of the American Chemical Society, Albany, NY, August 8–11, 1976

Chairman and Organizer of the Symposium on Adhesion Measurement of Thin Film, Thick Films, and Bulk Coatings, sponsored by the American Society for Testing and Materials, held at the ASTM Headquarters in Philadelphia, PA, November 2–4, 1976

Co-chairman and Organizer of the Symposium on Solution Chemistry of Surfactants held at the 52nd Colloid and Surface Science Symposium, Knoxville, TN, June 12–14, 1978

Chairman and Organizer of the Symposium on Surface Contamination: Genesis, Detection and Control held as a part of the ICCCS Meeting, Washington, DC, September 10–13, 1978

Co-chairman of the International Symposium on Solution Behavior of Surfactants: Theoretical and Applied Aspects held at the 10th Northeast Regional Meeting of the American Chemical Society, Potsdam, NY, June 30–July 3, 1980

Chairman and Organizer of the Symposium on Adhesion Aspects of Polymeric Coatings held at the Electrochemical Society Meeting, Minneapolis, MN, May 10–15, 1981

Chairman and Organizer of the International Symposium on Physicochemical Aspects of Polymer Surfaces held at the American Chemical Society Meeting, New York City, NY, August 23–28, 1981

Co-chairman of the International Symposium on Surfactants in Solution held in Lund, Sweden, June 27–July 2, 1982

Chairman and Organizer of the International Symposium on Adhesive Joints: Formation, Characteristics and Testing held at the American Chemical Society Meeting, Kansas City, MO, September 12–17, 1982

Program Chairman of the First Technical Conference on Polyimides: Synthesis, Characterization and Applications held under the auspices of the Mid-Hudson Section of the Society of Plastics Engineers, Ellenville, NY, November 10–12, 1982

Co-chairman of the 5th International Symposium on Surfactants in Solution held in Bordeaux-Talence, France, July 9–13, 1984

Chairman and Organizer of the Symposium on Surface and Colloid Science in Computer Technology held as a part of the 5th International Conference on Surface and Colloid Science, Clarkson University, Potsdam, NY, June 24–28, 1985

Co-chairman of the Symposium on Particles on Surfaces: Detection, Adhesion and Removal held at the 17th Annual Meeting of the Fine Particle Society, San Francisco, CA, July 28–August 2, 1986

Chairman and Organizer of the 6th International Symposium on Surfactants in Solution held in New Delhi, India, August 18–22, 1986

Co-chairman of the Symposium on Particles in Fluids: Detection, Characterization and Control held at the 18th Annual Meeting of the Fine Particle Society, Boston, MA, August 3–7, 1987

Chairman of the 2nd Symposium on Particles on Surfaces: Detection, Adhesion and Removal held at the 19th Annual Meeting of the Fine Particle Society, Santa Clara, CA, July 18–22, 1988

Co-chairman of the Symposium on Polymers in Information Storage Technology held at the American Chemical Society Meeting, Los Angeles, CA, September 25–30, 1988

Chairman of the 7th International Symposium on Surfactants in Solution held in Ottawa, Canada, October 2–7, 1988

Co-chairman of the Symposium on Metallized Plastics: Fundamental and Applied Aspects held at the Electrochemical Society Meeting, Chicago, IL, October 9–14, 1988

Chairman of the 2nd Symposium on Particles in Gases and Liquids: Detection, Characterization and Control held at the 20th Annual Meeting of the Fine Particle Society, Boston, MA, August 21–25, 1989

Co-chairman of the 2nd Symposium on Metallized Plastics: Fundamental and Applied Aspects held as a part of the Electrochemical Society Meeting, Montreal, Canada, May 6–11, 1990

Chairman of the 8th International Symposium on Surfactants in Solution held at the University of Florida, Gainesville, FL, June 10–15, 1990

Co-organizer of the Symposium on Acid–Base Interactions: Relevance to Adhesion Science and Technology in honor of the 75th Birthday Anniversary of Dr. F. M. Fowkes held at Lehigh University, Bethlehem, PA, June 18–20, 1990

Chairman of the 3rd Symposium on Particles on Surfaces: Detection, Adhesion and Removal held at the 21st Annual Meeting of the Fine Particle Society, San Diego, CA, August 21–25, 1990

Organizer of the Symposium on Silanes and Other Coupling Agents in honor of the 75th Birthday Anniversary of Dr. E. P. Plueddemann held at Dow Corning Co. in Midland, MI, April 3–5, 1991

Chairman of the 3rd Symposium on Particles in Gases and Liquids: Detection, Characterization and Control held at the 22nd Annual Meeting of the Fine Particle Society, San Jose, CA, July 29–August 2, 1991

Co-chairman of the 3rd Symposium on Metallized Plastics: Fundamental and Applied Aspects held as a part of the Electrochemical Society Meeting, Phoenix, AZ, October 13–18, 1991

Chairman of the Symposium on Contact Angle, Wettability and Adhesion in honor of Dr. Robert J. Good held at the American Chemical Society Meeting, San Francisco, CA, April 5–10, 1992

Chairman of the 9th International Symposium on Surfactants in Solution held in Varna, Bulgaria, June 7–12, 1992

Chairman of the 4th Symposium on Particles on Surfaces: Detection, Adhesion and Removal held at the 23rd Annual Meeting of the Fine Particle Society, Las Vegas, NV, July 13–17, 1992

Organizer of the International Symposium on Adhesion Measurement of Films and Coatings held under the auspices of Skill Dynamics, an IBM Company, in Cambridge/Boston, MA, December 5–7, 1992

Co-chairman of the 4th Symposium on Metallized Plastics: Fundamental and Applied Aspects held as a part of the Electrochemical Society Meeting, Honolulu, Hawaii, May 16–21, 1993

Co-organizer of 4th Symposium on Particles in Gases and Liquids: Detection, Characterization and Control held at the 24th Annual Meeting of the Fine Particle Society, Chicago, IL, August 24–28, 1993

Organizer of the International Symposium on Polymer Surface Modification: Relevance to Adhesion held under the auspices of Skill Dynamics, an IBM Company, Las Vegas, NV, November 3–5, 1993

Organizer of the International Symposium on Plasma Polymerization/Deposition: Fundamental and Applied Aspects held under the auspices of Skill Dynamics, an IBM Company, Las Vegas, NV, November 8–10, 1993

Chairman of the 10th International Symposium on Surfactants in Solution held in Caracas, Venezuela, June 26–30, 1994

Co-organizer of the Symposium on Polymer Surfaces and Interfaces: Characterization, Modification and Application held at the 53rd Annual Technical Conference of the Society of Plastics Engineers, ANTEC95, Boston, MA, May 7–11, 1995

Co-chairman of the 5th Symposium on Metallized Plastics: Fundamental and Applied Aspects held as a part of the Electrochemical Society Meeting, Los Angeles, CA, May 5–10, 1996

Chairman of the 11th International Symposium on Surfactants in Solution held in Jerusalem, Israel, June 9–13, 1996

Co-organizer of the Symposium on Particles on Solid Surfaces: Detection, Adhesion and Removal held as a part of the Fine Particle Society Meeting, Chicago, IL, August 6–9, 1996

Co-chairman of the 6th Symposium on Metallized Plastics: Fundamental and Applied Aspects held as a part of the 1997 International Meeting between the Electrochemical Society and the International Society of Electrochemistry, Paris, France, August 31–September 5, 1997

Organizer of the Symposium on Particles on Solid Surfaces: Detection, Adhesion and Removal held as a part of the Fine Particle Society Meeting in Dallas, TX, April 1–3, 1998

General Chairman of the 12th International Symposium on Surfactants in Solution held in Stockholm, Sweden, June 7–11, 1998

Co-organizer of the Symposium on Emulsions, Foams and Thin Films in honor of Dr. Darsh T. Wasan on his 60th Birthday, as a part of the 72nd ACS Colloid and Surface Science Symposium, held at Penn State, PA, June 21–24, 1998

APPENDIX II

Books

1. K. L. Mittal (Ed.), *Adsorption at Interfaces*, ACS Symposium Series No. 8. American Chemical Society, Washington, DC (1975).
2. K. L. Mittal (Ed.), *Colloidal Dispersions and Micellar Behavior*, ACS Symposium Series No. 9. American Chemical Society, Washington, DC (1975).
3. K. L. Mittal (Ed.), *Micellization, Solubilization, and Microemulsions*, Vol. 1. Plenum Press, New York (1977). Part of this book has been translated into Russian.
4. K. L. Mittal (Ed.), *Micellization, Solubilization, and Microemulsions*, Vol. 2. Plenum Press, New York (1977). Part of this book has been translated into Russian.
5. K. L. Mittal (Ed.), *Adhesion Measurement of Thin Films, Thick Films and Bulk Coatings*. American Society for Testing and Materials, Philadelphia, PA (1978). This book has gone into second printing. Also based on this book, K. L. Mittal was the recipient of the 1990 Charles B. Dudley Award of the ASTM.
6. K. L. Mittal (Ed.), *Solution Chemistry of Surfactants*, Vol. 1. Plenum Press, New York (1979).
7. K. L. Mittal (Ed.), *Solution Chemistry of Surfactants*, Vol. 2. Plenum Press, New York (1979).
8. K. L. Mittal (Ed.), *Surface Contamination: Genesis, Detection, and Control*, Vol. 1. Plenum Press, New York (1979).
9. K. L. Mittal (Ed.), *Surface Contamination: Genesis, Detection, and Control*, Vol. 2. Plenum Press, New York (1979).
10. K. L. Mittal and E. J. Fendler (Eds), *Solution Behavior of Surfactants: Theoretical and Applied Aspects*, Vol. 1. Plenum Press, New York (1982).
11. K. L. Mittal and E. J. Fendler (Eds), *Solution Behavior of Surfactants: Theoretical and Applied Aspects*, Vol. 2. Plenum Press, New York (1982).
12. K. L. Mittal (Ed.), *Physicochemical Aspects of Polymer Surfaces*, Vol. 1. Plenum Press, New York (1983).
13. K. L. Mittal (Ed.), *Physicochemical Aspects of Polymer Surfaces*, Vol. 2. Plenum Press, New York (1983).
14. K. L. Mittal (Ed.), *Adhesion Aspects of Polymeric Coatings*. Plenum Press, New York (1983).
15. K. L. Mittal and B. Lindman (Eds), *Surfactants in Solution (Lund Proceedings)*, Vol. 1. Plenum Press, New York (1984).
16. K. L. Mittal and B. Lindman (Eds), *Surfactants in Solution (Lund Proceedings)*, Vol. 2. Plenum Press, New York (1984).
17. K. L. Mittal and B. Lindman (Eds), *Surfactants in Solution (Lund Proceedings)*, Vol. 3. Plenum Press, New York (1984).
18. K. L. Mittal (Ed.), *Adhesive Joints: Formation, Characteristics and Testing*. Plenum Press, New York (1984).
19. K. L. Mittal (Ed.), *Polyimides: Synthesis, Characterization and Applications*, Vol. 1. Plenum Press, New York (1984).
20. K. L. Mittal (Ed.), *Polyimides: Synthesis, Characterization and Applications*, Vol. 2. Plenum Press, New York (1984).
21. K. L. Mittal and P. Bothorel (Eds), *Surfactants in Solution (Bordeaux Proceedings)*, Vol. 4. Plenum Press, New York (1986).
22. K. L. Mittal and P. Bothorel (Eds), *Surfactants in Solution (Bordeaux Proceedings)*, Vol. 5. Plenum Press, New York (1986).
23. K. L. Mittal and P. Bothorel (Eds), *Surfactants in Solution (Bordeaux Proceedings)*, Vol. 6. Plenum Press, New York (1986).
24. K. L. Mittal (Ed.), *Treatise on Clean Surface Technology*, Vol. 1. Plenum Press, New York (1987).
25. K. L. Mittal (Ed.), *Surface and Colloid Science in Computer Technology*. Plenum Press, New York (1987).

26. G. G. Fuller and K. L. Mittal (Eds), *Opportunities and Research Needs in Adhesion Science and Technology*, Proceedings of the NSF Workshop on Adhesion Science and Technology held in Lake Tahoe, CA, October 14–16, 1987, published in 1988.
27. K. L. Mittal (Ed.), *Particles on Surfaces 1: Detection, Adhesion and Removal*. Plenum Press, New York (1988).
28. K. L. Mittal (Ed.), *Particles in Gases and Liquids 1: Detection, Characterization and Control*. Plenum Press, New York (1989).
29. K. L. Mittal (Ed.), *Surfactants in Solution (New Delhi Proceedings)*, Vol. 7. Plenum Press, New York (1989).
30. K. L. Mittal (Ed.), *Surfactants in Solution (New Delhi Proceedings)*, Vol. 8. Plenum Press, New York (1989).
31. K. L. Mittal (Ed.), *Surfactants in Solution (New Delhi Proceedings)*, Vol. 9. Plenum Press, New York (1989).
32. K. L. Mittal (Ed.), *Surfactants in Solution (New Delhi Proceedings)*, Vol. 10. Plenum Press, New York (1989).
33. K. L. Mittal (Ed.), *Particles on Surfaces 2: Detection, Adhesion and Removal*. Plenum Press, New York (1989).
34. K. L. Mittal (Ed.), *Polymers in Information Storage Technology*. Plenum Press, New York (1989).
35. K. L. Mittal and J. R. Susko (Eds), *Metallized Plastics 1: Fundamental and Applied Aspects*. Plenum Press, New York (1989).
36. K. L. Mittal (Ed.), *Particles in Gases and Liquids 2: Detection, Characterization and Control*. Plenum Press, New York (1990).
37. K. L. Mittal and H. R. Anderson, Jr. (Eds), *Acid–Base Interactions: Relevance to Adhesion Science and Technology*. VSP, Zeist, The Netherlands (1991).
38. K. L. Mittal (Ed.), *Particles on Surfaces 3: Detection, Adhesion and Removal*. Plenum Press, New York (1991).
39. K. L. Mittal (Ed.), *Metallized Plastics 2: Fundamental and Applied Aspects*. Plenum Press, New York (1991).
40. K. L. Mittal and D. O. Shah (Eds), *Surfactants in Solution (Gainesville Proceedings)*, Vol. 11. Plenum Press, New York (1991).
41. K. L. Mittal (Ed.), *Silanes and Other Coupling Agents*. VSP, Zeist, The Netherlands (1992).
42. K. L. Mittal (Ed.), *Metallized Plastics 3: Fundamental and Applied Aspects*. Plenum Press, New York (1992).
43. K. L. Mittal (Ed.), *Particles in Gases and Liquids 3: Detection, Characterization and Control*. Plenum Press, New York (1993).
44. K. L. Mittal (Ed.), *Contact Angle, Wettability and Adhesion*. VSP, Zeist, The Netherlands (1993).
45. A. Pizzi and K. L. Mittal (Eds), *Handbook of Adhesive Technology*. Marcel Dekker, New York (1994).
46. M. Strobel, C. S. Lyons and K. L. Mittal (Eds), *Plasma Surface Modification of Polymers: Relevance to Adhesion*. VSP, Zeist, The Netherlands (1994).
47. K. L. Mittal (Ed.), *Particles on Surfaces: Detection, Adhesion and Removal*. Marcel Dekker, New York (1995).
48. D. S. Rimai, L. P. DeMejo and K. L. Mittal (Eds), *Fundamentals of Adhesion and Interfaces*. VSP, Zeist, The Netherlands (1995).
49. K. L. Mittal (Ed.), *Adhesion Measurement of Films and Coatings*. VSP, Zeist, The Netherlands (1995).
50. K. L. Mittal (Ed.), *Polymer Surface Modification: Relevance to Adhesion*. VSP, Zeist, The Netherlands (1996).
51. M. K. Ghosh and K. L. Mittal (Eds), *Polyimides: Fundamentals and Applications*. Marcel Dekker, New York (1996).
52. A. K. Chattopadhyay and K. L. Mittal (Eds), *Surfactants in Solution (Caracas Proceedings)*. Marcel Dekker, New York (1996).
53. K.-W. Lee and K. L. Mittal (Eds), *Polymer Surfaces and Interfaces: Characterization, Modification, and Application*. VSP, Zeist, The Netherlands (1997).

54. K. L. Mittal (Ed.), *Metallized Plastics: Fundamentals and Applications*. Marcel Dekker, New York (1998).
55. P. Kumar and K. L. Mittal (Eds), *Microemulsions: Fundamentals and Applied Aspects*. Marcel Dekker (in press).
56. K. L. Mittal and A. Pizzi (Eds), *Adhesion Promotion Techniques: Technological Applications*. Marcel Dekker (in press).
57. K. L. Mittal (Ed.), *Metallized Plastics 5&6: Fundamental & Applied Aspects*. VSP (in press).

APPENDIX III

Educational activities

- Taught a 16-week course on Surface and Colloid Chemistry at IBM, San Jose, CA, 1973
- Taught a 14-week course entitled 'Selected Topics in Surface Chemistry' at IBM, East Fishkill, NY, 1976
- Course Advisor and key instructor for the short course on Fundamentals of Adhesion: Theory, Practice, and Applications held under the auspices of the State University of New York at New Paltz, NY, November 8–10, 1976; October 31–November 2, 1977; October 25–27, 1978; October 22–24, 1979; and November 5–7, 1980. This course was well attended and highly acclaimed
- Taught a 3-day short course on Adhesion at the Universidade de Sao Paulo, Sao Paulo, Brazil, December 10–12, 1979; and at the Universidade Federal de Rio de Janeiro, Rio de Janeiro, Brazil, June 3–5, 1981, under the auspices of the Brazilian Chemical Society
- Taught a mini-course on Clean Surface Technology under the auspices of the International Society for Hybrid Microelectronics in Chicago, IL, October 12, 1981
- Taught a mini-course on Clean Surface Technology at Printed Circuits '82 in New York City, NY, May 11, 1982
- Course Co-Coordinator and key instructor for the short course on Surface Cleaning, Modification, and Adhesion given under the auspices of the Swedish National Technical Board in Soderköping, Sweden, June 7–11, 1982
- Course Director and key instructor for the short course on Adhesion Science and Technology given under the auspices of the Center for Professional Advancement, East Brunswick, NJ:

E. Brunswick, NJ, March 2–4, 1981
San Mateo, CA, Feb. 8–10, 1982
San Francisco, CA, March 14–16, 1983
Millbrae, CA, Oct. 10–12, 1983
Millbrae, CA, May 14–16, 1984
Chicago, IL, March 4–6, 1985
San Mateo, CA, March 3–5, 1986
San Mateo, CA, Sept. 22–24, 1986
Chicago, IL, March 30–April 1, 1987
San Mateo, CA, Sept. 28–30, 1987
Chicago, IL, May 16–18, 1988
E. Brunswick, NJ, Sept. 14–16, 1988
E. Brunswick, NJ, May 1–3, 1989
E. Brunswick, NJ, Oct. 2–4, 1989
Jamesburg, NJ, May 15–17, 1990
E. Brunswick, NJ, April 29–May 1, 1991
E. Brunswick, NJ, Jan. 11–13, 1982
E. Brunswick, NJ, Feb. 14–16, 1983
Des Plaines, IL, Sept. 26–28, 1983
E. Brunswick, NJ, April 25–27, 1984
E. Brunswick, NJ, Feb. 11–13, 1985
E. Brunswick, NJ, Feb. 10–12, 1986
E. Brunswick, NJ, Sept. 15–17, 1986
E. Brunswick, NJ, March 16–18, 1987
E. Brunswick, NJ, Sept. 14–16, 1987
E. Brunswick, NJ, April 25–27, 1988
Burlingame, CA, August 15–17, 1988
Chicago, IL, April 10–12, 1989
San Francisco, CA, Sept. 18–20, 1989
Chicago, IL, April 18–20, 1990
E. Windsor, NJ, Oct. 29–31, 1990
E. Brunswick, NJ, Oct. 28–30, 1991

E. Brunswick, NJ, May 13–15, 1992
Chicago, IL, June 7–9, 1993
E. Brunswick, NJ, June 13–15, 1994
E. Brunswick, NJ, June 21–23, 1995
E. Brunswick, NJ, May 20–22, 1996
E. Brunswick, NJ, June 23–25, 1997
E. Brunswick, NJ, July 13–15, 1998

- Instructor for a 1-day course on Adhesion of Polymeric and Metallic Thin Films under the auspices of the American Vacuum Society:

Detroit, MI, Oct. 15, 1980
Albuquerque, NM, April 17, 1981
Denver, CO, May 4, 1981
Detroit, MI, May 19, 1981
Seattle, WA, June 12, 1981
Anaheim, CA, November 6, 1981
Phoenix, AZ, February 1, 1982
Burlington, VT, September 27, 1982
Anaheim, CA, April 11, 1983
Reno, NV, December 7, 1984
New Orleans, LA, April 28, 1986
Dearborn, MI, May 23, 1986
Baltimore, MD, October 31, 1986
San Jose, CA, March 9, 1987
Boston, MA, June 19, 1987
Anaheim, CA, November 6, 1987
San Jose, CA, March 14, 1988
Rochester, NY, June 9, 1989
Newton, MA, June 15, 1989
Boston, MA, October 27, 1989
San Diego, CA, April 6, 1990
Toronto, Canada, October 8, 1990
Columbia, MD, May 7, 1991
Pittsburgh, PA, June 3, 1991
St. Paul, MN, June 13, 1991
Seattle, WA, November 11, 1991
San Jose, CA, March 9, 1992
Rochester, NY, May 12, 1992
Minneapolis, MN, June 24, 1992
Chicago, IL, November 12, 1992
Minneapolis, MN, June 10, 1993
Orlando, FL, November 19, 1993
Denver, CO, Aug. 25, 1995

- Course Director and key instructor of the course on Adhesion Science and Technology given under the auspices of the Center for Professional Advancement in Amsterdam, The Netherlands:

November 28–December 1, 1983
November 26–29, 1984
June 3–6, 1985
June 2–5, 1986
July 13–16, 1987
November 16–19, 1987 (The Hague)
June 20–23, 1988
November 14–17, 1988
July 3–6, 1989
December 11–14, 1989
June 25–28, 1990
November 19–22, 1990
November 25–28, 1991
March 23–26, 1992
September 7–10, 1992
September 20–23, 1993
October 17–20, 1994
November 20–23, 1995

- Taught a one-day course on Surface Contamination and Cleaning at the Institute of Environmental Sciences Meeting, San Jose, CA, May 4, 1987
- Taught a half-day course on Surface Contamination and Cleaning at the Fine Particle Society Meeting, Boston, MA, August 3, 1987; Santa Clara, CA, July 18, 1988; Boston, MA, August 21, 1989; San Diego, CA, August 22, 1990; San Jose, CA, July 30, 1991; Las Vegas, NV, July 16, 1992; Chicago, IL, August 24, 1993; New Brunswick, NJ, July 26, 1994; Chicago, IL, August 6, 1996
- Taught a one-day course on Surface Contamination and Cleaning at the ICCCS Meeting, Los Angeles, CA, September 26, 1988
- Taught a two-day course on Surface Contamination and Cleaning via the NTU Satellite Network, Professional Development Programs (Broadcast from Northeastern University, Boston, MA), October 18–19, 1988; October 10–11, 1989
- Taught a half-day course on Surface Contamination and Cleaning at CleanRooms '92, Atlantic City, NJ, April 13–16, 1992

- Taught a short (half-day) course on Adhesion and the Surface and Interface Techniques on October 2, 1993 as a part of the ECASIA 93 held in Catania, Italy, October 4–8, 1993
- Taught a half-day course on Surface Contamination and Cleaning at Clean Rooms '94, Philadelphia, PA, March 14, 1994
- Taught a half-day course on Surface Contamination and Cleaning at Precision Cleaning, Rosemont, IL, May 16, 1994; Rosemont, IL, May 15, 1995; Anaheim, CA, May 14, 1996; Cincinnati, OH, April 15, 1997; Rosemont, IL, May 19, 1998
- Taught a one-day tutorial on Adhesion Aspects of Films and Coatings at the International Conference on Metallurgical Coatings and Thin Films, San Diego, CA, April 23, 1995
- Taught a one-day course on Adhesion: Principles and Practices at ISHM '97, Philadelphia, PA, October 13, 1997
- Taught a one-day course on Adhesion Aspects of Coatings at the 5th Coatings for Aerospace Industry, Toronto, Canada, October 22, 1997
- Taught a one-day course on Adhesion: Principles and Practices at the International Systems Packaging Symposium, San Diego, CA, December 2, 1997

APPENDIX IV

Other publications

1. H. Singh and K. L. Mittal, Negative catalyst, *J. Chem. Ed.* **46**, 185–186 (1969).
2. J. O'M. Bockris, K. L. Mittal and R. K. Sen, Mechanism of activation in electron transfer in solution, *Nature Phys. Sci.* **234**, 118–119 (1971).
3. K. L. Mittal, Conceptual clarification of the terms used to describe emulsion behavior, *J. Soc. Cosmet. Chem.* **22**, 815–824 (1971).
4. J. McHardy, K. L. Mittal and J. O'M. Bockris, Electrocatalysis of oxygen reduction by the tungsten bronzes, Final Report, U.S. Army Mobility Equipment Command, Ft. Belvoir, Virginia. Contract No. DA44-009-AMC-469 (T), 73 pages, AD-747 766 (December 1971).
5. R. D. Vold and K. L. Mittal, The effect of lauryl alcohol on the stability of oil-in-water emulsions, *J. Colloid Interface Sci.* **38**, 451–456 (1972).
6. S. H. Kim, K. L. Mittal and J. O'M. Bockris, Ellipsometric studies of passivation, Final Report, Office of Saline Water, Washington, DC, Project No. 14-30-2801, 19 pages (April 1972).
7. R. D. Vold and K. L. Mittal, Differences in ultracentrifugal stability of various oil-in-water emulsions, *J. Soc. Cosmet. Chem.* **23**, 171–188 (1972).
8. R. D. Vold and K. L. Mittal, Determination of sodium dodecyl sulfate in the presence of lauryl alcohol, *Anal. Chem.* **44**, 849–850 (1972).
9. R. D. Vold and K. L. Mittal, Effect of age on the ultracentrifugal stability of liquid petrolatum–water emulsions, *J. Pharm. Sci.* **61**, 769–772 (1972).
10. K. L. Mittal, Determination of CMC of Polysorbate 20 in aqueous solution by surface tension method, *J. Pharm. Sci.* **61**, 1334–1335 (1972).
11. K. L. Mittal and R. D. Vold, Effect of the initial concentration of emulsifying agents on the ultracentrifugal stability of oil-in-water emulsions, *J. Am. Oil Chem. Soc.* **49**, 527–532 (1972).
12. J. O'M. Bockris, R. K. Sen and K. L. Mittal, On quantum electrochemical kinetics, *J. Res. Inst. Catal.* (*Hokkaido*) **20** (3), 153–184 (1972).
13. R. D. Vold and K. L. Mittal, Effect of temperature on the stability of Nujol–water emulsions, *J. Colloid Interface Sci.* **42**, 436–440 (1973).
14. D. F. O'Kane and K. L. Mittal, Plasma cleaning of metal surfaces, *J. Vac. Sci. Technol.* **11**, 567–569 (1974).

15. K. L. Mittal, Adherate: a new and general term, *J. Adhesion* **6**, 377–378 (1974).
16. K. L. Mittal, Ultracentrifugal technique in the study of emulsions, in: *Colloidal Dispersions and Micellar Behavior*, K. L. Mittal (Ed.), ACS Symposium Series No. 9, pp. 76–96. American Chemical Society, Washington, DC (1975).
17. K. L. Mittal, A critical appraisal of the methods for measuring adhesion of electrodeposited coatings, in: *Properties of Electrodeposits: Their Measurement and Significance*, R. Sard, H. Leidheiser, Jr. and F. Ogburn (Eds), Ch. 17, pp. 273–306. Electrochemical Society, Princeton, NJ (1975).
18. A. N. Saxena and K. L. Mittal, Optical and dielectric constants of hafnium and its anodic oxide films, *J. Appl. Phys.* **46**, 2788–2789 (1975).
19. K. L. Mittal, Surface chemical criteria for maximum adhesion and their verification against the experimentally measured adhesive strength values, in: *Adhesion Science and Technology*, L. H. Lee (Ed.), Part A, pp. 129–168. Plenum Press, New York (1975).
20. K. L. Mittal, Adhesion aspects of metallization of organic polymer surfaces, *J. Vac. Sci. Technol.* **13**, 19–25 (1976). Adapted in Italian in *Galvanotecnica* **28** (7), 113–115 (1977).
21. K. L. Mittal, Adhesion measurement of thin films, *Electrocomponent Sci. Technol.* **3**, 21–42 (1976).
22. K. L. Mittal and D. F. O'Kane, Vapor deposited silanes and other coupling agents, *J. Adhesion* **8**, 93–97 (1976).
23. K. L. Mittal, Reply to the Letter by W. A. Dukes, *J. Adhesion* **8**, 101–102 (1976).
24. R. P. Misra, S. Bhatia and K. L. Mittal, A study of surface degradation of polymeric high voltage insulating materials due to UV exposure, in: *1976 Annual Report, Conference on Electrical Insulation and Dielectric Phenomena*, pp. 325–331. National Academy of Sciences, Washington, DC (1978).
25. K. L. Mittal and P. Mukerjee, The wide world of micelles, in: *Micellization, Solubilization, and Microemulsions*, K. L. Mittal (Ed.), Vol. 1, pp. 1–21. Plenum Press, New York (1977).
26. K. L. Mittal, The role of the interface in adhesion phenomena, *Polym. Eng. Sci.* **17**, 467–473 (1977).
27. S. Bhatia, R. P. Misra and K. L. Mittal, Surface degradation of polymeric high voltage insulator materials due to environmental stresses, *Electrochemical Society Meeting, Extended Abstracts*, Vol. 77-1, pp. 58–60 (May 1977).
28. G. DiGiacomo and K. L. Mittal, Reaction of lead film in perfluoro-*n*-hexane (low conductivity) in the presence of added HCl and CH_3COOH, *Electrochemical Society Meeting, Extended Abstracts*, Vol. 77-2, pp. 294–296 (October 1977).
29. S. P. Bhatia, R. P. Misra and K. L. Mittal, Effect of UV exposure on leakage current of high voltage insulating materials, in: *1977 Annual Report, Conference on Electrical Insulation and Dielectric Phenomena*, pp. 360–364. National Academy of Sciences, Washington, DC (1979).
30. K. L. Mittal and H. R. Anderson, Jr., Interfacial chemistry and adhesion: recent developments and prospects, *Science-Ciencia* **6** (1), 19–20 (1978).
31. K. L. Mittal, Adhesion measurement: recent developments, unsolved problems, and prospects, in: *Adhesion Measurement of Thin Films, Thick Films, and Bulk Coatings*, K. L. Mittal (Ed.), pp. 5–17. American Society for Testing and Materials, Philadelphia, PA (1978). Translated in Italian in *Galvanotecnica* 139–145 (July/August 1979).
32. J. Ahn, K. L. Mittal and R. H. MacQueen, Hardness and adhesion of filmed structures as determined by the scratch technique, in: *Adhesion Measurement of Thin Films, Thick Films, and Bulk Coatings*, K. L. Mittal (Ed.), pp. 134–157. American Society for Testing and Materials, Philadelphia, PA (1978).
33. R. D. Vold, K. L. Mittal and A. U. Hahn, Ultracentrifugal stability of emulsions, in: *Surface and Colloid Science*, E. Matijevic (Ed.), Vol. 10, pp. 45–97. Plenum Press, New York (1978).
34. S. Bhattacharya and K. L. Mittal, Mechanics of removing glass particulates from a solid surface, *Surface Technol.* **7**, 413–425 (1978).
35. K. L. Mittal, Surface contamination: an overview, in: *Surface Contamination: Genesis, Detection and Control*, K. L. Mittal (Ed.), Vol. 1, pp. 3–45. Plenum Press, New York (1979).
36. K. L. Mittal, Factors affecting adhesion of lithographic materials, in: *Durability of Macromolecular Materials*, R. K. Eby (Ed.), ACS Symposium Series No. 95, pp. 371–391. American Chemical Society, Washington, DC (1979). Also reprinted in *Solid State Technol.* 89–95, 100 (May 1979).

37. A. U. Hahn and K. L. Mittal, Mechanism of demulsification of oil-in-water emulsion in the centrifuge, *Colloid Polym. Sci.* **257**, 959–967 (1979).
38. K. L. Mittal, Surface contamination: its genesis, detection and control, in: *Proceedings of the Nordic Association for Contamination Control Meeting*, held in Abo (Turku), Finland, May 21–23, pp. 4–9 (1979).
39. K. L. Mittal, Progress in adhesion science, *Adhesives Age* 36–37 (January 1980).
40. K. L. Mittal and R. O. Lussow, Adhesion measurement of polymeric films and coatings with special reference to photoresist materials, in: *Adhesion and Adsorption of Polymers*, L. H. Lee (Ed.), Part B, pp. 503–520. Plenum Press, New York (1980). Preprint published in *ACS Org. Coatings Plastics Preprints* **40**, 527–532 (April 1979).
41. K. L. Mittal, Interfacial chemistry and adhesion: recent developments and prospects, *Pure Appl. Chem.* **52**, 1295–1307 (1980).
42. N. K. Eib, K. L. Mittal and A. Friedrichs, Wettability studies of plasma deposited hexamethyldisilazane, *J. Appl. Polym. Sci.* **25**, 2435–2438 (1980).
43. K. L. Mittal, Clean surface technology, *Adhesives Age* 18–19 (January 1981).
44. K. L. Mittal, Adhesion measurement of polymeric coatings, *Electrochemical Society Meeting, Extended Abstracts*, Vol. 81-1, pp. 326–327 (1981).
45. K. L. Mittal, Adhesion aspects of polymeric coatings: an overview, *Electrochemical Society Meeting, Extended Abstracts*, Vol. 81-1, pp. 263–265 (1981).
46. K. L. Mittal, Microscopic aspects of adhesion: recent trends and prospects, in: *Microscopic Aspects of Adhesion and Lubrication*, J. M. Georges (Ed.), pp. 153–165. Elsevier, Amsterdam (1982).
47. T. Furukawa, K. L. Mittal and H. R. Anderson, Jr., Fundamental interfacial studies using inelastic electron tunneling spectroscopy and their relevance to polymer adhesion, in: *Proceedings of the IUPAC Macro 82*, held at the University of Massachusetts, Amherst, MA, July 12–16, 1982, p. 697.
48. T. Furukawa, N. K. Eib, K. L. Mittal and H. R. Anderson, Jr., Inelastic electron tunneling spectroscopic study of the silane coupling agents: I. Aminophenyltrimethoxysilane adsorbed on plasma grown aluminum oxide and effects of high humidity, *Surface Interface Anal.* **4**, 240–244 (1982).
49. T. Furukawa, N. K. Eib, K. L. Mittal and H. R. Anderson, Jr., Inelastic electron tunneling spectroscopic study of silane coupling agents: II. γ-Aminopropyltriethoxysilane adsorbed from vapor on aluminum oxide surface and effects of high humidity, *J. Colloid Interface Sci.* **96**, 322–338 (1983).
50. M. El-Shabasy, B. Szikora, G. Peto, I. Szabo and K. L. Mittal, Investigation of multilayer systems by the scratch method, *Thin Solid Films* **109**, 127–136 (1983).
51. D. Suryanarayana and K. L. Mittal, Effect of pH of silane solution on the adhesion of polyimide to silica substrate, *J. Appl. Polym. Sci.* **29**, 2039–2043 (1984).
52. N. K. Eib, T. Furukawa and K. L. Mittal, Digital low noise DC power supply for tunneling spectrometers, *Rev. Sci. Instrum.* **55**, 1161–1163 (1984).
53. H. R. Anderson, Jr., A. I. Baise, L. P. Buchwalter, P. M. Cotts, J. Greenblatt, R. H. Lacombe, H. Lee and K. L. Mittal, Adhesion aspects of organic insulators in semiconductor technologies: I. Polyimides, in: *Proceedings of the International Adhesion Conference*, held at the University of Nottingham, Nottingham, UK, September 12–14, 1984.
54. D. Suryanarayana and K. L. Mittal, Effect of polyimide thickness on its adhesion to silicon nitride substrate with and without adhesion promoter, *J. Appl. Polym. Sci.* **30**, 3107–3111 (1985).
55. K. L. Mittal, Selected bibliography on adhesion measurement of films and coatings, in: *Testing of Metallic and Inorganic Coatings*, ASTM STP 947, W. B. Harding and G. A. DiBari (Eds), pp. 343–362. American Society for Testing and Materials, Philadelphia, PA (1987).
56. K. L. Mittal, Opportunities for joint Indo-US research in materials relative to microelectronics, in: *Proceedings of the NSF Workshop on Indian Scientific Strengths: Selected Opportunities for Indo-US Cooperation*, held at SRI International, Arlington, VA, March 12–13, 1987, pp. 137–153.
57. K. L. Mittal, Selected bibliography on adhesion measurement of films and coatings, *J. Adhesion Sci. Technol.* **1**, 247–259 (1987).
58. K. L. Mittal, Interfacial phenomena in microelectronics: research needs, in: *Proceedings of the NSF Workshop on Interfacial Phenomena in the New and Emerging Technologies*, held at the University of Colorado, Boulder, CO, May 29–31, 1986, pp. 2-1 to 2-10. Published in 1987.

59. K. L. Mittal, A multidisciplinary approach to surface contamination control (interview), *Microcontamination* 16–24 (May 1988).
60. G. G. Fuller and K. L. Mittal, Workshop on adhesion science and technology: executive summary, in: *Opportunities and Research Needs in Adhesion Science and Technology*, G. G. Fuller and K. L. Mittal (Eds). *Proceedings of the NSF Workshop on Adhesion Science and Technology*, held in Lake Tahoe, CA, October 14–16, 1987. Published in September 1988, pp. VII–XII.
61. K. L. Mittal, Surface cleaning: recent developments and needs, *Proc. 34th Annu. Technical Conf. Soc. Vacuum Coaters*, pp. 13–17 (1991).
62. K. L. Mittal, Adhesion measurement of thin films: a critical assessment, *Proc. 5th Int. Web Coating Conf.*, pp. 201–212 (1991).
63. K. L. Mittal, Surface modification techniques: a summary report, *Proc. Surface Processing Workshop*, NREL/CP-412-5007, pp. 9-1 to 9-4 (Sept. 1992).
64. K. L. Mittal, Surface contamination control primer (interview), *CleanRooms* 10–14 (April 1993).
65. K. L. Mittal, Surface contamination concepts and concerns, *Precision Cleaning* 17–22 (January 1995).
66. K. L. Mittal, Adhesion measurement of films and coatings: a commentary, in: *Adhesion Measurement of Films and Coatings*, K. L. Mittal (Ed.), pp. 1–13. VSP, Zeist, The Netherlands (1995).

APPENDIX V

Professional services

Discussion Leader for one of the Sessions at the 1977 Gordon Research Conference on the Science of Adhesion, New Hampton, NH

Member, Colloid and Surface Science Symposium Committee of the Division of Colloid and Surface Chemistry, American Chemical Society, 1977

Member, Advisory Committee of the Division of Organic Coatings and Plastics Chemistry of the American Chemical Society, 1980

Member of the Panel on Manufacturing Problems Associated with Contamination in Semiconductor and Electronics Industries held at the Fine Particle Society Meeting, Miami Beach, FL, April 24, 1985

Member, Plastics Advisory Board, The Center for Professional Advancement, East Brunswick, NJ, 1985

Chairman, Session on Interfacial Phenomena in Microelectronics held as a part of the NSF Workshop on Interfacial Phenomena in New and Emerging Technologies, University of Colorado, Boulder, CO, May 29–31, 1986

Member, Advisory Board of the Center for Surface Science and Engineering, University of Florida, Gainesville, FL, 1986

Member, Executive Committee, Indian Society for Surface Science and Technology, 1986

Member, Executive Council of the Fine Particle Society, 1986

Participated in the NSF Workshop on Indian Scientific Strengths: Selected Opportunities for Indo-US Cooperation held at SRI International, Arlington, VA, March 12–13, 1987

Co-chairman, NSF Workshop on Adhesion Science and Technology held in Lake Tahoe, CA, October 14–16, 1987

Chairman, Contamination Control Division of the Fine Particle Society, 1987

Guest Editor for a special issue entitled 'Dispersion Technology in Microelectronics', *Colloids and Surfaces* **29** (3) (Feb. 1988)

Chairman, NSF Workshop on Adhesion Science and Technology held in Washington, DC, October 20–21, 1988

Co-chairman, NSF Workshop on Interfacial Phenomena and Rheology of Fluid Dispersions held in Washington, DC, January 22–23, 1991

Member, Subcommittee on Emerging Technologies of the International Society for Hybrid Microelectronics, 1991

Member, Canvassing Committee for the ACS Award in Colloid or Surface Chemistry (October 1, 1992–September 30, 1995)

Member of the Panel dealing with Adhesion at the Society of Vacuum Coaters Meeting, Dallas, TX, April 29, 1993

Guest Editor for a special issue on Adhesion of the *IBM Journal of Research & Development* **34**, No. 4 (July 1994)

Member, Committee of Visitors for the Interfacial, Transport and Separation Processes Program of NSF, Arlington, VA, March 22–23, 1995

Member, International Advisory Committee, International Symposium on Micelles, Microemulsions and Monolayers: Quarter Century Progress and New Horizons held at the University of Florida, Gainesville, FL, August 28–30, 1995

Member, International Advisory Committee, International Conference on Polymer–Solid Interfaces II: From Model to Real Systems held in Namur, Belgium, August 12–16, 1996

Member, International Advisory Committee, International Conference on Rubbers held in Calcutta, India, December 12–14, 1997

Co-opted Council Member of the International Association of Colloid and Interface Scientists, 1997–1998

APPENDIX VI

Conference papers and invited presentations (lectures and seminars)

Ultracentrifugal evaluation of the stability of oil-in-water emulsions and the effect of lauryl alcohol. Annual Meeting of the Society of Cosmetic Chemists, New York City, NY, May 1970 (with R. D. Vold)

Differences in ultracentrifugal stability of various oil-in-water emulsions. 161st American Chemical Society Meeting, Los Angeles, CA, March 1971 (with R. D. Vold)

Effect of temperature on the stability of Nujol–water emulsions. 162nd American Chemical Society Meeting, Washington, DC, September 1971 (with R. D. Vold)

Dependence of ultracentrifugal stability of oil-in-water emulsions on emulsifier concentration. American Oil Chemists Society Meeting, Atlantic City, NJ, October 1971 (with R. D. Vold)

Potentialities and limitations of the ultracentrifugal technique in the study of emulsions. 167th American Chemical Society Meeting, Los Angeles, CA, March 31–April 5, 1974

Cleaning of surfaces: characterization of cleanliness and the kinetics of contamination.* Joint Fall Meeting (Electronics Division and NICE) of the American Ceramic Society, Denver, CO, September 18–20, 1974

---

*Invited conference paper.

A critical appraisal of the methods for measuring adhesion of electrodeposited coatings.* 146th Meeting of the Electrochemical Society, New York City, NY, October 13–17, 1974

Adhesion measurement of thin and thick films.* Meeting of the ASTM Committee F-1 on Electronics, Dallas, TX, January 28–30, 1975

Surface chemical criteria for maximum adhesion and their verification against the experimentally measured adhesive strength values. Symposium on Science and Technology of Adhesion, 169th American Chemical Society Meeting, Philadelphia, PA, April 6–11, 1975

Cleaning of surfaces.* 18th Annual Meeting of the Institute of Printed Circuits, Washington, DC, April 20–24, 1975

Adhesion aspects of metallization of organic polymer surfaces.* 22nd National Symposium of the American Vacuum Society, Philadelphia, PA, October 28–31, 1975

Adhesion: theory and practice. Talk at the Mid-Hudson Chapter of the Society of Manufacturing Engineers, Rhinebeck, NY, January 14, 1976

Criteria for adhesive bonding and the nature of the interfacial region. Seminar at the Wright-Patterson Air Force Base, Dayton, OH, February 17, 1976

Adhesion testing of polymers. Lecture at the Sixth Annual Summer Conference in Polymer Science and Technology sponsored by the State University of New York (New Paltz), Mohonk Mountain House, New Paltz, NY, June 16, 1976

Clean surface technology. Adhesion: theory and practice. Lectures at IBM Corp., Owego, NY, June 25, 1976

The wide world of micelles. International Symposium on Micellization, Solubilization and Microemulsions, 7th Northeast Regional Meeting of the American Chemical Society, Albany, NY, August 8–11, 1976 (with P. Mukerjee)

Accelerated tests for emulsion stability: validity and limitations. American Chemical Society Meeting, San Francisco, CA, August 29–September 3, 1976 (with A. U. Hahn)

The role of the interface in adhesion phenomena.* Chemical Institute of Canada Symposium on Polymers at Interfaces: Focus on Adhesion, Montreal, Canada, October 17–19, 1976

A study of surface degradation of polymeric high voltage insulating materials due to UV exposure. 1976 Conference on Electrical Insulation and Dielectric Phenomena, Buck Hill Falls, PA, October 18–21, 1976 (with R. P. Misra and S. P. Bhatia)

Adhesion measurement: recent progress, unsolved problems and prospects. ASTM Symposium on Adhesion Measurement of Thin Films, Thick Films and Bulk Coatings, ASTM Headquarters, Philadelphia, PA, November 2–4, 1976

Hardness and adhesion of filmed structures as determined by the scratch technique. ASTM Symposium on Adhesion Measurement of Thin Films, Thick Films and Bulk Coatings, ASTM Headquarters, Philadelphia, PA, November 2–4, 1976 (with J. Ahn and R. H. MacQueen)

Study of the mechanism of demulsification of dilute O/W emulsions using coulter counter. 69th Annual Meeting of the American Institute of Chemical Engineers, Chicago, IL, November 28–December 2, 1976 (with A. U. Hahn)

Why things stick together? Talk at the Mid-Hudson Section of the Society of Plastics Engineers, Wappingers Falls, NY, April 6, 1977

---

*Invited conference paper.

Chemical criteria for maximum adhesion of polymer films to various substrates. Seminar at the Dwight P. Joyce Research Center of Glidden Durkee, Strongville, OH, April 15, 1977

Adhesion: theory and practice. Lecture at IBM Corp., Kingston, NY, May 2, 1977

Surface contamination: an overview. Lecture at IBM Corp., Manassas, VA, May 23, 1977

Mechanics of removing glass particulates from a solid surface. 51st Colloid and Surface Science Symposium, Grand Island, NY, June 19–22, 1977 (with S. Bhattacharya)

Surface degradation of polymeric high voltage insulator materials due to environmental stresses. Electrochemical Society Meeting, Philadelphia, PA, May 8–13, 1977 (with S. Bhatia and R. P. Misra)

Reaction of lead film in perfluoro-*n*-hexane (low conductivity) in the presence of added HCl and CH_3COOH. Electrochemical Society Meeting, Atlanta, GA, October 9–14, 1977 (with G. DiGiacomo)

Effect of UV exposure on leakage current of high voltage insulating materials. 1977 Conference on Electrical Insulation and Dielectric Phenomena, Colonie, NY, October 17–20, 1977 (with S. Bhatia and R. P. Misra)

Interfacial chemistry and adhesion: recent developments and prospects.* 9th Caribbean Chemical Conference, San Juan, Puerto Rico, December 8–11, 1977 (with H. R. Anderson, Jr.)

Recent advances in adhesion. Seminar at the IBM Watson Research Center, Yorktown Hts., NY, March 29, 1978

Interfacial chemistry and adhesion: recent developments and prospects. Seminar at the Aerospace Corp., El Segundo, CA, May 22, 1978

Interfacial chemistry and adhesion: recent advances and prospects. Seminar in the Materials Department, U.C.L.A., Los Angeles, CA, May 22, 1978

Recent advances in adhesion science. Seminar in the Department of Chemical Engineering, University of Southern California, Los Angeles, CA, May 23, 1978

Adhesion: theory and practice. Seminar at FMC Corp., Santa Clara, CA, May 24, 1978

Surface cleanliness and adhesion. Theories of adhesion. Various techniques for improving adhesion. Adhesion measurement techniques. Lectures at IBM Corp., San Jose, CA, May 25–26, 1978

Surface contamination and cleaning. Adhesion: theory and practice. Lectures at IBM Corp., Burlington, VT, August 22, 1978

Surface contamination: an overview.* Symposium on Surface Contamination: Its Genesis, Detection and Control held as a part of the 4th International Symposium on Contamination Control held under the auspices of the International Committee of Contamination Control Societies, Washington, DC, September 10–13, 1978 (plenary opening paper)

Factors affecting adhesion of lithographic materials.* American Chemical Society Meeting, Miami Beach, FL, September 10–15, 1978

Surface contamination and cleaning. Why things stick together? Lectures at IBM Corp., Boca Raton, FL, September 15, 1978

Adhesion aspects of metallized plastics. Seminar at the Optical Coatings Laboratory, Santa Rosa, CA, March 30, 1979

---

*Invited conference paper.

Adhesion measurement of polymeric films and coatings with special reference to photoresist materials. American Chemical Society/Japan Chemical Society Chemical Congress, Honolulu, Hawaii, April 1–7, 1979 (with R. O. Lussow)

Accelerated tests for emulsion stability. Seminar at the Institute of Industrial Science, University of Tokyo, Tokyo, Japan, April 9, 1979

Interfacial chemistry and adhesion. Seminar at the Science University of Tokyo, Tokyo, Japan, April 10, 1979

Accelerated tests for emulsion stability. Seminar at Kao Soap Co., Tokyo, Japan, April 11, 1979

Role of the interface in adhesion phenomena. Seminar at the National Physical Laboratory, New Delhi, India, April 19, 1979

Interfacial chemistry and adhesion: recent developments and prospects. Seminar at the Indian Institute of Technology, New Delhi, India, April 20, 1979

Role of the interface in adhesion phenomena. Seminar at Panjab University, Chandigarh, India, April 26, 1979

Interfacial chemistry and adhesion. Seminar at FOM Institute for Atomic and Molecular Physics, Amsterdam, The Netherlands, May 1, 1979

Adhesion: theory and practice. Surface chemical criteria for maximum adhesion. A critical review of methods for measuring adhesion of films and coatings. Lectures at the Laboratoire de la Profession, Limelette, Belgium, May 2, 1979

Role of the interface in adhesion phenomena. Seminar at the Technical University of Denmark, Lyngby, Denmark, May 18, 1979

Interfacial chemistry and adhesion. Seminar at the Abo Akademi (Turku), Finland, May 21, 1979

Surface contamination: its genesis, detection and control.* Nordic Association for Contamination Control Meeting, Abo (Turku), Finland, May 21–23, 1979 (opening paper)

Role of the interface in adhesion phenomena. Seminar at the Swedish Institute for Surface Chemistry, Stockholm, Sweden, May 22, 1979

Role of the interface in adhesion phenomena. Seminar at Linköping University, Linköping, Sweden, May 23, 1979

Adhesion measurement of films and coatings. Role of the interface in adhesion phenomena. Seminars at the Norges Tekniske Hogskole, Trondheim, Norway, May 25, 1979

Interfacial chemistry and adhesion: recent developments and prospects.* Third International Conference on Surface and Colloid Science, Stockholm, Sweden, August 20–25, 1979

Adhesion: theory and practice. Seminar at the Zenith Research Center, Elk Grove, IL, October 11, 1979

Surface contamination and cleaning. Adhesion: theory and practice. Lectures at IBM Corp., Rochester, MN, October 12, 1979

Measurement of adhesion of films and coatings. Adhesion aspects of metallized plastics. Seminars at the IBM Watson Research Center, Yorktown Heights, NY, November 8, 1979

Interfacial chemistry and adhesion: recent developments and prospects. Surface contamination and cleaning. Seminars at the Universidade Federal de Santa Catarina, Florianopolis, Brazil, December 13, 1979

---

*Invited conference paper.

Accelerated tests of emulsion stability. Seminar at the Universidad Central del Ecuador, Quito, Ecuador, December 15, 1979

Interfacial aspects in lithographic polymers. Seminar at the Virginia Polytechnic Institute and State University, Blacksburg, VA, January 9, 1980

ESCA analysis of plasma-deposited hexamethyldisilazane (HMDS). Adhesion Society Meeting, Savannah, GA, February 10–13, 1980 (with N. Eib and H. R. Anderson, Jr.)

Measurement of the adhesion of films.* 1980 Annual Congress of the Chemical Society/Royal Institute of Chemistry, Durham, UK, April 9–11, 1980

Surface contamination and cleaning. Seminar at the Laboratorio de Semiconductores, Ciudad Universitaria, Madrid, Spain, April 14, 1980

Interfacial chemistry and adhesion: recent developments and prospects. Seminar at the Centro Tecnologia Materie Plastiche, ANIC s.p.a., Milano, Italy, April 15, 1980

Accelerated tests of emulsion stability. Seminar at the Eidgenossische Technische Hochschule, Zurich, Switzerland, April 16, 1980

Adhesion and surface cleaning. Seminar at the Balzers Corporation, Balzers, Liechtenstein, April 17, 1980

Interfacial chemistry and adhesion: recent developments and prospects. Seminar at the Technische Universitat Wien, Vienna, Austria, April 18, 1980

Modern progress in adhesion science. Seminar at the Universität Stuttgart, Stuttgart, Germany, April 21, 1980

Adhesion: theory and practice. Recent progress in adhesion science. Seminars at the University of Lund, Lund, Sweden, April 22, 1980

A comparative analysis (using ESCA, FT-IR, Auger, plasma chromatography and contact angle) of amino and vinylsilanes as deposited by different means. 54th Colloid and Surface Science Symposium, Lehigh University, Bethlehem, PA, June 15–18, 1980 (with N. Eib and H. R. Anderson, Jr.)

Thermal studies of silanes deposited by various means. 54th Colloid and Surface Science Symposium, Lehigh University, Bethlehem, PA, June 15–18, 1980 (with N. Eib and H. R. Anderson, Jr.)

Interfacial chemistry and adhesion: recent developments and prospects. Seminar at the Centro de Investigacion y de Estudios Avanzados del IPN, Mexico City, Mexico, November 27, 1980

Modern progress in adhesion science. Seminar at the Universidad Nacional Autonoma de Mexico, Mexico City, Mexico, November 27, 1980

Why things stick together? Seminar at IBM Scientific Center, Mexico City, Mexico, November 28, 1980

Recent developments in adhesion science. Seminar at the University of Connecticut, Storrs, CT, February 9, 1981

Surface and bulk characterization of plasma deposited silanes. Adhesion Society Meeting, Savannah, GA, February 22–25, 1981 (with N. K. Eib, R. Lacombe and H. R. Anderson, Jr.)

Recent trends in adhesion science. Seminar at Drexel University, Philadelphia, PA, April 20, 1981

Adhesion aspects of polymeric coatings: an overview. Symposium on Adhesion Aspects of Polymeric Coatings, Electrochemical Society Meeting, Minneapolis, MN, May 10–15, 1981

---

*Invited conference paper.

Adhesion measurement of organic coatings. Symposium on Adhesion Aspects of Polymeric Coatings, Electrochemical Society Meeting, Minneapolis, MN, May 10–15, 1981

Surface contamination and cleaning. Adhesion measurement of films and coatings. Seminars at INTEVEP, Caracas, Venezuela, June 8, 1981

Series of lectures in the area of adhesion. IBM Endicott, September 8–9, 1981

Microscopic aspects of adhesion: recent trends and prospects.* Symposium on Microscopic Aspects of Adhesion and Lubrication held at the 34th International Meeting of the Société de Chimie Physique, Paris, France, September 14–17, 1981

Adhesion: theoretical and applied aspects. Photoresist adhesion. Seminars at the Institut für Technologie der Fasern, Akademie der Wissenschaften der DDR, Dresden, Germany, September 17, 1981

Recent developments in adhesion science. Seminar at the Indian Institute of Technology, New Delhi, India, September 29, 1981

Metallized plastics: interfacial and adhesion aspects. Lecture at the Mid-Hudson Section of the Society of Plastics Engineers Meeting, Wappingers Falls, NY, December 9, 1981

Clean surface technology. Presentation at the 11th IBM Corporate Contamination Control Seminar, Poughkeepsie, NY, May 18–20, 1982

Clean surface technology.* New England Combined Chapter of the American Vacuum Society Meeting, Danvers, MA, June 14, 1982

Adhesion: theory and practice. Lecture as a part of the 2-week course on Applied Materials Technology, M.I.T., Cambridge, MA, June 19, 1984

Fundamental interfacial studies using inelastic electron tunneling spectroscopy and their relevance to polymer adhesion. IUPAC Macro '82 Meeting held at the University of Massachusetts, Amherst, MA, July 12–16, 1982 (with T. Furukawa and H. R. Anderson, Jr.)

Clean surface technology. Adhesion science and technology. Lectures as a part of the IBM Materials Science Seminar held at UC Santa Cruz, Santa Cruz, CA, August 15, 1984

Adhesion aspects of organic insulators in semiconductor technologies: I. Polyimides. International Adhesion Conference, University of Nottingham, UK, September 12–14, 1984 (with H. R. Anderson, Jr., A. Baise, L. P. Buchwalter, P. Cotts, J. Greenblatt, C. Goldsmith, R. H. Lacombe and H. Lee)

Interface technology and adhesion. Seminar at Norfolk State University, Norfolk, VA, November 15, 1984

Surface contamination: genesis, characterization and control. Lecture at the Society for Analytical Chemists of Pittsburgh Meeting, Pittsburgh, PA, April 8, 1985

Adhesion aspects of thin films.* 28th Annual Technical Conference of the Society of Vacuum Coaters, Philadelphia, PA, May 3, 1985

Surface contamination and cleaning. Surface analysis techniques and applications. Theories or mechanisms of adhesion. Contact angle, wettability and adhesion. Adhesion aspects of thin films. Adhesion of organic coatings. Adhesion aspects of lithographic materials. Adhesion measurement of films and coatings. Series of Lectures at the Shanghai Jiao Tong University, Shanghai, PRC, on the invitation of the International Advisory Panel and the Chinese Review Commission (under the auspices of the World Bank) as a Project Specialist in the area of Materials Science, July 1985

---

*Invited conference paper.

Clean surface technology. Lecture in the IBM European Course on Contamination Control Technology, La Hulpe, Belgium, September 4, 1985

Clean surface technology. Lecture at IBM Poughkeepsie, NY, October 24, 1985

Surface contamination and cleaning. Lecture in the seminar on Contamination Control Technology, IBM Canada, Bromont, Canada, November 5, 1985

Adhesion.* Featured presentation at the Association of Industrial Metallizers, Coaters and Laminators Meeting, Newark, NJ, November 13, 1985

Adhesion measurement of coatings: some myths and facts.* ASTM Symposium on Testing of Metallic and Inorganic Coatings, Chicago, IL, April 14–16, 1986

Interfacial phenomena in microelectronics: research needs.* NSF Workshop on Interfacial Phenomena in the New and Emerging Technologies, University of Colorado, Boulder, CO, May 29–31, 1986

Seminar on some personal glimpses and experiences under the programme 'meet the great'. Hans Raj College, University of Delhi, Delhi, India, August 13, 1986

Interfacial phenomena in microelectronics and related technologies.* Workshop on Advanced Topics in Surfactant Solutions, Bhabha Atomic Research Centre, Bombay, India, August 24–30, 1986

Opportunities for joint Indo-US research in materials relative to microelectronics.* NSF Workshop on Indian Scientific Strengths: Selected Opportunities for Indo-US Cooperation held at SRI International, Arlington, VA, March 12–13, 1987

Contact angle analysis of polymers.* Polymer Surfaces Symposium, 3M Center, St. Paul, MN, April 2, 1987

Interfacial phenomena in microelectronics: research needs. Plenary Lecture at the Symposium inaugurating the Langmuir Center for Colloids and Surfaces, Columbia University, May 19, 1987

Surface contamination and cleaning. Lecture at IBM Sindelfingen, Germany, July 17, 1987

The wonderful world of adhesion science. Lecture at the Packaging Institute International's Adhesion Committee Meeting, New York City, NY, September 10, 1987

Adhesion aspects of coatings. Seminar at ICI Explosives Group Technical Centre, McMasterville, Quebec, Canada, January 30, 1989

Interfacial phenomena in microelectronics: research needs.* Symposium on Innovations in Materials Processing using Aqueous Colloid and Surface Chemistry at the AIME Annual Meeting, Las Vegas, NV, February 27–March 2, 1989

Adhesion aspects of thin films.* 2nd ASM International Electronic Materials and Processing Congress, Philadelphia, PA, April 24–28, 1989

Adhesion measurement of films and coatings. Seminar at Polaroid Corp., Cambridge, MA, August 21, 1989

Adhesion improvement of polymeric coatings to various substrates.* International Symposium on Surface Phenomena and Fine Particles in Water-Based Coatings and Printing Technology held as a part of the Fine Particle Society Meeting, Boston, MA, August 21–25, 1989

Adhesion measurement of films and coatings.* Materials Research Symposium, Corning Section of the American Chemical Society, Corning, NY, September 15, 1989

---

*Invited conference paper.

Interfacial phenomena in microelectronics: research needs. Seminar at SUNY-Binghamton, NY, November 16, 1989

Interfacial phenomena in microelectronics: research needs. Seminar, Department of Chemistry, Delhi University, India, March 23, 1990

Interfacial phenomena in microelectronics: research needs. Seminar at Université du Quebec à Trois-Rivieres, Trois-Rivieres, Quebec, Canada, May 9, 1990

Various mechanisms of adhesion. Seminar at Schering-Plough Research, Pembroke Pines, FL, June 14, 1990

Adhesion science and technology: recent progress, prospects and research needs.* Plenary lecture at 5emes Journée d'Etude sur l'Adherence, sponsored by the French Vacuum Society, held in Villeurbanne, France, June 18–22, 1990

Thin film adhesion measurement: current status, prospects and needs.* American Vacuum Society Meeting, Toronto, Canada, October 8–12, 1990

Adhesion measurement techniques for films and coatings. Adhesion aspects of thin films. Seminars at Centre d'Etudes de Chimie Metallurgique, Vitry/Seine Cedex, France, November 23, 1990

Adhesion measurement of films and coatings: A critical assessment. Seminar at the National Institute of Standards and Technology, Gaithersburg, MD, January 14, 1991

Surface cleaning: recent developments and needs.* Plenary Lecture at the Society of Vacuum Coaters Meeting, Philadelphia, PA, March 17–22, 1991

Adhesion measurement of thin films: current status, needs and prospects.* American Physical Society Meeting, Cincinnati, OH, March 17–22, 1991

Surface contamination and cleaning.* Clean Rooms '91, Washington, DC, April 17–19, 1991

Recent developments in adhesion science. Seminar at Alcoa Laboratories, Alcoa Center, PA, August 27, 1991

Ways to improve adhesion in polymer/metal and metal/polymer systems.* Plenary Lecture at the International Conference on Polymer–Solid Interfaces held in Namur, Belgium, September 2–6, 1991

Adhesion measurement of thin films: a critical assessment.* Fifth International Vacuum Web Coating Conference, Austin, TX, December 1–3, 1991

Surface modification techniques.* Plenary Summary Status Presentation at the Workshop on Surface Processing and Applications to Transportation and Utilities Technologies, sponsored by U.S. Department of Energy, Dearborn, MI, December 10–12, 1991

Surface contamination and cleaning.* Upstate New York Chapter of the Institute of Environmental Sciences Meeting, Auburn, NY, April 2, 1992

Ways to improve strength and durability of adhesively bonded joints.* International Symposium on Structural Adhesive Bonding, Morristown, NJ, May 5–7, 1992

Various theories of adhesion.* Various ways to improve adhesion of polymeric materials.* NATO ASI on The Interfacial Interactions in Polymeric Composites, Antalya, Turkey, June 15–26, 1992

Interfacial chemistry and adhesion: recent developments. Seminar at Shell Research Laboratorium, Amsterdam, The Netherlands, September 9, 1992

---

*Invited conference paper.

Contact angle, wettability and adhesion. Seminar at Courtaulds Strategic Research, Coventry, UK, September 10, 1992

Improved adhesion of polymeric coatings. Seminar at PPG Fiber Glass Research Center, Pittsburgh, PA, November 3, 1992

Mechanisms of adhesion.* Opening Lecture in the Symposium on Chemistry of Interfaces for Improved Wettability and Adhesion — Fundamental and Technical Aspects held in Gothenburg, Sweden, November 16–17, 1992

Mechanisms of adhesion. Seminar in the Department of Polymer Technology, Chalmers University of Technology, Gothenburg, Sweden, November 17, 1992

Adhesion science: some recent developments and unsolved problems.* Plenary Lecture at the 1992 Industrial Symposium on Surface Chemistry held in Gothenburg, Sweden, November 18–19, 1992

Adhesion science and technology: recent developments, prospects and research needs. Seminar at Clarkson University, Potsdam, NY, March 2, 1993

Adhesion science: recent developments and prospects.* Symposium on Fundamentals of Adhesion and Interfaces held at the American Chemical Society Meeting, Chicago, IL, August 22–27, 1993

Polymer surface modification and adhesion: an overview.* Workshop Presentation at Surfaces in Biomaterials '93, sponsored by the Surfaces in Biomaterials Foundation, Boston, MA, September 12–15, 1993

Adhesion measurement of films and coatings: some questions, concerns and suggestions.* Second ASME International Electronics Packaging Conference held in Binghamton, NY, September 29–October 2, 1993

Adhesion measurement of films and coatings: a critical assessment.* Invited Lecture at ECASIA 93, 5th Conference on Applications of Surface and Interface Analysis held in Catania, Italy, October 4–8, 1993

Why things stick together? Seminar in the Department of Physics, Rochester Institute of Technology, Rochester, NY, October 12, 1993

Adhesion measurement of films and coatings: a critical assessment.* Keynote Lecture at the Arizona Chapter AVS Symposium — Adhesion in Microelectronics, Mesa, AZ, May 10, 1994

Adhesion testing.* Federation of Societies for Coatings Technology Spring Week Seminar on Adhesion in Coatings: Technology and Characterization, Minneapolis, MN, May 12–13, 1994

Metallized plastics: fundamental and applied aspects.* 5th Society of Polymer Science, Japan International Polymer Conference (IPC 94), Osaka, Japan, November 28–December 2, 1994

Adhesion science: recent developments, unsolved problems and prospects. Seminar at Kansai University, Osaka, Japan, November 30, 1994

Adhesion science: recent developments, unsolved problems and prospects. Seminar at the City College of the City University of New York, New York, NY, December 8, 1994

Adhesion testing of coatings.* 20th Annual FOCUS Conference of the Detroit Society for Coatings Technology, Detroit, MI, April 4, 1995

Stability of emulsions. Seminar at Stepan Company, Northfield, IL, May 15, 1995

Importance of surface/interface chemistry in manufacturing processes.* Thomas D. Callinan Award Address at the Electrochemical Society Meeting, Reno, NV, May 21–26, 1995

---

*Invited conference paper.

Interfacial chemistry and adhesion: recent developments and future needs. Seminar at the University of Wollongong, Wollongong, Australia, June 5, 1995

Interfacial chemistry and adhesion: recent developments and prospects. Seminar at RAFAEL, Haifa, Israel, June 29, 1995

Interfacial chemistry and adhesion: recent developments and prospects. Seminar at Technion, Haifa, Israel, June 29, 1995

Interfacial chemistry and adhesion: recent developments and prospects. Seminar at PPG, Allison Park, PA, August 7, 1995

Surface cleaning: some recent developments and needs.* Symposium on New Cleaning Technologies since Chlorofluorocarbons, American Chemical Society Meeting, Chicago, IL, August 20–25, 1995

Ways to improve adhesion of organic coatings to different substrates. Seminar at the Paints and Lacquer Association in Stockholm, Sweden, September 5, 1995

Interfacial chemistry and adhesion: recent developments and prospects. Seminar at Maria Curie-Sklodowska University, Lublin, Poland, September 11, 1995

Interfacial chemistry and adhesion: recent developments, unsolved problems and prospects.* Opening Lecture at the Seminar of the Technology Development Centre (TEKES), Espoo, Finland, February 15, 1996

Interfacial chemistry and adhesion: recent developments and prospects. Seminar at Abo Akademi University, Abo, Finland, February 16, 1996

Surface cleaning: recent developments and needs.* Workshop on Ion-Assisted Technologies and Surface Engineering held in conjunction with the International Conference on Metallurgical Coatings and Thin Films, San Diego, CA, April 21, 1996

Particle adhesion and removal. Seminar at the University of Florida, Gainesville, FL, May 30, 1996

Adhesion science and technology: recent developments, unsolved problems and prospects. OCMR Distinguished Lecture at the Institute for Aerospace Studies, University of Toronto, Toronto, Canada, September 23, 1996

Clean surface technology. Presentation at the New Jersey Chapter of ASM International, Mountainside, NJ, April 29, 1997

All what you wanted to know about adhesion but were afraid to ask. After-Dinner Lecture at the 5th Coatings for Aerospace Industry Meeting, Toronto, Canada, October 23, 1997

APPENDIX VII

Editorial and review activities

Editor

- *Journal of Adhesion Science and Technology* (1987–)
- *Journal of Surface Science and Technology* (1988)

Member, Editorial Boards of

- *Journal of Coatings Technology* (1979)

---

*Invited conference paper.

- *Solid State Technology* (1979–1995)
- *Advances in Colloid and Interface Science* (1979–1990)
- *Progress in Organic Coatings* (1980–1985)
- *Adhesives Age* (1980–1981)
- *Journal of Polymer Materials* (1984–)
- *Southern Brazilian Journal of Chemistry* (1991–)
- *Particulate Science and Technology: An International Journal* (1993–)
- *Precision Cleaning* (1994–1995)

APPENDIX VIII

Book reviews (since 1978)

Review of L. M. Prince (Ed.), *Microemulsions: Theory and Practice*, *J. Am. Chem. Soc.*, p. 1385, February 28, 1979.

Review of E. Wolfram (Ed.), *Proceedings of the International Conference on Colloid and Surface Science, Volume I*, *J. Am. Chem. Soc.*, p. 2256, April 11, 1979.

Review of E. Wolfram and T. Szekrenyesy (Eds), *Proceedings of the International Conference on Colloid and Surface Science, Volume 2*, *J. Am. Chem. Soc.*, p. 2256, April 11, 1979.

Review of J. F. Johnson and R. S. Porter (Eds), *Liquid Crystals and Ordered Fluids*, *J. Met.* **3**, 146–147 (1979).

Review of D. R. Paul and F. W. Harris (Eds), *Controlled Release Polymeric Formulations*, *J. Coatings Technol.*, p. 106, October 1979.

Review of B. V. Derjaguin, N. A. Krotova and V. P. Smilga, *Adhesion of Solids*, *J. Electrochem. Soc.*, pp. 492c–493c, December 1979.

Review of *Adhesives for Industry*, *Polymer News*.

Review of L. Engel, H. Klingle, G. W. Ehrenstein and H. Schaper, *An Atlas of Polymer Damage*, *Polymer News*.

Review of D. A. King and D. P. Woodruff (Eds), *Clean Solid Surfaces*, *Ceramic Abstracts*.

Review of R. M. Latanision and J. R. Pickens (Eds), *Atomistics of Fracture*, *Colloids Surfaces* **9**, 293–294 (1984).

Review of D. M. Brewis and D. Briggs (Eds), *Industrial Adhesion Problems*, *J. Adhesion Sci. Technol.* **1**, 99 (1987).

Review of A. Pizzi, *Wood Adhesives — Chemistry and Technology*, *J. Adhesion Sci. Technol.* **1**, 185 (1987).

Review of M. M. Sadek (Ed.), *Industrial Applications of Adhesive Bonding*, *J. Adhesion Sci. Technol.* **2**, 237–238 (1988).

Review of G. Verchery and A. H. Cardon (Eds), *Mechanical Behavior of Adhesive Joints*, *J. Adhesion Sci. Technol.* **2**, 238–239 (1988).

Review of S. Mindess and S. P. Shah (Eds), *Bonding in Cementitious Composites*, *J. Adhesion Sci. Technol.* **4**, 853 (1990).

Review of F. D. Lemkey, S. G. Fishman, A. G. Evans and J. R. Strife (Eds), *High Temperature/High Performance Composites*, *J. Adhesion Sci. Technol.* **4**, 853–854 (1990).

Review of D. M. Mattox, J. E. E. Baglin, R. J. Gottschall and C. D. Batich (Eds), *Adhesion in Solids*, *J. Adhesion Sci. Technol.* **4**, 854 (1990).

Review of R. L. Patrick (Ed.), *Treatise on Adhesion and Adhesives, Vol. 6*, *J. Adhesion Sci. Technol.* **4**, 855 (1990).

Review of B. D. Ratner (Ed.), *Surface Characterization of Biomaterials*, *J. Adhesion Sci. Technol.* **4**, 854–855 (1990).

Review of R. Yosomiya, K. Morimoto, A. Nakajima, Y. Ikada and T. Suzuki, *Adhesion and Bonding in Composites*, *J. Adhesion Sci. Technol.* **4**, 856 (1990).

Journal papers and other reviews (since 1978)

J. Colloid Interface Science	59
Research Proposals	19
J. Appl. Polymer Science	12
J. Coatings Technology	7
Symposium Proceedings Volumes	12
J. Vacuum Sci. Technol.	6
J. Electrochem. Society	4
Ph.D. Thesis	2
J. Environmental Sciences	1
Surface and Interface Analysis	2
Thin Solid Films	1
Adhesives Age	1
J. Dispersion Sci. Technol.	1
Colloids and Surfaces	2
J. Physical Chemistry	2
Powder Technology	1
Polymer Eng. Sci.	2
Colloid and Polymer Science	1
J. Materials Research	2
Particulate Science and Technology	1
Book Chapters	5

APPENDIX IX

Societies and other organizations

Fellow:	Adhesion Society American Institute of Chemists Indian Chemical Society
Member:	American Chemical Society American Society for Testing and Materials (ASTM) American Vacuum Society Electrochemical Society Fine Particle Society Indian Science Congress Association

Indian Society for Surface Science and Technology
Indian Vacuum Society
Institute of Environmental Sciences (Senior Member)
International Association of Colloid and Interface Scientists
International Microelectronics and Packaging Society
IUPAC Affiliate
Materials Research Society
New York Society for Coatings Technology (FSCT)
Phi Lambda Upsilon
Sigma Xi
Society for the Advancement of Electrochemical Science and Technology (India)
Society of Plastics Engineers
Society of Vacuum Coaters

APPENDIX X

Awards, scholarships, and other honors

Panjab University Merit Scholarship, 1962
Panjab State Merit Scholarship, 1962
Government of India's National Merit Scholarship, 1962–1965
Government of India's Atomic Energy Scholarship, 1965–1966
Anderson Clayton & Co.'s Fellowship in Colloid Chemistry at the University of Southern California, CA, 1967–1969
Certificate of Appreciation from the Institute of Environmental Sciences, 1979
Recipient of First Invention Achievement Award from IBM Corp., 1982
Outstanding Contribution Award from the Mid-Hudson Section of the Society of Plastics Engineers, 1984
Certificate of Appreciation from the IBM Corporate Contamination Control Committee, 1985
Listed in *American Men and Women of Science*, 14th edition, 1979
Listed in *Who's Who in the East*, 18th edition, 1981–1982
Listed in *Community Leaders of America*, 12th edition, 1982
Listed in *International Who's Who of Intellectuals*, Volume V, 1982
Listed in *Personalities of the East*, 1st edition, 1982–1983
Listed in *Men of Achievement*, 9th edition, 1983
Listed in *Men and Women of Distinction*, 1983
Listed in *International Book of Honor*, 1st world edition, 1983
Listed in *International Who's Who of Contemporary Achievement*, 1984
Listed in *Who's Who in Frontier Science and Technology*, 1984
Listed in *Who's Who in Technology Today*, 4th edition, 1984–1985
Invited as a Project Specialist in the area of materials science by the International Advisory Panel and the Chinese Review Commission (under the auspices of the World Bank) to visit Shanghai Jiao Tong University, Shanghai, PRC, 1985
Recipient of Recognition Plaque from the International Surface and Colloid Science Community (prominent scientists from 51 countries) for Continued Leadership and Distinguished Professional Service presented at the 6th International Symposium on Surfactants in Solution held in New Delhi, India, 1986
VHP Award of Excellence for Outstanding Professional Achievements, presented by the Vishwa Hindu Parishad of America at the Third Southeastern Hindu Conference held at the University of North Carolina at Charlotte, NC, 1987
Interviewed in the May 1988 issue of *Microcontamination* (a technical monthly publication)

Recognition Plaque from the IBM Corporate Contamination Control Committee for Continued Leadership and Significant Technical Professional Contribution to the Field of Contamination Control Technology, 1989

1990 Recipient of the Charles B. Dudley Award of the American Society for Testing and Materials

Interviewed on the topic of adhesion by the *New York Times*. The interview was published on May 31, 1992

Interviewed in the April 1993 issue of *Clean Rooms* (a technical monthly publication)

Robert L. Patrick Fellow of the Adhesion Society, 1994

Recipient of the 1995 Thomas D. Callinan Award of the Division of Dielectric Science and Technology of the Electrochemical Society

Interviewed in the September 1995 issue of *Adhesives Age* (a technical monthly publication)

Recognized by the dedication of a special issue of *Particulate Science and Technology — An International Journal* (Vol. 13, Numbers 3–4, July–December 1995) on his 50th birthday

Recognized by the world-wide adhesion community by organizing of the 1st International Congress on Adhesion Science and Technology in his honor on his 50th birthday in Amsterdam, The Netherlands, October 16–20, 1995

Recipient of the *Adhesives Age* Award given by the *Adhesives Age* Magazine, 1997

Recipient of the Adhesives Award of Committee D-14 of ASTM, 1997

Recipient of the 1997 John A. Wagnon Technical Achievement Award of the International Microelectronics and Packaging Society

Interviewed in the June 1997 issue of *Adhesives Age* (a technical monthly publication)

Mittal Festschrift, pp. lxv–lxvii
W. J. Van Ooij and H. R. Anderson, Jr. (Eds)

First International Congress on Adhesion Science and Technology

Amsterdam, 16-20 October 1995

ADHESION AND THE DUTCH

Ladies and Gentlemen,

As the representative of the Royal Dutch Chemical Society, I would like to express my appreciation to your organizing committee for choosing Amsterdam as the venue for the First International Congress on Adhesion Science and Technology in honor of Dr. Kash Mittal on his 50th Birthday. For me it was an extra and pleasant surprise to hear from the organizers that nearly forty different countries were represented, making this an event that genuinely stresses the worldwide importance of adhesion.

Adhesion is a phenomenon that dominates our lives both in the metaphorical and literal senses of the word. The 'Amsterdammers', the people from Amsterdam, who pass by this hotel and see the sign 'Adhesion Technology' will have some notion of what is happening. They all more or less understand the word Technology, while Adhesion sounds pretty to them. If we should tell them that adhesion is another word for the force between bodies of different nature, most of them will — after some reflection — come to think of love. And, indeed, adhesion plays an important role in love: for the closer the love is expressed, the more the Dutch forces will master the adhesion phenomenon. For those in this audience who have never heard of Dutch forces, I can reveal that one of your plenary speakers, Professor de Gennes, calls van der Waals forces: The Dutch forces.

When you open this Amsterdammer's eyes and tell him that in technical jargon 'adhesion' is mostly used in connection with glue, he will shrug his shoulders, remembering some disturbing failures with 10-second glues that stuck his fingers together. For many simple mortals, glue has lots of magic and only with luck will it do what has been promised on the tube or can.

I just referred to strong bonds because this is what the man in the street expects from glue. Often, however, weak bonds are desired more than strong bonds. In order to explain this we go back to The Netherlands and to its inhabitants: the Dutch.

The western part of this country was built by an ingenious use of dikes. Not so much by constructing sea dikes, because we had dunes, but by building low dikes to reclaim land from swamps and marshy regions. Windmills pumped the water over the dikes into canals and the dikes protected the polderland against the inflow of water. It all started some 1000 years ago when dike bodies were constructed from clay reinforced with twigs and sprigs. Nowadays, we call these composite structures. It was not easy for Dutch dike builders in the year 1000 AD to find the right reinforcing materials and the right proportion between twigs and clay. After years of experimenting, they succeeded and learned to make solid, tough and malleable dikes based on the typical combination of interlocking mechanical forces, hydrogen bonds, electrical forces, van der Waals forces and friction forces. Also in those days, dike building was not a recipe but an art and Dutch dike builders from West Holland were invited to Schleswig Holstein to teach the locals the tricks. Nobody realized then that they were early adhesion technologists, but in the end it was adhesion that shaped the present Dutch landscape.

All this explains how remote the science of adhesion is to many people who are every day confronted with the increasing number of applications of adhesion.

Do we recognize that adhesion should not be seen as a loose agglomerate of a couple of different operations, each having its roots in another specialty? I wonder whether adhesion should not be considered as a multidisciplinary subject that should be part of a university curriculum in, for instance, polymer teaching? I mention polymers on purpose, as polymers play an ever increasing role in manufacturing steps involving adhesion applications. We see polymers as a matrix in fibre-reinforced composites, as films in laminates, and as interlayers in three-dimensional mechanical constructions. In the university environment of polymer chemists and physicists, it must be challenging to teach adhesion technology as one discipline combining different kinds of expertise. Such a multidisciplinary approach is already practiced in industrial labs. Why should we not adapt the same multidisciplinary approach for teaching Adhesion Technology, making it part of a program of study?

Adhesion Technology as a coherent science will be needed in the further development of such diverse fields as composites, electronics, coatings and diagnostics.

Present-day composites consist of glass fibres and some thermoset matrix. For many reasons we now see attempts to use natural fibres, such as jute, hemp and flax as reinforcing materials. Thermosets are facing competition from thermoplastics — for instance, saturated polyesters.

I have noticed that only about 10 papers to be presented at this congress deal with interfaces in composites and I ask myself whether this is proportionate to the interest from the chemical and plastics industry. The use of natural fibres in combination with certain polymers imparts unique properties and could form an intriguing challenge in the development of replacements for construction wood.

In electronics we observe a continuous drive for further miniaturization. The interface between electrical and optical components on an electronic device will have to undergo a further reduction in thickness. Another easy-to-understand feature is that the surface-to-volume ratio will increase, as electronic devices become progressively smaller. Relatively more surface means relatively more attention to adhesion technologies.

Looking at coatings, we see them being increasingly applied to plastic surfaces. Especially in the very demanding automotive industry, adhesion between the coating and polypropylene car body parts is of very great interest. There is a strong urge to understand the basics of the chemistry and physics that govern the bond between the coating and the layer underneath.

In pharmacy and diagnostics the interactions between antigen and antibody have to be selective and fast. This has consequences for labeling, monolayer coverage and functional reactive sites. Every specific demand opens up other aspects of adhesion forces and calls for a need to understand what is going on.

Ladies and gentlemen: these are just a few of the directions in which industry will move or already has moved. Industry expects fundamental knowledge from the scientific world so that their technologies can be steered in the desired direction.

The Royal Dutch Chemical Society represents 60% of all the professional chemists and chemical engineers in The Netherlands. Its members are equally divided between industry and universities. This continuously stimulates our awareness that bridges should be built and reinforced between science on the one side and product development and manufacturing on the other side. I fervently hope that this congress will make a substantial contribution to bringing the need and abilities together in the fascinating and promising land of 'Adhesion Technology'.

Thank you very much.

Jan J. M. Mulderink
Valkenierspad 6
6891 CH Rozendaal
The Netherlands

Part 1

Fundamental aspects of adhesion and general topics

Mittal Festschrift, pp. 3–20
W. J. Van Ooij and H. R. Anderson, Jr. (Eds)

The molecular surface science of adhesion

G. A. SOMORJAI*

Department of Chemistry and Materials Sciences Division, Lawrence Berkeley National Laboratory, University of California at Berkeley, Berkeley, CA 94720-1460, USA

Abstract—The discovery of new techniques has permitted molecular level investigations of chemisorption bonds, which lead to strong adhesion and physisorption, and of van der Waals bonds, that lead to weak adhesion and control motion at the interface (molecular tribology). The atomic force and scanning tunneling microscopes (AFM and STM), the surface force apparatus (SFA), and vibrational spectroscopic studies using sum frequency generation (SFG) by nonlinear laser optics have permitted *in situ* investigations of buried interfaces: solid–liquid and solid–solid. Strong adhesion leads to adsorbate-induced restructuring of the substrate, where the exothermic heat of chemisorption produces weakening of metal–metal bonds (an endothermic process) that is needed to induce the relocation of the substrate atoms. The work of adhesion and the coefficient of friction were measured by AFM in the elastic deformation regime (nanonewton forces) for hard carbon films with variable hydrogen content. Both the work of adhesion and the friction coefficient increase with increasing hydrogen content of the carbon film that is smooth on the atomic scale. Friction can be markedly decreased by a thin perfluoroether film. SFG studies permit detection of the restructuring of adsorbed molecules at the buried interfaces, under applied load.

Keywords: Surface science of adhesion; atomic force microscopy; strong adhesion; weak adhesion; adhesion and catalysis.

1. INTRODUCTION

Adhesion is one of the most important surface properties, but also one of the most difficult to understand. Its magnitude is measured by mechanical tests, expressed well by its thermodynamic definition [1]

$$W_{ad} = \gamma_{13} + \gamma_{23} - \gamma_{12}, \tag{1}$$

where W_{ad} is the reversible work of adhesion or energy change on separating two media; 1 (the AFM tip, for example) and 2 (a solid carbon surface, for example) in contact in the presence of a medium 3. Medium 3 represents air or dry nitrogen in our experiments. γ_{13} and γ_{23} are the energies gained when interfaces 1–3 and 2–3 are formed while γ_{12} is the energy required to break interface 1–2. The mechanical

*E-mail: somorjai@cchem.berkeley.edu

work needed to separate two bodies in adhesional contact is determined by force measurements; yet the nature of the chemical bonds at the interfaces is responsible for W_{ad}, and these bonds must be broken and new bonds formed as the interface is replaced. Thus, scientists engaged in research in this field must be equally able to understand the mechanical as well as chemical behavior of surfaces.

Adhesion may be divided into two classes: (1) Strong Adhesion and (2) Weak Adhesion. Strong chemical bonds, similar to those formed during chemisorption at the solid–gas interface or during epitaxial growth of metal films on the surfaces of other metals are responsible for Strong Adhesion. In this circumstance, the bonds that form at the interface are as strong as or stronger than bonds in the bulk of the two adjoining materials. For weak adhesion, van der Waals bonds join the two phases in contact. These often form on top of the strongly adherent surface layer. In this circumstance, bonding is temporary and the two surfaces can move relative to each other. These weakly adhesive interfaces are responsible for tribology, friction, and lubrication, and for the transport of molecules along interfaces — often leading to phase separation or selective diffusion.

In the past, most measurements of the work of adhesion used such large forces as to destroy the interface. For example, the scratch adhesion test [2] uses a diamond stylus and the applied load is increased to a critical value at which the coating breaks. The appearance of interfacial failure leads to determination of the work of adhesion.

One important conclusion of these studies was that the contact area between two solids held together by adhesion forces is often much smaller than the total available surface area. The high pressures at the microasperities in contact can induce the plastic deformation that leads to wear.

2. TECHNIQUES FOR MOLECULAR LEVEL STUDIES OF ADHESION

The forces leading to breaking of chemical bonds are in the range of 10^{-8} newton or micrograms in the earth's gravitational field. The atomic force microscope (AFM) [3] permits us to apply forces as little as 10^{-9} N. Using this technique, one can measure the attractive force as the tip — usually a few hundred angstroms in radius — approaches the surface of a solid. It can measure the range of forces that lead to elastic deformation and the force that induces the onset of plastic deformation. The force needed to detach the tip from contact with the surface yields the adhesion force [4]. Definitions of the forces that can be measured by the approach and detachment of the AFM are given in Fig. 1. An example of how the adhesion force changes upon the oxidation of a carbon surface [4] is shown in Fig. 2. If the end of the AFM tip is approximated as a sphere, the adhesion force F_{ad} measured by the AFM tip can be interpreted as the interaction force between a sphere and a flat surface, which is given by the Derjaguin approximation [5] as

$$F_{ad} = 2\pi R W_{ad}. \tag{2}$$

Thus, it yields the work of adhesion directly if the radius of curvature of the end of the AFM tip is known.

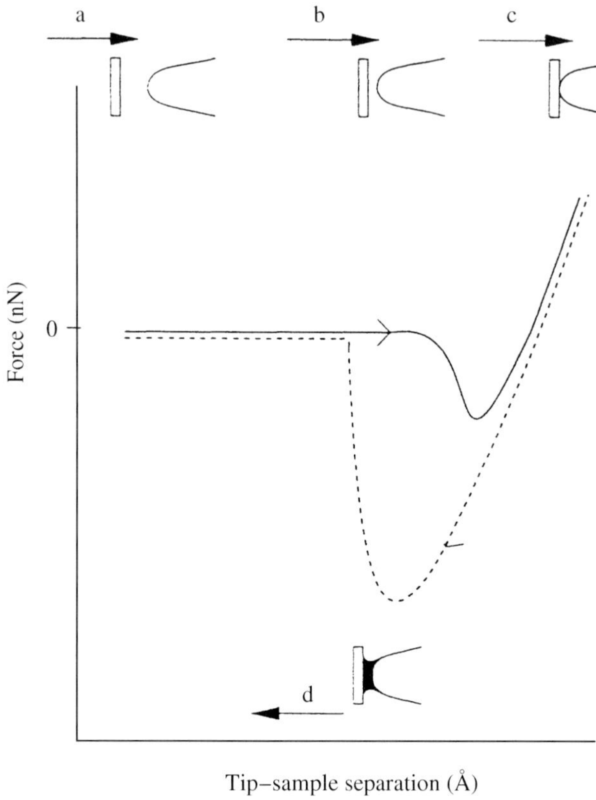

Figure 1. Schematic of the tip–surface interactions during measurement of force–distance curves. The solid line represents approach of the tip and surface as in schematic a, b and c, while the dotted line represents retraction of the tip and surface as in d. A = Adhesion force or negative force to pull the sample off the tip.

It can also be determined by measuring the electrostatic force between the tip (tip end is approximated as a sphere of radius R) and the sample (flat plane). When a voltage V (volts) is applied between a sphere of radius R (Ångstroms) and a flat plane at a distance S (Ångstroms), the Coulomb force can be approximated as [6]

$$F_{\mathrm{el}} = \frac{3RV^2}{S} \times 10^{-11}\ \mathrm{N}. \tag{3}$$

Another method uses a ridged strontium titanate surface, made of (100) and (101) crystal planes, with sharp features [7]. When the tip is rolled over in contact with this oxide surface, the tip profile is obtained.

It should be noted that when the AFM tip approaches the surface in air, capillary condensation occurs between the surface and the tip that drastically alters the forces at the contact. Thus, care should be exercised to reduce humidity or to use dry nitrogen atmosphere in atmospheric pressure studies. By applying a load parallel to the surface, the friction force can be measured (Fig. 3). The friction force divided

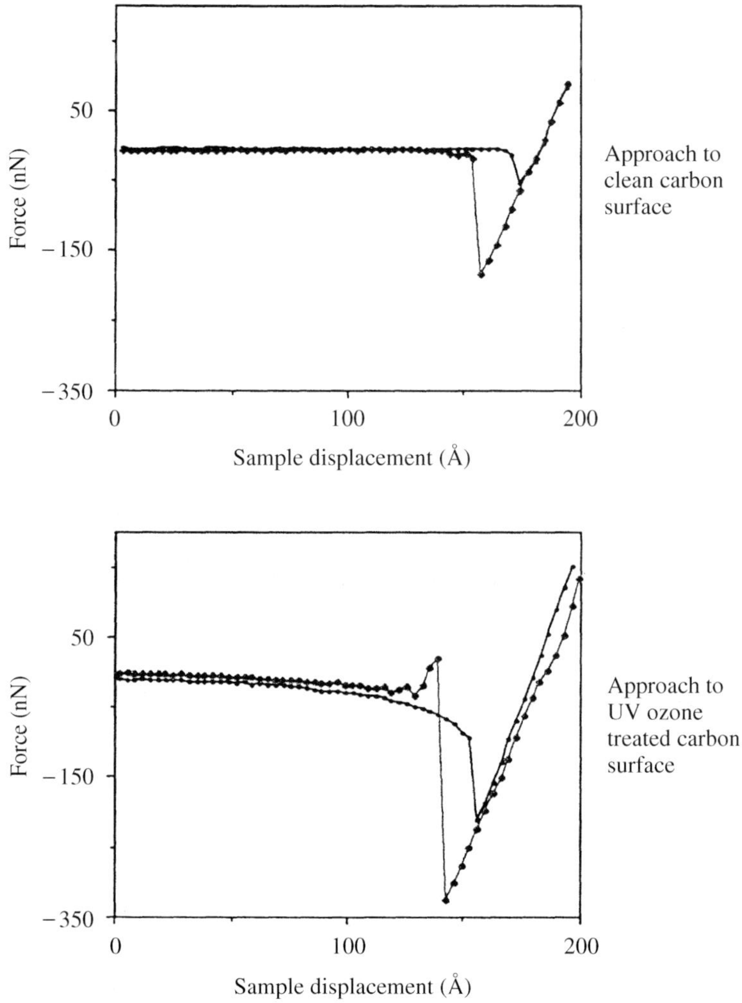

Figure 2. Force–displacement curves measured on a clean and oxidized amorphous carbon surface. For the oxidized surface, greater adhesion forces are observed.

by the load yields the friction coefficient. By moving the tip parallel to the surface, atomic resolution AFM images can be obtained under proper conditions. It is essential that the surface be as smooth as possible for these measurements. For this reason, mica or silicon crystal surfaces are often used [8]. The material to be studied (for example, platinum or gold) is deposited on the mica to obtain an atomically smooth metal surface [8].

The surface force apparatus (SFA) applies load to molecules sandwiched between two mica plates [9]. Unlike the AFM tip, the SFA configuration has a contact area in the 100 μm range, and loads applied are in the millinewton range. Because of the large contact area, spectroscopic studies can be carried out to determine the structure and orientation of molecules buried between the mica plates.

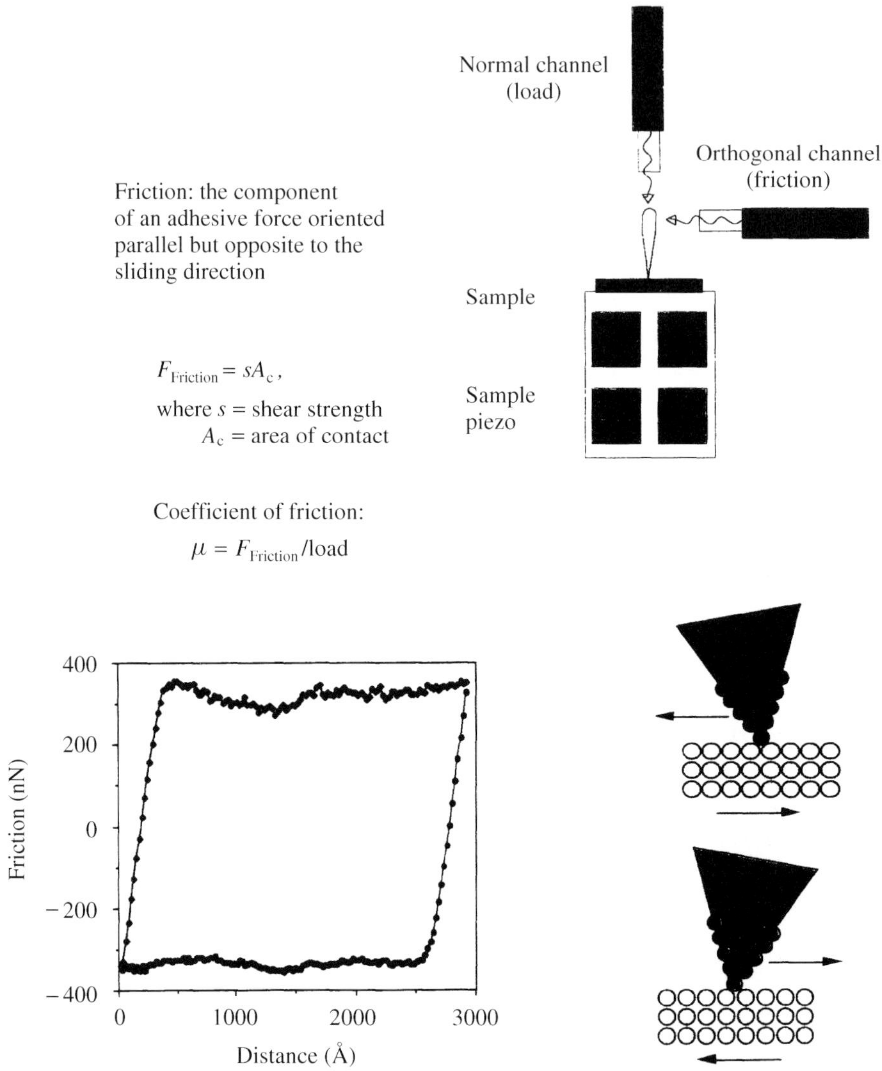

Figure 3. Monitoring friction forces with the atomic force microscope. A friction loop is shown as the tip moves relative to the surface. Initially the tip sticks, but then above a certain force the tip slides on the surface. When the tip is retraced, the same behavior is observed.

Spectroscopic studies can be performed by sum frequency generation (SFG) using nonlinear laser optics [10]. Two laser beams of frequency ω_1 (in the visible range) and w_2 (in the infrared range) are focused on the surface, and the sum frequency, $\omega_1 + w_2$ (in the UV range), is detected. This event, which involves ~ 1 in 10^{15} photons, occurs preferentially at interfaces because one requirement for its occurrence is the lack of inversion symmetry. By tuning the infrared frequency over a wide range, the vibrational spectrum of the molecules adsorbed at the interface is obtained. By changing the angle of incidence, the orientation of the molecules can be inferred. Attributes of SFG include submonolayer sensitivity, molecular specificity, ultra-fast time resolution, and interface specific (i.e. discriminates against the bulk of the sample).

3. STRONG ADHESION

Strong adhesion gives rise to the formation of chemisorption bonds — covalent or ionic — that are as strong as or stronger than the bonds between atoms in the solids we are trying to join by adhesional contact. Two metals in contact, for example, could transfer and intermix atoms at the interface. When such an adhering system is broken, the fracture is not likely to occur at the interface, but in the bulk of the material on one side of the joint or the other. This is responsible for cold-welding of clean metal surfaces. For example, when clean copper metal slides on clean steel in air, fragments of copper are found to be strongly welded to the steel surface. This phenomenon is not easily detectable in adhesional contact because metal surfaces are usually covered by thin oxide films or contamination layers.

The dominant molecular surface phenomenon that accompanies chemisorption is adsorbate-induced restructuring [11]. This can be best demonstrated by the STM studies displayed in Fig. 4. When the Pt (110) crystal face is exposed to atmospheric pressures of hydrogen, it forms a surface that has the missing-row surface structure. Thus, every second row of platinum atoms is displaced to produce a corrugated surface. When hydrogen is removed and replaced by oxygen at 1 atm, the Pt surface structure changes [12]. It develops microfacets of (111) orientation, as shown in Fig. 4. When oxygen is removed and substituted by carbon monoxide, the surface structure changes again; it becomes a smooth (110) crystal face. Thus, depending on which gas atoms or molecules chemisorb, the structure of the surface is altered. The surface assumes the structure that is thermodynamically most stable in the presence of the specific chemisorbed species. The heat of chemisorption (always exothermic) provides as the energy required for the weakening of metal–metal bonds (an endothermic process) that is needed to induce the relocation of the surface metal atoms.

STM images of the step structure of a Pt (111) crystal face in the presence of propylene and hydrogen during the hydrogenation reaction display marked changes that occur as the temperature is increased. When propylene decomposes (upon heating to 600 K) on such a surface, carbonaceous island deposits form [13]. As a result, the steps become kinked and the metal surface structure is markedly altered.

More quantitative information about adsorbate-induced restructuring is obtained from low-energy electron diffraction (LEED) surface crystallography. When carbon is chemisorbed on the (100) crystal face of nickel, it forms an ordered surface structure of 25% of a monolayer coverage. LEED surface crystallography studies revealed that carbon was located in four-fold sites [14]; however, the four nearest-neighbor metal atoms were rotated around the center carbon atom, and the nickel–nickel bond distance was somewhat enlarged (Fig. 5). This allows the chemisorbed carbon atom to sink deeper into the metal surface and achieve bonding to the nickel atom in the second layer directly under the carbon atom. Thus, the increased nickel–nickel bond

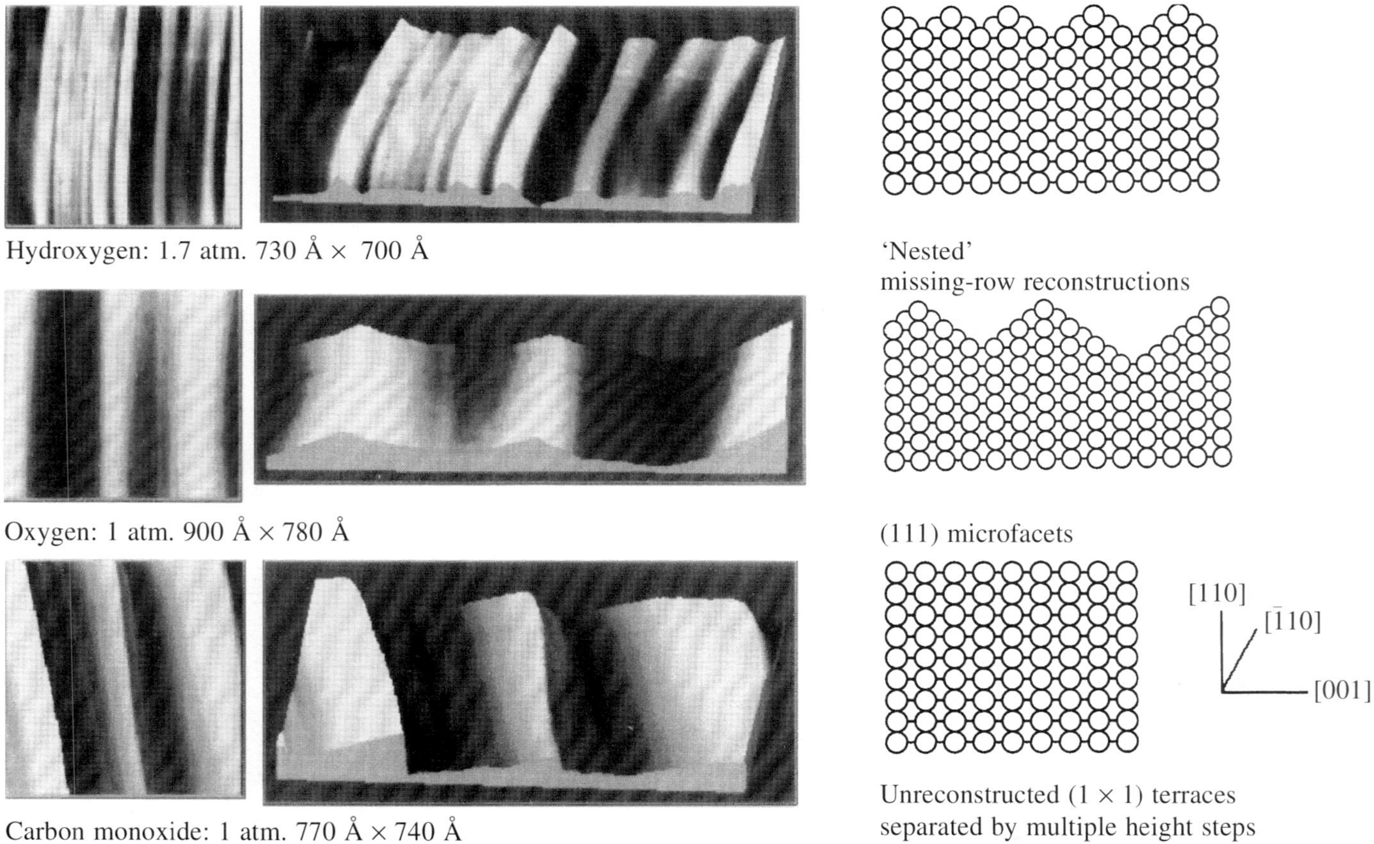

Figure 4. *In situ* high pressure scanning tunneling microscopy study of adsorbate-induced surface reconstructions on the Pt (110) crystal surface under atmospheric pressures of hydrogen, oxygen, and carbon monoxide.

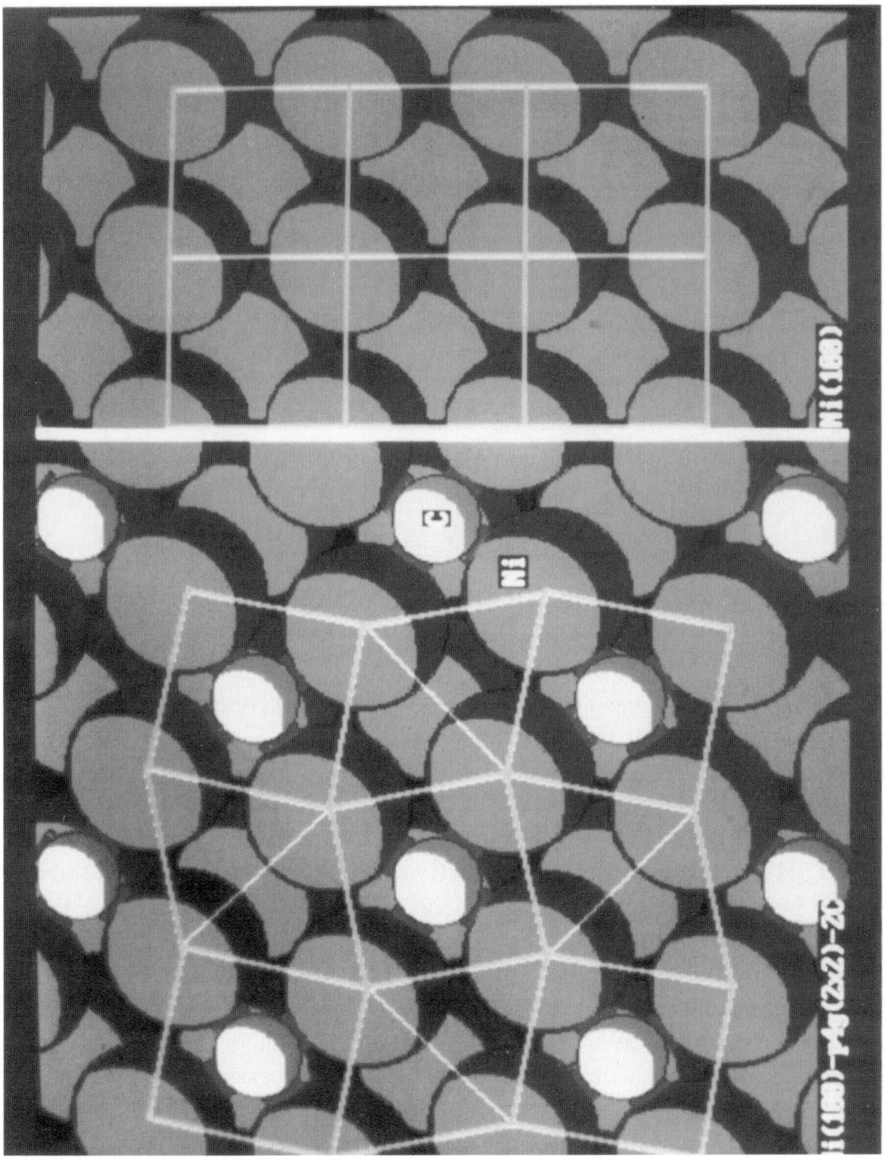

Figure 5. Carbon chemisorption-induced surface restructuring of the nickel (100) single-crystal surface.

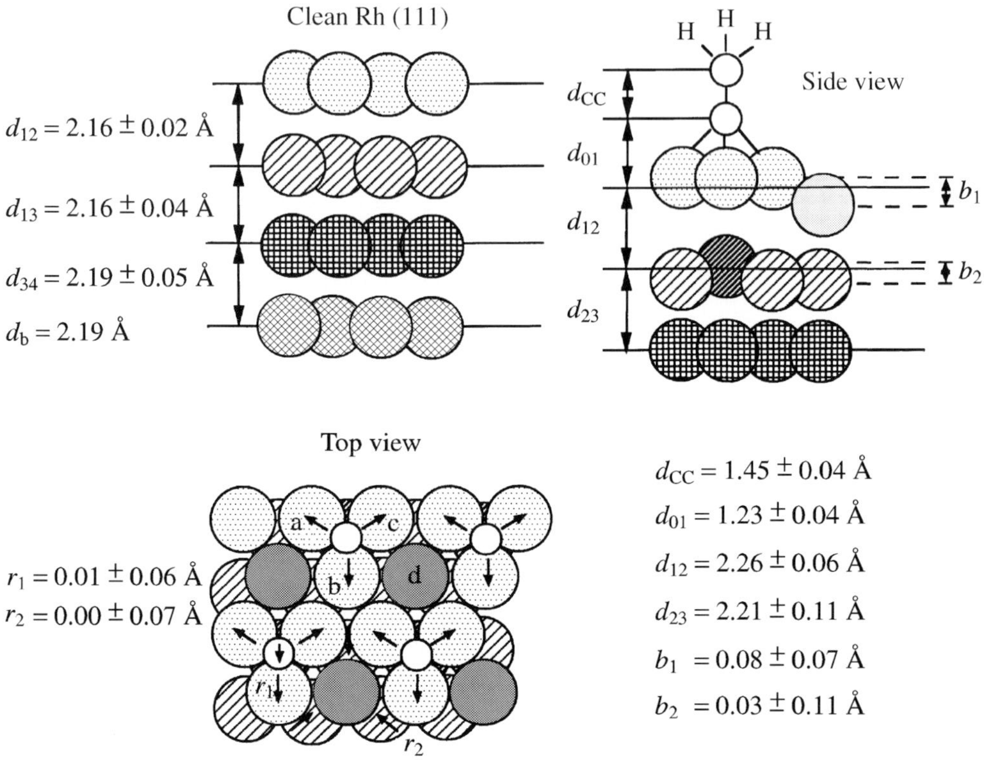

Figure 6. Ethylidyne chemisorption-induced restructuring of the rhodium (111) crystal face.

strength obtained this way provides the energy for the weakening of the metal–metal bonds that occurs because of bond expansion and rotation.

Another example of adsorbate-induced restructuring is ethylene chemisorption on the rhodium (111) crystal face at 300 K (shown in Fig. 6). Ethylene, C_2H_4, loses a hydrogen, and another hydrogen is transferred to the second carbon atom to form ethylidyne, C_2H_3. The carbon atom that is left without hydrogen bonds to a three-fold rhodium site [15]. The nearest-neighbor metal atoms move away from the bonding site, permitting the bonding carbon atom to sink deeper into the metal surface and partly bond to the rhodium atom directly underneath in the second layer. As a result, this atom moves up, closer to the carbon. The second nearest-neighbor rhodium atoms move down into the surface as compared to the nearest-neighbor metal atoms, and the surface becomes crumpled. Thus, chemisorption induces massive local restructuring, as shown by Figs 5 and 6.

The coadsorption of two molecules restructures metal surfaces differently from the adsorption of one or the other of the molecules [16]. For example, when benzene and carbon monoxide are coadsorbed on the Rh (111) crystal face (Fig. 7), the metal atoms move outward more under the benzene molecule than under the CO. Thus, the adsorbate–adsorbate interaction involves the local restructuring of the metal substrate as well. When sulfur alone adsorbs on a stepped platinum surface, two atom height

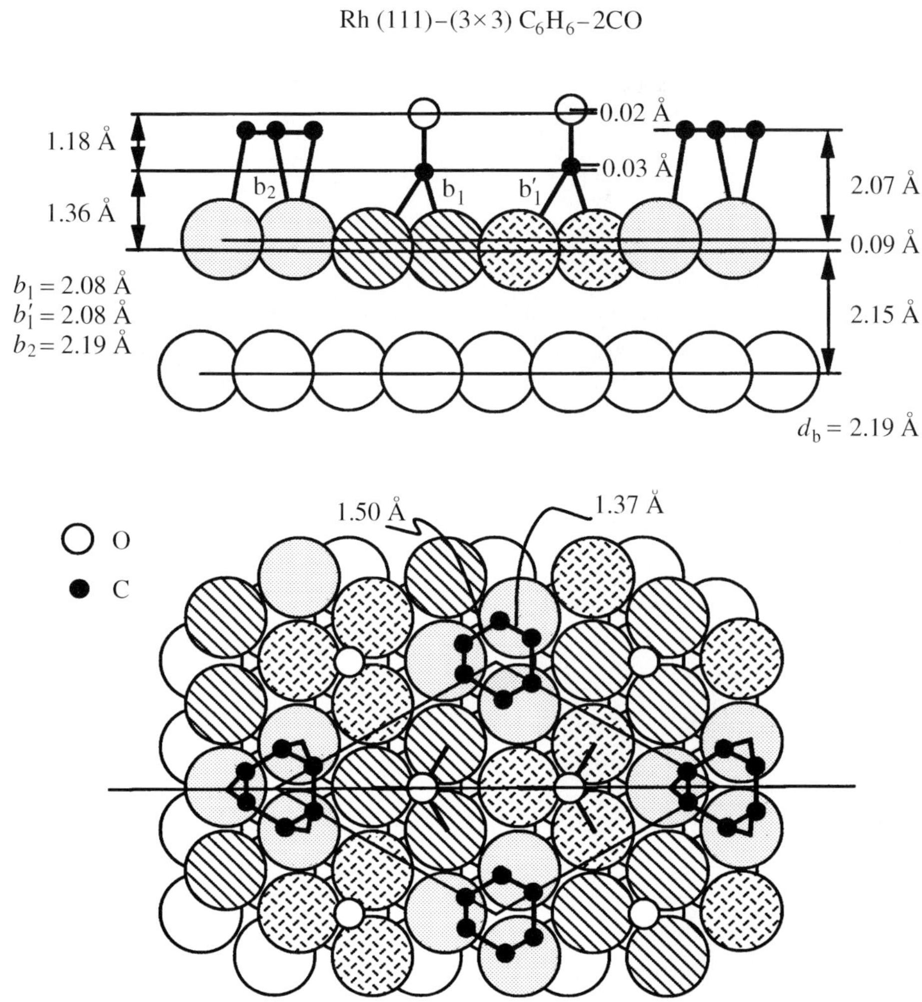

Figure 7. Adsorbate-induced restructuring under the influence of benzene and carbon monoxide coadsorption on the rhodium (111) crystal face.

steps are formed (Fig. 8a) [17]. When CO coadsorbs with sulfur on the same surface, the steps become one platinum atom in height, and sulfur and CO occupy alternating terraces, as shown in Fig. 8b.

The physical picture that emerges from these studies is one of a flexible surface; the surface rearranges its structure to optimize the chemisorption bond. The more open the surface, the lower the coordination of substrate atoms and the more readily chemisorption can induce surface restructuring [18].

There are two dominant surface phenomena that can be explained by adsorbate-induced restructuring: (a) cluster-like bonding of adsorbates on metal surfaces, and (b) the uniquely high activity of rough surfaces and surface defects (steps, kinks) in breaking strong chemical bonds (C—C, C—H, N≡N, etc.).

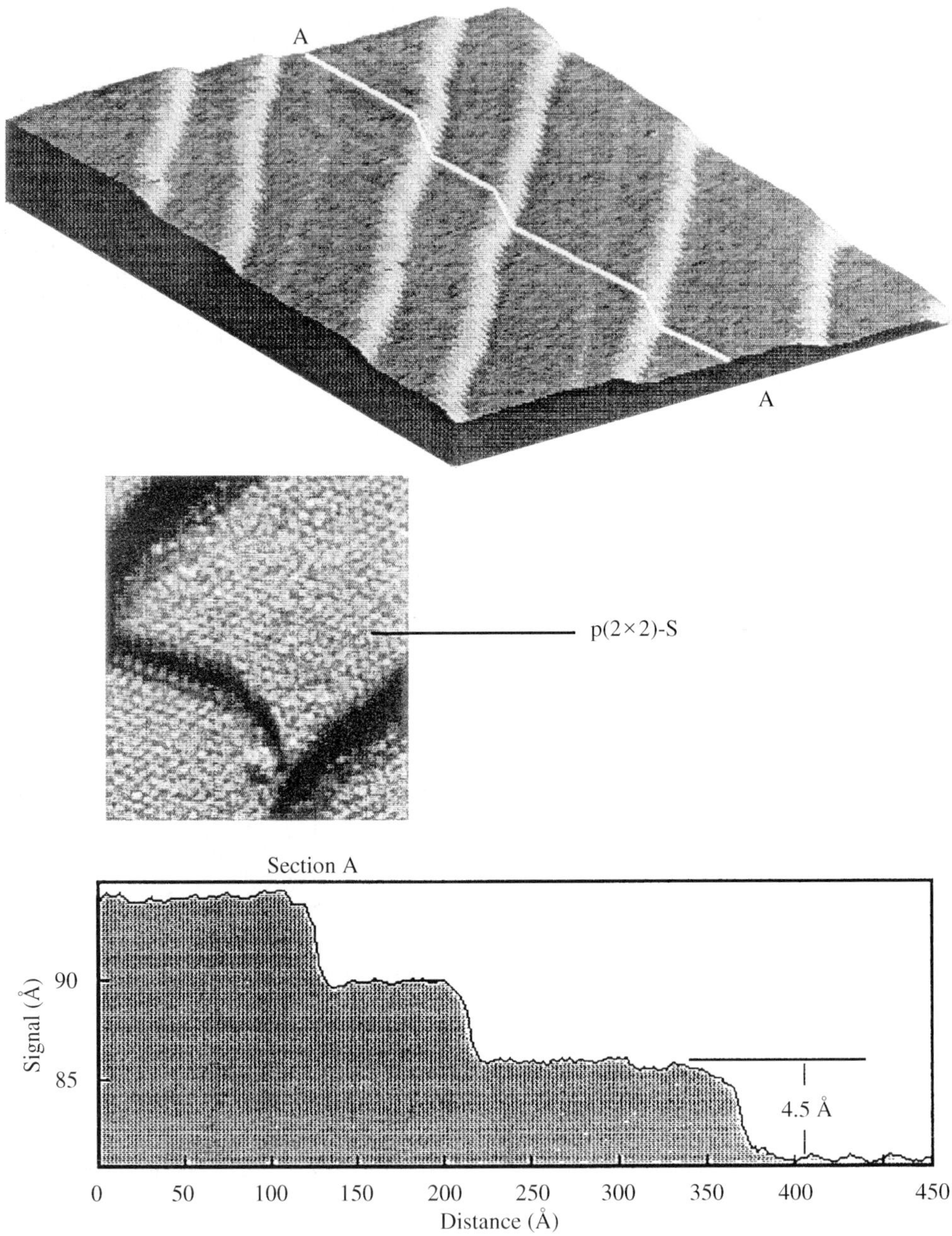

Figure 8a. STM study of the platinum step structure in the presence of a chemisorbed monolayer of sulfur. Note the double-height steps of 4.5 Å.

The vibrational spectra of many molecules adsorbed on metals are nearly identical for the vibrational spectra of organometallic species containing several metal atoms, and the same molecule as the adsorbate. The restructuring of the metal substrate around the adsorption site creates the cluster-like configuration [18].

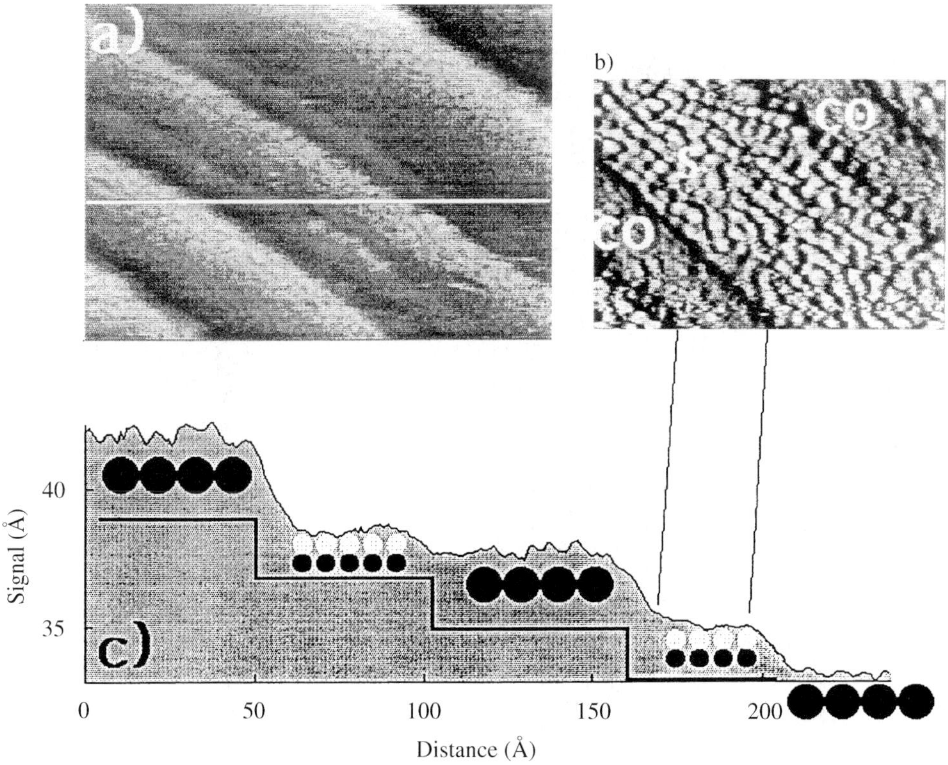

Figure 8b. Restructuring of the stepped platinum surface in the presence of coadsorbed sulfur and carbon monoxide. Note the single-height steps, and that sulfur and carbon monoxide occupy different terraces because of repulsive interactions.

The facile restructuring of metal atoms at low coordination sites breaks molecular orbital symmetries and helps in dissociating the molecules. For example, H_2–D_2 exchange was found to have three orders of magnitude higher reaction probability at stepped platinum surfaces than on the atomically smooth (111) crystal face [18].

4. WEAK ADHESION

In this circumstance, van der Waals or induced dipole forces control the interaction between the partners in the adhering system. These bonds are readily broken and replaced, thus permitting motion at the interface. This is the realm of tribology that gives rise to friction, and lubricants are able to modify it [19].

Using AFM, one can study weak adhesion and friction at the same interface using small enough nN forces to remain in the elastic regime (i.e. without causing damage to the surface). Much of the development of AFM was made necessary by the requirements of the magnetic hard disk used in information storage. The magnetic film is coated with a hard carbon film (smooth on the atomic scale) for mechanical protection, which is then lubricated with a high molecular weight fluoroether. The

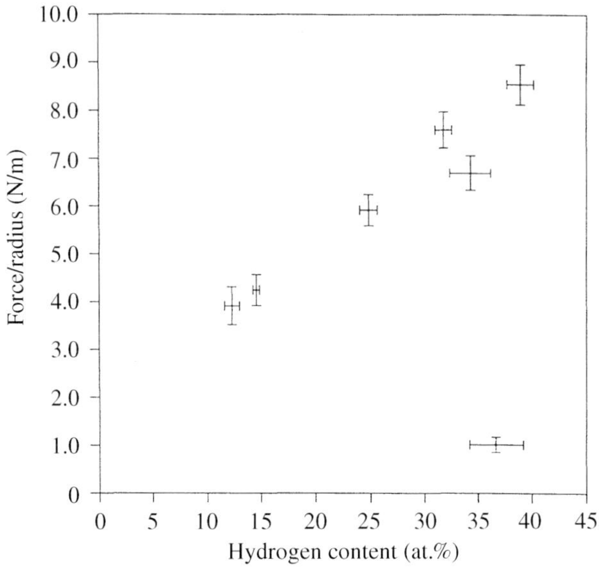

Figure 9. The adhesion force, which is proportional to the force over the radius, is plotted against the hydrogen content of the carbon films.

Table 1.

Atomic percent hydrogen	Adhesion force (nN)	Normalized adhesion force F/R (N/m)	Work of adhesion W_{ad} (mJ/m^2)
12.4 ± 0.7	153 ± 16	3.82 ± 0.46	608 ± 73
14.6 ± 0.3	640 ± 48	4.26 ± 0.33	679 ± 53
24.8 ± 0.8	890 ± 49	5.93 ± 0.34	944 ± 54
31.8 ± 0.8	1520 ± 75	7.60 ± 0.39	1210 ± 62
34.3 ± 1.9	476 ± 25	6.80 ± 0.43	1082 ± 68
36.7 ± 2.5	54.4 ± 8	1.09 ± 0.17	173 ± 27
39.0 ± 1.2	1280 ± 62	8.53 ± 0.47	1358 ± 75

carbon film is deposited in an argon plasma containing hydrogen, altering the hydrogen content of the film.

Figure 9 shows how the work of adhesion of the carbon film (that is, proportional to the adhesion force divided by the tip radius) varies with the hydrogen content of the film [4, 6]. It appears that the work of adhesion increases with hydrogen content. Figure 10 shows that the coefficient of friction also increases with the increasing hydrogen content of the carbon film [4, 6]. This is because an increase in hydrogen content increases the interfacial energy. This is related to the sp^2/sp^3-type carbon bond ratios and is not well understood. Table 1 summarizes the results. The large friction coefficient found for these carbon films has been reported by other carbon film studies as well [20]. Figure 11 shows that the work of adhesion is greatly decreased by a thin perfluoroether film, while a liquid lubricant film causes sudden detachment of the tip at a critical load due to capillary condensation.

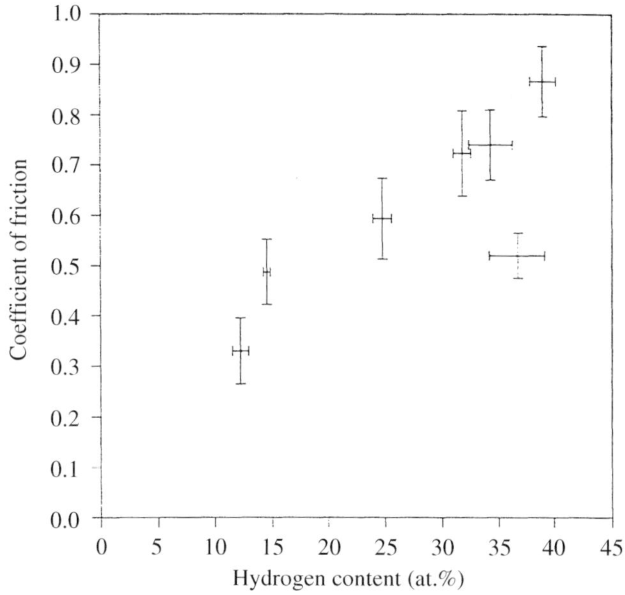

Figure 10. The coefficient of friction of the carbon films is plotted as a function of their hydrogen content.

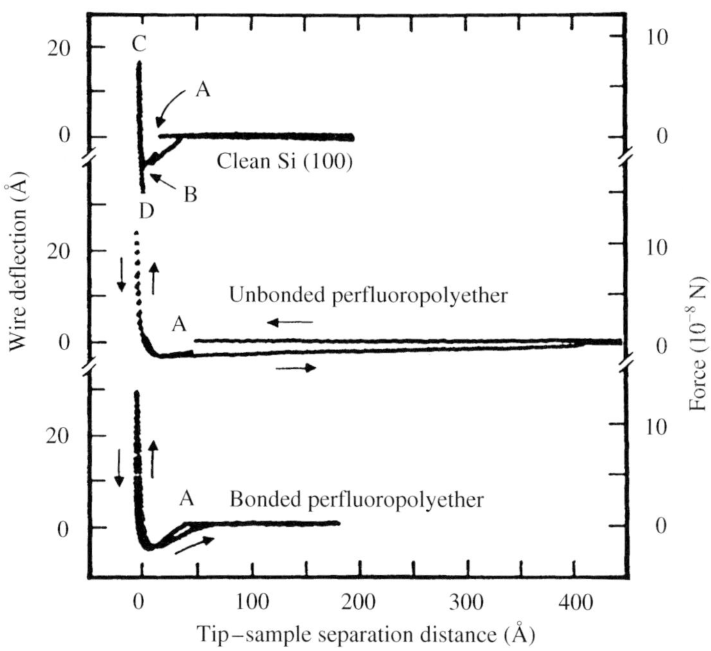

Figure 11. Load forces vs. distance curves comparing the behavior of clean Si (100), a 40 Å thick unbound, liquid perfluoropolyether film on Si (100), and a 15 Å thick bonded perfluoropolyether polymer film on Si (100) [23].

The presence of adhesion that is due to the surface energies of the contacting bodies changes the nature of the contact. There is an increase in contact area due to adhesion. It appears that in the presence of weak forces that cause only elastic deformation at the contact interface, the friction coefficient scales with the work of adhesion for the carbon surfaces studied. This relationship may be altered markedly at higher loads that cause plastic deformation, where the tip may plow into the film producing wear-track and debris. At the low loads applied in our experiments, a lubricant that greatly reduces that interfacial energy of the carbon surface thereby reducing the work of adhesion, also drastically reduces the friction coefficient.

5. FRICTION AND CATALYSIS

When there is strong chemical bond formation between contacting bodies, strong adhesion commences. The chemical equivalent of this mechanical phenomenon is chemisorption, which represents a stoichiometric reaction between the substrate and the adsorbing species. In the case of weak adhesion, the bonds between contacting bodies are readily broken to permit their relative motion (either parallel or perpendicular to the interface). The magnitude of the friction coefficient controls the motion parallel to the interface caused by continually breaking and making bonds between the two surfaces in relative motion. Walking would not be possible without friction, of course. The chemical equivalent of friction is heterogeneous catalysis, which involves adsorbed reactants that desorb as products after the surface chemical reaction to permit new reactant molecules to adsorb and repeat the process. If the chemical bonds between the reacting molecules and the surface are too strong, the process terminates in the chemisorption stage. When the reactant molecule–substrate surface bonds are too weak, there is no reaction. Only when the bonds are of intermediate strength will catalytic turnover occur.

As an example of how catalysis occurs exhibiting the type of bonding interaction responsible for friction, let us consider the hydrogenation of ethylene to ethane over platinum [21]. This reaction was studied using SFG to detect surface reaction intermediates by their vibrational spectra. The surface is covered with stagnant, strongly chemisorbed species (ethylidyne and di-σ ethylene) that turn over very slowly (making their hydrogenation rate unimportant). Weakly adsorbed π-bonded ethylene, however, that adsorbs only at atmospheric pressures and after all of the strongly chemisorbing surface sites are occupied, accounts for most of the ethane produced. The mechanism of this reaction is shown in Fig. 12.

6. STRUCTURE OF MOLECULES AT A BURIED INTERFACE

A versatile and successful technique uses the SFA in combination with SFG. An alkylthiol was adsorbed between the SFA plates, and its vibrational spectrum detected by SFG [22]. The end methyl groups were readily detectable, indicating their alignment perpendicular to the surface. As the load increased, the vibrational spectrum

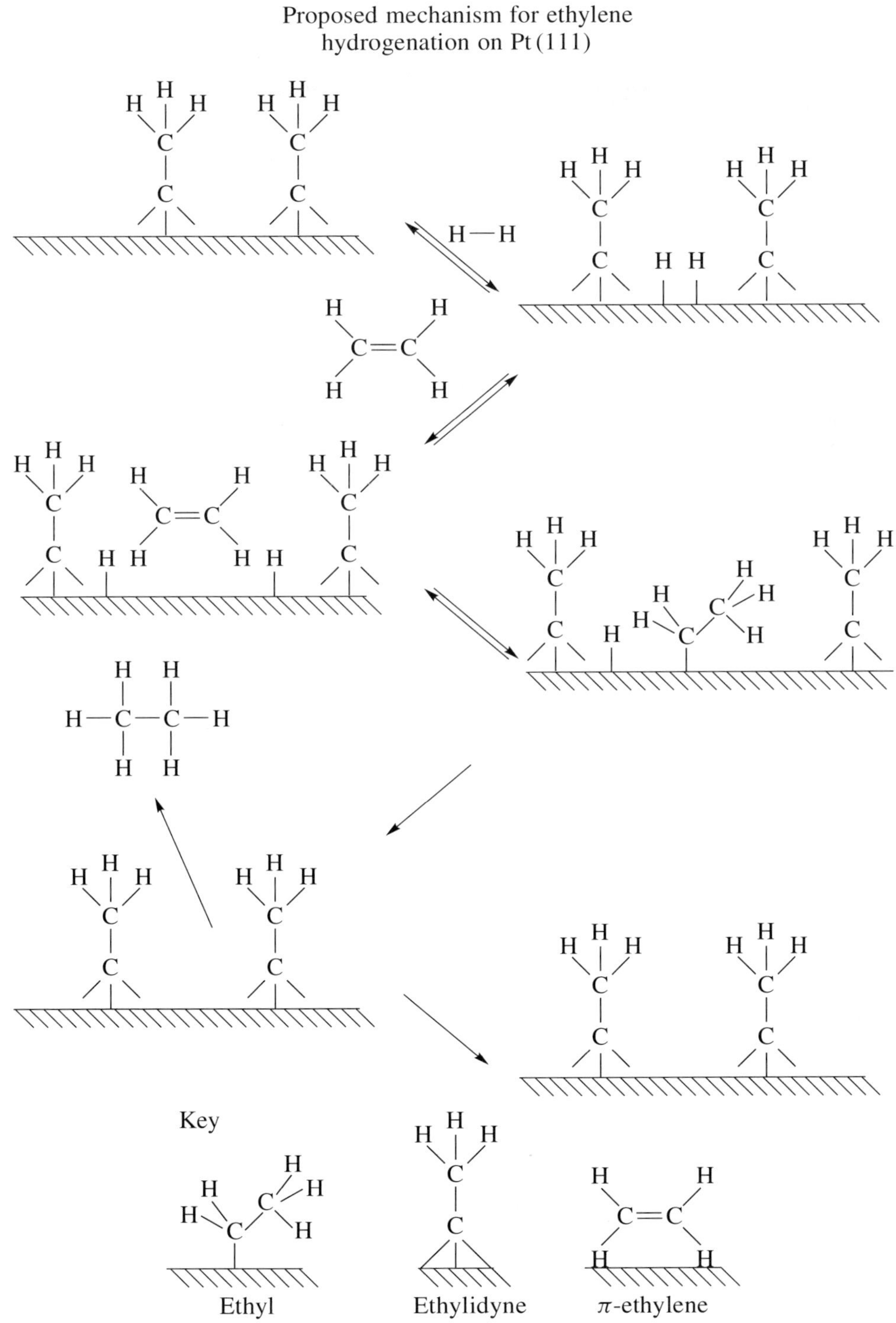

Figure 12. Mechanism of ethylene hydrogenation.

disappeared; however, it reappeared again reversibly as the load was decreased. The results could be interpreted by assuming that the alkyl chain bends under load, and the methyl groups freely rotate in this bent configuration; they straighten out and order again as the load is released. If this physical picture of chain bending and rotation under load is confirmed by further studies, we begin to uncover the mechanism of how molecules at a buried interface respond to the external force applied.

7. CONCLUDING REMARKS

There are exciting opportunities for molecular level studies of adhesional bonds using modern surface science techniques. Studies of polymer surfaces are underway, in my laboratory and in others, to determine the surface structure, identify areas of crystallinity and order, and also identify amorphous domains that exhibit different mechanical properties. Correlating the surface mechanical properties of polymers (hardness, chain alignment, crystallinity, friction coefficient) with their bulk mechanical properties will permit us to modify the surface in ways that will impart unique mechanical and chemical properties.

Acknowledgement

This work was supported by the Director, Office of Energy Research, Office of Basic Energy Sciences, Materials Sciences Division, of the U.S. Department of Energy under Contract No. DE-AC03-76SF00098.

REFERENCES

1. G. A. Somorjai, *Introduction to Surface Chemistry and Catalysis*. John Wiley & Sons, New York (1994).
2. A. J. Perry, *Thin Solid Films* **107**, 167 (1983).
3. E. Meyer, *Prog. Surf. Sci.* **41**, 1 (1992).
4. S. S. Perry, C. M. Mate, R. L. White and G. A. Somorjai, *IEEE Trans. on Magnetics* **32**, 115–121 (1995).
5. J. N. Israelachvili, *Intermolecular and Surface Forces*. Academic Press, London (1985).
6. S. S. Perry, G. A. Somorjai, C. M. Mate and R. L. White, *Tribol. Lett.* **1**, 233–246 (1995).
7. S. S. Sheiko, M. Moller, E. Reuekamp and H. W. Zandbergen, *Phys. Rev. B* **48**, 15 (1993).
8. M. Salmeron, G. Neubauer, A. Folch, M. Tomitori, D. F. Ogletree and P. Sautet, *Langmuir* **9**, 3600–3611 (1993).
9. J. N. Israelachvili, *Surf. Sci. Rep.* **14**, 109 (1992).
10. Y. R. Shen, *Ann. Rev. Phys. Chem.* **40**, 327 (1989).
11. U. Starke, M. A. Van Hove and G. A. Somorjai, *Prog. Surf. Sci.* **46**, 305–319 (1994).
12. B. J. McIntyre, M. Salmeron and G. A. Somorjai, *J. Vac. Sci. Technol. A* **11**, 1964 (1993).
13. P. S. Cremer, B. J. McIntyre, M. Salmeron, Y. R. Shen and G. A. Somorjai, *Catal. Lett.* **34**, 11–18 (1995).
14. Y. Ganthier, R. Baudoing-Savois, K. Heinz and H. Landskron, *Surf. Sci.* **251** (1991).
15. A. Barbieri, M. A. Van Hove and G. A. Somorjai, in: *4th Int. Conf. on the Structure of Surfaces*, X. D. Xie, S. Y. Tong and M. A. Van Hove (Eds), p. 201. World Scientific Publ. Co., Shanghai, China (1994).
16. A. Barbieri, M. A. Van Hove and G. A. Somorjai, *Surf. Sci.* **306**, 261 (1994).

17. J. Batteas, J. Dunphy, B. J. McIntyre, G. A. Somorjai and M. B. Salmeron, *Phys. Rev. Lett.* **77**, 534 (1996).
18. G. A. Somorjai, *Ann. Rev. Phys. Chem.* **45**, 721–751 (1994).
19. C. M. Mate, in: *Handbook of Micro/Nanotribology*, B. Bhushan (Ed.), pp. 167–181. CRC Press, Boca Raton (1995).
20. C. M. Mate, *Wear* **168**, 17–20 (1993).
21. P. S. Cremer and G. A. Somorjai, *J. Chem. Soc. Faraday Trans.* **91**, 3671 (1995).
22. Q. Du, X.-D. Xiao, D. Charych, F. Wolf, P. Frantz, Y. R. Shen and M. Salmeron, *Phys. Rev. B — Condensed Matter* **51**, 7456–7463 (1995).
23. G. S. Blackman, C. M. Mate and M. R. Philpott, *Phys. Rev. Lett.* **65**, 2270–2273 (1990).

Mittal Festschrift, pp. 21–47
W. J. Van Ooij and H. R. Anderson, Jr. (Eds)

Atomic force microscopy: Surface forces, adhesion and nanomechanics measurements*

RICHARD J. COLTON,† WILLIAM R. BARGER, DAVID R. BASELT, SEAN G. CORCORAN, DANIEL D. KOLESKE and GIL U LEE
Code 6177, Naval Research Laboratory, Washington, DC 20375-5342, USA

Abstract—This paper reviews work conducted at the Naval Research Laboratory to measure surface forces, adhesion and nanomechanical properties of materials using an Atomic Force Microscope (AFM). The article portrays the AFM as an enabling technology, evolving from a qualitative picturing-taking tool to a quantitative probe of materials properties. For example, surface forces can be used to reduce contact forces during imaging, improve image resolution, and change image contrast. Force–distance curves and maps can be used to correlate surface topography with the adhesion, surface energy, frictional and mechanical properties of materials. Applications presented include imaging phase-segregated Langmuir–Blodgett films, measuring molecular recognition forces between individual molecules, determining the absolute modulus of materials from nanoindentation, and studying the nucleation of dislocations. Finally, the evolution of the AFM towards dedicated micromachined instruments and sensors for on-line process control and sensing applications is introduced.

Keywords: Atomic force microscopy; surface forces; adhesion; nanomechanics; nanoindentation; molecular recognition; micromachined sensors; biosensors.

1. INTRODUCTION

The invention of the Scanning Tunneling Microscope (STM) [1], AFM [2] and other proximal probes (for an overview of recent literature, see [3]) has revolutionized the way we study the structure and properties of surfaces. These tools also enable the pursuit of more ambitious goals in nanotechnology, e.g. creating unique structures by manipulating individual atoms or molecules and developing new concepts in electronics, information storage, and sensor technology (for an overview of recent literature, see [4]). The study of 'tip–surface interactions' is our link between the macroscopic and nanoscopic worlds. These interactions permit atoms and molecules to be manipulated, provide contrast during imaging, allow different forces and properties to be measured, and in general, enable nanotechnology.

---

*Presented at the 1st International Congress on Adhesion Science and Technology in honor of Dr. Kash Mittal on his 50th birthday, Amsterdam, 16–20 October 1995.

†To whom correspondence should be addressed.

Our work at the Naval Research Laboratory (NRL) has focused on developing the AFM from a qualitative imaging tool to a quantitative probe of surface forces and the mechanical properties of materials. Experiments to elucidate the atomic- or molecular-scale mechanisms of adhesion, friction/wear, and deformation/fracture can now be done (for example, see the review articles [5, 6]). In addition, new theories and computational methods such as molecular dynamics simulations can be used to visualize the processes in order to understand the observed behavior ([7]; see the articles on nanotribology published in [8]).

1.1. Evolution of proximal probes

The AFM was invented to overcome a limitation of the STM, namely, its inability to image insulating materials [2]. While imaging is an important function of the AFM and perhaps the dominant mode in use today, several groups were quick to recognize other modes of operation. For example, Mate *et al.* configured the AFM to measure lateral forces and study atomic scale friction [9]; Burnham and Colton used force–distance curves to measure surface forces and mechanical properties of materials [10]; Wickramasinghe showed how other types of forces (capacitive, electrostatic) or physical properties (temperature) could also be measured or imaged [11]; and Hansma and coworkers performed early work in liquids which changed the way biological samples are studied [12]. The introduction of commercial AFM instruments that operated under ambient conditions found wide appeal with many potential applications.

This revolution in AFM instrumentation was soon followed by a revolution in nanometer-scale science and technology. Prior to the emergence of proximal probes, the discussion of nanometer-scale topics (with the exception of nanolithography) was uncommon. However, today, the literature is replete with articles on *nano*technology, *nano*electronics, *nano*mechanics, *nano*metrology, *nano*biology and *nano*tribology. In fact, many new journals with these names have appeared.

How the field of nanometer-scale science and technology evolves, only time will tell. Although the early expectation that an atomic-scale understanding of complex phenomena such as adhesion, friction and wear, etc., would lead to the development of better adhesives and lubricants is still unrealized, it is becoming clear that there are applications in metrology, boundary lubrication, and biology, for example, where nanometer-scale processes dominate and lend themselves to study by proximal probe techniques. The next evolutionary step in proximal probe technology appears to be the development of dedicated tools for information storage, fabrication, process control and sensing applications. Combining tunneling and force transducers with MicroElectroMechanical Systems (MEMS) (another discipline experiencing tremendous growth), is another area of expected growth.

1.2. Force–distance curves

As mentioned above, the AFM is used frequently as a tool to image the surfaces of insulating materials such as inorganic crystals, biomolecules and polymers. There are numerous examples of this approach in the literature. The goal is generally to image these surfaces with high (i.e. 'atomic') spatial resolution. However, this is very difficult to do and there are only a handful of examples of true atomically-resolved images obtained with the AFM [13–16]. The conditions needed to produce atomically resolved images are carefully controlled to reduce tip–surface interactions. Most high resolution AFM images show only nanometer-scale resolution. If an image appears to show atomic structure, it is usually an artifact showing the atomic periodicity of the lattice caused by large planes of atoms sliding against each other. Nevertheless, the AFM does not necessarily need to achieve atomic resolution to be useful. There are many important applications and phenomena that can be studied on the nanometer length scale.

The lack of understanding of the AFM imaging mechanism led us to develop the technique of measuring force–distance curves. Figure 1A shows one of our first force–distance curves obtained for a tungsten tip and graphite sample in air. Here, the force between the tip and surface (measured by the deflection of the cantilever) is plotted as a function of sample position. The sample position is the distance that the sample moves; it is *not* the actual tip–surface separation. The tip–surface separation is not known exactly because when the tip makes contact with the sample there will be a small, but unknown, amount of elastic deformation of the surfaces. As the sample approaches the tip, the initial interaction is usually attractive wherein the cantilever beam moves towards (negative force) the sample. When the tip finally contacts the surface only the leading end of the tip is in repulsive contact; however, the net force on the entire tip is still attractive. But as the sample is continually pushed against the tip, the repulsive force increases causing the cantilever to reverse direction and finally become net repulsive once the cantilever is pushed beyond its normal rest position. (The 'zero force' in the force–distance curve is defined as the cantilever rest position when the attractive and repulsive forces acting on the cantilever are equal.) Beyond the rest position, the sample is now under an applied load. From this quadrant of

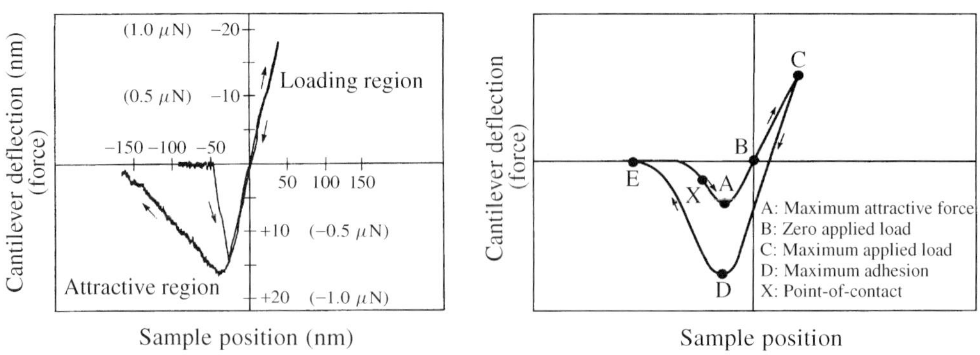

(A) (B)

Figure 1. (A) Early force–distance curve for a tungsten tip and graphite sample in air [10]; (B) schematic force–distance curve showing regions from which attractive, repulsive and adhesion forces are measured [18].

the graph the mechanical properties of the material can be studied. Reversing the sample's direction removes the loading force until contact is broken. However, in most cases a relatively large pull-off force is needed to break contact. The hysteresis in the curve is usually associated with adhesion between the tip and sample.

Today, many researchers use force–distance curves to measure surface forces, adhesion and the mechanical properties of materials. However, while the shape of the curves gives qualitative information about the tip–surface interaction (i.e. relative stiffness, adhesion, surface energy, etc.) more effort is needed to make the measurement quantitative. For example, there are several piezoelectric scanner artifacts (e.g. nonlinearity, hysteresis, creep, etc.) to correct and knowledge of the contact area and actual tip–sample separation is also desirable. These problems also contribute to imaging artifacts. Some control over these problems could be obtained by using non-contact, tapping or modulation modes of operation. However, it is still important to understand the nature of the tip–surface interaction which led us to the study of surface forces.

2. SURFACE FORCES

There are many types of forces that can occur between atoms, molecules and surfaces in close proximity [17]. Some of these forces are summarized in Table 1. Because they operate on the same scale as the AFM measurement [18], it may be difficult to distinguish between these forces if they are present together. One distinguishing property is, however, their interaction length, or the distance between surfaces over which the force can operate. Another distinguishing property is the functional relationship between force and distance.

There are basically three types of force — attractive, repulsive and adhesion — that we measure with the AFM. We will discuss each in turn.

Table 1.
Type and range of force interactions

Force	Example	Interaction distance (m)
Long range		
Magnetic	Magnetic domain	$\sim 10^{-7}$
Electrostatic	Tip–surface capacitance	$\sim 10^{-7}$
Van der Waals	Crossed cylinders (R≫D)	$\sim 10^{-7}$
	Tip–surface (R≪D)	$\sim 10^{-8}$
Hydrophobic	Molecular recognition & self-assembly	$\sim 10^{-9}$
Capillary condensation	Concave meniscus between tip–surface	$\sim 10^{-9}$
Short range		
Bond formation	Hydrogen and covalent bonds	$< 10^{-9}$
Adhesion	Jump-to-contact/chemical bonds	$< 10^{-9}$
Repulsion	Hard core, steric, hydration	$< 10^{-9}$

R = radius of curvature of the cylinder and tip; D = the distance between crossed cylinder or the tip and sample surface.

2.1. Attractive forces

There are many possible attractive force interactions between the tip and surface. Some long range forces include magnetic, electrostatic, van der Waals, hydrophobic, and capillary forces. Some short range attractive forces are adhesion (including the formation of chemical bonds) and charge regulation in electrolytes. As an exercise, imagine that all of these forces exist between the tip and surface. However, by choosing the proper conditions such as using non-magnetic, well-grounded conducting materials, in dry nitrogen or vacuum, one might be able to reduce or eliminate most of them leaving only one or two forces to consider. For example, if one wanted to measure the forces between individual molecules placed on a tip and surface, it may be necessary to eliminate the attractive forces between the two substrates (see Section 2.2).

For a sphere against a flat surface, the nonretarded van der Waals force, F_{vdw}, is

$$F_{\text{vdw}} = -\frac{AR}{6D^2} \quad \text{for } R \gg D, \tag{1}$$

where A is the Hamaker constant, R is the radius of the sphere, and D is the distance between sphere and flat. Israelachvili and Adams [19] used a Surface Force Apparatus (SFA) to measure the van der Waals force between two crossed cylinders of mica (geometrically equivalent to sphere against flat) in an aqueous electrolyte solution. Once the double-layer forces were removed, the SFA data agreed extremely well with equation (1) using a nonretarded Hamaker constant of 2.2×10^{-20} J. The data even revealed the onset of retardation at a separation of $D > 5$ nm. (Retardation effects occur when two atoms are far apart and the travel time of the electric field between them becomes comparable to the period of dipole oscillations. The dipoles do not oscillate in-phase causing the potential to drop off more rapidly. While these interactions between molecules are negligibly small, the interactions between macroscopic bodies, where the forces are significant at large separations, may show retardation effects.) This experiment validates the use of the SFA to measure van der Waals forces under well-controlled conditions.

This type of measurement with the AFM, while seemingly straightforward, has to our knowledge not been conducted (one group has made significant progress in this area — see for example [20]). Most AFM experiments are made in air where surface contamination and adsorbed water produce exaggerated pull-in and pull-off regions in the force–distance curve due to capillary forces [21, 22]. The capillary or meniscus force can be larger than the expected van der Waals force. However, control of the ambient conditions such as working under dry nitrogen, in vacuum, or in liquids should eliminate this problem.

The medium in which the two surfaces are immersed also affects the magnitude of the van der Waals force [23]. The van der Waals force between any two materials in vacuum is always attractive; the force between two identical materials is also always attractive; while the force between two different materials in a liquid medium may be repulsive. This effect can be used to our advantage in controlling the contact force and resolution of the AFM experiment (see Section 3).

The distance (D) dependence of the tip–surface force in an AFM experiment is usually longer range than the $1/D^2$ dependence of the van der Waals force [24]. This observation prompted us to consider other possible forces that could contribute to the measurement [25]. Among those considered (and not previously mentioned) are capacitance forces, and forces due to discrete electronic charges, surface layers of dielectric materials, permanent dipoles, and patch charges. Only the force associated with patch charges comes close to reproducing the experimentally observed distance dependence. Surface patch charges are due to different values of the workfunction on various regions of a surface. The regions may be inhomogeneous due to surface preparation, uneven distribution of adsorbates, crystallographic orientation, or variations in local geometry. These inhomogeneities give rise to differences in local surface charge density. It has been long recognized that patch charges significantly affect electron trajectories in field emission microscopy. However, until recently, the interaction of patch charges between two distinct bodies (i.e. tip and surface) had not been considered. Experimental confirmation of this effect is still needed.

2.2. *Repulsive forces*

Repulsive forces are very strong, short-range interactions between atoms and molecules due to electrostatic forces between like charges [17]. Repulsive forces are sometimes referred to as exchange, hard core, steric, or Born repulsion. Repulsive forces occur immediately upon tip–sample contact, i.e. at the 'X' in Fig. 1B. However, at this point the net forces are still attractive. Only between points B and C are the net forces repulsive. From this region of the curve, we are able to extract information about the mechanical response of the sample to load; the relative slope of the curve is related to the modulus of the material (see Section 5).

From the classical mechanics of Hertz (for a comparison of contact deformation theories, see [26]), we know that the contact area or the depth to which an indenter will penetrate depends both on the load imposed by the tip and the moduli of the indenter and sample, or

$$F_{\mathrm{Hertz}} = \frac{Ka^3}{R} = \left(K^2 R d^3\right)^{1/2}, \tag{2}$$

where F_{Hertz} is the applied load, R is the radius of the sphere, a is the radius of the contact area, d is the penetration depth or surface deformation, and K is the reduced modulus of the system defined as

$$\frac{1}{K} = \frac{3}{4}\left(\frac{1-\nu_1^2}{E_1} + \frac{1-\nu_2^2}{E_2}\right) \tag{3}$$

for Young's moduli E_1 and E_2 and Poisson's ratios ν_1 and ν_2 for the tip and sample, respectively. Hence the slope of a plot of load versus sample displacement or tip penetration (or explicitly, the tangent of this curve) will be proportional to the reduced modulus of the system.

In liquids, repulsive forces can be made to dominate the interaction between two surfaces (except at small separation where van der Waals forces, charge regulation, and adhesion may once again dominate). Surfaces in water or any liquid of high dielectric constant are usually charged and can be prevented from making contact by repulsive electrostatic forces. The magnitude of the repulsive electrostatic force depends upon the charging of surfaces caused by the ionization or dissociation of surface groups and by the adsorption of ions from solution. The ionic strength of the solution will also determine the distance dependence of the force through screening effects. At longer distances the repulsive electrostatic forces are due to double-layer interactions; at smaller distances, repulsive solvation or hydration forces dominate.

In solution, other short-range repulsive forces can also occur between atoms and molecules that are being pushed closely together. Attaching chain molecules or polymers to a surface, for example, causes parts of the molecule to extend into the solution depending on the nature of the solvent. When another surface approaches, the entropy of confining these dangling chains results in a repulsive entropic force known as steric repulsion [17]. The interaction will depend on the mobility and coverage of the surface layers and serves to prevent the surfaces from making adhesion contact.

Figure 2A demonstrates how some of these forces affect a force–distance curve acquired with a quartz sphere and quartz surface in a 10^{-5} M NaCl solution. At large distances, the force between the approaching surfaces is repulsive due to repulsive electrostatic forces. At smaller distances (< 3 nm), the net interaction becomes

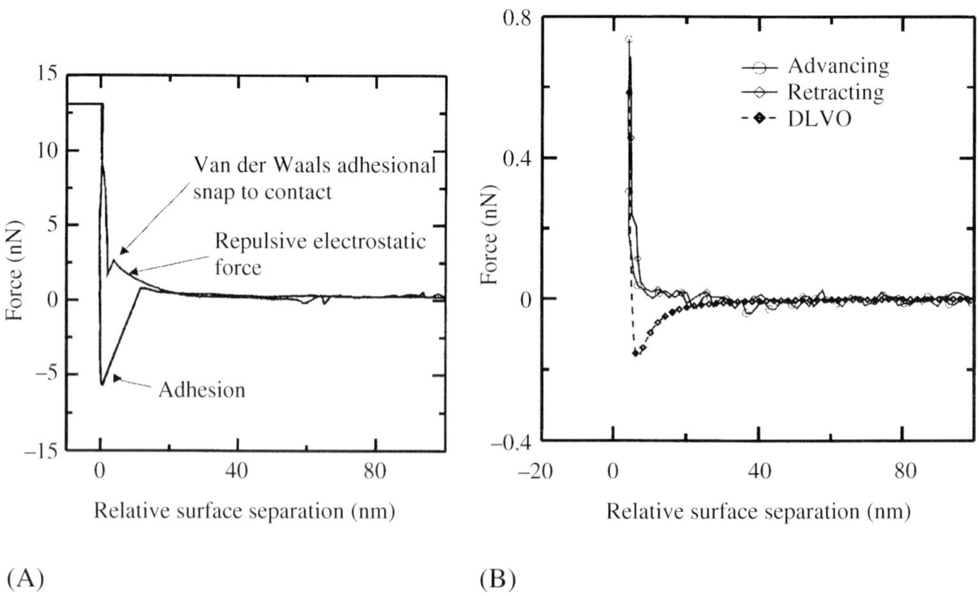

Figure 2. Force–distance curves illustrating the interaction between a quartz sphere ($R = 15\ \mu$m) and quartz surface (A) in a 10^{-5} M NaCl solution and (B) in pH 7.0 PBS solution but with the surface modified by the adsorption of a biomolecule which causes steric repulsion [27].

attractive due to van der Waals forces which eventually cause the surfaces to jump-to-contact. The surfaces are then loaded to a force of about 20 nN. The strong hysteresis of the curve when the surfaces are retracted is indicative of an adhesion interaction with a pull-off force of approximately 5 nN.

If the same two quartz surfaces are coated with bovine serum albumin (BSA) and immersed in a 10^{-1} M phosphate buffer saline (PBS) solution at pH 7, the interaction forces remain repulsive throughout the measurement and the adhesion forces are completely eliminated (Fig. 2B [27]). The theoretical Derjaguin–Landau–Verwey–Overbeek (DLVO) [17] force–distance curve (dashed line in Fig. 2B), which is the sum of the repulsive double-layer force and attractive van der Waals force, reveals that the long-range forces are strongly attenuated by the high ionic strength of the PBS. The measured force–distance curve also exhibits a short-range repulsive force that is not accounted for by the DLVO theory. This short-range repulsive force originates from steric or hydration forces associated with the BSA film.

2.3. *Adhesion forces*

There are as many ways to define adhesion as there are ways to measure it [28]. In atomic force microscopy we are generally concerned with the measurement of pull-off forces from the force–distance curve. The magnitude of the force will depend on the nature of the tip–surface interaction during contact. Although chemical bonds could form during tip–sample contact, we will focus on adhesion mechanisms based on classical mechanics.

Since attractive forces are proportional to contact area, adhesion depends on tip and sample deformation. Under load, both attractive and repulsive forces will increase with the amount of deformation or with contact area. At large applied loads, the repulsive force will dominate and we will observe Hertzian elasticity (as defined by equation (2)) or possibly plastic deformation. Hertzian contact mechanics does not consider the dependence of attractive forces on the contact area. However, there are existing models which do. The Johnson–Kendall–Roberts (JKR) fracture mechanics model [29] describes the effect of strong short-range interactions under conditions of relatively low elastic modulus and large radius of curvature; it neglects the fact that a part of the attraction acts over an area that lies outside the periphery of the contact region. It predicts an intrinsic jump-to-contact when the force gradient of the interaction exceeds the effective spring constant of the materials, with the establishment of a connective neck and a certain minimum contact area at separation. The JKR model predicts that the pull-off or adhesion force, F_a, is

$$F_a = -1.5\pi W_{12} R, \tag{4}$$

where W_{12} is the Dupré work of adhesion ($W_{12} = \gamma_1 + \gamma_2 - \gamma_{12}$, where γ_1 and γ_2 are the surface free energies of the tip and substrate, respectively, and γ_{12} is the interfacial free energy). The work of adhesion can be approximated by $2\sqrt{\gamma_1\gamma_2}$ when only dispersion forces are considered. From Hertzian contact mechanics we also know the geometrical relationship $R = a^2/d$ (where R, a and d are as defined above).

The Derjaguin–Muller–Toporov (DMT) long-range forces model [30] is appropriate for tip–surface combinations where the elastic modulus is high and the radius of curvature and surface energies are small. It neglects the additional non-Hertzian deformation that would be produced by an externally applied load alone, so that the contact area decreases towards zero as pull-off is approached. The DMT model predicts that the pull-off or adhesion force, F_a, is

$$F_a = -2\pi W_{12} R. \tag{5}$$

Both of these models reduce to the Hertzian model at high load.

Various AFM groups have had good success in using these continuum models to predict adhesion forces [31, 32] and contact areas [33, 34].

3. USING SURFACE FORCES

Despite the many problems associated with measuring surface forces, they can be exploited in various ways to reduce tip–sample loading force, improve image resolution or change image contrast.

3.1. Reducing contact force

With van der Waals forces, it is important to consider the medium in which the tip and sample are immersed, e.g. in a vacuum, air or liquid medium. According to Liftshitz's formalism the immersion medium should have a distinct effect on the van der Waals dispersion force between tip and sample in terms of the dielectric contribution of the medium [17]. (Van der Waals forces may be divided into three categories: (a) dipole–dipole, (b) dipole–induced dipole and ion–induced dipole, (c) induced dipole–induced dipole (also known as dispersion).) Highly polar liquids are known to modify van der Waals forces. In addition to the dispersion contribution such liquids exhibit a large entropic contribution to van der Waals force which includes induction and orientation forces between polar molecules. Polar liquids reduce the magnitude of van der Waals force interaction between the tip and sample with respect to the magnitude exhibited in vacuum. This was nicely illustrated by the work of Hartmann [23, 35, 36] who calculated the van der Waals forces between a sharp SiO_2 tip and a metal surface in polar media. Not only does the polar liquid reduce the force but it may also make the interaction repulsive. An estimate of the effect can be made using the principle of combining relations [17] used to obtain approximate values for unknown Hamaker constants from known ones, i.e.

$$A_{132} \approx \left(\sqrt{A_{11}} - \sqrt{A_{33}}\right)\left(\sqrt{A_{22}} - \sqrt{A_{33}}\right), \tag{6}$$

where A_{132} is defined as the nonretarded Hamaker constant for media 1 (tip) and 2 (sample) interacting across medium 3 (vacuum, air, liquid). Table 2 gives some examples; positive values indicate attractive forces and negative values indicate repulsive

Table 2.
Comparison of tip–surface interactions in different media

Tip	Surface	Hamaker constant ($A_{132} \times 10^{-20}$ J)		
		Air	Water	Ethanol
Tungsten	Gold	+25	+3	+12
Tungsten	Alumina	+15	−0.8	+9
Tungsten	Stearic acid film	+9	−2	+0.7
Tungsten	Poly(tetrafluoroethylene)	+3	−2	−0.2
SiO_2	Mica	+8	+1	+0.6
SiO_2	Gold	+14	−2	+2

Table 3.
Resolution as a function of Hamaker constant (van der Waals forces)

System	Hamaker constant (J)	Resolution (nm)
Si_3N_4/mica/air	1.4×10^{-19}	2.6
Si_3N_4/mica/ethanol	3.2×10^{-20}	1.6
Diamond/diamond/air	3.4×10^{-19}	1.9
Diamond/diamond/ethanol	1.9×10^{-19}	1.5

forces. The method should be used for guidance and is applicable only when dispersion forces dominate the interaction (which is generally the case except for small, highly polar molecules such as water). Nevertheless, the interaction force is predicted to be less when imaging in liquids compared to air (or vacuum). Weisenhorn *et al.* [37, 38] found the force between a silicon nitride tip and mica surface to decrease from 4×10^{-7} N in air to 5×10^{-9} N in water. This realization has had a profound effect on many imaging applications, particularly for biological samples.

Similarly, repulsive electrostatic forces can be used to control the interactions between surfaces, particularly in liquids. For example, in the work associated with Fig. 2 [27] we wanted to attach ligand-receptor molecules to the quartz surfaces and measure the binding or molecular recognition forces between them. However, the attractive and adhesion forces between the clean surfaces themselves (Fig. 2A) were larger than the expected molecular interaction. In order to overcome this problem, we placed the surface in a high ionic strength buffer solution and modified the surfaces with a protein film which contributed steric repulsion forces to the baseline interaction. Figure 2B shows that the baseline interaction was reduced to a level that eventually permitted the measurement of the molecular recognition forces.

3.2. Improving image resolution

Using the same principle, Hutter and Bechhoefer [39] manipulated van der Waals forces to improve image resolution. The relationship derived from the JKR theory is given as

$$\Re = \left(\frac{AR^2}{8KD_0^2}\right)^{1/3}, \tag{7}$$

where $\mathfrak{R}$ is the resolution, A is the Hamaker constant, R is the radius of the tip, K is the reduced modulus of the tip and sample and D_0 is the tip–sample separation. Table 3 summarizes their findings. This table illustrates that, even under the best conditions of reducing the interaction force by placing the tip and sample in ethanol and going to a hard material such as diamond, the resolution is still not atomic.

3.3. Changing image contrast

A limitation of both the STM and AFM is their inability to identify chemical composition. Nevertheless, it is exciting to see that phase-segregated organic films show both height and friction contrast that may allow the regions to be chemically differentiated. The Güntherodt group [40, 41] has done much of the pioneering work in this area. For example, Fig. 3 shows friction force microscope images of a monolayer film of a 1 : 1 mixture of partially fluorinated carboxylic ether acid ($C_9F_{19}C_2H_4{-}O{-}C_2H_4COOH$) and arachidic acid ($C_{19}H_{39}COOH$). Figure 3A shows the topography, and Fig. 3B shows the friction force map. They found that the friction on the fluorinated film was approximately four times higher than on the hydrocarbon film (the higher component in the topographic image). The films could also be manipulated producing the 'smiley faces' although the circular shape of the domains resulted from thermodynamic processes.

The image contrast was attributed to differences in tip–surface interactions. Several groups have shown that the surface force measured by the AFM is sensitive to changes in surface energy [31, 32]. For example, the surface energy of a $-CH_3$ terminated surface is vastly different from that of a $-CF_3$ terminated surface as illustrated in Table 4 from contact angle measurements. However, from surface energy considerations one would also anticipate that the $-CF_3$ terminated surface would show less adhesion, certainly, and presumably less friction [42].

Using equations (4) and (5), adhesion forces measured with the AFM can be related to surface energy measurements through the work of adhesion, W_{12}. The work of

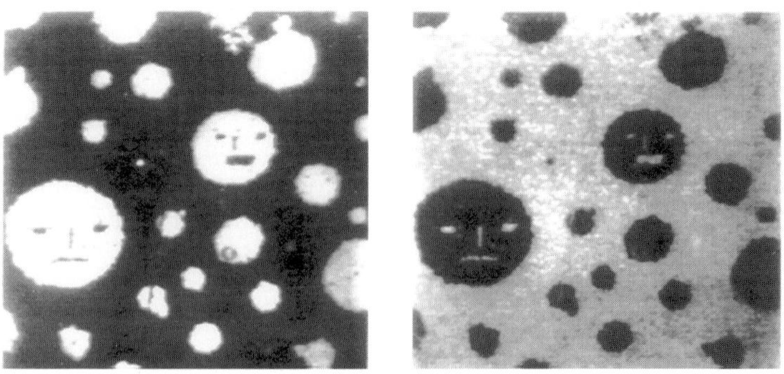

(A) (B)

Figure 3. Friction force microscope images of a monolayer film of a 1 : 1 mixture of partially fluorinated carboxylic ether acid and arachidic acid: (A) topography and (B) friction force map [41]. Light areas are high; dark areas are low.

Table 4.
Dispersion component of surface energies for different chemical groups

Chemical group	γ^{d} (mJ/m^2)
$-CH_3$	24
$-CH_2-$	31
$-CF_3$	6
$-CF_2-$	18

adhesion between the tip and sample can be approximated by

$$W_{ts} \approx 2\left(\gamma_s^{d}\gamma_t^{d}\right)^{1/2}, \tag{8}$$

where W_{ts} (or W_{12}) is the work of adhesion between the tip (t) and sample (s), and γ^{d} is the dispersion force contribution to the surface free energy.

3.4. Chemical force microscopy

The name 'chemical force microscopy' was coined by the Lieber group [43] to promote the idea that chemically sensitive imaging could be done with the AFM using chemically modified tips to measure differences in surface and frictional forces between chemically distinct regions of a surface. For example, a tip coated with $-COOH$ groups could be used to image a patterned surface containing films terminated with $-COOH$ and $-CH_3$ groups. The tip–surface interactions (particularly pull-off forces) are expected to be much stronger when the tip is over the $-COOH$ surface compared to when it is over the $-CH_3$ terminated surface. Hence, the image shows contrast in frictional force related to the functional group chemistry. While the method seems to work well in this example, it is unclear how general the effect will be. In addition, the method considers surface forces as the sole origin of contrast, whereas other tip–surface interactions, such as stiffness (see Section 5.1) may also contribute to or perhaps dominate the contrast mechanism. Because of these problems, the name chemical force microscopy has not received wide acceptance.

4. FORCE–DISTANCE CURVE MAPPING

We have tried to convey the importance of AFM force–distance curves to facilitate the quantitative analysis of surface forces and mechanical properties of surfaces with high spatial and depth resolution. The capability to 'probe' surfaces enhances the imaging mode of AFM and opens up new applications. The force–distance curve methodology can be further enhanced (in analogy to spectroscopic imaging in STM [44]) by collecting force–distance curves at every pixel of a topographic image. The data set can then be used to display force maps of adhesion, lateral force, and

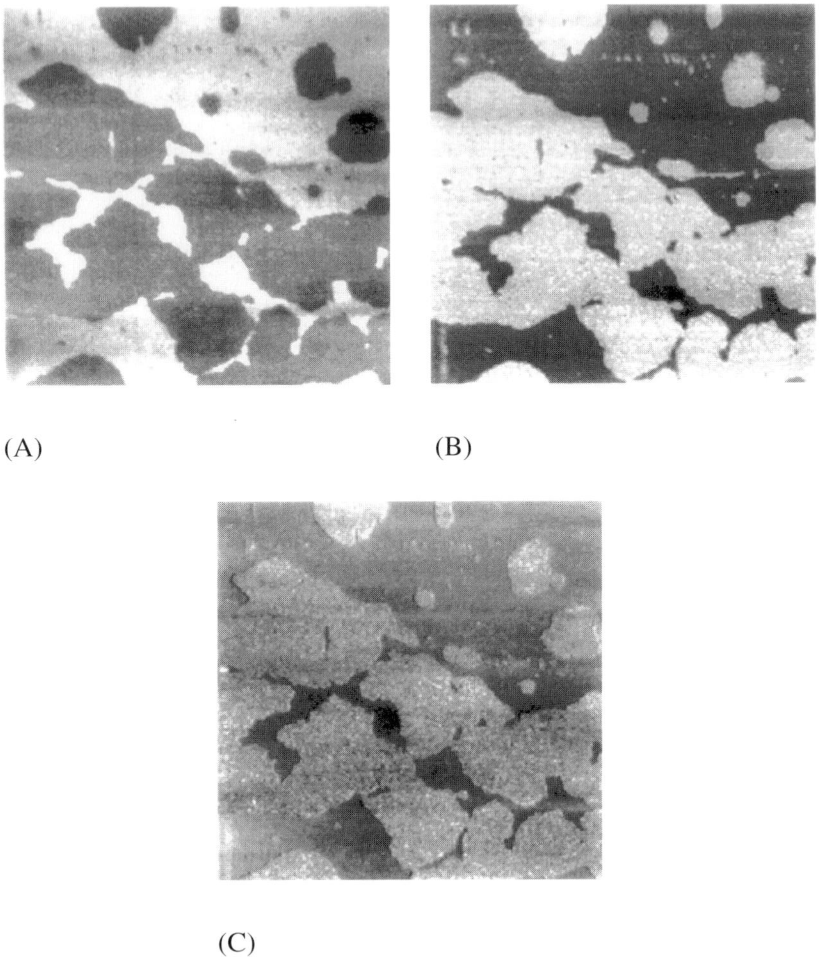

Figure 4. Force–distance curve maps obtained in air for a Langmuir–Blodgett monolayer film on mica consisting of a 1 : 1 mixture of palmitic (C16) and lignoceric (C24) acids [46]: the images show (A) topography, (B) adhesion, and (C) 'lateral' force maps.

stiffness/elasticity that are correlated with topography. An example is shown in Fig. 4 for a Langmuir–Blodgett (LB) film consisting of a 1 : 1 mixture of $C_{15}H_{31}COOH$ (C16) and $C_{23}H_{47}COOH$ (C24) fatty acids deposited as a monolayer film on mica [45, 46]. The images are constructed from 65 536 individual force–distance curves (256×256 array) which required about two hours to record. (An instrument with high thermal stability is needed to perform this measurement [47].) The film forms phase-segregated domains which differ in height. Figure 4A shows the topographic or height map. Figures 4B and C show adhesion and lateral force [48] maps, respectively. Adhesion and lateral forces are anti-correlated with topography: C24 regions have low adhesion (and friction), whereas C16 regions have high adhesion (and friction). The distribution of adhesion forces is shown in Fig. 5.

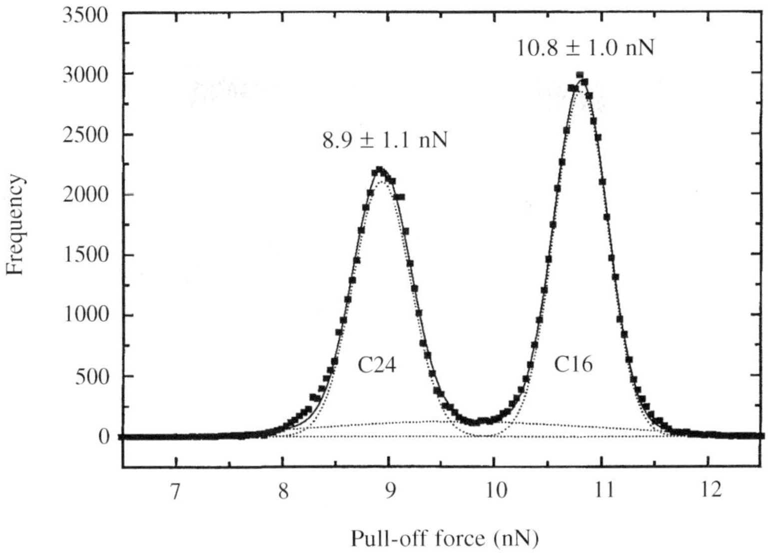

Figure 5. Histogram of the distribution of adhesion forces over the C16 and C24 region of the LB film.

5. APPLICATIONS DEALING WITH SURFACE FORCES AND ADHESION

5.1. Imaging phase-segregated Langmuir–Blodgett films

In Sections 3.3 and 3.4 we illustrated how the image contrast may depend on changes in surface energy (or van der Waals dispersion forces). The effect is quite common and occurs under a variety of experimental conditions with tips that have been chemically functionalized (although this is not absolutely necessary). Image contrast can be observed in topographic and frictional force images as well as in force curve maps. In all cases the tip–surface interaction gives a change in contrast allowing regions to be distinguished. To explore the contrast mechanism further we studied a series of mixed fatty acid films that varied in carbon chain length from C14 to C36 [45, 46]. Monolayer films were prepared on mica using the LB film method. Data for the C16 and C24 fatty acid films are shown in Fig. 4. The domain size and shape depend on the relative chain length of the fatty acids. However, we were most intrigued by the contrast in the frictional force images (not shown) and adhesion maps that occur despite the fact that the surface chemistry is the same, i.e. because both surfaces are terminated by $-CH_3$ groups, there should be no appreciable surface energy difference. The second surprising result occurs for the height difference measurement. The orientation, packing and thickness of these films are well known from X-ray diffraction studies [49]; hence the step height difference between different domains is also known. For the C16 and C24 system, the expected height difference measured by X-ray diffraction is 1 nm; however, a height difference of 2 nm is measured by the AFM.

To reconcile these differences we studied a series of mixed LB films that enabled us to formulate the model shown in Fig. 6 [45, 46]. With the LB film trough we can

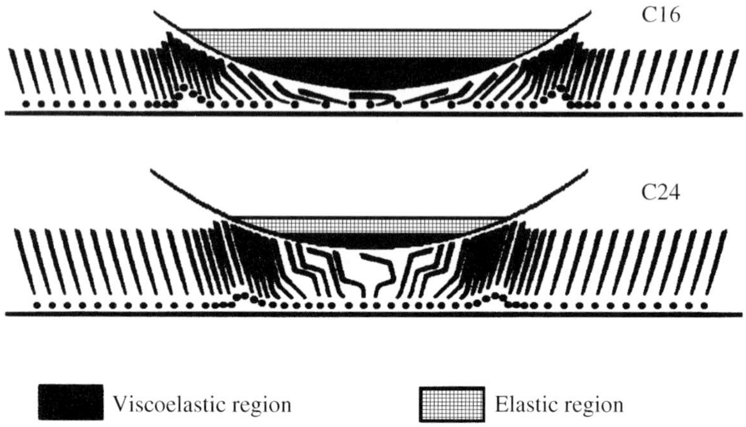

Figure 6. Model used to explain the image contrast in Fig. 4. Because the C16 film is much softer than the C24 film, the tip penetrates deeper into the C16 film causing more deformation. The lateral displacement of the film results in a viscoelastic response and concomitant increase in the measured adhesion or pull-off force due to the larger contact area.

measure directly the collapse pressure and compressional modulus (which measures resistance to compression in the horizontal plane, i.e. a 2-D rather than 3-D Young's modulus) of the pure and mixed films directly on the water surface. We find that the surface compressional modulus (and collapse pressure) increases with chain length. For example, the modulus of the C16 film deposited at a transfer pressure of 30 mN/m is 260 mN/m, whereas the modulus of the C24 film deposited at the same transfer pressure is 630 mN/m. Hence, the longer chain length film has a significantly higher compressional modulus. As a result, loading the film with the tip during contact imaging causes the C16 film to be compressed more than the C24 film. Under equivalent loads, the tip contact with the C16 film will be proportionally larger than the contact area with the C24 film. The local contact pressure also causes the films to move laterally; the shorter film moves more. Removing the load probably causes the films to respond viscoelastically, since the films are not usually permanently damaged. Hence at the time of pull-off, the contact area with the compliant film is higher and gives rise to a larger pull-off force than with the stiffer film. Because the films respond differently to load, we concluded that the image contrast is due to a different mechanical response. Therefore, in addition to surface energy effects related to chemical differences, the relative stiffness of a material must also be considered when interpreting the contrast in AFM images or force–distance curve maps. This becomes especially important when imaging compliant materials such as polymers and biological samples.

5.2. Biophysics: molecular recognition

The three-dimensional structure of complex biological molecules is determined not only by the strong covalent bonds between atoms, but also by relatively weak intramolecular forces such as electrostatic, van der Waals, hydrogen bonding, and hydrophobic forces. Similarly, when these forces act together between molecules and in a manner dependent on the specific 3-D structure of a molecule, they are collectively termed 'molecular recognition forces'. Understanding and controlling the nature of these forces will foster a deeper knowledge of biological processes and create new diagnostic tools with important medical and environmental applications.

Until recently, our understanding of molecular recognition was based on indirect physical and thermodynamic measurements such as X-ray crystallography, light scattering, nuclear magnetic resonance spectroscopy, and calorimetry. To measure forces directly requires a highly sensitive technique that can record both force and position with high resolution and accuracy. The AFM has both high force sensitivity (10^{-13} N or two orders of magnitude more sensitivity than the force of a single hydrogen bond) and position control (0.01 nm), and can probe areas as small as 10 nm^2 under physiological conditions. By attaching one type of molecule to the end of a force measuring cantilever beam and its complement to a flat substrate, we can measure directly the intermolecular forces between a single receptor–ligand complex, for example, when the cantilever first contacts the substrate and then pulls apart.

5.2.1. Receptor–ligand complexation. Streptavidin and biotin molecules comprise a receptor–ligand system that mimics an antibody–antigen interaction but with higher binding affinity. The streptavidin molecule contains four binding sites for biotin. By placing streptavidin on either the tip or sample and tethering biotin to the other surface using a short linker chain, we have measured the force necessary to rupture the intermolecular interaction once the surfaces are placed in contact and then pulled apart at a specific rate [27]. The measured rupture force is 0.34 ± 0.12 nN which is larger than the force of 0.06 ± 0.04 nN when the streptavidin sites are blocked with free biotin. Detailed analysis of the magnitude and distribution of the force indicates that it is possible to measure the intermolecular force between individual streptavidin–biotin molecules [27, 50]. However, we also find that the measured rupture force strongly depends on the rate at which the bond is ruptured.

5.2.2. DNA hybridization. Similarly, the two opposing surfaces can be functionalized with complementary strands of DNA and their binding interaction measured [51]. Using oligonucleotides consisting of five repeat units of four base-pairs, i.e. $(ACTG)_5$, we measured strong adhesion forces (up to 1.6 nN) between the surfaces only when the base sequences of the oligonucleotides were complementary, i.e. $(CAGT)_5$. Furthermore, the magnitude and distribution of the rupture forces depends on the extent of the Watson–Crick base pairing between 20, 16, and 12 base-pairs of individual molecules.

In Fig. 7 [51], we illustrate the case where noncomplementary oligonucleotides are immobilized on opposing surfaces and a complementary polynucleotide is placed between them. The addition of the homopolymer does not change the magnitude of the adhesion force but increases the distance (up to 400 nm in some cases depending on the length of the homopolymer) between the surfaces at which the bond rupture occurs. As the surfaces are separated, an exponentially increasing force is required to elongate the molecule resulting in a measure of the molecule's elasticity. Careful examination of the force curve reveals fine structure that may be interpreted as the

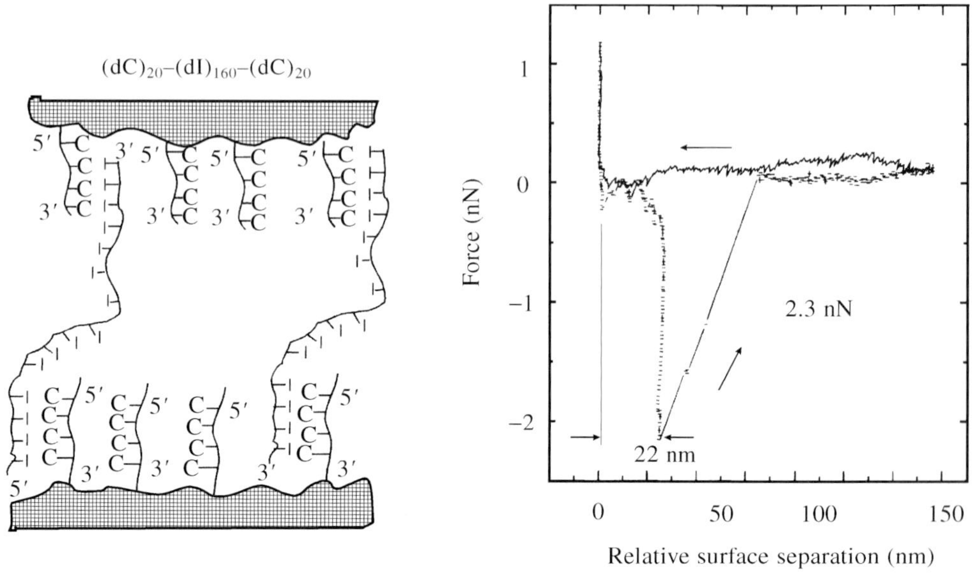

Figure 7. AFM measurement of the binding force between complementary strands of DNA [51]. The left panel shows two poly-inosine oligomers (dI) with a nominal length of 160 bases linked to two surfaces by base-pair interactions with its complement, namely 20-mers of cytosine (dC). The right panel shows the measured force–distance curve when one poly-inosine chain becomes attached between the two surfaces. The curve shows the force required to initially uncoil the single strand pulling it taut before finally breaking the specific base-pair interaction at a force of 2.3 nN.

force necessary to uncoil and elongate single stranded DNA. These results demonstrate that the AFM can measure *intra*molecular as well as *inter*molecular forces in complex macromolecular systems.

5.3. Single molecule sensors

We have used the force sensitivity and position control of the AFM to develop a molecular recognition tool capable of detecting the interaction between single molecules [27, 50, 51]. Furthermore, we have reduced this measurement to a new sensor concept. The sensor, which we call the Force Amplified Biological Sensor (FABS) [52, 53], uses an immunobead sandwich assay consisting of a cantilever force transducer to which antibodies have been bound. When the sample solution is introduced into the instrument, the antibodies bind any antigen — the protein, virus, or bacterium of interest — that may be present. Micrometer-sized, antibody-coated magnetic particles then bind to another part of the antigen, creating a (particle–)antibody–antigen–antibody(–cantilever) sandwich. After washing away excess particles, a magnetic field is turned on, which pulls on the particles causing the cantilever to bend. The amount of bending indicates the number of particles bound to the cantilever, and therefore, the amount of antigen present in the sample. Figure 8 is a photograph of the prototype sensor showing the Helmholtz coils with signal and reference Piezolevers^{TM} (Park Scientific Instruments, Sunnyvale, CA; see

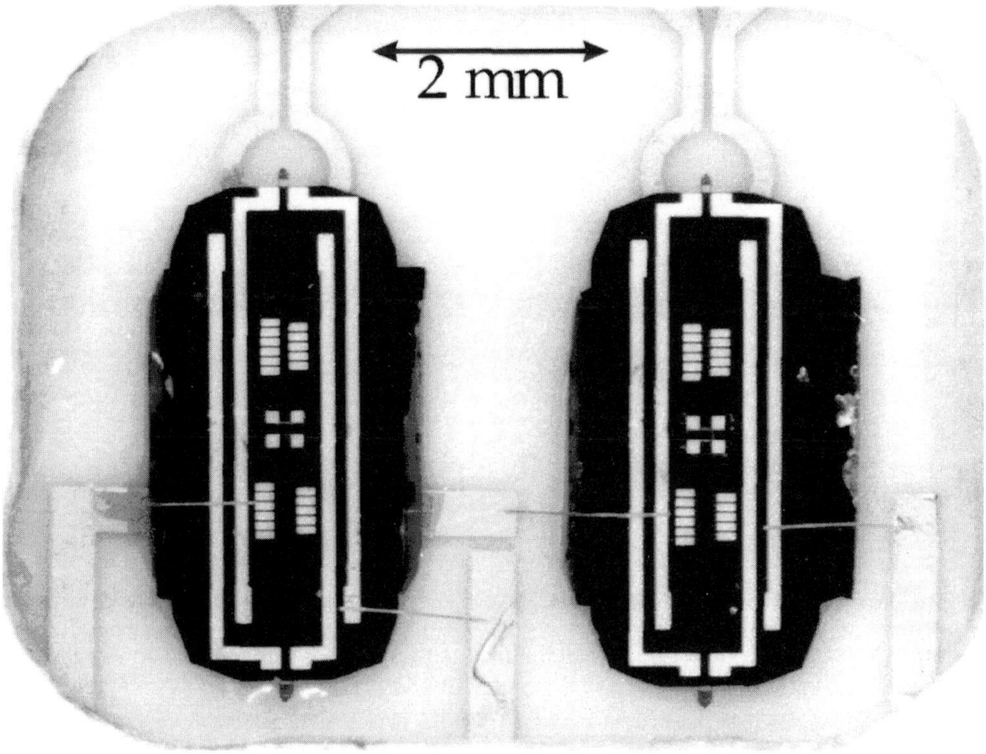

Figure 8. Photograph of the prototype FABS biosensor [53] showing the sample and reference piezoresistive cantilevers and one of the microfabricated coils used to generate the force gradient.

[54]) mounted onto a spacer. A second coil substrate (not shown) mounts above the cantilevers and seals off the liquid cell of the biosensor. The assay is expected to take about ten minutes to run.

There are other examples where technology derived from the STM and AFM has been used to develop new sensors [55–57], e.g. the tunneling accelerometer [58, 59], tunneling magnetometer [60–63], tunneling infrared sensor [64], tactile sensor [65], calorimeter [66], vapor detector [67, 68], radiation dosimeter [69], etc. This represents the next step in the proximal probe evolution, one where miniature sensors are used for process control.

6. NANOMECHANICS

The nanomechanical properties of materials is becoming an area of increasing importance as the scale of coatings, multilayer films, particles, components and devices approaches atomic dimensions. Microindentation methods have been used to measure the modulus and hardness of materials for many years. Nanoindentation methods have been developed to keep pace with developments in the materials arena. Commercial nanoindenters probe samples with indentation depths typically > 100 nm and applied loads > 1 mN. As a consequence, these instruments are limited to probing thin films with thicknesses greater than a few micrometers. Commercial atomic force microscopy, on the other hand, can perform indentations with depths < 1 nm and applied loads of order 1 nN. However, measurements suffer quantitatively because the contact area between the sample and tip (which has an inconsistent/unknown shape) is not known. In order to develop a quantitative nanoindenter with both high force and high depth resolution, the nanoindentation community has been improving the sensitivity and resolution of their measurements [70, 71] and the AFM community has been converting the AFM into a nanoindenter [10, 72] which employs stiffer springs (e.g. stiffness > 1000 N/m) or tips of known geometry (e.g. spheres). Other approaches include using the surface force apparatus [73] and various modulation or acoustic methods [74–76].

6.1. Qualitative measurements

The AFM can be used to determine the mechanical properties of materials. The force–distance curve (see Fig. 1) contains information about the mechanical properties in that portion of the curve where there is an applied load (top right hand corner). Even the earliest data contain information that gives the relative stiffness of the material. For example, Fig. 9 shows three force–distance curves that are drawn to emphasize

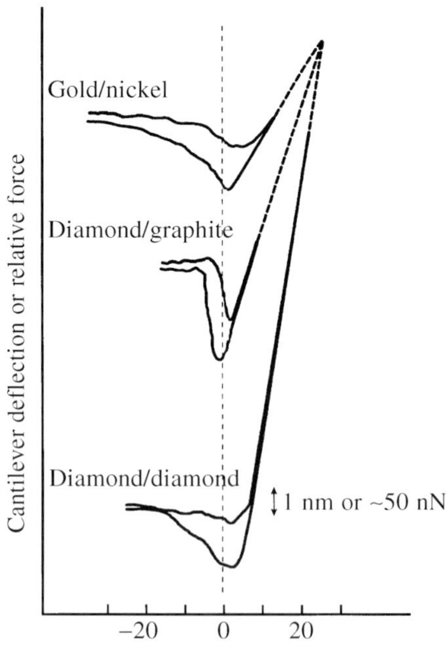

Figure 9. Comparison of force–distance curves for gold–nickel, diamond–graphite and diamond–diamond [25]. The plot shows both the loading and unloading curves and is drawn to illustrate the increasing slope of the loading/unloading curves (with applied load) with increasing modulus of the materials.

the relative slope of the loading part of the curve [25]; the reduced modulus of the tip and sample is directly proportional to the slope. From Fig. 9 it is easy to see that the gold–nickel system is more compliant than the diamond–diamond system. This should come as no surprise, but it is interesting to note that the force–distance curves give the relative order of the system compliance. Our goal is to determine how to quantify this measurement.

6.2. Quantitative measurements

The slope of the loading and unloading curves contains the modulus information. In order to interpret the curves as true indentation curves, it is desirable to convert the force–distance curve into a load–penetration depth curve as shown in Fig. 10. This is done by converting the cantilever deflection to force using Hooke's law and determining the difference between the sample and cantilever displacements to determine the tip penetration depth [77].

Using a pointed indenter with the AFM usually produces loading and unloading curves that are linear with respect to penetration depth. As a result we can use Sneddon's relationship [78] for a flat-ended punch (i.e. the indenter geometry for

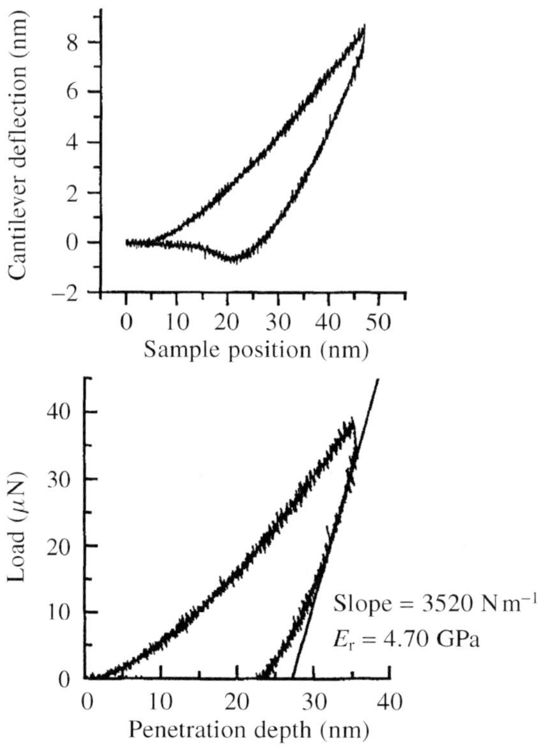

Figure 10. Nanoindentation experiment utilizing a glass sphere to indent a polystyrene film [84]. Top panel shows raw data. Bottom panel shows raw data converted to a plot of load versus penetration depth.

which load is linearly proportional to penetration depth) to determine the modulus from the slopes of these curves. Here,

$$E_r = \frac{\sqrt{\pi}}{2}\frac{Slope}{\sqrt{A}}, \tag{9}$$

where E_r is the reduced modulus of the sample and indenter, *Slope* is the slope of the loading curve (for a flat-ended punch indenter) or the initial slope of the unloading curve (for a spherical indenter), and A is the projected tip/surface contact area.

6.2.1. Instrumental considerations. Our earliest work showed that the slopes could not be measured reliably due to the drift and instability of the PZT (lead–zirconate–titantate) actuators [79–81]. The most severe problem occurs with piezo creep [79]. There are several approaches to overcome the nonlinearity, hysteresis, and creep problems. Some instrument manufacturers control the motion of the actuators using correction signals derived from additional strain gauges or capacitive sensors. In our laboratory we chose to replace the PZT ceramics with PMN (lead–magnesium–niobate) electrostrictive actuators which do not have this creep problem. As a result we began to record reproducible data where regions with different moduli (for example, the native oxides on metal surfaces and SiO_2 on Si as a function of oxide thickness [80, 81]) could be clearly distinguished. However, the results are only semiquantitative in that we could obtain reproducible slope measurements but could not convert them to modulus because we did not know the actual contact area.

6.2.2. Contact area. We have used both pointed and spherical indenters to probe the mechanical properties of a variety of materials — metals, semiconductors, polymers, etc. [80–84]. Pointed indenters offer the advantage that the contact area is limited and small, and that a high contact pressure can be achieved during loading, which is ideal for hardness measurements or for the study of nanoscale events such as dislocation nucleation and motion (see Section 7.2). However, on the nanometer scale, the apex of a pointed indenter is relatively blunt so that the contact area is difficult to determine unless an actual three-dimensional image of the indenter can be obtained. SEM micrographs only show the profile of the indenter and hence can only give an estimate of the actual contact area.

Spherical indenters, on the other hand, with well-defined contact geometries can be made from glass spheres attached to the end of metal shanks. Spherical indenters produce uniform stress fields and varying strain rate with indentation depth enabling the measurement of stress–strain relationships. However, spherical indenters are rather large compared to pointed indenters and, as a result, the lateral spatial resolution is limited. In addition, the thickness of a film that can be studied scales with the size of the indenter since the depth of the stress field increases with the diameter of the contact area. The best spheres for which there is excellent control of size and sphericity are made of low modulus ($\sim$75 GPa) glasses, limiting the type of materials that can be studied.

7. QUANTITATIVE NANOINDENTATION APPLICATIONS

7.1. Spherical nanoindentation of polystyrene films

Using the spherical indenters described in Section 6.2.2 we, together with collaborators from Dow Chemical Company and Wilson Instruments [84], studied the indentation of thin polystyrene films cast onto silicon wafers. Figure 11 shows a family of indentation curves. The data appear to fall into two groupings; the slope of the unloading curves (see Fig. 10) for the thicker films is not as steep as the slope of the unloading curves for the thinner films. Applying Sneddon's relationship [78], equation (9), to the initial region of the unloading portion of the indentation curve, we determine a modulus of 4.5 ± 0.4 GPa (one measurement shown as the solid line in Fig. 10) for the thicker films which compares favorably with reported literature values [85]. As the film thickness drops below 400 nm, the modulus begins to rise to a value approaching that of the soda-lime glass (~ 75 GPa). Hence the size of the indenter used here (i.e. 4 μm radius glass sphere) limits the measurement to film thicknesses greater than approximately 400 nm. Nevertheless, this example illustrates how the AFM was used successfully as a nanoindenter to measure the absolute modulus of a material. Similarly, spherical indenters were used to measure the bulk moduli of graphite and polycarbonate. The AFM nanoindenter using applied loads of the order of 10 μN and penetration depths around 20 nm is able to reproduce the bulk values measured with a commercial indentation instrument (UMIS 2000) which uses mN loads and penetration depths of 1 μm [84].

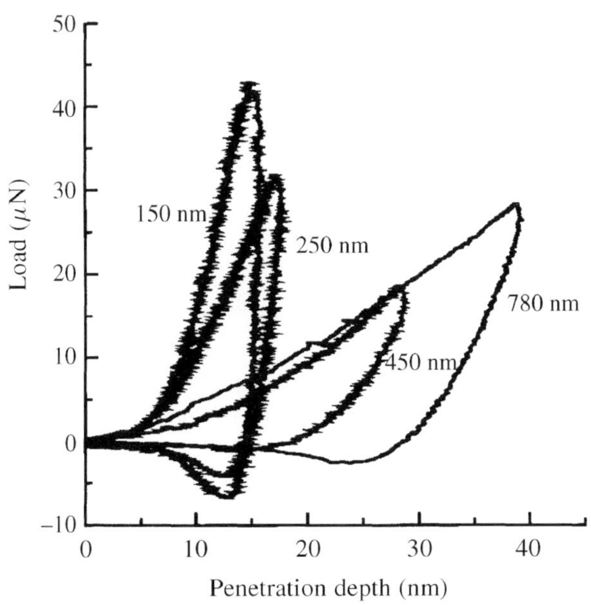

Figure 11. Typical indentation curves plotting load versus penetration depth for several polystyrene thin films [84]. The nominal film thickness is indicated in the figure.

7.2. Nanoindentation studies of yield point phenomena

Another exciting application of the AFM is the study of nanoscale phenomena such as the nucleation of dislocations. Nanoindentation studies often show an instantaneous displacement-excursion (i.e. usually one or sometimes two so-called 'pop-in' events) in the load–penetration depth curve. This anomaly is generally associated with a surface contamination effect, dislocation emission, phase transition, or oxide break-through event [86].

Using a commercial Hysitron, Inc. Picoindenter™ in conjunction with the Digital Instruments, Nanoscope™ III AFM/STM we performed indentations on Au (100), (110) and (111) single crystals [82, 83]. The Hysitron force transducer permits the direct imaging of the surface before and after the indentation. A Berkovich diamond indenter was used; its blunt end had a spherical radius of 205 nm determined by imaging the tip with an asperity on the Au surface. The Au single crystals were prepared by electrochemical polishing followed by flame annealing to red hot under a H_2 flame. The surfaces were then quenched and stored under triply distilled water and transferred into a dry N_2-purged glove bag.

Figure 12 shows a load–penetration depth curve for Au (100). The data are highly reproducible as long as the indentations are performed under dry N_2 in order to avoid surface contamination. The curve shows several yield displacements. The number of displacement 'steps' and their size varies between indentations but the total sum of all displacements is conserved, i.e. the average shape of the indentation curve is highly reproducible. This new phenomenon of 'staircase' yielding is observed for all of the Au single crystals and is unique because the plasticity appears to be confined to the instantaneous displacement excursions, i.e. the Au surface predominantly deforms elastically between each yield excursion. Since the contact area is more than an order

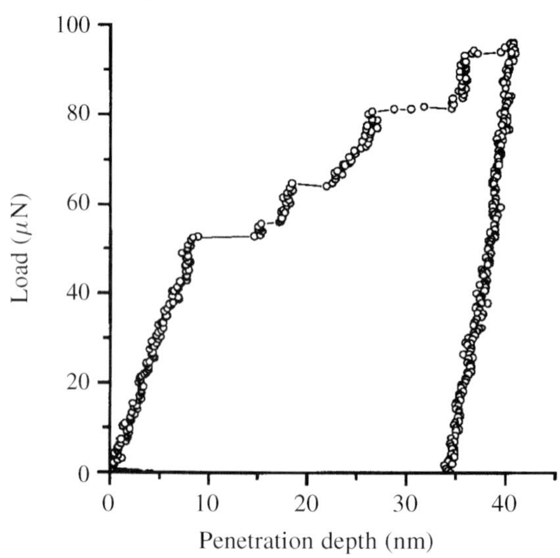

Figure 12. Nanoindentation curve for Au (100) showing the unusual 'staircase' yield point phenomenon [82, 83].

of a magnitude smaller than the average dislocation spacing expected for well annealed Au (~ 1 per μm^2), the excursions are explained in terms of multiple dislocation nucleation events on parallel slip bands.

8. CONCLUSIONS

This paper summarizes some of the work in the AFM area conducted by our group over the past eight years. The overall goal of our AFM programs has been to evolve the AFM from a qualitative imaging tool to a quantitative probe of surface forces and mechanical properties. To some extent we have accomplished this goal and are now pursuing the next step — developing new tools that could be used for on-line process control or novel sensing applications. The road ahead is still unknown but filled with adventure and discovery.

Some general conclusions of this work can be categorized in the following way. In imaging, we show that image contrast, in addition to being dependent on topography and surface energy, is also dependent on the mechanical response of the material. In nanomechanics, the AFM can be used to measure quantitatively the elastic modulus of materials and dislocation nucleation processes. In biophysics, the AFM can be used to measure the binding interaction between individual molecules. The latter work has led to the development of a novel biosensor concept based on the measurement of small forces.

Acknowledgements

This paper summarizes the work of many staff members and postdoctoral researchers (past and present) who have contributed to the AFM program at NRL over the years. Current members of the AFM group are listed as co-authors. Past members and/or other collaborators that we acknowledge include N. Burnham (EPFL), L. Chrisey (NRL), D. DiLella (Shepherds College), C. Draper (ST Research), Y. Dufrêne (NRL), K. Feldman (ETH Zurich), B. Gans (Damilic Corporation), K. Hansen (Geo-Centers Inc.), J. Harrison (US Naval Academy), S. Hues (Motorola), D. Krüger (Universität Münster), K. Lee (Geo-Centers Inc.), H. Pollock (University of Lancaster), D. Schaefer (Towson State University), S. Sinnott (University of Kentucky), K. Wahl (NRL), and L. Whitman (NRL). We also acknowledge the support of the Office of Naval Research.

REFERENCES

1. G. Binnig, H. Rohrer, Ch. Gerber and E. Weibel, *Phys. Rev. Lett.* **49**, 57–61 (1982).
2. G. Binnig, C. F. Quate and Ch. Gerber, *Phys. Rev. Lett.* **56**, 930–933 (1986).
3. Papers from *The Eighth International Conference on Scanning Tunneling Microscopy/Spectroscopy*, R. Hamers, M. Weaver, M. Weimer and P. Weiss (Eds) (*J. Vac. Sci. Technol. B* **14**, 1996).
4. Papers from *The Third International Conference on Nanometer-scale Science and Technology*, R. J. Colton and C. R. K. Marrian (Eds) (*J. Vac. Sci. Technol. B* **13**, 1995).
5. B. Bhushan, J. Israelachvili and U. Landman, *Nature* **374**, 607–616 (1995).

6. R. J. Colton, *Langmuir* **12**, 4574–4582 (1996).
7. U. Landman, W. D. Luedtke, N. A. Burnham and R. J. Colton, *Science* **248**, 454–461 (1990).
8. J. F. Belak, *Nanotribology*, Vol. 18 (*MRS Bulletin*, 1993).
9. C. M. Mate, G. M. McClelland, R. Erlandsson and S. Chiang, *Phys. Rev. Lett.* **59**, 1942–1945 (1987).
10. N. A. Burnham and R. J. Colton, *J. Vac. Sci. Technol. A* **7**, 2906–2913 (1989).
11. H. K. Wickramasinghe, *Scientific American*, pp. 98–105 (Oct. 1989).
12. B. Drake, C. B. Prater, A. L. Weisenhorn, S. A. C. Gould, T. R. Albrecht, C. F. Quate, D. S. Cannell, H. G. Hansma and P. K. Hansma, *Science* **243**, 1586–1589 (1989).
13. F. Ohnesorge and G. Binnig, *Science* **260**, 1451–1456 (1993).
14. F. J. Giessibl, *Science* **267**, 69–71 (1995).
15. Y. Sugawara, M. Ohta, H. Ueyama, S. Morita, F. Osaka, S. Ohkouchi, M. Suzuki and S. Mishima, *J. Vac. Sci. Technol. B* **14**, 953–956 (1996).
16. R. Lüthi, E. Meyer, M. Bammerlin, A. Baratoff, T. Lehmann, L. Howald, Ch. Gerber and H.-J. Güntherodt, *Z. Phys. B* **100**, 165–167 (1996).
17. J. N. Israelachvili, *Intermolecular and Surface Forces*. Academic Press, New York (1985).
18. N. A. Burnham and R. J. Colton, in: *Scanning Tunneling Microscopy and Spectroscopy: Theory, Techniques and Applications*, D. A. Bonnell (Ed.), pp. 191–249. VCH Publishers, New York (1993).
19. J. Israelachvili and G. E. Adams, *J. Chem. Soc. Faraday Trans. 1* **74**, 975–1001 (1978).
20. B. Gauthier-Manuel, *Europhys. Lett.* **17**, 195–200 (1992).
21. C. M. Mate, M. R. Lorenz and V. J. Novotny, *J. Chem. Phys.* **90**, 7550–7555 (1989).
22. G. S. Blackman, C. M. Mate and M. R. Philpott, *Phys. Rev. Lett.* **65**, 2270–2273 (1990).
23. U. Hartmann, *Phys. Rev. B* **43**, 2404–2407 (1991).
24. N. A. Burnham, R. J. Colton and H. M. Pollock, *Phys. Rev. Lett.* **69**, 144–147 (1992).
25. N. A. Burnham, R. J. Colton and H. M. Pollock, *Nanotechnology* **4**, 64–80 (1993).
26. R. G. Horn, J. N. Israelachvili and F. Pribac, *J. Colloid Interface Sci.* **115**, 480–492 (1987).
27. G. U Lee, D. A. Kidwell and R. J. Colton, *Langmuir* **10**, 354–357 (1994).
28. K. L. Mittal (Ed.), in: *Adhesion Measurement of Films and Coatings*, pp. 1–13. VSP, Zeist, The Netherlands (1995).
29. K. L. Johnson, K. Kendall and A. D. Roberts, *Proc. Roy. Soc. Lond. A* **324**, 301–313 (1971).
30. B. V. Derjaguin, V. M. Muller and Y. P. Toporov, *J. Colloid Interface Sci.* **53**, 314–326 (1975).
31. N. A. Burnham, D. D. Dominguez, R. L. Mowery and R. J. Colton, *Phys. Rev. Lett.* **64**, 1931–1934 (1990).
32. R. C. Thomas, J. E. Houston, R. M. Crooks, T. Kim and T. A. Michalske, *J. Am. Chem. Soc.* **117**, 3830–3834 (1995).
33. R. W. Carpick, N. Agraït, D. F. Ogletree and M. Salmeron, *J. Vac. Sci. Technol. B* **14**, 1289–1295 (1996).
34. E. Meyer, R. Lüthi, L. Howald, M. Bammerlin, M. Guggisberg and H.-J. Güntherodt, *J. Vac. Sci. Technol. B* **14**, 1285–1288 (1996).
35. U. Hartmann, *Ultramicroscopy* **42–44**, 59–65 (1992).
36. U. Hartmann, in: *Scanning Tunneling Microscopy III — Theory of STM and Related Probe Methods*, R. Wiesendanger and H.-J. Güntherodt (Eds), pp. 293–360. Springer-Verlag, Berlin (1993).
37. A. L. Weisenhorn, P. K. Hansma, T. R. Albrecht and C. F. Quate, *Appl. Phys. Lett.* **54**, 2651–2653 (1989).
38. A. L. Weisenhorn, P. Maivald, H.-J. Butt and P. K. Hansma, *Phys. Rev. B* **45**, 11 226–11 232 (1992).
39. J. L. Hutter and J. Bechhoefer, *J. Appl. Phys.* **73**, 4123–4129 (1993).
40. R. M. Overney, E. Meyer, J. Frommer, D. Brobeck, R. Lüthi, L. Howald, H.-J. Güntherodt, M. Fujihira, H. Takano and Y. Gotoh, *Nature* **359**, 133–135 (1992).
41. E. Meyer, R. M. Overney, R. Lüthi, D. Brobeck, L. Howald, J. Frommer, H.-J. Güntherodt, M. Fujihira, H. Takano and Y. Gotoh, *Thin Solid Films* **220**, 132–137 (1992).
42. H. I. Kim, T. Koini, T. R. Lee and S. S. Perry, *Langmuir* (1997) (in press).
43. C. D. Frisbie, L. F. Rozsnyai, A. Noy, M. S. Wrighton and C. M. Lieber, *Science* **265**, 2071–2074 (1994).
44. R. J. Hamers, R. M. Tromp and J. E. Demuth, *Phys. Rev. Lett.* **58**, 1972 (1986).

45. W. Barger, D. Koleske, K. Feldman, D. Krüger and R. Colton, *Polymer Preprints* **37**, 606–607 (1996).
46. D. D. Koleske, W. R. Barger, G. U Lee and R. J. Colton, *Mater. Res. Soc. Symp. Proc.* **464**, 377 (1997).
47. D. D. Koleske, G. U Lee, B. I. Gans, K. P. Lee, D. P. DiLella, K. J. Wahl, W. R. Barger, L. J. Whitman and R. J. Colton, *Rev. Sci. Instrum.* **66**, 4566–4575 (1995).
48. When the tip of the cantilever contacts the sample surface, the cantilever will have a tendency to twist during loading particularly if the tip is not aligned with the sample normal. It is in fact nearly impossible to have the tip perfectly aligned since the cantilevers are normally curved due to stress relaxation and the sample will have some tilt and surface roughness. The lateral force map shown here, therefore, may not be related to the frictional property of the material. However, in this case, we do find that the lateral force map does show the same features as frictional force images obtained by sliding the sample relative to the tip.
49. A. Schaper, L. Wolthaus, D. Möbius and T. M. Jovin, *Langmuir* **9**, 2178–2184 (1993).
50. E.-L. Florin, V. T. Moy and H. E. Gaub, *Science* **264**, 415–417 (1994).
51. G. U Lee, L. A. Chrisey and R. J. Colton, *Science* **266**, 771–773 (1994).
52. D. R. Baselt, G. U Lee and R. J. Colton, *J. Vac. Sci. Technol. B* **14**, 789–793 (1996).
53. D. R. Baselt, G. U Lee, K. M. Hansen, L. A. Chrisey and R. J. Colton, *Proc. IEEE* **85**, 672–680 (1997).
54. M. Tortonese, R. C. Barrett and C. F. Quate, *Appl. Phys. Lett.* **62**, 834–836 (1993).
55. M. F. Bocko, K. A. Stephenson and R. H. Koch, *Phys. Rev. Lett.* **61**, 726–729 (1988).
56. M. F. Bocko and K. A. Stephenson, *J. Vac. Sci. Technol. B* **9**, 1363–1366 (1991).
57. T. W. Kenny, W. J. Kaiser, J. A. Podosek, H. K. Rockstad, J. K. Reynolds and E. C. Vote, *J. Vac. Sci. Technol. A* **11**, 797–802 (1993).
58. S. B. Waltman and W. J. Kaiser, *Sensors & Actuators* **19**, 201–210 (1989).
59. T. W. Kenny, S. B. Waltman, J. K. Reynolds and W. J. Kaiser, *Appl. Phys. Lett.* **58**, 100–102 (1991).
60. J. H. Wandass, J. S. Murday and R. J. Colton, *Sensors & Actuators* **19**, 211–225 (1989).
61. R. A. Brizzolara, R. J. Colton, M. Wun-Fogle and H. T. Savage, *Sensors & Actuators* **20**, 199–205 (1989).
62. A. Witek and D. G. Onn, *J. Vac. Sci. Technol. B* **9**, 639–642 (1991).
63. D. D. DiLella, L. J. Whitman, R. J. Colton, T. W. Kenny, W. J. Kaiser, E. C. Vote, J. A. Podosek and L. M. Miller, *Sensors & Actuators* (1997) (submitted).
64. T. W. Kenny, W. J. Kaiser, S. B. Waltman and J. K. Reynolds, *Appl. Phys. Lett.* **59**, 1820–1822 (1991).
65. J. C. Jiang, V. Faynberg, R. C. White and P. K. Allen, *J. Vac. Sci. Technol. B* **11**, 1962–1967 (1993).
66. J. K. Gimzewski, Ch. Gerber, E. Meyer and R. R. Schlittler, *Chem. Phys. Lett.* **217**, 589–594 (1994).
67. T. Thundat, G. Y. Chen, R. J. Warmack, D. P. Allison and E. A. Watcher, *Anal. Chem.* **67**, 519–521 (1995).
68. T. Thundat, E. A. Watcher, S. L. Sharp and R. J. Warmack, *Appl. Phys. Lett.* **66**, 1695–1697 (1995).
69. T. Thundat, S. L. Sharp, W. G. Fisher, R. J. Warmack and E. A. Watcher, *Appl. Phys. Lett.* **66**, 1563–1565 (1995).
70. W. C. Oliver, R. Hutchings and J. B. Pethica, in: *Microindentation Techniques in Materials Science and Engineering*, Vol. STP 889, P. J. Bau and B. R. Lawn (Eds). American Society for Testing and Materials, Philadelphia (1986).
71. T. F. Page, W. C. Oliver and C. J. Hargue, *J. Mater. Res.* **7**, 450–473 (1992).
72. B. Bhushan and V. N. Koinkar, *Appl. Phys. Lett.* **64**, 1653–1655 (1994).
73. J. L. Loubet, M. Bauer, A. Tonck, S. Bec and B. Gauthier-Manuel, in: *Mechanical Properties and Deformation Behavior of Materials Having Ultra-Fine Microstructures*, M. Nastasi (Ed.), pp. 429–447. Kluwer Academic Publishers (1993).
74. N. A. Burnham, G. Gremaud, A. J. Kulik, P.-J. Gallo and F. Oulevey, *J. Vac. Sci. Technol. B* **14**, 1308–1312 (1996).
75. N. A. Burnham, A. J. Kulik, G. Gremaud, P.-J. Gallo and F. Oulevey, *J. Vac. Sci. Technol. B* **14**, 794–799 (1996).

76. N. A. Burnham, A. J. Kulik and G. Gremaud, in: *Procedures in Scanning Probe Microscopies*, R. Colton, A. Engel, J. Frommer, H. Gaub, A. Gewrith, R. Guckenberger, W. Heckl, B. Parkinson and J. Rabe (Eds). Wiley, Chichester (1997) (in press).
77. S. M. Hues and C. F. Draper, in: *Procedures in Scanning Probe Microscopies*, R. Colton, A. Engel, J. Frommer, H. Gaub, A. Gewrith, R. Guckenberger, W. Heckl, B. Parkinson and J. Rabe (Eds). Wiley, Chichester (1997) (in press).
78. I. N. Sneddon, *Intl J. Engng Sci.* **3**, 47–57 (1965).
79. S. M. Hues, C. F. Draper, K. P. Lee and R. J. Colton, *Rev. Sci. Instrum.* **65**, 1561–1565 (1994).
80. S. M. Hues, C. F. Draper and R. J. Colton, *J. Vac. Sci. Technol. B* **12**, 2211–2214 (1994).
81. C. F. Draper, D. M. Schaefer, R. J. Colton and S. M. Hues, in: *NATO ASI Series: Forces in Scanned Probe Methods, Series E: Applied Sciences*, Vol. 286, H.-J. Güntherodt, D. Anselmetti and E. Meyer (Eds), pp. 85–90. Kluwer Academic Publishers, Dordrecht (1995).
82. S. G. Corcoran, R. J. Colton, E. T. Lilleodden and W. W. Gerberich, *Mater. Res. Soc. Symp. Proc.* **436**, 159–164 (1997).
83. S. G. Corcoran, R. J. Colton, E. T. Lilleodden and W. W. Gerberich, *Phys. Rev. B, Rapid Commun.* **55**, R16057–R16061 (1997).
84. S. G. Corcoran, S. M. Hues, R. J. Colton, D. M. Schaefer, C. F. Draper, G. F. Meyers, B. M. DeKoven and S. C. Webb, *J. Mater. Res.* (1997) (in press).
85. I. Gilmour, A. Tainor and R. N. Haward, *J. Polym. Sci. Polym. Phys. Ed.* **12**, 1939–1940 (1974).
86. W. W. Gerberich, J. C. Nelson, E. T. Lilleodden, P. Anderson and J. T. Wyrobek, *Acta Mater.* **44**, 3585–3598 (1996).

Mittal Festschrift, pp. 49–61
W. J. Van Ooij and H. R. Anderson, Jr. (Eds)

Lifshitz–van der Waals, Lewis acid–base and electrostatic interactions in adhesion in aqueous media

C. J. VAN OSS,[1,2,*] W. WU[2] and R. F. GIESE[3]

[1]*Department of Microbiology,*
[2]*Department of Chemical Engineering and*
[3]*Department of Geology, State University of New York at Buffalo, Buffalo, NY 14214-3000, USA*

Abstract—Lifshitz–van der Waals (LW) and Lewis acid–base (AB), together with electrostatic (EL) forces are the non-covalent forces acting in adhesion in condensed phase media, such that the work of adhesion, $W^{adh} = W^{LW} + W^{AB} + W^{EL}$. In the case of serum albumin (SA) and glass surfaces or silica particles, on a *macroscopic* scale, $W^{LW} > 0$, $W^{AB} < 0$ and $W^{EL} < 0$, so that W^{adh} is negative, i.e. repulsive. Nonetheless, in aqueous media, at neutral pH, SA adheres to glass surfaces, as well as to silica particles. It may be hypothesized that on a *microscopic* level, negatively charged, electron-donating SA moieties, located on prominent sites with a small radius of curvature, can penetrate the macroscopic repulsion field and bind to electron-accepting cations imbedded in the glass surfaces (Ca ions) or in silica particles (Si ions). The correctness of the hypothesis is supported by the fact that all adhering SA can be desorbed from, say, silica particles with Na_2-EDTA. Furthermore, energy vs. distance diagrams demonstrate that the more prominently located SA sites with a small radius of curvature should indeed be able to overcome the macroscopic repulsion field and to adhere locally to microscopic cationic sites in the glass or silica. Thus, energy vs. distance balances of the extended DLVO type (including AB as well as LW and EL forces), combining macroscopic and microscopic interactions, can be used to predict adhesion in complex systems.

Keywords: Lewis acid–base interactions; electrostatic interactions; Lifshitz–van der Waals interactions; protein adhesion; protein adsorption; adhesion to glass; adhesion to metal oxides.

1. INTRODUCTION

It has long been known that the adhesion or adsorption of particles or macromolecules in aqueous media onto hydrophobic surfaces has an outcome that tends to be quantitatively and qualitatively different from their adhesion or adsorption onto hydrophilic surfaces [1]. However, the fact that forces acting at entirely different levels could be responsible for the diverging effects has so far rarely been hinted at and the actual forces acting at different levels have only recently been quantitatively estimated [2, 3].

---

*To whom correspondence should be addressed.

The fairly recent developments that made it possible to measure the polar and the apolar properties of hydrophobic as well as of hydrophilic surfaces [4–7] enabled us to calculate the net interaction energy between a clean glass surface and dissolved serum albumin (SA) molecules, in an aqueous medium at pH 7. It turned out that, under these conditions, on a macroscopic level, SA is repelled by clean glass at all distances [2, 3]. This at first glance should make it impossible for SA to adsorb onto glass, at pH 7. However SA definitely does adsorb onto glass under these circumstances [1, 8, 9]. This induced us to perform a closer analysis of the forces at work in the interaction between proteins and glass [2, 3] and, more in general, between proteins and various multivalent metal oxides immersed in water [10, 11]. In all these cases the attractive force acts at a microscopic level between discrete cationic sites (e.g. Ca^{2+} embedded in the glass) and net anionic, protruding peptide moieties with a small radius of curvature of the protein molecules [3].

An interesting aspect of the discrete-site, microscopic level mode of adhesion is that the ultimately adhering protein molecules have to overcome the macroscopic level repulsion. To that effect they must first undergo some degree of rotation, to become properly oriented in one of only a few possible alignments. Next, they can pierce the macroscopic repulsion field, allowing them to become adsorbed to small discrete sites on a microscopic level.

2. THEORY

The theory of macroscopic level interactions in aqueous media has been given elsewhere in some detail [4–7]. Briefly the work of adhesion can be expressed as

$$W^{\text{adh}} = W^{\text{LW}} + W^{\text{AB}} + W^{\text{EL}}, \tag{1}$$

where the superscripts signify LW — Lifshitz–van der Waals, AB — Lewis acid–base, and EL — electrostatic interactions. The work of interaction between two different macromolecules, particles, or surfaces, 1 and 2, immersed in water, w, is designated as $W_{1\text{w}2}$. Using Dupré's approach [4–7], the interfacial part of the work of adhesion, $W^{\text{IF}}_{1\text{w}2}$, may be expressed as

$$W^{\text{IF}}_{1\text{w}2} = \gamma_{1\text{w}} + \gamma_{2\text{w}} - \gamma_{12}, \tag{2}$$

where

$$W^{\text{IF}}_{1\text{w}2} = W^{\text{LW}}_{1\text{w}2} + W^{\text{AB}}_{1\text{w}2} \tag{3}$$

and where γ_{ij} indicates the interfacial tension between i and j. ($W^{\text{EL}}_{1\text{w}2}$ must be determined separately, from electrokinetic data [5, 7].)

Equation (2) may be separated into two parts (cf. equation (3)):

$$W^{\text{LW}}_{1\text{w}2} = \gamma^{\text{LW}}_{1\text{w}} + \gamma^{\text{LW}}_{2\text{w}} - \gamma^{\text{LW}}_{12} \tag{4}$$

and

$$W^{\text{AB}}_{1\text{w}2} = \gamma^{\text{AB}}_{1\text{w}} + \gamma^{\text{AB}}_{2\text{w}} - \gamma^{\text{AB}}_{12}. \tag{5}$$

2.1. The apolar or Lifshitz–van der Waals component of the interfacial tension

The LW or apolar component of the interfacial tension is expressed as

$$\gamma_{ij}^{\mathrm{LW}} = \left(\sqrt{\gamma_i^{\mathrm{LW}}} - \sqrt{\gamma_j^{\mathrm{LW}}}\right)^2. \tag{6}$$

In completely apolar condensed-phase systems, $\gamma_{\mathrm{S}}^{\mathrm{LW}}$ (S = solid) can be measured by contact angle (θ) determination with an apolar liquid (L) using a simple version of Young's equation [4–7]:

$$1 + \cos\theta = 2\sqrt{\gamma_{\mathrm{S}}^{\mathrm{LW}}/\gamma_{\mathrm{L}}^{\mathrm{LW}}}. \tag{7}$$

$\gamma_{\mathrm{S}}^{\mathrm{LW}}$ can also be measured by contact angle determination with a polar liquid, provided that γ_{L}, as well as $\gamma_{\mathrm{L}}^{\mathrm{LW}}$ is known ($\gamma_{\mathrm{L}} = \gamma_{\mathrm{L}}^{\mathrm{LW}} + \gamma_{\mathrm{L}}^{\mathrm{AB}}$, and $\gamma_{\mathrm{L}}^{\mathrm{AB}} = 2\sqrt{\gamma_{\mathrm{L}}^{\oplus}\gamma_{\mathrm{L}}^{\ominus}}$; see below):

$$(1 + \cos\theta)\gamma_{\mathrm{L}} = 2\sqrt{\gamma_{\mathrm{S}}^{\mathrm{LW}}\gamma_{\mathrm{L}}^{\mathrm{LW}}}. \tag{8}$$

W^{LW} decays with the square of the distance, l, for flat parallel surfaces, and with the distance, l, between interacting spheres.

2.2. The polar or Lewis acid–base component of the interfacial tension

The AB or polar component of the interfacial tension is expressed as

$$\gamma_{ij}^{\mathrm{AB}} = 2\left(\sqrt{\gamma_i^{\oplus}\gamma_i^{\ominus}} + \sqrt{\gamma_j^{\oplus}\gamma_j^{\ominus}} - \sqrt{\gamma_i^{\oplus}\gamma_j^{\ominus}} - \sqrt{\gamma_i^{\ominus}\gamma_j^{\oplus}}\right). \tag{9}$$

In polar condensed-phase systems, $\gamma_{\mathrm{S}}^{\mathrm{AB}}$ can also be measured by contact angle determinations, using an extended version of Young's equation [4–7]:

$$(1 + \cos\theta)\gamma_{\mathrm{L}} = 2\left(\sqrt{\gamma_{\mathrm{S}}^{\mathrm{LW}}\gamma_{\mathrm{L}}^{\mathrm{LW}}} + \sqrt{\gamma_{\mathrm{S}}^{\oplus}\gamma_{\mathrm{L}}^{\ominus}} + \sqrt{\gamma_{\mathrm{S}}^{\ominus}\gamma_{\mathrm{L}}^{\oplus}}\right). \tag{10}$$

Clearly, γ_i^{LW}, $\gamma_i^{\oplus}$ and $\gamma_i^{\ominus}$ are measured together so that when i = S (where S is a flat, smooth, solid surface) contact angle determinations should be done with three different liquids, L (of which two must be polar) to solve for the three unknowns in equation (10). Conversely, when i = L, contact angle determinations should be done on three different solids, S. The high-energy polar contact angle liquids for which $\gamma_{\mathrm{L}}^{\oplus}$ and $\gamma_{\mathrm{L}}^{\ominus}$ have been characterized [4–7] are water, glycerol, and formamide. The high-energy apolar contact angle liquids that are used most for the measurement of $\gamma_{\mathrm{S}}^{\mathrm{LW}}$ are diiodomethane and 1-bromonaphthalene (both of which are almost completely apolar). It is necessary to use high-energy liquids for contact angle determinations, because it is essential that $\gamma_{\mathrm{L}} > \gamma_{\mathrm{S}}$, for θ to be greater than 0°. A number of γ^{LW}, $\gamma^{\oplus}$ and $\gamma^{\ominus}$ values for various materials can be found in Chapter XIII of [7]. It is not possible to measure contact angles on rough surfaces with any accuracy [12], so that it is not advisible to measure contact angles on layers of particles, deposited on a flat surface. However, the γ^{LW}, $\gamma^{\oplus}$ and $\gamma^{\ominus}$ values of particles (e.g. glass powder, non-swelling clay particles, etc.) can be determined by measuring the rate of capillary rise of different liquids through flat beds of such particles, by means of thin layer wicking [6, 7, 13], which yields $\cos\theta$.

3. MACROSCOPIC SCALE INTERACTIONS

With a view to the measurement methods, described above, it is clear that only quantitative surface properties are obtained that are averaged over several μm^2 (for particles) to several mm^2 (for flat surfaces on which contact angles were measured). The macroscopic-scale properties that can be measured by contact angle determinations, and by electrokinetic approaches, are treated in this Section. Microscopic-level interactions are considered in Section 4, below. Excluding for the moment the electrostatic work of macroscopic interaction, the interfacial work of macroscopic interaction between two molecules or particles, 1, immersed in water, w, is expressed as

$$W_{1w1}^{IF} = 2\gamma_{1w}, \tag{11}$$

(cf. equation (2)), which can also be written (equation (3)) as

$$W_{1w1}^{IF} = 2\left(\gamma_{1w}^{LW} + \gamma_{1w}^{AB}\right). \tag{12}$$

Inserting the values for γ_{1w}^{LW} (equation (6)) and for γ_{1w}^{AB} (equation (9)) into equation (12):

$$\left[W_{1w1}^{IF}\right] = \overset{(a)}{2\left(\gamma_1^{LW} - \gamma_w^{LW}\right)^2} + 4\left(\overset{(b)}{\sqrt{\gamma_1^{\oplus}\gamma_1^{\ominus}}} + \overset{(c)}{\sqrt{\gamma_w^{\ominus}\gamma_w^{\oplus}}} - \overset{(d)}{\sqrt{\gamma_1^{\oplus}\gamma_w^{\ominus}}} - \overset{(e)}{\sqrt{\gamma_1^{\ominus}\gamma_w^{\oplus}}}\right), \tag{13}$$

where term (a) represents the Lifshitz–van der Waals or apolar contribution and terms (b–e) the polar contributions.

3.1. Attractive forces: Hydrophobic interactions

3.1.1. Completely apolar substances. For completely apolar substances, 1, immersed in water, term (a) of equation (13) is positive (i.e. attractive), but small, in practice varying only between 0 and 1 mJ/m^2 [6, 7].

For completely apolar materials, 1, immersed in water, terms (b, d and e) (equation (13)) are zero [7, 14]. However, term (c), which is always present and considerable, has the value of 25.5 mJ/m^2, at 20°C; it represents the hydrogen-bonding work of cohesion of water, such that $W_{ww}^{AB} = 102\ \mathrm{mJ/m^2}$, at 20°C. Thus for completely apolar materials, 1, immersed in water, and neglecting for the moment the small contribution of W_{1w1}^{LW}, it may be stated that

$$W_{1w1}^{IF} = W_{ww}^{AB} = 102\ \mathrm{mJ/m^2}. \tag{14}$$

The driving force for the hydrophobic attraction between two completely apolar molecules or particles immersed in water therefore corresponds for more than 96% with the hydrogen-bonding work of cohesion of water (W_{ww}^{AB}), which amounts to $102\ \mathrm{mJ/m^2}$, at 20°C. Thus, W^{AB}, in the attractive mode, is usually by far the strongest factor in noncovalent interactions in water.

3.1.2. Partly polar substances. For partly polar substances, 1, immersed in water, term (a) can, in practice, amount to 0 to $3.5\ \mathrm{mJ/m^2}$ (making W_{1w1}^{LW} from 0 to $13\ \mathrm{mJ/m^2}$). Term (c) (equation (13)), i.e. the omnipresent hydrogen-bonding work of cohesion of water remains $25.5\ \mathrm{mJ/m^2}$ (contributing $102\ \mathrm{mJ/m^2}$ to W_{1w1}^{AB}). Most partly polar substances have a zero or negligible $\gamma_1^{\oplus}$ (which eliminates terms (b) and (d) of equation (13)). However in the case of partly polar substances, term (e) of equation (13) is no longer negligible; the more polar (i.e. the more hydrophilic) substance 1 is, the higher the value of term (e) of equation (13) becomes, thus [on account of the negative sign of (e)], diminishing the work of hydrophobic attraction embodied in term (c) [7, 14]. In other words, whereas the work of hydrophobic attraction is attributable to the hydrogen-bonding work of cohesion of water, in the case of partly polar substances [term (c), equation (13)] this work of attraction is counterbalanced to a greater or lesser degree by a repulsion originating from the interaction between the electron-donors of substance 1 with the electron-acceptors of water [term (e) of equation (13)], which corresponds to the polar energy of hydration of substance 1.

For $W_{1w1}^{AB} > 0$, W_{1w1}^{AB} decays with distance, l, as $\exp[(l_0 - l)/\lambda]$, where $l_0 \approx 0.16$ nm and λ varies from 0.2 nm for short distances to (allegedly) more than 10 nm for $l > 10$ nm [7].

However, it now seems increasingly safe to assume λ to be equal to 1 nm for all distances l equal to or larger than 1 nm [15]. The reason for the exponential decay with distance of both attractive and repulsive AB interactions has to do with the change in water structure, density and/or orientation in close vicinity to the solid surface. This peculiarity of the structure of 'vicinal' water decays exponentially to the random state as a function of distance away from the solid surface [7, 16, 17]. In contrast to EL forces, ionic strength as such does not significantly affect AB forces. For the influence of multivalent counterions on polar surfaces, however, see Section 3.4, below.

3.2. Repulsive forces: Hydration pressure

Returning to equation (13), it is clear that W_{1w1}^{LW} always remains atractive [term (a)] but, as mentioned above it usually amounts to less than $13\ \mathrm{mJ/m^2}$. Term (c) is positive: it represents the underlying energy of hydrophobic attraction, amounting to 102 mJ/m^2 (the hydrogen-bonding energy of cohesion of water at 20°C) — this attractive energy is always present. Thus, to achieve a net total work of repulsion, an underlying work of attraction of 102 to 115 mJ/m^2 must be surmounted, usually in the guise of term (e) (equation (13)). Thus term (e) (equation (13)) must be greater than 1/4 of 102 to 115 mJ/m^2, i.e. greater than 25.5 to 28.75 mJ/m^2. To that effect, the electron-donor parameter of the polar surface tension component of substance 1, $\gamma_1^{\ominus}$, must have a minimum value from 25.5 (for $W_{1w1}^{LW} = 0$) to 32.4 mJ/m^2 (for $W_{1w1}^{LW} = 13$ mJ/m^2), to overcome the underlying omnipresent hydrophobic and van der Waals attraction [6, 7, 14]. All water-soluble proteins, soluble polysaccharides, and other polymers such as poly(ethylene oxide), poly(vinyl alcohol), etc., in the hydrated state, have $\gamma^{\ominus}$ values well over 30 mJ/m^2 [7]. Thus, in most cases where hydrophilic repulsion occurs, W^{AB} is also the predominant factor in the total noncovalent interaction in water.

For $W_{1w1}^{AB} < 0$, W_{1w1}^{AB} decays with distance, l, as $\exp[(l_0 - l)/\lambda]$, where $l_0 \approx 0.16$ nm and λ varies from 0.2 nm at close range to about 1 nm for other distances [7].

3.3. Reversal of hydrophobic attraction

As the work of hydrophobic attraction originates from the hydrogen-bonding work of cohesion of water, the hydrophobic attraction can be reversed by decreasing that hydrogen-bonding energy of cohesion among the water molecules. This can be done by admixing water-miscible organic solvents that have a much lower hydrogen-bonding energy of cohesion than water, thus lowering the total energy of cohesion among the liquid molecules and, in so doing, decreasing the hydrophobic attraction energy. In practice one uses, for example, ethylene glycol, ethanol, dimethyl sulfoxide or acetonitrile, as additives to water, to decrease or reverse the hydrophobic attraction [6, 7]. Admixtures of, for example, acetonitrile or ethylene glycol are used for reversing the hydrophobic attraction in the eluting step of reversed-phase liquid chromatography [7, 14].

3.4. Link between electrostatic and Lewis acid–base interactions

Electrostatically totally neutral surfaces can have a very high $\gamma^{\ominus}$-value, making them exceedingly hydrophilic and causing them to repel each other. Charged (say, negatively charged) surfaces also tend to have a high $\gamma^{\ominus}$-value, which contributes strongly to their hydrophilicity, and concomitantly to their mutual repulsion. The hydrophilic repulsion (hydration pressure) is usually even stronger than the repulsion engendered by opposing electric charges of the same sign. When measures are taken to decrease the electric surface (ζ) potential, the $\gamma^{\ominus}$-value is lowered at the same time, which then causes the formerly strongly charged surfaces to become hydrophobic. The decrease in ζ-potential brought about by the addition of plurivalent counterions, for example, plurivalent cations, in the case of negatively charged particles, causes such particles in aqueous suspension to attract each other, thus destabilizing the suspension and giving rise to flocculation. This phenomenon has long been known [18–20] and is traditionally ascribed to the van der Waals attraction becoming dominant as a consequence of the decrease in the electrostatic repulsion. In reality, it can be shown that the hydrophobizing effect caused by the decrease in ζ-potential is quantitatively much more important than the (small) van der Waals attraction [21, 22]. The hydrophobization (i.e. the decrease in $\gamma^{\ominus}$) of charged surfaces, macromolecules, phospholipids or particles, concomitant with a lowering of their ζ-potential, can be brought about by causing the pH of the medium to approach the isoelectric point (in the case of amphoteres) [23, 24]; by adding modest amounts of plurivalent counterions [21, 22, 25, 26]; or by raising the ionic strength with mono-monovalent salts [27, 28]; see also [6, 7]. Thus, the admixture of fairly low concentrations of $CaCl_2$ to phospholipid surfaces (e.g. in liposomal or cell-membranes) favors cell–cell adhesion and, subsequently, cell fusion, by enhancing hydrophobic interactions [14, 29].

3.5. Extended DLVO approach

The DLVO theory[1] does not work in water [6, 7, 27]. Briefly, this approach holds that it suffices to determine both van der Waals and electrostatic interactions, as a function of distance, to achieve a complete description of the interaction between macromolecules, particles, or other surfaces immersed in a liquid medium. However, as shown in Section 3.1, in water, hydrophobic attractions can account for more than 90% and up to 100% of the total interaction forces. In the case of biological materials, such as blood plasma proteins or blood cells, the ζ-potentials are relatively so low as to represent only a small proportion of the total interaction forces. For instance, the stability of blood cells in peripheral blood is for more than 90% due to hydrophilic (Lewis acid–base) and less than 10% to electrostatic repulsion forces, whilst the van der Waals attraction is exceedingly small [7, 33, 34]. The ζ-potentials of biological materials, such as proteins and cells, at physiological pH and ionic strength are rather small. At a maximum, for example, in the cases of serum albumin or erythrocytes, $\zeta \approx -25$ mV; for other blood plasma proteins and blood cells, $|\zeta|$ is much smaller. In systems with higher ζ-potentials, such as aqueous suspensions of clay particles, hydrophobic attractions [27,28] or hydrophilic repulsions [21, 22, 28] also play a crucial role. Thus, in aqueous media, Lewis acid–base forces usually are predominant and must be taken into account in an extended DLVO (XDLVO) approach (see Fig. 1).

4. MICROSCOPIC SCALE INTERACTIONS

4.1. Discrete-site EL and AB attraction

As was already mentioned in Section 1, notwithstanding an overall macroscopic-scale repulsion between serum albumin (SA) and clean glass, at pH 7.0 [2, 3], SA does adsorb onto glass, at pH 7.0, to a degree which is about 10% of that attainable in the adsorption of SA to hydrophobic surfaces [1]. For the adsorption of proteins to glass, this is due to the microscopic scale interaction between local sites comprising plurivalent cations embedded in the glass (e.g. Ca^{2+}) with negatively charged moieties on the protein [3].

---

[1]Named after Derjaguin and Landau [30], Verwey and Overbeek [31, 32].

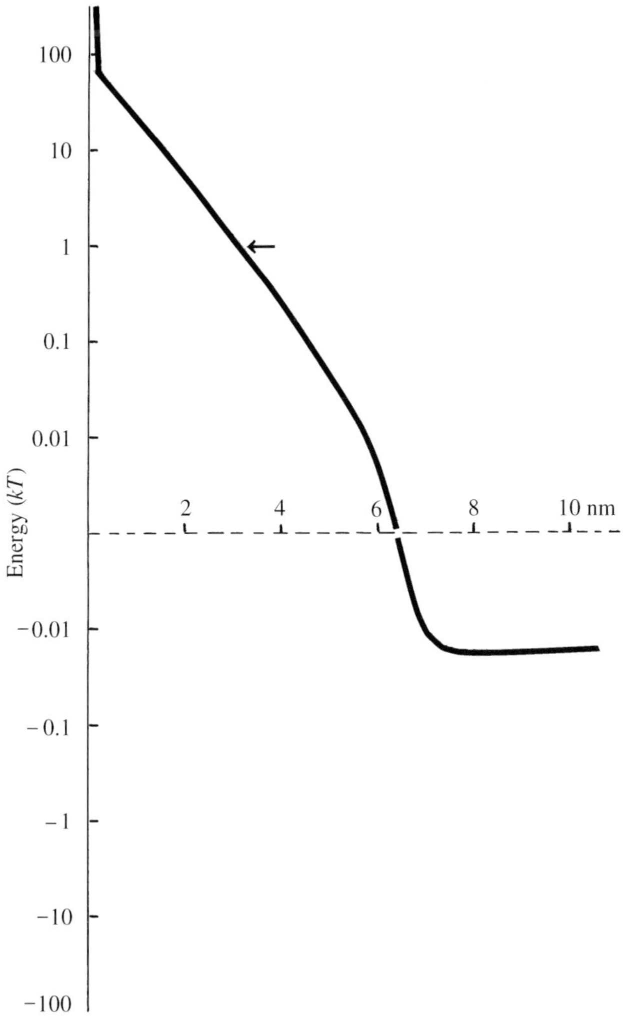

Figure 1. XDLVO diagram of the macroscopic interaction between a human serum albumin (HSA) molecule in aqueous solution and a clean glass plate [3]. The HSA molecule is oriented with one of its sites of small area of curvature ($R = 1.3$ nm) toward the glass surface, which allows the closest approach. Nonetheless, on a macroscopic level, the HSA molecule cannot approach the glass surface more closely than about 3 nm (at $\Delta G = +1\ kT$; see arrow). Note that $W = -\Delta G$.

4.2. *Combined macroscopic and microscopic XDLVO approach*

By combined microscopic and macroscopic XDLVO approaches it could be shown that it is only the prominent protein moieties with the smallest radii of curvature that can pierce the macroscopic repulsion field and bind to discrete cationic sites embedded in the glass surface [3] (see Fig. 2). It should also be noted that the cationic sites in the glass are also electron-acceptors, which can engage in a Lewis acid–base attraction with the electron-donor moieties on the proteins' periphery. It was subsequently shown that SA also readily adsorbs onto SiO_2, SnO_2, ZrO_2 and other metal oxide particles

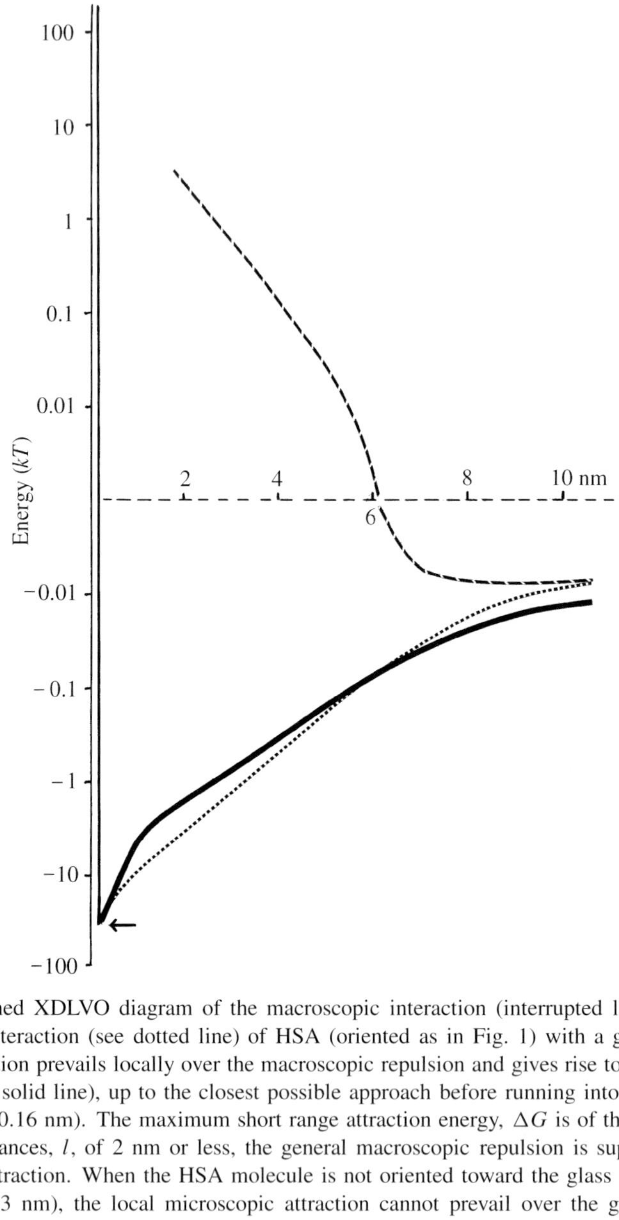

Figure 2. Combined XDLVO diagram of the macroscopic interaction (interrupted line, cf. Fig. 1) and the microscopic interaction (see dotted line) of HSA (oriented as in Fig. 1) with a glass plate [3]. The microscopic attraction prevails locally over the macroscopic repulsion and gives rise to a strong net attraction (see the thick solid line), up to the closest possible approach before running into the Born repulsion (vertical line at = 0.16 nm). The maximum short range attraction energy, ΔG is of the order of $-20\ kT$, see arrow. At distances, l, of 2 nm or less, the general macroscopic repulsion is superseded locally by the microscopic attraction. When the HSA molecule is not oriented toward the glass with a site of small curvature ($R = 1.3$ nm), the local microscopic attraction cannot prevail over the general macroscopic repulsion [3].

with increased SA adsorption in the order: $SiO_2 < SnO_2 < ZrO_2$. Of these metal oxide particles, SiO_2 is the most hydrophilic and ZrO_2 the most hydrophobic [11].

4.3. Reversal of microscopic-scale EL and AB adhesion

As a strong indication that the interaction between SA and these metal oxides may indeed be attributed to a metal cation–peptide chain interaction, it was observed that the microscopic-scale protein adsorption to these insoluble metal oxides was reversible upon the admixture of Na_2-EDTA [11].

4.4. Other microscopic-scale discrete-site modes of adhesion

4.4.1. Ion exchange driven adhesion. Although the mechanism of protein adsorption to ion exchange beads or other surfaces has the microscopic-scale electrostatic attraction in common with the type of discrete-site protein adsorption discussed in Sections 4.1 to 4.3 above, the ion exchange mechanism differs from that of the latter. Ion exchange offers only a weak macroscopic-scale repulsion, or often a net macroscopic-scale attraction, as a consequence of the much higher surface potential of the ion exchange beads' surface, with an overall sign of charge opposite to that of the protein. Thus, in most cases, a macroscopic-scale and a microscopic-scale cooperate in achieving protein adhesion to the ion-exchange material's surface. Reversal of protein adhesion is usually effected by changing the pH of the liquid medium, so as to change the sign of charge of the protein, but not of the ion exchanger.

4.4.2. Metal chelate protein adhesion. Metal chelate affinity binding of histidine-containing proteins [35], or immobilized metal-activity protein adhesion [36], works by means of the microscopic-scale attraction between a divalent metal (usually Zn^{2+} or Cu^{2+}), attached to a chelating gel (e.g. iminodiacetic acid coupled to a gel) and a histidine-containing protein. Here, the metal appears to act as a Lewis acid (electron-acceptor), which can bind fairly strongly to exposed histidine residues on a peptide chain of a protein, and somewhat less strongly to tryptophan residues [36]. The histidine moieties here act as Lewis bases (electron-donors). The adhesion of histidine-containing proteins (e.g. lysozyme, ribonuclease, albumin, myoglobin, protein A, avidin, interferon [36]) occurs at alkaline pH. Reversal of protein adhesion can be effected at nearly neutral pH by exchange with low molecular weight electron-donors such as imidazole [36].

4.4.3. Specific site bioadhesion. Specific site bioadhesion to various ligands (or receptors) is an important and very common phenomenon, which is at the origin of antibody–antigen, lectin–carbohydrate, enzyme–substrate, and other receptor–ligand interactions. The mechanism has much in common with the discrete-site type of adhesion of proteins to glass, metal oxides, etc., treated in Sections 4.1 to 4.3 above. An important difference is that in specific site bio-adhesion, the three-dimensional shapes of the specific sites on, e.g. antibody (Ab) and antigen (Ag) have to be complementary. The 'complementarity determining regions', or CDR of Ab and Ag are named *paratope* and *epitope*, respectively. The paratopes (most antibodies, or immunoglobulins have two identical paratopes) represent only 1%, or less, of the total immunoglobulin molecules, i.e. they usually consist of about 6 and rarely more than 10 amino acids each. Epitopes are, necessarily, about the same size; they can be part of Ag molecules which are proteins, or polysaccharides, or glycoproteins, or lipoproteins, or lipopolysaccharides, etc. Ag molecules can have many epitopes, which may all be identical in shape and structure, or several or all of the epitopes of a given Ag may be different in their primary structures.

The microscopic-scale attraction forces between epitope and paratope are fairly long-range, i.e. they have to be able to surmount an overall long-range macroscopic-scale repulsion field when still 2 to 4 nm removed from each other [37]. In the first stages of the epitope–paratope attraction, only electrostatic (EL) and/or Lewis acid–base (AB) (or hydrophobic) attractive forces play a role. At close range, more and more interstitial water of hydration becomes extruded, which increases the entropy of the system. When all or most interstitial water of hydration has been extruded, Lifshitz–van der Waals (LW) forces become dominant [37, 38].

Adhesion between lectins and carbohydrates, enzymes and substrates, and other receptor–ligand systems have similar mechanisms. In all cases of specific attraction, the closest conformation in shape between CDRs gives rise to the highest energy of adhesion, because the closer the three-dimensional conformation, the shorter the final distance between the two CDR surfaces. It should be recalled that AB and EL attraction energies increase exponentially with a decrease in distance, and LW energies decrease as the inverse square of the distance [5–7]; see also Sections 2.1 and 3.1, above.

It should be recalled that hydrophobic (AB) attractions can readily occur between a hydrophobic and a hydrophilic surface [6, 7], and that for proteins to be soluble in water, their overall surface must be hydrophilic [6, 7, 14]. Nevertheless, epitopes, which have hydrophilic surfaces [39], can bind hydrophobically to specific immunoglobulins, which also are hydrophilic (to remain soluble). This hydrophobic attraction is made possible because the hydrophobic patch of the paratope tends to be situated in a concavity or cleft of the immunoglobulin molecule [37, 38]. This facilitates the epitope–paratope attraction, while avoiding undue adhesion among immunoglobulins molecules or between immunoglobulins and other protein molecules.

Thus, microscopic-scale specific adhesion between discrete complementary sites on biopolymers can occur whenever required, without impairing the overall macroscopic-scale repulsion between biopolymers, which is a *conditio sine qua non* for their continued solubility and for the avoidance of large-scale formation of thrombi.

4.4.4. Semi-specific bioadhesion. Somewhat less specific for unique epitopes on specific proteins, some organic compounds are prone to bind a whole category of proteins, but do not bind to proteins outside of the category. An important case in point is the organic dye Cibacron Blue. When attached to a gel or other carrier, Cibacron Blue will bind, for example, to serum albumins and to proteolytic enzymes [40], but not, for example, to immunoglobulins.

5. CONCLUSIONS

By taking Lifshitz–van der Waals, electrostatic and Lewis acid–base interaction energies, as a function of distance (extended DLVO approach) into account, adhesion in aqueous media can be quantitatively depicted. For adhesion of polar polymers to high-energy polar surfaces (e.g. metal oxides), an analysis has to be made which combines extended DLVO plots on a macroscopic and on a microscopic level.

Adhesion of proteins to metal oxides such as silica appears to take place between a protein moiety of low radius of curvature and an ionized metal site, as this type of adhesion can be reversed by the addition of a complexing agent.

REFERENCES

1. F. MacRitchie, *J. Colloid Interface Sci.* **38**, 484 (1972).
2. C. J. van Oss, W. Wu and R. F. Giese, *Abstr. 207th Natl ACS Meeting*, San Diego, CA (March 1994); Abstr. No. 88, Div. Colloid and Surface Chem.
3. C. J. van Oss, W. Wu and R. F. Giese, in: *Proteins at Interfaces 1994*, T. E. Horbett and J. L. Brash (Eds), pp. 80–91. ACS Symposium Series 602, Washington, DC (1995).
4. C. J. van Oss, M. K. Chaudhury and R. J. Good, *Advan. Colloid Interface Sci.* **28**, 35 (1987).
5. C. J. van Oss, M. K. Chaudhury and R. J. Good, *Chem. Rev.* **88**, 927 (1988).
6. C. J. van Oss, *Colloids Surfaces-A* **78**, 1 (1993).
7. C. J. van Oss, *Interfacial Forces in Aqueous Media*. Marcel Dekker, New York (1994).
8. I. Michaeli, D. R. Absolom and C. J. van Oss, *J. Colloid Interface Sci.* **77**, 586 (1980).
9. D. R. Absolom, I. Michaeli and C. J. van Oss, *Electrophoresis* **2**, 273 (1981).
10. C. J. van Oss, B. Sébille, J. Piquion, C. Vidal-Madjar, W. Wu and R. F. Giese, *Abstr. 207th Natl ACS Meeting*, San Diego, CA (March 1994); Abstr. No. 109, Div. Colloid and Surface Chem.
11. C. J. van Oss, W. Wu, R. F. Giese and J. O. Naim, *Colloids Surfaces-B* **4**, 185 (1995).
12. M. K. Chaudhury, *Short-range and Long-range Forces in Colloidal and Macroscopic Systems*. Ph.D. Dissertation, SUNY, Buffalo (1984).
13. C. J. van Oss, R. F. Giese, Z. Li, K. Murphy, J. Norris, M. K. Chaudhury and R. J. Good, *J. Adhesion Sci. Technol.* **6**, 413 (1992).
14. C. J. van Oss, *Colloids Surfaces-B* **5**, 91 (1995).
15. J. Wood and R. Sharma, *Langmuir* **11**, 4797 (1995).
16. N. A. M. Besseling, *Statistical Thermodynamics of Fluids with Orientation-Dependent Interactions*. Ph.D. Dissertation, Agricultural University of Wageningen, The Netherlands (1993).
17. B. V. Derjaguin, in: *Theory of Stability of Colloids and Thin Films*, pp. 23–27. Consultants Bureau/ Plenum Press, New York (1989).
18. H. Schulze, *J. Prakt. Chem.* **25**, 431 (1882); **27**, 320 (1883).
19. W. B. Hardy, *Proc. Roy. Soc. London* **66**, 110 (1900); *Z. Phys. Chem.* **33**, 385 (1900).
20. J. Th. G. Overbeek, in: *Colloid Science*, H. R. Kruyt (Ed.), Vol. 1, p. 58. Elsevier, Amsterdam (1952).
21. W. Wu, R. F. Giese and C. J. van Oss, *Colloids Surfaces-A* **89**, 241 (1994).
22. W. Wu, *Linkage between ζ-Potential and Electron-Donicity of Charged Polar Surfaces*. Ph.D. Dissertation, SUNY, Buffalo (1994).
23. S. R. Holmes-Farley, R. H. Reamey, T. J. McCarthy, J. Dent and G. M. Whitesides, *Langmuir* **1**, 725 (1985).
24. C. J. van Oss, *J. Protein Chem.* **8**, 661 (1989).
25. S. Ohki, *Biochim. Biophys. Acta* **689**, 1 (1982).
26. Y. Guo, S. Ohki and C. J. van Oss, *Colloids Surfaces-B* (submitted).
27. C. J. van Oss, R. F. Giese and P. M. Costanzo, *Clays Clay Miner.* **38**, 151 (1990).
28. P. Chheda and D. Grasso, *Langmuir* **10**, 1044 (1994).
29. C. J. van Oss, M. K. Chaudhury and R. J. Good, in: *Molecular Mechanics of Membrane Fusion*, S. Ohki, D. Doyle, T. D. Flanagan, S. W. Hui and E. Mayhew (Eds), p. 113. Plenum, New York (1988).
30. B. V. Derjaguin and L. D. Landau, *Acta Physicochimica URSS* **14**, 633 (1941).
31. E. J. W. Verwey and J. Th. G. Overbeek, *Theory of the Stability of Lyophobic Colloids*. Elsevier, Amsterdam (1948).

32. J. Th. G. Overbeek, in: *Colloid Science*, H. R. Kruyt (Ed.), Vol. 1, p. 245. Elsevier, Amsterdam (1952).
33. C. J. van Oss, *Cell Biophys.* **14**, 1 (1989).
34. C. J. van Oss, in: *Biophysics of the Cell Surface*, R. Glaser and D. Gingell (Eds), p. 131. Springer, New York (1990).
35. J. Porath, J. Carlsson, I. Ilsson and G. Belfrage, *Nature* **258**, 598 (1975).
36. E. Sulkowski, *Trends Biotech.* **3**, 1 (1985).
37. C. J. van Oss, in: *Immunochemistry*, C. J. van Oss and M. H. V. van Regenmortel (Eds), p. 581. Marcel Dekker, New York (1994).
38. C. J. van Oss, *Molec. Immunol.* **32**, 199 (1995).
39. C. J. van Oss and J. O. Naim, *Immunol. Invest.* **21**, 649 (1992).
40. R. K. Scopes, in: *Protein Purification*, p. 125. Springer-Verlag, New York (1982).

Mittal Festschrift, pp. 63–80
W. J. Van Ooij and H. R. Anderson, Jr. (Eds)

Relationships between interfacial acid–base interactions and adhesive bond strength

DAVID W. DWIGHT

Composites Innovation Laboratories, Owens Corning Science and Technology Center, 2790 Columbus Road, Granville, OH 43023, USA

Abstract—Acid–base interactions across interfaces are shown to have predictable influences on adhesion. The history of this development, and methods to assay the acid–base character of solvents, polymers and a variety of powders and fibers are reviewed briefly. Recent studies are described that demonstrate directly how acid–base interactions influence both 'fundamental' and 'practical' adhesion.

Keywords: Acid–base; donor–acceptor; work of adhesion; adhesive bond strength; AFM; IFM; XPS; SEM.

1. INTRODUCTION

On the occasion of the First International Conference on Adhesion Science and Technology, it was an honor to be invited to present an overview of the role of acid–base interactions in adhesion. This paper captures the essence of that presentation. In the first section, the development of our understanding of intermolecular interactions is traced from the initial, gas phase studies to modern work on adhesive bonding. Due to the breadth of the subject and the availability of several excellent references, this section is little more than an annotated bibliography. The main focus is on descriptions of recent studies in which quantitative measurements of both acid–base surface chemistry and adhesive bond strengths have been correlated. Also, these studies span the range of both *fundamental* and *practical* adhesion as defined by Mittal [1]. Section 2 discusses fundamental studies at atomic–molecular dimensions, in which the Surface Forces Apparatus (SFA), Atomic Force Microscopes (AFM), and the Interfacial Force Microscope (IFM) have been used to measure uniquely strong attractive interactions and charge exchange as surfaces with different chemistry were brought into contact and separated. Section 3 summarizes some of our own work applying acid–base principles in a study of practical adhesive bonding. Approximate work-of-adhesion values were calculated from contact angle adsorption isotherms, and correlated with adhesive peel strength measurements and semi-quantitative fracture surface analyses using X-ray photoelectron spectroscopy (XPS).

1.1. Historical background

The development of understanding of forces of attraction and repulsion at the atomic, molecular and colloidal levels occurred during the period from 1870 to 1950. Understanding the forces involved in *separating* solids — the field of fracture mechanics — developed amongst ceramists and metallurgists from 1920 to 1950. In the 1960s and 1970s a wealth of new information became available about macroscopic solid surfaces, *via* thermodynamic, spectroscopic and microscopic methods. Many of those data were used to develop relationships between the composition and structure of interfaces, and the physical properties of bonded surfaces. At the same time, significant interaction began to develop between the surface science and fracture mechanics communities, setting the stage for the rapid expansion of adhesion science and technology that has occurred over the last 20 years.

The science of intermolecular forces was first established in the gas phase, where pair-wise, 'physical' interactions are dominant because of the relatively large intermolecular distances. The so-called non-polar or 'dispersion–force' interactions caused by the ubiquitous electronic fluctuations in all matter were clarified sequentially in the works of van der Waals [2], London [3] and Lifshitz [4]. Specific interactions in the gas phase were explained adequately by Keesom [5] and Debye [6] in terms of dipole–dipole attractions. Note that hydrogen bonds (electron donor–acceptor interactions) require specific bond angles and distances, and are therefore rare in the gas phase. It is important to realize that in the liquid and solid states, the high density of neighboring atoms renders dipole–dipole attraction ineffective, and the short bond lengths enhance specific interactions. In the case of water, for example, dipole–dipole interactions account for 76% of its cohesive energy density in the *vapor* phase, but less than 3% in the *liquid* phase.

Hildebrand [7] showed how to extrapolate the non-specific interaction theory to liquids in order to predict *endothermic*, liquid solubilities. The key equation in that 'solubility parameter theory' relates the interaction energy between liquids to the geometric mean of their individual cohesive energy densities. Girifalco and Good [8], and Fowkes [9] applied that concept to liquid–liquid interfacial interactions in non-polar systems. Many researchers have incorrectly extrapolated those concepts to systems with so-called 'polar' interactions. Consequently, geometric mean equations are found in the literature in various attempts to rationalize liquid and solid interfacial interactions. However, interfacial systems that have specific interactions of the electron donor–acceptor type display *exothermic* behavior, which requires a different theoretical framework to generate fundamental understanding.

The unique, 'monopolar' interactions predominant in condensed matter were termed hydrogen bonds by Brönsted [10]. Pearson's work on the hard–soft acid–base (HSAB) principle expanded the concept beyond proton transfer to include polarizability and ionization potential [11]. The research of Lewis [12] and of Mulliken [13] extended understanding of acid–base interactions to non-aqueous systems, and introduced the theory of electron donor–acceptor interactions. Drago [14] applied those principles in studies of the *exothermic* heats-of-solution of dilute solutions of Lewis acids and bases in non-interacting solvents. Accurate correlation with a four-parameter equation resulted in the assignment of 'E' (electrostatic) and 'C' (covalent) constants for each of 40 organic acids and bases. Relative hardness or softness is measured by the ratio, E/C, thus incorporating the HSAB principle. Gutmann [15] also used dilute solution calorimetry to determine a scale of donor numbers, DN. However, in the NMR spectroscopic shift method developed for Gutmann's scale of acceptor numbers, AN, no solvent was used. Subsequently, the van der Waals contribution to the chemical shift has been determined for 30 liquids, and a corrected list of acceptor numbers, AN^*, has been published [16].

Turning to the solid state, understanding of the forces of attraction and repulsion was first extended to colloidal particles in terms of van der Waals dispersion forces by Hamaker [17]. Subsequently, electrostatic forces were addressed by the Derjaguin–Landau–Verwey–Overbeek (DLVO) model [18]. The influence of chemical composition of surfaces — as well as the influence of the intervening medium — could now be computed and compared with experiment. In verifying those theories, however, measurements were made of surfaces coming into contact, not separating from each other after contact.

Quite independently, an energy balance approach was developed by Griffith [19], Orowan [20] and Irwin [21] in order to understand the fracture of brittle solids. According to the principles of fracture mechanics, the energy required to cause failure is a product of two terms: one for the surface energies of the *parting* surfaces, and the other for the energy-storage capacity of the solid, normalized by a geometrical factor accounting for stress intensity at crack tips. Fracture surface analysis became important for technological applications of fracture mechanics. Microscopic and spectroscopic data help to explain the influence of initial flaws, weak boundary layers, and inherent microstructure, as well as possible deviations from elastic behavior during fracture. The fracture mechanics approach was extended to polymeric solids through the Johnson–Kendall–Roberts (JKR) compliance model [22]. With specific focus on extending the fracture mechanics approach to the field of practical adhesion, the nature of the proportionality between the thermodynamic work of adhesion and adhesive bond strength was demonstrated by Gent and Schultz [23]. Good [24] considered the fracture mechanics variables in adhesive joints, and concluded that detailed study of *failure zone* would be necessary to understand fundamental mechanisms operating in *practical* adhesion. Theoretical and experimental elaboration of these concepts has been published [25].

Frederick M. Fowkes was one of the foremost pioneers of the acid–base approach to help understand adsorption and adhesion. Twenty two papers documenting the state of development of the field were collected in a 1991 Festschrift volume [26] by Mittal and Anderson, honoring Prof. Fowkes. This is the primary reference for theoretical background, analytical methods and many practical, applied studies; according to the editors, it "represents the cumulative wisdom and thinking of contemporary researchers engaged actively in unravelling acid–base interactions and how these can be utilized in understanding and tailoring adhesion." However, direct correlation between adhesive bond strength and some measure of acid–base interactions were rare. Applications of acid–base interactions usually demonstrated their influence upon adsorption and composite materials properties.

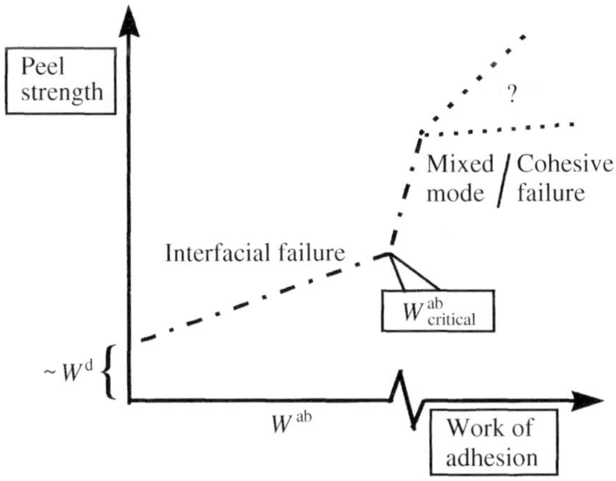

Figure 1. Schematic diagram illustrating the relationship between adhesive peel strength (practical adhesion) and thermodynamic work of adhesion, according to our hypotheses. Note that dispersion forces are assumed to contribute an essentially constant amount, W^{d}, to the peel strength. This is shown as the (non-zero) minimum value of peel strength on the ordinate. It is the acid–base component of the work of adhesion that can be varied from zero to values even greater than the dispersion contribution. In the low-strength regime, where failure occurs at the original interface, a monotonic increase in peel strength with work of adhesion is predicted. Above some value denoted $W^{ab}_{critical}$, the original interface does not fail, and thus predictions based upon equilibrium thermodynamics are no longer expected to hold. The right-hand side of the figure illustrates the qualitative prediction that peel strength will increase substantially above $W^{ab}_{critical}$. There are new fracture mechanisms in operation, and the question mark means that there is no known relationships to compute peel strength in cases of mixed modes of failure.

Part of the difficulty in predicting adhesive bond strength from data on the characteristics of the adherends and adhesives is the fact that the *formation* of adhesive bonds is governed by principles that are very different from the principles governing the *failure* of the bond. The schematic diagram in Fig. 1 illustrates this problem: quantitative predictions of mechanical strength based on surface thermodynamics can be valid only on the left-hand side, when (1) fracture occurs at the original interface, and (2) the joining and failure processes are reversible. In this regime, strength values usually are too low for practical applications. But it is only here that predictions can be made from surface chemistry or thermodynamics. Therefore, an important question that relates fundamental to practical adhesion for any specific materials system is: How can the surface or interphase be modified so as to exceed the critical value of the acid–base contribution to the work of adhesion, $W^{ab}_{critical}$? Then, as indicated on the right-hand side of Fig. 1, bulk (cohesive) failure mechanisms dominate, and produce much greater adhesive bond strengths. Among the new, energy-absorbing mechanisms available during mixed mode and cohesive failure of adhesively bonded joints are elastic and plastic deformation, crazing and high-surface-area, brittle cleavage cracking, many examples of which have been published [27].

1.2. Surface analytical tools and applications

Semi-empirical approaches for enhancing adhesive bond strength may be carried out if suitable methods are available to rank the acid–base character of the materials involved. The development of methods to characterize acidity and basicity of powders, fibers and films is an active area of research (and controversy). Most of the spectroscopic, thermodynamic and electrostatic techniques were described (and often applied) in the Fowkes Festschrift volume [26]. Developments in the myriad thermodynamic methods have been reviewed recently [28]. Investigators who are new to this field should proceed with caution in the use of published techniques, and especially in applying them in new and different materials studies. A good approach is to use at least two different techniques and develop corroborating information. Most methods rely upon measuring (and interpreting) the interaction of solid surfaces with small molecules or ions ('probes') of known acid–base character. Conclusions can depend upon (1) the choice of probes, (2) whether they are used in the gas phase, as a neat liquid or in dilute solution, and (3) the measurement method. For example, strongly interacting (chemisorbed) probes are useless in the usual inverse gas chromatography (IGC) experiment because they will not desorb unless the temperature is ramped up. On the other hand, when using probes in combination with XPS analyses, probe chemisorption is essential: weakly interacting probes will be volatilized in ultrahigh vacuum during the XPS experiment.

1.2.1. Fiber-reinforced polymer composites. In terms of practical application of acid–base concepts, perhaps fiber-reinforced polymer matrix composites have received the most attention. Several groups have characterized various surface treatments on glass, carbon and graphite fibers by XPS, IGC, contact angles, programmed thermal desorption (PTD) and flow microcalorimetry (FMC). Some studies have been successful in correlating fiber and matrix surface chemistry with measured adhesion (interfacial shear strength) and composite strength.

IGC and wetting methods have dominated the literature of fiber surface thermodynamic characterization. However, the specific approaches to selection of probes and theories of data reduction are still under development. Carbon fiber surface treatments (electrooxidation and plasma) were clearly distinguished from each other, even though the correlation to interfacial shear strength was not direct [29]. Wetting, IGC and PTD showed qualitative correlation with each other and with XPS spectra, indicating that treatments served primarily to increase acidity through the addition of hydroxyl and carboxyl groups [30].

Adsorption isotherms of silver and magnesium ions from solution, followed by XPS, quantified surface acidity and correlated with single fiber fragmentation results [31]. Aqueous streaming potential measurements confirmed the acidity of treated carbon fibers and correlation was shown between thermodynamic work of adhesion and interfacial shear strength with an epoxy resin [32]. Subsequently, correlation has been shown among Lewis acidity measured by IGC and by wetting (using water, dimethyl sulfoxide (DMSO), α-bromonaphthalene), Brönsted acidity measured by streaming potential, XPS hydroxyl content, and interfacial shear strength in an epoxy resin [33].

Using somewhat the reverse approach, four different matrixes — poly(vinyl chloride) (PVC), poly(methyl methacrylate) (PMMA), poly(ethyl methacrylate) (PEMA) and poly(methacrylonitrile) (PMAN) — were evaluated for 'polarity' and Lewis basicity by contact angles of formamide, glycerol, ethylene glycol, and water [34].

Plasma-treated carbon fibers were found to be acidic using XPS. Single filaments were imbedded in each matrix, and critical fiber lengths were determined. Correlation was found between interfacial adhesion (inversely related to critical lengths) and the degree of Lewis basicity (PVC < PMMA < PEMA < PMAN), but not with the 'polar component of surface tension.'

The role of Lewis acid–base interactions at the fiber–matrix interface in composites was studied with both glass [35] and Teflon® fibers [36]. In the glass fiber case, surface chemistry was modified with amino-, methacryloxy- and glycidoxy-silane coupling agents. Silane adsorption mechanisms as well as the properties of filament-wound, unidirectional epoxy and polyester composites were explained by a combination of XPS, scanning electron microscopy (SEM), and FMC. Heats of adsorption of pyridine and phenol proved that the coupling agents added acidic sites to the glass fiber surface as well as much stronger basic sites. The subsequent adhesion of the matrix polymers and the short beam shear strengths of composites were explained on this basis. The Teflon fibers were first etched with sodium naphthalene solutions, and then monofunctional hydroxyl (acidic) and ester (basic) surfaces were created by sequential reactions with diborane/peroxide and acetic anhydride. Composites of both functional fibers and controls were prepared with a chlorinated poly(vinyl chloride) (acidic) matrix. Those made with the acetylated fibers were superior in tensile properties, and SEM fractography showed PTFE fibrillation, indicative of good fiber–matrix adhesion and stress transfer. The hydroxylated fibers performed no better than untreated Teflon fibers. This study provided semi-quantitative evidence for the positive influence of acid–base interactions on adhesion in composite materials. Section 3 of this paper reports follow-up studies using Teflon FEP films with four different surface functional groups, and four model adhesives with different functional group acidity and basicity.

1.2.2. Polymer film surfaces. The sorption of acidic and basic 'probe' molecules and ions has been monitored by XPS and by contact angle changes, making it possible to quantify the donor–acceptor character of low-surface area specimens. A relative scale of polymer basicity has been obtained using XPS to measure the uptake of sodium cations (Na^+) by polymeric films cast onto glass slides containing sodium oxide [37]. The diffusion of Na^+ from the glass substrate to the polymers was interpreted in terms of acid–base interactions between Na^+ (Lewis acid) and a series of basic polymers; poly(styrene) (PS), PMMA, and poly(ethylene oxide) (PEO), and a series of acidic polymers; poly(vinyl butyral) (PVB), PVC, and poly(vinylidene fluoride) (PVdF).

The acid–base properties of PMMA and a commercial polyurethane have been studied by swelling of polymer films in solvent vapors having a range of Gutmann's donor and acceptor numbers. The solvent concentration in the swollen polymer determined by XPS described the Lewis base strength of the polymer [38]. The binding energies of Cl $2p$ and F $1s$ from the acidic solvents were shifted to low values because of the formation of the solvent-polymer adduct. The N $1s$ chemical shift of pyridine following interaction with PMMA was positive suggesting that pyridine was the Lewis base and PMMA was the Lewis acid species, i.e. PMMA is amphoteric [39]. Using chloroform as an acidic probe, a correlation was determined between the Cl $2p$ binding energies and ΔH^{ab} for a series of polymers (acetate, acrylates, ethylene oxide and urethane) [40]. These shifts in core orbital binding energies are small — 0.1 eV commonly — and that may limit broad applicability. Nevertheless, the approach was useful in understanding the adhesion of vacuum evaporated aluminum to polypropylene (PP) treated by ammonia plasmas. In the case of oxygen plasma treated PP, DMSO was employed as a probe to interact with the acidic sites on that surface [41]. Finally, acid–base interactions accompanying PMMA adsorption onto conducting polymers were studied by XPS and ToF–SIMS [42].

Two approaches have been published for determining the acid–base character of polymer films from contact angles of neat liquid probes that have previously determined quantitative acid–base parameters [43, 44]. Lee has reviewed the controversies among the various approaches (e.g. surface tension components and equation-of-state) [45]. Our approach is illustrated in the practical adhesion examples in Section 3 of this paper. We measure contact angles versus temperature of several concentrations of Lewis acid and base probe solutions in methylene iodide [46]. From these data adsorption isotherms and heats of adsorption are calculated.

2. FUNDAMENTAL ADHESION

Fundamental experiments using the SFA and the AFM have provided a new level of detail about the physics and chemistry of joining and separating solid surfaces. For example, simultaneous measurements of force–distance curves and charge exchange between *dissimilar* surfaces have provided convincing evidence of an electron donor–acceptor mechanism of interfacial bond formation. Comparison of contact and separation measurements made with various substrates in aqueous solution, humid air and dry nitrogen appear to point the way to a fundamental unification of dispersion, acid–base, electrostatic and frontier orbital energetics. Elegant measurements of the physics of interaction between model surfaces with controlled chemistry have led to interpretations of the data that bridge the gap between chemists' and physicists' descriptions.

In the pioneering SFA experiments, silica–mica and silica–silver surfaces (both asymmetric cases) 'jumped' into contact and transferred charge up to 10 mC/m^2 upon separation, whereas mica–mica pairs (symmetric case) showed weak attraction and little charge transfer [47]. Subsequent work showed pure van der Waals attraction of 0.1 J/m^2 in symmetric cases, and no attraction until after the first contact in the asymmetric case, whereupon 6–9 J/m^2 work was required for separation (i.e. in range of the fracture energies of ceramics) [48].

The connection between Derjaguin's electrostatic model of joint separation and Fowkes' acid–base model of joint formation was made in the interpretation of similar SFA experiments with silica and silica coated with a monolayer of aminopropyl silane [49, 50]. Force–distance curves obtained for symmetrical surface contacts in water were fitted very well by the classical DLVO theory of electrostatic repulsion between electrical double layers. However, long range electrostatic attraction characterized the interactions between the asymmetric pairs of surfaces. Contact electrification (2 mC/m^2) in dry nitrogen produced 3 J/m^2 separating energies, as well as electrical discharges, leaving the silica (silanols) negative and the silated surface (amines) positive. This confirms that the expected acid–base charge exchange occurred upon contact. Similar data were collected in an AFM with a silica sphere attached to the cantilever, and the results compared with the corresponding SFA experiments [51]. Again, the DLVO theory described symmetrical cases in aqueous solution, but silica/gold interactions became attractive within 2–3 nm of contact, and produced strong adhesion and instabilities in the cantilever.

The 'instability problem' has been overcome with a new, zero-compliance force sensor in an instrument called the Interfacial Force Microscope (IFM). Using this tool, artifact-free force–distance curves have been measured for self-assembled monolayers of methyl-, carboxyl- and amine-terminated alkanes [52]. The schematic diagram of those unique adhesion experiments is shown in Fig. 2. After evacuating to 10^{-4} mm Hg and backfilling with dry nitrogen, the first contact and separation produced the force–distance curves in Fig. 3. Inspection of Fig. 3 provides qualitative confirmation of the relative magnitudes predicted for van der Waals, hydrogen bonding and acid–base interactions. Not shown in the figure are data discussed in the paper on the asymmetric contact: methyl/carboxyl, which produced results equivalent to the methyl/methyl contact. This is an important experimental validation of the hypothesis of additivity of van der Waals and acid–base interactions.

The authors were unable to model the force–distance curves to their satisfaction. Nevertheless, it is interesting to note the values of the work of adhesion they calculated from the peak values of the force–distance curves. Using mechanics models developed for a sphere interacting with a flat surface, the work of adhesion equals 60, 100, 230 and 680 mJ/m^2 going from the top to the bottom of Fig. 3. Remarkably, there is quantitative agreement with expectations for the first three cases (methyl–methyl:

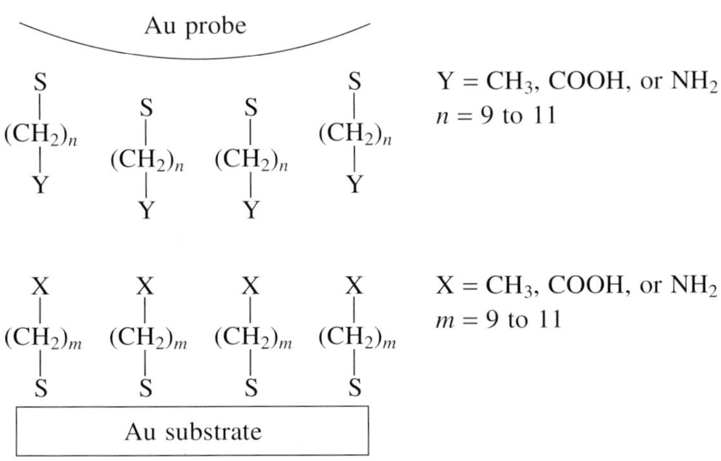

Figure 2. Schematic diagram illustrating the Interfacial Force Microscope (IFM) experiments using alkane thiol Self-Assembled Monolayers (SAMs) with methyl, amine, and carboxylic acid terminal groups. By bringing various pairs of functional groups together in this arrangement, direct measurements of dispersion–forces and donor–acceptor interactions at solid surfaces were made [52].

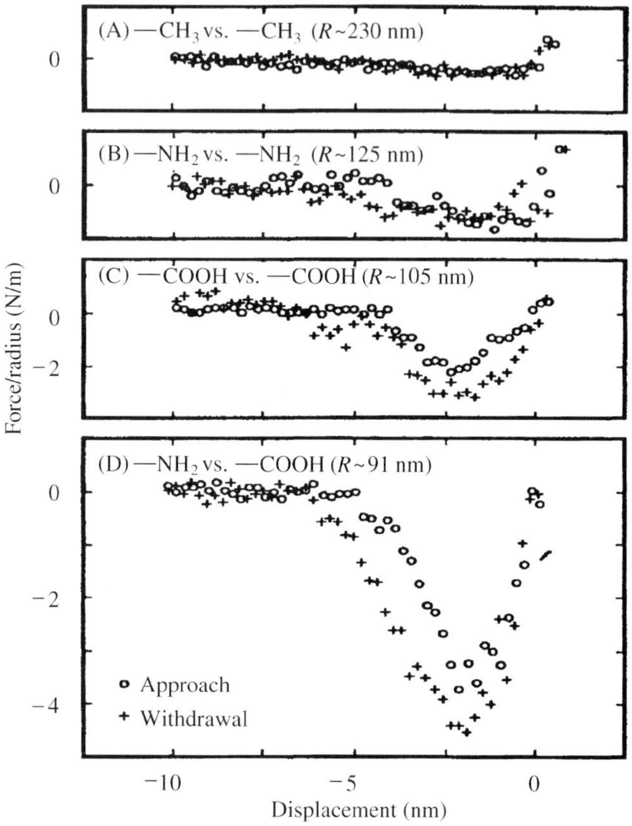

Figure 3. Representative IFM force–distance curves from the experiments sketched in Fig. 2 (R indicates the tip radius used in each case). Panels (A) and (B) show that dispersion–force interactions between amine groups are somewhat larger than those between methyl groups. Hydrogen bonding interactions between carboxylic acid groups in panel (C) are at least twice the magnitude of the dispersion forces, and the acid–base interaction in panel (D) appears greater by another factor of two [52].

van der Waals only; amine–amine: one hydrogen bond per molecule; carboxyl–carboxyl: two hydrogen bonds per molecule). The very high value obtained for the acid–base interaction (amine–carboxyl) was rationalized by the additional contributions of interfacial free energy and proton transfer. However, failure analysis might reveal that irreversible mechanisms also accompanied the detachment event in this case. Indeed, it appears that there is a unique hysteresis in the last force–displacement curve: upon unloading, much higher forces persist from about −2 to −7 nm displacements. The authors also mention the possibility of various extrinsic effects, e.g. the size and shape of the probe as well as adsorption of impurities.

3. PRACTICAL ADHESION

Previously, we demonstrated how the properties of Teflon FEP film were uniquely suited to high resolution surface analyses, surface modification and adhesion studies [53]. In a natural extension of that background, we began a comprehensive project to make quantitative correlations among: (1) the acid–base character of polymer film substrates and adhesives, (2) their adhesive peel strengths, and (3) fracture surface analyses [54]. The formation of four monofunctional surfaces, and their characterization has been described in detail [55]. The peel strength data and some of subsequent SEM and XPS fractographs are available in conference proceedings [56]; herein those results are contrasted and compared in terms of the schematic diagram in Fig. 1.

XPS and attenuated total reflection infrared (ATR-IR) spectroscopy were used to characterize the molecular structure of the surface region of four functional Teflon FEP film surfaces: two acidic (containing hydroxyl and carboxyl groups), and two basic (containing acetyl and dinitrobenzoate groups). Contact angle adsorption isotherms at two temperatures were determined using phenol as an acidic probe and tetrahydrofuran (THF) as a basic probe. The carboxylated surface had a higher molar heat of adsorption, ΔH^{ab} with THF than the hydroxylated surface, and neither surface had any interaction with phenol, the acidic probe. The acetylated surface behaved as a base, interacting with phenol but not with THF, while the dinitrobenzoyl surface had both acidic and basic characters.

At this point, the study adopted a technological style: first, it was assumed that the acid–base contribution to *any adhesive's work of adhesion* is directly proportional to the *specific probe's heat of adsorption*; second, adhesive acidity or basicity was 'selected' by comparing the primary functional group in the polymer repeat unit with Drago's E and C tables; and third, variations in adhesive viscoelastic response were not taken into account explicitly. Four polymers were chosen as model adhesives — two with acidic functional groups and two with basic groups: (1) a copolymer of polyethylene and 20% acrylic acid (PEAA, Scientific Polymer Products) with carboxylic acid groups and a softening point of 85°C, (2) post-chlorinated poly(vinyl chloride) (CPVC, B.F. Goodrich) with acidic hydrogen atoms on the chlorinated backbone carbon atoms and a softening point of 120°C, (3) PMMA (Polysciences) with basic carbonyl functional groups and a glass transition temperature of 105°C, and (4) a 1 : 1 copolymer of methyl vinyl ether and maleic anhydride (PMVE, GAF Chemicals) with basic methoxy and anhydride groups and a softening point of 205°C. Solutions of 5% by weight in the appropriate solvent (PEAA: decahydronaphthalene (decalin), CPVC: 1,2-dichloroethane, PMMA: toluene, PMVE: THF) were flow-coated onto 1×4 cm strips of functionalized Teflon substrate. After overnight evaporation of the solvents, films were bonded together for 5 min in a hydraulic press at 35 kPa and 100–150°C, depending upon the adhesive. Bondline thickness was about 2 μm. Adhesion tests were carried out at room temperature in the 180° peel mode at 0.5 cm/min using an Instron Universal Testing Machine.

The relationships between the work of adhesion calculated from the adsorption of the basic probe (THF), and the peel strengths measured for the two basic adhesives is illustrated in Fig. 4A. In that bar chart, the trends in strength from one type of surface treatment to another (indicated by the dashed lines) are similar to the trends in W^{ab}. Also, the CPVC adhesion data in Fig. 4B correspond well to the trend predicted by phenol adsorption. Note that the magnitudes associated with both phenol adsorption and acidic adhesive peel strengths range 50% or more lower than in the

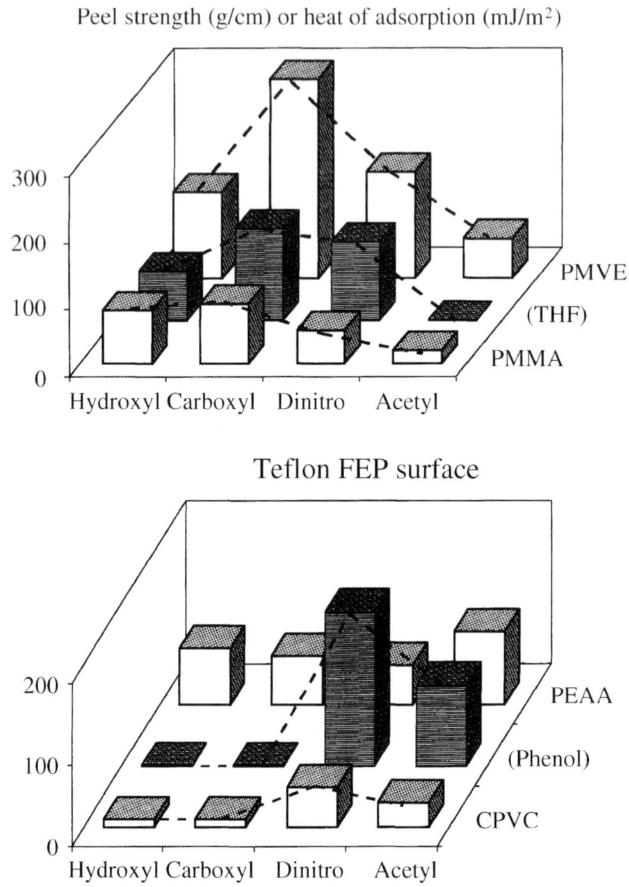

(A)

(B)

Figure 4. (A) Bar charts illustrating the similarity in trends between the thermodynamic heat of adsorption calculated from tetrahydrofuran probe-solution contact angle data with the trends in peel strength results for model adhesives poly(methyl methacrylate) (PMMA) and poly(methyl vinyl ether) (PMVE). It is pure coincidence that the values for work of adhesion are of similar magnitude as the peel strength values; the units are entirely different. (B) Bar charts comparing the thermodynamic heat of adsorption calculated from phenol probe-solution contact angle data with the peel strength of the two acidic model adhesives. Chlorinated poly(vinyl chloride) (CPVC) adhesion follows the trend predicted by probe adsorption, but all the poly(ethylene-co-acrylic acid) (PEAA) joints failed cohesively in the PEAA, and were insensitive to variations in substrate surface chemistry.

cases of THF and the basic adhesives. In terms of Fig. 1, this leads to the expectation that interfacial failures are more likely with the acidic adhesives. In fact, in the case of CPVC adhesive, we have published SEM and XPS fractographs showing pure interfacial failure on both acidic Teflon film surfaces, but on the basic surfaces mixed-mode failure (bulk Teflon plus interfacial failure) [55] was observed.

The schematic diagram in Fig. 5 illustrates the three loci-of-failure observed: (A) interfacial, and cohesive (B) in the substrate, or (C) in the adhesive. The following SEM and XPS fractographs on the peeled specimens provide a semi-quantitative under-

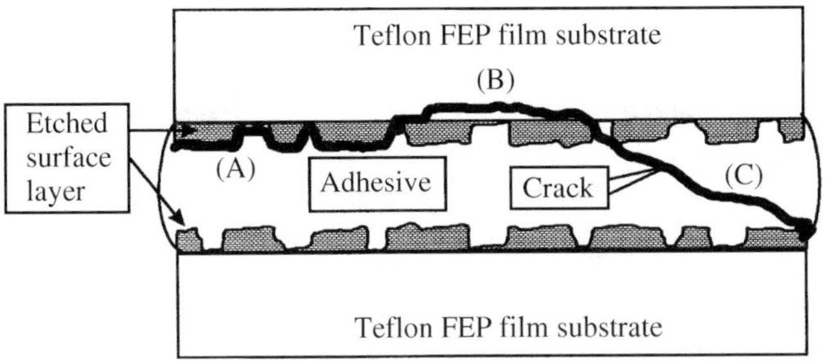

Figure 5. Schematic diagram summarizing the observations of SEM fractography in terms of types of failure observed on the peel strength test specimens: (A) at the original interface, (B) in the Teflon substrate, and (C) in the adhesive layer.

standing of the relationships between peel strength and locus-of-failure as indicated in Fig. 1. Figures 6A, B and D show the SEM evidence for interfacial failure. Figures 6C, 7B and D are examples of cohesive failures in the various adhesives, and Fig. 7C is a cohesive failure in the Teflon FEP substrate. At lower magnification in Fig. 7A, the patches of adhesive and substrate result from the mixed modes of failure caused by the peeling front moving back and forth from one substrate to the other through the adhesive layers, as illustrated in Fig. 5.

PEAA peel strength values appear to be essentially independent of the Teflon surface chemistry. Fractography explains this apparent anomaly: both XPS and SEM results indicate that all PEAA joints failed cohesively in the middle of the adhesive layer. Virtually no fluorine (from the Teflon FEP substrate) appears in the XPS spectra on any of the fracture surfaces, and SEM shows ductile polymer failure features covering both sides of the peeled surfaces (Fig. 6C). Apparently in these cases, the work of adhesion arising from dispersion forces alone is sufficient to drive the peeling crack tip into the ductile PEAA adhesive, where it is stable at the slow peel rate used. In fact, these peel specimens, especially when bonded with PEAA, provided interesting examples of the interplay among the interfacial, viscoelastic, and geometric factors involved in 'practical' adhesion failure. In general, very erratic results were obtained for bondlines thicker than about 4 μm. Reproducible data were obtained for 1–2 μm bondlines. Also, some PEAA-bonded specimens were peeled by hand at a rate much higher than attainable with the Instron. In these cases, SEM fractography shows interfacial failure (Fig. 6D) with both acidic substrates. However, PEAA cohesive failure was still observed with both basic substrates. We were unable to measure the peel strengths at high rates, but we would predict from Fig. 1 that the latter cases would be stronger. Thus PEAA would also fit the pattern seen in CPVC.

All the other fracture surfaces were much more complex, showing 100 to 500 μm patches of alternating bulk and interfacial failure; the SEM photomicrograph in Fig. 7A is one example. Attempts to derive quantitative information from the SEM photomicrographs were abandoned because of the wide distribution of shapes and areas of the three types of failure, as well as differences in their microscopic deformation and fracture processes. The latter were more pronounced in the cases of the basic adhesives, illustrated in Figs 7B, C and D. There are copious microscopic deformations that undoubtedly are responsible for increased energy absorption, and thus the

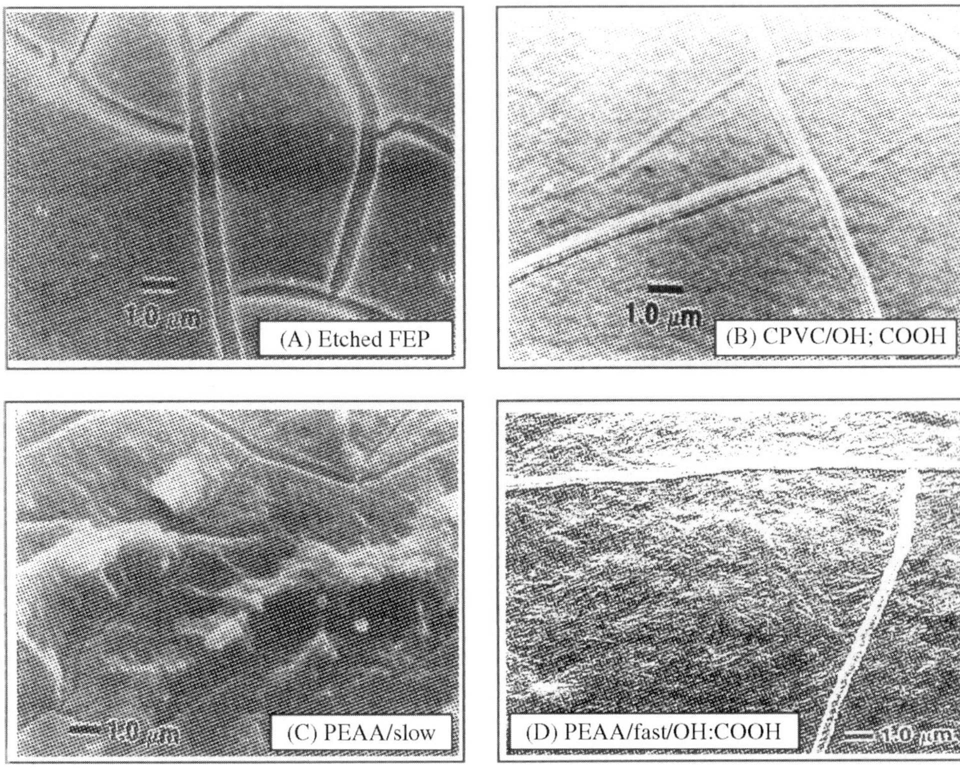

Figure 6. High magnification (5000×) SEM photomicrographs representative of several of the fracture surfaces selected from the acidic adhesive (CPVC and PEAA) peel strength test specimens. (A) Micro-cracked Teflon FEP substrate typical of all the surfaces after defluorination by the etching solution. Subsequent 'monofunctionalizing' reactions did not alter this topography which is characterized by myriad cracks about 1 μm wide and 0.5 μm deep. (B) Interfacial failure typical of CPVC bonded to either acidic substrate. Note the evidence of excellent wetting and interfacial contact. It is remarkable that the adhesive pulled cleanly from such fine surface features with 'mirror-image' replication of the substrate. (C) Typical failure surface from all the PEAA Instron tests, showing a layer of adhesive covering the etched surface. Some of the substrate's microcracks are still visible in places, partially filled with residual adhesive. All of both sides of the peeled surfaces were identical to this view, indicating ductile, cohesive failure within the PEAA. (D) PEAA failure surface typical of either acidic substrate, but obtained by high speed peeling by hand. Obviously the adhesive failed at the original interface, even pulling neatly out of the microcracks, very similar to the low-strength CPVC fracture in panel (B).

relatively high peel strength values. However, close inspection of the relative volume and structure of individual microscopic deformations indicates that they are characteristic of the specific adhesive polymer. From their radically different appearance, one would expect different contributions to the measured peel strength.

Our XPS data provide an average of the surface composition of approximately 3 mm^2 of the fracture surface. Furthermore, the XPS fluorine spectrum is a unique and sensitive indicator for the Teflon substrate. If failure were 100% in the Teflon FEP film substrates, the relative fluorine concentration would be 67%. The XPS fluorine concentrations from the fracture surfaces are listed in Table 1, and clearly dif-

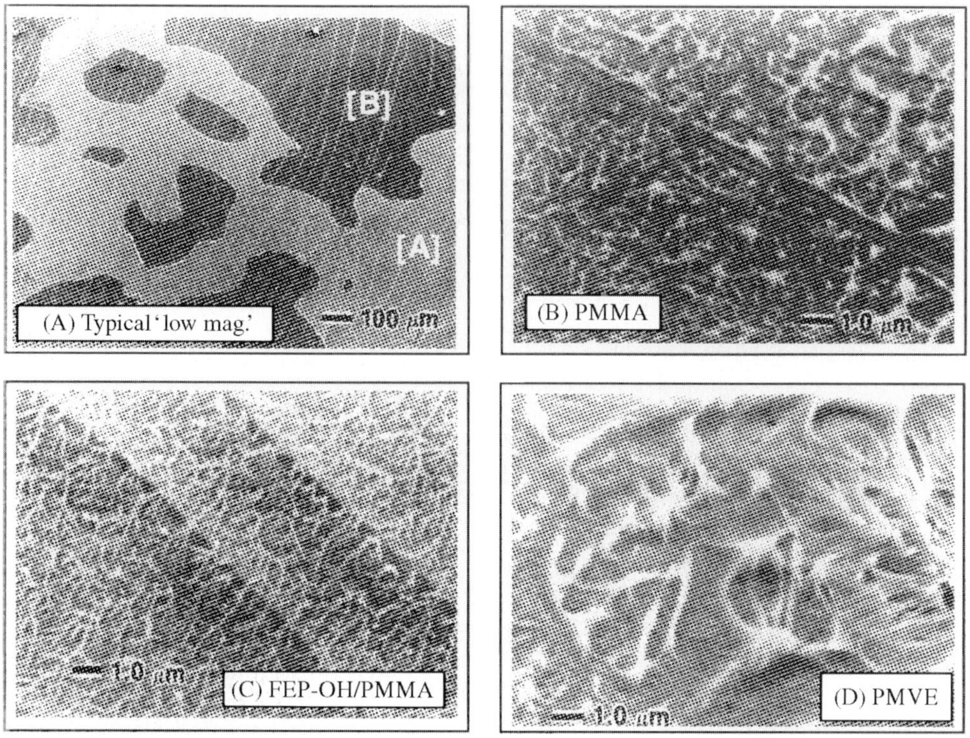

Figure 7. Fractographs representative of the basic adhesives (PMMA and PMVE). (A) Low magnification SEM photomicrograph of a fracture surface typical of many in this study. At the scale of about 0.1 to 1 mm, there are patches of adhesive [A] adhering to both substrates. There are series of roughly parallel cracks, separated by 10 to 100 μm, running in the etched surface layer [B], and sometimes propagating through the adhesive layer. At a smaller scale, it is apparent that varying amounts of the surface region of the etched and monofunctionalized Teflon FEP substrates remain attached to the patches of adhesive. Therefore XPS fluorine concentrations obtained from areas about five times this size were used as an index to quantify this complex type of locus-of-failure (see Table 1 and Fig. 8). (B) Interfacial failure in the PMMA case. (C) Teflon substrate failure region on a PMMA-bonded sample. Microfibrillation of the fluoropolymer created a significant amount of new surface area, and the plastic deformation processes obviously absorbed some strain energy. (D) PMVE interfacial failure surface, showing 'leathery' plastic deformation, along with crazing and micro-cracking that created a large amount of new surface area.

Table 1.
XPS fracture surface analysis from the sixteen different pairs of peeled surfaces. Tabulated are the average fluorine atom percentages detected, which indicates the degree to which the crack ran in the Teflon FEP substrate (Teflon contains about 67% fluorine)

Teflon surface	Model adhesives			
	PMVE	PMMA	CPVC	PEAA
Hydroxyl	40	50	2	2
Carboxyl	35	10	2	2
Dinitrobenzoyl	40	50	20	3
Acetyl	55	60	20	0

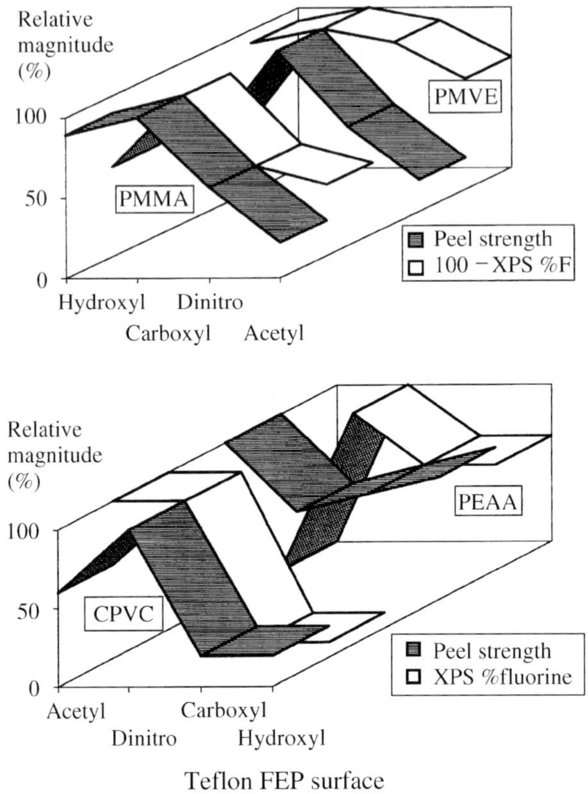

(A)

(B)

Figure 8. Ribbon plots comparing normalized peel strength values (dark) with the corresponding normalized function of the XPS fluorine concentrations (white) from the fracture surfaces of the peel strength test specimens. There is a similarity in trends, as predicted by Fig. 1, even though most of the failures were not interfacial. (A) For both basic adhesives, there is no interfacial failure on any of the samples, and peel strength increases when the crack prefers to run in the adhesive, indicated by the increase in the value of (100 − XPS %fluorine). (B) In the case of CPVC, interfacial failure corresponds to low strength, but on the basic (acetyl and dinitro) surfaces, peel strength increases as the crack runs to some degree in the Teflon substrate (indicated by the increase in XPS %fluorine).

ferentiate the acidic and basic adhesives. Failure in the Teflon FEP film substrate is predominant for the basic adhesives, but is almost absent in the acidic adhesives. The relationships between peel strength and locus-of-failure are illustrated in Figs 8A and B for the basic and acidic adhesives, respectively. Peel strengths are normalized to the highest value at 100%. For the basic adhesives, the XPS data are normalized by (100 − %fluorine), whereas for the acidic adhesives, the data were normalized by (%fluorine), illustrating generic differences in fracture mechanisms. In the cases of the basic adhesives, the failures were cohesive, alternating between the Teflon substrates and the adhesive. A greater proportion of failure in the adhesive promoted higher peel strength. In the CPVC case, strength increased when failure shifted from entirely interfacial to partially in the Teflon substrate. Even though most failures were of the complex, mixed-mode type, still there was a correspondence in the trends of locus-of-failure and peel strength. Referring again to Fig. 1, we conclude that the fundamental principles relating interfacial acid–base interactions to adhesive bond strength and locus-of-failure are correct, and that predictions made by probe adsorption measurements can be valid even when failure does not meet the interfacial criterion.

4. SUMMARY AND CONCLUSIONS

Beginning with gas phase studies, the evolution of the science of intermolecular forces created a rich base of theoretical and experimental tools. Applications of those concepts to solid–solid interfaces has its foundation in composite materials research, attempting to predict stress transfer between the matrix and the reinforcing fiber or powder. The relationships between performance and the relative acid–base character of the surfaces have been demonstrated in a variety of systems by correlation between surface chemistry and mechanical properties. Acid–base interactions are shown to play a significant role in both *fundamental* and *practical* adhesion. Several spectroscopic and thermodynamic methods are discussed for quantifying the chemical functional groups involved as well as the magnitude of their interactions. Systematic investigations with gas- and liquid-state 'probe' molecules provide quantitative tables that rank the acid–base character of many functional groups.

The acid–base concept for predicting the contribution of exothermic interactions at interfaces to adhesion has been validated in two recent studies. Direct measurement of force–distance curves between methyl-, carboxylic acid- and amine-terminated self-assembled alkane thiol monolayers using the new Interfacial Force Microscope quantified van der Waals, hydrogen bond, and acid–base forces of interaction between solid surfaces. Potentially, adhesion experiments with model substrates using the IFM could quantify many functional group contributions to the forces at interfaces. Such data could advance efforts in theoretical modeling of adhesion, and enhance the ability to predict the effects of different functional groups on the performance of fiber-reinforced composite materials and adhesive joints.

Our comprehensive study used a unique contact angle method to quantify the number and strength of acidic and basic sites on Teflon FEP films that were functionalized with hydroxyl, carboxyl, acetyl and dinitrobenzoate groups. Combinations of those four substrates with four adhesives containing chlorine, acrylic acid, methacrylate, and acrylic anhydride functional groups produced sixteen pairs of interfaces. Correlation between approximate calculations of the thermodynamic work of adhesion with measured adhesive peel strength, and locus-of-failure demonstrates the general and practical utility of optimizing acid–base interactions at interfaces.

Acknowledgements

I am greatly indebted to the late Professor Frederick M. Fowkes, from whom I learned most of what I know about acid–base interactions. Also, for all the experimental work in Section 3, thanks to Mary B. Kaczinski, who persisted throughout the birth of her first child and the untimely death of her principal advisor. Several reviewers are thanked for substantial contributions.

REFERENCES

1. K. L. Mittal, in: *Adhesion Science and Technology*, L.-H. Lee (Ed.), Part A, p. 129. Plenum Press, New York (1975).
2. J. D. van der Waals, Thesis, Leiden University (1872).
3. F. London, *Z. Phys. Chem.* **B11**, 222 (1930).
4. E. M. Lifshitz, *J. Expl. Theor. Phys. USSR* **29**, 94 (1955).
5. W. H. Keesom, *Physik Z.* **22**, 129; 643 (1921).
6. P. Debye, *Physik Z.* **21**, 178 (1920); **22**, 302 (1921).
7. J. H. Hildebrand and R. L. Scott, *Solubility of Nonelectrolytes*, 3rd edn. Reinhold, New York (1950).
8. L. A. Girifalco and R. J. Good, *J. Phys. Chem.* **61**, 904 (1957).
9. F. M. Fowkes, *J. Phys. Chem.* **66**, 382 (1962).
10. J. N. Brönsted, *J. Chem. Soc.* **119**, 574 (1921).
11. R. G. Pearson, *Hard and Soft Acids and Bases*. Dowden, Hutchinson and Ross, Stroudsburg, PA (1973).
12. G. N. Lewis, *Valence and the Structure of Atoms and Molecules*. Chem. Cat. Co., New York (1923).
13. R. S. Mulliken and W. B. Person, *Ann. Rev. Phys. Chem.* **13**, 107 (1962).
14. R. S. Drago, G. C. Vogel and T. E. Needham, *J. Am. Chem. Soc.* **93**, 6014 (1971).
15. V. Gutmann, *The Donor–Acceptor Approach to Molecular Interactions*. Plenum Press, New York (1978).
16. F. J. Riddle, Jr. and F. M. Fowkes, *J. Am. Chem. Soc.* **112**, 3259 (1990).
17. H. C. Hamaker, *Physica* **4**, 1058 (1937).
18. B. V. Derjaguin and L. Landau, *Acta Physicochim. USSR* **18**, 733 (1941);
    E. J. Verwey and J. Th. G. Overbeek, *Theory of the Stability of Lyophobic Colloids*. Elsevier, New York (1948).
19. A. A. Griffith, *Philos. Trans. Roy. Soc.* **A221**, 163 (1920).
20. E. Orowan, *Rep. Prog. Phys.* **12**, 185 (1948).
21. G. R. Irwin, *Welding J.* **31**, 450 (1952).
22. K. L. Johnson, K. Kendall and A. D. Roberts, *Proc. Roy. Soc. Lond.* **A324**, 301 (1971).
23. A. Gent and J. Schultz, *J. Adhesion* **3**, 281 (1972).
24. R. J. Good, *J. Adhesion* **4**, 133 (1972).
25. D. Maugis, *J. Mater. Sci.* **20**, 3041 (1985).
26. K. L. Mittal and H. R. Anderson, Jr. (Eds), *Acid–Base Interactions: Relevance to Adhesion Science and Technology*. VSP, Utrecht, The Netherlands (1991).
27. D. W. Dwight, M. E. Counts and J. P. Wightman, in: *Colloid and Interface Science*, M. Kerker (Ed.), p. 143. Academic Press, New York (1976);
    D. W. Dwight, E. Sancaktar and H. F. Brinson, in: *Adhesion and Adsorption of Polymers*, L.-H. Lee (Ed.), p. 141. Plenum Press, New York (1980);
    D. L. Kotzev, T. C. Ward and D. W. Dwight, *J. Appl. Polym. Sci.* **26**, 1941 (1981).
28. T. B. Lloyd and D. W. Dwight, in: *Characterization of Polymers*, H.-M. Tong, S. P. Kowalczyk, R. Saraf and N. J. Chou (Eds), pp.150–168. Butterworth-Heinemann, Boston (1994);
    T. B. Lloyd, *Colloids Surfaces* **A93**, 25 (1994).
29. J. Schultz, L. Lavielle and C. Martin, *J. Adhesion* **23**, 45 (1987).
30. S. P. Wesson and R. Allred, *J. Adhesion Sci. Technol.* **4**, 277 (1990).
31. C. A. Baillie, J F. Watts and J. E. Castle, *J. Mater. Chem.* **2**, 939 (1992).
32. E. Mader, K. Grundke, H. J. Jacobasch and G. Wachinger, *Composites* **25**, 739 (1994).
33. H. J. Jacobasch, K. Grundke, P. Uhlmann, F. Simon and E. Mader, *Composite Interfaces* **3**, 293 (1996).
34. P. D. Bourgeois and T. Davidson, *J. Adhesion* **45**, 73 (1994);
    P. D. Bourgeois, Ph.D. thesis, Univ. of Connecticut (1993).
35. F. M. Fowkes, D. W. Dwight and D. A. Cole, *J. Non-Crystalline Solids* **120**, 47 (1990).
36. D. W. Dwight, F. M. Fowkes, D. A. Cole, M. J. Kulp, L. J. Salvati, Jr. and T. C. Huang, *J. Adhesion Sci. Technol.* **4**, 619 (1990).

37. J. F. Watts, M. M. Chehimi and E. M. Gibson, *J. Adhesion* **39**, 145 (1992).
38. M. M. Chehimi, J. F. Watts, S. N. Jenkins and J. E. Castle, *J. Mater. Chem.* **2**, 209 (1992).
39. J. F. Watts and M. M. Chehimi, *J. Adhesion* **41**, 81 (1993).
40. M. M. Chehimi, J. F. Watts, W. K. Eldred, K. Fraoua and M. Simon, *J. Mater. Chem.* **4**, 305 (1994).
41. M. Tatoulian, F. Arefi-Khonsari, N. Shahidzadeh-Ahmadi and J. Amouroux, *Plasmas and Polymers* **1**, 27 (1996);
M. Tatoulian, F. Arefi-Khonsari, N. Shahidzadeh-Ahmadi and J. Amouroux, *Colloids Surfaces* **A105**, 277 (1995).
42. M.-L. Abel, M. M. Chehimi, A. W. Brown, S. R. Leadley and J. F. Watts, *J. Mater. Chem.* **5**, 845 (1995).
43. H. P. Schreiber, *J. Adhesion* **37**, 51 (1992).
44. R. J. Good, M. K. Chaudhury and C. J. van Oss, in: *Fundamentals of Adhesion*, L.-H. Lee (Ed.), p. 153. Plenum Press, New York (1991).
45. L.-H. Lee, *J. Adhesion Sci. Technol.* **7**, 583 (1993).
46. F. M. Fowkes, M. B. Kaczinski and D. W. Dwight, *Langmuir* **7**, 2464 (1991).
47. D. T. Smith, *J. Electrostatics* **26**, 291 (1991).
48. R. G. Horn and D. T. Smith, *Science* **256**, 362 (1992).
49. A. Grabbe, *Langmuir* **9**, 797 (1993).
50. R. G. Horn, D. T. Smith and A. Grabbe, *Nature* **366**, 442 (1993).
51. W. A. Ducker and T. J. Senden, *Langmuir* **8**, 1831 (1992).
52. R. C. Thomas, J. E. Houston, R. M. Crooks, T. Kim and T. A. Michalske, *J. Am. Chem. Soc.* **117**, 3830 (1995).
53. D. W. Dwight and W. M. Riggs, *J. Colloid Interface Sci.* **47**, 650 (1974);
D. W. Dwight, in: *Characterization of Metal and Polymer Surfaces*, L.-H. Lee (Ed.), p. 313. Academic Press, New York (1977);
D. W. Dwight, *J. Colloid Interface Sci.* **59**, 447 (1977).
54. M. B. Kaczinski, Ph.D. thesis, Chemistry Dept., Lehigh University, Bethlehem, PA (1993).
55. M. B. Kaczinski and D. W. Dwight, *J. Adhesion Sci. Technol.* **7**, 165 (1993).
56. D. W. Dwight, *Proc. Soc. Plast. Eng., ANTEC '95*, 2744 (1995);
M. B. Kaczinski, *Proc. Adhesion Society meeting*. Williamsburg, VA (1993).

Mittal Festschrift, pp. 81–108
W. J. Van Ooij and H. R. Anderson, Jr. (Eds)

The mechanical theory of adhesion — A seventy year perspective and its current status

D. E. PACKHAM

School of Materials Science, University of Bath, Claverton Down, Bath, BA2 7AY, UK

Abstract—The mechanical theory featured in some of the earliest scientific discussions of the mechanism of adhesion, published by McBain and Hopkins in 1925, and still finds a place in contemporary accounts of theories of adhesion. At different times during the intervening period, quite different assessments of its significance have been made. In this paper the development of the theory over this seventy year period and the reasons for its changes of fortune are discussed. Contemporary evidence for 'mechanical adhesion' is critically examined, and the meaning of the theory is assessed in the context of modern theory of adhesion.

Keywords: Adhesion theory; anodised aluminium; anodised titanium; FPL etch; mechanical theory; macro-roughness; microfibrous surfaces; polyimides; surface roughness; work of adhesion.

1. INTRODUCTION: SCOPE OF THE PAPER

The phenomenon of adhesion must have been known to our ancestors for millions of years and no doubt excited the curiosity of lively minds. We know that Newton was interested in it, and challenged 'experimental philosophers' to find out why things stuck [1]. It is only relatively recently, however, that the phenomenon has been the subject of sustained scientific study.

A key stage in the development of modern adhesion theories is usually considered to be the classic paper [2] '*On adhesives and adhesive action*' published by McBain and Hopkins in 1925. This claimed to be the first systematic study of the action of adhesives and popularised the concept of 'mechanical adhesion'. It is appropriate seventy years later to mark this publication by a review of the mechanical theory of adhesion and of its current status. This present paper sets out to do this. The intention is not to present a comprehensive review of 70 years of work. Such a review would be of inordinate length and of doubtful value as there have been many reviews over the years, some of mechanical adhesion in general, others of particular aspects of this field. Papers of Rinke [3] in 1945, of Reinhart [4] in 1954 and of Wake [5] in 1976, together with the references they cite, give good surveys of the situation at the times they were written. In the 1980s important reviews, mostly directed at polymer–metal adhesion, include those published by Venables [6], by Arslanov and Ogarev [7] and by Brockmann *et al.* [8]. More recent work by Davis [9] and by Critchlow and Brewis [10] should also be noted.

What this present paper aims to do is to give (in Section 2) an historical overview of the main stages of development of the mechanical theory up to the mid-1980s and then (Section 3) to survey in more detail work of the last ten years in order to assess the present status of the theory (Section 4). These last two sections can be seen as an updating of my own 1983 review, '*The adhesion of polymers to metals: the rôle of surface topography*' in the important collection edited by Kash Mittal [11]. In Section 2, I have also sought to minimise the overlap with my 1992 paper [12], written to mark the contribution of John Venables to the development of adhesion theory. A fuller assessment of his work on adhesion to porous surfaces on aluminium and titanium can be found there.

2. AN HISTORICAL PERSPECTIVE

2.1. Early adhesion theories — mechanical and specific

McBain and Hopkins considered that there were two kinds of adhesion — mechanical and specific. Specific adhesion involved 'some species of interaction' between the smooth non-porous surface and the adhesive. This interaction might be 'chemical or adsorption or mere wetting'. Specific adhesion has developed into the model we today describe in terms of the adsorption theory.

In contrast, mechanical adhesion was only considered possible with porous materials. It occurred 'whenever any liquid material solidifies *in situ* to form a solid film in the pores'. They cite as examples adhesion to wood, unglazed porcelain, pumice and charcoal. For McBain and Hopkins mechanical adhesion was very much a common sense concept: 'It is obvious that a good joint must result when a strong continuous film of partially embedded adhesive is formed *in situ*'.

When they consider whether mechanical adhesion can act as a mechanism of adhesion on its own, in the absence of adsorption effects, their position was somewhat ambiguous. On one hand they wrote that 'A liquid which wets a surface and is then solidified possibly always makes a joint'. Neglecting the equivocality of the 'possibly always', this seems to mean that 'whenever wetting occurs, some form of adhesion must result'. This would be accepted today. By contrast, elsewhere in the paper McBain and Hopkins talk of mechanical adhesion as being 'independent of adhesive action'. They found no evidence of adsorption of adhesive from solution onto filter paper, and concluded that the adhesive adsorption did not occur during the bonding of wood.

McBain and Hopkins distinguish adhesion to porous surfaces from adhesion to surfaces which are merely rough. Roughness, they report, has little influence on adhesion, 'provided that the roughening has been carried out without pitting the surfaces appreciably'. If metal surfaces are 'rather deeply furrowed' the joints are decidedly weaker than those obtained with smooth surfaces. This reduction may be a result of either an increase in average adhesive film thickness, or there being insufficient adhesive to bridge the gap between the surfaces.

It is interesting to note how many of the themes of contemporary discussions of adhesion mechanisms (Section 3) were broached in this classic paper in 1925.

2.2. Reaction against the mechanical theory

Despite its intuitive appeal, the mechanical theory of adhesion was subjected to increasingly hostile criticism. Browne and Truax [13] published a paper on wood adhesion soon after McBain and Hopkins. While stating that they had 'no serious disagreement with McBain and Hopkins' point of view', their main emphasis was that specific adhesion was more important than mechanical. For joints with wood, they obtained high adhesion to smooth surfaces and were unable to observe a correlation between adhesion and penetration of the glue into the wood. McBain and Lee replied by arguing that the penetration was on an *ultra*-microscopic scale and so could not be observed [14].

A survey of adhesion by Rinker and Kline [15] in 1945 advised that 'in order to obtain a strong joint, a smooth surface is more desirable than a roughened one', and glossed Browne and Truax's work [13] as 'this mechanical interpenetration cannot account for more than a small fraction of the joint strength' [15].

In an influential review published in 1954 Reinhart considered the question 'of specific *versus* mechanical adhesion' [4]. On the basis of high adhesion to smooth surfaces and lack of microscopic evidence supporting the mechanical theory, he concluded that 'mechanical adhesion occurs seldom, if at all'. The poorer adhesion often observed with rough surfaces was explained in terms of poor wetting leading to void formation, and the development of stress concentrations around projections on the rough surface. This conclusion was endorsed by Wake writing in 1965, 'theories that mechanical interlocking ... added to the strength of a joint have been largely discredited' [16].

2.3. Revival of the mechanical theory

Thus by the 1950s the mechanical theory was all but abandoned; yet in the 1970s it was again being taken seriously. The extent of the change can be judged again by quoting Wake, writing this time in 1976 in his book '*Adhesion and the Formulation of Adhesives*'. Wake concluded [5] that 'adhesive joints frequently possess an important mechanical component essential to the performance of the joint but this type of component cannot suffice as the sole mechanism whereby surfaces are joined. It must be enhanced by, just as it enhances, specific adhesion'.

Most of the new work from the 1960s cited by Wake falls into one of two categories. The first is associated with the electroless deposition of metals onto plastics such as ABS and polypropylene. In the process, the plastics must be etched in a way which produces pits on a micrometre scale. Such a topography had been shown to be a necessary but not sufficient condition for adequate adhesion.

The second category was concerned with adhesion to porous or microfibrous surfaces on metals. In 1969 Packham *et al.* [17] had demonstrated the importance of pore structure in the adhesion of polyethylene to anodised aluminium, showing that the polymer entered the pores on the anodised surface. Arrowsmith [18] worked with electroformed copper and nickel, and argued that mechanical adhesion was the main mechanism of adhesion of these surfaces to epoxide laminates. There were further reports of the importance of topography in adhesion, for example, to copper [19] and titanium [20] with needle-like oxides.

It was developments at this time in electron microscopy (scanning electron microscopy and scanning transmission electron microscopy) and in electron spectroscopy (Auger and X-ray photoelectron spectroscopies) that enabled the physical structure and chemical composition of surface layers to be established in detail previously impossible. This, together with the increasing need of the aerospace industry for strong, consistent and durable adhesive bonds was among the factors that stimulated considerable work on pore-forming surface treatments for aluminium and titanium published during the late seventies and into the early years of the eighties.

It is not within the scope of this paper to present a comprehensive critique of the work of this period. Some such reviews have already been mentioned [7, 11]; others, written from a European industrial perspective, can be found in references [8, 21]. Significant research was carried out in the United States, for example, by Boeing [22–25] which led that company to adopt a standard phosphoric acid anodising pretreatment for structural bonding of aluminium [26]. Another American example is provided by Venables and his colleagues at the Martin Marietta Laboratories who, from the mid-1970s, undertook a study of the influence of pretreatment variables on the structure and adhesion properties of the films formed [12, 27–31].

The broad consensus that comes from most of this work is that strong bonds, and more particularly bonds of high durability, tend to be associated with a highly porous surface oxide, providing, of course, that the values of viscosity and surface tension of the adhesive are such as to allow it to penetrate the pores [11]. Thus for aluminium, a sulphuric acid/dichromate etch, known as the FPL etch, produces a surface of low porosity and bond durability compared to phosphoric acid anodisation [6, 17, 22, 24]. Figures 1a and 1b [33] give very much simplified schematic representations which emphasises the contract between the surfaces mentioned. It must be emphasised that the actual structures depend upon the various pretreatment parameters and usually show features omitted here [6, 8, 30, 32, 33].

The importance of porosity was brought out strongly in a 1984 review by Venables [6]. He concluded that for aluminium and titanium 'certain etching or anodization pretreatment processes produce oxide films on the metal surfaces, which because of their porosity and microscopic roughness, mechanically interlock with the polymer forming much stronger bonds than if the surface were smooth'. This is as unequivocal a statement of the mechanical theory of adhesion as can be found in the original work of McBain and Hopkins.

Since this time, the acceptance of a 'mechanical theory' has not been seriously challenged, and it now has a generally accepted place within the canon of adhesion theories [9, 10, 34–36]. It is nevertheless of value to examine work of the past decade or so to see what advances have been made. What kind of adhesion phenomena are now being rationalised in terms of the theory? Has the theory itself undergone

(a)

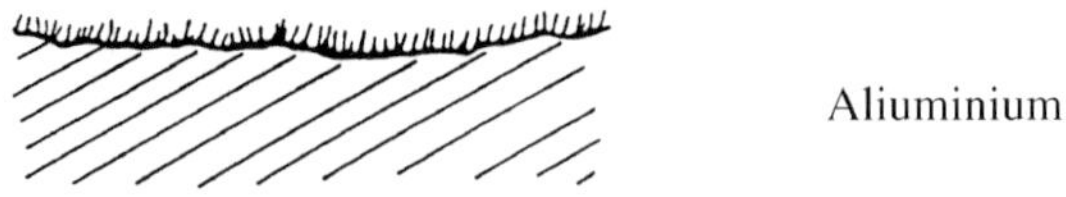

(b)

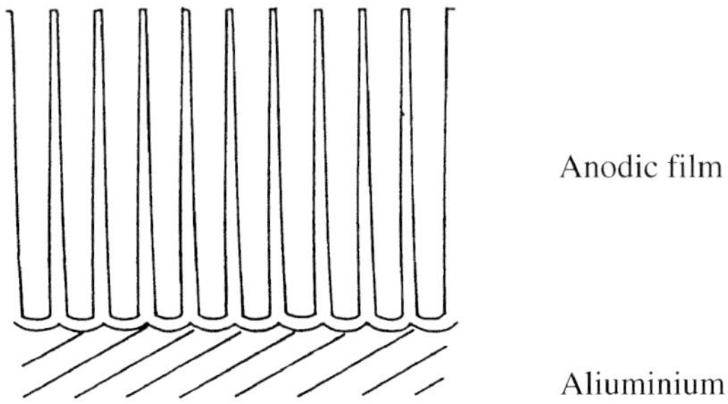

(c)

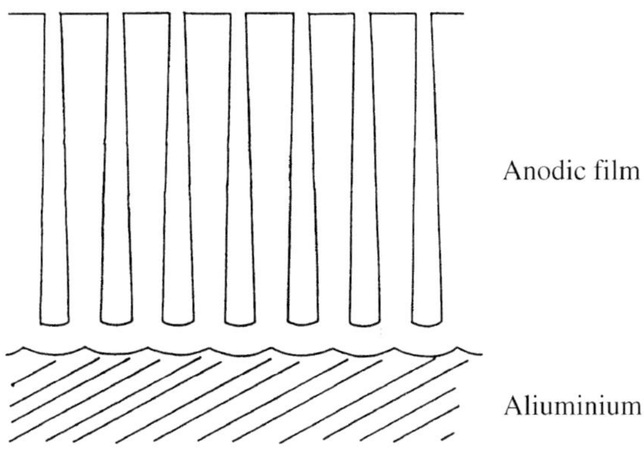

Figure 1. Simplified schematic representation of the porous surface produced on aluminium by the treatments (a) FPL etch, (b) anodising in phosphoric acid, (c) anodising in chromic acid. After [33].

significant development? If so, has the tendency been to reinforce or to undermine the orthodoxy of the eighties?

3. THE LAST DECADE

McBain and Hopkins considered that wood, unglazed porcelain and charcoal were substrates to which mechanical bonding occurred. What types of system are regarded today as ones where mechanical factors play a significant rôle in adhesion?

References to wood and ceramics still appear in the literature. For example, Bogner has reported correlations between bond strength and increased penetration of the adhesive into the wood [37]. Correlations between ceramic porosity and adhesion have been published by Packham and Johnston [38] for unglazed porcelain and by Kanai [39] for alumina prepared by a sol-gel process.

The scope of the mechanical theory has moved well beyond these materials. The main features of recent work may conveniently be surveyed under three headings dealing, in turn, with *metal*, *polymer* and *fibre* substrates.

3.1. Metal surfaces

Discoveries in this area were principally responsible for the re-establishment of the mechanical theory in the 1970s [5]. Microporous and microfibrous surface topographies can be produced to form a pre-treatment on many metals including magnesium, copper, steel, titanium and aluminium.

As was emphasised in the previous section, many of these pretreatments are important industrially, and give increased durability to the joint in humid conditions as well as giving adequate initial adhesion. Treatments for titanium and aluminium have been reviewed within the past decade, for example by Davis [9] and by Critchlow and Brewis [10]. Bishopp *et al.* [33] have specifically discussed the effect of pretreatment of aluminium for bond durability. Broadly speaking, their results confirm the consensus discussed in the previous section. For unprimed surfaces they report considerably better durability for phosphoric acid anodising than for FPL etching. They found poorer durability for a chromic acid anodised surface, which they attribute to inability of the adhesive to penetrate the narrow entry to the pores (see Fig. 1c). It should be noted that they were using a toughened epoxy film adhesive with a 120°C cure temperature which might have had a relatively high viscosity [11]. As their electron microscopy showed little damage to adhesive or substrate even after a fall in adhesion, they suggested subtle undermining of alumina film or disruption of physico-chemical bonds across the interface as possible mechanisms of degradation. This is in contrast to the view of Venables [6] who argues that conversion of the aluminium oxide to hydroxide is a significant factor in bond degradation in a humid environment.

This is still an active research area where the concepts of the mechanical theory of adhesion are commonly invoked in order to rationalise the results obtained. Three recent papers of Wightman's on bonding, respectively, to aluminium, steel and titanium illustrate current work.

Ko and Wightman [40] demonstrated the effectiveness of anodising aluminium–lithium alloys in either sulphuric acid or phosphoric acid. A porous surface oxide was produced into which a polysulphone adhesive penetrated giving a 'mechanical means of adhesion'. Bonds were obtained which were much more durable in high humidity than comparable ones made to surfaces without this high porosity.

Microfibrous surfaces on steel produced by a hydrothermal oxidation (cf. Fig. 2 [41]) were used by Hollenhead and Wightman in a study of bonding with a thermoplastic polysulphone [42]. Lap shear strengths for the grit blasted steel, made for comparison, were 40% lower than those for microfibrous surfaces pretreatments: by increasing the strength of the interfacial region, failure is directed into the polymer where dissipation of fracture energy leads to a higher fracture stress. Although bonds with both pretreatments deteriorated with exposure to humid environments, those with the microfibrous surfaces were the more durable.

Filbey and Wightman [43] have clearly demonstrated the applicability of analogous principles in the bonding of epoxies to titanium alloys. They found that anodising

Figure 2. Microfibrous oxide formed on steel by oxidation in moist nitrogen at 400 °C. Scanning electron micrograph, base + 5.5 μm [41].

either in chromic acid, based on a method outlined in Boeing patents [44, 45], or in sodium hydroxide was capable of producing a surface with pores of 40 to 50 nm diameter. Surface analysis showed the penetration of resin into these pores and wedge test results demonstrated the superior durability of these anodic pretreatments, compared with others which did not give a microporous surface. These results both confirmed and extended findings of Boeing [44, 45] and of Martin Marietta Laboratories [6]. In their recent review of 35 types of surface treatment for titanium, Critchlow and Brewis [10] conclude that microporous surfaces are 'widely recognised ... to produce the best durability results', an effect ascribed by many authors to an increased likelihood of mechanical keying.

The generally good adhesion with a range of adhesives, usually combined with good bond durability in a humid environment [46], is generally associated with entry of the adhesive into the pores on the substrate surface.

The pores on anodised aluminium may be as small as tens of nanometres in diameter, and at one time skepticism was expressed as to whether any polymeric adhesive could enter them. There is now a wealth of evidence from electron microscopy and surface analysis [6, 33, 43, 47, 48] backed up by theoretical calculations on idealised models [7, 11, 43, 49] which shows that polymeric adhesives and primers will often penetrate the microporous surface layers. The ease of ingress depends strongly on the polymer viscosity and time for which it is in a fluid state [11, 21, 49, 50], so a change of adhesive or of bonding conditions may significantly alter the extent of penetration.

Recent work has further confirmed the importance of adhesive penetration for good bond performance. Davies and Ritchie [51] in their work on bonding metal matrix composites to aluminium emphasise the importance of the 'micro-composite' interfacial region formed by penetration of the adhesive into the porous oxide on chromic or phosphoric acid anodised aluminium for the durability of the joints under stress and in humid environments. The 'micro-composite' region decreases the sharpness of the discontinuity in modulus between adhesive and substrate.

Sargent has recently published an investigation of bonding chromic acid anodised aluminium with epoxy and phenolic adhesives [48]. By variation of anodising conditions (in a way not revealed) a range of peel strengths were obtained. As the fine structure of the surface oxide increased, so did the peel strength (Fig. 3). Depth profiling using argon ion bombardment and XPS showed evidence of a greater proportion of primer within the oxide layer, to a depth of at least 40 nm, for high peel strength specimens than for low peel strength ones. This result should be contrasted with that of Bishopp *et al.* [33] discussed above.

With increasing realisation of the environmental damage caused by many synthetic chemicals, there have been moves to develop pretreatments for aluminium which avoid the use of chromates. The widely used phosphoric acid anodisation is commonly preceded by a chromate-containing FPL etch. As long ago as 1977, Russell [52] published results which suggested that an etch designated 'P' etch might be able to replace the FPL process, both as a treatment in its own right and as a preliminary to phosphoric acid anodising. The 'P' etch consisted of sodium sulphate and ferric sulphate dissolved in a sulphuric acid–nitric acid mixture.

A further development of this treatment is known as the 'P2' etch. It avoids using nitric acid and consists of ferric sulphate and sulphuric acid alone. The P2 etch produces an oxide topography comparable to that of FPL, and has shown some promise as a substitute for the FPL treatment [27, 53, 54]. It was even suggested as a replacement for the phosphoric acid anodising process itself [27]. More recent findings, such as the wedge test results shown in Fig. 4, confirm its effectiveness when followed by the anodising, but do not support its use as a substitute [54].

3.1.1. Macro-roughness. Although some polymers are able to enter the pores produced on aluminium by sulphuric acid anodising [41], they are much finer than those from anodising in phosphoric acid [32, 55, 56]. There has been an interest in opening up the structure of the sulphuric acid-formed films by a post anodising dip. Arrowsmith and Clifford [57] have shown that a controlled etch in phosphoric acid after anodising in either sulphuric acid or chromic acid results in improved bond strength and durability of adhesively bonded aluminium compared with standard etching and anodising treatments. Digby and Packham [58] found that the durability of bonds to phosphoric acid anodised aluminium could also be improved by controlled exposure to phosphoric acid after anodising (Fig. 5). Figure 6 shows the more open structure of the surface after such a treatment.

Plasma spraying [59] provides a method of producing a porous surface layer without the use of chemical solutions. The porosity is much coarser than that associated with anodising, being of the order of micrometres rather than tens of nanometres. Clearfield *et al.* [60] at Martin Marietta Laboratories compared anodising treatments for titanium alloy [Ti-6Al-4V] with plasma sprayed coating of the alloy. Durability results showed comparable performance. Alumina powder has been used on aluminium, titanium and

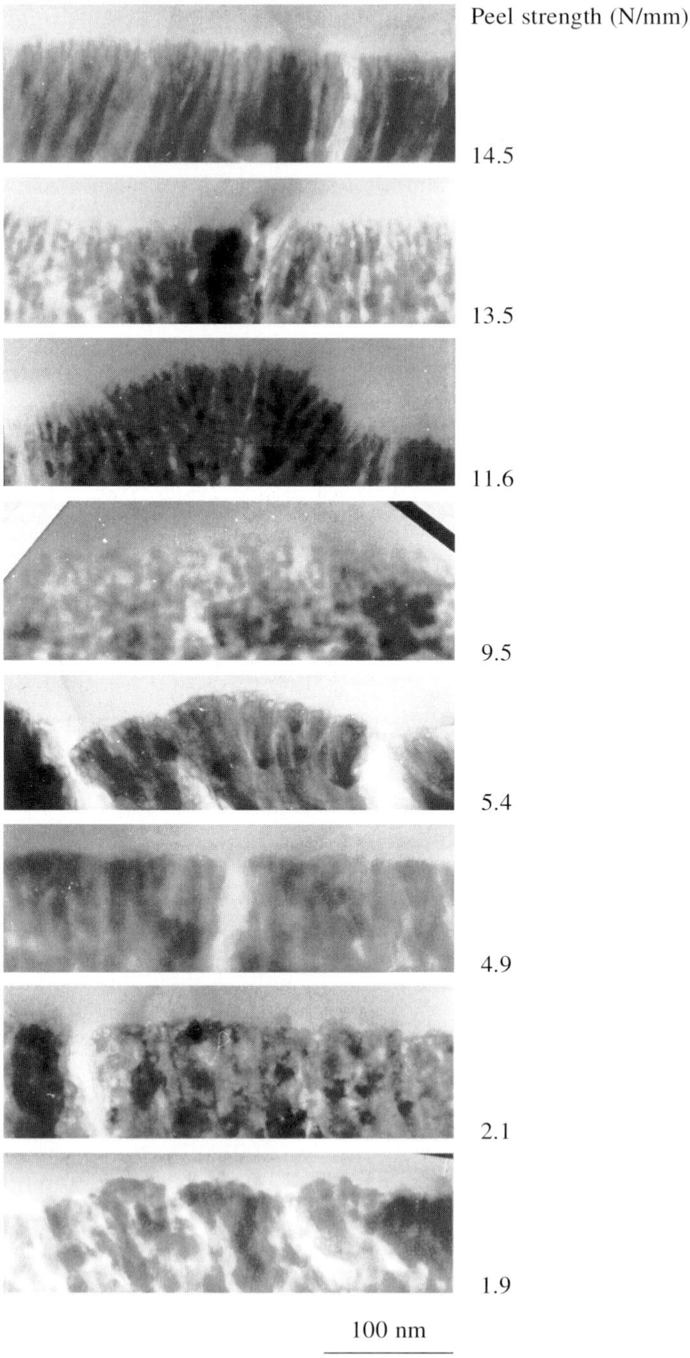

Figure 3. Adhesion of epoxy resin to aluminium anodised in chromic acid to give different oxide porosities. Increasing fine structure of the oxide corresponds with increasing peel strength [48].

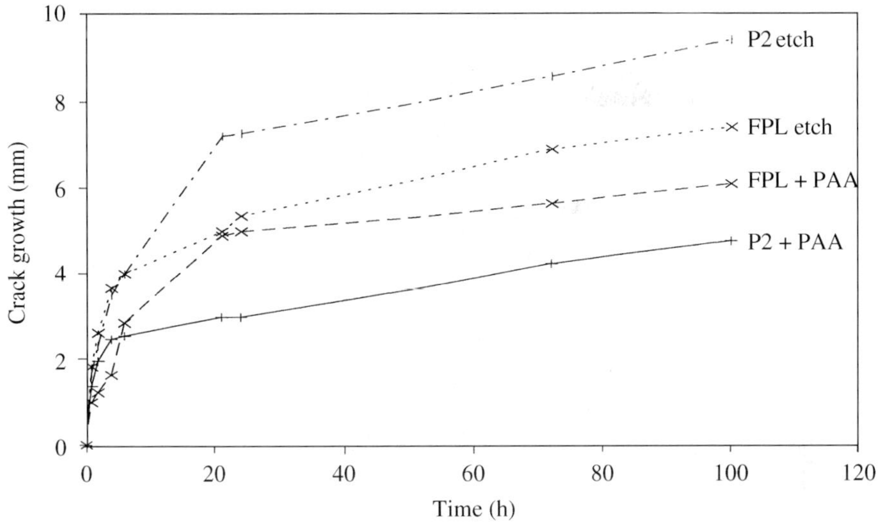

Figure 4. Crack growth on exposure to 96% relative humidity at 56°C for Boeing wedge test specimens of aluminium bonded with epoxy resin. The P2 pre-treatment is compared with the FPL etch used both alone and prior to phosphoric acid anodising (PAA) [54].

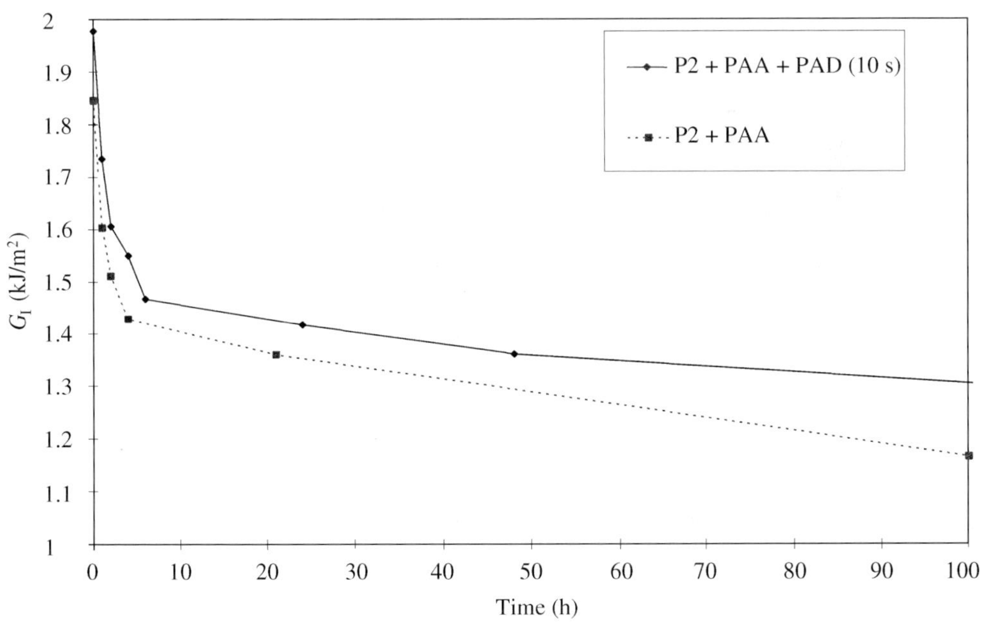

Figure 5. Change in strain energy release rate (G_I) on exposure to 96% relative humidity at 56°C for Boeing wedge test specimens of phosphoric acid anodised (PAA) aluminium bonded with epoxy resin. The effect of a 10 s phosphoric acid post-anodising dip (PAD) is shown [58].

steel [61]. More recent work from Martin Marietta by Davis *et al.* has reported on the potential of plasma sprayed coating as pretreatments for aluminium and titanium

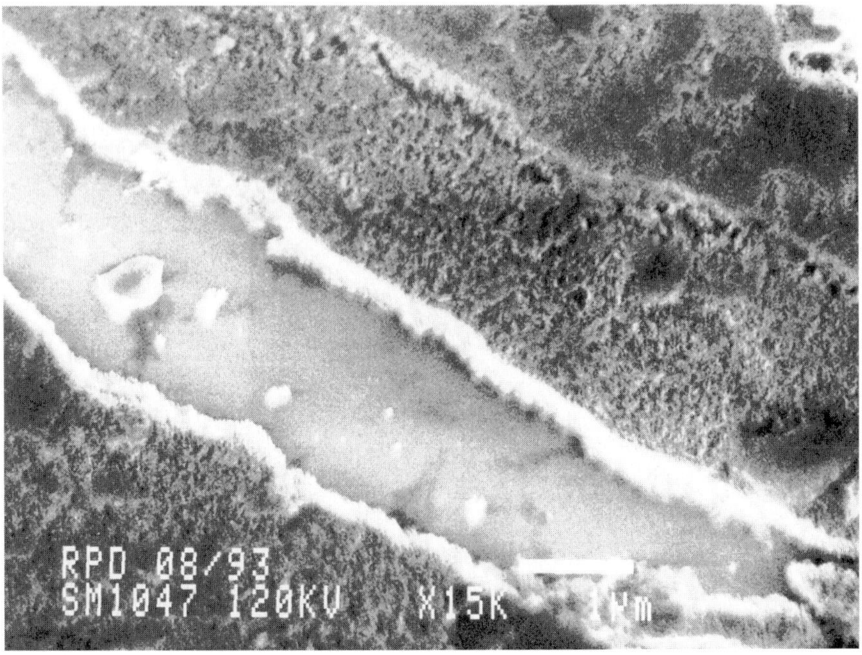

(a)

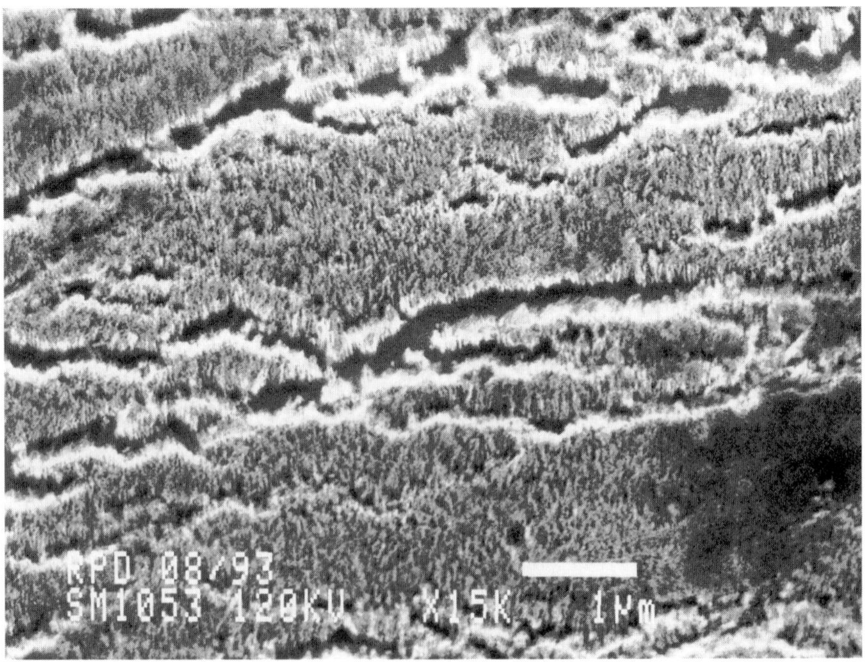

(b)

Figure 6. Scanning electron micrographs showing the change in surface structure of phosphoric acid anodised aluminium after a phosphoric acid post-anodising dip (PAD) for (a) 10 s, (b) 20 s, (c) 30 s [58].

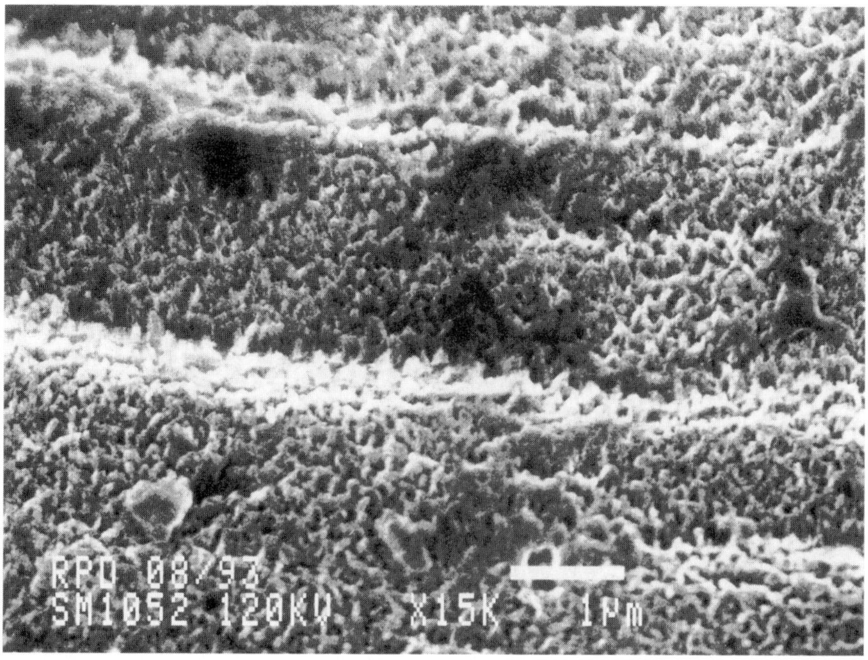

(c)

Figure 6. (Continued).

[62, 63]. The best performance on aluminium which they report is for a 40 μm thick coating of aluminium–silicon alloy mixed with polyester: crack growth in a wedge test was comparable to that for P2 and FPL.

There has been a renewed interest in this coarser, micrometre sized, porosity in recent years. Indeed Sargent [48] reported influences of micrometre scale roughness as well as those discussed above which were associated with a far finer scale of roughness. Another example is the work of Morris and Shanahan on sintered steel substrates bonded with a polyurethane adhesive [64]. They ascribed the much higher fracture energy for joints with sintered steel compared with those with fully dense steel to the mechanical interlocking of polymer within the pores. Extra energy was required to extend and break these polymer fibrils.

A picturesque example of providing a mechanical 'key' has recently been described by van der Putten who was concerned to bond copper directly to silicon in the context of integrated circuit technology [65].

Copper sticks poorly to silica but titanium tungstide sticks well. Using conventional lithographic techniques, islands of TiW 0.1 μm thick were sputtered onto the silica and the photoresist was removed (Fig. 7a).

Palladium acts as a nucleating agent for the electroless deposition of copper. By treating the surface with palladium(II) chloride in hydrochloric acid, a monolayer or so of palladium is deposited on the TiW surface. The palladium chloride solution also contains 1% of hydrofluoric acid which attacks the silica, undercutting the TiW islands (Fig. 7b). Copper is now deposited, nucleating on the palladium-covered TiW

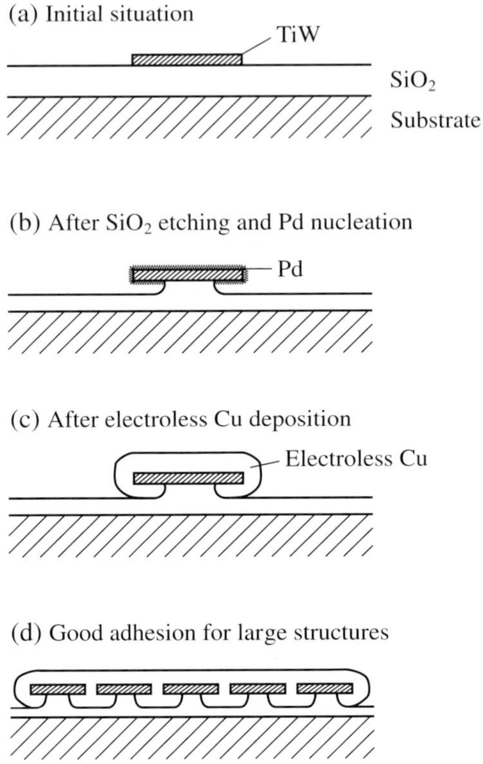

Figure 7. Adhesion of copper to a silica layer achieved via deposition of titanium tungstide islands [65].

and growing from it. It is thus mechanically anchored to the silicon surface (Fig. 7c and d).

It is important to point out that this work on macroroughness qualifies, rather than supersedes, the earlier work [3, 4, 16] discussed above, where such roughness was generally reported to have an adverse effect on adhesion. Critchlow and Brewis [66] have recently reported that use of different grit sizes in the grit blasting of aluminium has only a small (albeit significant) influence on the durability of epoxy bonds.

The effect of roughness on this scale is a subtle one, depending very much on the detailed characteristics of the surface [11]. This point has recently been eloborated by Keisler and Lataillade [67] for steel joints made with epoxy resin and tested at high strain rates. Differences in the surface topography affected the deformation mechanism of the joint (cf. [68, 69]). Excessive roughness combined with sharp asperities and narrowly spaced valleys gave rise to poor wettability and to the creation of microcracks near the interface. Stronger joints, especially at very high test rates, were observed for some 'laser roughened' steel where wide valleys and asperities with a high radius of curvature were produced.

It is clear then that porous surfaces and the concepts of the mechanical theory of adhesion play a large part in the surface treatment of metals for structural adhesive bonding. Davis [9] has recently summarised the situation with respect to aluminium and titanium: 'all high performance adhesive bonds require clean, microscopically rough substrate surfaces allowing wetting of polymer and complete filling of the pores of nooks and crannies ... the resulting mechanical bonds supplement or supplant the chemical bonds present'.

3.2. Polymer substrates

In contrast with metals, polymers have low surface energies and are consequently wetted less readily by adhesives. Both classes of material may have cohesively weak surface layers — with polymers they result from migration of additives or of low molecular weight polymer fractions to the surface [70]. A wide range of treatments have been described which enhance adhesion to polymers. Some involve aggressive chemical treatment. Strong oxidising solutions, such as chromic acid or permanganate have been employed with hydrocarbon polymers, and highly reactive combinations such as sodium in liquid ammonia have been used for fluorocarbons. Alternative pretreatments to 'wet' chemical treatment include surface flaming, the action of gas plasma or reactive ion etching [71–73]. (It is important to appreciate that the plasma treatment referred to here and described in references [71–73] is very different from the plasma *spraying* of a porous layer referred to above, and described in reference [59].)

Surface analysis has established in some detail the various functional groups that different treatments introduce into the surfaces of different polymers. These are often polar oxygen-containing groups. It is commonly reported that surface roughening accompanies these chemical changes, and a 'mechanical' effect is sometimes claimed or mentioned as a possible contribution to the adhesion. The importance of the roughening of rubber reinforced plastics, such as high impact polystyrene and ABS, is well established (see [5, 11]). Oxidising agents etch away the spherical rubber-rich phase where it comes close to the surface [74]. It may be easy to speculate about mechanical effects in other cases, but less easy to prove that they apply.

O'Kell *et al.* have recently reported the effects of treating polyethylene in a low energy (< 1 W) oxygen or nitrogen plasma [75, 76]. The changes induced were particularly well-characterised as the plasma cell was attached directly to the preparation chamber of an X-ray photoelectron spectrometer, so that spectra could be obtained immediately after treatment, as well as after exposure to air. Topographical changes were observed using atomic force microscopy. The plasmas introduced both chemical and topographical changes into the surfaces within the first few seconds of treatment. The authors consider the amount of oxygen functionality introduced to be the dominant parameter governing the level of adhesion of the polymer to aluminium. The importance of this conclusion would seem to be a little reduced, as bonds were made by melting the polyethylene against the aluminium. This might alter the surface chemistry and would almost certainly alter the topography.

Similar work treating polypropylene in oxygen and nitrogen plasmas has been published by Harth and Hisbt [77]. Again, rapid formation of functional groups was observed, but roughening was slower. Adhesion specimens were made by bonding the treated polymer to aluminium using a polychloroprene adhesive. As the peel strength increased before the roughening became significant, the authors again considered that the introduction of chemical groups was the main cause of enhanced adhesion. They speculated that the 'filigrees' of polymer formed after longer treatment times were weak, but unfortunately gave no information about the failure modes which might have enabled a less ambiguous conclusion to be reached.

Recent publications on treatment of styrene–butadiene rubber (SBR) poly-*p*-phenylene–vinylene and PTFE–polyperfluoroalkylvinylether (PFA) copolymer report evidence for mechanical interlocking which together with chemical interaction contributes to the adhesion. These are now discussed.

Pastor Blas *et al.* [78] chlorinated vulcanised SBR with trichloroisocyanuric acid solution. The treatment produced microcracks in rubber surface which, along with the formation of carbonyl, hydroxyl and ether functional groups, were considered to aid adhesion with a polyurethane adhesive.

Nguyen *et al.* studied the vapour deposition of aluminium on thin poly-*p*-phenylene–vinylene films [79]. Satisfactory adhesion to the aluminium was obtained when the polymer films were treated with oxygen by radio frequency sputtering. The roughness of the surface produced was considered important as it allowed deeper penetration of the deposited aluminium which subsequently formed an 'Al–O–C complex'.

Surface roughening was found by Marchesi *et al.* [80] to be a necessary stage in obtaining adequate adhesion of PFA to copper. The roughening was followed by chemical modification of the surface with sodium naphthalenide.

3.2.1. Polyimides. Bonding to polyimides has received attention in recent years because of their applications in the microelectronics industry and in high performance aerospace composites.

Buchwalter and Saraf [81] have treated polyimide surfaces prior to deposition of tantalum or chromium by argon sputtering, both on its own and in combination with reactive ion etching with carbon tetrafluoride. They have systematically characterised the chemistry and topography of the treated surfaces. The peel strength measured on peeling the metals from the optimally treated surfaces (*c.* 1000 J/m^2) was much higher than that of the virgin polyimide (*c.* 100 J/m^2). Buchwalter carefully examined the failure modes and concluded that the increased adhesion was not a consequence of the increased surface roughness *per se*, nor of chemical change but of weak boundary layer removal with the increased energy dissipation perhaps being associated with cracking of underlying polyimide, toughened by the pretreatment [82].

Inagaki *et al.* treated polyimide films in a range of gas plasmas, deposited a copper layer and measured the adhesion [83]. They obtained infra-red spectroscopic evidence for co-ordinate bond formation between copper and the treated polymer surface. The argon plasma, which gave the highest peel strength, produced a rough surface with fine, high protuberances. The authors are undecided as to whether to ascribe the enhanced adhesion to mechanical keying, to weak boundary layer removal or to the co-ordinate bond formation. Unfortunately, they did not publish micrographs of the failure surfaces which could have been compared with Buchwalter's to judge whether a similar explanation to hers applied here.

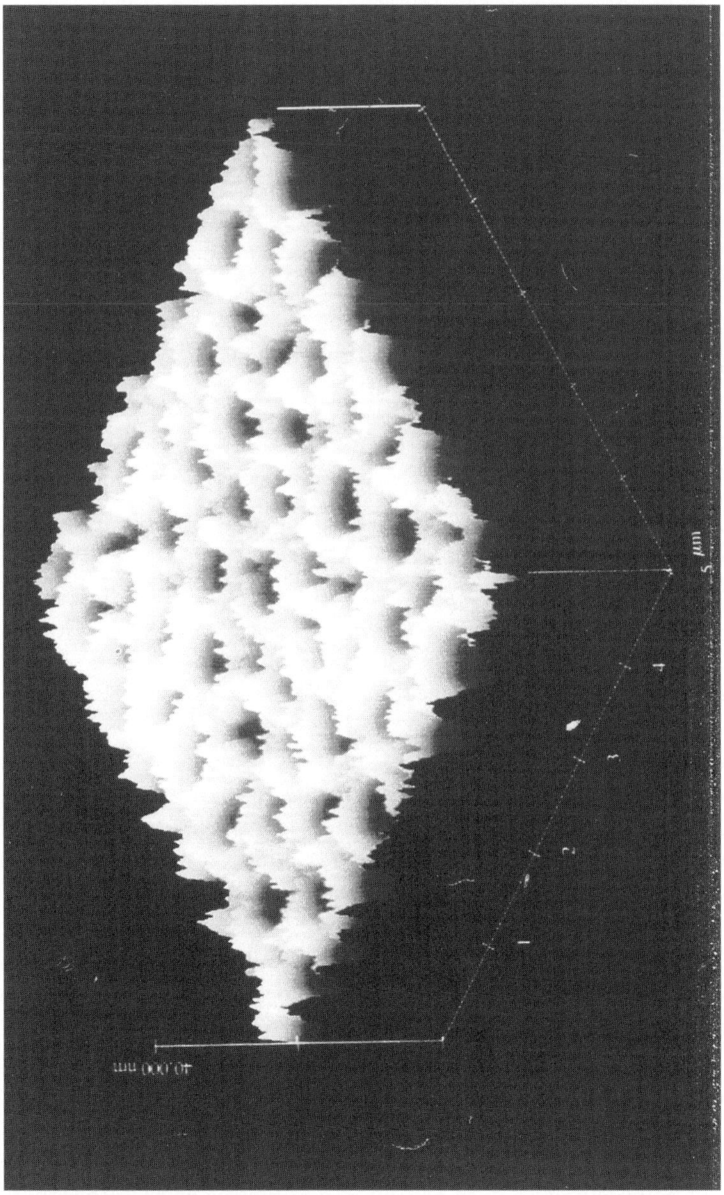

Figure 8. Example of the nanoscale roughness on polyimide produced by phase separation [84].

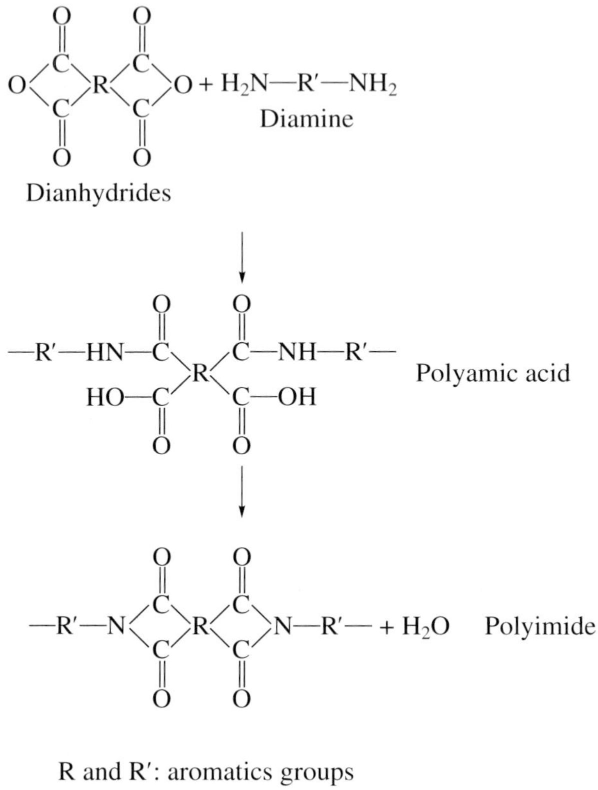

Figure 9. Formation of a condensation polyimide (after [85]).

A particularly elegant method of producing polyimide with a rough surface and enhanced adhesion has been described by Saraf *et al.* [84]. It is achieved by controlling the initial stages of formation of the polyimide from its dianhydride and diamine precursors, so no post-polymerisation treatment is required for adhesion. An example of the nanoscale roughness that can be produced is shown in Fig. 8.

In order to understand how the roughness is achieved, it is necessary to remember the basic chemistry of formation of condensation polyimides [85]. An appropriate acid derivative, such as a dianhydride or diacyl chloride, and a diamine are dissolved in a solvent and a 'soft bake', typically at 70°C, is employed to remove solvent. Condensation to form a polyamic acid (PAA) occurs, (Fig. 9). A subsequent 'hard bake' at around 400°C eliminates water, and leads to the formation of the polyimide itself.

The method of Saraf *et al.* [84, 86, 87] relies on the fact that exactly the same polyimide would be obtained on hard baking of a polyamic ester, rather than polyamic acid: i.e. the carboxyl groups in the acid shown in Fig. 9 would be esterified and the corresponding alcohol, rather than water, would be eliminated during the hard bake. The polyamic ester can be obtained by simple modification of the original anhydride or acyl chloride.

Saraf *et al.* [84] produced poly(pyromellitic dianhydride–oxydianiline) (PMDA-ODA) from both the polyamic acid (PAA) and the ethyl ester (PAETE). Precursors of both acid and ester were dissolved in solvent (*N*-methyl pyrrolidone, NMP), the solution spun on a suitable substrate and 'soft baked' at 70°C to remove the solvent. Depending on the proportions of the two precursors, polyimide films with different roughnesses on the scale of nanometres could be produced.

The rough surface is the result of two phenomena. As the NMP solvent evaporates during the soft bake, phase separation by spinodal decomposition occurs. The PAA-rich phase swells more than the PAETE-rich phase: this produces surface roughness on a fine scale.

Imidisation occurs during the hard bake, but this is some fifteen times faster for PAA than for PAETE. Thus at an intermediate stage, the former PAA phase has formed rigid polyimide while the PAETE is still essentially the softer precursor. As the PAETE is converted to polyimide, further modification of the topography is spontaneously achieved.

Saraf *et al.* [84] ascribe the poor adhesion of conventional, untreated polyimide films to an ordered surface layer and the alignment of the polymer chains parallel to the film surface, as well as to the absence of nanoscale roughness. They claim that the technique described overcomes all these problems. They quote peel strengths for vapour deposited copper which are nine times greater than those to untreated conventionally formed polyimide.

3.3. Fibres

A third area in which mechanical effects in adhesion are often reported deals with fibre–matrix adhesion in composites. Whereas adhesion to glass fibres is satisfactorily effected by use of silane coupling agents, carbon, Kevlar and high strength polyethylene fibres usually require chemical etching or gas plasma treatment. These modify the surfaces chemically, but many authors have reported a roughening which they see as contributing to the enhanced adhesion.

Treatments of carbon fibre in various plasmas have been reported to improve adhesion. For example, the formation of rough fibre surfaces and consequent improvement in adhesion to the matrix have been mentioned by Moyer and Wightman [88] for an oxygen plasma and by Zang [89] for oxygen and argon plasmas. In contrast, Yuan *et al.* [90] ascribed improved interfacial shear strengths in oxygen and argon plasma-treated fibres to the formation of hydroxyl, ether or aromatic groups, not to surface roughening.

Anodising has been shown by King *et al.* to produce surface roughness on pitch-based carbon fibre [91, 92]. They used XPS and chemical probes in order to better understand the reasons for the improved properties of an epoxy composite made from the treated fibres. They concluded that the increased interfacial shear strength resulted primarily from improvements in mechanical interlocking rather than chemical interactions between fibre and matrix.

An interesting example of mechanical interlocking was reported by Marshall and Price in composites made from treated carbon fibre [93]. They talk of graphite sheet

Table 1.
Epoxy resin reinforced with polyethylene fibres: effect of fibre treatment on inter-laminar shear strength (ILSS) and impact energy (after [94])

Treatment	ILSS (MPa)	Impact energy (kJ/m^2)
None	15	158
Chromic acid	20	—
Plasma etching	27	119

edges which 'increase the number of potential bonding sites', and also present a 'jagged surface profile (which may) increase mechanical keying'. It is curious in a paper in a primary research journal that they give no detail whatever of the surface treatment they used!

Ultra-high molecular weight polyethylene is now of interest as fibre reinforcement in composites. Bonding problems apply, similar to those addressed in their work on polyethylene sheet by O'Kell *et al.* [75, 76] (see above). Similar treatment methods have been applied.

Ward and Ladizeski [94] have used chromic acid treatment and plasma etching to achieve a range of values of both monofilament pull-out adhesion and inter-laminar shear strength (Table 1). This work provides an example of an adverse effect of high fibre–matrix adhesion. Enhanced adhesion resulting from oxygen plasma treatment prevented delamination during impact, and so reduced the impact strength [95].

Ward *et al.* [96] have also studied the effect of the oxygen plasma treatment time. At short treatment times, general oxidation of the polyethylene fibre surface occurred. With intermediate times, surface cross-linking predominated. Long treatments produced a rough surface 'which could give rise to a mechanical keying effect'.

Silverstein *et al.* have used chromic acid, potassium permanganate and hydrogen peroxide to treat polyethylene fibres [97, 98]. Of these three, they found only chromic acid to be effective in increasing the adhesion. They suggest the removal of weak boundary layers and surface roughening as well as surface oxididation as explanations for the better adhesion.

3.4. The last decade — summary

It can be seen then that the last decade has been one of steady progress and consolidation, during which the main features of the mechanical theory as it was re-established in the 1970s and 1980s have been confirmed in a wide range of experimental situations. The theory has proved a useful one in the sense that it has stimulated the development of new surface treatments for metals, polymers and fibres and has assisted in giving an understanding of their efficacy. There has perhaps been a tendency, now that the theory is again 'respectable', to invoke 'mechanical effects' somewhat uncritically wherever an increase in surface roughness has been observed.

4. DISCUSSION: CURRENT STATUS OF THE THEORY

4.1. Fundamental and practical adhesion

It is now seventy years since McBain and Hopkins' classic paper proposing the mechanical theory of adhesion was published. In the intervening years, it has suffered from the vicissitudes of scientific orthodoxy [12], but is now well regarded and is often invoked to rationalise results of adhesion experiments. In the light of contemporary models, how might we regard mechanical theory of adhesion and its relationship to the adsorption theory?

Before addressing this question directly, let us consider the relationship between the forces acting at the interface between the bonded phases and the 'adhesion' measured by some destructive test performed on the adhesive joint. This practical adhesion, as Mittal [99] points out, reflects the properties of the interfacial region and, indeed, of the joint as a whole. The adhesive joint is usefully regarded as a 'composite'. The interfacial forces are, of course, important, but the performance of the joint is a consequence of the interface, the bulk phases and the interactions between interface and bulk phases joined. The manner in which the components interact will depend upon the geometry of the joint, its physical form, and upon the way in which it is loaded.

This interdependent relationship can be expressed in simple mathematical form. Good did this when he applied the Griffith–Irwin theory of fracture to a joint comprising a bond between two phases [100]. The fracture stress, σ_f, is given by

$$\sigma_f = k\left(\frac{EG}{l}\right)^{1/2}, \tag{1}$$

where k is a constant, l is the length of the critical crack which leads to fracture and E is the modulus and G the fracture energy.

The modulus arises in the Griffith theory as its value determines the amount of elastically stored energy at a given strain. It is this stored energy which is released, providing the fracture energy, G, when fracture occurs. Within the adhesive joint, E and G are semi-local properties. Fracture will occur where the term EG/l is lowest, whether at or near the interface or within one of the bulk phases. Factors which alter E or G or l locally within the joint may alter its strength and other properties.

The fracture energy G will involve a term associated with the energy required for bond breaking at the interface when the joint fails. These may be primary or secondary bonds. This surface energy term will be the thermodynamic work of adhesion W_a or of cohesion W_c, depending on whether fracture occurs at the interface or cohesively within one phase [101]. To this surface energy term must be added a term ψ representing other energy absorbing processes — for example plastic deformation — which occur during fracture:

$$G = W_a \text{ (or } W_c) + \psi. \tag{2}$$

Usually, ψ is very much larger than W_a (or W_c). This is why practical fracture energies for adhesive joints are almost always orders of magnitude greater than works of adhesion or cohesion.

A similarly approach may be adopted towards the peel test, where the 'peel strength' is essentially a peel energy, the fracture energy per unit area. The peel energy, P, then comprises a surface work term and (according to circumstances) terms for plastic, viscoelastic and other losses which occur during fracture [102]:

$$P = W_a \text{ (or } W_c) + \psi_{\text{plast}} + \psi_{\text{ve}} + \ldots \tag{3}$$

Factors which enhance either W or ψ will increase the fracture energy (the adhesion) of the bond. Although again it is usual that

$$\psi \gg W_a \text{ (or } W_c)$$

a modest increase in W may result in a large increase in adhesion as ψ and W may be coupled. For some mechanically simple systems, where ψ is largely associated with viscoelastic loss, a multiplicative relation has been found:

$$P = W \times \phi(c, T), \tag{4}$$

where $\phi(c, T)$ is a temperature and rate dependent viscoelastic term [103, 104]. In simple terms, stronger bonds (increased W) may lead to much larger increases in fracture energy because they allow much more bulk energy dissipation (increased ψ) during fracture.

Theories of adhesion can be understood in terms of the effects they predict for local modulus (E), surface energy (W), energy dissipating capacity (ψ) or defect size l. Let us use these ideas first to consider the mechanical theory, and then to compare the mechanical with the adsorption theory.

4.2. Mechanical theory

Adhesion to a rough or porous surface may be enhanced. How might this work?

4.2.1. Roughness factor — increased area of contact. An obvious way in which a rougher surface might lead to higher adhesion than a smooth one results from the increased area of contact. A surface with a higher 'roughness factor', defined as the ratio of the true area to the nominal area, would be expected to give a higher effective work of adhesion. This point has often been made, and in careful experiments with rubber adhesion, Gent *et al.* have convincingly demonstrated the effect [105, 106]. With grit blasted steel they observed increases in adhesion by factors of 2 to 3 times which they ascribed to an increase in surface area.

Such a traditional roughness factor approach overcomes the complication of surface roughness, by treating the surfaces as if they were intrinsically the same as smooth ones, and applying a correction factor for the roughness.

4.2.2. Increased work of adhesion. But what if the surface is very much rougher? What about the porous surface of anodised aluminium shown in Fig. 3 or a microfibrous oxide on steel (Fig. 2)? As the scale of roughness becomes finer, use of a simple scaling factor becomes increasingly unrealistic and unconvincing. It becomes unconvincing not just because of increasing practical difficulty in measuring the 'true' area of such surfaces, it is unconvincing because the roughness itself is an essential characteristic of the surfaces.

These surfaces are not just the same as smooth surfaces that happen to be rough. As we approach molecular scale roughness — indeed long before we get there — the molecular environment of the surface molecules, and therefore their surface energy, is quite different from that of similar molecules on a nominally plane surface. Their energy is a consequence of the topological configurations they take up. Moreover, it has been argued that many engineering surfaces are fractal, possessing a similar structure over a range of scales [107]. The area ascribed to such surfaces is arbitary, depending entirely on the scale chosen.

Further, roughness may develop as a result of the reorganisation of two phases once brought into contact. It would then be an intrinsic property of the interface concerned. Using the surface forces apparatus [101], Israelachvili and his colleagues have demonstrated the rearrangement of a solid–solid interface after formation by reptation, reorientation and exchange of molecular species [108, 109]. Roughness is produced where it was not previously present.

Even if it were practicable, it would make no scientific sense to apply some sort of 'roughness factor' to allow for the roughness of such surfaces. The concept of work of adhesion is applied in a macroscopic context: the appropriate area to be used in calculating it in these cases should therefore be the macroscopic area (i.e. ideal, geometric area) of the interface.

We see then that the effect of producing many of the more intricate forms of surface topography, described in this paper, is to enhance the surface energy and work of adhesion. The result is not fundamentally different from that obtained by increasing the surface energy by chemical modification. Both have the potential to enhance the measured adhesion (see equations (2), (3) and (4)).

4.2.3. Surface roughness and stress distribution. Modification of the topography of a surface may alter the way that the stress is distributed when the joint is loaded. This can increase the energy dissipation, ψ, which occurs during fracture, and provide an important increase in adhesion.

A conceptually simple example of this was the adhesion of silica to copper discussed above [65]. Here the stress is directed away from the low W_a interface (silica/copper) towards the stronger silica/palladium interface by the topography produced (Fig. 7). This is in many ways a simply 'key' of the sort that McBain and Hopkins envisaged. The surface topography protects weak regions from a high stress field.

The suggestion has been made that the occlusion of cohesively weak material in the pores or fibres of microfibrous surfaces might contribute to the good adhesion often associated with such surfaces [110, 111]. This can be understood as another example of the direction of the applied stress away from weak material towards the stronger bulk polymer. Arslanov and Ogarev [7] argue that the diffusion coefficient is too low for this to occur with polyethylene on anodised aluminium. Even if this be true in the particular example, there would seem to be no reason for dismissing the mechanism altogether.

Another way in which microfibrous and microporous surfaces enhance adhesion is by causing stress discontinuities at the interface. Fibre tips, for example, act as stress concentrators causing plastic yielding, at first locally, and then further into the bulk of the polymer. Evidence for this effect has been found, not only with high ductility polymers, such as low density polyethylene and ethylene vinyl acetate copolymers, but also with epoxy resins [69, 112, 113]. By altering the interfacial topography, the energy absorbing capabilities of the bulk materials are exploited to a greater extent. In terms of equation (3), the energy losses, ψ_{plast}, are increased.

Gent and Lin have shown that large amounts of energy can also be involved in peeling an *elastic* material from a rough surface [105]. The energy is essentially used for the elastic deformation of embedded filaments: this energy is lost because when the filaments become free, they immediately relax.

Gent and Lin experimented with rubber bonded to aluminium plates with regular arrays of cylindrical holes. The peel energy was low for the plates in the absence of holes. An energy balance analysis gives the ratio of fracture energy for peeling from the material with cylindrical pores G'_a to that from a smooth substrate G_a as

$$\frac{G'_a}{G_a} = 1 + \frac{4\phi l}{a}, \tag{5}$$

where l is the pore length, a its radius and ϕ the ratio of pore area to total area of the plate [105]. Their experimental results demonstrated the essential validity of this relationship. Where pull-out alone occurred the work of detachment for their system increased by up to 20 times.

They further considered the situation where fracture of strands occurred. The extra work is proportional to depth of pores and for their system could be several hundred times the work of detachment from a smooth surface. This energy term dominates for deep pores.

There are obvious similarities between the polymer which has solidified within the pores of a microfibrous surface and fibres embedded in the matrix of a composite material. Standard treatments of fibre composites (e.g. [114]) draw attention to the significance of the critical length of fibre. When short fibres are stressed axially, shear failure at the fibre/matrix adhesion is considered to occur, and the fibres may be pulled out of the matrix. Fibres greater than the critical length, with a consequently larger fibre/matrix surface area, fail in tension, and only the broken ends are pulled out. This, of course, is one of the points that Gent and Lin were demonstrating. The fracture toughness of the composite may be enhanced by energy terms associated with fibre fracture, with fibre/matrix adhesion and with fibre pull-out. By assuming that the fibre is linearly elastic and equating the interfacial shear force to the tensile force for a fibre of critical length, l, it immediately follows that

$$\frac{2l}{a} = \frac{\sigma}{\tau}, \tag{6}$$

where a is the fibre radius, σ its tensile strength and τ the interfacial shear strength. As in equation (5), the l/a ratio is significant.

Arslanov and Ogarev [7] use equation (6) to argue that the critical length of an adhesive in a microporous anodic film is very small, so the adhesive filaments will fail in tension and the most of the pore length is irrelevant to adhesion. Application of the simple model of equation (5) to this situation shows that even with a short length of elastic adhesive a useful increase in peel strength might be expected. For polyethylene embedded in a film formed by anodising in phosphoric acid, a ratio G'_a/G_a of three to four times is obtained.

In a realistic situation the adhesive filament will not act as a perfect elastic body uniformly stressed up to fracture. Uneven stress distributions and plastic yielding would be expected to increase the energy dissipation observed beyond that calculated for the ideal elastic model.

While calculations like those discussed involve serious simplifications and idealisations, they do serve to show that surface roughness *per se* is capable of increasing the fracture energy of an adhesive joint by a significant amount.

The mechanical theory of adhesion is associated with adhesion to rough and porous surfaces. Adhesion to such surfaces may be effective because their intrinsically high surface energy gives rise to a high work of adhesion in the bond. When stressed, these surfaces may be able to redistribute the stress so as to increase energy dissipation during failure of the joint. These mechanisms are consistent with the theory discussed above, expressed in terms of equations (1) to (4).

4.3. Adsorption theory compared with the mechanical theory

McBain and Hopkins, like most commentators since them, tended to treat mechanical adhesion and specific adhesion (adhesion by way of the adsorption theory) as occurring by two entirely distinct mechanisms. Is this distinction really useful? Is there a danger of an unnecessary polarisation, where an integration might be preferable? Let us briefly examine adhesion as seen by the adsorption theory, considering adhesion to a 'smooth' surface.

If wetting is poor because of an inappropriate thermodynamic relationship between surface energies of adhesive and substrate, or because there is insufficient time for equilibrium to be reached before the adhesive sets, poor adhesion is likely to result because of a reduced area of contact and because of stress concentrations associated with interfacial voids. In terms of equations (1) and (2), W_a is low (part of the interface is air) and crack length l is increased. These considerations apply in the same way in the context of the mechanical theory. For thermodynamic or kinetic reasons poor penetration of adhesive into the interstices of the surface may result [11].

If wetting is complete, adsorption of the adhesive is considered to occur whether the surface is smooth or rough, but poor adhesion may result from low work of adhesion, a result of low surface energy. Adhesion may be enhanced by surface treatments some of which increase the surface energy and thence W_a. The treatment which raises W_a may be one which essentially alters the chemical groups in a smooth surface (although some topographical change will usually accompany it), or it may drastically alter the topography of the surface (although some chemical change will usually accompany it). Although W_a is increased the measured adhesion, G or P above, may be increased by a far greater extent (equation (4)). Adhesion fracture energies are commonly far higher than the energies that can be ascribed to interfacial bonds, even to strong primary bonds. The treatment which increases surface energy is effective because it enables much more energy to be absorbed during fracture: it increases ψ (equations (2) and (3)). This energy may be absorbed in regions near the interface or within the bulk phases present. This accounts for the widespread observation that high adhesion is associated with cohesive failure of the joint. Both adsorption and mechanical model work in a similar way, i.e. with an increase of work of adhesion leading to failure away from the interface.

Sometimes despite relatively strong interfacial forces, adhesion is low because stresses are strongly concentrated at the interface when the joint is loaded. This may be a result of a sharp discontinuity of modulus between adhesive and substrate. This is equivalent to introducing a critical defect, size l, close to the interface. Some surface treatments interpose a layer of intermediate modulus, lowering the stress concentrations and enabling larger volumes of material to be involved in energy dissipating viscoelastic or plastic deformation. The action of silane coupling has been described in such terms. The effect of an interfacial 'composite' region [51] where substrate and adhesive merge over a distance of nanometres or micrometres will similarly smooth out a sharp modulus change between the two phases.

5. CONCLUSIONS

It is important to recognise that the bias towards analysis, which is a characteristic of the scientific method, is not usually an end in itself, but a means of approaching the objective of achieving an overall rationalisation of phenomena [115]:

> 'In weakness we multiply distinctions, then
> Deem our petty boundaries are things
> That we perceive and not that we have made.'

Discussion of adhesion phenomena separately in the terms of adsorption and mechanical theories may sometimes help to a clarify our ideas, but 'adsorption' and 'mechanical' effects can rarely be completely isolated from one another. Each theory emphasises a different aspect of a more comprehensive model which, in principle, relates molecular dispositions in the region of the interface to macroscopic properties of an adhesive joint. It might be a mistake today to lay too much emphasis on the distinction between the two classical theories, although this has been valuable at various times during the last seventy years in stimulating the development of new concepts and in suggesting fruitful experiments.

REFERENCES

1. I. Newton, in: *Opticks*, 4th edn, Book 3, Part 1, Query 31 (1730).
2. J. W. McBain and D. G. Hopkins, *J. Phys. Chem.* **29**, 188 (1925).
3. R. C. Rinker and G. M. Kline, U.S. National Advisory Committee on Aeronautics, Tech. Note 989 (August 1945).
4. F. W. Reinhart, in: *Adhesion and Adhesives: Fundamentals and Practice*, p. 9. Society of Chemical Industry, London (1954).
5. W. C. Wake, in: *Adhesion and the Formulation of Adhesives*, pp. 65–71. Applied Science, London (1976).
6. J. D. Venables, *J. Mater. Sci.* **19**, 2431 (1984).
7. V. V. Arslanov and V. A. Ogarev, *Prog. Organic Coatings* **15**, 1 (1987).
8. W. Brockmann, O. D. Hennemann, H. Kollek and C. Matz, *Int. J. Adhesion Adhesives* **6**, 115 (1986).
9. G. D. Davis, *Surf. Interf. Anal.* **17**, 439 (1991).
10. G. W. Critchlow and D. M. Brewis, *Int. J. Adhesion Adhesives* **15**, 161 (1995).
11. D. E. Packham, in: *Adhesion Aspects of Polymeric Coatings*, K. L. Mittal (Ed.), p. 19. Plenum, New York (1983).
12. D. E. Packham, *J. Adhesion* **39**, 137 (1992).
13. F. L. Browne and T. R. Truax, *Colloid Symposium Monograph* **4**, 258 (1926).
14. J. W. McBain and W. B. Lee, *Ind. Eng. Chem.* **19**, 1005 (1927).
15. R. C. Rinker and G. M. Kline, *Modern Plastics*, p. 168 (Nov. 1945).
16. W. C. Wake, in: *Adhesion and Adhesives*, 2nd edn, R. Houwink and G. Salomon (Eds), Vol. 1, p. 405. Elsevier, Amsterdam (1965).
17. K. Bright, B. W. Malpass and D. E. Packham, *Nature* **223**, 1360 (1969).
18. D. J. Arrowsmith, *Trans. Inst. Metal Finishing* **48**, 88 (1970).
19. K. Bright, B. W. Malpass and D. E. Packham, *Brit. Polym. J.* **3**, 205 (1971).
20. K. W. Allen, H. S. Alsalim and W. C. Wake, in: *Faraday Special Disc. Chem. Soc. No. 2, Solid/Solid Interfaces*, p. 38. Chemical Society, London (1972).
21. O. D. Hennemann and W. Brockmann, *J. Adhesion* **12**, 297 (1981).
22. G. Farkas, *Surfaces* **14** (93), 37 (1975).
23. A. W. Bethune, *SAMPE J.* **11** (3), 4 (1975).
24. J. C. McMillan, J. T. Quinlivan and R. A. Davis, *SAMPE Q.* **7** (3), 13 (1976).
25. J. A. Marceau, *SAMPE Q.* **9** (4), 1 (1978).
26. Boeing Process Specification, *Phosphoric Acid Anodising of Aluminum for Structural Bonding*, BAC 5555, revision A (1975).
27. J. D. Venables, *J. Adhesion* **39**, 79 (1992).
28. J. M. Chen, T. S. Sun, J. D. Venables and R. Hopping, in: *Proc. 22nd SAMPE Symposium*, p. 25 (April 1997).
29. T. S. Sun, J. M. Chen and J. D. Venables, *Appl. Surface Sci.* **1**, 202 (1978).
30. J. D. Venables, D. K. McNamara, J. M. Chen, T. S. Sun and R. L. Hopping, *Appl. Surface Sci.* **3**, 88 (1979).
31. T. S. Sun, D. K. McNamara, J. S. Ahearn, J. M. Chen, D. Ditchek and J. D. Venables, *Appl. Surface Sci.* **5**, 406 (1980).
32. J. P. O'Sullivan and G. C. Wood, *Proc. Roy. Soc. A* **317**, 511 (1970).
33. J. A. Bishopp, E. K. Sim, G. E. Thompson and G. C. Wood, *J. Adhesion* **26**, 237 (1988).
34. A. Roche, *Vide Couches Minces* (257), 197 (1991).
35. K. W. Allen, *J. Physique IV* **3** (C7, Part 2), 1511 (1993).
36. K. W. Allen, *Int. J. Adhesion Adhesives* **13**, 67 (1993).
37. A. Bogner, *Holz als Roh- und Werkstoff* **49**, 271 (1991).
38. D. E. Packham and C. Johnston, *Int. J. Adhesion Adhesives* **14**, 131 (1994).
39. H. Kanai, *J. Mater. Sci.* **26**, 1815 (1991).
40. C. U. Ko and J. P. Wightman, *J. Adhesion* **24**, 93 (1987).
41. D. E. Packham, *Int. J. Adhesion Adhesives* **6**, 225 (1986).

42. J. B. Hollenhead and J. P. Wightman, *J. Adhesion* **37**, 121 (1992).
43. J. A. Filbey and J. P. Wightman, *J. Adhesion* **28**, 1 (1989).
44. Y. Moji and J. A. Marceau, Anodising titanium to promote adhesion, US Patent 3,959,091 (1976) (*Chem. Abs.* **85**, 70057 (1976)).
45. Y. Moji and J. A. Marceau, Anodising titanium to promote adhesion, US Patent 3,989,876 (1976) (*Chem. Abs.* **85**, 200102 (1976)).
46. R. G. Schmidt and J. P. Bell, *Adv. Polym. Sci.* **75**, 33 (1986).
47. B. W. Malpass, D. E. Packham and K. Bright, *J. Appl. Polym. Sci.* **18**, 3249 (1974).
48. J. P. Sargent, *Int. J. Adhesion Adhesives* **14**, 21 (1994).
49. S. A. Tschetter, N. E. Shephard and J. P. Wightman, in: *Proc. 18th Adhesion Society Annual Meeting*, p. 165. Hilton Head, SC (1995).
50. R. P. Digby, Ph.D. Thesis, University of Bath (1996).
51. R. J. Davies and M. D. Ritchie, *J. Adhesion* **38**, 243 (1992).
52. W. J. Russell, *J. Appl. Polym. Sci., Appl. Polym. Symp.* (32), 105 (1977).
53. N. L. Rogers, in: *Proc. 13th National SAMPE*, p. 640 (1981).
54. R. P. Digby and D. E. Packham, *Vide Couches Minces* (272 SS), 586 (1994).
55. F. Keller, M. S. Hunter and D. L. Robinson, *J. Electrochem. Soc.* **100**, 411 (1953).
56. G. C. Wood and J. P. O'Sullivan, *Acta Electrochimica* **15**, 1865 (1970).
57. D. J. Arrowsmith and A. W. Clifford, *Int. J. Adhesion Adhesives* **5**, 40 (1985).
58. R. P. Digby and D. E. Packham, *Int. J. Adhesion Adhesives* **15**, 61 (1995).
59. H. Reiter, in: *Handbook of Adhesion*, D. E. Packham (Ed.), p. 192. Longman (1992).
60. H. M. Clearfield, D. K. Shaffer, S. L. Vandoren and J. S. Ahearn, *J. Adhesion* **29**, 81 (1989).
61. R. A. Pike, V. H. Patarini, R. Zatoski and F. P. Lamm, *Int. J. Adhesion Adhesives* **12**, 227 (1992).
62. G. D. Davis, D. K. Shaffer, P. L. Whisnant, G. B. Groff and J. D. Venables, *3rd Int. Conf. Adhesion and Surface Anal.*, paper L13, Loughborough (April 1994).
63. G. D. Davis, P. L. Whisnant, D. K. Shaffer, G. B. Groff and J. D. Venables, *J. Adhesion Sci. Technol.* **9**, 527 (1995).
64. A. E. P. Morris and M. E. R. Shanahan, *Int. J. Adhesion Adhesives* **14**, 145 (1994).
65. A. M. T. van der Putten, *J. Electrochem. Soc.* **140**, 2376 (1993).
66. G. W. Critchlow and D. M. Brewis, *Int. J. Adhesion Adhesives* **15**, 173 (1995).
67. C. Keisler and J. L. Lataillade, *J. Adhesion Sci. Technol.* **9**, 395 (1995).
68. J. R. G. Evans and D. E. Packham, *J. Adhesion* **10**, 177 (1979).
69. P. J. Hine, S. El Muddaris and D. E. Packham, *J. Adhesion* **17**, 207 (1984).
70. L. A. Reeves, D. Kiroski and D. E. Packham, *Surf. Interf. Anal.* **23**, 299 (1995).
71. D. E. Packham (Ed.), *Handbook of Adhesion, passim*. Longman (1992).
72. R. A. Ryntz, *Prog. Organic Coatings* **25**, 73 (1994).
73. D. M. Brewis (Ed.), *Surface Analysis and Pretreatment of Plastics and Metals*. Applied Science, London (1982).
74. D. Hace, V. Kovacevic and D. Pajc-Liplin, *3rd Int. Conf. Polymer Characterisation*, Denton, Texas (Jan. 1995).
75. S. O'Kell, M. Aindow and C. Jones, *Vide Couches Minces* (272), 420 (1994).
76. S. O'Kell, T. Henshaw, G. Farrow, M. Aindow and C. Jones, *Surf. Interf. Anal.* **23**, 319 (1995).
77. K. Harth and H. Hisbt, *Surface Coatings Technol.* **59**, 350 (1993).
78. M. M. Pastor Blas, M. S. Sánchez Adsuar and J. M. Martín Martínez, *J. Adhesion Sci. Technol.* **8**, 1093 (1994).
79. T. P. Nguyen, K. Amgaad, M. Cailler and V. H. Tran, *J. Adhesion Sci. Technol.* **8**, 821 (1994).
80. J. T. Marchesi, H. D. Keith and A. Garton, *J. Adhesion* **39**, 185 (1992).
81. L. P. Buchwalter and R. Saraf, *J. Adhesion Sci. Technol.* **7**, 925 (1993).
82. L. P. Buchwalter, *J. Adhesion Sci. Technol.* **9**, 941 (1993).
83. N. Inagaki, S. Tasaka and K. Hibi, *J. Adhesion Sci. Technol.* **8**, 395 (1994).
84. R. F. Saraf, J. M. Roldan and T. Derderian, *IBM J. Res. Develop.* **38**, 441 (1994).
85. S. J. Shaw, in: *Handbook of Adhesion*, D. E. Packham (Ed.), p. 319. Longman (1992).
86. R. F. Saraf, *Macromolecules* **26**, 3623 (1993).

87. N. C. Stoffel, E. J. Kramer, W. Volksen and T. P. Russell, *Polymer* **34**, 4524 (1993).
88. D. J. D. Moyer and J. P. Wightman, *Surf. Interf. Anal.* **14**, 496–504 (1989).
89. B. Z. Jang, *Compos. Sci. Technol.* **44**, 333 (1992).
90. Y. L. Yuan, S. S. Shuy and J. Y. Lai, *J. Appl. Polym. Sci.* **42**, 2525 (1991).
91. T. R. King, D. F. Adams and D. A. Buttry, *Composites* **22**, 380 (1991).
92. T. R. King, D. F. Adams and D. A. Buttry, *Compos. Sci. Technol.* **44**, 351 (1992).
93. P. Marshall and J. Price, *Composites* **22**, 388 (1991).
94. I. M. Ward and N. H. Ladizeski, in: *Composite Interfaces*, H. Ishida and J. L. Koenig (Eds), p. 37. North Holland, New York (1986).
95. B. Tissington, G. Pollard and I. M. Ward, *Compos. Sci. Technol.* **44**, 197 (1992).
96. B. Tissington, G. Pollard and I. M. Ward, *Compos. Sci. Technol.* **44**, 185 (1992).
97. M. S. Silverstein and O. Breuer, *Compos. Sci. Technol.* **48**, 151 (1993).
98. M. S. Silverstein, O. Breuer and H. Dodiuk, *J. Appl. Polym. Sci.* **52**, 1785 (1994).
99. K. L. Mittal, *IBM J. Res. Develop.* **38**, 346 (1994).
100. R. J. Good, *J. Adhesion* **4**, 133 (1972).
101. D. E. Packham, *Int. J. Adhesion Adhesives* **16**, 121–128 (1996).
102. T. A. Hatzinikolaou and D. E. Packham, *3rd International Adhesion Conference*, paper 8, York (Sept. 1987).
103. A. N. Gent and A. J. Kinloch, *J. Polym. Sci. A2* **9**, 659 (1971).
104. E. H. Andrews and A. J. Kinloch, *Proc. Roy. Soc. A* **332**, 385 and 401 (1973).
105. A. N. Gent and C. W. Lin, *J. Adhesion* **32**, 113 (1990).
106. A. N. Gent and S. M. Lai, *Rubber Chem. Technol.* **68**, 13 (1995).
107. B. Bhushan, J. N. Israelachvili and U. Landman, *Nature* **374**, 607 (1995).
108. Y. L. Chen, C. A. Helm and J. N. Israelachvili, *J. Phys. Chem.* **95**, 10 736 (1991).
109. H. Yoshizawa, Y. L. Chen and J. N. Israelachvili, *J. Phys. Chem.* **97**, 4128 (1993).
110. D. E. Packham, K. Bright and B. W. Malpass, *J. Appl. Polym. Sci.* **18**, 3237 (1974).
111. A. Marceau, *SAMPE Q.* 1 (July 1978).
112. D. E. Packham, in: *Developments in Adhesives — 2*, A. J. Kinloch (Ed.). Applied Science Publishers (1981).
113. T. A. Hatzinikolaou and D. E. Packham, in: *Adhesion 9*, K. W. Allen (Ed.), p. 33. Elsevier Applied Science, London (1984).
114. D. Hull, in: *An Introduction to Composite Materials*, p. 142 *et seq.* Cambridge University Press (1981).
115. W. Wordsworth, *The Prelude* **II**, 217–219, text of 1805. Oxford University Press (1970).

Mittal Festschrift, pp. 109–125
W. J. Van Ooij and H. R. Anderson, Jr. (Eds)

Metal adhesion on the nanometer scale

U. DÜRIG* and A. STALDER
IBM Research Division, Zurich Research Laboratory, CH-8803 Rüschlikon, Switzerland

Abstract—Adhesion, wear and friction are closely related to the interactions and mechanical responses that occur between and within nanometer-sized surface structures. Thus, besides being interesting in its own right, the study of material properties on this length scale provides the basis for an understanding of phenomenologically well-known macroscopic features on a fundamental level. Experimentally, the nanometer regime has become accessible by the development of scanning tunneling microscopy. Interaction mechanics on the nanometer scale is addressed in this paper from this perspective both in terms of theoretical concepts and recent experimental work.

Keywords: Adhesion; interaction forces; plastic deformation; nanometer scale.

1. INTRODUCTION

Adhesion between bodies is more than just a force acting between them. Adhesion involves a subtle interplay between interaction forces and mechanical deformations that occur on a wide span of length scales because surfaces are never perfectly smooth. One must visualize them as corrugated landscapes having a myriad tiny protrusions. When two such surfaces are brought into contact, adhesion is governed by the interaction between asperities and their mechanical response, a fact that has long been known and emphasized [1, 2].

A common way to look at adhesion is to express the interaction energy in terms of surface free energies

$$W_{\mathrm{ad}} = A_{\mathrm{c}} \Delta\gamma_i, \tag{1}$$

where A_{c} is the contact area and $\Delta\gamma = \gamma_1 + \gamma_2 - \gamma_{12}$ is the the gain in surface free energy by bringing two surfaces with respective surface free energies γ_1 and γ_2 into contact, thereby forming an interface with a surface free energy γ_{12}. The concept of contact area and interface free energy is not well defined in a realistic adhesion problem, however, which involves asperity contacts and voids as depicted in Fig. 1. One can intuitively define the contact area A_{c} as the sum of the cross sectional areas A_{c}^i of the gap free bridges that connect the two bodies. Furthermore, a measure of the distance between the two interacting bodies, s, is introduced. The measure

*To whom correspondence should be addressed. E-mail: drg@zurich.ibm.com

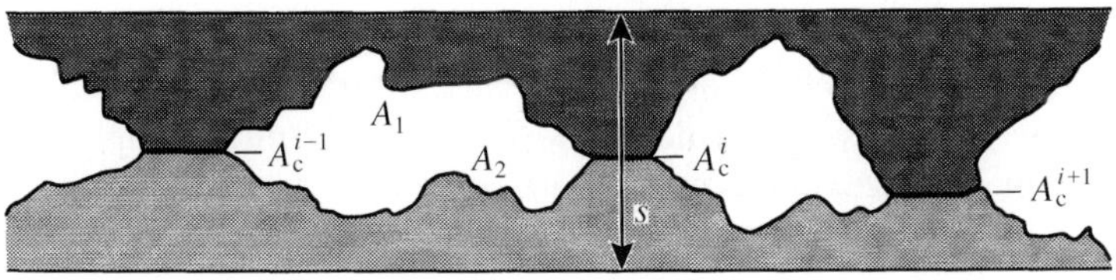

Figure 1. Schematic diagram of two bodies in contact. The interface is characterized by free surfaces $A_{1,2}$ and contact areas A_c^i where the two bodies touch. The distance between the bodies, s, is measured with respect to some convenient frame of reference.

is defined with respect to reference frames that are accessible to measurements and, consequently, that are typically far away from the interacting surfaces. Clearly, one cannot specify an adhesion energy here simply in terms of the energy difference between a contact and a noncontact state. A more useful approach is to ask how the total energy of the system depends on s.

Let us now consider a small change Δs, and assume that the bodies are pulled slightly apart. As a result, the connective bridges, henceforth termed necks, will be elongated. Accordingly, the contact cross section will decrease by ΔA_c, and an additional free surface area $\Delta A_{1,2}$ on the two bodies will be exposed. Exactly how neck elongation, contact cross section, and free surface area are related is governed by the elasto–plastic deformation mechanics of the necks. Irrespective of these details, the change of adhesion energy can be written (following the concepts in equation (1)):

$$\Delta W_{ad} = \gamma_1 \Delta A_1 + \gamma_2 \Delta A_2 - \gamma_{12} \Delta A_c, \tag{2}$$

which corresponds to an adhesion force of

$$F_{ad} = \frac{\Delta W_{ad}}{\Delta s} = \gamma_1 \frac{\Delta A_1}{\Delta s} + \gamma_2 \frac{\Delta A_2}{\Delta s} - \gamma_{12} \frac{\Delta A_c}{\Delta s}. \tag{3}$$

Equations (2) and (3), however, are not correct. The pitfall is that it is implicitly assumed that the free surfaces A_1 and A_2 are sufficiently far apart for their mutual interaction to be neglected. This presumption is justified only for short-range interactions, which decay exponentially over a distance that is small compared to the radii of curvature involved in the interface structure. However, for long-range interactions such as van der Waals (VdW) interactions, whose magnitude decays as a power law with distance, the situation is much more complex because no obvious length scale can be identified. Formally, one can correct for this deficiency by adding a long-range interface energy term $W_{lr}(s)$, which depends explicitly on distance and implicitly on the topographic structure of the free surfaces $A_{1,2}$. This long-range interface term gives rise to an additional adhesion force $dW_{lr}(s)/ds$, which is not contained in equation (3). Thus, the intrinsic mutual adhesional interaction consists of a global long-range force term plus a local capillary force term connected with the creation and annihilation of free surface area at contacts.

In order to separate two bodies, one must overcome not only the intrinsic adhesional interaction but also the mechanical forces involved in the deformation of contacts. The interplay between these forces depends on the topographic structure of the surfaces and their chemical properties which, in turn, strongly influence the characteristics of the adhesional interaction. Thus, adhesion is an extremely complex subject. The purpose of this paper is to discuss some of the fundamental microscopic issues involved in metal adhesion. In particular, attention is focused on the interaction characteristics between nanometer-sized asperities and their deformation mechanics. The coalescence and fracture of such asperity contacts is one of the key processes in friction and wear.

2. INTERACTIONS IN METAL ADHESION

The interaction between two metallic bodies arises from the quantum-mechanical description of the Coulomb interaction of the conduction electrons. For the moment, we consider the interaction between two parallel half-spaces separated by a distance s. At large distances, the electron wavefunctions do not overlap at the interface. The interaction is then dominated by VdW forces originating from quantum-mechanical fluctuations of the electron gas, which in turn give rise to space- and time-modulated dipole moments. The leading term of the nonretarded (we assume that the interaction distance is less than 100 nm) VdW interaction energy per unit area is given by the well-known Lifshitz formula [3]:

$$W_{\mathrm{ad}}(s) = \frac{\hbar\overline{\omega}}{16\pi^2 s^2}, \tag{4}$$

where $\overline{\omega}$ is related to the dielectric loss in the half-spaces. Loosely speaking, $\overline{\omega}$ corresponds to the sum of the peak absorption frequencies of independent dissipation channels, and typical values are of the order of 10 eV for most metallic elements.

The Lifshitz expression emerges from a local approximation of the dielectric function. The motion of electrons is strongly correlated in metals, however, so nonlocal screening effects cannot be neglected. These effects become appreciable at separations smaller than 0.6 nm [4]. The corresponding theory is rather elaborate. Numerical results have been obtained for simple metals such as aluminum in the regime of negligible electron exchange [5]. The most important difference is that for small distances the interaction is weaker than the Lifshitz formula would suggest. In particular, the divergence at $s \to 0$ is eliminated.

For sufficiently small gap widths, i.e. $s \leqslant 0.3$ nm, electron wavefunctions overlap appreciably, making it necessary to consider exchange and correlation effects. Rose *et al.* [6] calculated the adhesion energy for simple metals using a self-consistent local density formalism and a jellium model. They discovered that the interaction energy as a function of interface separation was fairly accurately represented by the Rydberg function with two scaling parameters, namely the maximum of the adhesion energy per unit area E_{ad} and a length scale l_{sc} approximately equal to the Thomas–Fermi screening length

$$W_{\mathrm{ad}} = E_{\mathrm{ad}}(1 + s/l_{\mathrm{sc}}) \exp(-s/l_{\mathrm{sc}}). \tag{5}$$

This theory was extended in a semi-phenomenological way to include all metallic elements, whereby the quantities E_{ad} and l_{sc} are inferred from cohesive properties of the metals [7].

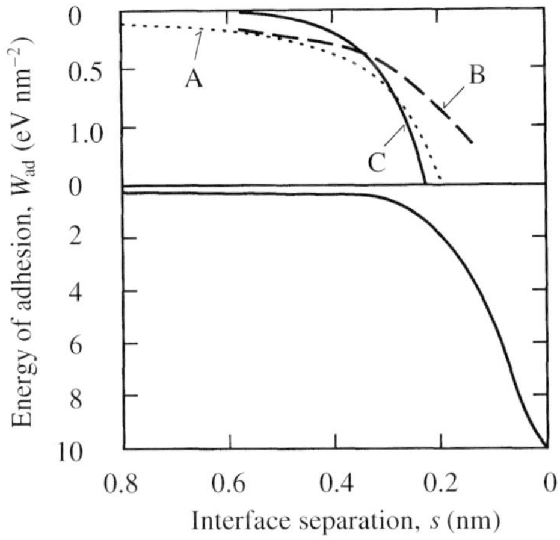

Figure 2. Adhesion energy of two Al half-spaces. Upper panel: (A) van der Waals interaction, Lifshitz result; (B) van der Waals interaction, Inglesfield result; (C) short-range adhesion interaction, Ferrante–Smith potential. Lower panel: Overall interaction potential obtained by interpolating curves (B) and (C). (Adapted from [9].)

As a representative example, the interaction potential for two Al half-spaces is shown in Fig. 2 using the following parameters: $\hbar\bar{\omega} = 8.2$ eV [8], $E_{ad} = 10$ eV nm^{-2} [6], and $l_{sc} = 0.066$ nm [7]. The upper panel shows the various contributions to the adhesion energy on an expanded scale. The dotted curve (A) corresponds to the nonretarded Lifshitz–VdW interaction, equation (4). The dashed curve (B), taken from [5], represents the VdW interaction corrected for nonlocal effects. The short-range metallic adhesion interaction, equation (5), is depicted by the solid line (C). The overall interaction potential obtained by interpolating the latter two curves is displayed in the lower panel. It becomes apparent from the figure that more than 90% of the adhesion energy in metals is due to short-range interactions.

Next we consider the interaction of a sharply pointed tip with a flat substrate surface. The situation corresponds to an extreme realization of a single surface asperity interacting with a counter surface in an adhesion problem. Ignoring many-body effects, the tip–sample force can be written as [9]:

$$F_{TS} = -\frac{dW_{ad}(s)}{ds}\sigma(0) + W_{ad}(s)\frac{d\sigma(0)}{ds} + \int_0^\infty W_{ad}(s+z)\frac{d^2\sigma}{dz^2}\,dz, \tag{6}$$

where $\sigma(z)$ denotes the tip cross section perpendicular to the surface normal, z, of the substrate. The tip–sample interaction is thus determined by the shape of the apex expressed by its cross section $\sigma(0)$, its radius of curvature $R_a \equiv d\sigma(0)/2\pi\,dz$, plus a correction term accounting for contributions of the tip shaft. In passing we note that the second term in equation (6) corresponds to the so-called Derjaguin approximation [10] that is generally used to describe particle adhesion.

Given that the surface of a real tip is corrugated (see Fig. 3) it is not immediately clear how to evaluate the tip parameters in equation (6). Matters are relatively simple for the short-range interaction, which is sensitive to the atomic structure only at the

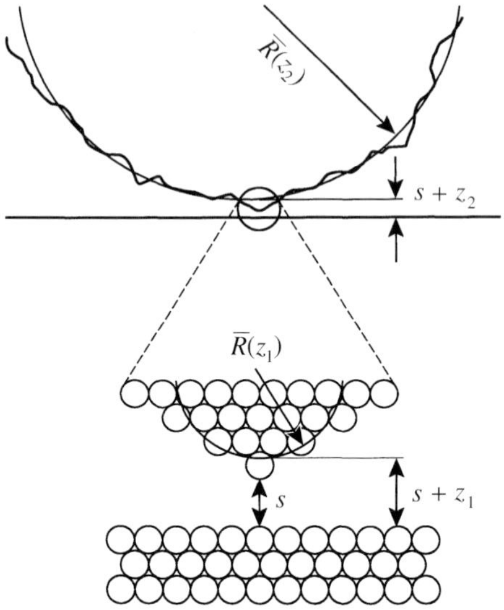

Figure 3. Schematic diagram of a corrugated tip apex. Effective VdW radii, $\overline{R}(z_1) = 0.8$ nm and $\overline{R}(z_2) = 10$ nm, corresponding to local maxima of the VdW force are indicated.

very apex of the tip. Thus, two parameters, namely the zeroth and the first moment of the exponentially weighted tip cross section

$$M_n = \int_0^{\infty} \sigma(z)\, z^n \exp\left(-z/l_{sc}\right) dz, \qquad n = 0, 1, \tag{7}$$

are sufficient to characterize fully the tip–sample force. The effective tip parameters are then expressed in terms of these moments as [9]:

$$\sigma(0) = \frac{2M_0 - M_1/l_{sc}}{l_{sc}} \tag{8}$$

and

$$R_a = \frac{M_1/l_{sc} - M_0}{2\pi l_{sc}^2}. \tag{9}$$

The situation is different for the VdW interaction for which no natural length scale exists. It is easy to see by substituting the Lifshitz expression, equation (4), for W_{ad} that the first, then the second, and finally the third term dominate in equation (6) with increasing tip–sample distance. Thus, the farther the tip is from the surface, the less important the details of the apex. In fact, it can be shown that details of the tip cross section are averaged out over a range $1/\sqrt{3}$ times the distance from the surface [9]. Therefore, it is useful to define an effective VdW radius of curvature which depends on the position, z, above the tip apex in the following way:

$$\overline{R}(z) = \frac{1}{2\pi} \frac{\sigma\left(z + (s+z)/\sqrt{3}\right) - \sigma(z)}{(s+z)/\sqrt{3}}. \tag{10}$$

The idea behind this construction is to replace the infinitesimal derivative in the definition of the radius of curvature, $2\pi R \equiv \mathrm{d}\sigma/\mathrm{d}z$, by the corresponding difference form, which conforms to the appropriate averaging interval for the VdW interaction. The VdW force corresponding to a particular effective tip radius is given by

$$F_{\overline{R}(z)}(s) = 2\pi \overline{R}(z) W_{\mathrm{ad}}(s+z) \tag{11}$$

and, restricting ourselves to the Derjaguin approximation [11], the leading term of the VdW tip–sample force will be dominated by that section of the tip that maximizes $F_{\overline{R}(z)}$. Hence, the effective tip radius is implicitly defined by equations (10) and (11) and the condition to select a value for z such that $F_{\overline{R}(z)}$ is maximized. Thus, the tip–sample force can be approximated by

$$F_{\mathrm{TS}}(s) = \max_{(z)} F_{\overline{R}(z)}(s). \tag{12}$$

This procedure will be illustrated by an example. We consider a corrugated spherical tip with a conical asperity at the apex as depicted in Fig. 3. The asperity shall be representative of the structure of the tip corrugation, which can be characterized by an rms height of ± 0.6 nm and a corresponding lateral correlation length of the order of 1 nm. The VdW forces are evaluated using the nonlocal interaction potential, curve (B) in Fig. 2, and we assume a tip–sample distance of $s = 0.2$ nm. A first maximum of $F_{\overline{R}(z_1)} = 0.25$ nN is obtained at $z_1 = 0.2$ nm, and the corresponding radius is $\overline{R}(z_1) = 0.8$ nm. This maximum corresponds to an effective VdW radius of the asperity at the tip apex. A second maximum of $F_{\overline{R}(z_2)} = 0.7$ nN evolves at $z_2 = 1$ nm. Here, the effective tip–radius $\overline{R}(z_2)$ converges towards the overall radius of curvature of the tip, assumed to be 10 nm, as corrugations are being averaged out by the smoothing procedure in equation (10). Thus, in terms of the VdW interaction, the tip–sample force is dominated by the large-scale structure of the tip apex. Note that, because of the large z offset, namely 1 nm, this contribution is only weakly dependent on the actual tip–sample distance s.

Having established the effective VdW tip parameters, we are now in a position to calculate the interaction force and force gradient as a function of the tip–sample distance. The result of the calculation is shown in Fig. 4. The interaction potential, including short-range interactions, is the same as that illustrated in Fig. 2. The dashed curve corresponds to the interaction of the isolated asperity with the flat substrate, and the solid curve is obtained if the entire tip structure is considered. As expected, the long-range VdW interaction due to the large-scale tip structure gives rise to a more or less constant, yet substantial, background force (difference between solid and dashed curves in Fig. 4a). However, because of the weak distance dependence of the background force it has only a small effect on the force gradient. Note also that short-range adhesion forces dominate the tip–sample interaction for $s \leqslant 0.3$ nm, and its magnitude is determined by the effective tip parameters given in equations (8) and (9).

Experimental investigations of the force laws of surface interactions have a long history. Long-range molecular attraction due to VdW forces was the first surface

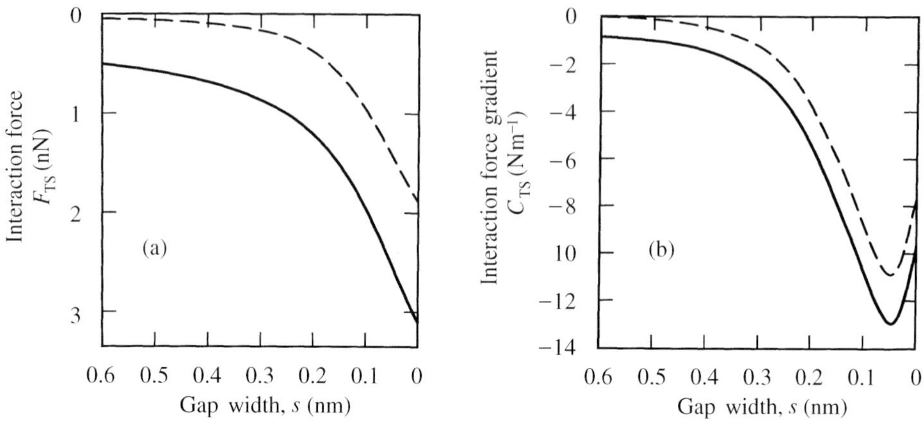

Figure 4. (a) Interaction force and (b) force gradient for a tip like the one shown in Fig. 3. Dashed lines correspond to the interaction of the isolated conical asperity at the tip apex with a flat substrate. This interaction is dominated by short-range forces for $s \leqslant 0.2$ nm. The total interaction including the large-scale tip structure is represented by the solid line. (Adapted from [9].)

interaction to be studied systematically [10]. Accurate control of the distance between the interacting bodies was one of the major experimental challenges. Ultimately, resolution is limited by the intrinsic roughness of the surfaces. To circumvent this problem, Tabor and Winterton [12] used a pair of orthogonally crossed mica cylinders as a sample. This technique allowed VdW interactions to be studied at distances approaching 1 nm [13]. The scheme was also successfully used to investigate solvation forces and adhesion in polymer layers [14].

Among the first to perform quantitative measurements of the interaction between metal surfaces were Derjaguin *et al.* [15]. However, lack of precise knowledge of the interaction geometry rendered interpretation of the data difficult. In particular, no information on the adhesion at distances close to contact could be obtained. On the other hand, as discussed above, short-range forces are responsible for most of the adhesion energy in metals, and therefore they are essential in metal adhesion. Because of their short range, namely of the order of one tenth of a nanometer, their direct experimental study is not a trivial subject.

The development of scanning tunneling microscopy (STM) and related local probe techniques, in particular scanning force microscopy [16], paved the way for a new class of experiments. These techniques allow the gap width between a probe tip and a sample surface to be controlled on a subnanometer scale, and simultaneously tip–sample forces to be measured with nano-Newton sensitivity. The theme of our own experimental work has been to characterize tip–sample interactions during normal metal–vacuum–metal tunneling operation of the STM. Here, the tunneling current provides a convenient parameter for controlling the tip–sample distance down to atomic dimensions, thus enabling us to study short-range metallic adhesion interactions involving electron wavefunction overlap [17, 18]. The interaction is probed by measuring the tip–sample force gradient. The latter can be deduced from shifts of the resonance frequency of a spring sensor to which either the tip or the sample is

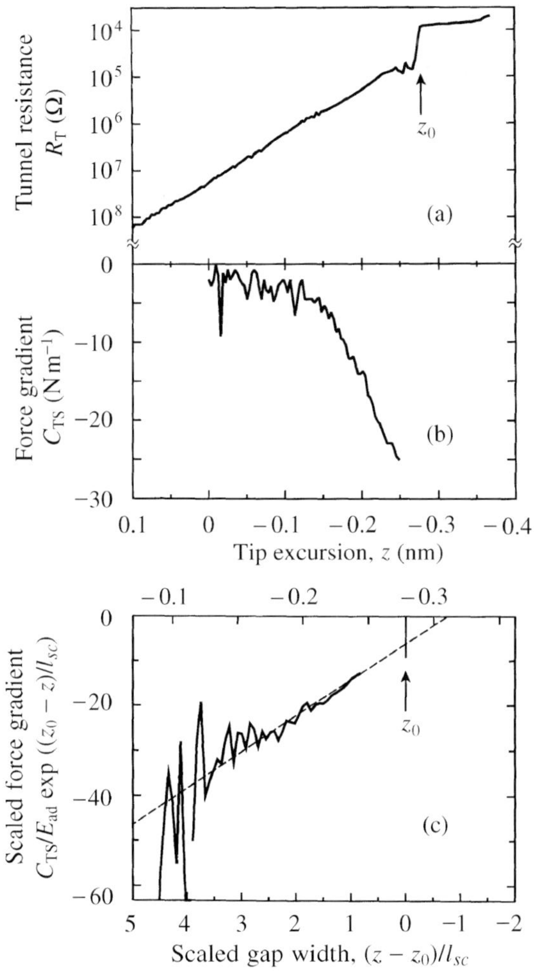

Figure 5. (a) Tunneling resistance and (b) interaction force gradient measured as a function of tip excursion using an Ir tip above an Ir sample. (c) Plot of the interaction force gradient normalized by $E_{ad} \exp[(z_0 - z)/l_{sc}]$ using $E_{ad} = 30$ eV nm^{-2}, $l_{sc} = 0.042$ nm, and $z_0 = -0.27$ nm. (Adapted from [17].)

attached. A detailed discussion of the experimental methodology has been published in [19].

Tunnel resistance and the interaction force gradient measured as a function of tip–sample distance using an Ir tip and an Ir substrate are shown in Fig. 5. The tunnel resistance decreases exponentially with decreasing gap width up to a transition point (marked by the arrow) at which the tunnel resistance abruptly falls to a value of the order of 10 kΩ. This transition can be identified as the making of electrical contact, viz. the vanishing of the energy barrier for electron conduction. As expected for an adhesional interaction, the force gradient is negative and increases rapidly in magnitude as the contact point is approached.

Substituting the short-range adhesion potential equation (5) for W_{ad} in equation (6), one obtains the following expression for the force gradient:

$$C_{TS} = E_{ad} \exp\left(- s/l_{sc}\right) \frac{2M_0 l_{sc} - M_1 - sM_0}{l_{sc}^4}. \tag{13}$$

It is proportional to an exponential term that reflects the chemical nature of the interaction and to a linear term that contains the geometrical information. A plot of the interaction force gradient divided by $E_{ad} \exp((z_0 - z)/l_{sc})$ should yield a straight line from which the moments $M_{0,1}$ can be calculated. Such a plot is shown in Fig. 5c using the onset of electrical conductance at $z_0 = -0.27$ nm to define contact. The scaling parameters are taken from the literature, namely $E_{ad} = 30$ eV nm^{-2} [20] and $l_{sc} = 0.042$ nm [7]. The effective tip radius deduced from the plot is of the order of 0.1 nm, thus indicating that the short-range interaction essentially sees only one atom at the very apex of the tip. Furthermore, a detailed analysis [9] reveals that the transition to contact involves a mechanical instability whereby the tip apex deforms by approximately 0.05 nm as it yields to strong tensile adhesion forces. Thus, we can say that a tiny adhesion neck consisting of only one atom has been formed. This represents the ultimate lower limit of an asperity contact. Furthermore, because the formation of the contact is coupled to a mechanical instability, the process is not adiabatically reversible, and hence it represents a friction mechanism on a truly atomic scale.

The adhesion properties change dramatically when the surface is covered with an organic overlayer. For example, positive interaction force gradients of the order of 200 N m^{-1} were measured on a (111)-oriented Au substrate covered with a self-assembled monolayer of alkane chain molecules (mercaptohexadecanol, $HS-(CH_2)_{16}-OH$) [21]. The results clearly show that the alkane molecules act as a chemically inert spacer for the tunneling tip, preventing it from coming into contact with the metallic Au substrate. In fact, the alkane layer was able to sustain compressive tip forces in excess of 50 nN, corresponding to a pressure of the order 5 GPa based on an experimentally determined radius of contact of the order of 2 nm. Hence, such alkane layers act as excellent lubricants for nanometer-scale structures by suppressing strong adhesional interactions that would lead to the spontaneous formation of adhesion necks and, thus, to wear and associated friction.

In summary, one can say that surface forces play two distinct roles in adhesion. First, they provide an intrinsic loading force that pulls two bodies together. This loading force is determined to a large degree by the surface topology and long-range forces. Second, short-range forces are mainly responsible for the coalescence of small-scale asperities as they come into contact. The mechanics of such contacts is another important factor in surface adhesion. This aspect will be discussed in more detail in the next section.

3. MECHANICS OF NANOMETER-SIZED CONTACTS

Metallic contacts have traditionally been studied in indentation experiments. These entail the measurement of stress–strain characteristics of a joint formed by a sharply pointed stylus, typically an etched tungsten tip, which is first pressed against and then retracted from a flat sample [2, 22–24]. Experimental results have been interpreted in terms of continuum deformation theory and semiempirical approaches to account for plastic deformation [23–25]. Neglecting the atomic structure may be a viable concept for analyzing micrometer-sized contacts. Here we are interested in the initial stage of contact formation, however, which involves the interaction of nanometer size asperities and their subsequent deformation which definitely is influenced by atomistic processes.

Spurred by the development of STM and scanning force microscopy, indentation experiments have been extended to the nanometer regime [26–28]. Ultimately, point contacts of atomic dimensions can be made, as has been convincingly demonstrated in a number of experiments [29–32]. A common feature of all these experiments is the observation of a jump to contact as the tip approaches the surface due to a mechanical instability driven by adhesion forces [18, 33, 34]. This jump manifests itself as an abrupt increase in the electrical conductance and in the adhesion force. The latter is also responsible for the subsequent spontaneous growth of nanometer-sized connective necks by means of plastic yielding. Such connective necks provide an ideal test ground for studying nano-mechanical material properties.

Important contributions in the field of nano-mechanics have been made by molecular dynamics (MD) simulations. Such simulations provide, in principle, a complete description of all mechanical parameters of the system. In their pioneering work, Landman *et al.* studied a variety of phenomena related to the field of nanoscience [35], in particular the consequences of dynamic tip–substrate interactions. The formation of the adhesion neck and its plastic deformation have been studied extensively for a Au tip indenting a Ni surface and vice versa [36, 37]. It was found that the deformation of the neck proceeded in a sequence of structural transformations between ordered and disordered states of a narrow region where the neck was thinnest [36, 37]. Accordingly, the tip–sample force exhibits a characteristic oscillation each time a new atomic layer is added to the neck in this process. Other MD investigations of the mechanics of nanometer scale metal contacts have followed; they also predict that yielding proceeds in a sequence of mechanical instabilities localized within the neck [38, 39].

Recently, mechanical instabilities were observed in an experimental study of the plastic deformation of Au contacts having a diameter of typically 3 to 6 nm [40]. These experiments, conducted at a temperature of 4 K, showed that the contacts became deformed in a repetitive sequence consisting of an elastic strain phase followed by a sharply defined yielding event. Depending on the neck diameter, qualitatively different yielding behavior was observed in our experiments, which were performed at room temperature [41, 42]. For necks having a diameter greater than ≈ 5 nm, fracture-like yielding prevailed. Here, the neck cross-section changes typically by one order of magnitude in catastrophic events. On the other hand, elongation of thin necks having a diameter of less than ≈ 5 nm occurred as a quasi-continuous plastic flow. Here, the cross-section shrinks exponentially with neck elongation in a repetitive sequence of small steps.

Au is selected as sample material in these contact studies because it is chemically inert and because it is highly ductile on a macroscopic scale. The latter property makes Au a promising choice for studying plastic deformation processes. However, one must expect that the yielding properties of hard and brittle materials might be different from the ones observed with Au, even on the nanometer scale. Consequently, this section is not meant to provide a comprehensive discussion of the topic. It merely gives a glimpse across one of the present frontiers in adhesion physics.

In short, experiments are conducted using a standard tunneling microscope complemented by a stiff force sensor (a 'nanoguitar' [43]) as sample stage. Contacts are made by indenting a Au tip in a (111)-oriented Au substrate by typically 1 nm. The deformation mechanics of the contact is studied by measuring the applied force and the electrical conductance as a function of relative tip–sample displacement. The preparation of tips is somewhat critical. They must be sharp in the sense that the measured force indeed corresponds to the genuine mechanical and short-range surface forces and that no corrections for long-range VdW forces need to be made (for details see [42]).

The conductance measurement allows one to estimate the contact area using the Landauer–Büttiker formula for ballistic quantum transport [44]. The conductance of the contact may be written as

$$G_c = \frac{2e^2}{h} \sum_{i=1}^{N_c} T_i, \tag{14}$$

where N_c is the number of conduction channels supported by the contact and T_i denotes the transmission probability of the i-th channel. The value of N_c depends on the electronic structure of the material and on the area of the contact, A_c. For a free-electron gas, one has $N_c \simeq 4A_c/\lambda_F^2$. As $\lambda_F \simeq 2/a$, where a is the mean interatomic distance, the number of channels corresponds approximately to the number of atoms that fit into the contact cross-section (for a valence of one). Theoretical investigations [38] and experiments [29–32] suggest that elastic interchannel scattering is of minor importance, and one may set $T_i = 1$ in equation (14) (which for large N_c is equivalent to the classical Sharvin conductance [45]). One then obtains $A_c \simeq A_0 g_c$, where $A_0 \simeq 0.08$ nm^2 is the mean area occupied by a Au atom and g_c is the conductance of the contact in units of $2e^2/h$ (henceforth, normalization of the conductance will be implicitly assumed).

By their very nature, indentation experiments are not strictly reproducible. However, characteristic features can be identified. Three representative examples of conductance and force data are shown in Fig. 6. During the approach of the tip towards the sample surface, an abrupt transition from tunneling to contact is seen when the tunnel conductance reaches values between 0.2 and 1. Contact manifests itself as a jump of the conductance to values of about 10 to 30, indicating that a connective neck between tip and substrate has been established. The spontaneous neck formation is driven by surface forces which, in turn, also give rise to the overall attractive tip–sample force observed. As the tip is indented into the substrate, compressive

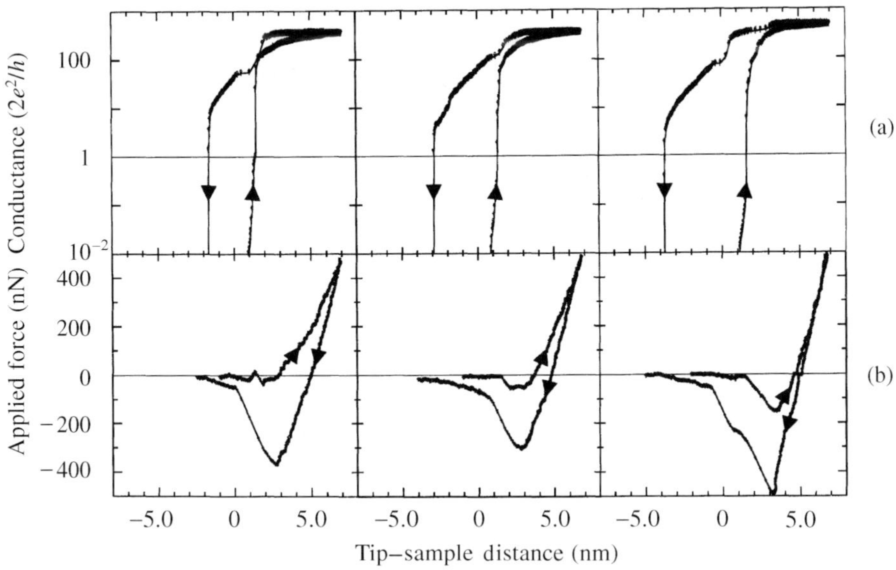

Figure 6. (a) Conductance on a logarithmic scale and (b) applied force versus tip–sample displacement. Data of three representative examples are shown. Approach and retraction are indicated by arrows. (Adapted from [41].)

deformation forces build up which eventually dominate, and the overall tip–sample force becomes repulsive.

Irrespective of the details of the indentation process, such as maximum loading force or contact area, three consecutive phases can be differentiated when the tip is retracted after indentation: buildup of tensile stress, incomplete fracture, and plastic flow. Unlike indentation, the neck diameter remains practically constant as the load is relieved during retraction. Thus, there is substantial hysteresis in terms of the shape of the neck. In energetic terms, however, loading and unloading are quasi-reversible in the compressive stress regime. The mechanical work executed during indentation is recovered to a large degree in the retraction phase because there is little hysteresis in the corresponding section of the force curves. As the tip is retracted further, tensile stress is built up, yet no significant change in the contact diameter is observed until the tensile stress exceeds a critical value of the order of 10 GPa. Thus, once a neck has been formed it acts like a cylindrical wire whose diameter remains approximately constant up to the point where yielding occurs. The yield point can be identified as the maximum of the tensile force the neck is able to support.

Yielding sets in smoothly as can be seen from the gradual decrease of the conductance. The shrinking rate of the neck accelerates rapidly, however, as the neck is strained. Simultaneously, the tensile force acting on the tip decreases at a somewhat lower rate than that at which the neck contracts. As a consequence, tensile stress escalates to the point at which this stress can no longer be sustained, and the neck abruptly contracts in a catastrophic event. In some cases a cascade of such events can be observed. The process can be considered a partial fracture, whereby large amounts of elastic strain energy are released in single events. Fracture yielding stops when the neck diameter falls below a critical value of the order of 5 nm (corresponding to a conductance of 200).

Adhesion necks of subcritical dimensions deform in a quasi-continuous manner, whereby (1) the conductance decreases exponentially as the neck is elongating, and (2) irrespective of the initial neck size, the adhesion neck breaks at a conductance value of either 3 ± 0.2 or 4 ± 0.2. Thus, the cross section of the smallest contacts that can be sustained comprises either three or four atoms. Furthermore, as the initial conditions have no effect on the final deformation process, one suspects that this process involves some kind of a local plastic flow within a limited range of the constriction.

In fact, molecular dynamics simulations performed by Landman *et al.* [37] suggest that plastic yielding of nanometer-sized contacts involves a series of subsequent structural transformations between ordered and disordered states of a small number of atomic layers adjacent to the constriction. In each transformation the neck shrinks as one atomic layer is added. In the spirit of the simulation results we consider yielding steps involving a transformation zone consisting of a fixed number n of atomic layers, which for simplicity is modeled as a cylindrical section [46]. Only atoms within this zone are able to migrate when the neck yields, whereas atoms outside this zone are assumed to be static.

We now consider a yielding step in which the neck elongates by a distance Δs, and, correspondingly, the cross section shrinks by ΔA_c. Hence, a volume $\Delta A_c n d_0$ is removed from the transformation zone, where d_0 denotes the spacing between atomic layers. This volume of material is incorporated into the rigid section of the neck adjacent to the transformation zone. Hence, because its width is constant, namely nd_0, a mass balance equation $nd_0 \Delta A_c = A_c \Delta s$ can be written. Integration of this equation yields that the cross section of the neck decreases exponentially as the neck is elongated, precisely as observed in the experiment. Measured decay constants $\kappa = 1/nd_0$ vary between 0.7 and 1 nm^{-1}, from which one concludes that plastic flow is confined within a region comprising five to six atomic layers (assuming 0.2 nm for the layer spacing).

The local order–disorder model of plastic flow stipulates that yielding of the neck proceeds in a sequence of discrete steps each time a new atomic layer is added to the neck. Indeed, the contact area decreases stepwise as evidenced by the plateaus found in the conductance data, and the conductance plateaus coincide with corresponding fluctuations of the applied force (Fig. 7). This behavior has also been observed by Rubio *et al.* [47] in Au contacts of similar dimensions. These observations provide clear evidence in support of the notion that the conductance plateaus are connected with discrete structural transformations of the neck, and are not solely a consequence of quantization. As expected for the local order–disorder model, the structure repeats itself quasi-periodically in intervals of $\Delta s \simeq 0.25$ nm, which is close to the lattice spacing of closed-packed planes.

In order to elongate the neck in the plastic flow regime, the apparent contact pressure, defined as the applied force divided by the contact area, must be increased constantly as the neck becomes longer and thinner. Typically, the apparent contact pressure is of the order of 10 GPa for a contact diameter of 2.5 nm ($g_c \approx 60$) and approaches values

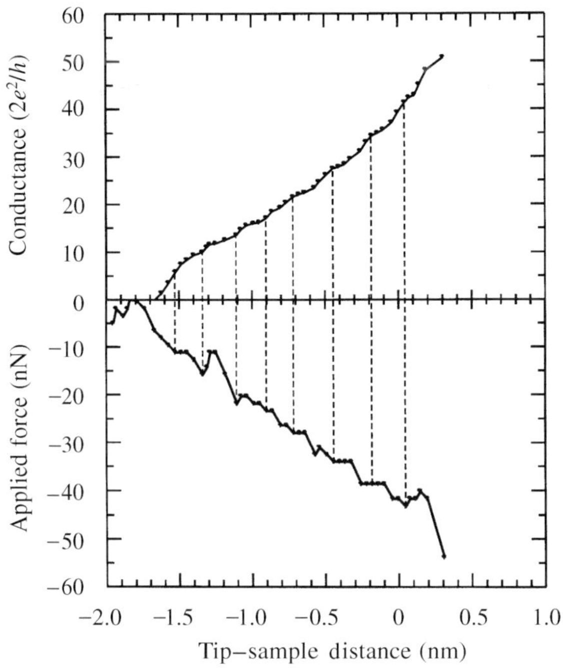

Figure 7. Conductance and applied force measured in the plastic flow regime shortly before the connective neck breaks. (Adapted from [41].)

of the order of 20 GPa shortly before the neck breaks. One is tempted to equate the apparent contact pressure with the yield strength of the neck. However, the applied force contains two contributions, namely a surface force acting at the periphery of the neck plus more or less evenly distributed elastic strain forces. The latter are decisive for yielding to occur. Therefore, to obtain the intrinsic yield stress σ_y, i.e. the force per unit area that must be applied in order to overcome the internal elastic strain forces at the yield point, the surface force contribution must be subtracted from the applied force. Surface forces are estimated to be of the order of $2\pi R \Delta\gamma$, where $R \simeq \sqrt{g_c A_0/\pi}$ is the contact radius and $\Delta\gamma \simeq 3$ nN nm^{-1} is the free energy associated with a virtual increment of the neck surface. The surface force contribution to the apparent contact pressure is significant. For a contact diameter of 2.5 nm one obtains a value of ≈ 5 GPa, which increases to ≈ 10 GPa at a diameter of 1 nm. Subtracting this surface force contribution we obtain an intrinsic yield strength of the order of 5 to 8 GPa, which does not depend on the contact diameter. For comparison, the macroscopic yield strength of Au is of the order of 200 MPa. Thus, the rigidity of these nanoscopic necks is more than one order of magnitude larger than that of macroscopic objects.

Nano-indentation experiments have revealed size dependences of the yield stress in the compressive stress regime for a number of materials [48, 49]. In particular, Pharr and Oliver [50] noted that the yield stress measured on silver substrates increased markedly at indentation depths smaller than 50 nm, and that the increase of the yield stress correlated with the absence of observable near-surface dislocation activity.

Hence, it was speculated that a different yielding mechanism must be in effect at these length scales. We have confirmed this notion in our experiments. In particular, it has been shown that the yield stress measured in atomic-scale connective necks reached values approaching the theoretical lattice limit, which is of the order of $E/2\pi \approx$ 30 GPa. Furthermore, these necks are clearly too small to accommodate dislocation loops in the classical sense. Hence, yielding must proceed in a fundamentally different way than in the bulk.

4. CONCLUSIONS

Adhesion involves a number of important parameters, such as surface morphology, interactions at various length scales and the deformation mechanics at these length scales. Interaction forces manifest themselves in two ways, namely as extremely short-range forces, which are sensitive to the chemical state of the surface, and long-range forces responsible for an overall attractive force, which is mainly influenced by the large-scale topographic structure. In this paper, the main focus has been on the processes that occur on a nanometer scale, where short-range forces are a major cause of the deformation and coalescence of small asperities. The deformation mechanics of such asperities is different from the mechanics at larger scales, say, greater than 1 μm. In particular, plastic deformation becomes important because it is intrinsically irreversible, and thus provides an internal mechanism for friction. The commonly accepted picture of plastic deformation involves the motion of dislocations. Their mobility is higher, the larger the range over which they extend. Thus one concludes that small structures are more resistant against deformation. This topological hardening is indeed observed in Au contacts made by indenting a sharp tip into a flat Au substrate. Furthermore, the observed yielding mechanics can be interpreted in terms of a local deformation mechanism proposed by Landman *et al.* [37] in which yielding proceeds by a sequence of order–disorder transitions of a small number of atomic layers.

Acknowledgements

We would like to thank E. Delamarche for preparing the Au samples. We gratefully acknowledge fruitful discussions with G. Dietler, A. Baratoff, J. Pethica, and U. Landman.

REFERENCES

1. D. Tabor, *J. Colloid Interface Sci.* **58**, 2–13 (1977).
2. S. K. Roy Chowdhury and H. M. Pollock, *Wear* **66**, 307–321 (1981).
3. E. M. Lifshitz, *Sov. Phys. JETP* **2**, 73–83 (1956).
4. J. Heinrichs, *Solid State Commun.* **13**, 1595–1598 (1973);
   J. E. Inglesfield and E. Wikborg, *J. Phys. F: Metal Phys.* **5**, 1475–1489 (1975).
5. J. E. Inglesfield, *J. Phys. F: Metal Phys.* **6**, 687–701 (1976).
6. J. H. Rose, J. Ferrante and J. R. Smith, *Phys. Rev. Lett.* **47**, 675–678 (1981);
   J. H. Rose, J. Ferrante and J. R. Smith, *Phys. Rev. B* **31**, 3427–3434 (1985).

7. J. H. Rose, J. R. Smith and J. Ferrante, *Phys. Rev. B* **28**, 1835–1845 (1983).
8. H. Krupp, *Adv. Colloid Interface Sci.* **1**, 111–239 (1967).
9. U. Dürig, *IBM J. Res. Develop.* **38**, 347–366 (1994).
10. B. V. Derjaguin, A. S. Titijeskaja, I. I. Abricossova and A. D. Malkina, *Disc. Faraday Soc.* **18**, 24–41 (1954).
11. The second (Derjaguin) term in equation (6) is the dominant term for the VdW interaction in most practical situations, viz. 100 nm $> s >$ 0.5 nm and $R >$ 1 nm.
12. D. Tabor and R. H. S. Winterton, *Nature* **219**, 1120–1121 (1968);
D. Tabor and R. H. S. Winterton, *Proc. Roy. Soc. A* **312**, 435–450 (1969).
13. J. N. Israelachvili and D. Tabor, *Nature* **236**, 106 (1972);
J. N. Israelachvili and D. Tabor, *Proc. Roy. Soc. A* **331**, 19–38 (1972).
14. For an overview, see e.g. J. N. Israelachvili, *Intermolecular and Surface Forces*, 2nd edn. Academic Press, London (1991).
15. B. V. Derjaguin, Y. I. Rabinovich and N. V. Churaev, *Nature* **265**, 520–521 (1977).
16. For a review, see e.g. *Scanning Tunneling Microscopy: Theory and Application*, D. Bonell (Ed.). VCH Publishers, Weinheim (1993);
D. Sarid, *Scanning Force Microscopy with Applications to Electric, Magnetic, and Atomic Forces*, Oxford Series in Optical Sciences, M. Lapp and H. Stark (Eds). Oxford University Press, Oxford (1991).
17. U. Dürig, O. Züger and D. W. Pohl, *Phys. Rev. Lett.* **65**, 349–352 (1990).
18. U. Dürig, O. Züger, L. C. Wang and H. J. Kreuzer, *Europhys. Lett.* **23**, 147–152 (1993).
19. U. Dürig, O. Züger and A. Stalder, *J. Appl. Phys.* **72**, 1778–1798 (1992).
20. W. R. Tyson and W. A. Miller, *Surf. Sci.* **62**, 267–276 (1977).
21. U. Dürig, O. Züger, B. Michel, L. Häussling and H. Ringsdorf, *Phys. Rev. B* **48**, 1711–1717 (1993).
22. H. M. Pollock, P. Shufflebottom and J. Skinner, *J. Phys. D: Appl. Phys.* **10**, 127–138 (1977).
23. M. D. Pashley, J. B. Pethica and D. Tabor, *Wear* **100**, 7–31 (1984).
24. M. D. Pashley and J. B. Pethica, *J. Vac. Sci. Technol. A* **3**, 757–761 (1985).
25. D. Maugis and H. M. Pollock, *Acta Metall.* **32**, 1323–1334 (1984).
26. J. K. Gimzewski and R. Möller, *Phys. Rev. B* **36**, 1284–1287 (1987).
27. N. A. Burnham and R. J. Colton, *J. Vac. Sci. Technol. A* **7**, 2906–2913 (1989).
28. S. R. Cohen, G. Neubauer and G. M. McClelland, *J. Vac. Sci. Technol. A* **8**, 3449–3454 (1990).
29. J. M. Krans, J. M. van Ruitenbeek, V. V. Fisun, I. K. Yanson and L. J. de Jongh, *Nature* **375**, 767–769 (1995).
30. J. I. Pascual, J. Mendez, J. Gómez-Herrero, A. M. Baró, N. Garcia, U. Landman, W. D. Luedtke, E. N. Bogachek and H.-P. Cheng, *Science* **267**, 1793–1795 (1995).
31. L. Oleson, E. Laegsgaard, I. Steensgaard, F. Besenbacher, J. Schiøtz, P. Stoltze, K. W. Jacobsen and J. K. Nørskov, *Phys. Rev. Lett.* **72**, 2251–2254 (1994).
32. N. Agraït, J. G. Rodrigo and S. Vieira, *Phys. Rev. B* **47**, 12345–12348 (1993).
33. K. L. Johnson, K. Kendall and A. D. Roberts, *Proc. R. Soc. Lond. A* **324**, 301–313 (1971).
34. J. R. Smith, G. Bozzolo, A. Banerjea and J. Ferrante, *Phys. Rev. Lett.* **63**, 1269–1272 (1989).
35. U. Landman, W. D. Luedtke and M. W. Ribarsky, *J. Vac. Sci. Technol. A* **7**, 2829–2839 (1989).
36. U. Landman, W. D. Luedtke, N. A. Burnham and R. J. Colton, *Science* **248**, 454–461 (1990).
37. U. Landman, W. D. Luedtke and E. Ringer, *Wear* **153**, 3–30 (1992).
38. T. N. Todorov and A. P. Sutton, *Phys. Rev. Lett.* **70**, 2138–2141 (1993).
39. R. M. Lynden-Bell, *Science* **263**, 1704–1705 (1994).
40. N. Agaït, G. Rubio and S. Vieira, *Phys. Rev. Lett.* **74**, 3995–3998 (1995).
41. A. Stalder and U. Dürig, *Appl. Phys. Lett.* **68**, 637–639 (1996).
42. A. Stalder and U. Dürig, *J. Vac. Sci. Technol. B* **14**, 1259–1263 (1996).
43. A. Stalder and U. Dürig, *Rev. Sci. Instrum.* **66**, 3576–3579 (1995).
44. M. Büttiker, Y. Imry, R. Landauer and S. Pinhas, *Phys. Rev. B* **31**, 6207–6215 (1985).
45. Y. V. Sharvin, *JETP* **48**, 984–985 (1965).

46. Because the constriction corresponds to an extremum of the neck cross section, the derivative of the cross section with respect to the neck axis vanishes at this point. Therefore, the cylinder approximation is correct to the first order.
47. G. Rubio, N. Agraït and S. Vieira, *Phys. Rev. Lett.* **76**, 2302–2305 (1996).
48. N. Gane and J. M. Cox, *Philos. Mag.* **22**, 881–891 (1970).
49. J. B. Pethica, R. Hutchings and W. C. Oliver, *Philos. Mag. A* **48**, 593–606 (1983).
50. G. M. Pharr and W. C. Oliver, *J. Mater. Res.* **4**, 94–101 (1989).

Mittal Festschrift, pp. 127–136
W. J. Van Ooij and H. R. Anderson, Jr. (Eds)

Application of fracture mechanics to adhesion observed between semiconducting surfaces in electric fields

K. KENDALL

Birchall Centre for Inorganic Chemistry and Materials Science, Keele University, Keele, Staffs ST5 5BG, UK

Abstract—Adhesion can be augmented by electric fields applied across contacting semiconducting surfaces. In the simplest example of contacting flat ceramic discs held at a potential difference of 100 V, a stress of around 100 kPa is necessary for separation. Early theories have suggested that this electrical adhesion stress should depend on the square of the voltage, as expected from the forces acting between capacitor plates. This paper shows that such theories must be incorrect because they do not take into account the cracking mechanism which applies to the fracture of most adhesive joints. When fracture mechanics is included in the theory, it is evident that the electrical adhesion must depend on voltage to the 2/3 power, as observed experimentally. The extension of these ideas to the electrostatic precipitation of powders and to the yield points of electrorheological fluids is discussed. In addition, the general nature of the fracture mechanics theory of adhesion in the presence of long-range forces, such as electrical, mechanical and magnetic is outlined.

Keywords: Electrical adhesion; fracture mechanics; semiconducting surfaces.

1. ADHESION DUE TO ELECTRIC FIELDS

Adhesion resulting from electrostatic attraction has been observed from ancient times, and has often been confused with the molecular adhesion resulting from van der Waals or other surface forces, which we now view as 'true adhesion'. The essential difference between electrical and molecular adhesion is the range of the force. Electrical forces appear to act from the centre of bodies, and obey an inverse square law of separation, falling off slowly as the solids are separated, whereas molecular forces are located very close to the surface and fall rapidly within nanometre separations.

Mathematically, we recognise the difference between electrical adhesion and molecular adhesion through the two equations below:

Electrical: $$\sigma = \frac{\varepsilon V^2}{8\pi z^2}, \tag{1}$$

where σ is the mechanical stress required to separate two plates at an electrical potential difference V a distance z apart where the permittivity of the gap is ε. This is derived from the potential theory of electrostatics.

$$\text{Molecular:} \qquad \sigma = \left[\frac{8EW}{\pi(1-\nu^2)d} \right]^{1/2}, \tag{2}$$

where σ is the mean contact stress needed to pull apart two semi-infinite solids of elastic modulus E and Poisson's ratio ν joined over a circle of contact diameter d with work of adhesion W. This is essentially a cracking equation similar to that of Griffith [1–3] derived from an energy balance theory of fracture mechanics, as shown later. The main differences between equations (1) and (2) stem from the uniformity of the stress pattern around the contact, together with the fall in stress with separation. In the electrostatic case the stress is uniform across the whole area and remains high as the surfaces are separated, whereas the molecular adhesion gives an infinite stress at the edge of the contact, with much lower stress in the centre and zero attraction once the surfaces crack apart. These effects have two consequences: firstly, the electrical adhesion is tougher, in other words it expends more energy on separation, whereas the molecular contact is brittle and cracks with little energy expense; secondly, there is no size effect for electrostatic adhesion because the failure stress is the same for all sizes, whereas there is a size effect for molecular adhesion, since smaller contacts appear to adhere more strongly, in stress terms, because of the inverse square root of d dependence of equation (2).

2. ELECTRICAL PLUS MOLECULAR ADHESION

There are situations in which both molecular and electrical adhesion act. An early example was noted by Edison who observed the influence of electrical fields at phonograph contacts [4]. Johnsen and Rahbek [5] studied this effect systematically using lithographic stone and obtained a pressure of 5 kPa when a potential difference of 400 V was applied between polished stone and metal plates. Balakrishnan [6] and Stuckes [7] used polished sintered magnesium titanate against a metal surface to obtain an adhesion of 25–100 kPa at 100 V depending on the surface smoothness.

The important practical applications of electrical adhesion enhancement are to electrostatic precipitation and to photocopying. McLean [8] studied the adhesion of dust cake collected on the passive electrode in an electrostatic precipitator, and found detachment stresses up to 0.18 kPa at fields of 1 MV/m. Robinson and Jones [9] measured the yield stress between the surface of a packed bed of particles and a metal plate under electric fields perpendicular to the plate, finding a change in the yield condition as the electric field was applied. Others [10, 11] explored the behaviour of fluidised beds under the influence of electric fields and found that the field was much more effective than expected in holding up the powder. Moslehi and Self [12], following [8] and [11] produced a theory of this effect based on single particle models, in which current constriction produced large fields at the contacts, but their analysis was flawed by poor contact mechanics.

More convincing theories have been produced by Martin, Ghadiri and colleagues [13–15] who not only made single contact measurements of the electrical adhesion between lead glass spheres, but also demonstrated that this effect could be used to control flow of powders from hoppers. This area of flow control of powders under electrical fields relates to the electrorheological fluids discovered by Winslow [16] who showed that dispersions of fine silica particles in oil could be gelled by the application of a strong field of around 2 MV/m. Such gelation was shown to stem from adhesion between the particles which were pulled together by the electric force to form strings which thickened the fluid. A description of these electrical stringing phenomena is given in a recent book [28]. Electrorheological fluids are under investigation as valves, dampers and clutches [17, 18].

The objective of this paper is to show how fracture mechanics theory can be applied to these situations in which both molecular and electrostatic forces act.

3. FRACTURE MECHANICS OF ADHESION

The theory which has been most successful in describing adhesion phenomena is that of fracture mechanics which was first systematically applied to adhesive joints around 1970 [19, 20], although it had been used to describe the peeling of paint films and adhesive tapes much earlier [21], but without realising the connection to the Griffith theory of cracks [1]. The basis of fracture mechanics is that adhesive joints fail by cracking, just like glass, because the molecular forces have such short range. At the tip of the crack, in a perfectly elastic material, stress rises mathematically to infinity so that the progress of the crack cannot be decided by a stress balance but must be dictated by an energy balance. This energy balance, according to Griffith, is between the potential energy in the load driving the crack, the stored elastic energy within the material, and the surface energy demanded by newly exposed molecules at the crack surfaces, quantified as the thermodynamic work of adhesion, W, the work required to separate unit area of contact reversibly. This work of adhesion is considered to be of such short range that there is no attraction across the opened crack faces, an assumption which is almost perfectly true for van der Waals and other surface forces.

Such fracture mechanics arguments have been extremely useful in describing how the forces required to separate surfaces depend on the geometry, elasticity and chemistry of adhesive joints [2, 3, 19, 22]. For example, the theory explains why large engineering objects never stick together without glue whereas small particles below 1 μm in diameter always stick together unless some anti-stick material is interposed at the contact. The theory also accounts for the widely different strengths of adhesive joints, showing why the same materials can give a strength of 0.1 MPa in the peeling configuration, but 100 MPa in the lap-shear configuration, merely as a consequence of geometry change in the fracture mechanics analysis [23]. Similarly, the theory can account for the large drop in adhesion brought about by the presence of water and other environmental agents or surfactants, because the work of adhesion W is reduced by the presence of additional molecules at the interface [24].

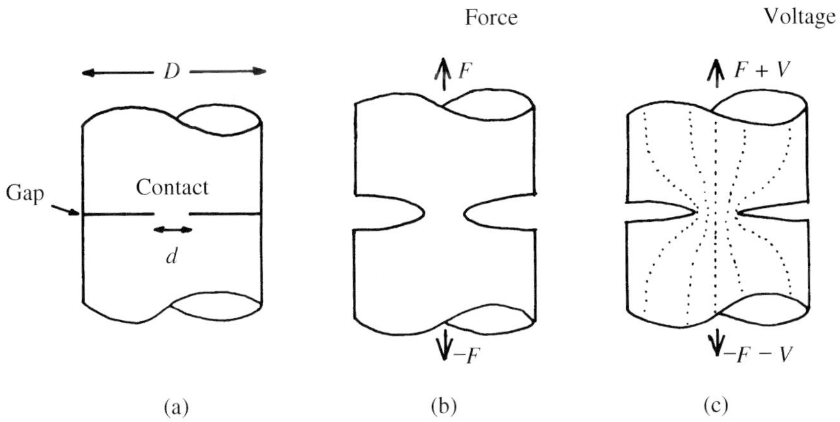

Figure 1. (a) Discs of diameter D making molecular contact over central region diameter d. (b) Application of tensile force F opens gap to form crack. Discs are held together by short-range molecular adhesion forces acting within the contact region. (c) Electrical attraction across the gap closes the crack due to voltage V applied to discs; the discs are also held together by short-range molecular forces in the contact region.

Consider the application of this theory to two flat discs of poorly conducting elastic material, diameter D, with flat polished ends brought together to make contact over a small central circular region of diameter d, as in Fig. 1a. The ends of the cylinders have been made nearly flat but do not make contact over the remaining surface outside the contact spot because there is a small gap, say 10 nm, too large for any molecular adhesion to develop. When a small force F is applied to pull the cylinders apart, Fig. 1b, this narrow gap between the surfaces opens up elastically, and closes when the force is removed. Essentially this gap is a crack, to which the Griffith energy balance argument can be applied.

4. THEORY OF ADHESION

Consider the potential, elastic and surface energy terms for Fig. 1b. The displacement δ of each applied force F due to the elastic opening of the gap is given by [2]

$$\delta = \frac{F(1-\nu^2)}{Ed}, \tag{3}$$

so that the total potential energy is $-2F\delta$. The stored elastic energy is half this and of opposite sign, giving a sum of $-F\delta$. The other energy term is the surface energy which is $-\pi d\, d\, W/4$. Therefore, the total energy U is given by

$$U = -\frac{F\,F(1-\nu^2)}{Ed} - \frac{\pi d\, d\, W}{4}. \tag{4}$$

Equilibrium is when

$$\frac{\mathrm{d}U}{\mathrm{d}d} = 0,$$

i.e. when

$$F = \left[\frac{\pi E d^3 W}{2(1-\nu^2)}\right]^{1/2}. \tag{5}$$

The nominal stress σ in the bulk of each disc resulting from this adhesion is

$$\sigma = \frac{4F}{\pi D D}.$$

Therefore

$$\sigma = \left(\frac{d}{D}\right)^2 \left[\frac{8EW}{\pi d(1-\nu^2)}\right]^{1/2}. \tag{6}$$

Equation (6) defines the adhesion stress for a small contact spot between large discs under a tensile loading. Putting in some numbers for ceramic material of elastic modulus $E = 200$ GPa, Poisson ratio $\nu = 0.3$ and work of adhesion $W = 0.1\ \mathrm{J\,m^{-2}}$, then a 1 mm diameter contact between 30 mm diameter discs would fracture at a modest stress of 8.3 kPa, based on the overall disc diameter. The maximum value of the elastic opening of the gap before fracture would be 53 nm — a small value, but larger than the original gap of 10 nm between the highly polished surfaces. This calculation shows that the elastic deformations are significant when considering the fracture of adhesive joints.

Now consider the application of an electrical potential across the contact, V on one side and $-V$ on the other as in Fig. 1c. Because the material is slightly conducting, with resistivity around 1 $\mathrm{G\Omega\,m}$, a small current will be constrained to flow through the contact spot. This has been shown to give a uniform field strength across the gap [25], leading to a constant attractive pressure σ^* closing the crack faces. This is a long-range force which adds to the short-range adhesion acting within the contact zone itself. Such pressure can be large because of the narrow gap, even at low voltage. The gap therefore closes significantly as a result of the applied field. A similar closing of the gap could be achieved by reducing the fluid pressure in the gap with a pump, or by introducing some other long-range force which can act across the gap between the surfaces, e.g. magnetic or electrostatic. Clearly, this argument will only be true for modestly conducting bodies. Too high a conductivity will result in heating effects, too low will lead to unpredictable voltages at the surfaces.

The electrical closure pressure is [25]

$$\sigma^* = \frac{\varepsilon}{8\pi}\left[\frac{VEd}{F(1-\nu^2)}\right]^2 \tag{7}$$

and this must be overcome by the external stress to cause fracture of the contact, giving the new stress σ for fracture of

$$\sigma = \frac{\varepsilon}{8\pi}\left[\frac{VEd}{F(1-\nu^2)}\right]^2 + \left(\frac{d}{D}\right)^2 \left[\frac{8EW}{\pi d(1-\nu^2)}\right]^{1/2}. \tag{8}$$

This result is interesting because it shows that the electrical pressure pulling the surfaces into contact dominates at high voltages, in which case the final equation for fracture stress is

$$\sigma = \left(\frac{\varepsilon}{8\pi}\right)^{1/3} \left[\frac{4VEd}{\pi D^2(1-\nu^2)}\right]^{2/3} \tag{9}$$

showing that the voltage dependence is a two thirds power law. By putting in values of $\varepsilon = 8.85$ pF/m, $V = 3$ kV, $E = 200$ GPa, $d = 1$ mm, $D = 30$ mm and $\nu = 0.3$, the fracture stress may be calculated to be 674 kPa, some 80 times larger than the molecular adhesion alone.

Obviously, if the surfaces are truly parallel, or if the contacts are not conducting, then equation (9) reverts to equation (1) and square law behaviour should be observed as for electrostatic chucks [27].

5. EXPERIMENTAL EVIDENCE

Equation (9) is compared with experimental results on polished magnesium titanate discs [7] which give a reasonable fit to the theory (Fig. 2). In these experiments, the tensile stress required to separate a disc 16 mm diameter from a steel plate was measured at a number of different voltages. Figure 2 shows that the results follow equation (9). The contact spot diameter d was not known for these experiments but was assumed to be 0.7 mm to give the best fit to the points. Figure 3 shows the results for the static torque needed to obtain slip between the surfaces at several voltages. These again fitted the 2/3 power law of equation (9).

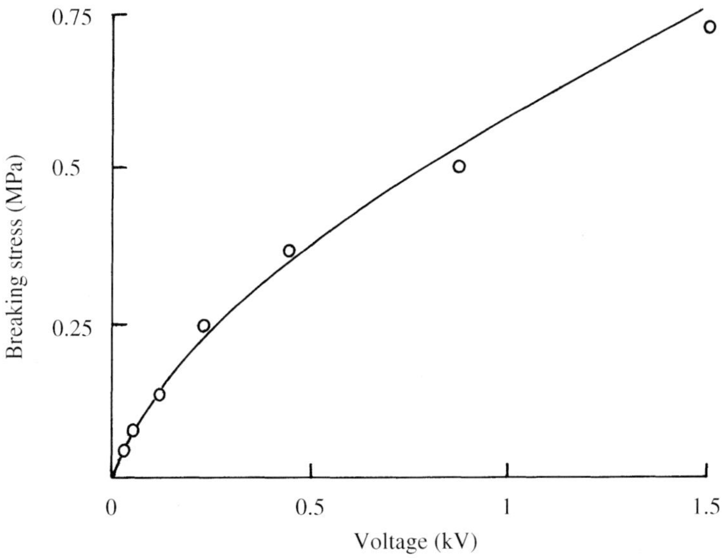

Figure 2. Experimental results [7] for the tensile stress required to separate a magnesium titanate disc of 16 mm diameter, 4 mm thickness from a steel surface at various voltages compared with equation (9).

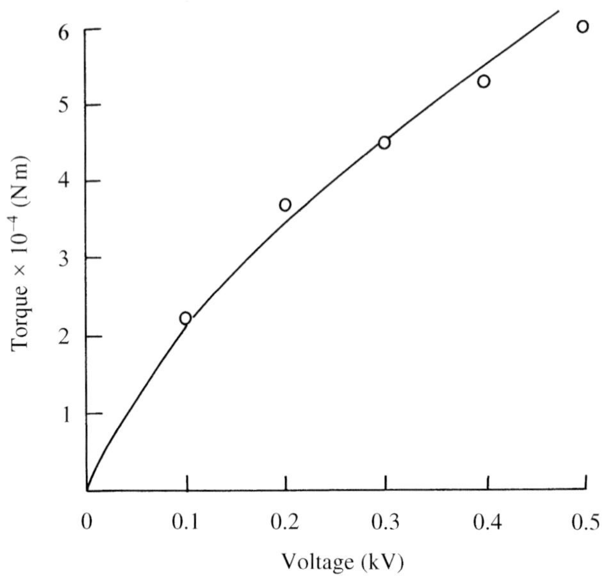

Figure 3. Results for static torque required to rotate the contact between magnesium titanate and steel as the voltage was increased.

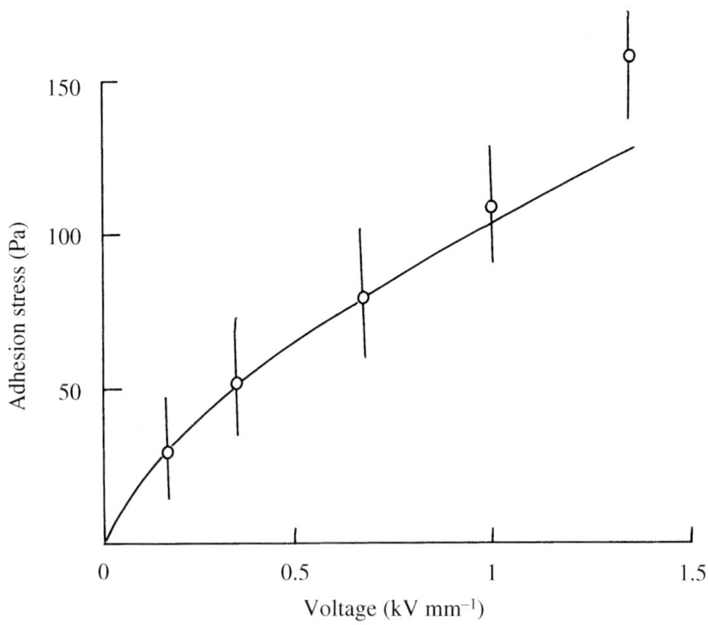

Figure 4. Adhesion of fly ash as function of voltage [8].

Figure 4 shows the results of McLean on the adhesion of fly ash as a function of electric field. There was much scatter but the voltage dependence was about 2/3. Adhesion stresses were lower than for flat plates, but this was to be expected because

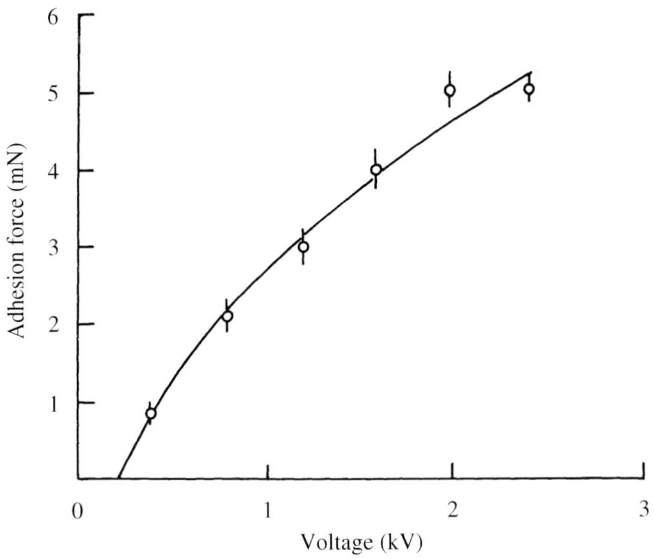

Figure 5. Single contact measurements on glass sphere 3.7 mm diameter [15].

the area of action of the electrical pressure must be less for spheres. Martin [15] has carried out adhesion tests on single spheres of lead glass 3.7 mm diameter. Such spheres were coated with a small amount of surfactant to give reproducible surface conductivity. However, the sphere surfaces were rough at the 1 μm scale. The small forces were measured by a nano-indenter device. The results shown in Fig. 5 indicate that the adhesion force increased with voltage to the 2/3 power but there was a small zero shift.

6. ELECTRICAL TOUGHENING

The general form of equation (8),

$$\sigma = \sigma^* + \frac{K}{c^{1/2}}, \tag{10}$$

where σ is the stress applied to the joint, σ^* is an attractive pressure which acts at long range between the open crack faces, K is the toughness and c is the characteristic dimension of the adhesive crack, shows how an adhesive joint may be made less brittle by electrical means. When σ^* is zero, such that there is no long-range electrical influence to pull the separate crack faces together, the joint is perfectly brittle and gives the classic Griffith behaviour, as shown by the broken curve in Fig. 6, the stress at failure falling rapidly as c increases. But as the electrical pressure becomes significant, the joint becomes less crack sensitive because the failure stress of the joint does not fall so quickly with increasing c. Therefore the joint is less brittle. It

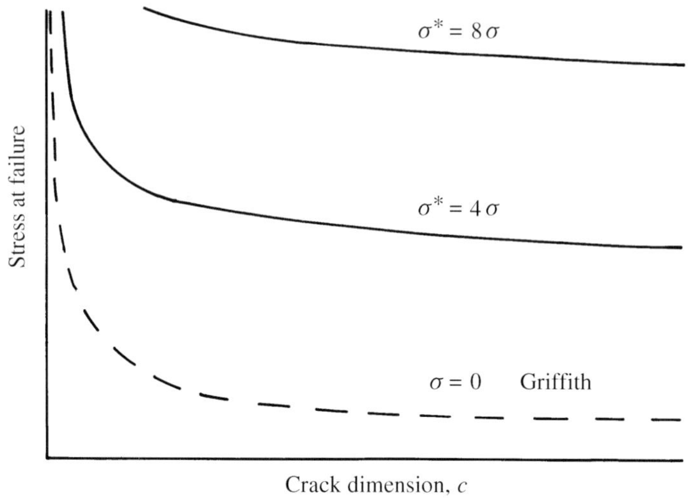

Figure 6. Effect of the electrical stress σ^* in reducing the crack sensitivity of an adhesive joint, thereby increasing toughness.

displays the so-called *R*-curve behaviour typical of toughened systems, in which the resistance to cracking appears to increase with crack length.

This same argument would apply to an adhesive material which did not crack cleanly, but left strings of adhesive material bridging the crack faces to provide a long range attraction across the crack. If these strings supported a stress σ^* between the crack faces then they would also toughen the joint according to equation (10). This is the mechanism which has been demonstrated for crazing of poly(methyl methacrylate) when stretched in methanol. The craze is a crack which is bridged by strands of polymer still able to support stress, thus reducing brittleness [26].

Therefore it is evident that the toughening effect of electrical adhesion forces is equivalent mathematically to the toughening observed in crazed polymers, in plastic metals, in carbon fibre reinforced plastics and so forth.

7. CONCLUSIONS

Electrical fields applied across solid contacts can increase the adhesion substantially. A theory based on fracture mechanics has been derived to explain the electrical effect in terms of the attractive forces developed across the crack faces near the contact spot. The theory shows that the adhesion can be expressed as the sum of two terms: one depending on the applied voltage which gives long-range attractions across the crack faces, and the second on the work of adhesion due to molecular forces acting in the contact spot. When the electrical term is much larger than the other, the adhesion stress is proportional to the 2/3 power of voltage and can readily attain a value of 1 MPa for flat plates. These results are consistent with measurements on ceramic discs. The same power law has also been observed with powders and on single spherical particles, though then the adhesion was smaller. It has been demonstrated that the electrical adhesion gives a toughening to the adhesive contact. These results are important in the fields of electrostatic precipitation, photocopying, electrostatic control of powder flow and electrorheological fluids.

REFERENCES

1. A. A. Griffith, *Phil. Trans. Roy. Soc. Lond.* **221A**, 163 (1920).
2. K. Kendall, *J. Phys. D: Appl. Phys.* **4**, 1186 (1971).
3. K. Kendall, *J. Adhesion Sci. Technol.* **8**, 1271 (1994).
4. T. A. Edison, Brit. Pat. 2909 (1877).
5. A. Johnsen and K. Rahbek, *J. Inst. Elect. Eng.* **61**, 713 (1923).
6. C. Balakrishnan, *Brit. J. Appl. Phys.* **1**, 211 (1950).
7. A. D. Stuckes, *Proc. Inst. Elect. Eng. Lond.* **103**, 125 (1956).
8. K. J. McLean, *J. Air Pollution Control Assoc.* **27**, 1598 (1984).
9. K. S. Robinson and T. B. Jones, *IEEE Trans. Ind. Appl.* **IA-20**, 1573 (1984).
10. T. W. Johnson and J. R. Melcher, *Ind. Eng. Chem. Fundamentals* **14**, 146 (1976).
11. P. W. Dietz and J. R. Melcher, *AI Ch. E. Symp.* **175**, 166 (1978).
12. G. B. Moslehi and S. A. Self, *IEEE Trans. Ind. Appl.* **IA-20**, 1598 (1984).
13. M. Ghadiri, C. M. Martin, J. E. P. Morgan and R. Clift, *Powder Technol.* **73**, 21 (1992).
14. M. Ghadiri and R. Clift, Eur. Pat. Appl. No. 87308304.2 (1987).
15. C. M. Martin, PhD thesis, University of Surrey, UK (1994).
16. W. M. Winslow, *J. Appl. Phys.* **20**, 1137 (1949).
17. H. Block and J. P. Kelly, *J. Phys. D: Appl. Phys.* **21**, 1661 (1988).
18. R. T. Bonnecaze and J. F. Brady, *J. Chem. Phys.* **96**, 2183 (1992).
19. K. L. Johnson, K. Kendall and A. D. Roberts, *Proc. Roy. Soc. Lond.* **A324**, 301 (1971).
20. M. L. Williams, *J. Appl. Polym. Sci.* **13**, 29 (1969).
21. R. S. Rivlin, *Paint Technol.* **9**, 215 (1944).
22. K. Kendall, *Science* **263**, 1720 (1994).
23. K. Kendall, *J. Phys. D: Appl. Phys.* **8**, 512 (1975).
24. M. K. Chaudhury and G. M. Whitesides, *Langmuir* **7**, 1013 (1991).
25. K. Kendall, *J. Phys. D: Appl. Phys.* **24**, 1072 (1991).
26. K. Kendall, W. J. Clegg and R. D. Gregory, *J. Mater. Sci. Letters* **10**, 671 (1991).
27. M. Nakasuji and H. Shimizu, *J. Vac. Sci. Technol.* **10**, 3573 (1992).
28. T. B. Jones, in: *Electromechanics of Particles*, p. 139. University Press, Cambridge (1995).

Mittal Festschrift, pp. 137–146
W. J. Van Ooij and H. R. Anderson, Jr. (Eds)

Current status and future directions of ultra large scale integrated circuits and the crucial role of adhesion science and technology

ARJUN N. SAXENA
International Science Company, 4217 Pomona Avenue, Palo Alto, CA, USA; and Rensselaer Polytechnic Institute, Troy, NY 12180, USA

Abstract—The entire world has been revolutionized since the advent, during the 1960s, of monolithic silicon integrated circuits (ICs), and their large volume manufacturing. There is hardly any activity that mankind needs for its survival and progress that does not use ICs in one form or another. The rapid advancement of the science and the manufacturing technologies have now made Ultra Large Scale Integration (ULSI) an indispensable commercial reality. The world market in 1995 for all the semiconductor components including ICs was over $ 100 billion, which then feeds into the world electronic systems market exceeding about $ 1 trillion! These markets are expected to grow larger still due to the omnipresence of the ICs in ever increasing numbers of applications. A brief review of the current status and future directions of ULSICs will be presented. Although adhesion science has played a crucial role from the very inception of ICs, it has not yet been appreciated in the literature. While it is impossible to provide a review of all the techniques in this paper, a novel technique to characterize adhesion quantitatively, especially in microminiature geometries used in ULSICs, is presented here for the first time along with a model to explain the results. Another unique feature of this quantitative technique is that it enables adhesion characterization both during and after normal and accelerated test conditions. Such data are of extreme importance for the yield, reliability assessment and improvement of ULSICs as well as the forthcoming Ultra Performance ICs (UPICs), which are the ultimate in ICs beyond ULSICs.

Keywords: ULSICs; UPICs; adhesion; multilevel films; reliability.

1. INTRODUCTION

The most widely used integrated circuits are fabricated on a single crystal silicon wafer. Various types of devices (transistors, diodes, resistors, capacitors) are fabricated and interconnected by multilevel metallization techniques to perform the desired circuit functions. They use a large number of metal and dielectric films, which must have extremely good adhesion to each other and to the various surfaces to achieve high reliability and yield. The number and variety of the materials used in these films is ever increasing with the complexity and the high performance demands of the Ultra Large Scale Integrated Circuits (ULSICs). So the role of adhesion in ULSICs is also ever increasing. But the crucial role of adhesion in the very inception of integrated circuits (ICs), has not been appreciated in the literature so far. Hardly any paper has been published in the entire field of ICs which gives even a qualitative measurement technique of adhesion, and certainly not a quantitative one. There is a large proliferation of techniques to measure adhesion, usually on large surfaces. However, there is still no satisfactory method to measure it quantitatively, in particular for the submicron geometries of metal and dielectric films used in ULSI. While it is impossible to provide a review of all the measurement techniques in this paper, a novel technique to characterize adhesion quantitatively, especially in microminiature geometries used in ULSICs, is presented here for the first time along with a model to explain the results. Another unique feature of this quantitative technique is that it enables adhesion characterization both during and after normal and accelerated test conditions. Such data are of extreme importance for the yield, reliability assessment and improvement of ULSICs, as well as the forthcoming Ultra Performance ICs (UPICs), which are the ultimate in ICs beyond ULSICs.

2. CURRENT STATUS OF ICs

ICs were invented independently by Noyce [1] and Kilby [2]. The US patent was granted first to Noyce on April 25, 1961, and then to Kilby on June 23, 1964, although the filing date of Kilby was a little earlier (May 6, 1959) than that of Noyce (July 30, 1959). Without going into details of these patents in this paper, in the opinion of the present author sought in the early discussions [3], one of the key reasons for the award of the IC patent to Noyce ahead of Kilby was due to the desired adhesion of the materials chosen to interconnect the devices on a monolithic chip. Since the advent of the ICs in 1961, there has been rapid progress in the fabrication of ever more complex IC chips having an ever larger number of devices. Moore [4] prescribed laws which predicted remarkably well the evolution of the technologies and

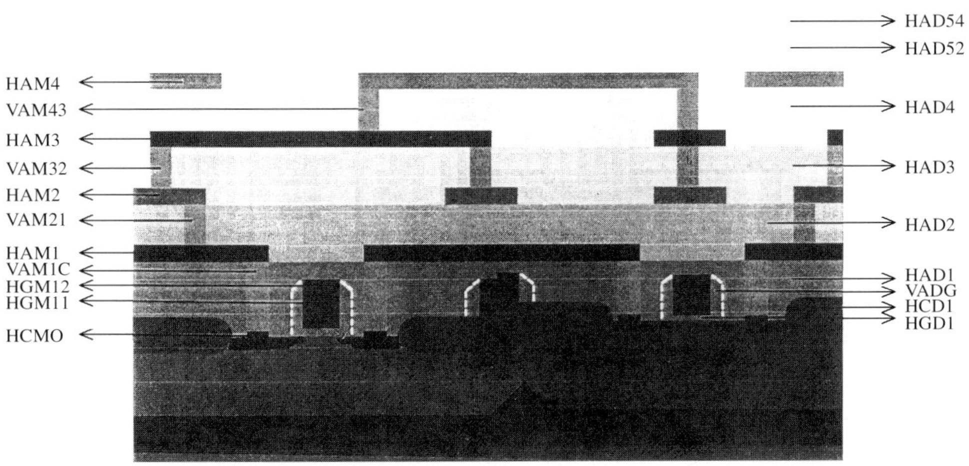

Figure 1. Cross section of CMOS with 4-level metallization.

the products in the entire multibillion dollar Si IC field. It can be stated quite categorically that Moore divined the growth and progress of this indispensable field. The current status of ICs can best be summarized in a practical way by pointing out that the commercially available chip in large volume is the Pentium chip [5], which has about 3.2 million transistors per chip interconnected with four levels of metallizations. It uses CMOS technology (Fig. 1) having about 0.5 μm geometries, and 'more than 20 complex layers' of various materials as films. More advanced chips, e.g. P-6 [6] and others, are being developed and put into manufacturing. (Figure 1 is a computerized drawing of the cross section of a CMOS structure with 4 levels of planarized horizontal and vertical interconnects. It is not drawn to scale, but it illustrates the various components of the CMOS transistors and the multilevel metallizations. For details, see [9].)

3. FUTURE DIRECTIONS

The incessant desire to put an ever increasing number of functions on a chip has provided the thrust to keep on scaling down the device dimensions, and achieve an ever larger number of devices on a chip. The laws of physics have been modified and extended by the innovative use of advanced technologies, such as phase shift masks, to achieve submicron features in large volume manufacturing. Also, to amortize the manufacturing cost per wafer, the wafer sizes have now grown to 200 mm diameter, and are expected to grow even larger. The economics of constructing a fab with all the equipment needed for microelectronics manufacturing, as well as the fundamental limits of materials and processes, appear to put a roadblock to the validity of Moore's Law. None other than Moore himself reviewed several of these factors recently [7], and predicted that his law, which has guided the destiny of the entire industry, is approaching the limits of its validity. However, a recent analysis [8] has shown new technology directions which can extend the validity of Moore's Law.

The roadmap of future monolithic ICs has been given by Saxena [9]. Figure 2 shows two roads to achieve the ultimate in ICs, viz. Ultra Performance ICs (UPICs). A user of IC does not care whether a particular chip is a CMOS, bipolar or BiCMOS chip, or how many devices or number of multilevel metallizations are on the chip. Its performance, cost, reliability, and a few other factors [9] are of paramount importance. The performance of the Si chip can be enhanced by invoking active devices in the 3rd dimension in addition to multilevel metallizations, and by fabricating compound semiconductor (e.g. GaAs) devices monolithically on Si. Thus, electronic and optical functions can be integrated, and UPICs can be fabricated, whose performance will be superior to that of the ULSICs. However, the number of films and materials used for UPICs increase significantly beyond those needed for Si ULSI. Implied in this is the increasing role of adhesion science and technology.

4. ROLE OF ADHESION SCIENCE AND TECHNOLOGY

As discussed in Section 2, adhesion played a key role in the award of the IC patent to Noyce ahead of Kilby. A key factor to explain [3] the superior adhesion of Al

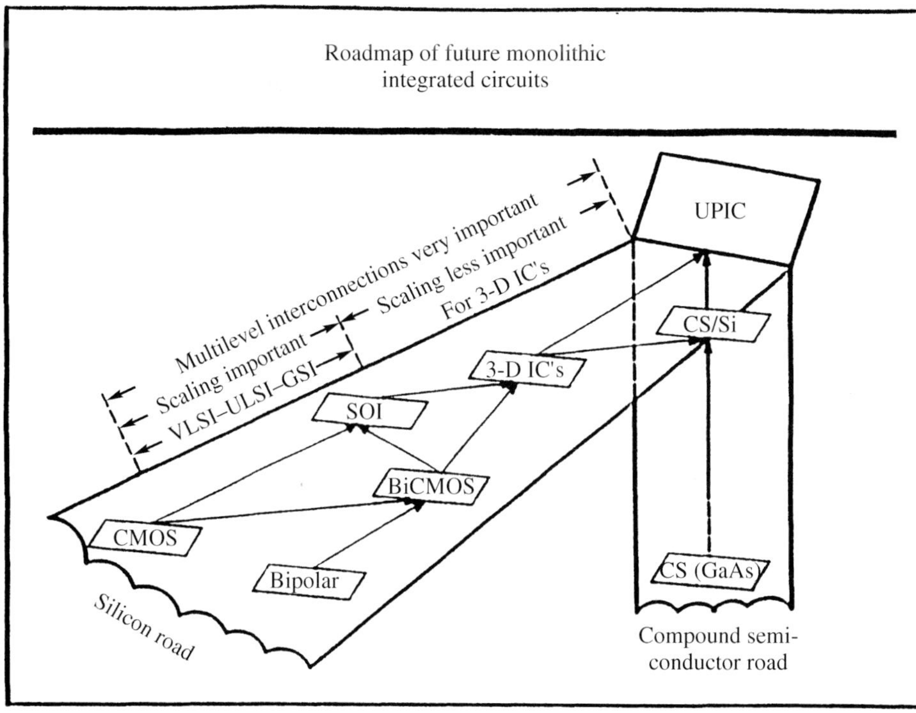

Figure 2. Roadmap of future ICs. The ultimate in ICs are the Ultra Performance ICs (UPICs).

to SiO_2, as compared with that of Au, was the formation of interfacial monolayer(s) of Al oxide contiguous to the underlying SiO_2 surface during the Al deposition process, which was absent in the case of Au deposition. This key factor also explains the superior adhesion of Ti, TiW, and TiN films used currently as underlayers to AlX (aluminum alloys with a suitable metal X, most common being Cu) interconnects in ULSI. Saxena obtained a patent [10] which used selective modification of this factor on SiO_2 in ICs, thereby achieving adhesion of metal films selectively only on the areas where needed for the interconnects without etching of the metal.

To get an appreciation of the role of adhesion science and technology in ULSI, Fig. 3 shows the number of layers/interfaces in a CMOS IC with a 4-level metallization. The detailed discussion of the various layers, materials and technologies is beyond the scope of this paper. However, some key information on them is as follows. All the layers are deposited by chemical vapor deposition (CVD) or physical vapor deposition (PVD), except for the gate oxide and that on the silicide, which are grown thermally. Barrier metals (BMs) are not specified, but the most common ones are Ti, TiW, and TiN. The inter-layer dielectrics (ILDs) are mostly SiO_2 based, some having small additions of boron and/or phosphorus. The Si_3N_4 films often used as cap layers on SiO_2 ILDs have not been shown.

Adhesion problems have been indicated mostly for the interfaces created during the deposition/formation of BMs, AlX, and silicides. The adhesion problems have been neglected for the interfaces created during the deposition of ILDs on metals, and

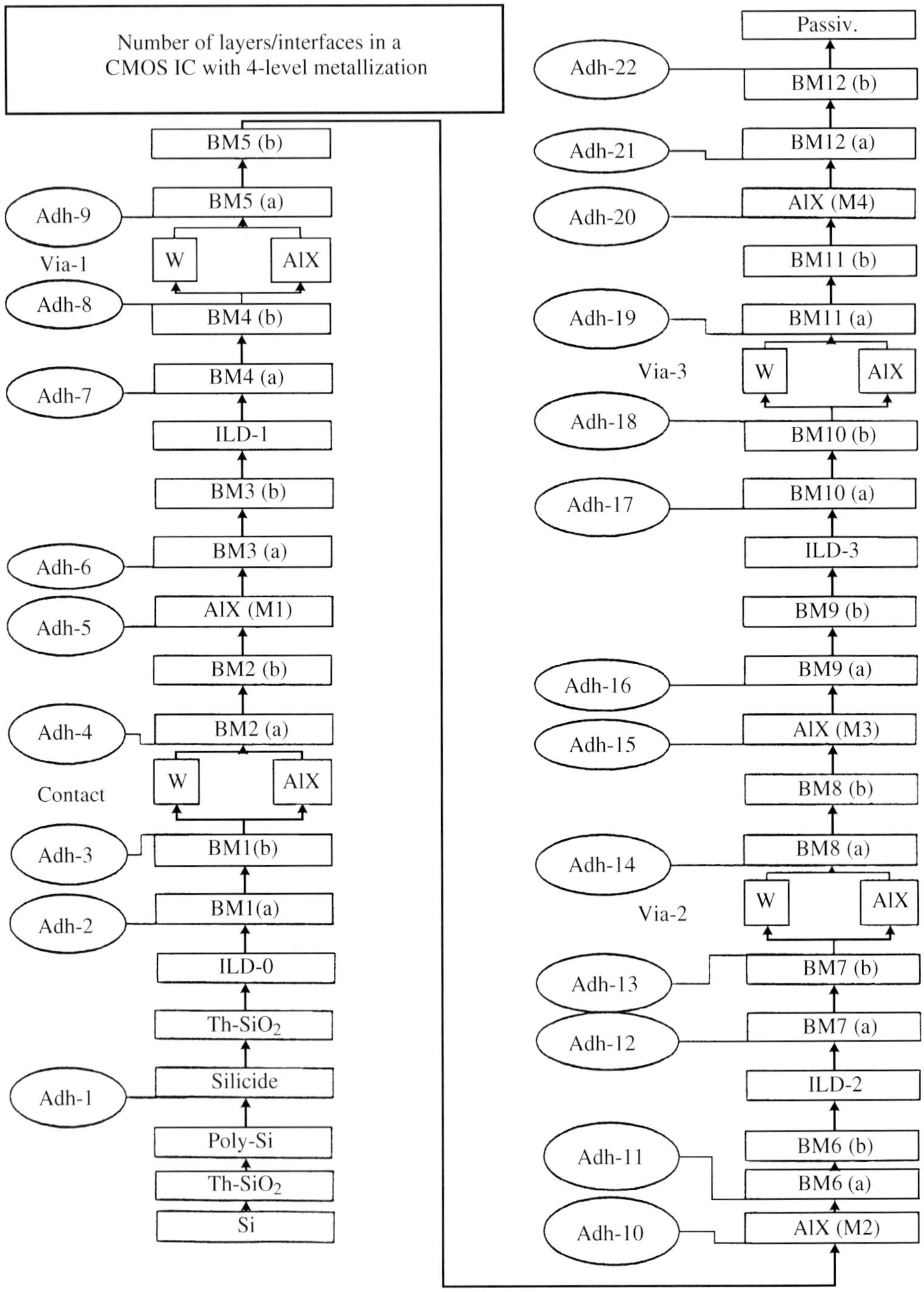

Figure 3. Number of layers/interfaces in a CMOS with 4-level metallization. There are potentially 22 interfaces in such a structure at which adhesion problems can occur.

for the intra-layer delamination of ILDs. Even with these simplifying assumptions, as shown in Fig. 3, there are potentially 22 interfaces in a 4-level CMOS IC at which adhesion problems can occur, if appropriate precautions in the processing are not taken. The current microelectronic manufacturing technologies have advanced enough to assure that the occurrence of the adhesion problems are minimized, if not eliminated. Nevertheless, in view of the very large number of interfaces at which adhesion problems can occur, extraordinary care has to be taken in the manufacturing processes for high yield and reliability. In practice, adhesion problems due to lack of appropriate process control do occur, which are manifested in the loss of yield, reliability failures of ICs, and high cost of manufacturing. Exact information on these are usually not available in the open literature, as they are considered proprietary by most of the corporations.

5. DIAGNOSTICS FOR ADHESION

Compounding the adhesion problems mentioned above in Section 4 is the lack of suitable diagnostics for adhesion, in particular for submicron geometries. Mittal [11] has recently given an excellent commentary on the adhesion measurement of films and coatings. There are more than 300 techniques to measure what is termed 'practical adhesion', which depends on the fundamental (interfacial) adhesion and many other factors. But these techniques are inadequate for measurements on films with submicron geometries as used in ULSICs, and to monitor adhesion as it changes during and after normal and accelerated electrical testing. As pointed out earlier in Section 3, the number of materials, layers, and interfaces are ever increasing with the advancement of ULSICs to UPICs. Thus, it is highly desirable to have a diagnostic technique which will characterize adhesion of films quantitatively in submicron geometries in addition to large areas, both during and after normal and accelerated electrical testing.

5.1. Novel technique to characterize adhesion

The novel technique proposed for the first time in the literature uses the measurement of the thermal conductivity, k, of films whose adhesion on a given surface needs to be measured. It is well known [12] that the k of thin films is lower than that of bulk materials k_0. Savvides and Goldsmid [13] gave a model to explain this, which was verified for the various types of SiO_2 films used in ULSICs [12]. According to this model, the difference in the thermal conductivities of the bulk and the film varies linearly with the reciprocal of the square root of film thickness, i.e. $1/t^{1/2}$. Writing this as a simple equation

$$\Delta k = s \cdot \frac{1}{t^{1/2}}, \tag{1}$$

where

$$\Delta k = k_0 - k. \tag{2}$$

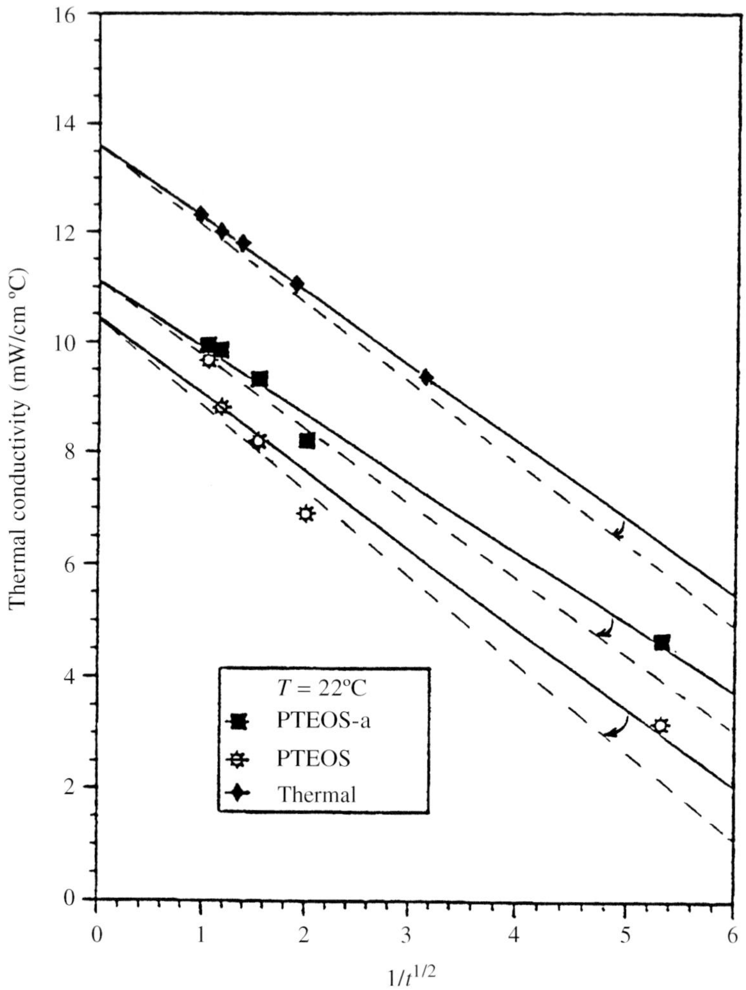

Figure 4. Thermal conductivity at 22°C of as-deposited PTEOS, annealed PTEOS and thermal oxides vs. inverse square root of oxide thickness. The dashed lines represent the larger slopes, s, for the respective films when their adhesion is poorer due to processing problems, or aging during subsequent thermal/chemical treatments and electrical testing. The fact that the slope for as-deposited PTEOS (PTEOS) films is larger than for the annealed PTEOS (PTEOS-a) films, provides an experimental confirmation of the model proposed in this paper. It is well known that the adhesion of PTEOS films improves on annealing; therefore the slope, s, of the former is higher than for the latter!

The slope s is related to the mean free relaxation times for Umklapp phonon scattering, τ_U, and for phonon scattering by point defects in the material of the film τ_{PD}. For details, see [12] and [13].

The effective bulk thermal conductivity, k_0, of each type of SiO_2 film was obtained from the experimental data for as-deposited and annealed films, by using microminiature test structures [14]. Figure 4 shows that the experimentally measured k varies linearly with $1/t^{1/2}$. For detailed discussion and analyses, see [12].

Adhesion of the films was not taken into consideration in the above work. Therefore, invoking adhesion of the film to affect its thermal conductivity, the model for the novel technique presented here is as follows.

First, we observe that the slope s of the experimental data in Fig. 4, and given in equation (1), is a constant independent of film thickness t. Next, we note that the effective bulk thermal conductivity of the film, k_0, is characteristic of the material of the film, but it is independent of its adhesion to the surface for the given set of k data. Now, neglecting the variation in τ_U for the bulk and the surface of the film, the mean free relaxation time for phonon scattering from point defects, τ_{PD}, is related to its bulk and surface components as

$$1/\tau_{PD} = 1/\tau_{PDB} + 1/\tau_{PDS}. \quad (3)$$

The τ_{PDB} is independent of the adhesion of the film; however, τ_{PDS} is dependent on the adhesion of the film to the surface. For the two cases of good and poor adhesion of the film to the surface, the following model is given.

(a) *Good adhesion* (denoted by subscript g henceforth): For good adhesion of the film to the substrate, there will be low surface scattering of phonons in the film, because of good coupling with the phonons in the substrate. Therefore, k_g will be higher than that for a film with poor adhesion, k_p (see below). Thus, Δk_g and s_g will be lower. This implies that τ_{PDSg} is larger.

(b) *Poor adhesion* (denoted by subscript p henceforth): For poor adhesion of the film to the substrate, there will be large surface scattering of phonons in the film, because of poor coupling with the phonons in the substrate. Therefore, there will be fewer phonons available to conduct heat through the film. Thus, Δk_p and s_p will be higher. This implies that τ_{PDSp} is smaller.

Thus, using the microminiature test structures, thermal conductivity, k, of films on desired substrates can be measured [12, 14]. By making such measurements as a function of film thickness, and using equation (1), the effective bulk thermal conductivity, k_0, of the film, and slope, s, can be calculated from the experimental data as shown in Fig. 4. Constant value of s verifies that the deposition process of the film of different thicknesses gives the same adhesion on the substrate. Thus, s is a quantitative parameter which characterizes adhesion for that particular film and process. As indicated in Fig. 4, the dashed lines drawn for the respective films represent the larger slopes, s, which would have been observed if the adhesion of these films was, or had become, poorer because of bad processing conditions and/or aging during subsequent thermal/chemical treatments. An experimental confirmation of the model proposed above is given by the fact that the slope for as-deposited Plasma Tetra Ethyl Ortho Silicate (PTEOS) films is larger than for the annealed PTEOS (PTEOS-a) films. It is well known that the adhesion of PTEOS films improves on annealing; therefore the slope, s, of the former is greater than for the latter!

When the detailed k measurements at different film thicknesses are not possible, in order to determine effective bulk thermal conductivity, k_0, of the film, and the slope s to characterize adhesion quantitatively, then measurements at only one film thickness can be used for a quantitative comparison of different degrees or levels of adhesion. This comparison is done for this technique by quantitatively referencing to a specific adhesion standard. For example, let $s_1, s_2, \ldots, s_i$, be the parameters characteristic of different qualities of adhesion. Then, for a film of given thickness, but deposited/processed differently to give different adhesion characteristics to the same type of substrate, equation (1) gives

$$\frac{\Delta k_1}{\Delta k_2} = \frac{s_1}{s_2}, \tag{4}$$

$$\frac{\Delta k_2}{\Delta k_3} = \frac{s_2}{s_3}, \tag{5}$$

and so on. Thus, relative adhesion characteristics can be assessed quantitatively by making k measurements at only one film thickness. The usefulness of this can be further appreciated when assessments of adhesion variations on normal and accelerated testing of ULSICs are required, which can be measured with this technique.

6. SUMMARY

Adhesion science and technology has played a key role from the very inception of ICs. ICs are omnipresent in their applications to almost every activity that mankind needs for its survival and progress. IC technologies have advanced to the ULSI level today, in which several million transistors are integrated on a chip, and are now available commercially. To achieve the ultimate in ICs, UPICs need to be developed, whose performance will be superior to that of the ULSICs. The number of films and materials used for UPICs increase significantly beyond those needed for Si ULSICs. Thus, the role of adhesion science and technology will be greater in UPICs than in ULSICs. However, to get an appreciation of the crucial role of adhesion science and technology in the simpler case of ULSI, the number of layers/interfaces in a typical CMOS IC with a 4-level metallization has been given. It was shown that there are potentially 22 interfaces at which adhesion problems can occur. More than 300 techniques are known which give parameters to assess adhesion of films on large area surfaces. But they are inadequate for measurements on films with submicron geometries as used in ULSICs. A novel technique to characterize adhesion quantitatively, especially in the microminiature geometries used in ULSI, has been presented for the first time in this paper along with a model to explain the results. Experimental data of the thermal conductivities of SiO_2 films have been given, which confirm the validity of the technique and the model proposed for characterizing adhesion in microminiature device geometries of ULSICs. Even though the discussion and the experimental data presented above are for dielectric (SiO_2) films, the technique and the model proposed can also be applied with minor modifications to characterize the adhesion of metal films. Further studies are necessary to evaluate the adhesion of various types of films being used in microelectronics and other fields. The novel technique proposed here can be used effectively in all these fields to characterize adhesion quantitatively.

Acknowledgements

Discussions with Dr. Gordon Moore on his law, factors for its limitations as viewed by him, and the various films/interfaces in ULSICs are gratefully acknowledged by the author. Thanks are also due to him and Dr. Sheldon Roberts for the discussions on the early issues of metallizations related to the Noyce and Kilby patents. Thanks are also due to Dr. Kash Mittal for discussions and his latest book on adhesion, and to Dr. Stan Wolf for discussions on ULSIC technologies.

REFERENCES

1. R. N. Noyce, US Patent 2,981,877 (1961).
2. J. S. Kilby, US Patent 3,138,744 (1964).
3. A. N. Saxena, unpublished (1961).
4. G. E. Moore, *Technical Digest, 1975 IEDM*, p. 11 (1975).
5. *Pentium Processor*, Intel Corporation (1992).
6. *P-6 Processor*, Intel Corporation (1995).
7. G. E. Moore, No exponential is forever — some approaching limits to VLSI technology. Keynote address, VLSI Multilevel Interconnection Conference (VMIC), unpublished (1993).
8. A. N. Saxena, Technologies to extend the validity of Moore's law, to be published.
9. A. N. Saxena, Roadmap of future monolithic ICs, VMIC State-of-the-Art Seminar, p. 8 (1989); A. N. Saxena, K. Ramkumar, S. K. Ghosh and M. A. Bourgeois (Eds), Technology and Reliability Issues of Multilevel Interconnects in Bipolar, BiCMOS and CMOS VLSI/ULSICs, *Proc. IEEE Bipolar/BiCMOS Circuits and Technology Meeting* (*BCTM*), Minneapolis, MN, p. 12 (1993).
10. A. N. Saxena, US Patent 3,687,722 (1972).
11. K. L. Mittal (Ed.), in: *Adhesion Measurement of Films and Coatings*, pp. 1–13. VSP, Utrecht, The Netherlands (1995).
12. A. N. Saxena and M. Bourgeois, *Proc. 11th International VMIC*, p. 126 (1994).
13. N. Savvides and H. J. Goldsmid, *Physics Letters* **41A** (13), 193 (1972).
14. Schafft, J. S. Suehle and P. Mirel, *Proc. IEEE International Conference on Microelectronics Test Structures*, IEEE Cat. #89CH2693-0, p. 121 (1989).

Mittal Festschrift, pp. 147–159
W. J. Van Ooij and H. R. Anderson, Jr. (Eds)

Adhesion problems in electronic devices

S. PIGNATARO

Dipartimento di Scienze Chimiche, Università di Catania, Viale A. Doria 6, 95125 Catania, Italy

Abstract—The fabrication of every manufactured article requires the solution of several adhesion problems among the interfaces between the various materials contained in the article. The situation becomes more and more delicate with the increasing technological level of the product considered. A semiconductor device is typical of such situation since it is constituted, from a chemical point of view, by a series of layers whose sequence and thickness is the result of a delicate balance among various needs which are often contradictory. The adhesion between the various layers is of crucial importance.

In this paper some adhesion problems currently encountered in the fabrication of a semiconductor device are reported. In particular, the interfaces between the molding resins used in the device package and the various materials which constitute the outermost part of the die are considered. In addition the adhesion between the silicon chip and the metallic frame, realized in the so-called 'back end' of the line with soft solder alloy, or that between polyimide and the materials of the 'front end' of the line is also discussed.

Keywords: Electronic devices; adhesion; interfacial chemistry; molding resin; polyimides; solder alloy; pad contacts.

1. INTRODUCTION

The micro and macroscopic properties of any given chemical system and in particular of its surfaces and interfaces are governed by the structure (electronic, atomic, molecular, geometric, and morphological) of the system, which ultimately depends upon the interatomic and intermolecular forces. This is true [1] for the surface of a fullerene molecule or the surface of the molecule of an enzyme as well as for the surfaces and interfaces of more complex systems like biosensors and biomaterials or other manufactured products like those shown in Fig. 1.

The performances of finished manufactured products, in turn, depend on the micro- and macroscopic properties of the chemical systems which constitute the product. Adhesion is one of the properties critical for the functionality (reliability) of an electronic device. But, what do we mean by adhesion? According to Mittal [2], for adhesion we may intend: (1) fundamental or intrinsic adhesion; (2) thermodynamic adhesion; or (3) practical adhesion.

(1) is defined [2] as the summation of all interfacial intermolecular interactions between the contacting materials.

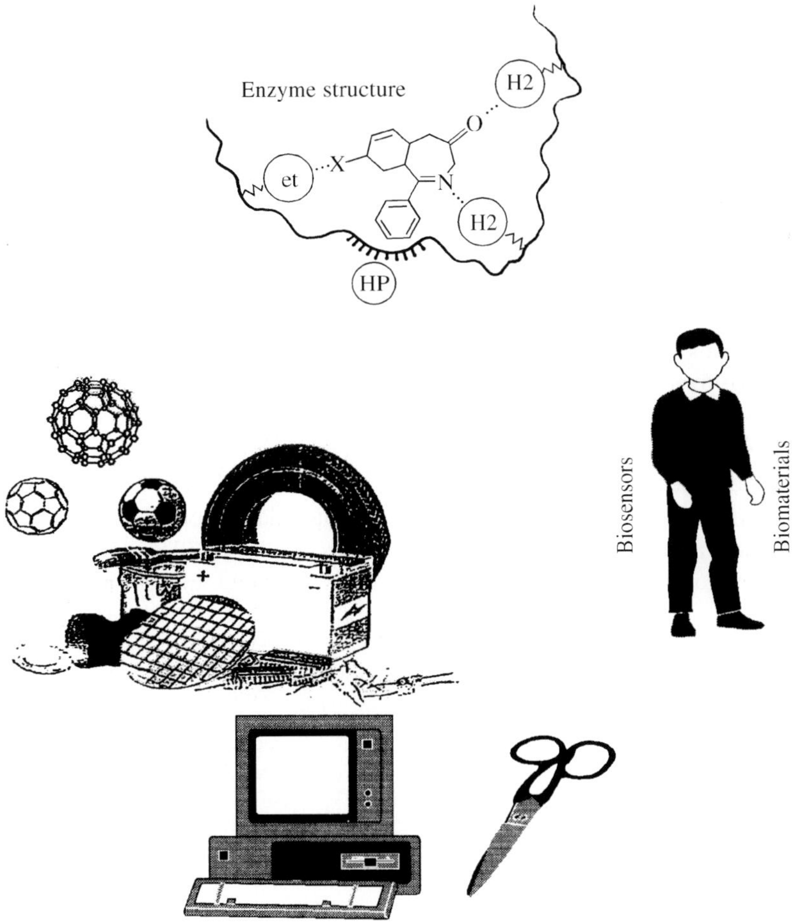

• Ski soles
• Tennis shoes
• Electronics cards, etc.

Figure 1. Some chemical systems and products encountered in daily life. Note that the molecules also have their surfaces and interfaces. Thus for molecules as well as the most complex products we have to study, design, prepare and control the surfaces and interfaces with nanometer definition in order to improve the ultimate performances. On the enzyme surface et is a cationic site, H2 are hydrogen bonding sites, HP is a hydrophobic pocket.

(2) signifies [2] the change in free energy when an interface is formed (or separated) and is expressed as $W_A = \gamma_1 + \gamma_2 - \gamma_{1,2}$ where W_A is the reversible work of adhesion and γ_1 and γ_2 represent the surface free energy of the substrate 1 and of the film 2. $\gamma_{1,2}$ is the interfacial free energy. For a series of metal/oxide systems a relation between the work of adhesion (mJ/m^2), interfacial free energy (mJ/m^2) and observed adhesion strength has been reported [3]. In the case of a liquid coating, $W_A = \gamma_{LV}(1 + \cos\theta)$, where γ_{LV} is the surface free energy of the liquid, and θ is the contact angle of the liquid coating with the substrate.

(3) signifies [2] the force or the work required to remove or detach a film or a coating from the substrate. It is a function of the fundamental adhesion and many other factors including intrinsic stress in the coating and the elastic properties of the materials involved.

Which then are the factors and processes governing adhesion in practice? Adhesion, like any other property, is governed by the initial 'structure' of the considered system. It is to be noted however that from a practical point of view we also have to consider the chemical evolution (reactions, diffusions, etc.) of the system with time. Table 1 summarizes most of the factors and processes influencing adhesion of a material to a substrate for a manufactured article. In particular, the chemical bond is at the base of intrinsic adhesion and it is governed, as we are learning more and more from the applications of tools like STM/AFM or from life science research [1], by the average force field created at the interfaces by the relevant interatomic and intermolecular forces. Also, interlocking is in some respect governed by these forces, since it depends on the interpenetration of the adhering layers at a macroscopic level and also on the interpenetration of the layers at an atomic or molecular level.

Contaminations and induced modifications, as well as diffusion in or out of the interface layer and reactions among the various chemical species populating the in-

Table 1.

Factors affecting adhesion of a material to a substrate of a manufactured article

Film / Substrate →	•Bond (governed by interatomic and intermolecular forces)	Type of bonds: van der Waals, electrostatic, covalent, metallic, hydrogen, acid–base interaction, etc.
	•Mechanical interlocking (depends on interpenetration at atomic and macroscopic scale)	Deals with: morphology, surface roughness, chain entanglement
	•Induced modification of a given interfacial structure	May occur by using: promoters, ion mixing, spallation, etc.
	•Interfacial contaminations	Contribute to define the interphase structure and the initial level of adhesion
	•Stability under working conditions (no chemical evolution like reactions, diffusion of not anchored species in or out of the interfacial region, etc.)	May be altered by: time, heat, mechanical stress, humidity (swelling, corrosion), etc.

terfaces, may concur to define (build) the 'structure' of the 'as prepared' interface. These last factors and processes cause the improvement or degradation of adhesion with respect to that due to the bonding and interlocking factors alone. This 'structure' and consequently the adhesion may be changed in practice later on if there is a chemical evolution (modification with time) of the system during its life under working conditions. It is important to realize that too good adhesion may sometimes be detrimental for the functionality of the considered system. This is true, for example, for the passivation of optoelectronic devices where such detrimental effects may be found as due to the interfacial bond with the passivation layer which alters the electronic structure of the semiconductor. In addition for medical purposes (adhesion of bacteria to surfaces of biomaterials [4a], electronic biochip [4b]), or for microoptical [4c] structure applications, one may want to have the minimum possible adhesion (abhesion). Now let us discuss how to measure adhesion of a film or coating. More than 300 techniques have been proposed [5] but what do they measure?

Macroscopic adhesion measurements give average values that depend on interactions occurring at a microscopic level. These are related to the 'as prepared' interface structure and its chemical evolution as well as to mechanical factors. These last factors include stored stress energy and elastic properties of the involved materials (in the peel test, the work consumed by inelastic deformation is crucial). Note that both 'chemical' and 'mechanical' factors depend ultimately on interatomic and intermolecular forces. As a last comment, I would like to emphasize that the practical adhesion may depend not only on the characteristic of the interface but also on the whole system, namely on the near interface layers which can critically affect the performance.

On the basis of the above considerations and statements one can appreciate why adhesion is a crucial property for electronic devices. An electronic device is constituted (Fig. 2), from a chemical point of view, of a series of layers (with their surfaces and interfaces) having different chemical nature [6]. The sequence and thickness of these layers is dictated by various requirements, which often are contradictory. The

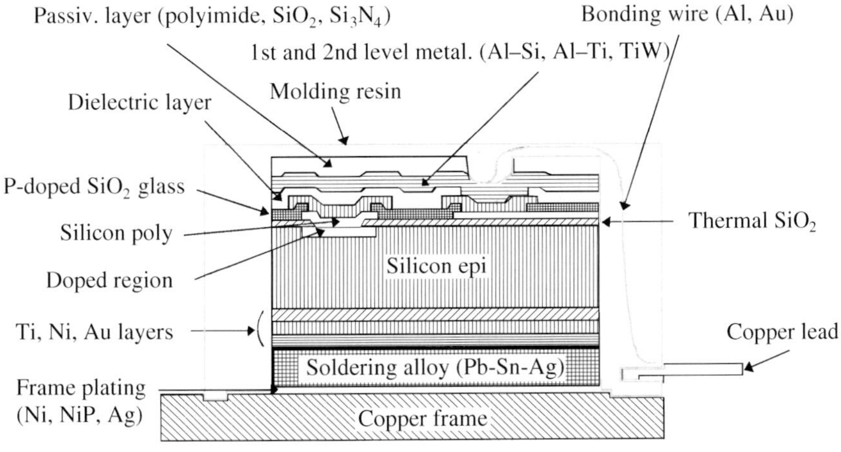

Figure 2. Scheme of the chemical nature of the layers of an electronic device with two metal level, showing the chemical complexity of the system.

adhesion between the various layers, in any case, is of crucial importance. The materials involved should be 'compatible' and in particular they should adhere to each other. The interfaces of the finished device as well as those built and destroyed in the steps of the production line, are all critical points in this respect. In the next sections of this paper, some adhesion problems currently encountered in the fabrication of a semiconductor device will be discussed.

2. ADHESION-RELATED PROBLEMS

2.1. Adhesion problems involving molding resins: The epoxy resin–metal-frame interface

Encapsulating resins are very frequently used in the packaging of electronic devices for silicon chip protection. The interfaces involving the resins are critical for the reliability of the devices [7]. Figure 3 summarizes in a scheme some observed chemical evolution phenomena (modification with time) at the resin–frame interface. The specific reported effects are discussed in detail in [8, 9]. Figure 3 shows that the interfacial layer is populated by a variety of chemical species and competition between the various phenomena involving these species may occur. It is therefore essential to understand the interplay of these phenomena. In a previous work [6] we reported the dependence of the pull strength on: (a) the chemical nature of the commercial resin and of the frame; (b) the degreasing treatment and the lifetime of the bath used in the production line for this degreasing treatment; (c) the physical design of the frame; (d) the pressure cooker treatment.

Figure 4 reports the data we obtained recently [10] on this problem by using the experimental scheme reported in Fig. 5. These data refer to the following processes (treatments of the frame surface): Process 1 is a soap degreasing process; Process 2

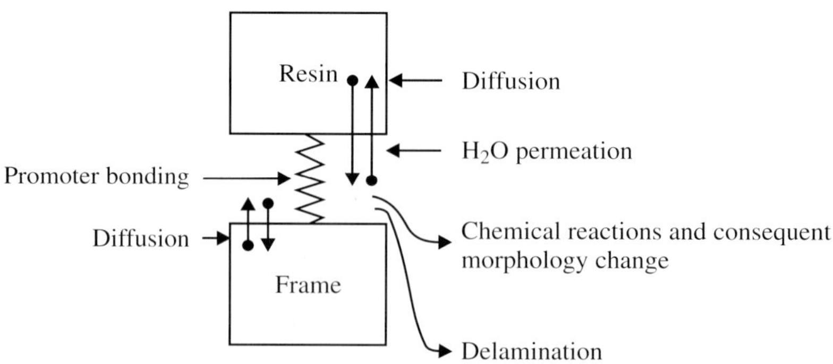

Figure 3. Some chemical modifications with time of an epoxy novolac resin–copper frame interface. Specific observed phenomena are: (a) the adhesion promoter diffuses to the interfacial layer (IL). Its bonding improves the adhesion level; (b) Cu ions diffuse out of the IL and may catalytically degrade the epoxy resin polymeric network; (c) phenolic moieties of resin additives diffuse into the IL reducing or retarding the oxidation of copper during aging or molding; (d) H_2O permeates through the resin and corrodes the metallic surfaces; (e) copper of nickel plated frames diffuses through nickel coating and reaches the IL where it reacts with brominated flame retardant additive of the resin.

Type of resin	Type of process	Pull strength (MPa)	% Delamination
A	1	5.1 ± 0.3	0
A	2	4.7 ± 0.3	0
A	3	5.8 ± 0.2	0
A	4	Close to zero	100
B	1	Close to zero	100
C	1	3.6 ± 0.7	12.5

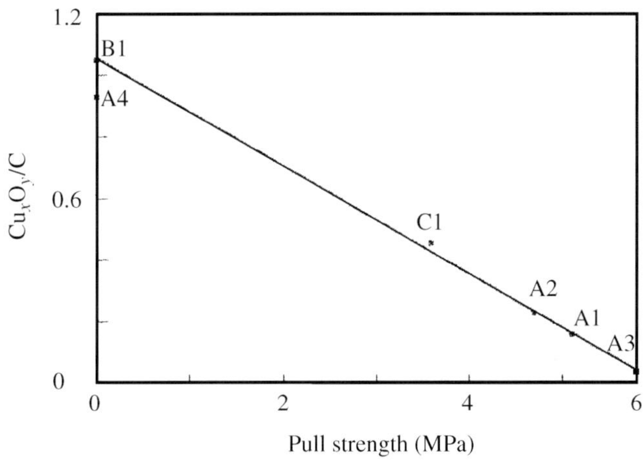

Figure 4. Resin–copper frame interface: pull strength values (MPa), % delamination and ESCA data (Cu $2p$ bands and Cu_xO_y/C ratio) for different commercial resins and frame surface treatment processes. The resins A and B are amine cured epoxy novolac resins while C is an anhydride epoxy novolac resin. The symbols A1–A4, B1, C1, indicate the particular resin–process couple.

= Process 1 followed by plasma treatment in 100% O_2; Process 3 = Process 1 followed by plasma treatment in O_2 + successive plasma treatment in 100% N_2; Process 4 = Process 1 followed by treatment in air at 225°C. These treatments were all followed by a treatment at 175°C in air for 20 s in the molder, in order to simulate more closely the surface used during molding. Process 4 was used in order to have a relatively stable initial (before molding) Cu(II) surface. Figure 4 shows the interdependence of the pull strength of the specimen with the type of the frame surface treatments, the type of resin, the Cu_xO_y/C ratio at the interface observed after the pull test, and the percentage of delaminated specimens. In particular, relatively high pull strengths are correlated with the absence of delamination (as measured by scanning acoustic microscopy) and with relatively low Cu_xO_y/C ratio observed by XPS (ESCA) on the metal side of the surface after the pull test. This result shows that thick oxide–hydroxide layers of Cu(II) (A4 and B1 specimens) induce cohesive failures within themselves and this leads to poor adhesion performance. The situation at the resin interface with Si_3N_4, SiO_2, P-doped SiO_2 glass, native AlO_x, polyimide are other critical points in the device and are currently being investigated in our laboratory again by using the experimental scheme reported in Fig. 5. The results will be reported elsewhere.

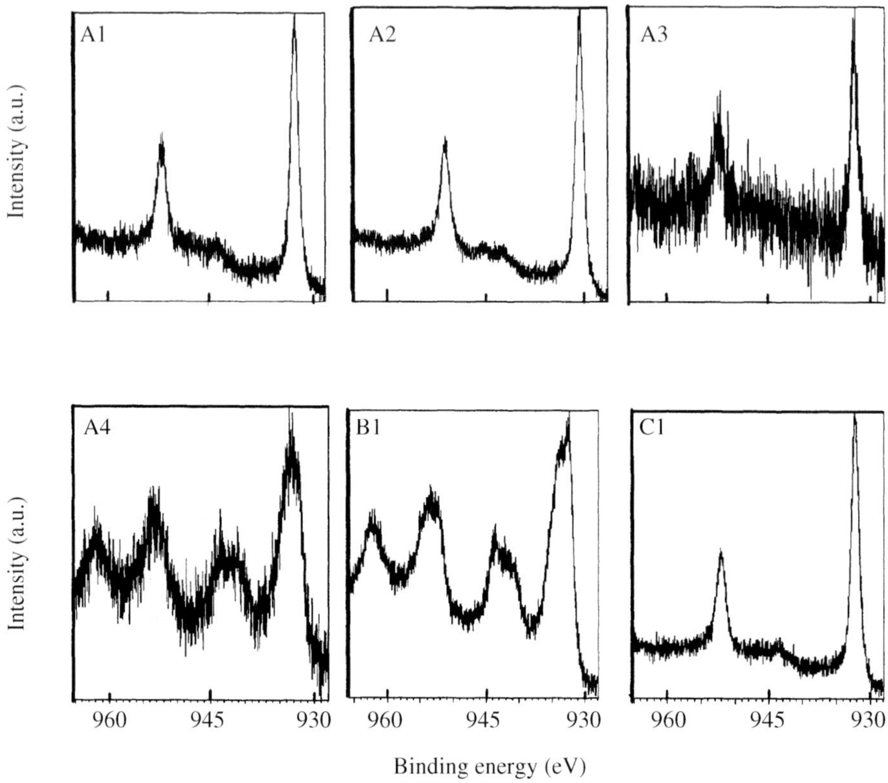

Figure 4. (Continued).

2.2. *Adhesion problems involving polyimide*

Polyimide (PI) is a material being used more and more in the electronics industry because of its desirable characteristics [11]. A number of studies deal with the PI interfaces with many materials. Good adhesion is, for example, reported with CrO_x [12, 13] as well as with epoxy resin using an adhesion promoter [14] or to polyimide itself after being treated in order to present an amorphous surface which may provide entanglement with the semicrystalline PI substrate [14]. No chemical interaction and therefore very poor adhesion is reported with Si/SiO_x surface [8]. Poor adhesion and in some cases decomposition of the PI is found with Cu/CuO_x [13, 16] and Ni/NiO_x [8] substrates. The chemical nature of the substrate oxide is quite important in determining the interfacial chemistry in these systems [8]. PI adhesion mechanisms to metal and metal oxides is still a debated problem even if, at least for Cr and possibly for Cu, bonding to the carbonyl group seems to be responsible for the observed adhesion [12, 17]. It should be pointed out that the confusion that exists in the literature on the adhesion mechanisms stems from the fact that the studied interfaces were not prepared in a well defined chemical state. For example, when one prepares a polymer coating on a metal substrate by spin coating, the real interface is, at best (no contamination layer present), between the polymer coating and the native

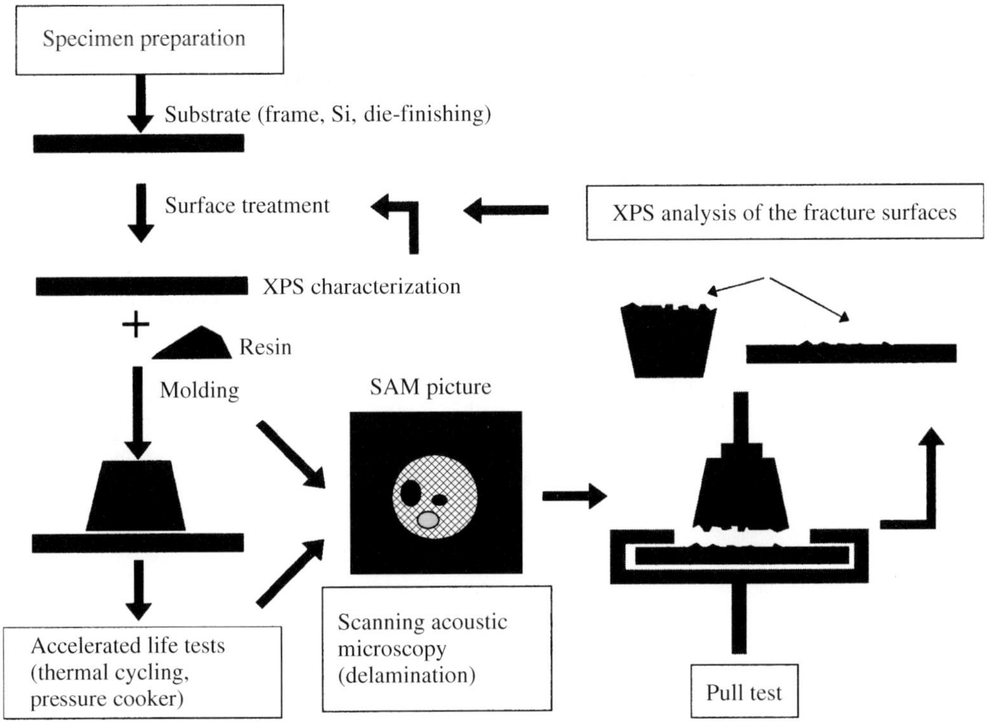

Figure 5. Sequence of experiments carried out to study and optimize the resin interface with the metallic frame or the materials used for the silicon die finishing.

oxide present on the 'metal' substrate. The enhancement of adhesion of PI to SiO_x or Si_3N_4 is often reached by using an Al chelate adhesion promoter which is then converted to Al_2O_3.

Figure 6 reports the peel strength of the PI coating on plasma enhanced chemical vapour deposited Si_3N_4, against the percentage of the chelate conversion to Al_2O_3 obtained by using various curing conditions of a Al acetoacetatediisopropylate type primer. The PI used was of isoindoloquinazolinedione type. Both polyamic acid solution (for the PI formation) and the chelate solution were deposited by spin coating on the Si_3N_4 substrate [1, 6]. The same graph reports the PI peel strength vs. the PI/Si_3N_4 ratio observed on the Si_3N_4 surface after peeling. This shows that the peel strength of the PI/Si_3N_4 system is governed by the percentage of chelate conversion in a basically independent way from the curing conditions of the primer (type of heating; used atmosphere: O_2, air, or N_2). In addition, higher peel strength is related to a greater amount of PI left on the Si_3N_4 surface after peeling. Again, as pointed out in the Introduction and shown in the resin–frame example, the interfacial chemistry is quite important. A pressure cooker treatment (121 °C, 2 atm., 100% humidity, 48 h) of the PI/silicon nitride interfaces causes a decrease of the peel strength. For the systems obtained by using B, C, and D processes (see caption of Fig. 6) this treatment causes a decrease of the peel strength from 52, 73, 81 g/cm to 30, 63, 63 g/cm, respectively. This should be due to the formation of $Al(OH)_3$ at the expense of Al_2O_3 [19].

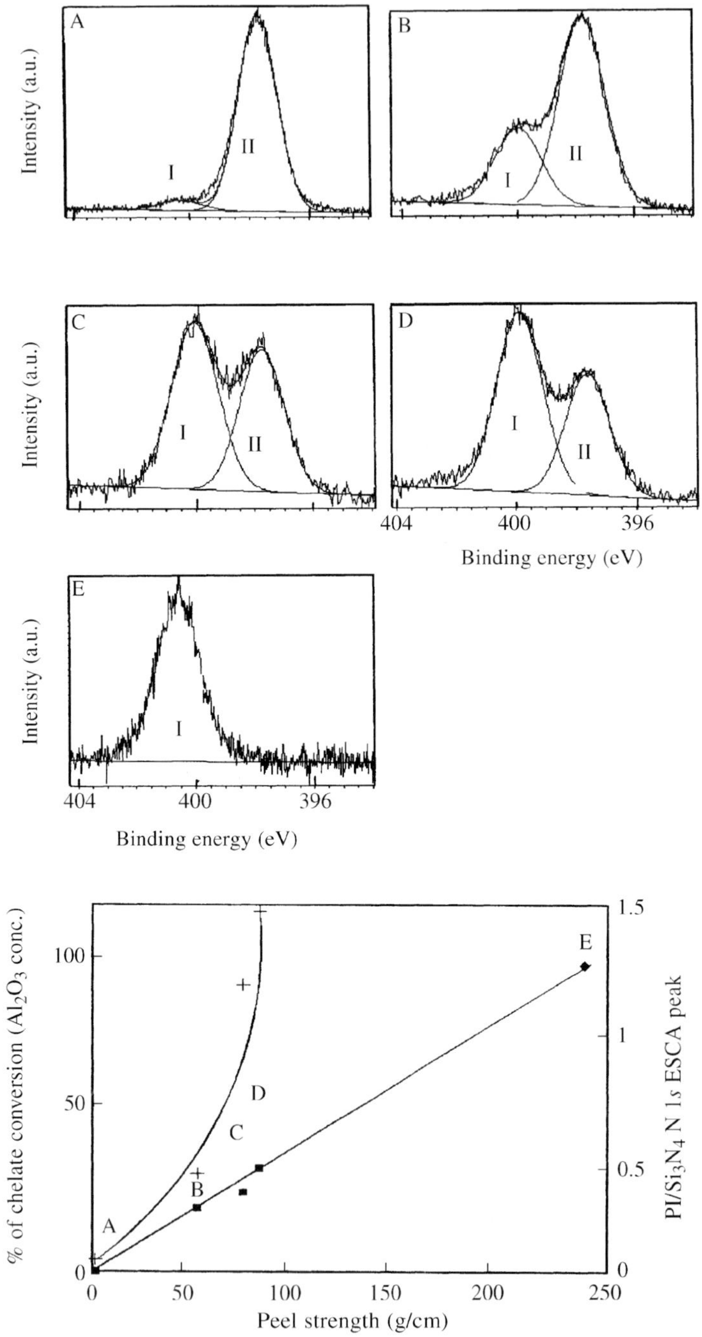

Figure 6. Peel strength of polyimide/silicon nitride system vs. % of Al chelate primer conversion to Al_2O_3 (■) and polyimide/silicon nitride ratio from ESCA N 1*s* peak areas obtained on the silicon nitride side after peeling (+). The top of the figure reports such peaks: (I) polyimide; (II) silicon nitride [18]. A = without primer; B = primer cured at 350°C, 1 h, under N_2; C = primer cured at 250°C, 100 s, in air; D = primer cured at 350°C, 1 h, under O_2; E = PI spun and cured on Al having native Al_2O_3 oxide.

2.3. Adhesion problems at metal/metal or metal/oxide interface

A variety of metal/metal and metal/oxide interfaces are present in a finished device or are formed during the fabrication steps of the device itself. Adhesion problems are of key concern here also. In this section just two examples are cited to show how important is the interfacial chemistry and how insidious are the contamination effects.

The first example refers to the wire bonding to leads. The Ni/Al bond was considered in the microelectronics community to be more reliable than the Cu/Al bond. This was due to the fact that Ni has a thinner overlay of oxides, hydroxides and carbonaceous contaminants than Cu. This thinner overlay is more easily broken in an ultrasonic bonder and the intermetallic compounds responsible for the adhesion formed in a cleaner and easier way. Both bonds are, however, heavily influenced by the interfacial chemistry as is shown by a strong dependence of the bond strength on thermal or pressure cooker treatments, especially in the presence of chlorine contaminants [20]. In agreement with these studies, a hydrogen plasma treatment of the metal surface is found to improve wettability, shear force and reproducibility of bond quality [21].

The second example refers to the 'pad' contacts. These show good bonding properties only when the oxide layer is no thicker than 4–8 nm [22]. This system is also greatly affected by the interfacial chemistry. The fluorine is ubiquitous in an electronic plant and it is often responsible for the observed loss of adhesion due to corrosion effects. In a recent paper [22], corrosion of Al bond pads and consequent loss of adhesion was attributed to fluorine coming from the outgassing of the wafer, contamination from the polypropylene plastic boxes used for storage and subsequent contamination of the Al pads. In another study, fluorine containing solutions used to etch the SiO_2 layer coating the pad were found [6] to be responsible for the 'spotted pads'.

The control of contaminants like those mentioned above is not trivial in most manufacturing situations and in the electronics industry in particular; yet it is critical for the formation of reproducibly strong adhering interfaces.

2.4. Adhesion involving soft solder alloy

Another interface traditionally studied in our laboratory is that of the soft solder alloy used to join the chip to the metallic frame. There is an increasing interest in electronics packaging for conductive epoxy adhesive [23] mainly because of the environmental demands, but still soft solder alloys are very much used in power devices.

Figure 7 shows the observed chemical evolutions of soft solder alloy of the Pb/Sn/Ag type. It is interesting to note that mechanical stress facilitates dealloying with growth of Pb-rich phase grains [24], while a melting process under inert atmosphere causes a reverse process inducing the alloying of Au into a Pb-rich phase [25]. Figure 8 shows that this last phenomenon is of such importance that a 100 nm thick Au layer, on top of which the solder alloy was allowed to melt, disappears at the edge of the lead alloy solder bump. The underlaying Ni is observed in the XPS (ESCA) map at this Au-Pb alloy edge. If the melting is performed under O_2, PbO_2 and SnO_2 are formed. These oxides, from the chemical point of view, have a surface

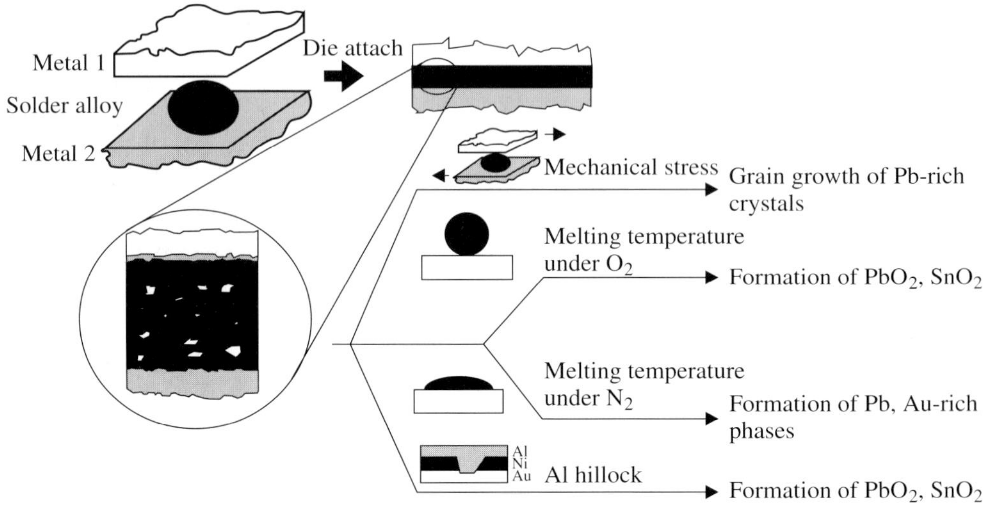

Figure 7. Observed chemical evolution of soft solder alloy of the Pb/Sn/Ag type under various conditions during die attachment or the life of the joint.

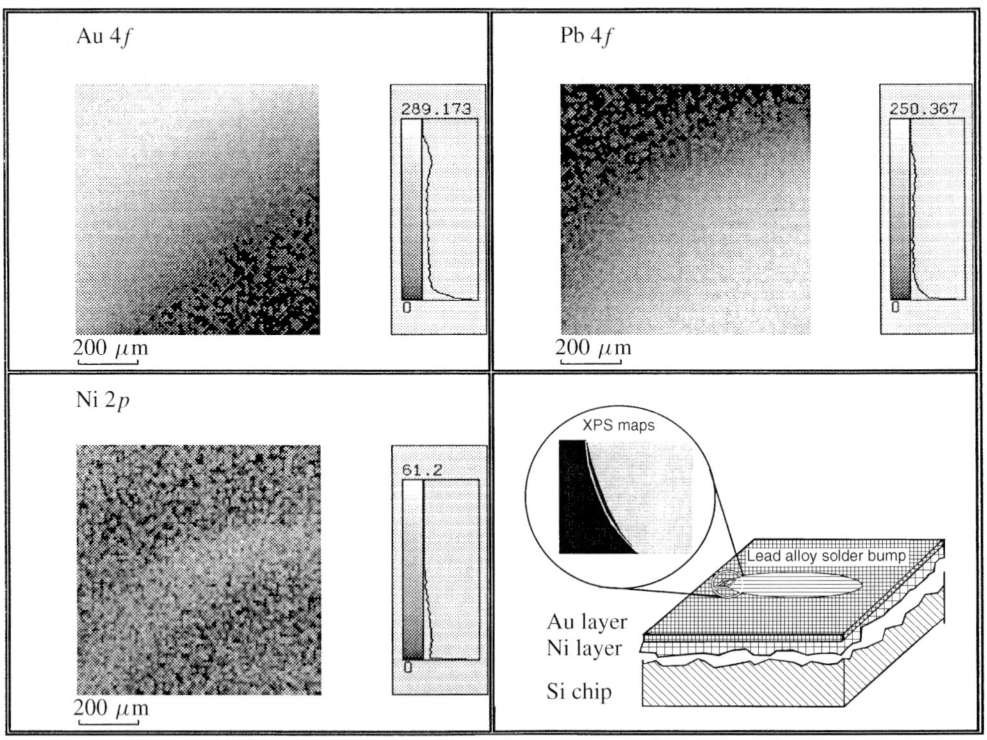

Figure 8. ESCA Ni $2p$, Pb $4f$ and Au $4f$ chemical maps and relative intensity colour scale at the edge of the lead solder alloy bump on top of a Au layer. The alloy was melted in N_2 on a 100 nm thick Au layer deposited on a Ni layer grown on top of a silicon wafer. The ESCA maps were obtained by using an AXIS spectrometer operating in the magnetic mode (lateral resolution 20 μm).

bond force field quite different from that of Au which does not oxidize. This different interatomic and intermolecular force field (see Introduction) causes a low wettability of the Au layer by the molded alloy and the lack of adhesion of the alloy itself with consequent failure of the joint. This behaviour seems to be in agreement with: (a) the finding that adhesion between couples with similar bond structure [3] like SiC and metals (Ni, for instance) is good, while the adhesion between SiO_2 or Al_2O_3 and metals is poor; (b) the qualitative observation [26] that adhesion between two metals is good when the solubility in the liquid phase is 100% and in solid state $>1\%$.

Catastrophic effects are also observed at a Au–metallic frame joint if hillocks (Fig. 7) are formed from an Al layer near the Au layer and reach the solder level. These hillocks bring oxygen to the solder alloy causing the formation of PbO_2 and SnO_2 and consequent failure of the joint. The hillock example is just an example showing the importance of a global approach in designing an interface of a given high-tech manufactured product and in particular of an electronic device. As pointed out in the Introduction, the near interface layers can critically affect the performance of a bonded system.

Acknowledgement

SGS-THOMSON, MURST and CNR are gratefully acknowledged for partial financial support. Thanks are due to A. Scandurra and G. F. Indelli for useful discussions and help in preparing the manuscript.

REFERENCES

1. S. Pignataro, *La Chim.e L'Ind. (Milan)* **78**, 583 (1996).
2. K. L. Mittal, *Electrocomponent Sci. Technol.* **3**, 21–42 (1976).
3. J. E. E. Baglin, in: *Materials and Processes for Surface and Interface Engineering*, Y. Pauleau (Ed.), pp. 111–149. Kluwer Academic Publishers, Dordrecht (1995).
4. (a) M. Collaud Coen, J. Kressler, H. Frey, F. Kropec, R. Mulhaupt and F. Daschner, Paper presented at the *ECASIA meeting*, Lausanne (1995);
   (b) P. I. Aebischer and S. Makohliso, Paper presented at the *ECASIA meeting*, Lausanne (1995);
   (c) P. Groning, A. Schneuwly and L. Schlapbach, Paper presented at the *ECASIA meeting*, Lausanne (1995).
5. K. L. Mittal, in: *Adhesion Measurement of Films and Coatings*, K. L. Mittal (Ed.), pp. 1–13. VSP, Utrecht (1995).
6. S. Pignataro, *Fresenius J. Anal. Chem.* **353**, 227 (1995).
7. S. Kim, presented at the *41st Electronic Components and Technology Conference*, Atlanta, Georgia (May 1991).
8. S. Pignataro, in: *Metallized Plastics 2: Fundamental and Applied Aspects*, K. L. Mittal (Ed.), pp. 179–188. Plenum Press, New York (1991) (and references therein).
9. S. C. Hong and T. C. Wang, *J. Appl. Polym. Sci.* **52**, 1339–1351 (1994) (and references therein).
10. A. Scandurra, R. Zafarana, A. Grasso, B. Pelligra and G. F. Indelli, presented at *Second European Conference on Electronic Packaging Technology*, Essen (1996).
11. (a) K. L. Mittal (Ed.), *Polyimides: Synthesis, Characterization and Applications*, Vols 1 and 2. Plenum Press, New York (1984);
    (b) P. Vasquez, G. Ferla and S. Pignataro, *Chimica Oggi*, 17 (1984).
12. N. Inagaki, S. Tasaka and K. Hibi, *J. Adhesion Sci. Technol.* **8**, 395–410 (1994) (and references therein).

13. K.-W. Lee, G. F. Walker and Viehbeck, *J. Adhesion Sci. Technol.* **9**, 1125–1141 (1995).
14. H. K. Yun, K. Cho, J. H. An, C. E. Park, S. M. Sim and J. M. Park, *J. Adhesion Sci. Technol.* **8**, 1395–1411 (1994).
15. K.-W. Lee, *J. Adhesion Sci. Technol.* **8**, 1077–1092 (1994).
16. L. P. Buchwalter, *J. Adhesion Sci. Technol.* **9**, 97–116 (1995).
17. A. Ouhlal, A. Selmani, A. Yelon and M. Andrews, *J. Adhesion Sci. Technol.* **8**, 485–499 (1994).
18. A. Scandurra, A. Santangelo and S. Pignataro, to be published.
19. (a) W. P. de Wilde, G. van Vinckenroy, L. Tirry and A. H. Cardon, *J. Adhesion Sci. Technol.* **9**, 149–158 (1995);
    (b) R. B. Comizzoli, R. L. Opila and Y. H. Wong, *Mater. Res. Soc. Symp. Proc.* **153**, 205–210 (1989).
20. (a) S. Pignataro, A. Torrisi, O. Puglisi, A. Cavallaro, A. Perniciaro and G. Ferla, *Appl. Surf. Sci.* **25**, 127–136 (1984);
    (b) A. Torrisi, A. Cavallaro, A. Licciardello, A. Perniciaro and S. Pignataro, *Surf. Interf. Anal.* **10**, 306–310 (1987).
21. N. Onda, E. Beck, A. Domman, V. Jaecklin, C. Luchinger and J. Ramm, presented at *ECASIA 95 meeting*, Lausanne (1995).
22. K. H. Ernst, D. German, R. Harvrt and E. Hollander, *Surf. Interf. Anal.* **21**, 691–696 (1994) (and references therein).
23. See, for example, O. Rusanen and J. Lenkkeri, *IEEE Trans. CPMT Part B* **18**, 320–325 (1995).
24. A. Scandurra, A. Porto and O. Puglisi, *Appl. Surf. Sci.* **89**, 1–10 (1995).
25. A. Scandurra and S. Pignataro, to be published.
26. Y. Pauleau (Ed.), in: *Materials and Processes for Surface and Interface Engineering*, pp. 475–527. Kluwer Academic Publishers, Dordrecht (1995) (and references therein).

Mittal Festschrift, pp. 161–178
W. J. Van Ooij and H. R. Anderson, Jr. (Eds)

Particle adhesion: Recent advances and current challenges

D. S. RIMAI,[1,*] L. P. DEMEJO[1] and R. C. BOWEN[2]

[1] *Office Imaging, Eastman Kodak Company, Rochester, NY 14653-6402, USA*
[2] *Imaging Research and Advanced Development, Eastman Kodak Company, Rochester, NY 14650-2151, USA*

Abstract—Over the past two decades great strides have been made in understanding the adhesion of particles to surfaces. These strides have been led by theoretical work as exemplified by the models of Johnson, Kendall, and Roberts; Derjaguin, Muller, and Toporov; and Maugis and Pollock. Experimental studies using microscopic and nanoindentor technologies have complemented these studies. More recently, there have been significant advances in technology which allow particle adhesion to be studied both experimentally and theoretically in significantly greater depth than could be achieved previously. Moreover, advances in applied areas, such as materials science and the fabrication of semiconducting devices, have driven the need to understand the adhesion of small particles to novel materials in much greater detail than was necessary previously. These advances have, to a great degree, verified many of the predictions of the appropriate theoretical models. They have also resulted in unexpected observations, which have raised questions concerning the fundamental understanding of particle adhesion. It is the goal of this paper to summarize recent advances in the field of particle adhesion and to describe some of the current challenges facing researchers in this field.

1. HISTORICAL PERSPECTIVE

While the concept of a particle adhering to a substrate is fairly simple for most people to visualize due to everyday experience, the interactions between that particle and substrate giving rise to the adhesional properties are far from simple. Prior to adhering, the particle must first be attracted to the substrate and impinge upon it. During this process, the particle must either convert its kinetic energy into elastically stored potential energy or into heat and sound, or the energy must be dissipated by the creation of plastic deformations [1–4].

At around the time that 'contact' is established, both the particle and substrate experience stresses associated with the forces of adhesion. These stresses cause the materials to deform and the deformations can be either elastic or plastic, depending on whether or not the yield strengths of any of the materials are exceeded. It should be noted that the forces leading to the relatively long range attraction and deposition of the particles may not be the same as the forces adhering these same particles to

*To whom correspondence should be addressed.

the substrate, as the ranges of the adhesion forces are often short compared to, for example, the range of electrostatic forces. However, some adhesion forces can still be considered long in the sense that the materials interact through them beyond their region of contact.

Modern research aimed at developing an understanding of particle adhesion was pioneered by individuals such as Bradley [5], Derjaguin [6], Hamaker [7], Lifshitz [8], and Krupp [9]. Derjaguin attempted to calculate the surface-force-induced contact radius between a particle and substrate by treating the system as a Hertzian indentor. This approach presupposed that the interactions were strictly elastic and compressive in nature. Moreover, Derjaguin assumed that the surface forces were caused by van der Waals interactions.

The relationship between van der Waals interactions and surface forces was originally explored by Hamaker, who related the forces to the number of atoms per unit volume. Lifshitz later related the van der Waals interactions to the dielectric properties of the materials and electronic density fluctuations within those materials. As discussed by Krupp, when all three theories are combined, it is found that the Derjaguin model predicts that the contact radius a depends on the particle radius R, the separation distance between the two materials z_0, and an interaction energetics term $\hbar\overline{\omega}$ by the relationship

$$a^3 = \frac{3\hbar\overline{\omega}}{32\pi z_0^2}\left(\frac{1-\nu^2}{E}\right)R^2, \tag{1}$$

where E and ν are the Young's modulus and Poisson ratio of the deforming material.

The first major advance in what may be termed 'modern' particle adhesion theory was made by Johnson *et al.* [10] (hereafter referred to as the JKR theory) in 1971 and further advanced by Kendall and Padget [11]. Basically, Johnson and coworkers observed, while studying the adhesion-induced deformations between homogeneous combinations of macroscopic gelatin and rubber spheres, that the contact radius predicted by the Derjaguin model was significantly smaller than that observed. Johnson *et al.* realized that tensile forces, as well as the compressive forces of the Derjaguin model, contributed to the size of the contact radius. Assuming that the deformations were strictly elastic and that all interactions occurred within the contact zone, the JKR model predicts that

$$a^3 = \frac{R}{K}\left\{P + 3w_{\mathrm{A}}\pi R + \left[6w_{\mathrm{A}}\pi RP + \left(3w_{\mathrm{A}}\pi R\right)^2\right]^{1/2}\right\}, \tag{2}$$

where w_{A} is the thermodynamic work of adhesion, P is any externally applied load, and

$$K = \frac{4}{3\pi\left(k_1 + k_2\right)}, \tag{3}$$

where k_1 and k_2 are related to the Young's moduli and Poisson ratios of the two materials by

$$k_i = \frac{1-\nu_i^2}{\pi E_i}. \tag{4}$$

The JKR model is significant, not only because it incorporates the tensile interactions, but because it allows parameters such as the thermodynamic work of adhesion between dissimilar solids to be determined and predicts the response of the particles to specific parameter changes. For example, in the absence of any applied load, equation (2) reduces to

$$a^3 = \frac{6w_A \pi R^2}{K}. \tag{5}$$

Alternatively, if adhesion did not exist, equation (2) reduces to that of the Hertzian indentor. Finally, if a negative load is applied to the particle, the contact radius decreased. However, it is required that the radical in equation (2) remain real. This implies that the separation of the particle from the substrate occurs, not when a vanishes, but, rather, when

$$a = 0.63a_0, \tag{6}$$

where a_0 is the contact radius under no externally applied load. Moreover, the force needed to effect separation P_s is given by

$$P_s = -\frac{3}{2} w_A \pi R, \tag{7}$$

which is independent of the Young's modulus.

An alternative to the JKR theory was proposed by Derjaguin *et al.* in 1975 [12]. This model, hereafter referred to as the DMT theory, approached the problem of particle adhesion from a molecular level and assumed that the shape of the adhesion-induced contact was Hertzian. As a direct consequence of this assumption, it was found that exactly half of the interaction force was within the contact region and half was outside. As shown by Tabor [13], these assumptions led to the JKR and DMT theories making radically different predictions. For example, although both models predict the same power law dependencies of the contact radius on particle radius and Young's modulus, the contact radius predicted by the JKR model under no load conditions is approximately 1.8 times greater than that predicted by the DMT model. Moreover, whereas the JKR model predicts that, upon application of a negative load, separation of the particle from the substrate occurs at a finite contact radius, the DMT model predicts that the contact radius will continually decrease and that separation occurs when $a = 0$. Finally, the force predicted by the DMT model to effect separation is $4/3$ as much as that predicted by the JKR model. As was the case with the JKR model, the DMT model predicts that the separation force is independent of Young's modulus.

The discrepancies between the JKR and DMT models was finally resolved by Muller *et al.* [14] (hereafter referred to as MYD) who showed that both theories have their own regions of validity, defined by the dimensionless parameter μ, defined by

$$\mu = \frac{32}{3\pi}\left(\frac{2Rw_A}{\pi E^* z_0^3}\right)^{1/3}, \tag{8}$$

where

$$\frac{1}{E^*} = \frac{1-\nu_1^2}{E_1} + \frac{1-\nu_2^2}{E_2}. \tag{9}$$

For the case where $\mu > 1$, corresponding to larger particles, higher surface energies, and lower Young's moduli, the JKR theory should appropriately describe the adhesion of a particle to a substrate. On the other hand, if $\mu < 1$, corresponding to smaller particles, lower surface energies, and higher Young's moduli, the DMT model would be appropriate. The JKR-DMT transition has recently been studied by Maugis [15]. However, it should be noted that, as a consequence of equation (8), the adhesion-induced deformations of most real materials should be describable by the JKR model.

There are occasions where the adhesion-force-induced deformations between a particle and substrate can be sufficiently large to exceed the yield strength of one of the contacting materials. In that instance a totally or partially plastic deformation will occur. This effect, originally postulated by Krupp [9], was described in detail by Maugis and Pollock [16] and is referred to, hereafter, as the MP theory. According to the MP theory, as either the externally applied load P or the adhesion force increases, the deformation goes from elastic to elasto-plastic and, finally fully plastic. Although no theory of particle adhesion exists for the elasto-plastic regime, when the deformation is fully plastic,

$$P + 2\pi w_A R = \pi a^2 H, \tag{10}$$

where the hardness H is related to the yield strength Y of the material by

$$H = 3Y. \tag{11}$$

By comparing equations (5) and (10) it is readily seen that elastic and plastic adhesion-induced deformations can be distinguished by the different power law dependence of the contact radius on particle radius. Moreover, deviations from the extrapolated curves of the contact radius versus the particle radius to the 1/2 or 2/3 power from the origin can be considered a measure of the significance of any externally applied load, such as those arising from electrostatic interactions.

Experimental studies related to the mechanics of particle adhesion have been performed for some time. Techniques used in these studies include, but are not limited to, the use of nanoindentor simulations of particle–substrate contact [17–21] and microscopic examination of particle–substrate interfaces [11, 22, 23]. More recently, equipment such as the surface force apparatus [24], high resolution and field emission scanning electron microscopes and the atomic force microscope have been used to more critically study particle adhesion [25–28]. These studies have led to the validation of many of the predictions of the JKR and MP theories. However, many of these studies have also indicated that there remains much in the area of particle adhesion to be explained. In addition, the authors are unaware of any conclusive observations indicating situations where the DMT model correctly predicts the behavior of particles in contact with surfaces. Presumably, this is due to the fact that, for real materials, the criterion that $\mu < 1$ implies that $R \ll 1$ μm. Such particles tend to exist in colloids rather than in a dry, isolated state. Theories presupposing interactions between dry particles and substrates would not be appropriate under such circumstances. It is possible, of course, that submicrometer-size aerosol particles that deposit on a substrate may behave according to the predictions of the DMT theory, but that is not known at the present time and is a topic for future research.

At this time, consider a quintessential JKR system: rigid particles (glass) having radii from about 1 μm to just over 100 μm contacting an elastomeric substrate (polyurethane). As discussed elsewhere in detail [29], the contact radius between the particles and the substrate can be both viewed and measured using a high resolution scanning electron microscope (SEM). In particular, by mounting the sample onto a cross-sectional stub and viewing it at an 88° tilt angle, the deformation of the polyurethane around the particle, caused by the forces of adhesion, can be directly viewed. Micrographs of such particle–substrate interfacial regions are shown in Fig. 1 for particles having nominal radii of 3.3 μm (Fig. 1A), 12.5 μm (Fig. 1B), 22 μm (Fig. 1C), and 65 μm (Fig. 1D). As can be seen from these micrographs, there is a finite contact area which, presumably, is the result of the surface-force-induced stresses between the two materials causing the polyurethane to deform. If this is, indeed, the case and there are no other significant forces, the contact radius should vary as the particle radius to the 2/3 power and the least squares fit line through the data should intersect the origin. This was found to be correct, as illustrated by the graph of the contact radius as a function of particle radius to the 2/3 power shown in Fig. 2. The work of adhesion was estimated, using equation (5) to be 0.17 J/m^2, which is a reasonable value for the combination of a polymer and soda-lime glass.

As also discussed in this paper, the JKR model predicts that, upon application of a negative load, separation occurs from a finite contact rather than when the contact radius vanishes. The accuracy of this prediction is also illustrated with soda-lime glass particles on polyurethane [29]. In this instance, the particles had radii of approximately 100 μm and were sufficiently massive that most fell off the substrate when the substrate was raised and tilted to the angle needed for viewing in the SEM. The average contact radius of the remaining particles, such as that observed in the micrograph shown in Fig. 3, was about $0.7a_0$. This suggests that the separation of the particle from the substrate occurs at a finite contact radius that is consistent with the predictions of the JKR formalism.

Examples of plastic response to the stresses associated with the adhesion of a particle to a substrate have also been reported in the literature [30–33]. The materials studied included both particles composed of polymers, such as polystyrene, and particles composed of inorganic materials, such as soda-lime glass. As an example, consider the case of polystyrene particles on a silicon wafer, as seen in the scanning electron micrographs shown in Fig. 4. In these micrographs most of each particle was imaged at a relatively low magnification. However, near the actual contact region, the magnification was increased. This allowed both the particle and contact radii to be determined for each particle in a single scan.

The power dependence of the contact radius on particle radius was found by regressing $\log a$ onto $\log R$. Accordingly, it was found that a varied as $R^{0.42 \pm 0.13}$, which is in agreement with the predictions of the MP theory of adhesion. Figure 5 shows

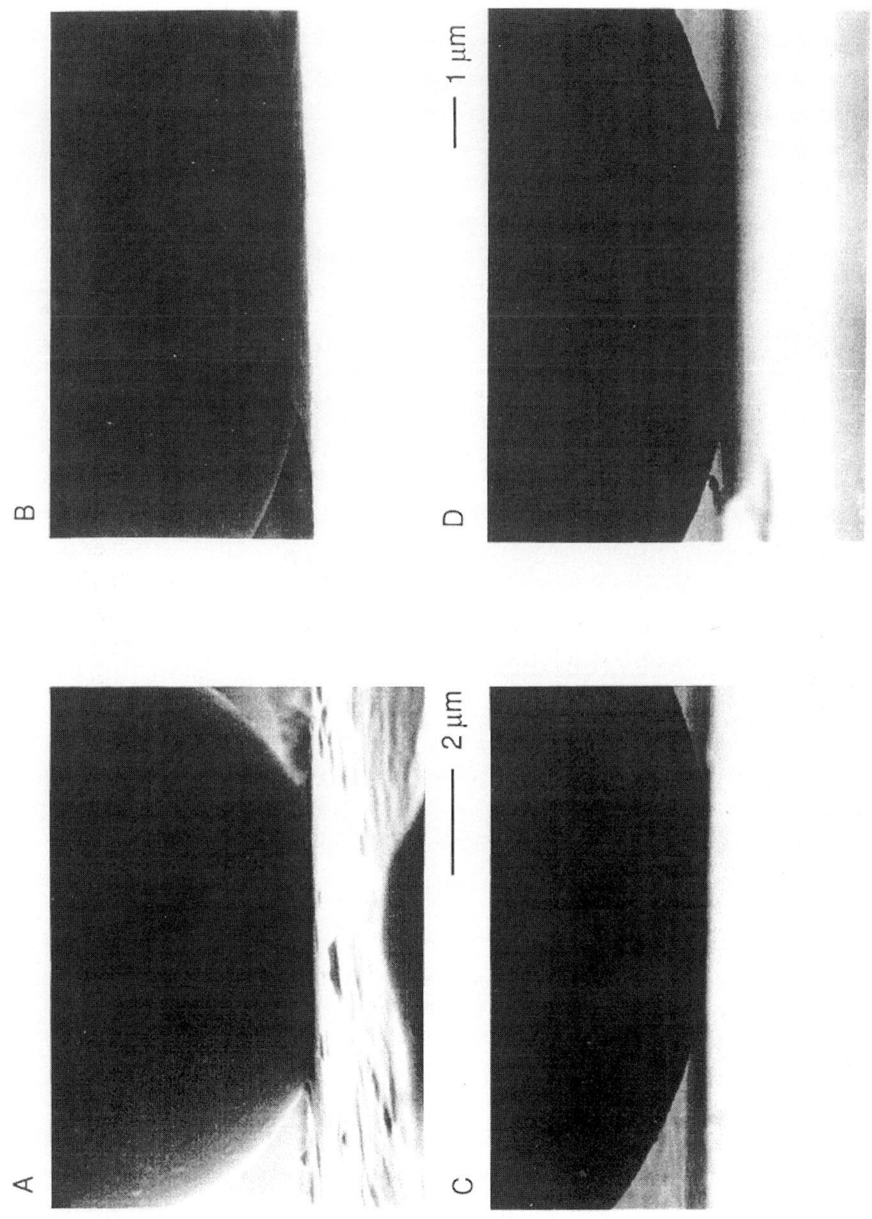

Figure 1. High tilt SEM micrographs of glass particles in contact with a polyurethane substrate. Magnification scales are drawn beneath each micrograph.

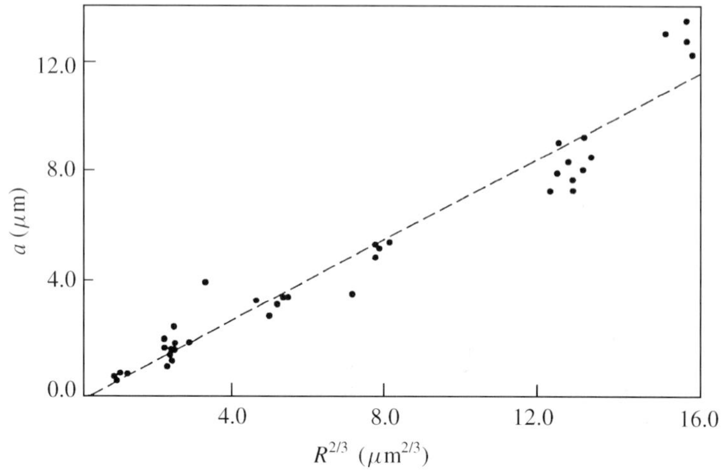

Figure 2. The contact radius as a function of the particle radius to the 2/3 power for glass particles on a polyurethane substrate. The intersection of the least squares fit curve with the origin suggests that the effects of external loads, such as those arising from the presence of electric charge or gravity, are negligible in this example.

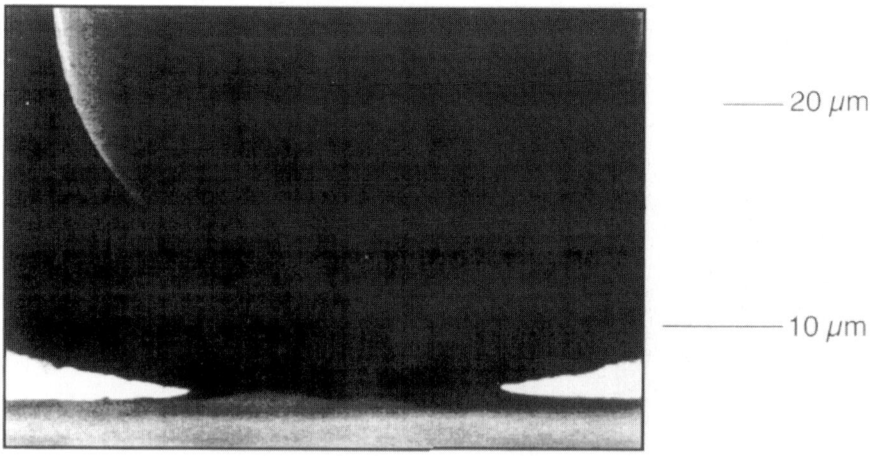

Figure 3. An SEM micrograph of a glass particle with a nominal 103 μm radius on a polyurethane substrate. The micrograph is split into two regions: a relatively low magnification region showing most of the particle and a high magnification region showing the contact region. The distention of the substrate is due to gravity pulling on this rather large particle.

a plot of the contact radius as a function of the square root of the particle radius. The extrapolated least squares fit line through the data comes close to intersecting the origin, suggesting that the effects of interactions other than surface forces are small. It is possible that the slight offset may be caused by some electrostatic charge on the particle. The work of adhesion, calculated using equation (10) and assuming that $P = 0$ and using the measured yield strength of the material (10.8 MPa) was found to be 0.32 J/m^2.

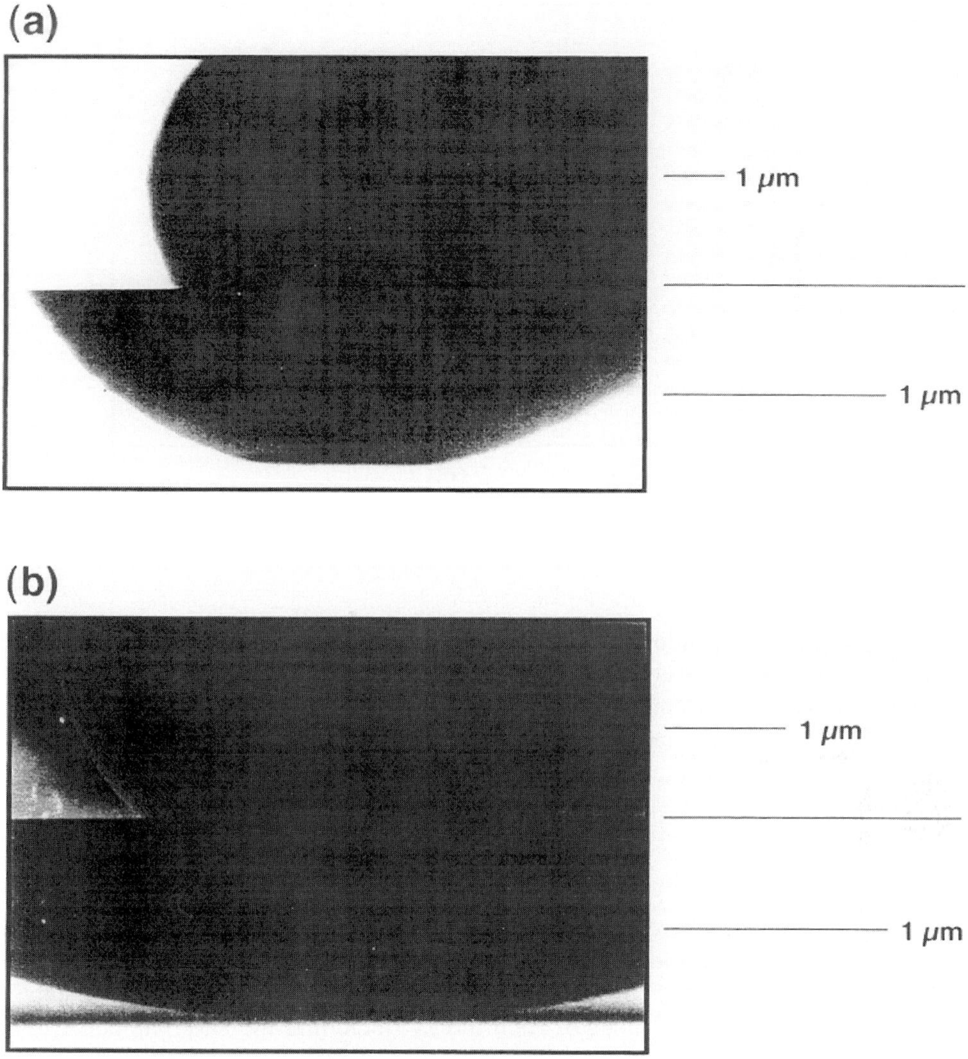

Figure 4. Polystyrene particles having nominal radii of 1.7 μm (a) and 5.1 μm (b) on a silicon substrate. Split magnification views are shown similar to that used in Fig. 3.

If, indeed, the deformations of the polystyrene spheres are plastic rather than elastic, the average pressure exerted on the particle should exceed its yield strength. Bowling showed [34] that, as a particle in contact with a substrate deforms, the force of adhesion becomes proportional to the contact radius squared divided by the separation distance cubed. Upon dividing by the average contact area, it is found that, with the correction for the increased force due to the finite contact radius [34], the pressure p exerted on the particle is given by

$$p = \frac{2w_A}{z_0}. \tag{12}$$

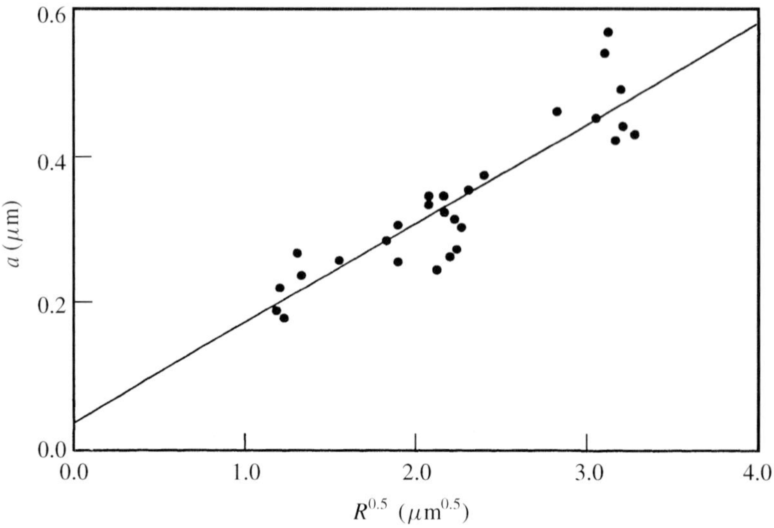

Figure 5. The contact radius of polystyrene particles on a silicon substrate as a function of the square root of the particle radius.

Upon substituting the stated values it was found that p is of the order of the Young's modulus of polystyrene and far in excess of its yield strength. It would be expected, then, that the stresses associated with the surface forces should cause the polystyrene to deform plastically and that the MP theory should predict the behavior of this system, as appears to be the case.

2. FUTURE OPPORTUNITIES

It may appear to the casual reader, after the discussion in the first part of this paper, that particle adhesion is fully understood. After all, quantitative theories describing particle adhesion in both the elastic and plastic regimes have been proposed and experimentally verified. The appropriate regimes of each theory (including the DMT model) can be readily distinguished and the adhesion responses predicted for given systems. Indeed, the level of understanding of particle adhesion is impressive. However, as is generally the case, solid scientific advances in the area of particle adhesion have both raised new questions and have provided the foundation upon which to build. The opportunities in the field of particle adhesion are far too numerous to be described in a single paper. Accordingly, the authors will try to highlight the area with a few current problems and hope that these discussions will lead to further advances in general.

2.1. What kind of interactions affect the adhesion of particles

It is frequently assumed that forces of adhesion are dominated by van der Waals interactions. This assumption has recently been questioned by Burnham *et al.* [35], who realized that the range of the attractive forces was too great for van der Waals interactions. They proposed that, due to local inhomogeneities in the work-function of a material, a microscopic charge transfer would occur, resulting in a dipole. If the work-function of two such 'patches' differed by an amount $\Delta\phi$, a particle would have a dipole moment $p = 4\pi\varepsilon_0 R^2 \Delta\phi$. If the centers of two particles are some distance r apart, the force on each particle will have a maximum value of

$$\frac{3p^2}{\pi\varepsilon_0 r^4} = 48\pi\varepsilon_0 \left(\frac{R}{r}\right)^4 (\Delta\phi)^2 \tag{13}$$

and will vary with the orientation of the dipoles. This model has been referred to as the 'patch charge' model.

The interactions between real particles and surfaces become more complex than merely distinguishing between patch charges and van der Waals interactions because real particles are not perfect spheres and generally possess an electric charge. For example, a 10 μm xerographic toner particle may have a typical charge of the order of 8 fC [36]. If this particle were spherical and the charge uniformly distributed over its surface, the Coulombic forces would be small compared to the van der Waals interactions for the case of the particle in contact with the surface. For example, for the Coulombic attraction between the described particle in contact with a grounded metal plate, the force would be given simply by

$$F_{\mathrm{I}} = \alpha \frac{q^2}{16\pi\varepsilon_0 R^2}, \tag{14}$$

where α is a factor to correct for the polarization of the particle, and q and R are the charge and radius of the particle, respectively. For a material with a dielectric constant of 4, $\alpha = 1.9$. For the ideal particle under consideration, $F_{\mathrm{I}} = 10$ nN.

The van der Waals force F_{vdw} between an undeformed spherical particle and a rigid substrate is given by

$$F_{\mathrm{vdw}} = \frac{\hbar\varpi}{8\pi z_0^2} R, \tag{15}$$

where $\hbar\varpi$ is approximately 2 eV for typical polymers and somewhat higher for more electrically conducting materials and z_0 is the separation spacing for a van der Waals bonded solid (approximately 0.4 nm). Accordingly, F_{vdw} is of the order of 1 μN or two orders of magnitude greater than F_{I}.

In the real world, things are not that simple. Particles are not perfectly spherical and, as shown by Hays [36], the charge distribution on such particles is not uniform and the more highly charged areas are the ones that tend to contact the surface of the substrate. This effect, which is referred to in the literature as a 'charged patch', as opposed to the 'patch charge' mentioned previously, would increase the effect on particle adhesion of the Coulombic attraction. Moreover, as a particle approaches a surface, higher order multipole terms become significant in determining the force of attraction [37]. The effect of these multipoles depends on the difference in the dielectric constants between the materials and the medium surrounding the particle.

Although both non-uniform charge distributions and polarization factors affect particle adhesion in principle, there is still a need to further develop that understanding before these contributions to particle adhesion can be quantitatively determined. Specifically, although the occurrence of non-uniform charge distributions appears to be based on firm experimental observations, these observations are more qualitative than quantitative. This is due to the difficulty in measuring specific charge levels at specific sites on a given particle and being able to quantitatively relate those charge levels to the particle geometry and to specific adhesional forces. The effect of polarization is also not fully understood. Although dipole effects can be estimated for particles and substrates separated by a finite distance, once the particle actually contacts the surface, the materials deform and, in some aspects, can be considered to be 'welded' together by the surface forces. Indeed, separation of the materials is often accompanied by a cohesive failure of one of the materials rather than a fracture along the interfacial plane. Since many common materials have dielectric constants that are fairly close, the effects of polarization in the regions of closest proximity would be small. Multipoles would be more pronounced in the separated regions but their effects on adhesion would be mitigated by the large separation distances. Solving the problem of multipole effects on particle adhesion is further complicated because of one's inability to draw a Gaussian surface around a particle in intimate contact with a substrate, especially where significant deformation of the materials has occurred.

Recently, Gady *et al.* [38] have attempted to distinguish the types of interactions governing particle adhesion. Specifically, they measured the gradient of the attractive force as a function of separation distance between a particle and a planar substrate. In their studies, they mounted polystyrene particles on the tip of an atomic force cantilever and oscillated them in a low vacuum near the surface of a substrate consisting of highly oriented pyrolytic graphite (HOPG). This allowed the power law dependence of the gradient of the force of attraction to be determined. The advantage of this technique over direct measurements of the attractive forces is that it gives very precise measurements. An example of their reported results for a 6 μm radius polystyrene particle is shown in Fig. 6. The solid line represents the anticipated van der Waals force gradient, whereas the line composed of alternating dashes and single dots represents the anticipated gradient of the attractive force due to a charged patch on the surface of the particle having a radius equal to that predicted by the JKR model. The line represented by the alternating single dash and triple dots is the sum of these contributions. As can be seen, the data are in good agreement with that predicted by adding the effects of these two curves, with the cross-over between electrostatically and van der Waals dominated attraction occurring at approximately 30 nm. It would appear from these results that the forces of adhesion between particles and substrates may not be due to one type of interaction, but rather two or more. Further exploration along these lines is clearly needed.

2.2. Molecular level understanding of particle adhesion

Present understanding of particle adhesion is based principally on the JKR, MP, and DMT/MYD models. The JKR and MP theories are based strictly on continuum mechanics and thermodynamics. The DMT and MYD models allow for some molecular

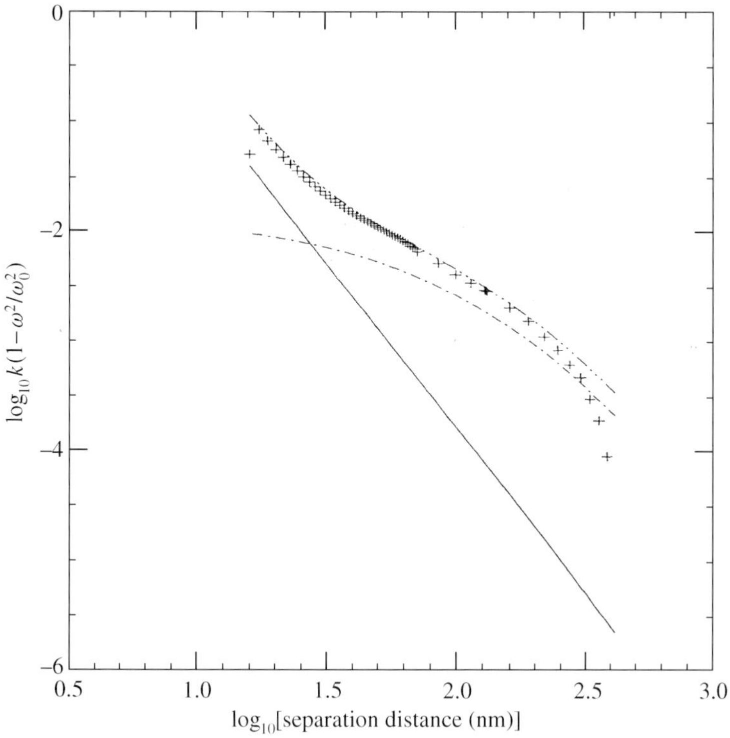

Figure 6. Force gradient as a function of particle–substrate separation distance. The solid line represents van der Waals interactions whereas the alternating dash-dot line represents interactions due to electrostatically charged patches. The line illustrated by three dots and a dash is the sum of the aforementioned curves. The experimental data are represented by the symbol 'x'.

interactions in that they assume the force of adhesion can be calculated by allowing the molecules in the adhering materials to interact via a Lennard-Jones potential and performing a pairwise summation of the forces between atoms in the two materials. However, the physical parameters of the individual materials, such as their Young's moduli, are assumed to be given.

More to the point, current particle adhesion models are able to predict the behavior of materials but do not address the question of why they behave in a specific manner. For example, the JKR model predicts that, upon application of a negative load, separation of the particle from the substrate occurs not when a vanishes, but when a is finite. This occurs from the requirement that the radical in equation (2) must be real. But what on a molecular level is giving rise to this behavior?

Intuitively, it would seem that the adhesion of a particle to a substrate should be determined by the same intermolecular potentials that give rise to other physical properties of the materials in question. Once the intermolecular potential for a material is defined, the elastic properties, melting point, yield strength, and surface energy of that material should also be fixed. If this is done, the problem of particle adhesion becomes one of solving Newton's equations for many molecules that tend to be clustered within two generally distinct bodies. Such analysis is now possible owing to ongoing advances in computer technology and is the subject of much molecular dynamic modeling. For example, Quesnel *et al.* [39] first calculated the elastic constants of a three-dimensional FCC lattice composed of several hundred atoms, assuming the atoms comprising that lattice interacted via a Lennard-Jones potential. They then calculated the surface energy of a similar, two-dimensional rectangular lattice and then studied the adhesion of a similarly composed particle to that lattice [40].

Although the model proposed by Quesnel is fairly simple and calculations have been limited by computational facilities, several interesting features have arisen. For example, the hysteresis between the attractive and removal forces between particles and substrates, such as those commonly observed using cantilever-force techniques, is seen to be associated with the decay of phonons, generated at the point where the particle and substrate snap together, within the materials. In other words, even if the deformations are elastic, the phonon decay allows for an energy loss mechanism that can account for the observed hysteresis. More interesting, perhaps, is the observation that, upon application of a negative force, the particle and substrate separate at a contact radius comparable to that predicted by the JKR model. This suggests that the origin of the finite separation radius predicted by the JKR model may be caused by the relatively short-range nature of van der Waals interactions, rather than by the size of the particle and the Young's moduli and surface energies of the materials, as suggested by the MYD model. Obviously, this area is ripe for further research.

2.3. Behavior not predicted by current theories

Although current theories appear to describe the adhesion of particles to substrates for many systems, recent studies have revealed a variety of phenomena that have not been predicted. These phenomena include anomalous power law dependencies of the contact radius on particle radius [41], engulfment of the particles by the substrate [42], viscoelastic creep [43], and the time dependence of adhesion [44]. These effects are most pronounced when at least one of the contacting materials is a polymer.

An example of viscoelastic creep and particle engulfment is shown in the micrographs presented in Fig. 7. These scanning electron micrographs show nominal 4 μm (Fig. 7A) and 11 μm radius glass particles on a plasticized polystyrene substrate. As can be seen, the smaller particles have become almost totally engulfed by the substrate, whereas the contact radius for the larger particles is slightly less than the particle radius. Intuitively, it would seem that engulfment might be feasible at some critical particle radius R_c given by

$$R_c = \frac{7\gamma_{12}}{E} \tag{16}$$

for the case of elastic deformations, where γ_{12} is the interfacial energy and E is the Young's modulus of the substrate. For the case of the substrate deforming plastically, the critical radius is related to the yield strength Y of the substrate by

$$R_c \approx \frac{2\gamma_{12}}{3Y}. \tag{17}$$

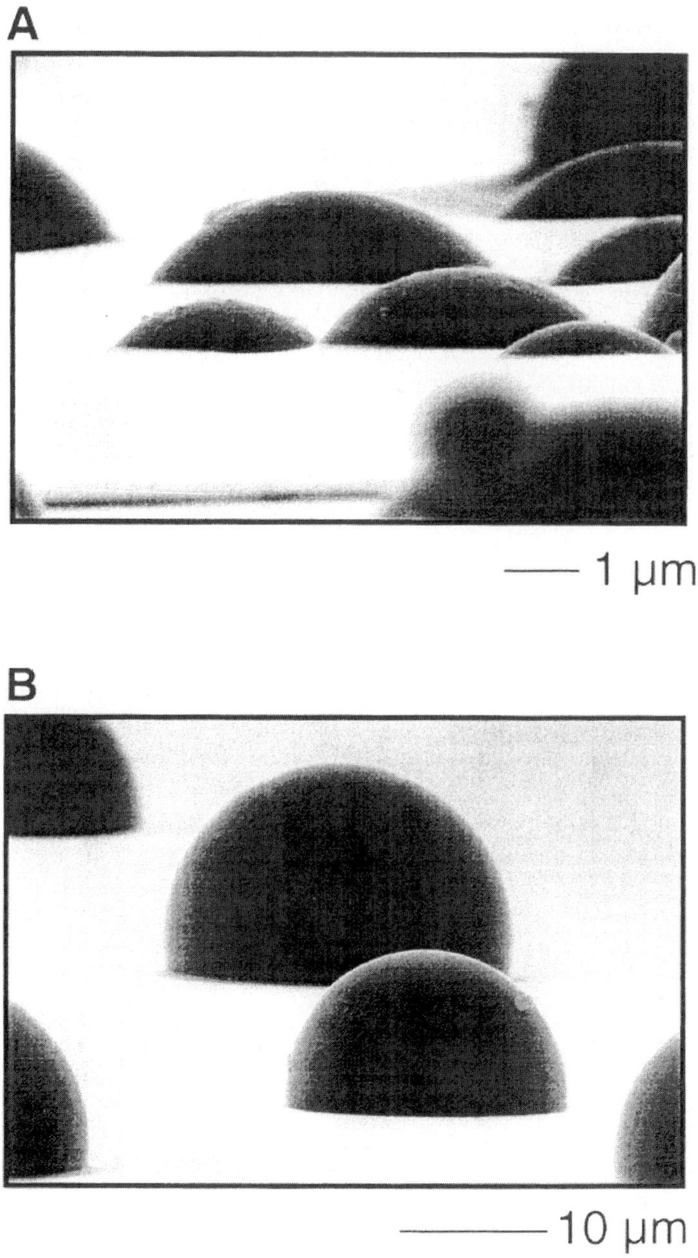

Figure 7. SEM micrographs of glass particles on a plasticized polystyrene substrate. Surface-force-induced engulfment of the particles by the substrate is observed.

Similar effects were observed to occur with other highly compliant and/or viscoelastic substrates, with time periods ranging from several days to several weeks. The micrographs presented in Fig. 7 were taken after the particles were in contact with the substrate for about one month and no further embedding had been observed.

Another interesting and unexplained observation is that, for highly compliant elastomeric substrates (e.g. $E \approx 0.2$ MPa) the contact radius was found to depend on the particle radius to the 3/4 power rather than the 2/3 power predicted by the JKR model. This relation was found to hold for particles having radii between approximately 1 and 100 μm for highly compliant elastomeric substrates [41], but was also found to be true for particles having radii between approximately 1 and 6 μm in contact with a more normal elastomeric substrate (i.e. one having a Young's modulus in the range of 3 MPa) [45, 46].

An example of the 3/4 power law dependence is shown in Fig. 8, which shows the surface-force-induced contact radius for glass particles on a plasticized polyurethane substrate ($E = 45$ KPa) [41]. The power law dependence was determined by linearly regressing $\log a$ on $\log R$. The slope of this curve was found to be 0.75 ± 0.02 to within the 95% confidence interval.

Tabor [13] has shown that, for materials following the JKR theory, the height of the adhesion-induced meniscus h around a particle is approximately given by

$$h = \frac{R\left(\frac{w_A}{2}\right)^2}{E^2}. \tag{18}$$

The work of adhesion w_A can be estimated by substituting the experimentally determined contact and particle radii into equation (5). Accordingly, it was found that the estimated values of w_A ranged between 0.040 and 0.047 J/m^2, depending on the size of the particle chosen. These are reasonable values for these materials.

The calculated and observed meniscus heights are plotted in Fig. 9. As can be seen, there is good agreement between the predicted and experimentally determined meniscus heights. These results suggest that, rather than needing a totally new theory,

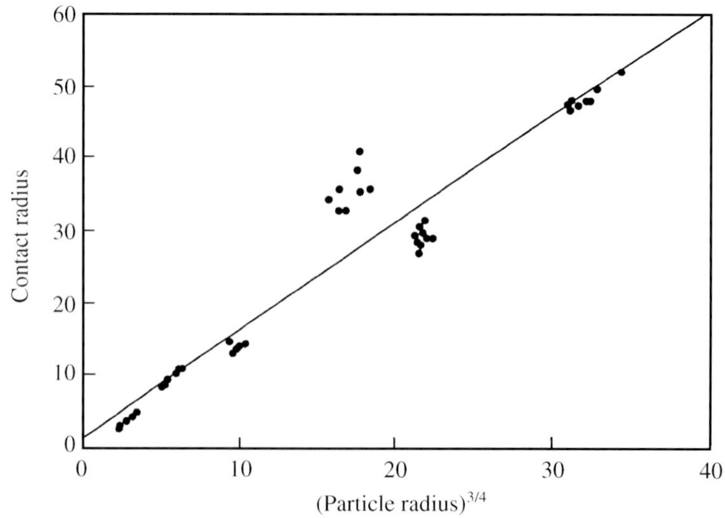

Figure 8. Contact radius as a function of particle radius to the 3/4 power for glass particles on a highly compliant polyurethane substrate.

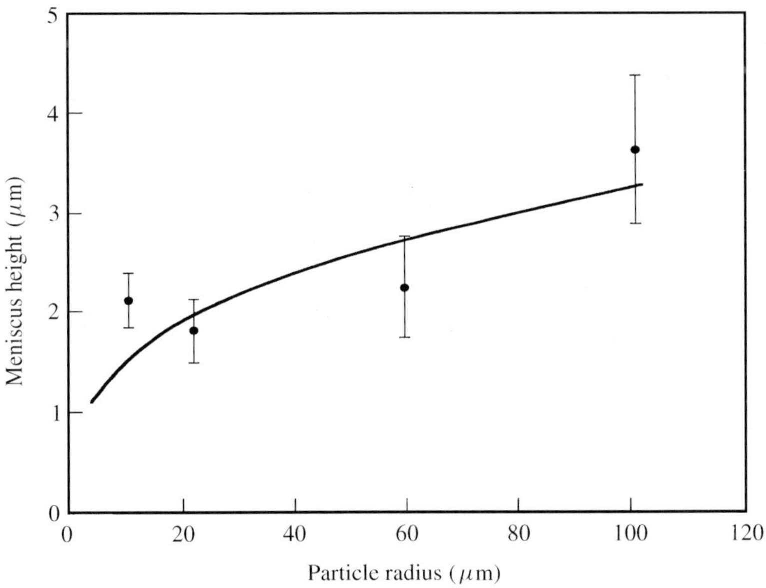

Figure 9. Meniscus height as a function of particle radius for the same materials as used in Fig. 8. The solid line represents the meniscus height as predicted by the JKR theory.

the anomalous power law dependence may be explainable if generalizations to the JKR theory are made.

One such generalization of the JKR theory was proposed by Maugis [47] who showed that the 3/4 power law dependence could be obtained if, rather than using a parabolic approximation for the profile of a spherical particle, the exact profile was used. Quesnel [48] showed that a similar power law dependence could be obtained by solving the adhesion equations in a self-consistent manner so as to take into account the increase in attractive force with increasing contact area. Other hypotheses advanced include the effects of viscoelastic creep, elastic-to-plastic transitions, and non-linear elastic contributions.

3. CONCLUSIONS

Particle adhesion is an important topic both scientifically and technologically. In accordance with its importance, it has been the subject of much investigation and much progress has been made in understanding this topic. Theories such as those advanced by Johnson, Kendall and Roberts; Derjaguin, Muller and Toporov; and Maugis and Pollock have given us a fundamental understanding of particle adhesion. These theories have been complemented by experimental studies which have shown both the ranges of validity of these theories as well as their limitations. Mechanisms giving rise to the forces of adhesion have also been recently elucidated. However, despite all the progress (or, perhaps because of all the advances), many more issues have come to light that have created great opportunity for further scientific advances.

The examples presented in this paper are merely meant to be illustrative, rather than all-inclusive, of the challenges in this field. One could equally well highlight topics such as the effect on adhesion of the particle impact energy, temporal effects on adhesion, the effects of thin films and coatings on bulk substrates, humidity effects, surface roughness, etc. In summary, particle adhesion research is, and should remain, a suitable subject for scientific research.

REFERENCES

1. L. N. Rogers and J. Reed, *J. Phys. D* **17**, 677 (1984).
2. S. Wall, W. John and S. L. Goren, in: *Particles on Surfaces 2: Detection, Adhesion, and Removal*, K. L. Mittal (Ed.). Plenum, New York (1989).
3. K. L. Johnson and H. M. Pollock, *J. Adhesion Sci. Technol.* **8**, 1323 (1994).
4. B. Dahneke, *J. Adhesion* (in press).
5. R. S. Bradley, *Philos. Mag.* **13**, 853 (1932); also, *Trans. Faraday Soc.* **32**, 1088 (1936).
6. B. V. Derjaguin, *Kolloid Z.* **69**, 155 (1934).
7. H. C. Hamaker, *Physica* **4**, 1058 (1937).
8. E. M. Lifshitz, *Soviet Physics JEPT* **2**, 73 (1956).
9. H. Krupp, *Advan. Colloid Interface Sci.* **1**, 111 (1967).
10. K. L. Johnson, K. Kendall and A. D. Roberts, *Proc. Roy. Soc. London, Ser. A* **324**, 301 (1971).
11. K. Kendall and J. C. Padget, *Intl J. Adhesion Adhesives* **2**, 149 (1982).
12. B. V. Derjaguin, V. M. Muller and Yu. P. Toporov, *J. Colloid Interface Sci.* **53**, 314 (1975).
13. D. Tabor, *J. Colloid Interface Sci.* **58**, 2 (1977).
14. V. M. Muller, V. S. Yushchenko and B. V. Derjaguin, *J. Colloid Interface Sci.* **77**, 91 (1980); also, V. M. Muller, B. V. Derjaguin and Yu. P. Toporov, *Colloids Surfaces* **7**, 251 (1983).
15. D. Maugis, *J. Colloid Interface Sci.* **150**, 243 (1991); also, D. Maugis and B. Gauthier–Manuel, *J. Adhesion Sci. Technol.* **8**, 1229 (1994).
16. D. Maugis and H. M. Pollock, *Acta Metall.* **32**, 1323 (1984).
17. H. M. Pollock, P. Shufflebottom and J. Skinner, *J. Phys. D* **10**, 127 (1977).
18. M. D. Pashley and D. Tabor, *Vacuum* **31**, 619 (1981).
19. H. M. Pollock, *J. Phys. D* **11**, 39 (1978).
20. R. Mouginot and D. Maugis, *J. Mater. Sci.* **20**, 4354 (1985).
21. D. Maugis and M. Barquins, *J. Phys. D* **16**, 1843 (1983).
22. M. M. Chaudhri and E. H. Yoffe, *Philos. Mag. A* **44**, 667 (1981).
23. M. K. Chaudhury and G. M. Whitesides, *Langmuir* **7**, 1013 (1991).
24. J. N. Israelachvili and D. Tabor, *Proc. Roy. Soc. London Ser. A* **311**, 19 (1972).
25. D. S. Rimai, L. P. DeMejo and R. C. Bowen, *J. Adhesion Sci. Technol.* **8**, 1333 (1994).
26. W. W. Merrill, A. V. Pocius, B. V. Thakker and M. Tirrell, *Langmuir* **7**, 1975 (1991).
27. D. M. Schaefer, M. Carpenter, R. Reifenberger, L. P. DeMejo and D. S. Rimai, *J. Adhesion Sci. Technol.* **8**, 197 (1994).
28. H. A. Mizes, K. G. Loh, M. L. Ott and R. J. D. Miller, in: *Particles on Surfaces: Detection, Adhesion, and Removal*, K. L. Mittal (Ed.), pp. 47–59. Marcel Dekker, New York (1995).
29. L. P. DeMejo, D. S. Rimai and R. C. Bowen, *J. Adhesion Sci. Technol.* **5**, 959 (1991).
30. J. K. Vrtis, C. D. Athanasiou, R. J. Farris, L. P. DeMejo and D. S. Rimai, *J. Adhesion Sci. Technol.* **8**, 929 (1994).
31. D. S. Rimai, R. S. Moore, R. C. Bowen, V. K. Smith and P. E. Woodgate, *J. Mater. Res.* **8**, 662 (1993).
32. D. S. Rimai, L. P. DeMejo and R. C. Bowen, *J. Appl. Phys.* **68**, 6234 (1990).
33. R. C. Bowen, L. P. DeMejo and D. S. Rimai, *J. Adhesion* **51**, 201 (1995); and R. C. Bowen, L. P. DeMejo and D. S. Rimai, in: *Advances in Particle Adhesion*, D. S. Rimai and L. H. Sharpe (Eds). Gordon and Breach, New York (1995).

34. R. Allen Bowling, in: *Particles on Surfaces 1: Detection, Adhesion, and Removal*, K. L. Mittal (Ed.), pp. 129–142. Plenum, New York (1988).
35. N. A. Burnham, R. J. Colton and H. M. Pollock, *Nanotechnology* **4**, 64 (1993); see also H. M. Pollock, N. A. Burnham and R. J. Colton, in: *Advances in Particle Adhesion*, D. S. Rimai and L. H. Sharpe (Eds). Gordon and Breach, New York (1995).
36. D. A. Hays, *J. Adhesion Sci. Technol.* **9**, 1063 (1995); see also D. A. Hays, in: *Fundamentals of Adhesion and Interfaces*, D. S. Rimai, L. P. DeMejo and K. L. Mittal (Eds), pp. 61–72. VSP, Utrecht (1995).
37. W. Y. Fowlkes and K. S. Robinson, in: *Particles on Surfaces 1: Detection, Adhesion, and Removal*, K. L. Mittal (Ed.), pp. 143–155. Plenum, New York (1988).
38. B. Gady, D. Schleef, R. Reifenberger, D. Rimai and L. P. DeMejo, *Phys. Rev. B* **53**, 8065 (1996).
39. D. J. Quesnel, D. S. Rimai and L. P. DeMejo, *Phys. Rev. B* **48**, 6795 (1993).
40. D. J. Quesnel, D. S. Rimai and L. P. DeMejo, *J. Adhesion Sci. Technol.* **9**, 1015 (1995); see also D. J. Quesnel, D. S. Rimai and L. P. DeMejo, *J. Adhesion* **51**, 49 (1995).
41. D. S. Rimai, L. P. DeMejo, W. B. Vreeland and R. C. Bowen, *Langmuir* **10**, 4361 (1994).
42. D. S. Rimai, L. P. DeMejo and R. C. Bowen, *J. Adhesion* (in press).
43. L. P. DeMejo, D. S. Rimai, J. H. Chen and R. Bowen, in: *Particles on Surfaces: Detection, Adhesion, and Removal*, K. L. Mittal (Ed.), pp. 33–45. Marcel Dekker, New York (1995).
44. S. Krishnan, A. A. Busnaina, D. S. Rimai and L. P. DeMejo, *J. Adhesion Sci. Technol.* **8**, 1357 (1994).
45. D. S. Rimai, L. P. DeMejo and R. C. Bowen, *J. Appl. Phys.* **66**, 3574 (1989).
46. L. P. DeMejo, D. S. Rimai and R. C. Bowen, *J. Adhesion Sci. Technol.* **5**, 959 (1991).
47. D. Maugis, *Langmuir* **11**, 679 (1995).
48. D. J. Quesnel, unpublished results (1994).

Part 2

Contact angle, wettability and surface energetics

Mittal Festschrift, pp. 181–197
W. J. Van Ooij and H. R. Anderson, Jr. (Eds)

A new approach for determining roughness by means of contact angles on solids

R. J. GOOD,* M. K. CHAUDHURY† and C. YEUNG‡

Department of Chemical Engineering, State University of NY at Buffalo, Buffalo, NY 14260, USA

Abstract—Contact angle hysteresis on rough surfaces is caused by the contortion of the liquid surface that must occur as the liquid front passes from one metastable configuration to another. We have combined the Wenzel equation for the effect of roughness on the contact angle, θ, with the well-known equation relating contact angles to the surface free energy of the solid and of the liquid, and with Good's hypothesis of a free energy barrier to liquid front motion. The method that is developed calls for measuring θ for a series of liquids and plotting $\cos\theta_a$ vs. $\sqrt{\gamma_l^{\mathrm{LW}}}/\gamma_l$ and extrapolating to the limit of $(1/\gamma_l) \to 0$. On a perfectly smooth, homogeneous surface, the intercept is -1 and the Wenzel ratio for a rough surface is given, approximately, by the negative of the value of the intercept. A shift of the γ_c value for the solid, due to roughness, is also predicted. Experimental data are presented for measurements with Teflon FEP.

Keywords: Contact angle hysteresis; roughness; Teflon FEP; Wenzel ratio.

1. INTRODUCTION

Surface roughness is one of the major causes of contact angle hysteresis [1]. The contact angle of a liquid on a solid is strongly affected by any deviations of the surface from ideality [2, 3], such as roughness. For a rigid, chemically homogeneous surface, Young's equation,

$$\gamma_{\mathrm{sv}} - \gamma_{\mathrm{s}l} = \gamma_{l\mathrm{v}} \cos\theta_{\mathrm{Y}}, \tag{1}$$

can be exact only if the surface is smooth§. Wenzel [4] identified a roughness ratio, r,

$$r = \frac{a}{A}, \tag{2}$$

---

*To whom correspondence should be addressed.

†Present address: Department of Chemical Engineering, Lehigh University, Bethlehem, PA 18015, USA.

‡Present address: Department of Chemical Engineering, University of Illinois, Urbana, IL 61801, USA.

§Strictly speaking, a rough surface cannot be perfectly homogeneous as regards local energy density. So the theory that we employ must be recognized as an approximation for real surfaces.

where a is the actual, microscopic area and A is the apparent area, the projection of the microscopic area on a plane. The Young–Wenzel equation is

$$\cos\theta_{\mathrm{W}} = \frac{r(\gamma_{\mathrm{sv}} - \gamma_{\mathrm{s}l})}{\gamma_l} \quad (3a)$$

$$= r\cos\theta_{\mathrm{Y}}, \quad (3b)$$

where θ_{W} is the contact angle on the solid whose roughness ratio is r.

It has been shown [5–8] that for a liquid on a homogeneous smooth solid, the following general equation holds:

$$\gamma_l(1 + \cos\theta_{\mathrm{Y}}) = 2\sqrt{\gamma_{\mathrm{s}}^{\mathrm{LW}}\gamma_l^{\mathrm{LW}}} - 2\left(\sqrt{\gamma_{\mathrm{s}}^{\oplus}\gamma_l^{\ominus}} + \sqrt{\gamma_{\mathrm{s}}^{\ominus}\gamma_l^{\oplus}}\right), \quad (4a)$$

where the superscript LW denotes the apolar (London–van der Waals) component of the surface free energy [3, 8] and the polar (acid–base) components of γ_{s} and γ_l are designated by $\oplus$ for Lewis acidity and $\ominus$ for Lewis basicity. For apolar solids such as polypropylene and Teflon, $\gamma_{\mathrm{s}}^{\oplus}$ and $\gamma_{\mathrm{s}}^{\ominus}$ are zero, so that

$$\gamma_l(1 + \cos\theta_{\mathrm{Y}}) = 2\sqrt{\gamma_{\mathrm{s}}^{\mathrm{LW}}\gamma_l^{\mathrm{LW}}} \qquad \text{if } \gamma_{\mathrm{s}}^{\oplus} = \gamma_{\mathrm{s}}^{\ominus} = 0 \quad (4b)$$

$$\cos\theta_{\mathrm{Y}} = -1 + 2\sqrt{\gamma_{\mathrm{s}}^{\mathrm{LW}}}\left(\frac{\sqrt{\gamma_l^{\mathrm{LW}}}}{\gamma_l}\right). \quad (4c)$$

Fox and Zisman [9] recommended plotting $\cos\theta$ vs. γ_l for a series of liquids, and extrapolating to the intercept with the line for $\cos\theta = 1$ ($\theta = 0$) to determine γ_{c}, the critical surface tension for the wetting of a solid. Good and Girifalco pointed out that a plot of $\cos\theta$ vs. $1/\sqrt{\gamma_l}$ gave a theory-based straight line, and hence (when extrapolation was required) a much more valid number for γ_{c}. Good [11] demonstrated mathematically the relation between the Fox and Zisman plot and the plot of $\cos\theta$ vs. $1/\sqrt{\gamma_l}$.

On the basis of the form of equations (4b) and (4c), Fowkes [10a, b] recommended including a point at ($\gamma_l^{-1} = 0, \cos\theta = -1$) in a plot of $\cos\theta$ vs. $\sqrt{\gamma_l^{\mathrm{LW}}}/\gamma_l$. This approach needs to be improved upon, by combining equations (4c) and (3b):

$$\cos\theta_{\mathrm{W}} = -r + 2r\sqrt{\gamma_{\mathrm{s}}^{\mathrm{LW}}}\left(\frac{\sqrt{\gamma_l^{\mathrm{LW}}}}{\gamma_l}\right). \quad (5)$$

Then in the graph of $\cos\theta$ vs. $\sqrt{\gamma_l^{\mathrm{LW}}}/\gamma_l$ the line should *not* be forced to go through the point $(0, -1)$. The intercept at $\gamma_l^{-1} \to 0$ should be a measure of the roughness.

We will show below how the extrapolation of the observed advancing and retreating $\cos\theta$ values can be used to obtain a measure of the Wenzel roughness ratio.

The critical surface tension for wetting, γ_c, should also be a function of the roughness. From equations (3) and (5), with $\cos\theta = 1$,

$$\gamma_c^{LW} = \gamma_s^{LW}\left(\frac{2r}{1+r}\right)^2, \tag{6}$$

provided that (as is commonly the case) the critical wetting liquid is apolar. If the true value of γ_s^{LW} has been determined, using a molecularly smooth sample, then for a rough sample an *apparent* critical surface tension, γ_c^*, may be determined using the contact angle of a series of liquids. The roughness may be obtained by means of equation (7), which is equation (5) for $\theta_W = 0$:

$$r = \left[-1 + 2\sqrt{\gamma_s^{LW}}\left(\frac{\sqrt{\gamma_c^{LW*}}}{\gamma_l}\right)\right]^{-1}. \tag{7}$$

2. HYSTERESIS THEORY

In 1952 [1], it was proposed that hysteresis on a chemically homogeneous, rigid solid could be explained as due to the free energy of activation, F_a or F_r, required in order for a liquid front to move from one metastable configuration to another while a constant *local* angle at the solid surface is maintained. Equation (8) is a slightly modified form of the equation in ref. [1]:

$$\gamma_l \cos\theta_a = r(\gamma_s - \gamma_{sl}) - F_a \tag{8a}$$

$$\gamma_l \cos\theta_r = r(\gamma_s - \gamma_{sl}) - F_r \tag{8b}$$

Figure 1 shows a model surface with a sinusoidal cross-section, and indicates the location of the liquid front such that the local angle is the Young angle and the shape of the liquid surface (e.g. for a drop whose volume is constant) is determined by the constancy of the Laplace curvature $(1/\rho_1 + 1/\rho_2)$, where the ρ's are principal radii of curvature, over the whole liquid surface. When the drop front is not at a (metastable) equilibrium location, the curvature must depart from constancy, and this will cost energy, e.g. F_a.

Johnson and Dettre [12] used a model of a surface having ridges in concentric rings, with a sinusoidal profile. A drop of liquid is placed at the center, and liquid is added or withdrawn so that the diameter changes slowly. The free energy of activation is due to the local increase (or decrease) in area of the liquid surface. Consider the liquid to be a drop whose volume is constant. If the liquid surface is a section of a sphere when the front is in a metastable configuration, then a small motion (say,

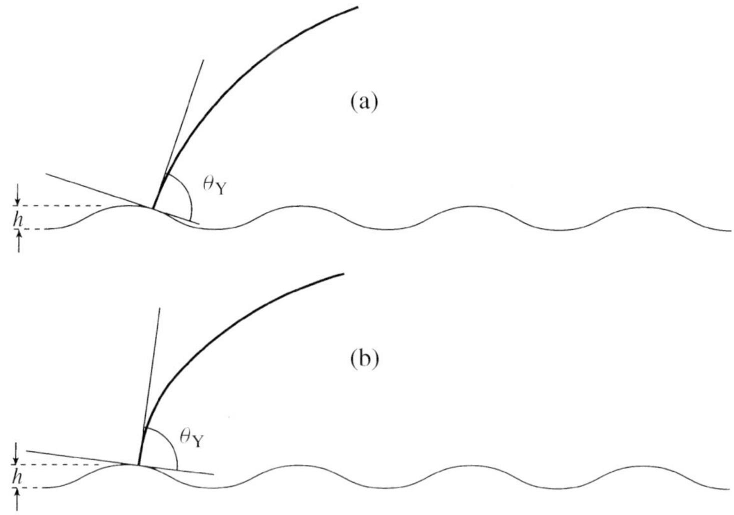

Figure 1. (a) A model surface with a liquid front at a metastable location. (b) Liquid front at an unstable location, to the left of the location in (a). Note the smaller radius of curvature near the solid.

from right to left) will require that the liquid/vapor area increase (or decrease) before returning to a minimized area when a new metastable configuration is reached.

If, instead of concentric rings, the surface consists of random hills and valleys, then in a metastable state the liquid front will wander across the surface in such a way as to keep the *local* contact angle as nearly constant as possible, with the free energy being minimized by allowing local contortion of the liquid surface near the solid, together with very local fluctuations of the contact angle away from the Young angle.

The magnitudes of the energy barriers should be a function of the surface tension of the liquid, as well as of the roughness ratio. For the advancing angle, we have hypothesized that

$$F_a = \gamma_l f_1(r). \tag{9a}$$

It should not be expected that F_r will be equal to F_a in general. It became evident from the experimental results that the form of the energy barrier in the retreating case, F_r, must be more complex than that of F_a. A possible form for F_r is

$$F_r = \gamma_l f_2(\gamma_l, r). \tag{9b}$$

Then equations (8a) and (8b) become

$$\gamma_l \cos\theta_a = r(\gamma_s - \gamma_{sl}) - \gamma_l f_1(r) \tag{10a}$$

$$\gamma_l \cos\theta_r = r(\gamma_s - \gamma_{sl}) - \gamma_l f_2(\gamma_l, r). \tag{10b}$$

The Wenzel angle is now taken, provisionally, to be the advancing angle, θ_a, or the retreating angle, θ_r, depending on the direction of the liquid front motion. Equation (8a) combined with equation (5) yields

$$\cos\theta_{\mathrm{a}} = -r + r\sqrt{\gamma_{\mathrm{s}}^{\mathrm{LW}}}\left(\frac{\sqrt{\gamma_l^{\mathrm{LW}}}}{\gamma_l}\right) - \gamma_l f_1(r) \tag{11a}$$

$$-r = \lim_{\frac{1}{\gamma_l}\to 0}\left[\cos\theta_{\mathrm{a}} + f_1(r)\right] \tag{12a}$$

$$\gamma_{\mathrm{c}}^{\mathrm{LW}} = \gamma_{\mathrm{s}}^{\mathrm{LW}}\left(\frac{2}{1 + r + f_1(r)}\right)^2. \tag{12b}$$

Also,

$$\cos\theta_{\mathrm{r}} = -r + r\sqrt{\gamma_{\mathrm{s}}^{\mathrm{LW}}}\left(\frac{\sqrt{\gamma_l^{\mathrm{LW}}}}{\gamma_l}\right) + \gamma_l f_2(\gamma_l, r) \tag{11b}$$

$$-r = \lim_{\frac{1}{\gamma_l}\to 0}\left[\cos\theta_{\mathrm{r}} - f_2(\gamma_l, r)\right] \tag{12c}$$

$$\gamma_{\mathrm{c}}^{\mathrm{LW}} = \gamma_{\mathrm{s}}^{\mathrm{LW}}\left(\frac{2}{1 + r + f_2(\gamma_l, r)}\right)^2. \tag{12d}$$

Thus, γ_{c} for a solid is a function of the roughness.

Now a data-handling method, for advancing angles, is to plot $\cos\theta_{\mathrm{a}}$ vs. $\sqrt{\gamma_l^{\mathrm{LW}}}/\gamma_l$ and extrapolate the regression line to the limit at the line, $\sqrt{\gamma_l^{\mathrm{LW}}}/l \to 0$. If the intercept is less than -1, denote it by r^*:

$$r = r^* - f_1(r), \tag{13}$$

$$f_1(r) = -\cos\theta_{\mathrm{a}} - r + 2r\sqrt{\gamma_{\mathrm{s}}^{\mathrm{LW}}}\left(\frac{\sqrt{\gamma_l^{\mathrm{LW}}}}{\gamma_l}\right). \tag{14}$$

For this purpose, the value of $\gamma_{\mathrm{s}}^{\mathrm{LW}}$ is needed. It can be determined using a sample of the solid that is actually smooth, with an apolar liquid such as CH_2I_2, and the following equation [based on equation (4)]:

$$\gamma_{\mathrm{s}}^{\mathrm{LW}} \cong \frac{\gamma_l^{\mathrm{LW}}(1 + \cos\theta_{\mathrm{a}})^2}{4}. \tag{15}$$

If a graph of $\cos\theta_{\mathrm{a}}$ vs. $\sqrt{\gamma_l^{\mathrm{LW}}}/\gamma_l$ for the smoothest available sample has indicated appreciable (but not too serious) roughness, e.g. according to equations (11a) and (12a), ignoring $f_1(r)$, then a second approximation can be estimated using

$$\gamma_s^{LW} = \gamma_l^{LW} \frac{\left[1 + \cos\theta_a + f_1(r)\right]^2}{4}. \tag{16}$$

In principle, the same general data-handling method applied to retreating angles should yield the same value of r as the advancing angle data. This turned out not to be the case.

Finally, we can compare samples of a particular solid that have the same surface composition, using the smoothest available surface as a reference sample. We write for the relative roughness, R,

$$R = \frac{\cos\theta_{rough}}{\cos\theta_{ref}} = \frac{r}{r_{ref}} \tag{17}$$

and replacing r by R in all the above equations. For the reference surface, $R_{ref} \equiv 1$.

3. EXPERIMENTAL

3.1. Materials

Teflon FEP sheet from DuPont was used as received (after rinsing with a light hydrocarbon) for the 'smooth' solid, and it was roughened by random abrasion with SiC paper of various grit sizes. Lucite PMMA sheet, also from DuPont, was used as received. The liquids employed were triply distilled water; formamide, 99% spectrophotometric grade, Aldrich Chemical Co.; ethylene glycol, $>99\%$, spectrophotometric grade, Aldrich Chemical Co.; diiodomethane, 99%, Aldrich Chemical Co.; bromobenzene, ACS certified grade, Fisher Scientific Co.; benzene, Spectro ACS grade, Eastman Chemical Co.; n-octane, $>99\%$, Aldrich Chemical Co.; activated alumina, Fisher Scientific Co.; and activated carbon, Darco G-60, 100 mesh, Aldrich Chemical Co.

The diiodomethane was purified with activated carbon and stored over pure copper shot in a darkened container. The rest of the organic liquids were purified by treatment with activated alumina.

3.2. Equipment

A Ramé Hart contact angle goniometer, model 100-00, modified by the insertion of a zoom lens (for control of magnification without loss of focus) was used. Images could be recorded with a system consisting of a Panasonic video monitor, model WV-CM110A, and a Mitsubishi video printer, model P71U. The drops were delivered onto the solid samples with a micrometer syringe, from Fisher Scientific Co., with a stainless steel needle. The needle tip was used to hold the liquid as a 'captive drop', with the needle well centered. In measuring the advancing angle θ_a and the retreating angle θ_r, it was observed (on examination at higher magnification) that after stopping the addition or withdrawal of liquid, the three-phase line continued in slight but detectable motion for as much as 1 min with ethylene glycol. With the lower-viscosity liquids, a wait of 30 s before recording a measurement appeared to be satisfactory.

4. RESULTS

Tables 1 and 2 show the measured contact angles on Teflon FEP. Figure 2 was part of an early series of measurements. The Teflon FEP was from a different sample from that used in the rest of the study. The figure shows the graph of $\cos\theta_{\mathrm{a}}$ and $\cos\theta_{\mathrm{r}}$ for liquids on Teflon FEP vs. $\sqrt{\gamma_l^{\mathrm{LW}}/\gamma_l}$ for the smoothest Teflon that was available. The extrapolated value of $\cos\theta_{\mathrm{a}}$, for abscissa $= 0$, was -1.05. The extrapolated value of $\cos\theta_{\mathrm{r}}$ was -0.7. The latter value is clearly not equal to $\cos\theta_{\mathrm{a}}$. The extrapolated γ_{c} is 19.0 ± 3 mJ/m^2 based both on θ_{a} and θ_{r}.

Figure 3 shows the $\cos\theta_{\mathrm{a}}$ results for 'smooth' Teflon FEP, second series of measurements.* The estimate of $\cos\theta$ for $1/\gamma_l = 0$ is 1.07 ± 0.15, in excellent agreement with Fig. 2. The error band was set at 95% confidence limits. The estimate of γ_{c} is 18.7 ± 3.0 mJ/m^2, in excellent agreement with the earlier work. Figures 4, 5, and 6 show the curves of $\cos\theta_{\mathrm{a}}$ vs. $\sqrt{\gamma_l^{\mathrm{LW}}/\gamma_l}$ for surfaces that had been roughened with 4000, 1000, and 600 grit papers, respectively.

Table 1.

Advancing contact angles (in degrees) of some liquids on the Teflon FEP surface roughened with SiC paper of various grit sizes

Liquid		Teflon FEP surface roughened with:		
		4000 grit	1000 grit	600 grit
Water	118.6 ± 1.9	125.8 ± 2.3	140.4 ± 3.7	155.6 ± 4.0
Formamide	102.7 ± 1.7	103.9 ± 2.3	120.2 ± 3.6	133.7 ± 3.0
Ethylene glycol	94.9 ± 0.8	93.8 ± 2.7	115.8 ± 3.0	132.2 ± 2.9
Diiodomethane	80.9 ± 1.1	95.7 ± 1.5	105.0 ± 2.6	112.4 ± 2.4
Bromobenzene	70.2 ± 0.5	75.4 ± 2.4	72.2 ± 2.9	80.0 ± 1.6
Benzene	58.0 ± 3.3	52.7 ± 3.0	54.5 ± 1.5	59.8 ± 1.8
Octane	34.6 ± 0.4	34.4 ± 1.7	0.0	0.0

Table 2.

Receding contact angles (in degrees) of some liquids on the Teflon FEP surface roughened with SiC paper of various grit sizes

Liquid		Teflon FEP surface roughened with:		
		4000 grit	1000 grit	600 grit
Water	105.2 ± 1.3	97.6 ± 1.0	93.8 ± 2.3	128.7 ± 3.7
Formamide	86.5 ± 0.3	74.8 ± 1.7	76.3 ± 3.5	56.6 ± 2.2
Ethylene glycol	79.8 ± 0.6	65.6 ± 1.1	45.0 ± 1.1	28.5 ± 5.9
Diiodomethane	78.8 ± 0.7	70.5 ± 2.3	45.9 ± 3.1	0.0
Bromobenzene	58.1 ± 1.3	48.0 ± 0.9	0.0	0.0
Benzene	46.4 ± 1.7	31.9 ± 3.4	0.0	0.0
Octane	19.6 ± 0.9	9.9 ± 0.5	0.0	0.0

*The first series was conducted by M. K. Chaudhury and the second series by C. Yeung.

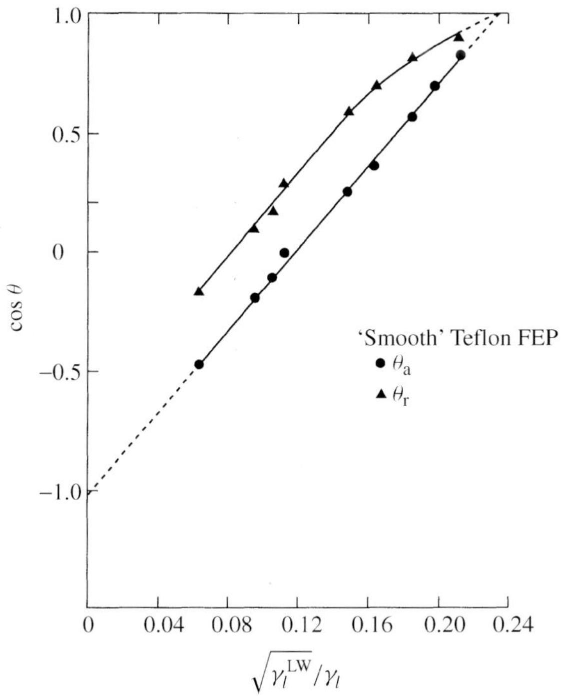

Figure 2. $\cos\theta_a$ and $\cos\theta_r$ for liquids on the 'smoothest' Teflon FEP, first series.

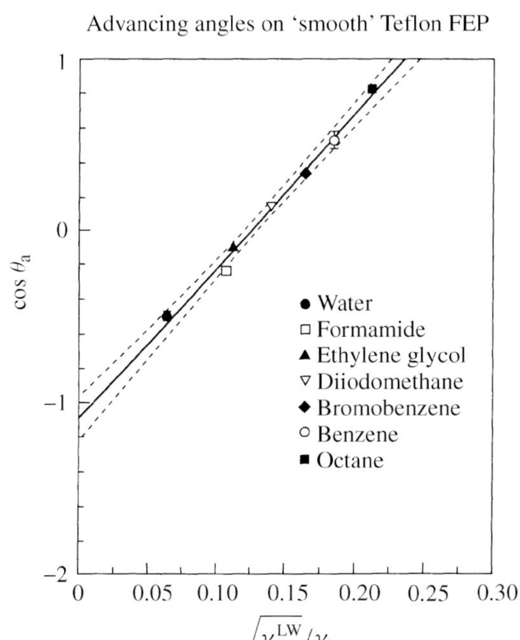

Figure 3. $\cos\theta_a$ for liquids on the 'smoothest' Teflon FEP, second series. The error band corresponds to 95% confidence limit.

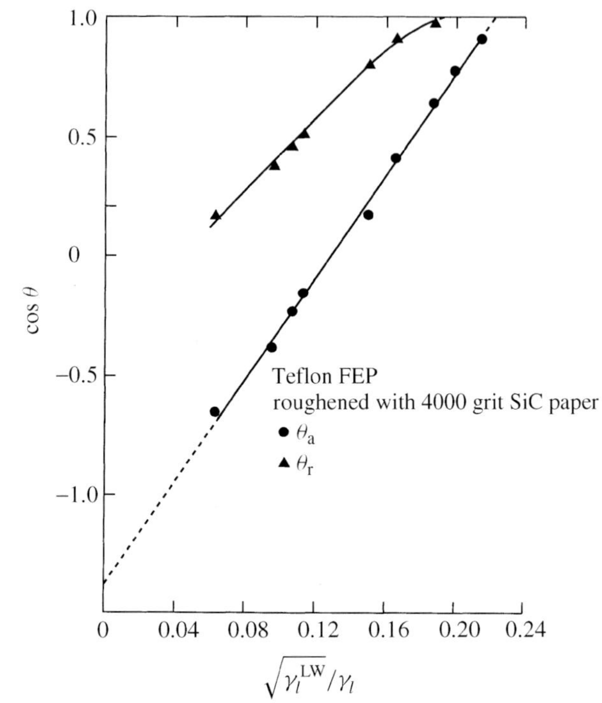

Figure 4. $\cos\theta_a$ and $\cos\theta_r$ for liquids on Teflon FEP roughened with 4000 grit SiC paper.

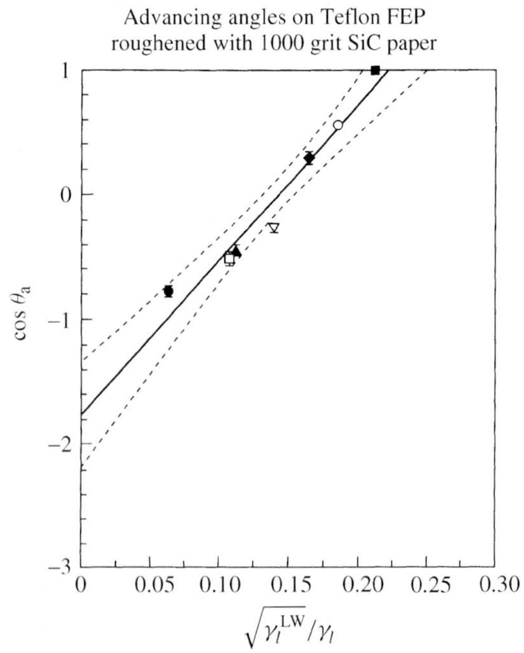

Figure 5. $\cos\theta_a$ for liquids on Teflon FEP roughened with 1000 grit SiC paper.

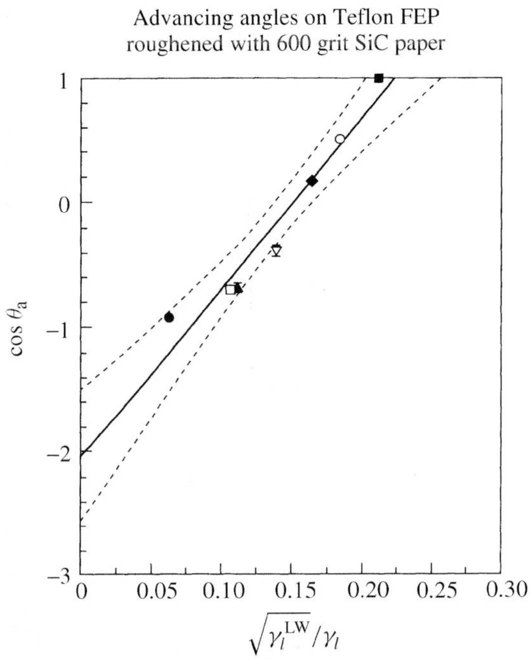

Figure 6. $\cos\theta_a$ for liquids on Teflon FEP roughened with 600 grit SiC paper.

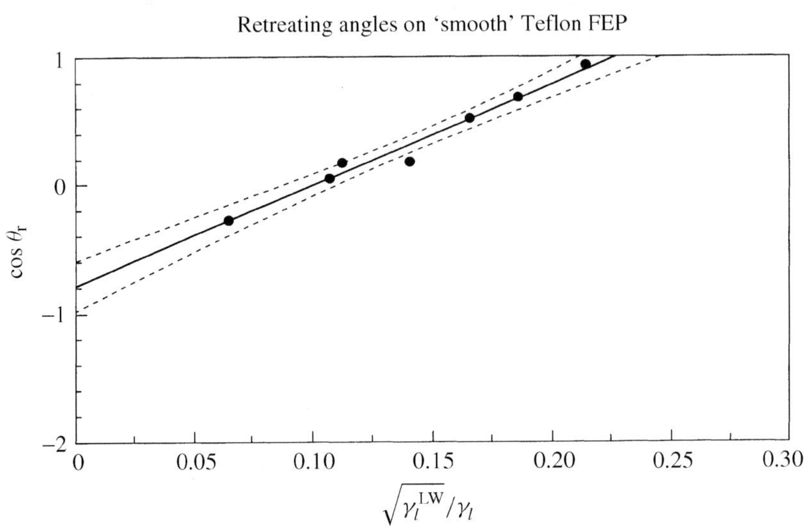

Figure 7. $\cos\theta_r$ for liquids on the 'smoothest' Teflon FEP.

Figures 7 and 8 show $\cos\theta_r$ for the smoothest Teflon and the 4000 grit-roughened solid. The $1/\gamma_l \to 0$ intercept is -0.8 ± 0.2 and -0.5 ± 0.3, both of which are physically impossible as a measure of r. So we reach the tentative conclusion that the theory, i.e. equations (9)–(14), may not apply to retreating angles. This conclusion

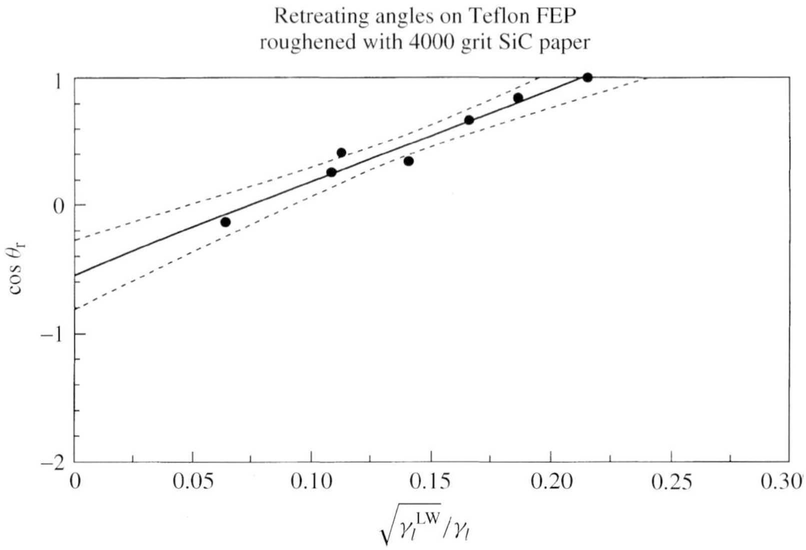

Figure 8. $\cos\theta_r$ for liquids on Teflon FEP roughened with 4000 grit SiC paper.

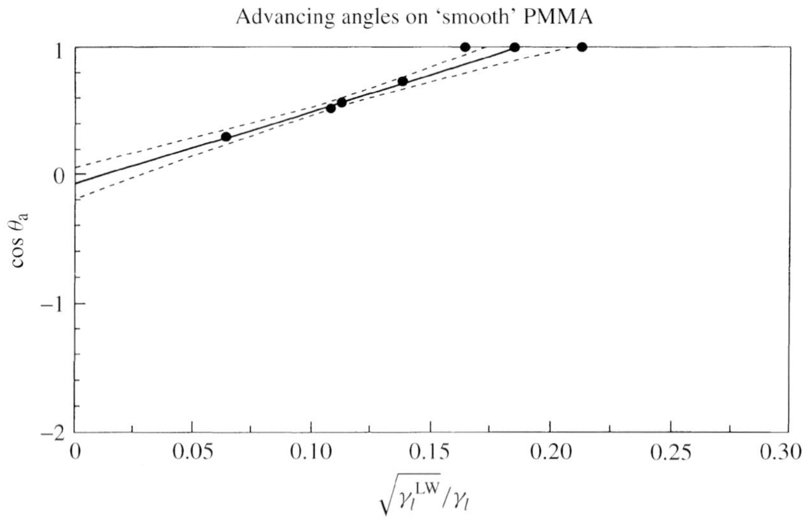

Figure 9. $\cos\theta_a$ for liquids on 'smooth' PMMA.

is supported by the fact that with retreating angles on Teflon FEP roughened with 1000 grit and 600 grit paper, γ_c was 37 mJ/m^2 on the former and 56 mJ/m^2 on the latter.

A series of θ_a measurements on roughened PMMA were also made (see Fig. 9 for the smoothest surface). The γ_c value was 37 ± 10 mJ/m^2, in fair agreement with Zisman's value of 39 mJ/m^2 [13]. The intercept at $1/\gamma_l \to 0$ was at -0.1 ± 0.15. For PMMA roughened on 4000 grit paper (Fig. 10), γ_c remained the same within experimental error, but the $\gamma^{-1} = 0$ intercept was -0.1.

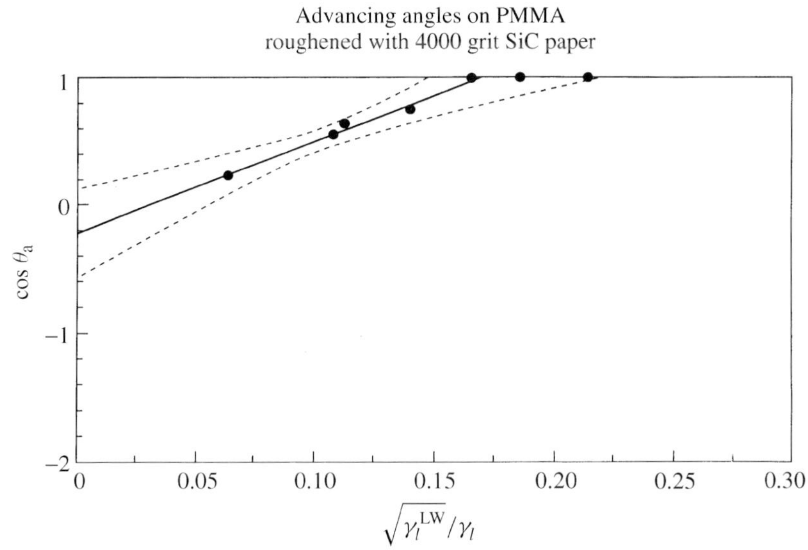

Figure 10. $\cos\theta_a$ for liquids on PMMA roughened with 4000 grit SiC paper.

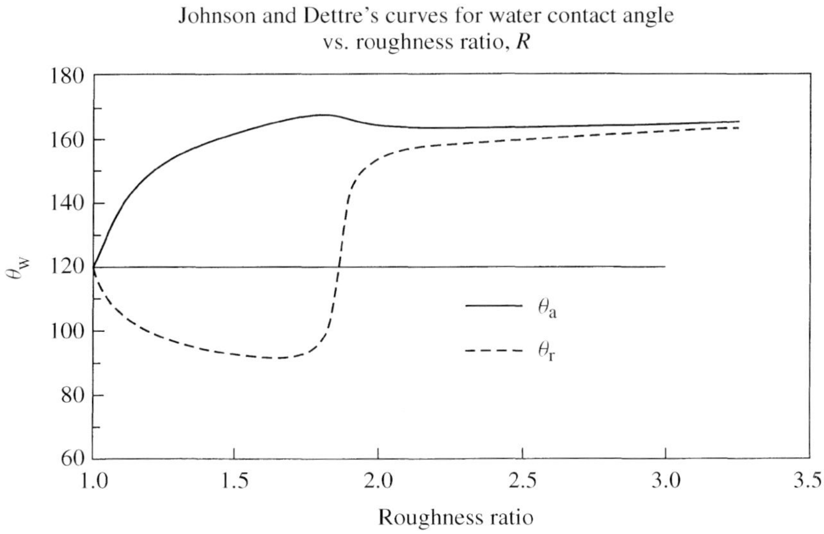

Figure 11. Johnson and Dettre's computations for the contact angle of water vs. the roughness ratio.

The combination of the Young–Wenzel equation with equation (4a) for a monopolar basic solid (such as PMMA) and a polar liquid is

$$\cos\theta_W = r\left[1 - 2\sqrt{\gamma_s^{LW}}\left(\frac{\sqrt{\gamma_l^{LW}}}{\gamma_l}\right) - 2\sqrt{\gamma_s^{\ominus}}\left(\frac{\sqrt{\gamma_l^{\oplus}}}{\gamma_l}\right)\right]. \tag{18}$$

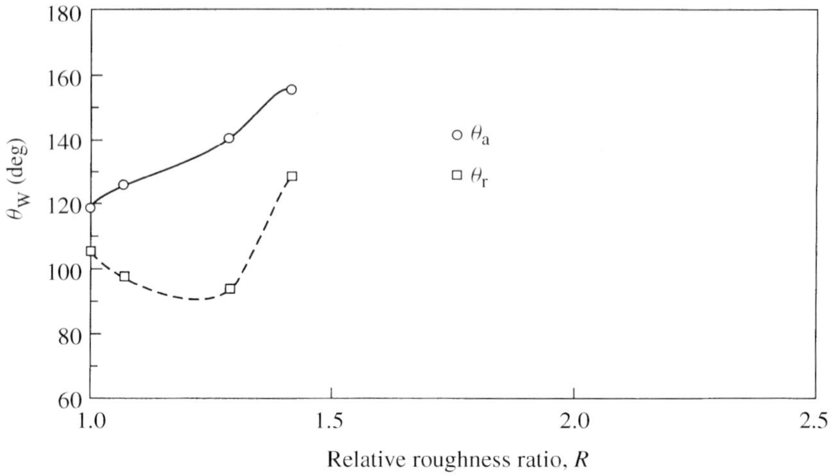

Figure 12. Observed contact angle of water on Teflon FEP vs. the relative roughness ratio R.

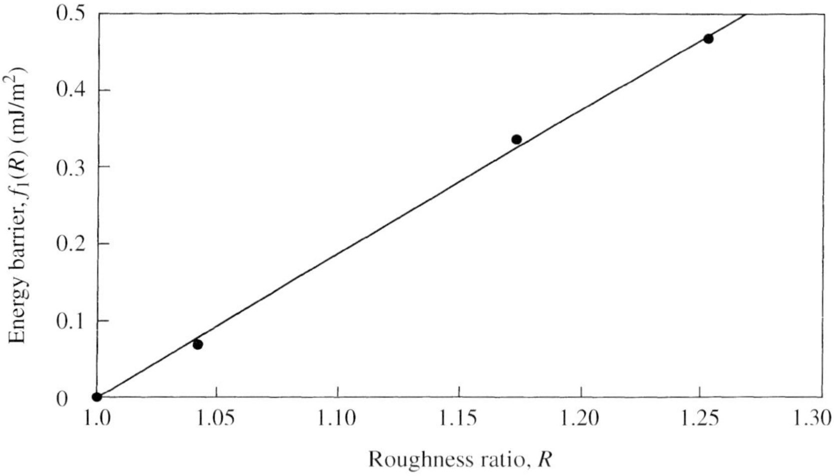

Figure 13. Energy barrier for liquid advance vs. the roughness ratio R on Teflon FEP.

It would not be expected, on the basis of equation (18), that the simple extrapolation of $\cos\theta$ vs. $\sqrt{\gamma_s^{LW}}/\gamma_l$ would yield very useful information.

Johnson and Dettre [12] computed the curve of advancing and retreating contact angles vs. the roughness ratio for their model surface (see Fig. 11). The sharp rise in $\cos\theta_r$, at about $r = 1.8$, was said to be due to the entrapment of air in the grooves on the surface. Our experimental results are shown in Fig. 12, which is in qualitative agreement with Fig. 11. The rise in $\cos\theta_r$, around $r = 1.4$, does correspond to the entrapment of air.

Table 3 shows the numerical results for roughened Teflon FEP. In this table, the value of 17.0 for γ_s^{LW} is probably valid because the roughness term ($r^* = 1.07$) is

Table 3.
Results for roughened Teflon FEP

Grit size	Intercept r^*	γ_c^{LW} (mJ/m^2) (apparent)	γ_s^{LW} (mJ/m^2) (uncorrected)	R	$f_1(R)$ (mJ/m^2)
—	1.07 ± 0.15	17.6 ± 3	17.0	[1.0]	[0.0]
4000	1.23 ± 0.4	17.6 ± 4	15.7	1.07	0.070
1000	1.76 ± 0.6	19 ± 4	12.4	1.29	0.34
600	2.05 ± 0.75	21 ± 4	10.7	1.42	0.47

Figure 14. Scanning electron micrograph of 'smooth' Teflon FEP. Magnification × 500.

close to 1. The apparent γ_c varies with the roughness. The values of γ_s^{LW} in column 4 are less than 17.0 mJ/m^2, and do not represent the smooth solid. These trends agree with expectations.

Figure 13 shows the energy barrier, $f_1(R)$, as a function of R from our data.

Figure 15. Scanning electron micrograph of Teflon FEP roughened with 4000 grit SiC paper.

We do not report the calculations based on the retreating angles. Values of r^* that were less than -1 indicated that the analysis was not applicable. We have not treated the data on PMMA, or other polar solids, for similar reasons.

Figures 14–16 show scanning electron micrographs of three of the surfaces that were studied.

5. CONCLUSIONS

We have developed and tested a method of determining the roughness of apolar solids from advancing contact angles. The method is validated for moderate roughness, e.g. up to a Wenzel roughness ratio of 1.4. The retreating angle, θ_r, appears to be less directly related to the Wenzel roughness than is θ_a. The energy barrier for liquid front advancing is about 0.5 mJ/m^2 for Teflon that has been abraded with 600 grit SiC paper.

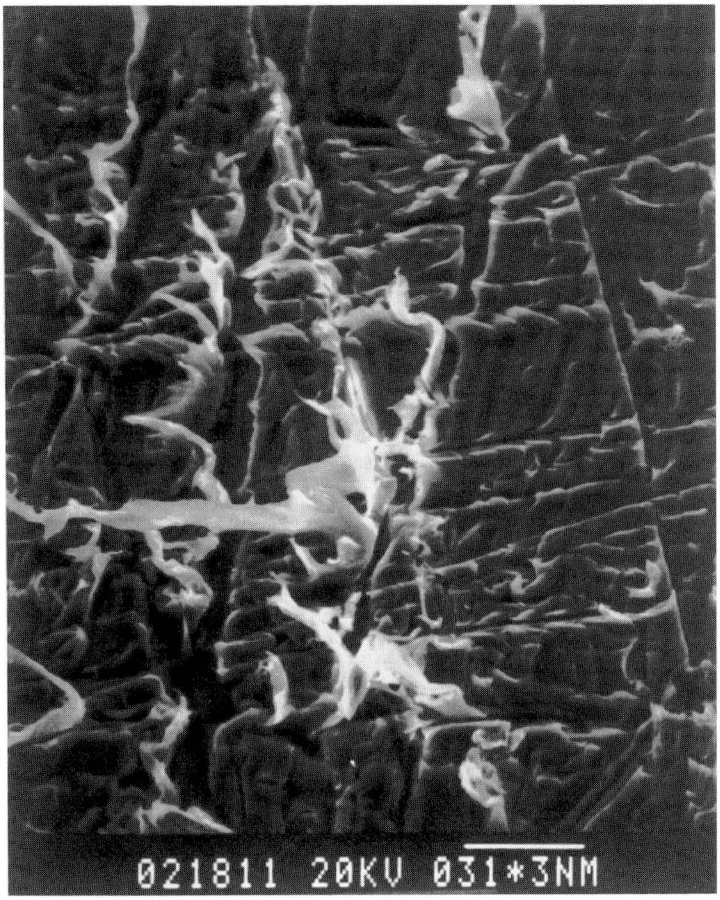

Figure 16. Scanning electron micrograph of Teflon FEP roughened with 1000 grit SiC paper. Magnification × 500.

Acknowledgements

We thank the National Science Foundation for support of M. K. Chaudhury, and the Industry/University Center for Biosurfaces at the State University of New York at Buffalo for the support of C. Yeung. This work constituted part of the Ph.D. thesis of Dr. Chaudhury (1986) and of the MS thesis of Mr. Yeung (1995).

REFERENCES

1. R. J. Good, *J. Am. Chem. Soc.* **74**, 5041 (1952).
2. R. J. Good, in: *Surface and Colloid Science*, Vol. 11, R. J. Good and R. R. Stromberg (Eds), pp. 1–29. Marcel Dekker, New York (1979).
3. R. J. Good, *J. Adhesion Sci. Technol.* **6**, 1269 (1992).
4. (a) R. N. Wenzel, *Ind. Eng. Chem.* **28**, 988 (1936);
   (b) R. N. Wenzel, *J. Phys. Colloid Chem.* **53**, 1466 (1949).
5. R. J. Good and L. A. Girifalco, *J. Phys. Chem.* **64**, 561 (1960).

6. C. J. van Oss, M. K. Chaudhury and R. J. Good, *Separation Sci. Technol.* **22**, 1 (1987).
7. C. J. van Oss, M. K. Chaudhury and R. J. Good, *Chem. Rev.* **88**, 927 (1988).
8. R. J. Good, M. K. Chaudhury and C. J. van Oss, in: *Fundamentals of Adhesion*, L. H. Lee (Ed.), p. 153. Plenum Press, New York (1991).
9. H. W. Fox and W. A. Zisman, *J. Colloid Interface Sci.* **7**, 428 (1952).
10. (a) F. M. Fowkes, *Ind. Eng. Chem.* **56** (12), 40 (1964);
(b) F. M. Fowkes and M. A. Mostafa, *Ind. Eng. Chem. Prod. Res. Dev.* **17**, 3 (1980).
11. R. J. Good, *J. Colloid Interface Sci.* **59**, 398 (1977).
12. (a) R. E. Johnson, Jr. and R. H. Dettre, in: *Contact Angle, Wettability and Adhesion*, Adv. Chem. Ser. No. 43, 112. American Chemical Society, Washington, DC (1964);
(b) R. H. Dettre and R. E. Johnson, Jr., in: *Contact Angle, Wettability and Adhesion*, Adv. Chem. Ser. No. 43, p. 136. American Chemical Society, Washington, DC (1964).
13. W. A. Zisman, *Ind. Eng. Chem.* **53** (10), 18–38 (1963).

Mittal Festschrift, pp. 199–237
W. J. Van Ooij and H. R. Anderson, Jr. (Eds)

Adhesion failure and dewetting of thin liquid films

ASHUTOSH SHARMA,[1,*] RAJESH KHANNA[1] and GÜNTER REITER[2]
[1]*Department of Chemical Engineering, Indian Institute of Technology at Kanpur, Kanpur-208016, India*
[2]*Institut de Chimie des Surfaces et Interfaces, CNRS, 15 rue Jean Starcky, 68057 Mulhouse, France*

Abstract—Theory, numerical simulations, and experiments on the stability, dynamics, morphology and dewetting of thin (< 100 nm) films on variously modified solid surfaces are presented. Results for two distinct types of thin films are presented in greater details: (1) films subjected to the long range van der Waals attraction and extremely short range Born repulsion, and (2) films subjected to the long range attraction combined with shorter range repulsion. In the first case, true dewetting (adhesion failure) occurs in the form of growing holes which eventually coalesce to form a giant polygonal structure. The number density, growth rates and the morphology of rims are characterized for growing holes. In the second case, saturation of the initial instability leads to a quasi-stable interconnected microstructure consisting of an array of small droplets in equilibrium with largely flat thin regions. This 'morphologically phase separated' (MPS) structure slowly coarsens by ripening (merger) of drops due to spatial gradients of curvature and disjoining pressure. Nonadsorbing polystyrene (PS) films on variously modified nonwettable silicon wafers, and on densely grafted PS 'brushes', show true dewetting. In contrast, PS and PS–PEO (a block copolymer) films adsorbed on UV-ozone treated wafers show the MPS and ripening.

Keywords: Thin fluid films; dynamics; dewetting, morphology; surface properties.

1. INTRODUCTION

Adhesion failure secondary to dewetting of thin (< 100 nm) fluid or fluidlike (e.g. viscoelastic) films on solid surfaces is usually initiated by a spontaneous deformation of the film surface, which is engendered by the excess intermolecular interactions (conjoining pressure) [1–7]. The conjoining pressure (negative of disjoining pressure, Π) is simply the excess pressure in a thin film as compared to the pressure in the corresponding bulk phase. Thus, the free surface of a thin film deforms spontaneously due to the flow from thinner to thicker regions whenever the intermolecular force field is such that the conjoining pressure increases with a decrease in the local film thickness ($-\mathrm{d}\Pi/\mathrm{d}h < 0$).

Depending on the nature of the (nonlinear) force field in the film, the evolution of the initial surface instability leads to one of the following two sequence of events: (A) formation, growth and coalescence of circular holes on the substrate leading to a

*To whom correspondence should be addressed.

true adhesion failure and dewetting of the film [8–12]; or (B) saturation of the initial instability leading to the formation of a (quasi) stable interconnected microstructure consisting of an array of small drops in equilibrium with thin films [13–15]. The latter scenario (roughening of the film), which can be thought of as a 'morphological phase separation' (MPS) [16–17] rather than a true dewetting, should usually occur in the presence of a short range repulsion in addition to the long range attraction causing the initial destabilization [7, 16, 17]. In some cases, the initial roughening (MPS) of the film is followed by the appearance of isolated circular holes [15].

While the conditions for the film instability and its initial stages of evolution are now well understood [1–7, 16–18], the long term evolution of the instability and details of the resulting microstructure have been rather difficult to analyze, because of several theoretical and computational problems. For example, in the case of true adhesion-failure, the initial formation of a hole is followed by the appearance of a rapidly retracting true three-phase contact line [8–12], where the thin film analysis breaks down due to the divergence of the stresses. It is possible theoretically [19] to remove the (nonphysical) singularity at the contact line by including a short range repulsion (e.g. Born repulsion), but it results in an extremely 'stiff' equation due to widely separated length (and time) scales, slopes and curvatures. This difficulty has so far prevented direct simulations of the dynamics of the hole-growth.

Another interesting but poorly understood problem concerns the fate of morphologically phase separated microstructures [16, 17], consisting of micro-drops joined by ultrathin nano-scale films. A slow coarsening (ripening) of the microstructure may be anticipated [11] due to the propensity of flow from smaller to larger drops because of Laplace pressure. However, this phenomenon has not been studied by direct hydrodynamic simulations, which would require a very long term tracking of a collection of drops.

The objective of this paper is to provide the theoretical and experimental understanding of various stages of dewetting and morphological phase separation in thin films on uncoated and coated substrates. The patterns of evolution of instability are related to the behavior of disjoining pressure isotherms. Ultrathin coatings are often applied to tailor the substrate properties including adhesion and wettability [8–12, 20–22]. Even in the absence of externally applied coatings, few substrates are really homogeneous right up to their surfaces. Surface reactions (Si $\rightarrow$ SiO_2, UV-ozone treatment, hydration, etc.), and contamination can give rise to pseudo-coatings with surface properties vastly different from that of the the underlying bulk substrate. Some fascinating aspects of wetting of coated (stratified) substrates by large drops have been studied previously [20, 23]. Here, the role of thin surface coatings in various stages of the *thin film instability* and *dewetting* are explored.

2. EVOLUTION OF INSTABILTY: THEORY

2.1. Thin film equation

The following 'long-wave' equation [4–6, 16] describes the 2-D dynamics and shape of a thin Newtonian fluid film on a solid (gravity effects and evaporation/condensation are neglected):

$$3\mu\frac{\partial h}{\partial t}+\frac{\partial}{\partial x}\left[\gamma_{\mathrm{f}} h^{3}\frac{\partial^{3}h}{\partial x^{3}}\right]+\frac{\partial}{\partial x}\left[h^{3}\frac{\mathrm{d}\Pi}{\mathrm{d}h}\frac{\partial h}{\partial x}\right]=0, \tag{1}$$

where $h(x,t)$ is the local film thickness, coordinate x is parallel to the substrate surface, t is time, μ is viscosity, γ_{f} is surface tension of the film (f), and Π is the disjoining pressure of the film which is also related to the total excess free energy (ΔG) of the film per unit area due to various types of intermolecular interactions:

$$\Pi=-\left(\frac{\partial\Delta G}{\partial h}\right). \tag{2}$$

For tangentially immobile films [5], the viscous coefficient 3 in equation (1) is replaced by 12. Equation (1) is the sum of three forces (per unit width of the film), which are (from left to right in equation (1)): viscous force, surface tension force due to the curvature at the free interface, and the sum of excess intermolecular forces. The *viscous* force merely *retards* the growth of instability, whereas the *interfacial force* (surface tension effect) has a *stabilizing* influence until the film ruptures. After the appearance of a hole, however, the surface tension assists in the growth of the hole. Finally, negative diffusivity ($\mathrm{d}\Pi/\mathrm{d}h > 0$) can engender the interfacial instability (spinodal decomposition) manifested as the growth of thicker regions of the film at the expense of the thinner regions. This is also verified by the solution of equation (1) after linearization around the mean thickness (h_0), viz.

$$h=h_0+\varepsilon\exp(\iota kx+\omega t),\quad \iota^2=-1, \tag{3}$$

which gives the linear dispersion relation [1–7] for the initial growth coefficient (ω) of a small amplitude ($\varepsilon \ll h_0$) disturbance with wavenumber k as

$$\omega=\left(\frac{h_0^3k^2}{3\mu}\right)\left[\left(\frac{\partial\Pi}{\partial h}\right)-\gamma_{\mathrm{f}}k^2\right]. \tag{4}$$

Thus, $\mathrm{d}\Pi/\mathrm{d}h > 0$ is the necessary condition for the spinodal decomposition ($\omega > 0$), and the sufficient condition is the presence of a long mode in the disturbance with wavenumber, $k^2 < (1/\gamma_{\mathrm{f}})(\partial\Pi/\partial h)$. The wavelength of instability is given by $\lambda = 2\pi/k$. The length scale of the fastest growing (dominant) initial wave in the linear approximation is obtained from $(\partial\omega/\partial k)=0$, which gives

$$\lambda_{\mathrm{m}}=2\sqrt{2}\pi\left[\gamma_{\mathrm{f}}\left(\frac{\partial\Pi}{\partial h}\right)^{-1}\right]^{1/2}. \tag{5}$$

2.2. Disjoining pressure and its relation to film stability

In the absence of electrical double layer effects, the total excess free energy (and disjoining pressure from equation (2) is usually the sum of the long range apolar Lifshitz–van der Waals (LW) interactions and the shorter range polar (e.g. acid–base) or other non-LW interactions. The non-LW interactions [24–26] are significant for hydrogen bonding substances like aqueous films [7, 16–18, 24–26] and for polymer films on grafted or adsorbed polymer chains [27, 28].

The LW interaction energy on a coated substrate is made up of attractive LW interactions among the molecules of the thin film and interactions among the molecules of the film with the coating and the underlying substrate (Fig. 1). The total excess LW interaction energy per unit area can be obtained by pairwise summation of molecular LW potentials (with r^{-6} decay) in the microscopic approach of London and Hamaker, which gives [12, 20, 29]:

$$\Delta G^{\mathrm{LW}} = -\frac{(A_{\mathrm{ff}} - A_{\mathrm{cf}})}{12\pi h^2} - \frac{(A_{\mathrm{cf}} - A_{\mathrm{sf}})}{12\pi (h+\delta)^2}, \tag{6}$$

where δ is the coating thickness, A_{ij} refers to the Hamaker constant, and subscripts f, s and c denote the film, substrate and coating materials, respectively. Clearly, the well known form $\Delta G^{\mathrm{LW}} = -A_{\mathrm{e}}/12\pi h^2$, is recovered in the absence of coating ($\delta = 0$, $A_{\mathrm{e1}} = A_{\mathrm{ff}} - A_{\mathrm{sf}}$), and also when the coating is thick ($\delta \to \infty$, $A_{\mathrm{e2}} = A_{\mathrm{ff}} - A_{\mathrm{cf}}$) and the effect of substrate is thus screened. The effective Hamaker constants of film material on the substrate (A_{e1}) and the coating (A_{e2}) materials are also related to the LW components of the spreading coefficients on the substrate ($S_{\mathrm{s}}^{\mathrm{LW}}$) and the coating ($S_{\mathrm{c}}^{\mathrm{LW}}$), respectively [7, 16–18, 24–26].

$$A_{\mathrm{e1}} = A_{\mathrm{ff}} - A_{\mathrm{sf}} = -12\pi d_0^2 S_{\mathrm{s}}^{\mathrm{LW}}, \tag{7}$$

$$A_{\mathrm{e2}} = A_{\mathrm{ff}} - A_{\mathrm{cf}} = -12\pi d_0^2 S_{\mathrm{c}}^{\mathrm{LW}}, \tag{8}$$

where d_0 is a cut-off distance, which for a large array of condensed phase materials equals 0.158 ± 0.008 nm [24, 25]. The LW component of the spreading coefficient of the film material on a surface ($k =$ s or c) is defined in terms of the LW components of various surface (γ_i^{LW}) and interfacial ($\gamma_{ij}^{\mathrm{LW}}$) tensions as

$$S_k^{\mathrm{LW}} = \gamma_k^{\mathrm{LW}} - \gamma_{\mathrm{f}}^{\mathrm{LW}} - \gamma_{k\mathrm{f}}^{\mathrm{LW}}. \tag{9}$$

While it is more difficult to determine the effective Hamaker constants, especially on ill-defined modified surfaces encountered in practice, the spreading coefficient can be obtained from facile measurements of macroscopic contact angles [16, 24, 25, 29]. In the absence of polar interactions, $S_k^{\mathrm{LW}} > 0$ implies perfect wettability of material k whereas $S_k^{\mathrm{LW}} < 0$ implies partial wetting (finite contact angle).

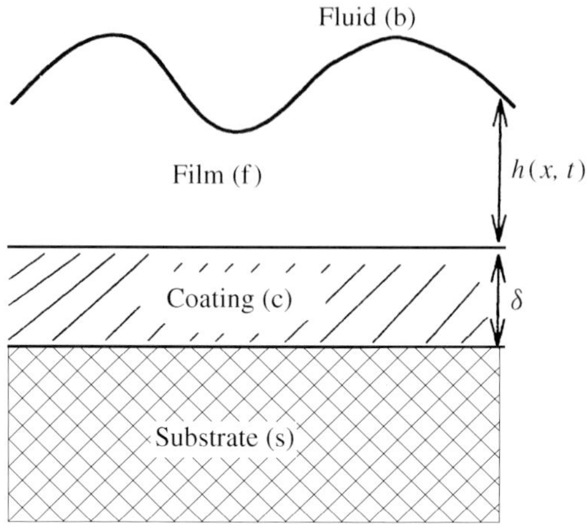

Figure 1. Schematic of a deforming unstable thin film on a coated substrate.

Combining equation (6) with equations (7), (8) gives

$$\Delta G^{\mathrm{LW}} = S_{\mathrm{c}}^{\mathrm{LW}} \frac{d_0^2}{h^2} + \left(S_{\mathrm{s}}^{\mathrm{LW}} - S_{\mathrm{c}}^{\mathrm{LW}}\right) \frac{d_0^2}{(h+\delta)^2}. \tag{10}$$

The LW component of the force (per unit volume) which governs the film stability in equation (1) is, from equation (2):

$$\frac{\mathrm{d}\Pi^{\mathrm{LW}}}{\mathrm{d}h} = -\frac{6S_{\mathrm{c}}^{\mathrm{LW}} d_0^2}{h^4} - 6\left(S_{\mathrm{s}}^{\mathrm{LW}} - S_{\mathrm{c}}^{\mathrm{LW}}\right) \frac{d_0^2}{(h+\delta)^4}. \tag{11}$$

The LW force field on coated substrates remains destabilizing ($\mathrm{d}\Pi/\mathrm{d}h > 0$) for all film thicknesses, when both the coating and the substrate are partially wettable ($S_{\mathrm{c}}^{\mathrm{LW}}, S_{\mathrm{s}}^{\mathrm{LW}} < 0$) [12, 29]. When either the coating or the substrate is perfectly wettable, but the other is not, the opposite signs of $S_{\mathrm{c}}^{\mathrm{LW}}$ and $S_{\mathrm{s}}^{\mathrm{LW}}$ give rise to a combination of attractive and repulsive force fields in the film [20, 23].

The free energy of the polar interactions (when present) usually show exponential decay [7, 16–18, 24, 25, 30].

$$\Delta G^{\mathrm{P}} = S^{\mathrm{P}} \exp\left[\frac{(l_0 - h)}{l_{\mathrm{p}}}\right], \tag{12}$$

where l_0 is the cut-off distance, l_{p} is a correlation length for polar interactions and S^{P} is the polar component of the spreading coefficient of film material on the surface, viz. $\Delta G^{\mathrm{P}} = S^{\mathrm{P}}$ at $h = l_0$ [7, 16]. S^{P} is almost always negative [7, 16, 24, 25] for aqueous films due to a large polar component of the cohesive energy of water which denotes hydrophobic attraction. A known exception [6] is water film on hydrophilic quartz surface, for which a strong 'hydration pressure' or 'hydrophilic repulsion' exists, signifying $S^P > 0$. S^P vanishes for films of apolar liquids and also when the film and the substrate materials have the same type of monopolarity.

In addition to the apolar (LW) and polar (e.g. acid–base) interactions, films of end-adsorbed amorphous polymers exhibit wetting autophobicity [15, 28, 31–35] engendered by the entropic effects associated with the connectivity of polymer chain repeat units [27, 28, 32]. Similarly, roughening as well as true dewetting of polymer films on a grafted brush of an identical or a different polymer are also observed [15, 36]. Simplified calculations of entropic effects [27, 28] show a short range repulsion, which causes a minimum in the free energy curve (see Fig. 2b discussed later in this section). The total free energy in such cases is a combination of the entropic effects and the net LW interactions with the substrate (equation (10)). Finally, in the absence of other short range repulsive forces, the extremely short ranged Born repulsion [24–26] provides a 'cut-off' distance, which also prevents the divergence of the LW interaction at contact.

The total free energy for non-equilibrium (unstable) thin films can display one of the three distinct extended DLVO types of behavior summarized in Fig. 2. The Born repulsion at $h = l_0$ is shown in the hard sphere approximation, since it climbs from zero to infinity very rapidly.

The total free energy of adhesion, $\Delta G(l_0)$ is identical to the total spreading coefficient [7, 16, 34]. Thus, the latter may be written as the sum of as many different components (e.g. S^{LW}, S^P, etc.) as the number of components of the free energy of interaction.

Type I behavior (Fig. 2a) is for a purely attractive force field, e.g. when both the apolar and polar components of the spreading coefficient are negative [20, 29]. In the case of coated substrates, it also implies that both the coating as well as the substrate materials are only partially wetted by the film fluid. The simplest potential which describes the Type I behavior corresponds to an apolar film on an uncoated nonwettable substrate (s)

$$\Delta G = -\frac{A}{12\pi h^2} + \frac{B}{h^8}, \quad A, B > 0. \tag{13}$$

The repulsive term (B/h^8) derives from the r^{-12} decay of intermolecular repulsion in the Lennard-Jones potential. The minimum of the free energy occurs at a distance l_0, where

$$\Pi(l_0) = -\frac{\partial \Delta G}{\partial h} = 0 \quad \text{and} \quad \Delta G(l_0) - \Delta G(\infty) = S^{LW} < 0.$$

Equation (13) can thus also be written as ($S^{LW} < 0$).

$$\Delta G = -|S^{LW}|\left(\frac{d_0^2}{h^2}\right) + \left(\frac{27}{256}\right)|S^{LW}|\left(\frac{d_0}{h}\right)^8, \tag{14}$$

where $d_0^2 = (4/3)l_0^2$, which is in accord with the best estimates [24] of d_0 (= 0.158 nm) and l_0 (= 0.137 nm). Since $(\mathrm{d}\Pi/\mathrm{d}h)$ remains positive for all thicknesses greater than

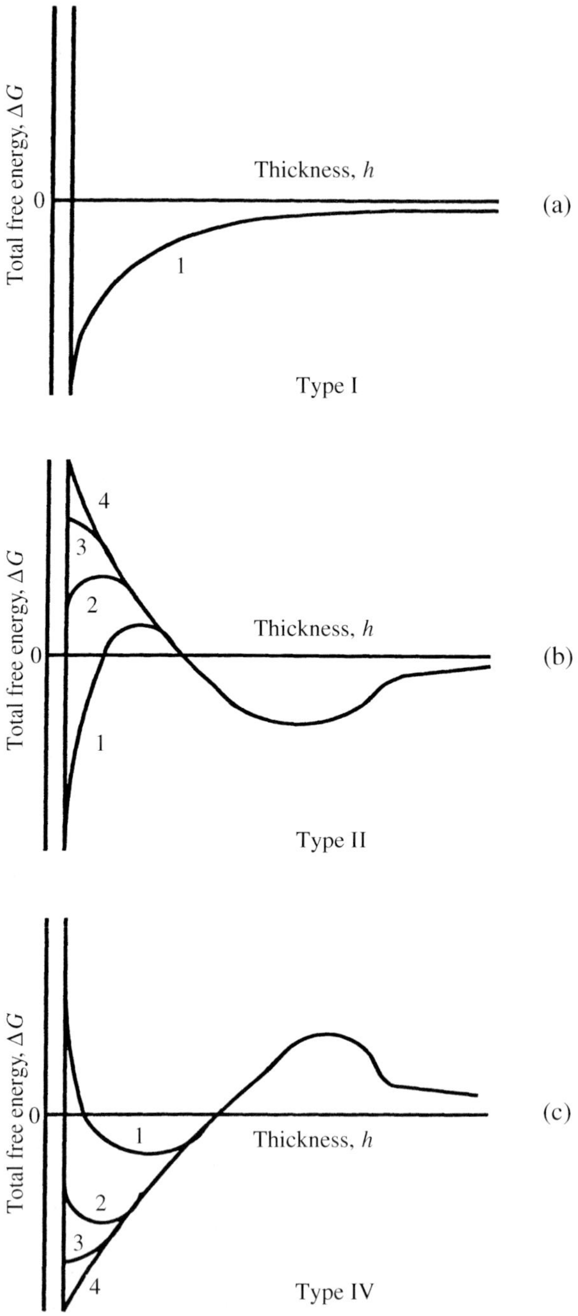

Figure 2. Different possible variations of the total free energy of interaction (ΔG). (a) Type I films with attractive force. (b) Type II films with a long range attraction combined with a short range repulsion. (c) Type IV films with a short range attraction combined with a long range repulsion. In (b) and (c), the shape of corresponding $-\mathrm{d}\Pi/\mathrm{d}h$ curve for curves 1–3 resembles curve 1 of ΔG curves, and the shape of corresponding $-\mathrm{d}\Pi/\mathrm{d}h$ curve for curve 4 parallels itself.

the cut-off (l_0), the growth of instability remains unhindered and a true dewetting in the form of retracting three phase contact line is the only possibility.

Type II behavior (curves 1–4 of Fig. 2b) is derived from a combination of a long range attraction with a shorter range repulsion. The free energy of adhesion (ΔG at l_0) is positive for curves 2–4, signifying a positive total spreading coefficient. Qualitative behavior of the function $(-\mathrm{d}\Pi/\mathrm{d}h) = \partial^2\Delta G/\partial h^2$ can be inferred from the curvature of ΔG curves [7]. For ΔG curves 1–3, the shape of $(-\mathrm{d}\Pi/\mathrm{d}h)$ is qualitatively similar to curve 1, since the sign of curvature changes only twice in the interval (l_0, ∞). For curve 4, Fig. 2b, however, $(-\mathrm{d}\Pi/\mathrm{d}h)$ remains qualitatively similar to the variation of ΔG (curve 4), since the sign of the curvature changes only once. Sufficiently thick Type II films are initially unstable $(-\mathrm{d}\Pi/\mathrm{d}h < 0)$ due to long range attraction. The initial instability, however, eventually saturates leading to MPS [7, 16, 17] when thinner portions of the film encounter a stabilizing force field, viz. $(-\mathrm{d}\Pi/\mathrm{d}h)$ becomes positive. The film 'roughens' and evolves into a collection of small drops which attain (quasi) equilibrium with the intervening largely flat ultrathin films [7, 16, 17]. While a true rupture of resulting ultrathin films is not possible for ΔG in curve 4 of Fig. 2b, it is certainly possible by nucleation for systems governed by curve 1, and possibly for curves 2 and 3 as well. This is because $(-\mathrm{d}\Pi/\mathrm{d}h)$ again becomes negative for very small thicknesses for curves 1–3. This situation is analogous to rupture of metastable 'black' soap films by nucleation, and to the slow coagulation of colloids over a potential barrier. Numerical simulations [16] also confirm rupture in later stages of evolution beyond the MPS by nucleation mechanism. In simulations [16], the small deformations needed for nucleation were automatically introduced by numerical inaccuracies (by Gibbs' phenomenon in the Fourier spectral technique used). A true rupture can thus be induced at the thinnest regions of the morphologically phase separated microstructures, when the barrier for further growth of instability is locally surmounted [16]. Further, it may be anticipated that the microstructure away from the sites of rupture should undergo slow coarsening due to the Laplace pressure induced flow from smaller to larger drops. All of the above qualitative anticipations are indeed confirmed in the simulations and experiments reported later in this paper.

Examples of Type II films include amorphous polymer films on end-adsorbed molecules and on grafted polymer 'brushes' [15, 17, 28, 32, 36]. The entropic effects in these films are described by curve 3 or 4 of Fig. 2b [27, 28]. Apolar films on coated substrates with $S_{\mathrm{s}}^{\mathrm{LW}} < 0$ (nonwettable substrate) and $S_{\mathrm{c}}^{\mathrm{LW}} > 0$ (wettable coating) also show Type II behavior (curve 4) [20, 23]. Thus, equation (10) provides a simple model for understanding the physics of roughening and ripening in Type II polymer films, without the accompanying possibility of rupture due to nucleation. Inspection of equation (10) shows that the coating properties ($S_{\mathrm{c}}^{\mathrm{LW}}$) dominate the free energy at locally thin regions ($h \ll \delta$) of the film, and thus true dewetting cannot occur when the coating is wettable ($S_{\mathrm{c}}^{\mathrm{LW}} > 0$).

Finally, Type IV behavior for ΔG (Fig. 2c) is a combination of a long range repulsion with a short range attraction. Type IV behavior can be viewed simply as an inversion (change of sign) of ΔG for Type II systems. Again, $(-\mathrm{d}\Pi/\mathrm{d}h)$ for curves 1–3 resembles the shape of curve 1 in Fig. 2c [7]. Also, the shape of $(-\mathrm{d}\Pi/\mathrm{d}h)$ for curve 4 parallels itself. Interestingly, Type IV thick films are stable ($-d\Pi/dh > 0$), and can be made unstable only by large amplitude external disturbance exceeding a critical size (nucleation) [37]. Films thinner than a critical thickness (where $d\Pi/dh$ changes sign) can either rupture (curve 4), or undergo MPS as in Type II films (curves 1–3) [7, 16, 17]. Polyethylene–propylene films on native-oxide covered silica surface show abrupt roughening (MPS) when their thickness is reduced [13], suggesting Type IV behavior. Polyethylene–propylene is a statistical copolymer with CH_3 branches randomly distributed along the chain. Another example [7, 16, 17] is water films on high energy surfaces ($\gamma_s^{LW} > 21.8$ mJ/m^2), where $S^{LW} > 0$ (long range LW repulsion) and $S^P < 0$ (shorter range polar attraction). $S^P < 0$ is almost always encountered for aqueous films due to a large polar cohesion energy of hydrogen bonded water. Finally, the LW energy of films on coated substrates with $S_c^{LW} < 0$ (nonwettable coating) and $S_s^{LW} > 0$ (wettable substrate) also displays the Type IV behavior (curve 4 of Fig. 2c) [20].

2.3. Equilibrium contact angle

The above discussion presents different possible patterns for the growth of instability and their relationship to the characterstics of the disjoining pressure isotherms. The equilibrium contact angle (θ) of relatively large (> 100 nm) drops that finally emerge after cessation of dewetting is given by the following extended Young–Dupre equation [31, 33, 34], which can also be obtained directly [34] from equation (1) at equilibrium ($\partial h/\partial t = 0$).

$$\cos\theta = 1 + \frac{\Delta G(h_e)}{\gamma_f}, \tag{15}$$

where h_e is the stable equilibrium film thickness corresponding to the minimum of the free energy, $\Delta G(h_e) < 0$. Thus, h_e is found from the conditions, $\Pi(h_e) = 0$ and $\partial^2\Delta G(h_e)/\partial h^2 > 0$. For the potential of the form given by equation (14), $h_e = l_0$ and the Young–Dupre equation (without the film pressure), viz.

$$\cos\theta = 1 + \frac{S^{LW}}{\gamma_f}, \tag{16}$$

is recovered [34]. In general, if the minimum of the free energy is due only to the Born repulsion (e.g. Type I films), the Young–Dupre equation without film pressure is always obtained, with S^{LW} in equation (16) replaced by an effective total spreading coefficient. Based on equations (10), (12) and (15) the equilibrium contact angle of the film fluid on a nonwettable coated substrate is thus given by ($S_c^{LW}, S_s^{LW} < 0$) [12, 29],

$$\cos\theta = 1 + \frac{1}{\gamma_f}\left[S_c^{LW} + S_c^P + \frac{l_0^2}{(l_0+\delta)^2}\left(S_s^{LW} - S_c^{LW}\right)\right]. \tag{17}$$

Since even monolayer coatings are usually in the nm range ($\delta \gg l_0$), the equilibrium contact angle in Type I systems is governed almost entirely by surface properties of the coating material, regardless of the nature of the substrate. The paramount influence of coating surface properties on the contact angle is also experimentally well documented [21, 22]. In contrast, equation (10) shows that the initial instability of a relatively thick film ($h \gg \delta$) is governed mostly by the substrate wettability (S_s) regardless of the coating wettability.

Inspection of Fig. 2b reveals that Type II systems display partial wettability due to presence of secondary minima in the free energy even when the total spreading coefficient is positive (curves 2–4 of Fig. 2b). In these cases, the drop attains mechanical equilibrium with a 'precursor' film of thickness, h_e ($\Delta G(h_e) < 0$). As discussed previously, thin films of free polymer on a grafted 'brush' lead to Type II behavior due to entropic repulsion engendered by stretching of grafted molecules. In such cases, the polymer melt should not completely wet a 'brush' of chemically identical but grafted molecules. This autophobic behavior of polymer films has been predicted theoretically [27, 28, 32], and also verified experimentally [15, 32, 36]. For curve 1 of Fig. 2b, two different equilibrium contact angles are possible, corresponding to primary and secondary minima in the free energy, at l_0 and h_{es}, respectively. Usually, the 'receding' equilibrium angle (with $\Delta G(h_{es})$ in equation (15) is smaller than the 'advancing' equilibrium contact angle (with $\Delta G(l_0)$), since the primary minimum is usually deeper, viz. $\Delta G(l_0) < \Delta G(h_{es})$.

Similarly, for Type IV systems, finite contact angles can result even when the total spreading coefficient is positive (curve 1 of Fig. 2c). In contrast to the autophobicity of polymer melts, the well known autophobic behavior of small molecules [33, 35] is usually due to their Type IV characterstics (S^{LW}, $S > 0$, but $S^P < 0$) [34]. Ideally, the equilibrium values of advancing and receding contact angles should be identical for Type IV systems. In reality, the retraction of a drop is likely to be very slow due to the presence of repulsion at greater thicknesses (which favors spreading). Thus, the slope of the retracting contact zone on macroscopic scales may well appear to remain very small for long times due to kinetic reasons. This idea, however, needs further exploration.

Finally, it may be noted that for very small (nanometer sized) drops (which are produced in the initial stages of MPS), the equilibrium contact angles are always smaller than the predictions of equation (17) for large drops [17]. This was found to be due to the strong curvature and disjoining pressure effects in nanodrops.

In what follows, we present models describing the dynamics of Type I and Type II films, and briefly discuss a methodology for their numerical solutions.

2.4. Nondimensionalization and numerical methods

The physics of dewetting in Type I films is studied by the use of the simplest Type I potential (equation (14)) together with the thin film equation (1). A compact representation of results is facilitated by introducing the following nondimensional scales ($S^{LW} < 0$) [16].

$$H = \frac{h}{h_0}, \qquad l = \frac{l_0}{h_0}, \qquad d = \frac{d_0}{h_0} = \sqrt{(4/3)}l, \qquad \nu = \frac{\mu}{\rho}, \tag{18}$$

$$C = \frac{3\rho\nu^2}{\gamma_f h_0}, \qquad B = \frac{-6S_s^{LW} d^2 h_0}{3\rho\nu^2}, \tag{19}$$

$$X = (|B|C)^{1/2}\left(\frac{x}{h_0}\right), \qquad T = B^2 C\left(\frac{\nu}{h_0^2}\right)t. \tag{20}$$

The above scalings reduce equation (1) combined with equation (14) to the following strongly conservative form:

$$\frac{\partial H}{\partial T} + \frac{\partial}{\partial X}\left[H^3 \frac{\partial}{\partial X}\left[\frac{\partial^2 H}{\partial X^2} - \frac{1}{3H^3}\left\{1 - \left(\frac{l}{H}\right)^6\right\}\right]\right] = 0. \tag{21}$$

The initial growth of the instability has been extensively studied [4, 5, 16, 18] based on the above equation after neglect of the Born repulsion, since $l \ll 1$ initially. The nondimensional length scale (Λ_m) of the fastest growing mode of instability is $2\sqrt{2}\pi$, and the nondimensional time for appearance of the first dry spot is $[4\ln(1/\in)]P_n$, where P_n (< 1) is a nonlinear factor which depends on the ratio ($\in$) of initial amplitude (nonhomogeneity) of disturbance and the mean film thickness. While the length scale of the instability and the film breakup time are almost unaffected by the extremely short range Born repulsion, its inclusion is necessary to prevent the nonphysical singularity of equation (21) at film rupture ($H \to 0$), and to study the process of hole-growth. The limit $h \to l_0$ denotes a dry spot (i.e. absence of real film), since l_0 is the equilibrium distance (gap) between molecules of the liquid and the solid. Dynamics of hole-growth based on equation (21) has not been studied previously, due possibly to numerical difficulties introduced by widely separated length and time scales.

As discussed previously, the simplest prototype of a Type II potential is given by equation (10) with wettable coating ($S_c^{LW} > 0$) and nonwettable substrate ($S_s^{LW} < 0$). The cut-off equilibrium film thickness is introduced by the short range LW repulsion due to the coating. The governing equation is nondimensionalized in this case also with the help of scalings (equations (18)–20)) with $S^{LW} = S_s^{LW}$ and the following additional nondimensional groups

$$R = \frac{S_c^{LW}}{S_s^{LW}} < 0 \quad \text{and} \quad D = \frac{\delta}{h_0}. \tag{22}$$

The nondimensional equation of evolution for a Type II film model is thus obtained as

$$\frac{\partial H}{\partial T} + \frac{\partial}{\partial X}\left[H^3 \frac{\partial}{\partial X}\left[\frac{\partial^2 H}{\partial X^2} - \frac{1}{3H^3}\left\{\left(1 + \frac{D}{H}\right)^{-3}(1 - R) + R\right\}\right]\right] = 0. \tag{23}$$

The fastest growing nondimensional wavelength from the linear instability analysis around $H = 1$ is

$$\Lambda_m = 2\sqrt{2}\pi\left[|(1 + D)^{-4}(1 - R) + R|\right]^{-1/2}. \tag{24}$$

Equation (23) with $R > 0$ also governs the instability of Type I apolar films when both the coating and the substrate are nonwettable (S_c^{LW}, $S_s^{LW} < 0$). Nonlinear stability and rupture of such films have been considered in detail elsewhere [29].

Equations (21) and (23) were solved with periodic boundary conditions over a large domain size (L), which was at least twice the size of the dominant wavelength. The periodicity interval was rescaled to 2π by defining $\eta = KX$, where K is the nondimensional wavenumber. The initial condition was taken to be a volume preserving random disturbance with a small maximum nondimensional amplitude of 0.01. The Fourier collocation technique provides greatly superior spatial resolution and efficiency for solution of thin film equations until the film breakup [16–18, 38]. However, we found that the Fourier collocation results in nonphysical oscillations (Gibbs' phenomenon) over largely flat portions of retracting films during the late stages of hole-growth. In view of this problem, space discretization of the film equations in their strongly conservative form was performed by successive central differencing, with dependent variable (and its functions) evaluated at half nodes. Half node interpolation was done by the second order Everett's method. The time integration of the resulting set of ordinary differential equations was performed by Gear's method which is suitable for stiff equations. While the grid convergence for a Type II film investigated here was obtained for about 50 grids per wavelength, tracking the growth of a single hole in a Type I film required about 4000 grid points. In general, grid convergence required increased grid density for thicker films.

3. EXPERIMENTAL

Hydrodynamic instability in thin films is best studied with amorphous polymer films because of their high viscosity and low vapor pressure. In the first set of experiments [8, 9], smooth thin (10–60 nm) films of polystyrene ($M_w = 28$ k) were deposited by spin coating on three different types of nonwettable coated silicon wafers. The thickness of various coatings (labelled A, B and OTS) and the initial roughness of interfaces were all less than 1 nm, as measured by X-ray reflectometry. Dewetting was induced by heating (135 to 170°C) the film samples above their glass transition temperature ($T_g = 103$°C). The morphology at various stages of dewetting was characterized by an optical phase interference microscope. All of these films showed true dewetting by formation of circular holes. X-ray reflectometery confirmed that no residual film was left behind on the substrate where dewetting occurred. The average contact angles of PS on wafers A, B and OTS wafer were estimated to be 22, 42 and 48 degrees, respectively.

In the second set of experiments, polystyrene films ($M_w = 9$ k) were deposited and heated on UV-ozone and 'Piranha' treated silicon wafers. A weak adsorption of polystyrene to the substrate hydroxyl groups resulted initially in roughening of the film, rather than true dewetting by circular holes.

The third set of experiments [15] probed the instability of methyl terminated polystyrene films ($M_w = 43$ k or 1120 k) on a polystyrene 'brush' ($M_w = 9.3$ k or 43.5 k) end grafted on UV-ozone treated silicon wafers. The polymer forming the 'brush' was terminated by a monochloro-silane group, which reacted with the hydroxyl groups on the substrate [15]. Films of free polymer on grafted 'brush' initially became rough by forming an interconnected structure of hills and valleys (MPS), which showed ripening and eventual dewetting. The latter, however, occurred only on densely grafted layers. On very densely grafted brushes, dewetting dominated over MPS.

The fourth set of experiments explored the instability of polystyrene films (M_w = 43 k) on a grafted polystyrene brush (M_w = 9 k) as described above, on which an additional film of end-functionalized long polystyrene molecules (M_w = 44 k or 126 k) was spin coated and annealed. The long grafted molecules of these bi-modal brushes could act as 'connectors' between the brush and film, thus slowing down the adhesion failure of the film considerably. The surface of films became rough and true dewetting could even be suppressed for large density of connectors.

In the fifth and final set of experiments, films of a symmetric block copolymer, polystyrene–polyethylene oxide (PS–PEO; M_w = 3 k + 3 k) were spin coated from its toluene solution containing some water onto UV-ozone treated silicon wafers, and annealed in air at 120 to 150°C. PEO contains some adsorbed water. UV-ozone treatment causes adsorption of PS and, even more easily, of PEO. The growth of instability resulted in a variety of 'bi-continuous' structures consisting of shallow low curvature regions and high curvature 'drops' of various shapes. True dewetting did not occur; an adsorbed layer about 20 nm thick was always present. Upon further annealing, the 'morphologically phase separated' structure coarsened by ripening of drops, i.e. bigger droplets 'ate' smaller droplets.

4. RESULTS AND DISCUSSION

In what follows, we present the various stages of evolution of instability and its physics for Type I and Type II films in greater details. A direct comparison of theory and experiments is also presented whenever possible.

4.1. Instabilities of Type I films

Figures 3a and 3b show various phases of evolution of 2-D structures in a Type I film (h_0 = 5 nm) on uncoated substrates (from equation (21)). Four distinct stages of evolution may be readily identified.

In the first stage (Fig. 3a), the initial random disturbances ($\in = 0.01$) are quickly reorganized on the (nondimensional) length scale of the dominant wave ($2\sqrt{2}\pi$), and the instability continues to evolve on this length scale until the appearance of dry spots. The length scale of the instability is indeed quite accurately predicted by the linear theory (e.g. equation (23) with $D = 0$). Since a lone hole appears in each wavelength, the number density of initial holes on a substrate of area a_s can be approximated by

$$N_H = \frac{a_s}{\lambda^2} = \frac{-3a_s d_0^2 S_s^{LW}}{4\pi^2 \gamma_f h_0^4}. \tag{25}$$

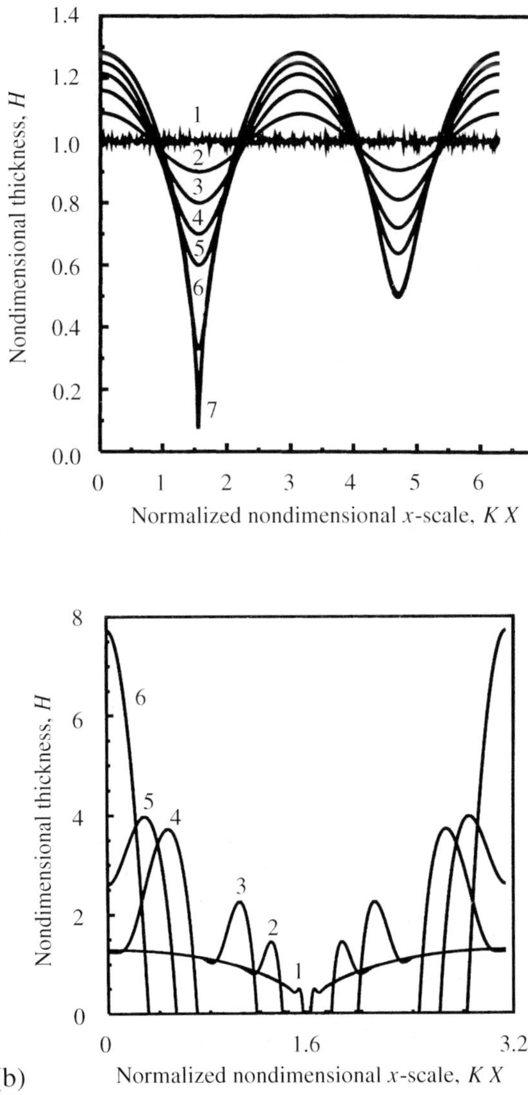

(a) (b)

Figure 3. Various stages of evolution of the instability in a Type I film (mean thickness = 5 nm, $L = 2\Lambda_m = 4\sqrt{2}\pi$). Normalized (nondimensional) x-scale is KX, which varies between 0 and 2π. In reality, dimensional x-scale is several orders larger than the mean film thickness and (dimensional) slopes are very small. (a) Evolution of the initial instability until the film ruptures with appearance of the first dry spot (phase 1). Curves 1–7 correspond to nondimensional time (T) = 0, 19, 21, 22, 22.6, 22.98 and 23, respectively. The second dry spot appears soon thereafter, since the growth rate at this point is very high. (b) Various stages of hole-growth. Curves 1–6 correspond to $T(\times 10^4)$ = 1.025, 1.059, 25.5, 64.5, 79.4, 120, respectively. The overlap between neighboring rims (phase 4) begins at $T \sim 70$ (curve 4), and their merger leads to a cylindrical thread (curve 6 shows half of the cross section of the thread). In the second (unsteady) and third (quasi-steady) phases of hole growth, the film beyond the rim remains undisturbed.

Further, for films that are initially much thicker than the coating ($D \ll 1$), the length scale of the instability is not affected by the coating properties (S_c^{LW} and S_c^{AB}). This conclusion is clear from equation (24) and has also been verified by nonlinear simulations [18, 29]. The initial density of holes is, therefore, governed essentially by the long range interactions with the substrate (S_s^{LW}), as shown by equation (25).

The micrograph in Fig. 4 shows a spontaneous formation of holes in Type I PS films on nonwettable substrates (silicon wafers) coated with a nonwettable coating (first type of experiments discussed earlier). In contrast to a rather uniform hole-spacing witnessed in 2-D simulations, experiments show a more random distribution of holes. This may be due to a combination of the hole–hole interaction, nonhomogeneity of initial defects, and other 3-D effects. While these complexities cannot be easily captured in simulations, predictions based on simple 2-D theory do match very well with the mean spacing of holes in experiments [12].

Open symbols in Fig. 5 present the data on average number of holes per $10^4\ \mu m^2$ of the surface, and the solid line is the best fit to the data using equation (25) with $(S_s^{LW}/\gamma_f) = 0.213$. Experiments also show $N_H \propto h^p$, where $p = -4.1 \pm 0.15$ agrees well with the predicted value of -4 from equation (25). The value of $(S_s^{LW}/\gamma_f) = 0.213$ is also realistic for a PS–silicon system, as indicated by the following properties. At the temperature of annealing ($\sim 150°C$), $\gamma_f(PS) = 32\ mJ/m^2$ [39]. Thus, the best fit value of $S_s^{LW} = -6.8\ mJ/m^2$ corresponds to an effective Hamaker constant (A_{el}) of 6.4×10^{-21} J (from equation (7)). Since the Hamaker constant A_{ff} for PS at the annealing temperature is calculated to be 6×10^{-20} J (from $A_{ff} = 24\pi d_0^2 \gamma_f$ [24]), the Hamaker constant for silicon substrate, A_{ss} should be about 4.8×10^{-20} J (from $A_{el} = A_{ff} - A_{sf}$, and $A_{sf} = \sqrt{A_{ss} A_{ff}}$). This magnitude of A_{ss} is in accord with the measured values of the Hamaker constant for fused quartz, which ranges from 5×10^{-20} to 6×10^{-20} J at room temperature [26]. In view of the fact that estimation of Hamaker constants from macroscopic parameters (γ_f, S_s^{LW}) involves some uncertainty ($\sim 10\%$) [24], the agreement is good.

The first phase of instability is completed with the film breakup (Fig. 3a). We now redefine T so that $T = 0$ corresponds to the instant of film breakup. In the second phase of the instability (Figs 3b and 6a for $T < 0.001$), the dry-spot (hole) expands with a rapid concurrent increase in the dynamic contact angle at the retracting three phase contact line (Fig. 6a). Small rims begin to surround the hole due to the removal of material from the expanding hole. The rate (velocity) of hole-expansion is maximum immediately after its appearance (Fig. 6b), which rapidly declines to an almost constant value in the third phase of the evolution ($0.002 < T < 0.01$). In the third phase, dynamic contact angle increases very slowly (Fig. 6). While the non-dimensional hole-radius R_H varies approximately as $T^{0.75}$ in the second phase, the variation changes approximately to $T^{0.85}$ in the third phase ($0.002 < T < 0.009$). The best linear fit is poor, and extrapolates to a finite hole radius at $T = 0$, which is unrealistic. Indeed, experiments [15] also suggest the same inadequacy in the linear fit of the data for hole-diameters. Experiments [10, 15] on dewetting of thin polymer films also show initial rapid retraction with $R_H \propto T^q$ ($q \approx 0.7$), and a slight increase in q at later times.

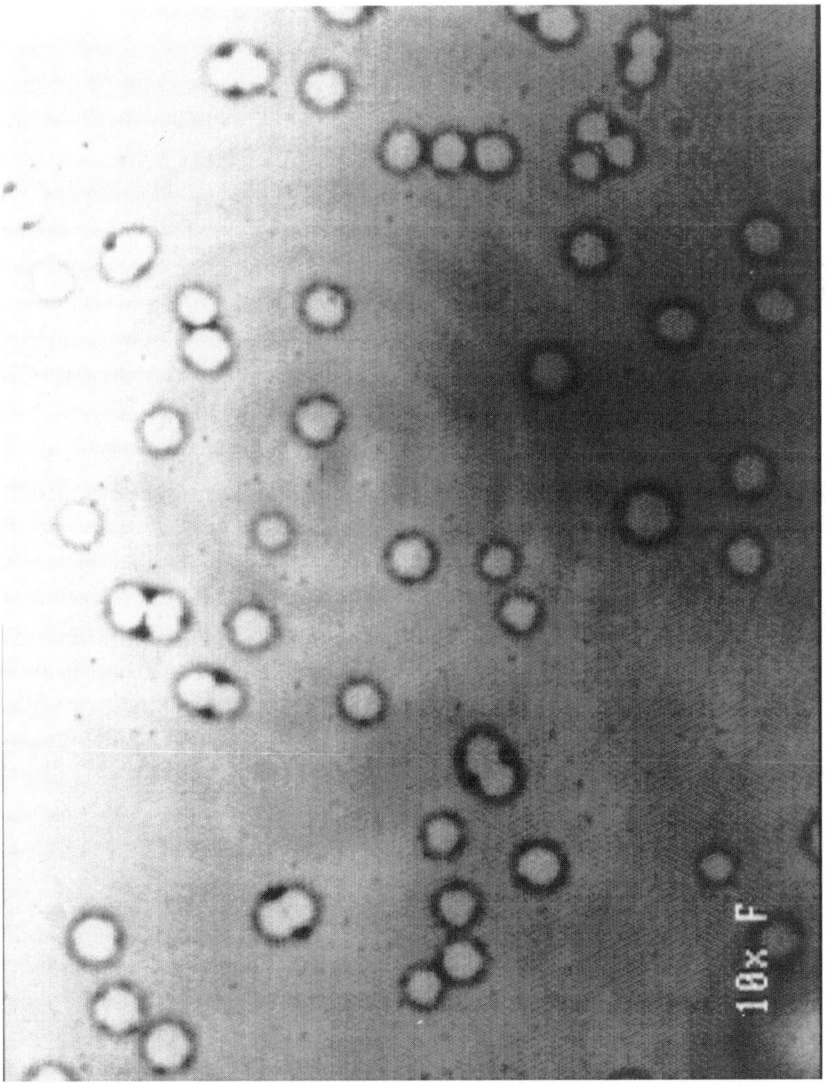

Figure 4. Optical micrograph of holes in a 16 nm thick polystyrene film ($M_w = 28$ k) on wafer A (Type I system). Area shown is 10^4 μm^2.

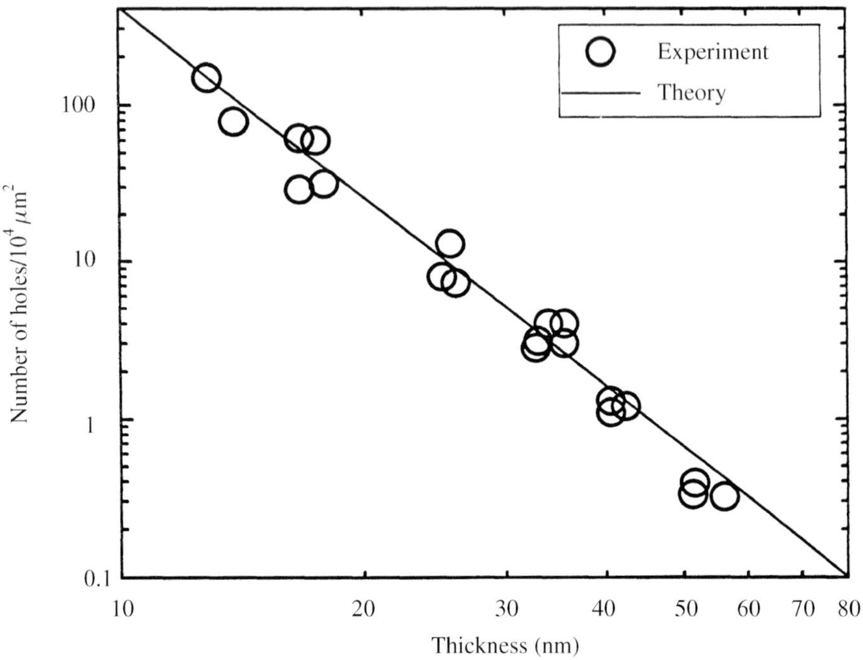

Figure 5. Double logarithmic plot of average number of initial holes per 10^4 μm^2 for PS films on wafer A. The line and open circles represent the theory (equation (25) with $S_s^{LW} = -8.1$ mJ/m^2) and experiment, respectively.

The second (transient) stage of dewetting therefore corresponds to initial rapid and unsteady expansion of the hole, followed by a (quasi) steady expansion in the third phase. Inspection of the Fig. 6 suggests that if the third phase can continue for longer times (as in an unbounded film with no other holes), the dynamic contact angle as well as the velocity of hole-growth may reach constant values asymptotically. However, in reality, a fully developed regime is only approximately reached when several holes are present. This is because the third stage of evolution lasts only until the growing and moving rims (disturbances) reach close to the 'cell' corner, and begin to feel the presence of the neighboring hole-rim.

The fourth and final phase of 2-D simulations now commences with a significant overlap of neighboring rims, which greatly slows down the hole expansion, and the dynamic contact angle rapidly climbs to its equilibrium value (Fig. 6, $T > 0.01$). Eventually, a cylindrical liquid thread forms by the merger of rims. The equilibrium (nondimensional) slope (H_X) of liquid thread from Fig. 6 is 17.5, which compares well with its theoretical value of 18.3 obtained by nondimensionalizing equation (16). The theoretical value is, however, for an infinite drop. For finite drops (as in curve 6 of Fig. 3b), the equilibrium slope should be smaller [17], as seems to be the case here in simulation. 2-D simulations cannot track the evolution of structure beyond this time. In reality [8, 9], drainage (normal to the plane of paper) from liquid threads also occurs leading to hole-coalescence, as in 3-D foams.

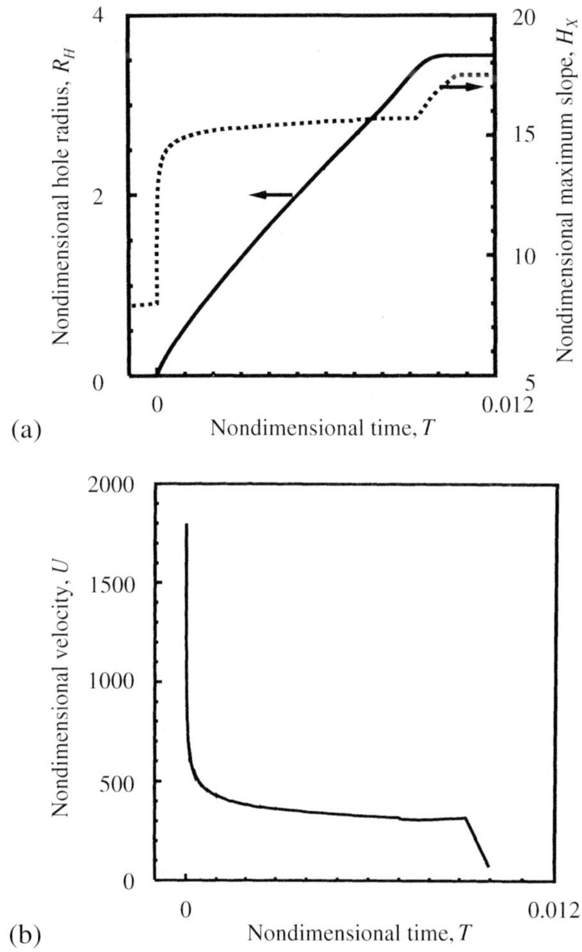

Figure 6. (a) Variations of hole-radius, R_H (solid line) and dynamic contact angle (broken line) with time (all are nondimensional). (b) Variation of nondimensional velocity, U $(= \mathrm{d}R_H/\mathrm{d}T)$. T represents the nondimensional time after appearance of the dry-spot ($R_H = 0$, $T = 0$). Dynamic contact-angle increases rapidly in the second phase (unsteady growth), becomes almost constant in the third phase (quasi-steady growth), and finally climbs to its equilibrium value in the fourth phase (overlap and merger of rims). Hole-radius scales as T^q in second ($q \sim 0.75$) and third ($q \sim 0.85$) phases. Velocity decreases rapidly in the second and fourth phases, but it is quasi-steady in the third phase.

While the simulation reported here is for a 5 nm thick film ($l^{-1} = 36.5, l_0 = 0.137$ nm), our simulations (not shown) for other thicknesses confirm that the dimensional hole-radius is not sensitive to the mean film thickness. This is expected since the hole growth is governed largely by a balance of forces close to the contact line. However, the hole-radius (or time) at the onset of the fourth, hindered, phase depends on the mean film thickness, since the mean hole-spacing is governed by h_0 ($\lambda \propto h_0^2$; equation (25)).

The following salient features of structure during dewetting also emerged from the simulation. The thickness of the film beyond the hole rim remains largely undisturbed,

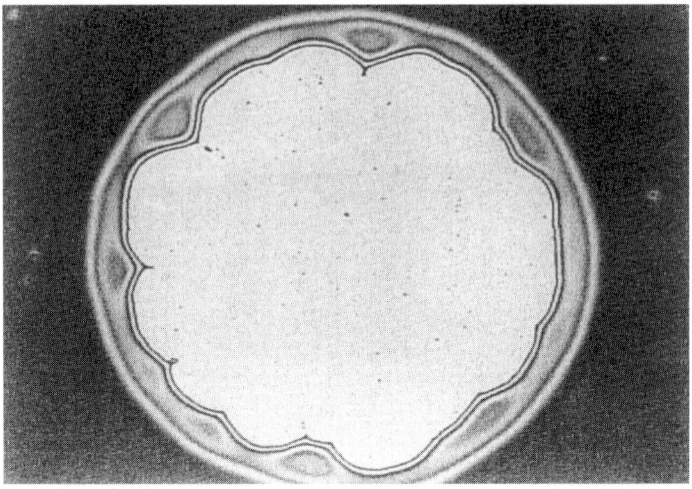

(a)

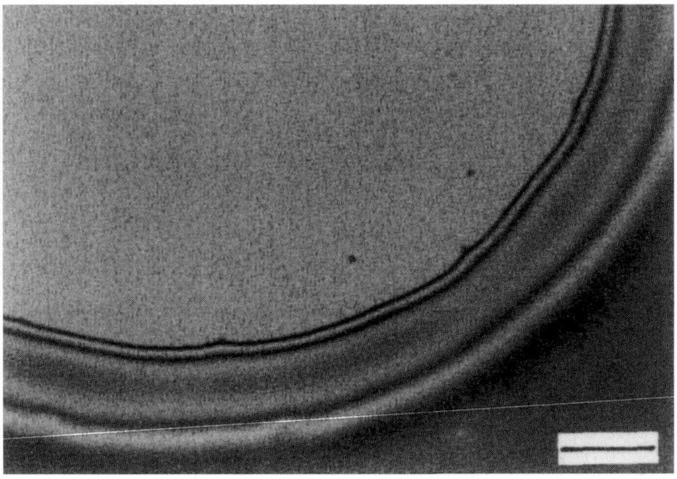

(b)

Figure 7. (a) Optical micrograph of a typical growing hole obtained in a ~100 nm thick PS (M_W = 43 k) film on a monochlorosilane terminated PS (M_W = 9.3 k) brush. Diameter of hole is ~230 μm. (b) Magnification of a portion of another hole in the above system indicating the asymmetric shape of the rim. The length of the bar is 10 μm.

since the material removed simply accumulates in the rim (Fig. 3b). Thus, roughening of the film away from the rim does not occur for Type I films. Further the rim profile is highly asymmetric, with higher slopes near the three-phase contact-line. Finally, the far side of the rim does not level-off to the height of the undisturbed film monotonically, but dips below the film level, causing a slight depression ahead of the moving rim. All of the above results are in accord with experimental observations [15]. A close-up view of a lone hole in Fig. 7 shows the depression before the rim merges with the film (dark surroundings). The spreading between two successive fringes is smaller towards the hole side, signifying a steeper rim. The rim also shows unevenness in circumferential direction, which is engendered by the Rayleigh instability of long threads due to their cross sectional curvature. The fingering instability of retracting rims, which was witnessed on low wettability substrates, is also related to the above consideration [9, 12]. This instability cannot, however, be captured in 2-D simulations.

It is interesting to note that the factors other than those considered here e.g. viscoelastic effects [40], surface diffusion [41] and polymer slippage [42] may also be instrumental in producing the rim asymmetry and its non-monotonic shape. However, as shown here for the first time, these features are already present even for viscous flow in Newtonian fluids.

An approximate expression for the velocity of hole-growth in the third phase with approximately constant dynamic contact angle is easily obtained from simulations by noting the relation between dimensional (u) and nondimensional (U) velocities (see scalings (equations 18–20)).

$$u \propto \frac{\left(|S_{\mathrm{s}}^{\mathrm{LW}}|\right)^{3/2} U}{\sqrt{\gamma_{\mathrm{f}}\mu}}.$$

Further, $|S_{\mathrm{s}}^{\mathrm{LW}}| = (1 - \cos\theta)\gamma_{\mathrm{f}}$ from the Young equation, which gives,

$$u \propto U\left(\frac{\gamma_{\mathrm{f}}}{\mu}\right)(1 - \cos\theta)^{3/2} \propto U\left(\frac{\gamma}{\mu}\right)\theta^3 \quad \text{(for small } \theta\text{)}.$$

Thus, $u \propto (\gamma_{\mathrm{f}}/\mu)\theta^3$ whenever a constant velocity regime is found for (nondimensional) velocity, U (as in Fig. 6 approximately for $T > 0.002$). This result (Tanner's law) has previously been obtained by a simplified analysis for the long time behavior of a lone expanding hole in a thick film [43, 44]. In this long time asymptotic regime, the hole-diameter is very large compared to the width of the rim. Further, the rim cross section was assumed to be circular. In the same analysis, an estimate for the dynamic contact angle was obtained as $\theta_{\mathrm{d}} = \theta/\sqrt{2}$. Our simulation shows slightly different relation in the third phase, viz., $\theta_{\mathrm{d}} = (0.85 \pm 0.01)\theta$. The cross section of the rim was also not found to be circular. The result, $u \propto (\gamma_{\mathrm{f}}/\mu)\theta^3$, however, appears to be rather independent of these details in the constant velocity regime, even for relatively thin films with many holes. Possibly, the approximate analytical result may be better for unhindered expansion of a hole in thicker films, as appears to be the case experimentally for holes created by nucleation, rather than by spinodal decomposition [44]. Expansion of a spontaneously formed hole (by spinodal decomposition) has not been previously studied theoretically. For such cases, Tanner's law holds only approximately in the quasi-steady regime, since U is not truly constant, but declines slowly as $U \propto T^{-0.15}$ (Fig. 6b).

As noted previously, experiments [10, 15] on growth of holes in thin polymer films show $R_H \propto T^q$ ($q \approx 0.7$ initially), and a slow increase in q with time. These observations are consistent with our results. It is interesting to note that if slippage of liquid (polymer) on the substrate is allowed, q is also predicted to be (2/3) in the strong slippage regime of hole expansion [42]. However, since all of the observations can also be reproduced without assuming slippage, further experiments need to be designed to assess the relative contributions of slip and viscous flow in hole expansion.

Due to scalings given by equation (20), dimensional radius of the hole, r_H is related to its nondimensional radius, R_H by (proportionality constant has units of length):

$$r_H \propto R_H \left(\frac{|S_s^{LW}|}{\gamma_f} \right)^{-1/2} \propto R_H (1 - \cos\theta)^{-1/2} \sim \frac{R_H}{\theta}.$$

Thus, it is clear from scaling arguments that the (dimensional) hole-radius (r_{Hc}) at the onset ($R_H \approx 0.06$) of third phase (quasi-steady growth) is related to the equilibrium contact angle by $r_{Hc} \propto \theta^{-1}$. Thus, the quasi-steady growth is realized for increasingly smaller holes on less wettable substrates (high θ). Similarly, from nondimensional scalings, it is easily shown that the contact angle at the time of film rupture, $T = 0$, is proportional to the equilibrium contact angle.

In what follows, we give a summary of stages of evolution beyond the fourth [8, 9, 12], which is the last phase before the onset of drainage and coalescence.

The micrograph in Fig. 8a shows the repeated coalescence and growth of holes, until a polygonal network structure of giant holes is produced (Fig. 8b). The polymer threads become long and the drainage (leading to further coalescence) slows down greatly since the time of coalescence in 2-D foams varies as D_p^n ($n = 5$–7), where D_p is the polygon (hole) diameter [12]. At this time, the Rayleigh instability engendered by the cross sectional curvature of long cylindrical threads becomes dominant and disintegrates polygon-edges into drops, which appear neatly arranged in polygonal patterns (Fig. 9).

The mean circle equivalent diameter of the final polygonal pattern (Fig. 8b) can be obtained from $\pi D_p^2/4 = N\lambda^2$, where N is the average number of holes that coalesce to form a polygonal cell. With the help of equation (25), D_p is obtained as

$$D_p = \left(\frac{4h^2}{d_0} \right) \left[\frac{-N\pi\gamma_f}{3S_s^{LW}} \right]^{1/2}. \tag{26}$$

Experiments [8, 9] indeed show that $D_p \propto h^m$, where the best fit value of $m = 1.99 \pm 0.07$ agrees closely with the theoretical value of 2. Interestingly, experiments (Fig. 10) also show that D_p is largely unaffected by wettability of the coating [9], which implies that the degree of coalescence (N), like the density of holes, also depends only on the substrate properties, regardless of the coating (Fig. 10). This is because the Rayleigh instability manifests only when the time scale for drainage ($\tau_D \propto D_p^n$) becomes comparable or larger than the characterstic time scale for the Rayleigh instability (τ_R). The condition $\tau_D \sim \tau_R$ for two different coatings with θ_1 and θ_2 implies $(D_{p1}/D_{p2}) \sim (\tau_{R1}/\tau_{R2})^{1/n}$, where n is large (5–7) [12]. However, τ_R is a weak function of the wettability (θ) of the coatings. Thus, $D_{p1} \sim D_{p2}$ is indeed expected [12] in the range of contact angles investigated here (22–48°).

Finally, while the density of holes and polygon diameter are unaffected by wettability (θ) of thin coatings, the diameter of final drops (Fig. 9) depend strongly on θ [8, 9, 12]. This is indeed expected because the equilibrium contact angle depends almost solely on wettability of coating (equation (17)), regardless of the substrate properties.

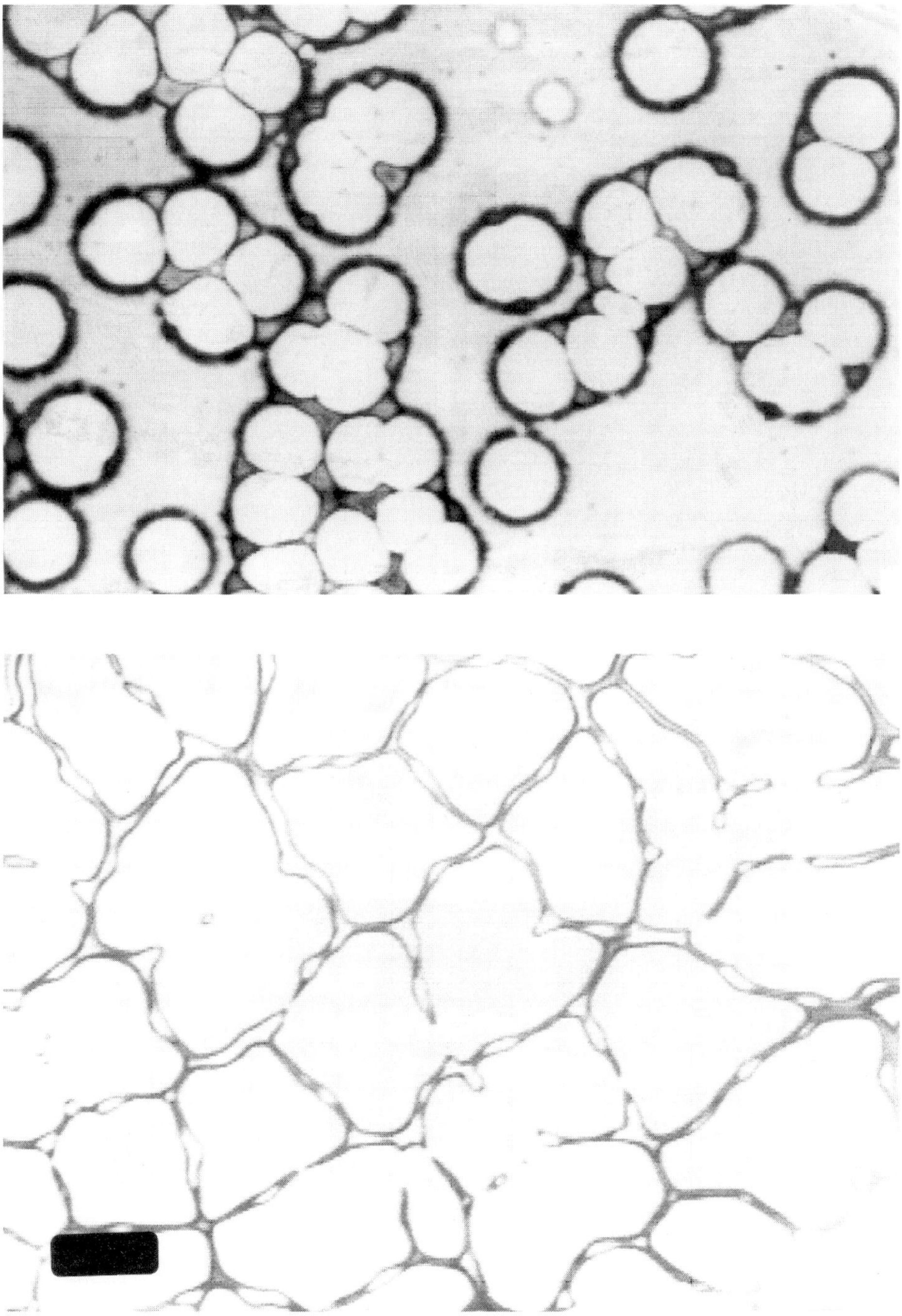

(a)

(b)

Figure 8. (a) Typical appearance of coalescing holes in PS films on wafer A (Type I system). (b) Typical appearance of polygonal pattern in PS films on wafer A (Type I system). The length of the bar is 10 μm.

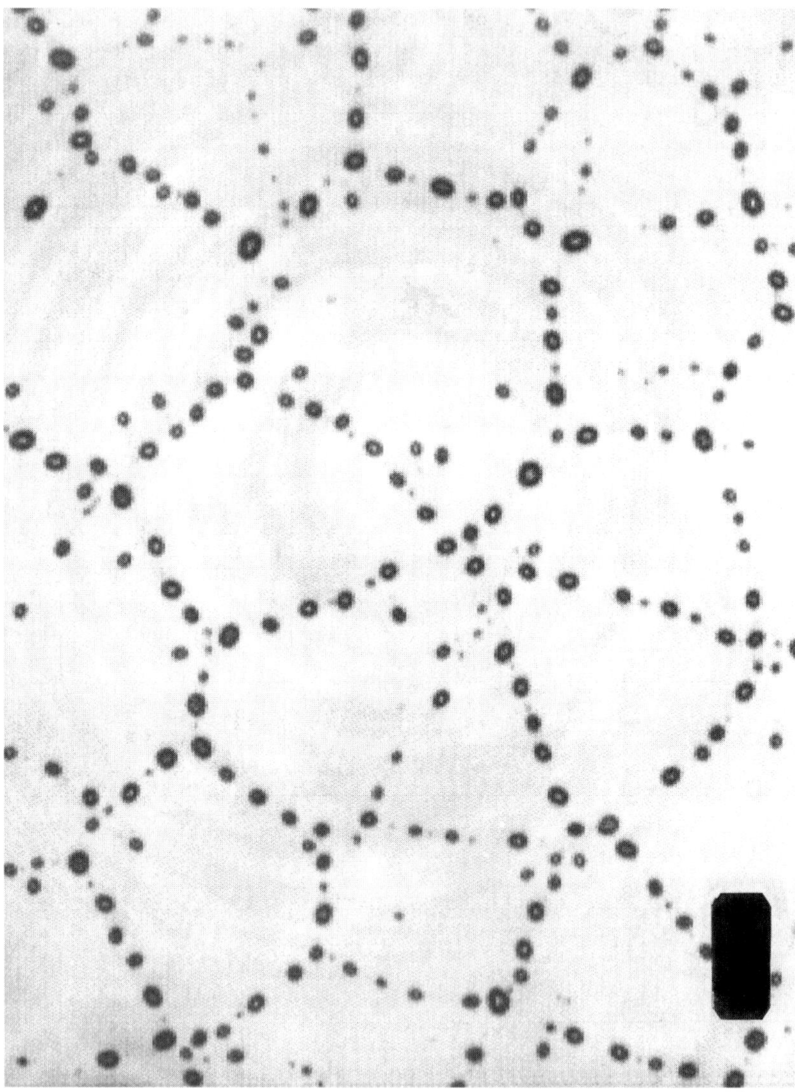

Figure 9. Typical appearance of spherical droplets produced by decay of polygonal rims due to Rayleigh instability. The length of the bar is 100 μm.

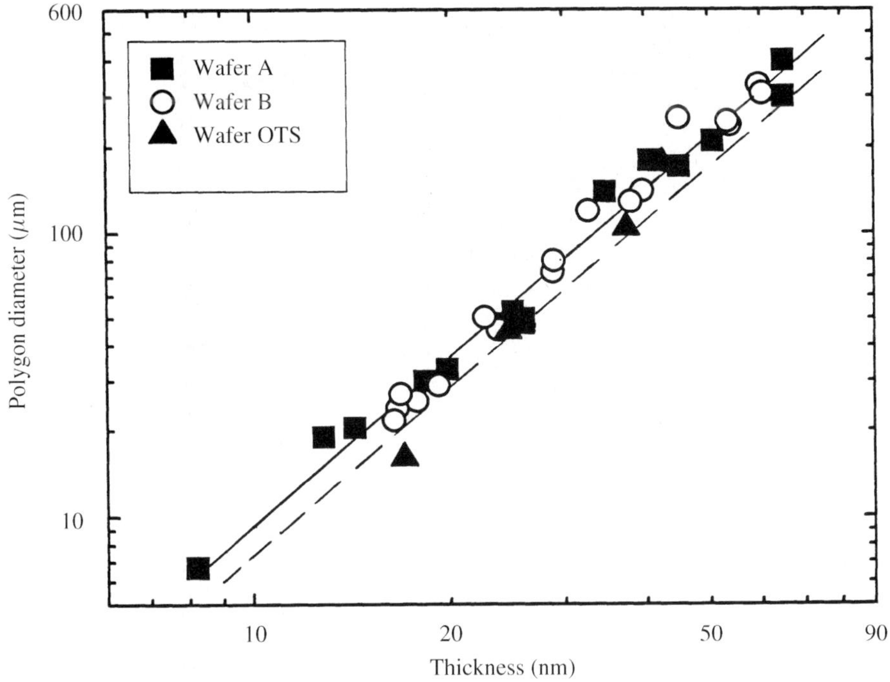

Figure 10. Double logarithmic plot of the average diameter of polygons for PS samples on wafer A (solid squares), wafer B (open circles) and OTS wafer (full triangles). The lines are the best theoretical fit to the data using equation (26) with $S_s^{LW} = -8.1$ mJ/m^2 and $N = 2.7$ for wafers A and B (solid line), and $N = 2$ for OTS wafer (broken line).

The low magnification micrographs of Fig. 11 show the later stages of dewetting in another Type I system — a PS film on a densely grafted PS brush (see Experimental section). Since free polymer on top is excluded from the region occupied by the dense brush, creation of a well defined brush-free polymer interface engenders 'autophobicity' leading to true rupture. Micrographs show the progression of hole-coalescences leading to bigger holes and longer polymer ridges. On low density, 'wet' brushes, which allow significant interpenetration of the free polymer, only MPS was observed.

The later stages of evolution of the instability including drainage and drop formation are discussed in detail elsewhere [9, 12]. Here, we have confined our attention largely to the earlier stages of dewetting before the onset of drainage and hole-coalescence.

4.2. Instabilities of Type II films

We consider the behaviour of a Type II prototype governed by equation (23) with wettable coating ($S_c^{LW} > 0$) and nonwettable substrate ($S_s^{LW} < 0$). The energy and force curves for the example chosen here are shown in Fig. 12. This behaviour corresponds to curve 4 of Fig. 2b. As discussed earlier, a true-adhesion failure of the film is not possible in this case due to the short range (LW) repulsion engendered by a wettable coating. This allows examination of roughening (MPS) and ripening in their pristine forms without the accompanying possibility of true rupture.

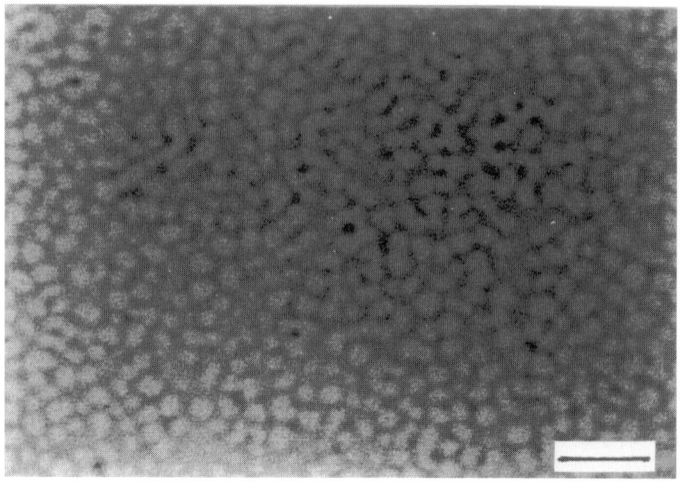

(a)

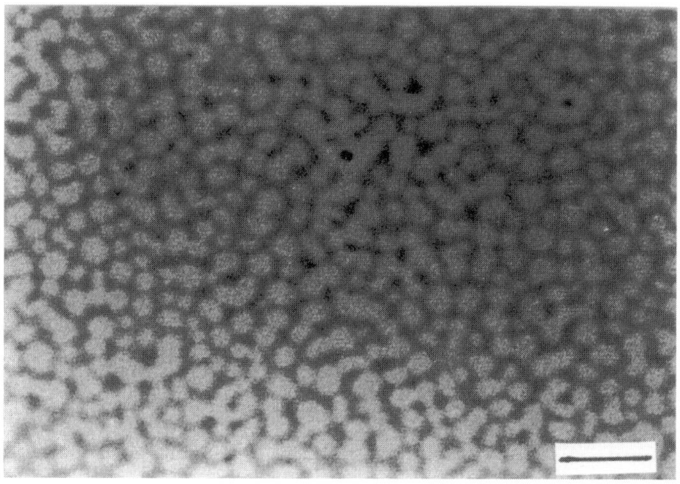

(b)

Figure 11. Dewetting and coalescence of holes in a 20 nm thick PS (M_w = 1120 k) film on a dense 12.5 nm thick PS (M_w = 43 k) brush: (a) after annealing for 28 h at 170°C, (b) after an additional 2.5 h of annealing, and (c) after an additional 6 h of annealing. The length of the bar is 10 μm. Long polymer ridges eventually disintegrate into drops (not shown).

Figures 13a–13c show theoretical evolution of the instability in a large 2-D domain ($4\Lambda_m$) starting with a small amplitude random disturbance (curve 1). In curve 2, instability evolves on the length scale (Λ_m) announced in equation (24), and a quasi-steady four drop structure (curve 3) emerges with saturation of initial instability. This stage may be referred to as the initial roughening (MPS) of the film without true rupture. This intermediate structure is, however, not truly stable unless all drops are *exactly* of the same shape and size. In reality, the flow occurs through the intervening films in the direction of decreasing curvature (due to Laplace pressure) and increasing disjoining pressure. This is witnessed in gradual merger of the two smaller drops in the left corner of Fig. 13a (curves 4–8) to produce a larger drop

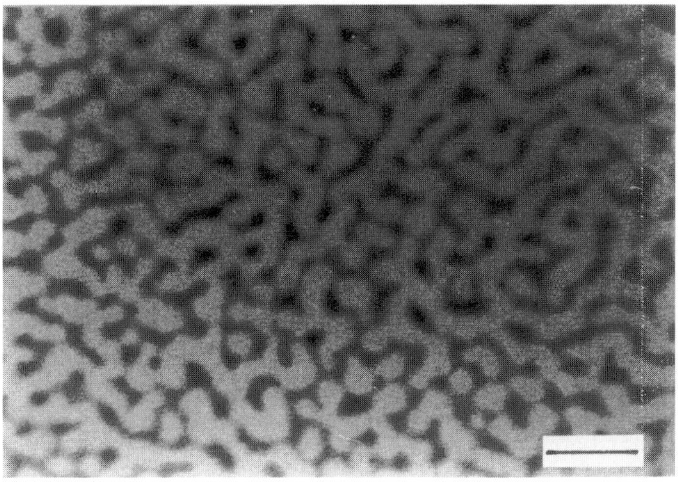

(c)

Figure 11. (Continued).

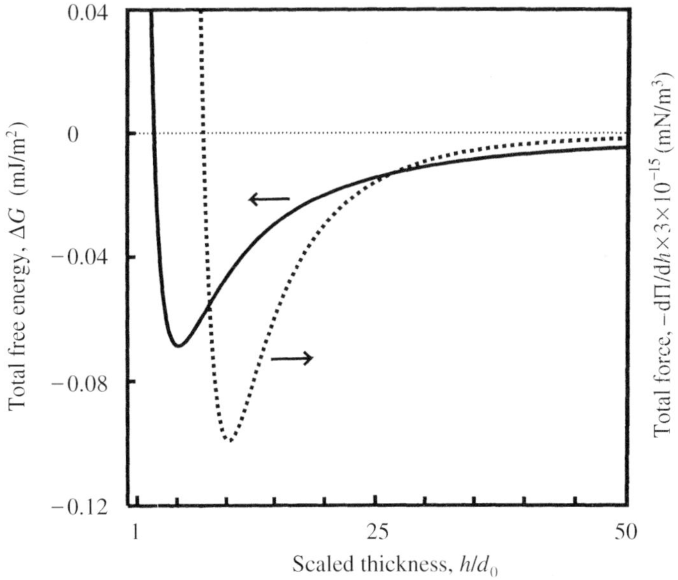

Figure 12. Variations of the free energy (ΔG in mJ/m^2; solid line) and force ($-\mathrm{d}\Pi/\mathrm{d}h \times 3 \times 10^{-15}$ in mN/m^3; broken line) for a Type II film on a coated substrate. d_0 is a cut-off distance for intermolecular forces (= 0.158 nm). Parameters are $S_\mathrm{s}^\mathrm{LW} = -15$ mJ/m^2, $S_\mathrm{c}^\mathrm{LW} = 1.5$ mJ/m^2 and $\delta = 1$ nm. h denotes the film thickness in nm.

(curve 8). Further ripening or coarsening of the structure continues in a similar fashion (Figs 13b and 13c). Thus, curves 4–12 depict the slow transformation (ripening) of initial four drop structure to a two drop structure, which in turn produces a single drop (curves 13–16 in Fig. 13c). When neighboring drops are of similar sizes, they appear to move towards each other and merge in the middle (Fig. 13a). When size

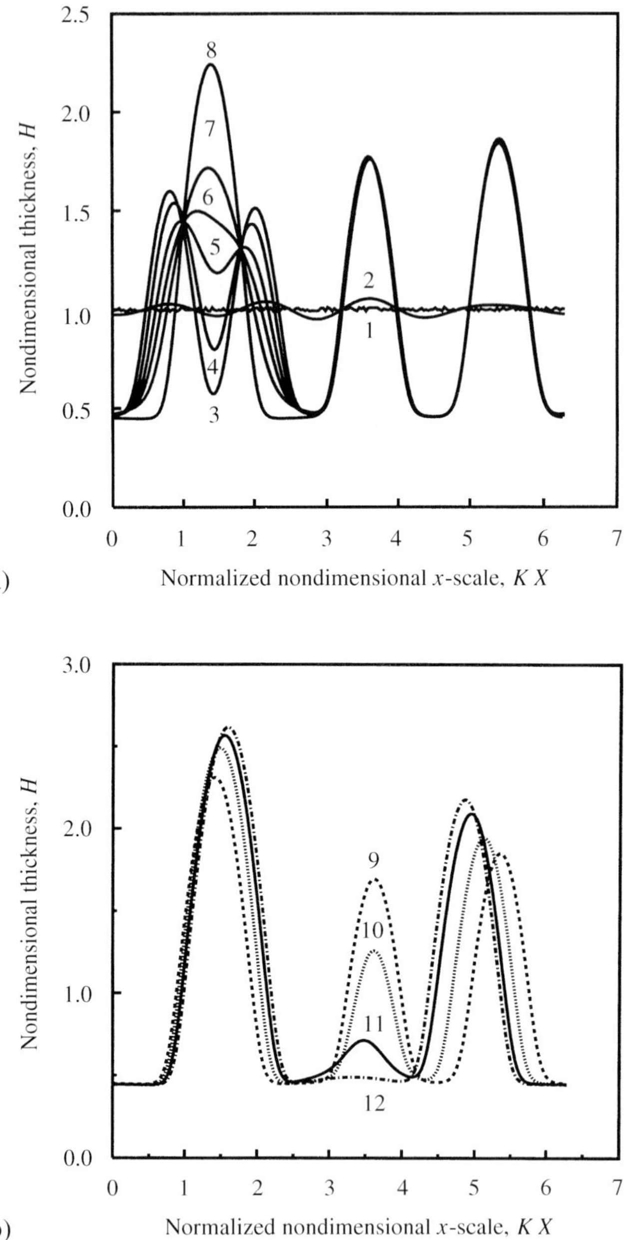

Figure 13. Different stages of roughening (curves 1–3) and ripening (curves 4–16) in a Type II film represented by $R = -0.1$ and $D = 0.5$. The periodicity interval (L) was chosen to be $4\Lambda_m$. Curves 1–16 correspond to $T(\times 10^{-3}) = 0$, 1, 15, 15.6, 15.8, 15.9, 16, 20, 30, 50, 55, 57, 60, 170, 180 and 200, respectively. Initial random perturbation (curve 1) evolves into a four drop structure (curve 3), which ripens into a three drop structure (curve 8) by merger of two smaller drops (a). Similarly, the three drop structure gives way to a two drop structure (curve 12 of (b)). Eventually, a single drop remains (curve 16 of (c)).

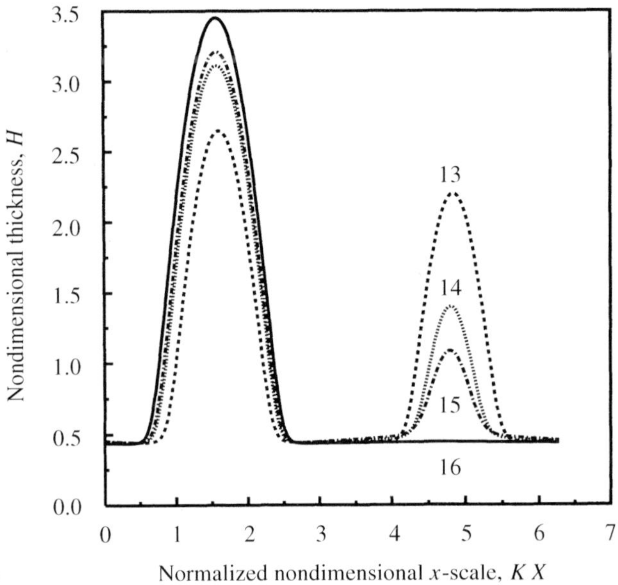

(c)

Figure 13. (Continued).

differences are significant (Fig. 13c), the center of mass of the bigger drop does not shift much but the smaller drop disappears, i.e. the bigger drop 'eats' the smaller one. It may be noted that the time scale for ripening (curves 4–16) is much longer than the time scale for the saturation of the initial instability leading to 'morphological phase separation' (roughening) in the form of quasi-stable coexistence of drops with intervening thin films (curve 3).

The physics of roughening and ripening can be better appreciated by examining the total (nondimensional) free energy including the surface tension and disjoining pressure contributions.

$$F = \int \left(\Delta G^* + \frac{H_X^2}{2} \right) \mathrm{d}x,$$

where ΔG^* is a nondimensional interaction energy defined as (see equation (23)).

$$\Delta G^* = \frac{-1}{6H^2} \left[\left(1 + \frac{D}{H} \right)^{-2} (1 - R) + R \right].$$

Figure 14 presents the reduction of F (solid curve) during initial roughening (regime labelled 4, $T < 2 \times 10^3$), and during later stages of ripening ($T > 10^4$). The intermediate structures with 4, 3 and 2 drops *appear* to be stable for long times, and for this reason they can be easily overlooked in simulations. The nondimensional root mean square roughness (R_F) of the film also increases in steps (broken curve). The abrupt transitions in the free energy and roughness provide the 'drop-level' picture of ripening in which only a single binary-merger is witnessed at one time. The change should be

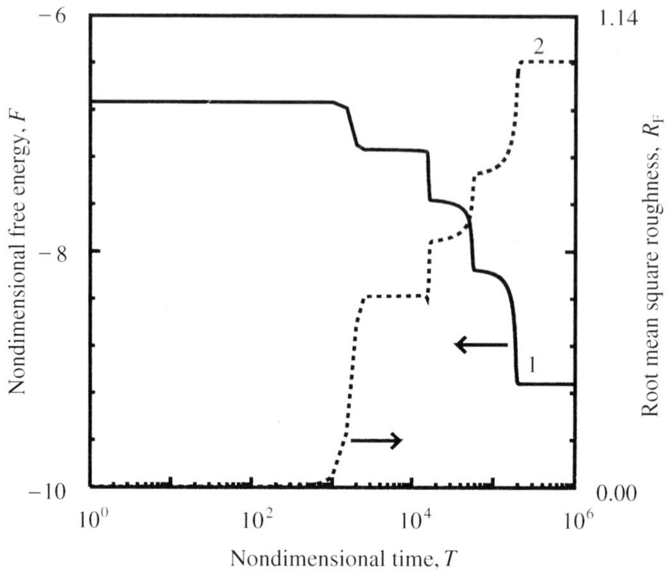

Figure 14. Reduction of the total (nondimensional) free energy (F) during initial roughening to produce a 4 drop structure, and during different stages of ripening to produce 3, 2 and 1 drops (solid curve). Increase in the root mean square 'roughness' (standard deviation of film thickness, R_F) is shown by broken line. The x-scale (T) is nondimensional time.

more continuous for a collection of large number of drops of varying sizes over a sufficiently large area where many 3-D 'drops' are undergoing ripening at the same time.

Rapid step like reductions in the free energy occur at the time of successive stages of ripening of the structure ($4 \to 3 \to 2 \to 1$ drops). Changes in F are confined largely to the last stages of merger of neighboring drops. It is for this reason that the intermediate structures may be termed as 'quasi-stable' or 'kinetically-stable' [45]. It is also clear that intermediate structures persist for longer times as drops grow larger (note that T is shown on log scale).

For an *infinite* drop on a coated surface, the equilibrium contact angle is obtained from equation (17). For the system simulated here, the (nondimensional) equilibrium slope, (H_x) equals 0.494 for an infinite drop. For comparison, the maximum average slope of the quasi-stable 4 drop structure (curve 3 in Fig. 13a) was found to be 0.2 and the slope of the final drop (curve 16) was 0.3. This is expected because smaller drops also display smaller contact angles [17]. Further ripening and the growth of the microstructure would continue to increase the contact angle until the asymptotic value for a macroscopic drop is attained.

Complete 3-D simulations to uncover the 3-D morphology and 2-D dynamics, currently under progress in our group, have shown that we can approximately construct the visual appearance of the 3-D structure by a random superposition of 2-D structures shown in Figs 13a–13c. Such a procedure is rigorous in linear regime [46, 47], or when the interaction between orthogonal waves (x and y directions) is minimal in the nonlinear regime. The latter appears to be a good approximation for Type II films undergoing roughening (MPS) and ripening, as shown in the experiments reported later in this paper. This approximation is also suggested by a closely related problem of patterns in spinodal phase separation and ripening of polymer blends [47].

We simulated the 3-D structure by superposition of 2-D profiles of Fig. 13, $f(KX)$, with random phase-shifts and orientations, viz. by the construction

$$H = C_n \sum_{j=1}^{M} f\left[K\left(X\cos\theta_j + Y\sin\theta_j\right) + \phi_j\right], \tag{27}$$

where M is the number of waves, θ_j is the (random) orientation from 0 to π, ϕ_j is (random) phase shift within the periodicity interval, and C_n normalizes the film thickness by the condition of conservation of mass in the film.

Figure 15 presents the film structure in different stages of roughening and ripening, starting with a small amplitude random disturbance ($M = 20$). The film-texture, and processes of roughening (MPS) and late coarsening are reminiscent of the concentration field in the spinodal decomposition (phase separation) and ripening in nonequilibrium polymer blends [46, 47]. This is not a coincidence, since the density in phase separation is also governed by an equation which is qualitatively similar to equation (1) for the film thickness.

Micrographs in Figs 16 and 17 present the roughening (MPS) in the form of interconnected structure of 'hills' and 'valleys' for some Type II PS films on UV-ozone treated silicon wafers (Fig. 16) and on bi-modal PS brushes (Fig. 17; see Section 3). In Fig. 17, a slow retraction (dewetting) of the film from the edge (where it was broken) is also visible. This behaviour is consistent with curves 1–3 Fig. 2b for the free energy (see Theory section). Thick films of this type show MPS, but true rupture (or dewetting) becomes possible by nucleation at the thinnest parts of the film. Once the energy barrier is locally breached by external means (or by nucleation), a true dewetting is initiated from the broken edges.

Micrographs in Fig. 18 show the sequence of ripening in a 'morphologically phase separated' thin film of PS–PEO on a UV-ozone treated Si wafer (see Section 3). The structure becomes coarser and the heights of polymer ridges increase because of the flow from the thinner (low curvature) to thicker (high curvature) regions (Fig. 18b). Further ripening (Fig. 18c) leads to shorter ridges of greater heights which resemble droplets. Finally, micrographs in Fig. 19 depict the continued ripening of droplets in the form of growth of bigger droplets at the expense of smaller ones. These observations are in accord with the simulations of Figs 13 and 15.

In contrast to Type II systems, the sequence of hydrodynamic instabilities stops in Type I systems with the emergence of final equilibrium disjointed drops (Fig. 9). Since drops are separated by intervening patches of largely dry surface, further direct ripening of drops by viscous flow on the surface is not possible in Type I systems. However, ripening on a very long time scale is still possible through vapor phase mediated transport (by evaporation/condensation as in Ostwald ripening). This form of ripening (not considered here) is engendered by the higher equilibrium vapor pressure of smaller drops (as predicted by the Kelvin equation). For low molecular weight liquids, slow ripening may also occur by yet another mechanism — by diffusion or

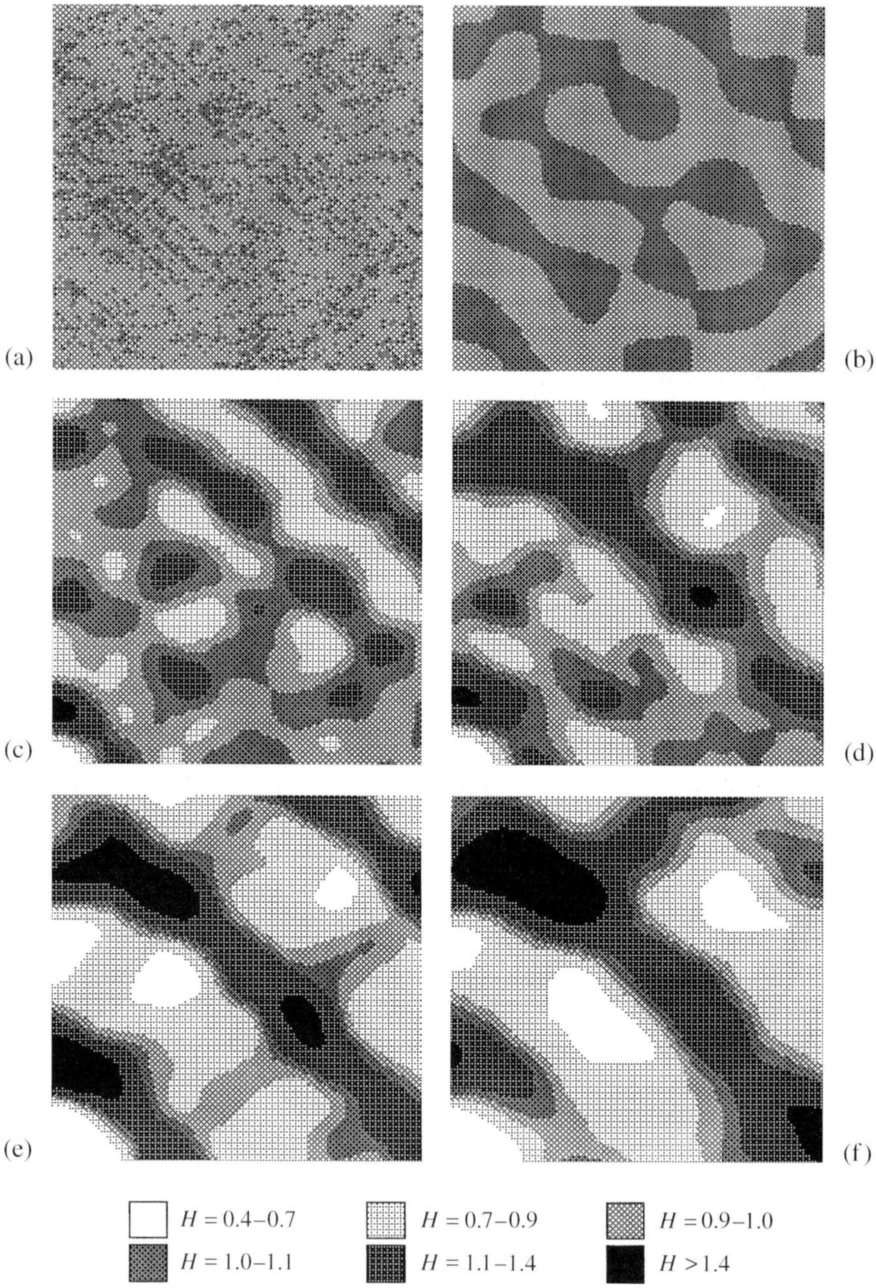

Figure 15. Gray scale appearance of 3-D microstructures in different stages of roughening and ripening. These structures were generated by random superposition of twenty 2-D profiles of Figs 13a–13c. Different shades represent different heights. Figures 15a–15f correspond, respectively, to $T = 0$ (random perturbation), 10^3 (intermediate stage of roughening), 1.5×10^4 (4 drop structure), 3×10^4 (3 drop structure), 6×10^4 (2 drop structure) and 2×10^5 (1 drop structure). Similarly, further ripening should continue among neighboring drops due to gradient of curvature.

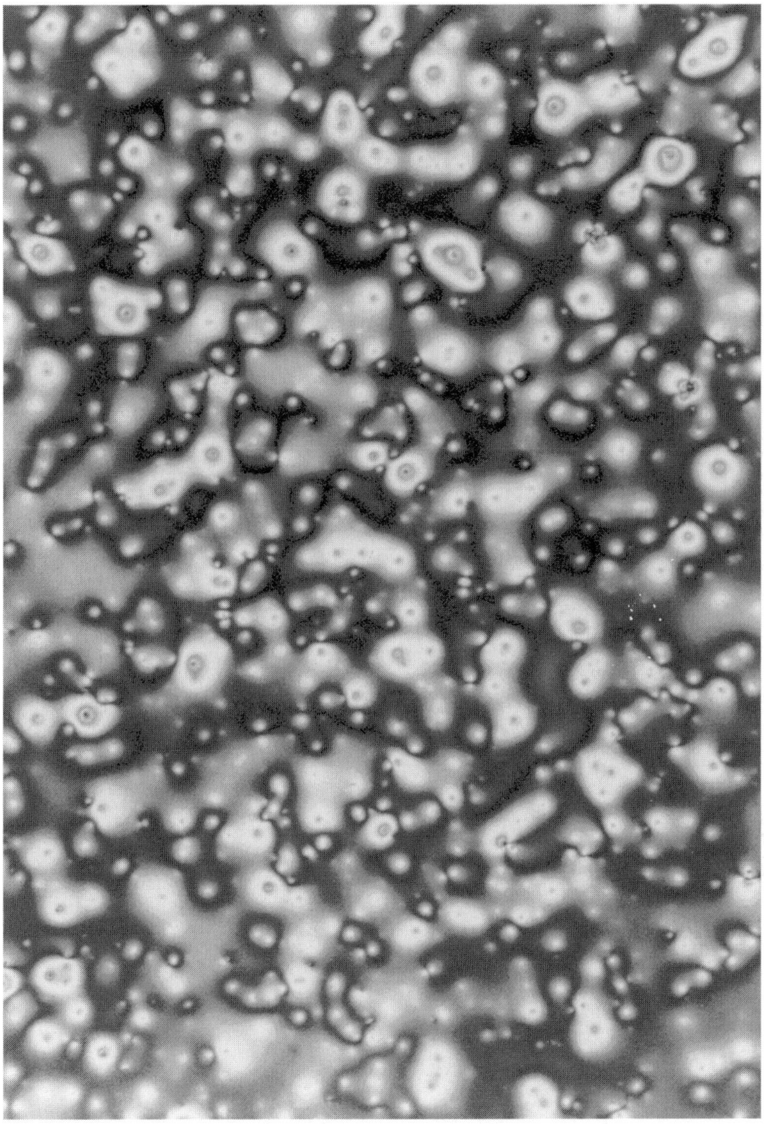

Figure 16. Appearance of a 110 nm thick PS ($M_w = 9$ k) film on a 'UV-ozone + Piranha' treated silicon wafer after annealing at 180°C for 30 min. The area shown is 4500 μm^2.

'hopping' of adsorbed submonolayer molecules on the substrate in the direction of decreasing chemical potential.

The mathematics of ripening by direct viscous flow through connecting films, and its analogy to spinodal decomposition, were also identified in previous studies [19, 45]. However, a Type I system was unfortunately chosen to illustrate the physics of ripening [19, 44], which applies really only to films undergoing roughening (MPS) in the form of an interconnected structure, without a true rupture (e.g. Type II behaviour investigated here).

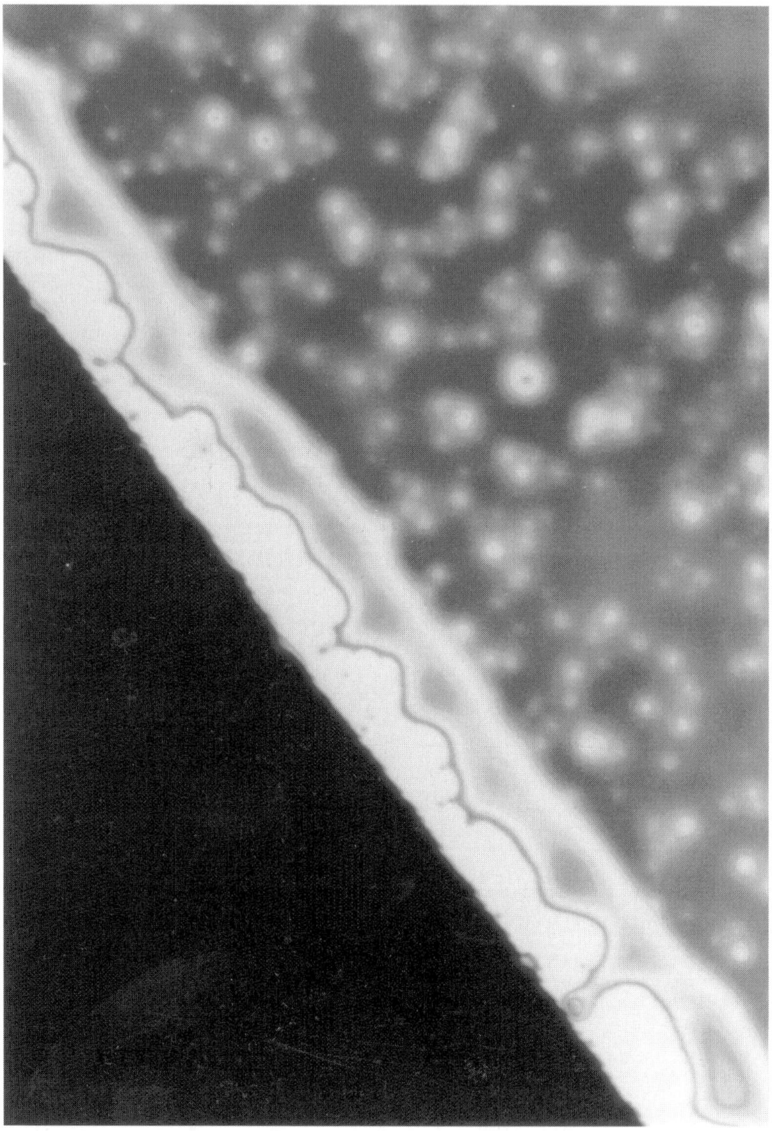

Figure 17. Appearance of a 90 nm thick PS (M_w = 43 k) film on a bimodal brush consisting of a PS (M_w = 9 k) brush (6.4 nm) plus additional 0.8 nm of grafted PS (M_w = 44 k). Annealing was done for 20 min at 180°C. In addition to roughness, retraction of the film from the sample edge is also visible.

Thus, the main difference in the behaviour of Type I and Type II films is the absence of roughening and ripening in the former, because the destabilizing (attractive) force grows stronger with the local film thinning leading to true rupture. In Type II films on the other hand, the (interconnected) structure indeed resembles the density field in spinodal phase separation, and viscous flow through the connecting films leads to the ripening.

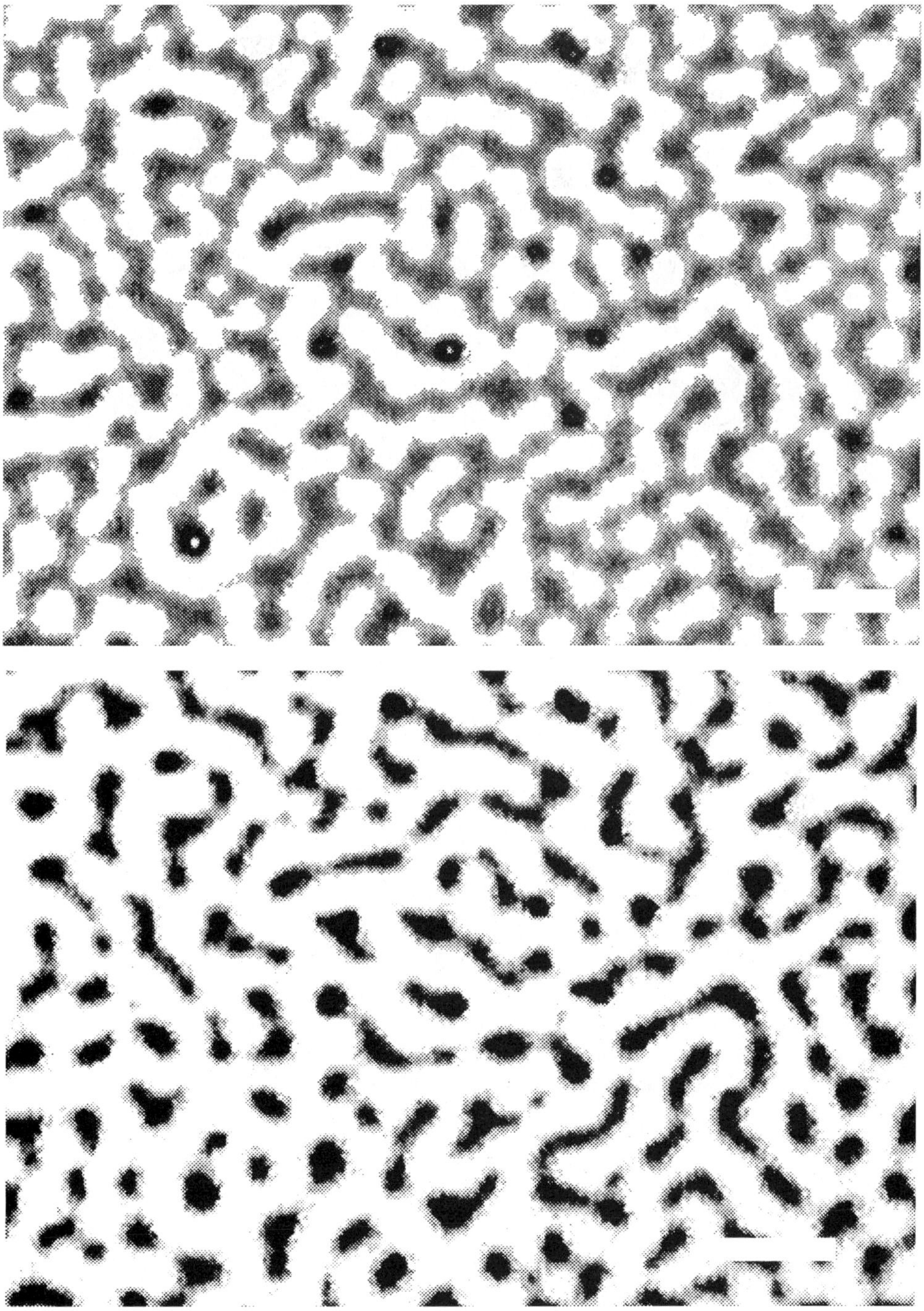

(a)

(b)

Figure 18. Ripening of a 'morphologically phase separated' microstructure in a 34 nm thick PS–PEO (M_w = 3 k + 3 k) block copolymer film on UV-ozone treated silicon wafer: (a) after annealing for 18 min at 122°C, (b) 21.5 min, and (c) 26.5 min. The length of the bar is 25 μm. The darker shades correspond to higher heights. White areas contain films about 20 nm thick.

(c) 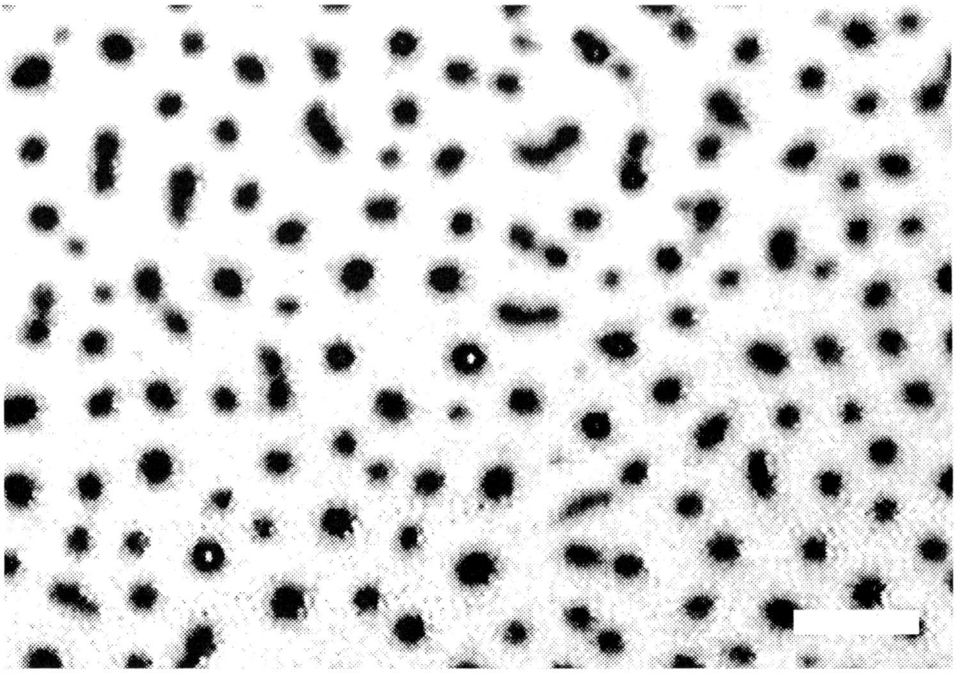

Figure 18. (Continued).

While we have focused on the role of disjoining pressure (and its variations as in Fig. 2) in the film instability, the dynamic behaviour of polymer films is also complicated by a host of other factors, e.g. rheology (visco-elasticity) [40, 41], possible slippage over the surface [10, 42], chain entanglements and chain pull-out [48]. Nevertheless, it appears that many qualitative and quantitative features of instability, adhesion failure and pattern formation in thin films can be uncovered by simulations based on a simple model (equation (1)). The theory and simulations presented here will improve the design of future experiments, and will also aid our understanding of the inverse problem of determination of disjoining pressure characterstics from the observations of film instability.

5. SUMMARY

The free surface of a thin film becomes unstable and deforms spontaneously due to the excess intermolecular interactions whenever the disjoining pressure increases with increase in the film thickness. Three distinct types of unstable thin films may be identified based on their disjoining pressure isotherms (Fig. 2): (Type I) films subjected to the long-range Lifshitz–van der Waals (LW) attraction and the extremely short range Born repulsion; (Type II) films subjected to the LW attraction and a shorter ranged repulsion, which may be derived from 'polar' interactions, adsorption

(a) 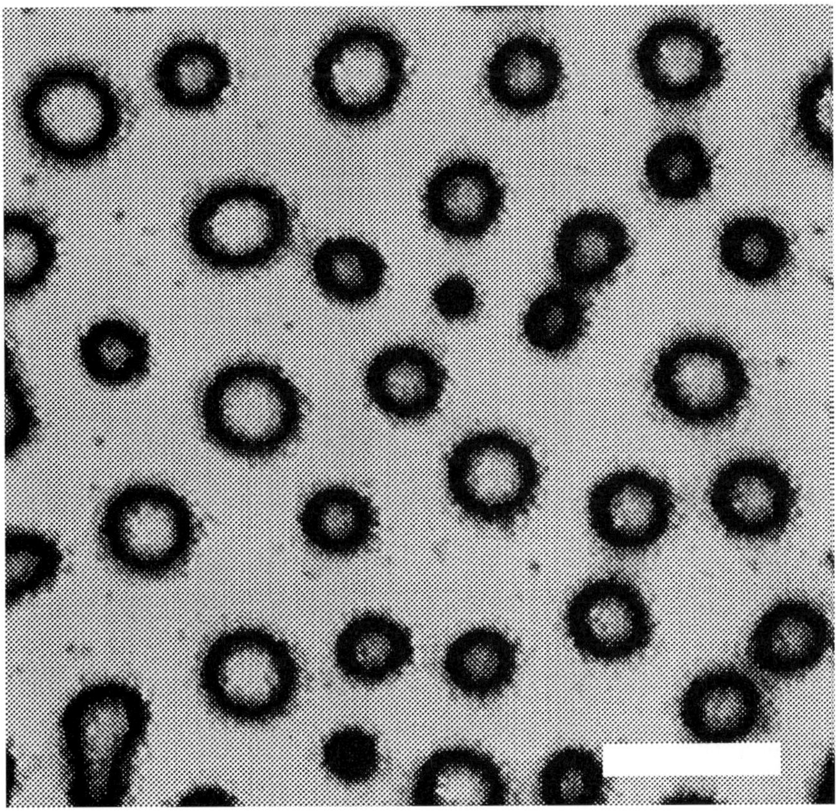

Figure 19. Ripening of droplets produced by Morphological Phase Separation in a 56 nm thick PS–PEO film: (a) after annealing for 16.7 min at 144.5°C, and (b) after 25 min. The length of the bar is 25 μm. The darker shades denote higher heights, except for the lighter shade which is repeated within the droplets, where height is greater than 200 nm.

or wettable coatings; and (Type IV) films subjected to the long-range LW repulsion and a shorter range attraction, which may be derived from polar 'hydrophobic' interaction, or a nonwettable coating.

For Type I films, the evolution of surface instability leads to true dewetting by the following sequence of events: growth of disturbance → hole formation → hole-growth → hole coalescence → polygonal pattern → drops (Figs 3, 4, 8, 9).

The initial length scale of instability determines the number density of holes ($N_H \propto h^{-4}$; Fig. 5). For relatively thick Type I films on coated substrates, N_H is governed only by the wettability (θ_s) of the substrate ($N_H \propto \theta_s^2$), regardless of the wettability of the coating. The polygon (Fig. 10) and drop diameters [12], however, depend on the wetting properties of the coating, as well as the substrate.

The dry-spot (hole) initially expands ($R_H \sim T^{0.75}$) with a rapid concurrent increase in the dynamic contact angle, followed by a long quasi-steady phase of expansion ($R_H \sim T^{0.85}$) where the dynamic contact angle is almost constant (Fig. 6). Eventually,

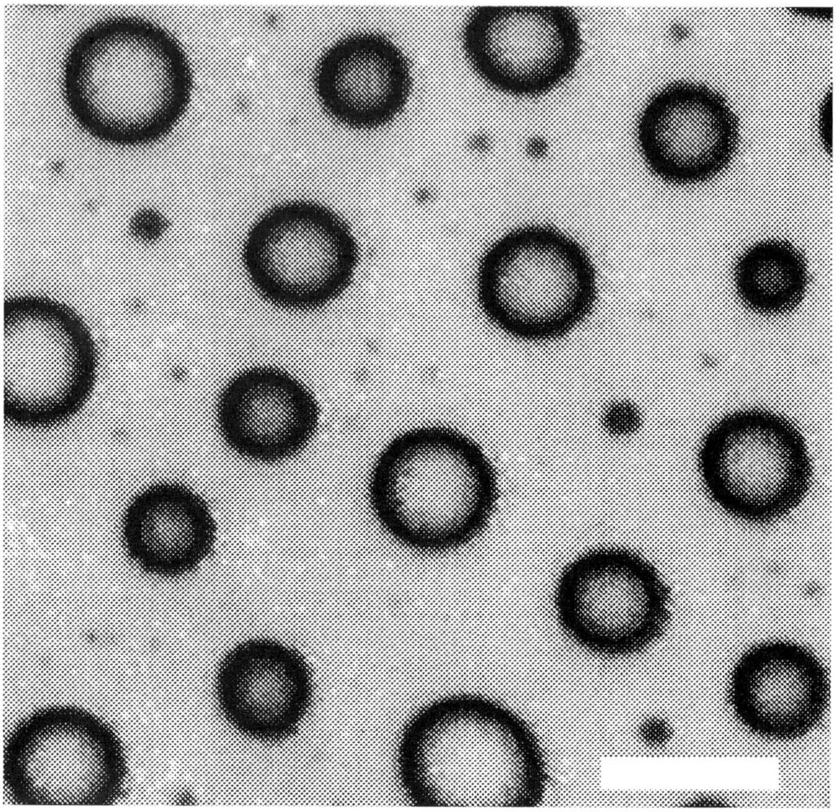

(b)

Figure 19. (Continued).

rims of neighboring holes merge to form a ridge (Figs 3b, 8), before hole coalescence occurs.

During the hole expansion, the film beyond the growing rim remains undisturbed (Figs 3b, 4). The rim is asymmetric and does not merge with the film monotonically, i.e. a depression is created ahead of the moving rim (Figs 3b, 7).

For Type II and Type IV films, the growth of instability eventually saturates without causing a true rupture, unless 'nucleation' can occur due to dust-particle, etc. The resulting 'morphologically phase separated' (MPS) microstructure (Figs 13–19) consists of an array of droplets (high curvature phase) in quasi-stable coexistence with largely flat thinner films (low curvature phase). The MPS structure slowly coarsens (Figs 15, 18, 19) by ripening (merger) of droplets (Figs 13, 19) due to spatial gradients of curvature and disjoining pressure. Ripening is accompanied by almost step like changes in the free energy and roughness (Fig. 14).

It is hoped that a combination of the theory, numerical simulations and experiments presented here will improve the design and interpretation of future experiments on stability, dynamics, morphology, dewetting and adhesion failure of thin films.

Acknowledgements

This work was partially supported by the Indo-French Centre for the Promotion of Advanced Research/Centre Franco-Indien Pour la Promotion de la Recherche.

REFERENCES

1. A. Sheludko, *Adv. Colloid Interface Sci.* **1**, 391 (1967).
2. A. Vrij and J. Th. G. Overbeek, *J. Am. Chem. Soc.* **90**, 3074 (1968).
3. E. Ruckenstein and R. K. Jain, *Faraday Trans.* **70**, 132 (1974).
4. M. B. Williams and S. H. Davis, *J. Colloid Interface Sci.* **90**, 220 (1982).
5. A. Sharma and E. Ruckenstein, *J. Colloid Interface Sci.* **113**, 456 (1986).
6. G. F. Teletzke, H. T. Davis and L. E. Scriven, *Chem. Eng. Commun.* **55**, 41 (1987).
7. A. Sharma, *Langmuir* **9**, 861 (1993).
8. G. Reiter, *Phys. Rev. Lett.* **68**, 75 (1992).
9. G. Reiter, *Langmuir* **9**, 1344 (1993).
10. C. Redon, J. B. Brzoska and F. Brochard-Wyart, *Macromolecules* **27**, 468 (1994).
11. G. Reiter, *Macromolecules* **27**, 3046 (1994).
12. A. Sharma and G. Reiter, *J. Colloid Interface Sci.* **178**, 383 (1996).
13. W. Zhao, M. H. Rafailovich, J. Sokolov, L. J. Fetters, R. Plano, M. K. Sanyal, S. K. Sinha and B. B. Sauer, *Phys. Rev. Lett.* **70**, 1453 (1993).
14. J. M. Guerra, M. Srinivasrao and R. S. Stein, *Science* **68**, 75 (1992).
15. G. Reiter, P. Auroy and L. Auvray, *Macromolecules* **29**, 2150 (1996); see also P. Auroy and L. Auvray, *Macromolecules* **29**, 337 (1996).
16. A. Sharma and A. T. Jameel, *J. Colloid Interface Sci.* **161**, 190 (1993).
17. A. T. Jameel and A. Sharma, *J. Colloid Interface Sci.* **164**, 416 (1994).
18. A. Sharma and A. T. Jameel, *Chem. Soc. Faraday Trans.* **90**, 625 (1994).
19. V. Mitlin, *J. Colloid Interface Sci.* **156**, 491 (1993).
20. F. Brochard-Wyart, P. G. de Gennes and H. Hervet, *Adv. Colloid Interface Sci.* **34**, 561 (1991).
21. W. A. Zisman, in: *Contact Angle, Wettability and Adhesion*, Adv. Chem. Ser. No. 43, p. 1. American Chemical Society, Washington, DC (1964).
22. C. W. Extrand, *Langmuir* **9**, 475 (1993).
23. P. Silberzan and L. Leger, *Phys. Rev. Lett.* **66**, 185 (1994).
24. C. J. van Oss, M. K. Chaudhury and R. J. Good, *Chem. Rev.* **88**, 927 (1988).
25. C. J. van Oss, *Colloids Surfaces A* **78**, 1 (1993).
26. J. N. Israelachvili, *Intermolecular and Surface Forces*. Academic Press, London (1985).
27. K. R. Shull, *J. Chem. Phys.* **94**, 5723 (1991).
28. K. R. Shull, *Faraday Discuss.* **98**, 203 (1994).
29. R. Khanna, A. T. Jameel and A. Sharma, *Ind. Eng. Chem. Res.* **35**, 3081 (1996).
30. V. S. Mitlin and M. M. Sharma, *J. Colloid Interface Sci.* **157**, 447 (1993).
31. F. Brochard-Wyart, J. M. di Meglio, D. Quere and P. G. de Gennes, *Langmuir* **7**, 335 (1991).
32. L. Leibler, A. Ajdari, A. Mourran and G. Coulon, in: *Ordering in Macromolecular Systems*, A. Teramoto, M. Kobayashi and T. Norisuje (Eds), pp. 301–311. Springer-Verlag, Berlin (1994).
33. G. J. Hirasaki, in: *Thermodynamics of Thin Films and Three Phase Contact Regions in Interfacial Phenomena in Petroleum Recovery*, N. R. Morrow (Ed.), pp. 23–27. Marcel Dekker, New York (1991).
34. A. Sharma, *Langmuir* **9**, 3580 (1993).
35. E. F. Hare and W. A. Zisman, *J. Phys. Chem.* **59**, 335 (1955).
36. E. Vitt and K. R. Shull, *Macromolecules* **28**, 6349 (1995).
37. A. T. Jameel, Ph.D. thesis, Indian Institute of Technology at Kanpur, India (1994).
38. A. Sharma, C. S. Kishore, S. Salaniwal and E. Ruckenstein, *Phys. Fluids A* **9**, 1832 (1995).
39. G. T. Dee and B. B. Sauer, *J. Colloid Interface Sci.* **152**, 85 (1992).

40. S. A. Safran and J. Klein, *J. Phys. II France* **3**, 749 (1993).
41. D. J. Srolovitz and S. A. Safran, *J. Appl. Phys.* **60**, 255 (1986).
42. F. Brochard-Wyart, P. G. de Gennes, H. Hervet and C. Redon, *Langmuir* **10**, 1566 (1994).
43. L. Leǵer and J. F. Joanny, *Rep. Prog. Phys.* **55**, 431 (1992).
44. C. Redon, F. Brochard-Wyart and F. Rondelez, *Phys. Rev. Lett.* **66**, 715 (1991).
45. V. Mitlin and N. V. Petviashvili, *Phys. Lett. A* **192**, 323 (1994).
46. J. W. Cahn, *J. Chem. Phys.* **42**, 93 (1965).
47. S. Reich, *Phys. Lett.* **114A**, 90 (1986).
48. G. Reiter, J. Schultz, P. Auroy and L. Auvray, *Europhys. Lett.* **33**, 29 (1996).

Mittal Festschrift, pp. 239–253
W. J. Van Ooij and H. R. Anderson, Jr. (Eds)

Retarded wetting and dewetting on elastomeric substrates

M. E. R. SHANAHAN[1] and A. CARRÉ[2]

[1]*Centre National de la Recherche Scientifique, Ecole Nationale Supérieure des Mines de Paris, Centre des Matériaux P.M. Fourt, B.P. 87, 91003 Evry Cédex, France*
[2]*Centre de Recherche Corning Europe, 7 bis, Avenue de Valvins, 77210 Avon, France*

Abstract—Many wetting processes involve interpretation using Young's equation to describe contact angle equilibrium on a solid surface. By assuming the solid to be rigid, no account is made of the component of the liquid surface tension perpendicular to the solid surface. It is shown that a wetting ridge must be formed and, although negligible for hard solids, this mesoscopic disturbance of the solid near the wetting front can have significant consequences on a soft solid. The theory of triple line displacement, taking into account viscoelastic dissipation in the wetting ridge, is developed both for wetting and dewetting processes. Experimental studies using tricresyl phosphate and two types of model solid — a rigid polymer and silicone elastomers — have been carried out. Both for wetting and dewetting, triple line motion is markedly slowed down on the soft solids as a result of viscoelastic dissipation near the triple line. Theoretical predictions and experimental findings are found to be in good agreement.

Keywords: Dewetting; dry patch; elastomers; spreading; surface tension; viscoelasticity; wetting (kinetics); wetting ridge.

1. INTRODUCTION

Wetting represents a topic of considerable importance both in adhesion and tribology, to name but two domains in the natural and industrial world. Most adhesives make contact with their substrate in the liquid state and thus relative surface and interfacial free energies must have suitable values to assure good wetting. Similarly, a liquid lubricant must adequately cover solid surfaces and intervene to impede solid/solid contact. Equilibrium wetting is essentially governed by the three interfacial tensions, γ_{ij}, where i and j represent solid, S, liquid, L, and vapour, V, and under ideal conditions is related to the contact angle, θ_o, by Young's equation [1]:

$$\gamma_{SV} = \gamma_{SL} + \gamma_{LV} \cos \theta_o. \tag{1}$$

(Vapour, V, may be replaced by a second, immiscible liquid but in the following we shall take the surrounding fluid as gaseous.)

Satisfaction of equation (1) with a low or zero value of θ_o is deemed necessary for good adhesion to ensue: even if final interfacial bonds are of a chemical nature, good initial contact between the two phases is a necessary, even if insufficient, condition for acceptable adhesion. However, if the adhesive, after application, hardens (by crosslinking, solvent evaporation, etc.) too rapidly, before spreading to equilibrium has been attained, insufficient contact may result. Thus the kinetics of wetting, or spreading, is of utmost importance. Similarly, spreading phenomena are of paramount importance in other contexts, such as tribology, the application of paints and inks and the production of composite materials.

The kinetics of wetting under 'classical' conditions is now well established, at least as far as the basic physics is concerned. In the case of an ideal solid (flat, smooth, homogeneous, isotropic and rigid) on which a (small — gravitational effects are ignored here) sessile drop is deposited, spreading rate to equilibrium is governed by a dynamic equilibrium. The lack of capillary balance furnishes potential energy which is consumed, at a diminishing rate, by viscous dissipation within the liquid [2]. In the present study, we relax the constraint of having an absolutely rigid solid. As will be seen below, if we allow the solid to be (sufficiently) deformable, a 'wetting ridge' forms near the L/S/V triple line, due to γ_{LV} $(= \gamma)$ 'pulling' on the solid. The height of this ridge is of the order of $\gamma \sin \theta / Y$ where Y is Young's modulus of the solid and θ is the contact angle, either θ_o at equilibrium or $\theta(t)$ before equilibrium is attained (a function of time, t). This solid deformation may have an influence on Young's equation, (1), under certain special conditions (soft solid and very small drop) [3], but of more importance is the action of the wetting ridge on the kinetics of wetting. The mesoscopic ridge can have macroscopic effects in spreading leading to retarded behaviour. These will be described below.

2. THE WETTING RIDGE

2.1. Statics

Consider Fig. 1 which represents the local region of the triple line of a drop of liquid on a deformable solid. The vertical component of the liquid surface tension must be balanced by a stress field in the vicinity of the triple line. We shall simplify by assuming a typical, average stress σ acting in a region of extent ω, the perturbed zone. Thus for equilibrium we have:

$$\gamma \sin \theta \approx \sigma \omega. \tag{2}$$

The local strain, ε, will be given, in order of magnitude, by:

$$\varepsilon \approx \frac{h}{\omega}, \tag{3}$$

where h is the height of the wetting ridge. Assuming the solid to be elastic, we have $\sigma = \varepsilon Y$ and using this with relations (2) and (3), we obtain:

$$h \approx \frac{\gamma \sin \theta}{Y}. \tag{4}$$

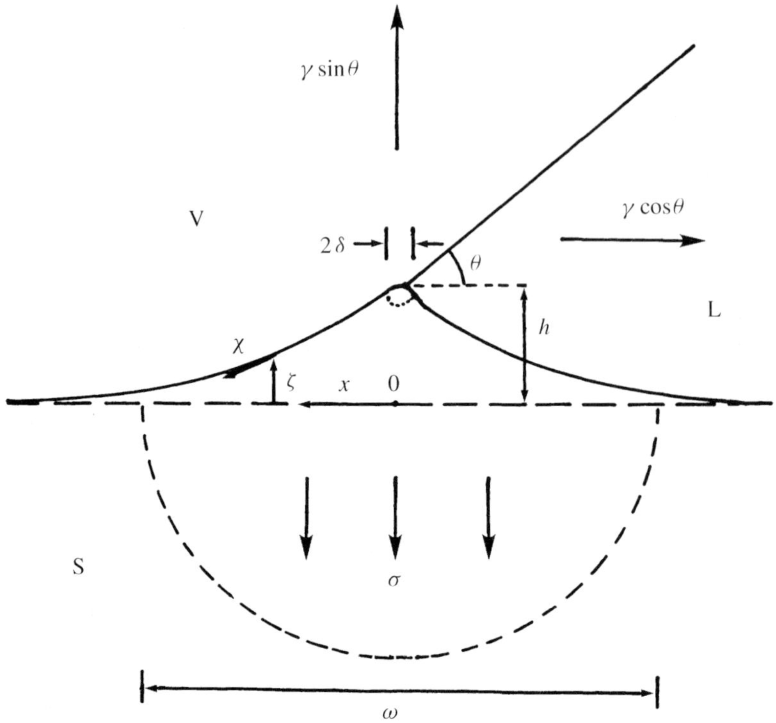

Figure 1. Schematic representation of the wetting ridge produced near the triple line and the nomenclature used in the text. $\gamma \sin\theta$ leads to vertical disturbance, ζ, and $\gamma \cos\theta$ to surface stretching, χ.

This very simple scaling law shows how the ratio γ/Y essentially governs the dimensions of the wetting ridge and that for hard solids, the effect becomes negligible. As an example, for steel $Y \approx 200$ GPa and for mercury, $\gamma \approx 500$ mN m^{-1}, thus leading to $h \approx 2.5 \times 10^{-12}$ m, assuming that $\theta = 90°$. Clearly this value of h is meaningless and we may take it that no deformation occurs. However, an elastomer may typically present a Young's modulus of *ca.* 1 MPa and with a liquid of $\gamma \approx$ 50 mN m^{-1} (the range of usual liquids), $h \approx 50$ nm, small yet non-negligible. It will be shown later how this *meso*scopic phenomenon can produce *macro*scopic effects.

The above gives a very simple demonstration of the order of magnitude but a more exact approach has been used to obtain the form of the wetting ridge [4, 5]:

$$\zeta(x) = \frac{\gamma \sin\theta}{2\pi\mu} \ln\left(\frac{d}{|x|}\right); \quad |x| > \delta, \tag{5}$$

where ζ is the (deformed) height of the ridge as a function of distance x from the triple line (see Fig. 1), μ is the shear modulus of the solid and d is a constant (datum depth) as usually obtained in two-dimensional problems of stress analysis. The cut-off at δ corresponds to the lower limit of linear elastic behaviour and is essential to prevent divergence of the logarithmic function. Equation (5) corresponds to an elastomer so that Poisson's ratio is equal to 0.5 (N.B. θ is still defined conventionally as being the angle subtended by tangents to the L/V interface at the triple line and the *undisturbed* solid surface).

In addition to the vertical deformation due to $\gamma \sin\theta$, the horizontal component, $\gamma \cos\theta$, leads to an effect of 'stretching' of the elastomeric surface, χ, on the vapour side of the triple line for $\theta < 90°$ and on the liquid side for $\theta > 90°$ [6]:

$$|\chi(x)| = \frac{\gamma|\cos\theta|}{2\pi\mu}\ln\left(\frac{d'}{|x|}\right); \quad |x| > \delta, \tag{6}$$

where d' is another constant.

2.2. Dynamics

Let us now consider the triple line and its associated wetting ridge in motion, at speed U. As a simple approach we may use scaling concepts as above. With average vertical stress σ in the zone of (linear) dimension ω below the solid surface, and influenced by the liquid surface tension, γ, we have a stored energy density of order σ^2/Y leading to an overall value of stored elastic energy, E_1, of *ca.* $(\sigma\omega)^2/Y$ (per unit length of triple line). Using relation (2), we find $E_1 \approx \gamma^2 \sin^2\theta/Y$. However this relationship is obtained assuming that the only contribution to local stress is vertical, σ, corresponding to $\gamma \sin\theta$. However, the horizontal component of stress should also be taken into consideration leading to $E_2 \approx \gamma^2\cos^2\theta/Y$ and thus the overall stored elastic energy should be of the form:

$$E \approx E_1 + E_2 \approx \frac{\gamma^2}{Y}. \tag{7}$$

As the triple line moves, the elastomer in the deformed region will undergo a strain cycle at speed U. A variety of frequencies is involved depending on the position of the deformed zone with respect to the triple line, ranging from U/ω, up to $\nu \approx U/\delta$ at the cut-off at $x = \delta$. The latter will be dominant since the region near the triple line is the most strained. As a result, the rate at which work is done, $\dot{E}$, is of the order $\gamma^2 U/(Y\delta)$. Defining the complex modulus as $Y^*(\nu) = Y'(\nu) + \mathrm{i}Y''(\nu)$ where Y' and Y'' are respectively storage and loss moduli of the elastomer and $\mathrm{i}^2 = -1$, the quantity of energy dissipated (per unit time and length of triple line) is:

$$\dot{E}\Delta \approx \frac{\gamma^2 U}{Y\delta}\cdot\frac{Y''(U)}{Y'(U)}, \tag{8}$$

where Δ represents $Y''(U)/Y'(U)$ and corresponds to the fraction of energy dissipated, or loss tangent of the elastomer at frequency $\nu \approx U/\delta$.

Equation (8) summarises the basic physics in simple terms but a more exact approach involves use of equations (5) and (6). As the triple lines moves at speed U, the work done in 'pulling up' and 'stretching' the elastomer ahead of the triple line is given by [6]:

$$\dot{E} = \gamma U\left(\left|\frac{\mathrm{d}\zeta}{\mathrm{d}x}\right|\sin\theta(t) + \left|\frac{\mathrm{d}\chi}{\mathrm{d}x}\right|\cos\theta(t)\right), \tag{9}$$

where both differentials are evaluated at $x = \delta$. Here we neglect the (small) region of elastomer contained in the immediate proximity of the three phase line (i.e. $|x| < \delta$). Evaluation of equation (9) using equations (5) and (6) leads to:

$$\dot{E} \approx \frac{\gamma^2 U}{2\pi\mu\delta}, \tag{10}$$

which is comparable to relation (8) (neglecting, for the moment, Δ). [In fact equation (8), derived on purely scaling arguments, suggests that the rate at which work is done is about twice that given by equation (10).]

The use of the loss tangent in equation (8) is really an oversimplification and, with the lack of any rigorous theoretical arguments relating the fraction of energy dissipated, Δ, to the range of strain rates covered in the disturbed region of the elastomer, we refer to an empirical relation initially used in the study of rate dependent adhesion of elastomers [7]. (It should perhaps be noted, even so, that a theoretical attempt to explain the relation has been made [8].) The following relation was proposed by Maugis and Barquins [7] to explain the adhesion of elastomers in various test configurations:

$$G - W_o = W_o\phi(a_T v), \tag{11}$$

where G is strain energy release rate, W_o is the intrinsic energy of adhesion, $\phi(a_T v)$ is a dimensionless function of separation rate, v, and a_T is the Williams, Landel and Ferry time–temperature shift factor [9]. It was shown that $\phi(a_T v)$ varies as $(a_T v)^n$ where the constant $n \approx 0.6$. We shall therefore write the rate of energy dissipation (per unit length of triple line) as:

$$\dot{E}\Delta \approx \frac{\gamma^2 U}{2\pi\mu\delta}\left(\frac{U}{U_o}\right)^n, \tag{12}$$

where U_o is a constant, or characteristic triple line speed.

3. DYNAMICS OF TRIPLE LINE MOTION

When a drop of liquid is placed on a solid surface, it is not usually at equilibrium initially. In fact, in general, its contact angle, θ, is greater than the equilibrium value, θ_o, and spreading ensues, as shown schematically in Fig. 2. This corresponds

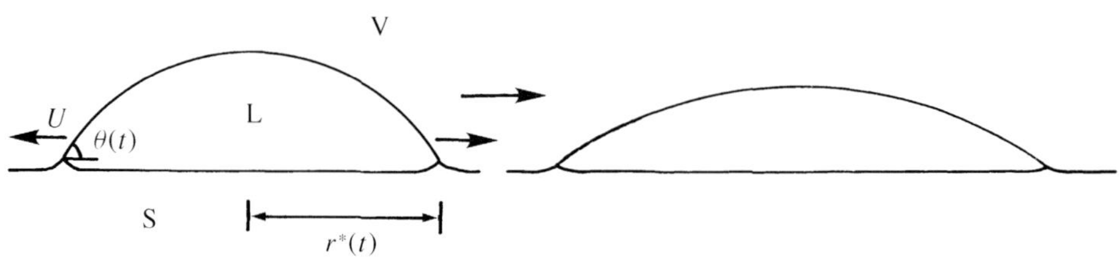

Figure 2. Spreading of a sessile drop accompanied by displacement of the wetting ridge.

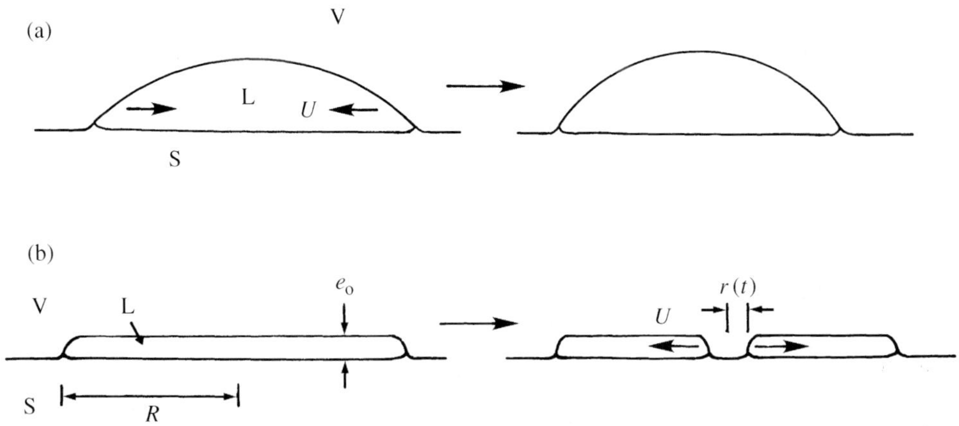

Figure 3. Dewetting in (a) the case of a sessile drop after 'splashing' (contact angle smaller than the equilibrium value) and (b) the case of dry patch nucleation and growth.

to wetting. However, in some circumstances, for example after letting a drop of liquid fall or splash, rather than being placed, onto a solid, it may actually have a contact angle less than θ_o, in which case retraction of the triple line, or dewetting, will occur, as shown in Fig. 3a. Dewetting may also take place when a thin film of liquid is forced to cover a solid, despite its unwillingness to wet the surface. Nucleation of a hole will occur and this dry patch will spread, as shown in Fig. 3b. In all of these cases, motion of the triple line is involved and thus, on an elastomeric or soft solid, displacement of a wetting ridge occurs. We shall discuss the spreading case of Fig. 2 and the patch formation process of Fig. 3b (dewetting by a drop is really a hybrid of these two).

3.1. Spreading of a drop on an elastomeric surface

By analogy with equation (11), which applies to adhesion, we may treat the spreading of a (small) drop of liquid on a viscoelastic substrate using the language of fracture mechanics. The first thing to observe is that equation (11) applies to an *opening* interface whereas in the present case, spreading corresponds to a *closing* interface, so the left hand member of equation (11) should be inverted to $W_o - G$. Since the excess capillary energy, or force, is due to non-satisfaction of Young's equation (the entire treatment below considers unit length of triple line), we have:

$$W_o - G = \gamma_{SV} - \gamma_{SL} - \gamma \cos\theta(t) = \gamma\big(\cos\theta_o - \cos\theta(t)\big). \quad (13)$$

It should perhaps be pointed out that in equation (13), the angles to be considered are indeed θ_o and $\theta(t)$. As can be seen from the schematic representation of the triple line zone shown in Fig. 1, the angle subtended in the liquid phase between the tangents to the liquid/vapour and solid/liquid interfaces at the triple line may be considerably greater than θ when a wetting ridge exists. Nevertheless, as demonstrated in [3], Young's equation still holds except when the drop is exceedingly small and thus θ, the angle measured with respect to the undeformed solid, is the correct angle to be used when evaluating excess capillary energy. The overall, macroscopic thermodynamics remains unaffected by the increase in angle subtended at a mesoscopic scale.

By definition (Dupré's equation [10]):

$$W_o = \gamma + \gamma_{SV} - \gamma_{SL} = \gamma\big(1 + \cos\theta_o\big), \tag{14}$$

resulting in:

$$G = \gamma\big(1 + \cos\theta(t)\big). \tag{15}$$

Thus the equivalent of a 'strain energy release rate' in the present case can be considered as an 'instantaneous' value of Dupré's energy of adhesion, with $\theta = \theta(t)$ instead of the equilibrium value! Clearly there is no stored elastic energy in the spreading liquid. Equation (13) may be considered as the 'motor' producing spreading, its power being expressed as $U(W_o - G)$ at triple line speed U.

As spreading ensues, the energy released by the 'motor' is dissipated by 'braking' effects. Under conditions of wetting on a rigid solid, virtually the totality of this dissipation occurs due to viscous shear in the liquid. A calculation of this rate of dissipation, $T\dot{S}$, has been previously given [2] and it can be expressed as:

$$T\dot{S} \approx \frac{3\eta U^2 \ell}{\theta(t)}, \tag{16}$$

where η is liquid viscosity, and ℓ represents the logarithm of the ratio of two lengths, the numerator corresponding approximately to drop contact radius and the denominator, a small distance from the contact line necessary to avoid divergence (a singularity of the flow-field). Equation (16) is a simplification based on the so-called lubrication approximation, in which flow is taken to be predominantly parallel to the (undeformed) solid surface, and thus only strictly applies to low values of θ. Nevertheless, it is worth noting that viscous dissipation is smaller for higher contact angles.

When the solid is a soft elastomer, we have a supplementary viscoelastic dissipation term, given by equation (12). Thus the dynamic energy balance takes the form:

$$U\big(W_o - G\big) \approx T\dot{S} + \dot{E}\Delta. \tag{17}$$

Large contact angles lead to low viscous dissipation, as pointed out above and, in addition, as found experimentally, $T\dot{S} \ll \dot{E}\Delta$. Thus to a good approximation we may reasonably neglect $T\dot{S}$ and equations (12) and (17) lead to:

$$W_o - G \approx \frac{\gamma^2}{2\pi\mu\delta}\left(\frac{U}{U_o}\right)^n. \tag{18}$$

The similarity between equation (18) and equation (11), describing adhesion, is clear. Making use of equation (13) and rearranging, we find:

$$U \approx U_o\left[\frac{2\pi\mu\delta}{\gamma}\big(\cos\theta_o - \cos\theta(t)\big)\right]^{1/n}. \tag{19}$$

Equation (19) describes the dependence of spreading speed of the triple line of a small liquid drop on the change in contact angle as it approaches its equilibrium value, under conditions in which viscous dissipation may be considered negligible [11, 12].

3.2. Dewetting of a liquid film on an elastomeric surface

We consider a circular 'puddle' of liquid L on solid S in the presence of vapour V (see Fig. 3b). Its radius and thickness are respectively R and e_o before a small hole nucleates at the centre and starts to grow with radius $r(t)$. This nucleation is the result of instability of the wetting film since the equilibrium contact angle, θ_o, is finite. We consider $r(t) \ll R$ and $U = \mathrm{d}r/\mathrm{d}t$ relatively small, anticipating viscoelastic braking effects. Under these conditions, film thickness, $e(t) \approx e_o$, and any thickening of the film in the vicinity of the triple line may be neglected, in contrast to dewetting on a rigid solid where triple line speed is sufficient to cause liquid build-up, or a rim, to occur [13].

As for drop spreading, we shall consider the energy balance for dewetting per unit length of triple line. The driving force for dewetting is given simply by $\gamma + \gamma_{SL} - \gamma_{SV}$, if, which seems reasonable, we take it that the contact angle at the receding triple line remains approximately constant (and close to the equilibrium value). Dewetting is more directly akin to the propagation of a fracture front in adhesion experiments, such as peel, and thus, by analogy with equation (11), we have:

$$G - W_o = \gamma + \gamma_{SL} - \gamma_{SV} = \gamma(1 - \cos\theta_o). \tag{20}$$

With the definition of the Dupré reversible energy of adhesion, W_o, [relation (14)], we obtain:

$$G = 2\gamma, \tag{21}$$

which may be compared to the value of the 'strain energy release rate' for drop spreading, given by equation (15). In this case, G is equivalent to the energy of cohesion of the liquid; equation (20) represents the dewetting 'motor'.

The energy balance of equation (17) also applies to dewetting, provided a sign change is effected for the left hand member, and so we can write:

$$G - W_o = k\eta U + \frac{\gamma^2}{2\pi\mu\delta}\left(\frac{U}{U_o}\right)^n. \tag{22}$$

In equation (22), the first term on the right hand side is equivalent to expression (16), but since contact angle is assumed essentially constant during dewetting, θ, as well as ℓ, are absorbed in the constant k.

Equations (18) and (22), respectively for spreading and dewetting, are directly comparable. Apart from the sign change accounting for 'crack closing' or 'crack opening', the only difference is that we have not neglected the viscosity term in the latter case, for reasons which will become clear below.

For a known value of the constant n, equation (22) can, in principle, be solved initially as an algebraic equation (although this may not be easy with expected values of

n; $n < 1$ but is not necessarily an easy number to handle) and then simply integrated to give the radius of the hole as a function of time, $r(t)$ (note: $U = \mathrm{d}r/\mathrm{d}t$). However, as the algebraic step may be delicate, we shall anticipate that the viscous effect will be small compared to the viscoelastic contribution to dissipation and thus neglect the former. Equations (20) and (22) then give, after simple integration:

$$r(t) \approx U_o \left[\frac{2\pi \mu\delta(1 - \cos\theta_o)}{\gamma} \right]^{1/n} t, \tag{23}$$

and thus we may expect that the triple line will recede at constant speed, as indeed is the case without viscoelastic dissipation, provided gravitational effects remain negligible [13, 14]. (Note that inclusion of the viscous contribution will not alter the constancy of triple line recession rate, only the actual value of the speed.)

Let us consider the hypothetical case in which the solid becomes rigid, all other factors remaining the same, and compare this to that in which viscoelastic dissipation dominates. Ignoring the second term on the right hand side of equation (22), we have the case of the rigid solid. Let U_1 represent recession rate with viscoelastic dissipation and U_2 the equivalent without it, such that $U_2 = mU_1$ with $m \geqslant 1$; then:

$$G - W_o = k\eta U_1 + \frac{\gamma^2}{2\pi\mu\delta}\left(\frac{U_1}{U_o}\right)^n, \tag{24}$$

$$G - W_o = k\eta U_2, \tag{25}$$

leading to:

$$(m - 1)k\eta U_1 = \frac{\gamma^2}{2\pi\mu\delta}\left(\frac{U_1}{U_o}\right)^n. \tag{26}$$

Substitution of relation (26) into (24) gives:

$$G - W_o = k\eta U_1 + (m - 1)k\eta U_1, \tag{27}$$

where the right hand terms are respectively the viscous and viscoelastic contributions to dissipation. Thus the ratio of viscoelastic to viscous energy dissipation is equal to $(m - 1)$. As the recession speeds vary in a ratio of m, the dissipation rates due uniquely to viscous effects for the two cases are in a ratio of m^2.

4. EXPERIMENTAL

4.1. Drop spreading

The liquid chosen for this study was tricresyl phosphate (TCP) (Aldrich Chemical Company, Inc.), representing a non-volatile product of reasonably low viscosity ($\eta = 0.070$ Pa s) and a fairly average surface tension ($\gamma = 40.9$ mN m^{-1}). Small sessile drops (2 μl) were placed on the substrates and their contact angles, $\theta(t)$, measured as a function of time until equilibrium was attained using a Ramé-Hart contact angle goniometer. The solids were placed in a glass box in which the atmosphere was saturated in TCP vapour, both to prevent evaporation of liquid from the sessile drops and to reduce possible water vapour adsorption from the atmosphere.

Results obtained with two solids are presented here, corresponding to a relatively rigid polymer (Teflon PFA, du Pont de Nemours — shear modulus $\mu \approx 250$ MPa) and a reasonably soft elastomer — a two component silicone rubber (RTV 630, General Electric Co. — $\mu = 1.5$ MPa), both polymers presenting a smooth, flat, horizontal, surface for drop deposition. The latter is prone to slight swelling by TCP and therefore samples were immersed in TCP for several days before the wetting experiment until equilibrium swelling had been attained. The elastomeric surface was carefully wiped before drop deposition to remove extraneous liquid.

Wetting experiments were conducted at a temperature of 20°C.

4.2. Dewetting

Dewetting experiments, at least in the early stages which are presently of interest, take place on a much shorter time scale than the spreading to equilibrium of sessile drops. For this reason, the formation of dry patches was studied using a video camera mounted on a low power optical microscope.

Again the liquid employed was TCP and the two solids were the rigid Teflon PFA and a two component silicone elastomer. However for reasons of convenience, the elastomer used for dewetting was RTV 615 (General Electric Co.) with a shear modulus of 0.8 MPa, whose transparency permits very clear observation by light transmission of the nucleation and growth of dry patches.

Circular puddles of TCP of 50 mm diameter were obtained by attaching a 0.1 mm thick ring of plasticized adhesive paper on the solids and filling the inside with liquid. A thin layer was made by drawing across the liquid surface with a microscope slide. Calibration at 0.1 mm thickness was ensured by the spacer. At this thickness, the liquid film is in an unstable state since the equilibrium thickness, $2K^{-1}\sin(\theta_0/2)$, where K^{-1} is the capillary length, is *ca.* 1.5 mm [14]. ($K^{-1} = \sqrt{\gamma/\rho g}$ where ρ is liquid density and g gravitational acceleration.)

In most cases, nucleation of dewetting was initiated spontaneously, probably due to surface defects or thermal agitation energy.

5. RESULTS AND DISCUSSION

Figure 4 gives results of sessile drops spreading in the form of contact angle, $\theta(t)$, and contact radius, $r^*(t)$, as a function of time, t, after drop deposition. It can be seen clearly that a small drop of TCP spreads rapidly to equilibrium on the harder solid, Teflon PFA, spreading being complete after *ca.* 15 s. By contrast, the same liquid on a soft elastomer spreads far more slowly and approaches equilibrium only after *ca.* 20 min. This strange behaviour is attributed to the existence of a wetting

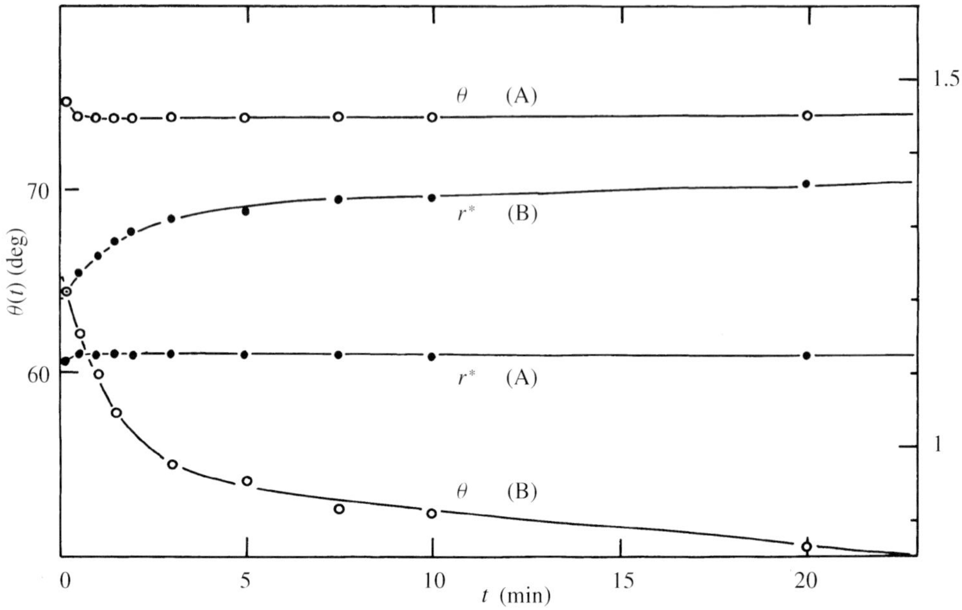

Figure 4. Contact angle, $\theta(t)$, and contact radius, $r^*(t)$, *vs.* time, t, for 2 μl drops of tricresyl phosphate on (A) Teflon PFA and (B) silicone elastomer RTV 630.

ridge effectively braking the spreading process. It is conceivable that this reduction in spreading rate is due to the fact that contact angles on the elastomer are smaller than on the rigid polymer, thus increasing the importance of the viscous dissipation term [see equations (16) and (17)], but results obtained with other hard solids for which θ is smaller than for the elastomer preclude this eventuality [15]. Even in the case of contact angles of *ca.* 20°, spreading is complete after *ca.* 15 s.

Figure 5 [line (A)] corresponds to an interpretation of the results of Fig. 4 for the elastomer making use of equation (19) in a rearranged form:

$$\log\left[\cos\theta_o - \cos\theta(t)\right] = n \log U + \log\left[\frac{\gamma}{2\pi\mu\delta U_o^n}\right], \tag{28}$$

where the spreading speed, U, was calculated from the variation of drop contact radius, r^*, with time, t. The linear relation shown has a correlation coefficient close to 1 and a gradient of 0.56. This latter figure is in good agreement with the value of 0.6 obtained for viscous dissipation in adhesion experiments [7], thus corroborating effects of a wetting ridge (as opposed to an 'adhesion ridge' — the two phenomena are clearly very similar).

Unfortunately, the spreading to equilibrium of TCP on Teflon PFA is too rapid for analysis of the results to be made. However, we expect the following equation to apply [from equations (13), (16), and (17)]:

$$\log\left[\cos\theta_o - \cos\theta(t)\right] \approx \log U + \log\left[\frac{3\eta\ell}{\gamma\theta(t)}\right]. \tag{29}$$

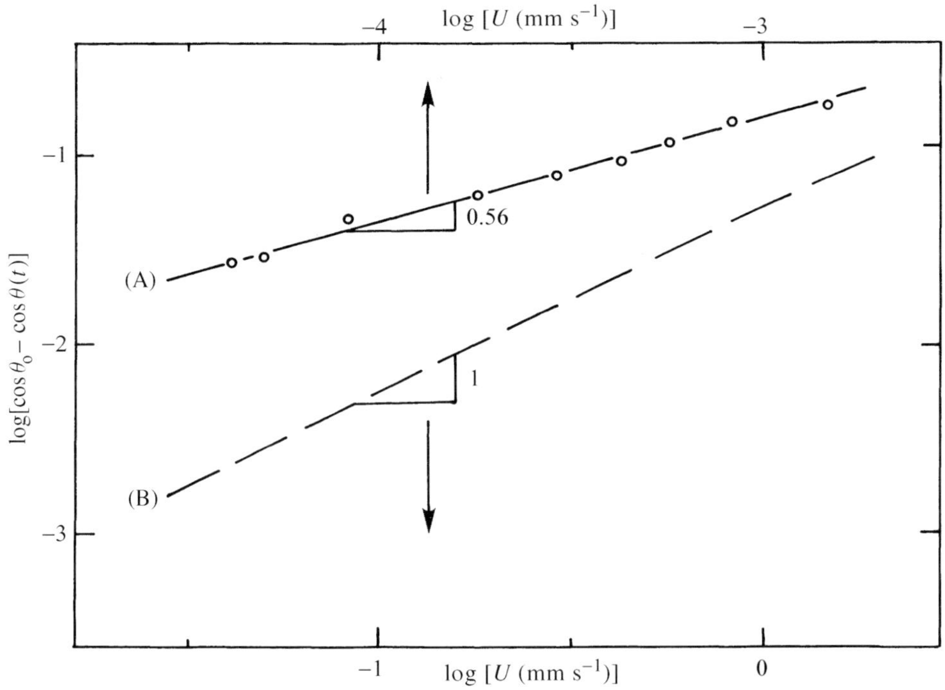

Figure 5. Log[cos θ_o − cos $\theta(t)$] vs. log U for (A) sessile drop of tricresyl phosphate on RTV 630 and (B) expected behaviour on Teflon PFA. Note that different log U scales are required for the two cases.

The value of ℓ is, of course, not strictly speaking a constant but it varies slowly. It may be taken as having the form $\ell \approx \ln(r^*\theta_0^2/a)$ where a is a molecular length [2]. Taking $r^* \approx 1$ mm, $\theta_0 \approx 75°$ and $a \approx 10^{-9}$ m, we obtain $\ell \approx 14$.

Figure 5 also shows the expected behaviour of spreading on Teflon PFA as given by equation (29) [line (B)]. Not only is the gradient greater, being virtually unity (the $\theta(t)$ term in the second right hand member has little effect), but it can be seen that for a given spreading speed, considerably less 'power' is required, as represented by [cos θ_o − cos $\theta(t)$]. Note, however, that use of relation (29) is somewhat approximate since the dissipation term given by expression (16) is really valid only for small values of θ (<1 rad.).

This clear difference in spreading behaviour has been found with other liquids and, in fact, it has been shown that wetting on an elastomer becomes viscosity independent [16], confirming the validity of equation (28).

Results of dewetting are represented in Fig. 6. Line (A) corresponds to the steady-state variation of the average radius of the dry patch growing in the unstable film of TCP on the elastomer RTV 615 as a function of time t. Transient behaviour for $t < 1$ s is here neglected. In agreement with the prediction of equation (23), the dry patch radius increases linearly with time. Each value shown represents the mean of five experiments and the average dewetting rate is found to be 0.39 ± 0.03 mm s^{-1}.

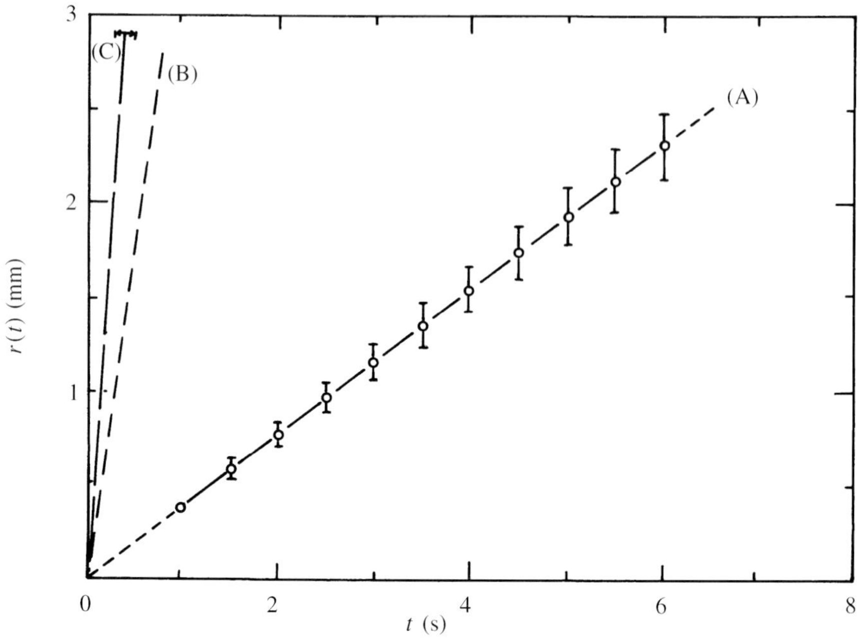

Figure 6. Dry patch radius, $r(t)$, vs. time, t, for tricresyl phosphate on silicone elastomer RTV 615 [line (A)]. The expected behaviour in the absence of a wetting ridge is given by line (B). Line (C) corresponds to the observed dewetting on Teflon PFA.

We may rewrite equation (25), corresponding to wetting without a viscoelastic contribution to dissipation from the wetting ridge, as:

$$U_2 \approx \frac{\gamma\theta_o(1 - \cos\theta_o)}{3\eta\ell}, \tag{30}$$

assuming equation (16) also applies to dewetting and where $k \approx 3\ell/\theta_o$. Taking $\ell \approx 14$ as above (the production of a rim under viscosity-controlled conditions is expected and this should be of the order of 1 mm across [14]) and $\theta_o = 47^\circ$, we obtain $U_2 \approx 3.6$ mm s^{-1}. Thus without any contribution from viscoelastic dissipation, the receding triple line would be expected to travel at *ca.* 10 times the speed. This is shown by line (B) in Fig. 6.

The kinetics of dewetting was also studied on Teflon PFA, representing a hard solid. It was found that the average speed of dewetting was 8 ± 2 mm s^{-1}. This is shown as line (C) in Fig. 6 and is found to be in good agreement with the value of *ca.* 7.6 mm s^{-1} deduced from equation (30) when taking a value of θ_o of 61° (apparent receding equilibrium contact angle). The dewetting rate is an order of magnitude higher on this solid presenting a shear modulus approximately 300 times greater than the silicone elastomer. We may therefore suppose that it is rigid enough for the solid surface to be insufficiently perturbed by the wetting triple line for there to be any consequences in wetting behaviour. (Another possibility, albeit unlikely, is that the wetting ridge is significant but that the dissipative properties of the Teflon PFA are low.) Thus, despite similar equilibrium wetting properties (θ_o in the range of 50–60°), the dynamic dewetting properties vary enormously as a result of the relative deformability of the substrates and the formation of a wetting ridge which must travel with the triple line.

In Section 3.2 we deduced that the ratio of dissipation rates due to viscous effects for the (equivalent) uniquely viscous and viscoelastically controlled cases was equal to m^2 where m is the dewetting speed ratio. We thus see from the data obtained with the silicone elastomer that m^2 in this case is of the order of 100.

6. CONCLUSIONS

Wetting phenomena, both static and dynamic, are of great importance in a multitude of domains including adhesion and related subjects. The dynamics of spreading on an ideal, rigid solid is controlled by an energy balance between excess capillary potential and viscous dissipation within the liquid. If the solid is sufficiently soft, the liquid surface tension, γ, causes a mechanical disturbance, in particular, a wetting ridge, or vertical protrusion, of the otherwise flat solid. The magnitude and the form of this perturbation are discussed here. When spreading occurs, this wetting ridge must accompany the triple line and thus a layer of solid near the surface undergoes a strain cycle. With a dissipative solid, this leads to an extra viscoelastic term, often outweighing the viscous contribution in the dynamic energy balance of spreading. Approximate and more exact equations for the process are derived using simplified fracture mechanics terminology. Both wetting and dewetting are treated.

Experimental studies have been carried out using tricresyl phosphate as the liquid and two types of solids as model surfaces — a relatively rigid polymer and two silicone elastomers. During wetting as represented by the spreading to equilibrium of small sessile drops, it was found that whilst equilibrium was reached after *ca.* 15 s on the rigid surface, over 20 min was required on a soft elastomer. The 'braking' effect has been quantitatively related to viscoelastic dissipation in the wetting ridge. Dewetting was studied by considering the spontaneous nucleation and growth of dry patches in unstable films of tricresyl phosphate. It was found that the growth rate of dry patches was an order of magnitude slower on the soft elastomer when compared to the rigid polymer, although in both cases triple line receding rate was constant (in the range considered).

It is believed that these results not only lead to a parallel in the understanding of elastomeric wetting and adhesion processes, but also open the way to controlling dynamic wetting phenomena by modifying solid, rather than liquid properties.

REFERENCES

1. T. Young, *Philos. Trans. R. Soc.* **A95**, 65 (1805).
2. P. G. de Gennes, *Rev. Mod. Phys.* **57**, 827 (1985).
3. M. E. R. Shanahan, *J. Phys. D: Appl. Phys.* **20**, 945 (1987).
4. G. R. Lester, *J. Colloid Sci.* **16**, 315 (1961).

5. M. E. R. Shanahan and P. G. de Gennes, in: *Adhesion 11*, K. W. Allen (Ed.), p. 71. Elsevier, London (1987).
6. M. E. R. Shanahan, *J. Phys. D: Appl. Phys.* **21**, 981 (1988).
7. D. Maugis and M. Barquins, *J. Phys. D: Appl. Phys.* **11**, 1989 (1978).
8. J. A. Greenwood and K. L. Johnson, *Phil. Mag. A* **43**, 697 (1981).
9. J. D. Ferry, *Viscoelastic Properties of Polymers*, 2nd edn, p. 314. Wiley, New York (1970).
10. A. Dupré, *Théorie Mécanique de la Chaleur*, p. 369. Gauthier-Villars, Paris (1869).
11. A. Carré and M. E. R. Shanahan, *C.R. Acad. Sci.* **317(II)**, 1153 (1993).
12. M. E. R. Shanahan and A. Carré, *Langmuir* **10**, 1647 (1994).
13. F. Brochard, J. M. di Meglio and D. Quéré, *C.R. Acad. Sci.* **304(II)**, 553 (1987).
14. F. Brochard, C. Redon and F. Rondelez, *C.R. Acad. Sci.* **306(II)**, 1143 (1988).
15. M. E. R. Shanahan and A. Carré, *Langmuir* **11**, 1396 (1995).
16. A. Carré and M. E. R. Shanahan, *Langmuir* **11**, 24 (1995).

Mittal Festschrift, pp. 255–263
W. J. Van Ooij and H. R. Anderson, Jr. (Eds)

Surface properties of silicone release coatings

MICHAEL J. OWEN
Dow Corning Corporation, Midland, MI 48686-0994, USA

Abstract—The Harkins relationship, that the spreading coefficient of an adhesive on a release coating is the difference between the work of adhesion of the materials and the work of cohesion of the adhesive, has been found to apply to a variety of silicone release coatings. The works of adhesion and cohesion were estimated from contact angle data using the Owens and Wendt approach. The prediction of the Harkins relationship is obeyed by almost all the combinations of pressure-sensitive adhesive and release coatings we have examined. Release occurs when the spreading coefficient is negative and does not when it is positive. The main exception to this general spreading coefficient rule is the failure of polytetrafluoroethylene to release polydimethylsiloxane-based pressure-sensitive adhesives. The cause is believed to be roughness of the polytetrafluoroethylene surface.

Keywords: Silicones; coatings; surfaces; contact angle; release.

1. INTRODUCTION

The phenomenon of release is of interest to adhesion scientists for a number of reasons. It is the opposite of adhesion and an understanding of the one aids in the understanding of the other. Deliberate inversion of thinking can help shed light on the adhesion behavior of polymers and coatings. Moreover, release substrates are vital to the processing of many polymers. They offer a convenient way of delivering adhesive products, particularly pressure sensitive adhesives (PSAs). Silicone products dominate this PSA release coating application.

Release agents function either by lessening intermolecular interactions between the two surfaces in contact or preventing such close contact [1]. Thus they can be either low surface tension materials based on aliphatic hydrocarbon or fluorocarbon groups or particulate solids. The major categories of materials used are waxes, fatty acid metal soaps, other long-chain alkyl derivatives, polymers, and fluorinated compounds. The most technologically familiar, low-surface-tension silicone polymer is polydimethylsiloxane (PDMS) which is widely used as a release agent, both as a liquid or emulsion and as a cross-linked coating. The former types of release agent are non-permanent and usually result in significant transfer of release agent to the released surface. The latter type are ideally permanent, or at least capable of multiple uses. This requires a cured film surface free from migratory species. Here we are concerned only with such crosslinked coatings. They are commonly used with

PSAs in the form of a release liner supported on various base substrates including papers such as glassine and kraft, polymer-coated papers and polymer films such as polyethylene, polypropylene, and polyesters.

Technical requirements for such coatings extend beyond good release characteristics [2]. They include such features as good thermal and dimensional stability and no impairment of subsequent adhesion of the adhesive but, nevertheless, the primary requirement is one of lower surface tension than the adhesive to be released. Surface tension values and composition of candidate adhesives and release materials are needed for characterization, but the question arises whether such data can be used for predictive purposes. This paper brings together some new and some previously published results, not all readily accessible, to address this question. Central to this possibility is the spreading coefficient, S [3]. For an adhesive spreading on a release coating it is given by

$$S = \gamma_R - \gamma_A - \gamma_{AR}, \tag{1}$$

where γ_A is the surface tension of the adhesive, γ_R that of the release substrate, and γ_{AR} the interfacial tension between them. Harkins [3] has also shown that

$$S = W_A - W_C, \tag{2}$$

where W_A is the thermodynamic work of adhesion, the change in energy per unit surface area when two interfaces come into contact [4], and W_C is the work of cohesion of the adhesive. This latter quantity is simply twice the surface tension as two such surfaces are produced on cohesive cleavage. The former quantity W_A is given by

$$W_A = \gamma_A + \gamma_R - \gamma_{AR}. \tag{3}$$

It follows from Harkins' relationship [equation (2)] that a negative spreading coefficient means that W_A is less than W_C and the failure will be at the interface, the desired situation in the release coating application. Conversely, a positive spreading coefficient implies a greater interfacial than cohesive strength which is clearly not conducive to easy release. The prospect of using such simple equations to predict release behavior is most attractive.

2. SURFACE TENSION CHARACTERIZATION

This study is based on contact angle measurements. Of the various equations available, the two which most readily permit calculation of the interfacial tension between adhesive and coating, γ_{AR}, are the Owens–Wendt [5] and Wu [6] equations. Now that directly measured work of adhesion values of silicone systems are possible through the application of the Johnson, Kendall and Roberts technique by Chaudhury and coworkers [7, 8], this is clearly the direction this type of study will take next. Both these equations use the simplification of separating the surface tension into the component arising from dispersion force interactions, γ^d, and the component arising from other polar forces, γ^{p}. The Owens–Wendt equation, also known as the geometric mean equation, is

$$\gamma_{SL} = \gamma_{SV} + \gamma_{LV} - 2\left(\gamma_{LV}^{d}\gamma_{SV}^{d}\right)^{1/2} - 2\left(\gamma_{LV}^{p}\gamma_{SV}^{p}\right)^{1/2} \quad (4)$$

and the Wu equation, also known as the harmonic mean equation, is

$$\gamma_{SL} = \gamma_{SV} + \gamma_{LV} - \frac{4\gamma_{LV}^{d}\gamma_{SV}^{d}}{\gamma_{LV}^{d} + \gamma_{SV}^{d}} - \frac{4\gamma_{LV}^{p}\gamma_{SV}^{p}}{\gamma_{LV}^{p} + \gamma_{SV}^{p}}. \quad (5)$$

The limitations of these equations have been known for over a decade and newer, more refined approaches have been established [9], yet they remain the most useful way of estimating γ_{SL} from contact angle data. We use the Owens–Wendt equation because it is straightforward to work with and readily accommodates the use of several liquids, has some theoretical justification, gives values that bear some resemblance to critical surface tension of wetting values and provides polar and dispersion force components of reasonable magnitude. Others favor the use of the Wu equation [10].

The contact angles were determined using a Ramé-Hart, Inc., NRL Model A-100 contact angle goniometer. Measurements were made several times on both sides of at least three drops. In those cases where a progressive change was observed, the value extrapolated to zero time was used. The justification for this is that time changes are more likely to be due to interaction with the substrate (softening, swelling, chemical reaction) than to orientation equilibrium adjustment for simple liquids. In each case the advancing contact angle over previously unwetted surface was used. The measurements were made at laboratory temperature (23 ± 1)°C and humidity without using the controlled environment chamber. Sufficient repeats were made to satisfy us that this was an acceptable procedure.

Contact angle test liquids were the best purity available. Liquids whose surface tension did not agree with accepted literature values, or did not maintain such values for several hours, were redistilled until agreement was obtained. A minimum of two liquids is needed to apply the Owens–Wendt equation. Water and methylene iodide are the two common choices. Sometimes we restricted ourselves to this pair, other times more liquids such as glycerol, α-bromonaphthalene and n-hexadecane were included.

The surface tensions for the various substrates used, calculated by equation (4) and the simple form of the Young equation, are given in Table 1. The silicone coating data have been previously published [11] but the organic PSA data have not been previously readily available. The acrylic PSA is GELVA® GMS-263 from Monsanto Co. and the SBR PSA is RESYN® 36-6045 from National Starch and Chemical Corp. Syl-off® 7044 and 23 are commercial paper coatings of Dow Corning Corporation. Syl-off® 7044 is a solventless material and Syl-off® 23 is a solvent-based product. The paper substrate is James River Paper Co. super-calendered 40 lb kraft. These adhesives and release coatings have been widely used in the industry and have been the subject of other surface studies [10]. They are included to illustrate the surface tensions of conventional organic PSA and PDMS release coatings.

Table 1.
Surface tension of adhesives and release coatings

Material	Polar component, γ^p (mN m^{-1})	Dispersion component, γ^d (mN m^{-1})	Total, γ (mN m^{-1})	Reference
Acrylic PSA	3.4	31.5	34.9	
SBR PSA	4.0	27.1	31.1	
Paper coating substrate	18.8	29.6	48.4	
Syl-off® 7044	1.7	20.3	22.0	11
Syl-off® 23	0.7	19.9	20.6	11
PDMS PSA	2.0	17.9	19.9	
PDPS PSA	0.5	23.4	23.9	
$CF_3(CF_2)_7(CH_2)_2M^a$	0.9	9.9	10.8	
$CF_3(CF_2)_3(CH_2)_2M^a$	1.4	13.1	14.5	
$CF_3(CF_2)_6(CH_2)_2SiO_{3/2}$	1.0	12.3	13.3	
$(CF_3)_2CFO(CH_2)_2SiO_{3/2}$	2.6	15.2	17.8	
PMNFHS^b	1.1	8.4	9.5	12
PMTFPS^b	2.8	10.8	13.6	12
PTFE^b	0.5	18.6	19.1	5

^{a}M = methacrylate group.
^{b}Using water and methylene iodide contact angle data only.

The PDMS PSA is DC® 280A from Dow Corning Corp., and the PDPS PSA is GE SR-574 from General Electric Corp. and contains polydiphenylsiloxane as well as PDMS. The presence of the aromatic phenyl ring raises the surface tension of this adhesive. The two fluoromethacrylate polymers were pure materials, a gift from Dr. R. H. Dettre of Du Pont's Jackson Laboratory. Also included in Table 1 are polymethyl(nonafluorohexyl)siloxane (PMNFHS) and polymethyl(trifluoropropyl)siloxane (PMTFPS) values from our previous studies [12] and polytetrafluoroethylene (PTFE) data from Owens and Wendt [5].

3. TEST OF THE HARKINS RELATIONSHIP

Works of adhesion and cohesion (of the adhesives) calculated from the surface tension data of Table 1 are given in Table 2 together with the spreading coefficient calculated from the Harkins relationship, equation (2).

Whether the separation of the adhesive tape from the substrate is classed as cohesive failure (C) or apparent adhesive (interfacial) failure (A) is based on visual observation together with some limited release force determination. A 2.54 cm wide (i.e. one inch) tape is peeled at 180° from the substrate at a rate of 30.5 cm (i.e. one foot) per min. Release forces less than 100 N m^{-1} with no obvious cohesive failure are classed as releasing, i.e. the desirable interfacial failure mode. This arbitrary standard peel speed is very slow compared to industrial practice. We also know that some

Table 2.
Harkins relationship quantities

System	Work of adhesion, W_A (mJ/m^2)	Work of cohesion, W_C (mJ/m^2)	Spreading coefficient, S (mJ/m^2)	
Acrylic PSA/paper	77.1	69.8	+7.3	C
SBR PSA/paper	73.9	62.2	+11.7	C
Acrylic PSA/Syl-off® 7044	55.4	69.8	−14.4	A
SBR PSA/Syl-off® 7044	52.1	62.2	−10.1	A
Acrylic PSA/Syl-off® 23	53.2	69.8	−16.6	A
SBR PSA/Syl-off® 23	49.7	62.2	−12.5	A
PDMS PSA/Syl-off® 7044	41.8	39.8	+2.0	C
PDPS PSA/Syl-off® 7044	45.4	47.8	−2.4	A
PDMS PSA/Syl-off® 23	40.1	39.8	+0.3	C
PDPS PSA/Syl-off® 23	44.4	47.8	−3.4	A
PDMS PSA/$CF_3(CF_2)_7(CH_2)_2M^a$	29.3	39.8	−10.5	A
PDMS PSA/$CF_3(CF_2)_3(CH_2)_2M^a$	33.9	39.8	−5.9	A
PDMS PSA/$CF_3(CF_2)_6(CH_2)_2SiO_{3/2}$	32.5	39.8	−7.3	A
PDMS PSA/$(CF_3)_2CFO(CH_2)_2SiO_{3/2}$	37.6	39.8	−2.2	A
PDMS PSA/PMNFHS	27.5	39.8	−12.3	A
PDMS PSA/PMTFPS	32.5	39.8	−7.3	A
PDMS PSA/PTFE	38.5	39.8	−1.3	C

^{a}M = methacrylate group.
C = cohesive failure (in adhesive) — undesired separation mode (≡ adhesion).
A = apparent adhesive failure (at interface) — desired separation mode (≡ release).

substrates, for example PMTFPS, release reasonably well from PDMS PSAs at this slow speed but are much less effective at higher speeds. Nevertheless, this slow speed relates more to the surface tension values which are typically quasi-equilibrium values, and is adequate for initial screening. At higher release speeds the viscoelastic response of both adhesives and coatings dominate the release behavior, a complexity we choose to avoid in these studies.

It follows from the Harkins relationship, equation (2), that a negative spreading coefficient means that W_A is less than W_C and the failure will be at the interface, the desired situation in the release coating application. Conversely, a positive spreading coefficient implies a greater interfacial than cohesive strength that will not offer easy release. We have found that this prediction of the Harkins relationship is obeyed by almost all the PSA/release coating combinations we have encountered. Examples are given in Table 2. The greatest exception to this general spreading coefficient rule is the lack of facile release of PDMS PSAs from PTFE.

There are several possible explanations for this discrepancy. The calculated spreading coefficient value is small and it may be that with a different choice of contact angle test liquids agreement could be found. This is a general problem with contact angle studies and remains speculative until future direct measurements of the work of adhesion are made. Silicon and fluorine have a strong affinity for each other and there may be specific acid/base silicon/fluorine interactions that are stronger in the PDMS/PTFE case than in the PDMS/fluorosilicone case. A more likely explanation lies in the very different physical state of a fluorosilicone such as PMNFHS and PTFE. For instance, the glass transition temperature, T_g, of the former polymer is 198 K [13], well below room temperature, whereas that of the latter polymer is 390 K [14], well above ambient temperature.

Ho and coworkers [15] in their study of minimally adhesive surfaces believe that in addition to low surface energy, a low T_g for the surface phase is desirable for minimizing mechanical locking of a prospective adherent to the surface. This may not be so important for the slow separation of a pressure-sensitive material but the high T_g might support a more structured, rough or porous surface that could contribute to higher adhesion by virtue of an increased real area of contact. Scanning electron micrographs of PMNFHS and PTFE of the type used in our qualitative release tests are shown in Figs 1a and 1b, respectively. They were taken with a JEOL JSM-6100 SEM. Samples were coated with 15 nm of Au/Pd prior to examination to prevent charging effects.

Although published studies of SEM of silicone release coatings have focused on defects and other inconsistencies [16, 17], effectively applied silicone release coatings are generally smooth and virtually featureless with an even coating following the contours of the original substrate. An example is given in Fig. 1a, which is a scanning electron micrograph of our PMNFHS-based coating, DC Syl-off 7785 on a Mylar substrate. The situation is quite different for PTFE, as can be seen in Fig. 1b where two kinds of heterogeneities are evident: roughly parallel chains of solid material several micrometers wide separated by aligned webs of fibrils of the order of one to ten micrometers. The long axes of the fibrils are approximately at right angles to the long axes of the solid, island-like regions. The PTFE shown is a general-purpose, over-the-counter product purchased from Ace Hardware and manufactured in Taiwan that we use for release testing.

Another possible explanation of the anomalous release behavior of PTFE has been advanced by Newby *et al.* [18]. They show that the peel adhesion values of a viscoelastic adhesive measured on surfaces prepared by grafting organofunctional silanes to silicon wafers do not conform with conventional expectation. The adhesion on the fluorocarbon surface was highest even though it has the lowest surface free energy. Lowest adhesion was found on a flexible PDMS surface. Remarkable differences in the meniscus instability patterns were also observed on these surfaces. These observations show that it is the interfacial slippage, not the surface free energy, that controls viscoelastic adhesion on low energy surfaces.

With these various complicating possibilities it is little wonder that no general quantitative relationship between the magnitude of the negative spreading coefficient and the release force is found. Only in particular, restricted circumstances [10] is this kind of dependency predicted by the Gent and Schultz theory [19] obtained. This theory holds that the energy required to separate a viscoelastic adhesive from

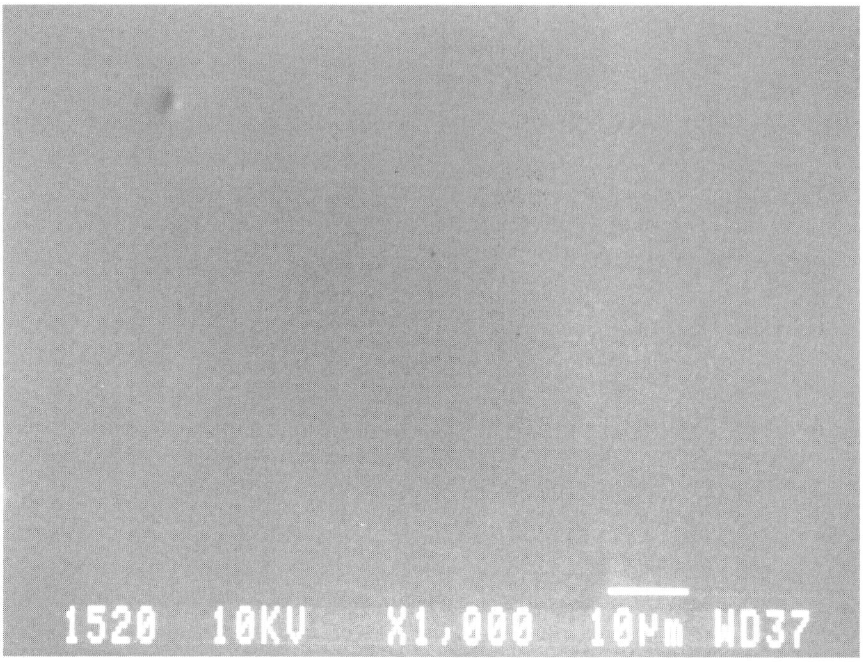

(a)

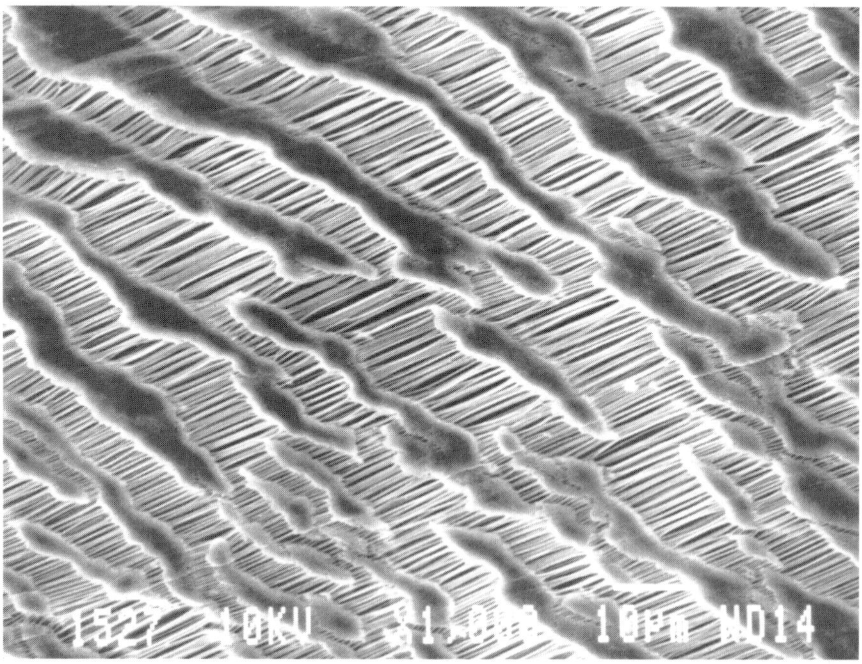

(b)

Figure 1. Scanning electron micrographs of release surfaces. (a) PMNFHS based coating and (b) PTFE tape.

a solid substrate is a product of the interfacial work of adhesion and a term proportional to the viscoelastic energy dissipation. For a given adhesive this implies adhesion strength will be proportional to the surface free energy of the substrate. Although this theory accounts for a large body of adhesion data it is evidently not always applicable for low energy surfaces. This lack of correlation of surface properties with release performance has been observed before with silicone release systems. For example, Hsu and coworkers [20] characterized siloxane-containing methacrylates by contact angle, ATR-IR and XPS. They showed that systems indistinguishable by all three methods gave significant differences in release performance. A further complicating factor in some cases is significant transfer of silicone material across this interface during the release process [21] despite efforts to avoid this.

Such observations underline the obvious feature that release forces depend very significantly on viscoelastic response in the bulk materials as well as on surface effects. The importance of bulk effects in the PSA release process is emphasized by the fact that one of the few established correlations with release force is with cross-link density in the release film [22]. The higher the cross-link density the lower the release force. This can be explained by a Lake and Thomas critical bond rupturing process [23] or interpenetration and polymer reptation effects (suction model) introduced by de Gennes [24]. This approach suggests a useful strategy for release coating development: use surface tensions to make go/no-go decisions on material suitability and then tailor the polymer structure to control the desired release force.

4. SUMMARY

Works of adhesion and cohesion for a variety of pressure-sensitive adhesive/release coating combinations, estimated from contact angle data using the Owens and Wendt approach, generally obey the prediction of the Harkins relationship. When the calculated spreading coefficient is positive then cohesive failure occurs in the adhesive, i.e. the undesired separation mode from a release perspective. When the spreading coefficient is negative, apparent interfacial failure, i.e. the desired release mode, occurs. The main exception to this predictively useful relationship is the failure of PTFE to release from PDMS-based pressure-sensitive adhesives. SEM micrographs reveal marked surface structure and roughness of the PTFE surface which is believed to be the cause of this anomaly.

Acknowledgements

I thank Dr. R. H. Dettre of Du Pont and Terry Swihart and Darrell Jones of Dow Corning for provision of materials. Jennifer Fritz and Jeff Janowicz took the electron micrographs, Diane Kelly measured the contact angles. Dave Thomas of our Barry, Wales, Laboratories first suggested the roughness explanation for the anomalous behavior of PTFE.

REFERENCES

1. M. J. Owen, in: *Encyclopedia of Polymer Science and Engineering*, 2nd edn, H. F. Mark, N. M. Bikales, C. G. Overberger and G. Menges (Eds), Vol. 14, pp. 411–421. Wiley-Interscience, New York (1988).
2. J. D. Jones and Y. A. Peters, in: *Handbook of Pressure-Sensitive Adhesive Technology*, 2nd edn, D. Satas (Ed.), pp. 601–626. Van Nostrand Reinhold, New York (1989).
3. W. D. Harkins, in: *Colloid Chemistry*, J. Alexander (Ed.), Vol. 1, pp. 192–264. The Chemical Catalog Co., New York (1926).
4. A. N. Gent and G. R. Hamed, in: *Encyclopedia of Polymer Science and Engineering*, 2nd edn, H. F. Mark, N. M. Bikales, C. G. Overberger and G. Menges (Eds), Vol. 1, pp. 476–518. Wiley-Interscience, New York (1985).
5. D. K. Owens and R. C. Wendt, *J. Appl. Polym. Sci.* **13**, 1741–1747 (1969).
6. S. Wu, *J. Polym. Sci., Part C* **34**, 19–30 (1971).
7. M. K. Chaudhury and G. M. Whitesides, *Langmuir* **7**, 1013–1025 (1991).
8. M. K. Chaudhury, *J. Adhesion Sci. Technol.* **7**, 669–675 (1993).
9. C. J. van Oss, *Interfacial Forces in Aqueous Media*. Marcel Dekker, New York (1994).
10. D. J. Gordon and J. A. Colquhoun, *Adhesives Age* **19** (6), 21–30 (1976).
11. M. J. Owen, *J. Coatings Technol.* **53** (679), 49–53 (1981).
12. M. J. Owen and H. Kobayashi, *Macromol. Symp.* **82**, 115–123 (1994).
13. H. Kobayashi and M. J. Owen, *Macromolecules* **23**, 4929–4933 (1990).
14. P. Peyser, in: *Polymer Handbook*, 3rd edn, J. Brandrup and E. H. Immergut (Eds), pp. VI/209–VI/277. Wiley-Interscience, New York (1989).
15. T. Ho, K. J. Wynne and R. A. Nissan, *Macromolecules* **26**, 7029–7036 (1993).
16. J. E. Wilson and H. A. Freeman, *Tappi* **64** (2), 95–97 (1981).
17. J. Stein, R. B. Bolon, M. L. Porta, C. D. Robertson and M. E. Grenoble, *Int. J. Adhesion Adhesives* **12** (2), 111–116 (1992).
18. B. Z. Newby, M. K. Chaudhury and H. Brown, *Science* **269**, 1407–1409 (1995).
19. A. N. Gent and J. Schultz, *J. Adhesion* **3**, 281–294 (1972).
20. T. Hsu, S. S. Kantner and M. Mazurek, *Polym. Mater. Sci. Eng.* **55**, 562–566 (1986).
21. L. A. Duel and M. J. Owen, *J. Adhesion* **16**, 49–59 (1983).
22. A. E. Bey, *Adhesives Age* **15** (10), 29–32 (1972).
23. G. J. Lake and A. G. Thomas, *Proc. Royal Soc. (London)* **A300**, 108–119 (1967).
24. P. G. de Gennes, *C.R. Acad. Sci. Ser. 2* **308**, 13–17 (1989).

Part 3

Surface modification: Relevance to adhesion

Mittal Festschrift, pp. 267–283
W. J. Van Ooij and H. R. Anderson, Jr. (Eds)

Pretreatments of fluoropolymers. A review of studies between 1990 and 1995

D. M. BREWIS* and I. MATHIESON
Institute of Surface Science & Technology, Loughborough University, Loughborough Leics LE11 3TU, UK

Abstract—The most effective commercial methods for pretreating fully fluorinated polymers at present are sodium in liquid ammonia and sodium naphthalenide in a solvent such as tetrahydrofuran. A number of other methods can produce large increases in bondability but are relatively inefficient. Plasma treatments of fully fluorinated polymers such as PTFE often introduce functional groups and/or increase wettability but result in moderate improvements at best in bondability. However, recent work with ammonia plasmas has resulted in big improvements in bondability.

Some batches of partially fluorinated polymers may be effectively bonded without a pretreatment. However, other batches possess surface regions of low cohesive strength. Partially fluorinated polymers such as poly(vinyl fluoride) may be rapidly and effectively treated with a conventional flame treatment. Where a flame treatment is unsuitable, as with complex shapes, potassium hydroxide solutions provide effective treatments for partially fluorinated polymers such as poly(vinyl fluoride) and poly(vinylidene fluoride).

Keywords: Potassium hydroxide; weak boundary layers; XPS; pretreatment; tetrabutylammonium bromide; surface modification; plasmas; flame treatment.

1. INTRODUCTION

Effective pretreatments for poly(tetrafluoroethylene) PTFE, were developed in the 1950s. These were sodium in liquid ammonia [1] and sodium naphthalenide in tetrahydrofuran [2]. Many other treatments for PTFE have since been investigated including plasma treatment [3], direct electrochemical reduction [4], treatment with an alkali metal amalgam [5] and reduction with benzoin dianion [6]. However, none of these has proved as effective as the original two methods which remain the main commercial pretreatments.

The sodium complex methods are also effective with other fluoropolymers. For example, large increases in adhesion have been observed with ethylene–chlorotrifluoroethylene copolymer, poly(vinyl fluoride) PVF and poly(vinylidene fluoride) PVdF [7].

*To whom correspondence should be addressed. E-mail: D.M.Brewis@lboro.ac.uk

It was found, however, that the treatment rate with these polymers was much slower than with PTFE.

Although the sodium in liquid ammonia and sodium naphthalenide treatments are very effective, especially with PTFE, they have undesirable environmental features. There is clearly scope for more environmentally-friendly alternatives to the two methods used commercially. In the last few years, interest has centred on plasma treatment but other work has included the treatment of partially fluorinated polymers with flames and Group I hydroxides. Work carried out in these areas over the last five years is now reviewed.

2. PLASMA TREATMENT

Although flames and corona discharges consist of plasmas, the term plasma treatment usually refers to a process carried out at reduced pressure, typically one torr. Much research was carried out in the 1960s and 1970s on the use of plasmas to pretreat fluoropolymers and other polymers [3, 8–10] and has continued to the present date. Until recently, the improvements in adhesion obtained with PTFE have been much less than those achieved with partially fluorinated and other polymers.

Some of the main plasma studies involving fluoropolymers carried out between 1990 and 1995 are summarised in Table 1 and then each are commented on in more detail.

Kasemura *et al.* [11] used a domestic microwave oven with an output of 560 W at 2450 MHz operating at a pressure of 0.2–1 torr to pretreat PTFE and FEP. As the pressure of the air was decreased from 1 to 0.2 torr, the efficiency of the treatment increased as measured by lower contact angles with various liquids. XPS indicated that a 'significant' amount of oxygen had been introduced into the surfaces after a 10 s treatment in air at 0.2 torr. Peel strengths using a pressure sensitive adhesive tape increased by about 50 and 200% for PTFE and FEP, respectively. However, it should be pointed out that even the highest peel strengths obtained (2 kg per 25 mm) are only moderate.

Morra *et al.* [12] showed that treatment of PTFE for 30 s in an oxygen plasma did not cause any significant topographical changes according to SEM, but treatment for 15 min produced a sponge-like surface. However, with argon no topographical changes were evident, even after 15 min treatment. In contrast to the topographical changes, chemical changes were more pronounced with an argon plasma as can be seen in Table 2.

It is interesting to note that the amount of defluorination and the amount of oxygen introduced into the surface decrease with treatment time. With an oxygen plasma treatment for 15 min, the amount of surface oxidation is less than that for the untreated PTFE.

Measurement of advancing and receding contact angles revealed an ageing effect which is summarised in Table 3. In the case of PTFE treated for 30 s in an oxygen plasma, it took about 80 h to reach equilibrium.

Table 1.
Plasma pretreatment of fluoropolymers

Polymers	Plasma frequency	Gases	Surface studies	Wettability study	Bonding study	Comments	Ref.
PTFE, FEP	2450 MHz	Air	XPS, SEM	Yes	Peel test	Domestic microwave oven used. Wettability and peel strength improved	11
PTFE	13.56 MHz	O_2, Ar	XPS, SSIMS, SEM	Yes	No	For a given treatment level, Ar caused a greater chemical change, but O_2 produced a greater topographical change	12
E-TFE copolymer	13.56 MHz	$O_2 + SF_6$, O_2, NH_3	XPS	Yes	Double lap shear	Up to 30-fold increase in joint strength. Treatment in an O_2 plasma for 15 min introduced 7–8 at.% of oxygen	13
FEP	700 kHz	O_2, Ar, air, H_2O	XPS	Yes	No	Marked decreases in contact angles after treatment for a few seconds. On ageing for a few weeks, the contact angles increased but not to their original level	14
PTFE	700 kHz	O_2, Ar, air, H_2O	XPS, SEM	Yes	No	Substantial increases in surface energy. Only small amount of O introduced. Main effects were abstraction of F and crosslinking	15
PTFE, PVdF	13.56 MHz	O_2/H_2O, H_2/H_2O, H_2/CH_3OH, H_2/CH_2O	XPS, FTIR, SEM	Yes	No	Hydrogen used as a fluorine scavenger. Up to 17 at.% O introduced into PTFE for 30 min treatment with H_2/methanol plasma	16
PVdF	13.56 MHz	Ar, O_2, N_2 and NH_3	XPS	Yes	180° peel	N_2 and Ar plasmas improved wettability and gave 7-fold increase in peel strength	17

Table 1.
(Continued)

Polymers	Plasma frequency	Gases	Surface studies	Wettability study	Bonding study	Comments	Ref.
PTFE	50 Hz	Ar, O_2	XPS	Yes	Scratch test	O_2 and Ar plasma treatments (10–20 s) caused increases in surface energy of PTFE and adhesion of vacuum deposited Au to PTFE	18
PTFE	13.56 MHz	NH_3	XPS	No	Tensile test	Improvement in bondability of PTFE after plasma treatment much less than after sodium naphthalenide treatment	19
PTFE, PFA	13.56 MHz	Ar	XPS	No	No	For both polymers, a moderate decrease in F/C ratio together with a moderate increase in O/C ratio occurred after very short exposures, after which the ratios remained constant	20
PFA	2450 MHz	N_2, O_2, H_2	XPS, SEM	No	Peel test	Efficiency decreased $N_2 > O_2 > (N_2 + H_2) > (O_2 + H_2) > H_2$	21
PTFE, FEP	13.56 MHz	NH_3	XPS, STM, AFM	Yes	No	Reorientation and chemical changes on ageing are demonstrated by contact angles and XPS measurements	22
PTFE, PVF	13.56 MHz	O_2, N_2, Ar, Air	XPS	No	Lap shear	With PVF, up to 12× increase in joint strength and substantial quantities of O introduced. With PTFE, much less O introduced and joint strength increases limited to 4×	7
PTFE	13.56 MHz	O_2, N_2, H_2, He, Ne, Ar, CF_4	XPS, AFM	No	No	The oxygen plasma caused the greatest change in topography but the smallest change in surface chemistry	23

Table 2.
XPS composition (atomic percentages) of oxygen and argon plasma-treated PTFE as a function of treatment time [12]

Treatment time (min)	C	F	O
Untreated	39.8	60.4	0.8
Oxygen plasma			
0.5	44.6	48.9	6.4
1.0	42.7	51.1	7.1
2.0	42.6	50.9	6.5
5.0	40.9	57.0	2.1
10.0	38.3	60.5	1.2
15.0	38.3	61.4	0.3
Argon plasma			
0.5	50.2	27.0	22.8
1.0	50.8	25.0	24.2
2.0	46.9	37.2	15.9
5.0	45.6	44.9	9.5
10.0	38.5	52.5	9.0
15.0	37.6	53.9	8.5

Table 3.
Water advancing and receding contact angles on oxygen and argon plasma-treated PTFE as a function of treatment time [12]

Treatment time (min)	Just treated (degrees)		Equilibrated (degrees)	
	Advancing	Receding	Advancing	Receding
Untreated			106	90
Oxygen plasma				
0.5	65	10	105	12
1.0	69	12	104	11
2.0	135	32	134	33
5.0	165	142	166	145
10.0	166	145	165	147
15.0	166	156	165	158
Argon plasma				
0.5	61	13	90	13
1.0	64	12	90	11
2.0	68	10	88	12
5.0	68	13	95	14
10.0	72	13	95	10
15.0	69	11	96	10

The bondability of a plasma-treated ethylene–tetrafluoroethylene (E-TFE) copolymer was investigated by Hansen *et al.* [13]. They found that $O_2 + SF_6$, O_2 and NH_3 plasmas increased the strength of double lap joints bonded with various epoxide adhesives by a factor of 20–30; this compared with an increase of 11-fold with the commercial etchant, Tetra-Etch.[1] However, the highest value obtained with a plasma treatment (2.3 MPa) was surprisingly low. The present authors have obtained lap shear strengths in excess of 20 MPa with various treatments of PVF, PVdF and PTFE and in excess of 16 MPa for an ethylene–chlorotrifluoroethylene copolymer [7].

Hansen *et al.* [13] used XPS to examine the E-TFE copolymer after it had been treated for 15 min in an oxygen plasma. They found 7–8 at.% oxygen had been introduced into the polymer surface and attributed this mainly to ester groups.

Griesser *et al.* [14] studied the treatment of FEP using argon, air, oxygen or water vapour plasmas; most of the work utilised argon. A custom-built oscillator operating at 700 kHz and 10 W was used at a pressure of 0.6 torr. XPS data showed that treatment for a few seconds in an argon plasma introduced about 4 at.% O, significantly reduced the F content and increased the C content. They found that water contact angles decreased markedly after treatment for a few seconds. This effect was partially reversed over a period of three weeks. The authors attributed this to a partial reorientation of chain segments such that oxygen-containing functional groups moved towards the bulk of the polymer. From the decreased F : C ratio, the authors concluded that some crosslinking had occurred and that this might account for the incomplete recovery in contact angles. In contrast to the contact angle data, Griesser *et al.* found that the oxygen content within the XPS sampling depth increased on storage over a few weeks and suggested that a possible mechanism could involve the relatively slow breakdown of peroxy radicals; these radicals could decompose into stable oxygen-containing groups plus carbon-centred radicals which, in turn, could react with oxygen.

The treatment of PTFE using air, oxygen, argon and water plasmas was studied by Youxian *et al.* [15]. After treatment for a few seconds, substantial increases in surface energy occurred mainly due to changes in the polar component. However, XPS showed that little oxygen was introduced by the treatment. A reduced F : C ratio on treatment indicated crosslinking.

By using mixtures of hydrogen and methanol, high levels of oxygen (*c.* 17 at.%) were introduced into PTFE [16]. It was found that H_2/O_2 mixtures caused major changes in topography to porous PTFE whereas no topographical changes were observed for H_2/H_2O or H_2/methanol mixtures when SEM was used at 10 000× magnification.

Anderson *et al.* [17] studied the use of plasmas and an acid etch to improve bonding to PVdF. All four plasmas examined (O_2, N_2, Ar and NH_3) caused substantial reductions in water contact angle after 10 s treatment. The best improvements in bondability with a plasma was a 7-fold increase in peel strength after a 45 s treatment with argon. A 12-fold increase in peel strength was achieved with an acid etch

[1] Tetra-Etch is a proprietary brand of a sodium naphthalenide solution [19].

consisting of acetic, phosphoric and nitric acids. However, the peel strength after the acid etch was 48 N/m which is surprisingly low.

The effects of oxygen and argon plasmas on the surface energy and bondability of PTFE were studied by Kinbara *et al.* [18]. The surfaces became considerably more water-wettable after a 20-s treatment in either an oxygen or argon plasma. The adhesion of a vacuum deposited gold film, as measured by a scratch test, showed a considerable increase after these treatments. The authors concluded from XPS data that an oxygen plasma introduced $-C-O-$, $-CH_2-$ and $-CHF-$ groups into the PTFE surface.

In a study comparing ammonia plasma treatment with a sodium naphthalenide etch, the latter was found to be much more effective at increasing the bondability of PTFE [19]; a joint strength of 12.9 MPa was obtained with the sodium naphthalenide treatment compared with 2.8 MPa for the best plasma treatment. The authors used XPS to study the pretreated surfaces and also the failure surfaces after the adhesive joints had been destroyed. Some of these results are given in Table 4.

It is apparent that, with both treatments, failure is largely cohesive within the PTFE. Kaplan *et al.* [19] concluded that the much lower joint strength obtained with the plasma treatment was due to a cohesively weak layer between the bulk PTFE and the chemically modified layer. On the other hand, they believe the naphthalenide treatment may give higher joint strengths due to considerable crosslinking.

Golub *et al.* [20] used XPS to study the argon plasma treatment of PTFE and PFA. Contrary to other published work, they found that with both polymers after very short treatments there was a moderate decrease in the F : C ratio and an increase in the O : C ratio to about 0.1; thereafter the ratios remained essentially constant. They attributed the levelling off in the F : C and O : C ratios as being due to a dynamic competition between degradation (chain scission, $C-F$ bond rupture), crosslinking and CF_3 formation. They explained the difference in their results with other work as being due to the use of hydrocarbon-free polymers and the use of a monochromatic X-ray source for XPS.

Table 4.
XPS elemental analysis (at.%) of pretreated PTFE and failure surfaces of PTFE bonded with an epoxide [19]

Sample description	C	S	Mg	Ca	N	O	F	Na
PTFE virgin	33	—	—	—	—	—	67	—
Tetra-Etch								
Treated surface	71	0.4	1.2	0.9	1.9	17.0	7.9	—
PTFE side	37	—	—	—	1.2	1.0	61	0.2
Bolt side	40	—	—	—	2.2	2.8	53	2.1
Plasma activation								
Treated surface	46	—	—	—	6.4	6.3	39	—
PTFE side	33	—	—	—	—	—	67	—
Bolt side	35	—	—	—	0.7	0.4	64	—

The bondability test involved bonding a bolt to the PTFE and then pulling off the bolt.

The adhesion of evaporated copper to PFA, as assessed by both a peel test and a scratch test, was much enhanced by various plasma treatments [21]. The effectiveness of the plasma treatments towards metal adhesion as assessed using the scratch test decreased in the order:

$$N_2 > O_2 > (N_2 + H_2) > (O_2 + H_2) > H_2$$

No topographical changes were apparent with SEM after the plasma treatments and the improved adhesion was attributed to cleaning, improved wetting and formation of Cu−O and Cu−N bonds. Some chemical changes caused by the different plasmas under a particular set of conditions are given in Table 5.

Gengenbach *et al.* [22] made a detailed study of the ammonia plasma treatment of PTFE and FEP. For FEP, the N : C ratio was determined as a function of ageing time at take-off angles (TOA) of 15 and 90° with respect to the sample surface. Initially the ratio was higher at 15°, indicating a shallow treatment. At both angles, the N : C ratio decreased substantially over a period of 325 days, the decrease being much more rapid in the first few days. In marked contrast, the O : C ratio increased from about 0.05 to about 0.28 (at 90° TOA) over 325 days, with most of the change taking place in the first four days. The rapid uptake of oxygen is attributed to reaction between atmospheric oxygen and radicals produced by the plasma treatment whereas the slow uptake is thought to be due to secondary reactions. The corresponding water contact

Table 5.
Atomic concentration ratios of F, O and N to C on PFA, after different plasma treatments, as determined by XPS (treatment conditions: 100 W and 60 s) [21]

Treatment	F/C	O/C	N/C
Untreated	1.86	0.004	—
O_2	1.78	0.014	—
N_2	1.58	0.031	0.032
$O_2 + H_2$	1.51	0.069	—
$N_2 + H_2$	0.50	0.13	0.19
H_2	0.35	0.064	—

Table 6.
Summary of changes in elemental composition following plasma modification of PTFE (20 W, 5 min) [23]

Gas	% C	% F	% O	% N
No treatment	33.3 ± 0.6	66.8 ± 0.6		
O_2	33.0 ± 0.2	65.4 ± 0.2	1.7 ± 0.4	
N_2	33.2 ± 0.0	64.0 ± 0.6	1.6 ± 0.3	1.3 ± 0.4
H_2	56.1 ± 1.1	37.7 ± 0.4	4.7 ± 0.7	1.5 ± 0.1
He	38.7 ± 0.4	55.8 ± 2.0	4.1 ± 1.4	1.5 ± 1.0
Ne	34.6 ± 1.3	62.0 ± 1.0	2.6 ± 0.4	0.9 ± 0.0
Ar	37.2 ± 0.5	56.8 ± 0.8	4.3 ± 0.2	1.8 ± 0.1
CF_4	32.8 ± 0.6	67.2 ± 0.6		

angles for the treated FEP increased markedly over the first few days of ageing, despite the increased uptake of oxygen, although not to the level of untreated polymer. This reduced wettability is attributed to both reorientation of polar groups towards the bulk of the polymer and chemical changes in the first few atomic layers.

Changes in topography and surface chemistry for a range of plasmas (O_2, N_2, H_2, He, Ne, Ar and CF_4) were studied by Ryan and Badyal [23]. The changes in chemistry are summarised in Table 6. It can be seen the smallest chemical changes occur with the O_2 and CF_4 plasmas. On the other hand, an oxygen plasma gave the greatest topographical change.

3. OTHER TREATMENTS

A wide range of pretreatments exists both for fully and partially fluorinated systems. In the period 1990–1995, plasma treatments have attracted the most attention but several other methods have been studied. These include sodium naphthalenide, flame treatment and the use of Group I hydroxides.

3.1. Sodium naphthalenide

Ha *et al.* [24] using IR-IRS and XPS concluded that the sodium naphthalenide treatment of PFA introduced unsaturation fairly evenly to a depth of 112 nm while oxygen-containing functional groups such as $-OH$, $>C=O$ and $-CO_2H$ were concentrated in the first few nm. They calculated that in the top 5 nm of treated surface, 1 in 20 carbon atoms was involved in a hydroxyl group and 1 in 60 in a carbonyl group.

The same group studied the effect of sodium naphthalenide on PTFE, FEP and PFA [25]. They found that the washing procedure had a marked effect on the surface chemistry of treated PFA. Washing by tetrahydrofuran alone left a surface with much sodium carbonate but few hydroxyl groups. The authors concluded that hydroxyl groups were produced during the normal washing procedure, i.e. with water.

Marchesi *et al.* [25] found that the surface compositions of PFA and PTFE as determined by XPS were broadly similar after a 1 h treatment (Table 7). The SEM study indicated that the etching treatment did not cause any changes in topography to PFA or FEP but cracks were apparent with skived PTFE.

Table 7.
A comparison of surface functionality as measured by XPS of PFA and PTFE after a one-hour treatment with sodium complex [25]

Group	PFA	PTFE
C=C, C≡C	1 per 2.9 carbons	1 per 3.7 carbons
−OH	1 per 20 carbons	1 per 18 carbons
−CO	1 per 60 carbons	1 per 100 carbons
−F	1 per 80 carbons	1 per 40 carbons

Brewis *et al.* [26] compared the effectiveness of sodium naphthalenide towards PTFE and PVF. They found that the increase in lap shear joint strength with a two-part epoxy was much more rapid with PTFE, as can be seen from Table 8.

In a later study [7], the same authors studied the treatment of PTFE, ECTFE, PVdF and PVF with sodium naphthalenide (Table 9). As can be seen, the rate of chemical modification is in the order PTFE > PVdF > ECTFE > PVF. With PTFE, almost

Table 8.
Joint strengths for 'Tetra Etch'-treated PTFE and PVF [26]

	Treatment time (s)	Joint strength (MPa)*
PTFE	0	2.1
	2	18.3
	10	21.4
	30	20.5
	60	21.3
	600	22.2
	1800	20.8
	3600	22.3
PVF	0	1.8
	2	3.6
	10	4.0
	30	5.0
	60	10.4
	600	15.9
	1800	13.4
	3600	14.1

*Error is ± 1 MPa.

Table 9.
Effect of 'Tetra Etch' treatment on PTFE, PVF, PVdF and ECTFE [7]

Polymer		Colour	XPS (at.%)				Failure load (N)*
			C	Cl	F	O	
PTFE	None	White	38.4	—	61.6	—	420
	10 s	Brown	87.6	—	0.8	11.6	4280
	1 min	Black	82.2	—	0.9	16.9	4260
PVF	None	Colourless	70.4	—	28.8	0.8	360
	10 s	Colourless	72.4	—	26.7	0.9	800
	1 min	Colourless	75.4	—	23.0	1.6	2080
	60 min	Colourless	87.3	—	11.4	1.3	3020
PVdF	None	Colourless	51.4	—	47.9	0.7	—
	1 min	Faint brown	77.4	—	12.9	9.7	—
	60 min	Faint brown	79.5	—	9.2	11.3	—
ECTFE	None	Cream	53.2	14.3	32.5	—	240
	1 min	Cream	72.5	3.7	17.7	6.0	3300

*For a bonded area of 20 mm wide × 10 mm long; a two-part epoxide was used.

complete defluorination occurred in 10 s whereas with PVF defluorination was still far from complete after 1 h.

3.2. Treatment with Group I hydroxides

Crowe and Badyal [27] treated PVdF overnight at room temperature with a saturated aqueous solution of lithium hydroxide. Very different surface chemistries were obtained depending on the washing procedure. If the treated PVdF was washed in water, two new peaks were evident in the XPS spectrum, the dominant one being at 285.2 eV (most likely $-C_xH_y-$) and 287.4 eV (probably $>\underline{C}=O$ or $>\underline{C}<(OH)_2$). About 56% of the carbon atoms were modified. No Li 1*s* signal was observed.

If the treated PVdF was washed in isopropanol, both covalent (peak at 688.1 eV, 46%) and ionic (peak at 685.4 eV, 54%) fluorine species were observed. The existence of strong Li 1*s* and O 1*s* peaks at 55.7 and 532.1 eV was given as evidence for a reaction between PVdF and LiOH at the solid–solution boundary. Shoulders amounting to 53% of the total C 1*s* signal also emerge towards the lower binding energy side of both the $-\underline{C}F_2-$ and $-\underline{C}H_2-$ peaks. The amount of modified C 1*s* for both types of wash (56 and 53% respectively) compares with the observed ionic fraction (54%) of the F 1*s* spectrum. The authors concluded that dehydrofluorination had probably occurred. The treatment was reported to give much improved adhesion in a tape test.

Other workers [28–30] have reported the treatment of PVdF with Group I hydroxides and the use of phase transfer catalysts to increase the reactivity of the reagent. However, they do not report adhesion studies of the treated surfaces.

Brewis *et al.* [31] carried out a detailed study of the effect of Group I hydroxides, especially potassium hydroxide, on the bondability and surface chemistry of PVF and PVdF. Some of these results for PVF are summarised in Tables 10–12 and Fig. 1 and those for PVdF in Table 13.

Treating PVF for 1 min with a 15 M solution at 80°C gave a very similar bond strength with epoxy as a 2 h treatment but the surface modification was much more pronounced in the latter case. Likewise increasing the temperature of a 15 M solution beyond 80°C (Table 10) did not result in further large increases in adhesion, but the chemical changes were much more pronounced.

Table 10.

The effect of solution temperature on the surface composition and bondability of PVF treated with aqueous KOH solution (15 M, 1 min) [31]

Temperature (°C)	Colour	XPS (at.%)			Failure load (N)*	Standard deviation
		C	F	O		
Untreated	None	69.5	29.4	1.0	350	70
80	None	68.4	29.9	1.7	2900	250
100	None	68.9	29.1	2.0		
120	None	73.7	18.6	7.7		
140	Faint brown	76.5	10.6	12.9	3100	283
140 for 30 min	Brown	81.3	1.0	17.7	4330	249

*For a bonded area of 20 mm wide × 10 mm long; a two-part epoxide was used.

Table 11.
Effect of treatment time on surface compositions and adhesion levels of PVF treated with aqueous and alcoholic KOH solutions (5 M, 80°C) [31]

Time	Colour	Failure load (N)*	Standard deviation	XPS (at.%)			
				C	F	O	K
Aqueous							
0	None	350	70	69.5	29.4	1.0	—
1 min	None	440	160	71.2	27.9	0.9	—
10 min	None	450	0	69.6	28.0	2.4	—
30 min	None	430	50	71.6	23.7	4.7	—
1 h	None	580	115	73.1	20.2	6.7	—
Alcoholic							
0	None	350	70	69.5	29.4	1.0	—
10 s	None	500	115	74.5	24.9	0.6	—
30 s	None	530	126	68.6	29.6	1.8	—
1 min	Faint brown	890	85	69.2	28.7	1.1	1.2
10 min	Faint brown	3020	192	69.1	28.0	2.1	0.8
30 min	Light brown	3600	200	69.2	23.8	4.8	2.2
1 h	Brown	3490	85	70.6	21.8	6.2	1.4

*For a bonded area of 20 mm wide × 10 mm long; a two-part epoxide was used.

Table 12.
Effect of temperature on the treatment of PVF using 5 M aqueous KOH + 0.15 g TBAB per 50 ml of solution [31]

Temperature (°C)	Time	Failure load (N)*	Standard deviation
50	10 s	410	76
	1 min	670	76
	1 hour	450	141
80	10 s	430	177
	30 s	310	115
	1 min	320	29
	1 h	4220 CF	157
100	30 s	890	114
	1 min	2060	746
	1 h	4470 CF	58
110 (boiling)	10 s	650	265
	30 s	3280	171
	1 min	3980 CF	171
	1 h	4350 CF	129

*For a bonded area of 20 mm wide × 10 mm long; a two-part epoxide was used.
CF = cohesive failure within PVF.

The results in Table 11 compare aqueous and alcoholic KOH treatments. These results show that a 5 M aqueous solution of KOH at 80°C gives a relatively modest increase of 65% in the adhesion of PVF after 1 h, whereas a corresponding alco-

Table 13.
The effect on adhesion level and surface composition for various KOH treatments of PVdF [31]

Treatment	Colour	Failure load (N)*	SD	XPS (at.%)		
				C	F	O
None	None	1300	294	51.0	49.0	—
Aq. KOH (5 M 80°C)						
10 s	None	4120	135			
30 s	None	4040	261	52.7	45.7	1.6
1 min	Faint brown	4010	699	53.9	43.1	3.0
10 min	Faint brown	4560	723	58.0	37.0	5.0
1 h	Light brown	4320	286	62.0	28.9	9.1
Aq. KOH (5 M 80°C) + TBAB† 0.15 g						
10 s	Light brown	4250	250	62.0	28.4	9.6
30 s	Light brown	3940	241	63.1	24.4	12.5
1 min	Brown	4860	152	69.7	16.7	13.6
10 min	Brown	4430	222	74.6	9.9	15.5
1 h	Dark brown	4740	297	77.7	4.3	16.8‡
Alc. KOH (5 M 80°C)						
10 s	Light brown	4150	129	59.3	34.7	6.0
1 min	Brown	4630	330	62.2	27.5	10.3
5 min	Dark brown	4340	329	65.7	21.7	12.6
10 min	Black	4150	71	67.0	18.7	14.3
1 h	Black	3230	472	74.0	8.9	17.1

*For a bonded area of 20 mm wide × 10 mm long; a cyanoacrylate adhesive was used.
†TBAB per 50 ml of solution.
‡N 0.7 and Na 0.5 at.% were also present.
SD = standard deviation.

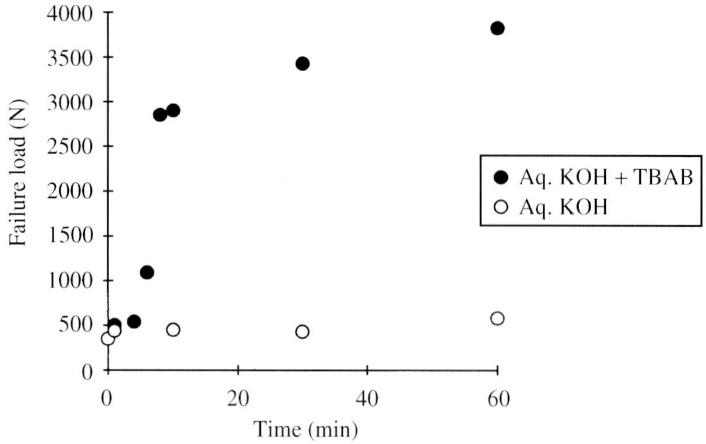

Figure 1. Bond strength with a two-part epoxide vs. time for PVF treated with 5 M aqueous KOH, and 5 M aqueous KOH + 0.03 g of TBAB per 50 ml of KOH solution at 80°C.

Table 14.
Flame treatment of PVF, PTFE and ECTFE [7]

Treatment (s)	XPS (at.%)				Failure load (N)*
	C	Cl	F	O	
PVF none	70.4	—	28.8	0.8	360
PVF 0.06	67.6	—	28.0	4.4	3240
PTFE none	38.4	—	61.6	—	420
PTFE 0.04	34.0	—	66.0	—	80
ECTFE none	53.2	14.3	32.5	—	240
ECTFE 0.06	68.6	8.0	17.2	6.0	2980

*For a bonded area of 20 mm wide × 10 mm long; a two-part epoxide was used.

holic solution results in a much larger increase (×10) for a given treatment time. However, the changes in surface composition shown by XPS are similar for the two pretreatments.

Figure 1 shows the effect of adding a small quantity of tetrabutylammonium bromide, TBAB, to aqueous KOH solution. The improvement in adhesion to an epoxide adhesive of PVF is much greater with the addition of TBAB to the solution although the surface chemical changes are similar to those obtained without this addition.

The effect of temperature on solutions containing TBAB to treat PVF is given in Table 12. Thus treatment times as low as 30 s can result in large increases in adhesion.

Table 13 summarises various KOH treatments of PVdF; the adhesive used was a cyanoacrylate namely Loctite Super Gel Xtra.

Clearly, very high joint strengths can be achieved much more rapidly than with PVF (see Tables 10–12). The changes in surface chemistry after a given treatment are also much greater with PVdF. In contrast to PVF, the chemical changes, for given treatment conditions, with alcoholic KOH and with KOH plus TBAB were much more pronounced than with the standard aqueous KOH solution.

3.3. Flame treatment

Mathieson *et al.* [7] showed that large increases in adhesion could be achieved by flame treating PVF and ECTFE (Table 14). With PTFE there was no surface modification and an actual decrease in bond strength. With the other two polymers a substantial quantity of oxygen functionality was introduced and this is probably responsible for the improved adhesion.

4. DISCUSSION

There are many methods that may be used to enhance adhesion to PTFE [1–6] but the two usually used commercially are sodium in liquid ammonia [1] and sodium naphthalenide [2]. They are both very effective but are environmentally undesirable.

The use of sodium and liquid ammonia is clearly hazardous during the pretreatment stage. Sodium naphthalenide solutions are less hazardous, and relatively stable proprietary products may be purchased. However, these are expensive and when the treated polymer is washed with water, naphthalene is liberated. The cleaning process results in an effluent typically consisting of naphthalene, tetrahydrofuran, methanol, sodium hydroxide and sodium fluoride.

The existence of a region of low cohesive strength on PTFE is well established [32]. To obtain good adhesion this layer must be removed, crosslinked to material of higher cohesive strength, or chemically modified so that it can be absorbed by the adhesive. It may also be necessary to chemically modify the surface of PTFE to improve wetting and also increase the interaction across the interface where wetting has occurred.

Until recently, plasma treatment of PTFE has led to only moderate improvements in bondability. Several studies have reported introduction of substantial quantities of oxygen-containing groups and/or substantial improvements in wetting but only small to moderate improvements in adhesion [7, 11, 19, 33]. This is consistent with the incomplete removal of the weak layer on PTFE. However, recently, Griesser [34] has reported much greater improvements in the bondability of PTFE after treatment with an ammonia plasma.

In contrast to fully fluorinated polymers such as PTFE, partially fluorinated varieties such as PVF and PVdF are moderately polar, although their total surface energies are similar to polyethylene [35]. It might be expected that good adhesion to the untreated polymers would be achieved with polar adhesives such as epoxides and this is sometimes the case [36, 37]. On other occasions poor adhesion is observed [7, 26, 31]. Poor and good adhesion to different batches of untreated PVF have been observed with the same adhesive [31, 37]. This is consistent with the existence of weak boundary layers on some batches of PVF.

From a practical viewpoint there are several options for treating difficult-to-bond partially fluorinated polymers. Sodium naphthalenide may be used but the treatment is quite slow and there is a significant effluent problem. Polymers such as PVF, PVdF, ECTFE and ETFE may be regarded as modified polyolefins and treated as such. For example, the flame treatment of PVF and ECTFE has been found to be very effective [7]. Recently, the use of potassium hydroxide has been found to be effective with PVF and PVdF. This treatment of PVF in 5 M aqueous solution of potassium hydroxide containing a small amount of tetrabutylammonium bromide for 30 s resulted in a 9-fold increase in shear strength and treatment of PVdF for 10 s in a 5 M aqueous solution of potassium hydroxide at 80°C resulted in a 3-fold increase to over 20 MPa [31].

The treatment of PVF with potassium hydroxide illustrates the need to remove potential weak boundary layers. The treatment of PVF for 1 h in a 5 M aqueous solution at 80°C resulted in only a relatively modest 65% increase in joint strength, despite a large increase in surface oxygen concentration. Conversely, treatment of PVF for 1 min in 15 M aqueous solution at 80°C resulted in an 8-fold increase in shear strength, but there was only a small change in surface chemistry (Table 10).

Clearly, much progress has been made in recent years in our understanding of the pretreatment of fluoropolymers and from the technological viewpoint a number of effective options for treating partially and fully fluorinated polymers exist. However, there are clearly gaps in our knowledge and there is room for further improvements in treatment methods.

5. CONCLUSIONS

- The use of sodium in liquid ammonia, and sodium naphthalenide in a suitable solvent are the most effective pretreatments for fully fluorinated polymers. However, recent work on ammonia plasmas may soon lead to an acceptable commercial alternative.
- Some batches of partially fluorinated polymers do not require a pretreatment. However, batches containing a weak surface layer may be effectively treated with plasmas (low pressure or flame) or with solutions of potassium hydroxide.

REFERENCES

1. R. J. Purvis and W. R. Beck, US Patent 2,789,063 to 3M Co. (16 April 1957).
2. A. A. Benderly, *J. Appl. Polym. Sci.* **6** (20), 221 (1962).
3. H. Schonhorn and R. H. Hansen, *J. Appl. Polym. Sci.* **11**, 1461 (1967).
4. R. H. Dahm, D. J. Barker, D. M. Brewis and L. R. H. Hoy, in: *Adhesion 4*, K. W. Allen (Ed.), p. 213. Applied Science, London (1979).
5. J. Jansta, F. P. Dousek and J. Riha, *J. Appl. Polym. Sci.* **34**, 3201 (1975).
6. C. A. Costello and R. J. McCarthy, *Macromolecules* **17**, 2941 (1984).
7. I. Mathieson, D. M. Brewis, I. Sutherland and R. A. Cayless, *J. Adhesion* **46**, 49 (1994).
8. H. Yasuda, *J. Macromol. Sci. Chem.* **A10** (3), 383 (1976).
9. R. R. Sowell, N. J. DeLollis, H. J. Gregory and O. Montoya, *J. Adhesion* **4**, 15 (1972).
10. B. W. Malpass and K. Bright, in: *Aspects of Adhesion 5*, R. J. Alner (Ed.), p. 224. University of London Press (1969).
11. T. Kasemura, S. Ozawa and K. Hattori, *J. Adhesion* **33**, 33 (1990).
12. M. Morra, E. Occhiello and F. Garbassi, *Surf. Interface Anal.* **16**, 412 (1990).
13. G. P. Hansen, R. A. Rushing, R. W. Warren, S. L. Kaplan and O. S. Kolluri, *Intl J. Adhesion Adhesives* **11** (4), 247 (1991).
14. H. J. Griesser, D. Youxian, A. E. Hughes, T. R. Gengenbach and A. W. M. Mau, *Langmuir* **7**, 2484 (1991).
15. D. Youxian, H. J. Griesser, A. W. H. Mau, R. Schmidt and J. Liesegang, *Polymer* **32** (6), 1126 (1991).
16. T. G. Vargo, J. A. Gardella, A. E. Meyer and R. E. Baier, *J. Polym. Sci. Part A* **29**, 555 (1991).
17. G. L. Anderson, D. A. Dillard and J. P. Wightman, *J. Adhesion* **36**, 213 (1992).
18. A. Kinbara, A. Kikuchi, S. Baba and T. Abe, *J. Adhesion Sci. Technol.* **7**, 457 (1993).
19. S. L. Kaplan, E. S. Lopata and J. Smith, *Surf. Interface Anal.* **20**, 331 (1993).
20. M. A. Golub, E. S. Lopata and L. S. Finney, *Langmuir* **10**, 3629 (1994).
21. M. K. Shi, A. Selmani, L. Martinu, E. Sacher, M. R. Wertheimer and A. Yelon, *J. Adhesion Sci. Technol.* **8**, 1129 (1994).
22. T. R. Gengenbach, X. Xue, R. C. Chatelier and H. J. Griesser, *J. Adhesion Sci. Technol.* **8**, 305 (1994).
23. M. E. Ryan and J. P. S. Badyal, *Macromolecules* **28**, 1377 (1995).
24. K. Ha, S. McClain, S. L. Suib and A. Garton, *J. Adhesion* **33**, 169 (1991).
25. J. T. Marchesi, K. Ha, A. Garton, G. S. Swei and K. W. Kristel, *J. Adhesion* **36**, 95 (1991).
26. D. M. Brewis, I. Mathieson, I. Sutherland and R. A. Cayless, *J. Adhesion* **41**, 113 (1993).
27. R. Crowe and J. P. S. Badyal, *J. Chem. Soc. Commun.* **958** (1991).
28. H. Kise and H. Ogata, *J. Polym. Sci., Polym. Chem.* **21**, 3443 (1983).

29. A. J. Dias and T. J. McCarthy, *Macromolecules* **17**, 2529 (1984).
30. A. J. Dias and T. J. McCarthy, *J. Polym. Sci., Polym. Chem.* **23**, 1057 (1985).
31. D. M. Brewis, I. Mathieson, I. Sutherland, R. A. Cayless and R. H. Dahm, *Intl J. Adhesion Adhesives* **16**, 87 (1996).
32. D. Briggs, in: *Industrial Adhesion Problems*, D. M. Brewis and D. Briggs (Eds), p. 15. Orbital Press, Oxford (1985).
33. N. Inagaki, S. Tasaka and H. Kawai, *J. Adhesion Sci. Technol.* **8**, 637 (1989).
34. H. J. Griesser, T. R. Gengenbach, L. Dai, S. Li and R. C. Chatelier, This Festschrift volume.
35. D. G. Rance, in: *Industrial Adhesion Problems*, D. M. Brewis and D. Briggs (Eds), p. 48. Orbital Press, Oxford (1985).
36. M. Levine, G. Ilkka and P. Weiss, *Polym. Letters* **2**, 915 (1964).
37. D. M. Brewis, *Progress Rubber Plastics Technol.* **1** (4), 1 (1985).

APPENDIX I. ABBREVIATIONS

PFA	poly(tetrafluoroethylene-co-perfluoroalkoxy vinyl ether), nominal formula $-(CF_2CF_2)_x-$ $(CF(OC_3H_7)CF_2)_y-$, $x/y \sim 49$
PTFE	poly(tetrafluoroethylene)
FEP	copolymer of tetrafluoroethylene and hexafluoropropylene
PVdF	poly(vinylidene fluoride)
ECTFE	ethylene–chlorotrifluoroethylene copolymer
ETFE	ethylene–tetrafluoroethylene copolymer
XPS	X-ray photoelectron spectroscopy
AFM	atomic force microscopy
STM	scanning tunelling microscopy
SEM	scanning electron microscopy
IR-IRS	infrared internal reflection spectroscopy

Mittal Festschrift, pp. 285–305
W. J. Van Ooij and H. R. Anderson, Jr. (Eds)

Oxygen plasma treatment of thermoplastic polyolefins: Relevance to adhesion

ANNA NIHLSTRAND,[1] THOMAS HJERTBERG[1] and KENTH JOHANSSON[2]
[1] *Department of Polymer Technology, Chalmers University of Technology, S-412 96 Gothenburg, Sweden*
[2] *Institute for Surface Chemistry, Box 5607, S-114 86 Stockholm, Sweden*

Abstract—Injection-moulded plates of four commercial thermoplastic polyolefins (TPOs) were subjected to oxygen plasma treatment. The modified surfaces were analyzed by water contact angle measurements and X-ray photoelectron spectroscopy (XPS), and the adhesion properties of the plates were evaluated by a 90° peel test after being lacquered with a two-component polyurethane lacquer. The study included treatments in two different plasma reactors operating at different frequencies. The influence of certain processing parameters, such as discharge power, flow rate and gas pressure, was investigated, as was that of frequency (using the same reactor). While the results revealed that oxygen plasma treatment indeed led to improved wettability, the degree of surface modification was not highly affected by changes in the processing conditions. In contrast, there was a great effect on the lacquer adhesion, in particular by changes in discharge power and gas pressure. The results also showed that the TPOs were sensitive in different ways towards changes in the processing conditions. It was also found that, regardless of the absolute peel force, the failures occurred in the substrate at some distance *below* the oxidized layer. These observations were attributed to a VUV-induced formation of radicals which, in the case of polypropylene-based materials, predominantly lead to β-scissions. As secondary radicals have a higher tendency to form crosslinks that can compensate for chain scission reactions, the difference in the sensitivity of the TPOs was proposed to be related to the amount and distribution of ethylene in the materials.

Keywords: Oxygen plasma; wettability; lacquer adhesion; thermoplastic polyolefins; chain scission; cohesive strength; VUV radiation.

1. INTRODUCTION

The number of application areas for polymeric materials is steadily increasing. Polyolefin-based plastics form a material group of special interest because of their excellent mechanical and processing properties in combination with their low cost. However, these materials are extremely difficult to lacquer, print and glue as a result of their inherently low surface energy. To overcome this problem and to further increase the applicability of polyolefin-based materials, some kind of pretreatment is necessary. Pretreatment methods can be divided into two categories: oxidation methods and the use of primers. Among the oxidation techniques, flame treatment [1, 2] and etching with chromic acid [3] are two of the most frequently used methods for altering the surface properties of components with irregular shape, while corona treatment and high-temperature extrusion are common ways to modify polymeric films or fibres. With any of these methods, a number of oxygen-containing groups are introduced onto the surface, which offers possibilities for stronger interactions than can be obtained by dispersive forces only. The other category — primers — is often used in the automotive industry for improving the paintability of thermoplastic polyolefins (TPOs) [4, 5]. Primers are usually based on a chlorinated polyolefin and, owing to the chlorine atoms, they are Lewis acids which, similar to oxygen-containing groups, offer possibilities for increased interactions. As demonstrated by Reihs [6], one important aspect is of course that the primer molecules are properly anchored to the substrate.

During the last decade, low-temperature plasma has attracted considerable interest as an alternative pretreatment method for plastics. Plasma treatment offers several advantages over the above-mentioned techniques. One of the most important features of the plasma technique is that neither solvents nor chlorinated products are involved. Instead, plasma utilizes a harmless gas, usually air or oxygen. Moreover, no hazardous by-products are formed and it gives, under optimized conditions, safe and uniform treatment to all kinds of object shapes. A vast number of studies have been carried out to gain more knowledge of the interaction between the plasma itself and polyolefin-based substrates. Most of these studies have focused on characterization of the treated surfaces by techniques such as contact angle goniometry, scanning electron microscopy (SEM), X-ray photoelectron spectroscopy (XPS), Fourier-transform infrared spectroscopy (FTIR), atomic force microscopy (AFM) and static secondary ion mass spectroscopy (SSIMS) (see for instance [7–11]). Studies including practical evaluation of the adhesion properties are far less common, although exceptions exist [12–14].

We have studied the adhesion between a lacquer and plasma-treated TPOs. Initial experiments (see Table 1) revealed that two commercial TPOs (TPO 1 and TPO 2) which were oxygen plasma-treated in exactly the same way (Reactor A) performed differently from an adhesion point of view. As measured by a 90° peel test, TPO 2 displayed a high level of lacquer adhesion (38 N/15 mm) with a clearly cohesive failure in the substrate (a thin layer of polymer was present on the lacquer side after peeling), while TPO 1 showed a mediocre peel force (11 N/15 mm) with an apparently adhesional failure. These results were surprising, since both TPOs displayed similar water contact angles and a similar degree of oxidation. The difference in lacquer adhesion obtained after plasma treatment in Reactor A became even more difficult to understand when it appeared that plasma treatment of the same materials in another reactor (Reactor B) led to excellent lacquer adhesion. In this case, both TPO 1 and TPO 2 showed peel forces of $\sim$ 37 N/15 mm, and the failures were clearly cohesive in the substrates.

An observation made in another study turned out to be vital for an understanding of the above results [15]. It was shown in that study that the lacquer adhesion obtained after plasma treatment of a TPO was not determined by the degree of surface modification, but rather by the cohesive strength of the near-surface region of the substrate. Independent of absolute peel force, the failure occurred in the substrate at some distance *below* the oxidized layer. Evidently, the lacquer adhesion was in that case determined by the cohesive strength of the near-surface region of the substrate,

Table 1.

Water contact angle data[a] (θ_A — advancing, θ_R — receding), at.% oxygen determined by XPS (% O), and peel forces (F, given as N/15 mm) after plasma treatment of two TPO materials (TPO 1 and TPO 2) using two different plasma reactors (A and B)

Treatment		Substrate	
		TPO 1	TPO 2
Untreated	θ_A	104	99
	θ_R	84	74
	% O	1.5	2.0
	F	~ 0	~ 0
Reactor A[b]	θ_A	51	55
	θ_R	< 5	< 5
	% O	17 ± 4	16 ± 4
	F	11 ± 3	38 ± 4
Reactor B[c]	θ_A	55	55
	θ_R	< 10	< 15
	% O	14 ± 2	15 ± 3
	F	37 ± 4	38 ± 4

[a]Standard deviation: 2–3°, measurements carried out one day after the plasma treatment.

[b]Treatment conditions, reactor A: 30 W, 20 sccm, 90 mTorr, 30 s, 125–375 kHz.

[c]Treatment conditions, reactor B: 50 W, 200 sccm, 460 mTorr, 60 s, 13.56 MHz.

which for some reason was seriously weakened. This weakening was proposed to be caused by the energetic vacuum–ultraviolet (VUV) radiation (< 200 nm) created during the plasma treatment, which is known to be strongly absorbed by hydrocarbons such as polyethylene (PE) and polypropylene (PP) [16, 17]. Moreover, as the VUV radiation is known to penetrate more deeply into the substrate than other active species, such as ions and atoms [18], radicals can be formed not only at the outermost surface but also in the near-surface region of the substrate.

The result of radical formation may be different depending on the nature of the radicals. Generally, if secondary radicals are formed, crosslinking is the expected result; whereas, if tertiary radicals are formed, β-scission is more likely to occur [19, 20]. Thus, if the lacquer adhesion is determined by the extent to which chain scission and crosslinking reactions occur in the near-surface region of the substrate, it is evident that the material composition is important, as PP forms predominantly tertiary radicals while PE forms secondary radicals. The variation in adhesion properties between TPO 1 and TPO 2 may thus be a consequence of differences in the material composition.

The objective of this work was to study how selected processing parameters affected the wettability and adhesion properties of four commercial TPOs after oxygen plasma treatments in two different plasma reactors. Part of the overall task was also to investigate whether the TPOs respond differently towards changes in the treatment conditions and, if possible, to relate such differences to variations in the material composition.

2. EXPERIMENTAL

2.1. Materials

TPOs generally consist of 60–75% of an ethylene-modified polypropylene matrix (ethylene content ~ 5–15%) and an ethylene–propylene rubber (EPR) or an ethylene–propylene–diene monomer (EPDM) rubber. Four commercial TPOs, supplied by Borealis, Norway (TPO 1–3) and Borealis, Belgium (TPO 4), were used in this study. TPO 1–3 were received as injection-moulded plates (melt temperature 230°C, mould temperature 40°C), while TPO 4 was received as pellets. TPO 1–2 (plate size 3 × 70 × 150 mm) were injection-moulded using a Netstal, model N150/75, and TPO 3 (plate size 2.7 × 69 × 120 mm) was injected-moulded using a Boy 50. TPO 4 was formed into plates (3 × 100 × 100 mm) by melt-pressing (2 min, 190°C, 10 ton), and then cold-pressed (5 min, 40°C, 10 ton) in order to avoid distortion of the plates. TPO 1–3 contain standard amounts of stabilizers whereas TPO 4 is marketed as a low-stabilized material. No further details about the materials were given, and for that reason selected bulk analyses were performed (see below).

2.2. Characterization of bulk materials

2.2.1. Extraction. Extraction was performed to estimate the rubber content of the TPOs. The plates were cut into small pieces (2–3 mm) and then ground into a fine powder. The ground powder (2 g) was heated to 75°C in 100 ml *p*-xylene for 6 h, whereafter the non-dissolved polypropylene (PP) was removed by filtering and the dissolved rubber phase was precipitated by pouring the solution into 500 ml of methanol. After the samples were rinsed, dried (vacuum-oven, 40°C) and weighed, the percentages of rubber were calculated as $m_R/(m_R + m_{PP})$, where m_R and m_{PP} are the weights of the rubber phase and the PP phase, respectively.

2.2.2. Precipitation. The rubber content was also estimated by a dissolution precipitation procedure. In this, the sample was dissolved in boiling *p*-xylene, whereafter the solution was cooled to room temperature, at which precipitation of the PP phase occurs. The two parts were isolated and handled according to the procedure described above, as was the calculation of the percentage of rubber.

2.2.3. Differential Scanning Calorimetry (DSC). DSC was used to estimate the material composition of the TPOs. The shape of the melting curves in the region 100–130°C provides information on how the ethylene is distributed in the materials (random or block), while a rough estimation of the rubber content can be achieved from the following formula: $1 - (\Delta H_{TPO}/\Delta H_{PP})$, where ΔH_{PP} and ΔH_{TPO} represent the energy (J/g) required to melt a reference PP and the TPO material, respectively. The reference PP had a crystallinity of 58% (calculated using 165 J/g as the heat of fusion for a 100% crystalline PP [21]). It should be noted that this formula can be used only under the assumption that the reference PP and the PP matrices in the TPOs have about the same degree of crystallinity, which was not controlled. Nevertheless, the percentages of rubber obtained from the DSC data should be comparable with each other, as they are related to the same reference PP. The samples were exposed to the following thermal history: heating 40–200°C, cooling 200–40°C and heating 40–200°C. All runs were performed at a temperature rate of 10°C/min, and the reported results were calculated from the second heating run. The typical sample weight was 10 mg. The measurements were carried out using a Perkin–Elmer DSC 7.

2.2.4. Size Exclusion Chromatography (SEC). SEC was used to compare the molecular weights (MW) and the molecular weight distributions (MWD) of the bulk materials, the PP phases and the rubber phases (obtained through the extraction procedure). The samples were dissolved in 1,2,4-trichlorobenzene (135°C) overnight. All sample solutions (5 g/l) were filtered prior to injection. The measurements were carried out on a Waters 150 CV system equipped with three columns — two Shodex AT-80 M/S and one MP 613 E column (particle size 10 μm, normal pore size distribution) — swelled in 1,2,4-trichlorobenzene. The calibration is based on narrow polystyrene standards and is recalculated for polyethylene via universal calibration [22].

2.2.5. Fourier-Transform Infrared Spectroscopy (FTIR). FTIR transmission measurements of thin films (50–100 μm) were carried out to identify differences in the distribution of ethylene in the TPOs. The runs were made at a resolution of 4 cm^{-1}, and 25 scans were recorded for each sample. The analyses were performed on a Perkin–Elmer FTIR 2000 system equipped with a triglycine sulphate (TGS) detector.

2.3. Plasma reactors

Two plasma reactors (located at the Institute for Surface Chemistry, Stockholm) were used in this study. The first reactor, R1 (*not* identical to Reactor B mentioned above), is a commercially available batch reactor (Technics Plasma, model 3027-E). The process chamber is made of aluminium and its inner dimensions (width × height × depth) are 650 × 600 × 700 mm. The chamber, originally designed for chemical etching of printed circuit boards, contains approximately 120 tubular electrodes, eight gas inlets and two gas channels with flow regulators. The power generator (ENI, model EGR 3200B) is tuned at 35 kHz with a maximum output up to 3000 W. The vacuum system includes a two-stage monoblock pump (Leybold-Heraeus, model DK 100) with an ultimate pressure of 15–20 mTorr and a working pressure of 100–1200 mTorr. The pressure in the chamber is controlled by a Leybold-Heraeus Thermovac TC gauge and a MKS Baratron capacitance manometer (Type 221A, Power Supply/Digital readout model PDR-C-1B). The samples treated in this reactor were always placed in the upper part of the reactor between the middle rows of electrodes. The system was pumped down to less than 50 mTorr before oxygen was introduced. Treatments were carried out at different discharge powers (300–2000 W) combined with various gas pressures (200–1200 mTorr). The treatment time throughout all experiments was 30 s. No measurement was made of the flow rate.

The second plasma reactor, R2 (identical to the above mentioned Reactor A), is an in-house construction built at the Institute for Surface Chemistry and consists of a glass vessel (inner diameter = 15 cm, length = 65 cm) connected to a double-stage rotary vacuum pump (Leybold-Heraeus D 65 B). Two externally wrapped, capacitively coupled copper electrode bands (3 cm wide, 10 cm apart) were powered by either a low radio-frequency (125–375 kHz) power generator (ENI, model HPG-2) or by a 13.56 MHz radio-frequency power generator (ENI, model ACG-3). In the former case, the discharge power was balanced by adjusting the frequency so as to give a minimum of reflected power, while an automatic impedance matching network was used in the latter case. The automatic matching network (ENI, model MW-5D) was controlled by a remote generator controller (ENI, model RFC-5MW). The oxygen flow rate was controlled by a mass flow controller (HITEC, Bronkhurst). The substrates were mounted on a glass petri dish and placed on top of a steel sample holder. The sample holder was kept at a floating potential at constant temperature (20°C) and was always positioned just below the most intense plasma zone (see Fig. 1). Prior to treatments, the chamber was pumped down to a base pressure of approximately 5 mTorr. Oxygen was introduced at the top of the reactor, and the flow rate was 20 standard cubic centimetres per minute (sccm), unless stated otherwise. The pressure during the plasma treatments was varied between 90–750 mTorr by adjusting the exhaust valve. The treatment time was always 30 s, and the applied discharge power was varied between 5–30 W.

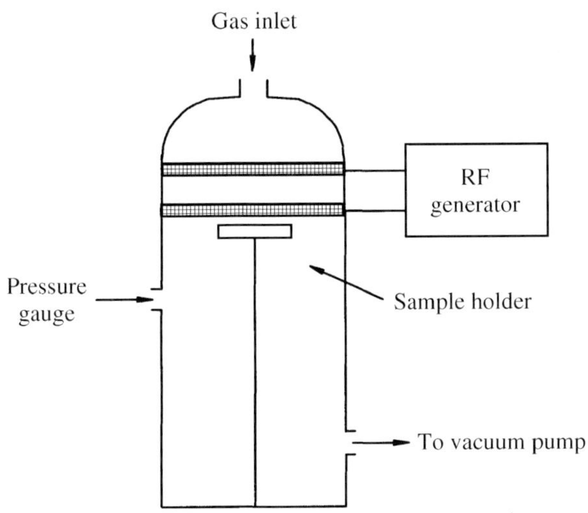

Figure 1. Schematic diagram of the lab-scale reactor (R2).

2.4. *Surface characterization*

2.4.1. Contact angle measurements (goniometry). To follow changes in wettability, water contact angles (both advancing and receding angles) were measured using a Ramé-Hart goniometer, model 100. Droplets (4–5 μl) of deionized water were placed at different locations on the sample surface, and a minimum of eight readings were made for each sample in order to determine average values. Owing to practical limitations, the contact angles were measured one day after the plasma treatment. Typical standard deviations were 2–3°.

2.4.2. X-ray Photoelectron Spectroscopy (XPS). XPS was used to determine the atomic composition of selected samples. The analyses were carried out using a Mg K_α (1253.6 eV) X-ray source at 250 W and 11 kV. The take-off angle was 45° with respect to the sample surface, which corresponds to a probing depth of $\sim$7 nm [23]. The pressure in the analysis chamber was $< 1 \times 10^{-9}$ Torr during data acquisition, and the analyzed area was 0.8 mm in diameter. The XPS analyses were performed on a PHI 5500 ESCA system from Perkin–Elmer, and the sensitivity factors used were according to the Perkin–Elmer manual (C = 0.296, O = 0.711, N = 0.477 and Si = 0.339).

2.5. *Lacquering and adhesion test*

The lacquer used throughout this study was a two-component polyurethane (PUR) lacquer delivered by Beckers, Sweden. The lacquer was composed of a hydroxyl-terminated polyester (Beckryflex TH-130 2601) and a hexamethylene–diisocyanate (HDI) prepolymer (Beckryflex TV-130). Lacquering was carried out immediately after performing the contact angle measurements. The test specimens used in the adhesion test were prepared as follows: a thin layer of the lacquer was manually applied onto the samples with a paintbrush. A polyester cloth (Monodur PES 71N, mesh width 71 μm) was placed in the lacquer as a reinforcement. The polyester cloth was plasma-treated prior to use to avoid failure between the textile cloth and the lacquer. Finally, another layer of lacquer was applied to ensure that the polyester cloth was completely soaked with the lacquer. The total thickness of the lacquer layer including the reinforcing cloth was 130 ± 20 μm. The lacquered samples were cured in an oven (90°C) for 40 min and then stored at atmospheric conditions (room temperature, normal humidity) for one week to complete the curing reaction. The lacquered plates were finally cut into 15-mm wide strips.

The lacquer adhesion was evaluated with a 90° peel test. The peel tests were performed using the TCT-5 model from Alwetron (located at Volvo Car Corporation, Gothenburg) equipped with a special fixture. The fixture includes a movable slide which enables the peel force to be measured at a constant angle of 90°. A detailed description of the peel test is given elsewhere [24]. The crosshead speed during the peel test was 600 mm/min, and the reported peel forces are average values for at least three strips peeled over a distance of 50 mm or more. The peel forces were registered as N/15 mm.

3. RESULTS

3.1. Characterization of the bulk materials

3.1.1. Rubber content. The results of the extraction, precipitation, DSC, SEC and FTIR experiments are summarized in Table 2. The table reveals that the 'rubber content' as determined by DSC correlates well with the results from the extraction and precipitation experiments. The highest rubber content is observed for TPO 1 and the lowest rubber content for TPO 2. It was not possible to estimate the amount of rubber in TPO 3 by either extraction or precipitation since the solution in this case was impossible to filter, indicating that this material may contain a reactor-modified TPO. Furthermore, it should be noted that neither of the methods gives an absolute measure of the rubber content, but rather a measure of the amount of amorphous material. This material can include the rubber phase as well as the amorphous part of the ethylene-modified PP matrix.

3.1.2. Ethylene modification. Figure 2 shows the DSC melting curves obtained for the TPOs. The figure reveals that the ethylene is differently distributed in the materials. Besides the melting peak observed in the region of 163–165°C (PP melting), the most

Table 2.
Material characteristics of TPO 1–4

Property	TPO 1	TPO 2	TPO 3	TPO 4
Rubber content (%)				
From extraction	43	29	NM	35
From precipitation	42	29	NM	34
From DSC[a]	46	29	42	34
Ethylene modification (DSC)	Random	Long blocks	Short blocks	Short blocks
Ethylene crystallinity (FTIR)	None	Much	Some	Some
$\overline{M}_n \times 10^{-3}$ bulk material	11.3[b]	23.8	18.5	16.6
$\overline{M}_w \times 10^{-3}$ bulk material	133	119	129	123
$\overline{M}_n \times 10^{-3}$ rubber phase[c]	26.7	24.8	20.7	17.5
$\overline{M}_w \times 10^{-3}$ rubber phase[c]	152	120	118	92.3
$\overline{M}_n \times 10^{-3}$ PP phase[c]	37.9	33.6	NM	NM
$\overline{M}_w \times 10^{-3}$ PP phase[c]	120	113	NM	NM
Carbon black[d]	Less	Less	Less	More
Other[e]	Talcum	—	Talcum	—

NM = not measured.
[a] Approximate rubber content determined as $1 - (\Delta H_{TPO}/\Delta H_{PP})$.
[b] The bulk material contained a low molecular weight fraction ($M_{PEAK} \sim 700$ g/mol) which could be found neither in the PP phase nor in the rubber phase.
[c] Obtained after extraction.
[d] Observed visually by comparing the colour of thin (50–100 μm) films.
[e] Identified by FTIR.

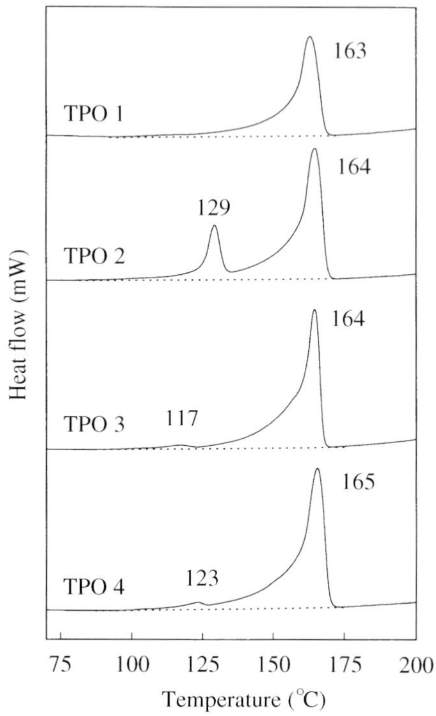

Figure 2. Melting curves of TPO 1–4 obtained by DSC. Heating rate 10°C/min.

obvious difference is found for TPO 2, which shows a pronounced melting peak in the interval of 125–130°C. This indicates that the ethylene is distributed in the form of long blocks that are able to crystallize in a way similar to high-density PE. In fact, it cannot be ruled out that this peak actually originates from 'pure' high-density PE, admixed as a means to further enhance the degree of ethylene modification. TPO 3 and TPO 4 both show signs of melting in the region of 100–125°C (melting region of low-density PE), which indicates that the ethylene in these materials is in the form of shorter blocks. TPO 1 shows no signs of melting peaks at all in the region of 100–130°C, implying that the ethylene in this case is randomly distributed.

Further evidence for the variation in ethylene distribution is achieved from the FTIR spectra shown in Fig. 3. The figure shows the in-phase $-CH_2-$ chain rock vibration. A peak at $\sim$722 cm^{-1} indicates an amorphous material with a minimum of two ethylene units in a row, while, if instead the peak is split into two peaks at 730 and 720 cm^{-1}, some crystallinity is present in the sample [25]. Again, the most obvious difference is found for TPO 2. The peak arising from the $-CH_2-$ rocking mode is clearly split into a doublet in this material, indicating that the ethylene is to some extent in the crystalline state. TPO 1 displays almost no sign of such a doublet, indicating an amorphous material with the ethylene units randomly distributed, while TPO 3 and TPO 4 show spectra intermediate to the ones obtained for TPO 1 and TPO 2. In the latter case, the $-CH_2-$ rocking peak is not clearly split into a doublet, but a shoulder on the left side of the peak reveals that some crystallinity is present in

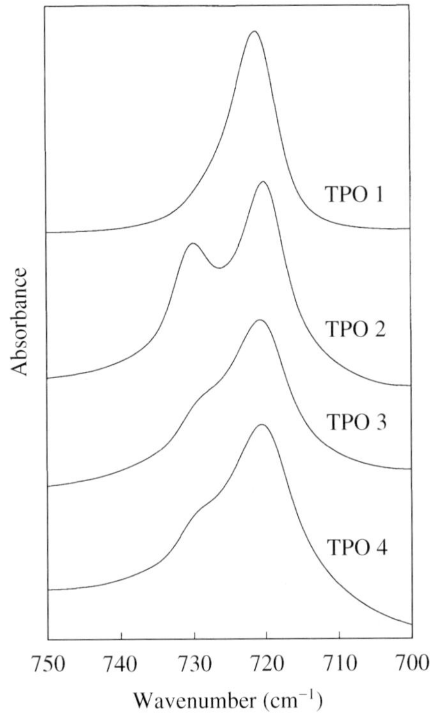

Figure 3. FTIR spectra of thin films showing the rocking mode of $-CH_2-$. A single peak indicates amorphous material (randomly distributed ethylene), whereas a split peak indicates crystallinity (longer sequences of ethylene units).

both samples. Apart from the information provided by the $-CH_2-$ in-phase rocking band, the FTIR spectra also revealed that TPO 1 and TPO 3 contained small amounts of talcum (peaks at 3677, 1018, 667 and 466 cm^{-1}).

3.1.3. Molecular weights and molecular weight distributions. The results obtained from the SEC analyses of the bulk materials, the PP phases and the rubber phases are presented in Table 2. The table shows that the bulk materials are similar with respect to the weight average molecular weight ($\overline{M}_w$). A lower number average molecular weight ($\overline{M}_n$) is obtained for TPO 1, however, which is caused by a low molecular (LM) weight fraction with a peak at $\sim$ 700 g/mol. This peak is not present in the other materials. No attempt was made to isolate and identify the material causing this peak but, according to its position, it may result from additives such as processing aids and stabilizers that appear in this region. The LM weight material is neither to be found in the PP phase nor in the rubber phase after the extraction procedure.

3.2. Results obtained after oxygen plasma treatment in R1

3.2.1. Influence of gas pressure. The influence of gas pressure on wettability and lacquer adhesion was studied using a constant power of 300 W. The results obtained

Table 3.
Water contact angle data (θ_A, advancing and θ_R, receding) and peel forces (F, given as N/15 mm) obtained after plasma treatment in R1, showing the influence of O_2 gas pressure. Plasma conditions: 300 W, 30 s

Gas pressure (mTorr)		TPO 1	TPO 2	TPO 3	TPO 4
200	θ_A	61	68	59	74
	θ_R	24	27	21	36
	F	11 ± 2	39 ± 4	43 ± 5	4 ± 1
400	θ_A	62	65	53	68
	θ_R	17	27	19	29
	F	18 ± 4	43 ± 2	50 ± 4	29 ± 6
600	θ_A	71	67	66	69
	θ_R	21	27	23	34
	F	23 ± 5	34 ± 5	41 ± 3	44 ± 6
800	θ_A	61	62	58	72
	θ_R	17	24	25	34
	F	34 ± 2	36 ± 1	46 ± 3	42 ± 4
1000	θ_A	74	75	63	72
	θ_R	19	24	23	20
	F	33 ± 3	39 ± 2	49 ± 1	23 ± 1
1200	θ_A	74	65	57	87
	θ_R	31	29	14	55
	F	14 ± 9	36 ± 4	43 ± 4	1 ± 1

Contact angles measured one day after the plasma treatment. Standard deviation: 2–3°.

after these treatments are presented in Table 3. The table reveals that changes in the gas pressure have limited influence on the degree of surface modification. This conclusion is based on the contact angles, which are seen to be generally independent of gas pressure and type of TPO. However, TPO 3 displays a weak tendency towards somewhat lower contact angles, and TPO 4 displays a tendency towards somewhat higher contact angles, as compared with the other TPOs. Moreover, the contact angles obtained for TPO 4 after treatment at 1200 mTorr deserve some extra attention. Both the advancing and receding contact angles are considerably higher than those observed after treatment using lower gas pressures. They are also considerably higher than the contact angles observed for the other TPOs treated at 1200 mTorr. Since the number of collisions in the plasma increases with increasing gas pressure (maintaining a constant power), the concentration of active species reaching the surface will be greatly reduced at higher pressures as compared with lower pressures. Hence, these high contact angles may simply be a consequence of inhomogeneous treatment and/or insufficient oxidation.

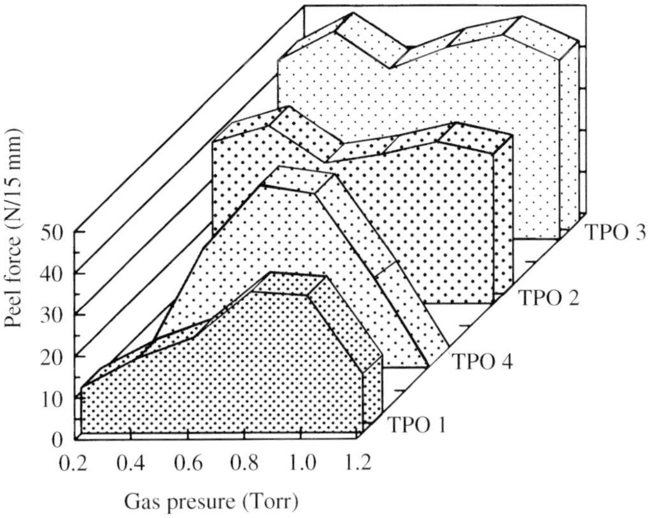

Figure 4. Influence of gas pressure on lacquer adhesion after O_2 plasma treatment in R1. Power 300 W; treatment time 30 s.

Although the wettability did not differ a great deal as a result of various gas pressures, the adhesion properties were highly affected. The peel forces obtained (displayed visually in Fig. 4) reveal that both TPO 2 and TPO 3 exhibit excellent lacquer adhesion, independent of the gas pressure used. All treatments of these materials led to clearly cohesive failures in the substrates. For TPO 1 and TPO 4, the gas pressure seems to be of greater importance. The trend is towards higher peel forces as the gas pressure increases. Starting from rather low values when 200 mTorr is used (11 N/15 mm for TPO 1 and 4 N/15 mm for TPO 4), peel forces comparable with those obtained for TPO 2 and TPO 3 are observed when 800 mTorr is used (> 30 N/15 mm). However, the improvement in lacquer adhesion with increasing gas pressure is limited for these TPOs and, at some level above 800 mTorr, the plasma treatment will no longer be as effective in improving the lacquer adhesion. In fact, a further increase in gas pressure to 1200 mTorr will lead to as low peel forces as those obtained after treatment at 200 mTorr. Keeping in mind the contact angle results discussed above, it would seem that the drop in lacquer adhesion is a manifestation of inhomogeneous treatment and/or insufficient oxidation.

3.2.2. Influence of discharge power. The discharge power is of great importance for the amount of active species in the plasma. If the discharge power is raised, the degrees of ionization, dissociation and electronic excitation will all be higher and the concentration of active species will thereby also be increased [26]. Hence, it can be assumed that the discharge power may influence both wettability and lacquer adhesion. To investigate whether this indeed is the case, plasma treatments were carried out at four different power levels: 300, 600, 1200 and 2000 W. The numerical results obtained after these treatments are summarized in Table 4, and the peel forces are presented visually in Fig. 5. Similar to what was found when the gas pressure was

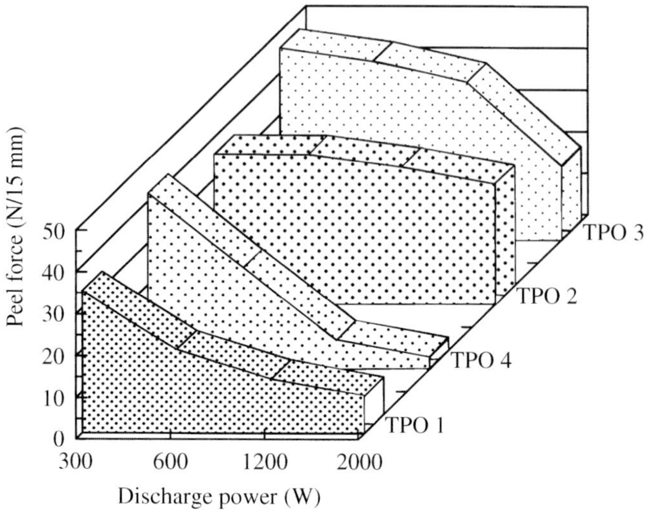

Figure 5. Influence of discharge power on lacquer adhesion after O_2 plasma treatment in R1. Gas pressure 800 mTorr; treatment time 30 s.

Table 4.
Water contact angle data (θ_A, advancing and θ_R, receding) and peel forces (F, given as N/15 mm) obtained after plasma treatment in R1, showing the influence of discharge power. Plasma conditions: 800 mTorr, 30 s

Discharge power (W)		TPO 1	TPO 2	TPO 3	TPO 4
300	θ_A	61	62	58	72
	θ_R	17	24	25	34
	F	34 ± 2	36 ± 1	46 ± 3	42 ± 4
600	θ_A	74	68	69	72
	θ_R	19	22	23	25
	F	20 ± 3	36 ± 1	43 ± 4	24 ± 2
1200	θ_A	66	71	67	72
	θ_R	23	26	25	29
	F	13 ± 2	33 ± 6	38 ± 8	7 ± 1
2000	θ_A	63	69	63	76
	θ_R	15	14	14	27
	F	9 ± 1	29 ± 8	18 ± 8	3 ± 1

Contact angles measured one day after the plasma treatment. Standard deviation: 2–3°.

varied, there is no great change in the contact angles as the discharge power is raised. Moreover, it is not possible to distinguish any clear trends among the materials except for a weak tendency towards higher contact angles in the case of TPO 4.

Logically, similar water contact angles would lead to comparable peel forces if the adhesion properties were determined by the degree of surface modification. However, contrary to what the contact angles suggest, the discharge power is of great importance for lacquer adhesion. At 300 W, all plasma-treated TPOs exhibit excellent lacquer adhesion (evident as cohesive failures in the substrates) but, when the discharge power is raised, the peel forces drop. The same trend is observed for all TPOs, although the deterioration is most pronounced for TPO 1 and TPO 4. When treated at 2000 W, the least sensitive material (TPO 2) maintains 81% (29 N/15 mm) of the peel force obtained after treatment at 300 W (36 N/15 mm). For the most sensitive material (TPO 4), the reduction is more pronounced: only 7% (3 N/15 mm) of the peel force obtained at 300 W (42 N/15 mm) is retained. TPO 1 and TPO 3 show more moderate drops, maintaining 26 and 39% of the peel strengths, respectively.

3.2.3. Influence of power-to-gas pressure ratio. According to the results presented so far, good lacquer adhesion is obtained when the plasma treatment is performed at a low power level in combination with a relatively high gas pressure, i.e. a low power-to-gas pressure ratio. This brings up the question of whether it is the absolute values of power and pressure that are important or whether it is the ratio between these two which determines the lacquer adhesion. Table 5 summarizes selected results from Tables 3 and 4 and it is evident from this table that a fixed power-to-gas pressure ratio (0.75 or 1.50) based on low absolute values results in as good lacquer adhesion as when the ratio is based on higher absolute values. Thus, it appears as though the important factor is the power-to-gas pressure ratio, rather than the absolute values.

3.3. Results obtained after oxygen plasma treatment in R2

3.3.1. Influence of gas pressure. To identify whether the influence of gas pressure is general and not specifically related to the use of R1, similar experiments were performed using R2. The main differences between R1 and R2 are the reactor size and geometry, the number and position of electrodes, and the discharge frequency. Using R2, the pressure was varied between 90 and 750 mTorr, and the power was 5 W. The differences in treatment conditions and reactor design of course make it impossible to directly compare the results obtained after treatment in these two reactors. However, the point of these experiments was not to compare the results in terms of absolute figures, but to see whether the same trends are found independent of

Table 5.
Peel forces (given as N/15 mm) obtained after plasma treatment in R1, showing the influence of various power-to-gas pressure ratios. Treatment time: 30 s

Power-to-gas pressure ratio (W/mTorr)		TPO 1	TPO 2	TPO 3	TPO 4
0.75	300/400	18 ± 4	43 ± 2	50 ± 4	29 ± 6
	600/800	20 ± 3	36 ± 1	43 ± 4	24 ± 2
1.50	300/200	11 ± 2	39 ± 4	43 ± 5	4 ± 1
	1200/800	13 ± 2	33 ± 6	38 ± 8	7 ± 1

Table 6.

Water contact angle data (θ_A, advancing and θ_R, receding) and peel forces (F, given as N/15 mm) obtained after plasma treatment in R2, showing the influence of gas pressure. Plasma conditions: 5 W, 20 sccm, 30 s, 125–375 kHz

Gas pressure (mTorr)		TPO 1	TPO 2	TPO 3*	TPO 4
90	θ_A	57	64	60	66
	θ_R	< 5	10	13	7
	F	16 ± 1	27 ± 6	46 ± 3	4 ± 1
500	θ_A	65	68	61	70
	θ_R	18	20	23	24
	F	32 ± 6	36 ± 2	47 ± 2	22 ± 2
750	θ_A	48	71	62	73
	θ_R	< 5	25	17	22
	F	22 ± 4	32 ± 2	47 ± 2	41 ± 5

*Treatment of TPO 3 at 500 mTorr and 750 mTorr were made at 30 W.
Contact angles measured one day after the plasma treatment. Standard deviation: 2–3°.

the reactor used. The results obtained after treatment in R2 are presented in Table 6. In accordance with the results reported in Table 3, it can be seen that the contact angles do not vary to any great extent with either gas pressure or among the TPOs.

As regards lacquer adhesion, it is evident from Table 6 that the peel forces are also greatly influenced by the gas pressure after treatment in this reactor and, again, the various TPOs perform differently. Similar to their behaviour after treatment in R1, TPO 2 and TPO 3 seem to be the materials least sensitive towards changes in the processing conditions. Independent of gas pressure, both show mainly cohesive failure in the substrates. It is also noteworthy that, in spite of the fact that TPO 3 was treated at 30 W (instead of 5 W) in the case of 500 mTorr and 750 mTorr, the failures are still clearly cohesive in the substrate. As it was shown above that a higher power level leads to a lower peel force, this indicates that TPO 3 can withstand tougher treatment conditions than, for instance, TPO 4. Furthermore, TPO 4 seems to be the material most affected by gas pressure. The peel force increases from 4 N/15 mm to 41 N/15 mm when the gas pressure is raised from 90 to 750 mTorr. TPO 1 also displays improved adhesion properties up to 500 mTorr but, in this case, the increase is followed by a decrease when the pressure is further increased to 750 mTorr. The decrease is unexpected, and it can only be concluded that it is not a result of insufficient oxidation, as the contact angles are still very low (48° and < 5°).

3.3.2. Influence of flow rate. It is clear from the above results that the gas pressure, independent of the reactor used, has an impact on the subsequent adhesion properties. However, while an increase in gas pressure using R2 can be carried out at a constant flow rate, the flow rate will automatically increase if the gas pressure is raised when treatments are carried out in R1. The strong dependence on gas pressure revealed after treatment in R1 may therefore be caused by variations in the flow rate. To

Table 7.
Water contact angle data (θ_A, advancing and θ_R, receding) and peel forces (F, given as N/15 mm) obtained after plasma treatment of TPO 3 in R2, showing the influence of gas flow rate. Plasma conditions: 30 W, 225 mTorr, 30 s, 125–375 kHz

		Gas flow rate (sccm)			
		5	20	50	100
TPO 3	θ_A	54	59	58	55
	θ_R	11	11	10	10
	F	13 ± 1	15 ± 2	18 ± 2	19 ± 3

Contact angles measured one day after the plasma treatment. Standard deviation: 2–3°.

obtain further information on the importance of the flow rate itself, a few additional plasma treatments (TPO 3 only) were performed for which the flow rate was varied between 5 and 100 sccm. The discharge power and the gas pressure were kept at values leading to mediocre lacquer adhesion ($\sim$ 15 N/15 mm) in order to see whether the lacquer adhesion would be improved or impaired when the flow rate was changed. The results are summarized in Table 7. No change in contact angles can be observed to result from different flow rates. The influence of flow rate on the peel force is also limited, although a slight increase can be distinguished when the flow rate is raised. Nevertheless, the flow rate must be considered to be of less importance for the adhesion properties than the gas pressure and discharge power.

3.3.3. Influence of discharge frequency. When the processing conditions are held constant, a change in the discharge frequency will, among other things, influence the electron energy distribution function (EEDF) [26, 27]. The energy and relative concentration of the active species will therefore also be affected and it is reasonable to assume that there will be an effect on both the degree of surface modification and the subsequent adhesion properties. The influence of frequency has been investigated previously, and was reported to be of minor importance for both wettability and the adhesion properties [15]. In that study, the comparison was made between radio-frequency (13.56 MHz) and microwave (2.45 GHz) discharges, whereas in this study the comparison is between 'low' radio-frequency (125–375 kHz) and 'normal' radio-frequency (13.56 MHz) discharges. Only one set of process parameters was tested in this case (5 W, 90 mTorr, 20 sccm, 30 s), and the results are summarized in Table 8. In agreement with the previously reported results, frequency does not appreciably affect wettability. Similar contact angles are obtained for both frequencies and no very large variation is found among the TPOs. On the other hand, frequency may significantly influence lacquer adhesion. TPO 1 and TPO 4 particularly show a clear increase in peel forces as the frequency is raised from kHz to MHz. The increase is less pronounced for TPO 2 and TPO 3.

3.3.4. Influence of discharge power. Obviously, a change in frequency (keeping the other processing parameters constant) can influence the subsequent adhesion properties. To verify that the influence of discharge power is independent of discharge

Table 8.
Water contact angle data (θ_A, advancing and θ_R, receding) and peel forces (F, given as N/15 mm) obtained after plasma treatment in R2, showing the influence of discharge frequency. Plasma conditions: 5 W, 90 mTorr, 20 sccm, 30 s

Discharge frequency		TPO 1	TPO 2	TPO 3	TPO 4
125–375 kHz	θ_A	57	64	60	66
	θ_R	< 5	10	13	7
	F	16 ± 1	27 ± 6	46 ± 3	4 ± 1
13.56 MHz	θ_A	56	62	55	69
	θ_R	6	16	12	20
	F	32 ± 4	30 ± 3	50 ± 4	25 ± 6

Contact angles measured one day after the plasma treatment. Standard deviation: 2–3°.

Table 9.
Water contact angle data (θ_A, advancing and θ_R, receding) and peel forces (F, given as N/15 mm) obtained after plasma treatment in R2, showing the influence of discharge power. Plasma conditions: 90 mTorr, 20 sccm, 30 s, 125–375 kHz

Discharge power (W)		TPO 1	TPO 2	TPO 3	TPO 4
5	θ_A	57	64	60	66
	θ_R	< 5	10	13	7
	F	16 ± 1	27 ± 6	46 ± 3	4 ± 1
30	θ_A	52	55	43	55
	θ_R	< 5	< 5	< 5	< 5
	F	11 ± 3	38 ± 4	19 ± 3	2 ± 1

Contact angles measured one day after the plasma treatment. Standard deviation: 2–3°.

frequency, treatments were also performed at different discharge powers using R2. Two power levels (5 and 30 W) were chosen for these experiments, and the plasma treatments were performed at 125–375 kHz. The results are summarized in Table 9. In accordance with what was found when using R1, the peel forces decrease when the discharge power is increased for all materials except TPO 2. The decrease is minor for TPO 1 and TPO 4 (both display low peel forces even after treatment at 5 W), but is more pronounced for TPO 3. The peel force drops to half its value when the discharge power is raised from 5 W to 30 W. The results obtained for TPO 2 are more surprising. The peel force is slightly lower after treatment at 5 W and, logically, this material should also reveal a decrease in peel force when the power is raised to 30 W. However, although the peel force is < 30 N/15 mm in the former case, the major part of this sample still displays a cohesive failure in the substrate.

4. DISCUSSION

The results presented above clearly show that the processing conditions during plasma treatment of TPOs must be taken into account when the adhesion properties are im-

Table 10.
Atomic composition (%), as measured by XPS, and peel forces (F, given as N/15 mm)

	XPS analysis				F
	% C	% O	% N	% Si	
Pure lacquer (cured facing air)	75.0	19.6	2.5	2.9	—
Untreated substrate					
TPO 1	98.1	1.5	0.4	0	$\sim 0^a$
TPO 2	98.0	2.0	0	0	$\sim 0^a$
TPO 3	97.0	2.6	0.4	0	$\sim 0^a$
TPO 4	98.2	1.8	0	0	$\sim 0^a$
Lacquer (cured facing untreated substrates)					
TPO 1	75.0	20.2	1.8	3.0	$\sim 0^a$
TPO 2	74.5	20.5	1.4	3.5	$\sim 0^a$
TPO 3	75.0	20.2	1.9	2.9	$\sim 0^a$
TPO 4	78.7	17.0	3.2	1.1	$\sim 0^a$
Lacquer (after peeling of plasma-treated[b] samples)					
TPO 1	98.5	1.5	0	0	14 ± 2
TPO 4	98.6	1.4	0	0	2 ± 1

[a]$\sim$ 0 N/15 mm means that the lacquer falls off the substrate as soon as the sample is touched.
[b]Plasma treatment in R2. Conditions: 30 W, 20 sccm, 90 mTorr, 30 s, 125–375 kHz.

portant. This is in contrast to the degree of surface modification as revealed by contact angle measurements. Apart from the highest oxygen pressure in R1, the processing conditions do not have any effect on the amount of polar groups introduced onto the surface, as judged by the wettability. Obviously, there is no relationship between the degree of surface modification and the lacquer adhesion. To understand the weak influence of surface chemistry on the adhesion, it is important to know the locus of failure. For the materials that generally showed high peel forces (TPO 2 and TPO 3), the failure was clearly cohesive in the substrate. This could be observed visually as a thin layer of substrate on the lacquer side after peeling. The samples showing peel forces of < 25 N/15 mm (mainly TPO 1 and TPO 4) did not reveal any substrate remaining on the lacquer side, as judged by visual inspection, which implies the failure to be adhesional. To determine the locus of failure for these samples as well, selected fracture surfaces were analyzed by XPS, as were untreated TPOs and corresponding lacquer surfaces (see Table 10). The table shows that a lacquer surface that was cured facing air consists of *ca.* 20% oxygen and a few percent nitrogen and silicon. Lacquer surfaces obtained by curing against untreated TPOs showed almost the same chemical composition as the pure lacquer surface. Evidently, lacquering untreated TPO plates does not leave any trace of polymer on the lacquer side after removal of the lacquer. The lacquer side of two peeled samples showing low peel forces were also analyzed. The peel forces were quite different (14 N/15 mm and 2 N/15 mm, respectively) but the composition in both cases was not that of the pure lacquer surface but equal to what was observed for untreated TPOs. Thus, regardless of the absolute peel force value, it can be concluded that the failures must have occurred *below* the oxidized surface layer for *all* materials and processing conditions, the only difference being the depth at which they occur. Conclusively, the lacquer adhesion is determined by the cohesive strength of the near-surface region of the substrate and, in accordance with the previously cited study, the varying cohesive strength is attributed to VUV radiation, as VUV is known to penetrate deeply into the substrate and to be strongly absorbed by PP and PE. Moreover, as TPOs are based mainly on PP, it is more likely that it is a weakening caused by excessive chain scission reactions rather than a decrease in the crosslinking density that is responsible for the varying cohesive strength of this region. This conclusion is further supported by a study concerning emissions from plasmas of selected gases in the VUV–visible region [28], in which it was reported that the radiation intensity in an oxygen plasma (observed as atomic oxygen lines) increased with increasing discharge power and decreased with increasing gas pressure, while no dependence on flow rate was found. As radiation intensity may be read as radical formation, this supports a chain scission mechanism, as these trends are exactly the opposite of what was found for the lacquer adhesion.

The above conclusion also explains why the degree of surface modification is negligible. As long as the surface has enough polar groups that can interact with the lacquer, there is no reason why the surface properties should be important if the lacquer adhesion is determined by the cohesive strength of a layer some distance below the oxidized surface. One exception to this general conclusion may exist, however. It was pointed out that TPO 4 showed a low peel force after treatment in R1 at 1200 mTorr. This was proposed to be the result of inhomogeneous treatment and/or insufficient oxidation and, in this particular case, the degree of surface modification may actually be the determining factor, as this sample had considerably higher contact angles than all other samples. Unfortunately, XPS analysis of the fracture surface was not carried out.

Moreover, it seems pointless to discuss the influence of eventual low molecular weight oxidized material (LMWOM) formed during the plasma treatment, and/or the presence of other low molecular weight material (LMWM) on the surface (e.g. stabilizers), as the failure obviously occurs in a region similar to a pure hydrocarbon. If LMWOM or LMWM is present, it is probably displaced and dissolved into the lacquer. In the case of TPO 1, however, it could be argued that the LMWM present in this material is a low molecular weight fraction of the polymer (i.e. a hydrocarbon) and, if this fraction is present on the surface during the plasma treatment, it would be the outermost surface of this material that would be oxidized. As this material has a lower cohesive strength than the bulk material, this could *per se* explain why low peel forces under certain conditions are obtained for TPO 1. On the other hand, it is known that the effect of etching increases with increasing discharge power [26]. Etching will remove, or at least reduce, the amount of LMWM present on the surface, which would thus lead to an increase in the adhesional bond strength. Since high peel forces are obtained at low powers, while considerably lower peel forces are obtained at high powers, it is unlikely that the low peel forces obtained for TPO 1 are caused by the presence of LMWM. It should be pointed out, however, that this LMWM, regardless of its nature, may influence the treatment 'lifetime' as, upon storage, it might diffuse towards the surface and reduce the plasma treatment effect.

Furthermore, ascribing VUV as the cause of low peel forces provides an increased understanding of why the TPOs display different sensitivity towards changes in the processing conditions. The reason is that the materials, owing to the variations in material composition, will have different tendencies for crosslinking and chain scission reactions. Thus, for example, the excellent adhesion properties and low sensitivity towards processing conditions obtained for TPO 2 can be related to the high ethylene content in the form of long blocks in this material, as these blocks favour crosslinking, while the high sensitivity of TPO 1 can be related to the random distribution of ethylene, as this distribution is unlikely to be favourable for crosslinking reactions. For TPO 3 and TPO 4, no such clear relations can be found. However, a possible explanation as to why TPO 3 shows a high resistance is that this material may contain a reactor-modified TPO, i.e. a PP matrix with a higher ethylene content than the PPs normally used in TPOs. This is supported by the fact that this material was impossible to filter after the extraction and dissolution precipitation procedures. Finally, a plausible explanation as to why TPO 4 has a high sensitivity is that the surface of these plates may be more rich in PP (and therefore more sensitive to chain scission) than the other TPOs as a consequence of this material having been melt-pressed into plates.

To summarize, the chemical composition of TPOs is definitely important for the adhesion properties obtained after plasma treatment. Exactly which factors are important cannot be concluded from this study, but it appears as though ethylene in the form of long blocks is preferable, as is probably also a high overall amount of ethylene. Moreover, other material parameters, such as the surface morphology and the type and amount of filler, may also be important factors. This is true of the latter, since it can influence the radiation intensity in the near-surface region of the substrate.

5. CONCLUSIONS

The most prominent findings from this work are summarized below:

- Regardless of the plasma reactor used, the degree of wettability is little influenced by variations in the processing conditions (within the intervals tested).
- A highly oxidized surface layer does not guarantee a high level of lacquer adhesion.
- The lacquer adhesion is highly dependent on both discharge power and gas pressure (the same trends were found regardless of reactor), while flow rate and discharge frequency seem to be of lesser importance (R2).
- Regardless of the absolute peel force, all failures are cohesive in the substrate at some distance *below* the oxidized surface layer. The cohesive strength of this region is proposed to be determined by the extent to which VUV-induced formation of radicals occurs.
- The TPOs have different sensitivity towards changes in the processing conditions. As crosslinking reactions in ethylene-rich regions can compensate to some extent for a weakening of the near-surface region caused by chain scissions, this is proposed to be related to the amount and distribution of ethylene in the materials.

Acknowledgements

The authors gratefully acknowledge The Swedish National Board for Industrial and Technical Development for financial support. The support of Beckers Industries, Volvo and Borealis is also acknowledged. Finally, the authors would like to thank Ms. Hélène Blixt and Mr. Anders Mårtensson for their assistance in the experiments.

REFERENCES

1. E. Sheng, I. Sutherland, D. M. Brewis and R. J Heath, *Surf. Interface Anal.* **19**, 151–156 (1992).
2. F. Garbassi, E. Occhiello and F. Polato, *J. Mater. Sci.* **22**, 207–212 (1987).
3. E. Sheng, I. Sutherland, D. M. Brewis and R. J Heath, *J. Adhesion Sci. Technol.* **9**, 47–60 (1995).
4. R. A. Ryntz, Q. Xie and A. C. Ramamurthy, *J. Coatings Technol.* **67** (843), 45–55 (1995).
5. W. L. Dechent, J. A. Giles, C. A. McMillan and J. O. Stoffer, *Polym. Mater. Sci. Eng.* **70**, 172–173 (1994).
6. K. Reihs, *Thin Solid Films* **264**, 135–140 (1995).
7. O. D. Greenwood, R. D. Boyd, J. Hopkins and J. P. S. Badyal, *J. Adhesion Sci. Technol.* **9**, 311–326 (1995).
8. N. Shahidzadeh-Ahmadi, F. Arefi-Khonsari and J. Amouroux, *J. Mater. Chem.* **5**, 229–236 (1995).
9. E. Occhiello, M. Morra, G. Morini, F. Garbassi and P. Humphrey, *J. Appl. Polym. Sci.* **42**, 551–559 (1991).
10. A. Holländer, J. Behnisch and H. Zimmermann, *J. Appl. Polym. Sci.* **49**, 1857–1863 (1993).
11. R. Foerch, G. Kill and M. J. Walzak, *J. Adhesion Sci. Technol.* **7**, 1077–1089 (1993).
12. D. Y. Wu, W. S. Gutowski, S. Li and H. J. Griesser, *J. Adhesion Sci. Technol.* **9**, 501–525 (1995).
13. J. F. Friedrich, W. Unger, A. Lippitz, T. Gross, P. Rohrer, W. Saur, J. Erdmann and H.-V. Gorsler, *J. Adhesion Sci. Technol.* **9**, 575–598 (1995).
14. M. Morra, E. Occhiello, L. Gila and F. Garbassi, *J. Adhesion* **33**, 77–88 (1990).
15. A. Nihlstrand, T. Hjertberg, H. P. Schreiber and J. E Klemberg-Sapieha, *J. Adhesion Sci. Technol.* **10**, 651–675 (1996).
16. R. H. Partridge, *J. Chem. Phys.* **45**, 1685–1690 (1966).
17. R. H. Partridge, *J. Chem. Phys.* **49**, 3656–3668 (1968).
18. D. T. Clark and A. Dilks, *J. Polym. Sci., Polym. Chem. Edn* **15**, 2321–2345 (1977).
19. H. F. Mark, N. M. Bikales, C. G. Overberger and G. Menges (Eds), in: *Encyclopedia of Polymer Science and Engineering*, 2nd edn, Vol. 4, pp. 418–436. John Wiley & Sons, New York (1986).
20. N. Grassie and G. Scott, in: *Polymer Degradation and Stabilisation*, pp. 103–105. Cambridge University Press, Cambridge (1985).
21. B. Wunderlich, in: *Macromolecular Physics*, Vol. 3, p. 63. Academic Press, New York (1980).
22. Z. Grubisic, R. Rempp and H. Benoit, *J. Polym. Sci., Part B* **5**, 753 (1967).
23. J. D. Andrade (Ed.), in: *Surface and Interfacial Aspects of Biomedical Polymers*, Vol. 1, pp. 175–184. Plenum Press, New York (1985).
24. Corporate Standard STD 1029, 5476, Volvo Car Corporation, Gothenburg (1994).
25. N. B. Coltrup, in: *Introduction to Infrared and Raman Spectroscopy*, pp. 220–233. Academic Press, New York (1975).
26. D. L. Flamm, in: *Plasma-Surface Interactions and Processing of Materials*, O. Auciello (Ed.), pp. 38–44. Kluwer Academic Publishers, Dordrecht (1990).
27. E. M. Liston, L. Martinu and M. R. Wertheimer, *J. Adhesion Sci. Technol.* **7**, 1091–1127 (1993).
28. A. Kruse, A. Fozza and M. R. Wertheimer, *Proceedings — 10th International Colloquium on Plasma Processes*, Antibes — Juan-les-Pins, 11–15 June (1995).

Mittal Festschrift, pp. 307–328
W. J. Van Ooij and H. R. Anderson, Jr. (Eds)

Plasma surface modifications for structural and biomedical adhesion applications

HANS J. GRIESSER,* THOMAS R. GENGENBACH, LIMING DAI, SHENG LI and RONALD C. CHATELIER

Division of Chemicals and Polymers, CSIRO, Private Bag 10, Rosebank MDC, Clayton 3169, Australia

Abstract—We discuss plasma surface modifications applied to perfluorinated polymers and polyolefins to achieve structural adhesive bonding or for biomedical purposes such as adhesion and proliferation of cells, and interfacial immobilization of biologically active molecules. We compare the properties of surface modifications performed in non-depositing plasma treatments with those of thin coatings produced in depositing plasma vapours (plasma polymerization), with particular emphasis on changes, on subsequent storage, to the properties and composition of the surface layers ('ageing'). Such changes usually proceed for extended periods of time after plasma processing. Polymer surfaces treated in non-depositing plasmas generally are unstable, showing an increase in the air/water contact angles over days and weeks due to surface reorientation motions. Concurrently, the composition of the surface layers is also affected by post-plasma chemical reactions: originating from trapped radicals, oxidative chain reactions lead to the production of substantial amounts of oxygen-containing groups. These reactions also convert some of the groups originally incorporated into the surface layers by the plasma treatment; for instance, amine groups are converted to amide groups as evidenced by shifts in the XPS N 1*s* binding energy. Plasma polymer coatings analogously undergo oxidative compositional changes with time, and are capable of some surface reorganization. Thus, the nature and densities of the chemical groups on plasma-treated surfaces and plasma polymer coatings can change considerably with time. The relative contributions by concurrent reorientation motions and oxidative reactions to the compositional changes vary markedly between different plasma-prepared surfaces, but usually both processes contribute to the ageing of a surface. The generally long time constants of the reorientation of plasma polymer surfaces suggest that their limited, slow mobility may be neglected when interpreting interactions with adsorbing proteins.

Keywords: Plasma surface modification; plasma polymers; surface reorientation; surface restructuring; oxidation; XPS; contact angles.

1. INTRODUCTION

Surface treatments can open up additional applications, for instance in the areas of adhesive bonding [1, 2] and biomedical materials [3, 4], for polymers whose bulk properties are suited to these applications but whose surface properties are inappropriate. Among the various surface modification techniques that can be applied to polymers,

---

*To whom correspondence should be addressed.

gas plasma (glow discharge) methods [5, 6] have attracted much interest as they are versatile. A wide variety of surface chemistries and properties can be obtained, and even polymers with very low chemical reactivity, such as perfluoropolymers, can be readily surface modified. Numerous workers have applied plasma surface treatments to polymers in order to substitute surface groups. The deposition of thin polymeric coatings by plasma polymerization is an alternative method for creating polymeric surfaces with various functional groups.

Our interests are in the design and development of polymer surfaces intended for structural adhesive bonding and biomedical applications. Although these may seem disparate areas of application, the ability to control the adhesion of biological entities (e.g. to promote colonization of polymer surfaces by cells or to reduce adhesion of bacteria) rests on the same issues of understanding and control of adhesional interactions across an interface. Adequate surface characterization is crucial for molecular interpretation of adhesional interfacial interactions and to unravel the effects of plasma processing (Fig. 1). The results of both surface-treating and plasma depositing processes depend not only on the substrate and the process gas, but also on a number of experimental variables. Detailed characterization of the complex chemical composition of plasma-prepared surfaces remains challenging even with state-of-the-art

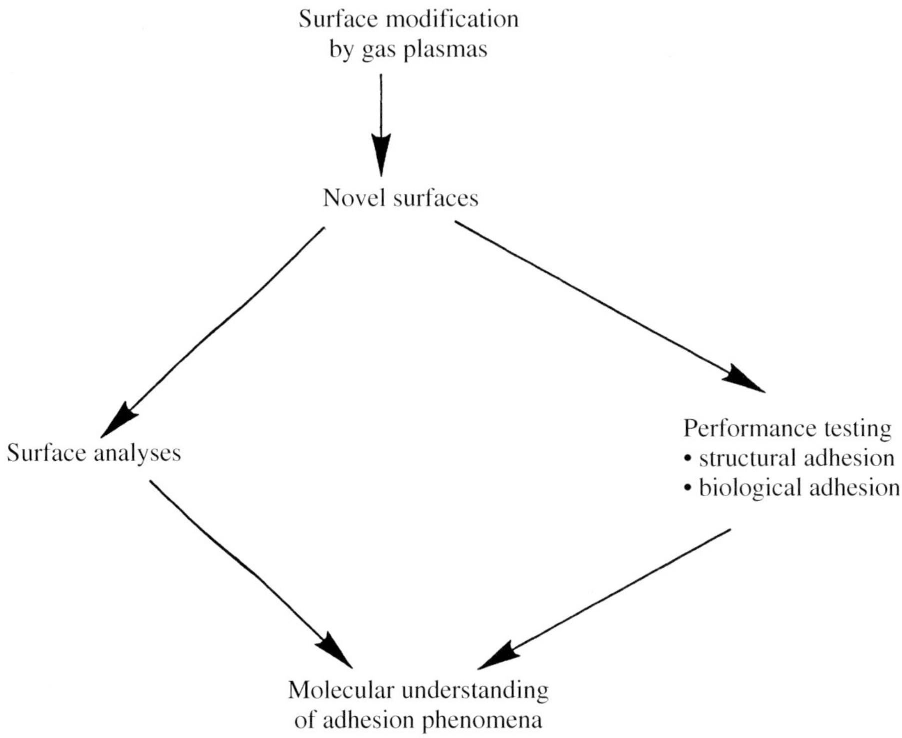

Figure 1. Surface modification by gas plasmas.

spectroscopic surface analytical methods, and industrial application often has preceded characterization of the chemical consequences of plasma processing.

An additional complication is that most modified polymer surfaces change ('age') considerably after treatment [7–10]. The most common manifestation is the 'hydrophobic recovery' (an increase in the air/water contact angles with time) of hydrophobic polymers stored in contact with air after they have been exposed to a plasma surface treatment that incorporated polar groups. This hydrophobic recovery is interpreted as resulting from 'surface reorientation' or 'surface restructuring', a term which describes the transport of polar groups away from the interface into deeper regions of the polymer with the concurrent migration to the air interface of unmodified, hydrophobic segments [7, 8]. The driving forces for this transport are the subject of ongoing study but appear to involve terms arising from interfacial enthalpy and translational entropy [10].

The ageing of modified polymer surfaces usually is, however, even more complex [9]: concurrent with the *physical* process of surface reorientation, which reduces the density of the polar groups in the surface layers, *chemical* changes may also proceed for extended periods of time after plasma processing, and they alter not only the relative densities but also the variety of chemical groups [11, 12]. Unfortunately, however, many studies performed surface analysis of plasma-prepared polymer surfaces only once and at unspecified times after fabrication, and did not consider possible time dependences of the surface compositions. Contact angle data may hide evidence of chemical changes superimposed on surface reorientation [9]. Multi-technique analyses of the ageing of a plasma-prepared polymer surface are therefore essential, including a surface analytical method that provides chemical information, to unravel which fundamental processes contribute to the ageing of the surface.

Detailed spectroscopic characterization of the ageing of plasma-prepared surfaces is required for understanding and optimizing surface chemistries and interfacial properties *at the time of intended use*. In structural adhesion and biomedical applications, variable periods of time may elapse between plasma processing and end-use of a polymer. However, while many studies have explored the effects of plasma parameters on surface chemistry and properties, and hydrophobic recovery is also well known, chemical ageing effects after plasma processing have received little attention. In this communication we discuss the importance of chemical and physical processes in the ageing of selected plasma-treated surfaces and plasma polymer coatings. Citing some previously published results and providing additional, new data, this contribution aims to present an overview and comparison of the diverse effects of ageing, in contrast to the detailed discussion of individual cases in our earlier reports [9–13]. Plasma processes intended to produce aminated surfaces will be given most attention; they are of applied interest [14–16] and also lend themselves to study of ageing effects better than many other surfaces. The examples selected illustrate issues of general relevance in the ageing of plasma-treated polymers and plasma polymer coatings, and these ageing issues must be considered when designing and interpreting structural and biomedical adhesive polymer surfaces.

2. EXPERIMENTAL

2.1. Materials and plasma methods

Plasma surface treatments and plasma polymer coatings were applied onto fluorinated ethylene–propylene copolymer tape (DuPont Teflon FEP 100 Type A), polytetrafluoroethylene (PTFE) tape (supplied by Dr. C. Lyons, 3M Company), and polyolefin sheet samples: high density polyethylene (HDPE), low density polyethylene (LDPE), and polypropylene (PP) (all from Cadillac Plastics) whose surfaces had been smoothed [17] to facilitate angle-dependent XPS analysis. The samples were cleaned as appropriate (typically by 5–10 min sonication in high purity ethanol, acetone, or toluene) and their surface cleanliness assessed by XPS [18].

Our plasma equipment [19, 20] and the general procedures we use for plasma surface treatments [9, 11, 15] and plasma polymerization [10] have been detailed previously. Briefly, lengths of FEP and PTFE tape were plasma-processed in a reactor that contained a transport system which moved the tape along the face of one of two vertical electrodes (90 × 18 mm; separated by 16 mm) [19]. This setup allowed the convenient fabrication of numerous identical specimens, so that a previously unanalyzed specimen could be used for the various analyses at various times after fabrication, thereby eliminating possible inaccuracies from accumulation of XPS radiation damage or contaminants in contact angle (CA) determinations. Specimens were stored under air in clean polystyrene tissue culture dishes at room temperature. To study the ageing of polyolefins, available in sheet form, the more laborious procedure was required of plasma-processing samples in a reactor that contained a horizontal disc electrode of diameter 100 mm on the bottom and a 6 mm rod electrode on the top, separated by 120 mm. Substrates were placed on the lower electrode. By adjusting the process gas flow, pressure, and the RF power, uniform plasmas across the disc electrode, except for the outermost 5–10 mm of the disc radius, were obtained, as indicated by uniform interference colouring of plasma polymer coatings and verified by XPS analysis.

Ammonia (Matheson) was supplied to the plasma chamber from a gas cylinder via a commercial flow controller. Water vapour and 'monomer' vapours were provided by evaporation at ambient temperature ($22 \pm 1\,^{\circ}C$) from the liquid contained in a thermostatted round bottom flask. Representative plasma parameters have been given in the earlier reports [9–11, 15]; additional experiments used similar conditions. While the magnitude of ageing depends somewhat on the processing conditions, the qualitative conclusions do not.

2.2. Surface analyses

Most XPS analyses were performed using a Vacuum Generators Escalab V unit with a non-monochromatic Al K_{α} source at 10 kV and 20 mA, and a hemispherical analyser operating in the fixed transmission mode. The base pressure in the XPS chamber, calibration of the binding energy scale, and correction for charging of the specimens under irradiation were the same as reported earlier [10]. Elements present were identified from survey spectra. High resolution spectra were recorded from individual peaks at 30 eV pass energy with a step width of 0.078 eV/channel. The elemental composition of the surface was determined as described previously [10]. The random error associated with quantitative elemental analysis was between 5 and 10% (usually 7 to 8%). The F/C ratio determined on PTFE tape was 1.94. The effects of sample decomposition under non-monochromatic X-rays were assessed as described earlier [9, 10] and found not to affect the accuracy of the measurements within experimental error as long as the total exposure time in angle-dependent XPS was < 30 min. As a further check, selected specimens were also analysed on a Surface Science Instruments SSX-100 spectrometer equipped with a monochromatic Al K_α source at NESAC/BIO, University of Washington, Seattle, WA. Corresponding values for the O/C ratio obtained on the two different instruments agreed to within 1 to 2%.

Contact angles were measured with triply distilled water using a modified Kernco Model G-II goniometer equipped with a syringe with a reversible, micrometer driven plunger, in order to enable determination of the advancing (θ_A), sessile (θ_S), and receding (θ_R) CAs [12]. Readings were taken from four drops and the average value calculated. The uncertainty associated with these measurements is estimated to be $\pm 3°$ in the case of θ_S and $\pm 4°$ in the case of θ_A and θ_R. Clean FEP surfaces ($\theta_A = 117°$, $\theta_R = 98°$ and $\theta_S = 107°$) were used to ascertain reproducibility of measurements and purity of the water.

3. RESULTS AND INTERPRETATION

3.1. Ammonia plasma treatments

3.1.1. FEP. Ammonia-plasma-treated FEP (FEP/NH_3) is the system we have studied most extensively because its surface reorients to a considerable extent, F is a convenient marker atom, and FEP is available with a much smoother surface topography [9] than PTFE and many other polymers, thus facilitating interpretation of angle-dependent XPS analyses. We have reported some of our work with FEP/NH_3 before [9]; in the following we will present additional, previously unpublished data and discuss this system in greatest detail as it demonstrates most clearly the issues that also apply to other polymer surfaces treated by NH_3 and other non-depositing plasmas. N-containing surfaces are particularly suited to separating the effects of surface reorganization and oxidation during ageing, and the ease of quantification by XPS of the element F further assists in the study of ageing effects.

Under various experimental conditions, surface modification of FEP in NH_3 plasmas produced surfaces with low (compared with other non-depositing process gases) air/water CAs [11]. However, when the treated surfaces were stored in contact with air, they were not stable with time; the CAs demonstrated typical hydrophobic recovery, increasing considerably over periods of a few days [9]. This recovery was interpreted as resulting from the transport of attached polar groups from the air interface into deeper regions of the polymer, with the simultaneous migration of untreated

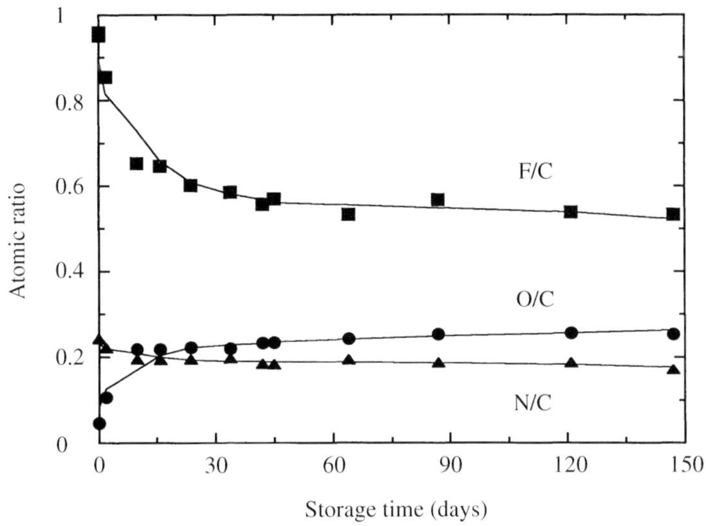

Figure 2. The composition of ammonia-plasma-treated FEP as a function of the storage time in air.

FEP chains to the interface (surface reorientation). The final CAs after extended equilibration showed values that were below the values characteristic of untreated FEP; thus, the recovery was incomplete.

The intensity of the XPS N 1*s* signal decreased slightly with time (Fig. 2), consistent with the reduced probability of detection of photoelectrons emitted from atoms that have moved some distance (but less than the XPS probe depth) into the polymer [9]. The magnitude of the reduction suggests a mean diffusional distance of the order of a nanometre; this distance is of a similar magnitude to the range of the interfacial forces effective in CA measurements [21] and thus accords with the view that interfacial forces drive surface reorganization. Angle-dependent XPS analyses of FEP/NH_3 samples at various times after fabrication have revealed changes in the elemental depth profiles upon storage. For freshly treated samples, the N depth profile is concentrated within the top 1 nm of the polymer [22]. In contrast, an aged sample was found to have a very different N depth profile (Fig. 3): the outermost 0.5 nm were relatively depleted in N, and the N distribution extended to depths of ~ 2 nm. This depth profile clearly is consistent with reorientation of the plasma-modified polymer surface layers.

The logarithmic nature of the above changes would have appeared to be consistent with a single process approaching an equilibrium situation. However, XPS analyses performed at various times after NH_3 plasma exposure not only showed the above-mentioned reduction in the intensity of the signal originating from N atoms but also revealed that the treated FEP surfaces underwent considerable changes by *chemical reactions during ageing* [9]. We note that the large majority of previous studies of the ageing of plasma-modified surfaces were based only on CA measurements and interpreted the data solely in terms of surface reorientation. Clearly, however, surface analysis by XPS or SSIMS is required since a logarithmic CA change may be deceptively simple, hiding marked chemical changes that also occur in the surface layers during ageing.

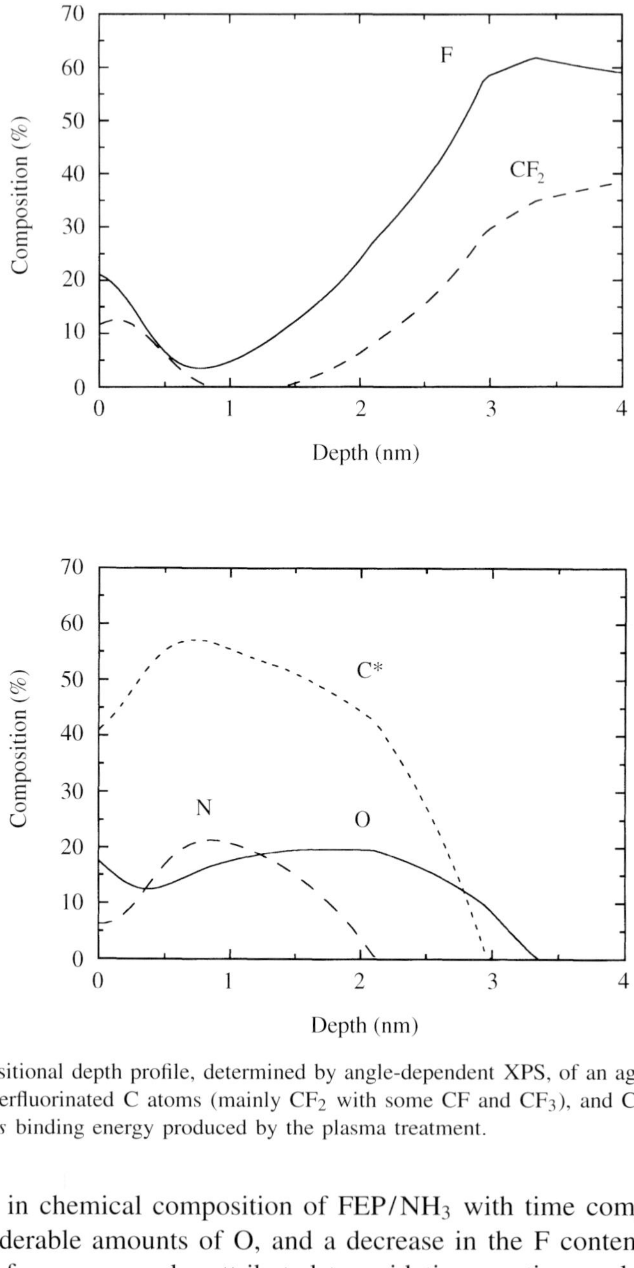

Figure 3. Compositional depth profile, determined by angle-dependent XPS, of an aged FEP/NH_3 sample. CF denotes perfluorinated C atoms (mainly CF_2 with some CF and CF_3), and C* denotes C atoms of lower XPS C 1*s* binding energy produced by the plasma treatment.

The changes in chemical composition of FEP/NH_3 with time comprised incorporation of considerable amounts of O, and a decrease in the F content (Fig. 2). The incorporation of oxygen can be attributed to oxidative reaction cycles. The plasma treatment is believed to create carbon-centered radicals (C•) in the surface layers by homolytic bond scissions. Evidently, some of the radicals were not able to dissipate by the time the reactor was vented, and in-diffusing, atmospheric oxygen reacted with these trapped C• radicals. The resulting peroxy radicals react further, and other types of oxygen-containing groups are produced. Some of the reactions partly recreate C• radicals for further oxygen addition. Moreover, some intermediates are metastable, thereby giving rise to delayed further reactions and hence an extended uptake of additional oxygen. This scenario is akin to the complex oxidative reaction pathways that are invoked for the radical-induced oxidation of polyolefins [23]. Like the plasma-incorporated N-containing groups, the oxidatively incorporated O-containing groups also are subjected to surface reorientation motions. The depth profile of Fig. 3 shows that after ageing, O was, like N, depleted within the outermost surface regions.

We have found that the oxidative reactions affected the binding energy of the XPS N 1*s* signal. This binding energy was 399.1 eV in a freshly treated FEP/NH_3 surface, a value which is in agreement with literature values for amine groups [24]. With time, however, the binding energy gradually increased, up to 400.3 eV, a value which indicated the presence of amide and possibly also some imide groups.

The *decrease* in the F content with time (Fig. 2) was of interest since we would have expected an *increase* in the F content from the migration to the surface of unmodified FEP chains and segments. This decrease may be due to several chemical reactions. The XPS F 1*s* signal recorded immediately after the plasma treatment consisted of *two* components [9]. The component located at 686 eV was assigned to fluoride ions (F^-); its intensity decreased rapidly within a few days and was no longer observable after 2 weeks. The F attached to C, located at 689.1 eV, also decreased in intensity with time. The mechanism by which the ammonia plasma produces F^- is not clear, although homolytic C$-$F bond scission followed by addition of a free electron from the plasma to the F• radical seems plausible. The reaction(s) by which F^- ions disappear are not understood; a speculative mechanism is the elimination of HF from $-CF_x-NH_3^+F^-$ structures. Some of the observed reduction in the fluorine intensity may also be related to the fact that the electronegativity of F will destabilize attached amine groups on FEP. Pentafluoroethylamine ($CF_3-CF_2-NH_2$) in solution is unstable, converting to $CF_3-C{\equiv}N$ [25, 26]. Thus, by analogy, on FEP/NH_3 (and PTFE/NH_3) surfaces HF may be liberated from $-CF_x-NH_2$ structures, yielding nitrile (from $x = 2$) and $-C{=}NH$ (from $x = 1$) groups. The plasma treatment may also produce secondary amines ($-CF_x-NH-CF_y-$) that would convert to $-CF_{x-1}{=}N-CF_y-$. Imines might then hydrolyze. Another possibility for the loss of F is the hydrolysis of acid fluorides produced by oxidation. Other, speculative mechanisms comprise disproportionation reactions of radicals such as C•, peroxy, and oxy radicals with elimination of volatile species such as $CF_2{=}CF_2$.

In summary, the ageing of NH_3-plasma-treated FEP surfaces is extremely complex and comprises both the physical process of surface reorientation and chemical reactions. Clearly, amine groups initially incorporated by the plasma treatment are largely removed from the surface within a few days; the XPS N 1*s* signal of aged surfaces indicates that little, if any, of the N is still present in the form of amines. *Amide* groups are a main species but other oxygen-containing groups are also present; however, extensive overlap of their spectral contributions presently prevents detailed analysis.

3.1.2. Environmental effects on reorientation. Figure 4 shows the dependence of the sessile contact angle, θ_S, on the time elapsed after plasma treatment, for FEP/NH_3 surfaces stored in air, pure nitrogen, and pure oxygen. Evidently the presence of

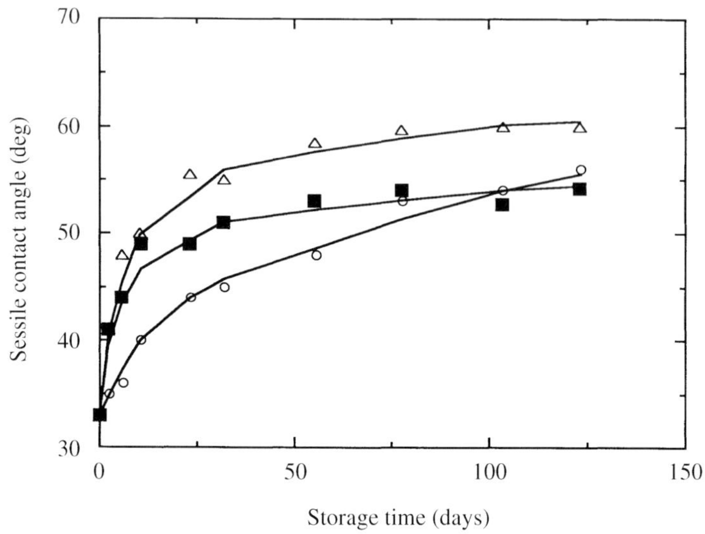

Figure 4. The sessile air/water contact angle of FEP/NH_3 surfaces stored in (△) pure oxygen, (■) air, and (○) pure nitrogen, as a function of time after the plasma treatment.

oxygen affects the reorientation of the surface. This observation points to an interplay between oxidation and surface reorientation. The slower surface reorientation in pure N_2 suggests that in the absence of O_2 the C· radicals that are present following the extinction of the plasma react with each other, thus forming additional inter- and intra-chain crosslinks and thereby reducing the mobility of the surface layers. On the other hand, when O_2 diffuses into the polymer, it adds to C· radicals. This addition quenches C· radicals and thus competes with the process of crosslink formation. Secondary reactions in the oxidative cycles may also influence the mobility of the polymer segments in the surface layers, by the formation of additional crosslinks and/or backbone chain scissions; however, Fig. 4 suggests that the initial reactions available to the C· radicals are the most influential upon the consequent surface reorientation. Unfortunately, current surface analytical methods do not provide a direct and reliable assessment of the crosslink density of thin surface layers to support this interpretation.

Surface reorientation is also affected by the medium in which the treated specimens are stored (Fig. 5). Of particular interest is the fact that the CAs increase even when the treated surfaces are stored in water. This trend towards a more hydrophobic surface with time when in contact with a polar medium may seem somewhat surprising as, at first glance, it appears to be energetically unfavourable; one might expect polar liquids to 'hold' the polar groups at the surface. However, we postulate a possible resolution for this paradox. In air, attractive or repulsive interfacial forces on surface groups are negligible; there is thus no significant net gain or loss of *interfacial enthalpy* when some of the polar surface groups are transported away from the interface. Also, the enthalpy of the migrating polar groups will not change much when comparing a surface location, contacting air, with a location 'inside' the FEP polymer. However, migration is driven by an increase in *translational entropy* which accompanies dilution

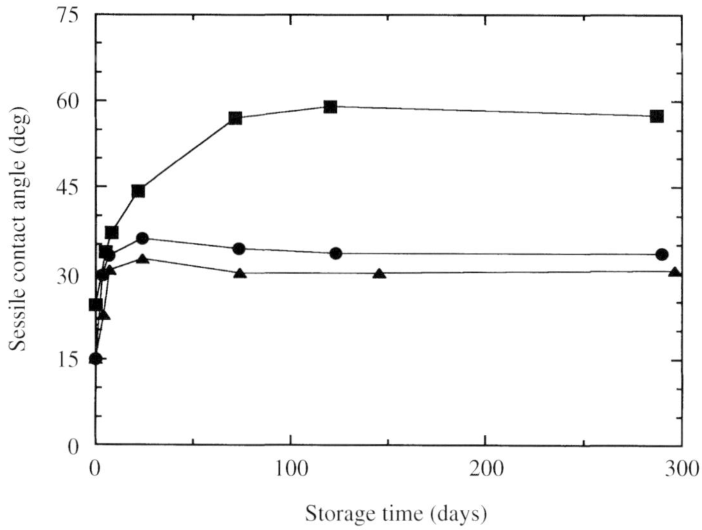

Figure 5. The sessile air/water contact angle of FEP/NH_3 surfaces stored in (■) air, (●) methanol, and (▲) water, as a function of time after the plasma treatment.

of the polar groups. When samples are stored in water, this entropic driving force still applies, but is partly counteracted by an attractive interfacial force (hydrogen bonding) between polar surface groups and water. It appears from Fig. 5 that the interfacial attraction between polar media and polar surface groups is not sufficient to completely overcome the entropic drive to (partial) migration. Thus, we contend that the increase in translational entropy that ensues upon inward diffusion of polar groups is a significant factor in surface reorientation, even in polar media.

3.1.3. Other polymers. The behaviour of PTFE/NH_3 is similar to that of FEP/NH_3 in showing analogous oxidation and an even larger extent of hydrophobic recovery [10]. Thus, while PTFE can be surface-aminated, the introduced amine groups are short-lived, becoming buried inside the polymer and/or converted to amides by post-treatment oxidative reactions.

These ageing effects are not unique to perfluorinated polymers although the presence of the strongly electronegative element F may promote some specific reactions in these materials. We have observed the same processes of surface reorientation and post-plasma oxidation to be involved in the ageing of LDPE, HDPE, and PP surfaces treated in NH_3 plasmas; qualitatively the same interpretations and mechanisms of ageing apply, although the magnitude of the changes with time of the CAs and the chemical composition varied significantly with the polyolefin. We suggest that the existence of concurrent processes of surface reorientation and post-plasma reactions should be assumed to be the rule rather than the exception when interpreting the ageing of plasma-treated surfaces. To illustrate the analogy with the interpretation of the ageing of FEP/NH_3 discussed above, we now present and interpret selected, representative data obtained with LDPE/NH_3; more extensive data for the entire set of polyolefins studied will be reported shortly elsewhere [17].

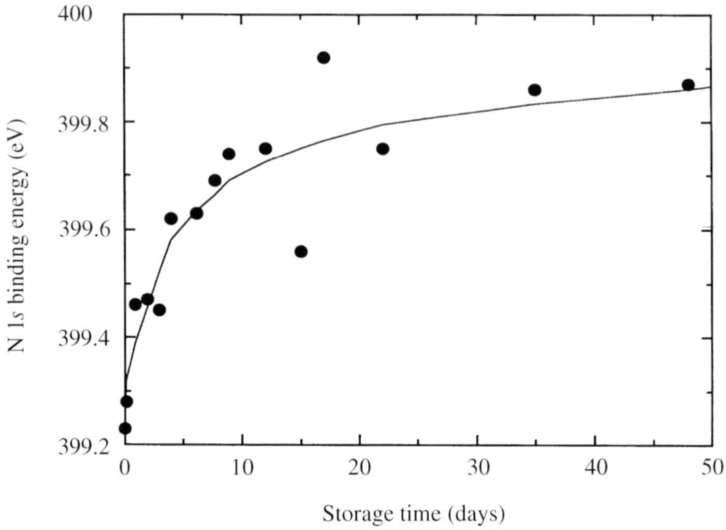

Figure 6. The binding energy of the N 1*s* signal of ammonia-plasma-treated LDPE as a function of time.

XPS spectra recorded at various times after NH_3-plasma-treatment revealed incorporation of oxygen and a shift in the binding energy of the N 1*s* signal, analogous to the observations with FEP/NH_3. Figure 6 displays the N 1*s* binding energy as a function of time elapsed since the plasma treatment. The binding energy value of freshly prepared surfaces is, by comparison with literature values [24], indicative of amine groups. With time, however, the N 1*s* binding energy increases towards a value indicative of amide groups (399.8 eV [24]). The oxygen uptake and the N 1*s* binding energy shift both are consistent with radical-induced oxidation on exposure of the treated samples to air. We note that it is not the N that becomes oxidized but the C that carries the amine group.

In addition to oxidative processes, surface reorientation motions also take place when LDPE/NH_3 samples are stored in air. Figure 7 shows the changes of the CAs with time. The observation that the CAs increase with time indicates that surface reorientation is more important than oxidative changes in defining the ageing of surface properties. The oxidative reactions are, however, important when seeking to interpret interfacial interactions in adhesive and biomedical applications because the chemical composition changes considerably during ageing; the nature and variety of the polar groups present after extended ageing may differ considerably from that of the groups on freshly treated surfaces.

The hydrophobic recovery of polyolefins treated by various plasmas has been known for many years but, surprisingly, oxidative changes with time have been largely neglected. Typically it appears to have been assumed that oxygen detected by XPS in N_2- or NH_3-plasma-treated polyolefin surfaces arose from contaminant vapours during the plasma process, although an interesting study by Gerenser [27] using a plasma reactor coupled directly to an XPS system showed that rapid oxygen incorporation resulted from the quenching of plasma-produced primary C· radicals. Our results confirm that

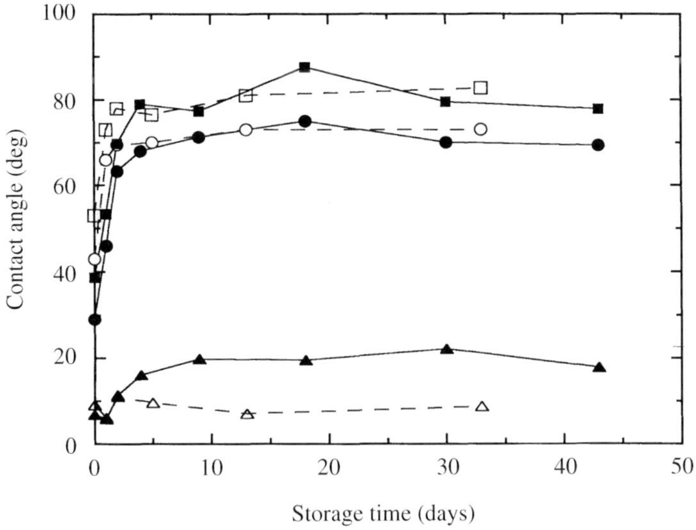

Figure 7. The advancing (squares), sessile (circles), and receding (triangles) air/water contact angles of ammonia-plasma-treated LDPE (filled symbols) and of water-plasma-treated LDPE (empty symbols) as a function of time.

plasma processing can be done with very little contamination by O-containing species, and that therefore the O detected in analyses of air-exposed samples is incorporated predominantly by post-treatment oxidative reactions consequent upon in-diffusion of atmospheric oxygen. Gerenser focused on short O_2 exposure times; the subsequent, slower, continuing oxygen uptake has not been characterized previously. Our results clearly show that the oxygen uptake continues much beyond the time required for quenching of the initially produced radicals. This is consistent with regeneration of radicals by oxidative reaction cycles.

Tatoulian *et al.* [28] have recently reported observation by XPS of an amide contribution on LDPE/NH_3. Our results provide an unambiguous explanation for the origin of the amide groups. Our results also suggest that the reasons why different research groups have reported substantially different O contents for particular polymer surface/plasma combinations may not only relate to the use of different plasma equipment configurations, but also to different exposure times to air prior to analysis. Unfortunately, the time elapsed between treatments and analyses typically has not been reported. Clearly, this time should be indicated, or preferably analyses at various times after fabrication should be performed, in order to provide an indication of the timeframe and magnitude of post-treatment changes.

3.2. Water vapour plasma treatments

Also included in Fig. 7 are the CAs as a function of time after treatment of LDPE in a plasma of water vapour. The changes are very similar in extent and rate to those observed with LDPE/NH_3, indicating that surface reorientation is not significantly different for these two surfaces under these plasma conditions. In contrast, the

CAs of FEP and PTFE samples modified in H_2O plasmas do not change on ageing in air [10]. XPS analyses showed that the elemental composition of these surfaces also remained invariant with time. FEP/H_2O and PTFE/H_2O surfaces thus undergo neither surface reorientation nor post-plasma oxidation. As this behaviour is comparatively rare, it is of considerable interest both for the study of fundamental polymer surface science issues and for practical applications. While typically the ageing of plasma-treated surfaces involves the concurrent processes of surface reorientation and post-plasma reactions, these two surfaces demonstrate that exceptions to this rule exist. For FEP/H_2O and PTFE/H_2O surfaces we have attributed the absence of surface reorientation to an unusually small entropic driving force inferred from the observation by XPS of a very shallow compositional gradient of O in the freshly treated surface [9].

3.3. Plasma polymers

3.3.1. Chemical changes during ageing. An alternative to plasma surface treatments, plasma polymerization [5, 6] enables the preparation of novel surface chemistries, by the deposition of ultrathin coatings that mask the underlying substrate. Whereas the effects of plasma surface treatments vary with the process gas, conditions, and the substrate, plasma polymerization is closer to a generic surface optimization method as it is easier to transfer to new substrates. However, the deposition reactions and the composition of the resultant coatings are complex, and the development of specific, controlled plasma polymer surface chemistries is still the subject of much research (e.g. [31]). Nevertheless, by appropriately adjusting plasma deposition parameters, most importantly the use of relatively low power levels, we have successfully produced amine-rich surfaces by the plasma deposition of alkylamine vapours [32, 33].

In the course of plasma polymerization, some of the radicals generated by the plasma and incorporated into the growing coating may not have dissipated by the time the reactor is vented. These trapped radicals then react rapidly with in-diffusing atmospheric O_2 and thereby start an oxidative reaction sequence. Accordingly, oxygen-containing groups are usually detected in plasma polymers deposited from monomers which do not contain this element [11, 32–36]. The increase with time in the amount of O in plasma polymer surfaces can continue for extended periods of time after plasma deposition and can be followed by XPS [11–13, 33]. In some cases compositional changes can still be observed more than a year after plasma deposition [12, 13], since the oxidative reaction sequences comprise peroxides and hydroperoxides [11], which are metastable and decompose slowly, regenerating radicals for further oxygen uptake. As we are interested in the fabrication of amine-containing surfaces, the question therefore arises whether, as for ammonia-plasma-treatments, oxidative processes occurring after the plasma deposition of amine plasma polymers would analogously remove amine groups, and if so, at what rates.

Figure 8 shows that post-deposition oxidative reactions also take place in the surface regions of an *n*-heptylamine plasma polymer deposited onto FEP (FEP/*n*HA). This oxygen uptake is accompanied by an increase in the N 1*s* binding energy, shown in Fig. 9. The initial value for the binding energy is 399.1 eV, which indicates that the

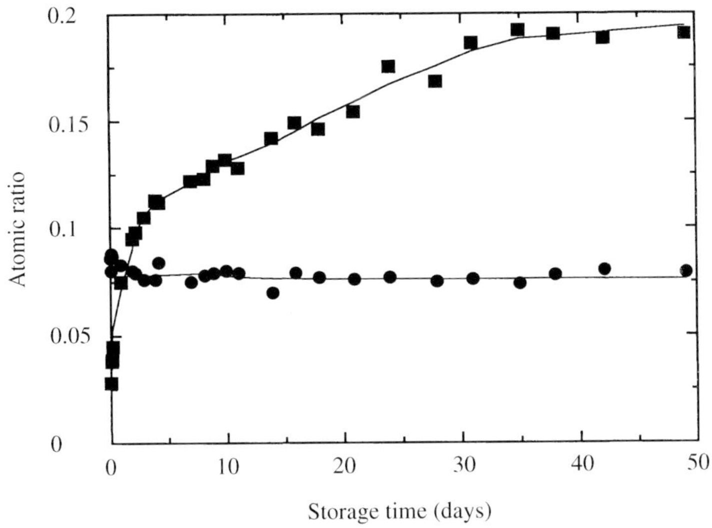

Figure 8. The elemental composition of an *n*-heptylamine plasma polymer deposited on FEP, as a function of time after deposition: N/C ratio (●) and O/C ratio (■).

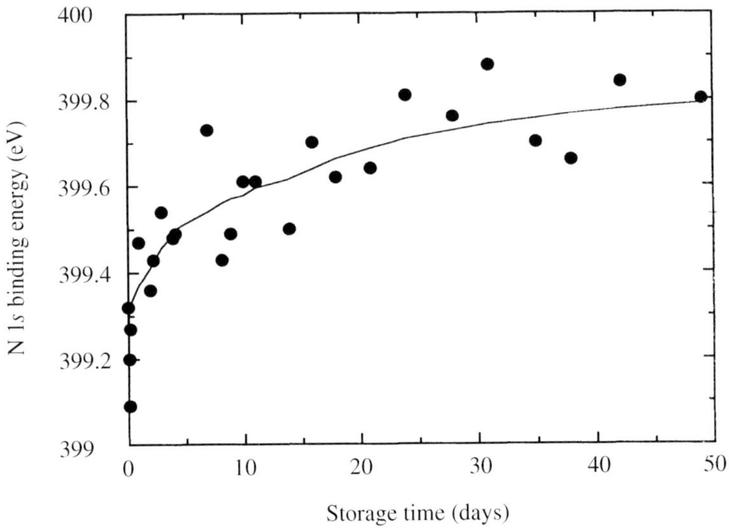

Figure 9. The binding energy of the N 1*s* signal of an *n*-heptylamine plasma polymer deposited on FEP, as a function of time after deposition.

N is predominantly in the form of amine groups. After a few weeks of storage in air, however, the N 1*s* binding energy is close to the literature value of 399.8 eV for amide groups [24]. Therefore, it appears that, as for FEP/NH_3 and LDPE/NH_3, the large majority of the amine groups incorporated by the plasma process are converted to amide groups. One might have expected oxidation of amines to produce nitroso, nitro, and nitrate groups but there is negligible intensity at the high N 1*s* binding energies [24] characteristic of these groups. Thus, the main chemical reaction that

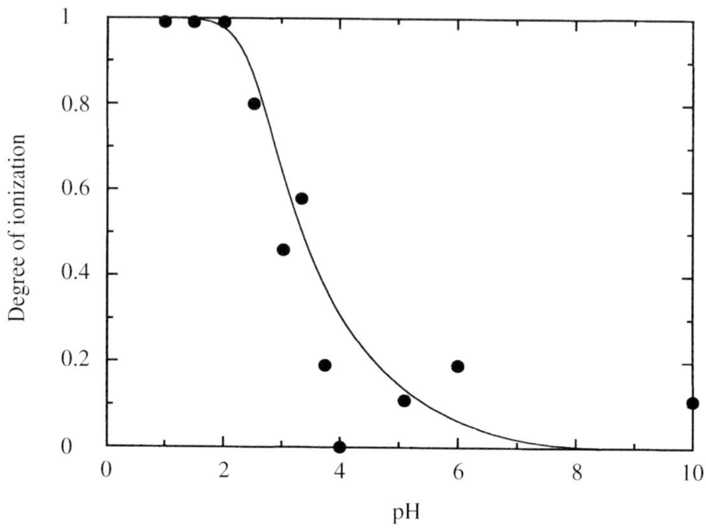

Figure 10. The degree of ionization of a freshly deposited *n*-heptylamine plasma polymer on mica, as a function of the pH of the probe solution. The circles represent values calculated from individual contact angle data points. The solid line represents the degree of ionization calculated via the pK_a obtained from nonlinear least squares fitting of the experimental contact angle data according to the procedure of [37].

removes the introduced amine groups is not the oxidation of the amine-N but the radical-induced oxidation of the C atoms to which the amine groups are attached.

Plasma polymers deposited from ethylene diamine (ED) or diaminopropane (DAP) process vapors have been found to undergo analogous chemical ageing [33, 40]. The much higher amine density of these plasma polymers apparently did not shift reaction mechanisms such as to prevent large-scale loss of amines. Thus, aminated surfaces have been found to be unstable towards ageing effects in all the cases studied in detail by us to date, namely, FEP/NH_3, PTFE/NH_3, LDPE/NH_3, and *n*HA, ED, and DAP plasma polymers on FEP and other substrates. Evidently it is not easy to produce surfaces that contain amine groups for extended periods of time after plasma processing.

The presence of substantial amounts of amine groups on the freshly plasma-prepared surfaces can also be documented by reaction with fluorescein isothiocyanate [9, 32] and by contact angle titrations [37]. Figure 10 shows results obtained by measuring air/water contact angles on a *n*-heptylamine plasma polymer surface as a function of the pH of the drop. The presence of chemical groups that are ionizable at low pH values is consistent with the putative presence of amine groups. The pK_a is lower than for amines in solution due to two factors: the placement of amine groups on a low dielectric medium such as a plasma polymer surface, and electrostatic repulsion between neighbouring charged groups [37]. On ageing, such contact angle titration effects are markedly reduced (not shown).

3.3.2. Surface reorientation. An interesting question is whether surface reorientation motions also contribute to the disappearance of amine groups from amine-plasma-polymer surfaces. The surface reorientation of plasma polymers is not as easy to study as is the reorientation of plasma-treated surfaces of conventional polymers. Plasma polymers are generally thought of as extensively crosslinked, rigid network structures; accordingly, one might assume negligible surface mobility but the few data available indicate that this is not necessarily so [12, 38–40]. The CAs of water on *n*-hexane [12] and *n*-heptylamine [40] plasma polymers *decrease* with time as the samples age in contact with air. This marked contrast to the usual CA *increases* with time indicates that the oxidative reactions incorporate substantially more polar groups into the surface regions than might be removed by reorientation motions. In fact, the CA data suggest that surface reorientation is quite limited compared with conventional polymers.

We have utilized angle-dependent XPS analyses to probe for compositional gradients that would indicate transport of polar groups away from the air interface. Such evidence was found with *n*-hexane [12] and *n*-heptylamine [40] plasma polymers: with time, the relative amount of oxygen detected at glancing emission became increasingly smaller than that detected at normal emission, indicating that some of the O-containing groups had migrated away from the air interface. The differences were, however, relatively small, and the time constant of reorientation was of the order of 10–50 days, which is considerably larger than that of plasma-modified conventional polymers [41].

An aminated plasma polymer surface can also be prepared by the polymerization of an alkane monomer followed by a subsequent surface treatment in an ammonia plasma. We have explored this route using *n*-hexane. The resulting *n*-hex/NH_3 surface is of interest because, in contrast to the *n*-hex and *n*HA surfaces discussed above, it was found to have CAs that *increased* with time as samples aged. This is shown in Fig. 11. Evidently, in this system the surface reorientation process is sufficiently extensive that it manifests itself clearly in the CA data, as for conventional polymer surfaces. The

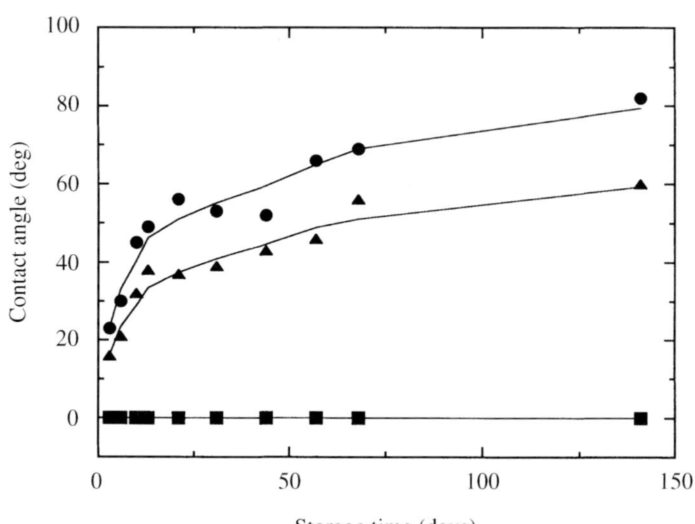

Figure 11. The advancing (●), sessile (▲), and receding (■) air/water contact angles of an ammonia-plasma-treated *n*-hexane plasma polymer as a function of storage time in air.

rate of this reorientation is, however, very slow; a time constant of $\sim$ 40–50 days can be estimated from Fig. 11.

4. DISCUSSION

We have shown that a number of plasma-treated polymer surfaces and surface layers of plasma polymer coatings change considerably as they age in contact with air, and these compositional changes can markedly affect their surface properties. More extensive data and discussion of the effects and mechanisms of ageing of some substrate/plasma combinations have been presented elsewhere [9, 10, 12, 40, 41]. Aged modified polymer surfaces are of interest because they can help elucidate fundamental aspects of polymer surfaces such as the rate and extent of mobility, and the thermodynamic factors (interfacial enthalpy versus translational entropy) [9, 10] that determine how a polymer surface adapts to changes in the contacting medium. In this communication, however, rather than focusing on ageing and its causes, we now wish to briefly discuss the *relevance* of surface ageing to the study and interpretation of interfacial adhesion in structural and biomedical applications.

As ageing alters the surface properties, it will affect adhesional interfacial interactions. However, to what extent can general conclusions be drawn given that the extent of ageing typically varies even when a particular polymer/process vapor combination is exposed to different plasma conditions? An example of this dependence of ageing on plasma conditions is FEP/NH_3 [22]. However, while the magnitude and relative importance of the processes causing ageing may vary with the process vapor, the plasma conditions, the substrate, and the equipment design, the fact that substantial surface reorientation and post-processing oxidation can be observed in a wide variety of plasma-treated surfaces and plasma coatings on ageing suggests that, in general, concurrent surface reorientation and post-plasma oxidative reactions should be considered when interpreting the ageing of plasma-prepared surfaces and their structural and biomedical adhesion properties.

It is particularly important to ascertain, by the study of ageing effects, that specific chemical groups originally placed onto a polymer surface are still present by the time the material is subjected to an adhesive bonding test to avoid erroneous interpretations. For example, amine groups on polymers are of interest for structural adhesive bonding [15] because they can react covalently with epoxy and cyanoacrylate adhesives. However, the interfacial bonding of aged, plasma-modified surfaces could, instead, be partly or entirely due to polar, *non*-covalent (acid–base) interactions. We have reported how NH_3-plasma-modification of polyolefins improves the durability of adhesive bonds in hot and humid environments [15]. However, the disappearance of amine groups from the surface layers of ammonia-plasma-treated and alkylamine-plasma-coated polymers may impose a time limit after fabrication for the utilization of such surfaces. Thus, the decrease in the bond strength of PP/NH_3 (Fig. 12) could be attributed to such effects. On the other hand, the markedly different behaviour of HDPE/NH_3 (Fig. 12) poses interesting questions that require detailed study by

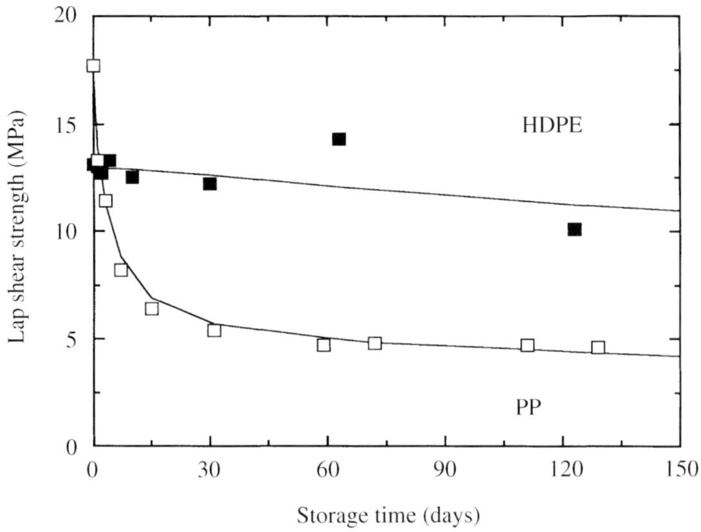

Figure 12. The lap shear strength of adhesively bonded, NH_3-plasma-treated HDPE and PP, a a function of storage time in air of the treated sample prior to bonding.

multi-technique surface analysis; this will be reported elsewhere. Clearly, interpretation of adhesion is contingent upon appropriate surface characterization data and understanding of ageing phenomena.

Closely related to such adhesive structural bonding is the use of reactive chemical surface groups for the covalent immobilization of biological macromolecules such as proteins or polysaccharides, with the intention of utilizing the attached layer of macromolecules to dictate biomedical adhesion effects. Ageing-induced compositional changes obviously are of major importance when plasma-prepared surfaces are intended for subsequent use in a second processing step, such as the covalent interfacial immobilization of a biologically active molecule by solution chemistry [16, 29, 30]. The disappearance within hours to days of amine groups by reorientation and oxidation imposes a limit to the time interval between plasma treatment and solution reactions. Use of freshly prepared amine surfaces is mandatory [30] but others have not specified the time elapsed between treatment and the covalent immobilization of proteins onto amine groups of PTFE/NH_3 surfaces [29]. Unfortunately, proteins adhere strongly to surfaces, and it is therefore difficult to differentiate between proteins coupled onto amine groups and proteins adsorbed onto surface amide groups. Ageing effects must be considered and appropriate surface analytical data included to verify that an intended interfacial covalent reaction has indeed taken place.

Understanding of the fundamental properties and behavior of polymer surfaces, such as rates of ageing, may also be of great value for the development of improved biomedical devices utilizing polymeric coatings rather than attached biological molecules (which, although offering the promise of specificity, entail additional costs and often problems in sterilization for instance). Plasma polymer coatings have become popular for creating novel biomaterials as they can be applied onto various commodity polymers to alter behavior such as cell adhesion and proliferation [3, 4].

Again, the surface properties of plasma polymers must be characterized in sufficient detail in order to understand how biological entities adhere. An issue of interest to us is how the mobility of a polymer surface affects the adsorption of proteins and the resultant adhesion of cells (adhesion of mammalian cells and bacteria to synthetic surfaces is dependent upon the prior adsorption of adhesive glycoproteins onto the bare synthetic material). Andrade [42] has proposed that the mobility of a polymer surface may interact with the process of protein adsorption: when a protein, which has some structural mobility, approaches a mobile polymer surface, interfacial forces arise and can rearrange the molecular structures *both of the protein and the polymer surface regions* to arrive at an interfacial energy minimum.

The very limited and extremely slow (in comparison with conventional polymers) surface reorientational motions of chemical groups on plasma polymer surfaces (although only very few systems have been studied in detail to date [12, 40]) is therefore of great interest to us as it may lessen the complexity of study of the interfacial forces involved in protein adsorption onto polymeric surfaces comprising a mixture of chemical groups. Interpretation of protein adsorption onto surfaces that are not capable of responding to the presence of altering interfacial forces (displacement of solvent molecules by an approaching protein molecule) within the timeframe of protein adsorption and denaturation reduces in complexity as only the protein can change its structure, and it may be less challenging to identify the contributions of the various interfacial forces.

'Immobile' (on the time scale of interfacial interactions) plasma polymer surfaces also present another advantage. To study the adhesion of proteins, the effective polymer surface composition in contact with a protein must be determined. However, the surface structure adopted by the polymer in contact with a protein may differ from the structure adopted in water and in freeze-hydration XPS sample preparation. Clearly, an 'immobile' polymer surface greatly reduces the requirements for specialized surface analysis methods; standard XPS sample preparation and analysis methods are sufficient.

Study of polymer surface ageing and the extent and rate of surface reorientation thus are valuable for assessing whether the mobility of a surface may be relevant to its biomedical interactions. The long time constants of the reorientation of the *n*-hexane and the aminated plasma polymer surfaces we have studied [12, 40] suggests that their limited and slow mobility may be neglected when interpreting interactions with adsorbing proteins.

Oxidative changes to the composition of plasma polymers are, however, clearly not negligible; the chemical nature and ionizability of plasma-prepared surfaces can change considerably with time, and this must be taken into account when interpreting biomedical interactions [14] and adhesive bonding performance [15]. The time elapsed between plasma deposition and evaluation must be stated, and chemical characterization must be performed at the same sample age as the performance testing. The rapid compositional changes that take place in many plasma polymers within the first few days make it difficult to obtain accurate correlations between surface chemistry and performance with fresh materials. We therefore test the biocompatibility of plasma polymers [14] that have been aged for some time (typically $> 2-3$ weeks) so that there are negligible chemical changes during the biocompatibility assay. On the other hand, when amine groups are used for further chemical reactions, such as the immobilization of biomolecules [16], then we use the plasma polymers within 4 h after fabrication. The same considerations apply to plasma-treated surfaces.

5. CONCLUSIONS

During storage in air, the surfaces of plasma-treated polymers and plasma polymer coatings usually are affected by surface reorientation (rotational and translational motions of treated and untreated polymer chains) and by post-plasma chemical reactions. These processes lead to considerable changes with time ('ageing') in the chemical composition and the properties of the surfaces. The relative importance of reorientation and oxidative reactions to the compositional changes varies markedly between different plasma-prepared surfaces, but generally both processes contribute to the ageing of a surface. However, exceptions such as water-plasma-treated FEP exist; as these materials age in air, these processes are absent and their surfaces remain invariant. Plasma-treated surfaces usually undergo 'hydrophobic recovery', in which the contact angle increases, indicating that surface reorientation is the main factor in defining the ultimate surface composition. However, this hydrophobic recovery can be accompanied by substantial oxidative changes to the surface composition. Plasma polymer surfaces, on the other hand, appear capable of only very limited and slow reorientational motions, although to date only a few examples have been characterized in sufficient detail.

The radical-induced oxidative reaction sequences that usually occur for extended periods of time after plasma treatment and deposition can interfere markedly with one's intention to produce specific surface chemistries for later use. Surfaces containing amine groups have been particularly unstable; XPS analysis indicated that the initially amine-rich surfaces were converted with time to surfaces containing amide and oxygen-containing groups. Thus, surfaces must be freshly prepared if one intends to perform a covalent interfacial reaction based on surface amine groups.

For adhesive bonding and biomedical applications, where there are delays between fabrication and end use, the preparation of polymer surfaces and plasma polymer coatings with predictable properties at the time of use requires detailed characterization of the ageing processes, since in many instances ageing is more important than the initial surface modification step in determining the composition and properties at the time of use of the surface. Therefore, the time elapsed between fabrication and testing must be specified, and surface analytical data must be recorded *at the time of performance evaluation*. The ability of surfaces to oxidize and reorient must also be taken into account when studying interfacial interactions. For biomedical applications, the limited and slow mobility of the surface layers of plasma polymers suggests that within the timeframe of adsorption and denaturation of a protein the plasma polymer surface is not capable of undergoing adaptive reorientational motions to minimize the interfacial energy. Thus, plasma polymers are particularly suited to the study of correlations between surface compositions and the adsorption behavior of proteins.

Acknowledgement

We thank Mr. Zoran Vasic and Mr. Paul Pasic for technical assistance. This work was partially supported by the Commonwealth Government's Co-operative Research Centres Scheme (Co-operative Research Centre for Eye Research and Technology).

REFERENCES

1. E. M. Liston, L. Martinu and M. R. Wertheimer, *J. Adhesion Sci. Technol.* **7**, 1091 (1993).
2. M. Strobel, C. S. Lyons and K. L. Mittal (Eds), *Plasma Surface Modification of Polymers: Relevance to Adhesion*. VSP, Zeist, The Netherlands (1994).
3. W. R. Gombotz and A. S. Hoffman, *Crit. Rev. Biocompat.* **4**, 1 (1987).
4. B. D. Ratner, A. Chilkoti and G. P. Lopez, in: *Plasma Deposition, Treatment, and Etching of Polymers*, R. d'Agostino (Ed.). Academic Press, San Diego (1990).
5. H. K. Yasuda, *J. Macromol. Sci.-Chem.* **A10**, 383 (1976).
6. R. d'Agostino (Ed.), *Plasma Deposition, Treatment, and Etching of Polymers*. Academic Press, San Diego (1990).
7. J. D. Andrade (Ed.), *Polymer Surface Dynamics*. Plenum Press, New York (1988).
8. F. Garbassi, M. Morra and E. Occhiello, in: *Polymer Surfaces — From Physics to Technology*, Ch. 2. Wiley, Chichester (1994).
9. T. R. Gengenbach, X. Xie, R. C. Chatelier and H. J. Griesser, *J. Adhesion Sci. Technol.* **8**, 305 (1994).
10. X. Xie, T. R. Gengenbach and H. J. Griesser, *J. Adhesion Sci. Technol.* **6**, 1411 (1992).
11. T. R. Gengenbach, Z. R. Vasic, R. C. Chatelier and H. J. Griesser, *J. Polym. Sci.: Part A: Polym. Chem.* **32**, 1399 (1994).
12. T. R. Gengenbach, Z. R. Vasic, R. C. Chatelier and H. J. Griesser, *Plasmas and Polym.* **1**, 207 (1996).
13. T. R. Gengenbach, R. C. Chatelier and H. J. Griesser, *Surf. Interface Anal.* **24**, 271 (1996).
14. H. J. Griesser, R. C. Chatelier, T. R. Gengenbach, G. Johnson and J. G. Steele, *J. Biomater. Sci., Polym. Edn* **5**, 531 (1994).
15. D. Y. Wu, W. S. Gutowski, S. Li and H. J. Griesser, *J. Adhesion Sci. Technol.* **9**, 501 (1995).
16. L. Dai, P. Zientek, H. A. W. StJohn, P. Pasic, R. C. Chatelier and H. J. Griesser, *Amer. Chem. Soc. Polym. Prepr.* **36** (1), 82 (1995).
17. S. Li and H. J. Griesser, in preparation.
18. H. J. Griesser, D. Youxian, A. E. Hughes, T. R. Gengenbach and A. W. H. Mau, *Langmuir* **7**, 2484 (1991).
19. H. J. Griesser, *Vacuum* **39**, 485 (1989).
20. H. J. Griesser and P. Zientek, *Polym. Mater. Sci. Eng.* **69**, 468 (1993).
21. C. D. Bain and G. M. Whitesides, *J. Amer. Chem. Soc.* **110**, 5897 (1988).
22. R. C. Chatelier, X. Xie, T. R. Gengenbach and H. J. Griesser, *Langmuir* **11**, 2585 (1995).
23. C. Vasile, in: *Handbook of Polyolefins: Synthesis and Properties*, 1st edn, C. Vasile and R. B. Seymour (Eds), p. 479. Marcel Dekker, New York (1993).
24. G. Beamson and D. Briggs, *High Resolution XPS of Organic Polymers. The Scienta ESCA300 Database*, 1st edn. Wiley, New York (1992).
25. R. C. Kumar and J. M. Shreeve, *J. Amer. Chem. Soc.* **102**, 4958 (1980).
26. W. Leidinger and W. Sundermeyer, *Chem. Ber.* **115**, 2892 (1982).
27. L. J. Gerenser, *J. Adhesion Sci. Technol.* **1**, 303 (1987).
28. M. Tatoulian, F. Arefi-Khonsari, N. Shahidzadeh-Ahmadi and J. Amouroux, *Intl J. Adhesion Adhesives* **15**, 177 (1995).
29. A. B. Fischer, R. Bizios, M. Sheu and I. H. Loh, *Trans. Amer. Soc. Biomater.* **17**, 21 (1994); and Verbal presentation, 20th Ann. Mtg Soc. Biomater., Apr. 5–9, 1994, Boston, MA, USA.
30. L. Dai, H. A. W. StJohn, J. Bi, P. Zientek, R. C. Chatelier and H. J. Griesser, *Surf. Interface Anal.* **24**, 611 (1996).
31. G. P. López and B. D. Ratner, *Langmuir* **7**, 766 (1991).

32. H. J. Griesser and R. C. Chatelier, *J. Appl. Polym. Sci.: Appl. Polym. Symp.* **46**, 361 (1990).
33. T. R. Gengenbach, R. C. Chatelier and H. J. Griesser, *Surf. Interface Anal.* **24**, 611 (1996).
34. H. Yasuda, M. O. Bumgarner, H. C. Marsh and N. Morosoff, *J. Polym. Sci., Polym. Chem. Ed.* **14**, 195 (1976).
35. A. M. Wrobel, *J. Macromol. Sci.-Chem.* **A22**, 1089 (1985).
36. R. A. Engelman and H. K. Yasuda, *Polym. Mater. Sci. Eng.* **62**, 19 (1990).
37. R. C. Chatelier, C. J. Drummond, D. Y. C. Chan, Z. R. Vasic, T. R. Gengenbach and H. J. Griesser, *Langmuir* **11**, 4122 (1995).
38. H. Yasuda, A. K. Sharma and T. Yasuda, *J. Polym. Sci., Polym. Phys. Ed.* **19**, 1285 (1981).
39. T. Yasuda, K. Yoshida, T. Okuno and H. Yasuda, *J. Polym. Sci.: Part B: Polym. Phys.* **26**, 2061 (1988).
40. T. R. Gengenbach, Z. R. Vasic, S. Li, R. C. Chatelier and H. J. Griesser, *Plasmas and Polym.* (in press).
41. R. C. Chatelier, X. Xie, T. R. Gengenbach and H. J. Griesser, *Langmuir* **11**, 2576 (1995).
42. J. D. Andrade, *Clinical Mater.* **11**, 19 (1992).

Mittal Festschrift, pp. 329–353
W. J. Van Ooij and H. R. Anderson, Jr. (Eds)

Adhesion, wettability and mechanical properties of ammonia- and helium-plasma-treated polypropylene

F. AREFI-KHONSARI,[1,*] M. TATOULIAN,[1] N. SHAHIDZADEH,[1] M. CHEHIMI,[2] J. AMOUROUX,[1] D. LEONARD[3] and P. BERTRAND[3]

[1] *Laboratoire de Génie des Procédés Plasmas, Université Pierre et Marie Curie ENSCP, 11 rue P. & M. Curie, 75231 Paris Cedex 05, France*

[2] *Institut de Topologie et Dynamique des Systèmes, Université Paris VII Denis Diderot, associé au CNRS (URA 34), 1 rue Guy de la Brosse, 75005 Paris, France*

[3] *Unité de Physico-chimie et de Physique des Matériaux, Université Catholique de Louvain, Batiment Boltzman, 1 place Croix du Sud, 1348 Louvain-La-Neuve, Belgium*

Abstract—The physicochemical modifications of ammonia-treated polypropylene (PP) films have been studied and characterized in terms of acid–base properties using the contact angle titration method and X-ray photoelectron spectroscopy in conjunction with a molecular probe technique using chloroform as a reference Lewis acid. These techniques have shown that PP surfaces that have been treated for between 0.7–1 s are basic in character. For longer treatment times, the basic character of the surfaces decreases, as shown by the above techniques and confirmed by time of flight–secondary ion mass spectroscopy (ToF–SIMS). On the other hand, for such treatment times, a degradation of the adhesion and mechanical properties was observed.

The ageing of an ammonia-plasma-treated PP was limited by a helium (He) plasma pretreatment known to crosslink the surface, stabilizing in this way the wettability, adhesion and mechanical properties. ToF–SIMS was performed on helium treated High Density Polyethylene (HDPE) in order to point out the structural modifications.

Keywords: Acid–base; plasma; ToF–SIMS; XPS; molecular probe; contact angle; emission spectroscopy; mechanical properties; ageing.

1. INTRODUCTION

Metallization of polymer substrates is used extensively in multilayer structures, packaging, automotive, decoration and microelectronic applications. Plasma treatment of polymer surfaces prior to metal deposition is a widely used technique to improve adhesion [1, 2]. The most important feature of the plasma techniques is that the surface properties of the treated material can be modified without changing their intrinsic bulk properties. It is well known that plasma treatments can simultaneously

---

*To whom correspondence should be addressed.

bring about many changes on the surface of polymers, such as degradation of polymer chain, formation of free radicals, chemical functionalization, etching and crosslinking [3, 4]. Because of the complexity of plasmas, the bulk of the research has been concentrated on establishing the dependence of the macroscopic and microscopic properties of the treated polymers on the major process variables such as the power, the nature of the plasma gas, the gas flow rate and the pressure. In order to obtain a better understanding of the plasma–polymer interactions, plasma diagnostic techniques such as Optical Emission Spectroscopy (OES), Actinometry Optical Emission Spectroscopy (AOES), or Mass Spectrometry (MS) have been coupled with surface analytical tools such as X-ray Photoelectron Spectroscopy (XPS), Static Secondary Ion Mass Spectroscopy (SSIMS), contact-angle measurements, Rutherford Back-scattering Spectroscopy (RBS), Scanning Electron microscopy (SEM), etc. [5–7].

In this study, polyaliphatics such as isotactic polypropylene (PP), 8 μm thick as well as high density polyethylene (HDPE), 14 μm thick, were plasma treated.

The reactor used in this study formed a low pressure plasma using asymmetrical geometry of electrodes (hollow electrode-grounded cylinder) with an excitation frequency of 70 kHz [2]. This special configuration provides the ability to simulate the continuous industrial surface treatment of polymer films, i.e. very short treatment times. In a previous work [8], the effect of a nitrogen (N_2) or ammonia (NH_3) plasma treatment on the adhesion of PP films to *in situ* deposited aluminium (Al) coatings was compared. It was pointed out that a higher rate of nitrogen uptake was observed in the case of the ammonia plasma as compared to the one obtained in a nitrogen plasma. This has been attributed to the high reactivity of the excited NH species identified in an ammonia plasma. As a result of this, the measured peel strengths were two times higher for NH_3-treated surfaces than for N_2-treated ones, thus revealing the important role of incorporation of nitrogen in improving the adhesion of thermally evaporated aluminium to PP. Moreover, the kinetics of the process by which nitrogen is incorporated in the PP film and the resulting adhesion improvement with respect to aluminium differ for low frequency (70 kHz) and RF (13.56 MHz) plasmas [8]. It was further pointed out in our previous study [8, 9] that overtreatments (corresponding to a few seconds in a low frequency discharge under our experimental conditions) very quickly gave rise to a degradation of the bulk (mechanical) properties of the modified surface layer leading to poor adhesion because of the formation of a weak boundary layer.

Since the stability of a plasma-treated surface is a very important issue for the industrial development of plasma processes, the ageing effects of ammonia-plasma-treated surfaces were studied [8, 10, 11]. It was shown that upon storage in a relatively non-polar medium such as air, the surface, rendered wettable by the plasma treatment, reverted to a less wettable state with time, probably due to a thermodynamically driven reorientation of polar moieties away from the surface into the sub-surface. This rotation of the lateral chains around the C—C bonds (β-relaxation) requires an energy of around 60 kcal/mol at room temperature; the thermal energy is sufficient to favour this mechanism on a small scale [12]. Moreover, it was hypothesized that the hydrogen bonds created between the different plasma-grafted groups can limit the destruction and elimination of the grafted functional groups and, secondly, the internalization of the chemical groups in contact with air. The equilibrium between these two opposing mechanisms gives rise to a stabilization of the surface chemistry over a few monolayers after a few days. We also found using the peel test that the adhesion also degraded when the samples were aged before metallization [8, 11]. It has been shown that the ageing phenomenon could be slowed down by stabilizing the surface layer by crosslinking using a noble plasma gas (e.g. helium or argon) [13]. The CASING (Crosslinking by Activated Species of Inert Gases) process consists of creating new bonds such as C=C, which improve the bond strength of the surface by forming a very cohesive and dense skin. On the other hand, Vogel and Schonhorn [14] showed that chemical bonds alone were not enough to achieve strong adhesion to polymers since, eventually, surface layers of low strength had to be eliminated by crosslinking and/or physical sputtering. Other authors have compared the crosslinking effect of HDPE caused by different plasma gases (argon, oxygen, nitrogen and hydrogen) [15]. They have shown that the thicker the crosslinked layer, the slower was the modification of the surface properties such as change in the contact angle measurements. The depth of crosslinking in polymers such as HDPE and poly(ethylene terephthalate) (PET) has been determined by different authors [13, 15, 16] by extracting the treated samples in solvents such as toluene or nitrobenzene in order to remove the non-crosslinked fraction, and the weight of the crosslinked gel which remained was then determined. In a previous paper [11], we have shown how the ageing phenomenon could be slowed down by stabilizing the surface layer of an ammonia-plasma-treated PP using a helium plasma pretreatment.

In this paper these modifications will be studied in terms of acid–base properties by using two different techniques, namely, contact angle titration method and XPS, in conjunction with the molecular probe technique. The applicability of contact angle measurements to determine the dispersive component of the surface energy of the solid as well as the acid–base component of the total work of adhesion in the case of plasma-treated polypropylene has been investigated in a previous paper [10]. On the other hand, it has recently been shown that XPS can indeed be used to quantitatively evaluate the acid–base properties of polymers (in the solid state) by monitoring the chemical shift experienced by an adsorbed molecular probe such as chloroform [17–20]. These results were then compared to the characterization of ammonia-plasma-treated PP by ToF–SIMS which is very suitable for studying modifications owing to their high mass resolution ($M/\Delta M = 3000$) under our conditions. It will be demonstrated in this paper how these methods can be used to assess the acid–base properties of modified polymer surfaces and to predict improved adhesion of these films with respect to thermally evaporated aluminium thin films.

ToF–SIMS was also employed in order to point out the structural modifications of the He-plasma-treated HDPE which is known to readily crosslink under plasma conditions. This method has been completed by mechanical tests such as elongation-at-break which is an indirect indication of the degree of crosslinking.

2. EXPERIMENTAL

2.1. Materials

PP films were 8 μm thick and showed a negligible amount of oxygen on the surface as judged from XPS analysis prior to plasma treatment and exposure to trichloromethane (TCM) or chloroform. The chloroform used as the molecular probe (Aldrich) was a high purity grade and stabilized with amylenes (alkenes, e.g. methyl-2-but-2-ene). Commercially available ultra-high purity gases (purity of $NH_3 \geqslant 99.96\%$, He $\geqslant$ 99.995%, tetrafluorocarbon (CF_4) $\geqslant$ 99.7%) were used without further purification.

2.2. Experimental set-up

Figure 1 shows the low-pressure plasma reactor used for the plasma treatment of polypropylene films. The electrode configuration is non-symmetrical. The reactor consists of a stainless-steel high-voltage, hollow electrode-grounded cylinder (diameter of the cylinder, 0.07 m). The power is coupled to the reactor from an RF generator with an excitation frequency of 70 kHz through a suitable impedance matching network. During the treatment, the polymer films (22 $\times$ 22 cm^2) were rolled onto the grounded cylinder. For a total plasma duration of 1 s, the calculated real treatment

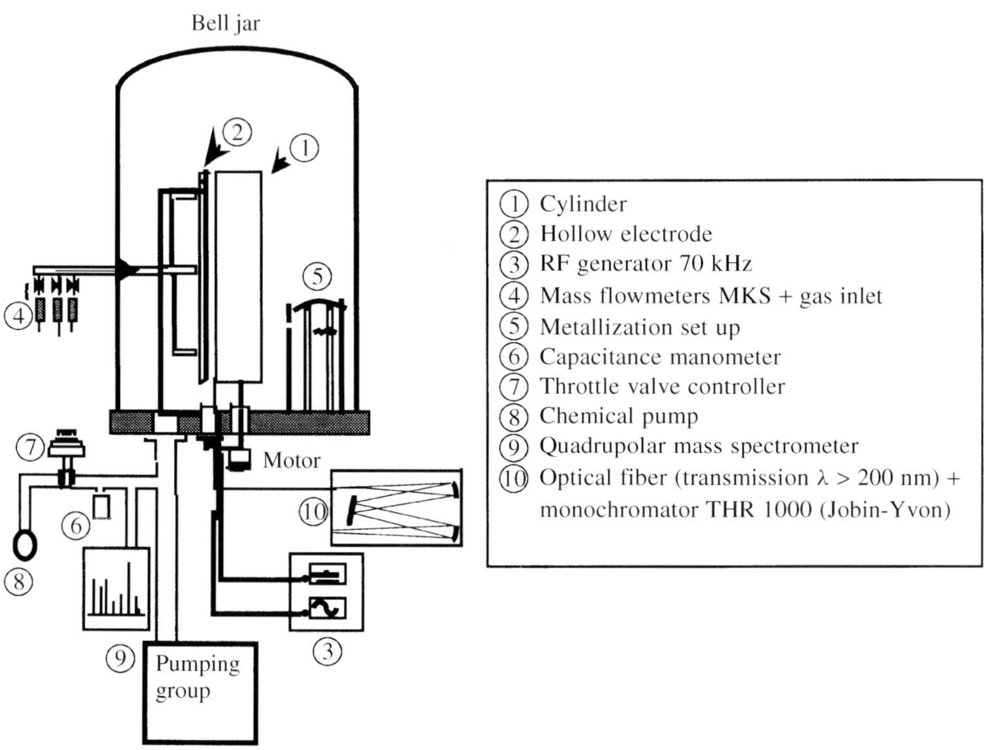

Figure 1. Schematic of the bell jar reactor used for the surface treatment of polypropylene films by a low-pressure plasma.

time is equal to 0.023 s. This system consequently allows the study of very short treatment times and simulates the industrial surface treatment of polymers. Gases were introduced through MKS mass flow controllers and the pressure was monitored with an MKS capacitance gauge. The main chamber was evacuated by a TPH 170 (Balzers) turbomolecular pumping system and a base pressure of 10^{-3} Pa was obtained. The operating pressure was maintained at 100–150 Pa by a 2012 AC chemical pump.

The electrical characteristics of the discharge were measured with a LeCroy 9400 digital oscilloscope with a sampling frequency of 100 MHz. Plasma emission was sampled by an optical fiber inserted into the plasma. The radiation transmitted was then focused by means of a quartz lens on the plane slit of a 1 m Jobin-Yvon HR monochromator equipped with a 3600 grooves/mm holographic grating and a Hamamatsu photomultiplier, for analysis of the 300–850 nm spectral domain.

In situ metallization of pretreated PP films was carried out by thermal evaporation in the same chamber, which avoided any contact of the pretreated polymer with air prior to deposition of the aluminium. Using this technique a thin aluminium film (20–30 nm) was deposited at 10^{-3} Pa.

2.3. Analytical techniques

2.3.1. X-ray photoelectron spectroscopy. XPS spectra were recorded using a VG Scientific ESCALAB MK I system interfaced to a Cybernetix Data Acquisition System based on a personal computer. The spectrometer was operated in the constant analyzer energy (CAE) mode at a pass energy of 20 eV to ensure good spectral resolution and a high signal-to-noise ratio. The electron take-off angle was 90° with respect to the surface. A non-monochromatic Mg K_α (1253.6 eV) source, operating at a power of 200 W, was used to record the spectra. Peak fitting of the high-resolution XPS spectra was done using software written at the ITODYS Laboratory, Université Paris VII. Quantification of the XPS data was achieved using peak areas and experimental sensitivity factors. The full width at half-maximum (FWHM) was constrained so that it was the same for all singlet peaks within a spectrum. A linear background was assumed and the peak shape was 100% Gaussian. Charge referencing was determined by setting the main C 1*s* component at 285 eV.

For the determination of acid–base properties of polymer surfaces [17, 18] by this technique, untreated and freshly ammonia-plasma-treated PP were exposed to chloroform or trichloromethane known to be a reference acid for a period of 10 min at ambient temperature. The polymer film was then allowed to outgas the excess of solute.* It was then transferred into the XPS analysis chamber for surface characterization. If the polymer–solute intermolecular interactions are strong enough (because of the existence of specific acid–base interactions) a residual amount of solute is detected and the molar ratio of solute/polymer repeat unit (in %) is then evaluated.

*In this context we consider the polymer and liquid vapour as a solution within the depth probed by XPS (approximately 5 nm). The polymer is consequently the solid solvent (phase in excess) and the liquid vapour the solute (the minor component).

2.3.2. Contact angle measurements. The apparatus developed in our laboratory is based on an image processing system. It allows the measurements of the contact angle of a drop of liquid on a flat surface and allows the contact angle values to be followed as a function of time. The apparatus has been described in detail elsewhere [21]. A monochromatic camera equipped with a zoom lens and a double focal extender produces a video image of the object (in this case, the drop of the liquid on the surface). With the help of an analog–digital interface the video image is digitized. We have written a software that allows the contour of the drop to be obtained and contact angle to be measured. Therefore, this system offers the advantages of being completely objective, reproducible (mean square deviation: 0.3 units for 100 responses) and rapid (0.6–1 s) as compared to the goniometer or photographic measurements.

The advancing angle is then measured each 0.6 s, so there is no need to extrapolate to time zero for the determination of the advancing contact angles. The humidity and the temperature of the chamber were not controlled. The latter varied between 20 and 25°C. The drop was considered to be a part of a sphere and the volume of the drop used was always 2 μl. Therefore, the deviation from sphericity due to gravitational perturbations was negligible.

Samples were cut to the size of 0.5×4 cm and attached by their back to a flat support using double-sided Scotch tape to keep the sample flat. All reported values are an average of seven measurements taken at different locations on the films and have a maximum error of $\pm 3^\circ$. The solid/liquid/vapour system is considered to be at thermodynamic equilibrium. Aqueous acidic and basic solutions with pH values ranging from 1 to 14 were prepared with double-distilled water with NaOH or HCl as the base or acid. The surface tension of the solutions was measured by the Wilhelmy technique. It showed that there was no difference between the surface tension of pure water and those of the acidic and basic solutions.

2.3.3. Time-of Flight SIMS. ToF–SIMS measurements were performed a few days after the treatment at Louvain-la-Neuve using a ToF–SIMS Trift spectrometer from Charles Evans & Associates [22].

The primary ion source was a pulsed ^{69}Ga ion source operated at 15 kV. The secondary ions were extracted perpendicularly to the sample surface. Since the sample surface is at ± 3 kV (according to the mode, i.e. positive or negative), a relatively high voltage (± 10 keV, according to the mode) was used for the extraction and acceleration of the secondary ions. The secondary ions were deflected in a 270° spherical sector consisting of three electrostatic analyzers to compensate for the initial energy distribution of ions with the same mass. The detector is a hybrid chevron stack of channel plates with resistive anode encoder (RAE). The 400 pA DC current was pulsed at 5 kHz repetition rate and with a 4 ns pulse width. The analyzed area was about 9.7×9.7 mm. The mass range was 0–5000 amu. The total primary ion fluence for a 5 min spectrum was limited to 7×10^{11} ions/cm^2. It is clearly within the so-called 'static' conditions [23, 24]. Charge compensation was performed with a pulsed electron gun operated at 20 eV electron energy [25]. The ToF–SIMS data were analyzed by the Cadence software from Charles Evans & Associates. This software allowed us to obtain calibration of mass from well-defined reference peaks (at least three). The peak integral was normalized systematically to the total intensity. The $H^{+/-}$ intensity was subtracted because it was too sensitive to slight adjustments of the setting of the spectrometer. The normalization to the total intensity appears to be the least arbitrary procedure.

3. RESULTS AND DISCUSSION

First we shall report on our study of the ammonia modified PP films as a function of the key parameter, treatment time.

The physicochemical modifications of ammonia-plasma-treated PP have been reported previously [10, 17]. In this paper these modifications will be clarified in terms of acid–base properties. The two techniques used are the contact angle titration method, and XPS in conjunction with a molecular probe. It was also interesting to look at the evolution of the different chemical functional groups detected at the surface as a function of the treatment time by a surface-sensitive technique such as ToF–SIMS.

3.1. Contact angle titration method

Contact angle measurements were made on the surfaces of non-treated and plasma-treated PP by using test liquids with different pH values [26–28]. The contact angles of aqueous solutions at the surface were measured immediately after the treatments. The use of measurements of contact angle (θ) of a liquid (l) with a surface tension γ_l on a solid (s) is a starting point for the calculation of the work of adhesion W_{sl}. Fowkes has shown that the total work of adhesion has two contributions [29]

$$W_{sl} = \gamma_l(1 + \cos\theta) = W_{sl}^{d} + W_{sl}^{ab}, \tag{1}$$

where W_{sl}^{d} defines the dispersion or London forces, and W_{sl}^{ab} the non-dispersive or acid–base interactions.

The measurements carried out on ammonia-plasma-treated PP for different treatment times are shown in Fig. 2. As can be seen in this figure, the behaviour of non-treated PP is independent of the pH of the test liquids (Fig. 2a). In this case, the substrate being non-polar, W_{sl} should be equal to W_{sl}^{d}, since the acid–base interactions do not exist [10]. It is noteworthy that for treatment times in the range of 0.7–1 s (Fig. 2b), the surface shows a clear basic character, i.e. the total work of adhesion decreases from 129 mJ/m^2 for acid pH values to 120 mJ/m^2 for basic pH ones. By increasing the treatment times ($t \geqslant 5$ s) one can show an increase in the total work of adhesion. However, in these cases (Figs 2c and 2d) the surface shows an amphoteric character with about the same total work of adhesion ($W_{sl} = 145$ mJ/m^2 for $t = 30$ s) for the strongly basic and strongly acidic aqueous solutions.

On the other hand, peak fitting of the N 1*s* and C 1*s* signals suggested that the grafted moieties on the surface were amine and amide groups among others [9, 10]. These chemical functional groups confer a predominant basic character to the surface for treatment times of the order of 1 s. Minor oxygen functional groups were also

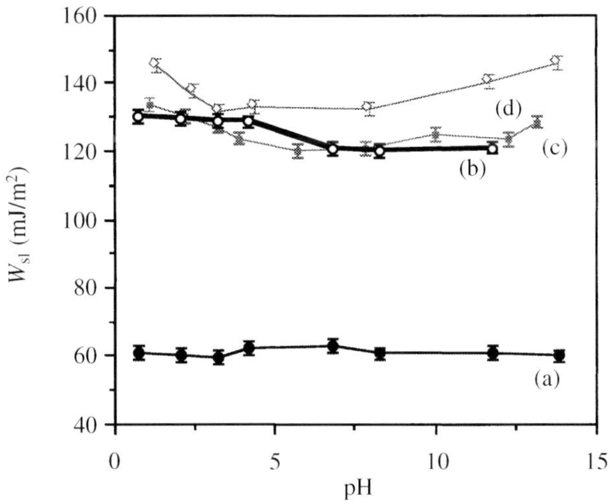

Figure 2. Variation of the total work of adhesion with different pH test liquids for different treatment times: (a) untreated PP; (b) 0.7 s; (c) 5 s; (d) 30 s. Experimental conditions: gas, NH_3; pressure, 150–200 Pa; Q, 150 sccm; power, 70 W; $d_{\text{inter-electrode}}$, 1cm; f, 70 kHz.

detected (O/C ≈ 4%) on the ammonia-plasma-treated surface [10]; their presence is due either to the interaction of residual oxygen atoms present in the reactor chamber with the surface or to oxygen-containing chemical functional groups grafted onto the surface during the transfer of the specimens from the plasma reactor to the XPS equipment [30, 31].

3.2. Assessment of the acid–base interactions of PP by XPS analysis

Since contact angle measurements demonstrated that ammonia-plasma-treated PP produces a Lewis basic character due to nitrogen-containing moieties, these films were exposed to TCM vapours known to be a reference Lewis acid. The XPS spectrum of the exposed PP exhibits a Cl $2p$ signal from chlorine whereas no chlorine is detected on the reference PP. This result is interpreted in terms of acid–base interactions between TCM and basic groups grafted onto the surface of the ammonia-plasma-treated PP.

For the ammonia-plasma-treated PP films, the Cl $2p3/2$ binding energy was in the range of 196.9–198 eV [32], which is much lower than the values obtained for TCM sorbed on reference basic polymers such as poly(methyl methacrylate) (PMMA) and poly(ethylene oxide) (PEO) [18]. This indicates the withdrawal of electron density from the acidic solute [18] and thereby shows a stronger acid–base bond.

For each treatment time, $\%S$, the percentage of TCM (solute) molecules sorbed per nitrogen atom, was calculated as follows:

$$\%S = \frac{\text{TCM}}{\text{N}} = \frac{\dfrac{I_{\text{Cl}\,2p}}{3\,s_{\text{Cl}\,2p}}}{\dfrac{I_{\text{N}\,1s}}{s_{\text{N}\,1s}}} \times 100\%,$$

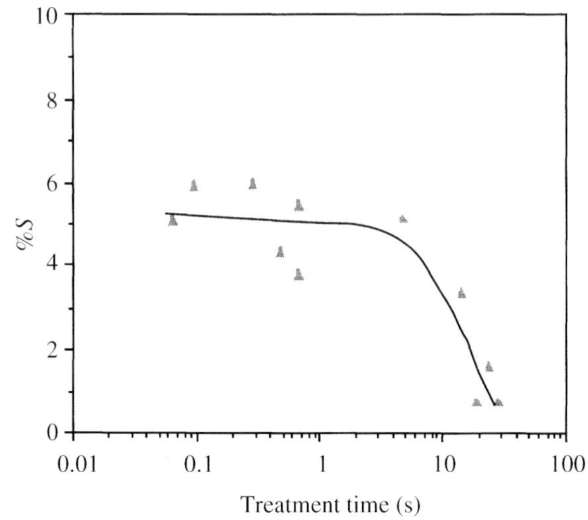

Figure 3. %*S*, number of TCM (chloroform) sorbed per nitrogen atom (in %) vs. the NH_3 plasma treatment time.

where the factor 3 was used because TCM contains three chlorine atoms. *I* and *s* are, respectively, the integrated peak area and the sensitivity for a given element. Figure 3 shows the change in %*S* vs. treatment time. Two distinct regions can be observed on this curve: the first one, i.e. for treatment times less than 1 s, the %*S* shows its upper value equal to 5%. In the second region, i.e. for treatment times longer than 1 s, the %*S* decreases sharply to 0.7% for a treatment time equal to 30 s. The %*S* starts to decrease from 1 s of treatment, although the N/C atomic ratio levels off and slightly increases for longer treatment times (30 s) [32]. This result is interpreted in terms of basic sites being less accessible. This could be due to a competition between TCM/ammonia-plasma-treated PP and lateral acid–base interactions of grafted groups at the surface, or a decrease of the specific groups with a basic character (cf. Section 3.3). The variation of %*S* vs. time is in good agreement with the results obtained from contact angle measurements (cf. Section 3.3), which showed a dominant basic character for treatment times in the order of 0.7 to 1 s whereas long treatment times (> 5 s) showed an amphoteric character.

The acid–base contribution to the enthalpy ΔH^{ab} for the polymer–TCM acid–base complex formation was calculated by using the published linear correlation [19, 33] between Cl $2p3/2$ BE and ΔH^{ab}, where the latter was determined by FTIR. The relationship is the following:

$$-\Delta H^{ab}\ (\text{kcal mol}^{-1}) = 228.8 - 1.1 E_B\ (\text{eV}). \quad (2)$$

The calculated values obtained from the above relationship vary between 3.1 and 4.3 kcal/mol and remain roughly constant with treatment time. The latter shows that the strength of the polymer–probe acid–base interactions does not vary even though the N/C ratio determined by XPS increases slightly for long treatment times (30 s). These values of ΔH^{ab} compare fairly well with the values of 4.3–4.9 kcal/mol obtained for TCM–amine adducts (Table 1) calculated using Drago's four parameter equation [33].

This provides strong supporting evidence for the specific interaction of TCM with the nitrogen-containing basic sites. However, ΔH^{ab} is generally smaller in the case of TCM/ammonia-plasma-treated PP with respect to TCM/amine complexes. This could be due to the possible steric effects at the surface of the polymer. Note that E and C values from Drago's tables were determined for experiments conducted in the liquid state, which is perhaps more favorable for acid–base contacts between opposing bodies. At this stage it was interesting to analyze the surface by ToF–SIMS which is much more sensitive to the topmost layers than XPS carried out with an electron take-off angle of 90°.

3.3. ToF–SIMS results

ToF–SIMS was performed in the negative as well as in the positive mode on the ammonia-plasma-treated PP films. The most important difference was observed in the negative mode. Figure 4a illustrates the 10–50 amu ToF–SIMS negative spectra of untreated PP, and PP treated in an ammonia plasma for $t = 0.5$ s and 30 s. The interesting ions are O^- (allowing the identification of the oxidation level of the sample through the value of the O^-/CH^- ratio) and nitrogen-containing ions such as CN^- and CNO^-. Table 2 presents the absolute values of the O^-/CH^-

Table 1.
ΔH^{ab} (kcal/mol) for several (TCM/base) adducts. E_A, E_B are Drago's electrostatic constants for an acid and a base, respectively. C_A, C_B are Drago's covalent constants for an acid or a base, respectively

	C_A	E_A	C_B	E_B	$-\Delta H^{ab}$ (kcal/mol)
TCM	0.159	3.02			
NH_3			3.46	1.36	19.64
CH_3NH_2			5.88	1.30	20.48
$(CH_3)_2NH$			8.73	1.09	19.64
$(CH_3)_3N$			11.54	0.81	17.97
$(CH_3CH_2)_3N$			11.09	0.99	20.06
*Treated PP 0.7 s					17.97
*Treated PP 25 s					12.95

*Measured values, others are calculated values.

Table 2.
O^-/CH^- ratio and normalized intensities of CN^- and CNO^- peaks obtained by ToF–SIMS in the negative mode for untreated PP, PP treated in NH_3 plasma (various times) and untreated Nylon 6

Ion	Untreated	0.5 s	1 s	5 s	10 s	30 s	Nylon 6
O^-/CH^-	0.004	0.5	0.4	0.5	0.75	0.5	1.2
CN^-	0	12	6	8.5	9	9	5
CNO^-	0	0.28	0.21	0.23	0.27	0.24	0.86

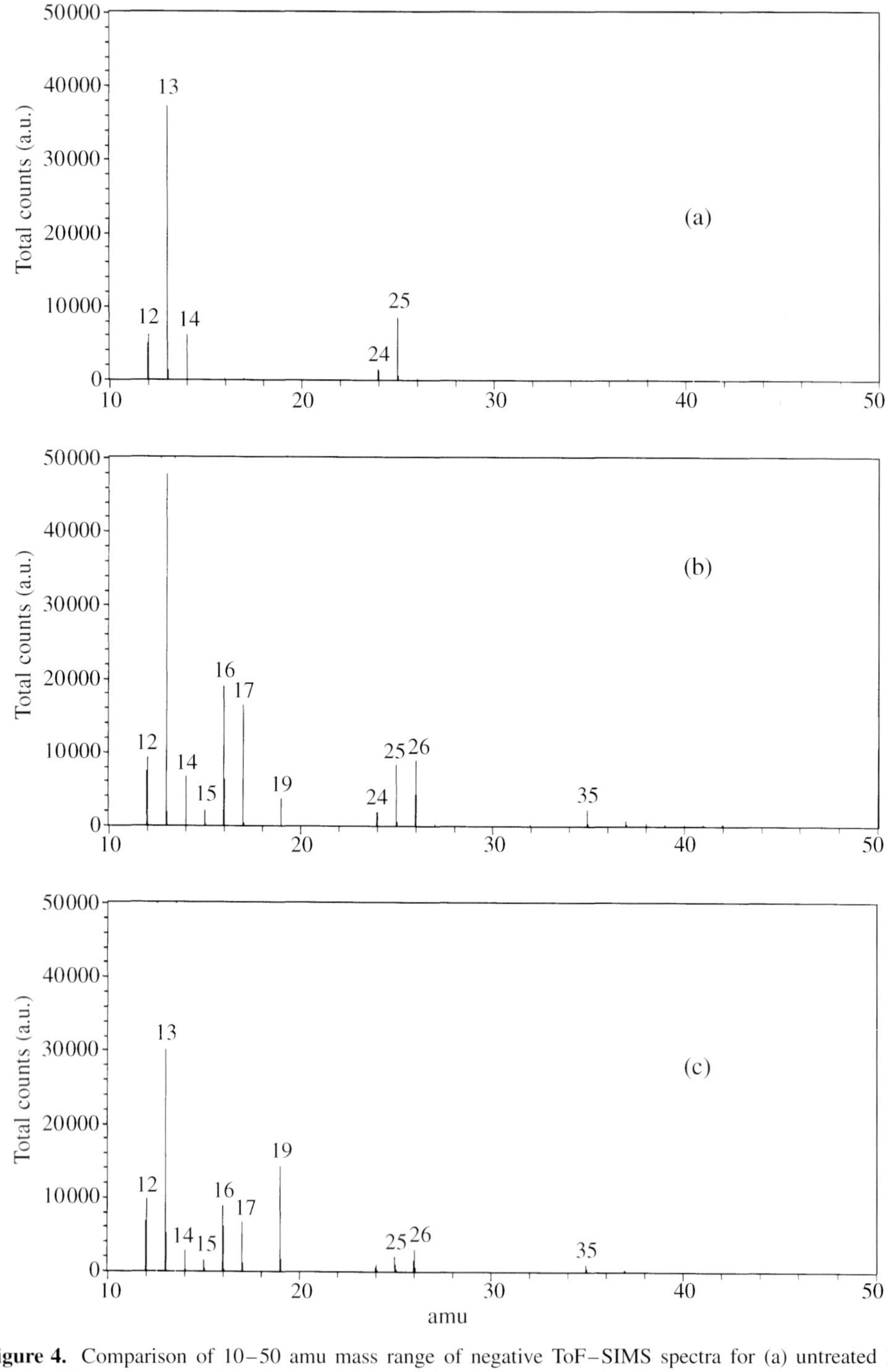

Figure 4. Comparison of 10–50 amu mass range of negative ToF–SIMS spectra for (a) untreated PP, (b) PP NH_3-plasma treated for 0.5 s and (c) PP NH_3-plasma treated for 30 s.

ratio and the normalized (see Experimental section) values of the intensity for CN^- and CNO^- ions as a function of the treatment time. Unfortunately, there is not a one-to-one correspondence between these ions and the content in amine and amide functional groups. This is illustrated by the normalized values observed in the case of the untreated Nylon 6 used here as a reference. In this case, the normalized intensity of the CN^- SIMS ion is certainly not negligible, even though there are only amide functional groups. A comparison with this reference allows us to conclude that:

(1) the oxidation could be partly related to the presence of amide functional groups;

(2) there are, logically, fewer amide functional groups than in the case of the Nylon;

(3) the very high intensity of CN^- groups is related to other functional groups containing only nitrogen such as amine functional groups.

Moreover, it is seen clearly that there is a decrease in the CN^-relative intensity without a decrease in the CNO^- relative intensity as a function of the treatment time. This can be related to an observed global decrease in basicity that was observed. It must indeed be noted that the basic character of the amines results from the electron pair on the nitrogen atom therefore classifying these species according to Lewis as n-EPD (non-bonding electron pair donors). In the case of amides, the delocalisation between the free electron pairs of nitrogen and carbonyl groups gives rise to a weaker base with respect to amines. However, an amide group can react as an n-EPD base by means of the free electron pair on nitrogen, and, in particular, that of carbonyl groups. These results could therefore explain why the basic character of the surface decreases for treatment times above 1 s in agreement with contact angle measurements and XPS technique in conjunction with the TMC acidic molecular probe.

Alternatively, the decrease of the basic character could be due to the cleavage effect of the ammonia-plasma treatment. Indeed, this leads to low molecular weight functionalized groups which can be mobile and hence can easily produce lateral acid–base interactions between the grafted groups, neutralizing these groups for further interactions.

It must be noted that there are also differences in the ToF–SIMS positive mode. Oxygen and nitrogen incorporation have been detected for treatment times as short as 0.5 s. They give rise to very complex ions. $^+C_nOH_{2n+1}$ and $^+C_nO_2H_{2n-1}$ ions (appearing at odd amu: 31, 45, 59, 73, 87, 101...) are some of the oxygen-containing ions, and $^+C_nN_mH_{2(n+m)-2}$ and $^+C_nN_mH_{2(n+m)}$ (appearing at even amu 28, 30, 42, 44, 56, 58...) are typical nitrogen-containing ions. Moreover, we have observed a positive ion containing C, H, N and O, i.e. the CH_2NO^+ ion (44.01 amu), as illustrated in Fig. 5, which represents the 43.5–44.5 amu ToF–SIMS positive spectra for the same samples as in Fig. 4. The intensity of the signal corresponding to this ion is very weak (the main component of the 44 amu peak is attributed to the $C_2H_6N^+$ ion at 44.03 amu), since it is the only one detected with C, H, N and O, and its deconvolution in the 44 amu peak is not evident (see the 30 s case). It is, however, not possible from positive SIMS results to draw conclusions on the relative concentration of amide and amine functional groups.

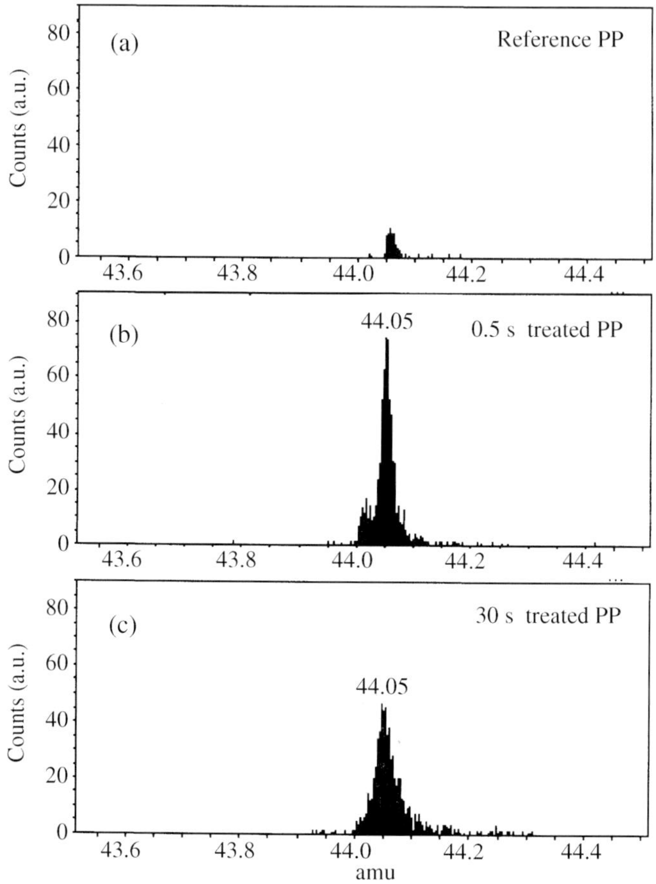

Figure 5. Comparison of ToF–SIMS in the positive mode of (a) untreated PP, (b) PP plasma treated for 0.5 s, (c) PP plasma treated for 30 s. In the case of short treatment time (t = 0.5 s), one observes a CH_2NO^+ peak (44.01 amu) which separates from the $C_2H_6N^+$ peak (44.01 amu).

3.4. Impact on the adhesion properties with respect to a thin aluminium layer

The ammonia-plasma-treated PP films were metallized *in situ* in the same chamber by thermal evaporation of aluminium [9]. The peel force was measured by a U-form peel test, developed in our laboratory, specially adapted to flexible substrates. After the peel test, the percentage of the metal peeled off was determined by an image processing system on the peeled off films. The evolution of this parameter as a function of the treatment time is shown in Fig. 6. Two distinct regions may be seen.

(1) For treatment times less than 1 s, the percentage of the peeled-off metal decreases from the very beginning of the plasma treatment (for 0.023 s only 4–5% of the metal was peeled off). Then it remained constant with increasing plasma treatment time. This good adhesion has been attributed to the presence of nitrogen [9]. However, no nitrogen was detected by AES at the interface, probably due to desorption of nitrogen under ion bombardment. On the other hand, Gerenser and Goppert-Berarducci [34] have observed the formation of Ag–N–C bonds

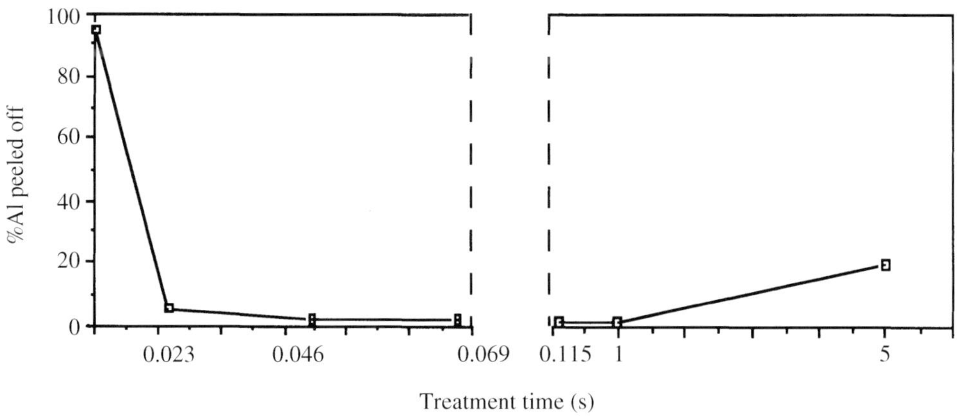

Figure 6. Percent aluminium peeled off as a function of the treatment time in the case of a NH_3 plasma. Experimental conditions: flow rate, 200 sccm; power, 500 W; pressure, 200 Pa; f, 70 kHz.

at low metal coverages by *in situ* XPS analysis in the case of N_2-plasma-treated polystyrene films and metallized by silver. This confirms the important role played by the nitrogen uptake of polymer surfaces.

(2) Above 1 s, the percentage of peeled-off metal increases, which shows an optimum value above which the metal–polymer adhesion starts to degrade. This is likely to be the formation of a weak boundary layer in the polymer because of an overtreatment leading to a cohesive failure in the polypropylene [11]. The depth of the latter was estimated to be about 50 nm for a treatment time of 5 s.

Furthermore, in-depth Auger Electron Spectroscopy (AES) carried out on the metallized ammonia-plasma-treated PP showed that the interface was oxidized, and Al–O–C type of bondings were detected [11]. Other authors have observed Al–O–C bonds by XPS analysis in the case of non-oxygen containing polymers such as poly-*p*-phenylene metallized at 10^{-6} torr [35]. So, in our experimental conditions, one can reason that there exists a high probability that the first monolayers of metal vapour reaching the polymer film are immediately oxidized. In this way, they are more likely to have a structure similar to alumina rather than aluminium, and hence one should consider the interactions of ammonia-plasma-treated PP with alumina.

Contact angle titration measurements carried out on a thin metallic aluminium layer covered by a natural passivating alumina layer (3 nm thick) shows an amphoteric behavior with an acidic tendency. The amphoteric character is shown by an increase in the total work of adhesion for strong acidic and basic solutions. The latter (W_{sl}) is higher for strong basic solutions than for measurements carried out with test liquids with pH $= 1$ to 2 ($\Delta W_{sl} \approx 10$ mJ/m^2) (Fig. 7). This explains its acidic tendency. On the other hand, the contact angle measurements are in good agreement with the classification of alumina, a trivalent oxide, as an acidic oxide in the literature [36], in contrast to bivalent oxides considered to have generally basic properties.

In fact, these results have been confirmed by comparing the adhesion improvement of O_2- and NH_3-plasma-treated polypropylene to aluminium. Oxygen treatment performed in the same experimental set-up used for the ammonia treatment gave rise to a

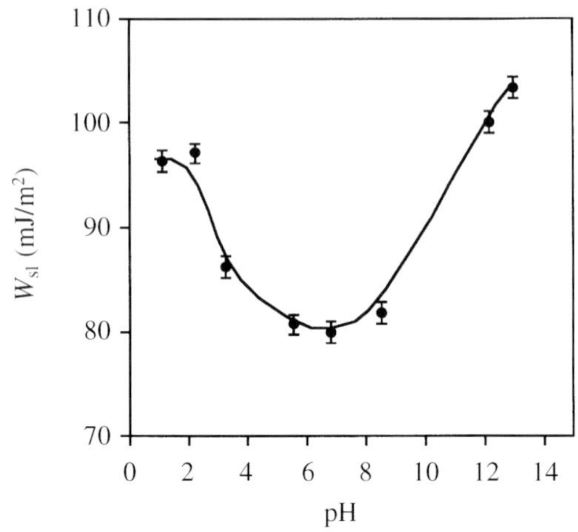

Figure 7. Total work of adhesion calculated by contact angle measurements carried out with different pH test liquids on a thin metallic aluminium layer covered by a natural passivating alumina layer.

predominantly acidic property of the surface [10, 37]. The adhesion of such a surface to aluminium was not improved as much as when ammonia treatment was performed under the same conditions [38].

We shall now discuss the ageing phenomena of the ammonia-treated PP and the role of the helium pretreatment.

In a previous paper, the ageing properties of NH_3-plasma-treated PP were studied in terms of its wettability and its adhesion to aluminium. A variation of 40% in the contact angle was found between a freshly treated PP and an aged sample and a plateau was obtained after 4 days [11]. In the following paragraphs, we shall consider the ageing effects in terms of acid–base interactions by using once again the contact angle titration method. We shall see that this ageing effect is not limited to a change in the wettability, but that the adhesion properties are also affected. Since helium pretreatment limits the variation or change of the wettability and the adhesion properties, this discharge was studied by emission spectroscopy. Finally, the structural modifications induced by this plasma were studied by ToF–SIMS and mechanical tests.

3.5. Study of the ageing phenomenon in terms of acid–base properties

Figure 2 showed that the freshly-prepared ammonia-plasma-treated polypropylene was found by the contact angle titration method to have a clear basic character: the total work of adhesion decreased from 130.7 mJ/m^2 for acid pH values to 120.6 mJ/m^2 for basic ones. The effect of ageing on the ammonia-plasma-treated PP is shown in Fig. 8c, indicating that the basic character is no longer clearly observed, i.e. the maximum value of W_{sl} decreases from 130 to 124 mJ/m^2. The surface shows an amphoteric character, as shown by the increase of W_{sl} for low (pH = 1) and high (pH = 13) pH values. The decrease of the basic character of the surface by ageing can be explained either in terms of neutralization of the groups possessing a basic character such as amine and amides by atmospheric contaminants such as CO_2, or attributed to

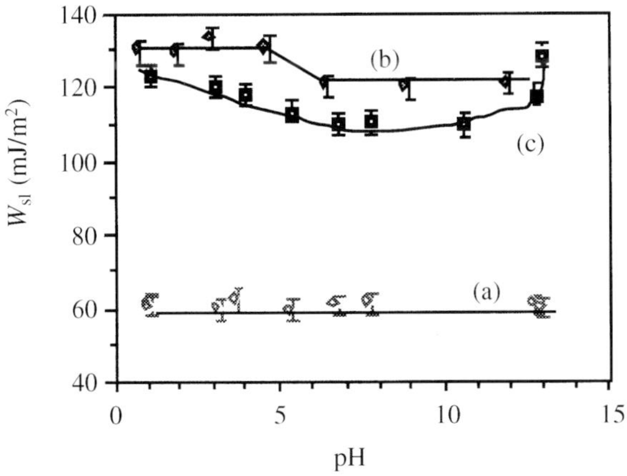

Figure 8. Total work of adhesion W_{sl} of NH_3-plasma-treated PP films (experimental conditions: t, 0.7 s; pressure, 100 Pa; power, 50 W; f, 70 kHz) as a function of pH value of test liquids; (a) nontreated PP; (b) 0.7 s; (c) aged 70 days.

a surface rearrangement through contact with air. The former has been shown to be a possibility by storing the treated films in a CO_2 atmosphere, which indeed showed a neutralization of the basic character by the contact angle titration method [38].

In order to remedy this problem, a helium pretreatment was carried out before the ammonia-plasma treatment; this limited the increase in contact angle to a great extent, i.e. it was less than 10% after 4 days and stayed constant thereafter [11].

3.6. Impact of the ageing on the adhesion properties

Figure 9 shows the effect of the helium-plasma pretreatment on the adhesion of NH_3-plasma-treated PP to aluminium. The metallization was carried out after storing the substrates in ambient air for different periods, and the adhesion strength was determined by the percentage of the metal peeled off after the 180°-U-form peel test was performed. This percentage increased by 30% after an ageing time of 70 days, compared to the sample that was metallized without exposure to air. The percentage decreased with a He pretreatment, reaching a maximum of 10% after 2 days (Fig. 9).

These results therefore show that He-plasma pretreatment, carried out before the ammonia-plasma, gives rise to a stable surface, thus avoiding the change of the surface energy and the degradation of the adhesion properties with ageing time. This stability could be explained by the fact that the polymer surface has been crosslinked by the different species existing in a helium plasma; for this reason the He discharge has been studied by Optical Emission Spectroscopy.

3.7. Study of the helium discharge by optical emission spectroscopy

The He discharge (Fig. 10) spectrum was recorded between 380 and 850 nm in order to characterize the different excited species responsible for the crosslinking effect of the polymer. It showed the lines corresponding to transition between the singlet and triplet He.

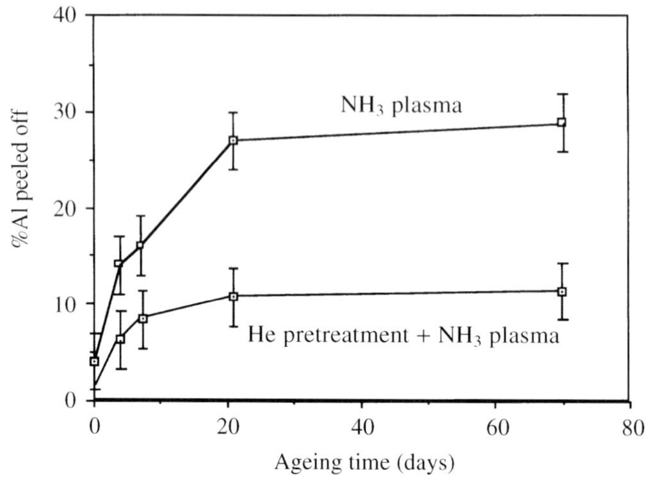

Figure 9. Effect of the helium pretreatment on the percentage of the metal peeled off as a function of the ageing time before metallization. Experimental conditions: pressure, 100 Pa; Q, 100 sccm; power, 5 W; f, 704 kHz; plasma gas, helium.

The triplet metastable, situated at about 19.8 eV, is the most abundant metastable species in a pure He discharge and can play an important role in the crosslinking effect. It may be formed by, in addition to direct electron pumping from the ground state, the relaxation of the singlet metastable He (2^1s) through collisions with electrons [39]. Its decay can be through diffusion to the walls, or most likely due to collisions with species derived from residual air in the reactor chamber. De Benedictis and Dilecce [40] have shown that N_2 and O_2 are very efficient quenchers of He (2^3s) and that the addition of N_2 to a He discharge reduces the He (2^3s) density and raises the quenching rates [40]. The process responsible for such strong quenching phenomenon is the Penning ionization of N_2 [41]:

$$\mathrm{He_M} + \mathrm{N_2}\ (\mathrm{X}^1\Sigma_g^+) \rightarrow \mathrm{He}\ (1^1s) + \mathrm{N_2^+}\ (\mathrm{B}^2\Sigma_u^+) + e^-. \tag{3}$$

The characteristic lines of atomic oxygen ($\lambda = 777.1$ nm and $\lambda = 844.6$ nm) may be due to charge transfer between He^+ species and O_2 molecules forming excited atomic oxygen according to reaction (4):

$$\mathrm{He^+} + \mathrm{O_2} \rightarrow \mathrm{He} + \mathrm{O^+} + \mathrm{O}\ (3^5p). \tag{4}$$

The other possibility for the consumption of helium ions is by ion bombardment of the surface of the polymer, which is always at a negative potential with respect to the plasma potential, giving rise to the formation of free radicals on the surface. It is important to note that helium ions have been detected by Optical Emission Spectroscopy, because the first excited energy level of the helium ion He^+ ($2s$) is situated at 40.8 eV [42], which is much higher than the energy of the electrons present in the discharge.

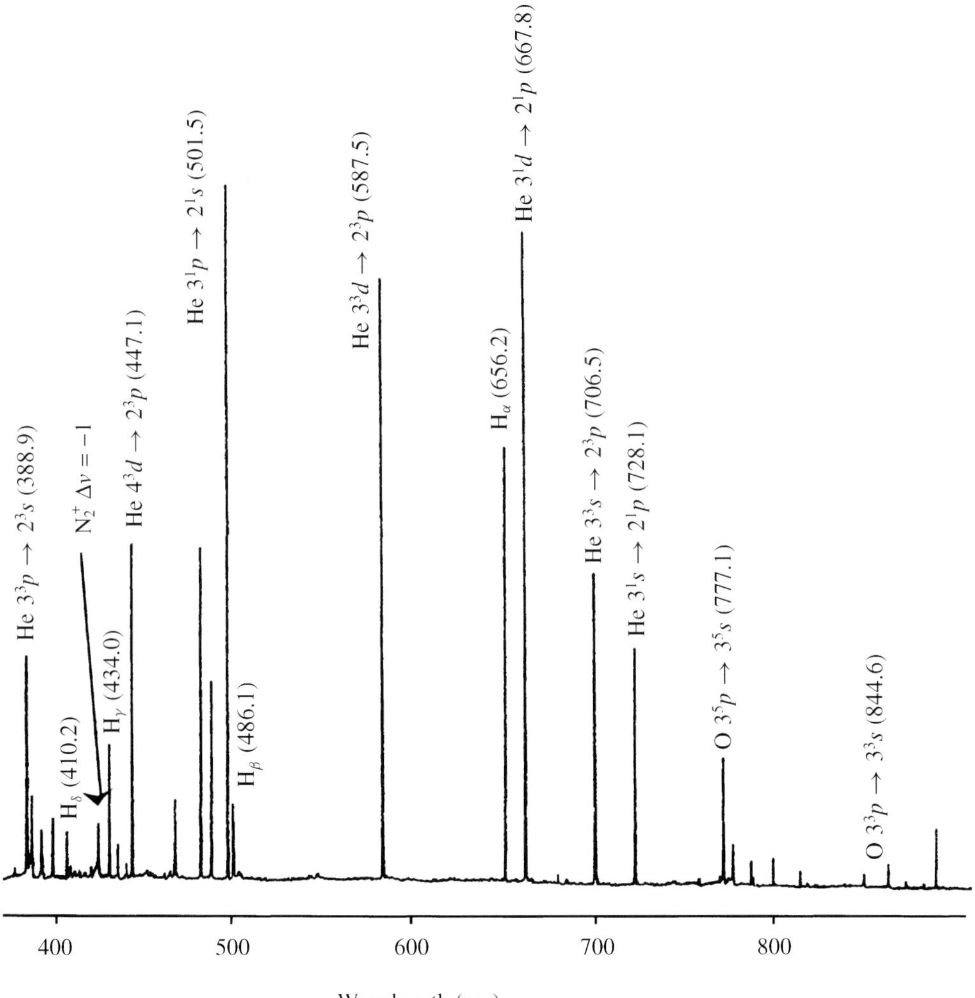

Figure 10. Emission spectrum of a He plasma in the 380–850 nm region. Experimental conditions: pressure, 100 Pa; power, 5 W; flow rate, 100 sccm.

The emissions of H_α (656.2 nm), H_β (486.1 nm), H_γ (434.0 nm) and H_δ (410.2 nm) clearly demonstrate the chain scission of the polymer surface; these results are in good agreement with mass spectrometric analysis of the effluent gases, which revealed that in the case of a helium discharge only hydrogen was generated [11]. This evidence about the dehydrogenation process of the polymer could explain the resulting crosslinking effect of this plasma. This could be caused by ions or the metastable He (2^3s) if it was not de-excited in the boundary layer. The initiation reactions can also be caused by extremely energetic vacuum-ultraviolet light (VUV). The absorption coefficient of PE, which is representative of polyaliphatics, is measured and varies in the range of 10^5-10^6 cm^{-1} atm^{-1} [43]. Some of the radiation may be re-radiated towards the polymer surface; VUV radiation can propagate through the boundary layer with little attenuation. An He plasma emits also very strongly at 59 nm ($2^1p \rightarrow 1^1s$). These photons have more than enough energy (59 nm = 2023 kJ/mol) to break C–C (345 kJ/mol) and C–H bonds (412 kJ/mol), and are therefore very effective in creating free radicals in the polymer. Once free radicals are created, they can react with other radicals on the surface in a pure He discharge to form a crosslinked surface, or react with air impurities to chemically modify the surface. The latter reaction is a competitive one and should be limited if one desires to crosslink the polymer.

Egitto and Matienzo [44] revealed the important role played by VUV photons and illustrated why different plasma gases were required to treat different polymers. Thus, they showed that the reason why an O_2 plasma did not efficiently chemically modify PTFE was that the 130.5 nm emission from an O_2 plasma was not strongly absorbed by PTFE and could not lead to its decomposition. On the other hand, an H_2-containing plasma radiates at 121.5 nm (the Lyman α radiation) and the corresponding photons can be absorbed by the surface of the PTFE, leading to initial breaking of the surface bonds.

3.8. Static SIMS study of the structural modifications. Effect of branching, crosslinking and unsaturation in the case of a helium treatment

In this study we have tried to point out the crosslinking effect of a He plasma, since this can increase the cohesive strength of polymer films and avoid problems associated with weak boundary layers [45]. For this study, high density PE (HDPE), which is known to readily crosslink under plasma conditions over thicknesses of a few micrometres was chosen. In the case of PP, the typical thicknesses of crosslinked surface layer reported were only 5–30 nm as determined by the gel content after solvent extraction [16, 46, 47]. That is why it can be even more difficult to demonstrate the crosslinking effect in the case of this polymer by ToF–SIMS.

It is possible to find in the literature quite different SIMS structural parameters calculated from the SIMS relative intensities. They are based on the comparison of unsaturated and saturated hydrocarbon homopolymer spectra. For branching (but also crosslinking, since it is quite impossible to separate both effects), the following SIMS structural parameters were proposed. They are all based on the pretext that the relative intensity increases for high mass hydrocarbon clusters because of the higher stability of ions with branched structures. They differ in the way these high mass clusters are involved and in the way they are normalized:

(1) C8/C2 (ratio of maximum peaks of clusters C8 and C2) [48];

(2) ΣC8/ΣC2 (ratio of the sum of the normalized intensities for each cluster) [49];

(3) ΣC(6–8) (ratio of the sum of all the peaks from clusters C6 to C8 to the sum of all the peaks from clusters C1 to C8) [50];

(4) ΣC8 (sum of the relative intensities of all the peaks from cluster C8) [51].

From our own measurements on reference samples (HDPE was compared to LDPE), it appeared that ΣC8 is the best one since it combines the least arbitrary reference and the highest sensitivity (the differences observed in similar calculations for the clusters C6 and C7 are smaller). The increase in the value of ΣC8 and also ΣC8/ΣC2 seen in Fig. 11 may thus be related to the development of crosslinking or branching at the surface of HDPE. However, one must be very careful. It was noted that there are active species other than helium species in our helium-plasma discharge. The active species related to the presence of residual air in the discharge can lead to various effects. Among these effects, chain breaking and introduction of heteroelements must be discussed. They can significantly influence the secondary emission of the hydrocarbon ions involved in the SIMS structural pointers. We must conclude that our results, while

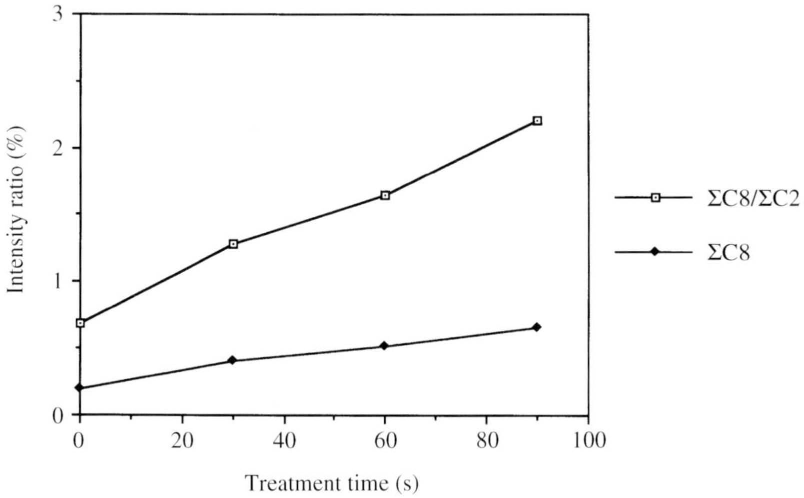

Figure 11. Variation of the $\Sigma C8/\Sigma C2$ ratio and the relative intensity of the $\Sigma C8$ as a function of the treatment time of HDPE in a helium plasma. Experimental conditions: pressure, 100 Pa; Q, 100 sccm; power, 5 W; f, 70 kHz.

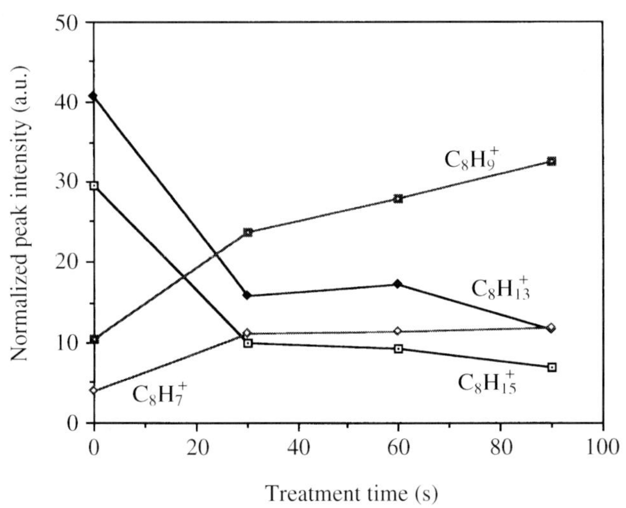

Figure 12. Evolution of some positive SSIMS hydrogen-containing peaks inside the C8 clusters as a function of the treatment time. Experimental conditions: pressure, 100 Pa; flow rate, 100 sccm; power, 5 W; f, 70 kHz.

compatible with the occurrence of crosslinking in the helium plasma, do not show a one-to-one correspondence (due to possible chain breakings and introduction of heteroelements).

On the other hand, other features must be noted that were not observed in the HDPE/LDPE comparison. There is a shift in the intensities of the C8 cluster peaks towards ions of lower masses, i.e. the less saturated fragments (cf. the relative intensity of the $C_8H_7^+$ ion as compared to that of the $C_8H_{15}^+$ ion shown in Fig. 12). Briggs [49] has also proposed that the distribution of the intensities within the C8 cluster could be used to fingerprint the unsaturation. This result is consistent with the fact that the surface of the polymer could be crosslinked during the He pretreatment, since the active sites are created on the surface by a dehydrogenation process confirmed by mass spectroscopy. On the other hand, the C2 cluster intensities shifted towards ions of higher masses in contrast to what was observed in the C8 cluster. The appearance of low-mass saturated fragments is not necessarily in contradiction to the former results, since this phenomenon could be attributed to the occurrence of branching in the polymer backbone. Therefore the comparative results of the plasma-treated samples with the HDPE/LDPE case point out that it is far more difficult to demonstrate the occurrence of crosslinking in the case of the former using ToF–SIMS. For this reason we attempted to verify the crosslinking effect of the helium plasma by carrying out indirect measurements as explained below.

3.9. Study of the mechanical properties of plasma-treated PP. Indirect study of the crosslinking effect induced by a helium discharge

As explained in Section 3.4, the adhesion properties of Al–ammonia-plasma-treated PP degraded when it was overtreated, probably due to the formation of a weak boundary layer. The latter could affect the mechanical properties of plasma-treated non-metallized polymers in different atmospheres (NH_3, He, CF_4).

In Fig. 13, the elongation-at-break of polypropylene is shown as a function of the treatment time. The role of the different plasmas in strengthening the topmost layers

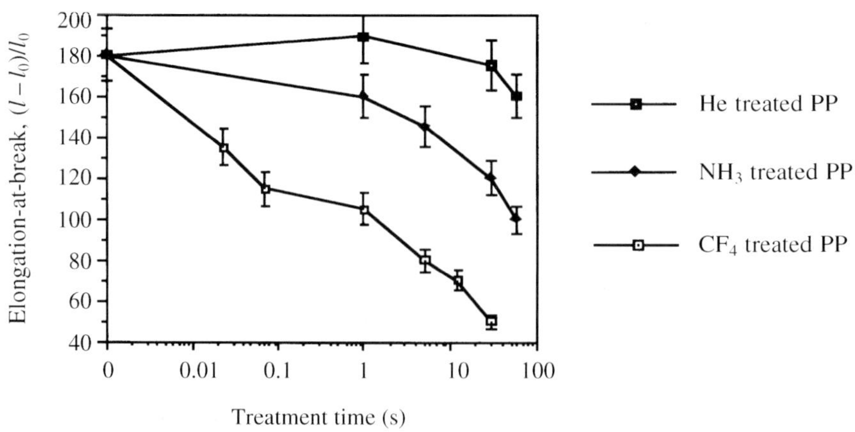

Figure 13. Elongation-at-break of PP treated in different plasma atmospheres as a function of the treatment time.

at the surface has been demonstrated based on the elongation-at-break measurement as well as the tensile strength. We should emphasize that the plasma treatment does not modify the stress–strain behaviour of the polymer, since this is an intrinsic property of the polymer bulk (Young's modulus is the same for the different plasma treatments). Nevertheless, the position of the break point on the stress–strain curve [11] depends on the gas. Moreover, by increasing the treatment time, the break point is shifted to the left, i.e. to lower tensile strength and elongation; in our case, the crack starts at the surface and then propagates into the bulk of the polymer when the latter has been weakened. The presence of a weak boundary layer gives rise to a premature break point even though the trends of the stress–strain curves remain unchanged.

In the case of a NH_3-treated PP, the elongation at break point decreases from 180% to 160% for a treatment time of 1 s, and reaches 145% for an overtreated sample ($t = 5$ s). On the contrary, the elongation properties of helium-treated PP remain relatively unchanged with treatment time, and are even improved for a treatment time of 1 s, which supports the hypothesis of a cohesive strengthening of the polymeric surface by the formation of a thin crosslinked layer that mechanically stabilizes the surface.

Since fluorinated monomers (e.g. CF_4, SF_6) are extensively used for the etching and surface modification of a large number of polymers as well as other materials, it was interesting to look at the influence of such plasmas on the mechanical properties of the polymers. The results show that in a CF_4 plasma, the PP undergoes degradation; whereas in inert gases only a few polymeric cleavage products were observed by mass spectrometry. Therefore the polymer is more crosslinked in the case of the latter. The degradation is enhanced by traces of oxygen present in the plasma reactor, which is well known to etch polymeric surfaces. The CF_4 plasma favours formation of cracks at a polymeric surface, which causes a loss of tensile strength at the break point. Besides, by using Rutherford Backscattering Spectroscopy (RBS), several authors have detected fluorination depths of different polymers treated by a CF_4 plasma of up to 100 nm [52, 53]. The alteration of mechanical properties of PP is due to the diffusion of fluorine through the surface layer, which is why a drastic decrease of the mechanical properties of CF_4-treated PP is observed with treatment time (Fig. 13). The failure mode of the peeled metallized plasma-treated films observed by SEM analysis was generally a cohesive one in the polymer, since the interfacial strength was usually higher than the cohesive strength of the polymer [11, 54].

Another indirect indication of the formation of a reinforced surface layer induced by the crosslinking effect caused by a helium plasma was the study of the in-depth Auger profiles of the metallized PP films [11, 54]. In the case of an ammonia-plasma treatment with a helium pretreatment, a sharp interface was observed showing almost no intermixing of Al with the carbon of the reinforced polymer substrate. In contrast, without a He pretreatment the interface seems to be much more diffuse, revealing an intermixing of Al in the fragmented zones of the overtreated PP films [11]. In the case of the CF_4-plasma-treated PP, this interface is much broader than that found in the case of an ammonia-plasma treatment. This indicates that PP is weakened over a greater depth in the former case, probably due to the diffusion of fluorine atoms or ions into the polymer [11, 54, 55], which in fact has been detected throughout the broad interfacial region.

4. CONCLUSIONS

The modifications of the ammonia-plasma-pretreated PP have been studied by a range of complementary methods of analysis: Contact Angle Titration method and XPS in conjunction with molecular probe and ToF–SIMS. Three different features have been observed as function of the treatment time:

(1) For treatment times less than 1 s, the NH_3-treated PP exhibits a dominant basic character; this is compatible with the amphoteric character with acidic tendency of alumina, which constitutes the first monolayer that arrives at the surface when aluminium is evaporated under a vacuum of 10^{-3} Pa.

(2) Above 1 s, two synergetic phenomena contribute to the decrease of adhesion: the basic groups are no longer accessible at the topmost surface and the mechanical properties are degraded by the formation of a weak boundary layer.

(3) ToF–SIMS performed on NH_3-treated PP confirmed the decrease in the general basic character of the surface; this was particularly exhibited by the decrease in intensity of CN^- signal, probably due to the competition between functionalization and ablation processes which take place in the plasma.

The ageing of ammonia-plasma-treated PP leads to a degradation of the adhesion to aluminium, which has been attributed to a decrease in the dominant basic character of NH_3-treated PP films. The use of an He pretreatment before NH_3 treatment stabilized the surface and limited the degradation. This was attributed to greater crosslinking of the surface induced by the different species present in a He discharge, i.e. ions, metastables as well as VUV radiation. ToF–SIMS results on He-plasma-treated HDPE can be interpreted in terms of crosslinking development as a function of the treatment time. However, there is not a one-to-one correspondence between SIMS structural parameters and crosslinking, since possible chain breaking and the introduction of heteroelements during the He plasma treatment can influence the secondary ion emission of the SIMS peaks involved in the parameter calculations. For this reason, indirect measurement such as elongation-at-break were performed on the He-treated samples. The results showed relatively unchanged elongation properties compared to those observed with the NH_3-treated ones, which supports the hypothesis of a cohesive strengthening of the polymeric surface by the formation of a thin crosslinked layer that mechanically stabilizes the surface.

REFERENCES

1. M. Strobel, C. S. Lyons and K. L. Mittal (Eds), *Plasma Surface Modification of Polymers: Relevance to Adhesion*. VSP, Zeist, The Netherlands (1994).
2. F. Arefi, V. André, P. Montazer-Rahmati and J. Amouroux, *Pure Appl. Chem.* **64**, 715–723 (1992).
3. J. M. Pochan, L. J. Gerenser and J. F. Elman, *Polymer* **27**, 1058–1062 (1986).

4. S. Saphieha, M. Vereault, J. E. Klemberg-Saphieha, E. Sacher and M. R. Wertheimer, *Appl. Surface Sci.* **44**, 165–169 (1990).
5. P. Montazer-Rahmati, F. Arefi and J. Amouroux, *Surf. Coating Technol.* **45**, 369 (1991).
6. Y. Khairallah, F. Arefi-Khonsari and J. Amouroux, *Pure Appl. Chem.* **66**, 1353 (1994).
7. Y. Khairallah, F. Arefi-Khonsari, J. Amouroux, D. Léonard and P. Bertrand, *J. Adhesion Sci. Technol.* **8**, 363–381 (1994).
8. M. Tatoulian, F. Arefi-Khonsari, N. Shahidzadeh-Ahmadi and J. Amouroux, *Int. J. Adhesion Adhesives* **15**, 177–184 (1995).
9. F. Arefi, M. Tatoulian, V. André, J. Amouroux and G. Lorang, in: *Metallized Plastics 3: Fundamental and Applied Aspects*, K. L. Mittal (Ed.), pp. 243–256. Plenum Press, New York (1992).
10. N. Shahidzadeh-Ahmadi, F. Arefi-Khonsari and J. Amouroux, *J. Mater. Chem.* **5** (2), 229–236 (1995).
11. M. Tatoulian, F. Arefi-Khonsari, I. Mabille-Rouger, J. Amouroux, M. Georgiu and D. Bouchier, *J. Adhesion Sci. Technol.* **9**, 923–934 (1995).
12. M. Morra, E. Occhiello and F. Garbassi, in: *Polymer-Solid Interfaces*, J. Pireaux, P. Bertrand and J. L. Bredas (Eds), pp. 407–428. IOP Publishing (1992).
13. H. Schonborn and R. H. Hansen, *J. Appl. Polym. Sci.* **11**, 1461 (1967).
14. S. L. Vogel and H. Schonhorn, *J. Appl. Polym. Sci.* **23**, 495 (1979).
15. Yaoguang Yao, Xeushu Liu and Yu Fen Zhu, *J. Adhesion Sci. Technol.* **7**, 63–75 (1993).
16. S. Wu, *Polymer Interface and Adhesion*. Marcel Dekker, New York (1982).
17. M. M. Chehimi, *J. Mater. Sci. Lett.* **10**, 908 (1991).
18. M. M. Chehimi, J. F. Watts, W. K. Eldred, K. Fraoua and M. Simon, *J. Mater. Chem.* **4**, 305 (1994).
19. M. M. Chehimi, J. F. Watts, S. N. Jenkins and J. E. Castle, *J. Mater. Chem.* **2**, 269 (1992).
20. J. F. Watts and M. M. Chehimi, *Int. J. Adhesion Adhesives* **15**, 91–94 (1995).
21. P. Montazer-Rahmati, F. Arefi, R. Borrin, A. Delacroix and J. Amouroux, *Bull. Soc. Chim. France* **5**, 811–816 (1988).
22. B. W. Scheuler, *Microsc. Microanalys. Microstruct.* **3**, 119–139 (1992).
23. D. Briggs, in: *Practical Surface Analysis*, 2nd edn, Vol. 2, *Ion and Neutral Spectroscopy*, D. Briggs and M. P. Seah (Eds), pp. 367–423. John Wiley, Chichester (1992).
24. A. Benninghoven, *Surf. Sci.* **300**, 246–260 (1994).
25. B. Hagenhoff and D. van Leyen, *J. Vac. Sci. Technol. A.* **7**, 3056–3064 (1989).
26. K. J. Hüttinger, S. Höhmann-Wien and G. Krekel, *J. Adhesion Sci. Technol.* **6**, 317 (1992).
27. K. J. Hüttinger, S. Höhmann-Wien and M. Sciferling, *Carbon* **29**, 449 (1991).
28. G. M. Whitesides and P. E. Laibinis, *Langmuir* **6**, 87 (1991).
29. F. M. Fowkes, in: *Acid–Base Interactions: Relevance to Adhesion*, K. L. Mittal and H. R. Anderson, Jr. (Eds), pp. 93–115. VSP, Utrecht (1991).
30. E. Occhiello, M. Morra, G. Morini, F. Garbassi and H. Humphry, *J. Appl. Polym. Sci. Symp.* **46**, 527 (1990).
31. F. Clouet, M. K. Shi, R. Prat, Y. Holl, P. Marie, D. Leonard, Y. de Puydt, P. Bertrand, J. L. Dewez and A. Doren, *J. Adhesion Sci. Technol.* **8**, 329–361 (1994).
32. N. Shahidzadeh, M. M. Chehimi, F. Arefi-Khonsari, J. Amouroux and M. Delamar, *Plasmas & Polymers* **1**, 27–45 (1996).
33. M. M. Chehimi, E. Pigois-Landureau, M. Delamar, J. F. Watts, S. N. Jenkins and E. M. Gibson, *Bull. Soc. Chim. France* **129**, 137 (1992).
34. L. J. Geresner and K. E. Goppert-Berarducci, in: *Metallized Plastics 3: Fundamental and Applied Aspects*, K. L. Mittal (Ed.), pp. 163–178. Plenum Press, New York (1992).
35. T. P. Nguyen, H. Ehaik, S. Lefrant and G. Leising, in: *Polymer–Solid Interfaces*, J. J. Pireaux, P. Bertrand and J. L. Bredas (Eds), pp. 253–256. IOP Publishing (1992).
36. J. Bernard, *Oxidation of Metals, I, Fundamental Processes*. Gauthier–Villars & C. Editor, Paris (1962).
37. N. Shahidzadeh-Ahamadi, M. M. Chehimi, F. Arefi-Khonsari, N. Foulon-Belkacemi, J. Amouroux and M. Delamar, *Colloids and Surfaces A: Physicochem. Eng. Aspects* **105**, 277–289 (1995).
38. N. Shahidzadeh, P. & M. Curie Thesis (Jan. 1996).
39. R. Deloche, R. Monchicourt, M. Cheret and F. Lambert, *Phys. A* **13**, 1140 (1975).

40. S. De Benedictis and G. Dilecce, *Plasma Sources Sci. Technol.* **4**, 212–223 (1995).
41. W. Lindinger, A. L. Schemeltekopf and F. C. Fehsenfeld, *J. Chem. Phys.* **61**, 2690 (1974).
42. P. Labori, J. M. Rocard and J. A. Rees, *Section Efficace Electronique, Coefficient Macroscopique*, Vol. 1. Dunod Editions (1968).
43. E. M. Liston, in: *Polymer–Solid Interfaces*, J. Pireaux, P. Bertrand and J. L. Bredas (Eds), pp. 424–441. IOP Publishing (1992).
44. F. D. Egitto and L. J. Matienzo, *Polym. Degrad. Stab.* **30**, 293 (1990).
45. H. Schonhorn and R. H. Hansen, *J. Appl. Polym. Sci.* **18**, 235 (1974).
46. C. A. L. Westerdahl, J. R. Hall, E. C. Schramn and D. W. Levi, *J. Colloid Interface Sci.* **47**, 610 (1974).
47. M. Hudis and L. E. Prescott, *J. Polym. Sci.* **B10**, 179 (1972).
48. W. J. van Ooij and R. H. G. Brinkhuis, *Surf. Interface Anal.* **11**, 430–440 (1988).
49. D. Briggs, *Surf. Interface Anal.* **15**, 734 (1990).
50. L. Lianos, C. Quet and T. M. Duc, *Surf. Interface Anal.* **11**, 430–440 (1988).
51. A. Delcorte, L. T. Weng and P. Bertrand, *Nucl. Instrum. Meth. Phys. Res. B.* **100**, 213–216 (1995).
52. G. A. Takacs, V. Vukanovic, D. Tracy, J. K. Chen, F. D. Egitto, L. J. Matienzo and F. Emmi, *Polym. Degrad. Stab.* **40**, 73–81 (1993).
53. Y. Khairallah, F. Arefi and J. Amouroux, *Thin Solid Films* **241**, 195–300 (1994).
54. M. Tatoulian, F. Arefi-Khonsari, I. Rouger-Mabille and J. Amouroux, *Proceeding of IUPAC, ISPC 12*, Minneapolis, August 21–25, 1995, pp. 45–50.
55. Y. Du and J. A. Gardella, Jr., *J. Vac. Sci. Technol. A* **13**, 1907–1912 (1995).

Mittal Festschrift, pp. 355–371
W. J. Van Ooij and H. R. Anderson, Jr. (Eds)

Novel metal pretreatments for improved adhesion of paints using plasma methods

W. J. VAN OOIJ,* K. D. CONNERS and P. J. BARTO

University of Cincinnati, Department of Materials Science and Engineering, Cincinnati, OH 45221-0012, USA

Abstract—Cleaning of a Cold-Rolled Steel (CRS) substrate by a plasma of a non-reactive gas and the deposition of a plasma-polymerized film (PP-film) of hexamethyldisiloxane (HMDS) as a pretreatment system to improve the corrosion of painted metals was investigated. Electrochemical Impedance Spectroscopy (EIS) was used to characterize the degree of corrosion protection achieved by this pretreatment system. An accelerated corrosion test was conducted to verify the EIS results. The EIS data indicated that the degree of corrosion protection achieved with this pretreatment system was comparable to that found using zinc phosphating as a pretreatment. The results of the accelerated environmental tests and the EIS results show that a plasma cleaning is required to achieve satisfactory corrosion protection of plasma treated/E-coated CRS.

Keywords: Plasma cleaning; plasma-polymerized films; pretreatment; electrochemical impedance spectroscopy; accelerated corrosion testing.

1. INTRODUCTION

Polymer coatings and paints are used extensively to protect metals from corrosion. The primary property that gives these coatings their corrosion protection capability is the ability to inhibit the diffusion of water and oxygen from the atmosphere to the metal [1]. However, all coatings allow a finite amount of water and oxygen to diffuse through to the metal surface [2]. When this occurs, corrosion is initiated. For corrosion to proceed, new metal surfaces must be exposed to the water and oxygen. This exposure takes place upon delamination of the coating from the surface due to stresses produced at the coating/metal interface as a result of corrosion product build up. Poor adhesion of the coating to the metal allows the spread of corrosion to proceed at an accelerated rate compared to a coating with good adhesion. Therefore, since it is not possible to completely inhibit the diffusion of water and oxygen to the metal surface, it is imperative that there be good adhesion of the coating to the metal.

*To whom correspondence should be addressed. E-mail: wvanooij@uceng.uc.edu

If the adhesion is good, especially in wet conditions, corrosion will not propagate after initiation and the metal is essentially protected.

The primary goal of the International Consortium for the Surface Engineering of Materials (ICSEM) at the University of Cincinnati is the development of novel pretreatments that are engineered to improve the adhesion of polymer coatings to a variety of materials. Among the treatments being investigated is the use of plasma processing for the pretreatment of cold-rolled steel (CRS) to improve the adhesion of a cathodic electrocoat primer and other paints. This research is being conducted in an attempt to replace the current technology of phosphating and chromating with environmentally compliant processes. The United States Environmental Protection Agency (EPA) has mandated that the amount of toxic materials used in industrial processes be reduced sharply. Processes carried out in a vacuum, i.e. plasma processes, using non-toxic gases, produce little or no toxic waste by-products requiring disposal. For that reason, these processes are considered environmentally friendly, and therefore may be suitable substitutes for phosphating and chromating. For a pretreatment to be considered as a replacement for current technology, several criteria must be met. Of these, it seems reasonable that the most important factor to be considered deals with the corrosion protection aspect. It is imperative that the same degree of corrosion protection afforded to the phosphated or chromated material be maintained by any pretreatment that replaces the current standard.

Plasma-polymerized films, also known as plasma-deposited polymers, or more simply, plasma polymers, are known to have a high degree of crosslinking, and excellent adhesion to a variety of substrates [3]. These two properties suggest that these films would provide excellent corrosion protection for metal substrates. The authors have shown in previous work that plasma polymerized films (PP-films) of hexamethyldisiloxane (HMDS) provide a degree of corrosion protection for cold-rolled steel (CRS) for short periods of time in an aggressive environment [4, 5].

It has also been shown that plasma polymers will adhere to a variety of substrates [6]. This added property suggests that these films can be used as a primer in a complete paint system. As is normally the case, the cleanliness of the metal substrate is likely to play a vital role in the adhesion of the plasma polymer film to the metal substrate.

A plasma treatment using a non-film forming gas, such as oxygen, has been found to be effective in removing organic carbon contamination from the surface of CRS [4]. The carbon contamination is thought to be a major contributor in the lack of adhesion between polymer coatings and metals [7]. Unfortunately, this oxygen treatment has the effect of producing a thickened oxide layer [4]. To compensate for this effect, i.e. reduce the thickness of this artificially produced oxide layer, the oxygen cleaning step was in this study followed by plasma cleaning with a mixture of Ar/H_2 [4]. Therefore, if the metal substrate is plasma cleaned to remove these contaminants, and a plasma–polymer film is subsequently deposited in the same reactor, this film can act as a corrosion protective coating, and also as a primer for subsequent coatings. Thus, the entire system of the plasma-cleaned surface, covered by a plasma polymerized coating and an E-coat should provide adequate adhesion and corrosion protection characteristics.

These films can be characterized for their chemical properties by most of the techniques commonly used for polymers and polymer coatings. These would include Infrared Spectroscopy (FTIR), X-ray Photoelectron Spectroscopy (XPS), Time-of-Flight Secondary Ion Mass Spectroscopy (TOFSIMS), and Nuclear Magnetic Resonance (NMR). The morphology of the films can easily be investigated using Scanning Electron Microscopy (SEM), Scanning Tunneling Microscopy (STM), or Atomic Force Microscopy (AFM). Other techniques, such as Cyclic Voltametry, have also proved to be useful to investigate certain diffusion characteristics of plasma polymer films [7].

In this work, the primary objective was the investigation of the degree of corrosion protection that could be achieved using the plasma processing steps described above. Electrochemical Impedance Spectroscopy (EIS) has been used extensively over the past decade to investigate the corrosion protection capabilities of polymer coatings on metals [8–11]. Historically this technique, which measures the current response to a fluctuating applied potential as a function of the AC frequency, has been used to measure the deterioration of a polymer coating with time [12]. This is best achieved by comparing data from a series of tests taken over an extended period of time. The deterioration of the coating is measured by the change in the elements of an Equivalent Circuit Model (ECM), which emulates the performance of the coating/metal system when a potential is applied. The difficulty with this type of analysis is the design of the ECM. If the elements of the ECM do not have a physical similarity to some property of the coating/metal system being investigated, the analysis will produce incorrect, and often misleading results.

The arrangement of the ECM is best determined using a graphical analysis of the EIS data in the form of a Bode plot [13]. In this format, the impedance and the phase angle of the coating/metal system are plotted as a function of the frequency of the applied potential. Changes in the slope of the impedance curve, and local maxima and minima in the phase angle curve indicate the existence of several simple electrical elements or diffusion phenomena, consisting most commonly of a resistor in parallel with a capacitor. This pair of elements is referred to as a ‘time constant’ in the ECM. The particular arrangement of these time constants will vary according to the system being investigated.

In addition to the deterioration of a coating system, the pre-assessment of a coating can also be investigated using EIS. This is again best achieved by fitting the collected EIS data to the performance of the ECM. Comparison of the EIS analysis with data from accelerated environmental corrosion performance tests is commonly performed and used for verification of the conclusions drawn from the EIS analysis.

As stated above, EIS is more commonly used to investigate coatings and paints on metals. However, if the parameters of the coatings are held constant, any differences in the performance of the prepared samples can be attributed to the metal pretreatment used before the application of the polymer coating.

In this work, EIS was used to evaluate the corrosion protection performance of a plasma cleaning/plasma polymer primer system. In one portion of the work, the conditions of the plasma polymerization were kept constant, and the plasma cleaning conditions were varied. In the other portion of the work, the plasma cleaning conditions were kept constant, and the plasma film properties were varied by introducing a small amount of oxygen carrier gas during the deposition, and also using a post-treatment by a plasma of oxygen. The objective of this second set of experiments was to enhance the barrier properties of the plasma polymer film by introducing a thin layer of SiO_x in the polymer film. The post-treatment by oxygen was performed in an attempt to increase the adhesion of the paint to the plasma polymer.

Accelerated corrosion tests were also performed to evaluate all samples in this research. Both EIS and accelerated corrosion tests were performed after painting with a cathodically deposited electrocoat system. Additionally, correlations were sought between these results and those obtained from EIS experiments.

2. EXPERIMENTAL

2.1. Materials and equipment

The metal substrate was automotive grade low carbon type 1010 CRS sheet of 0.5 mm thickness obtained from Armco Research and Technology, Middletown, Ohio. The non-film-forming gases used for the cleaning step were research grade argon, hydrogen, and oxygen acquired from Wright Bros., Cincinnati, Ohio. The HMDS was 99.7% pure, obtained from Aldrich Chemical Co., Milwaukee, Wisconsin, and was used without further purification. The cathodically electrodeposited primer, referred to as the 'cathodic E-coat' was type ED-5000 and was applied at PPG Industries, Allison Park, Pennsylvania. The phosphate control panel was a CRS panel phosphated in a modern fine-grained automotive type manganese-containing zinc phosphate bath and was obtained from ACT in Hillsdale, Michigan.

The CRS panels were first cut into 10×15 cm size. They were then ultrasonically cleaned in acetone for 20 min, blown dry with clean air and placed in the plasma reactor. This reactor was custom-built, used a DC power source, and had a triple parallel plate arrangement for sample placement in the reactor chamber. The sample substrate acted as the cathode, and was placed between two 22×22 cm cathodes made of bright-annealed stainless steel of type 434. In this arrangement, both sides of the sample could be cleaned, and deposition of the plasma polymer could take place on both sides of the sample simultaneously. The chamber itself was a glass bell jar whose diameter was 18 cm, with a center height of 18 cm. The glass bell jar sat on an 18 cm diameter stainless steel feed-through collar. The feed-through collar, in turn, sat on a stainless steel base plate. The vacuum pumping system consisted of a Varian DWP-250 diffusion pump, and a Welch 1397 mechanical backing pump. The gases were introduced into the chamber through 4 Brooks 5850-C mass flow controllers. The power supply was capable of delivering up to 1500 V at 200 mA.

2.2. Sample preparation

2.2.1. Samples with varied plasma cleaning. The samples with which the plasma cleaning was varied were cleaned in a plasma of oxygen, and/or a plasma of an argon/hydrogen mixture. The details of the plasma cleaning processes are given in Table 1. The plasma-polymerized HMDS was deposited for 10 min at a pressure of 250 mtorr, a flow rate of 2.5 sccm (standard cubic centimeters per minute) and a current of 24 mA at 1100 V.

2.2.2. Samples with varied plasma polymerization deposition conditions. Steel substrates of 17 × 22 cm in size were plasma cleaned similarly to sample G of the varied plasma cleaning set. The panels were then placed in the reactor chamber and held under vacuum. Plasma cleaning was carried out in two steps. First, the samples were cleaned using oxygen (O_2) for 10 min at a pressure of 200 mtorr, a power of 24 W, and a flow rate of 7.4 sccm. A 1 : 2 mixture of argon and hydrogen (Ar/H_2) was used in the next cleaning step for 10 min at a pressure of 125 mtorr, a power of 18 W, and a flow rate in the ratio 2/4 sccm, respectively. CRS panels were prepared as follows: one with a PP-FILM of HMDS (H), one with a PP-FILM of HMDS with an O_2 post-treatment (J), one with a plasma-polymerized laminate structure (PP-LAM) consisting of PP-film of HMDS layer + SiO_x layer + PP-film of HMDS layer (K), and PP-LAM with an O_2 post-treatment (M). The SiO_x layer was introduced into the PP-film by admitting O_2 into the reactor during part of the deposition process. The exact conditions of the deposition step for this experiment are given in Table 2. It is to be noted that in this experiment the PP-films were deposited at lower pressure and lower power conditions than in the cleaning experiment of Table 1. Lowering both pressure and the power in DC polymerization of HMDS results in films which are softer and have a lower crosslink density [3]. These conditions also effectively reduce the deposition rate, resulting in thinner PP-FILMS.

Table 1.
Conditions for plasma cleaning

Sample	Gas	Flow rate (sccm)	Pressure (mtorr)	Power (mA/V)	Time (min)
CRS	—	—	—	—	—
Phos. CRS	—	—	—	—	—
A	—	—	—	—	—
B	O_2	7.4	200	50/450	5
C	Ar/H_2	2/4	200	50/400	5
D	O_2	7.4	200	50/450	20
	Ar/H_2	2/4	200	50/375	10
E	O_2	7.4	200	50/450	10
F	O_2	7.4	200	50/450	10
	Ar/H_2	2/4	200	50/375	5
G	O_2	7.4	200	50/450	10
	Ar/H_2	2/4	200	50/375	10

Table 2.
Conditions for plasma polymer deposition experiment

Sample	Monomer	Carrier gas	Pressure (mtorr)	Power (W)	Flow rate (sccm)	Process step	Time (min)
H	HMDS	—	125	20	2.5	HMDS deposition	10
J	HMDS	—	125	20	2.5	HMDS deposition	10
	—	O_2	200	22	7.4	O_2 post-treatment	2
K	HMDS	—	125	20	2.5	HMDS deposition	10
	HMDS	O_2	200	20	2.5/3.8	SiO_x deposition	5
	HMDS	—	125	20	2.5	HMDS deposition	5
M	HMDS	—	125	20	2.5	HMDS deposition	10
	HMDS	O_2	200	20	2.5/3.8	SiO_x deposition	5
	HMDS	—	125	20	2.5	HMDS deposition	5
	—	O_2	200	22	7.4	O_2 post-treatment	2

2.3. Analysis and characterization of the PP-films

2.3.1. Depth profiling by X-ray photoelectron spectroscopy. XPS was performed on a Model 5400 small-spot Perkin–Elmer instrument to obtain elemental depth profiles of the PP-films of Table 2. It was equipped with a differentially pumped ion gun which was operated at 4 kV beam voltage and 25 mA beam current. Beam rastering in the X plane was maintained at 30% with a Y/X ratio of 1. Mg K_α radiation at a power of 400 W was used to obtain the spectra. The system base pressure was 6.7×10^{-6} Pa. The sputter rates were calibrated using a 100 nm thick Ta_2O_5 standard on Ta metal. For this standard, the sputtering rate was 3.7 nm per min. After each sputtering cycle of 3 min, narrow scans of all elements of interest were recorded. Peak areas of these elements were measured using a Shirley background correction. The areas were converted to elemental percentages using the internal sensitivity factors of the instrument.

2.3.2. Electrochemical Impedance Spectroscopy. The Electrochemical Impedance Spectroscopy data were collected using a Gamry CMS300 system with a Stanford Research Company 1825 Frequency Response Analyzer. The corrosion cell used was a three-electrode type (so-called flat cell). The counter electrode was a graphite rod. An Ag/AgCl reference electrode was used. Both of these were in contact with the working electrode, the sample being tested, through a 1 M NaCl solution electrolyte. The data were collected over a frequency range of 10^{-3} to 10^5 Hz.

A graphical analysis of the EIS data was performed to determine the arrangement of the ECM to be used for data fitting. The correct ECM was then used to fit the EIS data and determine quantitative values for the parameters of the ECM. If the ECM was correct, a good fit of the data will give values that can then be compared between samples to qualitatively determine which coating or coating system will provide the best corrosion protection.

2.3.3. Accelerated Environmental Corrosion Performance Tests (AECPT). The AECPT used was a test designed in our laboratory and was based upon test parameters found in similar corrosion tests such as the GM9511P and the Ford APGE tests. The samples were first scribed diagonally across the face of the sample. The scribe must penetrate through the E-coat and plasma polymer to the metal surface. The samples were then placed in a rack with the scribed faces oriented approximately 45° to horizontal. They were subsequently placed in a humidity/temperature chamber and exposed to 60°C and 85% RH for 23 h. Once removed from the chamber, the samples were submerged in a solution of 5% NaCl for 15 min. After completion of soaking, they were removed from the solution and allowed to air dry for 45 min. This procedure defines one cycle. The test protocol found in the GMP9511, also known as the GM-scab test, requires the samples be subjected to a minimum of 25 cycles for valid test results. This protocol was followed in this experiment for consistency of results.

Evaluation of the exposed samples was accomplished by measuring the length of delamination of the coating away from the scribe. Six measurements are taken along the scribe face. The results of these measurements are averaged and referred to as the average delamination value (ADV). If the average delamination values vary only slightly between samples after completion of 25 cycles, additional cycles may be added.

3. RESULTS AND DISCUSSION

3.1. XPS depth profiles of samples with varied plasma polymerization deposition

Figures 1–4 show the XPS elemental depth profiles of the PP-films with and without modification by adding the oxygen carrier gas (Table 1). It can be concluded (Fig. 1) that the bulk of the film had an elemental composition not much different from that of the monomer. The Si content was slightly increased, which may have been the result of preferential sputtering effects. The oxygen concentration at the interface with the CRS substrate is slightly higher than in the film, due to the presence of a residual iron oxide. The effect of the oxygen plasma post-treatment (Fig. 2) manifests itself as a higher surface concentration of oxygen. The thickness of the modified surface layer is approximately 7 nm. Both C and Si have a lower concentration in this surface region. The film with the intermediate SiO_x film (Fig. 3) has indeed a layer with increased Si and O and decreased C concentration. The layer thickness is about 70 nm and is situated around 80 nm below the surface. The film with the combined treatment (Fig. 4) shows a combination of these two effects.

3.2. EIS

3.2.1. Samples with varied plasma cleaning. Representative Bode plots of the EIS tests for the samples which had the plasma cleaning process varied are shown in Figs 5 and 6. The spectra of the remaining samples were very similar to the spectra

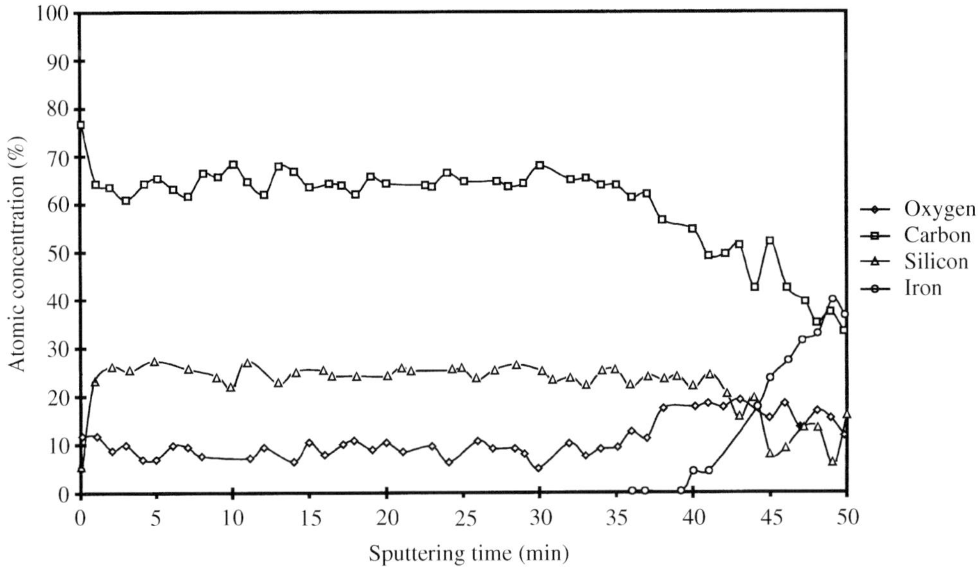

Figure 1. XPS depth profile of sample H.

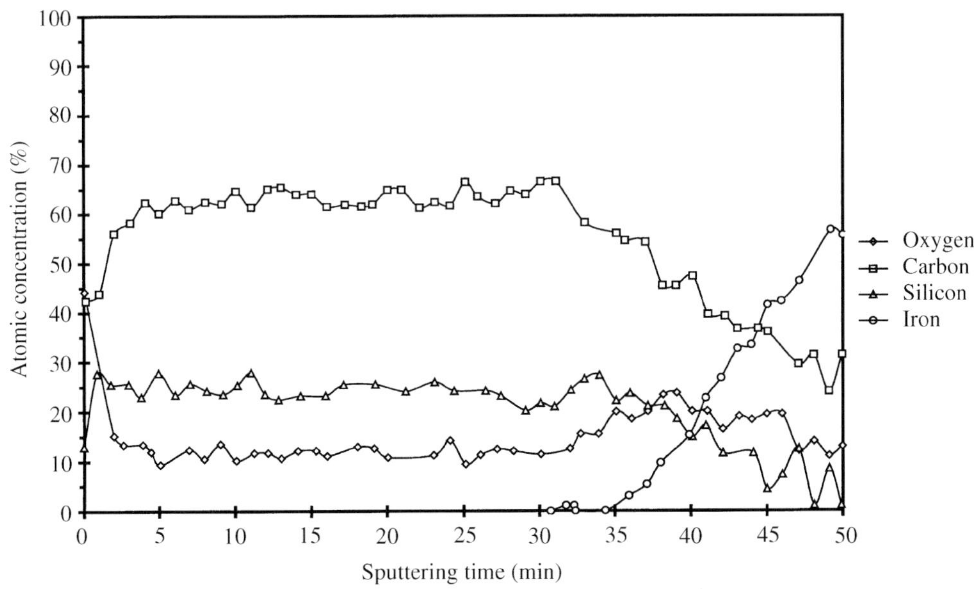

Figure 2. XPS depth profile of sample J.

presented in Fig. 6. An inspection of the shape and the position of both the phase angle and the impedance curves showed that there was very little difference in samples B–G. This is also seen in the similarity of the R_p and C_{dl} values given in Table 3. Sample A, whose spectrum is shown in Fig. 5, is seen to have a modulus and phase

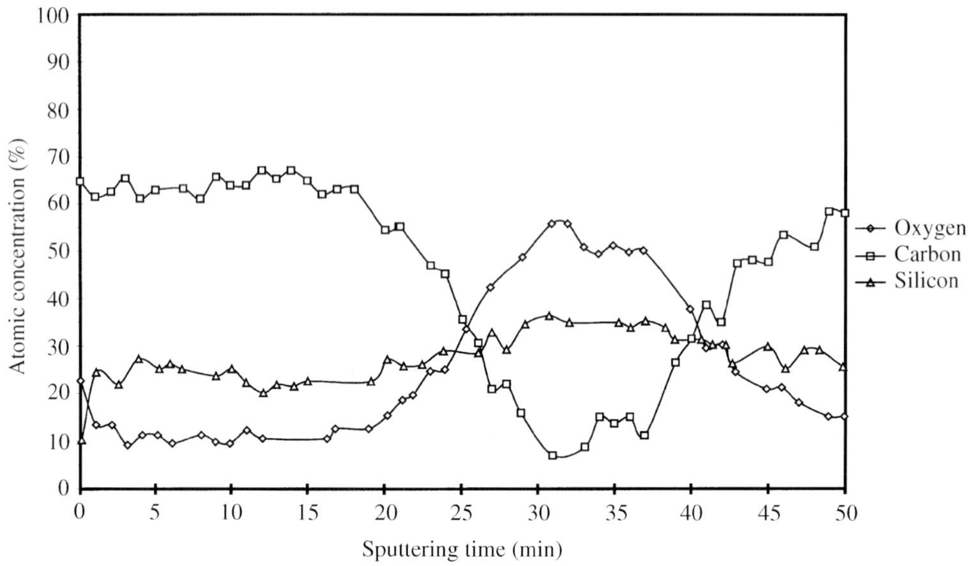

Figure 3. XPS depth profile of sample K.

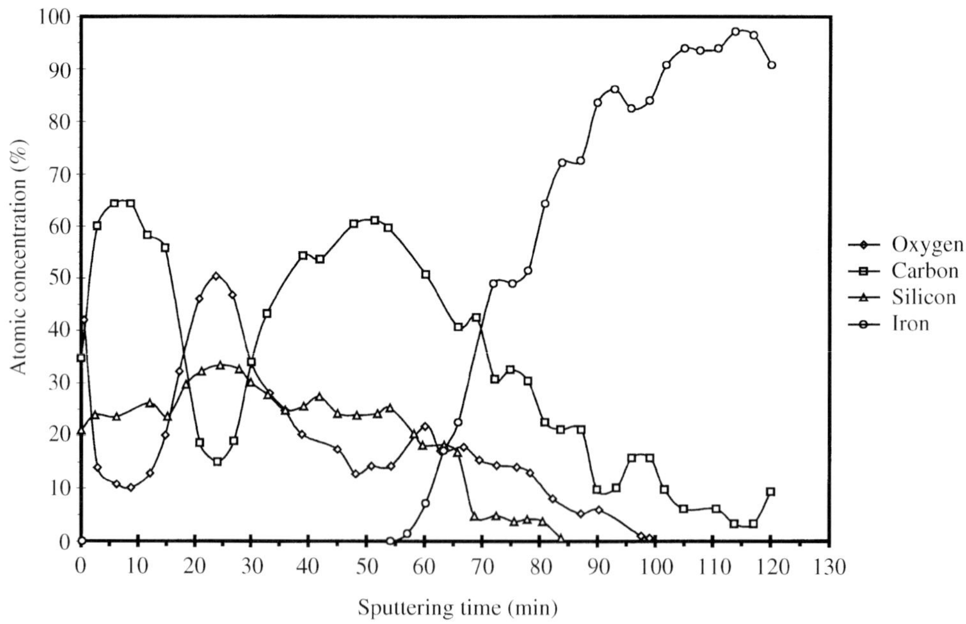

Figure 4. XPS depth profile of sample M.

angle curve very much different than that presented for sample B (Fig. 6). We again see in Table 3 that the R_p value is much lower than for the other samples, and the C_{dl} value is much higher than the other samples. From this preliminary analysis, we can conclude that sample A is clearly inferior to the rest of the sample set.

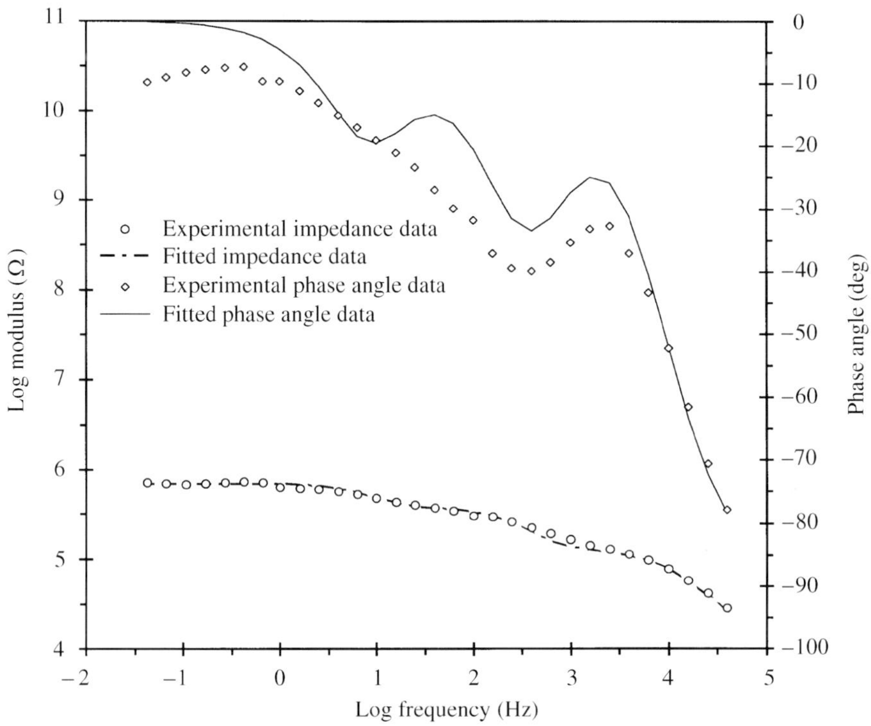

Figure 5. Bode plot of sample A; in this and all subsequent figures, the panel had been coated with a cathodically electrodeposited primer (ED-5000) and then immersed in a 1 M aerated NaCl solution for 24 h.

Table 3.
ECM parameter values for plasma cleaning experiment

Sample	R_{po} (Ω)	C_c (F)	R_{pl} (Ω)	C_{pl} (F)	R_p (Ω)	C_{dl} (F)
CRS	9.9×10^6	1.5×10^{-10}	9.6×10^8	1.4×10^{-11}	3.3×10^{11}	4.6×10^{-11}
Phos. CRS	4.1×10^5	1.8×10^{-10}	7.7×10^8	6.3×10^{-11}	1.4×10^{11}	8.6×10^{-11}
A	1.3×10^5	1.6×10^{-10}	2.5×10^5	3.1×10^{-9}	3.1×10^5	8.2×10^{-8}
B	8.2×10^6	1.7×10^{-10}	7.8×10^8	1.6×10^{-11}	1.0×10^{11}	5.4×10^{-11}
C	2.2×10^7	1.7×10^{-10}	1.0×10^9	1.7×10^{-11}	1.5×10^{11}	8.2×10^{-11}
D	4.4×10^6	1.7×10^{-10}	6.3×10^8	2.1×10^{-11}	1.5×10^{11}	5.8×10^{-11}
E	9.6×10^7	1.7×10^{-10}	2.3×10^9	3.8×10^{-11}	7.8×10^{11}	5.1×10^{-11}
F	9.2×10^6	1.6×10^{-10}	5.1×10^8	1.7×10^{-11}	3.4×10^{11}	1.0×10^{-10}
G	1.5×10^7	1.7×10^{-10}	1.5×10^9	1.7×10^{-11}	2.7×10^{11}	6.3×10^{-11}

From the graphical analysis of these plots, the ECM that best depicts the performance of the coating systems was designed, and is shown in Fig. 7. In this ECM, there was a resistor that represents the resistance of the solution (R_s). For analysis purposes, since the resistance of the solution was constant for all samples tested, this

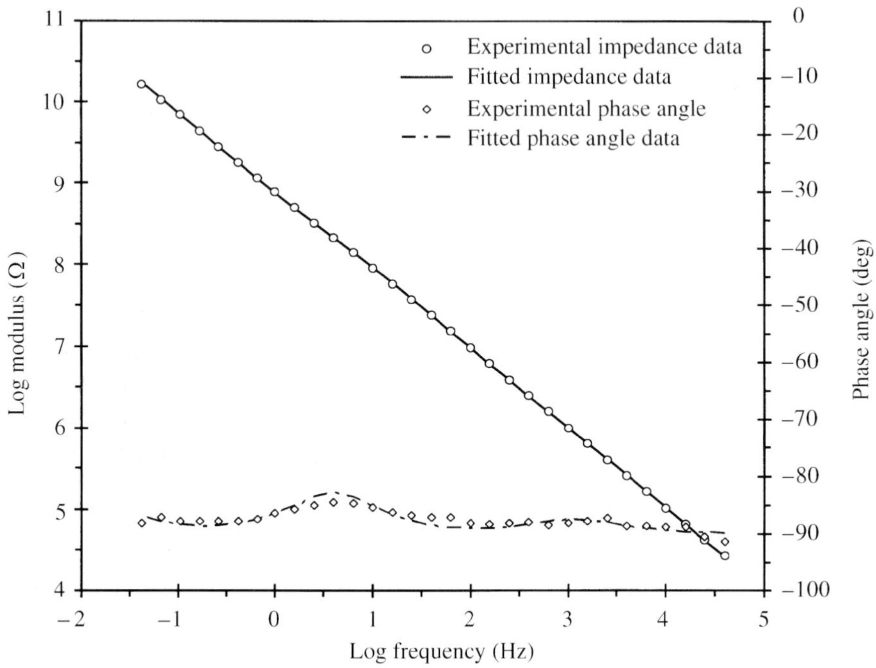

Figure 6. Bode plot for sample B.

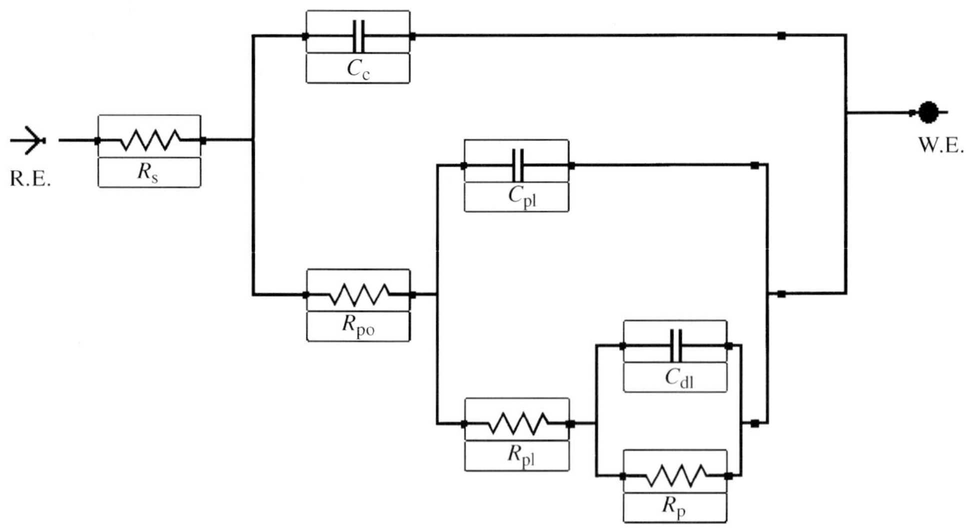

Figure 7. Equivalent circuit model of E-coat/plasma film/plasma-cleaned system; for explanation of elements, see text.

resistance was maintained at a constant value of 10 Ω. There were also three standard time constants, each containing a resistor and capacitor in parallel, each representing a different element in the coating system.

The first time constant consists of R_{po} and C_c. R_{po} represents the Pore Resistance of the E-coat. This value is indicative of the inherent defects and ionic or molecular pathways in a polymer coating. It is through these defects and pathways that diffusion of water and oxygen takes place. C_c represents the capacitance of the coating. This capacitance is dependent on the dielectric constant of the coating, as well as the amount of water present in the coating. Therefore, this parameter will change during the initial stages of exposure to a humid or aqueous environment. The value of this parameter will eventually stabilize at a constant value once the coating is completely saturated by a solution or vapor, i.e. once the coated metal has come to a pseudoequilibrium state with its environment.

The second time constant is related to the plasma polymer layer. In the samples which do not have a plasma polymer layer (the samples CRS and the phosphated CRS), this time constant is related to the oxide layer and the phosphate layer, respectively. The resistance (R_{pl}) is similar in nature to the pore resistance of the E-coat, and the capacitance (C_{pl}) is similar in nature to the coating capacitance of the E-coat.

In the third time constant, which represents the polymer/metal interface, R_p represents the polarization resistance. Traditionally, in DC polarization tests, R_p is used to measure the corrosion rate of corroding surface. This resistance gives an indication of the ease with which current can pass between the cathodic and anodic sites during the corrosion process. Thus higher the R_p, the lower the corrosion rate due to the inhibition of electrical transport between anode and cathode. If one thinks of the possible mechanisms that could inhibit this transport in the situation of a polymer coating on a metal surface, the absence of adjacent anode and cathode sites seems to be the only one that could be varied, if all other factors remain constant from sample to sample. One measure of adhesion is the density of chemical bonds between two substances: the greater the density of bonds, the greater the strength of adhesion. A high density of chemical bonds would also eliminate adjacent anodic and cathodic sites available for involvement in the corrosion reaction. Thus if all available anodic and cathodic sites are involved in chemical bonding between the polymer coating and the metal, they cannot participate in the corrosion reaction. Therefore, using this argument, one can infer that in tests where EIS is used for the assessment of the corrosion performance a coating system will provide, R_p gives an indication of the quality of adhesion of the coating to the metal surface.

The capacitor in the third time constant represents the Double Layer Capacitance (C_{dl}) which is set up at a saturated polymer/metal interface. A C_{dl} can only exist if there is a solution/metal interface. Since most researchers using EIS allow a penetration time for the solution to completely saturate the coating down to the metal surface before collecting the spectra, the value of C_{dl} gives an indication of the amount of solution present at the polymer/metal interface. If the same argument used in the case of R_p is used in this case, extensive chemical bonding at the interface would prevents large areas of solution/metal interface to form. Therefore, C_{dl} would also give an indication of the quality of the polymer–metal adhesion in EIS tests.

The fitted values of these parameters for the samples in which the plasma cleaning was varied, the CRS and the phosphated CRS, are listed in Table 3. The values of R_{pl}

and C_{pl} show that prior to exposure and initiation of corrosion, all samples, with the exception of sample A, have a stable plasma polymer layer, i.e. high resistance and low capacitance. Sample A, which did not have any plasma cleaning prior to deposition of the plasma polymer, clearly does not have a stable plasma polymer layer. The breakdown of the plasma polymer of this sample is thought to have occurred during the E-coating step. This was probably due to poor adhesion of the plasma polymer to the metal. The plasma polymer seems to have been partially removed during E-coating. This was corroborated by the 'orange peel' appearance of the E-coat of this sample. None of the other samples showed this effect.

Since the coatings on the samples are all essentially identical, any differences in the values that represent characteristics of the plasma coating or the E-coat must be due to minor experimental differences that cannot be easily controlled in the environments in which the plasma polymer or the E-coat are deposited. Therefore, the variations in R_{po}, C_c, R_{pl}, and C_{pl} were not considered in the analysis for corrosion protection in this study.

The two values that were considered are the polarization resistance, R_p, and the double layer capacitance, C_{dl}. Differences in these parameters have been related to adhesion properties of the system consisting of the polymer on a metal substrate prior to exposure to an aggressive environment [9]. Table 3 lists the fitted values obtained in the analysis using the ECM described previously.

As stated previously, the values of R_p were very similar for all the plasma cleaned samples, with sample E having the highest value, followed by samples F, CRS, G, C, D, Phos. CRS, B, and A, respectively. This ranking is thought to be associated with the thickness of the oxide layer that is produced during the oxygen plasma cleaning step [4]. The Ar/H_2 plasma cleaning, when used following the O_2 cleaning, has the effect of reducing the thickness of the oxide layer slightly. The Ar/H_2 plasma cleaning is also thought to render the surface of the metal (or oxide) very reactive. This creates excellent adhesion of the PP-FILM to the metal or the oxide layer. This would result in relatively similar initial values of R_p as long as a plasma cleaning was used. Thus, for sample A, which did not receive any plasma cleaning prior to PP-FILM deposition, we see a very much reduced initial value of R_p.

The reason for the high value of R_p seen for CRS is not fully understood, but may be the result of the uncoated metal having a lower resistance during the initial stages of the E-coat process due to the absence of the PP-FILM. Thus there is a high degree of adhesion for this initial layer of the E-coat, i.e. a denser film.

The values of C_{dl} for all samples in this set follow the same trend as seen in R_p. The higher value of sample A is again assumed to result from the lack of a plasma cleaning, which resulted in poor adhesion of the PP-FILM to the metal substrate. The slightly higher value seen for sample F is not clearly understood.

3.2.2. Samples with varied plasma polymerization deposition conditions. The shape and position of the impedance spectra for these samples were very similar to those for sample B of the plasma cleaning varied samples shown in Fig. 6. The modulus curves for this set of samples were slightly lower on the scale. The phase angle curves showed the same number of maxima and minima, but were located at slightly different frequencies than the plasma cleaning varied set of samples. A representative spectrum is shown in Fig. 8.

In this part of the research, it is important to observe the differences in the coating characteristics. These differences result from the inclusion of an SiO_x layer in the plasma coating, which should be more resistive than the plasma polymer without the SiO_x layer. The values for the elements in the ECM of these samples are given in Table 4. The parameter values in the ECM do not seem to correlate clearly with the variation in the deposition conditions. The addition of the SiO_x barrier film in the plasma polymer does not manifest itself as an increase in barrier properties as evidenced by the similarities of the R_{po} values of Table 4. Since there is no clear trend in Table 4,

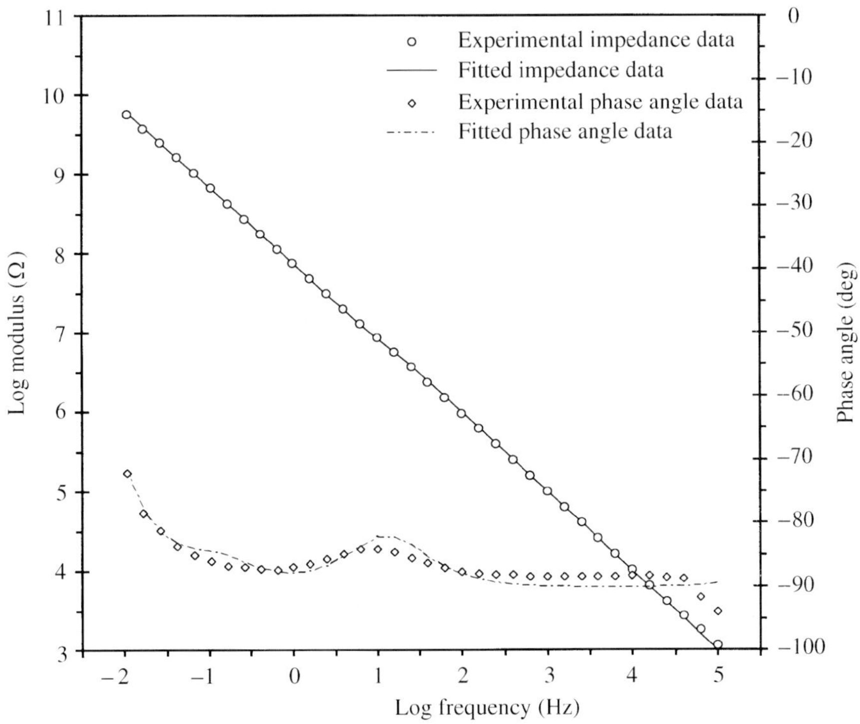

Figure 8. Bode plot for HMDS + O_2 etched sample I.

Table 4.
ECM parameter values for plasma polymer deposition experiment

Sample	R_{po} (Ω)	C_c (F)	R_{pl} (Ω)	C_{pl} (F)	R_p (Ω)	C_{dl} (F)
H	2.7×10^7	4.4×10^{-10}	1.1×10^8	1.3×10^{-9}	6.4×10^{10}	3.4×10^{-10}
J	2.7×10^7	5.2×10^{-10}	3.7×10^9	1.6×10^{-9}	1.6×10^{10}	4.4×10^{-10}
K	2.8×10^6	8.2×10^{-10}	8.2×10^8	1.5×10^{-9}	4.4×10^9	3.4×10^{-9}
M	2.7×10^5	1.7×10^{-10}	4.6×10^7	1.5×10^{-9}	3.1×10^{10}	5.6×10^{-10}

we must conclude that the modification of the plasma film probably also had an effect on the properties of the E-coat. These complex effects are the subject of further study.

3.3. Accelerated corrosion testing

3.3.1. Samples with varied plasma cleaning. The samples which had varied plasma cleaning provided a satisfactory degree of corrosion protection in the aggressive environment. The ADVs for these samples are listed in Table 5. Table 5 also compares the EIS results with the ADVs. Due to the similar values determined for R_p for all the plasma cleaned samples, a good correlation between the results of the two tests is hard to discern. One correlation that exists between the two tests is the difference in R_p values and ADVs between the samples that were plasma cleaned (samples B–G) and the sample that was not plasma cleaned (sample A). Sample A performed significantly worse than the plasma cleaned samples in the corrosion tests, as was expected from the lower values of R_p determined in the EIS tests. It should also be noted that in the EIS test, the data did not show any significant difference in the data from the CRS and phosphated CRS samples and the plasma cleaned samples. In this test, there is a distinct difference in the performance of the plasma-cleaned samples and

Table 5.
R_p and ADV in plasma cleaning varied samples

Sample	R_p (Ω)	ADV (mm)
CRS	3.3×10^{11}	3.36
Phosphated CRS	1.4×10^{11}	1.54
A	3.1×10^{5}	6.27
B	1.0×10^{11}	3.04
C	1.5×10^{11}	3.34
D	1.5×10^{11}	2.38
E	7.8×10^{11}	2.91
F	3.4×10^{11}	2.24
G	2.7×10^{11}	1.34

Table 6.
R_p and ADV for plasma polymer deposition experiment

Sample	R_p (Ω)	ADV (mm)
H	6.4×10^{10}	>10
J	1.6×10^{10}	2.0
K	4.4×10^{9}	7.2
M	3.1×10^{10}	3.5

the CRS and phosphated CRS panels. The phosphated CRS and sample G had lower ADVs than any of the other plasma-cleaned samples. It is reasonable to assume from these results that there is an optimum time for the oxygen cleaning step. Too short a period of time apparently does not efficiently remove the carbon contamination, and good adhesion of the PP-film to the metal substrate cannot take place. Too long a time, although it may completely remove the carbon contamination, produces a chemical change in the surface of the metal that does not promote good adhesion. The exact nature of this change is not yet fully understood, but may be caused by excessive growth of an iron oxide film. The CRS sample had an ADV similar to that of the plasma-cleaned samples. Sample A, which was not plasma cleaned, had the highest ADV. These results show that plasma cleaning is a necessary step prior to plasma polymer deposition for satisfactory corrosion protection performance, and further that the exact conditions of the cleaning process are important.

3.3.2. Samples with varied plasma polymerization deposition conditions. The results of this test for these samples were not conclusive. Although samples J and M had an ADV comparable to the better samples of the set with varied plasma cleaning, the reasons for the poor performance of samples H and K are not clear. However, since all other conditions of deposition were kept strictly constant, it appears that depositing the films under lower conditions of crosslinking (Table 2) seriously affected the corrosion performance. Such films can be expected to have a lower barrier action and higher diffusion rates than the films of Table 1, which were deposited at higher pressure and power [3].

4. CONCLUSIONS

The conclusions that can be drawn from the results presented here are summarized as follows.

- Plasma cleaning of CRS substrates, followed by the deposition of a homogeneous plasma polymer coating, when used as a primer system for a cathodic E-coat, provides corrosion protection comparable to that of state-of-the-art zinc phosphating.
- EIS can be used as a quick assessment technique for the degree of corrosion protection a cleaning/primer system can provide, as well as for investigating coating performance. However, if the pretreatment conditions are similar in nature in the actions on the metal substrate, initial EIS values may not be sufficiently different to discern which of these treatments is better than the others. For this reason, accelerated corrosion tests should be used in conjunction with EIS for verification of results. The ranking in the two types of test is not always identical.
- Using an O_2 plasma cleaning step seems to result in better corrosion protection by the plasma polymer film/E-coat system than when other gases are used in the plasma cleaning step. However, the duration of the O_2 plasma treatment needs to be optimized. The corrosion performance falls off at longer and shorter treatment times.

- The incorporation of an SiO_x layer in the plasma polymer does not seem to improve the corrosion protection for this system significantly. In fact, it seems that unless an O_2 post-treatment is used on these layered plasma polymers, the corrosion protection performance is actually adversely affected by the incorporation of this SiO_x layer. The mechanism of this effect appears to be very complex and may involve modification of the film in the electrodeposition process.

Acknowledgements

The authors would like to acknowledge Ralph Gray at PPG Industries, Inc., for his cooperation in the E-coating of the samples and Jerry Kallay of Weirton Steel Corporation for performing the XPS depth profiling analyses.

REFERENCES

1. H. H. Uhlig and R. W. Winston, *Corrosion and Corrosion Control: An Introduction to Corrosion Science and Engineering*. John Wiley (1985).
2. J. E. O. Mayne, *Official Digest* **24**, 127 (1957).
3. N. M. Morosoff, in: *Plasma Deposition, Treatment, and Etching of Polymers*, R. D'Agostino (Ed.). Academic Press, San Diego, CA (1990).
4. K. D. Conners, W. J. van Ooij, S. J. Clarson and A. Sabata, *J. Appl. Polym. Sci.: Appl. Polym. Symp.* **54**, 167 (1994).
5. K. D. Conners and W. J. van Ooij, in: *CORROSION '95 Proceedings*, Paper #379. NACE International, Houston, TX (1995).
6. M. R. Wertheimer and J. E. Klemberg-Sapieha, *Thin Solid Films* **115**, 109 (1984).
7. R. P. Wenz, in: *Organic Coatings Science and Technology*, G. D. Parfitt and A. V. Patsis (Eds), Vol. 6, p. 373. Marcel Dekker, New York (1984).
8. F. Mansfeld, *Corrosion* **35**, 301 (1981).
9. M. Kendig, F. Mansfeld and S. Tsai, *Corrosion Sci.* **23**, 317 (1983).
10. H. P. Hack and J. R. Scully, *J. Electrochem. Soc.* **138**, 33 (1991).
11. R. S. Lillard, J. Kruger, W. S Tait and P. J. Moran, *Corrosion* **51**, 251 (1995).
12. E. P. M. van Westing, G. M. Ferrari and J. H. W. de Wit, *Corrosion Sci.* **36**, 979 (1994).
13. K. D. Conners, W. J. van Ooij, D. J. Mills and G. P. Bierwagen, *Corrosion* (Dec. 1995).

Mittal Festschrift, pp. 373–385
W. J. Van Ooij and H. R. Anderson, Jr. (Eds)

Glow discharge oxidation of plasma polymerized organosilicon layers

J. L. C. FONSECA, C. P. BARKER and J. P. S. BADYAL[†]

Department of Chemistry, Science Laboratories, Durham University, Durham DH1 3LE, England

Abstract—Thin films of organosilicon plasma polymers derived from tetramethylsilane and hexamethyldisilane monomers have been deposited onto polyethylene substrates. Glow discharge oxidation of these overlayers produces an SiO_xH_y-rich skin. The level and rate of oxidative attack at the coating surface is found to be dependent upon the type of carbosilicon precursor originally employed for plasma polymerization.

Keywords: Organosilane; plasma polymerization; plasma oxidation.

1. INTRODUCTION

Conventional polysilane materials offer many attractive attributes for utilization in microlithography [1, 2]. These polymers are normally coated onto a substrate from solution. Their unique —Si—Si— backbone is σ-conjugated, where the σ–σ^* electronic transition is readily excited by UV laser irradiation, and can lead to local photoablation of the polysilane polymer [3]. Subsequent exposure to a non-equilibrium oxygen glow discharge results in preferential etching of exposed underlayer material, whereas any remaining regions of polysilane polymer become etch-resistant by oxidation to SiO_xH_y, together with a concomitant loss of volatile carbon and hydrogen oxides [4]. Such self-developing resists are highly desirable, since the resist image is generated in one step, thereby eliminating the need for solvent or plasma developers.

Plasma polymerized organosilicon [5] and organosiloxane [6, 7] films are promising alternatives to polysilane polymers for use in dry resist applications. Organosilicon plasma polymers can be synthesized at room temperature with a large number of unsaturated centres, and a high silicon content [8, 9]. These materials offer the additional benefit of enabling completely dry photolithographic processing: comprising of glow discharge carbosilicon polymerization, followed by UV laser imaging, and finally oxygen reactive ion etching (RIE) for image transfer into the underlying substrate.

Plasma polymerization of tetramethylsilane [$(CH_3)_4Si$] and the structurally related hexamethyldisilane [$(CH_3)_3SiSi(CH_3)_3$] results in the formation of extended networks

[†]To whom correspondence should be addressed.

of $-[CH_3]_m SiH_n-$ linkages [10, 11]. In this study, the susceptibility of these carbosilicon coatings towards oxygen glow discharge modification has been investigated by X-ray photoelectron spectroscopy (XPS), attenuated total reflection Fourier transform infrared spectroscopy (ATR-FTIR), and UV emission spectroscopy.

2. EXPERIMENTAL

Tetramethylsilane (TMS, 99.9%, Aldrich Chemical Company) and hexamethyldisilane (HMDS, 98%, Aldrich Chemical Company) were further purified by multiple freeze–pump–thaw cycles. These reagents were found to have a sufficiently high vapour pressure at room temperature, and therefore did not require any external warming. Low density polyethylene film (LDPE, Metal Box Company, UK) was used as the substrate; it was rinsed in isopropyl alcohol and dried prior to use.

Glow discharge experiments were carried out in a cylindrical glass reactor (4.5 cm diameter, 490 cm^3 volume) enclosed in a Faraday cage [12]. It was fitted with a monomer inlet, a Pirani pressure gauge, and a 47 l/min two-stage rotary pump attached to a liquid nitrogen cold trap. A 13.56 MHz radio frequency (RF) source was inductively coupled to the reactor through a matching network and a copper coil (4 mm diameter, 9 turns) spanning 11.0–18.5 cm from the monomer inlet. All joints were grease-free. A fixed substrate position of 6.5 cm from the reactor inlet was used for all experiments. Calculation of monomer and leak mass flow rates assumed ideal gas behaviour [13].

A typical experimental run comprised scrubbing the reactor with detergent, rinsing with isopropyl alcohol, and oven drying. This was followed by a high power (50 W) air plasma cleaning treatment lasting 60 min to ensure complete removal of any contaminants from the walls. Next, the reaction chamber was opened up to atmosphere and a piece of polymer placed onto the substrate holder. Then, the reactor was pumped down to its base pressure of 1×10^{-3} Torr, and a leak rate less than 2.6×10^{-10} kg/s. Subsequently, the monomer was introduced into the reaction chamber at 0.1 Torr pressure, and mass flow rates (F_M) of 3.6×10^{-8} kg/s and 4.3×10^{-8} kg/s for TMS and HMDS respectively (i.e. at least 99.3% of F_M is the organosilane precursor). Following 5 min of purging, the glow discharge was ignited at 0.8 W for 10 min. Upon termination of deposition, the RF source was switched off, the monomer reservoir closed, and the reactor pumped back down to its original base pressure for at least 5 min. High purity oxygen (BOC, 99.6%) was then introduced into the reaction chamber at 0.1 Torr pressure. The deposited organosilicon layer was treated with an oxygen glow discharge for 1 min. Finally, the system was vented to atmosphere, and the appropriate analytical measurement undertaken.

X-ray photoelectron spectroscopy (XPS) characterization was carried out on a Kratos ES200 X-ray photoelectron spectrometer operating in the fixed retarding ratio (FRR, 22 : 1) mode. Argon ion depth profiling studies were performed on a Vacuum Generators ESCALAB instrument, using constant analyser energy mode photoelectron detection (CAE, 50 eV pass energy), and a cold cathode ion gun (Vacuum Generators AG21, 1–2% energy spread) for substrate etching. Magnesium K_α X-rays were used as the photoexcitation source in both spectrometers, with an electron take-off angle of 30° from the substrate normal. A constant flux of 3 keV Ar^+ ions was maintained during depth profiling experiments by keeping the ion beam current fixed at 1.0 μA. Argon ion sputter depth profiles were carried out as a function of the ion beam current multiplied by sputtering time (henceforth referred to as current-time). An IBM PC computer was used for XPS data accumulation and component peak analysis (assuming linear background subtraction and Gaussian fits with fixed FWHM). All binding energies are referenced to the hydrocarbon ($-\underline{C}_xH_y-$) component at 285.0 eV. Sensitivity factors were instrumentally determined for unit stoichiometry using a range of standard polysiloxane polymers.

Attenuated total reflection (ATR) infrared absorbance spectra were collected on a FTIR Mattson Polaris instrument. Plasma polymer coated polyethylene film was mounted onto a variable ATR cell fitted with a KRS-5 crystal. An incident beam angle of 45° resulted in 14 internal reflections [14]. Typically, 100 scans were acquired at a resolution of 4 cm^{-1}. The carbosilicon overlayer was found to be thinner than the overall sampling depth of the ATR-FTIR technique [15].

A home-built UV emission spectrometer based upon a Czerny-Turner type monochromator was used for plasma glow analysis. A computer was used to rotate the grating using a stepping motor, and also to accumulate the counts from the photomultiplier tube detector. This instrument could scan continuously from 180 to 500 nm at 0.5 nm resolution.

3. RESULTS

3.1. XPS

Plasma polymerization of tetramethylsilane (Si : C ratio in monomer = 0.25) and hexamethyldisilane (Si : C ratio in monomer = 0.33) at 0.8 W yields organosilicon coatings with Si : C ratios of 0.35 ± 0.01 and 0.50 ± 0.01, respectively (as determined by XPS). Angle-resolved XPS studies reveal a small amount of oxygen at the surface (less than 4%): this is probably due to trapped free radicals in the plasma polymer layer reacting with the ambient atmosphere during sample transport from the deposition chamber to the XPS spectrometer. The main C 1*s* peak centred at a binding energy of 285.0 eV can be assigned to $\underline{C}-H$, $\underline{C}-C$ and $\underline{C}-Si$ environments [16]. The Si 2*p* core level at 100.7 ± 0.1 eV for TMS and 101.2 ± 0.1 eV for HMDS is characteristic of $\underline{Si}-H$ and $\underline{Si}-C$ bonds [16].

Plasma oxidation of each organosilicon film was studied as a function of glow discharge power (Figs 1 and 2). This treatment was found to be very rapid down to the depths accessible by XPS (0.5 to 2 nm) [3]. A plateau is reached well below 5 W corresponding to 20% carbon, 43% silicon, and 38% oxygen for the TMS plasma polymer, and 20% carbon, 35% silicon, and 45% oxygen for the corresponding HMDS material (Figs 3 and 4). Clearly this analysis excludes any hydrogen content, since it cannot be detected by XPS. The Si 2*p* core level shifts to higher binding energy during plasma oxidation, whereas the C 1*s* peak was attenuated, together with the emergence

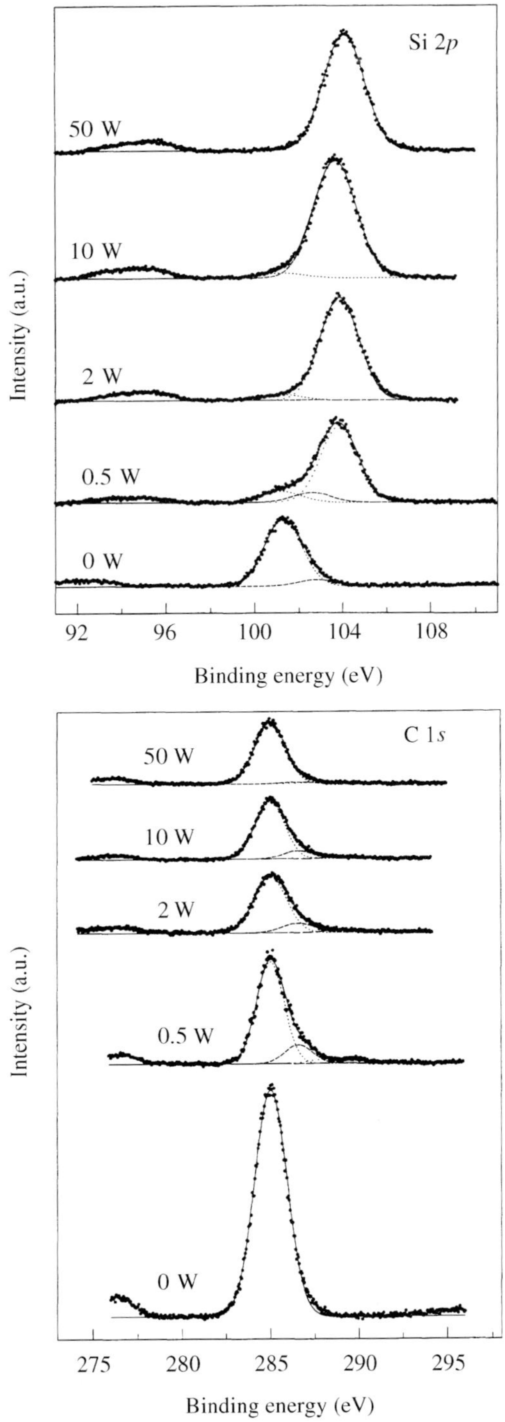

(a)

(b)

Figure 1. XPS spectra of oxidized TMS plasma polymer coating as a function of oxygen glow discharge power: (a) Si 2*p* and (b) C 1*s*.

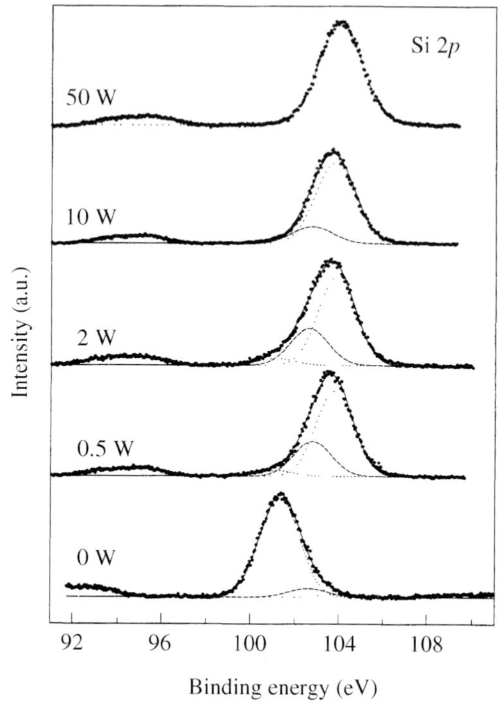

(a)

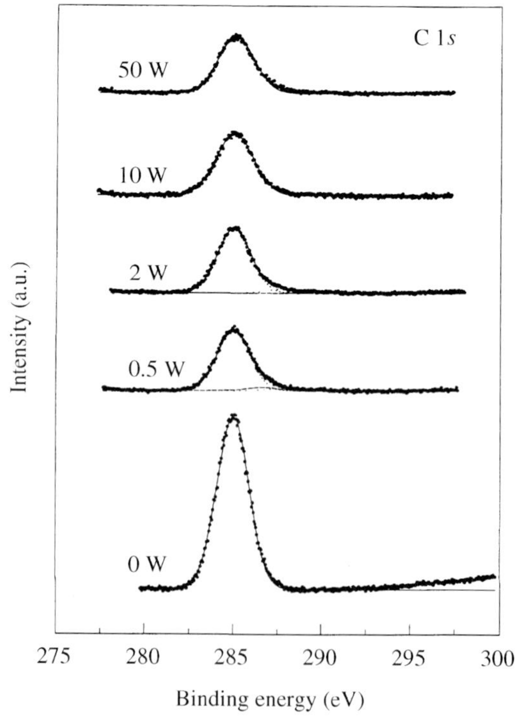

(b)

Figure 2. XPS spectra of oxidized HMDS plasma polymer coating as a function of oxygen glow discharge power: (a) Si 2*p* and (b) C 1*s*.

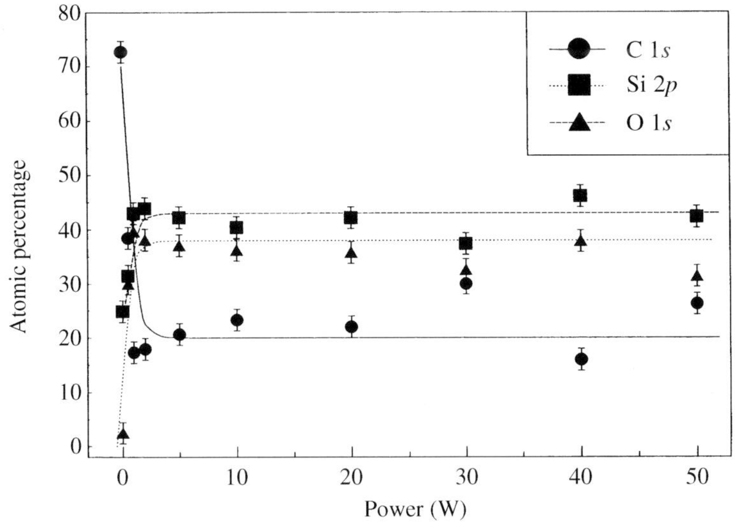

Figure 3. Compositional variation of TMS plasma polymer coating as a function of oxygen glow discharge power.

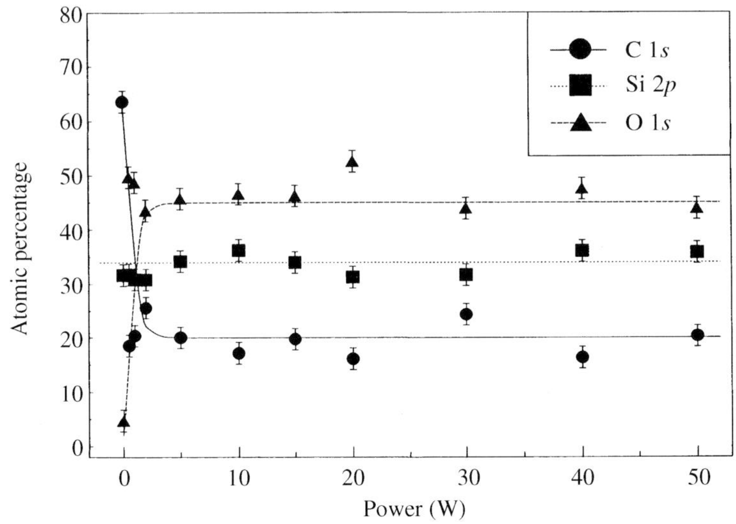

Figure 4. Compositional variation of HMDS plasma polymer coating as a function of oxygen glow discharge power.

of a weak oxidized component at high binding energy, which can be taken as a convolution of $\geqslant\underline{C}-O-$ (286.6 eV), $>\underline{C}=O/-O-\underline{C}-O-$ (287.9 eV), $-O-\underline{C}=O$ (289.0 eV), and $-O-\underline{C}O-O-$ (290.4 eV) peaks [17]. The Si 2p envelope can be fitted to $\underline{Si}-H/\underline{Si}-C$ (101.2 eV), $\underline{Si}-(OC_xH_y)_n$ (102.6 eV), and $O-\underline{Si}-O$ (104.0 eV) silicon environments [18]. The TMS plasma polymer oxidizes much more readily than its HMDS counterpart (Figs 1 and 2).

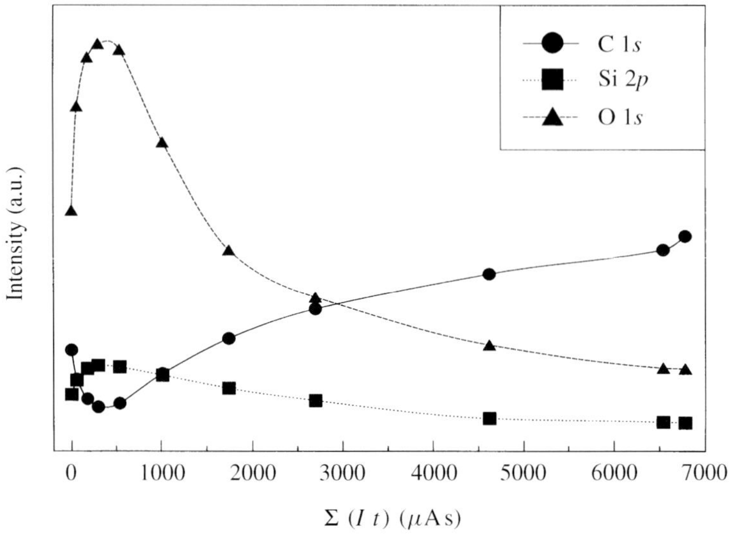

Figure 5. Ar^+ ion depth profile study of oxygen glow discharge treated (50 W) TMS plasma polymer coating.

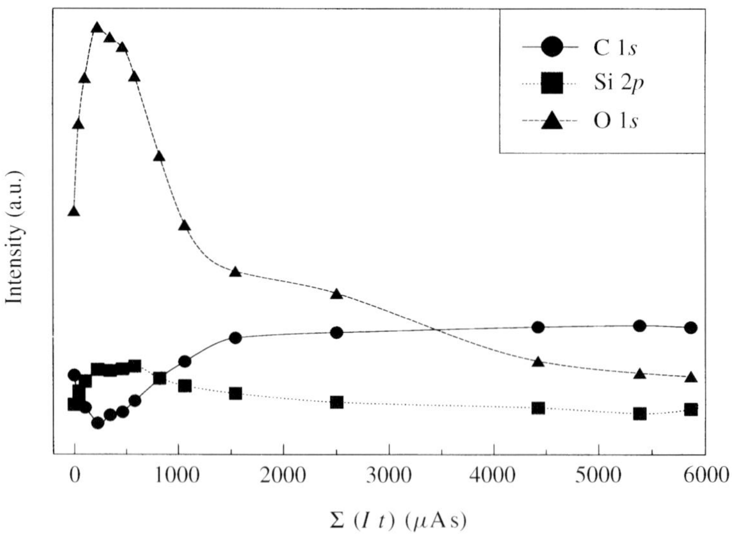

Figure 6. Ar^+ ion depth profile study of oxygen glow discharge treated (50 W) HMDS plasma polymer coating.

XPS analysis in conjunction with Ar^+ ion depth profiling of the oxidized organosilicon layers confirmed the build-up of oxygenated silicon species in the near-surface region (Figs 5 and 6).

Passage of oxygen over the organosilicon layer in the absence of a glow discharge resulted in negligible surface oxidation.

3.2. ATR-FTIR spectroscopy

Characteristic infrared absorption bands for clean polyethylene film can be assigned as follows [15]: 2915 cm^{-1} (asymmetrical C—H stretching in CH_2), 2847 cm^{-1} (symmetrical C—H stretching in CH_2), 1472 cm^{-1} (CH_2 bending), 1462 cm^{-1} (CH_2 bending), 1366 cm^{-1} (CH_2 wagging and/or symmetrical CH_3 movements), 729 cm^{-1} (CH_2 rocking, this is dependent upon the crystallinity and orientation of the polyethylene), and 720 cm^{-1} (mainly CH_2 rocking from the amorphous regions, with a minor contribution from crystal field splitting).

Both the freshly deposited organosilicon coating and the underlying polyethylene substrate contribute to the CH_3 and CH_2 stretching bands occurring at 2951 cm^{-1} and 2895 cm^{-1} (Figs 7 and 8). CH_3—C linkages can also be identified by peaks at 1460 cm^{-1} (methyl asymmetric bending in CH_3—C), and 1375 cm^{-1} (methyl symmetric bending in CH_3—C). Silicon related infrared absorptions are assignable as follows [15]: 2110 cm^{-1} (Si—H stretching), 1408 cm^{-1} (CH_2 symmetrical scissoring in Si—CH_2), 1250 cm^{-1} [CH_3 symmetric bending in $Si(CH_3)_n$], 1026 cm^{-1} [CH_2 wagging in Si—$(CH_2)_n$—Si], 833 cm^{-1} [CH_3 rocking in $Si(CH_3)_n$, $n = 2, 3$], 791 cm^{-1} [CH_3 rocking in $Si(CH_3)_n$, $n = 1, 2$], and 685 cm^{-1} (Si—C stretching) (Figs 7 and 8). All of the silicon related absorbances are much stronger for the HMDS plasma polymer since it contains a higher silicon content.

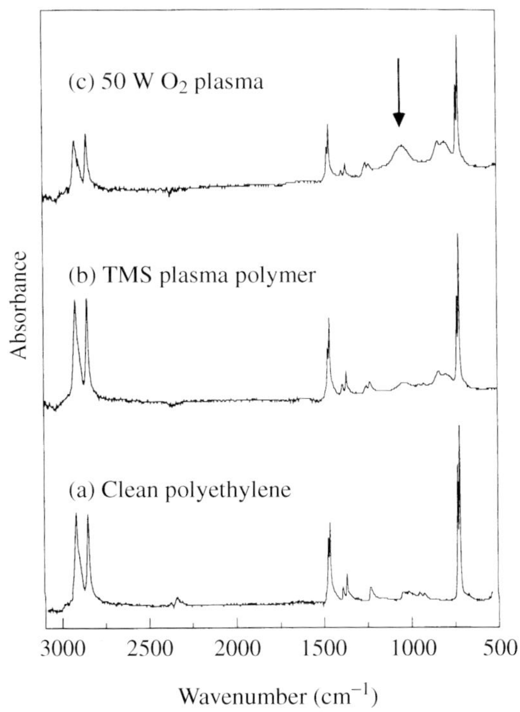

Figure 7. ATR-FTIR spectra of: (a) clean polyethylene; (b) TMS plasma polymer deposited on polyethylene; and (c) O_2 glow discharge oxidized TMS plasma polymer deposited on polyethylene.

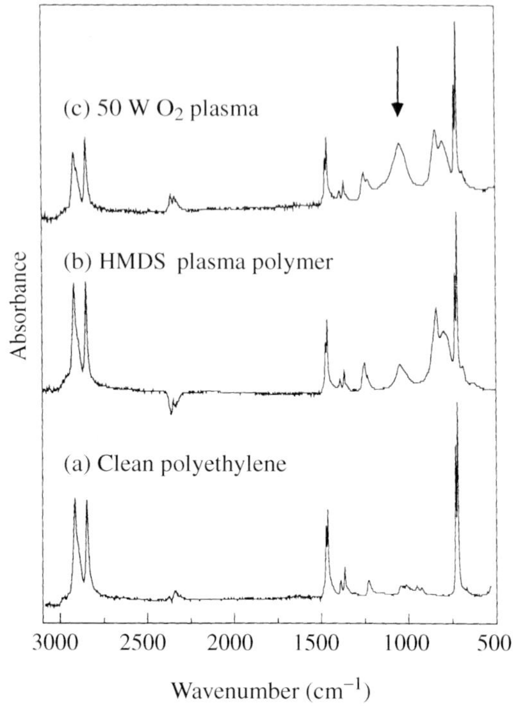

Figure 8. ATR-FTIR spectra of: (a) clean polyethylene; (b) HMDS plasma polymer deposited on polyethylene; and (c) O_2 glow discharge oxidized HMDS plasma polymer deposited on polyethylene.

ATR-FTIR analysis following glow discharge oxidation of both organosilicon plasma polymer layers shows a much stronger 1026 cm^{-1} [Si—O—Si and/or Si—O—C asymmetric stretching and/or CH_2 wagging in Si—$(CH_2)_n$—Si] absorbance (arrows in Figs 7 and 8). However it is clear that at the depths accessible to ATR-FTIR analysis ($\sim 0.1-1.0\ \mu m$), a significant amount of the virgin plasma polymer is still detectable. Therefore, the SiO_xH_y species are localized in the outer surface region and must be acting as a reactive ion etch barrier [3].

3.3. UV emission spectroscopy

The emission spectra taken during the oxygen glow discharge treatment of clean polyethylene film, TMS and HMDS plasma polymers are shown in Fig. 9. Their assignments are summarized in Table 1, and were checked by running glow discharges of each of the constituent gases (i.e. H_2, H_2O, CO, CO_2, etc.), Fig. 10. The main lines are due to transitions related to the presence of CO, OH and atomic hydrogen. For the carbosilicon plasma polymers, there is a strong attenuation in the CO line intensities with respect to clean polyethylene film, and also the appearance of an additional line (x) at 437.6 nm characteristic of the $H^3\Sigma_u^- \rightarrow X^3\Sigma_g^-$ transition in the Si—Si molecule. All of these observations are consistent with the formation of a $Si_xO_yH_z$-rich surface layer during the glow discharge oxidation of organosilicon plasma polymers. Other strong lines are due to the $C^3\Pi_u \rightarrow B^3\Pi_g$ transition for a

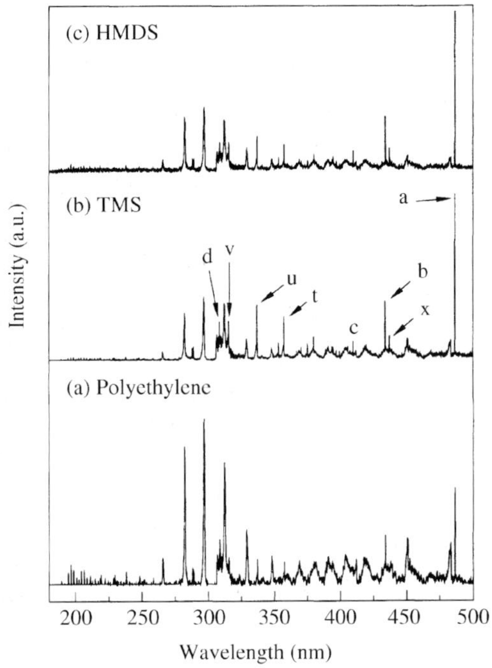

Figure 9. Emission spectra taken during oxygen glow discharge treatment (0.1 Torr, 10 W) of: (a) clean polyethylene; (b) TMS plasma polymer coating; and (c) HMDS plasma polymer coating.

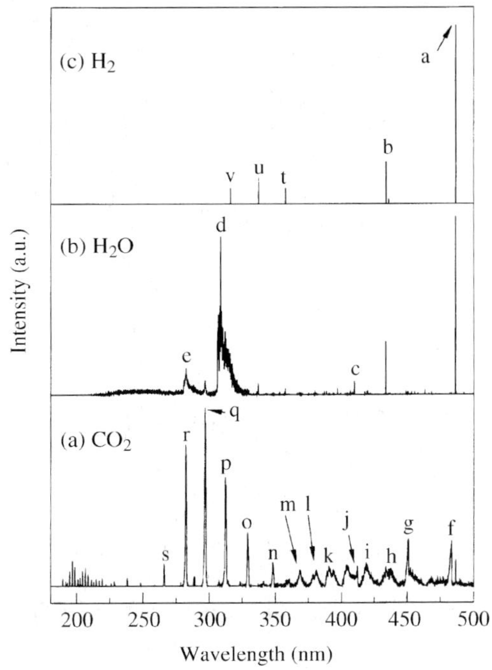

Figure 10. Emission spectra of glow discharges (0.1 Torr, 10 W): (a) CO_2; (b) H_2O; and (c) H_2.

Table 1.
Assignment of glow discharge emission lines (the alphabetical labels used in Figs 9 and 10 are indicated in parentheses) [19, 20]

Transition	Line/band (nm)	Vibrational assignment
Balmer lines (H_β, H_γ, H_δ)	486.1 (a)	—
	434.0 (b)	—
	410.2 (c)	—
OH ($A^2\Sigma^+ \rightarrow X^2\Pi$)	308.9 (d)	(0, 0)
	282.9 (e)	(1, 0)
CO ($C^1\Sigma \rightarrow A^1\Pi$)	438.0 (h)	(0, 3)
	412.5 (j)	(0, 2)
	389.3 (k)	(0, 1)
	368.1 (m)	(0, 0)
CO ($b^3\Sigma \rightarrow a^3\Pi$)	349.3 (n)	(0, 4)
	330.6 (o)	(0, 3)
	313.4 (p)	(0, 2)
	297.7 (q)	(0, 1)
	283.3 (e)	(0, 0)
	266.5 (s)	(1, 0)
CO ($B^1\Sigma \rightarrow A^1\Pi$)	483.5 (f)	(0, 1)
	450.9 (g)	(0, 0)
	439.3 (h)	(1, 1)
	412.4 (j)	(1, 0)
CO^+ ($B^2\Sigma \rightarrow A^2\Pi$)	420.9 (i)	(0, 1)
CO^+ ($A^2\Pi \rightarrow X^2\Sigma$)	379.6 (l)	(4, 0)
N_2 ($C^3\Pi_u \rightarrow B^3\Pi_g$)	315.9 (t)	(1, 0)
	337.4 (u)	(0, 0)
	357.7 (v)	(0, 1)
Si_2 ($H^3\Sigma_u^- \rightarrow X^3\Sigma_g^-$)	437.6 (x)	(2, 4)

nitrogen molecule: 315.9 nm (vibrational quantum number of final state: $\upsilon' = 1$, and vibrational quantum number of initial state: $\upsilon'' = 0$), 337.1 nm (0, 0) and 357.7 nm (0, 1). It is important to take into consideration that these nitrogen lines are weaker by an approximate factor of 2000 compared to their normal intensity in a pure nitrogen plasma, and therefore the actual amount of molecular nitrogen in the reactor during electrical discharge treatment must be less than 0.05%.

4. DISCUSSION

Plasma polymerization of the structurally related tetramethylsilane (TMS) and hexamethyldisilane (HMDS) molecules result in extended organosilicon networks [10, 15], comprising mainly $Si{-}CH_2$, $Si{-}(CH_2)_n{-}Si$, and $Si(CH_3)_n$, moieties. Solid state NMR and ATR-FTIR experiments have shown that the former contains a higher proportion of organic entities, and possesses a significant degree of unsaturation: this is consistent with its lower Si $2p$ core level binding energy (i.e. it is more electron-rich) [10, 11, 21]. These differences in plasma product have been correlated with the mode of precursor decomposition within the non-isothermal glow discharge. TMS has to undergo a relatively greater extent of fragmentation in order to generate solid forming moieties. In contrast to this, HMDS decomposes quite readily due to its weak $\geqslant$Si$-$Si$\leqslant$ bond.

Unsaturated hydrocarbon polymers undergo glow discharge oxidation much more readily than their saturated counterparts [22]. This explains why the TMS plasma polymer is oxidized at much lower electrical discharge powers (as seen by the shift in Si $2p$ envelope towards higher binding energies (Fig. 1a), and also by the more pronounced shoulder on the high binding energy side of the main C $1s$ peak (Fig. 1b). A faster rate of oxidation eventually results in a greater build-up of silicon content within the SiO_xH_y-rich layer due to the preferential loss of CO_2 and H_2O from the surface. Hence, despite there being a lower Si : C ratio in the as-deposited TMS plasma polymer layer, it exhibits a greater susceptibility towards glow discharge oxidation than its HMDS counterpart.

5. CONCLUSIONS

Coatings derived from the plasma polymerization of tetramethylsilane and hexamethyldisilane are excellent candidates for application as oxygen reactive ion etch barriers. A thin skin of SiO_xH_y-rich layer is formed at the surface of the organosilicon coating during oxygen glow discharge treatment, which restricts any further subsurface oxidation. Although the deposited tetramethylsilane plasma polymer layer is less Si-rich, it oxidizes at a much faster rate to yield a higher SiO_xH_y content. This behaviour can be accounted for on the basis that the TMS plasma polymer contains a greater degree of unsaturation, which in turn makes it more susceptible to oxidative attack.

Acknowledgements

J. L. C. F. thanks Brazil's Conselho Nacional de Desenvolvimento Científico e Tecnológico for financial support and EPSRC for provision of equipment during the course of this work.

REFERENCES

1. R. D. Miller, *Chem. Rev.* **89**, 1359–1410 (1989).
2. R. D. Miller, *Angew. Chem. Int. Ed. Engl. Adv. Mater.* **28**, 1733–1740 (1989).
3. M. A. Hartney, D. W. Hess and D. S. Soane, *J. Vac. Sci. Technol.* **B7**, 1–13 (1989).
4. F. Watanabe and Y. Ohnishi, *J. Vac. Sci. Technol.* **B4**, 422–425 (1986).
5. T. W. Weidman and A. M. Joshi, *Appl. Phys. Lett.* **62**, 372–374 (1993).
6. M. A. Hartney, J. N. Chiang, D. W. Hess and D. S. Soane, *Appl. Phys. Lett.* **54**, 1510–1512 (1989).
7. M. Morra, E. Occhiello and F. Garbassi, *J. Appl. Polym. Sci.* **48**, 1331–1340 (1993).

8. V. S. Nguyen, J. Underhill, S. Fridman and P. Pan, *J. Electrochem. Soc.* **132**, 1925–1932 (1985).
9. S. Y. Park, N. Kim, S. I. Hong and H. Sasabe, *Polym. J.* **22**, 242–249 (1990).
10. J. L. C. Fonseca, D. C. Apperley and J. P. S. Badyal, *Chem. Mater.* **4**, 1271–1275 (1992).
11. J. L. C. Fonseca, D. C. Apperley and J. P. S. Badyal, *Chem. Mater.* **5**, 1676–1682 (1993).
12. A. G. Shard, H. S. Munro and J. P. S. Badyal, *Polym. Comm.* **32**, 152–154 (1991).
13. C. D. Ehrlich and J. A. Basford, *J. Vac. Sci. Technol.* **A10**, 1–17 (1992).
14. R. T. Graf, J. L. Koenig and H. Ishida, *Fourier Transform Infrared Characterization of Polymers*. Plenum Press, New York (1987).
15. J. L. C. Fonseca and J. P. S. Badyal, *Macromolecules* **25**, 4730–4733 (1992).
16. R. C. Gray, J. C. Carver and D. M. Hercules, *J. Electron Spectrosc. Rel. Phenom.* **8**, 343–357 (1976).
17. J. Hopkins and J. P. S. Badyal, *Macromolecules* **27**, 5498–5503 (1994).
18. P. Laoharojanaphand, T. J. Lin and J. O. Stoffer, *J. Appl. Polym. Sci.* **40**, 369–384 (1990).
19. G. Herzberg, *Atomic Spectra and Atomic Structure*. Dover Publications, New York (1944).
20. R. W. B. Pearse and A. G. Gaydon, *The Identification of Molecular Spectra*. Chapman and Hall, London (1976).
21. J. L. C. Fonseca, D. C. Apperley and J. P. S. Badyal, *Macromolecules* **29**, 1705–1710 (1996).
22. A. G. Shard and J. P. S. Badyal, *Macromolecules* **25**, 2053–2054 (1992).

Mittal Festschrift, pp. 387–405
W. J. Van Ooij and H. R. Anderson, Jr. (Eds)

UV treatment of surfaces with excimer lasers and its effect on adhesion properties

H. DODIUK,[1,*] A. BUCHMAN,[1] M. ROTEL[2] and J. ZAHAVI[2]

[1] *Materials and Processes Dept., RAFAEL, P.O. Box 2250, Haifa 31021, Israel*
[2] *Israel Institute of Metals, Technion — Israel, Institute of Technology, Haifa 32000, Israel*

Abstract—The quality of adhesion is dependent upon the ability to apply proper surface treatment to the adherends. Both chemical modification and mechanical interlocking induced by surface treatments affect the strength and durability of the adhesive joint. Excimer laser UV irradiation presents a new technology for preadhesion surface treatment for various adherend materials. This technology presents an alternative to the use of environmentally unfriendly chemicals involved in conventional etching and abrasive treatments.

The effect of UV radiation at 193 nm on polymers, composites, metals and ceramics was investigated using Scanning Electron Microscopy (SEM), Fourier Transform Infra-red (FTIR) Spectroscopy, and X-ray Photoelectron Spectroscopy (XPS). The effect on adhesional strength and durability was tested on various joint geometries with and without exposure to hot/humid environment. Experimental results indicated that UV laser surface treatment improved significantly the adhesional shear strength, compared to untreated substrate, and was similar to or higher than that for conventionally treated substrates. Optimal UV laser treatment parameters (intensity, repetition rate and number of pulses) were defined depending on the substrate material and its chemical nature. The mode of failure changed from interfacial to cohesive as the number of laser pulses or energy increased during UV treatment until specific threshold values typical of the substrates were reached. Improved adhesion correlated with changes in morphology (uniform roughness) as indicated by SEM, with chemical modification and removal of contaminations as indicated by Auger, XPS and FTIR spectroscopies. It can be concluded that ArF UV excimer laser technology has potential as a clean and simple predhesion surface treatment for a wide range of adherend materials.

Keywords: UV; excimer laser; adhesional strength; adherends; surface treatment.

1. INTRODUCTION

Adhesive joint strength is affected by various factors such as nature of the substrate, its cleanliness, the nature of the adhesive, its curing cycle and stresses in the joint and mainly by interfacial interactions (physical and chemical). In order to improve joint strength, the common methods used are mechanical roughening, chemical modification, removal of weak boundary layers and stresses, and proper wetting. Application

---

*To whom correspondence should be addressed. E-mail: hannad@rafael.co.il

of a proper surface treatment to the adherend is among the decisive factors with respect to the final quality and durability of an adhesive joint. Many treatments have been devised for preparing materials surfaces for adhesive bonding, coating and the like. The general purpose of these preparation procedures is to modify the original surface of the adherend material: (a) to promote development of interfacial bonds with adhesives, and (b) to enhance the environmental resistance to moisture and humidity effects. Various materials (metals, plastics, composites and ceramics) require different surface treatments, some examples of which are detailed below.

The present processes for prebond surface preparation of thermoplastic, composite thermoplastic, and ceramic adherends involve the use of abrasive, welding, chemical and plasma treatments, which are, to a certain extent, destructive, poorly controlled and introduce undesirable changes in the morphology and composition of the surface such as cracks, uncontrolled pitting and contamination.

The prebond surface treatments which are commonly used for metals such as aluminum are [1]: chromate conversion coating, chromic acid anodization (with or without sealing), sulfuric acid anodization (with and without sealing) (SAA), phosphoric acid anodization (PAA), and chromic sulfuric acid etch (FPL — Forest Product Laboratory). The treatment for copper is chemical etching or sand blasting. The treatment for thermoplastic polymers and composites is sand blasting or etching with oxidizing acids. For simple structures welding, fusing, or plasma treatments are applied. The plasma treatment demands special environment and thorough cleaning of the surface following the treatment. Surface treatment for metal/ceramic bonds involves alumina blasting or brazing.

All these treatments involve the use of sand 'powders', acids (sulfuric, nitric, hydrochloric), strong bases or hexavalent chromium compounds. New OSHA (Occupational Safety and Health Administration) and EPA (Environmental Protection Agency) regulations ban such chemicals in industrial operations.

In recent years, the intensified development of ultraviolet (UV) laser systems has led to an increasing tendency to use these systems in various process applications. The main uses include cleaning of surfaces, precise cutting of materials, photopolymerization, ablation, coating in microelectronic circuits, and surgical applications. An outstanding feature of UV laser radiation is its ability to break chemical bonds or cause chemical reactions on the radiated surface due to photochemical effects in addition to a thermal heating effect. The range of wavelengths of UV lasers is within the energy of almost all organic chemical bonds. The ability to use specific coherent energies can result in specific chemical reactions which are otherwise difficult to achieve. The selection of wavelength can also determine the amount of heat effect compared with photochemical effect. Based on these effects, laser irradiation may provide a new method for surface preadhesion treatment for a wide range of materials.

Recently, excimer laser was used for preadhesion surface treatment of metals [2], thermoplastics [3, 4] and ceramics [5, 6].

Sancaktar *et al.* [2] used KrF (248 nm) excimer laser for surface modification and ablation of Al and Ti using laser energy of 0.5 $J/P\,cm^2$ with 5 and 20 pulses. They found changes in surface topography and increased oxidation and concluded that properly tailored ridge patterns on metal substrate were expected to improve the adhesion with polymer adhesives.

Heitz *et al.* [3] increased significantly the adhesion force of thin alloy films on PET [poly(ethylene terephthalate)] foils by excimer laser irradiation of the polymer surface prior to metal deposition. The foils were irradiated with 248 and 308 nm excimer laser in vacuum followed by metal deposition. The fluences used were below the ablation threshold.

Watanabe and Takata [4] compared irradiation of PET fiber or film with XeCl excimer laser or lamp in air (308 nm). They found that excimer laser irradiation showed a good adhesion property in comparison to excimer lamp irradiation because, above the threshold value, 340 $mJ/P\,cm^2$ irradiation with excimer laser caused an ablation due to the cleavage of chemical bonds, ripple shapes were produced and O/C ratio was reduced. In contrast, irradiation with excimer lamp at the same energy level did not cause surface changes because it was not intense enough to cause ablation and photochemical reaction.

Hourdakis and Hontzopoulos [5] used KrF excimer laser (248 nm) to treat α-alumina bulk material and Al_2O_3 plasma coated onto a metal. They found reduced surface roughness and reduced porosity and density of microcracks after irradiation at certain conditions. In addition, excimer laser surface treatment increased microhardness of Al_2O_3 coating, changed its orientation and increased its surface electrical conductivity. Processing of Al_2O_3, Si_3N_4, SiC and AlN with 308 nm XeCl excimer laser [6] proved to be a very precise technique because of the moderate thermal stresses applied to the material during processing.

Treatments with excimer lasers should produce desirable effects on the treated surfaces under proper laser conditions in order to provide better adhesion to the adherend. Laser treatment provides a clean and rather simple method of surface preparation and reduces the extent of damage to the treated surfaces. In our present and previous papers the application of ArF excimer laser (UV range at 193 nm) for surface treatment of thermoplastic Ultem and Lexan [7, 11] and carbon fiber reinforced thermoplastic (PEEK) composite [8], aluminum [9] and sealed anodic aluminum coating [11], copper, invar and fused silica is demonstrated. Thus, a new technology is presented for treatment of a variety of materials based on varying laser parameters according to the type of material treated.

2. EXPERIMENTAL

2.1. *Laser treatment*

The laser used during the course of this investigation was an ArF excimer laser model EMG 201 MSC (Lambda Physik, Germany). The laser parameters: repetition rate (5–30 Hz), energy density (0.9–4 $J/P\,cm^2$) and number of pulses (1–5000) were varied. All experiments were conducted at ambient temperature and in an air environment. The samples were irradiated by scanning the surface with the laser beam. The movement of the sample was governed by an $X-Y$ computerized table.

2.2. Materials

Table 1 lists the adherends and adhesives tested during the course of the investigation.

2.3. Testing

The various substrates were treated with the excimer UV laser using varying parameters such as intensity, repetition rate, beam area and number of pulses. The optimal laser treatment for each material was defined when the maximal shear strength of the corresponding bonded joint was achieved. Joint properties were determined using Single Lap Shear (SLS) test according to ASTM D-1002-72. Specimens were tested 4 days after preparation in an Instron machine, model 1185, at a cross-head speed of 2 mm/min. The mode of failure was determined visually. Surface morphology following laser treatment and following shear fracture was analyzed by means of Scanning Electron Microscope (SEM) model JSM-840 (JOEL, Japan). Surfaces were Au/Pd sputtered prior to analysis. Chemical changes at laser irradiated surfaces and at fracture surfaces were investigated by means of Fourier Transform Infra-red Spectroscopy (FTIR) using a Nicolet 5DX spectro-photometer in an external specular mode, equipped with a horizontal stage in a near-to-normal incidence and a gold-coated mirror as reference.

Surfaces of the laser-treated compared to untreated adherends were studied for chemical changes due to laser irradiation by X-ray Photoelectron Spectroscopy (XPS). A model PHI 555 spectrometer was used, with an Al K_α X-ray source, at 10 kV, 40 mA and a pressure of 3×10^{-8} torr. Microhardness tests as well as contact angle measurements were also performed for some adherends following laser treatment.

Two types of traditional surface treated references were used for each set of experiments. For the thermoplastics and thermoplastic composites, the references were untreated and/or abrasively treated with SiC (36 mesh). For the aluminum and sealed anodized aluminum adherends the references used were untreated aluminum and unsealed chromic acid anodized aluminum according to MIL-B-8625, respectively. For the copper, the references were samples with black conventional treatment (abonite) or sand blasting; for the glass epoxy fiberglass, the reference was a sample with sand blasting; and for the invar and alumina, the references were samples abraded with alumina blasting or silane primed.

3. RESULTS AND DISCUSSION

Table 2 summarizes the optimal laser treatment parameters for each of the materials tested. As can be seen from the results, different conditions are required for the different adherends tested. Pure polymers which are easy to activate require the least intensity for laser treatment (~ 0.1 J/P cm^2), composites which are more inert require higher energy (~ 0.2–1 J/P cm^2) and metals and ceramics which have strong chemical bonds require high energy (> 2 J/P cm^2).

Table 3 summarizes the maximal adhesive shear strength achieved for the various laser treated and bonded adherends after applying the optimal laser parameters. The

Table 1.
Adherends and adhesives tested

Adherend	Commercial name (producer)	Adhesive (producer), curing conditions
Polycarbonate	Lexan 9023-112 (General Electric)	Two-component polyurethane adhesive (Hexcel, USA), RT*, 48 h
Polyetherimide	Ultem 1000 (General Electric)	Two-component polyurethane adhesive (Hexcel, USA), RT, 48 h
Composite: PEEK (polyaryletheretherketone) reinforced with continuous carbon fibers	APC-2/AS-4 (ICI)	Structural epoxy adhesive • FM 300 2K (American Cyanamid) 2 h, 120°C, 2.6×10^3 torr • AF 163–2 OST (3M), 1 h, 120°C, 2.6×10^3 torr
Aluminum	2024-T3	• Rubber modified epoxy adhesive (RAFAEL), RT, 48 h • Structural epoxy adhesive FM73 (American Cyanamid) 2 h, 127°C, 1.8×10^3 torr
Sealed anodized aluminum	2024-T3, treated with • Chromic acid anodization • Sulfuric acid anodization	Rubber modified epoxy adhesive (RAFAEL), RT, 48 h
Copper	Electrolyte copper	• Rubber modified epoxy (RAFAEL), RT, 48 h • EA9394 (HYSOL), RT, 5 days
Fiberglass epoxy/copper	F-4	• Rubber modified epoxy (RAFAEL), RT, 48 h • Acrylic adhesive F0011
Copper coated fiberglass epoxy/polyimide	F-4/Kapton	• Rubber modified epoxy (RAFAEL), RT, 48 h • Acrylic adhesive F0011
Invar/fused silica		RTV159

*RT — room temperature.

Table 2.
Optimal laser parameters for the various adherend treatments

Adherend	Laser fluence ($J/P\,cm^2$)	Repetition rate (Hz)	No. of pulses
Polycarbonate	0.08	10	12
Polyetherimide	0.08	10	200
Composite PEEK (APC2/AS4)	0.19 or 1	5 5	100 10
Aluminum alloy	0.19	30	2000
Sealed anodized aluminum alloy	0.8–1.9	30	100–1000
Copper	2.7	10	50
Fiberglass epoxy/copper	0.18 (F4) 2.1 (Cu)	30 30	100 50
Copper/polyetherimide	0.18 (Kapton) 0.18 (Cu)	30 30	100 1000
Invar/fused silica	2.2	30	10

results indicate that ArF excimer laser treatment is effective for all different adherends treated and bonded. It should be stated that the laser parameters for the surface treatment are more dependent on the nature of the adherend than on the kind of adhesive used for bonding. Improvements in adhesive shear strength by 200–600% were achieved for laser treated adherends compared to non-treated ones and similar or higher compared to conventionally treated ones. Best results were achieved for the thermoplastic composite materials. Visual inspection of the failure surfaces shows clearly that laser treatment causes the mode of failure to change from interfacial in non-laser treated adherend to mostly cohesive at optimal laser operating conditions indicating that the interfacial adhesion was significantly improved.

SEM micrographs of laser treated adherends revealed morphological changes depending on the adherend material, laser energy and the number of pulses. The thermoplastic adherends' surfaces exhibited conic and rounded granules spread all over the surface (Figs 1 and 2). The carbon reinforced PEEK composite exhibited the same granules accompanied by partial exposure of the fibers (Fig. 3), at laser intensities of 0.1–0.2 $J/P\,cm^2$, while at higher laser intensities (above 1 $J/P\,cm^2$) the laser irradiated surface was smooth with randomly spread cracks (Fig. 4). SEM micrograph of the Al adherend after laser treatment showed no morphological changes at low laser intensities (0.18–0.2 $J/P\,cm^2$). Increasing laser intensity (0.7 $J/P\,cm^2$) reveals a fine microstructure on the treated surface demonstrating arrays of cracks about 1 μm wide and small holes (Fig. 5).

Irradiation of the sealed anodized aluminum samples at low laser intensities (0.2 $J/P\,cm^2$) showed no change in surface morphology even after 1000 pulses. At 0.7 $J/P\,cm^2$, changes in morphology include open bubbles, resulting probably from evaporated water. Some spherical droplets of Al_2O_3 due to splashing (caused by laser ablation of Al_2O_3) and cracks can be observed (Fig. 6). Irradiation of copper at low

Table 3.
Adhesive bond strengths[a] and failure modes[b] of the various studied bonded joints

Adherend/adhesive	Surface treatments		
	Untreated SLS[a] (failure mode)[b] (MPa)	Conventional treated SLS[a] (failure mode)[b] (MPa)	Laser treated SLS[a] (failure mode)[b] (MPa)
Polycarbonate/PU	3.5 (A)[b]	5.0 (M) SiC	7.5 (C)
Polyetherimide/PU	2.5 (A)	5.0 (M) SiC	5.5 (C)
Composite PEEK (APC2/AS4)/			
Structural epoxy FM 3002 K	6.1 (A)	14.7 (M) SiC	27.8 (M)
Structural epoxy AF 163–2	21.7 (A)	34.0 (M) SiC 20.0 (M) Plasma	45.4 (C)
Aluminum alloy/epoxy	2.0 (A)	10.2 (C) Unsealed anod	14.3 (C)
Aluminum alloy/structural epoxy	12.8 (C)	42.9 (C) Unsealed anod	34.4 (C)
Sealed anodized aluminum/epoxy	4.5 (A)	10.2 (C) Unsealed anod	11.0 (C)
Copper/modified epoxy	6.1 (A)	14.9 (C) Sand	14.3 (C)
Fiberglass epoxy/(rubber modified epoxy)/copper	—	31.7 (C) Sand	25.6 (A/M)
Fiberglass epoxy/(acrylic adhesive)/copper	—	8.5 (C) Sand	15.6 (C)
Copper/(acrylic adhesive)/polyimide	—	21.0 (M) Sand*	25.0 (A)*
Invar/fused silica/(RTV)	—	3.6 (A) Alumina	5.6 (C)

[a] ± 5% standard deviation (5 samples for each test).
[b] A — interfacial, M — mixed, C — cohesive mode of failure (in adhesive).
*These represent peel strength values in lb/inch.

energy showed only color changes due to oxidation. At higher energies morphological changes were observed showing round spheres spread evenly on the entire surface (Fig. 7).

Irradiated fiber glass showed the same morphology as composite PEEK except that the globes were surrounded by round circles which seemed like diffraction patterns (Fig. 8). Irradiated invar showed a change of color at low energies and, at higher energy, changes in morphology indicated as 'Moon'-like ridges and canals. At very high intensities the ridges formed into smooth shapes (Fig. 9). The fused silica showed no morphological change at low intensities and showed identical morphology as sealed anodized Al at high intensity. Kapton showed features similar to cones sticking out from the surface (Fig. 10).

All these morphologies increase the roughness of the surface enabling mechanical interlocking of the adhesive. It should be noted that the roughness is very uniform and evenly spread on the surface, which presents an advantage over abrasive treatments.

The enhanced adhesion due to mechanical interlocking is clearly revealed in the SEM micrographs for nearly all UV laser treatments and adherends. The cohesive

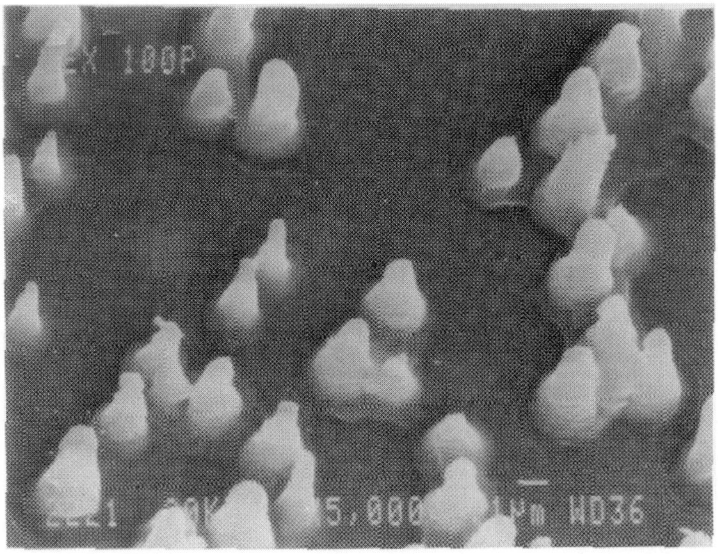

Figure 1. SEM micrograph of polycarbonate treated with excimer laser (0.085 J/P cm^2, 100 pulses).

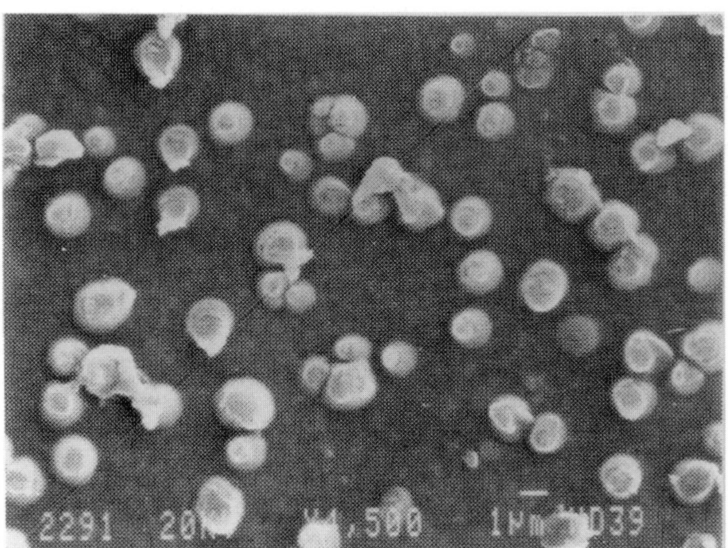

Figure 2. SEM micrograph of polyetherimide treated with excimer laser radiation (0.085 J/P cm^2, 100 pulses).

mode of the failure is indicated by an adhesive layer that can be found on each adherend showing a replica of the granules or bubbles formed during laser treatment. A far better adhesion (partly cohesive failure) can be observed compared to the inter-

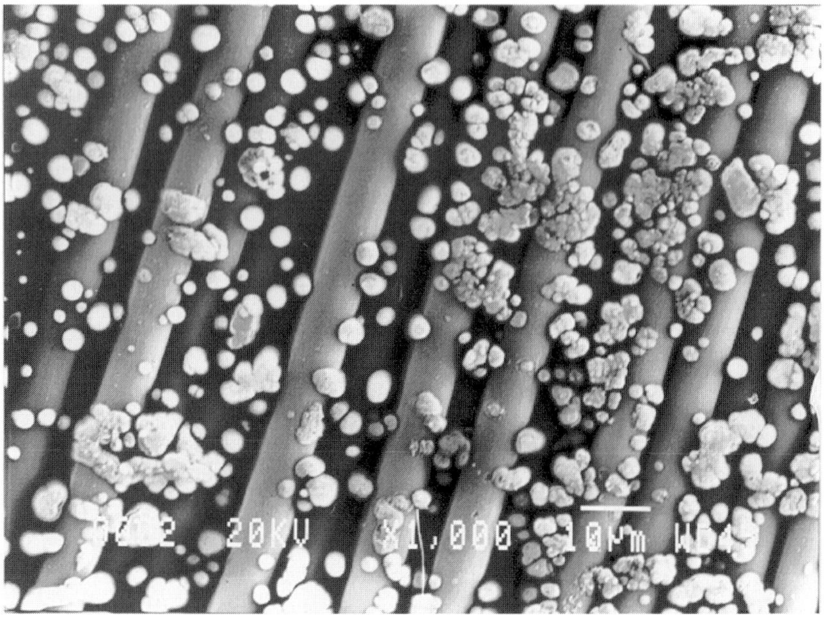

Figure 3. SEM micrograph of composite PEEK treated with excimer laser (0.18 J/P cm^2, 100 pulses).

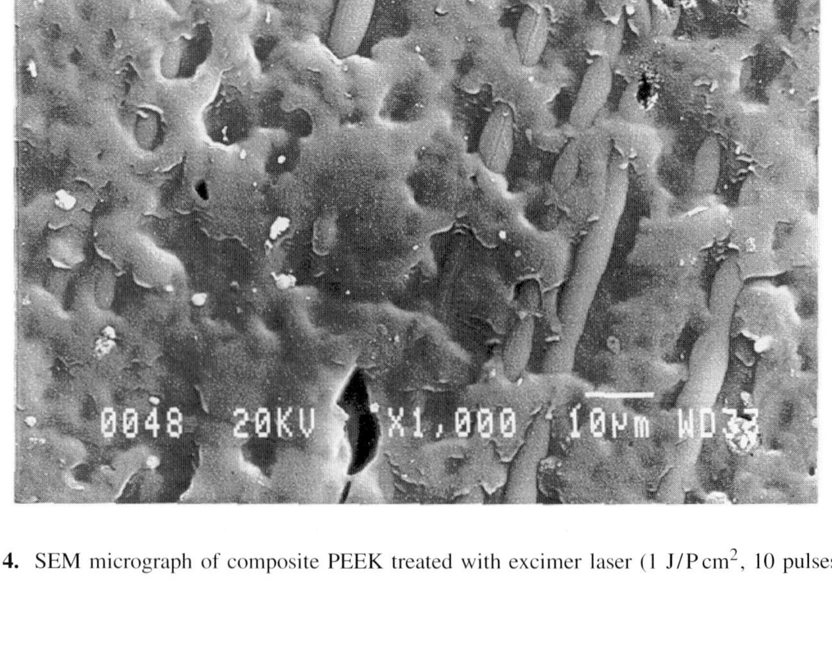

Figure 4. SEM micrograph of composite PEEK treated with excimer laser (1 J/P cm^2, 10 pulses).

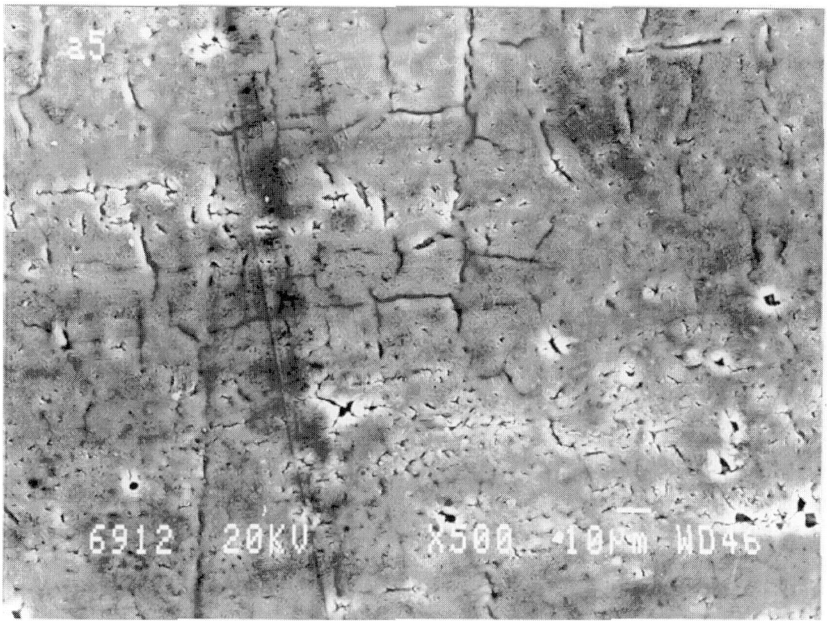

Figure 5. SEM micrograph of aluminum 2024-T3 treated with excimer laser (0.7 J/P cm^2, 200 pulses).

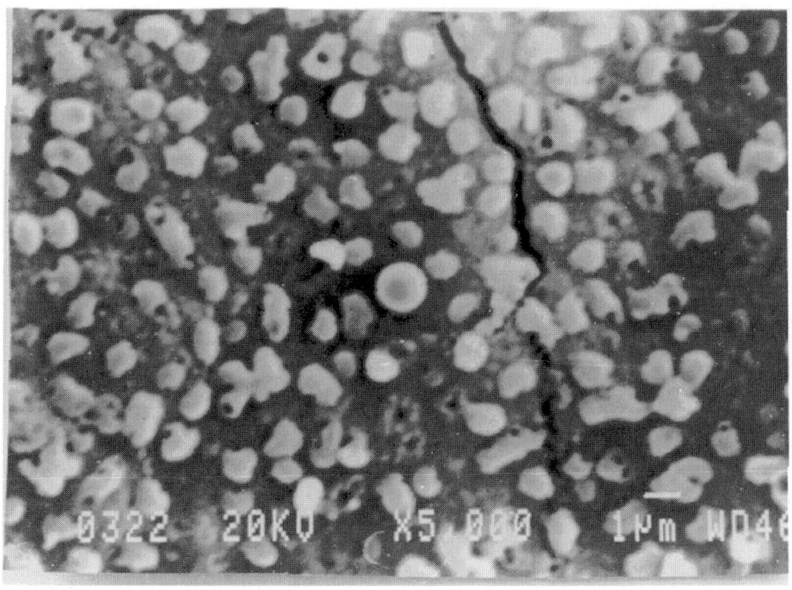

Figure 6. SEM micrograph of sealed anodized aluminum treated with excimer laser (0.7 J/P cm^2, 10 pulses).

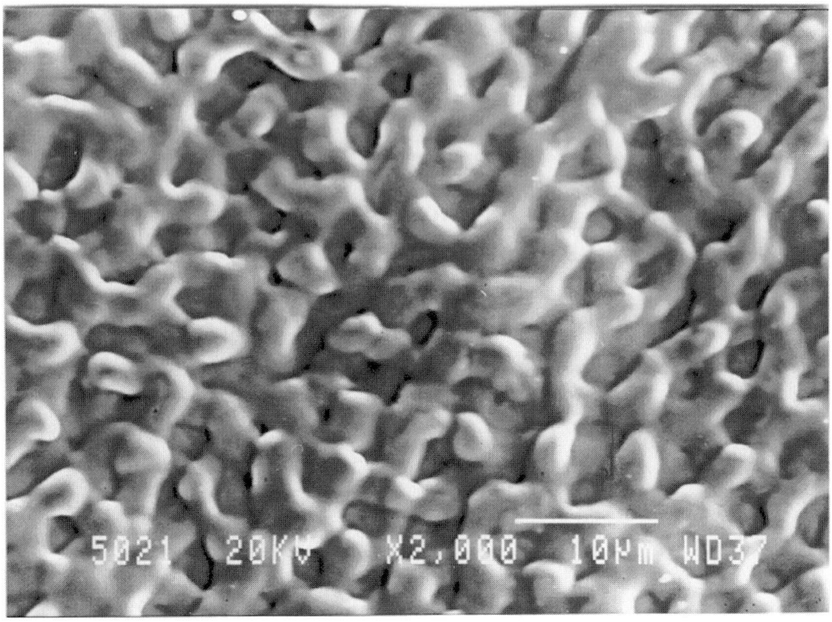

Figure 7. SEM micrograph of electrolyte copper treated with excimer laser (2.7 J/P cm^2, 50 pulses).

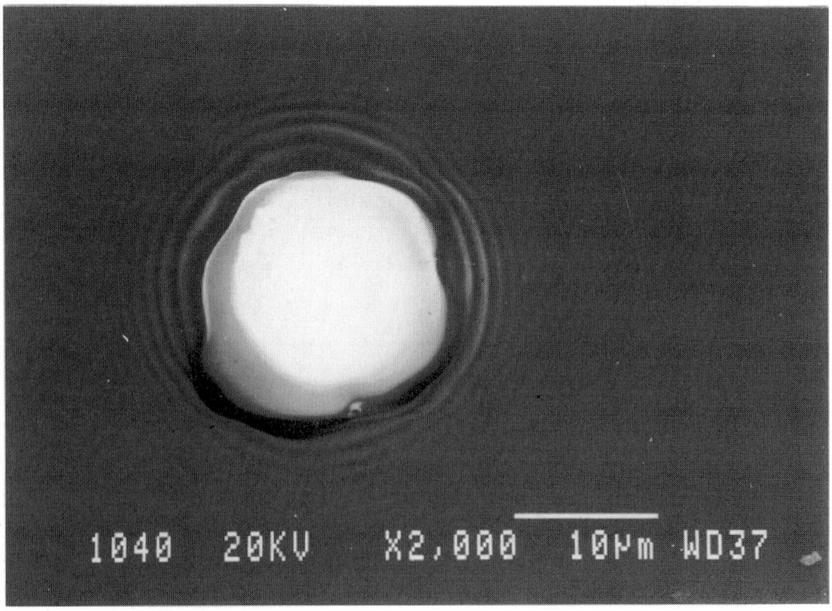

Figure 8. SEM micrograph of epoxy fiberglass treated with excimer laser (0.28 J/P cm^2, 500 pulses).

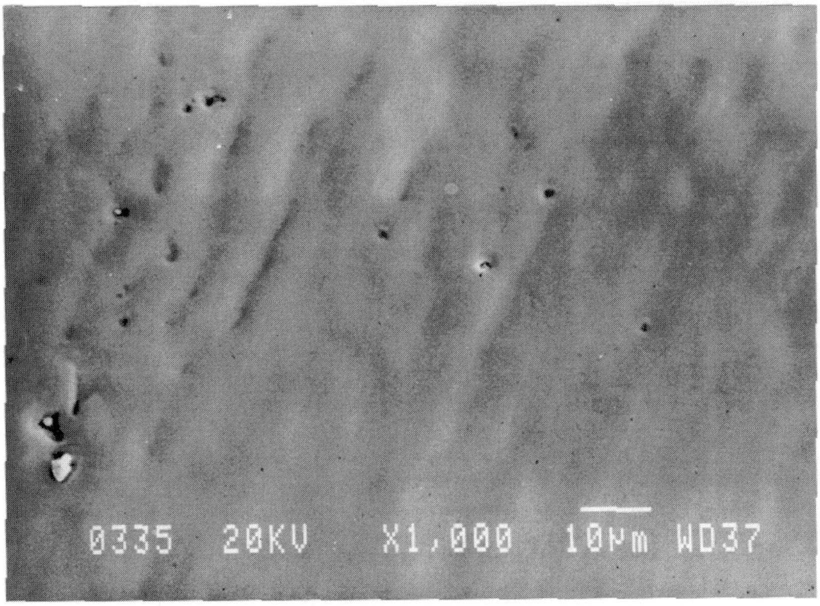

Figure 9. SEM micrograph of Invar treated with excimer laser (2.2 J/P cm^2, 100 pulses).

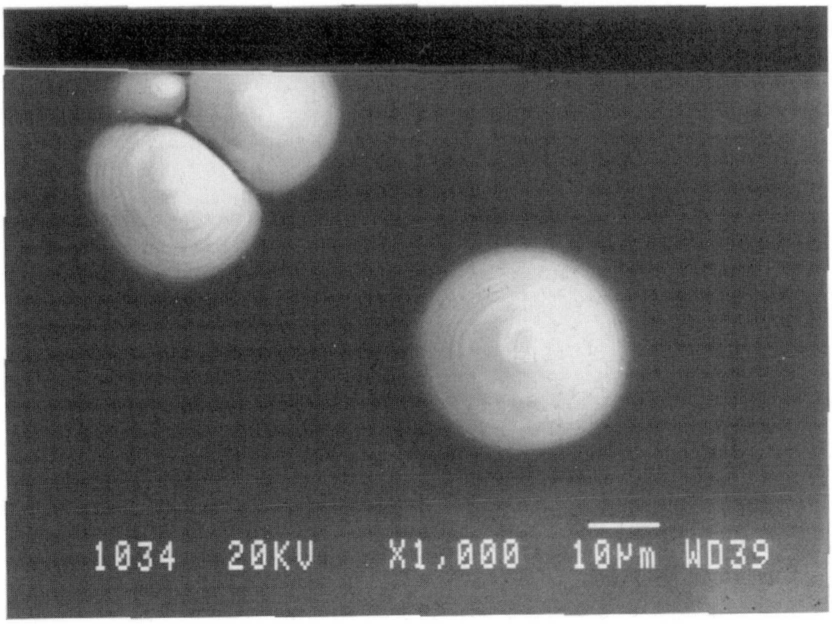

Figure 10. SEM micrograph of Kapton® treated with excimer laser (0.18 J/P cm^2, 500 pulses).

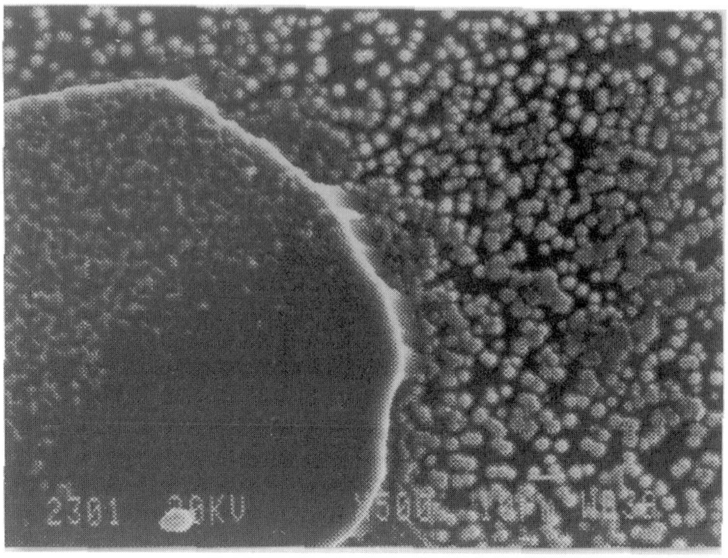

Figure 11. SEM micrograph of failed joint surfaces (cohesive failure) of polyurethane–polyetherimide treated with excimer laser radiation (0.085 J/P cm^2, 200 pulses).

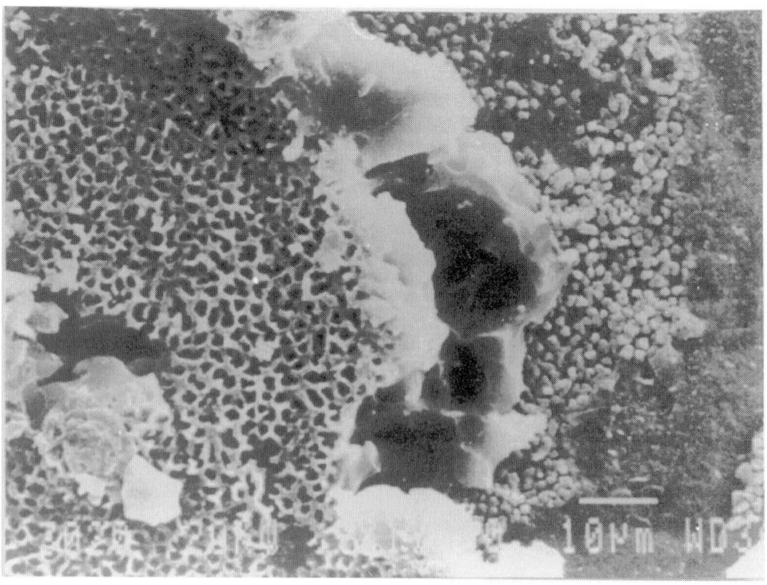

Figure 12. SEM micrograph of failed joint surfaces of sealed anodized Al–epoxy adhesive treated with excimer laser radiation (0.8 J/P cm^2, 100 pulses).

facial failure of the untreated adherend. Figures 11 and 12 represent the micrographs showing the replica formed during failure of thermoplastic and anodized Al adherends. The same phenomenon is also observed in PEEK composite, Cu, fiberglass, invar and fused silica. It should be noted that even if visually the failure seems partly interfacial, high magnification showed that a thin layer of adhesive was always present on the laser treated adherend.

At high laser intensity or number of pulses, a weakening of the treated layer of the adherend is observed due to ablation. This is shown by loosely connected granules at the thermoplastic surface, deformed and exposed fibers of the composite, molten dendritic areas of the Al and anodized Al surface, loose spheres on the Cu surface, and molten areas in the fused silica irradiated areas.

In addition to the pronounced morphological modification, chemical changes also take place following UV laser treatment of the adherends. These changes are depicted by FTIR and XPS. An increase in surface polarity of the polycarbonate occurs due to the scissioning of carbonate bonds to form hydroxyls and carboxyls [7, 11]. These changes probably increase the surface reactivity resulting in increased adhesional strength. A more energetic treatment causes massive degradation. This degradation results in the weakening of the outer layer as observed by SEM and probably causes a reduction in the adhesion strength.

FTIR analysis of polyetherimide reveals fewer chemical changes which occur mainly at the imide ring, to form polyamic acid, which improves the surface chemical activity [7, 11].

FTIR analysis of the PEEK composite reveals formation of hydroxyls and aldehydic groups from the scissioned carbonate bonds, which probably modify the adherend surface. It also reveals massive crosslinking of the outer surface layer resulting in a cohesively stronger surface with no weak boundary layers. High laser intensities result in a total peak reduction due to degradation and ablation (Fig. 13). XPS analysis

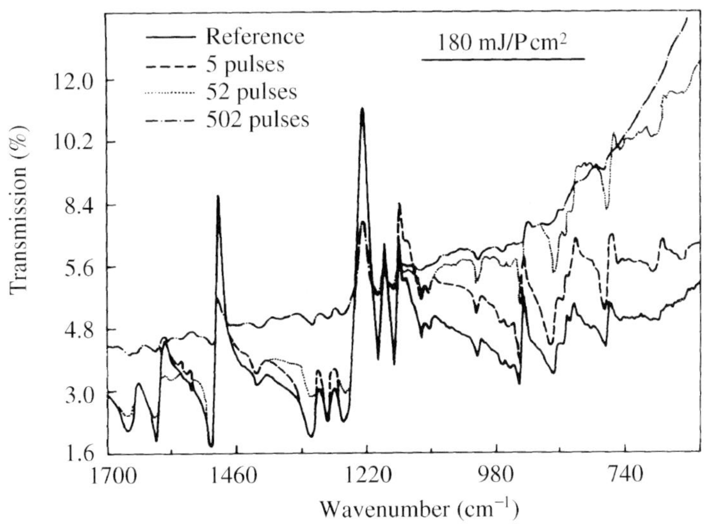

Figure 13. FTIR spectra of carbon reinforced PEEK before and after laser treatment.

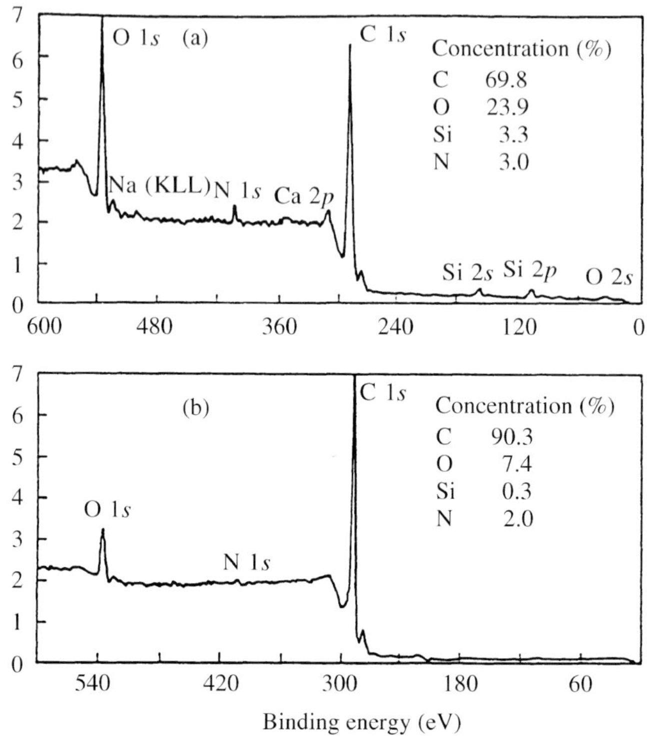

Figure 14. XPS spectra of polyetherimide before (a) and after (b) laser treatment.

(Figs 14–16) confirms the FTIR results and in addition reveals another important advantage of laser treatment, i.e. cleaning of the adherend surface by evaporating the contaminants and the weak boundary layers from the adherend surface at the initial stage of the treatment (Figs 14 and 15). From polyetherimide surface, Si was removed and from the PEEK composite surface, Mg and Si were removed at 0.18 $J/P\,cm^2$, 100 pulses.

FTIR spectra of anodized Al reveal the removal of adsorbed and structural water molecules from the anodic film and dehydration of the hydroxide to Al oxide [10]. A growth of a new Al oxide layer on the laser cleaned surface is observed by FTIR for the pure Al alloy [4]. Removal of organic carbon from the Al surface and formation of a new oxide was observed by Auger analysis and FTIR [9].

Measurement of the microhardness of the metal adherends which were polished prior to laser treatment showed a slight increase in hardness for both Cu and Al metals (of about 4–8%). This increase in hardness is typical of other laser processes as well (welding, cutting, printing, etc.) and results probably from oxidation and melting and solidification (aging).

Wetting properties were monitored by measuring water contact angles on Al adherends and composite PEEK adherend following laser treatment. The Al adherend showed a decrease in contact angle (better wetting) after laser treatment due to chemical modification of the surface (oxidation). PEEK composite showed increase in

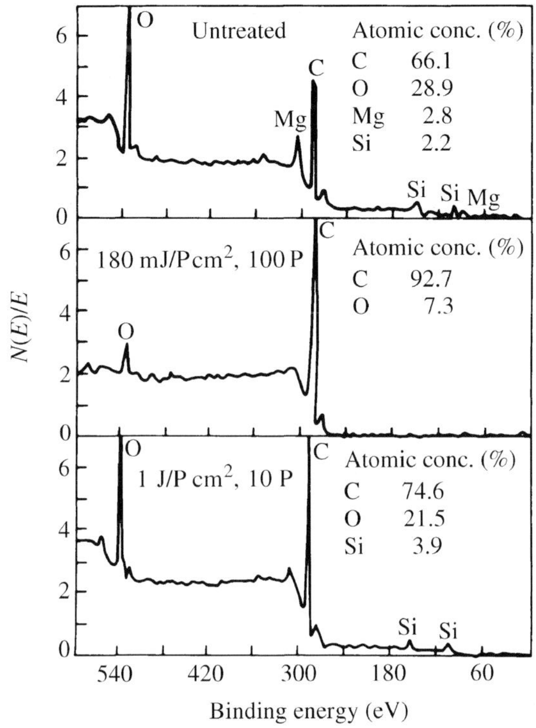

Figure 15. XPS spectra of carbon reinforced PEEK before and after laser treatment.

Table 4.
Contact angles of water on untreated and laser treated adherends and comparison to adhesive bond strengths (SLS)

Adherend	Surface treatment	Contact angle (deg)	SLS (MPa)
APC-2/AS-4	None	34	5.9
	Laser 0.18 J/P cm^2, 100 P	110	27.2
	Laser 1 J/P cm^2, 10 P	59	27.8
Al-2024	None	90	12.8
	Anodized	*	43.0
	Laser 0.18 J/P cm^2, 2000 P	41	34.0

*Anodized samples react with water.

contact angle due to increased crosslinking which lowers the polarity and increases the hydrophobic character (a positive effect of rejecting humidity from the surface) (Table 4).

Durability tests were conducted on Al 2024-T4 and on PEEK composites treated with UV laser at 60°C and 95% RH for up to 60 days. No degradation was observed during this period for either joint. An example of durability results is given in Table 5.

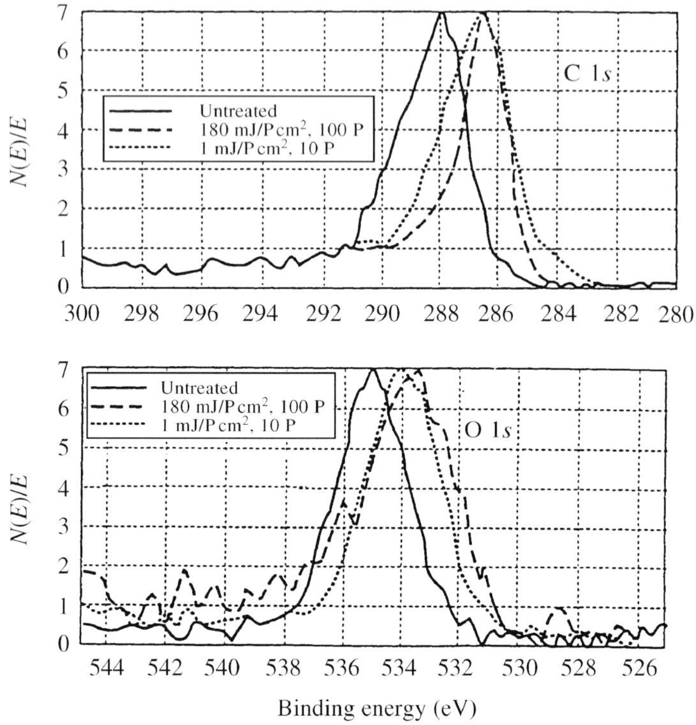

Figure 16. Comparison of XPS spectra of C 1*s* and O 1*s* of carbon reinforced PEEK before and after laser treatment.

Table 5.
Lap shear strength of APC-2/AS-4 with various surface treatments bonded with FM 300-2K and exposed to hot/wet environment

Time (days) at 60°C/95% RH	Lap shear strength (MPa)[a] Failure mode[b]		
	No treatment	Abrasive (SiC)	Laser treatment 0.18 J/P cm^2, 100 P
0	6.1 ± 0.8 (A)	14.7 ± 1.9 (A)	27.2 ± 0.8 (M)
10	2.6 ± 0.1 (A)	16.6 ± 1.5 (M)	38.6 ± 2.2 (M)
30	3.9 ± 0.4 (A)	13.2 ± 0.6 (A)	32.9 ± 6.0 (M)
60	4.5 ± 0.2 (A)	12.8 ± 1.5 (A)	29.2 ± 4.3 (M)

[a] ±5% standard deviation.
[b] A — interfacial, M — mixed.

4. CONCLUSIONS

The use of ArF excimer laser as a preadhesion surface treatment for various substrates (plastics, ceramics, composites and metals) has been investigated. The optimal laser treatment for each material to achieve the maximal shear strength was found to be dependent on the material. Deviating from optimal laser conditions may result in decomposition, melting and ablation, which should be avoided.

The UV irradiation as a surface treatment proved to be as effective as, or better than other conventional treatments such as chemical etching, abrasive blasting and even plasma treatments, and UV lamp irradiation.

The general phenomena observed due to UV laser treatment were: surface cleaning by removal of contaminations and weak boundary layers through evaporation; modification of surface chemistry by incorporating polar groups such as oxide derivatives and hydroxides; surface crosslinking and hardening (as observed by Auger, XPS and FTIR spectroscopies); and change of surface morphology by introduction of uniform roughness as observed by SEM. These phenomena were also observed by contact angle changes: its decrease for metal adherends indicating an enhancement of critical surface energy (γ_c) and its increase for PEEK composite due to hydrophobic effect, i.e. reduction of polar groups on the surface, and increase in hardness due to crosslinking (recombination) and removal of weak boundary layers.

Adhesional strength was improved compared to non-treated adherends and was similar to, or higher than conventional treatments. In addition, change in failure mode from interfacial to cohesive due to an increase in chemical and physical interactions, and mechanical interlocking occured in most cases.

It can be concluded that all four important considerations affecting adhesional strength were met using UV laser treatment: cleanliness, mechanical interlocking, chemical attraction, and wetting. The potential of using UV laser for preadhesion surface treatment was shown for a wide variety of substrates, each requiring its optimal laser parameters for a successful treatment.

UV laser treatment has many advantages such as: enhancement of adhesional strength and durability, limited surface effect without any deleterious effect on the bulk properties of the adherend treated. The process is effective in air and at room temperature, and is an effective, clean, environmentally friendly, precise, and safe process.

REFERENCES

1. J. D. Venables, *J. Mater. Sci.* **19**, 2431 (1984).
2. E. Sancaktar, S. V. Babu, G. S. D'Couto and H. Lipshitz, in: *Proc. of the Adhesion Society Meeting*, p. 203 (1993).
3. J. Heitz, E. Arenholz, T. Kefer, D. Bauerle, H. Hibst and A. Hagermeyer, *App. Phys. A* **55**, 391 (1992).
4. H. Watanabe and T. Takata, *J. Adhesion Sci. Technol.* **8** (12), 1425 (1994).
5. G. Hourdakis and E. Hontzopoulos, in: *Excimer Lasers and Applications III*, SPIE Vol. 1503, p. 249 (1991).
6. M. Geiger, N. Luz and S. Bierman, in: *Excimer Lasers and Applications III*, SPIE Vol. 1503, p. 238 (1991).
7. A. Buchman, H. Dodiuk, M. Rotel and J. Zahavi, *Intl J. Adhesion Adhesives* **11**, 144 (1991).
8. M. Rotel, J. Zahavi, A. Buchman and H. Dodiuk, *J. Adhesion* **55**, 77 (1995).
9. A. Buchman, H. Dodiuk, S. Kenig, M. Rotel, J. Zahavi and T. J. Reinhart, *J. Adhesion* **41**, 93 (1993).

10. Z. Gendler, A. Rosen, M. Bamberger, M. Rotel, J. Zahavi, A. Buchman and H. Dodiuk, *J. Mater. Sci.* **29**, 1521 (1994).
11. E. Würzberg, A. Buchman, E. Zylberstein, Y. Holdengraber and H. Dodiuk, *Intl J. Adhesion Adhesives* **10**, 254 (1990).

Mittal Festschrift, pp. 407–425
W. J. Van Ooij and H. R. Anderson, Jr. (Eds)

Adhesion activation of Twaron® aramid fibre

J. MAHY,[1,*] L. W. JENNESKENS,[2] L. L. T. VERTOMMEN,[1] P. J. DE LANGE[1] and O. GRABANDT[3]

[1]*Akzo Nobel Central Research, P.O. Box 9300, 6800 SB Arnhem, The Netherlands*
[2]*Debye Institute, Dept. of Phys. Org. Chem., Utrecht University, Padualaan 8, 3584 CH Utrecht, The Netherlands*
[3]*Akzo Nobel, Aramid Products, P.O. Box 9300, 6800 SB Arnhem, The Netherlands*

Abstract—Twaron aramid fibres require a surface treatment process in order to optimize their adhesion to polymer matrix systems. In this process, a finish containing epoxy reaction products is applied to the fibre surface and subsequently heat treated. The resulting fibre surface yields improved interfacial adhesion to, for example, epoxy and rubber matrices. In order to assess the role of the chemically and topographically modified fibre surface, detailed analysis of the interphase is a prerequisite. Well-defined starting materials and model components combined with surface analysis techniques are used. The reaction products in the epoxy-based finish are determined using a model formulation. The reaction chemistry, referred to as 'maturation' of the finish, and its relationship to an important process parameter, the pH of the formulation, is established. Quantitative information on the dimensions and the chemical composition of the interphase, the presence of functional groups at the fibre surface and the correlation between the interphase structure and the measured adhesion level to an epoxy matrix is studied.

Keywords: Interphase; adhesion; aramid fibre; epoxy; surface treatment.

1. INTRODUCTION

Twaron aramid fibre has exceptional physical properties, such as high specific tensile modulus, tenacity and dimensional stability, and chemical properties, such as chemical inertness and thermal/oxidative stability. These exceptional properties are associated with the highly ordered, crystalline structure of poly-(*p*-phenylene terephthalamide) (PPTA) [1]. Therefore, Twaron is now widely used as a reinforcing material in fields of application including tires, mechanical rubber goods (MRG), ropes, cables, optical fibre cables, friction material, protective fabrics and advanced composites.

The composite applications require optimal mechanical properties of fibre and matrix as well as optimal interfacial adhesion. However, for fibre reinforced thermoset, thermoplastic and rubber polymer matrix systems, interfacial adhesion may not pro-

*To whom correspondence should be addressed.

vide optimal load transfer between fibre and matrix. Notably for polyester fibre [2] and aromatic polyamides, such as aramid fibre [3, 4], it is known that the adhesion to, for example, polyvinylchloride, polyethylene and rubber is too low, and an 'adhesion activation' process is required. Ways have been developed to improve the aramid fibre–matrix adhesion, while at the same time retaining the excellent bulk physical and mechanical properties of the fibre.

Even when limited to aramid fibres made from PPTA, much work has been published on fibre surface modification and analysis, determination of the dominant adhesion mechanisms, micromechanical composite testing and modeling of the interface/interphase. Among the applied fibre surface modification techniques are reactive gas (e.g. fluorine) [5] and plasma [6] (e.g. ammonia) treatments, grafting and chemical attachment of functional groups, such as isocyanates and epoxides using wet-chemical or vapour phase techniques. Other surface modifications are obtained using metalation reactions, acidic or basic hydrolysis of the surface, etc. [7–9].

Surface analysis is aimed at detection and quantification of chemical groups at the fibre surface and their distribution and orientation as well as the characterization of thin (organic) films and contaminants. Among the techniques used to this end are X-ray Photoelectron Spectroscopy (XPS), Static Time-of-Flight Secondary Ion Mass Spectrometry (SSIMS, ToFSIMS) [10, 11] and infrared spectroscopy mapping [12]. Also, physical changes at the fibre surface, e.g. roughness and porosity, crystallinity, as well as composite failure mechanisms can be investigated using techniques such as electron microscopy (SEM, TEM) [13], surface force microscopy (AFM) and adsorption-based surface area determination (BET). Inverse Gas Chromatography (IGC) offers information on both the dispersive (i.e. nonpolar) characteristic and the polarity (acidity or basicity) of the surface [14]. Surface interactions can also be studied using thermogravimetrical analysis and microcalorimetry [15].

Adhesion theories, used to explain the (surface) *structure*–(adhesion) *property relationship* require detailed information at both the microscopic and the macroscopic levels. Theories such as *adsorption* (both physisorption, such as van der Waals and acid–base interactions, and chemisorption through the formation of, for example, covalent bonds), *electrostatic* interactions, *diffusion* theory of interpenetration in the interphase, and *mechanical interlocking* are used to describe the phenomena observed at the fibre–matrix interface/interphase [16].

As pointed out by Kendall [17], 'adhesion', or in practice 'adhesion strength' [16], involves both microscopical molecular interactions at the interface as well as macroscopic mechanical deformation required to separate the bonded materials. As indicated by K. L. Mittal on numerous occasions, e.g. at the Opening Ceremony of the 1st International Conference on Adhesion Science and Technology, this paper being a contribution to the Festschrift in his honour, there are several hundred 'adhesion tests', all of which measure a different (often convoluted) aspect of adhesion, usually for traditional or pragmatic reasons. However, by observing relative changes in the thus measured 'adhesion level', and correlating these changes to structural changes detected with other means, insight in the structure–property relationships can be obtained. Micromechanical analysis of fibre and (model) composites complements surface analysis in determining the interfacial properties under various load applications, for example by using *in situ* micro-Raman spectroscopic analysis of these systems [18].

There are different views as to the relative importance of the various mechanisms involved in the aramid-to-polymer matrix adhesion. For example, some workers emphasize the significance of covalent bonding as a necessary condition for improving interfacial adhesion of aramid fibre to, for example, epoxy matrix systems [12, 19, 20]. Other groups show evidence that covalent bonding, diffusion or interlocking phenomena are not operative in the aramid–epoxy system [21, 22]. In an earlier report, some of the present authors have presented evidence for a physisorption-dominated adhesion mechanism between Twaron aramid fibre and various epoxy systems [23]. This conclusion is based on XPS analysis of Twaron, treated with model epoxy components, before and after extraction with dichloromethane or acetone, in agreement with recent results reported elsewhere [15].

Generally, cleaning of the fibre surface prior to surface modification is essential in order to remove finishes, sizes and other processing or end-use aids as well as inorganic and organic contaminants from the surface [22, 24, 25]. However, in many cases, the surface structure is irreversibly modified [11, 26], either chemically or morphologically (porosity, crystallinity), which may additionally impair the physical properties of the fibre as a whole [27]. This is because the species present at the fibre surface may have a chemical composition consisting of the same elements (usually C, O, N) as that of the fibre, and are therefore difficult to detect, and also they may form a layer that is only a few nanometers thick, the removal of which is critical without affecting the underlying fibre. Moreover, the compatibility of the intended surface modification with some of the finish or size components present at the surface may be essential for the successful implementation of the modification in an industrial process.

It is therefore important to study not only the surface properties, e.g. wettability and adhesion, and the surface chemical composition resulting from a specific modification process, but also to take into account the possible changes in the bulk physical properties, as well as the surface chemistry and crystallinity of the starting polymeric fibre. Crystallinity at the surface is still a largely unknown feature. This 'surface crystallinity' means the crystalline order of the polymer chains in the uppermost nanometers of the fibre, and not the skin/core crystallinity differences, present in aramid, which extend over several hundreds of nanometers.

Aramid fibres have been studied as substrate for many surface modification techniques. However, only in very few cases was a thorough analysis made of the starting materials surfaces. Using XPS, it was observed by Sheu and co-workers on Kevlar 149 fibers (washed and dried) that the starting material had a very high surface oxygen concentration relative to the theoretically expected stoichiometry, which was attributed to surface oxidation during processing [11]. In a report by Rebouillat and co-workers on 'finish-free' Kevlar 29 yarns, similar high surface oxygen concentrations were noticed, but not explained [15].

Many interesting and advanced surface treatment processes have the disadvantage of being time-consuming, expensive, difficult to tune and control or chemically hazardous [27]. For major industrial applications, these aspects should be avoided when considering a fast, cheap, extremely reproducible and preferably on-line surface modification process, which moreover complies with occupational health and environmental regulations with regard to toxicity, processability and safety. There are few surface treatment processes which cover most, if not all, of the aforementioned constraints. An example of such an industrial surface treatment process, leading to improved interfacial adhesion properties of Twaron aramid fibre reinforced polymeric matrix systems is considered in this work.

The commercial Adhesion Activation (AA) process for Twaron aramid fibre (to rubber and synthetic matrix systems) is described in a patent [28]. It consists of the application to the fibre surface of an epoxy-containing finish, and subsequent curing of the finished fibre. This process can be carried out on-line, and yields an excellent adhesion level, both for rubber and epoxy-composite applications [4, 10]. The finish, apart from processing components, e.g. for yarn lubrication during spinning, contains the reaction product of an epoxy-based component and alkaline curing agent in the presence of one or more catalysts [28].

Prior to application of the finish to the fibre surface, the components present in the finish solution are expected to undergo different chemical reactions. This is the so-called maturation or 'ripening' of the aqueous formulation. In a previous report by some of the present authors, an account was presented of the effect of a commercial adhesion activation process on the interphase chemical structure of Twaron–epoxy composite systems [10]. This investigation is further elaborated, and now includes the determination of the maturation chemistry, the 'quantification' of the interphase, and the effect of the interphase structure on the properties of the adhesion activated (AA) Twaron fibre. The AA-process is aimed at optimizing the adhesion level of Twaron to different matrices. In this work, use is made of a 'bundle pull out' test for the adhesion of Twaron to thermoset matrix systems. A simulation of this AA-process was carried out, with well-defined starting materials and process conditions, to study the effect of these conditions on the surface structure and chemistry in a multi-analytical approach.

The adhesion activation process is monitored by the parameters describing the maturation chemistry and the thermal treatment of the fibre on which the adhesion active and maturated finish is applied. For the maturation chemistry, these parameters are the reaction time and temperature, the concentration and ratio of epoxy and hardener (curing agent) as well as the finish components present in the formulation. This process is monitored by the pH of the aqueous formulation as a function of time. The time-dependent pH-curve has, for specific temperature and concentration, a characteristic shape, which was not systematically investigated.

The thermal treatment of the finished fibre is characterized by the temperature and residence time, the amount of aqueous formulation applied to the fibre surface and the drying process prior to thermal curing of the finish.

This work contributes to a better understanding of the surface structure–property and structure–process relationships. This understanding serves two purposes: gaining a fundamental insight into microscopic interfacial phenomena and designing an optimal surface treatment process for the desired adhesion property of the fibre/composite.

2. EXPERIMENTAL

2.1. Starting materials

Akzo Nobel's Twaron aramid yarn (1680 dtex, i.e. 0.1680 kg per km of the bundle, with 1000 filaments per bundle) was used, without finish, size, pretreatment or processing oil and without twisting the bundle. Finish-free aramid fibre is not commercially available, since the static and frictional properties lead to difficult processing and handling of the yarn. However, this material can be spun, yielding the same physical properties (e.g. modulus, strength) and structure (crystallinity) as the commercial (finished) yarn. Equally important is the fact that the starting yarn has not received mechanical treatments such as (extreme) twisting, bending or abrasion, as in knitted fabrics, chopped fibres and aramid pulp (e.g. for friction material), since this influences the bulk and surface structure (morphology) and consequently the mechanical properties of the system due to, for instance, fibrillation or the formation of kinkbands in the filaments. The surface of the thus-obtained Twaron yarn was screened for the presence of adventitious contaminants and processing components using XPS.

2.2. Adhesion activation: model process

Because of the complex nature of the system, and the number of components and process steps involved, it was decided to investigate a model system of the adhesion activation process. In this model system, each relevant process step was analyzed separately using appropriate starting materials. For instance, the epoxy-reaction chemistry during maturation was investigated using a model epoxy component (glycidol) and piperazine in water.

The application to the fibre surface and subsequent curing behaviour was investigated using a commercial di-epoxy, diglycidylether of glycerol [29] and piperazine dispersed in a water/ethanol mixture. Preparation of the aqueous solution was performed by mixing a solution of the epoxy component in ethanol with a solution of piperazine in water at 20 or 30°C in a closed vessel under continuous stirring. The ethanol/water concentrations are 50/50 on mass fraction, while the epoxy/piperazine concentration is 9/1 on mass fraction, in a 1.5% dilution of the active components. The pH curve was monitored as a function of time.

Application and curing was carried out in a laboratory set-up with a calibrated application pump to treat the yarn surface with a known amount of finish solution. Application of 0.3% of the starting components (epoxy and piperazine) on the yarn, expressed as percentage of the bundle's mass, corresponds to a calculated average layer thickness of around 10 nm, assuming uniform coverage of the filaments by the finish components. The yarn with the applied finish was dried at 70°C before curing at 240°C in order to remove ethanol and water. The drying step was carried out in a hot air oven, while curing was accomplished by drawing the finished yarn over a hot metal plate for 5 s. During these stages, part of the finish is thermally and mechanically removed from the surface. Therefore, the cross-linked (cured) finish is expected to have a layer thickness below 10 nm.

2.3. Adhesion measurement

Adhesion measurements were carried out using a 'bundle pull-out' test for evaluation of the adhesion of a bundle of filaments to thermoset polymer systems, such as epoxy [30]. A twisted bundle (50–100 turns per meter) of aramid yarn is threaded through a hole in a metal plate. A pre-tension of around 1–2 N is applied to the bundle, which forms a tightly packed and nearly circular arrangement of filaments. A layer of 2–4 mm of epoxy is cast in a mould around the hole, and cured at room temperature for 72 h. The bundle itself is impregnated with the resin in order to provide a rigid ensemble. The bundle is cut off just above the resin surface at one end of the tablet. The bundle is then drawn from the cured resin tablet, using a tensile tester. The force per millimeter thickness of the thermoset matrix required to pull the bundle out of the matrix is calculated and normalized relative to the embedded length of the bundle, expressed in N/mm. The resin is a cold curing Novolak-epoxy/hardener system (XB 3052 A/B from Ciba).

2.4. Analysis

The maturation chemistry was analyzed using NMR, XPS and ToFSIMS. The starting epoxy/amine mass ratio was the same (9 : 1) in all experiments. Reactions were carried out in NMR tubes using deuterated solvents. The tubes were placed in a water bath at 30°C for 18 h. The reaction mixtures could be analyzed directly by ^{1}H- and ^{13}C-NMR, as a function of maturation time. For molecular weight determination, the reaction solvents first were removed before size exclusion chromatography (SEC) solutions could be prepared: this was done by freeze-drying the aqueous mixture.

The curing process, notably the interaction mechanism between the epoxide reaction products and the aramid fibre surface as well as the characterization of the interphase, was analyzed using XPS (in combination with extraction). The experiments were carried out on a VG Scientific Escalab 220i, using monochromatic Al K_α radiation and partly also on a VG Scientific Escalab MkII, using non-monochromated Al K_α radiation from a dual Al–Mg anode. Static secondary ion mass spectrometry was carried out on a VG Ionex ILX 23 Time-of-Flight SIMS, equipped with a 30 kV liquid gallium ion source and a Poschenrieder parallel mass detector. These instruments were described in detail elsewhere (see [10] and refs therein).

3. RESULTS AND DISCUSSION

3.1. Maturation reaction mechanism

The reaction mechanism of glycidol and piperazine in water (30°C) was studied as a function of maturation time. The piperazine curing agent reacts quantitatively (1 : 2) with glycidol, forming the epoxy ring opening adduct, as shown in Fig. 1a. The unreacted excess (9 : 1) undergoes an amine-catalyzed epoxy ring opening leading to oligomer formation, e.g. glycerol, as shown in Fig. 1b. In addition, in aqueous

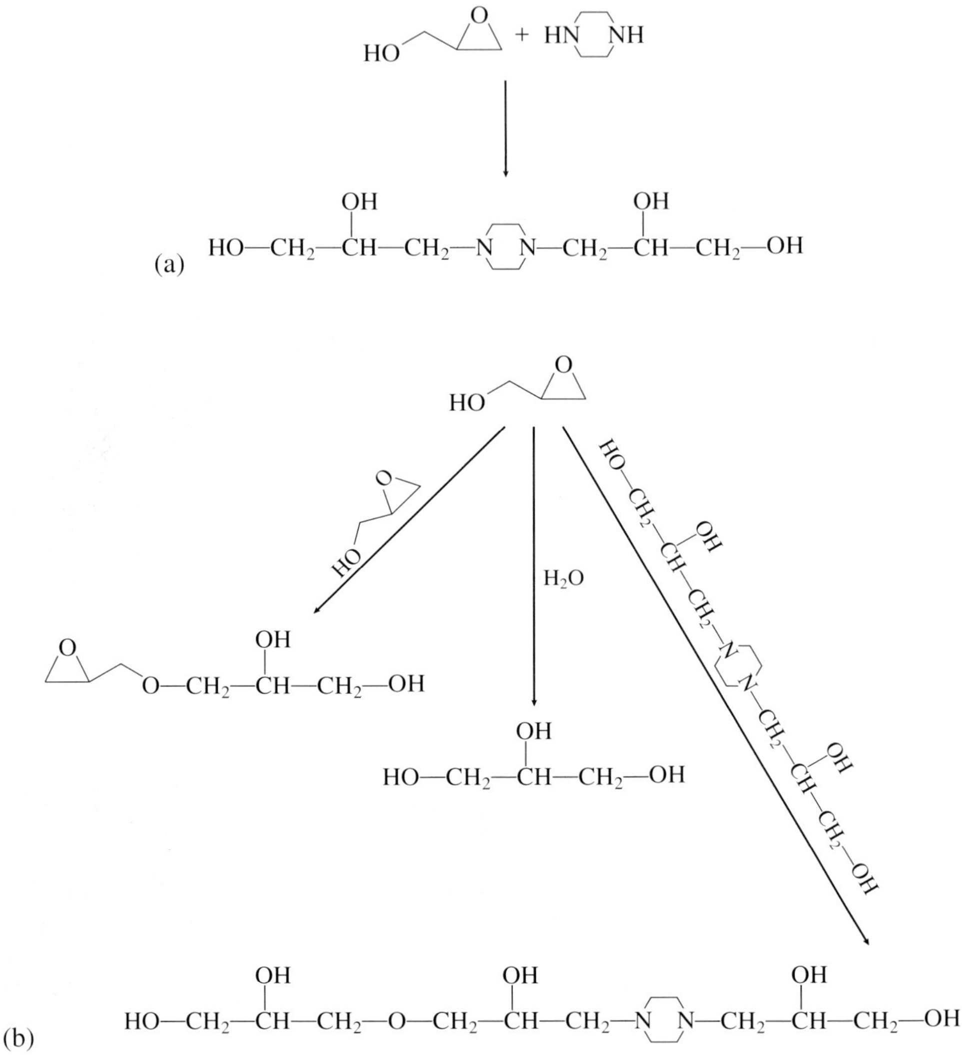

Figure 1. Schematic representation of the main reaction mechanisms observed in the epoxy-based finish model system (aqueous solution of glycidol/piperazine in 9/1 concentration), showing (a) adduct formation by stoichiometric reaction of piperazine with two glycidols, (b) oligomer formation by epoxy ring opening polymerization catalyzed by tertiary amines, formed according to (a), and hydrolysis of epoxy to vicinal diol due to presence of aqueous medium.

environment, the epoxy groups are also partially hydrolyzed to their corresponding vicinal diol. These reaction products were confirmed by NMR.

Whereas the first reaction (adduct formation) is relatively fast, the second reaction (amine-catalyzed epoxy ring opening) is much slower. Parallel to these observations, the pH behaviour as shown in Fig. 2a, is seen to consist of a fast decrease, and a subsequent slower increase. The decreasing pH can be explained by the conversion of the secondary amine into a tertiary amine (Fig. 1a), which leads to decreased basicity

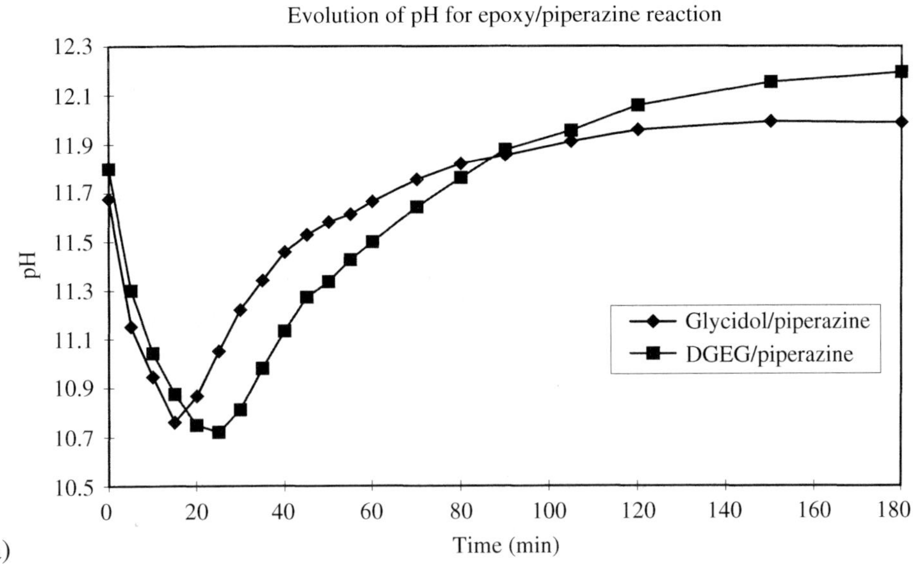

(a)

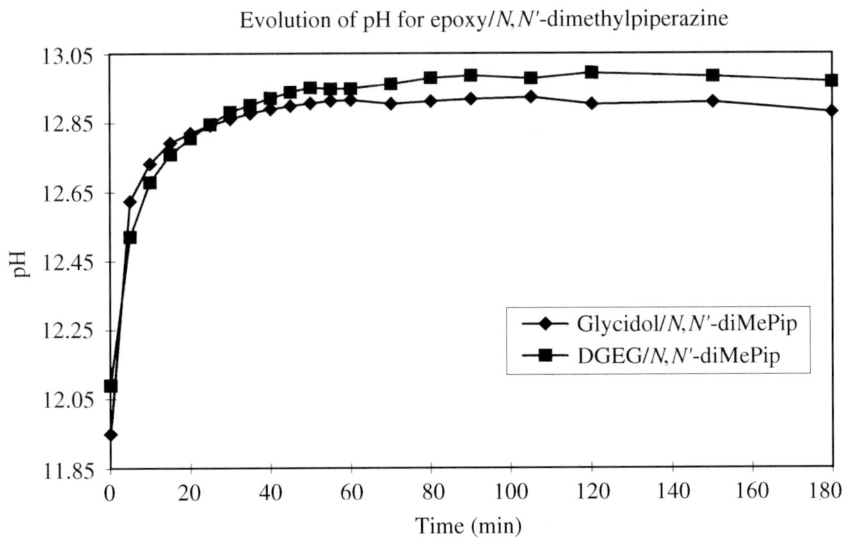

(b)

Figure 2. Time dependent behaviour of the pH for the aqueous epoxy : piperazine solution (a) and the aqueous epoxy : *N*,*N*′-dimethyl piperazine solution (b) at 30°C (both in 9 : 1 mass ratio). Results are similar for mono-epoxide glycidol and bis-epoxide diglycidylether of glycerol (DGEG).

of the system. The subsequent slow increase of the pH occurs simultaneously with the tertiary amine catalyzed epoxy ring opening reaction. However, the resulting reaction products alone (Fig. 1b) cannot be invoked to explain the high basicity of the system. In order to determine the role of the tertiary amine, a similar set of experiments was carried out using *N*,*N*′-dimethyl piperazine.

The stoichiometric reaction of glycidol with this tertiary amine is not observed in this case, and the epoxy ring opening reaction products of Fig. 1b are formed from the beginning. Moreover, no pH decrease was observed (Fig. 2b). This difference with the piperazine based system can be explained from the observation that the transformation from secondary amine in piperazine to tertiary amine in the adduct (Fig. 1a) leads to decreased basicity of the system. *N*,*N*′-Dimethyl piperazine, a tertiary amine, acts as catalyst from the beginning, leading to glycerol-type oligomer reaction products (Fig. 1b).

The slow pH increase with time, observed after the adduct formation with piperazine is completed and not observed with *N*,*N*′-dimethyl piperazine, cannot be explained by the epoxy ring opening reaction. It is proposed that a strong base is formed, which is responsible for this increase. A reaction product containing quaternary nitrogen could be considered as a strong base with a hydroxyl group as counter-ion. Evidence for the presence of such a quaternary nitrogen containing moiety was first observed in XPS. Subsequently, the molecular structure of this moiety was determined by ToFSIMS. Similarly, for *N*,*N*′-dimethyl piperazine a quaternary component was also detected.

The detection and characterization of the quaternary nitrogen containing component was carried out as follows. By gradually adding glycidol to a solution of piperazine in water, the pH starts to decrease after adding 2 equivalents of glycidol. It was found that the pH did not increase even after 20 h of reaction time (at 30°C). By adding more glycidol to the aqueous piperazine solution, up to the final ratio of 9 glycidol to 1 piperazine (mass ratio), the pH gradually increased again. By taking samples of the solution at different glycidol content, and after freeze-drying these samples, the solid could be transferred to a sample holder for subsequent analysis using XPS and ToFSIMS. As shown in Fig. 3, the N 1*s* signal obtained in this way from freeze-dried samples at different pH (different glycidol : piperazine ratio) shows an increasing contribution at higher binding energy (403 eV), associated with a charged nitrogen species (quaternary nitrogen). This contribution occurs only at ratios higher than 2 : 1 in agreement with the observation of the adduct formation, which at the 2 : 1 ratio contains only tertiary amines.

Figure 4 shows the ToFSIMS results obtained from the relative peak intensities of different molecular fragments as a function of the glycidol : piperazine ratio. Signals correspond to the molecular mass of the adduct with two or more glycidol units bonded to it. The relative peak intensities of oligomers with more than two glycidol units added to the diamine, i.e. the quaternary species, increase with increasing glycidol : piperazine ratio. Conversely, the signal associated with the mass of the adduct formed in the glycidol : piperazine = 2 : 1 ratio, i.e. the tertiary species with two glycidols, is seen to decrease with increasing glycidol/piperazine ratio.

The ToFSIMS peak intensities do not straightforwardly correspond with the actual (surface) concentration, but rather reflect the secondary ion yield of the corresponding mass fragment. However, since no other ion yield influencing mechanisms are assumed to be operative differently for the various adduct molecular ions, Fig. 4 is a qualitative indication of the relative surface concentration of the oligomers. The proposed oligomer structures, shown in the inset of Fig. 4, are not unambiguous however. The quaternary nitrogen containing structures, represented as R_3NH^+ type oligomers,

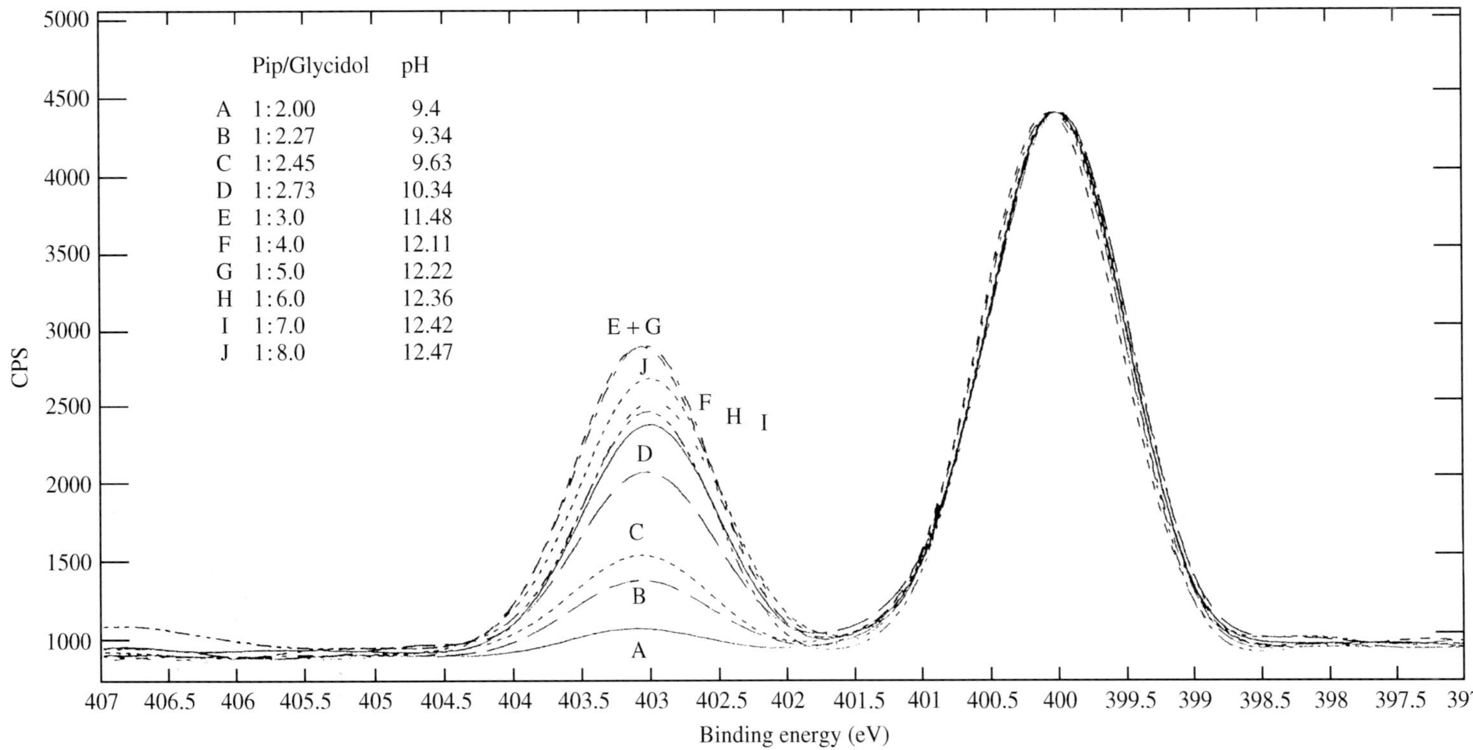

Figure 3. N 1*s* signal (XPS) obtained from freeze-dried samples with different glycidol/piperazine ratios. With increasing glycidol content, an increasing contribution at 403 eV is observed, which is attributed to N^+ species.

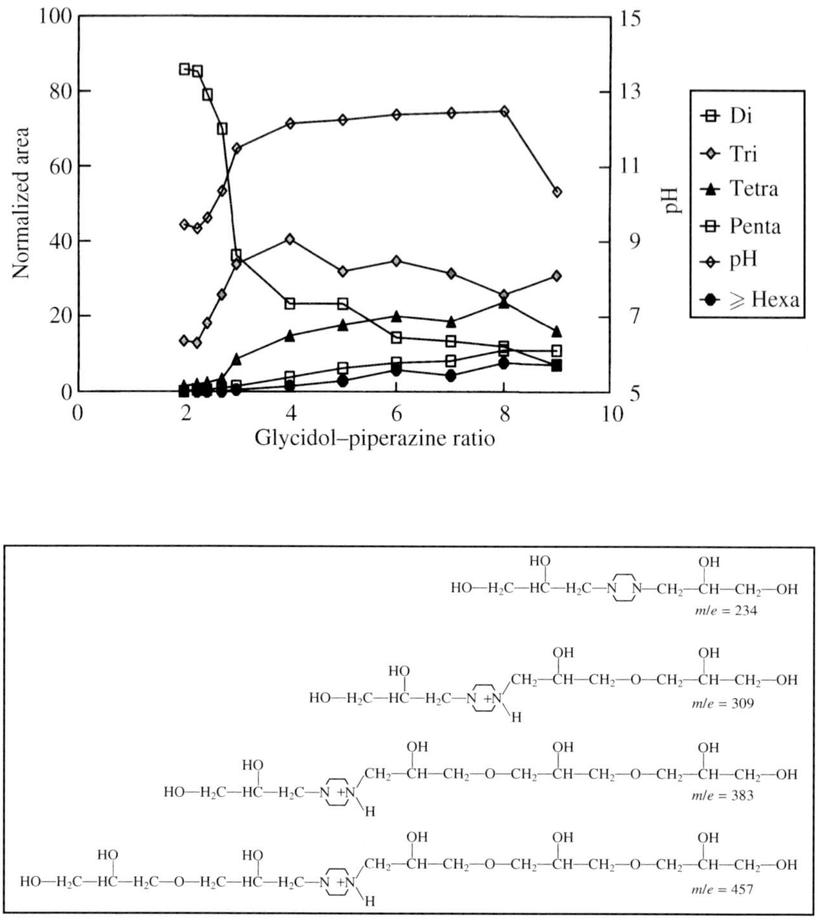

Figure 4. Plot showing the normalized static time of flight secondary ion mass spectrometry (S-ToFSIMS) peak areas for the molecular ion fragment signals of the adduct oligomers shown in the inset. Signal intensities are presented for different glycidol : piperazine ratios, as obtained from freeze-dried samples. These oligomers are possible reaction products of glycidol and piperazine, formed in the water/ethanol solution and desorbed by primary ion bombardment from the sample on a silver substrate as molecular ion (with or without Ag-cationization). The pH of the corresponding aqueous solution is also shown.

can also be represented as R_4N^+ type oligomers, i.e. components having the same m/e ratio. Only the latter can explain a high pH of the finish formulation (with OH^- as counter-ion).

These observations are in agreement with the assumption that the formation of the adduct with only two glycidol units (i.e. tertiary amine, uncharged species), which is responsible for the decrease in pH during maturation, is followed not only by the catalyzed epoxy ring opening polymerization, described previously (Fig. 1b), but also by the formation of oligomer adducts containing one quaternary nitrogen (i.e. charged species with OH^- counter-ion). These oligomers are supposedly mainly of the R_4N^+ type, and have been confirmed with NMR. The formation of the latter oligomers is responsible for the observed increase of the pH.

3.2. Curing mechanism and interphase characterization

The curing mechanism and the resulting fibre surface chemical composition and interphase structure have been studied using a different model system to that used in the maturation chemistry. A difunctional epoxide is required for the formulation of the finish, as described in the patent [28]. To this end, a commercial diglycidyl ether of glycerol was used with piperazine. For reasons of solubility and good wetting, a water/ethanol mixture was preferred. It was established in a similar way as described in the previous section that the maturation chemistry and notably the relation between the reaction product formation and the time-dependent behaviour of the pH are similar for both the aqueous glycidol and the difunctional epoxide.

Evaluation of the curing mechanism was carried out by monitoring the surface chemical composition of the treated Twaron fibre using XPS. Apart from the relative surface concentrations of the elements involved (mainly C, O, N), also the binding states of these elements can be quantitatively detected. Moreover, by analysing the detection angle resolved relative signal intensities and knowledge of the inelastic mean free paths of the core electrons, calculation of the surface layer thickness, homogeneity and the composition gradients can be performed. To this end, the treated Twaron fibres

Table 1.
XPS surface elemental concentrations for theoretical aramid stoichiometry for 'naked', unfinished Twaron aramid fibre, and for adhesion activated Twaron before (−) and after (+) extraction with ethanol. Concentrations measured at different detection angles: low information depth (30°) and high information depth (90°). Sample codes refer to (Non-)Ripened, (Non-)Cured system

Sample	Detection angle (deg)	Extraction	Concentration (at.%)				
			C	N	O	Na	Cl
Theory	—	–	77.8	11.1	11.1	—	—
Twaron	90	–	76.7	11.7	11.5	—	—
NRNC	90	–	73.5	7.4	18.0	0.2	0.9
RNC	90	–	74.0	8.6	16.6	0.2	0.6
NRC	90	–	75.5	9.4	14.7	0.1	0.3
RC	90	–	71.9	7.2	20.1	0.4	0.5
NRNC	90	+	76.4	10.7	12.7	—	0.1
RNC	90	+	76.2	10.3	13.4	—	0.1
NRC	90	+	73.3	7.6	18.1	0.4	0.5
RC	90	+	75.4	9.1	15.2	—	0.4
NRNC	30	–	70.2	5.4	21.9	0.1	2.4
RNC	30	–	72.0	7.0	19.3	0.2	1.5
NRC	30	–	68.9	5.5	23.4	0.3	2.0
RC	30	–	69.3	5.6	23.2	0.3	1.7
NRNC	30	+	76.4	10.1	13.4	—	0.2
RNC	30	+	76.3	9.3	14.2	—	0.3
NRC	30	+	73.6	8.2	17.5	—	0.6
RC	30	+	72.1	7.9	18.8	—	1.1

were analyzed both before and after extraction using ethanol. This was done after application of a freshly prepared epoxy-formulation as well as a formulation after maturation. Both systems were analyzed both with and without the subsequent curing step. The results for the chemical surface composition are shown in Table 1.

In general, the [C] : [O] : [N] concentrations deviate from the theoretical bulk stoichiometric concentrations for aramid, i.e. 78 : 11 : 11 at.%, respectively, notably prior to extraction. However, the starting material, unfinished Twaron fibre, shows an XPS surface chemical composition in excellent agreement to that of the theoretical bulk stoichiometry for *p*(phenylene terephthalamide) (Table 1). This shows that the material is very well defined, contrary to data in [11] and [15], where high surface oxygen concentrations indicate the presence of adventitious contaminants or functionalities. The C 1*s* signal for the starting material is presented in Fig. 5, and clearly shows the contribution of the aromatic, amide- and carbonyl-type carbon atoms, present in the PPTA polymer backbone.

The concentrations differ most from the theoretical bulk composition of aramid for the cured yarns treated with the formulation after maturation (RC), and show the smallest deviation for the non-cured yarns treated with the freshly made formulation (NRNC). Also, the results before and after extraction show the largest difference for NRNC, and the smallest for RC systems. This indicates that the extraction process

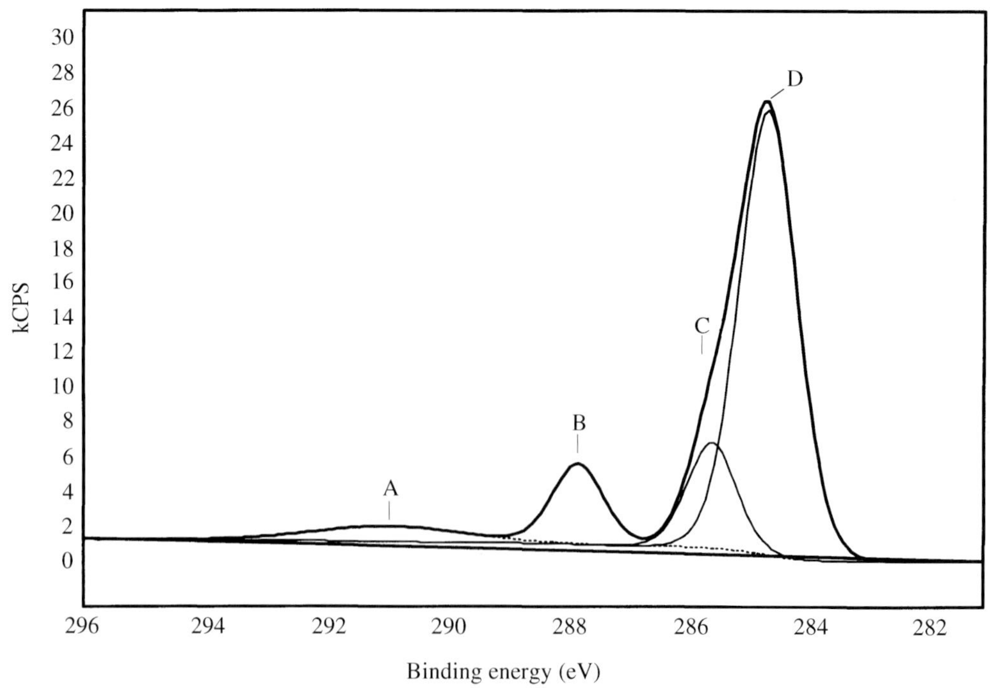

Figure 5. C 1*s* signal (X-ray Photoelectron Spectroscopy, XPS) obtained from the non-finished Twaron aramid fibre surface. Peak decomposition is in quantitative agreement with theoretical bulk stoichiometry for aramid, showing aromatic (A, D) and carbonyl (B) as well as amide bonded (C) carbons (see also Table 1 for XPS element concentrations).

removes part of the epoxy-reaction products from the Twaron surface less effectively when the finish formulation is mature ('ripened') and/or cured. The results also show that the presence of a higher amount of epoxy groups (highest concentration in freshly made formulation) at the yarn surface after application of the finish is not decisive for good interaction with the yarn, since this formulation can be more easily removed from the yarn than the maturated one, containing mainly oligomers of the glycerol-type (Fig. 1).

The [N] : [O] concentration ratio before extraction decreases with decreasing detection angle. This behaviour is not so clearly observed after extraction. It shows that the dominant part of the nitrogen signal is associated with the material deeper below the surface, while the oxygen is associated with the upper layers of the sample. This is expected since the nitrogen signal can largely be attributed to the Twaron, while the oxygen can for an important part be attributed to the epoxy-reaction products. The piperazine in the finished formulation is present only in a 1 : 9 mass ratio relative to the epoxy. Thus, the contribution to the nitrogen signal associated with the piperazine (adduct, oligomers) is much smaller than that associated with the aramid polymer backbone itself.

This conclusion is further supported by observation of the O 1*s* peak shape. This has been demonstrated earlier [10]: the oxygen signal associated with the ether- and hydroxyl-containing epoxy-reaction products shows a chemical shift in the O 1*s* signal which differs from that of the carbonyl moieties in the aramid structure. The N 1*s* peak shape, on the other hand, shows clearly a contribution from a quaternary nitrogen moiety, as expected from the maturation process (Fig. 3). From the O 1*s* peak shape contributions, i.e. the ether- and hydroxyl contribution and the carbonyl contribution, a calculation of the layer thickness of the epoxy formulation on the Twaron fibre can be made. An appropriate layer model and information about the peak shape and relative intensity of the elements present in both layer and fibre are required as input for this iterative calculation. In the calculation, the difference between the theoretically expected and measured intensities is minimized with the layer thickness as a variable. Both the 90° and 30° detection angle data can be used as starting points for the calculation; the difference in the layer thickness result is a measure for the reliability of the model used.

Alternatively, a similar procedure can be set up using the measured [C], [O], [N] concentrations from the C 1*s*, O 1*s* and N 1*s* signals, respectively. Now, the elemental concentrations of the layer and fibre are required as input. These concentrations are derived from the epoxy reaction products and the aramid bulk stoichiometry, respectively. Using the same layer model as in the previous approach, the same thickness should be obtained from the iteration, yielding a check of the reliability of the model used.

Assuming as layer model that the Twaron aramid fibre is completely covered with a uniform layer of the epoxy-reaction products, the results of the calculation are presented in Figs 6 and 7, before and after extraction with ethanol, respectively. The inelastic mean free paths are denoted in the figures, and elemental concentrations used for the calculation were obtained from Table 1. A detailed account of the calculation of the layer thickness is provided elsewhere [31].

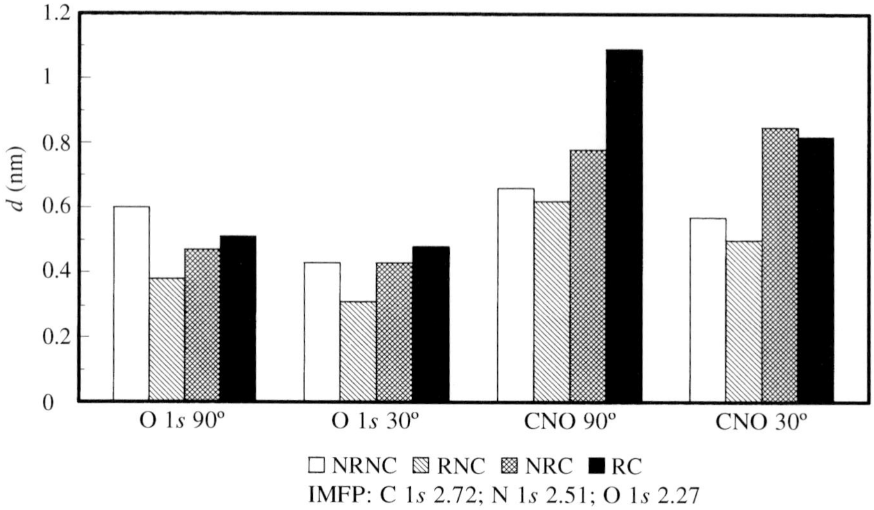

Figure 6. Interphase layer thickness calculations obtained from the XPS-derived surface concentrations of the elements C, N and O (Table 1) and peak decomposition of the O 1*s* binding states, assuming uniform coverage of the fibre surface with epoxy-reaction products. Results are obtained from as-received finished fiber samples for (non)ripened (N)R and (non)cured (N)C formulation of epoxy and piperazine at 30° and 90° XPS detection angles, and using inelastic mean free paths (IMFP) of electrons estimated for an organic matrix.

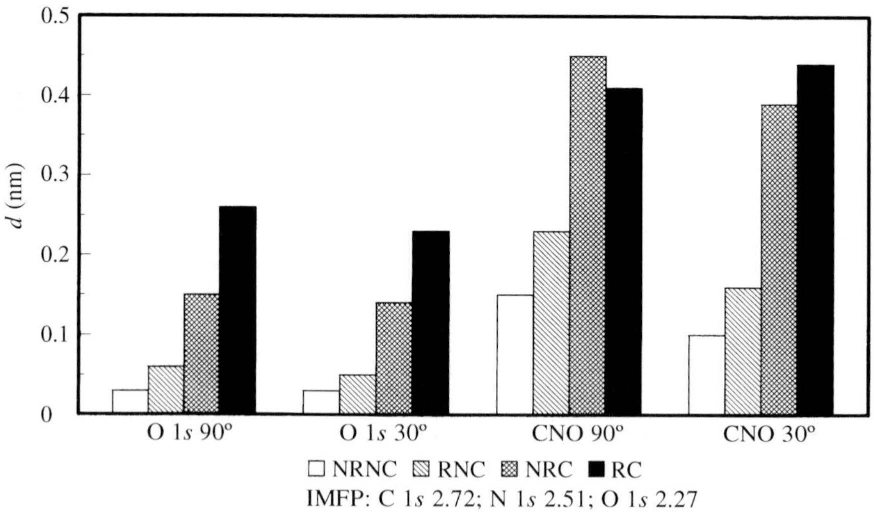

Figure 7. Same as Fig. 6, but obtained from finished fiber samples after Soxhlet extraction with ethanol.

Similar trends are present in all four data sets: thicknesses range between 0.4 and 1.0 nm with a general tendency for increasing thickness upon curing of the mixture. This tendency is less clear in the O 1*s* data than in the elemental concentration data. This may be due to the fact that carbonyl moieties may not entirely be attributable to the Twaron aramid fibre substrate as assumed in the model, but may also result from the curing process, thus leading to a lower calculated epoxy-reaction product layer thickness than in case of the elemental concentration data sets. Investigation of the curing process using infrared spectroscopy showed the presence of a carbonyl moiety at elevated temperatures. On the other hand, the good agreement between the 30° and 90° data sets suggests that the epoxy-reaction product layer is present as a well-defined layer at the aramid surface and is not penetrated (dissolved) into the fibre.

However, the results do not exclude a well-defined non-uniform layer consisting of thin domains of the reaction product, separated by regions of 'bare' aramid fibre. The iterations can also be carried out starting from a 'layer' model with incomplete coverage. For coverages between 60% and 100% of the analyzed surface, the model shows convergence. The resulting calculated layer thicknesses presented in Figs 6 and 7 must therefore be considered as minimum values and should roughly be doubled for non-uniform half coverage of the fibre surface.

This model of the interphase is supported by imaging ToFSIMS results reported earlier [10], where nanometer thick epoxy-based domains, extending typically 10–20 μm over the fibre surface were seen to be separated by regions of 'bare' Twaron fibre. The calculated subnanometer layer thickness suggests that only a fraction of the applied amount of the active species (calculated from the components to be 0.3% by mass, or approx. 10 nm thickness) is present at the fibre surface.

The same conclusions apply to the model system after ethanol extraction shown in Fig. 7. Here the tendency for increasing thickness with increasing maturation/curing is more obvious. The smaller overall layer thickness, i.e. from around 0.1 nm to below 0.5 nm corresponds with the extraction of reaction products from the fibre surface, which is more pronounced for not-ripened (less oligomeric) and not-cured (less cross-linked) materials than for the 'RC' reaction products. From these calculations it follows that both maturation and curing are necessary in order to obtain a layer which is cross-linked and which cannot be extracted from the fibre surface using ethanol.

This indicates that the 'RC' system adheres well to the fibre surface, although no evidence for diffusion of the components into the yarn has been observed using microcalorimetry. Based on these and earlier findings [10, 23], which excluded covalent and mechanical keying-based adhesion mechanisms, it can be concluded that there is a physisorption interaction mechanism between the aramid fibre and the epoxy-based finish. This physisorption is enhanced by the presence of the ether- and hydroxyl-groups from the reaction products described earlier (Fig. 1), as well as the possibility to form hydrogen bonds between the aramid and cured finish oligomers as originally proposed in [23].

3.3. Adhesion evaluation using Bundle Pull Out test

Obviously, a correlation between the observed surface structure (layer thickness and composition) with the adhesion strength to an epoxy matrix remains to be evaluated. The bundle pull out (BPO) adhesion results for the untreated reference Twaron (no finish), the commercial adhesion activated (AA) Twaron as well as the model system without/with maturation of the finish and without/with curing of the finished fibres are presented in Fig. 8. Clearly, the adhesion level increases from 'NRNC' (which is

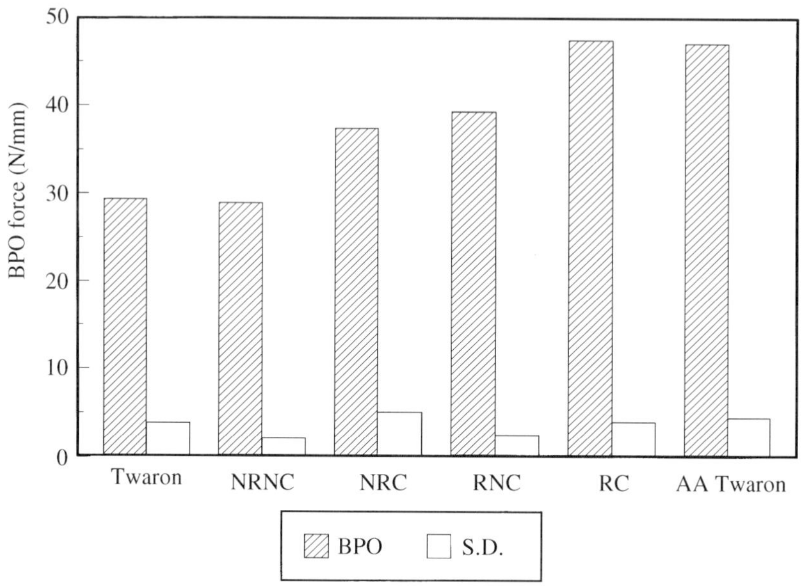

Figure 8. Bundle Pull Out Force (= practical adhesion) levels and their standard deviation (S.D.) expressed in N/mm, obtained from unfinished (Twaron) and commercial adhesion activated (AA Twaron) aramid bundles, as well as aqueous model finish of DGEG epoxy and piperazine on Twaron, (non)ripened ((N)R) and/or (non)cured ((N)C), pulled out of an epoxy matrix.

not significantly different from the untreated Twaron within experimental accuracy) to 'RC' (not significantly different from the AA-Twaron) by more than 50% from 30 N/mm to around 50 N/mm. The variability of these BPO experiments being rather large, as expressed by the relatively high standard deviation, the BPO levels for the differently treated Twaron bundles correlate within experimental accuracy with the interphase layer thicknesses calculated above (Fig. 6). It also shows that a very thin interphase (calculated to be around 1 nm), having an appropriate chemical composition, may be responsible for the large difference in the observed practical adhesion enhancement, as measured by BPO.

Using micro-Raman spectroscopy, it is possible to study the failure mechanism on a micrometer scale *in situ* and as a function of the applied macroscopic strain on the composite. Strain-induced Raman band shifts can be used to follow the distribution of stress or strain along individual fibres in a resin matrix, from which the interfacial shear stress can be derived at various loading levels (see e.g. [18] and refs therein). Low levels of interfacial shear stress (ISS) indicate a need to improve the adhesion properties of the fibres. It was observed that the maximum ISS for sized aramid fibres (Kevlar 49) differs from the ISS of de-sized ones, although the latter were found by XPS to have a surface composition largely different from that expected for the 'pure' aramid [32].

Recent experiments using Twaron aramid fibre both without finish and the adhesion activated material (commercial epoxy-based finish) have shown that the maximum ISS for AA-Twaron single filament epoxy composite systems differs significantly from that of the untreated Twaron, even at low matrix strains. At higher (> 1%)

matrix strains, the maximum ISS remains at a high level for the AA-Twaron, whereas the untreated Twaron maximum ISS decreases rapidly to around 10 MPa for 2.5% matrix strain [18]. The maximum ISS at high strains for the AA-Twaron system is at 40 MPa, close to the shear yield stress of the matrix epoxy resin. This indicates good interfacial adhesion along the filament, whereas at higher matrix strains for the untreated Twaron filaments debonding occurs. More detailed investigations of the mechanical deformation and failure for aramid fibres with and without an adhesion activation treatment will be reported soon.

4. CONCLUSIONS

The commercial adhesion activation process for Twaron aramid fibre consists of the application of an epoxy-based finish to the fibre surface, and subsequent curing thereof. The reaction products, which are being formed in the finish, have been determined ('maturation chemistry') and consist of oligomers of the epoxy–amine hardener adduct. An important process parameter for this 'maturation', the pH of the aqueous finish formulation, and its time-dependent behaviour have been explained in terms of the formation of these reaction products. The presence of quaternary nitrogen-containing oligomers, responsible for the high basicity of the finish, has been experimentally observed.

After application of the finish to the aramid fibre surface, an adhesion active interphase is created only after thermal treatment of the (mature) finish. This interphase has been characterized with respect to its chemical composition, the average layer thickness has been calculated and the layer (non)uniformity on a microscopic level has been established. A nanometer thin layer of epoxy-reaction products is held responsible for a significantly enhanced adhesion level to an epoxy matrix system, relative to the untreated fibre system, as measured using a Bundle Pull Out type test. From this and earlier [10, 23] work, a physisorption adhesion mechanism is invoked between fibre and matrix. The adhesion activated Twaron displays a micromechanical behaviour in epoxy composites, which is dramatically different from that of composites made with untreated Twaron [18].

Acknowledgements

The authors wish to thank their colleagues at Akzo Nobel Central Research: Mr. M. G. Simons for recording the XPS spectra and assistance with the development of the layer thickness calculations, Mrs. H. Henderiks for her help in the characterization of the maturation chemistry, Mr. M. E. F. Biemond for his help in the interpretation of the ToFSIMS results and Mr. P. Wessels for the finish formulation, application, curing and bundle pull out experiments.

REFERENCES

1. M. G. Northolt and J. J. van Aartsen, *J. Polym. Sci. Letters Ed.* **11**, 333–337 (1973).
2. T. Takeyama and J. Matsui, *Rubber Chem. Technol.* **42**, 159–256 (1969).

3. T. S. Solomon, *Rubber Chem. Technol.* **58**, 561–576 (1985).
4. H. Janssen, *Kautsch. Gummi Kunststoffe* **48**, 622–624 (1995).
5. M. A. Doherty, B. Rijpkema and W. Weening, Presented at the ACS Meeting, Rubber Division, Philadelphia, PA, May 2–5 (1995).
6. See e.g. M. Strobel, C. S. Lyons and K. L. Mittal (Eds), *Plasma Surface Modification of Polymers: Relevance to Adhesion*. VSP, Utrecht, The Netherlands (1994).
7. J. Kalantar and L. T. Drzal, *J. Mater. Sci.* **25**, 4186–4193 (1990).
8. Y. Wu and G. C. Tesoro, *J. Appl. Polym. Sci.* **31**, 1041–1059 (1986).
9. L. S. Penn, T. J. Byerley and T. K. Liao, *J. Adhesion* **23**, 163–185 (1987).
10. J. Mahy, L. W. Jenneskens and O. Grabandt, *Composites* **25**, 653–660 (1994).
11. G. S. Sheu and S. S. Shyu, *J. Adhesion Sci. Technol.* **8**, 531–542 (1994).
12. A. M. Mavrich, H. Ishida and J. L. Koenig, *Appl. Spectrosc.* **49**, 149–155 (1995).
13. B. C. Begnoche and R. L. Keefe, *Rubber Chem. Technol.* **60**, 689–704 (1986).
14. S. Rebouillat, M. Escoubes, R. Gauthier and A. Vigier, *Polymer* **36**, 4521–4523 (1995).
15. S. Rebouillat, M. Escoubes, R. Gauthier and A. Vigier, *J. Appl. Polym. Sci.* **58**, 1305–1315 (1995).
16. L.-H. Lee (Ed.), in: *Adhesive Bonding*, pp. 1–30. Plenum Press, New York (1991).
17. K. Kendall, *Science* **263**, 1720–1725 (1994).
18. M. C. Andrews, R. J. Young and J. Mahy, *Compos. Interf.* **2**, 433–456 (1995).
19. C. T. Chou and L. S. Penn, *J. Adhesion* **36**, 125–137 (1991).
20. E. G. Chatzi, S. L. Tidrick and J. L. Koenig, *J. Polym. Sci. B, Polym. Phys.* **26**, 1585–1593 (1988).
21. Y.-L. Hsieh, M. Wu and D. Andres, *J. Colloid Interface Sci.* **144**, 127–144 (1991).
22. A. Garton and J. H. Daly, *J. Polym. Sci.: Polym. Chem. Ed.* **23**, 1031–1041 (1985).
23. J. Mahy, L. W. Jenneskens, O. Grabandt, H. E. C. Schuurs, F. A. J. Geurts and J. Bijnen, in: *Interfacial Phenomena in Composite Materials '91*, I. Verpoest and F. Jones (Eds), p. 45. Butterworth-Heinemann, Oxford (1991).
24. E. M. Liston, L. Martinu and M. R. Wertheimer, *J. Adhesion Sci. Technol.* **7**, 1091–1127 (1993).
25. L. T. Weng, C. Poleunis, P. Bertrand, V. Carlier, M. Sclavons, P. Franquinet and R. Legras, *J. Adhesion Sci. Technol.* **9**, 859–871 (1995).
26. G. S. Sheu and S. S. Shyu, *Composites Sci. Technol.* **52**, 489–497 (1994).
27. P. Lee-Sullivan, K. S. Chian, C. Y. Yue and H. C. Looi, *J. Mater. Sci. Lett.* **13**, 305–309 (1994).
28. S. Willemsen, W. E. Weening and A. Steenbergen, 'Adhesive-coated multifilament yarn of an aromatic polyamide and a method for the manufacture thereof', European Patent Specification No. 0107887 (1987.01.21).
29. Nagase information sheet on Denacol® epoxy compounds.
30. F. Elkink and J. H. M. Quaijtaal, *Testing of Fibre–Matrix Adhesion*, Proc. Polymer Meeting at Rolduc on Integration of Fundamental Polymer Science and Technology, Kerkrade, The Netherlands (April 1988).
31. P. J. de Lange and M. G. Simons, *Surf. Interface Anal.* (submitted).
32. M. C. Andrews and R. J. Young, *J. Raman Spectrosc.* **24**, 539–544 (1993).

Mittal Festschrift, pp. 427–441
W. J. Van Ooij and H. R. Anderson, Jr. (Eds)

Surface photografting of polymer films and fibers for adhesion and adsorption control

BENGT RÅNBY

Department of Polymer Technology, Royal Institute of Technology, S100 44 Stockholm, Sweden

Abstract—Since the invention of an efficient one-step 'surface photografting' in our laboratories and its development as a *batch process* and a *continuous process* 10 years ago, the method has been applied extensively for adhesion and adsorption control of commercial film and fiber samples. In the batch process, the substrate is irradiated with UV light in an atmosphere of photoinitiator and monomer vapor in an inert gas. In the continuous process, the substrate is presoaked in a solution of photoinitiator and monomer which forms a thin liquid layer. The substrate is then UV-irradiated on line. In both processes, radicals are formed on the substrate surface by hydrogen abstraction. The surface radicals add monomer and form grafted chains as a thin surface layer which is analyzed by ESCA and reflection IR spectroscopy and by microtitration. A grafted layer 2 to 8 nm thick of a functional monomer, e.g. acrylic acid, acrylamide, vinyl pyridine or glycidyl acrylate, on an inert polymer surface, e.g. of a polyolefin or linear polyester, decreases the contact angle of water to complete wetting, increases the adsorption of textile dyes as much as 15 times and the adhesion to epoxy resin 3 to 6 times. Grafting with epoxyacrylates enhances surface bonding of heparin and polypeptides to the substrate. Secondary reaction of the epoxy groups with poly(ethylene glycol) makes the polymer surface hydrophilic, which prevents thrombosis of proteins.

Keywords: Surfaces; photografting; fibers; adsorption; proteins; adhesion; contact angle.

1. INTRODUCTION

Most commercial polymer materials have an inert surface with low adsorption of dyes and low adhesion to other materials. A simple test is the wetting of the surface with a drop of water. On a clean surface of polyethylene, polypropylene and polystyrene, the contact angle is about 90° while the surface of poly(methyl methacrylate) or a linear polyester, e.g. poly(ethylene terephthalate), shows lower values of about 70°. A contact angle of less than 20°, measured in advancing mode, can be obtained by surface modification using photografting.

The most inert polymer surfaces are silicones, e.g. poly(dimethylsiloxane), and Teflon®, poly(tetrafluoroethylene), with surface tensions of 20 to 25 mN/m. Commercial polymers like polyethylene, poly(vinyl acetate), poly(methyl methacrylate), and polystyrene have surface tension values of 35 to 40 mN/m.

This paper is largly a review of our recent work on the modification of polymer surfaces. After a short account of the background, it describes the surface photografting process, discusses the most recent studies on the mechanism of the surface photografting reactions, summarizes the available data on the structure of the grafted surfaces and presents the effects on the surface wetting, the adsorption of dyes and the adhesion of epoxy resins and commercial Scotch tape to the grafted surfaces. Some biomedical applications related to heparin, polypeptides and proteins are described. Later publications and patents by other authors are critically reviewed.

2. SURFACE PHOTOGRAFTING PROCESS

The one-step 'surface photografting' method was invented in our laboratories in 1984 [1]. Other grafting attempts described previously in the literature are either two-step processes or elaborate experiments which require long irradiation times (often hours) and give large amounts of homopolymer, i.e. they have low grafting efficiency [2–4]. We found that with benzophenone as a photoinitiator, with acrylic monomers like acrylic acid and acrylamide, with acetone as a solvent and high pressure UV lamps for irradiation, high yields of polymer and high grafting efficiency were obtained after short irradiation times [5].

In the *batch process* of surface photografting, the substrate is irradiated in a cell with an atmosphere of initiator and monomer vapor from a solution carried by an inert gas, e.g. nitrogen (Fig. 1). Because the vapor phase has low density, the transfer of initiator and monomer to the substrate is slow, and the process requires rather long irradiation times (minutes).

In the *continuous process* for surface photografting, the substrate is a filament, a yarn or a strip of film which is 'pre-soaked' by dipping in a solution of initiator and monomer and pulled on line into a reaction chamber where it is irradiated in a saturated atmosphere of the pre-soaking solution (Fig. 2). Because the transfer of initiator and monomer is in liquid phase on the substrate surface, only short irradiation times are required for grafting (seconds).

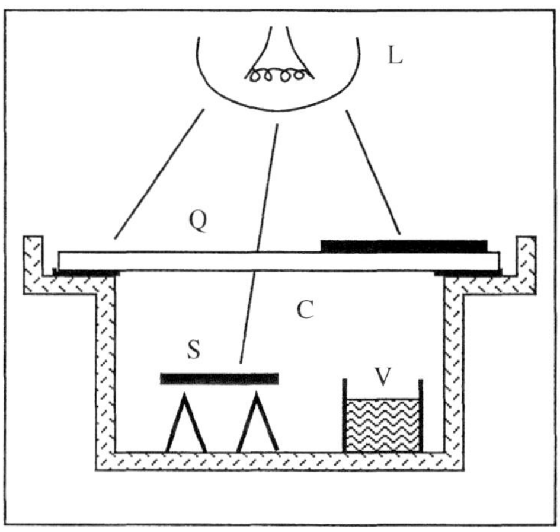

Figure 1. Experimental device for surface photografting with the batch process: specimen (S), solution of photoinitiator and monomer (V), reactor (C), quarts window (Q) and UV lamp (L).

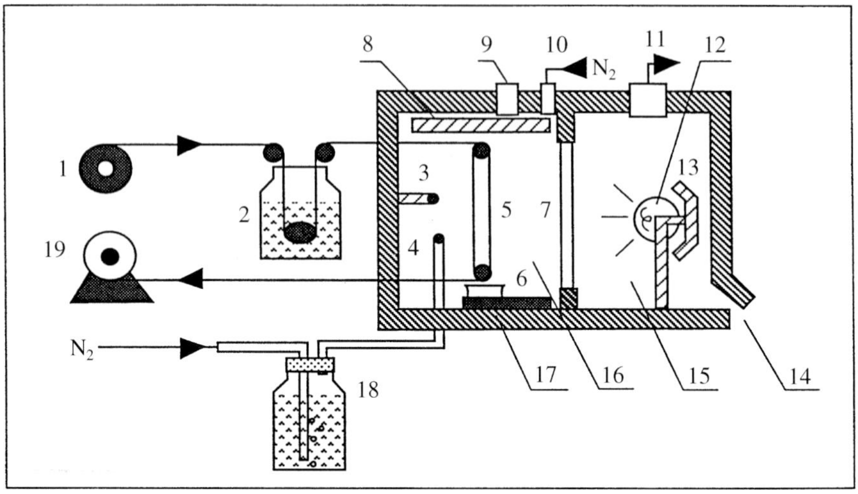

Figure 2. Experimental device for surface photografting with the continuous process: substrate on roll (1); pre-soaking solution of photoinitiator and monomer (2); temperature control (3); inlet for nitrogen (4) bubbled through solution (18) the same as in (2); rollers for specimen (5); electric heater (6); quartz window (7); water cooler (8); outlet (9); nitrogen inlet (10); air outlet (11); UV lamp with reflector (13); air inlet for cooling (14); lamp house (15); reaction chamber (16) and vessel for solution drops (17).

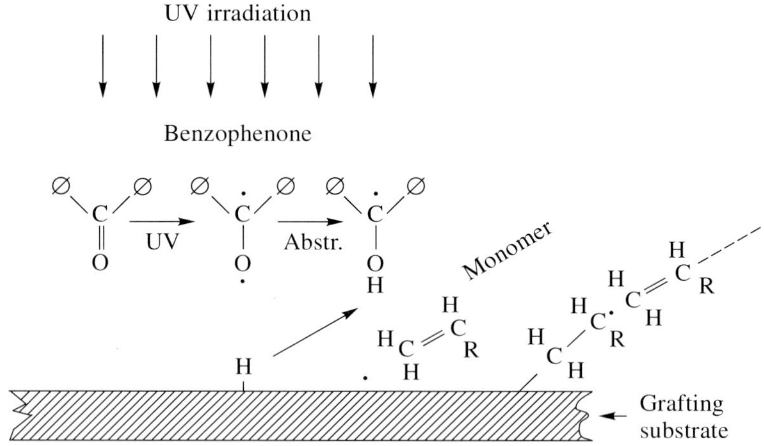

Figure 3. Mechanism of surface photografting of a vinyl monomer onto a substrate with benzophenone as photoinitiator.

In both processes, the reaction occurs at the substrate surface. The photoinitiator, usually benzophenone (BP) is excited to a singlet state by a ($\pi-\pi^*$) transition which forms a ($\pi-\pi^*$) triplet state by intersystem crossing (ISC). We have found recently that the ISC energy (singlet to triplet state) is dissipated by a non-radiative process to the substrate surface [6]. The BP triplet state formed abstracts hydrogen from the excited substrate surface and forms a surface radical ($P^\bullet$) and a ketyl radical ($BPH^\bullet$). The surface radical initiates graft polymerization by adding monomer. Most of the grafted chains are terminated by adding ketyl radicals (equation (1) and Fig. 3).

Acetone is an effective solvent in surface photografting. It completely wets the surfaces of most polymer substrates. Furthermore, acetone absorbs UV quanta

$$\begin{aligned} &BP \xrightarrow{h\nu} BP^{*S} \xrightarrow{ISC} BP^{*T} \\ &BP^{*T} + PH \longrightarrow BPH^{\bullet} + P^{\bullet} \\ &P^{\bullet} + nM + BPH^{\bullet} \longrightarrow P\text{—}M_n\text{—}BPH \end{aligned} \quad (1)$$

of short wavelength and takes part in the photo-initiating process by transferring energy to the substrate surface. Other aliphatic ketones have similar functions but acetone is the most convenient to use because of its excellent wettability, photoactive properties and high vapor pressure which makes it easy to remove.

3. PROPERTIES OF THE GRAFTED POLYMER SURFACES

A grafted functional surface layer applied onto an inert polymer substrate is most easily identified by measuring the water contact angle. For a surface of polystyrene film (PS), grafted with acrylic acid (AA), the contact angle decreases from 90° to approximately 20° after 2 min UV irradiation in the batch process [7]. Measured in the receding mode, the contact angle is virtually 0° after 40 s irradiation (Fig. 4). ESCA and reflection IR measurements give more specific information about the appearance and properties of the grafted polymer layer and the relative thickness of the layer (Fig. 5) [7].

An absolute value of the average thickness of the grafted layer is obtained by microtitration of the grafted functional monomer, e.g. AA on polyolefin fibers [8]. This analysis is possible only for substrates of very large surface area per weight, e.g. yarns of very thin fibers, which are surface photografted using the continuous process. An example is AA grafted onto polypropylene yarn of 444/58 denier/filament with a fiber diameter of 37.4 μm and a calculated surface area of 0.006286 m^2 per m yarn. The yarn before grafting has a titrated amount of 0.017 mmol COOH groups per m^2 and after grafting with AA 0.118 mmol COOH, which is the maximum

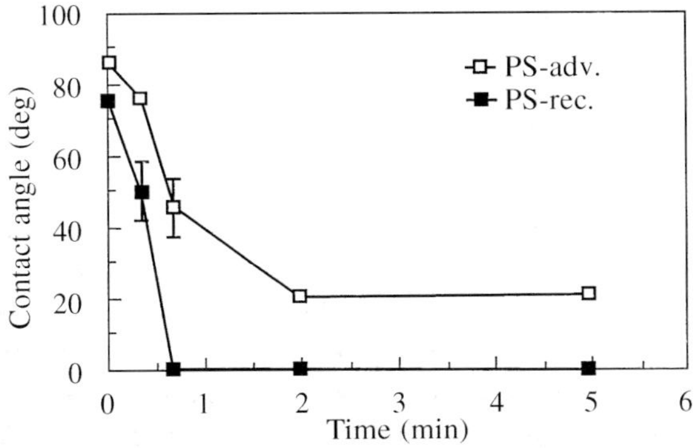

Figure 4. Contact angle of water on polystyrene before and after surface photografting with acrylic acid with the batch process under different irradiation times, measured in both advancing and receding modes.

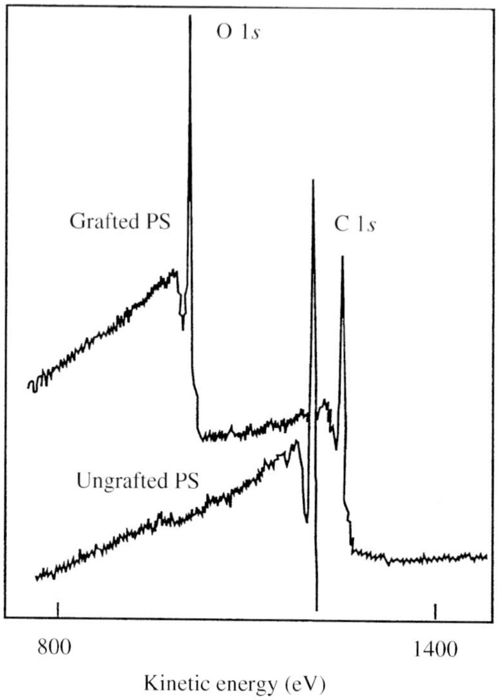

Figure 5. ESCA spectra of pure polystyrene and polystyrene grafted with acrylic acid. Irradiation time is 5 min for the grafted sample. The grafted PS spectrum is shifted by 45 eV.

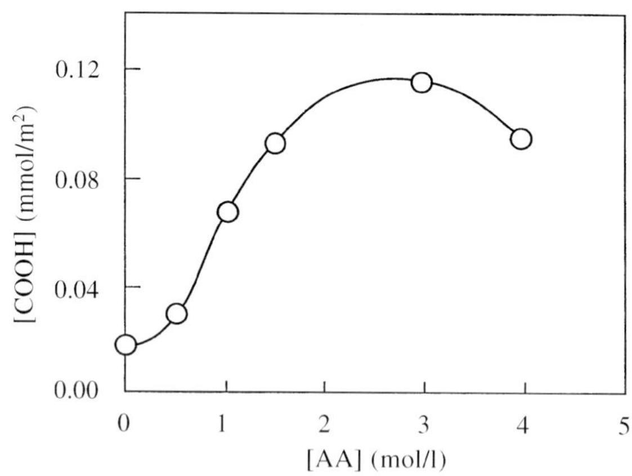

Figure 6. Titrated amount of COOH groups in mmol per m^2 of fiber surface of polypropylene fibers grafted at increasing acrylic acid concentration. [AA] = 0 refers to ungrafted fibers. Continuous process. Irradiation time 20 s, acetone as solvent.

obtained (Fig. 6). This means a calculated average thickness of the grafted AA layer of 7.9 nm [9]. Such a thin layer of grafted polymer cannot be analyzed by weighing. It has, in this case, a calculated weight of only 0.04% of the fiber weight but its presence has a profound effect on the surface properties of the fibers.

4. ADHESION TO GRAFTED POLYMER SURFACES

Molded round plates of LDPE (diameter 100 mm, thickness 3 mm) and film strips of PET [poly(ethylene terephthalate) 7 mm wide and 0.124 mm thick] were surface photografted with functional monomers using the batch process. The adhesion to commercial Scotch tape (19 mm wide) was measured in 180° shear test (Fig. 7) and 90° peel test (Fig. 8). The results of the shear tests for the PET surfaces grafted with acrylic acid (AA), glycidyl acrylate (GA), acrylamide (AM), and 4-vinylpyridine (4-VP) are

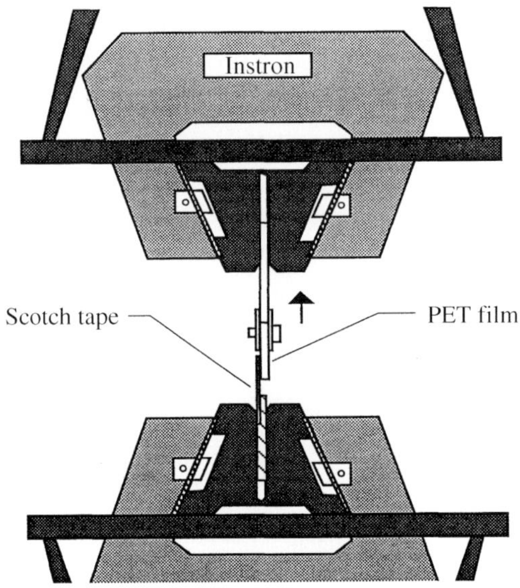

Figure 7. Device for shear test at 180° for the adhesion of Scotch tape to poly(ethylene terephthalate) film (PET).

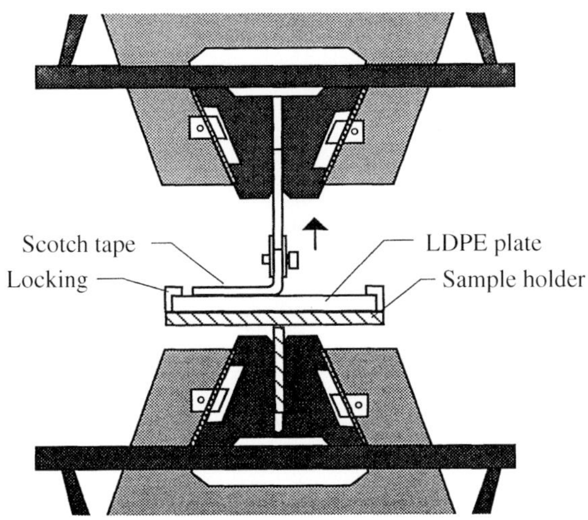

Figure 8. Device for peel test at 90° for the adhesion of Scotch tape to an LDPE plate.

given in Table 1. The increases are only 10 to 30%. The interpretation is that the test measures mainly the shear strength in the soft adhesive layer of the Scotch tape which is more than 1000 times thicker than the grafted layer.

The peel tests of the grafted LDPE plate surfaces show much larger effects. The Instron recordings for a blank surface and a surface grafted with AA are shown in Fig. 9. The increase in 90° peel strength is 5 to 8 times for the four grafted monomers (Table 2). The separation between the grafted LDPE surface and the Scotch tape appeared to be at the interface but no surface analysis was made.

The adhesion of photografted fibers to epoxy resin is measured with a device in the Instron instrument as shown in Fig. 10. A single filament is mounted in a hole of a brass block and the hole is filled with an epoxy resin which is then cured.

Table 1.
Results of 180° shear test for the adhesion of Scotch tape to PET film grafted with the monomers acrylic acid (AA), glycidyl acrylate (GA), acrylamide (AM) and 4-vinyl-pyridine (4-VP) using the batch process

Grafted monomer	Shear strength (N/m)
Blank	3600
AA	4543
GA	3838
AM	4219
4-VP	4257

PET sample size: 60 × 7 × 0.124 mm.
Scotch tape is commercially available.
Crosshead speed: 20 mm/min; full scale: 200 N.

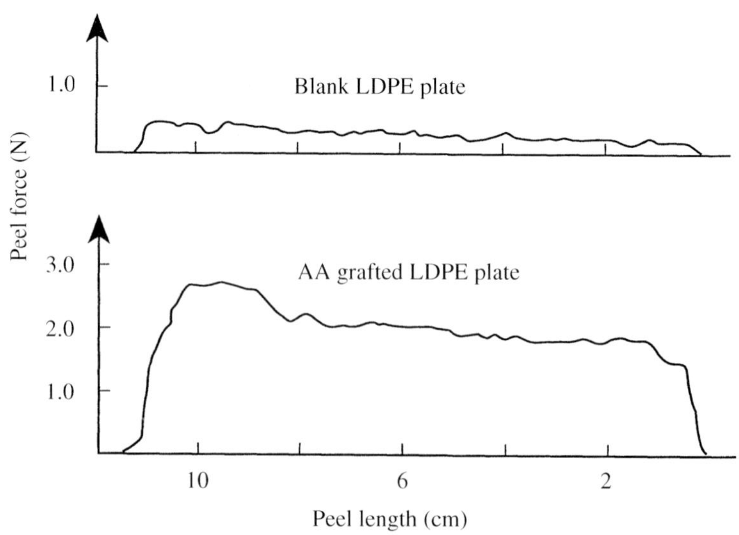

Figure 9. Recorded peel force measured at 90° for the adhesion of Scotch tape to LDPE plates, blank and grafted with AA using the batch process.

Table 2.
Results of measured peel strength (N/m) for the adhesion of Scotch tape to LDPE plates, blank and grafted with the monomers GA, AM, AA, and 4-VP using the batch process

Grafted monomer	Peel strength (N/m)	Increase in peel strength (times)
Blank	14.4	1.0
GA	69.5	4.8
AM	90.5	6.3
AA	96.8	6.7
4-VP	115.8	8.0

Sample width: 19 mm.

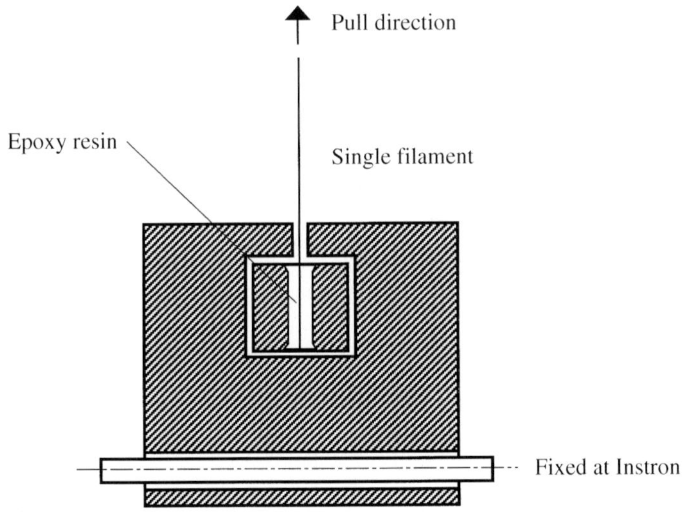

Figure 10. Device for pull-out test of single filaments, surface photografted with the continuous process, to measure the adhesion to epoxy resin.

The adhesion is measured in a pull-out test and the adhesion calculated in Newton per mm^2 of filament surface. High-strength polyethylene fibers (Spectra 900 from Allied-Signal Corp.) and polypropylene fibers (from Amoco Fabrics) were grafted with acrylamide (AM) using the continuous process. The relative degree of grafting is evaluated from ESCA spectra as the intensity ratio N $1s$/C $1s$ and the adhesion is measured by the pull-out test (Fig. 10). The results show an increase in adhesion by a factor of about 3 for the PP fibers and about 6 for HDPE fibers (Fig. 11). The initial grafting gives only small increases in adhesion indicating an early reaction of an inaccessible surface layer of the fibers. A remarkable result is that the extensive grafting of the HDPE fibers with UV irradiation times from 17 to 156 s has no measurable effect on the tensile properties of the HDPE fibers (Table 3). Tensile strength, elongation-at-break, and Young's modulus remain unchanged in the grafting process. For the photografted PP fibers (Fig. 12) the photografting has a measurable effect on the tensile properties (Table 4). The tensile strength decreases by a few

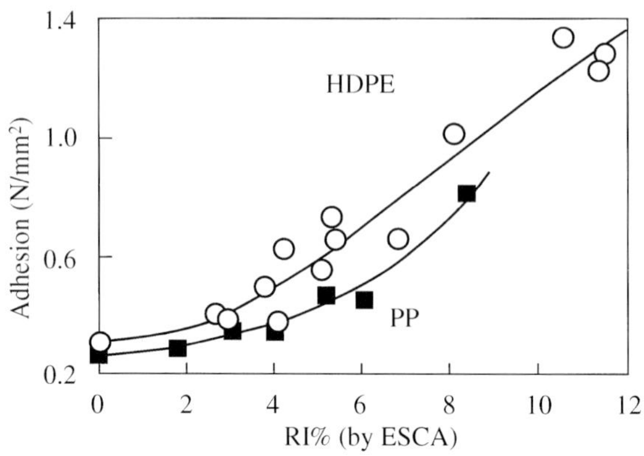

Figure 11. Results of pull-out test for adhesion to epoxy resin of high-strength single filaments of HDPE and textile fibers of PP after surface photografting with acrylamide (AM) to increasing degree, measured by RI = N 1*s*/C 1*s* in ESCA spectra.

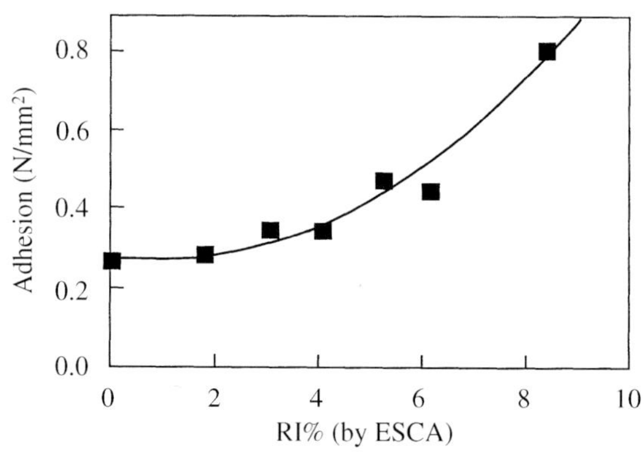

Figure 12. Adhesion to epoxy resin of single filaments of PP grafted with acrylamide (AM) with increasing degree of grafting (RI).

Table 3.
Bulk tensile properties of surface grafted HDPE filaments with increasing degree of grafting, measured by ESCA after different UV irradiation times

Grafted monomer	Time (s)	RI% (ESCA)	Tensile strength (GPa)	Elongation-at-break (%)	Young's modulus (GPa)
Blank	0	—	2.6	5.0	52
AM	156	4.0	2.5	4.5	57
AM	18	5.3	2.8	4.7	59
AM	17	5.9	2.7	4.5	59
AM	84	8.1	2.7	5.3	52
AM	30	11.4	2.8	5.4	52
AM	126	11.5	2.8	5.0	56
AA	14	16.8	2.7	4.9	55

The bulk tensile properties were measured for single filaments at 20°C. All data in this table are given as three sample averages.

Table 4.
Bulk tensile properties of surface grafted PP filaments with increasing degree of grafting with acrylamide, measured by ESCA

RI% (by ESCA)	Tensile strength (GPa)	Young's modulus (GPa)
Blank	0.404	5.1
1.8	0.399	5.7
4.3	0.395	6.0
6.1	0.398	5.8
7.5	0.392	5.8

AM as monomer, BP as photoinitiator, and acetone as solvent.

percent while the tensile modulus increases by 10 to 15% (Table 4). This may indicate a minor degradation of the PP material. PP is known to be more sensitive to photodegradation than polyethylene. The measured effects may also be related to the acetone which is the solvent for the photoinitiator and the monomer in the process. The changes in tensile properties of the PP fibers during the photografting are small and do not impair the use of the grafted PP fibers as textile material.

5. ADSORPTION OF TEXTILE DYES ONTO PHOTOGRAFTED FIBERS

The adsorption of dyes onto textile fibers is greatly enhanced by photografting with functional monomers. An example is PP fibers grafted with acrylic acid (AA) to increasing degree of grafting measured as the ratio RI = O 1*s*/C 1*s* in ESCA spectra (Fig. 13). The grafting increases the adsorption of a basic dye (Maxilon Blau from Ciba-Geigy) by approximately 6 times. The grafting has no effect on the uptake of a disperse dye (Dispersol Red B-2-3 C1 from ICI). The super-stretched HDPE fibers have high crystallinity and a very inert surface. These fibers are very difficult to dye. After grafting with AA to RI-values (O 1*s*/C 1*s*) of 5 to 15% the adsorption of Crystal Violet (a basic dye) is substantially increased (Fig. 14). The initial grafting (RI $<$ 5%) has only a small effect on the dye adsorption.

From the large experimental data on dying of grafted fibers with commercial textile dyes, some significant results are given in Table 5. Grafting PP fibers with an acid monomer (AA) increases the adsorption of a basic dye (Maxilon Blau) approximately 8 times. Grafting PVA fibers with a basic monomer (4-VP) increases the adsorption of an acid dye (Telon Echtrot Acid Red from Bayer) approximately 15 times. Grafting of HDPE fibers with glycidyl acrylate (GA) increases the adsorption of an acid dye (Irgalan Orange Dye from Ciba-Geigy) 3–4 times. Grafting polyester fibers (PET) with acrylamide (AM) increases the adsorption of a Direct Dye (Sirius Lichtbordo Azo Metal dye from Bayer) approximately 6 times. All these results are interpreted as being due to salt formation (acid/base), chemical reaction (epoxy groups reacting with acids) and strong polar interaction (amide/azo groups).

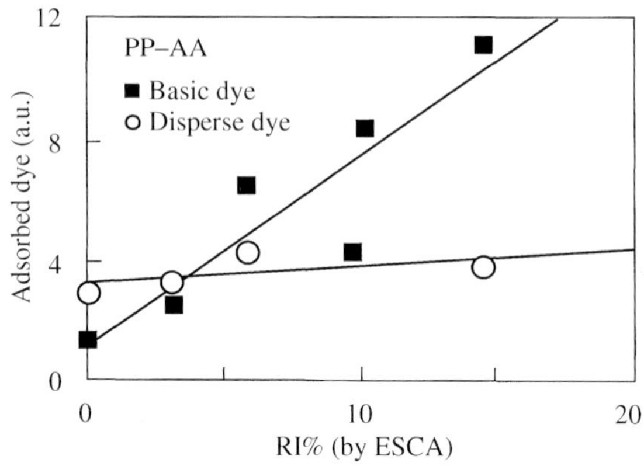

Figure 13. Adsorption of a disperse dye and a basic dye onto polypropylene fibers grafted with acrylic acid with increasing RI values (= O 1*s*/C 1*s*), using the continuous process, with benzophenone as photoinitiator and acetone as solvent.

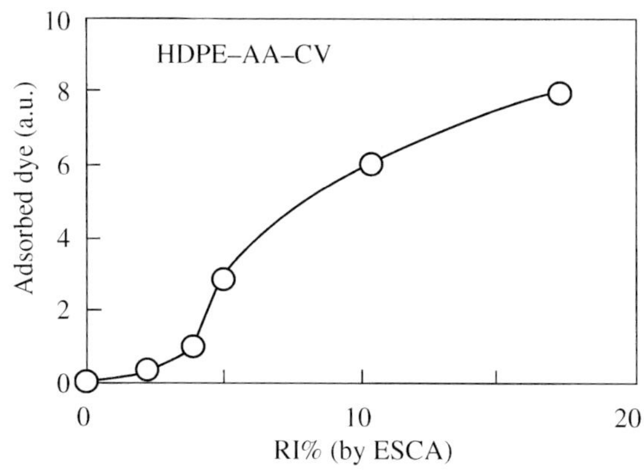

Figure 14. Adsorption of Crystal Violet (CV, a basic dye) onto HDPE fibers grafted with acrylic acid with increasing RI values (= O 1*s*/C 1*s*), using the continuous process, with benzophenone as photoinitiator and acetone as solvent.

Table 5.
Increase of dye adsorption onto synthetic fibers, surface grafted with different monomers to optimal degree using the continuous process, compared with the adsorption onto ungrafted fibers

Fiber/monomer/dye	Increase of dye adsorption (times)
HDPE fiber/GA/Acid Orange Dye IO	3.4
PP fiber/AA/Basic Dye MB	7.9
PET fiber/AM/Direct Dye SL	6.1
PVA fiber/VP/Acid Dye TE	15.3

6. APPLICATIONS OF PHOTOGRAFTED POLYMER SURFACES AS BIOMATERIALS

An important problem is the prevention of blood coagulation on synthetic polymer surfaces. A sheet of LDPE is photografted with glycidyl methacrylate (GMA) using the batch process and then reacted with heparin. This surface modification prevents thrombosis formation in contact with blood (Fig. 15). The heparin is chemically bonded to the grafted surface and cannot be washed away [13].

Another method of preventing coagulation of proteins on a polymer surface is to cover it with a hydrophilic coating [13]. A polyethylene surface is photografted with glycidylacrylate (GA) using the batch process and then reacted with poly(ethylene glycol) (PEG) containing a strong base (NaH) (Fig. 16). Through this reaction, PEG chains are bonded to the PE surface and form a swollen gel surface in water. The hydrophilic surface layer prevents adsorption of a human blood protein (transferrin) almost completely (Fig. 17).

A review of the surface photografting processes and their applications has been given in a lecture [14].

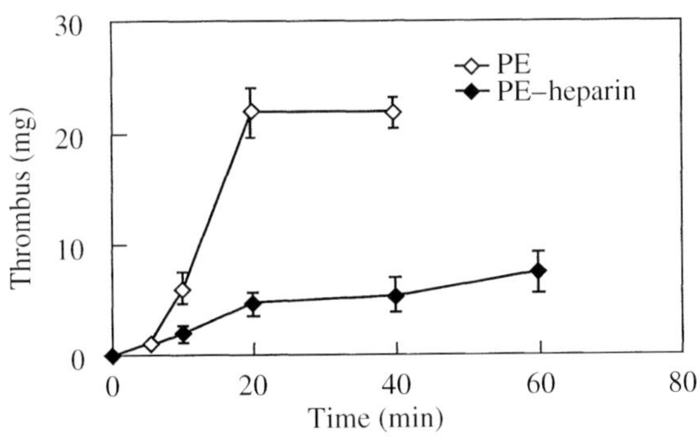

Figure 15. Thrombus formation *in vitro* from blood in contact with untreated polyethylene and on polyethylene with heparin grafted onto the surface.

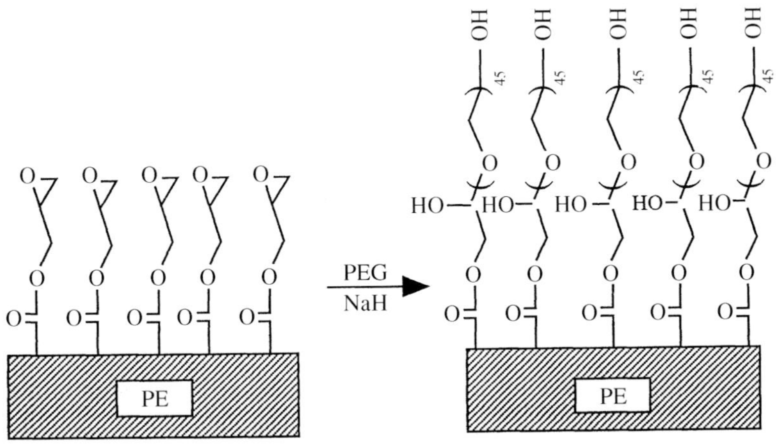

Figure 16. Attachment of poly(ethylene glycol) (PEG) to a polyethylene surface (PE) grafted with glycidyl methacrylate (GMA).

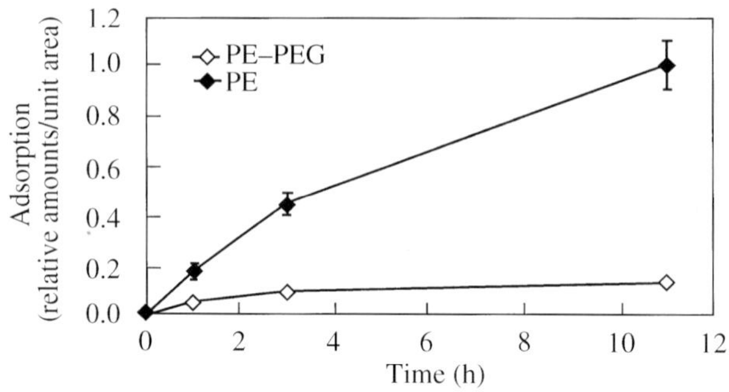

Figure 17. The adsorption of human transferrin on the surface of untreated polyethylene (PE) and polyethylene with poly(ethylene glycol) (PEG) grafted onto the surface. Relative adsorption values after 11 h contact with protein solution are shown.

7. SUBSEQUENT WORK ON 'SURFACE PHOTOGRAFTING'

Since the invention of efficient methods for 'surface photografting', first published in 1986 [1] and as a full paper in 1988 [5], photografting experiments have been described repeatedly by other authors without reference to our papers.

Ikada's group has reported extensive studies of surface modification of synthetic polymers for biomedical applications, e.g. for poly(ethylene terephthalate) (PET) using 'a novel method' [15, 16]. PET film samples were immersed in an aqueous solution of acrylamide (10 wt%) and $NaIO_4$ (10^{-4} to 10^{-2} M) in a Pyrex glass ampoule with no initiator added and without degassing and then UV-irradiated with a high pressure mercury lamp for 1 to 4 h. The treatment decreased the water contact angle from 65 to 20° and gave 20 to 60% conversion of monomer to polymer with only low amounts of grafted polymer. This method for photografting is apparently a rather inefficient method for surface modification compared to the 'batch process' described in [1] and [5].

The continuous process for 'surface photografting' was reinvented in 1993 by another group of authors, Riahi *et al.* [17], and applied to HDPE fibers for surface modification to increase adhesion. Some of the authors have also obtained a patent for the continuous surface photografting process [18] which in principle is identical to our continuous process from 1986.

From the content of the paper and the patent, it is obvious that the authors know our description of the continuous 'surface photografting' processes. They refer to a number of related papers but not to our papers. It is most surprising that the US Patent Office has granted a patent for a process which was described and published in easily available US publications several years earlier.

8. CONCLUSIONS

1. 'Surface photografting' is a new method for modification of polymer surfaces which was invented in our laboratories 10 years ago and developed into a *batch process* and a *continuous process*.

2. In surface photografting, a substrate of commercial polymer is UV irradiated in an atmosphere of initiator and monomer (the batch process) or with a thin liquid layer of initiator and monomer solution (the continuous process). The photoexcited initiator, e.g. benzophenone, abstracts hydrogen atoms from the substrate surface. The radicals formed on the substrate surface add the functional monomer to short grafted chains forming a thin surface layer. With reactive monomers, e.g. acrylic acid (AA), acrylamide (AM), glycidyl acrylate (GA) and 4-vinylpyridine (VP) the grafting reaction is fast and covers the substrate surface after a few seconds of UV irradiation.
3. The batch process has been applied to plates, sheets and films of polyolefins and polyesters. Functional monomers like AA, AM, GA, and VP grafted to the substrate surface as a thin layer (2 to 8 nm) decrease the contact angle for water from approximately 90° to values below 20° and increase the adhesion to Scotch tape by a factor of 5 to 8 in 90° peel tests and by 30 to 40% in 180° shear tests.
4. The continuous process has been applied to strips of film, filaments and yarns of commercial polymers like polyolefins, polyesters and poly(vinyl alcohol). With functional monomers grafted to the substrate surfaces, the adhesion to epoxy resin is increased 3 to 6 times in pull-out tests and the adsorption of commercial textile dyes by a factor of 3–4 to 15 times.
5. Surface photografting has been shown to be effective in biomedical applications. Heparin can be bound to an inert polymer surface like polyethylene, when grafted with glycidyl acrylate (GA). This prevents thrombosis from occurring when blood comes into contact with the surface. Surface photografting can also bind a surface layer of short chains of poly(ethylene oxide), which forms an aqueous surface gel preventing adsorption of human blood protein, e.g. transferrin.
6. Surface photografting is promising as a commercial method for adhesion and adsorption control on inert polymer surfaces.

Acknowledgements

The research work reviewed in this paper has been supported by grants and fellowships from the Swedich National Board for Technical Development (STU) and NUTEK, the Carl Trygger Foundation for Scientific Research and the Wenner-Gren Foundations, which is gratefully acknowledged.

REFERENCES

1. B. Rånby, Z. M. Gao, A. Hult and P. Y. Zhang, *Polymer Preprints (ACS)* **27** (2), 38 (1986).
2. G. J. Howard, S. R. Kim and R. H. Peters, *J. Soc. of Dyers and Colorists* **85**, 468 (1969).
3. S. Tazuke, T. Matoba, H. Kimura and T. Okada, *ACS Symp. Ser.* **121**, 217 (1980). See also S. Tazuke and H. Kimura, *Makromol. Chem.* **179**, 2603 (1978).
4. H. Kubota, K. Kobayaski and Y. Ogiwara, *Polym. Photochem.* **7**, 379 (1986). See also Y. Ogiwara, M. Takumi and H. Kubota, *J. Appl. Polym. Sci.* **27**, 3743 (1982).
5. B. Rånby, Z. M. Gao, A. Hult and P. Y. Zhang, *ACS Symp. Ser.* **364**, 168 (1988).
6. W. T. Yang and B. Rånby, *Polym. Bulletin* **37**, 89 (1996).

7. K. Allmér, A. Hult and B. Rånby, *J. Polym. Sci., Polym. Chem. Edn* **26**, 2099 (1988).
8. P. Y. Zhang and B. Rånby, *J. Appl. Polym. Sci.* **40**, 1647 (1990); **43**, 621 (1991).
9. Z. G. Feng, *Photoinitiated Surface Grafting of Synthetic Fibers*, Inaugural Dissertation, May 17, 1991, Stockholm, Sweden;
   Z. G. Feng and B. Rånby, *Polym. Adv. Technol.* **5**, 829 (1994).
10. B. Rånby, *J. Adhesion Sci. Technol.* **9**, 599 (1995).
11. Z. G. Feng and B. Rånby, *Angew. Makromol. Chem.* **195**, 17 (1992).
12. Z. G. Feng and B. Rånby, *Angew. Makromol. Chem.* **199**, 33 (1992).
13. K. Allmér, J. Hilborn, P. H. Larsson, A. Hult and B. Rånby, *J. Polym. Sci., Polym. Chem. Edn* **28**, 173 (1990).
14. B. Rånby, *Makromol. Chem., Macromol. Symp.* **63**, 55 (1992).
15. E. Uchida, Y. Uyama and Y. Ikada, *J. Appl. Polym. Sci.* **41**, 677 (1990).
16. E. Uchida, Y. Uyama and Y. Ikada, *J. Appl. Polym. Sci.* **47**, 417 (1993).
17. G. Riahi, G. Wood, H. Nguyen and A. Poursantip, *Compos. Interf.* **1** (1), 55 (1993).
18. H. X. Nguyen, A. Poursantip, G. Riahi, R. C. Bennett and G. M. Wood, US Patent 5,039,549 (1991).

Mittal Festschrift, pp. 443–460
W. J. Van Ooij and H. R. Anderson, Jr. (Eds)

Poly(ethylene glycol) grafting as a way to prevent protein adsorption and bacterial adherence

K. HOLMBERG[1] and J. M. HARRIS[2]
[1]*Institute for Surface Chemistry, P.O. Box 5607, S-114 86 Stockholm, Sweden*
[2]*Department of Chemistry, University of Alabama, Huntsville, AL 35899, USA*

Abstract—Grafting of poly(ethylene glycol) (PEG) is an effective way of reducing adsorption of proteins and bacteria to hydrophobic surfaces. The paper discusses and compares two different routes of attaching PEG chains to surfaces: adsorption of block copolymers of ethylene oxide and propylene oxide (EO–PO block copolymers) and grafting via use of an anchoring polymer, poly(ethylene imine) (PEI). An overriding goal is to achieve a dense packing of PEG chains. The best effect in terms of protein and bacteria rejection, judged from short-term experiments, is obtained by adsorbing a pre-formed copolymer of PEG grafted to PEI on a negatively charged surface. Using PEGs of molecular weight 1500 g/mol or higher, protein adsorption is reduced to a few percent of the amount adsorbed at an untreated surface. The block copolymer adsorption route is less effective, mainly due to protein-induced desorption of the hydrophilizing agent. Bacterial adherence is also minimal when the PEI–PEG route is used. Branched PEGs are slightly less effective than linear PEGs of the same molecular weight. The difference in performance between linear and branched PEGs is discussed in terms of difference in entropy change when the hydrophilic surface-bound layer is compressed by an approaching protein. Branched PEGs, having smaller exclusion volumes and less freedom of motion, will lose less entropy on compression. The effects exerted on protein adsorption by PEG attached to a surface parallel its effect on particle mobility in electrophoresis. Similar molecular properties seem to be responsible for both protein and bacteria rejection and reduction of electrokinetic effects.

Keywords: Poly(ethylene glycol); PEG; protein adsorption; bacteria adherence; protein rejection; bacteria rejection; biofouling.

1. INTRODUCTION

Adsorption of proteins and adherence of bacteria and other microorganisms to solid surfaces is a major problem in many biotechnological applications. Following pioneering work by the groups of Nagaoka, Merrill, Andrade and others, grafting of poly(ethylene glycol) (PEG) to surfaces has attracted attention as a way to minimize such biofouling [1–4]. It is by now a well-established fact that a dense and firmly attached PEG layer at the surface is a useful route for obtaining low protein adsorption and low cell adhesion characteristics. For instance, a PEG coating has been reported to give a marked suppression of plasma proteins and platelets at hydrophobic surfaces, leading to reduced risk of thrombus formation, as demonstrated both *in vitro* and *in vivo* [5, 6]. (It should be noted, though, that a critique of the antithrombogenicity of PEG has later been published by Llanos and Sefton [7].)

The surface grafted PEG layer is strongly hydrophilic, yet devoid of any charged groups. Provided the grafted layer is thick enough, both electrostatic and hydrophobic attraction forces between the surface and proteins in solution should consequently be small. The PEG chains, when exposed to water, are highly mobile and oriented away from the surface. The conformational freedom of the grafted chains would be severely restricted by a close approach by a large molecule or a particle [8]. The resulting repulsive interaction between surface and dissolved molecule or particle can be seen as an example of steric stabilization and is analogous to stabilization of dispersed systems by polyoxyethylene-based surfactants [9].

During several years we have studied the protein- and bacteria-rejecting effect of PEG attached to a variety of solid surfaces. In this work special attention has been paid to the following:

- developing ways to achieve a dense packing of PEG chains;
- exploring the possibility of using block copolymers of ethylene oxide and propylene oxide (EO–PO block copolymers) instead of PEGs for surface grafting;
- establishing a correlation between molecular weight of PEG chains and protein repellency;
- comparing linear vs. branched PEGs from the point of view of protein repellency;
- development of electroosmosis measurement as a means of analysis of PEG surfaces.

This paper addresses the above issues and provides a summary of our work on PEG grafting to prevent protein adsorption and bacterial adherence. Previously published results by this research group and others are put into a broader perspective and complemented by several unpublished findings.

2. METHODS

2.1. PEG derivatives

Reactive PEG derivatives were glycidyl ethers of PEG (PEG epoxides) prepared by reaction with epichlorohydrin in the presence of solid sodium hydroxide, as has been described earlier [10]. Linear PEGs of MWs 250, 500, 1000, 1500, 4000, and 19 000 and branched PEGs of MWs 500, 1000, 1700, and 6000 were prepared. Branched, reactive PEGs were tetrafunctional PEG epoxides prepared by ethoxylation of di(trimethylol)propane (Perstorp, Sweden) using KOH as catalyst followed by treatment with epichlorohydrin.

Three EO–PO block copolymers were used: one linear, one tetrabranched, and one dodecabranched. The linear copolymer was Pluronic PE 6200 from BASF (Germany). It has a molecular weight of 2660 g/mol and consists of a central poly(propylene oxide) block with oxyethylene units on both sides. The EO to PO weight ratio is 20 : 80

according to the supplier. The tetrabranched and the dodecabranched copolymers were synthesized by consecutive additions of PO and EO to di(trimethylol)propane and to a polyglycerol consisting of 10 condensed glycerol units (80 mol PO + 55 mol EO and 130 mol PO + 25 mol EO, respectively). Hence, also the branched copolymers consist of a central polyoxypropylene block with oxyethylene units stretching out.

Radiolabeling of the block copolymers was done in the ethoxylation step by use of ^{14}C ethylene oxide. Counting of radioactivity was done in a 1215 Rackbeta liquid scintillation counter (Pharmacia, Sweden) after adding scintillation liquid to the vials.

A branched poly(ethylene imine) (PEI) (Polymin SN, BASF, Germany) was used for anchoring PEG or EO–PO block copolymer to the surface. The PEI used had a MW of 500 000 g/mol and primary, secondary, and tertiary amino groups in an approximate ratio of 1 : 2 : 1 [11].

The PEI–PEG adduct was made by reaction of the diglycidyl ether of PEG with PEI as described elsewhere [12]. The adduct between the EO–PO block copolymer and PEI was prepared analogously from the tetraglycidyl ether of the copolymer [13].

PEG molecular weights, M_n, were determined by gel permeation chromatography (GPC) using 10 PEG standards from 106 to 23 000 g/mol for calibration. Polydispersities (M_w/M_n) for the PEGs used in this work were all less than 1.05, so M_w and M_n values are similar. GPC was performed with a 30 cm Waters Ultrahydrogel 250 column using a Biorad pump and a refractive index monitor. Molecular weights were also determined by NMR spectroscopy using the $-OH$ to $-CH_2CH_2O-$ ratio and the amount of ethylene oxide consumed relative to the amount of base initiator. NMR spectra were obtained with a Bruker 200 MHz instrument. The different methods of molecular weight determination gave the same values within experimental error.

2.2. *Proteins and bacteria*

Human fibrinogen (Grade L) was purchased from Pharmacia (Sweden) and normal swine IgG fraction was obtained from Dakopatts (Denmark). Other proteins and chemicals used in the protein adsorption studies, namely, peroxidase-conjugated goat immunoglobulin to human fibrinogen, peroxidase-conjugated rabbit immunoglobulin to swine IgG and 1,2-phenylenediamine dihydrochloride (OPD), were obtained from Dakopatts.

The oral bacterium used was *Streptococcus mutans* (strain LK, serotype c) which was labeled with ^3H [14].

2.3. *PEG-grafted surfaces*

Polystyrene surfaces were activated by oxidation with $KMnO_4$ in concentrated H_2SO_4 (1 g/l) for 30 s. Grafting of PEG to the activated surface was carried out by either a one-step or a two-step procedure. PEI was used as anchoring polymer in both routes.

In the one-step procedure, the oxidized surface was treated with a 10% aqueous solution of the PEI–PEG adduct for 2 h at 40°C. The pH was subsequently raised to 9.5–10.0, and the treatment was continued overnight at room temperature to ensure that all oxirane rings had been cleaved. The samples were then thoroughly rinsed with water and dried.

In the two-step procedure, previously described by Kiss *et al.* [15], the oxidized surface was reacted with a 5% aqueous solution of PEI for 30 min at room temperature. The substrates were rinsed with water, treated with a solution containing 10% PEG epoxide in carbonate buffer, pH 9.5, and reacted at 45°C for 20 h. Finally, the surfaces were rinsed with water and dried.

Grafting with EO–PO block copolymer was performed by the one-step procedure. The block copolymer consisting of 4 branches, all carrying terminal epoxy groups was reacted with PEI and the adduct formed was subsequently adsorbed at the oxidized surface [13].

Non-oxidized polystyrene was also hydrophilized by direct adsorption of non-reactive block copolymer, normally using a 0.1% aqueous solution and 30 min adsorption time at room temperature. The dodecabranched copolymer had very low water solubility and was used as a 0.25% solution of a mixture of water and isopropanol 85 : 15.

2.4. *Surface characterization*

Electron Spectroscopy for Chemical Analysis (ESCA) was performed on a Kratos AXIS-HS and a Hewlett Packard 5950 A instrument.

Ellipsometry was performed on an automated Rudolph thin-film instrument, type 436, controlled by a personal computer. A xenon lamp, filtered to 401.5 nm, was used as the light source. A thorough description of the experimental set-up is given by Landgren and Jönsson [16]. Before studying the adsorption process, the complex refractive index of the surface was measured. The refractive index of the oxidized surface was also determined at 30, 35, 40, and 45°C. The values obtained were used in the experiments with the EO–PO block copolymer, where the adsorption process was studied in the temperature interval 25–45°C. The polymer sample was then added to the cuvette at experimental conditions and the relative phase difference, Δ, and the amplitude ratio, $\tan\psi$, were obtained from the ellipsometer readings. Measurements were made under constant stirring at 300 rpm.

The average refractive index, n_f, and the mean optical thickness, d, of the grafted polymer layer were calculated numerically from the values of Δ and $\tan\psi$ using Drude's equation. The adsorbed amount of polymer, Γ, can be calculated from the d and n_f values [17]. The ellipsometry measurements were repeated several times. The accuracy of the numbers given is ±15%.

2.5. *Measurement of protein adsorption by ELISA*

A modified Enzyme-Linked Immuno Sorbent Assay (ELISA) was used to determine protein adsorption. The procedure given below for fibrinogen was used for other proteins as well. Details of the method can be found elsewhere [10].

PEG-coated polystyrene tubes of tissue culture grade were soaked in citrate–phosphate buffered saline (CPBS) for 2 h at 37°C. Human fibrinogen (0.4 mg) was added.

After 2 h at 37°C the tubes were rinsed by dilution–displacement with 100 ml CPBS. Two ml of a 0.1 M phosphate buffer (pH 7.0, containing 1.0 M NaCl and 0.2% Tween 20), containing 0.01 mg/ml peroxidase-conjugated goat immunoglobulin to human fibrinogen and 0.05 mg/ml rabbit immunoglobulin to human fibrinogen was added to each test tube. After 1 h at room temperature, the test tubes were rinsed extensively with water. Color was developed by addition of 2 ml of a solution containing 0.67 mg/ml 1,2-phenylenediamine dihydrochloride in 0.01% hydrogen peroxide. The reaction was terminated after 5 min with 2 ml of 1 M sulphuric acid, and the absorbance read at 490 nm. The color produced, after subtraction of a control, was proportional to the amount of adsorbed fibrinogen. Fibrinogen adsorption experiments with 96-well plates instead of test tubes were carried out analogously using 0.25 ml solution per well.

2.6. Measurement of bacterial adherence

Adherence of radiolabeled *Streptococcus mutans* (5×10^8 cells per ml) on polystyrene tube walls was investigated with and without surface hydrophilization. The dodecabranched EO–PO block copolymer was applied by adsorption from a 0.25% solution of a mixture of water and isopropanol 85 : 15. Coupling involving PEI as anchoring polymer was performed by the one-step and two-step procedures described above in the Section 2.3. The surfaces were incubated with the bacterial suspension in phosphate buffered saline (PBS) for 1 h at room temperature. During the incubation the tubes were kept in an upright position. Bacterial binding was evaluated by determining the radioactivity of the tube walls, as has been described elsewhere [14, 18].

3. RESULTS AND DISCUSSION

3.1. Comparison of PEG grafting methods

Two principal modes of attachment of PEG to polystyrene surfaces were compared. The simplest route comprised adsorption of EO–PO block copolymers to untreated polystyrene. Linear triblock polymers of EO–PO–EO type were used as well as branched copolymers also with a central polyoxypropylene block. All copolymers gave efficient hydrophilization as judged from contact angle determination (complete wetting by water). However, none of the copolymers was very effective in preventing fibrinogen adsorption. As can be seen from Fig. 1, the effectiveness seemed to increase with increasing degree of branching. The relatively poor performance of the copolymers in this test is likely to be due to protein-induced desorption of the hydrophilizing agent. Tests with radiolabeled material showed that a substantial portion of the block copolymers desorbed from the surface during incubation with the dilute fibrinogen solution. As can be seen from Table 1, there is an inverse correlation between protein rejection efficiency and amount of radiolabeled copolymer in solution.

The other general route of PEG grafting comprised oxidation of the polystyrene surface by $KMnO_4/H_2SO_4$ to introduce negative charges, followed either by adsorption

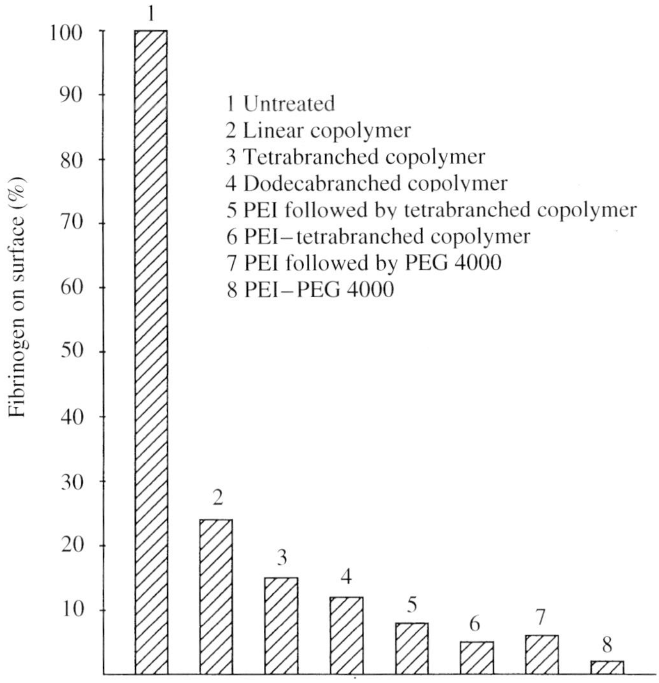

Figure 1. Adsorption of fibrinogen on hydrophilized polystyrene surfaces. Surface treatment was done either by adsorption of linear or branched EO–PO block copolymers to unmodified polystyrene or by using PEI as anchoring polymer for PEG or EO–PO block copolymer. The latter procedure used oxidized polystyrene and was done either by the one-step or the two-step route illustrated in Fig. 2.

Table 1.
Radiolabeled copolymer in solution after 30 min incubation of copolymer-treated polystyrene plates with a 0.1% aqueous solution of human fibrinogen. The values are given as % of total amount (surface-bound + dissolved) of radiolabeled material. Also given are fibrinogen adsorption values as % of adsorption at untreated polystyrene (values taken from Fig. 1)

Hydrophilizing agent	Radioactivity (%)	Protein adsorption (%)
Tetrabranched copolymer	9	15
Dodecabranched copolymer	6	12
PEI followed by tetrabranched copolymer	1	8
PEI–tetrabranched copolymer	< 1	5

of a ready-made PEI–PEG adduct (one-step grafting) or by adsorption of PEI followed by addition of PEG epoxide (two-step procedure). The two routes are illustrated in Fig. 2. The procedure was also used with the tetrabranched block copolymer as hydrophilizing agent.

Oxidation by $KMnO_4/H_2SO_4$ gives a mixture of negatively charged groups including carboxylate, sulfate and sulfonate, as has previously been shown by Gölander and Eriksson [19]. Our ESCA spectra showed the increase in O and decrease in C,

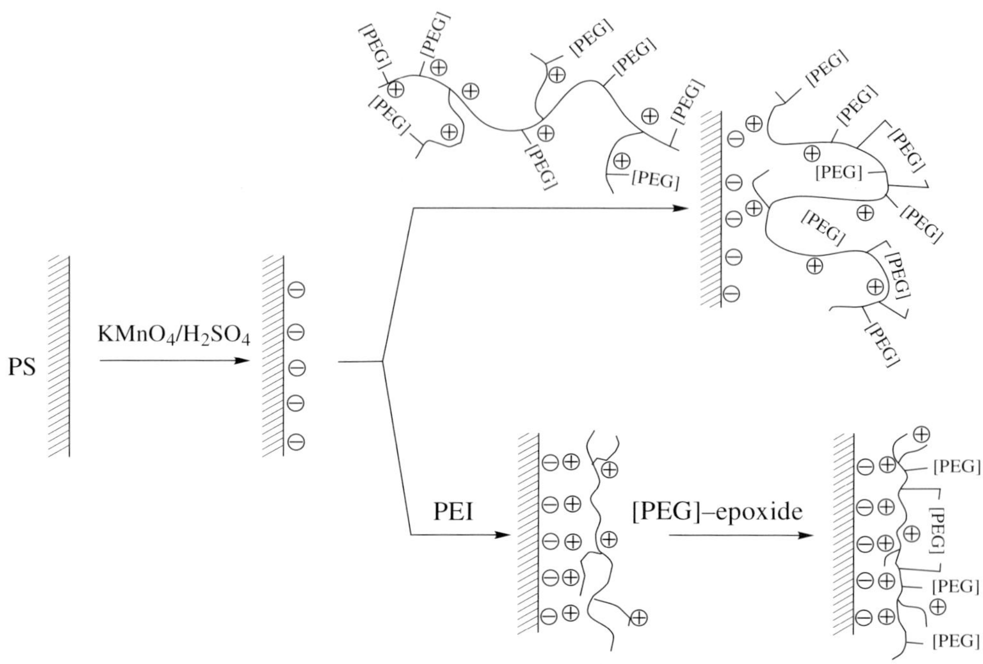

Figure 2. Schematic illustration of the PEG-grafted surface made either by adsorption of a PEI–PEG adduct or by consecutive surface treatments by PEI and PEG epoxide. Ellipsometry indicates an arrangement of PEI in loops on the surface in the former case and a flat adsorption of PEI in the latter case.

Table 2.
Surface composition (atomic %) measured by ESCA on polystyrene (all measurements were performed on a minimum of two samples)

Sample	C (%)	O (%)	N (%)	S (%)
Untreated	96–98	1–3	Trace	—
Oxidized sample	62–63	33	<1	2–3
PEI on sample	71–74	15	10–11	1–2
PEI followed by PEG 4000	70–71	25–26	4	—
PEI–PEG 4000	66	28	6	—
PEI–tetrabranched copolymer	57–59	36	5–7	—

characteristic of a surface oxidation. A small S peak was also present after the $KMnO_4/H_2SO_4$ treatment (Table 2) [10].

ESCA was also used to monitor PEI adsorption to the oxidized polystyrene surface as well as PEG grafting (using PEG epoxide) to the PEI layer. The results obtained are in accordance with expectations. ESCA did not, however, reveal any major differences between the one-step and the two-step methods of Fig. 2. Ellipsometry, on the other hand, showed that the grafted layer was much thicker when applied by the one-step than by the two-step procedure. Most likely, the one-step route gives a pronounced

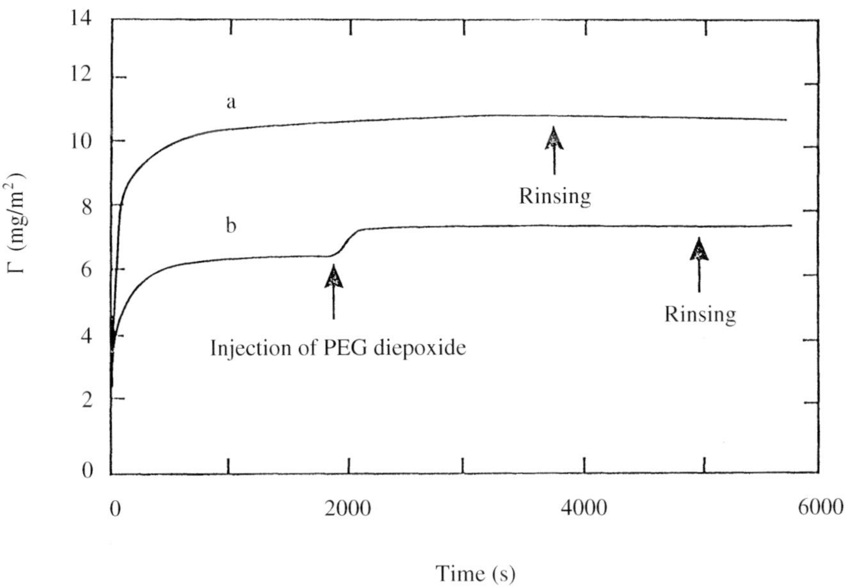

Figure 3. Adsorption of the PEI–PEG 4000 adduct on oxidized polystyrene (curve a), and adsorption of PEI followed by treatment with PEG 4000 epoxide (curve b).

Table 3.
Ellipsometry measurements on a series of oxidized polystyrene samples. Γ is total amount of adsorbed polymer and d is thickness of adsorbed layer

Sample	Γ (mg/m^2)	d (nm)
PEI on polystyrene	6.5	4.0
PEI followed by PEG 4000	7.5	4.9
PEI–PEG 4000	11	50
PEI–tetrabranched copolymer	6.8	23

loops-and-trains conformation of the PEG-grafted PEI (the schematic drawing of Fig. 2 is meant to illustrate the difference in conformation of the PEI–PEG adduct obtained by the two routes). The ellipsometry data are shown in Fig. 3 and Table 3 [12]. These experiments illustrate a severe shortcoming of ESCA for such investigations: being a dry vacuum technique, it does not give information about conformation of grafted chains and orientation into aqueous solution.

As can be seen from Table 3, the coating obtained from adsorption of PEI followed by reaction with PEG epoxide is rather thin: 4.9 nm, out of which PEI accounts for 4.0 nm. The small contribution by PEG to the layer thickness is an indication that the chains lie rather flat on the surface, which, in turn, implies that the PEG grafting is not very dense (a fully extended PEG 4000 should be able to reach well over 20 nm). The coating obtained from the pre-formed PEI–PEG adduct (the one-step route) gives a very thick layer, as measured by ellipsometry. A tentative explanation of this behavior is that adsorption of the PEI backbone of the PEI–PEG adduct in flat configuration on the polystyrene surface is prevented by unfavorable interaction between grafted PEG chains. Flat adsorption of PEI on the surface would require all PEG chains to be oriented away from the surface, towards the bulk water phase. Steric repulsion (steric stabilization) puts a limit to the packing of the PEG chains. The PEI–PEG copolymer can avoid unfavorable packing arrangements by forming loops above the surface. Hence, the large dimensions of the surface-bound PEI–PEG adduct obtained by the one-step route are likely to be due to two effects:

(1) a dense packing of PEG chains which leads to a high degree of chain elongation, and

(2) a loops-and-trains arrangement of the PEI backbone on the surface.

As shown in Fig. 1, the one-step route of attaching the PEI–PEG adduct to the surface is more efficient than the two-step route in preparing a fibrinogen-rejecting surface. Similar results have been obtained with other proteins such as IgG [12] and albumin.

Bacterial adherence was studied using the oral bacterium *Streptococcus mutans*. This organism is believed to be associated with caries and its adhesion characteristics is therefore of considerable interest. Adhesion of the bacterium to PEG-treated polystyrene followed the same pattern as discussed above for protein adsorption, although both copolymer adsorption [20] and the one-step and the two-step routes using PEI as anchoring polymer [21] gave low values of bacterial adherence. As can be seen from Fig. 4, virtually no *S. mutans* could be detected on the surfaces carrying PEG grafted to PEI after the 1 h incubation time used. Adsorption of the highly branched block copolymer is somewhat less efficient, possibly due to desorption of the hydrophilizing agent, as discussed above in the case of protein adsorption.

One may anticipate some kind of correlation between PEG grafting density and protein repellency. To date, conclusive evidence to support this hypothesis seems to be lacking. Ellipsometry can quantify the amount of surface-bound PEG. Provided the molecular weight of the PEG chains is known, a rough number of packing density, e.g. expressed as number of nm^2 per grafted PEG chain, can then be obtained. The two-step procedure applied to an oxidized silicon wafer has given a value of 1.5 nm^2 per molecule for PEG 1900 [22]. Other ways of introducing amino groups, such as silanization with aminopropyltrimethoxysilane, and plasma polymerization of 1,2-diaminocyclohexane, gave considerably lower PEG grafting density [22]. However, no systematic study of correlation between PEG chain packing and protein repellency seems to have been performed.

Electroosmosis applied to flat plates is a potentially useful technique for studying PEG grafting density on solid surfaces. Ongoing work in our group using a home-built electrophoresis cell has given promising results [23]. Electroosmotic fluid flow is measured at flat plates in a rectangular particle electrophoresis chamber. The technique provides a direct measure of surface density of charged groups. By using reactive PEG derivatives which transform basic amino groups on the surface (e.g. PEI amino groups) into non-basic linkages (e.g. amide or carbamate) one is able to directly estimate PEG grafting density. Thus, the technique eliminates the approximations in grafting density determinations involved in ellipsometry experiments with PEGs of broad molecular weight distributions.

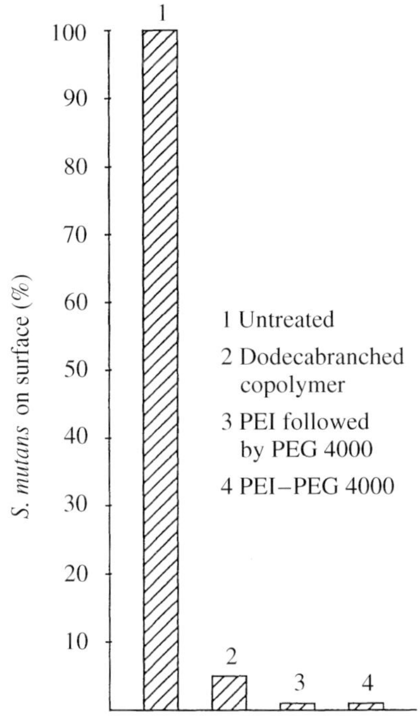

Figure 4. Adhesion of *Streptococcus mutans* to polystyrene treated with dodecabranched copolymer, with PEI followed by treatment with PEG 4000 epoxide and with the PEI–PEG 4000 adduct. The first surface treatment was performed on unmodified polystyrene, the two treatments involving PEI on oxidized polystyrene.

3.2. Views on the use of adsorbed EO–PO block copolymers for surface hydrophilization

In the use of EO–PO block copolymers as hydrophilizing agents to prevent protein adsorption, a compromise will have to be made between driving force for adsorption and surface hydrophilicity. The smaller the EO to PO ratio in the copolymer, the stronger the driving force for adsorption and the better the packing at the interface. However, surface treatment with copolymers having a large polyoxypropylene block surrounded by smaller polyoxyethylene segments is likely to result in a surface which is not entirely hydrophilic since the polyoxypropylene portion may show through the relatively short polyoxyethylene chains. Thus, proteins may be attracted by hydrophobic interactions to such a surface.

EO–PO block copolymers, like other ethylene oxide-containing polymers and surfactants, show a reverse temperature dependent phase behavior and a lower consolute temperature. The lower temperature phase boundary is often referred to as the cloud point (CP). The EO to PO ratio is the main governing factor of the CP; the higher the ratio, the higher the CP. Also, polymer molecular weight and cosolutes such as electrolytes and alcohols affect CP. By varying these parameters, phase separation temperatures (in water) ranging from less than 0°C to more than 100°C can be achieved.

In principle, an EO–PO copolymer is hydrophilic and water-soluble below and hydrophobic and water-insoluble above the CP. When attached to a surface, the copolymer can be expected to impart protein-repellency only when it is hydrophilic; i.e. below CP. The behavior of an ethoxylated cellulose ether adsorbed at a surface has been studied by the surface force technique. Hydrophobized mica was used as substrate and the surfaces were modified by adsorption of the amphiphilic, water-soluble polymer from aqueous solution [24]. By allowing two such modified surfaces to approach each other and measuring the forces that appear as the adsorbed polymer layers start to interact, one obtains information about the conformation of the adsorbed layers and the type of interaction, attractive or repulsive, that arises between the surfaces. It was found that as the temperature was raised towards the CP, the adsorbed polymer layers became more dense. The net force acting between the surfaces was repulsive, not only below the CP but also some 5–10 degrees above the CP. Net repulsion also above the CP is an example of the so-called enhanced steric stabilization and has been experienced with several amphiphilic, water-soluble polymers [25]. Enhanced steric stabilization can be explained by an orientation of hydrophilic groups towards the solution, which lowers the local Flory–Huggins χ parameter in the interaction region. An additional contribution to the enhanced repulsive force is the high polymer concentration in the adsorbed layer at higher temperatures. This is likely to result in poor interpenetration of the polymer layers on compression, and, consequently, in a volume restriction [26].

The behavior of the tetrabranched EO–PO block copolymer anchored to polystyrene via PEI (one-step route) was investigated by ellipsometry [13]. The layer thickness was measured at constant amount on the surface (6.8 mg/m^2). The temperature was varied between 25 and 45°C. The copolymer had a CP (in 0.1% aqueous solution) of 42°C. The variation in refractive index of the underlying surface with temperature was accounted for. Figure 5 illustrates the effect of temperature on average thickness, d, of the grafted copolymer layer. As can be seen from the figure, the thickness of the layer decreases in a close to linear fashion from approximately 23 nm at 25°C to less than 15 nm at 45°C. The theoretical length of the fully extended copolymer is approximately 30 nm, and it has previously been found that the polyoxyethylene layer obtained with adsorbed EO–PO block copolymers has a dimension close to the diameter calculated for the corresponding free PEG chain in solution [27]. The contraction of the grafted chain upon increasing the temperature, which is analogous to the above-mentioned findings using the surface force technique, can be seen as an effect of the decrease in solvency as the temperature increases [28]. As with the surface force measurements, there is no discontinuity in the curve around the CP. The behavior was completely reversible and repeated cycles gave identical values of d [13].

Adsorption of IgG on the same block copolymer-grafted surface was studied both by ellipsometry and ELISA [13]. The results are shown in Fig. 6. As can be seen, the adsorbed amount increases strongly with increasing temperature. The ellipsometry results show that the amount of IgG adsorbed at the surface increases from 1.6 mg/m^2 at 25°C to 3.7 mg/m^2 at 45°C (i.e. 3 degrees above CP). It can be noted that the

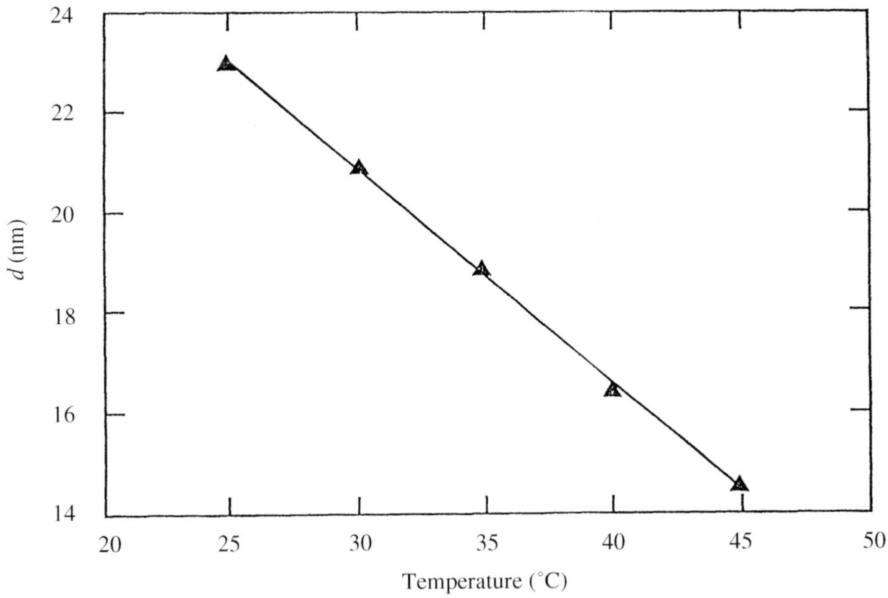

Figure 5. Mean thickness, *d*, of PEI–tetrabranched copolymer on polystyrene as a function of temperature. Adsorbed amount of copolymer was constant, 6.8 mg/m^2 [13].

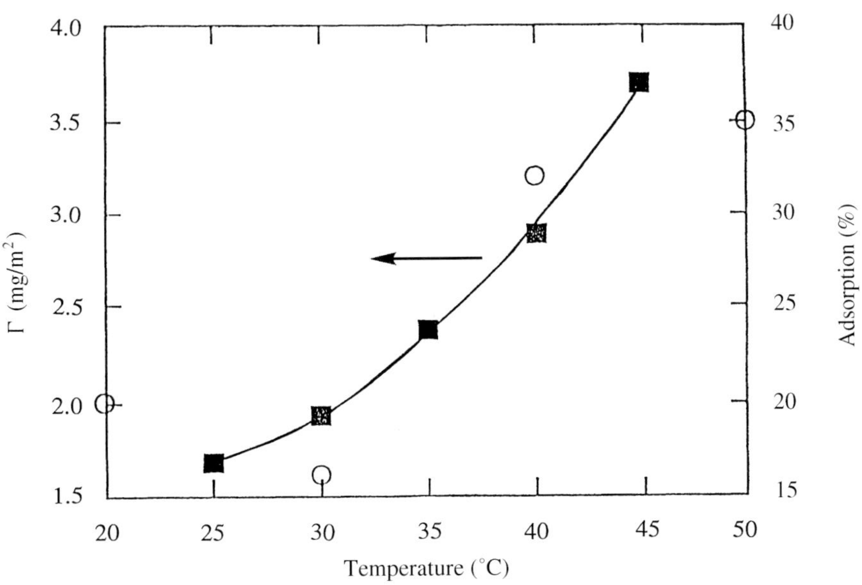

Figure 6. Adsorbed amount of IgG on polystyrene coated with PEI–tetrabranched copolymer after 1 h incubation at pH 7. Filled squares and open circles represent data from ellipsometry (left axis) and ELISA (right axis), respectively. The ELISA values are given as percentages of the amount adsorbed on untreated polystyrene under the same conditions [13].

amount of IgG adsorbed even at 25°C is high compared to adsorption values obtained at surfaces grafted with PEG homopolymers. The general trend of increasing adsorption of IgG with increasing temperature is confirmed by the ELISA measurements, although the correlation is not quite clear in this case (Fig. 6). However, the temperature control during the ELISA measurements is not as good as during the ellipsometry experiments. The ELISA technique is also less sensitive and accurate than ellipsometry in determining adsorbed amount of proteins on surfaces.

3.3. Influence of PEG molecular weight

Table 4 shows fibrinogen adsorption data related to two test series with a range of linear and tetrabranched PEGs of varying molecular weight. Protein adsorption was measured by ELISA which provides relative values rather than absolute amounts of protein on the surface. We have previously shown that 100% fibrinogen adsorption under the exact conditions used here corresponds to adsorption of 100 ng/cm^2 [10].

As can be seen from the table, reduction in molecular weight of linear PEG below 1500 g/mol is accompanied by an increase in fibrinogen adsorption. This effect becomes especially pronounced below 1000 g/mol, with the 250 g/mol material showing no protein rejection ability at all. The same general trend is found with the series of branched PEGs (a discussion of efficiency of linear vs. branched PEGs for protein rejection follows in the Section 3.4).

Several factors may contribute to the observed increase in fibrinogen adsorption for linear PEGs of low molecular weight. First, the small PEGs may be ineffective in hiding the exposed surface from approaching protein; i.e. the PEG layer produced

Table 4.
Effects of linear (L) and branched (B) PEGs bound to polystyrene on fibrinogen adsorption. Surface grafting was made by the two-step route using PEI as anchoring polymer. All measurements are in triplicate. Error limits are standard deviations

	Fibrinogen adsorption (%)	
PEG	Study 1 [10]	Study 2 [28]
Untreated	100 ± 11	100 ± 6
L-19 000	5 ± 1	—
L-4000	3 ± 2	5 ± 2
L-1500	5 ± 1	9 ± 1
L-1000	—	17 ± 3
L-500	—	52 ± 5
L-250	—	100 ± 1
B-6000	6 ± 5	—
B-1700	11 ± 3	14 ± 10
B-1000*	—	46 ± 17

*Smaller B-PEGs could not be employed due to insolubility in water.

by small PEGs may be so thin that the protein can sense the surface through the PEG layer. Alternatively, PEG packing density could be lower for the small PEGs so that the protein is able to find an exposed, unprotected portion of the surface (as discussed above, there are as yet no reliable, direct ways to measure packing density of grafted PEG chains). Evidence against this mechanism is that ESCA spectra for all the PEG-coated surfaces, regardless of the PEG molecular weight, are virtually identical [29]. In every case the ratio of —C—O— to —C—C— carbon in the C 1*s* peak is approximately 8 (within experimental error).

A second contributing factor derives from the theory that PEGs reject proteins because of the unfavorable entropy change that results upon compression of these conformationally-random, heavily hydrated chains by protein adsorption [10, 22, 30–32]. Adsorption could either be by interpenetration to provide contact with the surface or it could be directly on top of the bound PEG layer. The events are visualized in Fig. 7. Although the chains are randomly oriented, there is considerable ordering of water around these chains, with 2–3 water molecules closely associated with each oxyethylene unit [33, 34]. The negative entropy change upon PEG compression can be compensated by a positive entropy change associated with loss of waters of hydration from the PEG chain, although there would be a large endothermic enthalpy change opposing this loss. Thus, if water is lost, protein adsorption is enthalpically unfavorable, and if water is not lost, protein adsorption is entropically unfavorable; in either case, the overall free energy change is positive and unfavorable [29]. This thermodynamic explanation of reduction in protein adsorption by the PEG coating can be related to the observed loss of effectiveness with low molecular weight PEGs. With the short chains the unfavorable entropy effect may not be large enough to overcome attractive forces involved in the adsorption process.

It is interesting that the influence of PEG molecular weight on protein rejection seems to parallel the effect on particle mobility in electrophoresis [35, 36]. Evidently, similar molecular properties are responsible for both protein and bacteria rejection and reduction of electrokinetic effects. This effect of a PEG coating shows that the polymer layer is sufficiently thick to move the surface of shear beyond the electrical

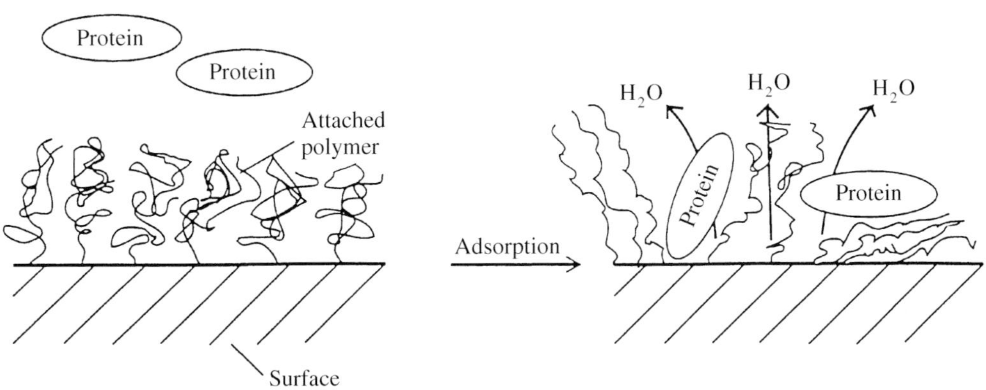

Figure 7. Compression of PEG layer by protein adsorption. A negative entropy change accompanies compression of random, surface-bound PEG chains.

double layer. For instance, coupling of PEG 5000 monoaldehyde to PEI adsorbed at polystyrene latex gives a particle with zero mobility in the pH range of 4–12 [37].

In a recent work, PEG was compared with the uncharged polysaccharide dextran as a coating material to prevent fibrinogen adsorption to polystyrene and also for effect on electrophoretic mobility of latex particles [38]. Two modes of coupling dextran to the surface were used: end-on and side-on. End-on coupling was performed by direct reductive amination to the aminofunctional PEI surface whereby the polysaccharide binds via the reducing end group. In order to couple dextran side-on, the polysaccharide was first oxidized by periodate and then bound by reductive amination. PEG was used as the PEG 5000 monoaldehyde which was attached end-on by the same coupling procedure. Clearly one would expect that dextran attached side-on would give a thinner layer than the two end-on attached polymers and this was confirmed by particle electrophoresis experiments. For side-on dextran the surface charge showed through at low pH. For the other two coatings the charges were totally hidden at all pH-values, indicating a thicker hydrophilic layer. End-on dextran is, of course, likely to give a thicker grafted layer than end-on PEG, since the former has a much higher molecular weight but, evidently, the layer thickness obtained with PEG 5000 was enough to move the hydrodynamic plane of shear beyond the effective electrokinetic layer.

ESCA showed that the amount of hydrophilizing polymer on the surface increased in the order end-on PEG < end-on dextran < side-on dextran, indicating a very dense packing of dextran when coupled side-on [38]. Protein adsorption measurements by ELISA showed that end-on PEG and side-on dextran gave very little fibrinogen on the surface (around 2% of the amount adsorbed on untreated polystyrene). More protein (around 11%) was obtained with end-on dextran. The obvious interpretation of these results is that packing density is a more critical parameter than chain extension, although certainly both must be important. Protein adsorption seems to be critically dependent on the availablility of bare surface, not covered by hydrophilic, non-charged polymer. Apparently, proteins can avoid long, extended tails grafted to the surface as long as they are ‘dilute’. This conclusion is supported by the observation (Table 4) that PEG coatings do not become more effective as the polymer chain length is increased above 1500 g/mol.

It should be pointed out that our findings of a PEG molecular weight of around 1500 g/mol being sufficient to give maximum protein repellency is not entirely in harmony with reports from other groups. Considerably longer PEG chains have been proposed to achieve a proper effect [31, 39, 40]. We believe that the discrepancy is due to the differences in chain packing densities which will inevitably be obtained when different grafting methods are used. Thus, without quantitative information about packing density, a statement about sufficient PEG molecular weight should be regarded as somewhat specific to the grafting method.

Prime and Whitesides have observed low degrees of protein adsorption for self-assembled monolayers with very short oligooxyethylene chains [41]. This result is very different from that observed by ‘conventional’ PEG grafting. As seen in Table 4, a PEG chain length of around 1500 g/mol, corresponding to 30–40 oxyethylene units, is needed to achieve a protein-repellent surface. Most probably, the extremely tight packing that can be achieved by the self-assembly technique reduces the need for a thicker coating. In fact, it is likely that the mechanisms of protein rejection are slightly different for the two modes of PEG grafting. Whereas conventional PEG coupling yields a very mobile, hydrogel-type surface with entropy-driven steric stabilization being a main mechanism by which protein adsorption is prevented (see discussion above), the dense and rigid grafted layer obtained by the self-assembly technique may function by totally eliminating all enthalpy contributions from the interaction between protein and surface.

3.4. Linear vs. branched PEGs

Table 4 also shows the effect of tetrabranched PEGs of varying molecular weights in preventing fibrinogen adsorption to polystyrene. As can be seen, the values obtained are slightly higher than those obtained with linear PEGs. Branched PEG 500 was also synthesized but, unfortunately, could not be used due to poor water solubility. The observed reduction in effectiveness of the branched, small PEGs, relative to the small, linear PEGs, lends credence to the theory that an unfavorable change in entropy accompanying protein adsorption is a major factor in the mechanism by which PEGs reject proteins. A branched PEG, compared with a linear PEG of the same molecular weight, should have fewer degrees of freedom, and thus compression of the polymer upon protein adsorption should not be as unfavorable entropically as for the more randomly oriented linear chains.

If branched PEGs are more restricted in their motion than their linear counterparts then one might expect the branched PEGs to have smaller exclusion volumes and to

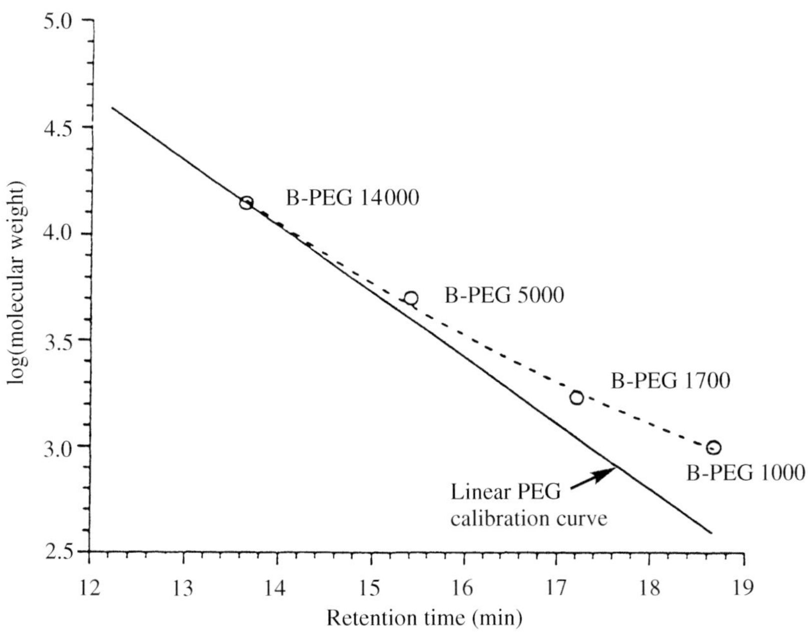

Figure 8. GPC analysis of branched vs. linear PEGs [29].

produce thinner layers when surface-bound. Figure 8 shows plots of log M_n vs. retention time for linear and branched PEGs [29]. As can be seen, with the branched PEGs there is a deviation from the straight line of the linear PEGs, which is particularly obvious for the small branched polymers. The branched PEGs 1700 and 1000 appear as if their molecular weights were 1100 and 400 g/mol, respectively. This result shows that placing a branching point in the polymer backbone does restrict conformational randomness and extension of the polymer chains for the small branched PEGs. For the larger branched PEGs this effect is apparently negligible.

Due to smaller exclusion volume, branched PEGs would, when compressed by protein adsorption, produce a smaller entropy change than would linear PEGs of the same molecular weight. Also, since the branched polymers have lower hydrated volume than their linear counterparts, they will produce a thinner surface layer with reduced ability to mask surface charges. Both effects will work in the direction of reduced effectiveness in preventing protein adsorption.

4. CONCLUSION

The results obtained indicate that grafting of PEG chains to a solid surface using PEI as anchoring polymer gives a densely packed layer of PEG grafts. Such a modified surface gives minimal protein adsorption and bacterial adherence. The efficiency in terms of protein rejection increases with PEG molecular weight up to a value of around 1500, above which it remains constant. Surface modification by adsorption of EO–PO block copolymers is less effective in reducing protein adsorption.

REFERENCES

1. S. Nagaoka, Y. Mori, H. Takiuchi, K. Yokota, H. Tanzawa and S. Nishiumi, *Polym. Prepr.* **24**, 67–68 (1983).
2. E. W. Merrill and E. W. Salzman, *J. Am. Soc. Artif. Internal Organs* **6**, 60–72 (1983).
3. J. H. Lee and J. D. Andrade, in: *Polymer Surface Dynamics*, J. D. Andrade (Ed.), pp. 119–136. Plenum Press, New York (1988).
4. S. W. Kim and J. Feijen, *Critical Reviews in Biocompatibility* **1**, 229–260 (1986).
5. S. Nagaoka, *Trans. Am. Soc. Internal Organs* **10**, 76–78 (1988).
6. S. Nagaoka and A. Nakao, *Biomaterials* **11**, 119–121 (1990).
7. G. R. Llanos and M. V. Sefton, *J. Biomater. Sci. Polym. Edn.* **4**, 381–400 (1993).
8. D. W. J. Osmond, B. Vincent and F. A. Waite, *Colloid Polym. Sci.* **253**, 676–682 (1975).
9. T. F. Tadros, in: *The Effect of Polymers on Dispersion Properties*, T. F. Tadros (Ed.), pp. 1–40. Academic Press, New York (1982).
10. K. Bergström, K. Holmberg, A. Safranj, A. S. Hoffman, M. J. Edgell, A. Kozlowski, B. A. Hovanes and J. M. Harris, *J. Biomed. Mater. Res.* **26**, 779–790 (1992).
11. D. Horn, in: *Polymeric Amines and Ammonium Salts*, E. J. Goethals (Ed.), pp. 333–355. Pergamon Press, Oxford, UK (1980).
12. C. Brink, E. Österberg, K. Holmberg and F. Tiberg, *Colloids Surfaces* **66**, 149–156 (1992).
13. F. Tiberg, C. Brink, M. Hellsten and K. Holmberg, *Colloid Polym. Sci.* **270**, 1188–1193 (1992).
14. J. Olsson, A. Carlén and K. Holmberg, *J. Dent. Res.* **69**, 1586–1591 (1990).
15. E. Kiss, C.-G. Gölander and J. C. Eriksson, *Prog. Colloid Polym. Sci.* **74**, 113–119 (1987).
16. M. Landgren and B. Jönsson, *J. Phys. Chem.* **97**, 1656–1660 (1993).

17. J. A. de Feijter, J. Benjamins and F. A. Veer, *Biopolymers* **17**, 1759–1772 (1978).
18. J. Olsson, M. Hellsten and K. Holmberg, *Colloid Polym. Sci.* **269**, 1295–1302 (1991).
19. C.-G. Gölander and J. C. Eriksson, *J. Colloid Interface Sci.* **119**, 38–48 (1987).
20. K. Holmberg, K. Bergström, C. Brink, E. Österberg, F. Tiberg and J. M. Harris, *J. Adhesion Sci. Technol.* **7**, 503–517 (1993).
21. B. Lassen, K. Holmberg, C. Brink, A. Carlén and J. Olsson, *Colloid Polym. Sci.* **272**, 1143–1150 (1994).
22. C.-G. Gölander, J. N. Herron, K. Lim, P. Claesson, P. Stenius and J. D. Andrade, in: *Poly(ethylene glycol) Chemistry*, J. M. Harris (Ed.), pp. 221–245. Plenum Press, New York (1992).
23. N. L. Burns, K. Emoto, K. Holmberg, J. M. van Alstine and J. M. Harris, *J. Biomed. Mater. Res.* (in press).
24. M. Malmsten, P. M. Claesson, E. Pezron and I. Pezron, *Langmuir* **6**, 1572–1578 (1990).
25. J. W. Dobbie, R. Evans, D. V. Gibson, J. B. Smitham and D. H. Napper, *J. Colloid Interface Sci.* **45**, 557–565 (1973).
26. D. H. Napper, *Polymeric Stabilization of Colloidal Dispersions*, p. 198. Academic Press, London (1983).
27. J.-T. Li, K. D. Caldwell and N. Rapoport, *Langmuir* **10**, 4475–4482 (1994).
28. M. Malmsten, F. Tiberg, B. Lindman and K. Holmberg, *Colloids Surfaces A* **77**, 91–100 (1993).
29. K. Bergström, E. Österberg, K. Holmberg, A. S. Hoffman, T. P. Schuman, A. Kozlowski and J. M. Harris, *J. Biomater. Sci. Polym. Edn.* **6**, 123–132 (1994).
30. S. I. Jeon, J. H. Lee, J. D. Andrade and P. G. de Gennes, *J. Colloid Interface Sci.* **142**, 149–158 (1991).
31. Y. Mori, S. Nagaoka, H. Takiuchi, T. Kikuchi, N. Noguchi, H. Tanzawa and Y. Noishiki, *Trans. Am. Soc. Artif. Internal Organs* **28**, 459–463 (1982).
32. W. R. Gombotz, W. Guanghui, T. A. Horbett and A. S. Hoffman, in: *Poly(ethylene glycol) Chemistry*, J. M. Harris (Ed.), pp. 247–261. Plenum Press, New York (1992).
33. K. P. Antonsen and A. S. Hoffman, in: *Poly(ethylene glycol) Chemistry*, J. M. Harris (Ed.), pp. 15–28. Plenum Press, New York (1992).
34. N. B. Graham, in: *Poly(ethylene glycol) Chemistry*, J. M. Harris (Ed.), pp. 263–281. Plenum Press, New York (1992).
35. N. L. Burns, MSc Thesis, University of Alabama, Huntsville, AL (1993).
36. J. M. van Alstine, N. L. Burns, J. A. Riggs, K. Holmberg and J. M. Harris, *Colloids Surfaces A* **77**, 149–158 (1993).
37. E. Österberg, K. Bergström, K. Holmberg, J. A. Riggs, J. M. van Alstine, T. P. Schuman, N. L. Burns and J. M. Harris, *Colloids Surfaces A* **77**, 159–169 (1993).
38. E. Österberg, K. Bergström, K. Holmberg, T. P. Schuman, J. A. Riggs, N. L. Burns, J. M. van Alstine and J. M. Harris, *J. Biomed. Mater. Res.* **29**, 741–747 (1995).
39. J. H. Lee, P. Kopeckova, J. Kopecek and J. D. Andrade, *Biomaterials* **11**, 455–464 (1990).
40. H. Elwing, S. Welin, A. Askendal, U. Nilsson and I. Lundström, *J. Colloid Interface Sci.* **119**, 203–210 (1987).
41. K. L. Prime and G. M. Whitesides, *J. Am. Chem. Soc.* **115**, 10714–10721 (1993).

Mittal Festschrift, pp. 461–494
W. J. Van Ooij and H. R. Anderson, Jr. (Eds)

Surface treatment of styrene–butadiene rubber with carboxylic acids

NURIA PASTOR-SEMPERE,[1] JUAN C. FERNÁNDEZ-GARCÍA,[1] A. CÉSAR ORGILÉS-BARCELÓ,[1] M. MERCEDES PASTOR-BLAS,[2] JOSÉ MIGUEL MARTÍN-MARTÍNEZ[2,*] and JOHN G. DILLARD[3]

[1] *INESCOP, Footwear Research Institute, Polígono Industrial Campo Alto, Apartado de Correos 253, 03600 Elda, Alicante, Spain*

[2] *Adhesion & Adhesives Laboratory, Department of Inorganic Chemistry, University of Alicante, 03080 Alicante, Spain*

[3] *Center for Adhesive and Sealant Science, Department of Chemistry, Virginia Technical College, Blacksburg, VA 24061-0201, USA*

Abstract—Solutions of carboxylic acids (fumaric acid, maleic acid, acrylic acid, succinic acid, malonic acid) in ethanol were used for surface treatment of two very different formulations of vulcanized synthetic styrene–butadiene rubbers to improve their adhesion to a polyurethane adhesive. The increase in the T-peel strength of adhesive joints formed between polyurethane and rubber surfaces treated with carboxylic acids depends on the rubber compounding and the characteristics of the carboxylic acid (acidity, solubility in ketones). The chemical composition of the treated surfaces was characterized by means of infra-red spectroscopy, and XPS (ESCA) analysis. The increase in the adhesion properties of styrene–butadiene rubbers treated with carboxylic acids is attributed to the elimination of abhesive substances on the surface of the rubber and to the modification of the failure mode of the joint (from interfacial to interfacial + cohesive failure of rubber). The effectiveness of the surface treatment of rubber with carboxylic acids is ascribed to a migration of the carboxylic acid into the polyurethane adhesive layer once the adhesive joint is formed. It is postulated that the carboxylic acid acts as a compatibilizer between rubber and polyurethane. The difference in effectiveness of the carboxylic acids studied is ascribed to the difference in the extent of rubber–adhesive interfacial interaction which, with increased interaction, allows varying degrees of cohesive failure at greater depths in the rubber.

Keywords: Surface treatments; carboxylic acid; polyurethane adhesive; styrene–butadiene rubber; T-peel strength; ESCA (XPS).

1. INTRODUCTION

Adhesion of rubber to polymers and metals is a key technology in the automotive, tire, aeronautical and footwear industries. Due to incompatibility between rubber and some adhesives (e.g. polyurethane adhesives), and with other adherents, a surface treatment

*To whom correspondence should be addressed. E-mail: JM.Martin@ua.es

is necessary to obtain adequate adhesion. Roughening [1] and halogenation [2–5] are common surface treatments for rubbers. Although roughening is effective, producing improvement of mechanical adhesion and elimination of abhesive substances from the rubber surface, its effectiveness is time-limited because these substances might migrate to the surface creating a weak boundary layer that lessens the strength of the adhesive joints [6]. Halogenation prevents the formation of these weak boundary layers in rubbers and generally produces a noticeable improvement in adhesion. However, halogenation of rubber presents some limitations: (i) Careful handling is required because of the oxidizing character of the halogenating solutions and their toxicity (chlorine emissions can be produced); (ii) The reaction takes at least 6 h for the optimum effectiveness of the surface treatment to be reached; (iii) The stability of the halogenating solutions is limited; and (iv) An excess of the halogenating agent can produce weak surface boundary layers [7].

The drawbacks of such surface treatments for rubber (roughening, halogenation) show the necessity of finding new surface treatments for elastomers. In this paper, the use of carboxylic acid solutions is proposed as a new surface treatment for synthetic vulcanized rubber.

Carboxylic acids have been used as additives in formulations of polyurethane (PU) adhesives [8–11] and cyanoacrylate adhesives [12] to improve their adhesion properties. It has been demonstrated [13] that the ageing properties of cyanoacrylate adhesives can be improved by adding small amounts of some selected organic anhydrides. On the other hand, the interaction between polyolefins and glass fiber in composite materials is generally improved by adding maleic anhydride or acrylic acid to the polyolefin [14] which results in a chemical interaction between C=O and amine group of silanes on the glass fiber. Grafted polymers [15] and polymers surface-treated with acrylic acid [16] show a surface orientation of the carboxylic groups in the presence of water which produces an increase of surface energy; this in turn, improves the compatibility between non-polar polymers and polar solvents. Grafting of polypropylene with maleic anhydride produces a noticeable improvement of adhesion to aluminum [17] due to the migration of the anhydride in the polyolefin surface to the aluminum surface. Grafted EVA (ethylene–vinyl acetate) with acrylic acid, maleic anhydride and their mixtures exhibit improved adhesion to several adherents [18].

The surface treatment of rubbers with fumaric acid solutions has been studied previously [19–21]. The increase in the adhesion between polyurethane adhesives and styrene–butadiene rubbers treated with fumaric acid has been attributed to an enhancement in the surface energy (due to the elimination of abhesive substances from the surface of the rubber and to the formation of C=O and C–OH groups by surface oxidation). However, carboxylic acids other than fumaric acid (FA) have not yet been studied so, in this work, five carboxylic acids were selected to analyze their effectiveness as a potential surface treatment of synthetic sulfur-vulcanized styrene–butadiene rubber (SBR) (40–60% styrene). The influence of the composition of SBR rubber, the characteristic of the carboxylic acid and the effects produced by the surface treatments are assessed. A migration of the carboxylic acid on the rubber surface to the adhesive may explain the effectiveness of the treatment, the extent of which depends on the formulation and nature of the rubber and on the characteristics of the carboxylic acid.

2. EXPERIMENTAL

2.1. Materials

Table 1 shows some properties of the carboxylic acids used in this study. This table contains the IUPAC nomenclature, first acid constant and melting point of the carboxylic acids. Because all the carboxylic acids selected in this study are quite soluble in ethanol, this solvent was used to apply most of them to the SBR surface, except for the fumaric and acrylic acid solutions in which a 1 : 1 v/v mixture of ethanol and 2-butanone was used.

The two synthetic vulcanized styrene–butadiene rubbers (R1, R2) used in this study, about 5 mm thick, were prepared using a moulding process (carried out at 150°C for 50 min) after open-mill mixing, using sulfur as the vulcanizing agent. They have different °Shore A hardness (93 and 72, respectively), different oils and plasticizer contents (7 phr for R1 and 12 phr for R2), different fillers (precipitated silica, carbon black), and different stearic acid and zinc oxide contents. Compositions of both rubbers are given in Table 2. The SBRs were surface-treated by roughening (0.5 mm thickness was eliminated) followed by treatment with carboxylic acid solutions to improve their adhesion properties. Rubbers were roughened (in a scouring machine at 2800 rpm using a sandpaper No. 60 of 200 mm diameter) to remove the most external rubber layer. A blank sample was prepared by wiping the roughened R2 rubber surface with a 2-butanone + ethanol mixture (1 : 1 in volume); 2-butanone was added because it is the solvent for the polyurethane adhesive used to determine the T-peel strength of rubber joints. The surface treatment with solutions containing different amounts (0.5–10 wt%) of carboxylic acid was carried out by rubbing the surface with a brush. Because the acrylic acid is liquid at room temperature, different diluted solutions (10–100 wt%) were prepared by adding the 2-butanone + ethanol mixture (1 : 1 in volume). The time that elapsed between the consecutive applications of carboxylic acid solutions and the PU adhesive to the surface of the rubber (treatment time) was varied between 30 min and 48 h. In Fig. 1, the T-peel strengths are given for R1 and R2 rubbers treated with 2 wt% malonic acid (MaA) — taken as a typical example — as a function of the treatment time. T-peel values included in Fig. 1 are the average of five experimental determinations (relative error was always lower than 5%). There were no variations in T-peel strength between 30 min and 48 h. One hour was selected as a suitable treatment time in this study.

To determine the T-peel strength a thermoplastic polyester–urethane (Desmocoll 510) manufactured by Bayer (Leverkusen, Germany) was used. This polyurethane contains around 5.8 wt% talc or fumed silica (estimated from thermogravimetric measurements). This one-component polyester–urethane has a very high crystallization rate (near 1 min at room temperature) and short open time. The adhesive was prepared by dissolving 18 wt% PU in 2-butanone, and a Brookfield viscosity of 1.7 Pa s (23°C) was obtained.

The adhesive joints were prepared by applying approximately 100 mg of adhesive to each one of the identically treated rubber surfaces. After allowing the solvent to evaporate for 30 min, the dry adhesive films were melted at 80°C, and immediately

Table 1.
Some characteristics of carboxylic acids

Nomenclature	Ordinary name	IUPAC name	Chemical formula	1st acid constant[a]	Melting point (°C)[a]
FA	Fumaric acid	*trans*-1,2-Ethylene dicarboxylic acid	HOOC−CH=CH−COOH	9.3×10^{-4}	230
MA	Maleic acid	*cis*-1,2-Ethylene dicarboxylic acid	HOOC−CH=CH−COOH	1.4×10^{-2}	139
AA	Acrylic acid	2-Propenoic acid	CH_2=CH−COOH	5.6×10^{-5}	14
MaA	Malonic acid	Methane dicarboxylic acid	HOOC−CH_2−COOH	1.2×10^{-3}	135
SA	Succinic acid	1,4-Butane dicarboxylic acid	HOOC−CH_2−CH_2−COOH	6.9×10^{-5}	119

[a]Data taken from *Handbook of Chemistry and Physics*. CRC Press, Boca Raton, FL (1994).

Table 2.
Compositions of R1 and R2 vulcanized synthetic rubbers

Component	Percentage (phr)	
	R1	R2
SBR 1620	71	—
SBR 1502	18	100
Carbon black HSR 1904	35	—
Precipitated silica	25	42
Hydrocarbon resin	3.5	5.0
Sulfur	1.8	2.0
N-Cyclohexyl-2-benzothiazole-sulphenamide	1.1	2.0
Tetramethylthiuram disulfide	0.2	—
Stearic acid	0.8	2.4
Zinc oxide	3.8	1.5
Poly(ethylene glycol)	1.1	—
Phenolic antioxidant	0.8	0.5
2-Mercaptobenzothiazole-disulfide	—	2.5
Microcrystalline paraffin wax	—	0.8
Hexamethylene tetramine	—	1.0
Fatty acid zinc salts	—	5.4

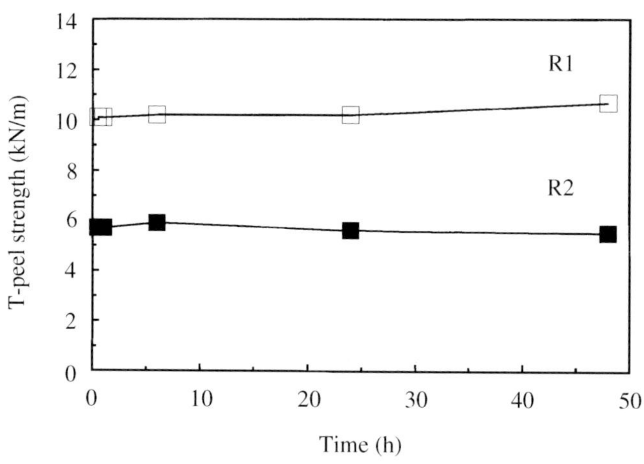

Figure 1. T-peel strengths (kN/m) of PU adhesive joints of roughened R1 and R2 rubbers treated with 2 wt% malonic acid (MaA) in ethanol. Influence of the treatment time.

brought into contact under a pressure of 0.8 MPa. The adhesive joints were conditioned for 72 h at 23°C and 50% relative humidity before undergoing the T-peel test.

2.2. *Experimental techniques*

2.2.1. T-peel strengths. The strength of the adhesive joints was determined using a T-peel test (European Standard: PREN 1391) in an Instron 1011 test instrument, with a peel rate of 100 mm/min. Five experimental determinations for each analyzed experimental variable were carried out, the experimental error obtained being less than 5%.

2.2.2. Scanning Electron Microscopy (SEM). The surfaces of R1 and R2 rubbers treated with carboxylic acids were analyzed using a scanning electron microscope JEOL SEM JSM 840. Both rubber surfaces were analyzed before and after T-peel tests.

2.2.3. Fourier Transform Infra-red Spectroscopy (FTIR). The IR spectra of R1 and R2 rubbers treated with different amounts (0.5–10 wt%) of carboxylic acids in ethanol or ethanol/2-butanone mixture were obtained to determine the nature of the chemical modifications produced by the surface treatment of the rubbers (before and after carrying out the T-peel test). The attenuated total multiple reflection method was used (ATR), with a thallium bromoiodide crystal (KRS-5) (for R2 rubber) or a germanium crystal (for R1 rubber); in this case, the germanium crystal reduces the absorption caused by carbon black. More details on this technique have been presented earlier [5].

2.2.4. X-ray Photoelectron Spectroscopy (XPS). XPS was used to determine the modifications produced in the most external (5–10 nm) rubber-treated surface. XPS analysis was performed using a Perkin–Elmer PHI 5400 spectrometer with a Mg K_α achromatic X-ray source (1253.6 eV), operating at 15 keV and 400 W with an emission current of 30 mA. Pressure inside the analysis chamber of the instrument was kept below 5×10^{-7} torr (6.6×10^{-5} Pa) during the course of the analysis. To avoid outgassing problems in rubber samples during the XPS experiments, the treated rubbers were outgassed for 1 h into the XPS analysis chamber before carrying out the experiments and, further, the samples were cooled at 0°C throughout the experiment. Rectangular sample pieces (20 × 30 mm) were used, although the dimensions of the analyzed areas on the samples were 1 × 3 mm. For each sample, a survey scan encompassing the region of 0–1100 eV was first taken. Multiplex scanning of all observed photopeaks (carbon, oxygen, zinc, nitrogen, sulfur, silicon) in the survey scan was then carried out, and typically ten scans for each element were done. Binding energies of all photopeaks were referenced to the C—C and C—H species at 285 eV. Atomic concentration calculations and curve fitting were carried out on an Apollo 3500 computer, using PHI software version 4.0. In general, one measurement was taken for each sample.

3. RESULTS AND DISCUSSION

3.1. T-peel strength

Figure 2 shows, as typical examples, the T-peel strengths measured on adhesive joints of R1 and R2 roughened rubbers after surface treatment with solutions containing

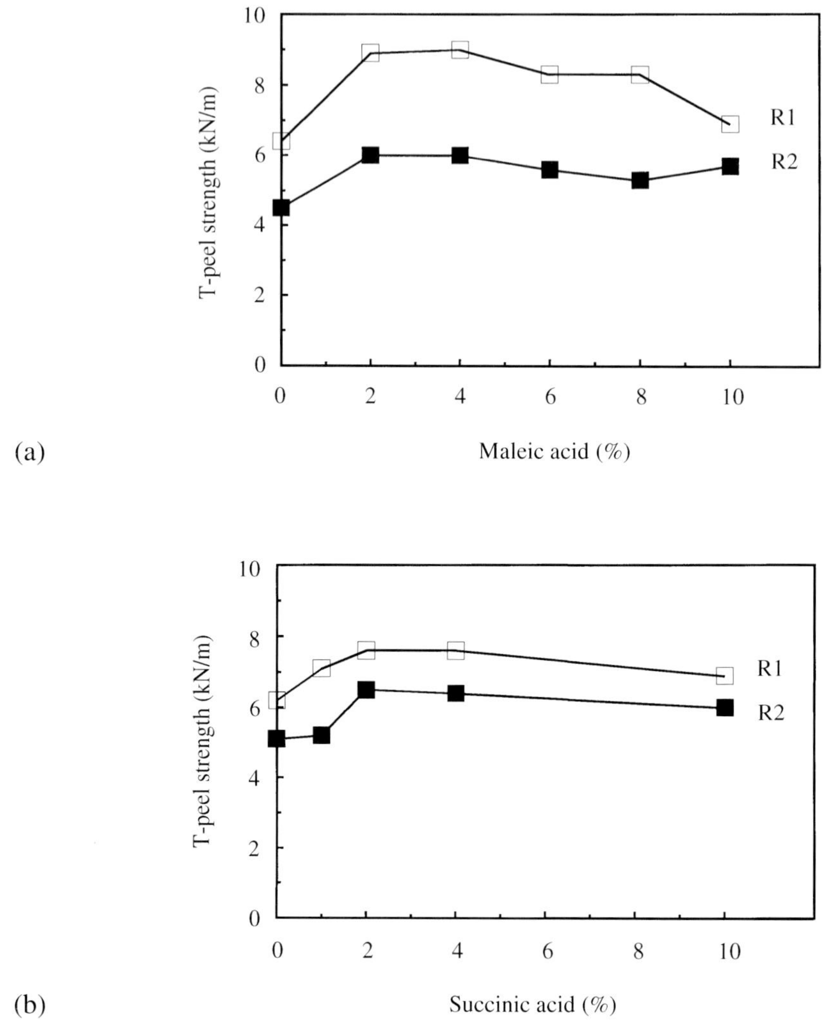

Figure 2. T-peel strengths (kN/m) of PU adhesive joints of roughened R1 and R2 rubbers treated with (a) maleic (MA) and (b) succinic (SA) acid solutions in ethanol.

different amounts of MA and SA in ethanol. Surface treatment with carboxylic acid improves the T-peel strength of rubber joints, although its effectiveness depends on the rubber compounding (more effective in R1). The highest T-peel strength was obtained with 2–4 wt% MA or SA; amounts of MA or SA larger than 4 wt% produced somewhat smaller T-peel strength (always higher than the strength of blank rubber joints).

A comparison of the effectiveness of the surface treatment with the different carboxylic acids is given in Fig. 3, where the percentage of maximum increase in T-peel strength (i.e. the ratio of T-peel strength obtained in adhesive joints of carboxylic acid surface treated roughened rubber with respect to that of only roughened rubber) is given for all carboxylic acids. In general, the treatment of rubbers with carboxylic

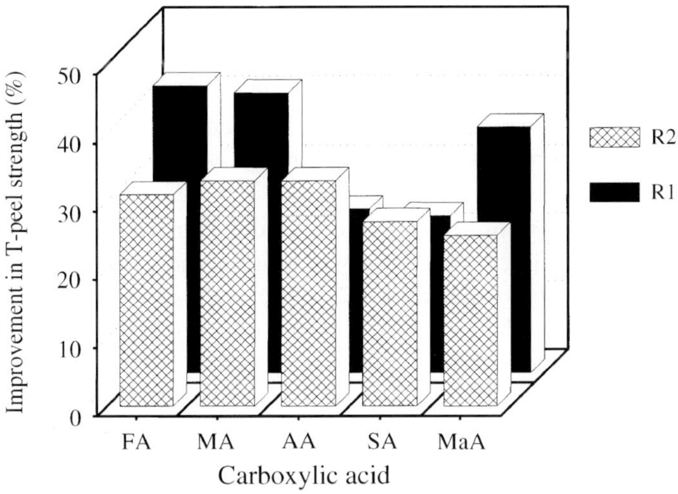

Figure 3. Improvement in T-peel strength (%) of PU adhesive joints of roughened R1 and R2 rubbers treated with solutions of different carboxylic acids in ethanol. The percentages were calculated as the ratio between the T-peel strength value of carboxylic acid-treated rubber joints with respect to the value for solvent-wiped roughened rubber joints.

acids increases the T-peel strength of adhesive joints by at least 25%, this increase being in general more pronounced in R1 rubber (possibly because R2 rubber contains a higher amount of abhesive substances, principally paraffin wax and zinc stearate). The treatment of R1 rubber with FA, MA and MaA produces the highest increase of T-peel strength, whilst for R2 rubber the highest increase corresponds to FA, MA and AA. The different degrees of effectiveness of the surface treatment of SBR rubbers with carboxylic acids (Fig. 3) can be arranged, in general, in two groups: (i) FA, MA and AA (only for R2 rubber) provide the highest increase of T-peel strength; (ii) SA and MaA provide the lowest increase of T-peel strength. Differences in chemistry of the treated surface (IR spectroscopy and XPS studies) and in the surface morphology (SEM micrographs) of the treated rubbers with carboxylic acids may explain the distinct effectiveness of the carboxylic acids.

3.2. Characterization of surface treated rubbers with carboxylic acids

The chemical modification produced on the rubber surface by treatment with carboxylic acids was followed by IR spectroscopy. In Fig. 4 the IR spectra of R2 rubber treated with FA, MA, AA, SA and MaA are shown. The features of the IR spectra of R2 rubber treated with the different carboxylic acids are very similar. Thus, the treatment of the R2 rubber surface with solutions of carboxylic acids produces the following effects:

- The band at 1535 cm^{-1} corresponding to zinc stearate (abhesive substance in the formulation of the R2 rubber) does not appear in the IR spectra of the carboxylic acids-treated rubber surface.

- Some bands corresponding to the carboxylic acids (SA: 1313, 1727 cm^{-1}; MA: 1218, 1260, 1429, 1636, 1704 cm^{-1}; AA: 1298, 1408, 1723 cm^{-1}; MaA: 1227, 1325, 1710 cm^{-1}; FA: 1229, 1275, 1314, 1410, 1667 cm^{-1}) are present.
- Poorly-defined bands between 1300 and 1400 cm^{-1}, and between 1500 and 1700 cm^{-1} appear, corresponding to C=O groups, which can probably be attributed to the carboxylic acids.

Figure 5 shows the SEM micrographs of R2 rubber treated with the different carboxylic acids. In general, the treatment with carboxylic acids does not modify the degree of roughness of the roughened R2 rubber surface. Treatment of R2 rubber with FA, MA, MaA and SA leads to the formation of crystallites of the carboxylic acids on the surface; the geometrical shape and size are different for each acid. For the roughened R2 rubber surface-treated with AA, no crystallite can be detected. In fact, according to Table 1, the melting point of AA is 14°C, and thus no crystallite formation can be expected at room temperature. The existence of carboxylic acid crystallites on the R2 rubber surface may explain the C=O bands of carboxylic acids in the IR spectra (Fig. 4), and no great differences in surface topography among the surfaces treated with different carboxylic acids has been found (AA surface-treated roughened R2 rubber is an exception).

The effects of surface treatment of R1 rubber with some carboxylic acids were also investigated with IR spectroscopy (Fig. 6) and SEM (Fig. 7). The IR spectra of

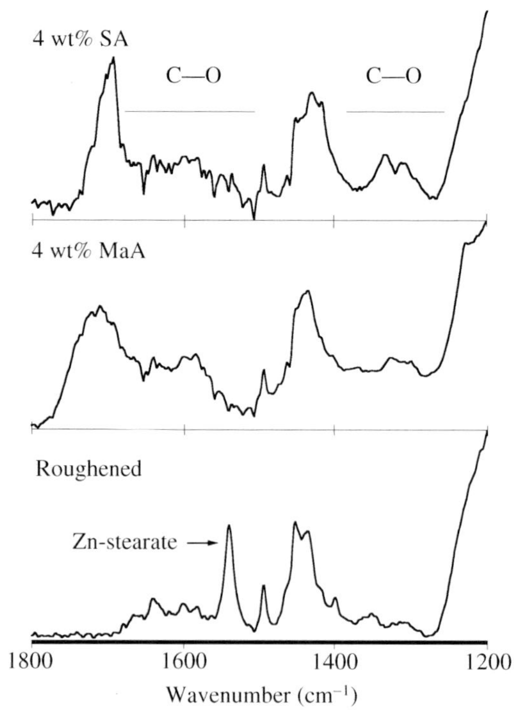

(a)

Figure 4. IR spectra of roughened R2 rubber treated with ethanol solutions of different carboxylic acids. (a) Roughened, 4 wt% MaA, 4 wt% SA; (b) 2 wt% FA, 4 wt% MA, 10 wt% AA.

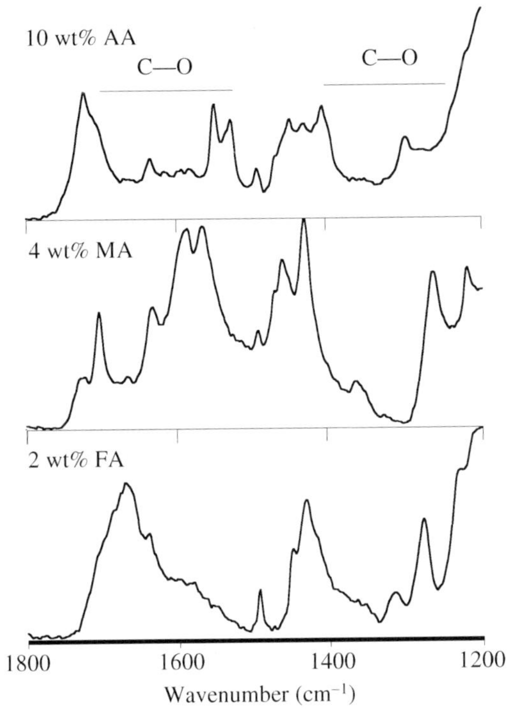

(b)

Figure 4. (Continued).

roughened R1 rubber (Fig. 6) show the presence of zinc stearate on the surface, which is completely removed by the treatment with carboxylic acids. This treatment also creates C=O bands (1300–1400 cm^{-1}, 1500–1700 cm^{-1}) on the rubber surface, the relative intensity of which is higher than for the R2 rubber (Fig. 4). On the other hand, all IR spectra (Fig. 6) show bands of carboxylic acids (the position of the bands were the same as indicated above) which correspond to the crystallites formed on the rubber surface (SEM micrographs of Fig. 7). The treatment of R1 rubber with AA produces a layer of carboxylic acid which covers all of the surface. Thus, in general, the same features can be found in the surface-treated R1 and R2 rubbers (although the use of carboxylic acid as surface treatment is more effective in R1 rubber): removal of zinc stearate, creation of C=O groups, and no modification of surface topography after treatment. However, no great differences exist between the surfaces treated with the different carboxylic acids (except no solid crystallites are found on the surface-treated rubbers with AA).

On the other hand, it might be expected that the carboxylic acid crystallites on the R1 and R2 rubber surfaces act as weak boundary layers, once the adhesive joint is formed, and thus, no improvement in T-peel strength should be observed. However, Figs 2 and 3 show an improved adhesion for carboxylic acids surface-treated rubbers. Therefore, IR and XPS spectra, and SEM micrographs of debonded rubber test pieces (after carrying out the T-peel test) were obtained.

4 wt% SA

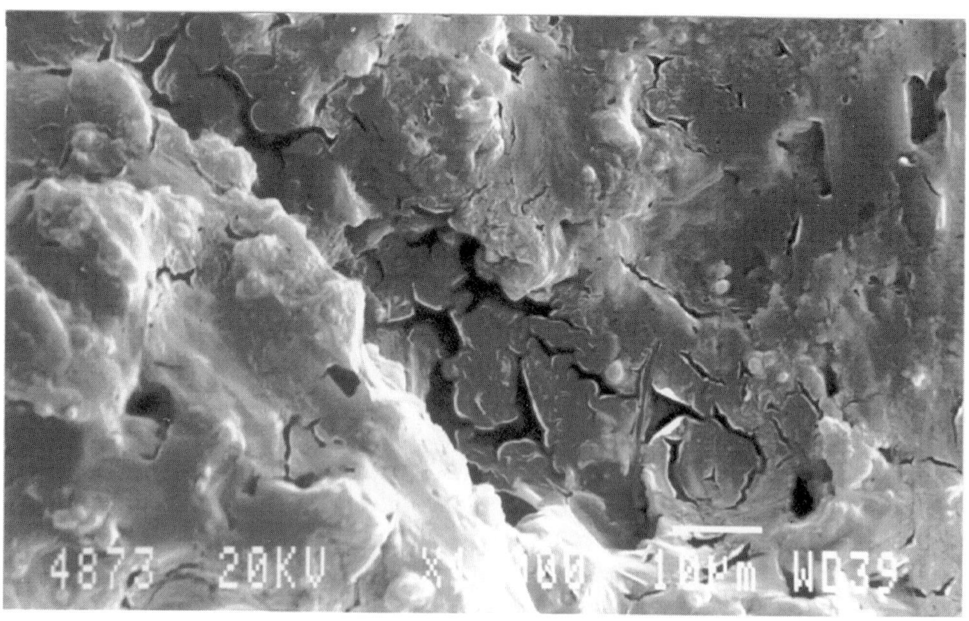

4 wt% MaA

Figure 5a. SEM microphotographs (×1000) of roughened R2 surfaces treated with ethanol solutions of different carboxylic acids. Roughened, 4 wt% MaA, 4 wt% SA.

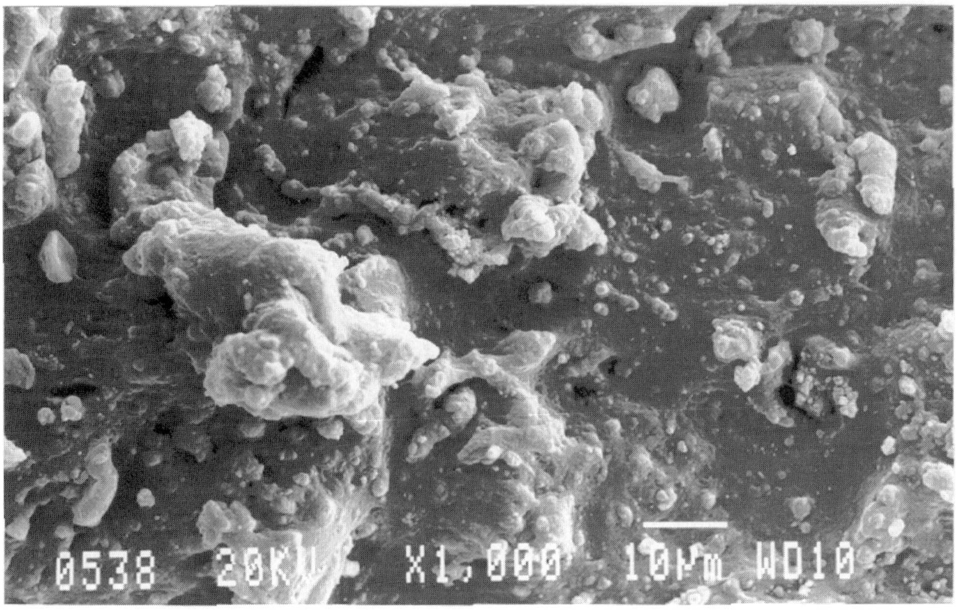

Roughened

Figure 5a. (Continued).

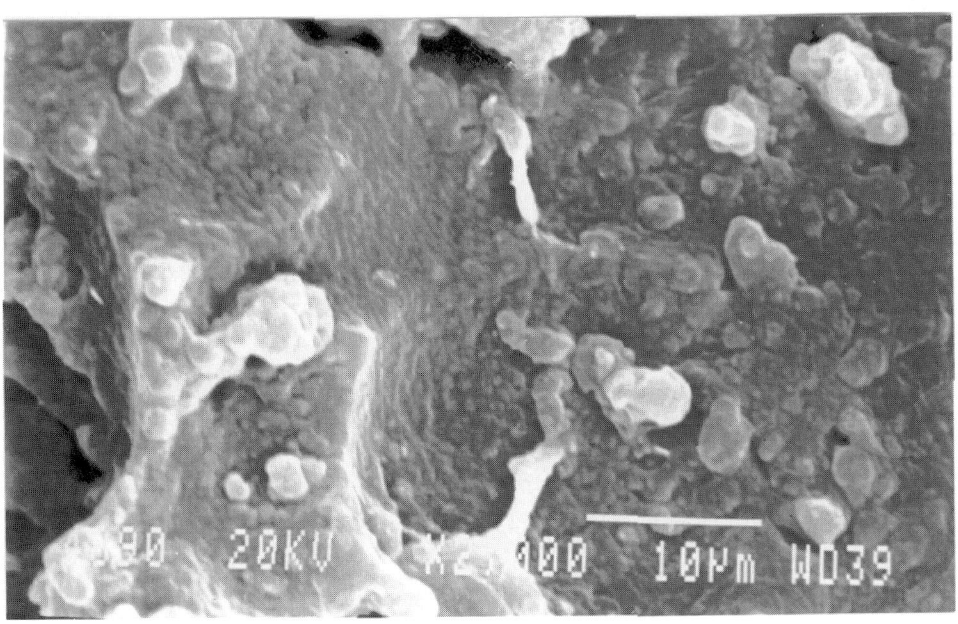

10 wt% AA

Figure 5b. SEM microphotographs (×1000) of roughened R2 surfaces treated with ethanol solutions of different carboxylic acids. 2 wt% FA, 4 wt% MA, 10 wt% AA.

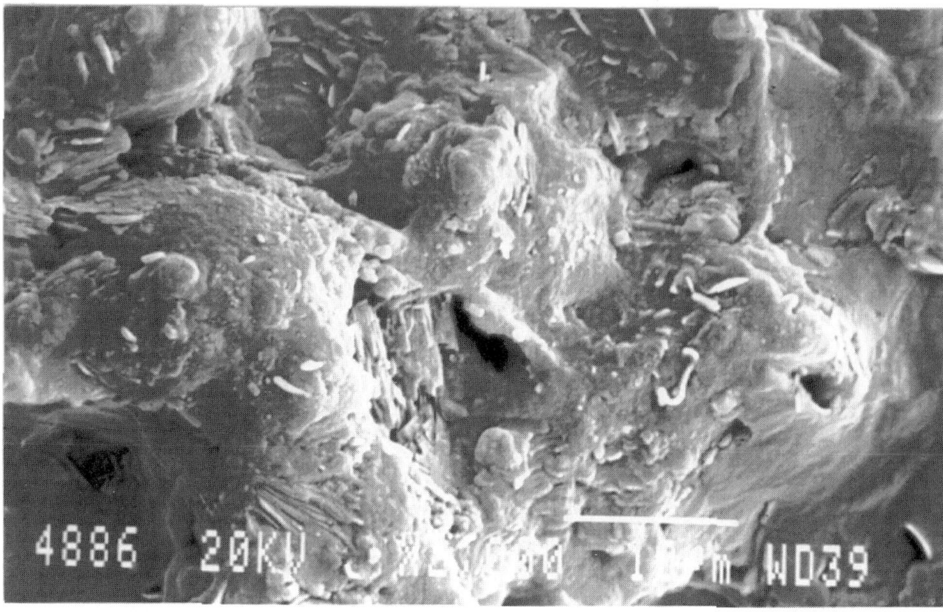

4 wt% MA

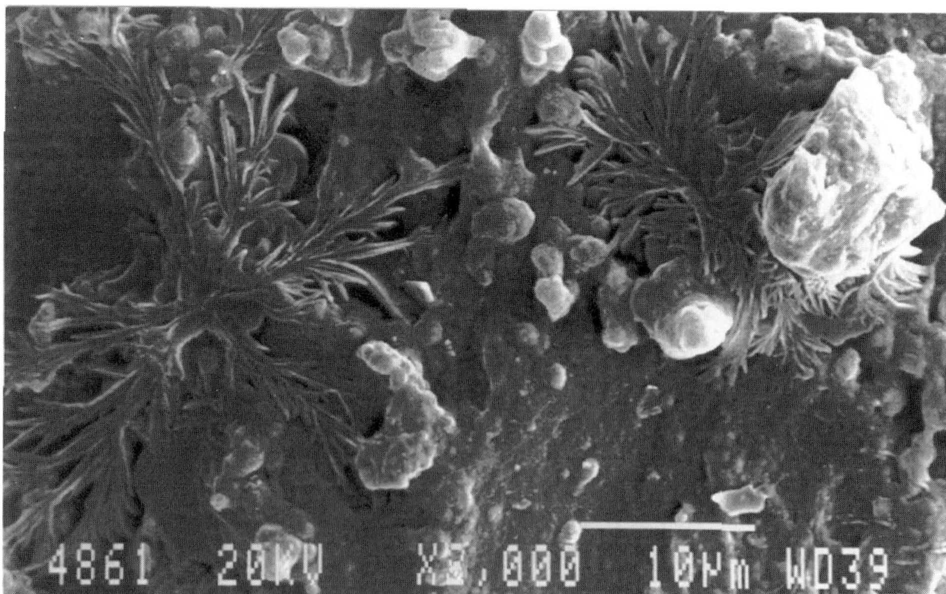

2 wt% FA

Figure 5b. (Continued).

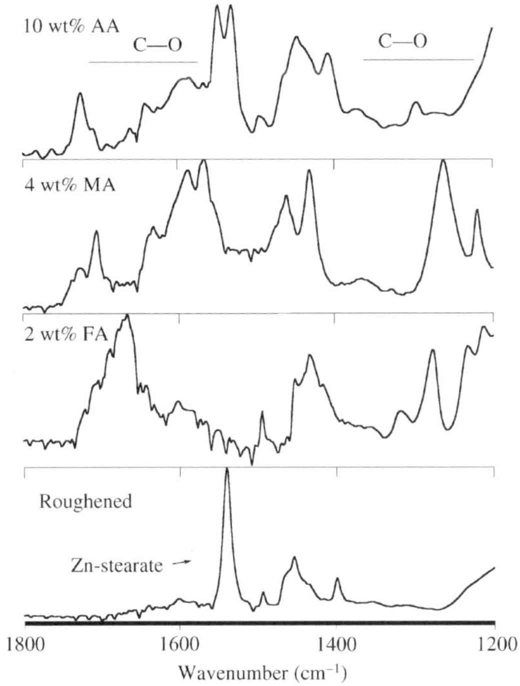

Figure 6. IR spectra of roughened R1 rubber treated with ethanol solutions of different carboxylic acids.

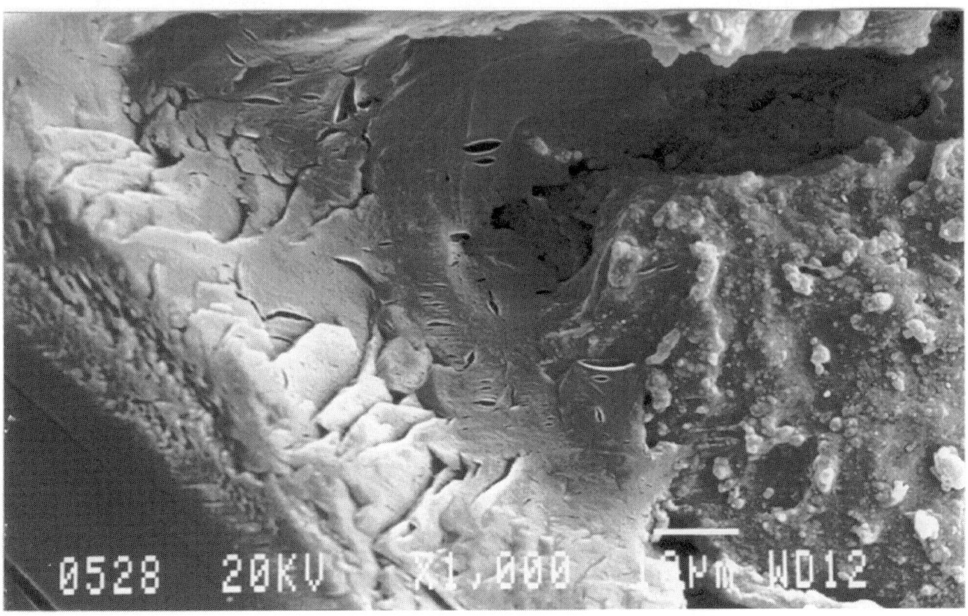

10 wt% SA

Figure 7a. SEM microphotographs (×1000) of roughened R1 rubber surfaces treated with ethanol solutions of different carboxylic acids. Roughened, 10 wt% MaA, 10 wt% SA.

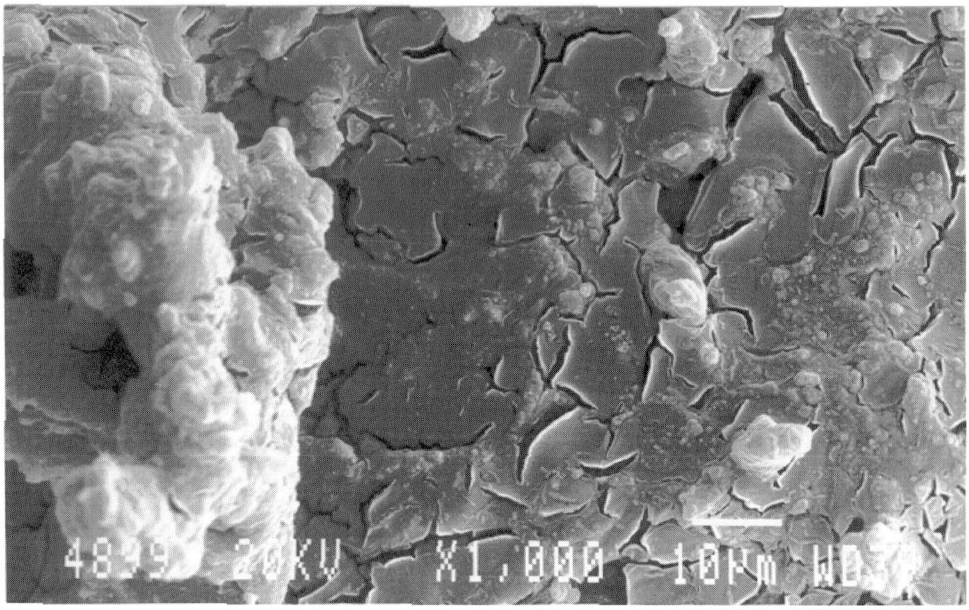

10 wt% MaA

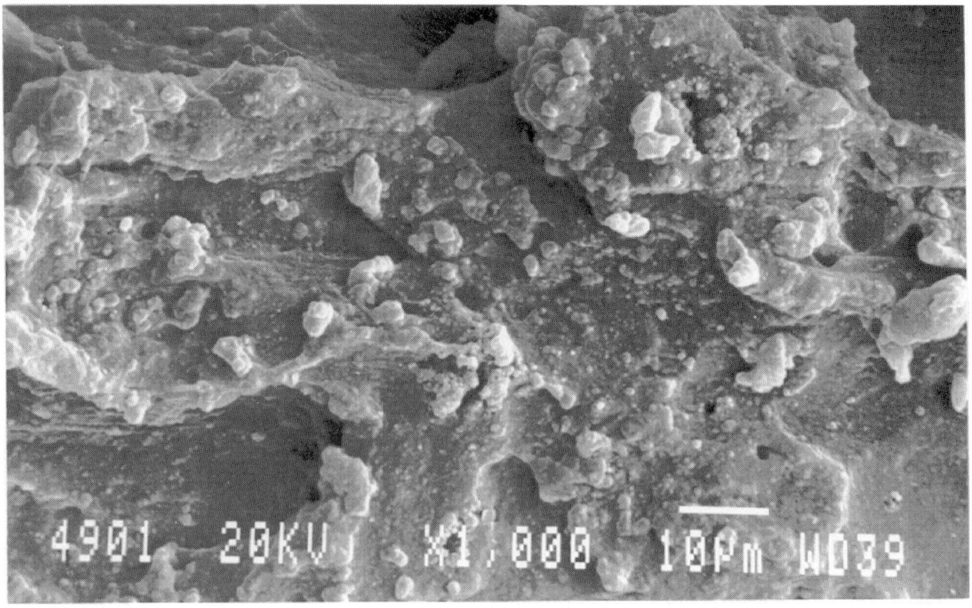

Roughened

Figure 7a. (Continued).

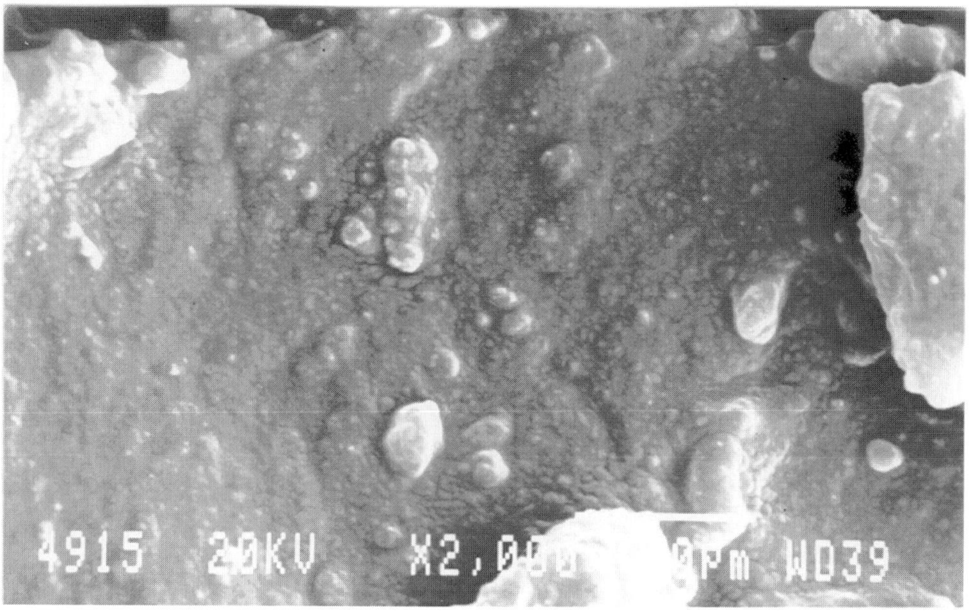

10 wt% AA

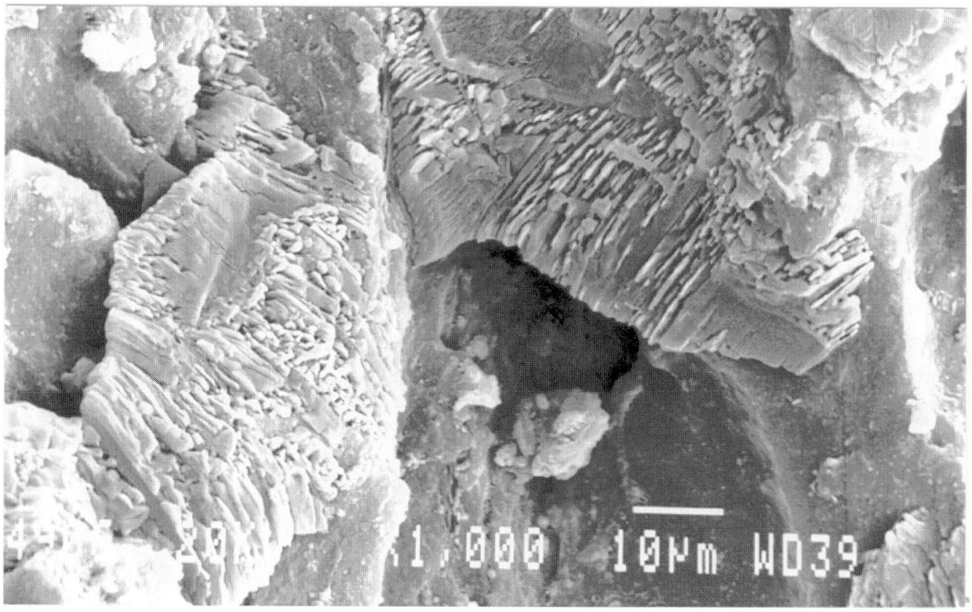

4 wt% MA

Figure 7b. SEM microphotographs (×1000) of roughened R1 rubber surfaces treated with ethanol solutions of different carboxylic acids. 2 wt% FA, 4 wt% MA, 10 wt% AA.

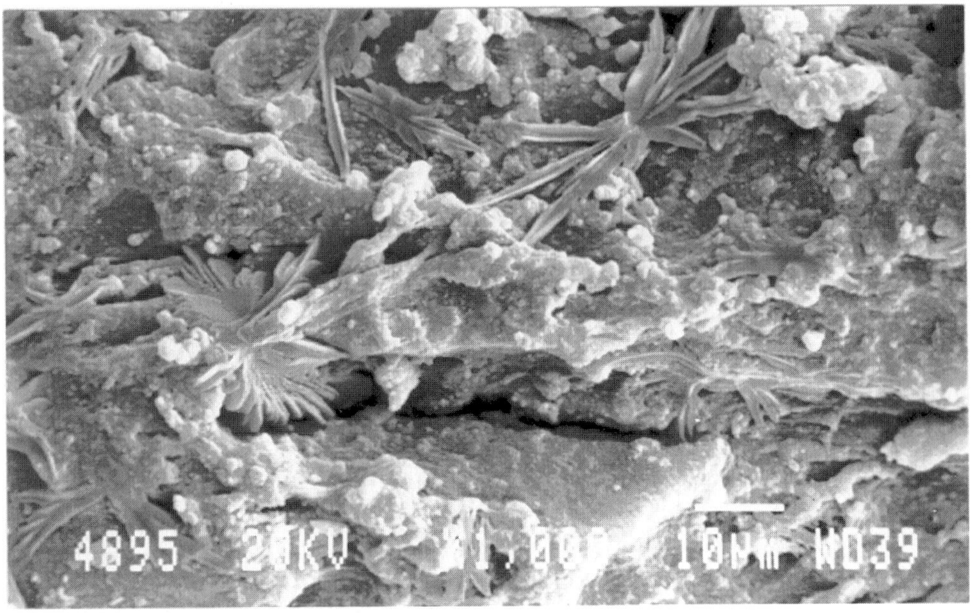

2 wt% FA

Figure 7b. (Continued).

3.3. Characterization of surface-treated debonded rubbers with carboxylic acids (after T-peel test)

Figure 8 shows the IR spectra of debonded R2 rubber surfaces (always the rubber side was analyzed) treated with carboxylic acids (i.e. after T-peel tests were carried out). The IR spectrum of debonded solvent-wiped roughened R2 rubber shows a band due to zinc stearate (1538 cm^{-1}) which is more intense than in the non-bonded roughened R2 rubber (Figs 4a and 8a). The increase in concentration of zinc stearate in debonded roughened R2 rubber is probably produced by migration of zinc stearate from the bulk to the rubber surface during the time that elapsed after formation of the joint [7]. Figure 9a shows the SEM micrograph of debonded solvent-wiped roughened R2 rubber. No great differences in surface topography are noted (with respect to the non-bonded roughened R2 rubber — see Fig. 5a), except for the fragments of polyurethane adhesive remaining on the surface. Chemical analysis of debonded roughened solvent-wiped rubber surfaces (both adhesive and rubber sides) confirm that zinc stearate (0.9–1.2 at.%) exists on both the rubber and adhesive sides of debonded solvent-wiped roughened R2 rubber joint (Table 3). According to Table 3, the chemical composition of the rubber and adhesive sides of debonded joints are relatively similar, except that more oxygen (—O—C=O) exists on the adhesive side. Therefore, the adhesive joint of blank roughened R2 rubber fails interfacially and in the adhesive (cohesive failure of adhesive).

The analysis of debonded roughened R2 rubber surfaces treated with carboxylic acids shows that this treatment avoids the migration of zinc stearate to the sur-

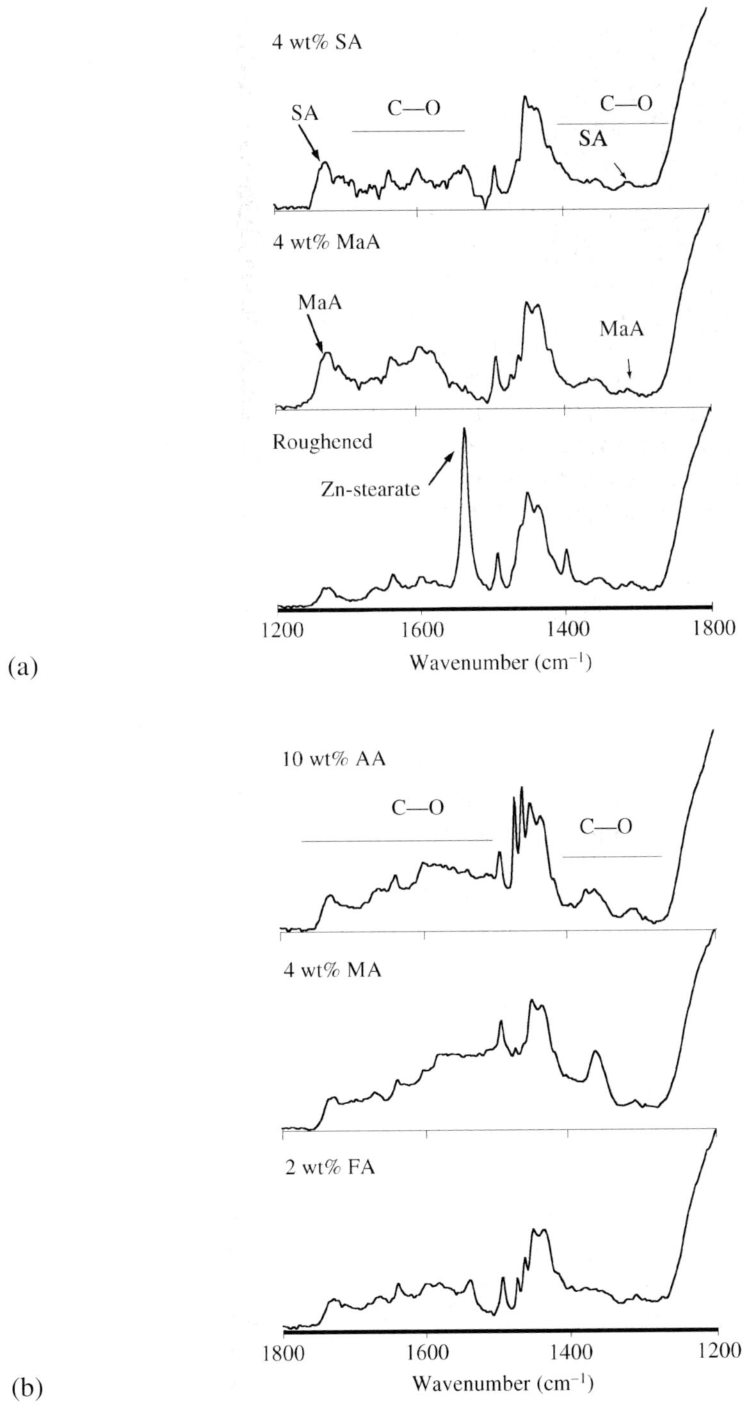

Figure 8. IR spectra of roughened R2 rubber treated with ethanol solutions of different carboxylic acids after carrying out the T-peel test. (a) Roughened, 4 wt% MaA, 4 wt% SA; (b) 2 wt% FA, 4 wt% MA, 10 wt% AA.

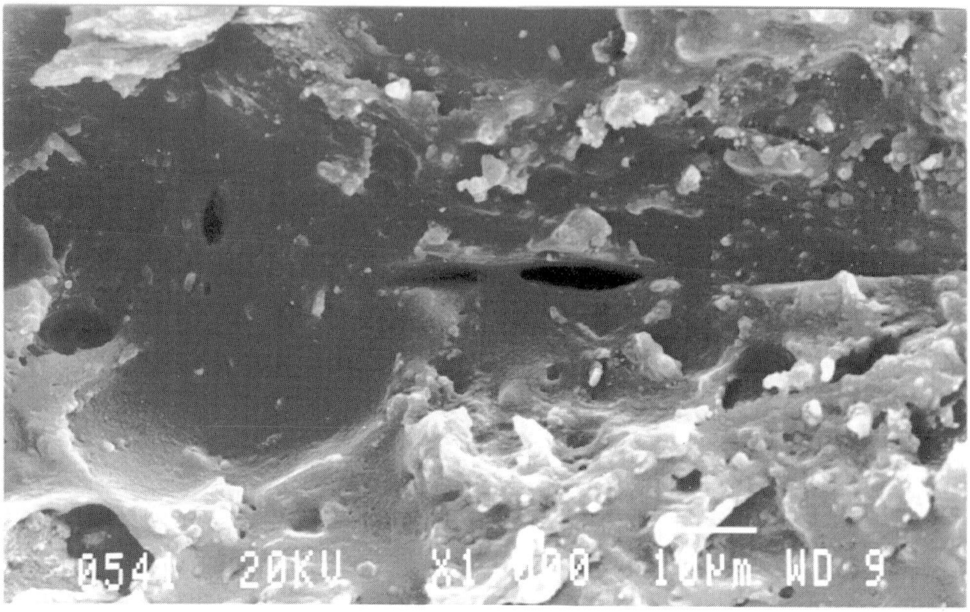

4 wt% SA

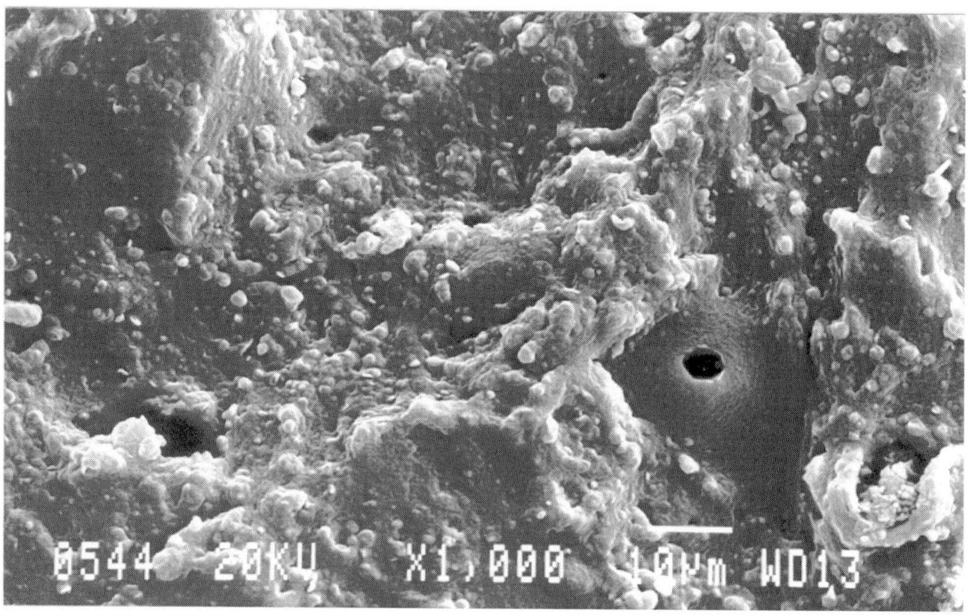

4 wt% MaA

Figure 9a. SEM microphotographs (×1000) of roughened R2 rubber surfaces treated with ethanol solutions of different carboxylic acids after carrying out the T-peel test. Roughened, 4 wt% MaA, 4 wt% SA.

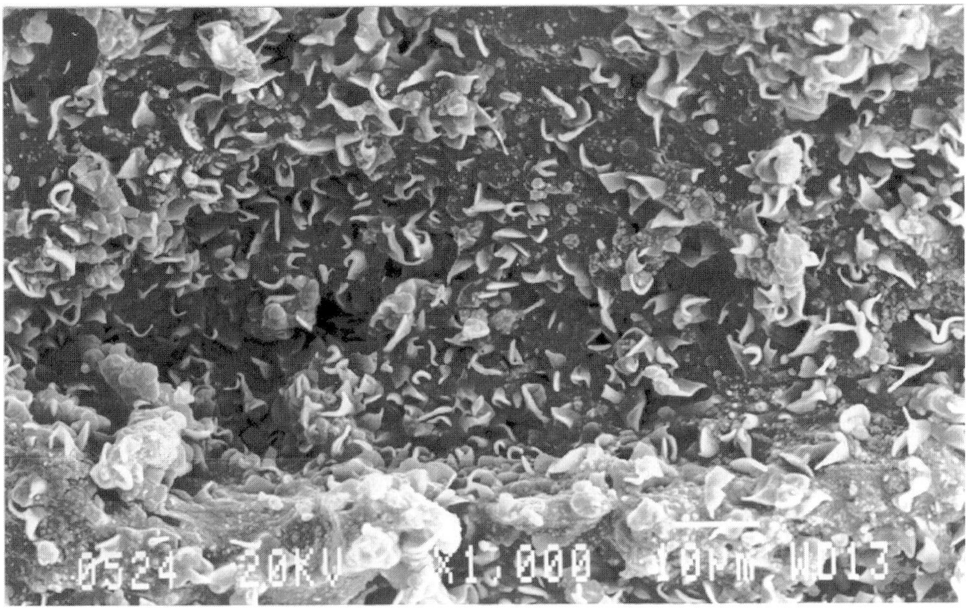

Roughened

Figure 9a. (Continued).

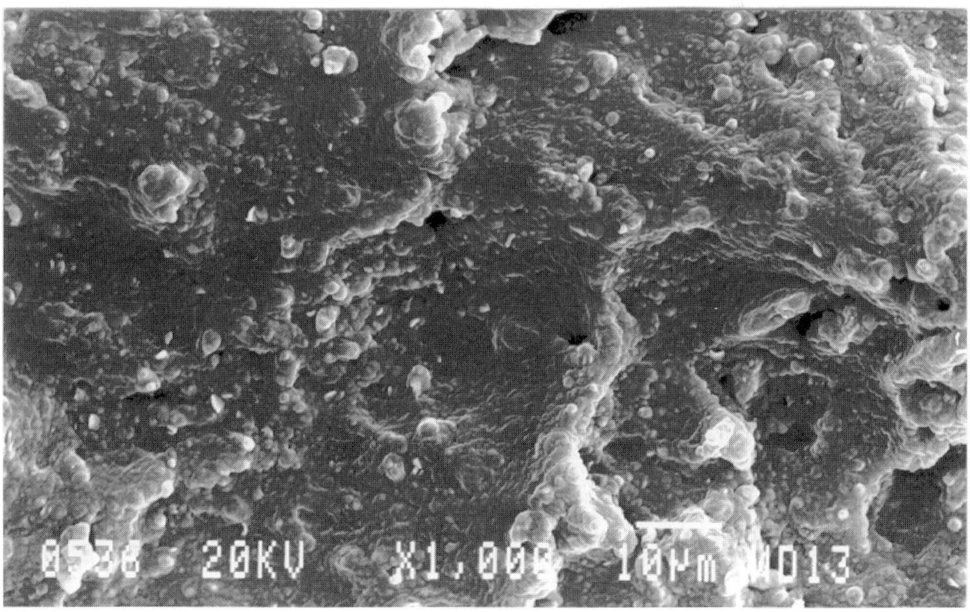

10 wt% AA

Figure 9b. SEM microphotographs (×1000) of roughened R2 rubber surfaces treated with ethanol solutions of different carboxylic acids after carrying out the T-peel test. 2 wt% FA, 4 wt% MA, 10 wt% AA.

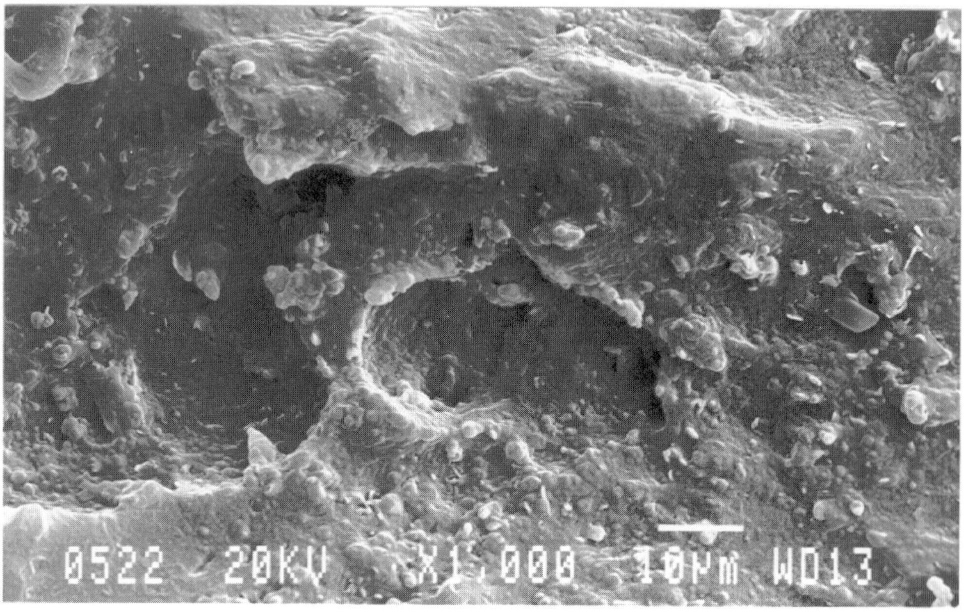

4 wt% MA

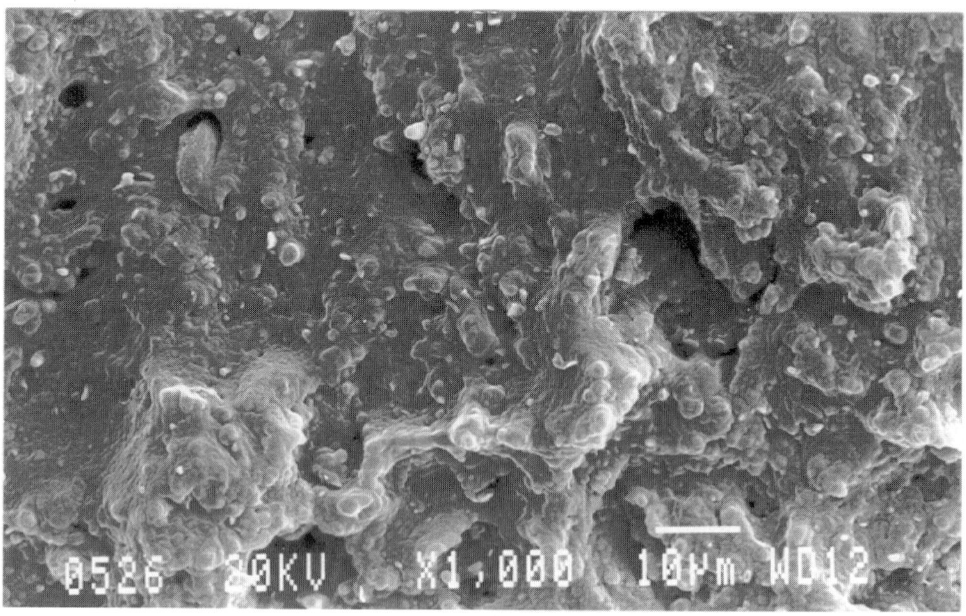

2 wt% FA

Figure 9b. (Continued).

Table 3.
XPS chemical composition (at.%) of debonded surface-treated roughened R2 rubber joints

Element	Non-bonded		Solvent treated		2 wt% FA		4 wt% SA	
	Roughened R2	PU	Rubber	Adhesive	Rubber	Adhesive	Rubber	Adhesive
C	96.8	73.1	91.7	88.2	97.2	87.7	98.0	90.8
O	2.0	20.4	5.0	8.5	2.0	9.9	1.2	7.1
N	ND[a]	0.7	0.6	0.4	ND	0.6	ND	0.3
Si	0.9	5.8	0.7	1.0	0.8	1.8	0.5	1.7
Zn	0.1	ND	0.9	1.2	ND	ND	ND	ND
S	0.2	ND	1.1	0.7	0.1	ND	ND	0.1

[a]ND = not detected; < 0.1 at.%.

face (Fig. 8 and Table 3). IR spectra show weak bands due to C=O groups at 1300–1400 cm^{-1} and 1500–1700 cm^{-1} on the rubber side of the debonded surface (Fig. 8b); these bands cannot be seen in the IR spectra of carboxylic acids themselves, but they are present in the IR spectrum of the polyurethane adhesive. On the other hand, SEM micrographs of the rubber side of debonded surface-treated roughened R2 rubber surfaces (Fig. 9b) do not show crystallites of carboxylic acids or polyurethane adhesive fragments on the surface, and the topography of these surfaces remains relatively similar to that of roughened R2 rubbers before the formation of the adhesive joint. Therefore, the failure of the surface-treated roughened R2 rubber joints occurs mainly at the interface, according to SEM results, but because some polyurethane adhesive bands can be seen in the IR spectra, a cohesive failure in the adhesive should be produced during the T-peel test.

On the other hand, the removal of crystallites of carboxylic acids from the rubber surface during the 'curing' of polyurethane adhesive in the joint should be related to the modification of the failure mode of adhesive joint (Figs 5 and 9). Thus, a migration of the crystallites of carboxylic acids from the roughened R2 rubber surface to the polyurethane adhesive may occur once the joint is created. To confirm this hypothesis, the following experiment was carried out: 2 wt% solid FA was added to the one-component solvent based polyurethane adhesive and an adhesive joint was formed with roughened R2 rubber; the T-peel test was carried out and the debonded roughened R2 rubber surfaces were analyzed with IR spectroscopy. Figure 10 shows the IR spectra of the rubber side of debonded roughened R2 rubber surfaces which were in contact with the polyurethane adhesive without and with 2 wt% FA. During the 'curing' of polyurethane adhesive which does not contain FA, a migration of zinc stearate to the roughened R2 rubber surface is produced (Fig. 8a, Table 3); such migration is completely avoided when the adhesive contains 2 wt% FA (Fig. 10).

Figure 11 shows the IR spectra of the rubber side of debonded roughened and (roughened + 2 wt% FA) treated R2 rubber surfaces; in this case, the polyurethane adhesive does not contain FA. IR spectra of roughened R2 rubber debonded surfaces given in Figs 10 and 11 are very similar and, consequently, similar effects are produced by surface treatment of roughened R2 rubber when the carboxylic acid is added to polyurethane adhesive or to the rubber surface. Therefore, a migration of the carboxylic acid is produced at the interface once the adhesive joint is formed. In

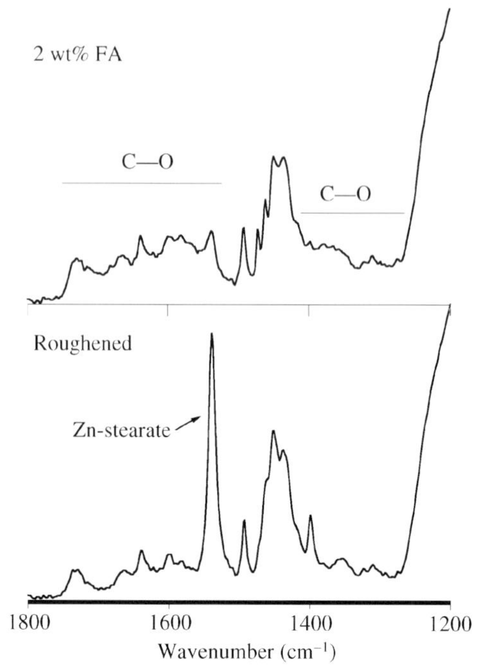

Figure 10. IR spectra of roughened R2 rubber test pieces obtained after carrying out the T-peel test with a polyester–urethane adhesive without and with 2 wt% fumaric acid.

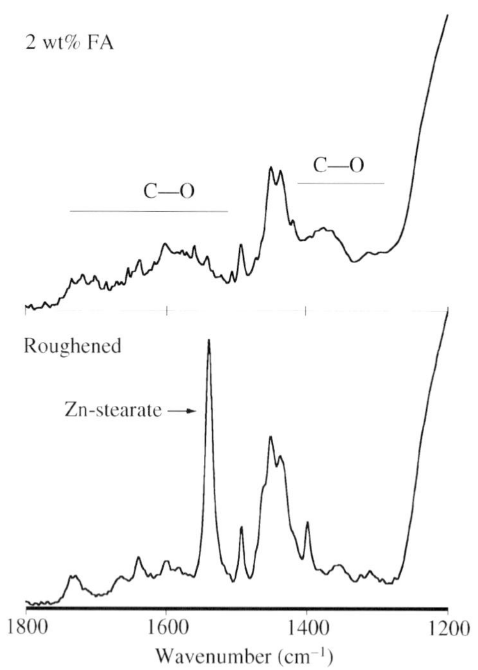

Figure 11. IR spectra of roughened and (roughened + 2 wt% fumaric acid) surface treated R2 rubber after carrying out the T-peel test.

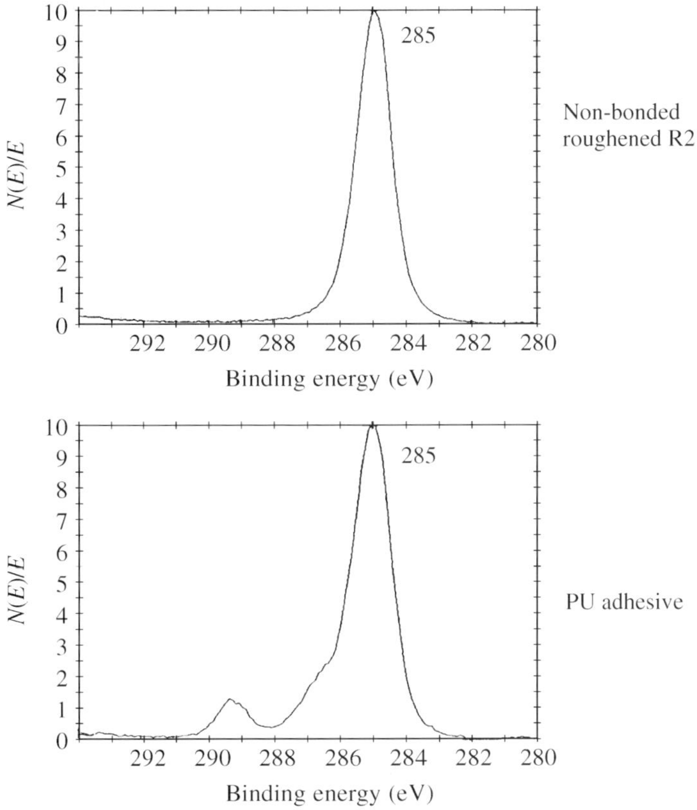

Figure 12a. C 1*s* photopeaks of roughened R2 rubber debonded joints. Non-bonded roughened R2 rubber; polyurethane adhesive.

other words, the carboxylic acid acts as a compatibilizer for the rubber–polyurethane interface.

The differences in the effectiveness of the surface treatment of roughened R2 rubber with different carboxylic acids may depend on their different tendency to migrate to the polyurethane adhesive once the adhesive joint is formed and/or on a different failure mode of the joints. XPS analysis of debonded roughened R2 rubber surfaces (Table 3) show some differences in chemical composition of the rubber and adhesive surfaces of debonded joints. The study was carried out with only two carboxylic acids, one of each group that can be distinguished in Fig. 3, i.e. the fumaric (FA) and the succinic (SA) acids. The atomic composition of the surface of the rubber side of debonded carboxylic acid-treated roughened R2 rubber joints is consistent with the removal of zinc stearate (there is no zinc — Table 3 — and the COO^- peak at 289.1 eV in the C 1*s* photopeak — Fig. 12b — is absent). The atomic composition of the surface of the adhesive side of the surface of debonded carboxylic acid-treated roughened R2 rubber joints shows higher oxygen and silicon concentrations at the surface than in the rubber side and also a small amount of nitrogen (which is only present in the polyurethane adhesive). Furthermore, the atomic compositions

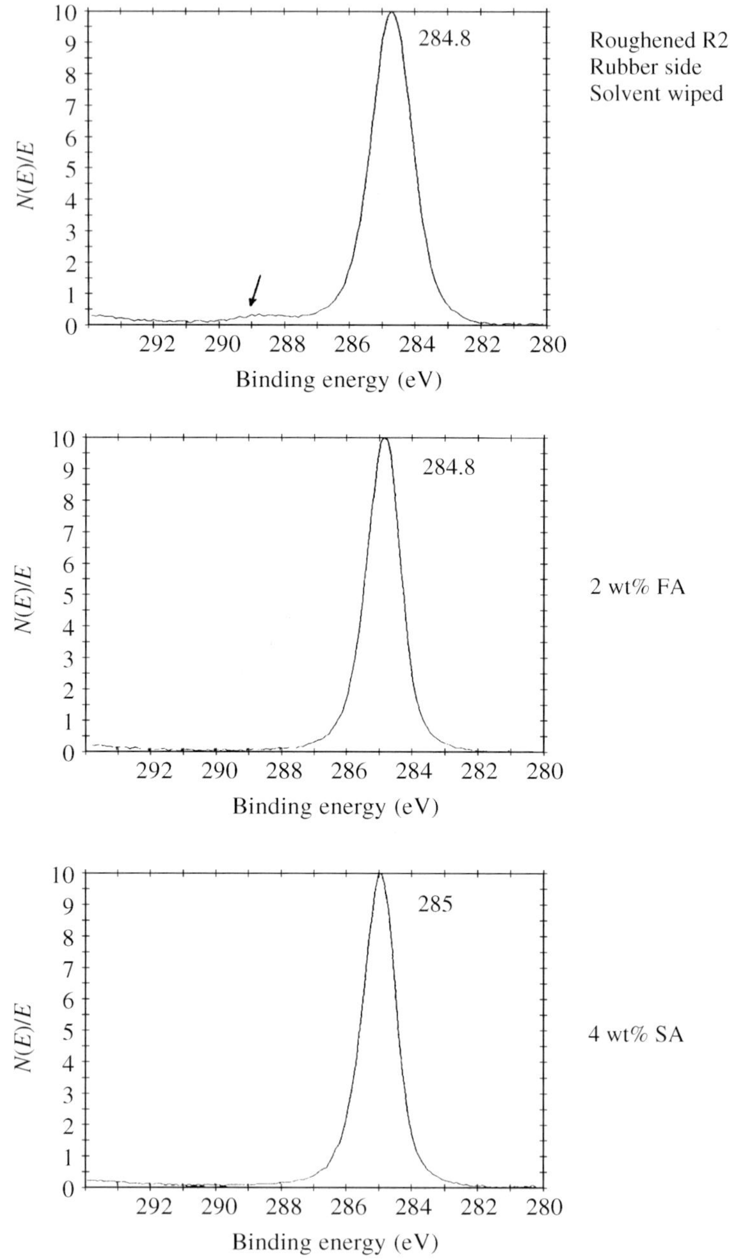

Figure 12b. C 1*s* photopeaks of roughened R2 rubber debonded joints. Rubber side of solvent-wiped, 2 wt% FA, 4 wt% SA.

of the polyurethane adhesive surface (Table 3) and the surface of the adhesive side of debonded joints do not completely agree, and therefore a mixed failure mode (interfacial + rubber cohesion) of the joints must be considered. However, depending on the carboxylic acid, the extent of interfacial and cohesive failure of the rubber is dif-

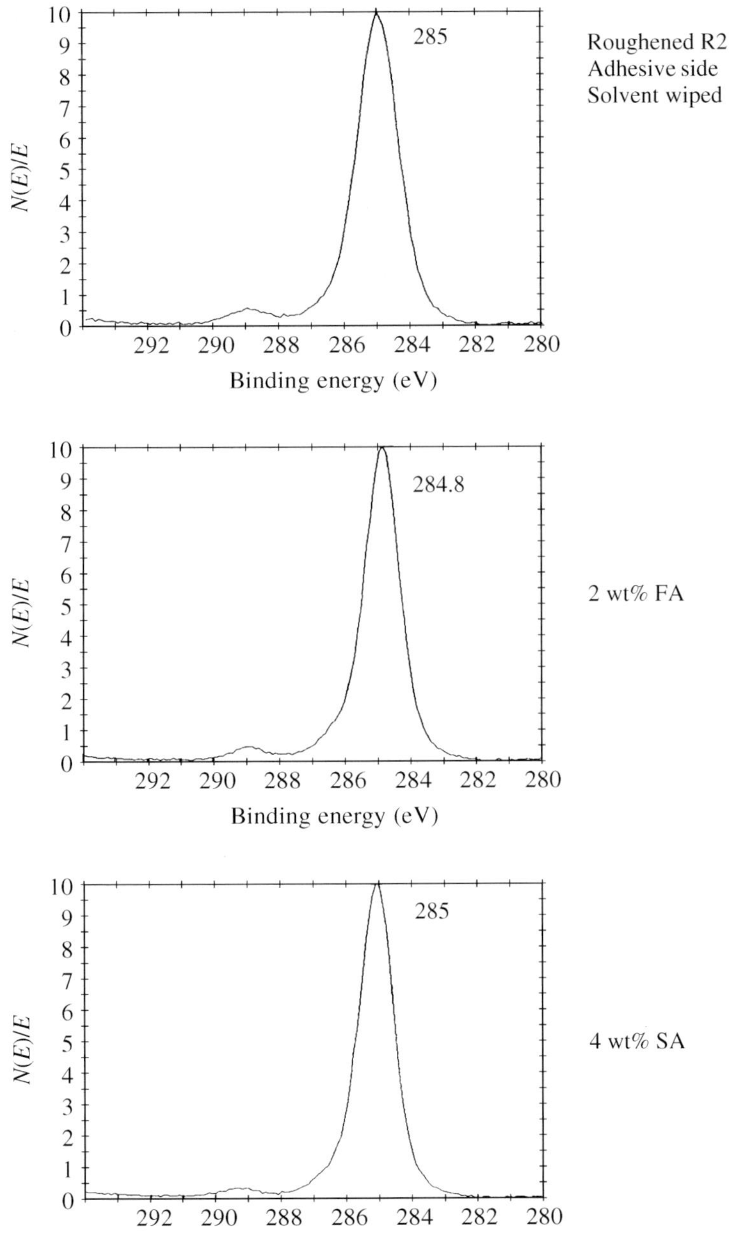

Figure 12c. C 1*s* photopeaks of roughened R2 rubber debonded joints. Adhesive side of solvent-wiped, 2 wt% FA, 4 wt% SA.

ferent. Table 3 shows a decreasing amount of oxygen (on both rubber and adhesive sides of debonded joints) from solvent-wiped to fumaric acid and to succinic acid surface-treated roughened R2 rubber. Figures 12 and 13 show the C 1*s* and Si 2*p*

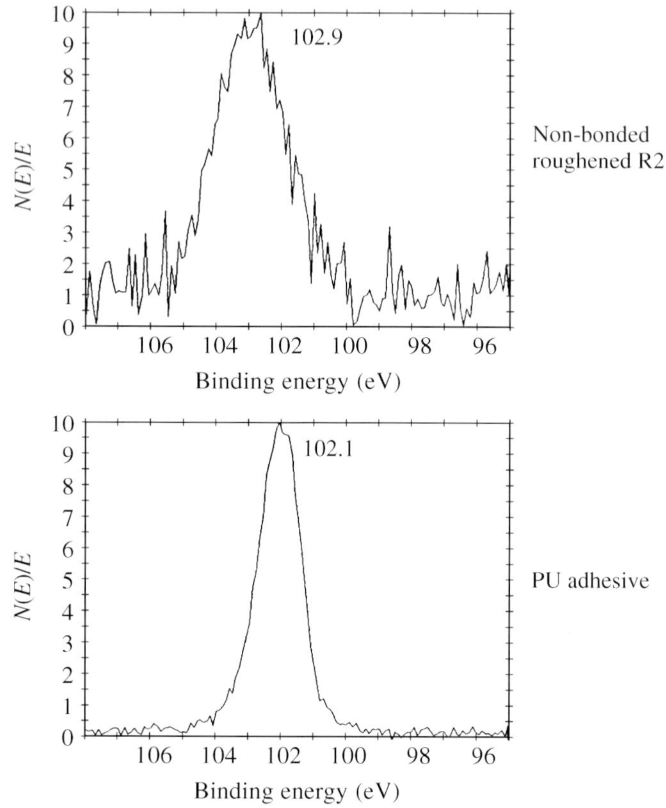

Figure 13a. Si 2*p* photopeaks of roughened R2 rubber debonded joints. Non-bonded roughened R2 rubber; polyurethane adhesive.

photopeaks, respectively, of both sides of debonded joints. C 1*s* photopeaks of the rubber surface of debonded joints are consistent with the composition of non-bonded roughened R2 rubber (Fig. 12a) (except in the solvent-wiped roughened R2 rubber joint, where the migration of zinc stearate to the surface is responsible for the COO^- peak at a binding energy of 289.1 eV), whereas the C 1*s* photopeaks of the surface of adhesive side of debonded joints are partially consistent with the polyurethane adhesive composition (Fig. 12a). Considering the results given in Table 3 and Figs 12b and 12c, both rubber residues and adhesive exist on the adhesive side of debonded joints and, further, the treatment with carboxylic acids produces a reduction of C=O (BE = 289.1 eV) and C—0 (BE = 286.9 eV) groups existing on the adhesive side of debonded joints. This reduction is more marked for SA than for FA surface-treated roughened R2 rubber. On the other hand, Figs 12b and 12c show that the treatment of roughened R2 rubber with carboxylic acids produces a change in the carbon–oxygen functionality. In the solvent-wiped roughened R2 rubber specimen, COO^- is the dominant carbon–oxygen group, whereas in the 2 wt% FA- and 4 wt% SA-treated specimens, C—O is noted. This may correspond to the C—O groups found in the IR spectra of debonded roughened R2 joints (Fig. 6).

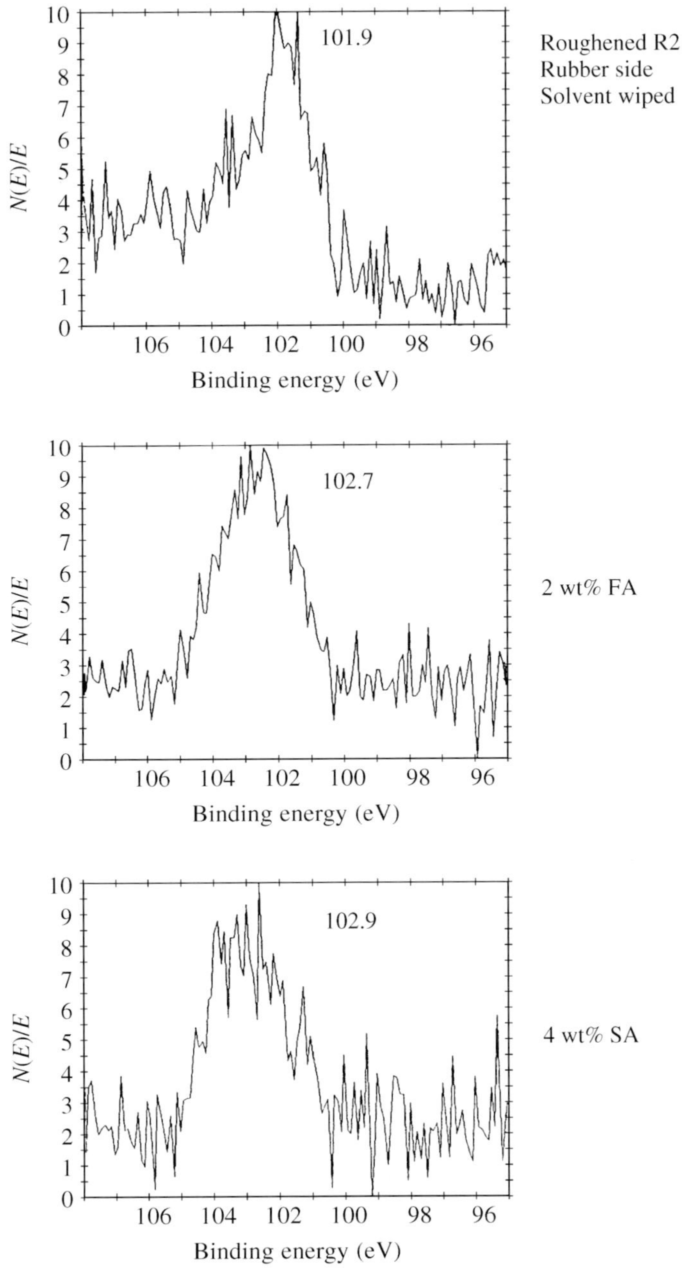

Figure 13b. Si 2*p* photopeaks of roughened R2 rubber debonded joints. Rubber side of solvent-wiped, 2 wt% FA, 4 wt% SA.

The same conclusions pertain when the evolution of Si 2*p* photopeaks is considered. The non-bonded R2 rubber and the polyurethane adhesive contain silica of different nature. Thus, in Fig. 13a, a Si 2*p* photopeak at 102.9 eV corresponds to silica gel

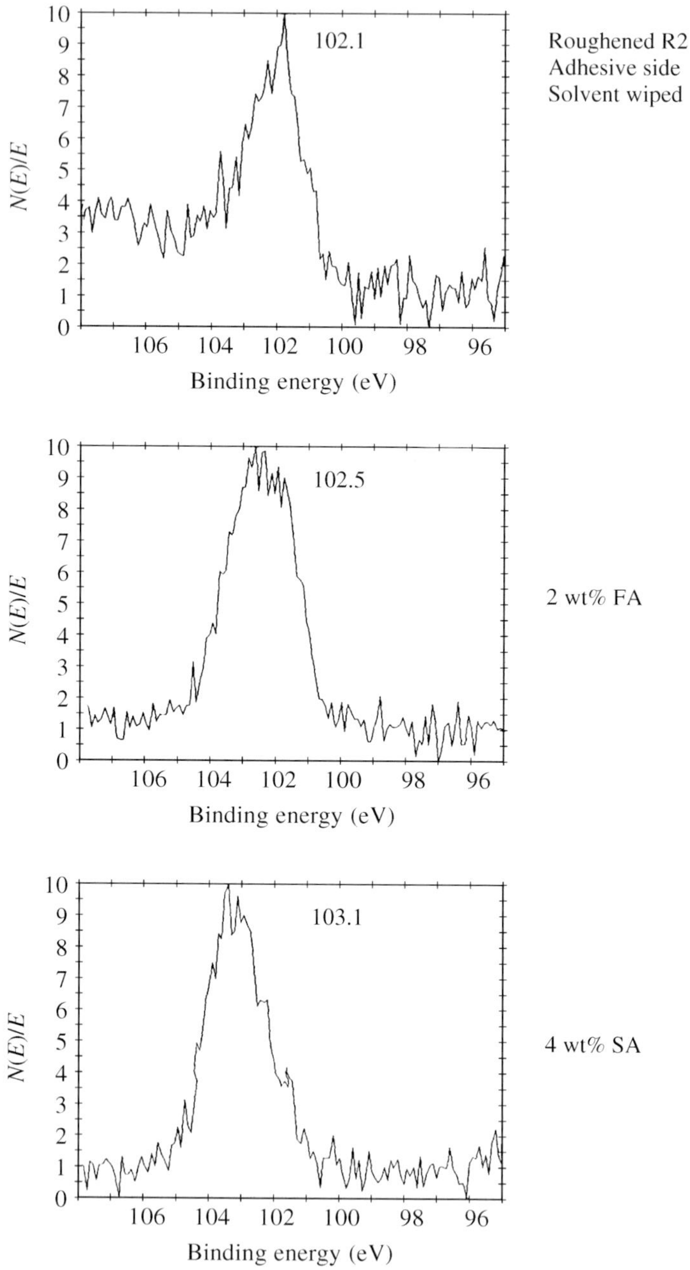

Figure 13c. Si 2p photopeaks of roughened R2 rubber debonded joints. Adhesive side of solvent-wiped, 2 wt% FA, 4 wt% SA.

(precipitated silica) in the roughened R2 rubber, whereas the polyurethane adhesive contains fumed silica or a silicate (BE = 102.1 eV). The surface of the adhesive side of debonded joints shows a continuous increase in the Si 2p binding energy in the series solvent-wiped to FA and to SA surface-treated roughened R2 rubber because the cohesive failure of rubber is more favoured in the SA surface-treated roughened R2 rubber.

The different strength of the interaction between the polyurethane adhesive and the surface-treated roughened R2 rubber can be related to the acid constant of carboxylic acids (Table 1). SA is less acidic than FA, and a weak acid–base interaction with the polyurethane adhesive (a base) can be produced. On the other hand, the solubility of carboxylic acids in ketones (2-butanone is the solvent for the polyurethane adhesive) is also relevant. FA and MA exhibit relatively good solubility in ketones, whereas SA and MaA do not. Thus, when the solvent-based polyurethane adhesive is applied to the rubber surface which was treated with FA, a partial solubility of the carboxylic acid may be produced facilitating its migration through the adhesive; the degree of solubility of SA in 2-butanone will be smaller and its migration to the adhesive is not facilitated.

The rubber and adhesive sides of debonded roughened R1 rubber joints were also characterized with XPS. Table 4 shows the atomic composition of the surfaces of the rubber and adhesive sides of some representative debonded joints, and Fig. 14 shows the C 1*s* photopeaks for the debonded joints. A comparison with the roughened R2 rubber debonded samples (Table 3) shows that the non-bonded roughened R1 rubber

Table 4.
XPS chemical composition (at.%) of debonded surface-treated roughened R1 rubber joints

Element	Non-bonded		Solvent treated		2 wt% FA		4 wt% SA	
	Roughened R1	PU	Rubber	Adhesive	Rubber	Adhesive	Rubber	Adhesive
C	87.0	73.1	84.7	80.9	86.5	80.0	87.1	78.7
O	9.1	20.4	10.1	15.6	9.3	17.0	8.5	18.9
N	0.7	0.7	0.9	0.9	0.5	0.5	ND	1.1
Si	1.0	5.8	2.8	1.6	2.5	1.7	2.9	0.7
Zn	1.8	ND[a]	1.2	0.9	0.7	0.8	0.9	0.5
S	0.4	ND	0.4	0.1	0.7	ND	0.7	0.1

[a] ND = not detected; < 0.1 at.%.

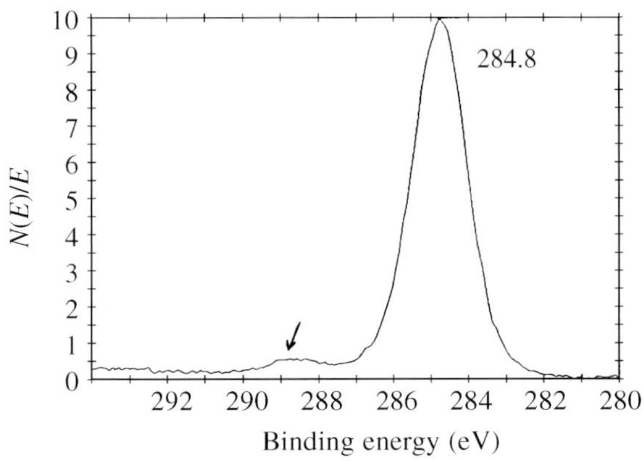

Figure 14a. C 1*s* photopeaks of roughened R1 rubber debonded joints. Non-bonded roughened R2 rubber.

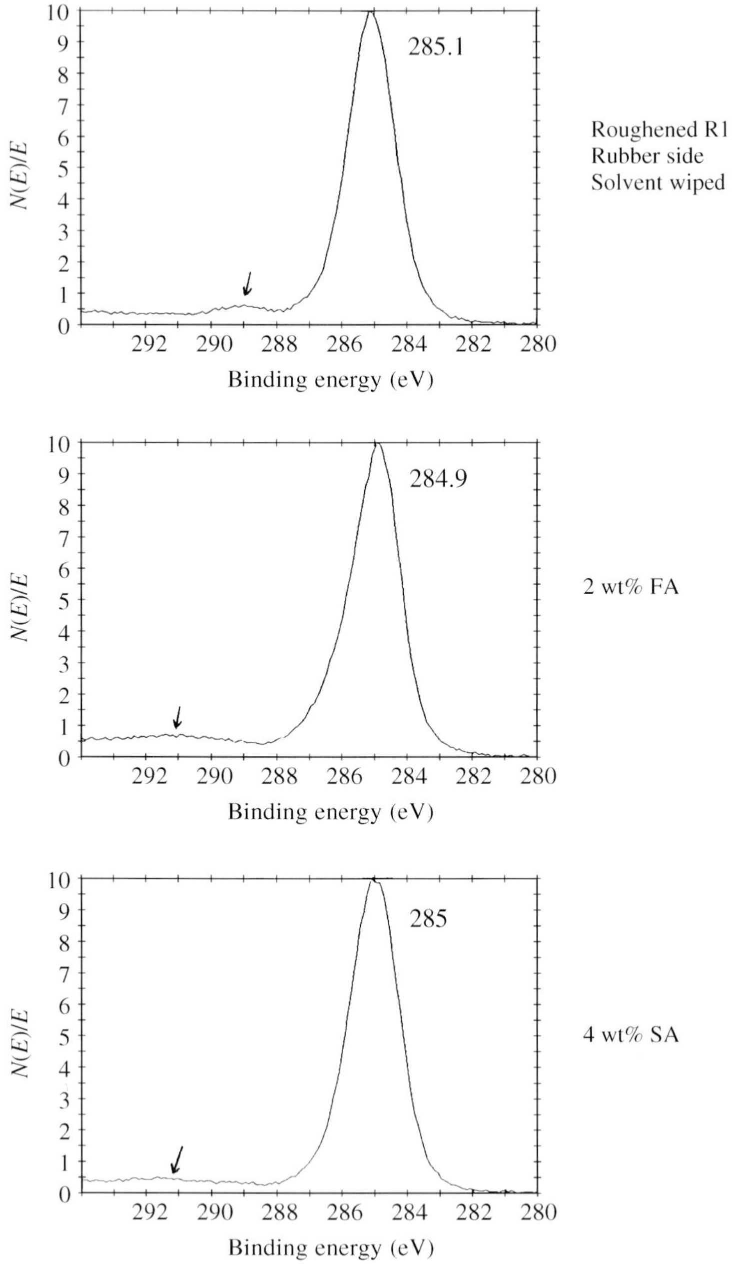

Figure 14b. C 1*s* photopeaks of roughened R1 rubber debonded joints. Rubber side of solvent-wiped, 2 wt% FA, 4 wt% SA.

contains lower carbon, and higher oxygen and silicon atomic concentrations. The respective atomic compositions on the rubber surface of all debonded joints (solvent-wiped, surface treated with carboxylic acids) are very similar, although the amount

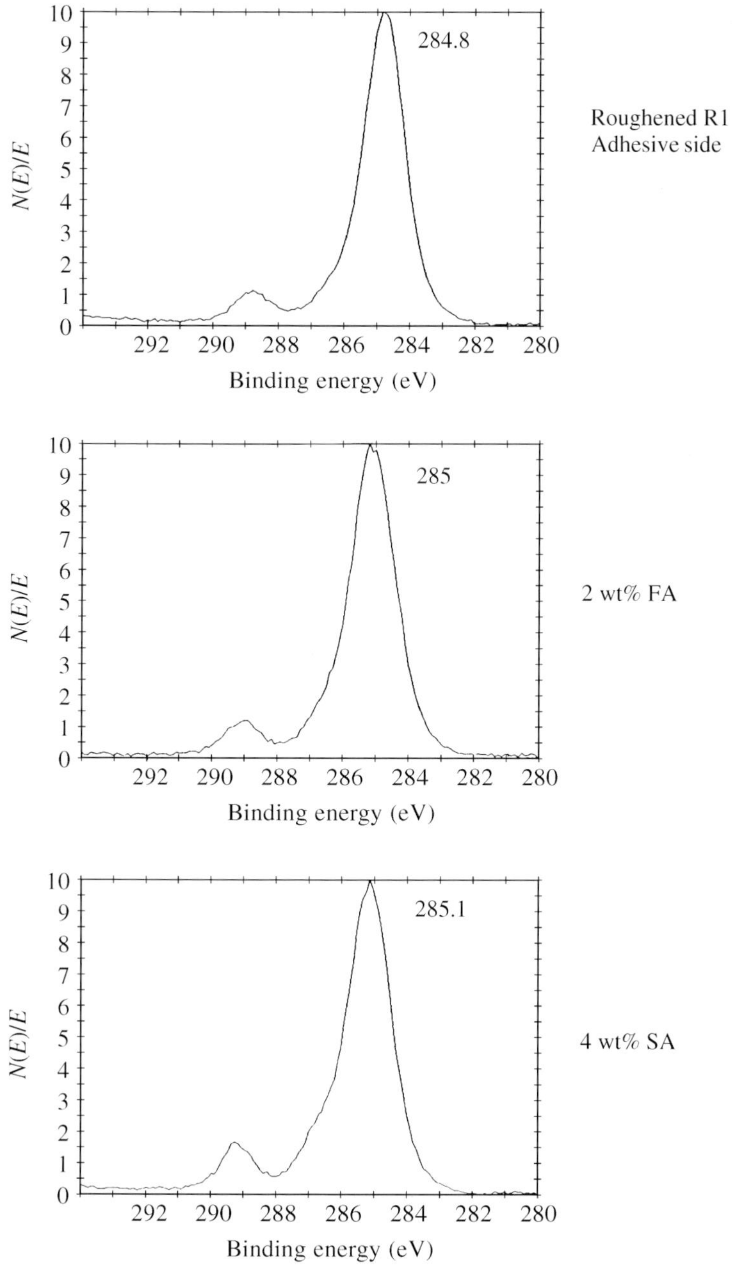

Figure 14c. C 1*s* photopeaks of roughened R1 rubber debonded joints. Adhesive side of solvent-wiped, 2 wt% FA, 4 wt% SA.

and kind of carbon–oxygen species are different. Thus, according to Fig. 14, the surface of the rubber side of solvent-wiped roughened R1 rubber joint shows a peak at 289.1 eV (carboxylic acid functionalities; 3.2 at.%). The surfaces of the rubber side of FA or SA treated roughened R1 rubber joints exhibit a broad carbon–oxygen peak at 291.4 eV, although in different atomic compositions (2.5 at.% in FA, and 1.2 at.% in SA surface-treated roughened R1 rubber). The shape and the binding energy of this broad carbon–oxygen peak is assigned to a $\pi-\pi^*$ transition associated with styrene moieties on the rubber side of debonded joints. The presence of styrene in the surface of carboxylic acid-treated samples but the absence of styrene in the polyurethane adhesive and in the solvent-wiped surface is indicative of a difference in failure mode. On the other hand, the atomic composition of the surface of the adhesive side of the debonded roughened R1 joints is consistent with the atomic concentration of polyurethane adhesive, but zinc is also present, and thus an interfacial + cohesive failure of the rubber is produced during the T-peel test. However, the atomic composition of carbon, silicon and zinc on the surface of adhesive side of the joint is reduced, and the atomic composition of oxygen and nitrogen is increased in the series solvent-wiped to FA and to SA surface-treated roughened R1 rubber, and the interfacial failure seems to be more facilitated in the carboxylic acid-treated surfaces.

4. CONCLUSIONS

Surface treatment of styrene–butadiene vulcanized rubbers with solutions of carboxylic acids in ethanol is effective, as a result of the elimination of abhesive substances and the modification of failure mode of the joints (from cohesive failure of the adhesive to interfacial + cohesive failure of rubber). These modifications contribute to the improvement of adhesion in these rubbers. The nature and formulation of the rubber determines the effectiveness of the surface treatment. FA and MA provide better T-peel results than MaA and SA, although a noticeable increase of T-peel strength (with respect to the solvent-wiped rubber) is always produced. The different effectiveness of the carboxylic acids can be ascribed to differences in acidity and solubility in 2-butanone (the solvent of polyurethane adhesive) which facilitate the removal of crystallites of carboxylic acids from the rubber surface during the formation of the joint.

On the other hand, a migration of the carboxylic acid is produced from the rubber surface to the interface, the carboxylic acid acting as a compatibilizer for the rubber surface and the polyurethane adhesive.

Acknowledgements

Financial support from CICYT (Project MAT92/0522) and Generalitat Valenciana (grant for J. M. Martín-Martínez to visit The Center for Adhesive and Sealant Science in USA) is gratefully acknowledged. Authors thank F. Cromer (Surface Analysis Laboratory, Virginia Technical College) for his collaboration in XPS measurements.

REFERENCES

1. D. Pettit and A. R. Carter, *J. Adhesion* **5**, 333 (1973).
2. J. C. Fernández-García, A. C. Orgilés-Barceló and J. M. Martín-Martínez, *J. Adhesion Sci. Technol.* **5**, 1065 (1991).

3. C. W. Extrand and A. N. Gent, *Rubber Chem. Technol.* **61**, 688 (1987).
4. D. Hace, V. Kovacevic, D. Manoglovic and I. Smit, *Angew. Makromol. Chem.* **176**, 161 (1990).
5. J. M. Martín-Martínez, J. C. Fernández-García, F. Huertas and A. C. Orgilés-Barceló, *Rubber Chem. Technol.* **64**, 510 (1991).
6. M. M. Pastor-Blas, M. S. Sánchez-Adsuar and J. M. Martín-Martínez, *J. Adhesion Sci. Technol.* **8**, 1093 (1994).
7. M. M. Pastor-Blas, M. S. Sánchez-Adsuar and J. M. Martín-Martínez, *J. Adhesion* **50**, 191 (1995).
8. Patent 1391722 (UK). *Polyurethane Adhesives Modified with Carboxylic Acids to Improve Strength* (Isar-Rakoll Chemie GmbH, 1971).
9. Patent 2113631 (Germany). *Klebstoffe auf Basis von Polyesterurethane für die Verklebung von Kautschuck-Materialen Unit Aderen Wekstoffen* (Isar-Rakoll Chemie GmbH, 1991).
10. N. Pastor-Sempere, A. C. Orgilés-Barceló and J. M. Martín-Martínez, in: *Proc. Adhesion '93 (Fifth International Conference)*, held in York, UK, pp. 268–271 (September 1993).
11. N. Pastor-Sempere, J. C. Fernández-García, A. C. Orgilés-Barceló and J. M. Martín-Martínez, in: *Proc. Int. Adhesion Science Symposium*, held in Yokohama, Japan, Paper B-26 (November 1994).
12. V. Vijayalakshmi, J. N. Rupanavi and N. Krishnamurti, *J. Adhesion Sci. Technol.* **6**, 781 (1992).
13. W. F. Thomsen, *Adhesives in Manufacturing*. Marcel Dekker, New York (1983).
14. H. A. Rijsdijk, M. Constant and A. J. M. Peijs, *Composite Sci. Technol.* **48**, 161 (1993).
15. J. Schultz, A. Carré and C. Mazeau, *Int. J. Adhesion Adhesives* **4**, 163 (1984).
16. A. Baszkin, M. Nishino and L. Ter-Minassian-Saraga, *J. Colloid Interface Sci.* **59**, 272 (1973).
17. J. Schultz, L. Lavielle, A. Carré and P. Comien, *J. Mater. Sci.* **24**, 4363 (1989).
18. K. Grundke, H. J. Jacobash, F. Simon and St. Schneider, *J. Adhesion Sci. Technol.* **9**, 327 (1995).
19. N. Pastor-Sempere, A. C. Orgilés-Barceló, R. Torregrosa-Maciá and J. M. Martín-Martínez, in: *Proc. Symposium on Hot Melt Adhesives*, held in Port-Jerome, France, pp. 53–59 (November 1993).
20. N. Pastor-Sempere, A. C. Orgilés-Barceló, J. C. Fernández-García, R. Torregrosa Maciá and J. M. Martín-Martínez, in: *Proc. 17th Adhesion Society Meeting*, held in Orlando, FL, pp. 43–45 (February 1994).
21. N. Pastor-Sempere, R. Torregrosa-Maciá, J. M. Martín-Martínez, J. C. Fernández-García and A. C. Orgilés-Barceló, *J. Adhesion* **50**, 25 (1995).

Mittal Festschrift, pp. 495–506
W. J. Van Ooij and H. R. Anderson, Jr. (Eds)

Starch thin film modified polystyrene surface for direct coupling of a synthetic GRGDSPK cell attachment peptide

E. C. ONYIRIUKA,* L. S. HERSH and M. D. BRYHAN

Science & Technology Division, Corning Inc., Sullivan Park Campus, SP-FR-03, Corning, New York 14831, USA

Abstract—A thin film of dialdehyde starch on polystyrene tissue culture ware as an alternative to the chemical surface oxidation modification by either corona-discharge or oxygen plasma treatment was characterized by X-ray photoelectron spectroscopy (XPS), infrared (IR) spectroscopy and scanning probe microscopy (SPM). The starch binding sites were quantitatively estimated using a 2,4-dinitrophenylhydrazine assay. Direct chemical coupling of the synthetic low molecular weight cell attachment peptide, GRGDSPK, to the starch reactive sites via reductive amination was demonstrated. The hybrid synthetic peptide–starch coated polystyrene surface is a viable alternative to the standard plasma oxidized, fibronectin (extracellular matrix glycoprotein) coated surface for the culture of primary anchorage-dependent endothelial cells.

Keywords: Starch thin film; cell culture ware; polystyrene; adhesion; cell attachment peptide.

1. INTRODUCTION

The binding of mammalian cells to biomaterials occurs by the interaction of the integrin adhesion receptors on the cell surface and the cell recognition domain in the adhesion protein pre-coating layer on the biomaterial surface [1]. The adhesion proteins, or the extracellular matrix (ECM) glycoproteins, which include fibronectin, vitronectin, collagen, and laminin listed in Table 1 are derived from animal sources, such as blood serum. Each of the ECM glycoproteins contains the tripeptidyl amino acid sequence: glycine–arginine–aspartic acid (RGD) as the main cell recognition site [2]. Fibronectin is considered to be the most effective ECM protein for cell attachment; hence it is the most widely used for cell culture applications. In the standard tissue culture procedure, a hydrophilic surface containing hydroxyl, carbonyl and carboxyl functional groups is usually generated by either corona-discharge or oxygen plasma oxidation. The standard tissue culture process starts with the application of a fibronectin pre-coating onto the chemically oxidized polymer surface before the cells are introduced. An alternative procedure involves the introduction of the ECM

*To whom correspondence should be addressed.

Table 1.
Cell binding domain sequences of ECM proteins

ECM protein	Cell recognition site
Fibronectin	RGDS
Collagen	RGDT
Laminin	LRGDN
Vitronectin	RGDV

R = Arginine; G = Glycine; D = Aspartic acid; S = Serine; T = Threonine; L = Leucine; N = Asparagine; V = Valine.

components together with the cells. In either case, a chemically modified hydrophilic polymer surface is required to anchor the ECM protein successfully to the polymer matrix.

It is desirable in industrial tissue culture (TC) applications to replace the standard procedure of chemical surface oxidation and ECM proteins with a hybrid attachment factor, such as a synthetic peptide containing the RGD cell binding site coupled directly to the biomaterial via a reactive molecule graft on the polymer matrix. The benefits of this novel approach are numerous. First, the number and density of RGD sites are more easily controlled and can be increased as desired depending on the application. The ability to manipulate the density of RGD sites on the substrate is especially important because certain cells do require multiple binding sites and thrive only when these sites are available. Obviously, fibronectin is unsuitable for these cells requiring multiple binding sites. The reason is that fibronectin is a large molecule (estimated molecular weight = 273 715) containing 2476 amino acids and has only one RGD cell recognition site found between amino acid residues 1615 and 1618 [3]. Secondly, complete purification of the ECM proteins is expensive and difficult with many complications, and the potential risk of contamination with animal viruses and bacteria is enormous. Thirdly, use of serum in cell cultures can be avoided simply by eliminating the ECM-fibronectin from the culture medium. The serum-free approach is also a simpler and relatively inexpensive process than the current practice. Selectivity is perhaps the most important attribute of the synthetic RGD approach. With this novel approach, it is possible to design a system which prevents the adhesion of one cell type and at the same time enhances the adhesion of others. In a recent review paper, Hubbell offered vascular grafts as an example where this selectivity is desired [4]. In vascular grafts, the ingrowth of endothelial cells which line the blood vessels is desired, whereas the binding and spreading of blood platelets is not because it could lead to occlusion of the graft by a thrombus. The selectivity advantage can also be employed in some cases where it is desirable to discriminate between proteins by just resisting the permeation of these materials based on size differences or class of protein.

Polystyrene (PS) culture surfaces have been successfully modified in our laboratory with a natural starch derivative, polymeric dialdehyde starch (DAS) graft to which we have coupled an RGD-containing synthetic peptide Glycine (Gly, G)–Arginine (Arg, R)–Glycine (Gly, G)–Aspartic acid (Asp, D)–Serine (Ser, S)–Proline (Pro, P)

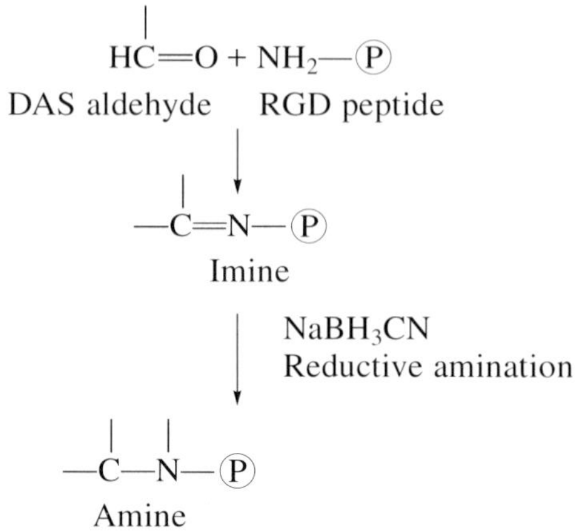

Scheme 1. Standard tissue culture procedure (left) versus the proposed hybrid GRGDSPK peptide/DAS graft approach.

–Lysine (Lys, K) (GRGDSPK). The RGD tripeptidyl sequence is also present in fibronectin as the main cell recognition site. The hybrid RGD-DAS approach is a viable alternative to the standard plasma oxidation followed by fibronectin-ECM procedure shown in Scheme 1. This paper describes the DAS coating on polystyrene and its reactivity toward the synthetic RGD molecule assayed using 2,4-dinitrophenylhydrazine (DNPH) as the probe molecule. The hybrid RGD-DAS approach is proposed as an alternative to the fibronectin coated polymer surface in the culture of anchorage-dependent cells.

2. EXPERIMENTAL

2.1. Materials and sample preparation

The molecular structures of the DAS, DNPH and GRGDSPK compounds are shown in Scheme 2. The commercially available DAS starting material derived from the oxidation of anhydroglucose natural starch contains two reactive aldehyde groups in each glucose unit. The DAS coating process is simple, fast and involves refluxing in water for 1 h. The resulting clear DAS suspension was poured into the untreated 24-well PS Petri dish (Corning Inc.) for 5 min to allow the DAS sufficient time to adsorb onto the PS dish. The solutions were then decanted and the adsorbed DAS coating was cured by either heating in an oven at 65°C for 10–15 min or by exposure to infrared lamp for 5 min.

The surface concentration of the aldehyde binding sites on the DAS-coated PS was determined using a DNPH assay. For the assay, DNPH was dissolved in 1 M HCl at a concentration of 0.25 mg/ml (~ 1 mM) and was allowed to stay in contact with the DAS-coated PS dish for 30 min at room temperature in the dark. The DNPH solution was decanted and the dish was washed several times with de-ionized water. The reacted DNPH was eluted with 0.1 M NaOH. The pH of the solution

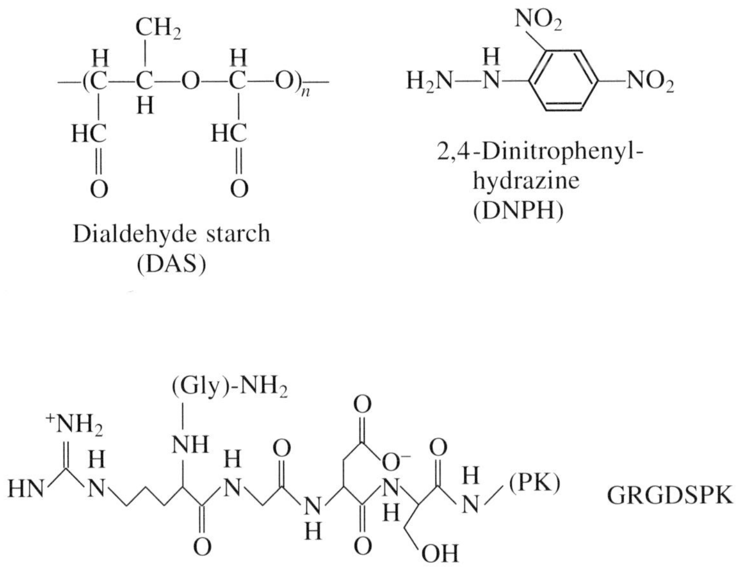

Scheme 2. Molecular structures of DAS, DNPH and GRGDSPK peptide sequence.

containing the DNPH elute was adjusted to 2 with 6 M HCl. The absorbance of the solution containing the DNPH extract was measured at 360 nm employing a Lamda 4C spectrophotometer. The amount of DNPH adsorbed on the DAS which is estimated from the absorbance value is proportional to the concentration of the reactive sites present on the surface of the DAS coated PS substrate.

The synthetic oligopeptide, GRGDSPK which contains the RGD cell recognition site was prepared by dissolution in 10 mM MOPS (3-[*N*-morpholino]propanesulfonic acid) buffer at pH of 7, in the presence of 20 mM sodium cyanoborohydride reducing agent. The coupling reaction is completed by exposing the DAS-coated PS Petri dish to the solution containing the GRGDSPK for 16 h at room temperature. The RGD/DAS-coated polystyrene dish was then dried at 30°C for 1 h.

2.2. *Characterization*

The surfaces were characterized by X-ray photoelectron spectroscopy (XPS) and infrared (IR) spectroscopy. The XPS experiments were carried out on a Physical Electronics (PHI) Model 5500 Multi-Technique XPS/SAM spectrometer using a standard Mg anode X-ray source operating at 400 W/27.5 mA.

The IR experiments were performed on a double beam Perkin–Elmer 1800 infrared spectrophotometer equipped with a DTGS (deuterated triglycine sulfate) detector. Infrared spectra were obtained by multiple attenuated internal reflection infrared spectroscopy using a Harrick Attenuated Total Reflectance (ATR) accessory with a ZnSe crystal and operated at 8 cm^{-1} resolution.

Surface topographic data were obtained using the TappingMode™ AFM from Digital Instruments Dimension 3000 Multimode NanoScope IIIa Scanning Probe Microscopy (SPM) system. Three-dimensional surface plots and roughness analysis were obtained from a 4 μm section. The average roughness (R_a) value was taken as being most representative of the mean surface roughness in all samples.

3. RESULTS

3.1. XPS

The absorbance of the DNPH extract solution measured at 360 nm as a function of soak time is plotted in Fig. 1. The graph shows that the maximum absorbance value of DNPH which is proportional to the amount of adsorbed DAS from a 0.1 mg/ml DAS solution is achieved after 5 min. Figure 2 displays the XPS survey spectra of the untreated and DAS-coated ware and it shows high surface oxygen concentrations after the DAS grafted to the PS. The high resolution spectrum of the C 1*s* region for the DAS coating shows the presence of an ether, C—O—C (286.5 eV), and an aliphatic aldehyde carbonyl, >C=O (288.2 eV), as expected for a DAS molecule. There was no discernible shift in the peak energies for either the DAS alcohol, C—OH, or the aldehyde carbonyl, RHC=O, functionalities when the DAS starting powder material was compared to the DAS thin film after application.

The survey spectra (not shown) after the reaction of both DNPH and RGD with DAS showed the presence of carbon, oxygen and nitrogen indicating that N had been incorporated onto the surface of the coating. Both of the DAS characteristic C—OH/C—O—C functional groups were observed in the C 1*s* high resolution spectra obtained for the DNPH-DAS and RGD-DAS reaction products. The XPS peak previously assigned to the DAS C=O functional group shifted by 0.5 eV to lower binding energy for both reaction products. The new energy value suggests that a new C=N bond was formed as a result of the reaction and hence the lower binding energy was observed. In Table 2 are listed the binding energies measured for the N 1*s* peak

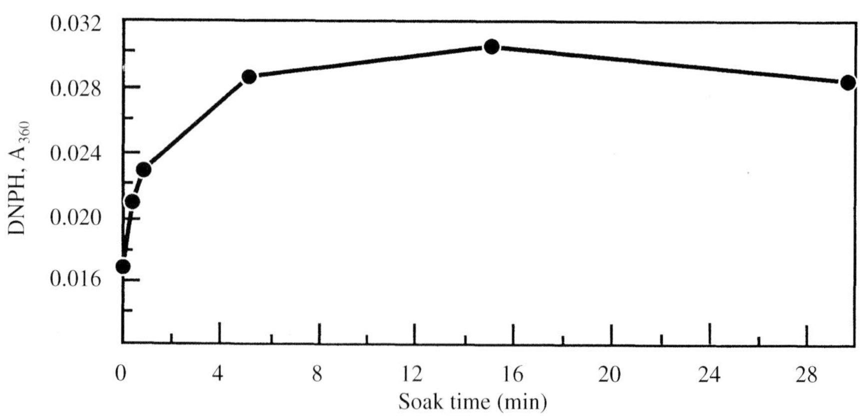

Figure 1. DAS (0.1 mg/ml) adsorption determined by 2,4-dinitrophenylhydrazine (DNPH) assay as a function of soak time in water.

before and after the respective reactions of DNPH and RGD with DAS. The data show that two new N peaks at 400.8 and 400.1 eV emerged after the reaction of DNPH and RGD molecules, respectively, with DAS.

The amount of the bound-RGD peptide was estimated from the nitrogen content determined by angle-dependent XPS (ADXPS) of reaction products from peptide solution concentrations ranging from 50–400 μg/ml. We found that the amount of bound-RGD increased with RGD solution concentration reaching a plateau at 400 μg/ml. The ADXPS results indicated that a 200 μg/ml peptide solution was sufficient to obtain high coverage of the reactive sites present on the DAS coating surface by the bound-RGD. Furthermore, at 90° take-off angle ($\sim$7 nm sampling depth), substantial

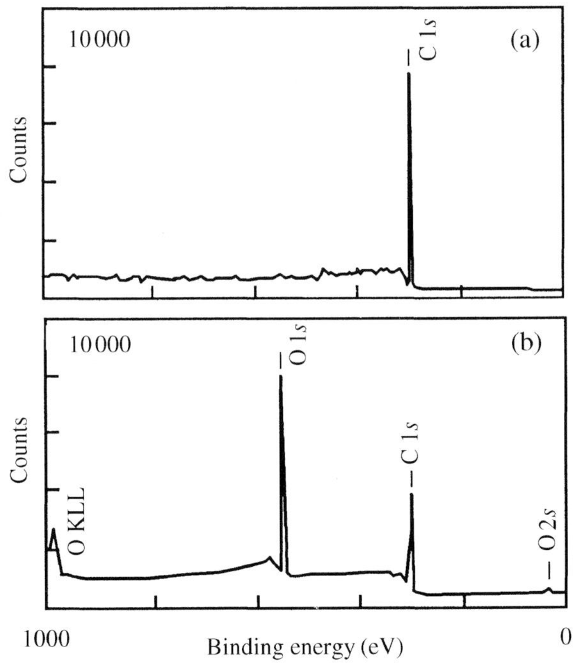

Figure 2. XPS survey spectra of (a) untreated polystyrene; and (b) dialdehyde starch (DAS) grafted onto polystyrene.

Table 2.
Nitrogen 1*s* peak energies measured by XPS before and after DNPH and GRGDSPK molecules are coupled to DAS

	N 1*s* binding energies (eV)		
DNPH	399.5	—	406.2
DNPH-DAS complex	—	400.8	406.3
GRGDSPK	399.4	—	—
GRGDSPK-DAS complex	—	400.1	—

amounts of bound-RGD reflected by the N content was still detectable. Thus, the ADXPS results clearly indicate multilayer binding of RGD to DAS.

3.2. Infrared spectroscopy

The identity of the DAS coating was verified by IR spectroscopy. A strong carbonyl stretching band at 1732 cm^{-1} was detected indicating that DAS coating layer had been adsorbed onto the surface of polystyrene. The band position observed above is well within the range typically observed for carbonyl groups in aliphatic aldehydes [5]. After the coupling reaction with DNPH, the IR spectrum showed a three-fold increase in the C—O/C=O ratio indicating that the aldehyde C=O groups had reacted to form a DAS-DNPH complex [6].

3.3. Atomic force microscopy

Table 3 compares the mean roughness (R_a) values of polystyrene surfaces determined by AFM. The data show that the R_a of DAS is the same as the values measured for plasma-oxidized and bound-RGD surfaces but is lower than that of untreated PS surface. The AFM micrograph of untreated PS was characterized by the presence of spheroids reflecting the particulate nature of the polymer material. This surface microstructure was still visible after the DAS thin film was applied except that the resulting surface was smoother. In contrast, with the plasma oxidized PS, the spheroids which were characteristic of PS were not visible. The three-dimensional AFM micrographs comparing DAS coating surfaces without and with bound-RGD are shown in Figs 3a and 3b, respectively. The micrographs clearly show an almost uniform coverage of the particulate structure of the surface of the DAS thin film by the bound RGD molecule.

3.4. Cell growth studies

The GRGDSPK/DAS surface was evaluated against the standard plasma oxidized and ECM-fibronectin coated surface by observing the performance of various cells grown on both surfaces. The major difference in performance was observed for primary cells (defined as cells that are still in their natural state) which showed 50–80 percent increase in growth when compared to the control surface. In contrast, with secondary cell lines (cells that are no longer in their natural state), the growth performance of

Table 3.
TappingMode™ AFM mean roughness (R_a) comparative analysis of polystyrene (PS) surfaces

Surface	Mean roughness (nm)
Untreated PS	2.152
Plasma-oxidized PS	1.739
DAS coated PS	1.702
RGD/DAS/PS	1.675

(a)

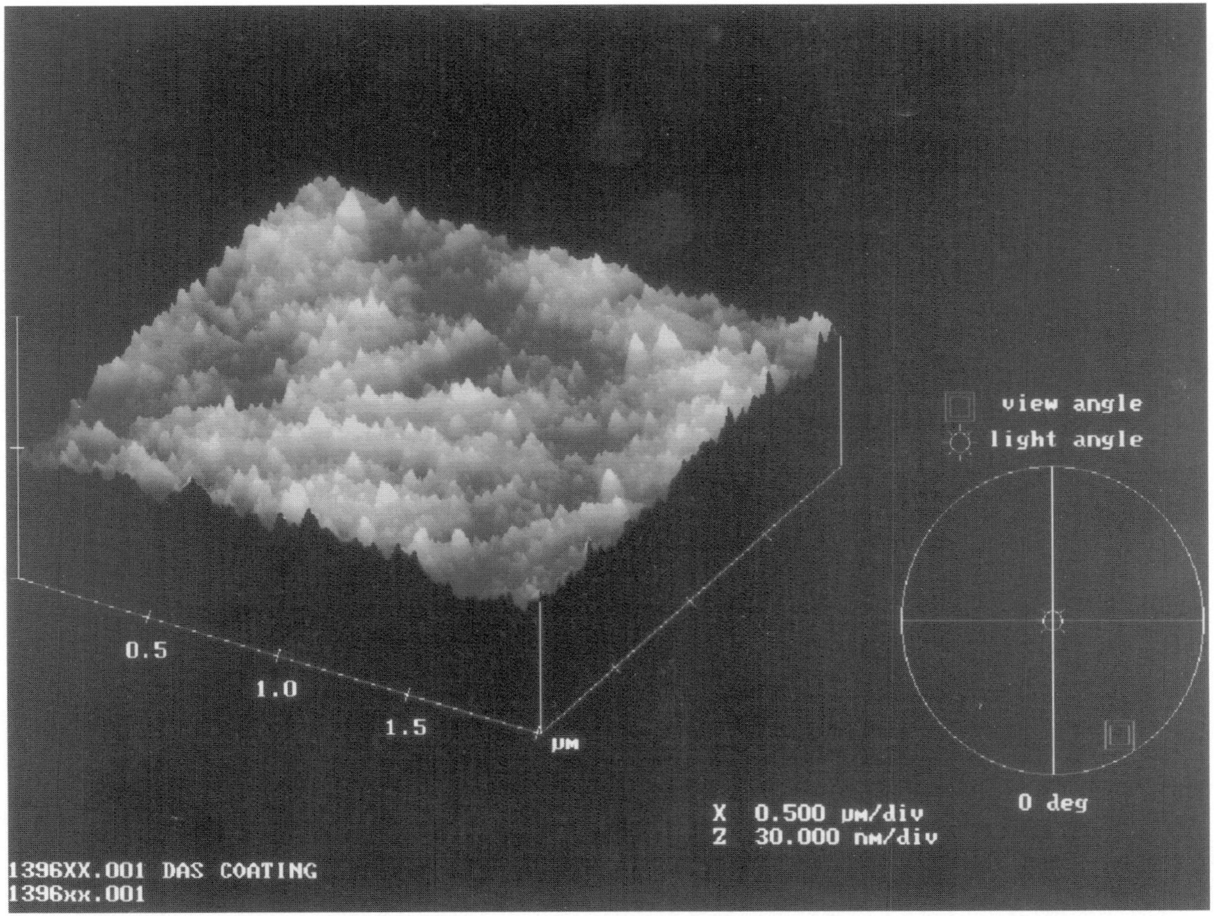

Figure 3. Comparison of the 3-dimensional AFM images of DAS coating surfaces (a) without, and (b) with bound-RGD peptide.

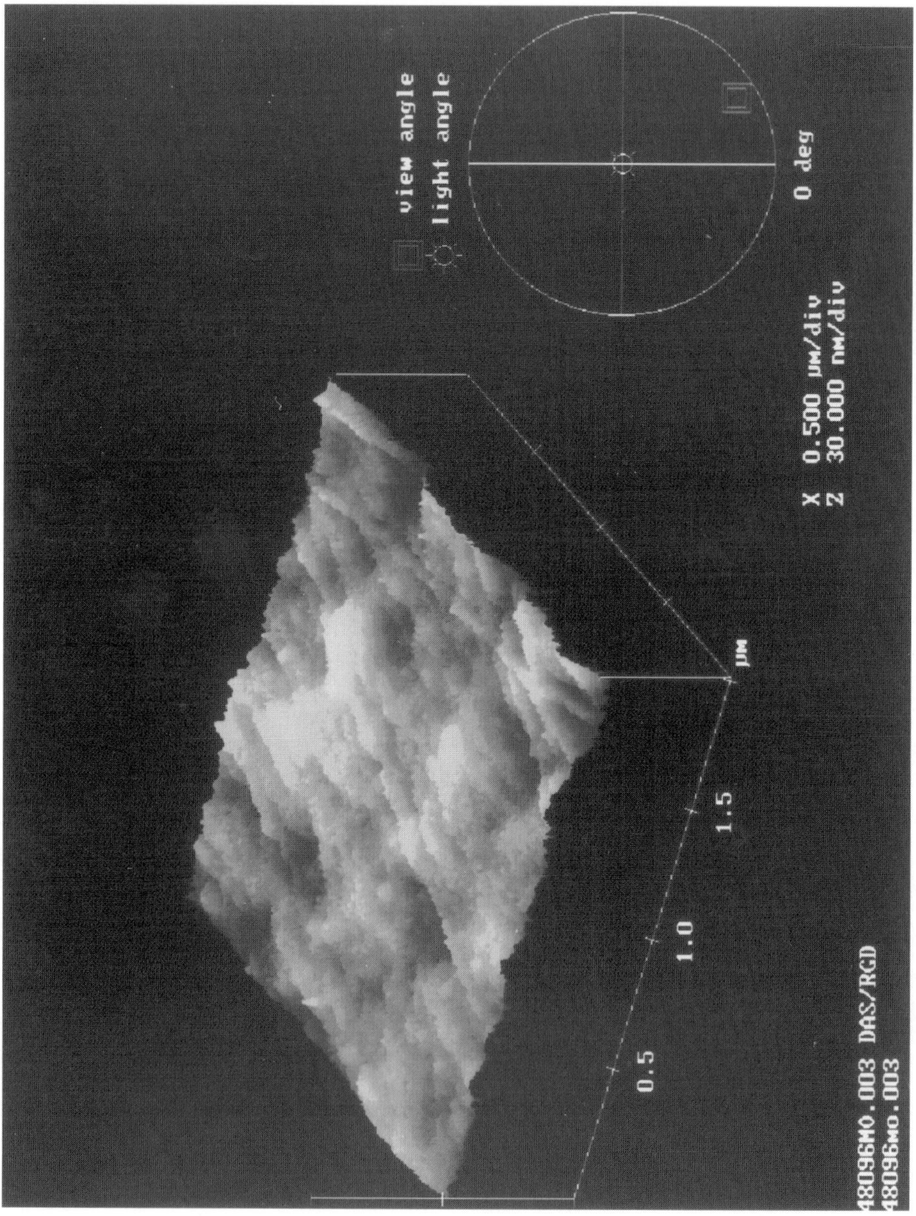

(b)

Figure 3. (Continued).

the hybrid GRGDSPK/DAS surface was actually lower than the standard fibronectin surface under similar conditions [7, 8].

4. DISCUSSION

The interaction of DAS with polystyrene surface is by physical rather than covalent chemical bonds. Chemical coupling of DAS to PS would most certainly be indicated by shifts in the XPS binding energies for the C—OH/C—O—C and C=O groups of adsorbed DAS coating. Such shifts were not observed, indicating that the adsorption of DAS to PS is physical in nature and occurs by a combination of hydrophobic interactions and/or London forces. We observed that without heating, the DAS coating was easily eluted upon exposure to aqueous media. This would indicate that the adsorbed DAS is cured via crosslinking by the heat treatment. The detection of the DAS characteristic alcohol/ether, C—OH/C—O—C functional groups within the XPS sampling depth (< 10 nm) after the DNPH coupling reaction clearly indicates that the DAS coating surface is not covered completely by the DNPH molecule. The observation suggests that only the DAS aldehyde reactive sites are involved in the coupling reaction with the amine groups provided by the DNPH molecule. The presence of the DNPH nitro group at 406.3 eV and a 400.8 eV peak assigned to an imine nitrogen, —HC=N—NH—, in the DNPH reaction indicates the formation of stable carbon–nitrogen covalent bond between DAS and DNPH. Similarly, the shift of the N 1*s* line to a lower binding energy of 400.1 eV observed for the RGD-DAS complex clearly indicates the formation of a stable covalent amine, $-H_2C-N-NH-$, bond. The reaction of RGD with DAS is consistent with reductive amination in a Schiff Base type reaction as shown in Scheme 3. However, with DNPH, XPS results strongly indicate that the reductive amination reaction is terminated at the usually unstable imine. Thus, the intermediate imine complex is stabilized by the phenyl group in the DNPH molecule and hence it was observed in the N 1*s* XPS spectrum of the DNPH-DAS reaction product.

The AFM results clearly indicate that the DAS coating alters the surface topography by smoothing of the polymer matrix and hence the decrease in the mean roughness. The uniform surface texture obtained for the DAS coating is identical to the texture observed for either the plasma or corona-discharge modified polymer surface. However, the surface roughness remained virtually unchanged for the bound-RGD indicating that uniform surface texture and binding were also obtained with RGD peptide. The uniform surface texture and subsequently the uniform binding of the synthetic RGD cell attachment peptide does influence how mammalian cells bind, spread and grow on these surfaces. It is therefore not surprising that anchorage-dependent primary cells would do better on these surfaces than on the ECM-fibronectin coated surface.

In fact, cell study results obtained from our laboratory clearly indicate that the growth performance is strongly dependent on surface preparation conditions and cell types. Hence we were not surprised to find that for anchorage-dependent primary cells, such as the human umbilical vein endothelial, rat aotic, and human tongue, the hybrid synthetic RGD-DAS surface was superior to the standard plasma-oxidized and ECM-fibronectin coated surface for polystyrene cell culture product.

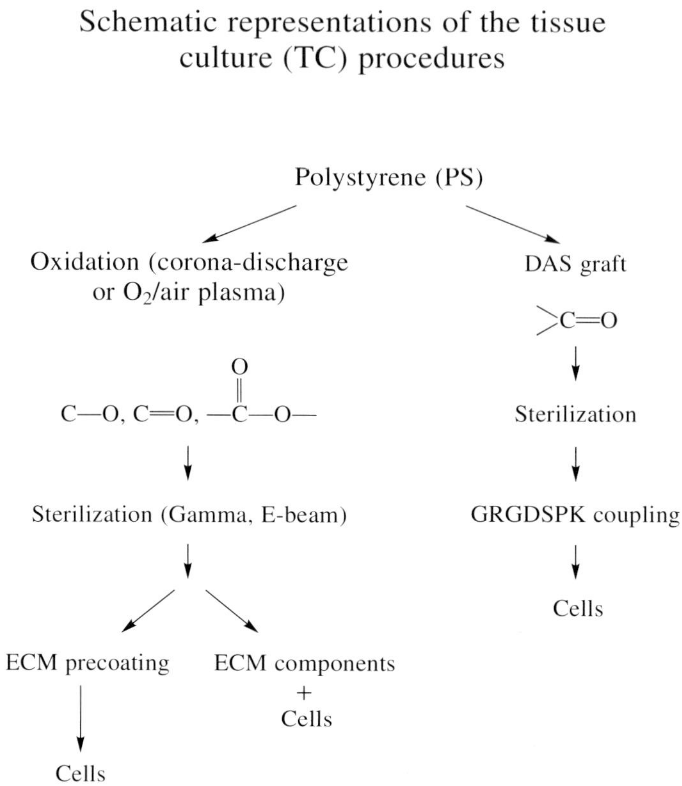

Scheme 3. Proposed mechanism for peptide coupling to DAS.

5. CONCLUSION

We have demonstrated an easy method of applying a thin film of DAS onto polystyrene on which the RGD-containing synthetic GRGDSPK peptide can be coupled. The binding sites on DAS-coated PS can be estimated quantitatively using a DNPH chromophore based assay. The GRGDSPK-DAS/PS surface is a superior alternative to the standard fibronectin-coated PS surface in the culture of primary anchorage-dependent cells.

Acknowledgements

We thank Brian Wheaton for carrying out the surface topography analysis employing the NanoScope Multimode TappingMode AFM instrumentation and Lyle D. Kinney for preparing some of the illustrations.

REFERENCES

1. B. D. Ratner, *Surf. Interface Anal.* **23**, 521–528 (1996).
2. Y. Yamada and H. K. Klienman, *Curr. Opinion Cell Biol.* **4**, 819 (1992).

3. S. Ayad, R. P. Boot-Handford, M. J. Humphries, K. E. Kadler and C. A. Shuttleworth, *The Extracellular Matrix Facts Book*. Academic Press, London (1994).
4. J. A. Hubbell, *Bio/Technology* **13**, 565–575 (1995).
5. N. B. Colthup, L. H. Daly and S. E. Wiberley, *Introduction to Infrared and Raman Spectroscopy*. Academic Press, Boston (1990).
6. E. C. Onyiriuka, *J. Adhesion Sci. Technol.* **10**, 617–633 (1996).
7. M. D. Bryhan, L. S. Hersh and F. M. Smith, US Patent 5,281,660 (1994).
8. J. Holland, L. Hersh, J. Myers, M. Bryhan, E. Onyiriuka and L. Ziegler, *Biomaterials* **17**, 2147–2156 (1996).

Part 4

Adhesives and adhesive joints

Mittal Festschrift, pp. 509–529
W. J. Van Ooij and H. R. Anderson, Jr. (Eds)

Electronically conductive adhesives: Conduction mechanisms, mechanical behavior and durability

E. SANCAKTAR* and Y. WEI

Center for Advanced Materials Processing, Department of Mechanical & Aeronautical Engineering, Clarkson University, Potsdam, NY 13699-5729, USA

Abstract—The objective of this paper is to add to the fundamental understanding of the conduction mechanisms and mechanical behavior of electrically conductive adhesives and their applications. Consequently, the research work to be presented involves two parts: (a) analysis of mechanical behavior, and (b) analysis of electrical behavior of conductive adhesives. Electrical and mechanical behaviors of both bulk conductive adhesives and single lap joints bonded with a conductive adhesive are studied. For bulk conductive adhesives, a model is established to predict the adhesive conductivity. The effect of film thickness on the conduction behavior of conductive adhesives is also presented, and it is shown that in thin film adhesives the resistivity values in planar directions are much larger than those encountered in 3-D measurements for the same adhesive material. The thermal mismatch stress induced at the interface between the adhesive and the substrate is studied in closed form and by finite element analysis, and the mechanical properties of the filled adhesive composite are obtained for this purpose. Calculations for a typical adhesive interconnection reveals the possibility of failure initiation due to thermal mismatch. Silver fillers are considered in discussions and calculations. Consequently, the silver migration phenomenon and its dependence on environmental temperature is also studied. The migration time is shown to decrease with increasing voltage, and high temperature ambient conditions are shown to speed up the migration.

Keywords: Electronically conductive adhesives; silver migration; contact resistance; thermal stresses; filled polymer properties; percolation theory; 2-D conduction; random particle dispersion; silver particle dispersion.

1. INTRODUCTION

Electronically conducting adhesives differ from regular adhesives due to the fact that they contain conductive metal particles. The properties of such filler particles are considerably different from those for the polymer adhesive matrix to which they are added. Consequently, the composite material which results acquires complex mechanical and electrical conduction properties which are likely to be affected by processing as well as environmental and service conditions.

*To whom correspondence should be addressed. Current address: The University of Akron, Department of Polymer Engineering, Akron OH 44325-0301, USA.

Even though conductive adhesives have the excellent potential of being a more efficient and less costly alternative to solder joining or brazing in electronic components, they still possess a number of problems in the areas of durability and design to fit specific needs [1–8]. Both of these issues require analysis, understanding and modeling of the physical and chemical mechanisms which govern electrical conduction and durability of the conductive adhesives.

In order to understand the fundamentals, the conducting microstructure needs to be modeled theoretically to explain the conduction mechanism. In order to do so, a computer simulation will be utilized. The contact behavior between the particle fillers in the adhesive is also a very important issue in determining conductivity, and a pressure dependent conduction model will be developed based on Holm's equation [9] to describe contact resistance in terms of pressure dependent constriction and tunneling resistances.

1.1. Migration of silver particles in conductive adhesives

Use of silver particles as conductive fillers presents a major problem due to silver aggregation and migration of silver from the anode to cathode in the presence of electric potential. With the presence of electric current, the reaction

$$\mathrm{Ag} - \mathrm{e}^- \longrightarrow \mathrm{Ag}^+ \tag{1}$$

takes place dissolving silver from the anode. At the cathode, the silver ions are discharged with the reaction

$$\mathrm{Ag}^+ + \mathrm{e}^- \longrightarrow \mathrm{Ag} \tag{2}$$

and silver is deposited. This process is basically what takes place in the Thum (Balbach) method of producing high purity silver [10]. In the Thum process, the cell voltage is 2.5 to 2.8 V with a current density of about 50 A per square foot of anode area. The current efficiency achieved is 95% [10]. In electrically conductive adhesives containing silver fillers it has been reported that the migration time increases with reduction in the level of voltage across the conducting path [11]. In fact, a 'threshold voltage' level below which silver migration is not expected to occur has been also predicted [12]. For example, Bolger *et al.* report that a reduction from (approximately) 20 to 5 V increased the silver migration time in their conductive adhesives from (approximately) 10 to 20 s. The shape of their voltage vs. migration time curve is nonlinear with the possibility of asymptotic behavior even though the relevant experiments were not performed to convincingly illustrate it. They also showed that the ionic purity level of the adhesive did not affect the migration time for silver. Bolger *et al.* report that the necessary conditions for silver migration to occur are the presence of electric potential, and moisture in the system [12].

The silver ions which form during this process also react with OH^- and form silver oxides [13] as described by the reaction equation

$$2\mathrm{Ag}^+ + 2\mathrm{OH}^- \longrightarrow \mathrm{Ag}_2\mathrm{O} + \mathrm{H}_2\mathrm{O} \tag{3}$$

Note that silver oxides are electrically conductive although their formation may change the tunnelling resistance properties of conductive adhesives.

1.2. Equivalent mechanical properties of the filler/matrix composite

It is known that particle size, shape and volume fraction all affect the mechanical behavior of filled polymers [14]. Using the generalized approach of Eshelby, Chow derived relations for the bulk modulus, transverse and longitudinal Young's moduli and, the two effective shear moduli, G_{12} and G_{13} of the filled polymer containing aligned ellipsoidal particles at finite concentration [14, 15]. Chow also derived relations for the linear and volumetric thermal expansion coefficients for filled polymers containing aligned ellipsoidal inclusions at finite concentrations [16]. He concluded that in such filled systems, the volumetric expansion varies only slightly with the aspect ratio of the ellipsoid while longitudinal and transverse linear expansions show strong dependence on the particle shape.

1.3. Adhesive/substrate thermal expansion mismatch problem

Due to the differences in the adhesive, substrate, and filler mechanical constants, conductive adhesive joints acquire mismatch stresses and strains. In order to be able to predict the durability of conductive adhesive joints, calculations of these thermal mismatch stresses and strains are needed. This analysis will help in characterization of the interfaces in conductive adhesive joints.

1.4. Objectives

The specific objectives of this current research project on electronically conductive adhesives are the following:

1. To show that the electrical resistance of powders made of homogeneous particles is dependent on the external pressure applied on them, and to model this behavior based on the representation of contact resistance as the sum of constriction resistance and tunnelling resistance.
2. To develop a computer simulation model to predict the average contact number and conduction path factor (relating the conductive path length to the unit element size) as functions of particle filler volume fraction in the adhesive matrix, and to assess any difference between the conduction behaviors in the 3-D and 2-D (thin film) cases.
3. To assess the effect of thermal mismatch on the stress state of joints made using electrically conductive adhesives, and to obtain mechanical properties for the filled adhesive composite for this purpose.
4. To assess any temperature dependence of the silver migration phenomenon of conductive adhesives containing silver particles.

2. MODEL FOR CONDUCTIVITY

Since conductive paths (or chains) are formed by metal particles, these conductive particles must initially make intimate contact (a general contact which includes the physical contact and the tunneling) under the action of some force so that, ultimately they transport electrons. Therefore, it is necessary to define a connection force for the metal particles, understand its nature and relate it to electric conductivity in the adhesive. For this purpose, a comparative model, based on homogeneous powder materials, is proposed for evaluating the correlation between this connection force and conductivity. Using the concepts developed initially with this model, we can then attempt to predict the nature of conductivity in conductive adhesives.

In order to be able to establish a relation between the compression force applied on metal particles in contact and the resulting conductivity, we first need to relate the external pressure, q, to the interparticle contact force. For this purpose we first make the following assumptions:

(i) The conducting metal particles are spherical in shape.

(ii) The metal particles are distributed with hexagonal packing arrangements.

(iii) The force is uniformly transferred from the top layer to the next layer in the axial direction.

In the presence of interparticle contact forces Q_e, the metal spheres will deform and a contact area is formed. Using elasticity theory [17], the radius, a, of the circular contact area can be calculated as [18]

$$a = 1.02\sqrt[3]{\frac{qr^3(1-\nu^2)}{E}}, \tag{4}$$

where ν is the Poisson's ratio, and E is the Young's modulus of particles.

For electronically conductive adhesives, the *in situ* contact resistance (R_s) is the sum of constriction resistance (R_c) and tunnel resistance, (R_t),

$$R_s = R_c + R_t, \tag{5}$$

$$R_c = \frac{\rho_1 + \rho_2}{4a}, \tag{6}$$

where ρ_1 and ρ_2 are the resistivities of spheres 1 and 2, respectively, and [9],

$$R_t = \frac{\rho_\sigma}{\pi a^2}. \tag{7}$$

In equation (7), ρ_σ ($\Omega\,m^2$) defines the tunnel resistivity.

3. THE DEPENDENCE OF ELECTRIC CONDUCTION ON FILM THICKNESS

We consider a cuboid domain with three side lengths of $L_x \times L_y \times L_z$. We assume that there are k_i paths ($i = x,\ y,\ z$). k_x is defined as the number of conductive paths along the x axis. k_y and k_z are defined similarly. Suppose the conductive path lengths are S_i ($i = x,\ y,\ z$). Here we have S_i as the average length in the i ($i = x,\ y,\ z$) direction.

With the volume fraction, φ, given, we can calculate the total particle number (N_T) in the chosen cuboid domain,

$$N_\mathrm{T} = \frac{6L_xL_yL_z}{\pi D^3}\varphi, \tag{8}$$

where D is the diameter of the spherical filler.

Suppose ρ_x, ρ_y, and ρ_z are resistivities along the $x,\ y,\ z$ axes. Let us consider the x-axis first. By the definition of resistivity, we have

$$\rho_x = R_x\frac{A_x}{L_x} = R_x\frac{L_yL_z}{L_x}, \tag{9}$$

where

$$R_x = \frac{1}{k_x}r_x \tag{10}$$

is the total resistance in the x direction and r_x is the resistance of each (parallel) path in the x direction, which is assumed to be the same for each path. Let us assume that the contact resistance between two particles is R_s which includes tunnel and constriction resistances as given by equation (5). Then, using the series resistance assumption,

$$r_x = \left(\frac{S_x}{D} - 1\right)R_\mathrm{s}, \tag{11}$$

where S_x is the average length of conductive paths. Since $(S_x/D) \gg 1$, $(S_x/D-1) = S_x/D$. Therefore,

$$\rho_x = \frac{1}{k_x}\frac{S_x}{D}R_\mathrm{s}\frac{L_yL_z}{L_x}. \tag{12}$$

The average length of conductive paths can be defined as $S_x = CL_x$, where C is a conductive path factor, which is a function of volume fraction, i.e.:

$$C = f_c(\varphi). \tag{13}$$

Statistically, the conductive path factor is inversely proportional to volume fraction φ. This implies that the average conductive path length increases with decreasing volume fraction. Note that, similarly, we have $S_y = CL_y$, and $S_z = CL_z$.

All particles in conductive paths should have at least two contact numbers for each. Therefore, the total contact numbers of all particles in conductive paths should be equal to $m(N_T/2)$, where m is the average contact number. Therefore, we have:

$$k_x\frac{S_x}{D}+k_y\frac{S_y}{D}+k_z\frac{S_z}{D}=m\frac{N_T}{2}. \tag{14}$$

Now solving for $\rho_x=\rho_y=\rho_z=\rho$ we get,

$$\rho=\frac{C^2\pi DR_s}{m\varphi}, \tag{15}$$

where D is the diameter of spherical filler, R_s is contact resistance [equation (5)], φ is volume fraction of fillers. In order to obtain an estimate of the resistivity using equation (15) we first need to obtain an expression for the terms m and φ. This can be done by using a computer simulation program, details of which are given in [18]. Using this computer simulation we get:

$$m=0.1625+5.7323\varphi-5.1716\varphi^2 \quad (1.5<m<2.0). \tag{16}$$

As discussed earlier, conductivity is established by the metal conductive paths through which the electrons move. These paths are developed in 3-D, referred to as 3-D paths. The average length (S) of the paths follow the statistical formula:

$$S=C\times L, \tag{17}$$

where L is the side length of a unit cuboid along the direction considered and C is the conductive path factor with values $1.3<C<1.6$.

For thin adhesive films (i.e. $z<150$ μm), the development of conductive particle paths is approximately in two dimensions, instead of the three dimensions discussed

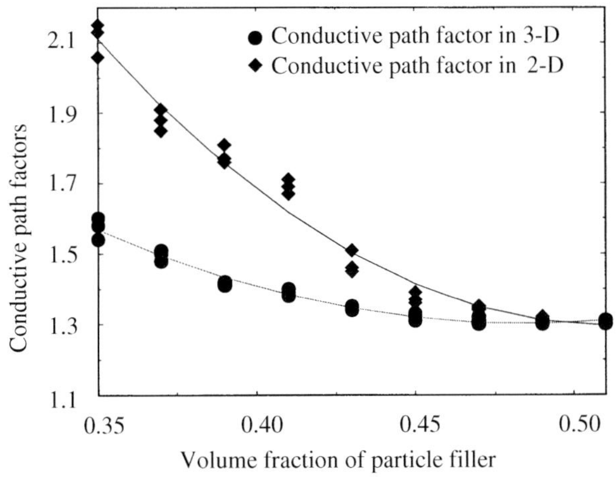

Figure 1. Conductive path factors C and C'.

above. The unit cuboid side length L_z considered infinite in 3-D becomes finite compared with the size of particle fillers. In this case, the conductive metal paths develop in a 2-D planar direction rather than in 3-D. By using a computer simulation we can show that the conductive path factor C' in a thin layer adhesive film differs from that in the three dimensional simulation. As shown in Fig. 1, at lower volume fraction of metal-filler, the difference between the C in three dimensions and the C' in two dimensions is very large. In other words, the conductivity in the planar directions is much lower. At higher particle volume fractions the difference between C and C' becomes smaller. This implies that with high volume fraction of fillers, the average length of conductive path is about the same for both three dimensional and two dimensional conductive adhesive applications.

4. CALCULATION OF EQUIVALENT MECHANICAL PROPERTIES FOR THE FILLER/MATRIX COMPOSITE, AND THERMAL MISMATCH STRESSES FOR ADHESIVE JOINTS

4.1. Chow's theoretical analysis of mechanical properties

As mentioned earlier, Chow [14, 15] derived relations for the shear (G_{ij}), bulk (k), and longitudinal (E_p) and transverse (E_t) Young's moduli using the generalized approach of Eshelby. All the modulus calculations developed by Chow are based on the following assumptions. (i) The filler particles are well bonded to the matrix; consequently, the isostrain condition at the particle/matrix interface can be used to develop the method. (ii) The shapes of particles are uniform ellipsoids with the aspect ratio C_ρ. (iii) The metal particles are distributed with their corresponding axes aligned.

The formulas to calculate the composite adhesive's shear moduli G_{12} and G_{13}, Young's moduli E_p and E_t in directions parallel and transverse to the particles' longitudinal axes, respectively, as well as bulk modulus k are summarized below:

$$G_{ij} = \frac{\left(\frac{G_1}{G_2} - 1\right) G_2 \varphi}{1 + 2\left(\frac{G_1}{G_2} - 1\right)(1-\varphi) S_{ijij}} + G_2, \quad ij = 12,\ 13, \tag{18}$$

$$E_p = \frac{\left[\left(\frac{k_1}{k_2} - 1\right) Y_1 + 2\left(\frac{G_1}{G_2} - 1\right) K_1\right] E_2 \varphi}{2K_1 Y_3 + Y_1 K_3} + E_2, \tag{19}$$

$$E_t = \frac{\left[\left(\frac{k_1}{k_2} - 1\right) Y_3 + \left(\frac{G_1}{G_2} - 1\right)(Y_3 A + K_3 B)\right] E_2 \varphi}{2K_1 Y_3 + Y_1 K_3} + E_2, \tag{20}$$

$$k = \frac{\left(\frac{k_1}{k_2} - 1\right)(Y_1 + 2Y_3) k_2 \varphi}{2K_1 Y_3 + Y_1 K_3} + k_2, \tag{21}$$

$$K_i = 1 + \left(\frac{k_1}{k_2} - 1\right)(1 - \varphi)\alpha_i \quad (i = 1,\ 3), \tag{22}$$

$$Y_i = 1 + \left(\frac{G_1}{G_2} - 1\right)(1 - \varphi)\beta_i \quad (i = 1,\ 3), \tag{23}$$

where G_1 and G_2, E_1 and E_2, as well as k_1 and k_2 are the elastic constants of filler and the matrix with the subscript 2 representing the matrix properties; φ is the filler volume fraction. The other terms in equations (10) through (23) are defined as follows:

$$\beta_1 = S_{1111} + S_{1122} - 2S_{3311}, \tag{24}$$

$$\beta_3 = S_{3333} - S_{1133}, \tag{25}$$

$$\alpha_i = \sum_{L=1}^{3} S_{LLii} \quad (i = 1,\ 3), \tag{26}$$

$$A = \frac{K_1}{1 + 2\left(\frac{G_1}{G_2} - 1\right)(1 - \varphi)S_{1212}}, \tag{27}$$

$$B = \frac{1 + \left(\frac{G_1}{G_2} - 1\right)(1 - \varphi)\left(S_{1111} - S_{3311}\right)}{1 + 2\left(\frac{G_1}{G_2} - 1\right)(1 - \varphi)S_{1212}}. \tag{28}$$

The components for Eshelby's tensor are given as follows:

$$S_{1111} = S_{2222} = \left(\frac{3}{4}P_1 + P_2\right)I - \pi P_1 C_\rho^2, \tag{29}$$

$$S_{1122} = S_{2211} = \left(\frac{1}{4}P_1 - P_2\right)I - \frac{1}{3}\pi P_1 C_\rho^2, \tag{30}$$

$$S_{1133} = S_{2233} = \frac{4}{3}\pi P_1 C_\rho^2 - \left(P_1 C_\rho^2 + P_2\right)I, \tag{31}$$

$$S_{3311} = S_{3322} = 4\pi\left(\frac{1}{3}P_1 - P_2\right) - \left(P_1 - 2P_2\right)I, \tag{32}$$

$$S_{3333} = 4\pi\left[\frac{1}{3}P_1\left(1 - 3C_\rho^2\right) + P_2\right] + 2\left(P_1 C_\rho^2 - P_2\right)I, \tag{33}$$

$$2S_{2323} = 2S_{1313} = 4\pi\left[\frac{1}{3}P_1\left(1 + C_\rho^2\right) + P_2\right] - \left[P_1\left(1 + C_\rho^2\right) + P_2\right]I, \tag{34}$$

$$2S_{1212} = S_{1111} - S_{1122}, \tag{35}$$

where

$$P_1 = \frac{3}{8\pi\left(1 - \nu_2\right)\left(1 - C_\rho^2\right)}, \tag{36}$$

$$P_2 = \frac{1 - 2\nu_2}{8\pi(1 - \nu_2)}, \tag{37}$$

and

$$I = \frac{2\pi C_\rho}{(1 - C_\rho^2)^{\frac{3}{2}}} \left[\cos^{-1} C_\rho - C_\rho (1 - C_\rho^2)^{\frac{1}{2}} \right] \quad \text{for } C_\rho < 1, \tag{38}$$

$$I = \frac{2\pi C_\rho}{(C_\rho^2 - 1)^{\frac{3}{2}}} \left[C_\rho (C_\rho^2 - 1)^{\frac{1}{2}} - \cosh^{-1} C_\rho \right] \quad \text{for } C_\rho > 1. \tag{39}$$

The aspect ratio C_ρ in the equations above is defined as the ratio of c/d in Fig. 2 and ν_1, ν_2 are the Poisson's ratios for the filler and the matrix, respectively.

Based on the above formulas given by Chow, we know that for the metal-filled composite materials, such as conductive adhesives, the elastic constants are functions of the aspect ratio C_ρ, volume fraction of filler φ, as well as the material properties of each constituent. Based on these equations, we can calculate the elastic constants for the filler/matrix composite for different C_ρ and φ values with examples shown in Figs 3 and 4 for relative Young's modulus and shear modulus, respectively. The relative moduli are defined as the ratio of composite modulus to the matrix modulus. These results are consistent with the well known rule of mixtures used in long fiber reinforced composites. When the aspect ratio of the particle is $C_\rho > 50$, the moduli are not a function of C_ρ. Therefore, for larger C_ρ, the composite behavior is more like that of long fiber reinforced composite. For intermediate aspect ratio cases, i.e. $1 < C_\rho < 50$ we propose an approximation which utilizes the method used for randomly oriented short fibers in predicting the behavior of particle filled adhesives so that the composite material can be treated as isotropic. Consequently, we propose to use the following formulas suggested by Agarwal and Broutman [19], in conjunction with equations (20), (22) and (23) given above

$$E_{\text{random}} = \frac{3}{8} E_{\text{p}} + \frac{5}{8} E_{\text{t}}, \tag{40}$$

$$G_{\text{random}} = \frac{1}{8} G_{\text{p}} + \frac{1}{4} G_{\text{t}}. \tag{41}$$

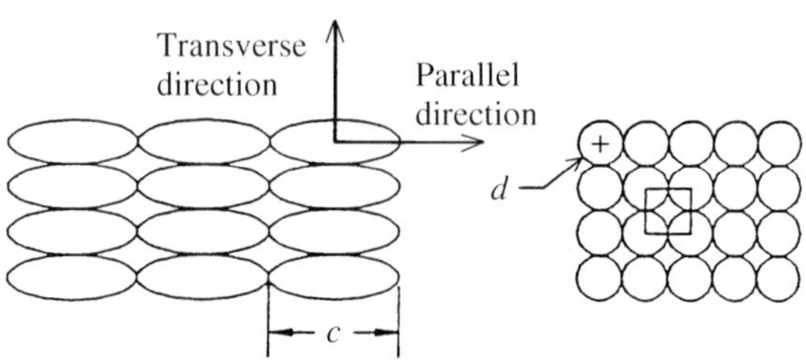

Figure 2. Uniform ellipsoids with the aspect ratio $C_\rho = c/d$.

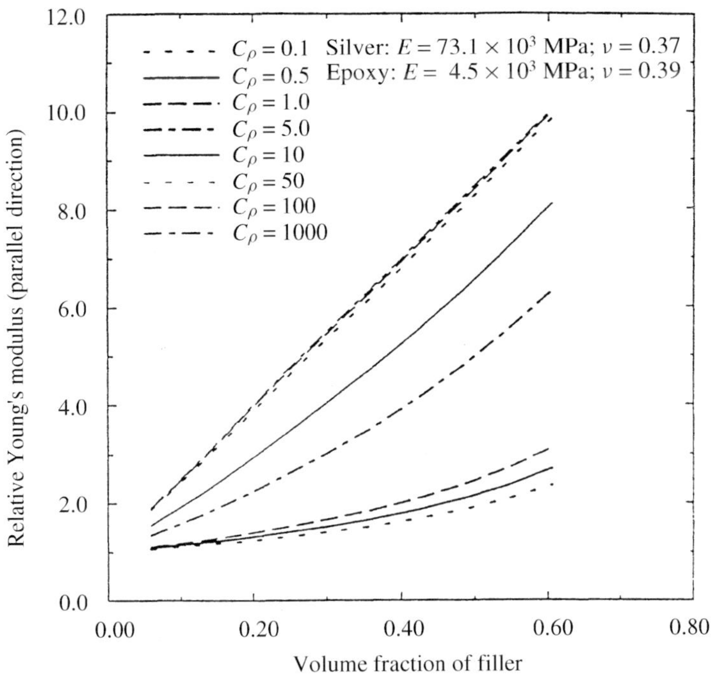

Figure 3. The effect of ellipsoid aspect ratio C_ρ and filler volume fraction φ on relative Young's modulus (parallel direction).

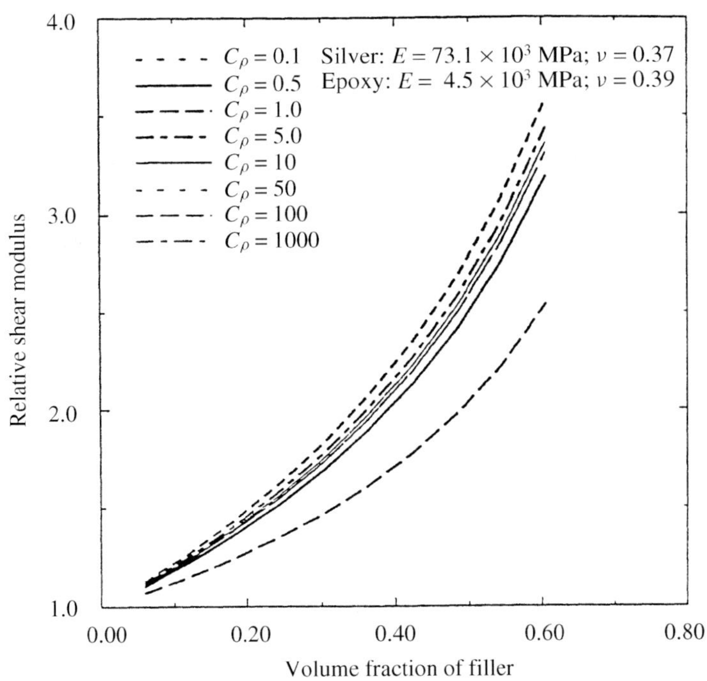

Figure 4. The effect of ellipsoid aspect ratio C_ρ and filler volume fraction φ on relative shear modulus.

The longitudinal linear thermal expansion coefficient, α_{33}, for the composite adhesive is given by Chow [16]:

$$\alpha_{33} = \alpha_2 + \frac{k_1}{k_2} \frac{(\gamma_1 - \gamma_2) Y_1 \varphi}{2K_1 Y_3 + Y_1 K_3}, \tag{42}$$

where γ_1 and γ_2 are the volumetric thermal expansion coefficients for the filler and the matrix, respectively, and α_2 is the linear thermal expansion coefficient for the matrix. Using a similar expression for the transverse linear thermal expansion coefficients $\alpha_{11} = \alpha_{22}$, Chow derives the volumetric thermal expansion coefficient, γ, as [16]:

$$\gamma = \gamma_2 + \frac{k_1}{k_2} \frac{(\gamma_1 - \gamma_2)(Y_1 + 2Y_3)\varphi}{2K_1 Y_3 + Y_1 K_3}. \tag{43}$$

If we want to treat the conductive adhesive as isotropic for ease of calculations, then an effective expression for the linear thermal expansion coefficient can be obtained as:

$$\alpha = \frac{\gamma}{3}. \tag{44}$$

4.2. Overlap joint stresses induced by thermal mismatch between the conductive adhesive and the substrate

Since the conductive adhesive and the substrate usually have different thermal expansion coefficients, thermal mismatch stresses develop at their interface. Finite element analysis and closed form solutions [20] can be utilized to illustrate these thermal mismatch stresses as they occur in an overlap configuration subjected to thermal variation such as the curing process followed by cool down.

4.3. Theoretical approximation of the shear stress introduced at the interface

We consider a bonded layer structure representing a single lap joint cured at an elevated temperature and subsequently cooled (Fig. 5). Suhir [20] gives the distribution

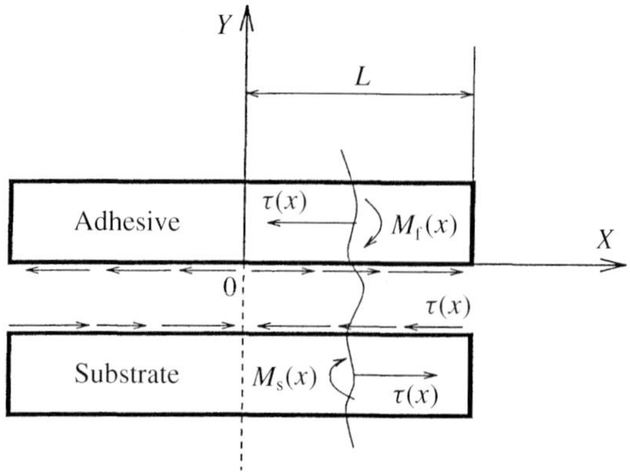

Figure 5. Shear stress analysis model.

of the shear stress at the substrate/adhesive interface resulting from the thermal mismatch as

$$\tau(x) = -\frac{\Delta\alpha\Delta T}{k(k_a + k_s)}\frac{\sinh(kx)}{\cosh(kL)}, \tag{45}$$

where

$$k^2 = \frac{1}{k_a + k_s}\left[\lambda_a + \lambda_s - \frac{3(h_a + h_s)^2}{E_a^0 h_a^3 + E_s^0 h_s^3}\right], \tag{46}$$

$\Delta\alpha = \alpha_a - \alpha_s$, where α_a and α_s are thermal expansion coefficients for the adhesive and substrate materials, $\lambda_a = (E_a^0 h_a)^{-1}$ and $\lambda_s = (E_s^0 h_s)^{-1}$ are coefficients of axial compliance for the adhesive and substrate, $E_a^0 = E_a/(1 - \nu_a)$ and $E_s^0 = E_s/(1 - \nu_s)$ are generalized Young's moduli of the materials, h_a and h_s are thicknesses of the adhesive and the substrate,

$$k_a = \frac{2(1 + \nu_a)h_a}{3E_a} \tag{47}$$

and

$$k_s = \frac{2(1 + \nu_s)h_b}{3E_s} \tag{48}$$

are coefficients of interfacial compliance, E_a and E_s are Young's moduli, ν_a and ν_s are Poisson's ratios for the adhesive and substrate materials, and ΔT is the temperature differential.

4.4. A case study

In order to illustrate the presence of thermal mismatch shear stress for an example case of electronic leads (plated copper) bonded with silver filled conductive thermoset adhesive, we can present the following calculations.

The adhesive's properties are: $\alpha_a = 46 \times 10^{-6}/°\text{C}$, $E_a = 27.6$ GPa, $\nu_a = 0.36$. The substrates' properties are: $\alpha_s = 16.6 \times 10^{-6}/°\text{C}$, $E_s = 117$ GPa, $\nu_s = 0.34$. The coefficient of thermal expansion mismatch is $\Delta\alpha = (46\text{–}16.6) \times 10^{-6}$, and the change in environmental temperature is: $\Delta T = 80°\text{C}$. The thicknesses of the adhesive layer and the substrate are $h_a = 0.11$ mm and $h_s = 0.254$ mm, respectively. Using these data and equation (45) the variation of the thermal shear stress along the 0.5 mm interface (half of the overlap length) can be calculated as:

$$\begin{aligned}\tau(x) &= -\frac{29.4 \times 10^{-6} \times 80}{6.41 \times 10^3 \times 5.55 \times 10^{-15}}\frac{\sinh(6413.7x)}{\cosh(3.2)} \\ &= -5.3809 \times 10^6 \sinh(6413.7x).\end{aligned} \tag{49}$$

At the overlap ends where $x = 0.5$ mm, the maximum thermal shear stress has the value, $\tau_{max} = -65.9$ MPa. Figure 6 shows the variation of the thermal shear stress along the interface as calculated by ANSYS 5.0 finite element anlaysis. The shear stress value of $\tau = -48.9$ MPa shown on this plot represents improvement in the state of stress which would be obtained if adhesive fillets were used at the overlap

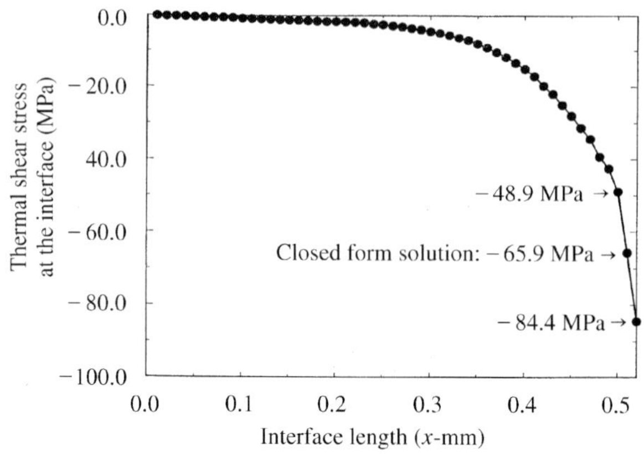

Figure 6. Thermal shear stress at the adhesive–substrate interface.

ends. However, even this value is above the adhesive failure stress level in shear, about 41.3 MPa (absolute value).

Suhir [20] also gives an expression for the distribution of the peel stress, $\sigma(x)$, created as a result of thermal mismatch

$$\sigma(x) = -\frac{1}{2} E_a^0 h_a^2 \Delta\alpha \Delta T k^2 \frac{\cosh(kx)}{\cosh(kL)}. \tag{50}$$

We note that the maximum values of thermal shear and peel stresses can easily be related:

$$\sigma_{max} = \frac{1}{2} k h_a \tau_{max}. \tag{51}$$

The results of our calculations based on typical values for electronic applications reveal that localized failure (plastic deformation or crack initiation) should be expected at overlap edges of bonded joints which undergo thermal cure or other similar thermal loading, especially if it is cyclic in nature.

Hence we believe that the thermal stress created during the curing process as well as subsequent thermal cycling may be a crucial factor affecting the durability of conductive adhesive joints and the curing schedule as well as in-service cooling must be considered carefully for this purpose. The use of slow, step-cooling subsequent to cure, and adhesive/substrate materials with coefficients of thermal mismatch close in value should help reduce such thermal stresses.

5. SILVER MIGRATION IN CONDUCTIVE ADHESIVES

5.1. Theory

Using Faraday's law, the relationship between charge transferred through a gap and the amount of silver ions involved can be concisely related as

$$m_{silv} = \frac{sMIt}{nF}, \tag{52}$$

where m_{silv} is the mass of the silver, s is the stoichiometric coefficient of the species, M is the atomic or molecular weight, I is the current, and t is the time elapsed. This equation is valid for a constant current process. In other cases the current must be integrated over time to calculate Q, the total charge passed:

$$Q = \int_0^t I \, dt. \tag{53}$$

Faraday's law basically depicts silver migration through a water medium. In this process we have ion conduction. Ion conduction differs fundamentally from electrical conduction in metals. Typically, the maximum concentration of charged species that can be obtained in a liquid is much lower than in a metallic conductor. Moreover, the mobility of large ions is much lower than that of electrons. As a consequence, the conductivity of metals is much higher than that of ions in liquid. Since the mechanisms of conductivity are different, the temperature dependencies of conductivity display different functional relationships. For ions, conductivity increases with increasing temperature. For the electrical conductivity of metals, the temperature coefficient of conductivity is negative and about an order of magnitude lower.

5.2. Experimental

In a typical silver ion migration experiment through a water filled gap, there are three ranges depending on the characteristics of conductivity, as shown in Fig. 7. During $0 < t < t_1$, the gap conductivity is ionic (period 1), and ion conduction dominates the process. The potential applied to the gap drives ions from the higher potential (anode) to the lower one (cathode), where the ions receive electrons, reduce to silver, and deposit. In the transition range ($t_1 < t < t_2$), gap conductivity is a mixture of ionic conduction, metal conduction and electron tunnelling conduction, with metal conduction replacing ionic conduction. This period is very short. During the final period ($t > t_2$), silver deposit occurs to bridge the gap. At that time the gap conduction

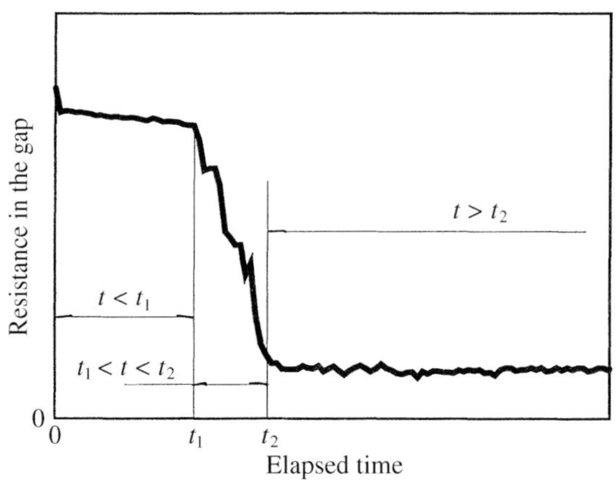

Figure 7. Change in conductivity during the different silver ion migration periods.

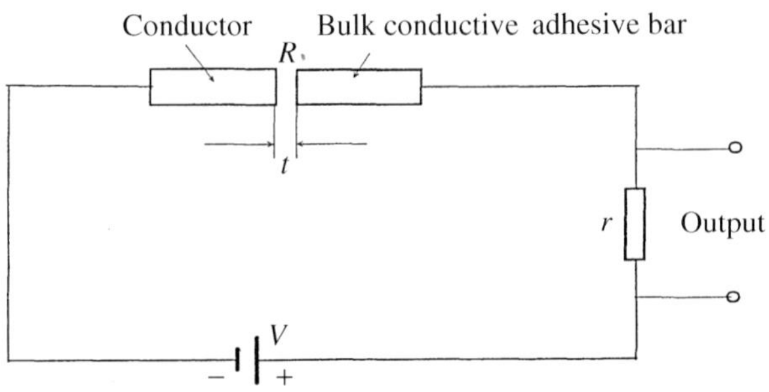

Figure 8. Test system for silver ion migration.

Table 1.
The properties of model adhesives A, B, C, and D

Samples	Resistivity (Ω cm)	Viscosity (cps) at 25°C	Cure temperature (°C)
Adhesive A	6×10^{-5}	55 000	130–150
Adhesive B	$< 5 \times 10^{-4}$	200 000	80
Adhesive C	1×10^{-4}	220 000	130–150
Adhesive D	1×10^{-4}	200 000	170–210

is totally by metal conduction, which is much higher than that in period 1, which is by ionic conduction.

The experimental device built for silver migration tests is shown in Fig. 8. In this device t_h is the thickness of the gap between the conductor and the bulk conductive adhesive bar. In our experiments, t_h is equal to 0.127 mm. In the associated circuit, R is the resistance of the gap, which can be used as a signal to identify a completed total silver migration experiment when it reduces to a few ohms; r is a standard resistor used to measure the current through the gap. A potential V is applied to the system with conductive adhesive on the anode side.

The experimental conditions used for the silver migration tests are as follows:

(1) Environmental temperatures used are $T = 25°C$ and $70°C$.

(2) Four groups of commercial conductive adhesives are tested: Adhesive A, Adhesive B, Adhesive C, and Adhesive D (see Table 1).

(3) The applied potential is 1.5 V.

(4) The gap distance is 0.127 mm.

(5) The migration medium is water with pH value of 7.

6. RESULTS AND DISCUSSION

Figure 9 shows the different stages of migration phenomenon. In stage (a), silver starts to deposit at the cathode side. In this stage, conduction between the cathode

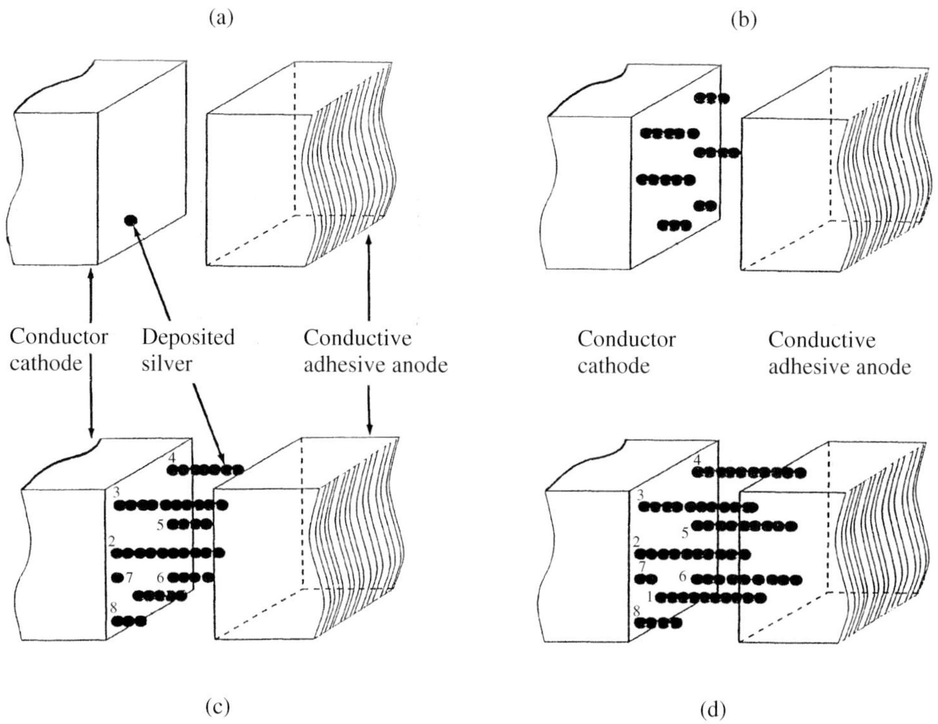

Figure 9. Different stages of silver ion migration.

and the anode is by ionic conduction. In stage (b), more silver ions deposit. In stage (c), silver ions continue to deposit and start to bridge the gap between the anode and the cathode. In stage (d), the bridging is accomplished and the gap between the anode and the cathode becomes a metallic (silver) conductor.

6.1. Migration results for four commercial conductive adhesives

In our experiments, we used four different conductive adhesives which are commercially available. The properties of these conductive adhesives are shown in Table 1. Both the cathode and the anode are made of the conductive adhesives. Experimental results shown in Fig. 10 reveal that the rate of migration differs significantly for different commercial conductive adhesives. For instance, the bridging time threshold for Adhesive A is approximately half of that for Adhesive B, which means that Adhesive B has higher resistance to silver ion migration.

6.2. DC potential consideration

As mentioned earlier, the presence of a DC potential is a necessary condition for silver migration. This potential should be high enough to have the silver in the adhesive lose electrons, drive the resulting ions (Ag^{+}) into and through the water medium, and deposit on the cathode. In this process, there are, principally, three resistances which

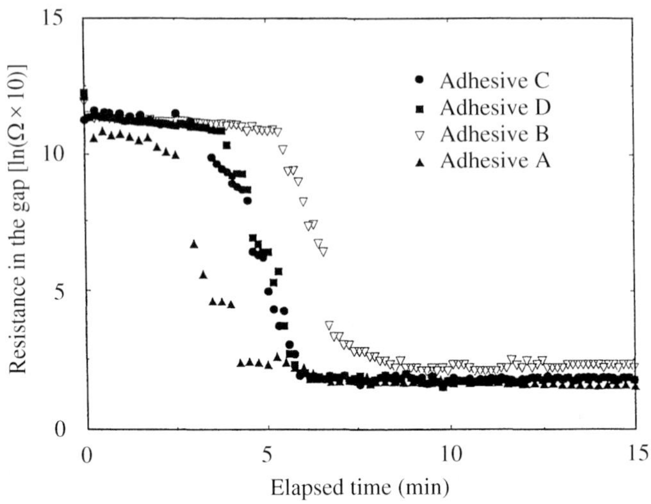

Figure 10. Experimental migration results for four commercial conductive adhesives.

balance the imposed potential energy: kinetic resistance (R_k) of Ag to lose electrons and to dissolve in the water medium; mobility resistance (R_m) to ions moving from the anode to the cathode; and deposition resistance (R_d) on the cathode. Let E_p be the potential energy and assume that all the energy needed in this process is for silver migration. Then the following equation must be satisfied as the minimum potential requirement:

$$E_p \geqslant \sum_i E_i, \qquad (54)$$

where $i = \text{k, m, d}$ and E_i are the energies to overcome R_k, R_m, and R_d resistances, respectively. Unfortunately, there are little experimental data available for E_i and it is very difficult to predict them quantitatively. But during our experiments we did notice a threshold potential level of 1 V was required to have any silver migration within 10 h. In practice, one volt potential between the conductors of electronic devices is very common and, consequently, very likely to cause silver migration problem in the circuits where silver filled conductive adhesives are used. The rate of silver migration increased exponentially with increasing voltage potential applied to the gap.

6.3. The effect of temperature on silver migration

In order to assess the effect of the environmental temperature on the silver migration behavior of four different commercial conductive adhesives, silver migration experiments were performed at 25 and 70°C conditions. The results shown in Fig. 11 reveal that the rate of silver migration is higher at higher temperature as depicted by shifts in the transition periods ($t_1 < t < t_2$) for all four commercial conductive adhesives. Adhesive B displays a higher sensitivity to temperature change by means of a large shift in its transition period although its threshold time for silver migration to start is longer than others.

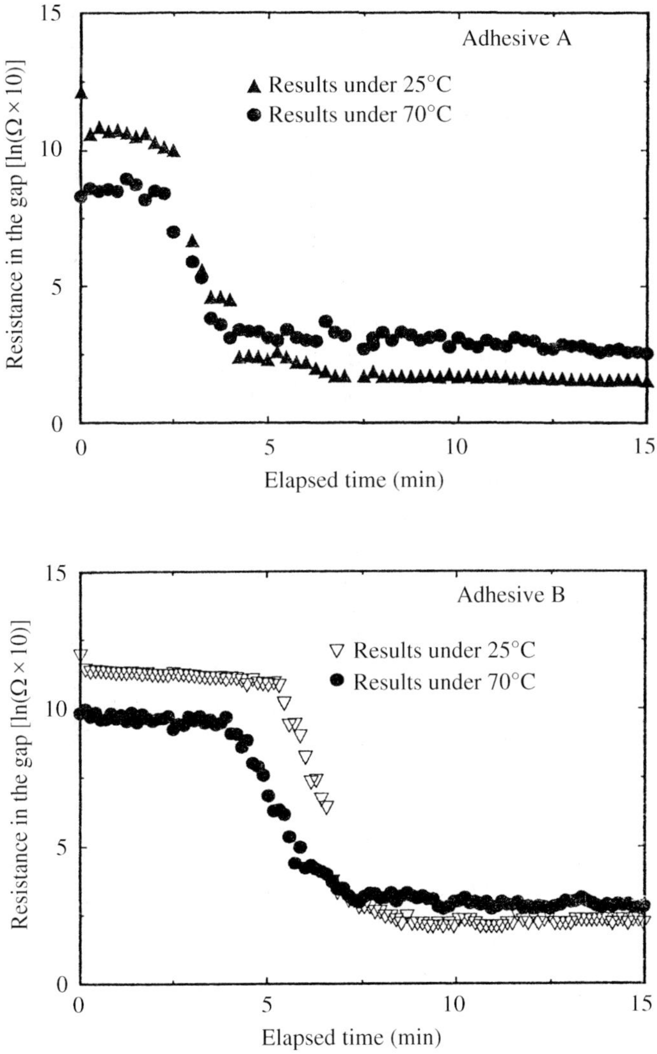

Figure 11. Silver ion migration results at 25 and 70°C.

7. CONCLUSIONS

Theoretical methodology was developed in order to understand the effect of external pressure on the resistivity of electronically conductive adhesives. The fact that the conductivity of metal particles increases with increasing compressive force was combined with Holm's equation, which considers total resistance in two parts: constriction resistance, and tunnelling resistance.

The analysis presented reveals that particle filled electrically conductive adhesive's resistivity depends on a parameter, S_i, representing the average length of conductive adhesive paths between the conductive particles incorporated in the adhesive matrix. The computer simulation utilized reveals that the S_i values in the planar direction

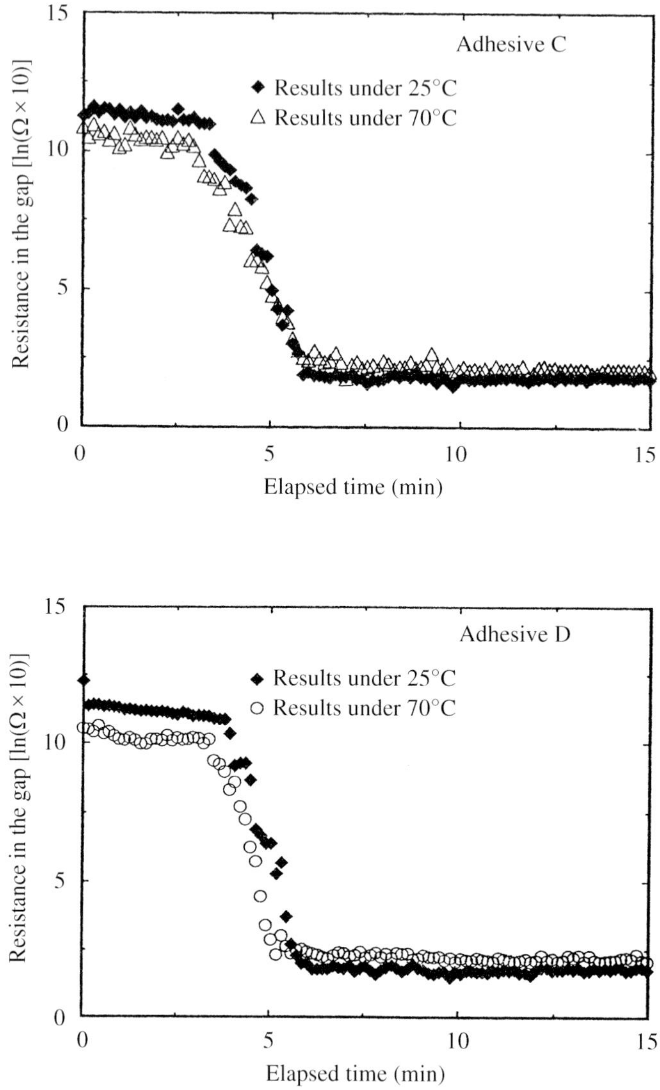

Figure 11. (Continued).

of 2-D thin film adhesives is much larger than that for the 3-D case. This leads to anisotropic electric conduction in thin film adhesives where the resistivity values in planar directions are much larger than those encountered in 3-D measurements for the same adhesive material [18, 21].

The equivalent mechanical properties for the filler/matrix composite, namely Young's, bulk and shear moduli, were calculated for the filled adhesive. All three of these moduli were shown to increase with increasing volume fraction of the filler. By using the theory developed for polymer based composites containing randomly oriented short fibers, analytical results were extended to approximate isotropic cases.

Thermal mismatch of the adhesive and substrate mechanical constants introduces stresses near the adhesive/substrate interface. Use of the closed form solution presented provides acceptable estimations about the magnitudes of these stresses and illustrates the possibility of failure initiation due to thermal mismatch. Consequently, thermal mismatch must be considered for proper design and function of conductive adhesive joints.

The occurrence of silver migration was illustrated by applying voltage to bulk adhesive samples separated with a migration gap of 0.127 mm and immersed in deionized water at different temperatures. The migration time was shown to decrease with increasing voltage. The temperature does affect the migration of silver ions. High temperature ambient conditions speed up the migration.

Acknowledgements

Financial support by IBM Corporation and New York State Science and Technology Foundation through the Center for Advanced Materials Processing at Clarkson University is gratefully acknowledged. We also would like to acknowledge an ISHM Educational Foundation Research Grant to Y. Wei for 1994–1995.

REFERENCES

1. A. O. Ogunjimi, D. Boyle, D. C. Whalley and D. J. Williams, *J. Electronics Manufacturing* **2**, 109–118 (1992).
2. B. Alpert and A. Schoenberg, *Electronic Packaging and Production* **31**, 1130–1132 (1991).
3. J. Bolger, J. Sylva and J. McGovern, *Surface Mount Technology* **6**, 66–70 (1992).
4. R. Pound, *Electronic Packaging and Production* **25**, 86–90 (1989).
5. G. Nguyen, J. Williams and F. Gibson, in: *Proceedings of the 1992 International Symposium on Microelectronics*, SPIE Vol. 1847, pp. 510–517. Int. Soc. Hybrid Microelectronics, Reston, VA (1992).
6. J. Honore, H. Rubin and M. Zierold, in: *Technical Program, National Electronic Packaging and Production Conference*, Vol. 3, pp. 1372–1380. Cahner Exposition Group, Des Plaines, IL (1992).
7. K. Brendfelt, J. Dral and B. Wells, in: *Proceedings of the Technical Program, Nepcon West '89*, Vol. 1, pp. 496–506. Cahner Exposition Group, Des Plaines, IL (1989).
8. K. Gileo, A. Chu and D. Vieau, in: *Proceedings of Technical Program, National Packaging and Production Conference*, Vol. 1, pp. 193–212. Cahner Exposition Group, Des Plaines, IL (1992).
9. R. Holm, *Electric Contacts*. Springer-Verlag, New York (1967).
10. M. C. Sneed, J. L. Maynard and R. C. Brasted, *Comprehensive Inorganic Chemistry*, Vol. II. Van Nostrand, New York (1954).
11. N. J. De Lollis, *Adhesives, Adherends, Adhesion*. Robert E. Krieger Publishing Company, Huntington, New York (1980).
12. J. C. Bolger, M. J. Herberg and C. T. Mooney, in: *Proceedings of the 1st International SAMPE Electronics Conference*, p. 616 (1987).
13. D. D. Chang, J. A. Fulton, H. C. Ling, M. B. Schmidt, R. E. Sinitski and C. P. Wong, in: *Proceedings of the 43rd Electronic Components and Technology Conference*, pp. 836–841. IEEE (1993).
14. T. S. Chow, *J. Appl. Phys.* **48**, 4072–4075 (1977).
15. T. S. Chow, *J. Polym. Sci.: Polym. Phys. Edn.* **16**, 959–965 (1978).
16. T. S. Chow, *J. Polym. Sci.: Polym. Phys. Edn.* **16**, 967–970 (1978).
17. S. P. Timoshenko and J. N. Goodier, *Theory of Elasticity*, 3rd edn. McGraw-Hill, New York (1970).
18. Y. Wei, *Electronically Conductive Adhesives: Conduction Mechanisms, Mechanical Behavior and Durability*. Ph.D. Dissertation, Clarkson University, Potsdam, New York (1995).

19. B. D. Agarwal and L. J. Broutman, *Analysis and Performance of Fiber Composites*. John Wiley, New York (1990).
20. E. Suhir, *J. Appl. Mech.* **55**, 143–148 (1988).
21. Y. Wei and E. Sancaktar, in: *Proceedings of the Electronic Components and Technology Conference*, J. Lau (General Chairman), pp. 701–706. IEEE Piscataway, NJ (1995).

Mittal Festschrift, pp. 531–542
W. J. Van Ooij and H. R. Anderson, Jr. (Eds)

Catalytic activation of wood adhesive polycondensation by the adhesion forces to a cellulosic substrate

A. PIZZI

ENSTIB, Ecole Nationale Superieure des Technologies et Industries du Bois, Université de Nancy 1, B.P. 1041, 88051 Epinal, France

Abstract—The reactions of polycondensation of phenol–formaldehyde resins, melamine–formaldehyde resins, and diisocyanates with wood residual water to polyureas, and of autocondensation of aldehyde-free tannin adhesives all have lower energies of activation on a cellulosic substrate than when such resins harden alone. This lower energy of activation has been identified as being due to (1) the catalytic self-condensation induced by the adherend carbohydrates such as crystalline and amorphous cellulose, (2) the diffusion hindrance exercised by the substrate on the hardening reaction of the resin, and (3) the formation of resin–substrate covalent bonds. The first of these is shown to be by far the major cause of the lowering of the activation energy of the adhesive polycondensation and hardening, with the second exercising some contribution and the third most often a negligible contribution under the conditions pertaining to thermosetting wood adhesive applications.

Keywords: Catalysis; wood adhesives; cellulose; adhesion forces; polycondensation; hardening; diffusion hindrance.

1. INTRODUCTION

A complete understanding of the phenomenon of adhesion of a formaldehyde-based resin, or any other resin, to a lignocellulosic substrate is fundamental to the evaluation of bond strength and bond strength development in wood products. The influence that the substrate exercises on the kinetic and curing behavior of the resin and the reasons for such an influence are not immediately obvious. For instance, it has been repeatedly established that the activation energy of the polycondensation reaction of resins such as phenol–formaldehyde (PF), urea–formaldehyde (UF), and melamine–formaldehyde (MF) is markedly influenced by the presence of wood [1–7]. In essence, the activation energy is considerably lowered [1–7]. This trend infers that resin polycondensation and crosslinking proceed at a much faster rate when the resin is in molecular contact with one or more of the main polymeric wood constituents. This phenomenon has, since its initial discovery, been ascribed to the formation of covalent bonds between the synthetic resin and the main constituents of the wood matrix. While there is no doubt that covalent bonds can be formed [8] linking PF resols (resins carrying hydroxybenzyl alcohol groups), UF resins, and

MF resins to lignin and wood carbohydrate, it is also important to consider the conditions under which they are formed. Prolonged heating and high temperatures are conducive to resin–substrate covalent bonds being formed [8], as demonstrated by certain applications of these resins to natural textiles [8]. What needs to be addressed, then, is to what extent are resin–substrate covalent bonds formed under the standard application conditions for thermosetting wood adhesives. For example, the maximum temperature in the strength-determining center of the core of a particleboard only reaches a maximum of 105–120°C at the pressing times and temperatures characteristic of such a product wood-bonding process [9]. Pressing times are fast, varying between 7 and 15 s per mm thickness of the panel. Thus, temperatures are not extreme, while times for curing are very short. Numerous investigations have been conducted to demonstrate the covalent bonding concept of resin to wood substrate. The majority of these studies have been by differential scanning calorimetry (DSC) or thermogravimetric analysis (TGA) [1–3, 10]. Infrared (IR) and ultraviolet (UV) spectrophotometry have also been used [6, 11]. The DSC and TGA investigations conducted are subjected to time and temperature limitations: for instance, scans of 10–20°C/min to a maximum of 250–300°C. Even considering the main curing exotherm of a PF resin as occurring approximately at 130–150°C, this curing exotherm will be reached in not less than 5–10 min after heating has started. A particleboard core instead remains at a maximum of 100–120°C for only 20–60 s and reaches this maximum temperature in 80–120 s. Some indications that covalent bonds do not form or form to a very small extent even in conditions much more drastic than these have also been reported [12–14]. The IR and UV investigations used even more extreme conditions: 14 h at 190°C for the IR [11, 15] and 2 h at 120°C for the UV investigations [6]. As a consequence, the formation of covalent bonds was demonstrated, but the extreme conditions used also implied that the situation might be very different at the much lower temperatures and shorter times characteristic of wood adhesive applications. A controversial but acute review [16] of the subject concluded that the evidence presented clearly indicates that no covalent bonds are formed under the conditions characteristic of thermosetting wood adhesive applications. This conclusion might also be too extreme. Molecular mechanics studies [17–21] have given a grounding to the theory that strong wood adhesive bonding can be equally well, and completely, explained by the sum of a multitude of resin/substrate secondary forces' interactions, without any need for interfacial resin to substrate covalent bonds. In this, the molecular mechanics conclusions are in agreement with the most accepted theories of adhesion in fields other than wood adhesives [22]. However, the undeniable experimental fact remains that a lignocellulosic substrate facilitates the curing of formaldehyde-based and other resins by lowering the apparent energy of the polycondensation.

2. EXPERIMENTAL [23–26]

2.1. *Phenol–formaldehyde (PF) resin*

Phenol (94 g, 1 mol) was mixed with 32 g of water and 8 g of methanol. The mixture was maintained at 40°C for 30 min under mechanical stirring during which two amounts of 5 g each of 33% sodium hydroxide (NaOH) were added. Paraformaldehyde (55 g, 1.8 mol) was added and the temperature of the reaction mixture was increased to reflux over a period of 30 min, during which another two aliquots of 5 g each of 33% NaOH were added. The reaction was maintained at reflux for 1 h and then cooled and stored. The PF resin obtained had a resin solids content of 59%.

2.2. *Preparation of samples*

2.2.1. Diluted PF resin and wood flour. PF resin (2.725 g) was diluted in water (6.652 g), affording a 2 : 5 resin to water solution. The diluted resin was then adsorbed onto wood flour (1.584 g) [*Pinus radiata* dry sapwood ground to a flour of 200 mesh (0.75 μm)].

2.2.2. PF resin and wood flour. PF resin (5.370 g) was adsorbed onto wood flour (0.88 g).

2.2.3. Saligenin on filter paper. Saligenin (*o*-hydroxybenzyl alcohol) (0.558 g) was dissolved in water (5.050 g), to which was added a 30% NaOH solution (0.667 g) to give a solution of pH 12.8. The saligenin solution was then adsorbed onto filter paper (Whatman Number 1 Chromatography paper) and allowed to dry in a desiccator. The concentration of saligenin on filter paper was 0.890 M.

2.2.4. p-Hydroxybenzyl alcohol on filter paper. *p*-Hydroxybenzyl alcohol (0.510 g) was dissolved in water (5.221 g) and a 30% NaOH (0.877 g) solution added to afford a solution of pH 13.8. The *p*-hydroxybenzyl alcohol was then adsorbed onto filter paper and allowed to dry in a desiccator. The concentration of *p*-hydroxybenzyl alcohol on filter paper was 0.787 M.

2.2.5. Saligenin and kraft lignin. Saligenin (undissolved) was crushed with either bagasse or kraft lignin using a pestle and mortar. A ratio of 50 : 50 saligenin to lignin was used.

2.2.6. Saligenin and wood flour. A 50 : 50 mixture by mass of saligenin (undissolved) and wood flour was crushed using a pestle and mortar.

2.2.7. Saligenin and cotton wool. A 50 : 50 mixture by mass of saligenin and cotton wool was crushed in a pestle and mortar. Cotton wool is composed of virgin cotton fibers of a degree of crystallinity of over 93%, the crystallites presenting the structure of cellulose I.

2.2.8. p-Hydroxybenzyl alcohol and kraft lignin. p-Hydroxybenzyl alcohol (undissolved) was crushed with either bagasse lignin or kraft lignin using a pestle and mortar. A ratio of 50 : 50 p-hydroxybenzyl alcohol to lignin was used.

2.2.9. p-Hydroxybenzyl alcohol and wood flour. A mixture of 50% p-hydroxybenzyl alcohol and 50% wood flour was crushed together in a pestle and mortar.

2.2.10. p-Hydroxybenzyl alcohol and cotton wool. A 50 : 50 mixture by mass of p-hydroxybenzyl alcohol and cotton wool was crushed in a pestle and mortar.

2.3. Sample testing

The samples were tested by placing a small amount into a sample crucible. The sample crucibles were then placed in a high precision DuPont thermal analyzer equipped with both DSC and TGA modules and heated at different rates, namely 5, 10, 15, 20, 25, and 30 K per min until a maximum of 573 K. In the case of the model compounds above, further heating rates were also used, namely 7, 8, 18, 18.5, 19, and 19.5 K, for a total of 12 different heating rates. The activation energy of the curing reaction of the PF resin alone, of the PF resin on different substrates, and of o-hydroxybenzyl alcohol (saligenin) and p-hydroxybenzyl alcohol both alone and on different substrates was calculated on the basis of the variation of the temperature of the maximum of each exotherm (T_{max}) of the DSC scan as a function of the temperature rate increase of the DSC scan, using the Kissinger equation [27]. Thus, ln(rate of temperature increase/T^2_{max}) was plotted as a function of $1/T_{max}$ and from the Kissinger equation, the energy of activation corresponding to each identified exotherm was obtained. Also, the onset and final temperatures of each exotherm were recorded. The same reactions were repeated and followed by TGA at the same temperature increment rates as for the DSC scans. TGA traces and first derivative traces were obtained as an initial step to distinguish which exotherms in the DSC scans corresponded to any type of polycondensation. Furthermore, DSC and TGA scans of all the substrates alone were obtained for the same purpose. All the scans were conducted in a nitrogen atmosphere. Each DSC and TGA scan was repeated using a Mettler DSC 20 of lower precision, to control the position of the main exotherm of the reactions.

2.4. Melamine–urea–formaldehyde (MUF) resins, isocyanate resins, and tannins autocondensation

The same experiments as those outlined above were carried out for melamine–urea–formaldehyde (MUF) resins [24], for polymeric MDI (4,4′-diphenylmethane diisocyanate) [25], and for polyflavonoid tannin autocondensation reactions [26].

3. RESULTS AND DISCUSSION

In UF and MF resin polycondensation, water is released at each condensation step; for a DSC analysis to have any meaning, it must always be coupled with a TGA analysis. This procedure is used to confirm which DSC exotherms correspond to distinct

TGA first derivative loss of mass by water evaporation and thus likely to belong to a condensation. DSC or TGA, if used in isolation, are techniques that are recognized to give results that are, at best, difficult to interpret [28, 29]. In this investigation, only DSC exotherms that could be clearly identified by TGA as relating to any polycondensation were considered. The model compound DSC scans yielded a series of several, discrete exotherms. This result is expected because hydroxybenzyl alcohols condense in a discrete series of steps: a monohydroxybenzyl alcohol first self-condenses to dimers and these self-condense then to ever higher oligomers in a series of discrete steps. These steps are well defined and separate and can be clearly observed in the early stages of the total condensation. The PF resin, instead, already a mixture of a variety of higher oligomers, gave only a series of three exotherms. The final energies of activation for each exotherm for the two model compounds and for a PF resin are reported in Tables 1–3. The temperature maxima and the range of the exotherms of the PF resin show correspondence with those of the model compounds. The first PF exotherm corresponds to the first exotherm of the *p*-hydroxybenzyl alcohol model compounds. The second PF exotherm corresponds to the sum of the second and third exotherms of the *p*-hydroxybenzyl alcohol model compound, or possibly only to the third exotherm of the *p*-hydroxybenzyl alcohol, and to the first exotherm of the saligenin. The third exotherm appears to be a wider example of the fourth exotherm of the *p*-hydroxybenzyl alcohol model compound. The main exothermic peak for PF and the *p*-hydroxybenzyl alcohol model compound appears clearly to be the same with a maximum at 135–140°C at the lower temperatures (up to 160°C) of the scan, and saligenin at a slightly lower temperature. From Tables 1–3, it appears that all the main constituents of the wood substrate, at some stages of the reaction, contribute to the decrease in the energy of activation of the whole polycondensation, or of important steps in it. The data in Tables 1–3 cannot, of course, tell directly if the decrease in the energy of activation is due to covalent co-reaction between the resin and substrate, or to a catalytic activation of the resin self-condensation induced by the wood constituents. An interesting aspect is, however, noticeable, and some conclusions can be drawn: any real exotherm that is not present in the self-condensation of the model compound alone is very likely due to reactions between model compound oligomers and the substrate. This conclusion is reached because any reaction between the model and substrate will be a different condensation reaction to the self-condensation of the model alone; as a consequence, it will give a differently positioned exotherm specific to, and characteristic of, that particular reaction. In this respect, Tables 1 and 2 make it obvious that covalent bonds between the resin and lignin appear to form and occur at a variety of stages in the polycondensation. It is also evident that some covalent bonds between the model compounds (and model resin) and cellulose form at the higher temperatures. What is also evident, however, is that the energy of activation of the self-condensation exotherms of the model compounds on some substrates, particularly cellulose at the lower temperatures, also strongly decreases. This type of exotherm cannot be ascribed to resin–substrate covalent bonds being formed, because they are characteristic self-condensation exotherms exclusively specific to the models alone. This characteristic implies that the substrate catalytically activates the self-condensation of the model and model resin. Of greater interest to wood adhesive

Table 1.
Activation energies of exotherms of *p*-hydroxybenzyl alcohol polycondensation and curing obtained by DSC and the Kissinger equation, corresponding to TGA mass losses

	Activation energy (kcal/mol)						
p-Hydroxybenzyl alcohol	—	24.58	31.18	—	—	26.83	—
p-Hydroxybenzyl alcohol + lignin	25.24	72.06	25.89	30.14	20.09	12.62	—
p-Hydroxybenzyl alcohol + cellulose	—	—	12.67	—	21.56	23.39	36.71
p-Hydroxybenzyl alcohol + wood	29.75	60.56	30.48	—	—	21.31	15.01
Average peak temperature (°C)	98.4	110.7	139.7	149.9	194	247.6	267.2
Peak temperature range (°C)	93.1–103	107.6–115.7	127–155.9	145.8–152	180.6–210.3	231.7–261	256.5–280
Main exotherm peak			Main				

Table 2.
Activation energies of exotherms of *o*-hydroxybenzyl alcohol (saligenin) polycondensation and curing obtained by DSC and the Kissinger equation, corresponding to TGA mass losses

	Activation energy (kcal/mol)				
Saligenin	[a]	[a]	20.64	16.97	—
Saligenin + wood flour	28.52	—	—	18.23	20.29
Saligenin + lignin	12.2	14.74	25.97	—	—
Saligenin + cellulose	11.83	—	29.82	17.77	—
Average peak temperature (°C)	115.5	161.2	203.9	232.6	269
Peak temperature range (°C)	111.8–151	156.6–165.7	184.5–218	207.5–264	259.7–

[a] A multitude of small peaks which are impossible to define in DSC. In TGA, a single, very wide step.

Table 3.
Activation energies of exotherms of a PF resin polycondensation and curing obtained by DSC and the Kissinger equation, corresponding to TGA mass losses

	Activation energy (kcal/mol)		
PF resin	55.61	23.85	29.15
PF resin + lignin	—	15.82	21.37
PF resin + cellulose[a]	6.91	16.73	20.84
PF resin + cellulose[b]	20.34	20.21	26.29
PF resin + wood flour	14.53	12.04	8.44
Average peak temperature (°C)	90.1	135.4	161.7
Peak temperature range (°C)	65–104	102.8–159	147.9–179.9
Main exothermic peak		Main	

[a] Cotton wool.
[b] Filter paper.

applications, both lignin and particularly cellulose, and crystalline cellulose at the lower temperatures, appear to activate PF self-condensation and curing. It is important to note the differences between *p*-hydroxybenzyl alcohol and the *ortho*-isomeride (saligenin). While for the *p*-hydroxybenzyl alcohol a close correlation with the PF resin trends can be observed, this trend is more difficult to observe in the case of saligenin. This tendency is the consequence of the significantly greater number and type of higher molecular weight isomerides that can be formed by progressive self-condensation of saligenin than by that of *p*-hydroxybenzyl alcohol. This higher number of isomerides causes a multitude of smaller exothermic peaks that smooth the curve at the more important lower temperature range, rendering more difficult or impossible the calculation of the energy of activation of saligenin self-condensation. Deductions become, as a consequence, incomplete and more fragmentary. The question that remains is: what is the cause of the resin-accelerated self-condensation?

The extents and intensities of the exotherms, or, better, the areas enclosed under the exotherm peaks, give an indication of the answer to the first question. Thus, the main exotherm of the PF resin self-condensation (139°C max., Table 3) encloses well over 95% of the total area enclosed under all the lower temperature exotherms. The situation is similar but not as extreme for the two hydroxybenzyl alcohol models [23–29], an indication that in a pre-condensed PF resin, lower mobility and diffusion problems, due to lower substrate wettability, limit even further the possibility of covalent bonding. In conclusion, the DSC part on PF resins of this study indicates that the proportion of exotherms due to resin–substrate covalent bonding is small in relation to the surface-activated self-condensation exotherms of the resin. Furthermore, TGA traces indicate that mass loss is small in the endotherms corresponding to a preceding resin–substrate covalent bonding exotherm; it is, instead, considerable for the endotherms following self-condensation of the resin.

To explain what is the cause of this acceleration effect, Table 4 presents the energies of interaction between different PF oligomers and the surface of an elementary cellulose crystallite obtained by molecular mechanics calculations [17–21]. In the case of *o*- and *p*-hydroxybenzyl alcohol, the maximum energies of interaction with the surface of crystalline cellulose are 19.7 and 13.3 kcal/mol, respectively. Such secondary force attraction between the synthetic oligomer and substrate, if mostly concentrated on a single bond such as the C—O of the hydroxybenzyl alcohol group, would weaken it by up to 25%. In theory, the reaction of two PF oligomer molecules activated to such an extent would yield a condensation rate up to 600 times faster than the same condensation when in the absence of the substrate. This relationship means, for example, that a PF condensation of rate constant 3.3×10^{-2} l/mol per s (Table 4) at 95°C could with surface activation have a rate constant of up to 19.8 l/mol per s under identical reaction conditions. Calculation of the decrease in energy of activation, at parity of collisional factor (which is not likely), yields a decrease in the energy of activation from 23.85 kcal/mol (Table 3) to 19.1 kcal/mol: approximately the same order of magnitude observed in Table 3 for cellulose surface activation of

Table 4.
Molecular mechanics-calculated sum of secondary forces binding PF oligomers to the surface of an elementary crystal of cellulose I

	Interaction energy (kcal/mol)	
	Average	Minimum
Trihydroxybenzyl alcohol	−35.9	−50.0
o,o-Dihydroxydiphenylmethane	−14.5	−20.8
o,p-Dihydroxydiphenylmethane	−14.2	−17.9
o-Hydroxybenzyl alcohol	−12.2	−19.7
p,p-Dihydroxydiphenylmethane	−11.8	−14.7
o,p-Dihydroxybenzyl alcohol	−10.5	−15.0
p-Hydroxybenzyl alcohol	−10.6	−13.3
o,o-Dihydroxybenzyl alcohol	−5.3	−9.3

the main curing exotherm. Expected variations of the collisional factor and diffusion considerations can account for the remaining difference (Table 3).

Similar trends to those observed above for PF resins are also observed for aminoplastic resins (UF, MF, MUF). In this context, the case of MUF resins is presented to describe the situation also for UF and MF resins. In Table 5, the apparent activation energies of the hardening exotherms of a MUF copolymer resin alone and in the presence of cellulose, lignin, and wood are presented, when an acid catalyst is not present (hence in an alkaline environment) as well as the same in the presence of 2% ammonium chloride, thus in an acid environment [24]. The exotherms' apparent activation energies of a MF and a UF resin, prepared as closely as possible to the MUF resin, are also presented as controls. The MUF resin when alone presents four main exotherms, in common with the MF and UF resins (Table 5).

The apparent energy of activation of the ammonium chloride-induced hardening of the MUF resin under acid conditions also decreases when the resin system is in the presence of cellulose and wood [24]. The decrease is as evident as that observed under alkaline conditions and is again even more strictly limited to the first two exotherms (melamine to melamine) [24]. The evident increase in apparent energy of activation of the subsequent three exotherms, and often their complete disappearance, in the ammonium chloride-hardened resins is an indication that the importance of melamine-to-melamine crosslinking has increased to such an extent as to deny, or to render considerably more difficult, any of the three other crosslinking reactions which follow. This crosslinking trend does not mean to say that a decrease in the

Table 5.

Apparent activation energies of exotherms of MUF resin polycondensation and curing by DSC and the Kissinger equation

	Activation energy (kcal/mol)				
MF	*32.2*[a]	*17.1*	[b]	[b]	*55.7*
UF	[b]	[b]	[b]	*10.9*	[b]
MUF	*30.0*	50.4	*12.0*	*10.6*	*12.6*
MUF + cellulose	← *12.7* →		[b]	10.8	[b]
MUF + lignin	23.6	62.5	16.1	13.4	*10.6*
MUF + wood	← *9.0* →		19.1	56.2	*82.0*
MUF + NH_4Cl	*32.0*	*19.9*	*9.4*	[b]	[b]
MUF + NH_4Cl + cellulose	← *8.5* →		*92.1*	[b]	[b]
MUF + NH_4Cl + lignin	[b]	*13.4*	[c]	[b]	*41.0*
MUF + NH_4Cl + wood	← *1.8* →		64.7	*108.5*	*37.2*
Peak max temperature range (°C)	83–102	89–109	113–150	123–168	168–204
Crosslink type	MM	MM	MU	UU	MM

[a]Italicized results are those where linearity of correlation exists (no diffusion hindrance).
[b]Peak does not exist or it is not noticeable.
[c]Peak exists but energy is not calculable.

apparent energy of activation of UF resins in the presence of cellulose and wood does not occur, as it has already been shown that it does [24]. It only means that in MUF systems the melamine reactivity is such that a decrease is only observed for the melamine groups, strongly depressing any effect on the crosslinking of other groups.

It is important to discuss to what extent the decrease in apparent energy of activation is due to the catalytic effect of the substrate surface, or rather to a diffusion hindrance effect. It is well known in heterogeneous catalysis that the correlation of $\ln \beta/T^2$ vs. $1/T$ (where β is the rate of temperature increase) from the Kissinger equation [27] shows a marked deviation from linearity when either diffusion hindrance on the catalyst surface or the occurrence of a low degree of surface covering exists [30]. The situation of a low degree of surface covering does not present itself under the conditions used. As a consequence, any deviation from linearity of the correlation would indicate a diffusion hindrance effect. In Table 5, all the apparent energies of activation in which the data give strictly linear results are italicized. From these, it can be seen that in the initial crosslinking reactions of the resin, those between melamine and melamine groups, the decrease in the energy of activation is mostly due to a substrate-induced catalytic effect and not to diffusion hindrance. That diffusion hindrance becomes important in the subsequent reactions, due possibly to the considerable increase in molecular mass of the resin, confirms the deductions of other authors [31, 32]. That the diffusion hindrance which is important is that exercised by the substrate onto the growing, networking resin can be seen by the fact that the Kissinger equation plots of the resin alone (or even with just ammonium chloride) are linear for all the exotherms, indicating that in the hardening of the resin alone, diffusion problems can only occur at the very end of the reaction. The result that linearity of the plots is lost at a much earlier stage in the reaction when the resin is cured in the presence of the substrate indicates that it is mainly the resin–substrate hindrance rather than the resin–resin hindrance which is the main contributor. This reaction trend also confirms what was found for paper–phenolic resin composites by other authors [31, 32], in which diffusion appeared to play an important role only at the higher degrees of resin conversion. It also shows that within-resin diffusion does not appear to play a role, as in MF, UF, and MUF resins in the absence of the substrate the correlations are all linear. This reaction trend has also been found for PF resins where the decrease in apparent energy of activation has been attributed to diffusion and to substrate-induced catalysis for the main hardening exotherm of the resin.

It is also of interest to discuss what could be the mechanism of the diffusion hindrance. The effect in the experiments is observed under more favorable conditions than what occurs on solid wood surfaces, eliminating possible wood anatomical interferences (if any), because a fine powder (200 mesh, 0.75 μm) wood flour had to be used as the substrate. The effect, then, might result from both portions of large and growing resin molecules chemically adsorbed on the lignocellulosic substrate surface, as well as the polymeric high-molecular-weight constituents of the substrate hindering relative molecular movement and consequently hindering and slowing down contact between resin reactive sites.

As for formaldehyde-based polycondensations, other reactions also appear to present similar acceleration effects on cellulosic substrates. Thus, the reactions of autocondensation of flavonoid tannins when alone, which are also an example of polycondensation, also present the same acceleration effect by surface catalysis by the cellulose substrate [26]. Another example of a reaction in which polycondensation acceleration of a non-formaldehyde-based resin by the same type of surface catalysis occurs has been shown to be the reaction of diisocyanate with water to form polyureas, also in the case of application of such a resin to wood adhesives [25].

In conclusion, acceleration of the self-condensation of resins by a catalytic effect of the lignocellulosic substrate appears to be a fairly common and diffuse phenomenon.

REFERENCES

1. H. Mizumachi, *Wood Sci.* **6**, 14 (1973).
2. H. Mizumachi and H. Morita, *Wood Sci.* **7**, 256 (1975).
3. R. Zakrzewski and S. Proszyk, *Rocz. Akad. Roln. Poznaniu* **117**, 91 (1979).
4. K. Horioka and M. Gamo, in: *IUFRO Conference on Wood Glueing*, R. Gillespie (Ed.), pp. 1–40. Madison, WI (1974).
5. I. Abe, K. Tsutsumi and M. Yoshimura, *J. Jpn. Wood Res. Soc.* **20**, 277 (1976).
6. S.-Z. Chow, *Wood Sci.* **1**, 215 (1969).
7. M. Cherubim and F. Henn, *Mitt. Dtsch. Ges. Holzforsch.* **57**, 165 (1971).
8. B. A. Kottes-Andrews, R. M. Reinhardt, J. G. Frick, Jr. and N. R. Bertoniere, in: *Formaldehyde Release from Wood Products*, B. Meyer, B. A. Kottes-Andrews and R. M. Bertoniere (Eds), pp. 52–66. American Chemical Society Symposium Series No. 316, Washington, DC (1986).
9. T. Maloney, *Modern Particleboard and Dry-Process Fiberboard Manufacturing*. Miller Freeman, San Francisco (1977).
10. M. V. Ramiah and G. E. Troughton, *Wood Sci.* **2**, 120 (1970).
11. S.-Z. Chow and H. N. Murai, *Wood Sci.* **4**, 202 (1972).
12. G. C. Allan and A. N. Neogi, *J. Adhesion* **3**, 13 (1971).
13. H. Poblete and E. Roffael, *Holz Roh Werkst.* **43**, 57 (1985).
14. G. E. Myers, in: *Wood Adhesives in 1985: Status and Needs*, A. W. Christiansen, R. Gillespie, G. E. Myers and B. H. River (Eds), pp. 119–158. Forest Products Research Society, Madison, WI (1986).
15. F. Mora, F. Pla and A. Gandini, Ecole Superieure de Papeterie, University of Grenoble, France, unpublished results (1987).
16. W. E. Johns, in: *Wood Adhesives Chemistry and Technology*, A. Pizzi (Ed.), Vol. 2, pp. 75–96. Marcel Dekker, New York (1989).
17. A. Pizzi and N. J. Eaton, *J. Adhesion Sci. Technol.* **1**, 191 (1987).
18. A. Pizzi and G. de Sousa, *Chem. Phys.* **164**, 203 (1992).
19. A. Pizzi and S. Maboka, *J. Adhesion Sci. Technol.* **7**, 81 (1993).
20. A. Pizzi, *J. Adhesion Sci. Technol.* **4**, 573–588 (1990).
21. A. Pizzi, *J. Adhesion Sci. Technol.* **4**, 589 (1990).
22. A. J. Kinloch, *Adhesion and Adhesives: Science and Technology*. Chapman & Hall, London (1987).
23. A. Pizzi, B. Mtsweni and W. Parsons, *J. Appl. Polym. Sci.* **52**, 1847–1856 (1994).
24. A. Pizzi and L. A. Panamgama, *J. Appl. Polym. Sci.* **58**, 109–115 (1995).
25. A. Pizzi and N. A. Owens, *Holzforschung* **49**, 269–272 (1995).
26. A. Pizzi, N. Meikleham and A. Stephanou, *J. Appl. Polym. Sci.* **55**, 929–933 (1995).
27. H. E. Kissinger, *Anal. Chem.* **29**, 1702 (1957).

28. M. E. Brown, *Introduction to Thermal Analysis*. Chapman & Hall, London (1988).
29. M. E. Brown, private communication (1993).
30. G. C. Bond, *Heterogeneous Catalysis, Principles and Application*. Clarendon Press, Oxford (1987).
31. E. Kiran and R. R. Iyer, *J. Appl. Polym. Sci.* **51**, 353 (1994).
32. E. Kiran and R. R. Iyer, *TAPPI* **76**, 128 (1993).

Mittal Festschrift, pp. 543–555
W. J. Van Ooij and H. R. Anderson, Jr. (Eds)

Characterization of the interface between a cured epoxy resin and aluminum

K. NAKAMAE,* T. NISHINO and H. YOSHITANI

Department of Chemical Science and Engineering, Faculty of Engineering, Kobe University, 1-1 Rokkodai, Nada, Kobe 657, Japan

Abstract—The mechanical stress residing at the interface between a cured epoxy resin and aluminum was analyzed by the non-destructive X-ray diffraction method. The molecular mechanism by which residual stress is generated was further studied by means of attenuated total reflection Fourier-transform infrared spectroscopy, X-ray photoelectron spectroscopy, and contact angle measurements. It was revealed that the residual mechanical stress is attributed to the predominant adsorption of the curing agent from the epoxy formulation onto the surface of aluminum substrate through an acid–base interaction.

1. INTRODUCTION

The events that occur at the interface between a matrix and a filler are believed to play an important role in the total performance of composite materials. Wetting and adhesion have been often discussed in relation to interfacial stress transmission, which may have great influence on the mechanical properties of composites. Residual mechanical stress and interfacial erosion are also implicated in loss of the mechanical properties and the stability of composite materials. The residual stress is generally attributed to the difference in the heat expansion coefficient between the matrix and the filler. However, only limited knowledge is available for understanding the molecular mechanisms by which residual stress is generated. Further investigations on the mechanism will enable us to develop strategies for reducing unfavorable residual stress.

In this study, we focus on residual mechanical stress that resides at the filler/matrix interfaces. The composite comprising an epoxy resin and an aluminum (Al) substrate was employed as a typical model of composite materials. The aluminum-filled epoxy system is widely used in industries, for example, as a protective coating for metals and semiconductors in electronics and as a thermosetting adhesive in aerospace technology, because of the excellent electrical and mechanical properties of an epoxy resin. The problem of residual stress, however, remains unsolved. When epoxy resin is

*To whom correspondence should be addressed.

cured against a substrate at a high temperature and cooled down to room temperature, residual stress is expected to arise because of substantial difference in thermal expansion coefficients of the resin and adherend [1–5]. The residual stress may reduce interfacial adhesion strength and eventually induce cracking in the composites.

In this study, residual mechanical stress was analyzed by means of the non-destructive X-ray diffraction method. To elucidate the mechanism by which residual stress is generated, the interface was studied by independent adsorption experiments of an epoxy resin and a curing agent. In addition, the interfacial structure was analyzed by attenuated total reflection Fourier-transform infrared spectroscopy (FTIR-ATR), X-ray photoelectron spectroscopy (XPS), and contact angle measurements. From these analyses, we attempt to explain the molecular events which undergo at the interface, producing residual mechanical stress, namely in terms of preferential localization of the curing agent at the interface.

2. EXPERIMENTAL

2.1. *Materials*

The liquid epoxy resin, a diglycidyl ether of bisphenol A (Epikote 828), was the product of Shell Chemical Co., USA. The resin had a number average molecular weight of 380 and an epoxy value of 190 ± 5 g/equivalent. An aromatic diamine, 4,4′-diaminodiphenyl-methane (DDM) used as a curing agent was purified by recrystallization from methanol solution. The molecular formulae of the curing agent and the resin are given in Fig. 1. The stoichiometric molar ratio of the epoxy resin to DDM was 2 : 1.

A plate and film made of Al with a purity of 99.9% were used as adherends. The surfaces of the adherends were polished using an abrasive and then cleaned ultrasonically in acetone for 5 min. After being washed with deionized water, the polished surfaces were treated with chromic acid, washed with deionized water, and then heated at 300°C for 3 h under nitrogen (N_2) atmosphere to remove water thoroughly. Surface roughness of the adherends, measured using a surface roughness meter, was shown to be less than 0.2 μm of peak-to-valley distance. Surface analysis of the Al plate and film with XPS indicated that the Al $2p$ spectra were resolved into a doublet, consisting of peaks at binding energies of 75 and 73 eV. These correspond to the oxidized form of aluminum, —Al—O—, and pure Al, respectively. A small peak was detected in the region of the C $1s$ core level spectrum, probably due to contamination of the surfaces.

Aluminum oxide (Al_2O_3) particles were used as a substrate in the adsorption experiment mentioned below.

2.2. *Curing of epoxy resin*

After mixing with DDM at 100°C, the epoxy resin containing 0 or 15 wt% of Al particles was immediately poured onto Al plates. The resin was subjected to precuring

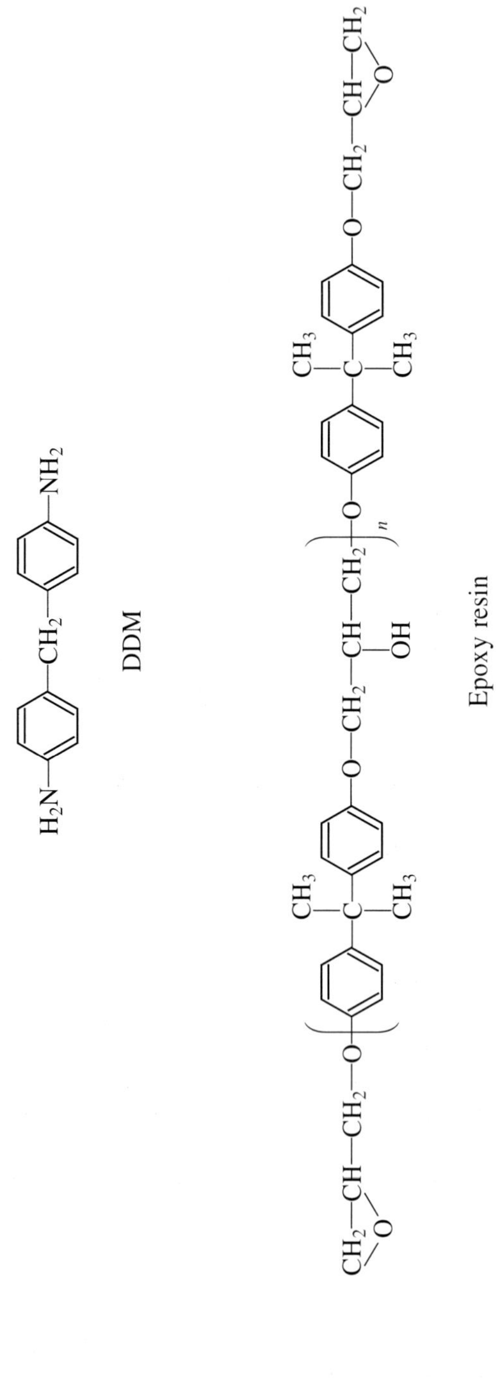

Figure 1. Molecular formulae of DDM and epoxy resin.

at 80°C for 2 h and subsequently temperature was elevated to 180°C and maintained for 6 h. After complete curing of the resin, the specimen composed of Al plate and the cured epoxy resin was cooled down gradually to room temperature and was kept in a desiccator for more than 24 h.

2.3. X-ray diffraction measurements

X-ray diffraction patterns of the Al plates bonded to cured epoxy resin matrix were recorded using an X-ray diffractometer (Rigaku Denki, RAD-B system; X-ray source: Cu K_α radiation; divergence slit width: 0.5°) operated at a tube voltage of 40 kV and a tube current of 20 mA. A lattice spacing of the (422) plane of Al crystals was determined from the shift of the diffraction angle and compared to the intrinsic spacing for the plane ($2\theta = 137.4°$). The temperature of specimens was monitored by attaching a thermocouple to the surface of specimen and the lattice spacing was converted to a value at 23°C. The accuracy of a peak shift measured is evaluated ordinally to be less than 0.003° in the angle 2θ.

2.4. Interface analyses

In order to analyze the interface between a cured epoxy resin and an Al substrate, some of the specimens were immersed in 0.1 M aqueous HCl solution to dissolve the Al adherend. The residual epoxy resin surface which had been in contact with the adherend was analyzed by XPS (Kratos XSAM800) using an X-ray source of Mg K_α radiation. The binding energy measured was referenced by setting the C—H peak maximum in a C 1*s* spectrum to 285.0 eV. The take-off angle was fixed at 0° or 60°. The pressure in the XPS chamber and the power output of the X-ray tube were at 5×10^{-6} Pa and 150 W, respectively.

Dynamic contact angle measurements on the surface of cured epoxy resin were carried out using a water droplet; the detailed procedure has been described elsewhere [6]. Equation (1) was used to determine the contact angle θ from the advancing contact angle θ_a and the receding contact angle θ_r:

$$\cos\theta = \frac{\cos\theta_a + \cos\theta_r}{2}. \quad (1)$$

Surface free energy was determined according to the Owens and Wendt equation [7] using contact angles of water and methylene iodide.

Interfacial tension measurements were performed with a duNoüy-type Shimadzu Interfacial Tension Meter.

2.5. Adsorption experiments

Competitive adsorption experiments of epoxy resin and DDM to Al_2O_3 particles were carried out from xylene solution. Xylene is a good solvent for both the epoxy resin and DDM. The solution was prepared by dissolving 1 g of the epoxy resin/DDM mixture at a stoichiometric molar ratio in 100 ml of xylene and then Al_2O_3 particles were added to the solution. After reaching equilibrium, adsorption of each component was assessed from a decrease in the bulk concentration which was determined from the infrared (IR) spectra of supernatant (Shimadzu FTIR-4200 spectrophotometer, Kyoto, Japan), by monitoring the absorbance peaks assigned to the each component, i.e. the C—O—C peak at 1040 cm^{-1} of the epoxy resin and the C—N peak at 1620 cm^{-1} of DDM.

3. RESULTS AND DISCUSSION

3.1. Residual stress

In order to estimate residual stress, we attempted to determine strain resident at the interface between cured epoxy resin and the Al adherend, using the non-destructive X-ray diffraction method.

The Al crystal is a cubic system ($a = 4.0597$ Å at 23°C) with a thermal expansion coefficient (2.386×10^{-5} K^{-1}) much smaller than that of the epoxy resin (6.698×10^{-5} K^{-1}). When an epoxy resin is cured at high temperature and cooled down to room temperature, Al crystals of the adherend will be subjected to a residual stress because an epoxy resin adheres strongly to Al, which causes strain in the Al crystals. The strain of Al crystal leads to a shift of diffraction angle 2θ, so the residual stress can be determined quantitatively [8–10]. The strain ε of Al crystals can be estimated by the use of following relation,

$$\varepsilon = \frac{\Delta d}{d_0}, \tag{2}$$

where d_0 denotes the initial lattice spacing and Δd is the change in the lattice spacing induced by residual stress.

Moreover, the information about the direction of residual stress can be obtained by employing the $\sin^2 \Psi$ method. Figure 2 schematically illustrates the method where the strain ε was measured at different inclination angles Ψ of an incident X-ray beam.

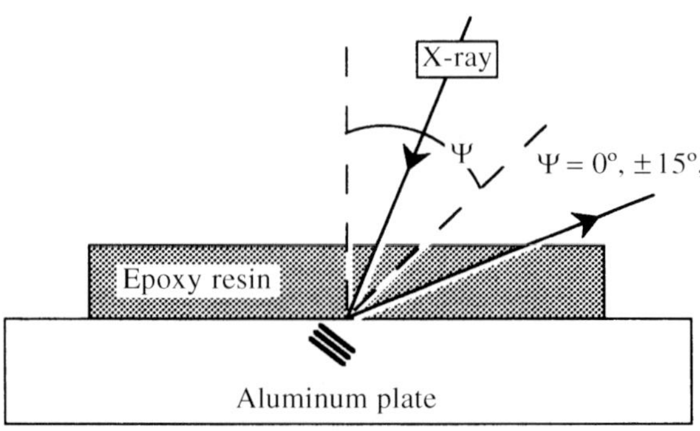

Figure 2. Schematic illustration for the measurement of residual stress by the non-destructive X-ray diffraction method.

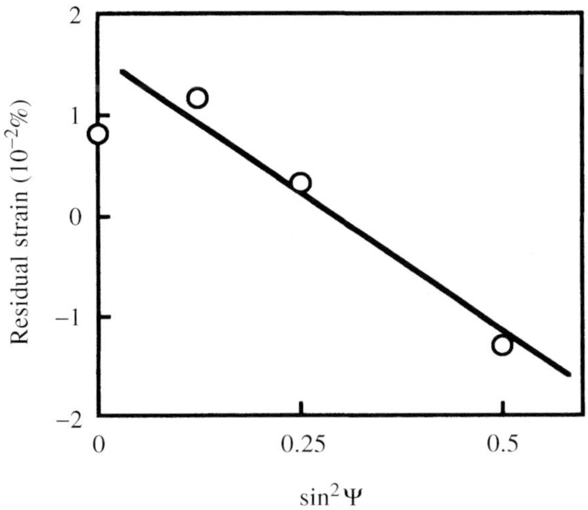

Figure 3. Relationship between strain of the Al plate and $\sin^2 \Psi$.

Uniaxial residual stress σ can be determined as a function of inclination angles Ψ, according to the following equation [10],

$$\varepsilon = \frac{(1+\nu)\sigma}{E} \sin^2 \Psi, \tag{3}$$

where E and ν are the elastic modulus (75.5 GPa) and the Poisson ratio (0.33) of the Al crystal, respectively.

In Fig. 3, the strain ε of Al crystal measured by the X-ray diffraction method is plotted against $\sin^2 \Psi$. It is clearly shown that residual strain was present in the Al substrate. The polarity of strain was found to depend on inclination angle Ψ; the Al substrate compressed ($\varepsilon < 0$) in the direction parallel to the surface, whereas expanded ($\varepsilon > 0$) in the direction perpendicular to the surface. From the slope of the straight line in Fig. 3, it is found that Al plate is subjected to a uniaxial compressive stress of 29.5 MPa parallel to the adherend surface. These findings may be explained in terms of the following phenomenon. When excellent adhesion is established between epoxy resin and Al substrate and the composite is cooled down from curing temperature to room temperature, the epoxy resin having a linear thermal expansion coefficient β_{ep} of 6.698×10^{-5} K^{-1} tends to shrink. Since β_{ep} of Al (2.386×10^{-5} K^{-1}) is much lower than that of the epoxy resin, the Al surface is compelled to contract, and a compressive stress is brought about at $\Psi = 90°$, while Poisson's deformation of Al crystals yields a strain $\varepsilon > 0$ at $\Psi = 0°$.

Figure 4 summarizes the strain residing in both phases. Details of the determination of residual stress and strain in an epoxy resin phase were given elsewhere [11]. It can be seen that the epoxy resin and the Al substrate are subjected to tensile and compressive residual stresses, respectively. This opposite tendency suggests that the residual stresses in the two phases obey the law of action and reaction. Experimental results, however, show that the absolute values are not the same in each phase. This

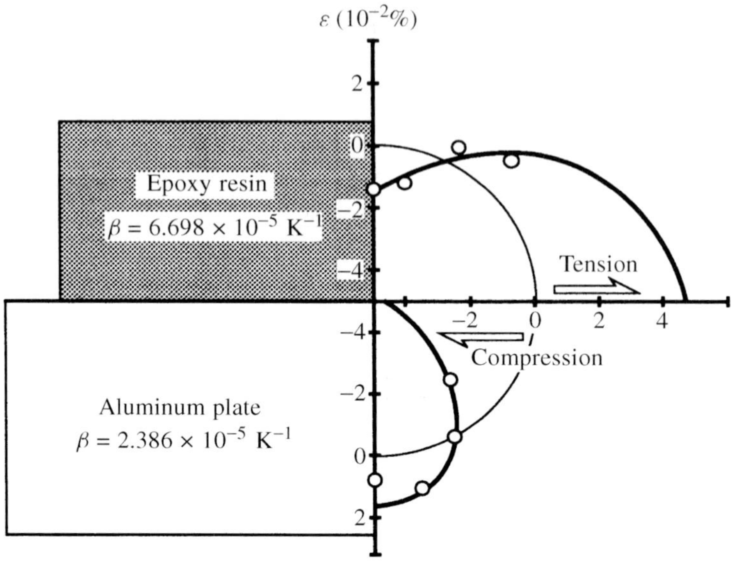

Figure 4. Strain ε residing in both the epoxy resin and the Al plate. β represents thermal expansion coefficient.

is probably due to the difference in the systems in that one contains the embedded Al particles in the epoxy resin and the other does not [11].

3.2. Interface analyses

As discussed in the previous section, residual stress distributes complementarily in each phase. This section will be devoted to discussing an interfacial molecular mechanism through which the difference in the thermal expansion coefficients generates residual stress at the epoxy resin/Al interface.

Figure 5 shows the results of competitive adsorption experiments. It is seen that adsorption of DDM onto the Al_2O_3 particles predominates over that of epoxy resin in the initial stage of the adsorption process. It seems that adsorption of the epoxy resin starts after the Al_2O_3 particles are covered with DDM.

IR spectra were also obtained from independent adsorption experiments in which each component was allowed to adsorb from a solution in xylene of each component separately. The absorbance ratios after 8 h are represented in Fig. 5 (dotted lines). After 3 h, the adsorption of DDM from the one-component solution was almost identical to that from the epoxy/DDM mixed system. However, the adsorption of epoxy resin after 8 h from the mixed system was approximately three times higher than that from the single component solution. This result signifies that epoxy resin did not adsorb directly onto the bare Al_2O_3 particles but onto the Al_2O_3 particles covered with DDM. These findings suggest that DDM has a stronger affinity for the Al surface than the epoxy resin.

Figure 6 shows the nitrogen content at the cured epoxy resin/Al interface, calculated from N 1*s* XPS spectra. Prior to the experiment, it was confirmed that the nitrogen

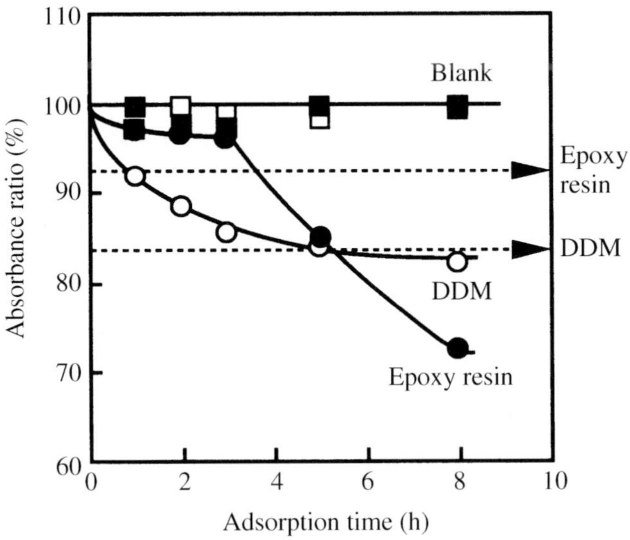

Figure 5. Adsorption of the epoxy resin and DDM from xylene solution on Al_2O_3 particles as a function of adsorption time; (●) epoxy resin; (○) DDM; (■) epoxy resin blank; (□) DDM blank. Dotted lines indicate the IR absorbance ratio obtained from the experiments in which each component was adsorbed from the single component solution.

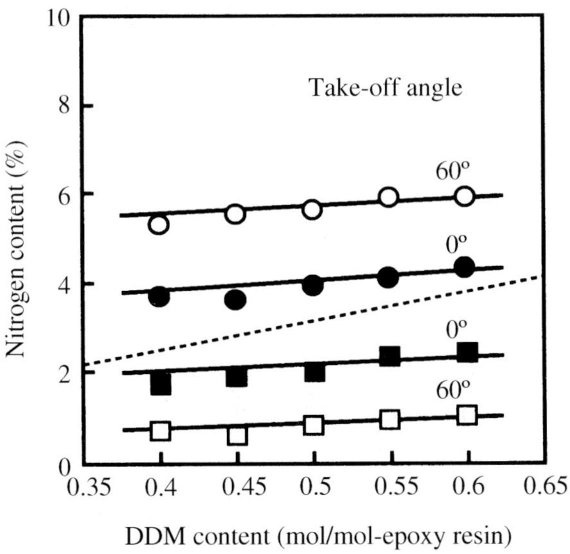

Figure 6. Nitrogen content at the interface for various epoxy resin/DDM ratios; (●, ○) epoxy resin/oxidized aluminum interface; (■, □) epoxy resin/air interface. Dotted line indicates N content in the bulk.

content of the cured epoxy resin/air interface did not change when the cured epoxy resin was immersed in HCl solution. As shown in Fig. 6, the nitrogen content did not change appreciably at the various DDM/epoxy resin ratios; however, there was clear evidence that the nitrogen content at the cured epoxy resin/Al interface was higher

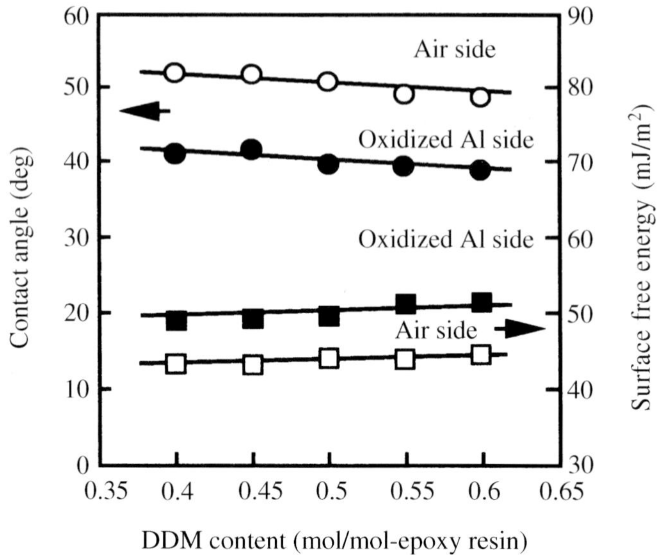

Figure 7. Water contact angle and surface free energy of the epoxy resins cured at various epoxy resin/DDM ratios; (●, ■) epoxy resin/oxidized aluminum interface; (○, □) epoxy resin/air interface.

than that at the cured epoxy resin/air interface and that in the bulk (dotted line). The same tendency was observed in the angle-dependent XPS analysis in which sampling depth was varied by changing the take-off angles (6 and 3 nm depths for take-off angles of 0° and 60°, respectively [12]). It was found that their nitrogen content was higher at the cured epoxy resin/Al interface compared to that far from the interface.

The results of contact angle and surface free energy measurements for various DDM/epoxy resin ratios are presented in Fig. 7. The θ_a and θ_r on the cured epoxy resin/air interface remained unchanged by immersing the cured epoxy resin in HCl solution. The contact angle of water and surface free energy were almost constant over various DDM/epoxy resin ratios. This result implies that the surface and interface structure was different from that expected from mixing ratios. It is clearly seen that the contact angle at the surface of cured epoxy resin which had been in contact with an Al plate was smaller than that at the cured epoxy resin/air interface. The surface free energy showed the opposite tendency. These observations are in good agreement with the results of XPS measurements that DDM units adsorb preferentially onto the Al surface, implying again that epoxy resin units localize at the interface between the cured epoxy resin and the air phase, whereas DDM units localize at the cured epoxy resin/Al interface. Thus, all the results obtained by FT-IR, XPS and contact angle measurements confirm that DDM interacts preferentially with the surface of Al adherend.

To investigate further the interactions between cured epoxy resin and Al, interfacial tension measurements were made between water and xylene solutions, and XPS analysis was carried out on thin cast films of DDM on Al and Al plates with adsorbed DDM. The interfacial tension results are given in Fig. 8. It is seen that the interfacial tension between water and xylene solution of epoxy resin is considerably lower

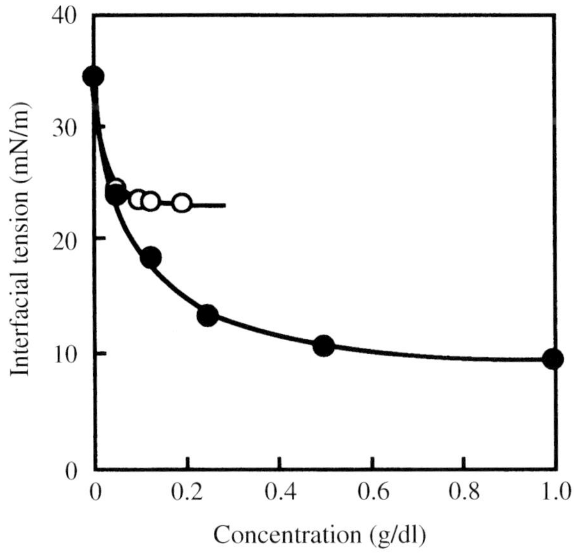

Figure 8. Interfacial tension measurements between water and xylene solutions of the epoxy resin (●) and DDM (○).

than that between water and xylene solution of DDM. The result is inconsistent with the effect expected from the consideration that DDM is more hydrophilic than the epoxy resin. This result could not explain the previous findings that DDM adsorbed preferentially onto the Al particles and that DDM units localized at the cured epoxy resin/Al interface. Generally, an oxidized Al surface is hydrophilic in nature. If the oxidized Al attracted any component of epoxy resin through a hydrophilic interaction, epoxy resin would localize at the interface. Accordingly, we propose the hypothesis that the interaction between DDM and oxidized Al is caused by factors other than a hydrophilic interaction.

The interfacial tension between water and epoxy resin or DDM solution was further measured as a function of pH (data not shown). It was found that the interfacial tension did not change appreciably at high pH. On the other hand, at low pH, while the interfacial tension between water and epoxy resin solution hardly changed, that between water and DDM solution could not be measured because DDM dissolved in the water phase. This phenomenon is interpreted as follows: ionization of DDM takes place under acidic conditions and ionized DDM has a strong interaction with acidic substances.

Figures 9 and 10 show the Al 2*p* and N 1*s* spectra, respectively, from the XPS analyses of (A) a DDM thin film cast on the surface of an Al substrate, and (B) DDM adsorbed onto an Al substrate from xylene solution. The Al 2*p* spectrum could not be measured on the cast film, indicating that the film thickness was larger than the sampling depth in the XPS analysis. In this case, the N 1*s* spectrum was assigned to the nitrogen involved in the bulk phase. On the contrary, for the adsorbed film, a clear Al 2*p* peak could be detected; i.e. the adsorbed film was thinner than the sampling depth. In this case, the N 1*s* spectrum includes information from the DDM/Al

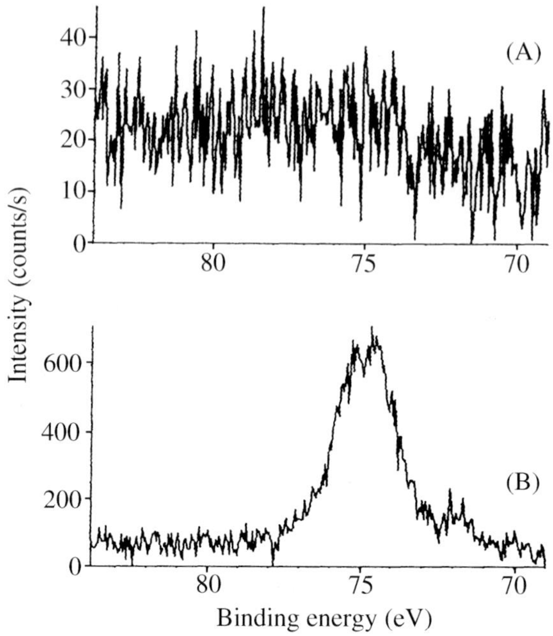

Figure 9. Al 2*p* spectra of the oxidized Al surfaces (A) covered with a DDM cast film and (B) with adsorbed DDM.

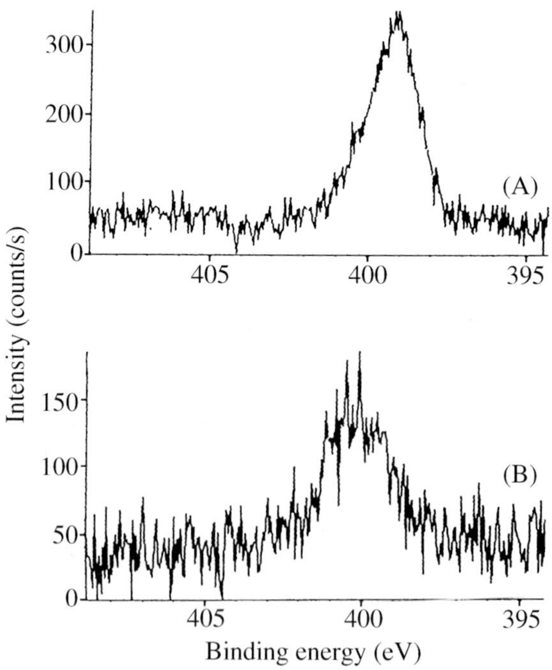

Figure 10. N 1*s* spectra of the oxidized Al surfaces (A) covered with a DDM cast film and (B) with adsorbed DDM.

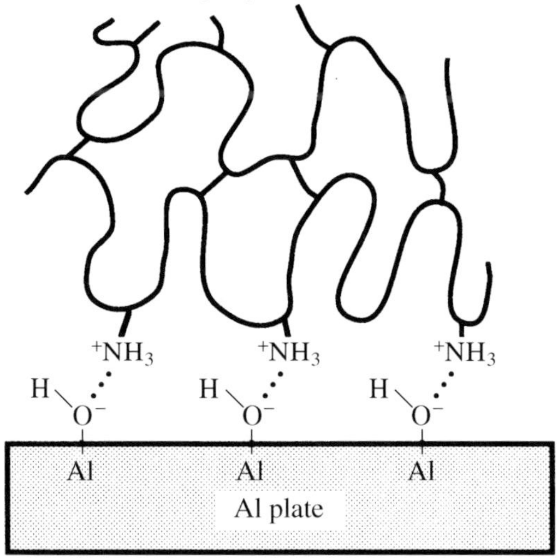

Figure 11. Schematic representation of the structure of the cured epoxy resin/oxidized Al interface.

interface. The most striking result seen in Fig. 10 is that the N 1*s* peak from the adsorbed film shifted to a higher binding energy, suggesting the ionization of the DDM adsorbed on the Al surface. Generally, the isoelectric point (IEPS) of an oxidized Al surface ranges from 7.5 to 9.1, implying that it has both acidic and basic characters. DDM has a strongly basic character, while the acid–base character of the epoxy resin is plausibly weaker than that of DDM. It is probable that the oxidized Al surface has many anionic sites and only a few cationic sites in the presence of a basic substance like DDM. Accordingly, electron exchange between the oxidized Al surface and DDM can easily occur, promoting acid–base interactions between the amino groups of DDM and acidic sites on the oxidized Al. The most likely form of the interaction is $H_3N^+ \cdots {}^-O{-}Al$. The involvement of acid–base interaction in adhesion has been discussed in the literature [13].

Figure 11 summarizes schematically the structure of the cured epoxy resin/Al interface, deduced from the findings in this study. When the epoxy resin is cured with DDM with the application of heat, molecules of both DDM and epoxy resin are able to diffuse freely in the system. DDM molecules preferentially adsorb onto the Al surface and form acid–base interactions between the amino groups in DDM and the acidic sites on the Al surface. These findings are in agreement with a previous study of Affrossman *et al.* [14]. It is likely that this interaction leads to tenacious adhesion at the interface, eventually producing mechanical residual stress, at the interface.

REFERENCES

1. M. Shinbo, M. Ochi and Y. Shigeta, *J. Appl. Polym. Sci.* **26**, 2265 (1981).
2. N. J. DeLollis and O. Mantoya, *Appl. Polym. Symp.* **19**, 417 (1972).
3. M. Shinbo, M. Ochi and K. Arai, *J. Coating Technol.* **56**, 45 (1984).

4. E. Alftham, A. DePwo and M. Rigdahl, *Intl. J. Polym. Mater.* **7**, 163 (1979).
5. Y. Inoue and Y. Kobatake, *Kolloid-Z.* **159**, 18 (1958).
6. T. Matsumoto and K. Kashiwa, *Polym. Sci. (Jpn)* **17**, 590 (1960).
7. D. K. Owens and R. C. Wendt, *J. Appl. Polym. Sci.* **13**, 1741 (1969).
8. H. P. Cheskis and R. W. Heckel, *Met. Trans.* **1**, 1931 (1970).
9. P. Predecki and C. S. Barrett, *J. Composite Mater.* **13**, 61 (1979).
10. B. D. Cullity, in: *Elements of X-ray Diffraction*, 2nd edn. Ch. 16. Addison-Wesley, Reading, MA (1980).
11. K. Nakamae, T. Nishino and X. Airu, *J. Appl. Polym. Sci.* **40**, 2231 (1990).
12. M. P. Seah and W. A. Dench, *Surf. Interf. Anal.* **1**, 2 (1979).
13. K. L. Mittal and H. R. Anderson, Jr. (Eds), *Acid–Base Interaction: Relevance to Adhesion Science and Technology*. VSP, Zeist, The Netherlands (1991).
14. S. Affrossman, J. M. R. MacAllister, R. A. Pethrick, B. Thomson, N. W. D. Brown and B. J. Meenan, *Polym. Surf. Interf.* **1**, 107 (1987).

Mittal Festschrift, pp. 557–571
W. J. Van Ooij and H. R. Anderson, Jr. (Eds)

Effects of surface passivation treatments of stainless steels on anaerobic cure activity: Part III, surface characterization by XPS

D. BILLY YANG*

Surface Science Laboratory, Loctite Corporation, 1001 Trout Brook Crossing, Rocky Hill, CT 06067, USA

Abstract—X-ray photoelectron spectroscopy (XPS) has been used to investigate the correlation between composition of 316L stainless steel (SS316L) surface films formed by different surface treatments and their catalytic effects on surface activated anaerobic cure. The four surface treatments studied were: (1) machined only, (2) machined and buffed, (3) machined and passivated, and (4) machined, buffed and passivated. The passivation was performed by immersion of SS in HNO_3 and EDTA solutions. It was found that the surface activity for anaerobic cure using Loctite 642 adhesive was significantly affected by the various treatments. The completely processed passivated 316L substrate, which includes machining, buffing and nitric acid/EDTA treatments, provided the greatest surface reactivity over the other treatments. The reason for this increase in surface activity for anaerobic cure appears to be due to alteration of surface composition. The results of XPS analysis indicate that layers of iron or chromium hydroxide with an overlayer of oxides were formed after machining. This metal hydroxide formation was due to metal contact with the cutting fluid, which contains the strong base ethanolamine. These iron/chromium hydroxide top layers were then partially converted to iron/chromium oxides after air exposure. The buffing apparently roughened the surface and removed minute iron particles and scale, but did not remove the oxide/hydroxide layers. The passivation process by HNO_3/EDTA, however, causes the surface metal to form a well-defined new phase by reducing the thickness of the topmost oxide layers and enriching the top layer with metal hydroxide and/or oxyhydroxide. The new phases are more active in catalyzing the anaerobic cure than the original existing metal oxides. Surface reactivity is interpreted on the basis of the catalytic activity of metal hydroxides or oxyhydroxides for anaerobic polymerization. The correlation between the composition and chemical state of the passive film and cure mechanism is discussed.

Keywords: Anaerobic adhesive; XPS; surface activation; passivation; nitric acid; stainless steel; ethanolamine; cutting fluid; iron oxide; iron hydroxide; iron oxyhydroxide.

1. INTRODUCTION

In the previous two papers of this series of anaerobic cure chemistry studies, the effects of transition metals, either by direct addition [1] or by contact with metal sur-

*To whom correspondence should be addressed. Current address: Avery Dennison Corporate Research Center, 2900 Bradley Street, Pasadena, CA 91107-1599, USA.

face [2], on cure kinetics have been investigated using a real time FT-IR (RT/FT-IR) technique [3]. Other studies to emphasize the effect of ligand type attached to the transition metal [4] as well as the synergistic effect of the co-initiators on anaerobic cure kinetics have also been reported [5]. Recently, 316L stainless steel (SS316L) was selected as the construction material of dispenser valves for our anaerobic adhesive automatic dispensing system. However, it was found that the surface reactivity of SS316L increases after a passivation treatment with HNO_3 and tetrasodium ethylenediaminetetraacetic acid (Na_4-EDTA). The increase of surface activity of the SS316L caused the adhesive to cure prematurely leading to the plugging of the dispenser valve. The purpose of passivating the SS316L is to improve the corrosion resistance and make the surface inert. However, contrary to expectations, the opposite effect was observed. Complete inertness of the metal used for the dispensing valve is essential to prevent unwanted instability of the adhesive.

Regarding metal passivation, there has been a strong interest in iron-containing alloys for many years. Several studies have been devoted to the development of the passivation process [6–9]. In order to investigate the properties of the passive layer, XPS studies have been carried out to investigate the composition, chemical structure and bonding state of the passive layer formed by different passivation treatments. The reagents used for stainless steel (SS) passivation include: HNO_3 [10–14], HCl [15–17], H_2SO_4 [18–22], H_3PO_4/H_2SO_4 [23], potassium or sodium hydroxides [24–26] and water [27, 28]. XPS studies of pure iron and chromium oxide, hydroxide, and related compounds have also been reported [29–33]. However, no study has been carried out to correlate the surface properties of the passive film with surface activated anaerobic polymerization. Presented here is an investigation of why surface composition or chemical state change, induced by HNO_3 passivation and other treatments, affects the cure of the anaerobic adhesive.

2. EXPERIMENTAL

2.1. Adhesive

A commercially available anaerobic adhesive (Loctite 642) was used. It contains a polyurethane methacrylate resin (*c.* 50%), methacrylic monomer (*c.* 40%), acrylic acid (*c.* 4%), cumene hydroperoxide (*c.* 2%), an amine (*c.* 1%), a saccharin accelerator (*c.* 1%) and stabilizer (*c.* 1%). Primers were not used in any of the anaerobic cure tests.

2.2. Surface preparations and treatments for stainless steel valve metal

The surface activity test for valve metal used for the present study involved 316L (low carbon) stainless steel with various surface treatments. The stainless steel test coupons were fabricated as a round disc 25 mm in diameter and 10 mm in thickness. All the stainless steel parts had previously been exposed to cutting oil (WS 5050, ITW Fluid Products Group, St. Louis, MO) during the fabrication process before buffing and passivation treatments. This cutting oil contains 4 to 5% of ethanolamine in a mixture of petroleum distillate and water. Buffing treatment refers to the exposure of the parts to the 220 grit silicon carbide impregnated in Nylon (Beartek, Norton Corp., Worcester, MA) and Scotch Brite® (3M, St. Paul, MN) buffing wheel to remove any loose particles or scales before passivation treatment was applied. The complete surface treatment procedures were as follows: (1) The parts were first degreased with perchloroethylene, rinsed with tap water, passivated in 45% nitric acid in water for 45 min at 49°C and rinsed again with tap water. (2) The final treatment included a distilled water rinse, a 4 h immersion at room temperature in 15% water in methanol mixture containing 2.6 wt% Na_4-EDTA, another rinse with distilled water and oven drying. Passivation of the stainless steel surface by nitric acid was performed by Harconn Chrome, Inc. (West Hartford, CT). The Na_4-EDTA treatment was performed in our laboratory. Stainless steel coupons with the following surface conditions were used for our studies: (A) as-received machined sample, no buffing or passivation; (B) no buffing but with passivation; (C) machined, followed by passivation; and (D) machined, buffed, followed by passivation (complete processing). All SS316L samples were cleaned by methanol and dichloroethane before use.

2.3. X-ray photoelectron spectroscopy (XPS) analyses

The XPS analyses were performed by Evans East Material Characterization Laboratory (Plainsboro, NJ, USA) using a Perkin–Elmer Model 5000LS photoelectron spectrometer with a monochromatic Al K_α X-ray radiation operated at a power of 400 W. This instrument was equipped with an ion pump and a titanium sublimation pump which was capable of a 120 l/s pumping rate. All measurements were made at an electron take-off angle of 45°. Conditions for the measurements were: pressure $< 1 \times 10^{-8}$ Torr, X-ray pass energy for survey and high resolution spectra: 178.95 and 35.75 eV, respectively, total number of scans: 100, and scanning rate: 1 eV/s. Time for the entire energy range was about 8–10 min. All spectra were corrected for shifting using the C 1*s* line at 284.6 eV as the reference. The atomic composition of each constituent was calculated based on the peak area ratio for those samples having relatively prominent elemental content on the surface. Sensitivity factors for peak area integration (Fe = 2.957, Cr = 2.427, Ni = 4.044, O = 0.711, N = 0.477 and C 1*s* = 0.296), obtained from the same spectrometer on other samples were used in the calculation of the compositions.

3. RESULTS AND DISCUSSION

3.1. Analysis of XPS survey spectra

The chemical structure of the films formed on the SS316L in different surface treatments was evaluated by XPS analysis. In general, the observed spectra consisted of five major peaks of carbon, oxygen, iron, chromium and nitrogen. The results of the passivation slightly increases the intensity of the chromium and nitrogen peaks, but

Table 1.
Atomic concentration (%) of elements detected on treated 316 SS samples, surface cleaned by MeOH and CH_2Cl_2

Sample	O	N	C	S	Si	Fe	Cr	Mn	Mo	Cr/Fe
Machined only	37.1	1.5	46	0	0.8	9.5	3.6	2	0.2	0.38
Machined + buffed	39.7	1.9	44.5	—	—	10.4	2.2	1.1	0.2	0.21
Machined + passivated	40.3	1.8	47	—	—	5.3	4.3	1	0.4	0.81
Machined + buffed + passivated	37.4	2.4	47.8	0.5	—	5.7	4.8	1.2	0.2	0.84

decreases the iron peak indicating the passivated film was enriched in chromium and nitrogen. This observation was supported by surface atomic concentration, as shown in Table 1, calculated from integrated peak areas of four tested SS substrates. Thus, the Cr and N content increased from 2.2 and 1.9 to 4.8 and 2.4%, respectively, and the Fe content decreased from 10.4 to 5.7% as indicated by the increase of Cr/Fe ratio from 0.21 to 0.82 for the surface of the machined/buffed substrate, without and with passivation, respectively. Similar results were also obtained for the machined-only and machined/passivated substrates. Thus, the Cr and N content increased from 3.6 and 1.5 to 4.3 and 1.8%, respectively, and the Fe content decreased from 9.5 to 5.3% for machined-only substrates, without and with passivation, respectively. The source of the trace amount of N, S and Cl of the unpassivated substrates appears to be due to their contact with the cutting fluid, which contains strong base, ethanolamine, and chlorinated and sulfurized mineral oils. The cutting fluid was used to reduce the surface temperature during machining as well as to interact with the cutting surface for corrosion inhibition and lubrication purposes. For passivated surfaces, the observed increase in N content could be due to their contact with the ethanolamine and sodium-EDTA. For detailed chemical structure, the oxidation state and nature of the bonding of the surface film constituents were analyzed by high resolution XPS for the elements of O 1*s*, Fe 2*p*3/2, N 1*s* and C 1*s* and the corresponding results are shown in Figs 1–4. For easy comparison, each figure represents the stacked plots obtained from four SS substrates with different treatments.

Before the XPS spectra are interpreted, some characteristic binding energy (BE) assignments of iron and chromium compounds should be mentioned. The BE assignments of the iron oxides (FeO, Fe_2O_3, Fe_3O_4), hydroxides ($Fe(OH)_2$ and $Fe(OH)_3$) and oxyhydroxide (FeOOH) have been reported by McIntyre and colleagues [31, 32], Asami and Hashimoto [29] and Allen *et al.* [33]. The O 1*s* peak of the metal–O bond for FeO, $-Fe_2O_3$ and $-FeO_3$, is located at 530.0, 529.8 and 530.0 eV, respectively. For –FeOOH, it is expected that the O 1*s* spectrum should contain approximately equal contributions from two types of oxygen, hydroxyl (OH bond) and oxide (OM bond) oxygen, which should be located at 530.3 and 531.4 eV, respectively, as expected from stoichiometry. The single O 1*s* peak of metal–O–H bond for $Fe(OH)_2$ and $Cr(OH)_3$ is known to appear at high BE side at 531.3 and 531.2 eV, respectively [29, 31].

3.2. Analysis of O 1s spectra

Analysis of the O 1*s* spectra, shown in Fig. 1, indicated that the surface film of the four SS substrates contained several different types of oxygen depending on the type of surface treatment. The O 1*s* spectra obtained from two unpassivated films are significantly different from the spectra of the two passivated films. The spectra of unpassivated samples show two major peaks centered at 529.6 and 531.5 eV that are indicative of the oxygen in metal–oxygen (OM) bond found in metal oxide and in metal—O—H (OH) type bonds found in metal hydroxide, oxyhydroxide or oxygen with adsorbed water molecules. The O 1*s* spectra of two passivated samples show only one major peak centered at 531.4 eV and a weak broad shoulder peak at 529.5 eV. As indicated in Fig. 1, the major O 1*s* peaks of the two unpassivated samples have similar intensities which indicates that the film thickness attributed to OM and OH type oxygen layers is very similar. Also, the relative O 1*s* peak intensity of the machined-only sample is similar to the machined-plus-buffed sample. This indicates that buffing treatment does not change the surface composition. The buffing treatment seems only to roughen the surface, as well as removing the scale or debris on the surface as indicated in a previous scanning electron microscopic (SEM) investigation [2].

After passivation by HNO_3/EDTA, a large portion of the OM peak at the low BE side decreases and becomes a shoulder on the OH peak. Thus, after passivation, the OM oxygen was reduced significantly and the OH oxygen still remained at its original level. These results indicate that the passivation reduces the oxide film by nearly one-half while leaving the metal hydroxide film intact. These data also indicate that the surface film formed after machine fabrication has a two-layer structure. The outer

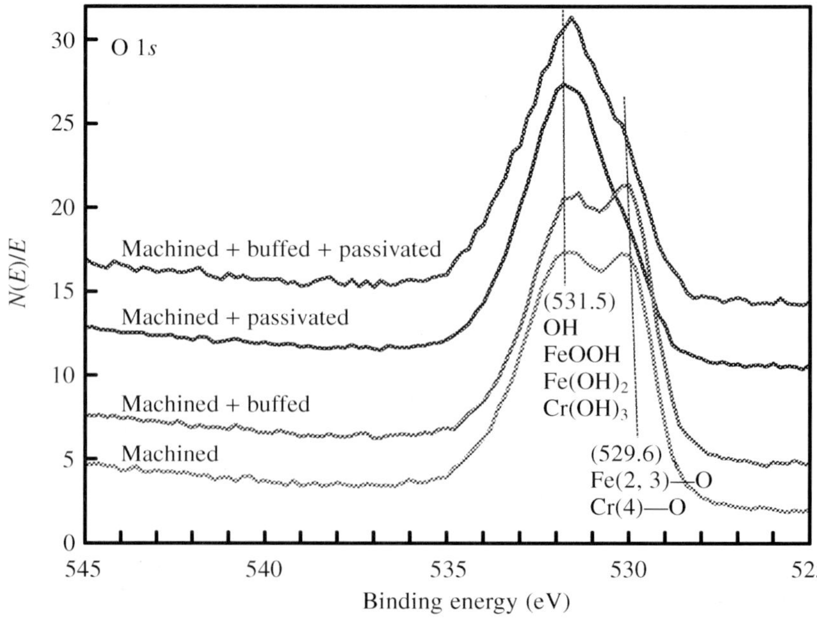

Figure 1. High resolution O 1*s* spectra of surface treated 316L stainless steel.

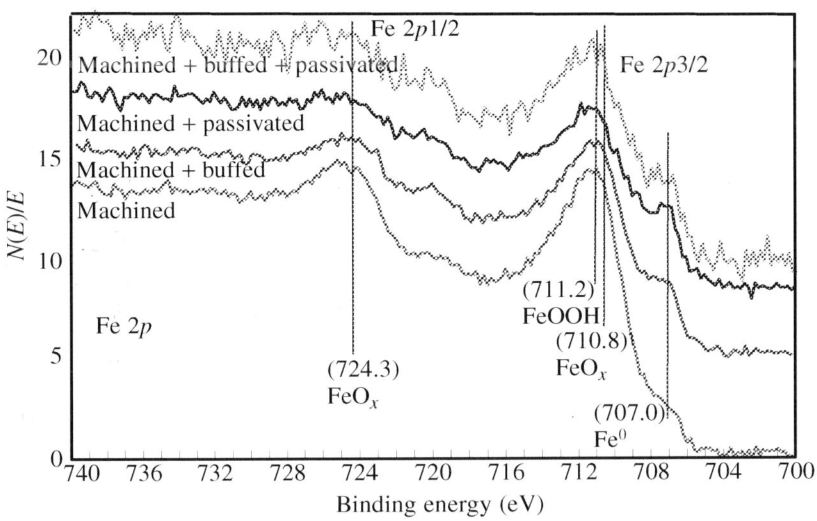

Figure 2. High resolution Fe $2p$ spectra of surface treated 316L stainless steel.

layer contains iron and chromium oxide and the inner layer contains hydroxide and/or oxyhydroxide. This outer film appears to be relatively inert, dense and non-porous and, therefore, immersion in HNO_3 for 45 min at 49°C reduces only the topmost oxide layer, while the lower hydroxide and/or oxyhydroxide remains unetched. Also, the spectra obtained from the passivated films show that the inner hydroxide layer contains some FeOOH and CrOOH. The O $1s$ line for the OM and OH bonds for —FeOOH should be located at 530.3 and 531.4 eV, respectively. These two kinds of oxygen could not be easily separated from the major O $1s$ of iron or chromium hydroxide spectra. Thus, the OH oxygen peak of oxyhydroxide may be buried underneath the major $Fe(OH)_2$ or $Cr(OH)_3$ peak near 531.5 eV and the OM oxygen peak of the remaining oxide at 529.6 eV as indicated in Fig. 3.

3.3. Analysis of Fe 2p spectra

It has been reported that the Fe $2p$ spectra of —Fe_2O_3, —FeOOH and —FeOOH are almost the same except for some differences in peak position and spin-orbit splitting [29, 31]. The BEs of Fe $2p3/2$ peak for metallic iron, FeO, —Fe_2O_3, —FeOOH and —FeOOH are known to appear at 706.82, 709.4, 710.97, 711.44, and 711.60, respectively. Figure 2 shows the comparison of four Fe $2p$ spectra obtained from four tested SS surfaces. Based on these XPS spectra, it may be concluded that the iron of the surface film is mainly present in three states: Fe^{3+} and Fe^{2+} (oxides and hydroxide forms) and metallic Fe^0. The most interesting result of these stacked plot spectra is the observation of a decreased intensity of iron oxide and/or hydroxide peaks at 710.8 and 711.2 eV and an increase in metallic iron peak at 707 eV on the surface upon the HNO_3 passivation treatment. Brundle *et al.* [30] reported that the broad satellite centered at 719.8 eV is characteristic of an Fe^{3+} species, independent of the oxides and the coordination number. The broad satellite at 715 eV is characteristic

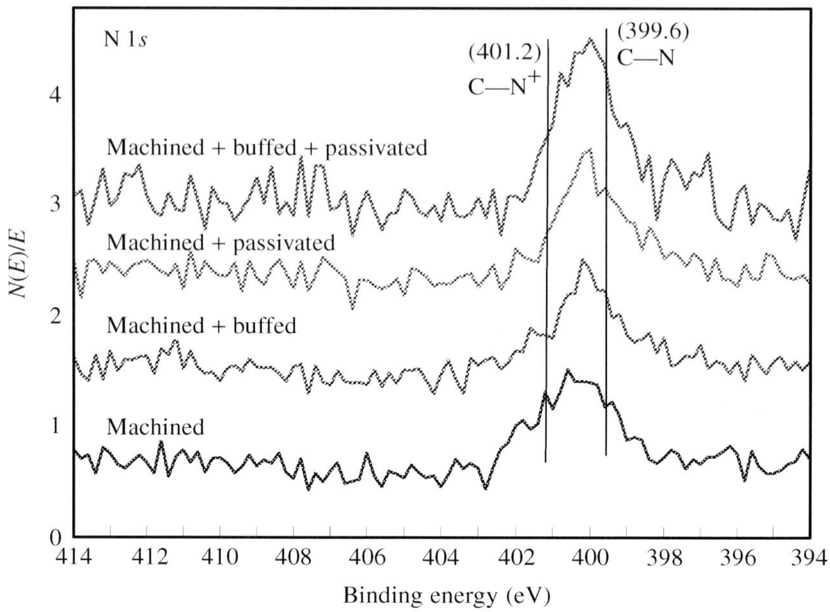

Figure 3. High resolution N 1*s* spectra of surface treated 316L stainless steel.

of an Fe^{2+} species. The combined presence of the main $2p3/2$ line and its satellite and the $2p1/2$ lines provide a reliable identification of Fe^{3+} and Fe^{2+} species, even in mixtures. Following the HNO_3/EDTA passivation treatment, the intensity of Fe^0 peak increases, but the FeO_x and/or $Fe(OH)_x$ peaks decrease, as shown in Fig. 2. This result strongly indicates that the outer oxide layers were removed exposing more metallic iron to the X-ray sampling depth. It should be noted that both $Fe(OH)_2$ and $Fe(OH)_3$ are soluble in acids [33, 34]. Since the treatment of the machined SS surface with the HNO_3 did not completely remove the iron or chromium hydroxide, it may be concluded that these hydroxide layers are still buried underneath other protective iron oxide thin layers.

3.4. Analysis of N 1s spectra

Figure 3 shows the comparison of four N 1*s* spectra obtained from these SS samples. A nitrogen peak from tertiary amine at 399.6 eV is detected for all four samples. Nitrogen from quaternary amine at 401.7 eV is also clearly detected for the machined and machined/buffed treated surfaces, indicating it may be due to contact with the ethanolamine type compound in the cutting fluid. The passivation treatment removed this type of nitrogen.

3.5. Analysis of C 1s spectra

Figure 4 shows the comparison of C 1*s* spectra obtained from four samples. Overall, the relative intensity of the C 1*s* peak for four tested samples shows insignificant differences except that the organic carboxylate containing carbon slightly increases

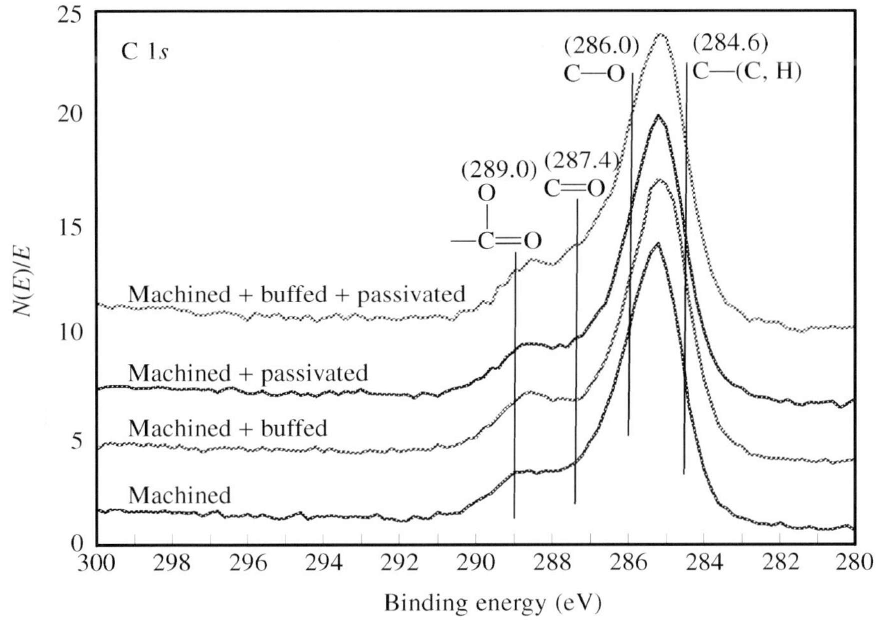

Figure 4. High resolution C 1*s* spectra of surface treated 316L stainless steel.

upon the passivation treatment. The source of the carboxylate appears to be from the acetate groups of the EDTA. This result seems to indicate that the Fe or Cr-EDTA type of compound, which cannot be removed by MeOH or methylene chloride wash, deposits on the surface and becomes one of the components in the passive film. This interpretation is also supported by the observed increase in nitrogen content after passivation as shown in Table 1. Since there is no significant change in overall hydrocarbon carbon, it seems to indicate that most of the carbon is adventitious carbon formed during the exposure to atmospheric air after solvent cleaning. It should be noted that we have also investigated the effectiveness of the cleaning solvents. If the SS substrates were cleaned by methanol only, the atomic concentration of carbonaceous carbon on these four samples can be as high as between 54.9 to 85%, as compared to between 44.5 to 47.8% for surfaces cleaned by both methanol and dichloroethane. Therefore, all four tested SS samples reported here were cleaned by both methanol and methylene chloride before the XPS analysis.

3.6. Analysis of Cr 2p spectra

The high resolution XPS spectra for Cr $2p$ spectra were not collected. However, based on a direct comparison of the Cr $2p$ peaks obtained from the survey spectra, detectable differences in relative intensity and chemical shift before and after passivation can be observed. Figure 5 shows the expanded four Cr $2p$ survey spectra indicating that before the passivation the chromium is mainly present in a Cr^{3+} (Cr_2O_3) state. After passivation, the shape of the Cr $2p3/2$ peak is not remarkably different but its intensity

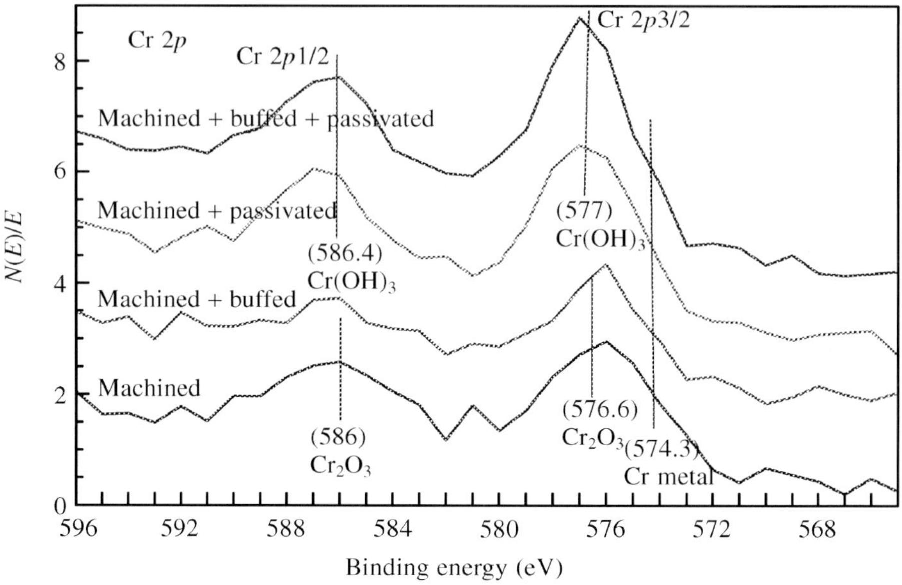

Figure 5. Expanded Cr 2p survey spectra of surface treated 316L stainless steel.

increases and BE shifts considerably indicating that the passivation depleted the iron oxide layer and caused the remaining layer to become enriched in Cr^{3+}. The BE shifts from 576 eV, mainly representative of Cr_2O_3, to 577 eV which is representative of $Cr(OH)_3$ [29]. However, because this is not a high resolution spectrum, this assignment is not very reliable. Also, it should be mentioned that the O 1s peaks for Cr_2O_3, $Cr(OH)_3$ and CrO_3, located at 529.98, 530.85 and 530.01 eV, respectively, may also co-exist with iron oxide and hydroxide peaks shown in Fig. 1.

3.7. Formation of metal hydroxide and oxyhydroxide

Barr [35, 36] reported that when an elemental metal was exposed to air at STP, natural passivation occurred by which most elemental metals formed a relatively regular, thin film of oxides through progressive oxidation and terminated by a near monolayer film of hydroxides. The formation of a hydroxide and/or carbonate layer is a continuous process, forming a very thin film that is constantly forced to the outer surface by any remaining oxygen with the ability to produce the generally more stable, oxide materials. The greater presence of oxides on most typical air-exposed elemental metal systems is apparently a result of the enhanced lattice forming ability of the oxide. The $Fe(OH)_2$ is also formed upon contact of a strong base with Fe^{2+} in the solution [34]. Based on our XPS data, the surface of the unpassivated surface was dominated with almost equal amounts of metal oxide and hydroxide and/or oxyhydroxide layers. We believe that the formation of such a passivated layer is due to the SS surface in contact with the cutting oil which contains the strong base ethanolamine. The natural passive iron oxide film on SS will react with ethanolamine to form the $Fe(OH)_2$ and

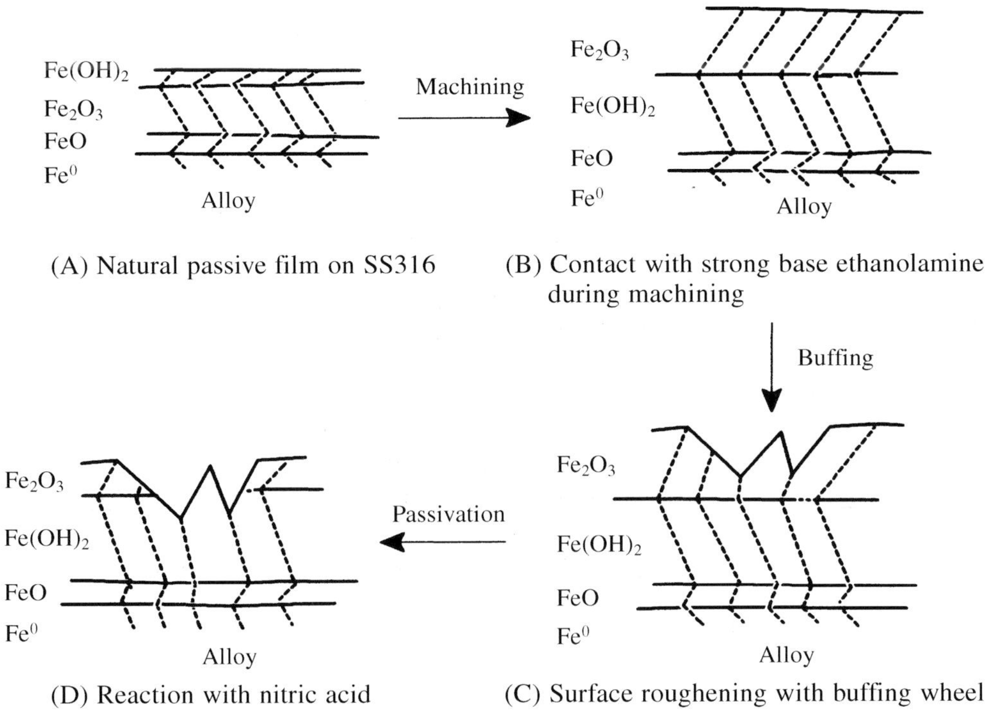

Figure 6. Schematic diagram of possible surface composition change after each treatment. (For clarity, the chromium layers are not displayed.)

$Fe(OH)_3$ according to the following possible reactions.

$$\begin{aligned}(HOCH_2CH_2)_3N + H_2O &\longrightarrow (HOCH_2CH_2)_3NH^+ + OH^- \\ Fe &\longrightarrow Fe^{2+} + 2e^- \\ Fe^{2+} + 2OH^- &\longrightarrow Fe(OH)_2 \\ 2Fe(OH)_2 + 1/2O_2 + H_2O &\longrightarrow 2Fe(OH)_3\end{aligned}$$

Banas *et al.* [28] also reported that in dilute aqueous electrolytes, water molecules participate in the passivation process of metals as a source of oxygen. In these media, an oxide–hydroxide film is formed on the metal surfaces as a result of the anodic reaction between metal atoms and water molecules.

$$Fe + 2H_2O \longrightarrow Fe(OH)_2 + H_2$$

It was also known that oxidation of Fe^{2+} in a basic solution, or of $Fe(OH)_3$ by air can lead to the formation of γ-FeO(OH) and Fe_3O_4 [34].

$$Fe(OH)_3 \longrightarrow FeO(OH) + H_2O$$

This topmost iron/chromium hydroxide is rapidly oxidized in air and eventually gives $Fe_2O_3 \cdot nH_2O$.

$$2Fe(OH)_2 \longrightarrow Fe_2O_3 + H_2O + 2H^+ + 2e^-$$

An observation, by Haupt *et al.* [26], of passive films of Fe and Fe–Cr alloys formed by NaOH treatment also showed a similar two-layer oxide structure with an inner FeO, $Fe(OH)_2$ and an outer Fe_2O_3. Passivation by HNO_3 appears to lessen the topmost oxide layers according to the following possible reaction:

$$6HNO_3 + Fe_2O_3 \longrightarrow 2Fe^{3+} + 6NO_3^- + 3H_2O$$

This process causes the surface to form a definite new phase which consists of a thin iron oxide outer layer and a thick Fe and Cr hydroxide and/or oxyhydroxide inner layer. Mansfeld *et al.* [11] studied the effect of passivation of the 304 stainless steel by HNO_3, after 60 min immersion at 22°C. They reported that the passivation by nitric acid apparently removed partial oxide layers formed by air exposure. The estimation of the thickness of the original air-formed oxide layer was about 1.7 nm. Passivation in HNO_3 reduces the oxide film thickness to about 1.2 nm. The passive film has a layer structure, which is made up of three distinguishable parts. The alloy–oxide interface is apparently covered with a layer of Cr^{3+} which is iron depleted, and above this is a region of mixed oxides with iron being present primarily as Fe^{2+}. The outer surface of the passive layer which is exposed to air has iron present principally as Fe^{3+}. Diekmann *et al.* [12] also reported that the passive oxide film formed by HNO_3 treatment on austenitic stainless steel consisted of SiO_2 on top of a Cr oxide layer containing Mo in both the 4 and 6 valence states, but no Ni was detected and the metal layer in contact with the oxide film was enriched with Cr. The depth of the oxide layer on passivated stainless steel was estimated to be 2–5 nm after 5 min exposure and 3–5 nm after 2 h in contact with 98% HNO_3. No formation of the metal hydroxide or oxyhydroxide film was reported in this work. However, Ohkubo *et al.* [14] reported a high oxygen atomic concentration on the SS passive film formed by HNO_3 treatment, and as a result, speculated that the FeOOH was formed. It should be noted that the iron or chromium hydroxides are soluble in acids and strong bases [34]. If the topmost layer of the passive film is enriched in hydroxide, the HNO_3 treatment should deplete this hydroxide layer. However, the XPS data show otherwise. Based on the relative intensity of our XPS spectra, the thickness of the passive film of our SS samples appears to be less than 5 nm assuming the analysis depth would be about 5 nm at an electron take-off angle of 45°. As pointed out during the analysis of Fe and Cr $2p$ spectra, three very interesting observations from the spectra obtained after passivation are: (1) a decrease in intensity of iron oxides and/or hydroxides peaks at 710.8 and 711.2 eV; (2) an increase in metallic iron peak at 707 eV; and (3) enrichment of the passive film with Cr. These results strongly suggest that the thickness of total passive film lies approximately between 2 to 5 nm. The final sodium-EDTA treatment may remove free metal ions on the surface, or form metal–EDTA complexes and deposit them on the surface. The use of EDTA as a stabilizer for anaerobic adhesive has been reported [37]. The overall results of the surface composition change after each surface treatment can be explained as shown in Fig. 6.

3.8. Effect of surface treatment on anaerobic cure mechanism

An anaerobic formulation typically contains an initiator, such as hydroperoxide, an amine, which functions as a reducing agent, and an accelerator such as saccharin or amine. Thus, the anaerobic system is cured by redox-initiated free radical polymerization. Oxygen inhibits the free radical and terminates the polymerization process. The proposed cure mechanism has the steps shown in Fig. 7 [38].

As shown in this scheme, both transition metal catalyst and saccharin/amine reducing agents play important roles in polymerization initiation and propagation. The role of the saccharin/amine and transition metal in anaerobic cure has been discussed previously [1, 2, 39]. The cure reactions are believed to have the redox pathways as labeled in Fig. 7: (1) the reaction of hydroperoxide with lower oxidation state metals results in the formation of an active free radical for polymerization initiation; (2) the saccharin/amine complex reacts with metals in their higher oxidation states to reduce the metal oxidation state. The overall result is to accelerate the redox process as well as to generate active free radicals; (3) in the presence of the oxygen, the active free radical will be quenched to form an inactive $ROO^{\bullet}$ free radical; and (4) in the absence of the oxygen, the active free radical will attack the monomers containing an active vinyl functional group to initiate the polymerization. In general, transition metals, which can undergo a one-electron transfer redox reaction with an oxidative and a reducing agent, are known as active catalysts for anaerobic cure acceleration. The results of the cure profile studies of anaerobic adhesive 642 cured on the SS316L surface with different surface treatments have been reported previously using a real time FT-IR technique [2]. The results indicated that the SS sample with complete surface treatments exhibits a higher surface activity for anaerobic curing than those samples having either no passivation, or no buffing. For example, after 4 h, the degrees of anaerobic cure on SS surface treated by machined only, machined/passivation and machined/buffed/passivation are 52, 23 and 39%, respectively. Thus, we found that

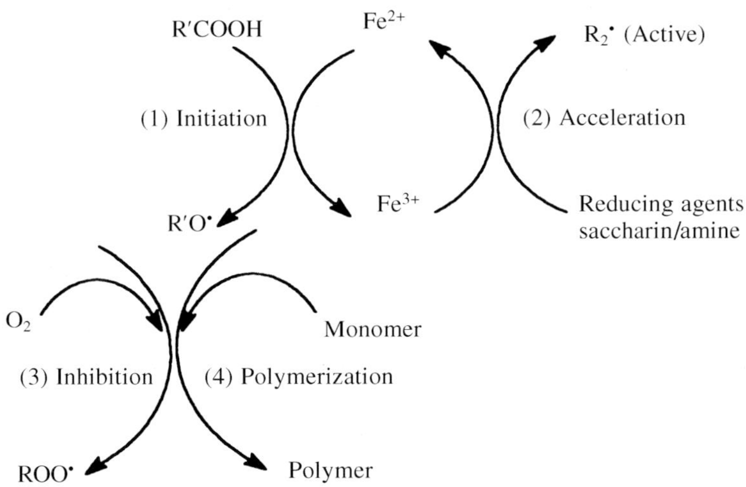

Figure 7. Proposed mechanism of anaerobic cure catalyzed by a transition metal ion.

the nitric acid/EDTA passivation treatment contributes a significant role in enhancing surface activity for anaerobic cure. These results clearly indicated that the cure acceleration was strongly associated with the composition and chemical state of the passive film.

On these SS surfaces, iron, not chromium, appears to be the most important source for surface activated polymerization. However, the availability or compatibility of the iron ion to participate in the redox cycle appears to be the key factor for cure acceleration. In the present XPS study, it has been demonstrated that the HNO_3 passivation treatment has depleted the catalytically inert iron oxide layers and caused the surface film to become enriched in iron hydroxide and/or oxyhydroxide. The formation of the basic hydroxide/oxyhydroxide layers also promotes the acid–base interaction [40] between the surface functional groups and the components in the anaerobic formulation. Thus, the iron hydroxide or oxyhydroxide could enhance interaction with an acidic curing agent such as saccharin to extract metal ions into the anaerobic cure system to accelerate the decomposition of hydroperoxide initiator. The observed surface activity enhancement of passivated SS surface is attributed to the increased availability of active iron ions for catalytic anaerobic polymerization. Also, our XPS analyses indicate that the buffing treatment does not change the surface composition. It seems only to roughen the surface, as well as remove the scale or debris on the surface as indicated by previous SEM study [2]. The XPS study performed by Asami *et al.* [41] also demonstrated that mechanically polished iron–chromium alloys showed no enrichment of chromium or iron in the oxide layer or the alloy surface. The mechanical polishing may also affect the surface stress levels [32] and this ultimately affects the density and distribution of defects present on the surface which may enhance the interaction with HNO_3 during the passivation treatment. Polishing to different surface roughnesses has been found to have an effect on the oxide film composition for Ni–Fe alloys [20], Ni–Cr–Fe alloy [42] and Cu–Ni alloy [43]. The roughened surface has a higher surface area contacting with adhesive than the smooth surface. Thus, the observed surface activity enhancement may be attributed to the increased exposure of active transition metal ions available for catalytic anaerobic polymerization. An increase of contact surface area with the adhesive will ultimately increase metal catalyst concentration for decomposition of the hydroperoxide initiator.

4. CONCLUSIONS

The surface film of machined SS316L substrate was found to consist mainly of Fe/Cr hydroxide/oxyhydroxide (i.e. $Fe(OH)_2$, $Fe(OH)_3$, $Cr(OH)_3$, FeOOH or CrOOH) with overlayers of Fe/Cr oxides (i.e. FeO, Fe_2O_3, Cr_2O_3, CrO_3). The origin of Fe and Cr hydroxide/oxyhydroxide on the surface of the machined SS316L is due to its contact with the cutting fluid. The cutting fluid (pH $\sim$ 9) contains strong base ethanolamine which could react with Fe/Cr under the high temperature fabrication environment. The presence of chemically bonded nitrogen, as indicated by XPS, indicated the occurrence of this type of interaction. Evidence of metal-carboxylate formation was observed as a result of the interaction of surface metal or metal oxides with EDTA.

The mechanical polishing also affects the stress level and ultimately affects the density and distribution of defects present on the surface which may enhance the interaction with HNO_3 and EDTA during the passivation treatment. In summary, the increase of metal ions of Fe/Cr oxyhydroxide or hydroxide on the stainless steel surface during the initial machining treatment is responsible for enhancing the anaerobic cure activity. Polishing by mechanical buffing roughens the surface and removes debris and scale thereby causing an increase of the contact surface area with the adhesive and, ultimately, increasing the metal catalyst concentration for decomposition of the initiator. The passivation treatment by HNO_3/EDTA basically removes the Fe/Cr oxide top layers and enriches the surface in Fe/Cr hydroxides and/or oxyhydroxide. The Fe/Cr oxyhydroxide or hydroxide also can increase surface basicity which could result in an enhanced interaction with acidic curing agent, such as saccharin, resulting in an accelerated cure.

Acknowledgements

I would like to thank Mr. Mark Holmes for arranging commercial surface treatments of stainless steel samples and to Ms. T. Wakamatsu for handling surface cleaning of the tested samples and for Japanese literature translation. Many thanks to Drs Paul Clark and John Woods for much interesting discussions and helpful suggestions. Thanks are also due to Ms. Maggie Orszak and Ms. Marian Glynn for conducting a literature search and to Ms. Judy Lombardo for proofreading the manuscript. Special thanks go to Mr. Jeff Schallenberger of Evans East Material Characterization Laboratory (Plainsboro, NJ) for performing the XPS experiments and helpful discussions in spectral interpretation. Thanks are also due to Loctite Corporation for permission to publish this paper.

REFERENCES

1. D. B. Yang, *J. Adhesion* **43** (4), 273–287 (1993).
2. D. B. Yang, D. Wolf, T. Wakamatsu and M. Holmes, *J. Adhesion Sci. Technol.* **9** (10), 1369–1379 (1995).
3. D. B. Yang, *J. Polym. Sci. Part A* **31**, 199–208 (1993).
4. D. B. Yang (Loctite), US Patent 5,126,416 (1992).
5. D. B. Yang, *Appl. Spectrosc.* **47** (9), 1425–1429 (1993).
6. U. R. Evans, *Metallic Corrosion, Passivity and Protection*. Longmans, Green & Co., New York (1948).
7. U. R. Evans, *Introduction to Metallic Corrosion*, 2nd edn. Edward Arnold, London (1963).
8. S. Spring, in: *Surface Preparation and Finishes For Metals*, J. A. Murphy (Ed.), p. 50. Soc. of Manuf. Eng., McGraw-Hill, New York (1971).
9. N. D. Tomashov and G. P. Chernova, *Passivity and Protection of Metals Against Corrosion*. Plenum Press, New York (1967).
10. R. P. Haak and T. Smith, *Intl J. Adhesion Adhesives* **3** (1), 15–23 (1983).
11. F. Mansfeld, J. V. Kendel and A. B. Harker, in: *Passivity of Metals*, R. P. Frankenthal and J. Kruger (Eds), pp. 752–757. Proc. Int. Symp. 4th Natl Asoc. Corros. Eng., Houston, TX (1978).
12. H. Diekmann, H. Graefen, R. Holm, E. M. Horn and S. Storp, *Z. Werkstofftech.* **9** (2), 37–45 (1978).
13. J. B. Lumsden and R. W. Staehle, in: *Proc. USA–JPN Seminar*, R. W. Staehle and H. Okada (Eds), pp. 75–76. Natl Assoc. Corros. Eng., Houston, TX (1976).

14. K. Ohkubo and M. Miki, *Fushoku Boshoku Bumon Iinkai Shiryo* (*Nippon Zairyo Gakkai*) **27** (3), 57–64 (1988).
15. K. Hashimoto, K. Asami and K. Teramoto, *Corrosion Sci.* **19**, 3–14 (1979).
16. B. Elsener and A. Rossi, *Metallurgy and Foundry Engineering* **18** (2), 175–186 (1992).
17. K. Hashimoto, M. Naka, K. Asami and T. Masumoto, *Corrosion Sci.* **19**, 165–170 (1979).
18. K. Asami, K. Hashimoto and S. Shimodaira, *Corrosion Sci.* **18**, 151–160 (1978).
19. K. Asami, K. Hashimoto and S. Shimodaira, *Corrosion Sci.* **16**, 387–391 (1976).
20. M. M. El Gomati, D. C. Peacock, M. Prutton and R. Roberts, *Scanning Electron Microsc.* **3**, 1091–1099 (1983).
21. B. Stypula, *Metallurgy and Foundry Engineering* **18** (2), 301–315 (1992).
22. P. Marcus and I. Olefjord, *Corrosion* **42**, 91 (1986).
23. J. Long and A. Fan, *Material Protection* **27** (8), 24–28 (1994).
24. A. Rossi, C. Calinski, H.-W. Hoppe and H.-H. Strehblow, *Surf. Interface Anal.* **18**, 269–276 (1992).
25. S. Haupt and H.-H. Strehblow, *Langmuir* **3**, 873–885 (1987).
26. S. Haupt, C. Calinski, U. Collisi, H. W. Hoppe, H.-D. Speckmann and H.-H. Strehblow, *Surf. Interface Anal.* **9**, 357–365 (1986).
27. J. E. Castle and C. R. Clayton, *Corrosion Sci.* **17**, 7–26 (1977).
28. J. B. Banas, B. Stypula, B. Mazurkiewicz, *Metallurgy and Foundry Engineering* **18** (2), 291–299 (1992).
29. K. Asami and K. Hashimoto, *Corrosion Sci.* **17**, 559–570 (1977).
30. C. R. Brundle, T. J. Chuang and K. Wandelt, *Surface Sci.* **68**, 459–468 (1977).
31. N. S. McIntyre and D. G. Zetaruk, *Analytical Chem.* **49** (11), 1521–1529 (1977).
32. N. S. McIntyre and T. C. Chan, in: *Practical Surface Analysis*, 2nd edn, D. Briggs and M. P. Seah (Eds), pp. 485–529. John Wiley, New York (1995).
33. G. S. Allen, M. T. Curtis, A. J. Hooper and P. M. Tucker, *J. Chem. Soc., Dalton Trans.* **14**, 1525 (1974).
34. F. A. Cotton and G. Wilkinson, in: *Advanced Inorganic Chemistry*, 5th edn, pp. 711–712. John Wiley, New York (1988).
35. T. L. Barr, *J. Vac. Sci. Technol.* **14** (1), 660–665 (1977).
36. T. L. Barr, in: *Modern ESCA: The Principles and Practice of X-ray Photoelectron Spectroscopy*, Ch. 8. CRC Press, Boca Raton, FL (1995).
37. E. Frauenglass and J. W. Gorman, US Patent 4,262,106 (1981).
38. C. W. Boeder, in: *Structural Adhesives: Chemistry and Technology*, S. R. Hartshorn (Ed.), pp. 217–247. Plenum Press, New York (1986).
39. F. J. Boerio, P.P. Hong, P. J. Clark and Y. Okamoto, *Langmuir* **6**, 721–727 (1990).
40. Y. C. Huang, F. M. Fowkes and T. B. Lloyd, in: *Acid–Base Interactions: Relevance to Adhesion Science and Technology*, K. L. Mittal and H. R. Anderson, Jr. (Eds), pp. 363–380. VSP, Zeist, The Netherlands (1991) and references therein.
41. K. Asami, K. Hashimoto and S. S Himodaira, *Corrosion Sci.* **17**, 713–723 (1977).
42. J. Russelet, G. Moulin and A. M. Huntz, *Reactive Solids* **5**, 1 (1988).
43. D. E. Dobb, J. P. Storvick and G. K. Pagenkopf, *Corrosion Sci.* **26**, 525 (1986).

Mittal Festschrift, pp. 573–583
W. J. Van Ooij and H. R. Anderson, Jr. (Eds)

Aspects of properties in polymer assemblies bonded by polyurethane adhesives

A. ZIANI,[1] H. P. SCHREIBER[1,*] and HELEN PAK-HARVEY[2]

[1] *CRASP, Department of Chemical Engineering, Ecole Polytechnique, PO Box 6079, Stn. Centre Ville, Montreal, QC H3C 3A7, Canada*

[2] *Reichhold Chemical Co., Box 13582, Research Triangle Park, NC 27709, USA*

Abstract—Adhesives based on a polyurethane blended with a series of styrene–butadiene rubber (SBR) latexes have been used to bond joints of vulcanized rubber (R) and of acrylonitrile–butadiene–styrene (ABS). It was shown that the bond strength generated was strongly dependent on the acid level and the gel content of the SBR additive. Bond characteristics were time-dependent. Aging in an inert environment raised bond strengths, apparently due to the ability of phase-separated polyurethanes to restructure at an interface. In aggressive humid media bond strengths decreased, the effect being inversely related to initial bond strengths. An estimate may be made of initial bond strengths needed to resist aging losses in the aggressive medium.

Keywords: Polyurethane–SBR blends; adhesives; rubber; ABS; initial bond strength; dry aging; humid aging.

1. INTRODUCTION

The use of adhesives has risen markedly in the past decades. Polyurethane-based (PUr) adhesives, in particular, find use in structural and semi-structural applications, as well as in the automotive, appliance and furniture industries. In certain cases, economics dictates that PUr be blended with other polymers, for example, latexes of the styrene–butadiene rubber (SBR) family. Whether used in their pure form or as components of a blend, an essential requirement for these formulations is their ability to meet stringent criteria of initial adhesion and resistance to aging. These criteria have motivated research designed to define the factors which determine the adhesion performance of PUr adhesive systems. The present paper is part of that research endeavor. It examines the strength of bonded assemblies in which rubber and acrylonitrile–butadiene–styrene (ABS) are the substrates joined by a series of PUr/SBR blends. An attempt is made to relate initial and aged bond characteristics to composition variables of the adhesive blends. The ultimate aim of work along these

---

*To whom correspondence should be addressed.

lines is to provide guidelines useful for the formulation of adhesives best suited to meet performance targets imposed by stated applications.

2. EXPERIMENTAL

2.1. Materials

A listing of the materials used in this study is given in Table 1. It identifies the PUr as a typical [1] linear, toluene di-isocyanate based polyester, with low-crystallinity polyol segments of M_n between 1500 and 2000. The 10 SBR latex components involved in formulating adhesive systems are coded by letters A–J; these codes will also be used to identify PUr/SBR blends. The copolymers vary in gel content, as determined by toluene extraction. Since swelling measurements indicated the latexes to be significantly crosslinked, the gel content may be used as an approximate index of molecular weight. The range in gel content from 15 to 71% thus indicates an analogous substantial range in molecular weights, and this should lead also to substantial variations in the mechanical properties of the materials. The latex copolymers also vary in the amounts of acrylic acid added to them, and this is expressed as the acid level in Table 1. Their capability of interacting with substrates through polar as well as dispersion forces therefore also should vary significantly. The glass transition values entered in the table were obtained from differential scanning calorimetric measurements. A Perkin–Elmer DSC 7 apparatus was used for the purpose, at a temperature ramp of 10°C/min.

In the major portion of this work, adhesive formulations were prepared by blending equal volumes of PUr and aqueous SBR latex; in preliminary experiments, however, the quantity of SBR added was varied, so as to relate adhesion performance with PUr content and thus to justify the choice of the 1 : 1 formulation for more intensive study. Blends of PUr and SBR were placed in a Rotovap apparatus where carrier liquids

Table 1.
Characteristics of experimental materials

SBR component			
Sample	Acid level (%)	Gel content (%)	T_g (°C)
A	0.2	23	−11
B	0.2	29	−11
C	0.5	26	−11
D	1.0	35	−20
E	1.5	43	−11
F	2.0	40	−11
G	3.0	47	−9
H	1.0	15	−14
I	1.0	31	−11
J	1.0	71	−7

Polyurethane matrix adhesive (PUr): TDI-based polyester PUr, with polyol segments of $M_n = 1500–2000$, of low crystallinity.

were evaporated under partial vacuum. Partially dried specimens were cut into small pieces and further dried under vacuum at 60°C for 24–30 h. This was sufficient to attain invariant weights, as shown by sequential gravimetric analyses. Dried adhesives were again cut into smaller pieces and stored in desiccators for further use.

A vulcanized SBR rubber, supplied by Slocum Chemical, Lynchburg, VA, was used as one of the substrates in this work. The other was an ABS compound from General Electric Co, identified as Cycolak AR 3501.

2.2. Characterization of adhesives

Characterization of formulated adhesives involved measurements of their mechanical properties and, in selected cases, of their surface energies.

Stress–strain measurements were carried out on a JJ Instruments M30K tester equipped with an ASYST data-acquisition system. The extension rate was 10 mm/min. From 3 to 5 specimens of each formulation were evaluated. The principal parameter drawn from the data was the total energy to rupture, ε_r, obtained by integrating under the stress–strain curve. Values of the parameter are reported in Table 2. Repeat determinations of ε_r were reproduced to better than $\pm 7\%$. Dog-bone shaped samples for these measurements were prepared to conform with ASTM D-638-89. Adhesives were placed between compression molding platens, covered by sheets of polyetherimide, to ease separation following molding. Three stage molding operations were developed for the formation of bubble-free sheets. The stages were (i) 2 min at 150°C and 2 metric tons applied pressure; (ii) 2 min at 150°C and 4 metric tons applied pressure; (iii) 10 min at room temperature and 4 tons applied pressure.

The range in ε_r, from 11.7 to 18.6 N/mm^2, is adequate to test an expected relationship between gel content and the mechanical strength of the adhesives. The relationship is not unique as indicated by inspection of data in Tables 1 and 2. For

Table 2.
Mechanical strength and surface energies of adhesives

Adhesive code	ε_r (N/mm^2)	γ_s^d (mJ/m^2)	γ_s^{nd} (mJ/m^2)
A	16.5		
B	18.0	33.8	1.4
C	17.7		
D	18.6	34.7	2.4
E	16.0		
F	15.1	36.0	4.0
G	13.5		
H	11.7	35.5	2.7
I	13.6	33.2	2.7
J	16.7	36.8	3.6
Rubber (average value)		33.5	1.0
ABS		36.8	6.6

Adhesives are 1 : 1 by weight formulations of PUr and SBR, identified by SBR sample code (see Table 1).

systems at acid levels $\geqslant 1.0\%$, a linear interdependence exists with a correlation coefficient of 0.961. A second linear relationship, with a correlation coefficient of 0.979, is generated for systems with acid levels $< 1\%$. Both extrapolate to a mechanical strength intercept (at zero gel content) at 12 ± 1 N/mm^2.

In certain cases, the static contact angle method was applied for surface energy determinations on molded adhesive specimens. A Rame-Hart goniometer was used with n-decane, tetrahydrofuran, glycerol, tricresyl phosphate, and water as the contact fluids. Following the method described elsewhere [2], contact angles were recorded as a function of contact time, the extrapolated datum at zero contact time being used for surface energy calculations. The harmonic mean equation, reported to be well-suited for the purpose [3], was used to obtain both the dispersive (γ^{d}) and the non-dispersive (γ^{nd}) component of the surface energy. Results of these evaluations are included in Table 2. Not surprisingly, the dispersive energies vary only narrowly, but the varying acid content of the SBR component results in substantial variations in the non-dispersive, or polar, contribution to the surface energy.

2.3. Bond strength of assemblies

An empirical but internally consistent peel test was used to measure the bond strength of rubber and ABS assemblies joined by each of the 10 adhesive formulations. Sample preparation differed slightly between the two substrates. For rubber, the substrate was cut into 10×2.5 cm rectangles. The central portions of these were coated with adhesive, leaving 2.5 cm free at either end. Coated samples were air-dried for 90 min followed by 4 min heating at 55°C. Immediately following this step, samples were overlaid and pressed in a Carver press for 20 s at 80°C under 700 kg load. ABS samples were cut to 5×2.5 cm, and the length covered by adhesive was restricted to 2 cm. Bonding followed the procedure noted above, but following the formation of the assembly, one of the free ends was heated gently with an open flame, raising the uncoated ABS to above its T_{g}. The heated ends were then bent to form a T-shaped specimen, the sample was cooled and then used for peel test determination, again with the stress–strain apparatus noted above. The deformation rate in these tests was 5 mm/min. From 3 to 5 specimens were tested for each system. Reported are integral peel strengths normalized to a contact area of 1 cm^2. The reported peel strengths were reproducible to within 11%.

Bond strengths evaluated on freshly-formed specimens are referred to as initial bond strengths. Assemblies were also examined following one of two accelerated aging sequences. An 'inert aging' protocol involved placing the samples in an oven set at 50°C, under an atmosphere of dry nitrogen for 72 h. On withdrawal, samples were equilibrated under ambient conditions for 24 h and re-tested. In an 'aggressive aging' sequence, samples were exposed to 100% RH air, under the same temperature/time settings as above.

3. RESULTS AND DISCUSSION

A first series of experiments, utilizing systems G and J, inquired into the variation of initial bond strength with the content of SBR diluent. Rubber was the substrate

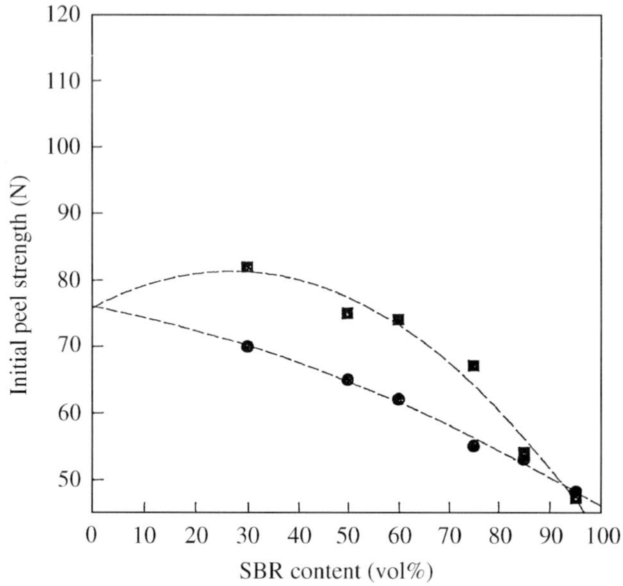

Figure 1. Initial integral peel strength of rubber joints produced by PUr adhesives with varying SBR content. (●) System G; (■) system J.

selected for this portion of work. Adhesives G and J were formulated so as to contain 30–95 vol% of the SBR latex. The results of this inquiry are shown in Fig. 1. A decrease in bond strength with diminishing PUr concentration (i.e. increasing SBR content) is to be expected, and generally confirmed. The mechanically weak system G (see Table 2) displays a systematic drop in peel strength with rising SBR content. However, the addition of the tougher SBR in formulation J has very little deleterious effect on bond strength when the SBR volume % is less than about 40. Speculatively, this may be due to a limited ability of the SBR to interpenetrate the rubber substrate, carrying some of the PUr adhesive into regions of the substrate removed from the surface, thereby reinforcing the interphase of the rubber. At high SBR concentrations, the quantity of PUr transferred may be too small to create the effect; indeed, removal of PUr from the surface region could, in these cases, transfer the failure locus closer to the rubber/adhesive interface. Given our primary objective, to identify the lead variables determining the bond characteristics generated by a 'typical' PUr/SBR blend, the remainder of the research focussed on the 1 : 1 composition.

One of the first relationships to be sought was that between peel strength of assemblies and the balance of surface energies for substrate and adhesive. Earlier documentation [4] had suggested that optimum adhesion would be obtained when the surface energies of the contacting materials were closely matched. Accordingly, the γ^{d} and γ^{nd}, entered in Table 2, were summed and the difference between those of adherend and adhesive expressed as $\Delta\gamma$. Absolute values of this difference and corresponding initial bond strengths of rubber and ABS assemblies have been entered

Table 3.
Surface energy differences of adherends and adhesives and initial peel strengths of rubber and ABS assemblies

Adhesive	$\|\Delta\gamma\|$ (mJ/m^2)		Initial peel strength (N)	
	Rubber	ABS	Rubber	ABS
B	0.7	6.2	50	111
D	2.6	6.3	63	90
F	5.5	3.4	79	135
H	3.7	5.2	17	48
I	1.4	7.5	65	131
J	5.9	3.0	97	146

in Table 3. An inspection reveals no discernible relationship between the parameters. Thus, although the rubber/adhesive F and J systems have large differences in their surface energies, their bond strengths are near the maximum for this substrate. In contrast, the close match between the surface energies of rubber and adhesive blend B produces a joint with rather feeble bond strength. For ABS substrates, the situation is somewhat modified. Here the high peel strengths for systems J and F also correspond to the lowest $\Delta\gamma$, inferring that polar interactions at the adherend/adhesive interface exert appreciable effects. The relationship between the parameters, however, is not a simple one, as is made evident by cases D and I. Other factors must be expected to make sizeable contributions to the adhesion performance of these assemblies.

3.1. SBR properties and bond strength

The most obvious of the variables alluded to above are the mechanical strength and acid content of the SBR diluent. Each of these was found to have an important effect on the initial bond strengths of assemblies. The relationship between peel strength and the adhesive gel content, signifying variation in the molecular weight of the SBR component, is displayed in Fig. 2. Albeit non-linear, the regression curves fitted to experimental data have high correlation ratios (0.987 for the rubber and 0.969 for the ABS substrate). Evidently, the mechanical strength of the adhesive is a strong factor in dictating bond strength. The displacement of curves, however, recalls the summation of Table 3, where the bond strengths with ABS uniformly exceed those with rubber. Since ABS is a more polar substrate, it would be expected that bond properties would relate to the specific, or polar, interaction potential of the adhesive. This would be set by the SBR acid level, and thus calls for a search of relationships between these factors. Figure 3 is the result of that inquiry. Once again well-defined, but non-linear relationships have been obtained, with the displacement of bond strengths between rubber and ABS again in evidence.

An inference drawn from Figs 2 and 3 points to the existence of a function combining the gel and acid contents of SBR which would generate a simpler and statistically more significant relationship with bond characteristics. The matter was approached empirically, in the absence of available theoretical guidance. One of the best fits to experimental measurements of bond strength was obtained from the logarithmic

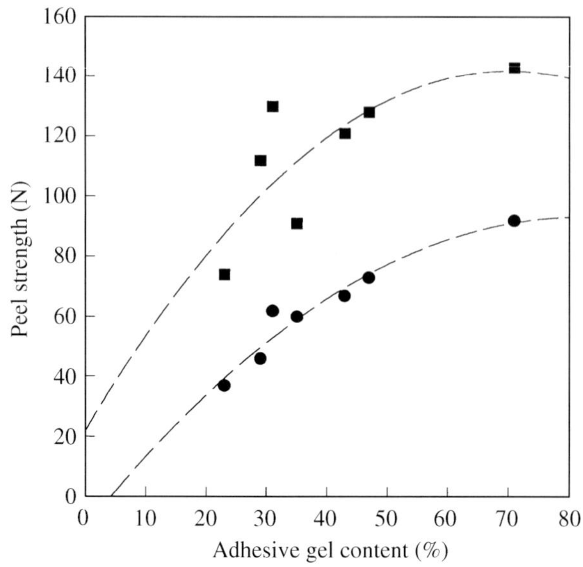

Figure 2. Initial integral peel strength of joints as function of adhesive gel content. (●) Rubber substrate; (■) ABS substrate.

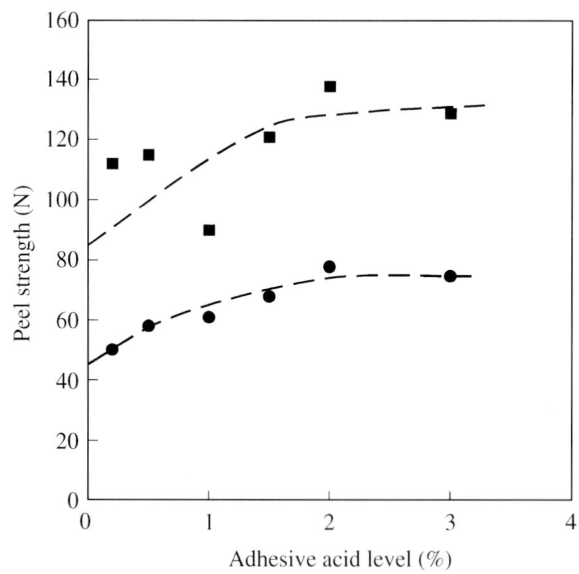

Figure 3. Initial integral peel strength of joints as function of adhesive acid level. (●) Rubber substrate; (■) ABS substrate.

product of gel and acid levels, as shown in Fig. 4. For clarity, shown are the results for the rubber joints, but a fully analogous relationship can be documented for the ABS systems. The first-order regression for data in Fig. 4 has a correlation coefficient

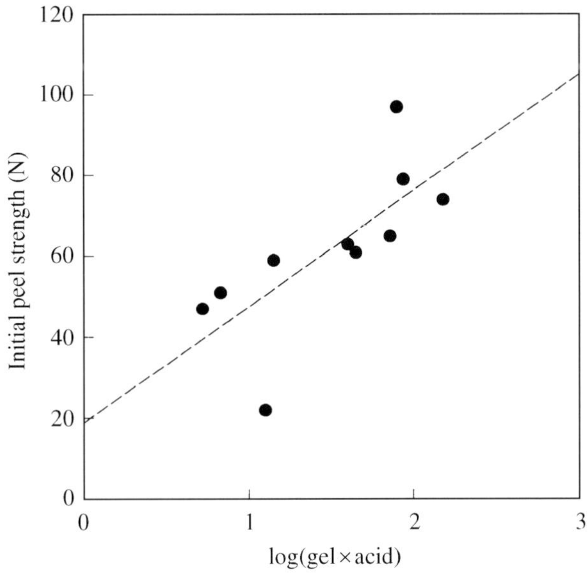

Figure 4. Initial integral peel strength of rubber joints vs. combined gel and acid contents of adhesive formulation.

of 0.980, rising to 0.992 if the point for system H is neglected. This omission is not unreasonable, since the experimental variation at these low levels of bond strength was greater than elsewhere. While a long extrapolation of the suggested line may not be warranted, the shown relationship nevertheless serves as a useful guideline to the optimization of an initial bond strength with substrates similar to the vulcanizate involved in this example.

3.2. Effects of aging on bond properties

Recent studies in our laboratories have shown [5, 6] that properties of joints bonded by PUr can display complex time dependencies. In many cases, these arise from the ability of PUr, among other polar-group containing polymers, to restructure when in contact with surfaces able to exert strong orienting forces [7, 8]. The restructuring capability of PUrs may be attributed to their phase-separated soft and hard segment components [1, 9], which allow the surface region of the polymer to be occupied preferentially by constituents which minimize surface or interfacial free energies. In the present work, the ABS would be a much stronger orienting substrate than rubber, as reflected by the non-dispersive surface energy component of these materials (cf. Table 2). The application of aging protocols described in the preceding section produced data summarized in Table 4.

Aging in the ‘inert’ nitrogen environment resulted in higher bond strengths than those observed on freshly prepared specimens. The effect is generally more pronounced with ABS than with rubber substrates, consistent with the argument above. The changes in bond characteristics thus may be due, in part at least, to the gradual

Table 4.
Bond aging characteristics

Adhesive	Substrate			
	Rubber		ABS	
	$(PS)_N/(PS)_i$	$(PS)_W/(PS)_i$	$(PS)_N/(PS)_i$	$(PS)_W/(PS)_i$
A	1.06	0.46	1.26	0.31
B	1.07	0.50	1.27	0.37
C	1.10	0.62	1.19	0.33
D	1.18	0.60	1.20	0.51
E	1.13	0.57	1.25	0.55
F	1.16	0.65	1.19	0.57
G	1.22	0.57	1.21	0.41
H	1.00	0.29	1.23	0.29
I	1.18	0.63	1.27	0.52
J	1.22	0.71	1.23	0.60

Inert aging in nitrogen (N), 72 h, 50°C. Aggressive aging (W) at 100% RH, 72 h, 50°C.

change in the composition of the adhesive at the substrate surface, leading to the establishment of equilibrium and enhanced interface strength. Aggressive aging data, on the other hand, show a distinct decrease in bond strength, and this also is more evident in ABS assemblies. The decreased bond strength may be attributed to the generally accepted notion of water diffusing along the interface, in the process weakening the bond. As water is intensely polar, it would follow that a greater tendency to displace the PUr adhesive would take place at the more polar ABS surface than at the essentially non-polar rubber vulcanizate. Of course, the ability of water to interpenetrate should be dependent on the initial bond strength, with the expectation that initially stronger bonds would offer greater resistance to the diffusing front.

This is borne out in fact, as indicated in Fig. 5. Assuming the linear functions to persist beyond the limits of current measurements, it would appear that a rubber/PUr–SBR/rubber joint would be fully resistant to water intrusion (in terms of the aging experiment actually carried out) if the initial bond strength were about 140 N. The corresponding initial bond strength in ABS assemblies would have to be in excess of 300 N, again reflecting the much greater difficulty in stopping ingress of water to the polar substrate. A second property of the adhesive involved in the aging process should be its mechanical strength. Arguably, the access to the interface of water would be favored by a flexible, less entangled structure, typical of lower molecular weight polymers (e.g. of the ABS family), or less highly crosslinked rubber vulcanizates. The concept was tested by current data, with reasonable success, illustrated in Fig. 6. Plotted in this manner, the discrimination between ABS and rubber data is greatly and expectedly reduced. The regression analysis applied in this case suggests, however, that control of adhesive gel content (and thus of molecular weight) alone will not succeed in generating adhesives fully tolerant to the present aging procedure.

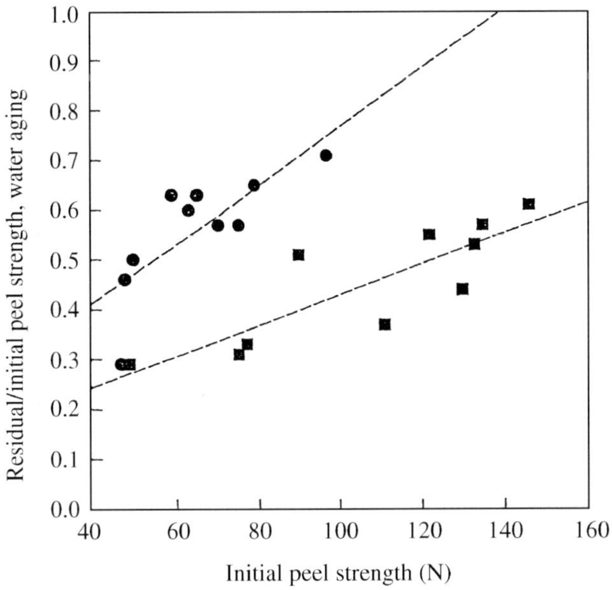

Figure 5. The residual integral peel strength of joints following aging in aggressive environment as a function of the initial peel strength. (●) Rubber substrate; (■) ABS substrate.

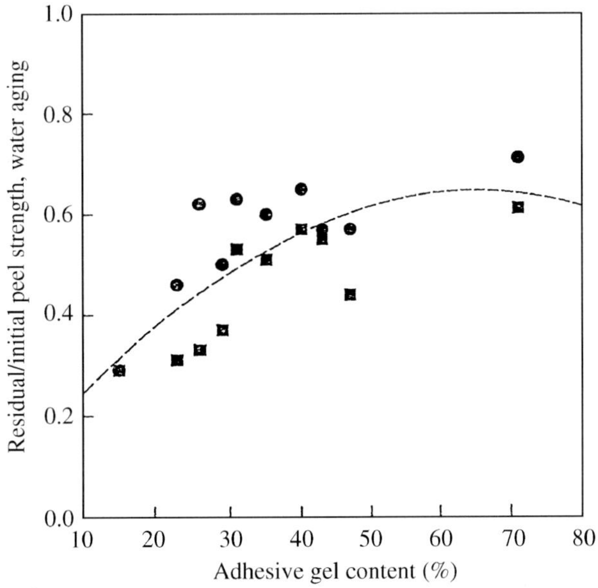

Figure 6. The residual integral peel strength of joints following aging in aggressive environment as a function of adhesive gel content. (●) Rubber substrate; (■) ABS substrate.

4. CONCLUSIONS

The following conclusions may be drawn:

- Peel strength evaluations have been used to characterize the initial and aged bond strengths of assemblies in which vulcanized rubber or ABS was bonded by a variety of adhesives based on PUr and containing SBR latex additive of varying composition.
- Initial bond strengths were not strongly correlated with the surface energies of adhesives and substrates.
- Substantial correlations were found between the initial bond strengths of joints and the gel content of SBR components, which may be assumed to correlate with the molecular weight of the SBR. Similarly, bond strengths were found to rise with increasing acid content of the SBR.
- Bond characteristics were shown to be time-dependent. Aging in dry nitrogen resulted in increased bond strengths, an effect attributed to the ability of phase-separated PUr to restructure when in contact with orienting surfaces.
- Significant bond strength losses were incurred during aging in a water-rich environment. The effect was attributed to the intrusion of water along the interfaces and was shown to be inhibited by an increase in the initial bond strength and by higher mechanical strength of the adhesive.
- The established, empirical correlations allow for the prediction of bond strengths needed to generate full resistance to aggressive aging, thereby serving as guideline to the formulation of superior adhesive blends.

Acknowledgements

We thank Dr. I. Ulkem, now at PAPRICAN, Canada for assistance in early portions of this work. Our thanks go to Drs Jan Grossman, Yakov Freidson and Martin Navratil, all of the Reichhold Company, for assistance in the preparation of experimental materials and for stimulating discussions.

REFERENCES

1. D. M. Hoffman, F. M. Kong and I. L. Chiu, *Polym. Plast. Technol. Eng.* **26**, 95 (1987).
2. H. P. Schreiber, *J. Adhesion* **37**, 51 (1992).
3. S. Wu, in: *Polymer Interface and Adhesion*, Ch. 3. Marcel Dekker, New York (1982).
4. G. A. Dyckerhoff and P. J. Sell, *Angew. Makromol. Chem.* **21**, 169 (1972).
5. A. Sengupta and H. P. Schreiber, *J. Adhesion Sci. Technol.* **5**, 947 (1991).
6. A. Migdal and H. P. Schreiber, *J. Adhesion* **52**, 1 (1995).
7. J. H. Chen and E. Ruckenstein, *J. Colloid Interface Sci.* **135**, 496 (1990).
8. J. Schultz and L. Lavielle, in: *Inverse Gas Chromatography*, ACS Symposium Series 391, D. R. Lloyd, T. C. Ward and H. P. Schreiber (Eds), Ch. 14. American Chemical Society, Washington, DC (1989).
9. I. Ulkem and H. P. Schreiber, *J. Appl. Polym. Sci.* **52**, 1857 (1994).

Mittal Festschrift, pp. 585–596
W. J. Van Ooij and H. R. Anderson, Jr. (Eds)

Improved evidence for chaos during peeling of pressure sensitive adhesive tape from a metal substrate

L. SCUDIERO, S. C. LANGFORD and J. T. DICKINSON*

Physics Department, Washington State University, Pullman, WA 99164-2814, USA

Abstract—Electrical currents produced when a pressure sensitive adhesive is peeled from a copper substrate provide a sensitive probe of the dynamics of interfacial failure. This current is produced when charged adhesive pulls away from the metal. Time-series analysis of the current fluctuations shows that the peel dynamics displays fractal behavior, with a correlation dimension of about 2.3. The small size of this dimension suggests that the dynamical behavior of the peel front could be described with a relatively simple model (3–5 degrees of freedom). Estimates of the Lyapunov exponent, λ, indicate a large, positive exponent (≈ 100 bits/s), showing that fluctuations along the peel front become inherently unpredictable on time scales of tens of ms (λ^{-1}). The presence of a positive Lyapunov exponent demonstrates conclusively that the current fluctuations (and thus adhesive detachment) exhibit deterministic chaos.

Keywords: Pressure sensitive adhesive; interfacial failure; contact electrification; fractal; chaos; time-series analysis.

1. INTRODUCTION

In previous work, we have shown that peeling a pressure sensitive tape from a copper substrate generates substantial electrical currents [1, 2]. These currents are produced when charged adhesive is drawn away from the metal surface. At peel speeds below about 1 cm/s, the current measured from the metal shows negative current spikes superimposed on an overall positive current. These negative current spikes are produced by electrical discharge events [1]. At higher peel speeds, electrical discharges are largely suppressed, and the electrical currents closely reflect the peel dynamics [1, 3]. For instance, simultaneous current and load measurements on thin strips of tape (where local advances along the peel front can significantly change the load) show strong correlations. Preliminary analyses of the current fluctuations at high peel speeds showed strong evidence for fractal-like fluctuations, but were inconclusive with regard to the presence of deterministic chaos [2].

In this work, we report improved current measurements which yield considerably more definitive results, including a determination of the largest Lyapunov exponent, λ,

*To whom correspondence should be addressed.

with units of information (dimensionless) per unit time. A positive exponent indicates that the system is extremely sensitive to initial conditions. Given a well specified initial state, systems with $\lambda > 0$ become unpredictable over time scales of λ^{-1}. Our data yield positive Lyapunov exponents $\lambda \approx 100$ bits/s for all three peel speeds tested, from 5 to 19 cm/s. Thus the underlying dynamics of peeling is conclusively chaotic. A dimensional analysis of this same data yielded correlation dimensions $D_c \approx 2.3$ over the same range of peel speeds. The small size of D_c suggests that a fairly simple dynamical model, with 3–5 degrees of freedom, could adequately characterize the chaotic behavior of the peel front.

The extremely nonlinear, viscous behavior of pressure sensitive adhesives during peeling is strongly suggestive of chaos. Adhesive fibrils and void fingers form and disappear in an irregular fashion. The energy dissipated in stretching these fibrils represents a large fraction of the energy required to detach the tape [4]. Thus these fluctuations are of fundamental importance in the design and application of pressure sensitive tapes. A simple model of the dynamics of peeling a pressure sensitive tape from a roll has been constructed by Hong and Yue, where the angular velocity of the roll displayed chaotic fluctuations at peel velocities near the transition from cohesive to interfacial failure [5]. Although their model was limited to the average behavior of the entire peel front, similar instabilities may be involved in the small-scale fingering observed experimentally. The sensitivity of current measurements to detachment along the peel front allows for study of the dynamical behavior of these small-scale fluctuations in real time.

2. EXPERIMENTAL

A schematic diagram of the peel geometry for current measurements appears in Fig. 1. A 180° peel geometry was employed to maximize the sensitivity of the current signal to events along the peel front. In this geometry, only the adhesive near the peel front has a significant component of velocity perpendicular to the substrate. The

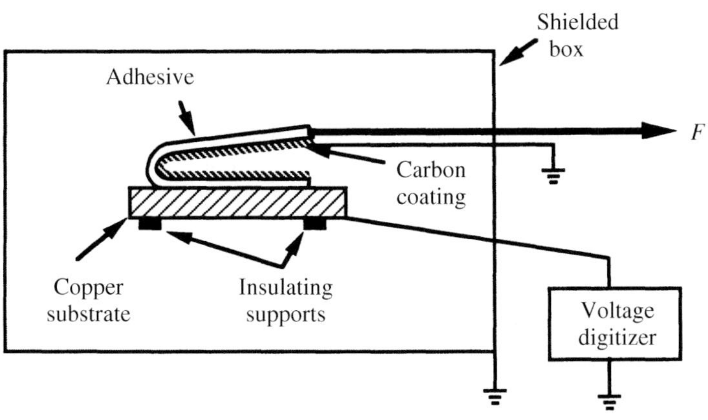

Figure 1. Diagram of experimental geometry for 180° peeling from copper.

motion of the rest of the adhesive (which remains charged) is parallel to the substrate and contributes little to the detected current. The sensitivity of the current signal to events in the peel zone was further enhanced by coating the tape backing with a thin conducting film (EMI-RFI Shield from GC Electronics). This carbon coating electrically shielded the copper substrate from charge carried by previously detached adhesive. These carbon films are quite thin ($<10\ \mu$m) and flexible, and thus should have little effect on the mechanical properties of the tape backing. The use of a carbon coating represented a significant improvement over the moving copper shield employed in [2].

This work employed 3M Scotch Brand Magic Tape (No. 810) 3/4 in. (19 mm), which was peeled from a rectangular copper sheet ($58 \times 27 \times 0.5$ mm^3). Prior to each experiment, the copper sheet was polished to a mirror finish with Simichrome polish (obtained from Competition Chemicals, Iowa City, Iowa, USA). The copper was then cleaned in acetone. Before applying the tape to the copper, the residual electric charge on the adhesive surface was neutralized with a Zerostat®, a commercial device used to neutralize static electric charge on phonograph records. Faraday cup measurements of the residual charge on the adhesive showed negligible values (± 10 nC/m^2). After applying the tape, a carbon film was sprayed onto the tape backing. The edges of the tape backing were masked to prevent the carbon film from grounding the backing to the substrate. The films were allowed to dry for at least one hour prior to peeling.

Peeling was performed in air in a shielded, grounded copper box at speeds ranging from 5 to 19 cm/s. To eliminate the electrical and mechanical noise introduced by mechanical loading devices, peeling was performed by hand. The average peel speed was determined from the length and duration of peel. The current signals were digitized at 10-μs intervals without additional amplification using a LeCroy 6810 transient waveform digitizer with an 1 MΩ input impedance and a 10-bit resolution. These improvements reduced the signal-to-noise ratio by a factor of 70 over our previous measurements. This improvement was critical in obtaining unambiguous results in the Lyapunov exponent calculations.

3. RESULTS AND DISCUSSION

3.1. Current signals and their power spectra

Figure 2 displays current measurements at three peel speeds between 5 and 19 cm/s. In the absence of peeling, the detected signal is negligible on these time and current scales. Thus all the fluctuations in the data seen on these scales are induced by peeling and not extraneous noise. The average current is roughly proportional to peel speed. (Note that different vertical scales are employed in Figs 2a–2c). The amplitude of the fluctuations also increases markedly as the peel speed is raised from 5 cm/s to 11 cm/s. The larger fluctuations, most evident in Fig. 2c, reflect the growth of low frequency components in the current signal. Significantly, the current signal at 19 cm/s is 'smoother' than the current signals at the lower peel speeds, indicating

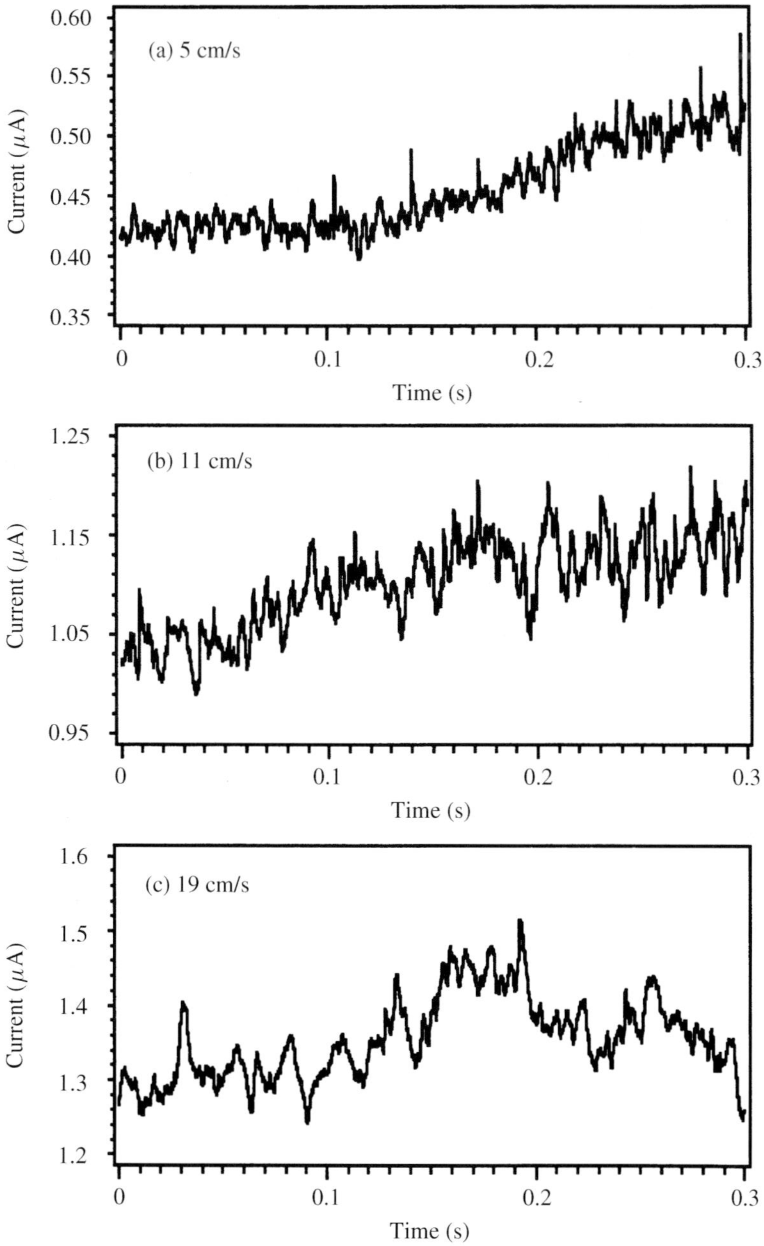

Figure 2. Current signals measured at peel speeds of (a) 5 cm/s, (b) 11 cm/s, and (c) 19 cm/s.

a drop in the amplitude of high frequency fluctuations. The raw data pass the first 'test' for chaos — they look chaotic!

The power spectra (computed from the Fourier transform) of the three current signals in Fig. 2 are shown in Fig. 3 on log–log plots. Also shown is the power spectrum of the background 'noise', acquired under identical conditions, but with no peeling.

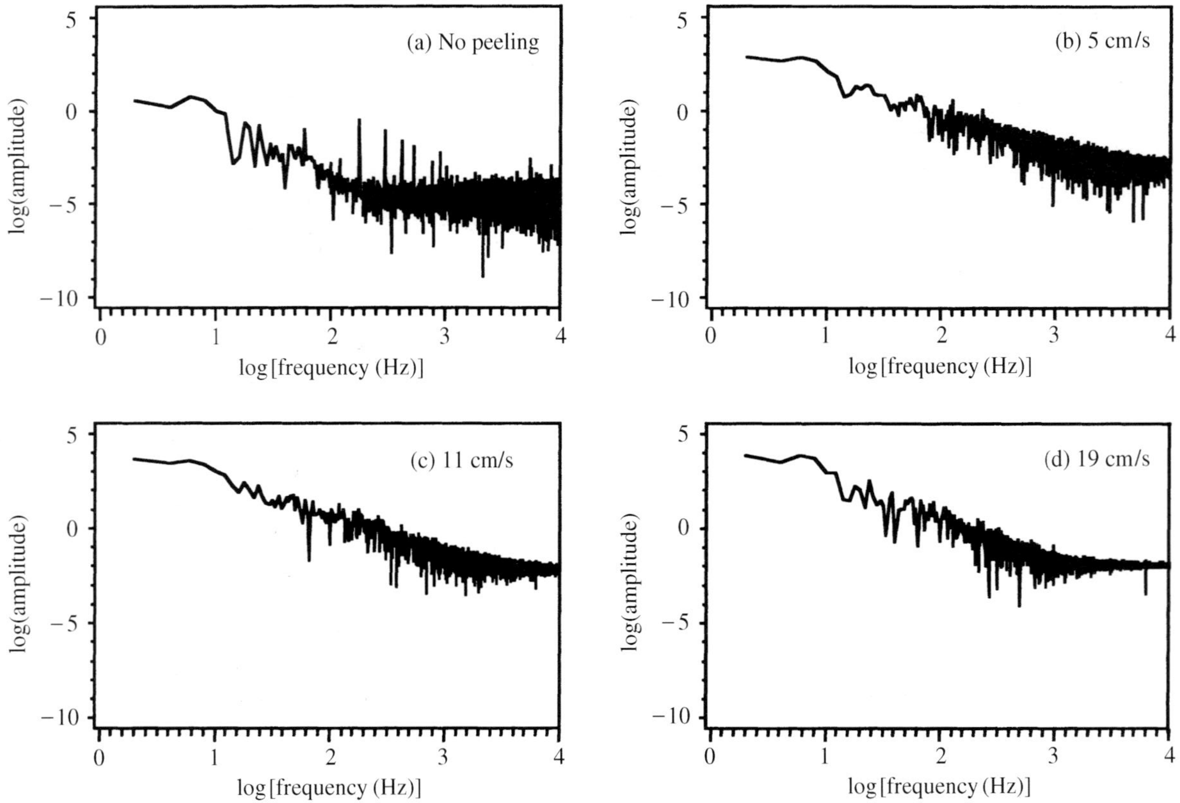

Figure 3. Power spectra of (a) noise, and current signals measured at peel speeds of (b) 5 cm/s, (c) 11 cm/s, and (d) 19 cm/s plotted on log–log scales.

At all frequencies, the average 'power' of the current signals during peeling is at least two orders of magnitude above the noise. In particular, the high frequency current fluctuations are well above the experimental noise, indicating that the signals are capable of representing fluctuations in the course of peeling on time scales less than 1 ms. The micromechanical events responsible for the highest frequency fluctuations are necessarily localized in nature, e.g. fibril advance and retraction along the peel front. By way of comparison with the power spectra of many non-chaotic processes, we note that these power spectra are very broad, with no obvious peaks or spikes that would indicate periodicity. Although not unique to chaotic processes, broad power spectra often accompany chaotic processes [6a].

Two main trends are observed in the power spectra with increasing peel speed. First, the spectra show a definite increase at the lowest frequencies (<10 Hz) as the peel speed is raised. Second, the power spectra flatten out at the highest frequencies, where the onset of the flat portion occurs at lower frequencies as the peel speed is raised. 'Flat' portions (where the spectral intensity is independent of frequency) typically reflect the operation of random processes (e.g. white noise). These high frequency components are quite weak relative to the low frequency components, and have little effect on the dynamical analysis described below. However, these components suggest the operation of a random process that becomes more important at higher peel speeds.

Portions of these power spectra (10–500 Hz) are replotted on linear scales and displayed in Fig. 4. (The components below 10 Hz are very large and have been omitted to emphasize the higher frequency components.) On a linear scale, the increase in spectral power as the peel speed is raised from 5 cm/s to 11 cm/s is especially clear — the lowest frequency components increase by almost a factor of ten. For simple periodic signals, the 'power' at a given frequency varies as the square of the amplitude. Thus if the amplitude of the fluctuations were simply proportional to peel speed, one would expect a factor of four change in the 'power' as the peel speed is raised from 5 cm/s to 11 cm/s. At the low frequencies, the fluctuations grow much faster, consistent with the operation of a nonlinear process. Nonlinearity is an essential feature of chaotic systems.

As the peel speed is increased further to 19 cm/s, the spectral power increases at frequencies below 40 Hz, but decreases at frequencies between 100 and 500 Hz. This is especially clear in the insets of Figs 4b and 4c. The weakness of the high frequency fluctuations is consistent with the smoother appearance of the current signal at 19 cm/s (Fig. 2c). Since the 'white noise' portion of the power spectra does not become weaker as the peel speed is raised, the decreased power between 100 and 500 Hz may contribute to the prominence of the white noise plateau at 19 cm/s.

3.2. Correlation dimension

Most dynamic systems (and all chaotic ones) are characterized by dimensions greater than 2. Since most experimental measurements (e.g. a measurement of a single experimental variable vs. time) are inherently one-dimensional, higher dimensional 'probes' are required to understand the behavior of these systems. The complex motion of a particle in a turbulent fluid, for instance, might be described by position

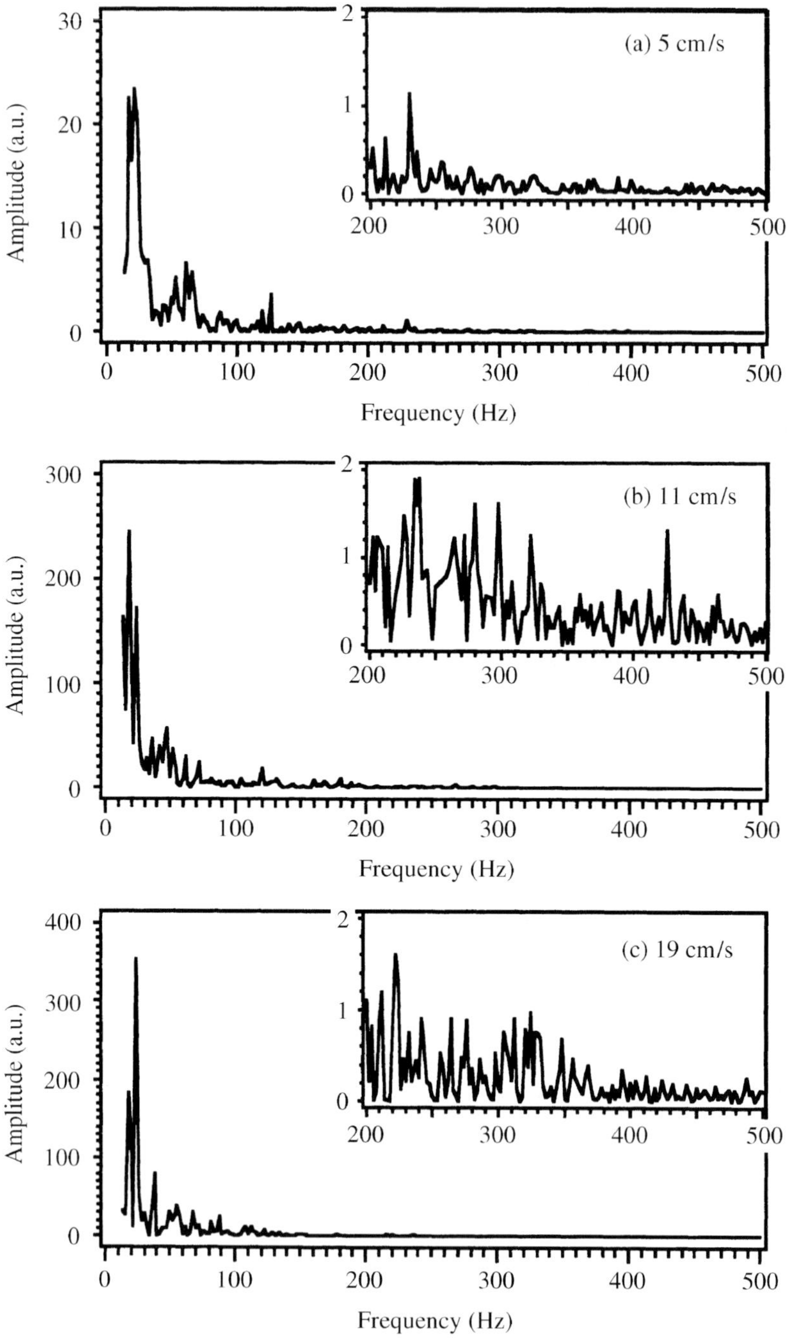

Figure 4. Power spectra of current signals measured at peel speeds of (a) 5 cm/s, (b) 11 cm/s, and (c) 19 cm/s plotted on linear scales.

or velocity measurements in three dimensions. However, in many systems, this can be provided by 'reconstructing' a vector-based time-series from measurements of a single variable [6b]. Given a set of N current measurements $\{I_j\colon\ j = 1 \ldots N\}$, we define a sequence of D-dimensional vectors $\mathbf{X}_j$ using current measurements made at intervals Δt (the 'delay time'); Δt is taken as a multiple of experimental sampling interval. (For instance, we employ $\Delta t = 5$ ms with current measurements made at 10 μs intervals.) The resulting vectors have the form

$$\mathbf{X}_j = \left\{I_j, I_{j+n}, I_{j+2n}, \ldots, I_{j+(m-1)n}\colon\ j = 1 \ldots (N - nD)\right\}. \tag{1}$$

Provided the measurements $\{I_j\}$ adequately reflect the system dynamics and that the dimension D is sufficiently high, the sequence of vectors $\{\mathbf{X}_j\}$ will have same topological form and dimension as the 'true' system using physically relevant variables. For example, a particle moving in a circle in a two-dimensional (x, y)-space will have a 'true trajectory' of the form $\{x = \cos(\omega t),\ y = \sin(\omega t)\}$, while a trajectory reconstructed from x-measurements will have the form $\{x_1 = \cos(\omega t),\ x_2 = \cos[\omega(t + \Delta t)]\}$. Except for certain 'anomalous' choices of Δt, the reconstruction is an ellipse, which has the same topology and dimension ($D = 2$) as a circle. Anomalous Δt values can generally be detected by trying several values. Similar reconstructed trajectories have been used to characterize the dynamics of various chaotic processes such as turbulence [7].

In particular, the physical object defined by $\{\mathbf{X}_j\}$ will have a (possibly fractal) dimension D_f. For D greater than D_f, the dimension of the set of $\mathbf{X}_j$ values is a property of the system. For instance, a circle has a dimension $D_\mathrm{f} = 2$ in a coordinate system of any dimension $D \geqslant 2$. Since the calculation of a physical dimension (D_f) is often computationally intractable, other measures of dimension have been developed. A particularly useful measure has been derived by Grassberger and Procaccia [8]. They define a correlation integral, $C(r)$, as

$$C(r) = \lim_{N\to\infty} \frac{1}{(N^2)} \times \left\{\text{number of pairs } (x_i, x_j) \text{ such that } |x_i - x_j| < r\right\}. \tag{2}$$

In systems with a well defined dimension, $C(r)$ scales as a power law in r. Thus one defines a correlation dimension D_c, such that

$$D_\mathrm{c} = \lim_{r\to 0} \frac{\ln[C(r)]}{\ln(r)}, \tag{3}$$

where D_c equals the slope of C vs. r when plotted on a log–log plot. In many systems, D_c is numerically equal to the conventional fractal dimension, and it can be shown that in all cases $D_\mathrm{f} \leqslant D_\mathrm{c}$.

Figure 5 shows a log–log plot of $C(r)$ vs. r for the current signal at a peel speed of 19 cm/s. The slopes (and thus dimensions) are consistent for $3 < D < 7$, as the superposition of curves indicates. (The quantity of data required to faithfully represent the system dynamics increases with increasing D; more data would be required

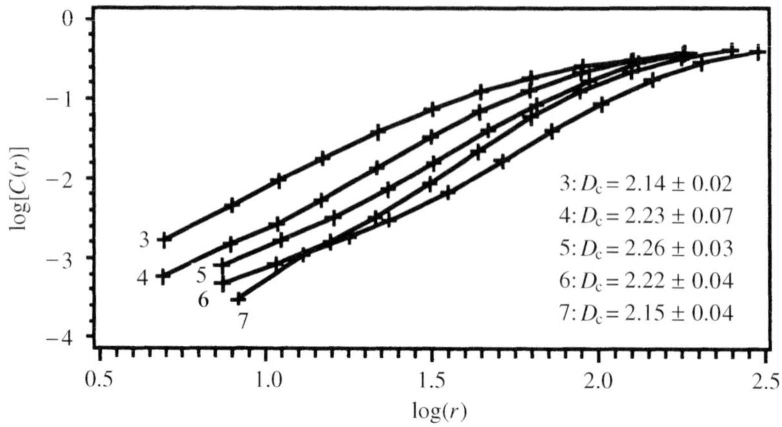

Figure 5. log–log plots of $C(r)$ vs. r for embedding dimensions $D = 3$–7 for the current signal measured at a peel speed of 19 cm/s.

for $D > 7$.) Least squares fits yield $D_c \approx 2.3$ for $3 < D < 7$. Noninteger correlation dimensions are generally associated with fractal structures and are consistent with the presence of low dimensional chaos. Analysis of the current signals at the lower peel speeds yielded very similar values of D_c. The fact that $D_c \ll D$ for the higher dimensions rules out 'noise' as the source of complex behavior in the current signals. Noise 'fills space' in any dimension, so that $D_c = D$ for all D's, at least to a good approximation [9].

The ability of the D_c calculation to rule out stochastic processes was verified by modeling the current data in terms of a rather general stochastic process (an autoregressive series). An autoregressive series can be constructed whose power spectrum is statistically consistent with any given set of data [10]. Model data sets generated in this fashion provide a strong test of procedures designed to distinguish random processes from deterministic ones. We note that 'model data' generated in this way do not have the same general appearance as our current signals, consistent with the conclusion that the current fluctuations are not due to truly random processes. Further, D_c calculations for the autoregressive series yielded effectively infinite values — i.e. $D_c \approx D$ for $D \leqslant 8$ (the highest dimension tested). Therefore, the D_c calculation distinguishes between our data and a known random process with a similar power spectrum. The low dimension of the current signals cannot be attributed to random, noise-like fluctuations.

The dimension D_f ($\leqslant D_c$) gives a rough idea of the number of physical variables which would reasonably be needed to describe the system. This corresponds to the dimension of the 'phase space' used to describe the system, or alternatively, to the number of coupled differential equations required in a mathematical model. The minimum number of variables is simply the first integer above D_f, while the maximum number is the first integer above $2D_f$ [11]. (Truly chaotic systems require at least 3 variables.) With $D_f \sim 2.3$, the number of variables required for a physical description of the current signals lies between 3 and 5. In many systems, the number of required variables is equal to the minimum value [12]. This suggests that a fairly simple dynamical model could adequately characterize the chaotic behavior of the peel front.

We note that D_c for these signals is significantly lower than that obtained in previous measurements ($D_c \approx 5.5$) [2], where currents from non-coated tape (no carbon coating) were measured with an electrometer. We attribute the high dimension obtained in previous work to higher levels of 60 Hz (and harmonics) interference — which can increment the measured dimension by integral amounts. Another potential source of higher dimensionality in previous work was breakdown (small electrical discharges) during peeling; these were further minimized in this work by the carbon coating and by faster peeling. As noted above, the higher dimension measurements in the previous work cannot be attributed to random processes (i.e. most 'noise' signals), which effectively prevent the convergence of the computation ($D_c \approx D$ for all D's).

3.3. Lyapunov exponent

The most definitive diagnostic for dynamical chaos is the determination of a positive Lyapunov exponent, λ. Simplistically, one can imagine that nearby orbits (paths traced out by the successive vectors $\mathbf{X}_i$) diverge or converge at a rate proportional to $\exp(\lambda t)$. Following convention, we express λ in units of bits per s, in which case the appropriate expression is $2^{\lambda t}$. If nearby orbits diverge exponentially, as in chaotic systems, the largest Lyapunov exponent will be positive. When nearby orbits diverge in this manner, predicting future behavior soon becomes problematic. For instance, if one has measured the initial state of a system with an accuracy of N bits (e.g. with an N-bit digitizer), the best possible accuracy in predictions of future behavior will decrease to $(N-1)$ bits after a time $\tau = \lambda^{-1}$. One bit of accuracy (information) is lost in every time interval τ.

The largest Lyapunov exponent was determined using an algorithm developed by Wolf *et al.*, because it is fairly robust in the presence of noise [13]. This estimate becomes

$$\lambda = \frac{1}{t_f - t_0} \sum_{i=0}^{P \times \Delta m} \log_2 \left(\frac{|\mathbf{X}_{i+\Delta m} - \mathbf{X}_{j+\Delta m}|}{|\mathbf{X}_i - \mathbf{X}_j|} \right), \tag{4}$$

where $|\mathbf{X}_i - \mathbf{X}_j|$ is the distance between points $\mathbf{X}_i$ and $\mathbf{X}_j$ (where $\mathbf{X}_j$ is deliberately chosen to lie close to $\mathbf{X}_i$) and $|\mathbf{X}_{i+\Delta m} - \mathbf{X}_{j+\Delta m}|$ is the separation of the successive points at a time $\Delta m \cdot \Delta t$ later. Because nearby trajectories cannot diverge indefinitely, the nearby point $\mathbf{X}_j$ is replaced at regular intervals. (The series $\{\mathbf{X}_m\}$ is searched for a point suitably close to $\mathbf{X}_i$.) P is the number of these replacements and t_f and t_0 are the times at the beginning and end of the series $\{\mathbf{X}_m\}$, respectively.

Figure 6 shows plots of the Lyapunov exponent estimate vs. the evolution time $t = \Delta m \cdot \Delta t$ for the three data sets in Fig. 2. A valid estimate will be stable over a range of evolution times. These estimates show a plateau of stable estimates between 5 and 15 ms. The corresponding Lyapunov exponents are independent of the peel speed over this range and are equal to $\approx$100 bits/s. These estimates were quite stable with respect to the choice of delay time Δt, and reasonably stable with respect to the embedding dimension D. (Raising D from 4 to 5 would typically raise λ

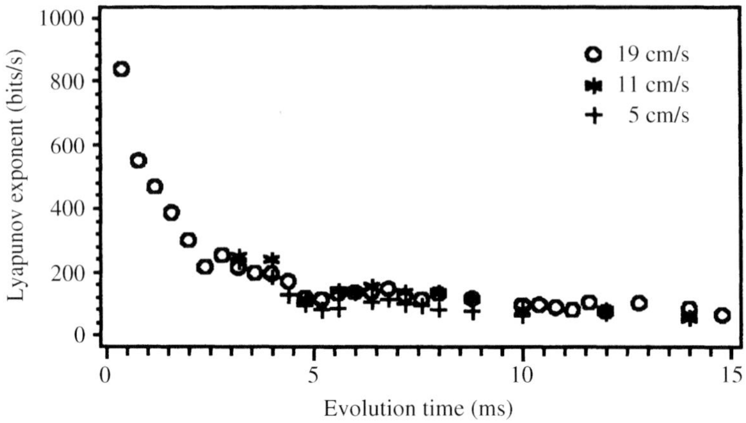

Figure 6. Lyapunov exponent estimates determined from the current signals. Estimates employed data vectors $\{\mathbf{X}_i\}$ of dimension $D = 4$, where the delay time $\Delta t = 5$ ms.

about 10%.) This stability provides assurance that the large, positive value of λ is not due to trajectories that appear to be close to each other when D is too low, but are actually far apart in the 'true' phase space of higher dimension; such trajectories often diverge rapidly and yield a false-positive exponent. (Similar problems can arise if Δt is too small.)

The positive value of λ indicates that this system is definitely chaotic in the deterministic sense. If the initial state of this system is known with an accuracy of N bits, the ability to predict its future state will be degraded at a rate of 1 bit every 10 ms ($\tau = 1/\lambda$). Although it is not clear how λ should be related to the physical properties of the adhesive–substrate system, we note that 10 ms is a very reasonable time scale for the onset of the rubber-to-glass transition in the adhesive phase at room temperature [14].

4. CONCLUSIONS

We have shown that the current signals generated during peeling of carbon-coated pressure sensitive tape in air exhibit chaotic behavior. Using the current signals to 'reconstruct' the behavior of the underlying dynamical system, we calculate a correlation dimension $D_c \approx 2.3$. Thus the dynamics of peel is fractal in character. Further, the small size of D_c suggests that a fairly simple dynamical model, with 3–5 degrees of freedom, could adequately characterize the chaotic behavior of the peel front. Analysis has also verified the presence of a positive Lyapunov exponent of $\lambda \approx 100$ bits/s — conclusive evidence for deterministic chaos. Since the current is produced by the motion of charged adhesive accompanying interfacial failure, we attribute this behavior to the chaotic progress of failure along the adhesive/substrate interface. Therefore, fluctuations in progress of the peel front under the conditions of this experiment become inherently unpredictable on time scales of tens of ms.

Additional information on the dynamics of this system may be obtained from the power spectra of the current fluctuations, although further work is required to identify the specific mechanical processes involved. The fluctuations at all frequencies below 10 kHz grow significantly as the peel speed is increased from 5 cm/s to 11 cm/s. As the peel speed is further increased to 19 cm/s, the spectral power in the 100–1000 Hz region actually decreases. This response is consistent with a stiffening of the adhesive at high strain rates. At the high frequency end of the power spectra (1–10 kHz), 'white noise' behavior becomes especially prominent at the high peel speeds. The power in this region is well above background, indicating the presence of low-amplitude random fluctuations. The small-scale fluctuations responsible for this random behavior are too small and too fast for convenient observation by photographic techniques. Also, the corresponding fluctuations in peel force would be quite small, since the fraction of the total load supported by the small, rapidly moving adhesive fibrils along the peel front is small. Novel techniques, such as the current measurements presented here, are therefore essential for their study.

Acknowledgements

This work was supported by the National Science Foundation under Grant CMS-9414405 and by the 3M Company.

REFERENCES

1. S. Lee, L. C. Jensen, S. C. Langford and J. T. Dickinson, *J. Adhesion Sci. Technol.* **9**, 1 (1995).
2. L. Scudiero, J. T. Dickinson, L. C. Jensen and S. C. Langford, *J. Adhesion Sci. Technol.* **9**, 27 (1995).
3. J. T. Dickinson, L. C. Jensen, S. Lee, L. Scudiero and S. C. Langford, *J. Adhesion Sci. Technol.* **8**, 1285 (1994).
4. R. J. Good and R. K. Gupta, *J. Adhesion* **26**, 13 (1988).
5. D. C. Hong and S. Yue, *Phys. Rev. Lett.* **74**, 254 (1995).
6. (a) F. C. Moon, *Chaotic and Fractal Dynamics* p. 55. John Wiley, New York (1992).
   (b) F. C. Moon, *Chaotic and Fractal Dynamics* p. 355 John Wiley, New York (1992).
7. A. Brandstäter, J. Swift, H. L. Swinney, A. Wolf, J. D. Farmer, E. Jen and J. P. Crutchfield, *Phys. Rev. Lett.* **51**, 1442 (1983).
8. P. Grassberger and I. Procaccia, *Physica* **9D**, 189 (1983).
9. A. Ben-Mizrachi and I. Procaccia, *Phys. Rev. A* **29**, 975 (1984).
10. S. Haykin and S. Kesler, in: *Nonlinear Methods of Spectral Analysis*, S. Haykin (Ed.), pp. 9–72. Springer, Berlin (1979).
11. H. D. I. Abarbanel, R. Brown, J. J. Sidorowich and L. Sh. Tsimring, *Rev. Mod. Phys.* **65**, 1331 (1993).
12. J. P. Crutchfield and B. S. McNamara, *Complex Systems* **1**, 417 (1987).
13. A. Wolf, J. B. Swift, H. L. Swinney and J. A. Vastano, *Physica* **16D**, 189 (1985).
14. D. H. Kaelble, *J. Adhesion* **1**, 102 (1969).

Mittal Festschrift, pp. 597–613
W. J. Van Ooij and H. R. Anderson, Jr. (Eds)

Electrical currents produced by peeling a pressure sensitive adhesive: Role of electrostatic forces in adhesion

L. SCUDIERO, S. C. LANGFORD and J. T. DICKINSON*
Physics Department, Washington State University, Pullman, WA 99164-2814, USA

Abstract—We report further measurements of current generated by peeling a pressure sensitive tape from a copper substrate where greater care was taken to minimize electrical breakdown and charge recombination across the failed interface. The highest currents, produced by peeling gold-coated tape in vacuum, correspond to surface charge densities $\geqslant 2.8$ mC/m^2. With typical displacements of 5 μm, this charge density would increase the work required for interfacial failure by 2 J/m^2; this is much greater than the thermodynamic work of adhesion (≈ 33 mJ/m^2) but much less than the observed peel energy (200–300 J/m^2). These currents can be manipulated by applying a bias voltage across the substrate and coating. The good correlation between peel force and current measurements under these conditions suggests that the peel force would be 30–60% lower in the absence of electrostatic effects. This result is discussed in terms of the irreversible nature of the charge separation during peeling.

Keywords: Pressure sensitive adhesive; interfacial failure; contact electrification; peel energy; work of adhesion.

1. INTRODUCTION

The role of electrostatic forces in adhesion is a matter of some controversy [1–4]. Considerable experimental effort has focused on measurements of the charge distribution near the interface, where electrical double layers can contribute to the work of adhesion. In many systems, however, the fracture energy is orders of magnitude higher than any plausible work of adhesion. This difference is due to irreversible processes accompanying failure, e.g. viscosity or plasticity in deformable phases. These dissipative processes depend strongly on the force required for interfacial failure; with significant charging across the interface, this force can be significantly greater than that required for 'reversible' bond breaking at the interface. In this work, we report improved measurements [5, 6] of charge separation across the interface between a metal and a pressure sensitive adhesive — showing that the total charge involved can be considerable. Correlations between the measured charge and the peel force suggest that electrostatic effects play an important role in enhancing deformation. Further, we

*To whom correspondence should be addressed.

argue that this enhanced peel energy does not necessarily imply a large electrostatic contribution to the thermodynamic work of adhesion.

In previous work [5, 6], we reported measurements of the current generated during the peeling of pressure sensitive tapes from copper substrates. This work was a direct extension of earlier measurements of currents generated during failure of epoxy/metal interfaces [7], as well as additional studies of electrical activity during peeling [8–12]. Peeling a pressure sensitive tape in air at speeds below 1 mm/s yielded coincident current spikes, flashes of light, and radio-frequency emissions associated with electrical breakdown events. Breakdown and tunneling limit the charge densities that can be supported on the separating surfaces. Thus charge density measurements based on current measurements (as presented here) are necessarily conservative lower bounds. In this work, three strategies were employed for minimizing charge recombination: (1) a thin conductive coating was applied to the tape backing to reduce the electric field between the adhesive and the substrate [13]; (2) peeling was performed in vacuum to raise the electric field required for breakdown; and finally, (3) peeling was performed at relatively high velocities to reduce the time available for charge recombination. With these precautions, charge densities as high as 2.8 mC/m^2 were observed.

The interpretation of simultaneous current measurements and peel energies is complicated by the fact that many treatments which reduce the charge density (e.g. raising the humidity) also affect other adhesion forces (e.g. van der Waals forces). In this work we show that applying a small negative bias voltage to the conductive backing can significantly reduce the currents measured during peeling. Although this phenomenon is not well understood, the lower currents correlated with significantly lower peel energies. Similarly, tapes provided with a thin conductive coating (reducing the charge on the substrate during peeling) displayed 10–15% lower peel forces than non-coated samples. Taken together, we conclude that at least in selected adhesive/substrate combinations, electrostatic forces can significantly enhance the energy dissipated during peeling by increasing the interfacial forces responsible for polymer deformation.

2. EXPERIMENTAL

A schematic diagram of the peel geometry for the current measurements appears in Fig. 1. This work employed 3/4 in. (19 mm) strips of 3M Scotch Brand Magic Tape (No. 810), which consists of a long-side-chain alkyl acrylate adhesive on a cellulose acetate backing. The substrates included copper sheet ($58 \times 27 \times 0.5$ mm^3) and glass slides (Gold Seal microscope slides from Becton, Dickinson and Company, Lincoln Park, NJ, USA). The copper was first smoothed with sandpaper through three successive grades and polished to a mirror finish with Simichrome polish (Competition Chemicals, Iowa City, Iowa, USA). Before each experiment, the copper block was electropolished at a potential of 6 V in concentrated phosphoric acid for 15–20 s and thoroughly rinsed with distilled water and acetone. The residual

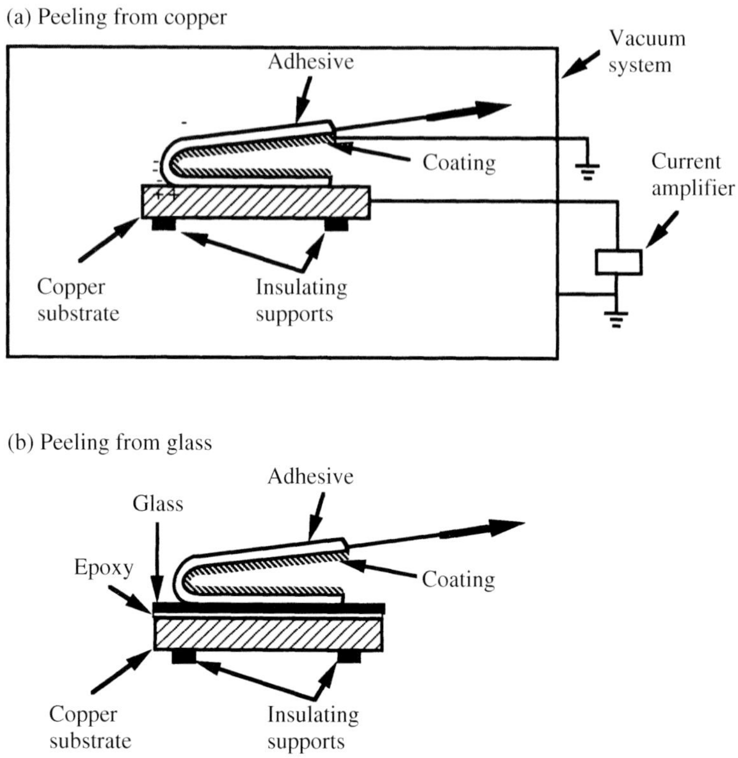

Figure 1. Diagram of experimental geometry for 180° peel (a) from copper and (b) from glass.

electric charge on the adhesive surface of the tape was neutralized prior to attachment with a Zerostat®, a commercial device used to neutralize static electric charge on phonograph records. Negligible residual charge (<100 nC/m^2) was detected in Faraday cup measurements. Vinyl gloves were worn during attachment and subsequent manipulation to avoid humidity effects due to the slight dampness of the hands.

After attachment, conductive coatings were applied to the tape backing to minimize charge recombination. Thin gold films (20 nm thick) were sputter deposited onto the backing under low vacuum (10 Pa Ar) and displayed a resistance of about 0.5 kΩ over a 10-cm length of tape. Alternatively, conductive carbon films were sprayed onto the tape backing (EMI-RFI Shield from GC Electronics); these films were much easier to prepare, but displayed considerably higher resistances (2.5–5.0 kΩ over a 10-cm length of tape). The conducting film was connected to an electrical ground prior to peeling. Some samples were peeled in air at relative humidities less than 40%. (Peeling at higher humidities yields significantly smaller currents.) Many of the samples were peeled in a vacuum system at pressures $<1 \times 10^{-4}$ Pa; these samples experienced vacuum conditions for periods ranging from 1 h to 1 week prior to peeling.

The current delivered to the copper substrate during peeling was measured with a Keithley Model 427 current amplifier, which has a rise/fall time response of about 50 μs at the sensitivity employed in this study. The output of the current amplifier was digitized at 50 μs intervals using a LeCroy 6810 transient waveform digitizer. Peel speeds ranged from 0.5 to 6.5 mm/s.

The effect of applying an electrical potential to carbon-coated tapes was also investigated in a series of experiments in air. The coating was allowed to dry for 2–10 h before applying the voltage. Current measurements were made with the voltage applied to the coating. The data described below where taken ~24 h after the application of the voltage. However, experiments performed within minutes after applying the adhesive show comparable features.

The magnitude of the corresponding surface charge density, σ, was determined from the observed current, I, by the relation

$$\sigma = \frac{Q}{A} = \frac{I}{vw}, \tag{1}$$

where v and w are the peel rate and the width of the tape, respectively.

We note that these current measurements reflect only charge remaining on the separated interfaces after failure and do not detect charge flowing directly from the charged adhesive to the copper substrate. As this direct recombination cannot be eliminated, equation (1) is an underestimate of the surface charge density. Other processes, which might lead to an overestimate of the surface charge, are expected to contribute very little to the observed currents. Current transport through the backing (in either direction) is limited by its high resistance, $>10^{10}\ \Omega$ for the total sample area. Similarly, free charge on the copper due to contact potential differences should be negligible. Contact potential differences (and other uncontrolled potential differences) are far less than 5 V. For a typical sample capacitance of 400 nF/m^2, this would increase the measured charge density by 0.002 mC/m^2 ($\sigma = QV/A$). This is less than 10^{-3} of the maximum charge densities detected in this work. Thus we are confident that the charge densities inferred from the detected currents represent conservative lower bounds.

To verify the work of adhesion in these adhesive/substrate systems, contact angle measurements were made by the sessile drop technique. Under rather general assumptions about the materials interactions, the thermodynamic work of adhesion, W_{ad}, is related to the surface free energy of the adhesive, γ_{adh}, and the equilibrium contact angle, θ_e, at the adhesive–substrate triple point by the Young–Dupré relation [14]:

$$W_{ad} = \gamma_{adh}(\cos\theta_e + 1). \tag{2}$$

Average values of θ_e were 38 and 27° for adhesive drops deposited on clean copper and glass plates, respectively. The surface energy of the adhesive, $\gamma_{adh} \approx 18$ mJ/m^2 (at 20°C) was determined by similar contact angle measurements where drops of distilled water were applied on the flat strip of tape with the adhesive facing upward ($\gamma_{water} = 72.8$ mJ/m^2 at 20°C). The observed contact angles for the adhesive on copper and glass then correspond to $W_{ad} = 33$ and 35 mJ/m^2, respectively.

3. RESULTS

3.1. Current measurements in air

Figure 2 shows signals generated by non-coated and gold-coated tape peeled in air at a rate of 3.2 mm/s. Both tapes yield positive currents with numerous negative spikes due to electrical discharges [5]. However, the average current from the coated tape is an order of magnitude greater than the current from the non-coated tape (38 nA vs. ~1 nA). The corresponding charge densities are 0.79 and 0.02 mC/m^2, respectively. Without the coating, a great deal of charge flows directly from the copper to the adhesive (direct recombination), and this charge is not detected by the electrometer. The grounded coating essentially provides an alternative, low resistance path for charge transport from the copper to the backing, and this current is detected by the electrometer. Therefore the coating allows for the detection of a much larger fraction of the initial surface charge. The 'detected' charge density from the gold-coated tape

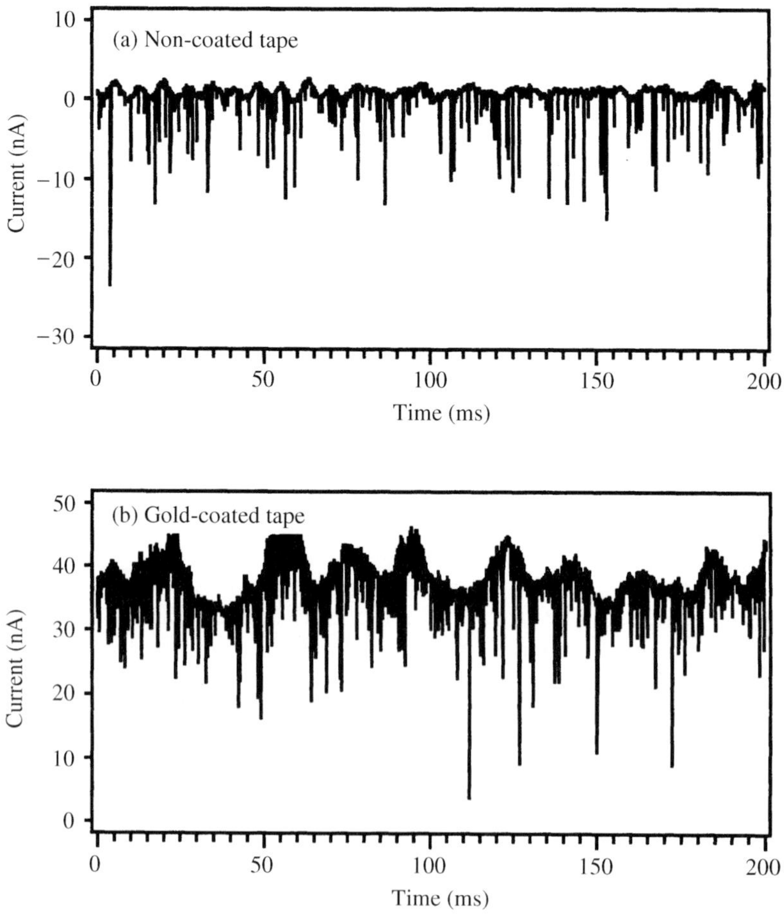

Figure 2. Current signals generated by peeling of (a) non-coated tape and (b) gold-coated tape in air at a speed of 3.2 mm/s.

($0.79\ \mathrm{mC/m^2}$) is similar to that reported by Johnson and Wallner, who measured the current generated by peeling coated tape from aluminum ($0.26\ \mathrm{mC/m^2}$) [13].

The number of negative spikes observed during peeling in air is similar for the coated and non-coated tapes, despite the much higher currents obtained from the coated tape. In previous work, we have shown that negative current spikes correspond to electrical breakdown events [5]. Electrical breakdown involves a direct transfer of charge from the copper to the adhesive, and again this charge is not detected by the electrometer. In addition, ions produced during breakdown adsorb onto the charged adhesive and metal surfaces, further reducing the surface charge densities. These effects significantly reduce the surface charge densities before they can contribute to the measured current.

3.2. Current measurements in vacuum

To minimize the frequency and intensity of electrical breakdown events, peeling was performed in vacuum. Current signals during peeling of non-coated and gold-coated tape in vacuum at 3.2 mm/s are shown in Fig. 3. The magnitudes of the average currents, 2 and 154 nA, respectively, correspond to surface charge densities of 0.05 and $2.8\ \mathrm{mC/m^2}$. By peeling in vacuum, the detected current has increased another factor of four in the gold-coated sample. Thus charge recombination due to breakdown during peeling in air is considerable. Because it is highly unlikely that charge recombination has been eliminated by peeling in vacuum, the calculated interfacial charge density for the coated sample ($2.8\ \mathrm{mC/m^2}$) is considered a conservative *lower bound* in this system.

Note that in vacuum, the current signal from the non-coated sample shows large numbers of positive current spikes. (Negative current spikes are observed in air.) In vacuum, breakdown produces positive current spikes due to the high secondary electron efficiency (>1) of oxidized copper [5]. When free electrons produced by electrical discharge events strike the positively charged (oxidized) copper surface, they stimulate electron emission — and the number of electrons emitted from the copper is greater than the number of electrons striking it. Nevertheless, even in vacuum, the average current detected from non-coated tape is orders of magnitude less than from the coated tape. In the non-coated tape, direct charge transfer, presumably due to surface currents and tunneling, strongly reduces the charge detected by the electrometer.

The charge density inferred from current measurements in vacuum is a fairly weak function of peel speed, as shown in Fig. 4. Between 0.5 and 4 mm/s, this current density varies no more than a factor of two. At high peel speeds, the adhesive is drawn quickly away from the copper and less time is available for direct charge recombination; all other things being equal, this would raise the detected current. However, charge flow from the copper to the coating is also characterized by a characteristic time (here $\tau = R_{\mathrm{load}} C_{\mathrm{sample}} \approx 100\ \mathrm{k\Omega} \times 1\ \mathrm{nF} \approx 100\ \mu\mathrm{s}$), and peeling at very high peel speeds may not allow sufficient time for this charge flow. Thus we expect that the highest measured charge densities will occur at a peel speed that minimizes direct recombination without seriously compromising charge flow to the coating.

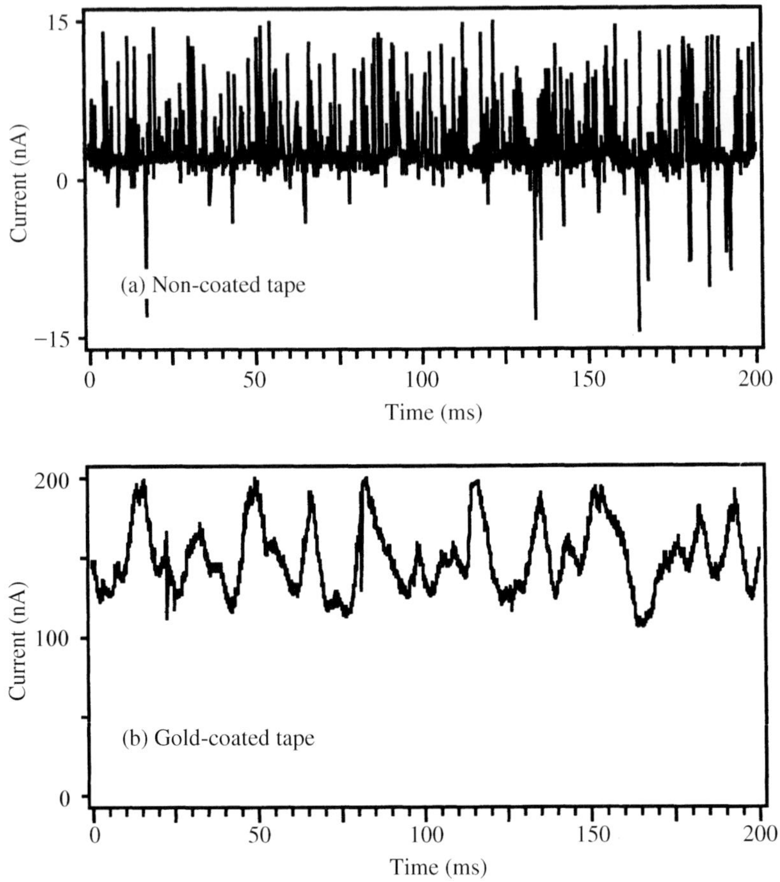

Figure 3. Current signals generated by peeling of (a) non-coated tape and (b) gold-coated tape in vacuum at a speed of 3.2 mm/s.

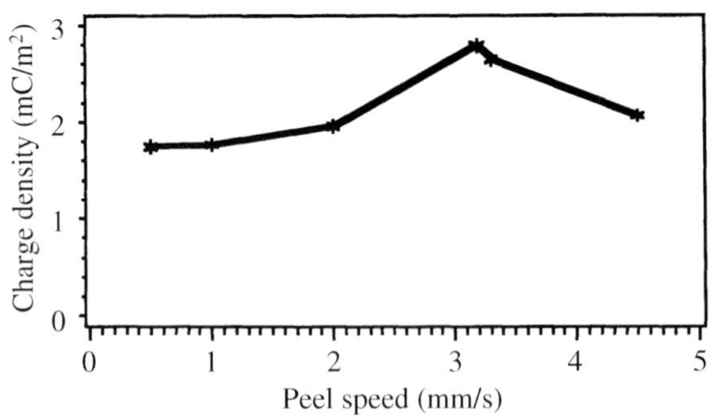

Figure 4. Surface charge density vs. peel speed for gold-coated tape peeled in vacuum.

Measurements with carbon-coated samples peeled in vacuum yield somewhat lower currents. Figure 5a shows a typical current signal generated by peeling carbon-coated tape from copper in vacuum at a peel speed of 3.5 mm/s. The average current from

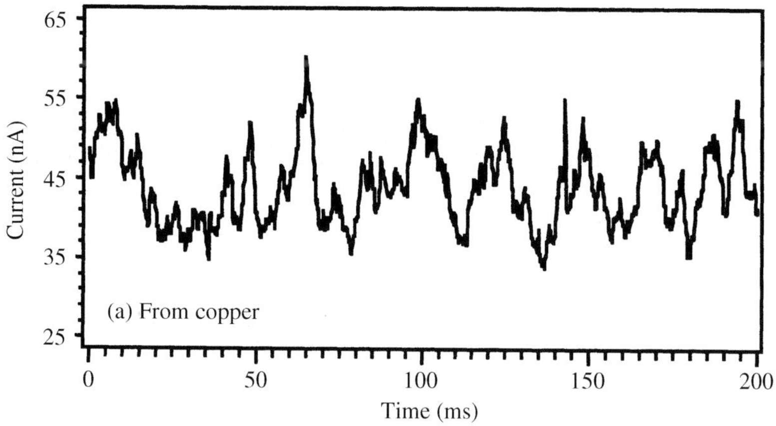

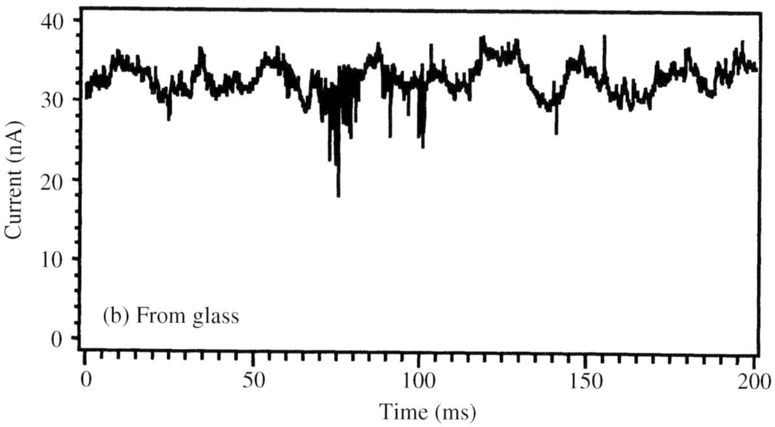

Figure 5. Current signals generated during peeling of carbon-coated tape from (a) copper and (b) glass in vacuum at a peel speed of 3.5 mm/s.

the copper substrate is 45 nA, corresponding to a charge density of 0.980 mC/m^2. This is almost a factor of three lower than the current detected from gold-coated tape under similar conditions. We attribute the lower current from the carbon-coated sample to the high ohmic resistance of the carbon films (2.5–5 kΩ) relative to the gold films (0.5 kΩ). The high resistivity of the carbon hinders charge flow from the copper to the coating and thus is not as effective in preventing direct recombination from the copper to the adhesive.

To assess the potential for similar charge density measurements involving insulating substrates, current measurements were made with a glass slide interposed between the copper and the adhesive. (The arrangement of the copper electrode and the glass substrate is indicated in Fig. 1b.) Here, the adhesive was peeled from the glass, and the measured current reflects charge separation along the adhesive/glass interface. A typical current signal is shown in Fig. 5b. The average current of 31 nA corresponds to a surface charge density of 0.87 mC/m^2. Although the current during peeling from copper shows little evidence for current spikes (Fig. 5a), peeling from glass shows a significant number of small, negative current spikes. The presence of the glass between the copper electrode and the carbon coating significantly lowers the sample capacitance and increases the potential required for charge transfer to the tape backing ($V = Q/C$). This higher potential appears across the separating surfaces, leading to breakdown and enhancing other recombination processes. Therefore, we expect that the charge density on the glass is significantly underestimated.

3.3. *Effect of coating on peel force*

Although the coating significantly improves the fraction of the initial surface charge detected by the electrometer, charge removed from the copper substrate no longer contributes to the electrostatic force acting between the copper and the failed portions of adhesive. The effect of the backing on the resulting peel force is illustrated in Fig. 6, where we show simultaneous force and current measurements (to the copper)

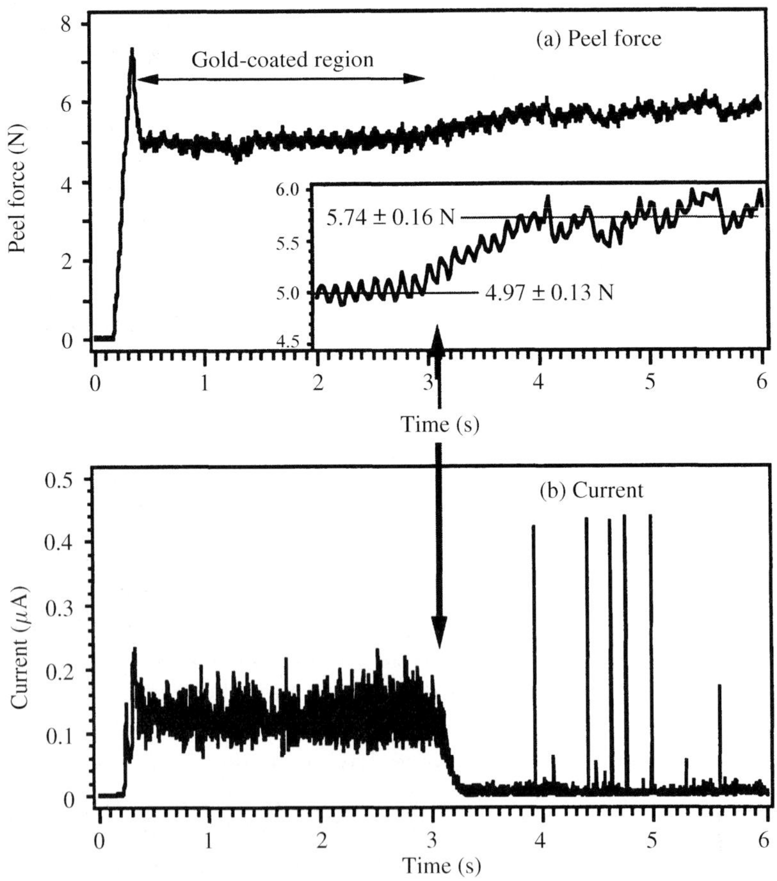

Figure 6. (a) Peel force and (b) current measured during peeling in vacuum (6.5 mm/s). Only the first half of the tape (to the left of the arrow) is coated with gold. The second half of the tape (to the right of the arrow) was left non-coated. The inset in (a) shows the peel force on a more sensitive vertical scale.

where a gold coating was applied to the first half of the tape only. Peel was performed in vacuum at a peel speed of 6.5 mm/s. The grounded gold-coated portion of the tape was peeled during the first 3 s of data collection, and the non-coated portion was peeled during the last 3 s of data collection. As expected, the current during the peeling of the non-coated portion drops to very low levels; however, the peel force for the non-coated portion rises about 15%. The peel force during peeling of non-coated tape in vacuum was typically 10–15% higher than from gold-coated tape ($\approx$300 N/m for the non-coated tape vs. $\approx$260 N/m for the coated tape).

Since the gold itself is extremely thin (30 nm), we cannot attribute this difference to a change in the mechanical properties of the adhesive/backing system. Rather, the change in peel force is due to a change in electrical properties of the system. Without the coating, charge remaining on the adhesive immediately after detachment is attracted to their image charges in the copper. With the coating, the image charges are soon transferred to the coating (through the electrometer) and the electrical attraction is now to the coating and not the substrate. This does not affect the electrostatic forces along intact portions of the interface, which can contribute to the strength of the interface. Since direct recombination in non-coated tape dramatically reduces the surface charge densities, a 10–15% higher peel force is remarkable. This suggests that the (presumably much larger) electrostatic forces acting across intact portions of the interface may strongly affect the mechanics of peeling.

3.4. Effect of substrate bias

Although removing the conductive coating can change the surface charge densities during peeling, the lack of a comparable charge density measurement for the non-coated tape prevents quantitative comparisons between charge density and peel force. In an attempt to modify the surface charge during peeling without physically altering the coating, small voltages were applied to the backing. Because of the large number of samples involved, these measurements employed carbon-coated tape (which is easier to prepare) and peeling was performed in air. The effect of varying the applied bias on the average current during peel is shown in Fig. 7 for three experimental runs. Figure 7a shows preliminary current measurements in which the peel force was not measured. Figures 7b and 7c show later measurements where the current and peel force were measured simultaneously. Surprisingly, a pronounced drop in current is observed when the carbon backing is held at -3 ± 0.3 V relative to the grounded copper substrate. In contrast, at higher and lower biases the detected current is roughly equal to the current produced when the carbon backing is grounded (0 V bias). Experiments where peeling is performed within minutes after applying the adhesive show currents quite comparable to those displayed above, where contact had been established for 24 h prior to peeling.

We emphasize that the direct contribution of the bias voltage to the free charge along the copper–adhesive interface is entirely negligible; furthermore, it would not be expected to show a ‘resonance’. The magnitude of the free charge associated with the bias voltage is given by the product of the applied voltage and the capacitance of the coating/tape/substrate structure. For a typical sample capacitance of 400 nF/m^2 and

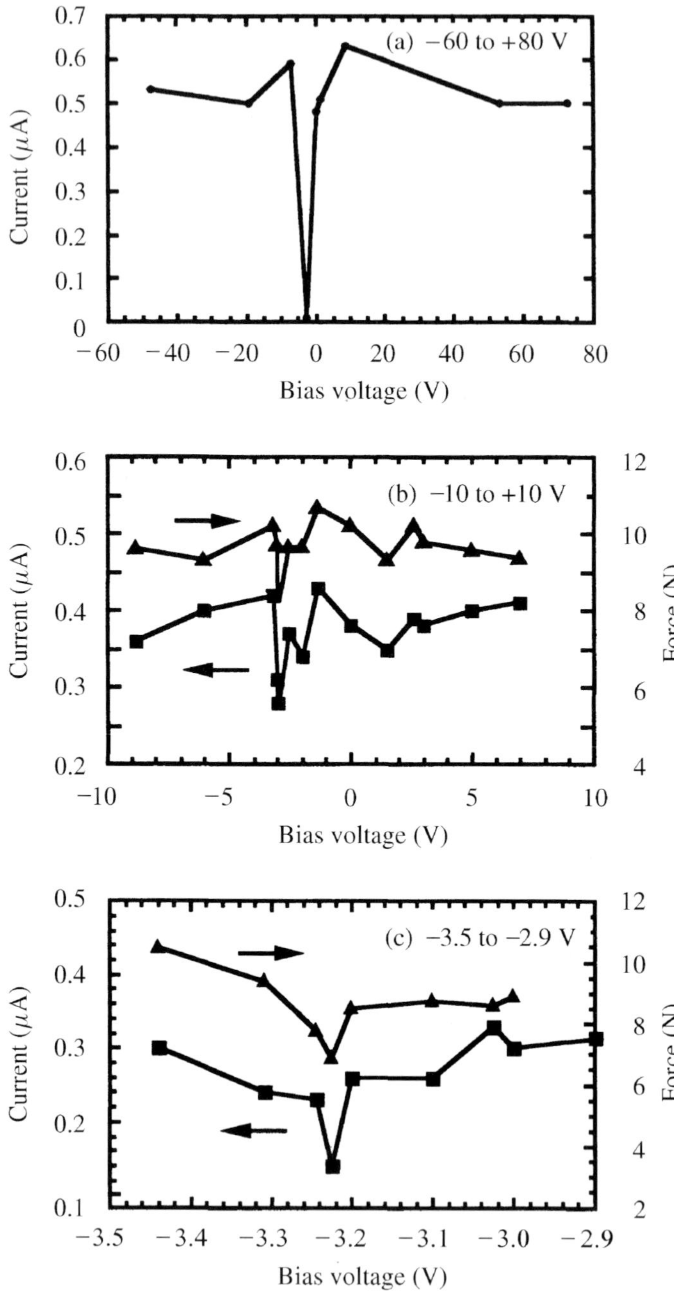

Figure 7. Average current and force measured during peeling of carbon-coated tape from copper in air as a function of bias voltage applied to carbon coating: (a) bias ranging from −60 to +80 V — peel speed 9.0 cm/s (force measurements not performed); (b) bias ranging from −10 to +10 V — peel speed 4.4 cm/s; (c) bias ranging from −3.5 to −2.9 V — peel speed 4.4 cm/s.

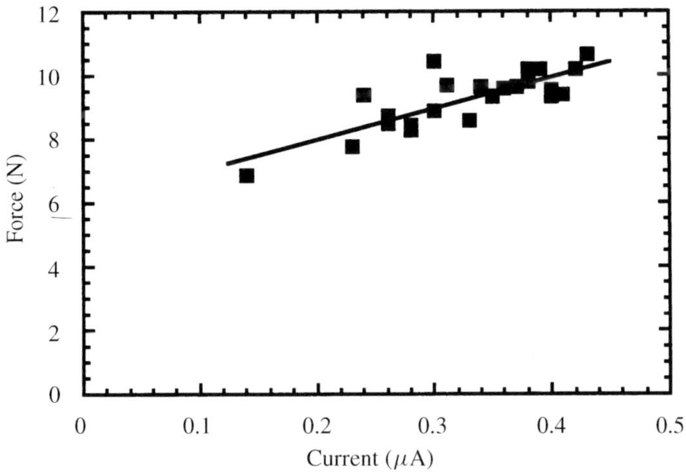

Figure 8. Peel force plotted as a function of peel current from the data of Figs 7b and 7c.

a bias of 100 V (the highest voltage applied in this work), the corresponding charge density is only 0.04 mC/m^2. This charge density would account for less than 5% of the observed current. At −3 V, the free charge associated with the bias voltage is about 0.001 mC/m^2. A similar argument rules out significant charge densities due to the contact potential difference between the copper substrate and the conductive coating, which is at most one or two volts. The electrical double layer along the adhesive–copper interface could produce comparable potential differences with similarly small effects on the observed current. Below we argue that an appropriate bias voltage may enhance charge transfer (re-neutralization) via tunneling as the two surfaces are drawn apart.

The current and peel force measurements in Figs 7b and 7c are correlated. Figure 8 displays the peel force as a function of the measured current. The correlation between peel force and current is significant (correlation coefficient 0.80). Extrapolating the peel force to zero current suggests that electrostatic effects are responsible for about 40% of the maximum peel force. Similar analyses of subsets of this data are consistent with enhancements in the 30–60% range. This shows convincingly that electrostatic effects can substantially enhance the peel force.

4. DISCUSSION

4.1. Electrostatic contribution to the work of adhesion

Historically, the contribution of electrostatic forces to adhesion has been treated in terms of reversible formation and failure of the interface. In this context, the work required to form the interface from two isolated surfaces is a well-defined thermodynamic state variable, the work of adhesion, W_{ad}. The contribution of electrostatic forces to W_{ad} is then the work required to produce the appropriate charge distribution, where we assume that the charges on the metal phase are concentrated at the interface and that charges in the adhesive are distributed into the bulk an average distance d from the interface. This is the geometry of an electrical double layer. For planar interfaces, the work required to establish this charge distribution is equal to the electrostatic force times the average separation. The electrostatic contribution to the work of adhesion, W_{el} then becomes [15]:

$$W_{el} = \frac{\sigma^2 d}{2\varepsilon_0 \varepsilon_r} = \sigma V_r, \tag{3}$$

where σ is the surface charge density, ε_0 is the dielectric permittivity of free space, ε_r is the relative dielectric permittivity ($\varepsilon_r = 2$–3 for typical polymers), and V_r $[=(\sigma d)/(2\varepsilon_0\varepsilon_r)]$ is the average potential of trapped charge in the adhesive relative to the metal.

The electrostatic contribution to the thermodynamic work of adhesion can therefore be estimated from the surface charge density and either d or V_r. Previous assessments of the contribution of electrostatic forces to W_{ad} have generally involved estimates or measurements of one or both of these quantities. For instance, $V_r \approx 1$ eV and $\sigma \approx 3$ mC/m^2 would yield $W_{el} \approx 3$ mJ/m^2, or about 10% of our measured work of adhesion for the copper/adhesive interface. Measurements of d typically show a complex dependence on material thickness and time of contact, as well as the identity of the conducting substrate [16–18]. A representative value of $d \approx 50$ nm yields $W_{el} \approx 10$ mJ/m^2, a somewhat larger fraction of the total work of adhesion.

4.2. Electrostatic forces acting on the intact interface

The total energy dissipated during peeling is typically orders of magnitude larger than W_{ad}, due to extensive deformation in the adhesive phase. The magnitude and duration of the forces that deform the adhesive phase are limited by the strength of the interface. Therefore W_{ad} and other factors which affect interfacial strength are critically important. In principle, electrostatic forces can contribute substantially to the interfacial strength.

The electrostatic forces acting across the intact interface can be approximated by the force attracting two parallel, charged plates:

$$\frac{F_{el}}{A} = \frac{\sigma^2}{2\varepsilon_0}. \tag{4}$$

At a charge density of 3 mC/m^2, F_{el}/A amounts to 500 kPa. (For planar surfaces, this force is independent of separation.) This is significantly smaller than the dispersion forces expected under these conditions. Assuming that the entire measured reversible W_{ad} for the adhesive–copper interface (33 mJ/m^2) is due to dispersion forces, the dispersion force per unit area at an interfacial separation of 1 Å comes to ~330 MPa [19]. However, the effect of the dispersion forces on interfacial strength is limited by their short range. (For planar surfaces, dispersion forces scale as $1/r^2$, where r is the distance between planes.)

4.3. Work done against electrostatic forces during interfacial failure

After interfacial failure, the newly formed surfaces are drawn apart. If these surfaces are charged, work is done against electrostatic forces. In the planar geometry, the electrostatic force does not depend on the separation and we can estimate the work done as the force times the displacement. Predictions of the maximum charge displacement based on typical criteria for electrical breakdown in air yield values in the range of a few μm [4]. When a conducting backing is electrically connected to a conducting substrate in vacuum, charge displacement is limited by charge flow to the backing. In a planar geometry the net charge displacement (before charge flows from the substrate to the backing) would be on the order of $l\kappa \approx 6$ μm, where l is the adhesive thickness (≈ 20 μm) and κ is the dielectric constant of the adhesive (≈ 3) — similar to the displacement in the case of breakdown-limited charge separation. (If the conducting coat is to effectively limit breakdown and other direct charge recombination processes, charge displacement with the coating must actually be somewhat less than displacement without the coating.) Displacing the measured surface charge density of 3 mC/m^2 merely 5 μm in a parallel plate geometry requires an energy of about 2 J/m^2, i.e. much more than the measured thermodynamic work of adhesion ($W_{ad} \approx 33$ mJ/m^2), but much less than the measured peel energies (200–300 J/m^2).

Although the work done against electrostatic forces is small relative to the total peel energy, it should make a major (if not dominant) contribution to the work required to separate the two interfaces. To incorporate the work done against electrostatic forces into fracture mechanics descriptions of peeling, a more generalized measure of interfacial energy (including irreversible work) is required. This generalized work of adhesion could then be incorporated into fracture mechanics descriptions of peeling that employ W_{ad}, such as that by Good and Gupta [20].

Under the conditions probed here, the total work required for interfacial failure appears to be on the order of 2 J/m^2, which corresponds to an interfacial peel force of 2 N/m. This represents a generalized force sustained by the interface (including electrically charged material on nearby portions of the failed interface) during dynamic peeling. This force must be applied by adhesive, which typically forms long, viscoelastic fibrils joining the backing and the interface. These fibrils can stretch to lengths of hundreds of microns, i.e. the displacement of the point where the fibrils attach to the backing is orders of magnitude greater than the displacement of interfacial charge (typically a few microns). Assuming that the forces at both ends of the fibrils are of similar magnitudes, the work done by the backing on the fibrils would thus be orders of magnitude higher than the work done by charge separation, as expected.

As described above, the presence of a conducting backing limits the net charge displacement during peeling. When the charge has been displaced a substantial fraction of the adhesive thickness, charge on the metallic substrate will flow to the backing. This effect actually reduces the electrostatic work performed during peeling in vacuum. In Fig. 6a, the presence of the coating reduced the peel energy (equal to the peel force divided by the tape width) by about 40 J/m^2. This change is much larger than the total work against electrostatic forces (expected to be on the order of 2 J/m^2). This again is consistent with the hypothesis that a relatively small change in the work performed against electrostatic forces (< 2 J/m^2) can produce a relatively large change in the work performed against adhesive deformation (≈ 40 J/m^2).

4.4. Role of coating–substrate bias

Applying small bias voltages across the coating–substrate gap has a strong effect on the observed current over a limited range of biases. The strong correlation between the measured peel forces and currents suggests that they share a common source, i.e. charge separation across the interface. Experiments where peeling is performed within minutes after applying the adhesive show currents quite comparable to those displayed above, where contact had been established for at least 24 h prior to peeling. Thus charge separation following interface formation appears to be a fairly rapid process. Under these experimental conditions, interfacial charging would be essentially complete before the application of a bias voltage. Therefore, the bias voltage must influence the process of charge recombination in the early stages of interfacial failure.

Although the mechanism for enhanced recombination at applied biases near -3 V is not clear, one possible explanation involves tunneling of charge carriers from the adhesive to the metal substrate during interfacial failure. Tunneling processes can be strongly enhanced in the presence of electron states with the same momentum and energy on both sides of the gap. Although metals provide a large number of unfilled states that would accommodate electrons tunneling from the adhesive, few of these states have the optimum momentum (near zero). If the applied bias is chosen so that the energy of the adhesive states equals the energy of an empty, zero-momentum copper state, tunneling could be strongly enhanced early in the failure process, when the interface experiences high stresses. This would both weaken the interface and reduce the observed currents. This mechanism suggests how recombination might be enhanced, should this be desired, e.g. by providing states favorable for tunneling by selection of an appropriate substrate or by introducing high densities of defects.

The correlation between peel force and current (Fig. 8) suggests that electrostatic effects account for 30–60% of the measured peel energy (assuming a linear relation). Since the conductive backing somewhat reduces the peel force, the actual contribution may be somewhat larger. As noted above, much of this enhancement is due to increased deformation in the adhesive phase.

4.5. Consequences of electrical discharges in special applications

Failure along highly charged interfaces can have serious consequences (e.g. electrical discharges) in explosive atmospheres and in sensitive electronic applications. We note that one generally successful strategy for minimizing interfacial charge has employed conductive polymers in the adhesive phase [21]. This does not necessarily eliminate or even minimize charge separation during attachment. However, it strongly promotes charge recombination by simple conduction and strongly suppresses discharges. Since most metals and semiconductors are coated with a thin oxide, we expect that tunneling and/or conduction through an insulating layer may limit charge redistribution on these substrates. Clearly, minimizing breakdown can involve trade-offs with mechanical properties, such as peel energy.

Electrostatic effects on macroscopic surfaces are strongly limited at high relative humidities. Similarly, charge separation during interfacial contact is a strong function of the choice of substrate. Although these factors affect other adhesion forces as well, we suggest that electrostatic effects are a significant source of variability in adhesive performance during peeling.

5. CONCLUSIONS

Electrical currents generated during peeling can provide a direct measure of the charge densities along the adhesive–substrate interface *if* charge recombination can be minimized. Recombination currents and electrical breakdown prevent detection of all of the interfacial charge. In this work, the electric fields along the zone of detachment have been reduced by coating the tape backing with an electrically conductive film. Electrical breakdown events have been further minimized by peeling in vacuum. With these precautions, electrical currents corresponding to charge densities up to 2.8 mC/m^2 have been observed. The electrostatic work required for interfacial failure is expected to be on the order of 2 J/m^2. This is orders of magnitude higher than the reversible work for interfacial failure (≈ 33 mJ/m^2). As undetected current cannot be entirely eliminated from our measurements, these values are lower bounds.

The good correlation between peel force and current when a bias is applied to the coating suggests that the peel force would be 30–60% lower in the absence of electrostatic effects. The importance of electrostatic effects is qualitatively supported by comparisons between gold-coated and non-coated tapes peeled under identical conditions. Since the charge separation responsible for the observed current is a distinctly irreversible process, electrostatic effects need not contribute significantly to the thermodynamic work of adhesion. In principle, the work done against electrostatic forces could increase peel energies. It is our opinion, however, that the primary role of charge separation is to increase the interfacial strength, which in turn enhances dissipative deformation in the polymer.

Acknowledgements

This work was supported by the National Science Foundation under Grant CMS-9414405 and the 3M Company.

REFERENCES

1. L.-H. Lee (Ed.), in: *Adhesive Bonding*, pp. 1–30. Plenum, New York (1991).
2. B. V. Deryagin, N. A. Krotova and V. P. Smilga, *Adhesion of Solids*. Consultants Bureau, New York (1978).
3. A. D. Roberts, *J. Phys. D* **10**, 1801 (1977).
4. W. Possart, *Int. J. Adhesion Adhesives* **8**, 77 (1988).
5. S. Lee, L. C. Jensen, S. C. Langford and J. T. Dickinson, *J. Adhesion Sci. Technol.* **9**, 1 (1995).
6. L. Scudiero, J. T. Dickinson, L. C. Jensen and S. C. Langford, *J. Adhesion Sci. Technol.* **9**, 27 (1995).
7. K. A. Zimmermann, S. C. Langford and J. T. Dickinson, *J. Appl. Phys.* **70**, 4808 (1991).
8. J. T. Dickinson, M. K. Park, E. E. Donaldson and L. C. Jensen, *J. Vac. Sci. Technol.* **20**, 436 (1982).

9. E. E. Donaldson, J. T. Dickinson and X. A. Shen, *J. Adhesion* **19**, 267 (1986).
10. J. T. Dickinson and E. E. Donaldson, *J. Adhesion* **24**, 199 (1987).
11. Z.-Y. Ma, J.-W. Fan and J. T. Dickinson, *J. Adhesion* **25**, 63 (1988).
12. E. E. Donaldson and J. T. Dickinson, *J. Adhesion* **30**, 13 (1989).
13. E. G. Johnson and A. J. Wallner, *J. Colloid Interface Sci.* **32**, 377 (1970).
14. A. W. Adamson, *Physical Chemistry of Surfaces*, 5th edn, p. 379. John Wiley, New York (1990).
15. J. Lowell and A. C. Rose-Innes, *Adv. Phys.* **29**, 947 (1980).
16. K. P. Homewood, *J. Phys. D* **17**, 1255 (1984).
17. T. J. Fabish, H. M. Saltsburg and M. L. Hair, *J. Appl. Phys.* **47**, 930 (1976).
18. F. Nordhage and G. Bäckström, in: *Static Electrification*, Institute of Physics Conf. Ser. **27**, 84 (1975).
19. J. H. Israelachvilli, in: *Intermolecular and Surface Forces with Applications to Colloidal and Biological Systems*, pp. 137–140. Academic Press, London (1985).
20. R. J. Good and R. K. Gupta, in: *Adhesive Bonding*, L.-H. Lee (Ed.), pp. 47–93. Plenum, New York (1991).
21. G. Gutman, S. Yau, R. Goetz, S. C. Koehn and D. E. Swenson, in: *Electrical Overstress/Electrostatic Discharge Symp. Proc. 1994*, pp. 230–236. EOS/ESD Association, Rome/New York (1994).

Mittal Festschrift, pp. 615–640
W. J. Van Ooij and H. R. Anderson, Jr. (Eds)

Fracture mechanics analyses of the behavior of adhesion test specimens

K. L. DEVRIES* and P. R. BORGMEIER
College of Engineering, University of Utah, Salt Lake City, Utah 84112, USA

Abstract—The strength of an adhesive joint is not a property of the adhesive alone but is a system property depending on adherends, the adhesive, the joint geometry, and service (or test) conditions. Most standard test methods specify giving the results as an average stress at failure. Use of this average stress is not very useful in the design and prediction of the strength of other adhesive joints that differ in even slight detail from the geometry of the test specimen. On the other hand, fracture mechanics methods have shown real promise in quantitatively explaining how differences in geometry can affect joint strength. The effects of modifications of lap specimens, similar to those described in ASTM D3165 'Strength Properties of Adhesives in Shear by Tension Loading of Laminated Assemblies' have been investigated. An earlier paper discussed changes in strength resulting from tapering of the adherends over the overlap region. Here this type of analysis is extended to lap joints in which large debond regions were introduced in the adhesive joint. In both cases, the finite element/fracture mechanics predictions were in good agreement with experimental observations on steel-epoxy lap specimens. Next, a study of failure in cleavage specimens is discussed. ASTM D3433 'Fracture Strength in Cleavage of Adhesives in Bonded Joints' is an ASTM standard whose basis lies in fracture mechanics. This standard makes use of specimens composed of 'double cantilever beams' (DCB) bonded together with the adhesive to be evaluated. The equation provided in ASTM D3433, to calculate the adhesive fracture toughness G_{Ic}, is based on the equations from mechanics of materials for flexure stress and vertical beam shear stress and assuming ideal cantilever end conditions. This equation, therefore, neglects any energy stored in the adhesive or through rotation of the cantilever end and/or stored in the beam beyond the crack tip (assumed cantilever point). This equation was carefully analyzed using both classical (from the literature) and numerical methods. One would, of course, anticipate that the original assumptions would be most valid for long slender beams and very thin adhesives. Careful finite element/numerical analysis indicates, however, that when using dimensions well within those suggested in ASTM D3433, the energy release rate or G_{Ic} determined including these factors can differ by more than 50% from that obtained using the recommended equation. A quasi-elastic adhesive was used with aluminum sheets to manufacture DCB specimens. A group of these specimens were tested to failure and the results used to determine a reference G_{Ic}. This G_{Ic} was then used with finite element methods to predict performance for test specimens with different geometries, i.e. the adhesive thickness and/or adherend thicknesses were altered. These test results confirmed the validity of the fracture mechanics/finite element approach. The calculated failure loads for the geometries typically differed by no more than 8% from the experimentally determined loads. The results for one sample lot differed by slightly more than 12% from the reference value but, as explained in the text, this difference might be attributed to mode dependence. On the other hand, use of the standard equation with the test results to calculate G_{Ic} yielded values that differed by as much as

*To whom correspondence should be addressed. E-mail: larry.devries@dean.eng.utah.edu

50% between the different sample geometries. The finite element methods were also used to calculate the energy release rate for various assumed crack paths through the adhesive. It was hypothesized that the fracture locus should follow paths of maximum energy release rate. Experimental observations of crack paths for a number of different geometries of DCBs were consistent with these predictions in every case.

Keywords: Adhesive failure; finite elements; fracture mechanics; locus of failure; prediction of joint failure; double cantilever beam specimen; experimental test results.

1. INTRODUCTION

Most engineering designs require that component parts be connected. Increasingly, adhesive joints are being used in place of mechanical connectors such as bolts, pins, rivets, screws, or nails. The reasons for this choice range from aesthetics to increasing the strength and integrity by distributing the transmitted load over larger areas or reducing the stress concentrations associated with the bolt or rivet holes. This latter aspect might be particularly important for fatigue considerations.

The stress analysis for any type of joint (mechanical, adhesive, or weld) is not a simple problem. Depending on the type of joint, these analyses may involve contact stresses, stress concentrations, or interfacial stresses across bondlines. Performance can be affected by time (creep, visco-elastic plastic behavior, corrosion, and other environmental effects) when the joint is expected to function under sustained loads. The designer is therefore critically in need of techniques for predicting the strength and durability of adhesive joints. It also would be useful to have techniques that could be used to optimize geometry and explore the effects of other factors to enhance joint performance. While these problems are not unique to adhesive joints, this paper will concentrate on adhesive joint failure.

What might appear as slight changes in joint geometry can dramatically influence the stresses in an adhesive joint. A wide variety of standard tests are available to measure 'strength' and other properties of adhesive joints. What is not so obvious is how to interpret the results from such tests or how to use the results to predict performance of practical joints that differ in any detail from the test configuration. Many standard test procedures (such as those for lap, tensile, and butt specimens) specify reporting results as the load at failure divided by the bonded area. As noted previously, however, the stress distribution for adhesive joints is complex and a strong argument can be made that failure is likely to be more closely related to the maximum stresses than to such 'average' stresses. These maximum stresses are highly dependent on the joint geometry and the properties of the materials making up the joint. Indeed, elastic analysis of many joints exhibit singularities where the predicted stresses become unbound.

This paper will discuss the analysis of and comparison with experimental results of three seemingly unrelated adhesion problems: (1) Lap shear joints for which the adherends are tapered over the overlapped region. (2) Lap shear joints in which relatively large regions of debond were introduced along the usual bonded region. (3) Double cantilever beam (DCB) specimens, in which adherend thicknesses and adhesive thickness were independently varied. It will be shown in the last case that the particular path followed by 'adhesive cracks' is dependent on these parameters.

The common thread in these outwardly very different tests is that while the experimentally observed phenomena cannot be explained by 'average stress criteria' they are consistent with predictions based on fracture mechanics. Computer based finite element analysis methods (FEA) make almost any joint geometry amenable to such fracture mechanics analysis.

2. MATERIALS AND SAMPLE GEOMETRIES

Techniques are being developed to handle nonlinear and nonelastic materials. It was felt, however, that the complexities associated with analyzing the test geometries considered here were in themselves rather difficult to resolve. Furthermore, the issues addressed in this paper could be just as well resolved with materials that behave in an approximately linearly elastic manner. Therefore, tests were conducted on a number of commercial adhesives seeking candidates that behaved relatively linear elastic in the time, loading rates, etc., used in the experiments.

For the single lap joint samples, Hysol® 934-NA epoxy was used to bond the steel adherends. This adhesive was selected because it had a relatively linear stress–strain response similar to that shown in Fig. 1, which is for the adhesive used in the DCB specimens. The adherends were hardened A614 or 4340 steel plates (22–25 R_c). The yield stress for steel of this hardness is well above any nominal stress induced during the loading to failure of the samples.

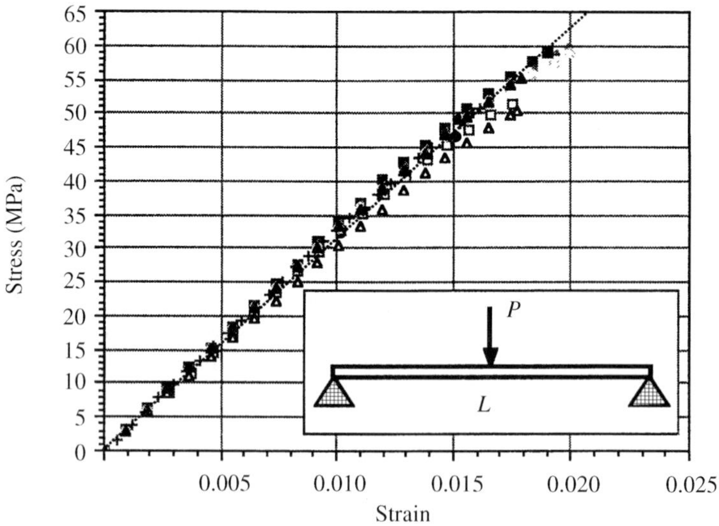

Figure 1. Stress versus strain plots for ten Hysol® 907 adhesive bars subjected to three-point bending. The plots depicted by open symbols had a length of 50.8 mm and were tested at a crosshead speed of 0.51 mm/min. The plots depicted by solid symbols had a length of 50.8 mm and were tested at a crosshead speed of 5.08 mm/min. The plot depicted by the + symbol had a length of 63.5 mm and were tested at a crosshead speed of 5.08 mm/min.

The DCB specimens were bonded with Hysol® 907 adhesive. The stress–strain behavior for this material at several different strain rates is shown in Fig. 1. The stress–strain behavior for Hysol® 934-NA was nearly as linear. The adherends were 6061-T6 aluminum alloy, which had a yield stress greater than 300 MPa as determined by 3-point bending tests. Again, the stresses in the aluminum were well below the yield stress during all the testing.

Both adhesive sample types were prepared by joining metal plates together with the adhesive, curing, and then cutting into strips (with a band saw or mill) to form the final specimens. The samples were manufactured using the appropriate ASTM standard as a guide (per ASTM D3165 for lap joints and D3433 for the double cantilever specimens, respectively) except with the modifications noted below.

Surface preparation of the steel adherends (4340 or A614 at 22 R_c) involved acetone solvent wiping followed by sand blasting with 200 grit zirconium silicate. The samples were then placed in an acetone degreasing solution for 2 h followed by another sand-blast and an additional 30 min in an acetone bath. Compressed 'dried' air (100 psi) was then blown across the surface to remove any dust left from the sand blasting operation and to facilitate rapid drying. The adherends were then bonded together using Hysol® 934-NA epoxy in configuration similar to that described in ASTM D3165. Spacers were used to accurately control adhesive thickness. The subsequent cutting was controlled so as to eliminate these spacers from the samples. Curing conditions were 25°C and 50±5% relative humidity for one week. More rapid high temperature cures were avoided in an effort to reduce cure/thermal residual stresses.

For the aluminum samples, initial preparations were made using a sulfuric acid–sodium dichromate treatment as outlined in ASTM 2651. It was found, however, that a somewhat less involved treatment was adequate. The residual stresses from sand blasting (even rather mild blasting) were found to severely distort the plates. As a substitute, the following treatment was used. The plates were initially solvent cleaned with acetone. This was followed by a surface roughening and cleaning operation with an electric drill mounted wire wheel. The adherends were again cleaned in an acetone bath and air dried at room temperature. The samples were bonded using Hysol® 907 adhesive applied 1 h after final degreasing. The cure was for 48 h at 23°C and 50 ± 5% relative humidity. As evidence for a good bond, it was noted that every specimen failed in the adhesive, not at the metal surface. Visual inspection revealed that even those samples (described later) in which the crack appeared to run along the interface exhibited a thin contiguous layer of adhesive on the adherend. This was also true for the steel samples.

The uniformity of adherend spacing/adhesive thickness was maintained by use of plastic spacers placed at strategic locations in the bonded area. In the subsequent sample cutting operation, these spacers were excluded from the specimens to be tested. From visual inspection and measurements before and after testing, it is estimated that the method controlled the thickness to better than 0.2 mm.

For the first series of lap joints (sample group 1), both adherends were tapered over the bond region as shown schematically in Fig. 2. Morton Thiokol Incorporated provided us with the second set of specimens. These samples were prepared by their personnel, also following procedures ASTM D3165, except intentional debond regions

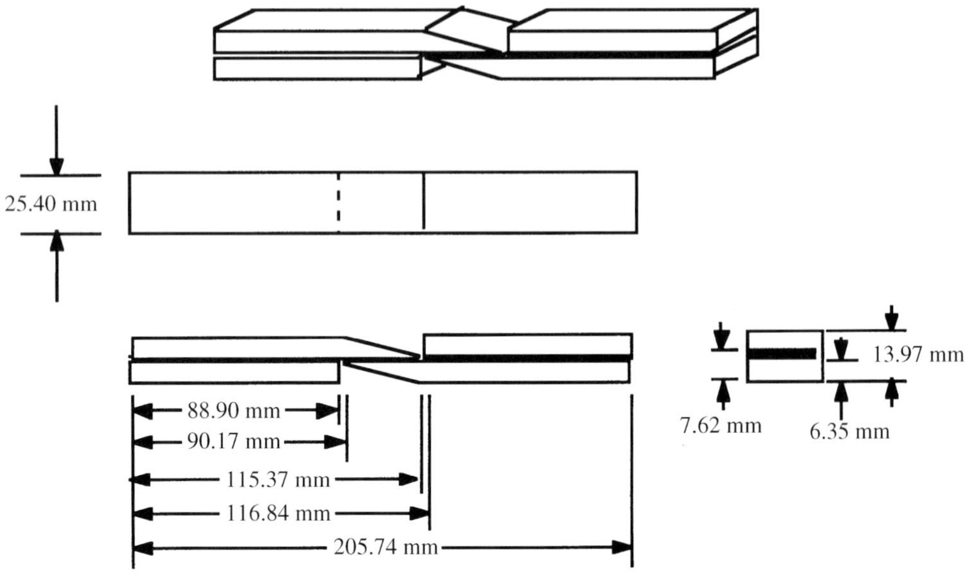

(a)

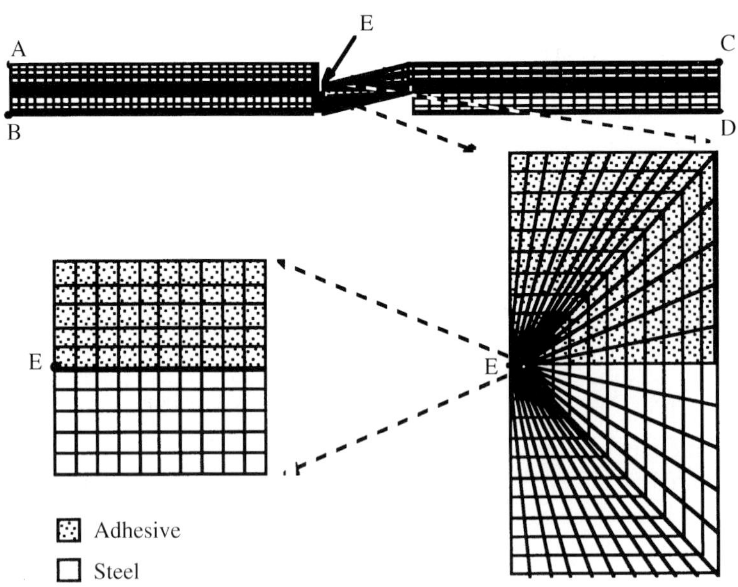

(b)

Figure 2. (a) The geometry and dimensions of specimen group 1. (b) The finite element grid used for the analysis of group 1 with expanded views showing areas of particular interest.

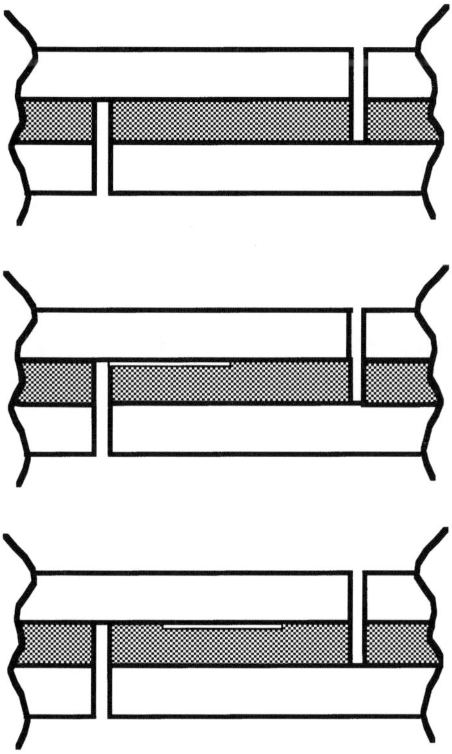

Figure 3. Schematic of specimen group 2 showing locations at which intentional large initial debonds, by using Teflon® tape, were introduced in specimens 2B and 2C.

were introduced. The debonds were formed with Teflon® tape (< 0.08 mm thick) at the locations shown in Fig. 3. It is recognized this tape would form a rather blunt starter crack but as will be discussed later a smaller, experimentally determined, 'inherent flaw' size was used in the analysis.

For the DCB specimens, ASTM D3433 suggests drilling 6.35 mm holes for insertion of pins to apply the loads. With the thin adherends for the DCB specimens used in this study, use of filament tape tabs was much more convenient and proved experimentally to be adequate. Two Scotch® Brand Filament Tapes (8-8700) were cut to a length of 355 mm. These were attached to the underside of each beam as outlined in Fig. 4 and folded back over themselves to form tabs that were gripped in standard Instron grips. The lap specimens were pulled to failure at a crosshead speed of 0.5 mm/min (0.02 in/min) and the DCB specimens at 51 mm/min (2 in/min).

2.1. Computer hardware and software used in the study

Most of the numerical analysis in this study used the ANSYS finite element program developed by Swanson Analysis System Incorporated. The finite element codes were run on a UNIX based Sun SPARC workstation. Although the FEA software could be used to perform 3-D analyses, this study used 2-D analyses assuming plane stress or plane strain conditions.

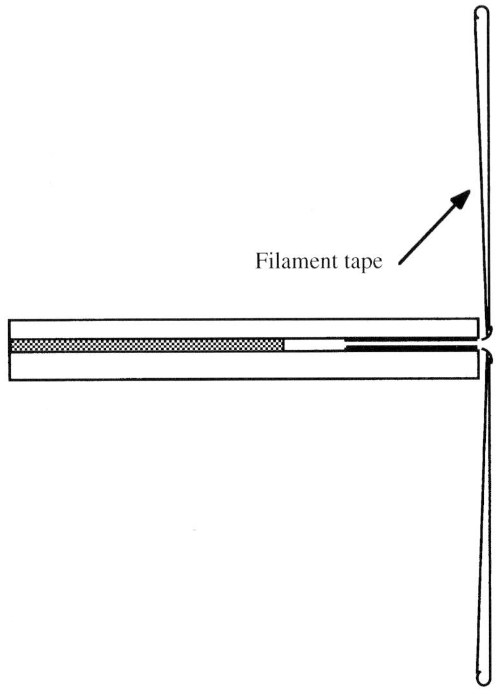

Figure 4. Schematic side view of DCB specimen demonstrating the use of filament tape to form tabs.

Details of the methodology developed for calculation of energy release rates, G, are given in a prior paper [1]. Suffice it here to note that two methods were used to calculate G [1–3]. (1) The compliance method (CM) where G is based on the difference in stored energy in the sample before and after incremental crack growth. This change in energy is equivalent to the area between the compliance curves before and after the assumed growth. (2) The crack closure integral method (CCI) is based on the work required to close a crack that has been assumed to increase in size an incremental amount, da [2, 3]. For a linear elastic system, this work is equal to the energy 'lost' when the crack propagates. If performed correctly, the two methods must yield the same result for G for an elastic body. In this way they provide a check on the numerical analysis part of the study. The CM method was, in general, simpler and more straightforward to use. It required significantly less program effort and computer time. On the other hand, the CCI method facilitated partitioning the energy release rate into two components, i.e. that associated with crack opening (Mode I) and that associated with shear displacement perpendicular to the crack front (Mode II). As will be discussed in the Results part of this paper, these components can have important ramifications for some adhesive problems.

2.2. Experimental results for modified lap joints

A prior paper [1] compared the predictions of a numerically aided fracture mechanics analysis on the effects of tapering the adherends of lap joints with experimental results. Tapering of the adherends reduces the rate at which the shear stress increases as the bond termini are approached. This results in a more uniform distribution of the shear stresses over the overlap region of the specimen. Tapering has, therefore, been proposed as a means of strengthening lap joints [4–7]. Indeed, a comparison of the FEA predicted shear stress distribution of tapered lap joints with conventional lap joints indicates that tapering reduces the magnitude of the shear stresses near the bond termini. If failure was solely dependent on the value of the shear stresses, one might anticipate that tapering would significantly enhance the strength of the joint. Shear stress magnitude, however, is not the complete story. Static equilibrium dictates that the integral of the shear stress over the 'plane area' of the joint in the direction of the load must equal the applied load. The stresses on such a plane differ significantly from the average value. The shear stresses are accompanied by induced normal stresses whose resultant's components perpendicular to the line of action must be zero. However, the values of these normal stresses may be very large locally as the bond termini are approached. There is strong evidence that joint failure is more closely related to these induced cleavage stresses than to the shear stresses [5, 8–14]. Tapering the adherends not only reduces the magnitude of the maximum shear stresses but is generally accompanied by an increase in outward deflection ('cleavage') of the region near the bond termini. As a result, the Mode I energy release rate is increased for the tapered samples and the total energy release rate for the two sample types is not markedly different. Use of a stress criterion to predict the strength of a joint is no small task. For reasons given above, it is difficult to exactly quantify the stresses. Use of fracture mechanics largely circumvents this problem.

Experiments on conventional and tapered lap joints indicated that tapering resulted in only a slight increase in strength ($< 15\%$), which was in general agreement with the fracture mechanics predictions [1]. The fracture mechanics analyses using both the CM and the CCI methods with a critical energy release rate determined from tests on conventional specimens predicted average failure loads for the tapered samples within 4% of the average of the experimentally determined values. Note that Kinloch and Kodokian also reached the conclusion that tapering of the adherends was not a very effective means of increasing joint strength [15]: they based their conclusion on somewhat different analysis and experiments.

In a second series of tests, samples that were otherwise similar to those referred to above as conventional, had intentional large regions of debond introduced along the bond line between the adhesive and the adherend. The location of these debonds are shown schematically in Fig. 3. (Note the difference between sample lots 2A, 2B and 2C.) If there is any validity at all to the average shear stress criterion, introducing debonds that reduce the bond area by 50% should decrease joint strength by a like amount.

These lap joint samples were tested in an Instron Machine at a loading rate of 0.5 mm/min. Samples 2A (no intentionally introduced debonds) failed at an average load of 4715 N with a standard deviation (SD) of 1210 N. Specimen group 2B failed at an average load of 3631 N with a SD of 1210 N. Specimen group 2C had an average failure load of 4564 N with a SD of 1454 N. Note that while the bonded areas for sample groups 2B and 2C were decreased by 50%, the average failure loads were only reduced by 13% and 3%, respectively. There were five to seven specimens in each sample lot. Clearly, an average shear stress criterion would not be very enlightening in the analysis of joint failure for such specimens.

On the other hand, fracture mechanics analysis is very informative. An 'inherent' flaw size of 0.075 mm was used in this determination. This quantity was determined by extrapolation from tests in which a series of smaller and smaller finite cracks had been introduced. The fracture mechanics analysis revealed several interesting features about the joints. While the energy release rates at the locations of the original bond termini are moderately increased, that at the end(s) of the intentionally introduced debond region (crack) is significantly larger for a given load. However, the CCI analysis indicates that while the calculated G value for sample lot 2A was almost pure Mode I, that at the end(s) of the 'debond regions' for sample lots 2B and 2C was more than 90% (some as high as 99%) Mode II. Several researchers have reported that the critical energy release rate in polymeric adhesives is mode dependent, with that for Mode II being significantly higher than that for Mode I [16–24]. The experimental results reported above are completely consistent with a fracture mechanics analysis using an energy release rate that is the weighted average of G_{Ic} and G_{IIc}, where G_{IIc} is taken as $2\,G_{\mathrm{Ic}}$. This latter relationship is reasonable based on the results reported in the above references.

It should be noted that at least one alternative (stress) explanation has been proposed to explain the observation that removing 50% of a bond area might not reduce strength proportionately [25]. Kinloch [25] notes that the method commonly used to create the debond areas (Teflon® tape) may create cracks that are rather blunt. This bluntness might, in some cases, even reduce the stress state, near that point, to a value below the stress concentrations of the original samples cut 90° corner.

2.3. Locus of crack path in DCB specimens

This section explores fracture in the adhesive in more detail, e.g. the fracture path as well as the load associated with joint failure. A question of long interest and considerable speculation relates to the underlying causes of the observation that 'good joints' rarely fail at the interface.* Failure in which the crack travels along or near the adhesive–adherend boundary such that the adherend surface appears to be relatively free of adhesive is often called an 'adhesion failure.' Failure in which there is an obvious layer of adhesive on the adherend is often termed 'cohesive failure.' These terms are somewhat arbitrary since a surface that appears clean may have, in fact, a microscopic layer of adhesive on it. Furthermore, almost all real surfaces are rough on a microscopic scale as well as having oxide layers, adsorbed layers of vapors and gases from the surroundings, etc. which can participate in the fracture process. Failure of practical adhesive joints in apparent 'adhesion failure' is often cited as evidence for environmental (particularly water) or corrosive attack [5]. Cracks running well away from the boundary are so common that they are considered as the norm. Chemical,

*Use of the term interface is not meant to imply that there is a well-defined plane between the adhesive and adherend(s) along which the crack can proceed.

statistical, and other reasons have been proposed to explain this behavior but none has been universally accepted nor conclusively proven to be correct.

From a chemical point of view, Good and Girifalco [26–27] related the different surface energies of the adhesive and the adherend in an analogous way to the thermodynamic work equations governing the separation of two mixed phases. They developed a molecular adhesion equation in the form (for the simplest case)

$$\gamma_{12} = \gamma_1 + \gamma_2 - 2\sqrt{\gamma_1\gamma_2}, \tag{1}$$

where γ_1 and γ_2 are the surface free energies of the two materials and γ_{12} is the interfacial free energy between them. From this equation, the work of adhesion W_a would be

$$W_a = 2\sqrt{\gamma_1\gamma_2}. \tag{2}$$

This quantity is bound by the surface free energies of the two materials and mathematically can be viewed as the geometric mean of γ_1 and γ_2. Similar arguments have been developed by others with slightly different approaches [28–31]. Although these equations may be oversimplified, they demonstrate the desired principle. Since the introduction of these ideas, many studies have broadened and refined these theories and have recently been reviewed by Allen [32]. An early proponent of the concept that statistically a crack would likely not run along an interface was Bikerman [33].

Other reasons proposed for crack paths avoiding the so-called interface are more mechanical in nature. As noted previously, several studies of adhesive joints have reported that G_{IIc} values are often higher than G_{Ic} values. This would mean that the total energy release rate (ERR) value ($G_c = G_{Ic} + G_{IIc}$) cannot be used as a material constant because of its dependence on the different modes. However, it can be used in this sense if the modal percentages are known along with the dependence of G_c on mode.

A basic premise assumed in this study is that, other things being equal, cracks will follow paths where the ERR is highest for a given load. Total ERR values were used for predicting behavior of DCB adhesive joints loaded to failure. The desired fracture planes were determined to be 100% Mode I for all except two of the geometries studied. In these two cases, the modal percentages did not converge but were near 96% and rising very slowly when the convergence studies were stopped. Therefore, all of the G values were, in effect, based on the G_I component. Except for one geometry (Group 3J), the FEA-Fracture Mechanics method used successfully predicted failure loads that were within approximately 5% of the average of the actual experimentally determined failure loads.

In the analysis, adhesive cracks were assumed to follow various paths through the adhesive. By varying the sample geometry (thickness of adhesive, different thicknesses of adherends, etc.) the path with maximum ERR could be varied from adhesive center to near the interface. In this way, it is possible to predict changes in failure load as well as identify the likely path the crack should follow. Experimentally, it was observed that the cracks 'kinked' away from the initial debond location (Teflon® tape or razor cut) and sought out and followed the predicted fracture plane (i.e. that with the highest G_I).

Akisanya and Fleck [34] and Fleck *et al.* [35] have studied cracks that kink away from bimaterial interfaces. They reported that cracks experienced both modes of fracture along these kink paths. They also postulated that these paths tended to lead to the planes that had the highest percentage of Mode I. These planes are hypothetically sought out because the energy required to drive a Mode II crack is greater than for a Mode I crack. It is theorized that, other things being equal, the path of least resistance has the minimum amount of Mode II. In composite systems, this 'Mode II effect' has been reported to drive a crack through the interface and into the adherend material [36, 37].

The measured failure loads of the specimens considered here actually reflect failure loads for cracks that were advancing in a mixed-mode environment. Because of this, the G_c values determined above are probably falsely high for a Mode I failure. This would mean that using a G_c value determined in this manner to predict the failure load of a crack that did not kink would overpredict its strength. This (where the calculated G was not G_I but mixed mode) is entirely consistent with the observations for the Group 3J geometry where the failure load was overpredicted by 5.3%.

The fracture paths observed for the experimental samples were compared to the predictions. None of the specimens had cracks that advanced exactly at an interface even though 'interfacial' failure was predicted for some of the geometries. The reasons why this type of failure does not occur in 'good joints' likely stems from the chemical, statistical, and mechanical reasons discussed previously. There is another micro-mechanical viewpoint worthy of note. From a mechanical point of view, there is a stiffness mismatch between the adhesive and the adherend. If the adherend and adhesive are tightly bonded and subjected to a loading they will have the same strain at their interface. The adhesive is typically more compliant than the adherend. Therefore, the adhesive at and very near the interface will not deform as much as the rest of the adhesive due to the reinforcing effect of the more rigid adherend. As a result, it is as if the adhesive material at and near the interface is different from the same material further away. This effect has been termed 'elastic stress shielding' [35]. This shielding effect has a diminishing influence on the adhesive as the distance from the interface is increased and may result in lower effective stresses in the adhesive immediately adjacent to the adherend. As a consequence, the locus of failure could move away from the boundary. Another reason for the lack of 'adhesional failure' could stem from the increase in surface area of the interface region due to surface roughness [38, 39]. This surface area has been estimated to be upwards of 10 000 times the area of a hypothetical plane section at the interface.

The samples studied were modifications of the geometry described in ASTM D3433 shown schematically in Fig. 2. Modifications involved changing adhesive and adherend(s) thicknesses. Figure 5 shows a schematic of possible crack paths. Figure 6 shows the deformed FEM grids used for an analysis of one geometry. The analyses of the DCB geometry appear deceptively simple. Indeed, some very enlightening information can be obtained using simple beam theory assuming the only elastic energy is that stored in the two adherend beams. This is the assumption used in the derivation of the equations for G_{Ic} given in Section 11 of ASTM D3433. This approach neglects at least two terms: (1) the energy stored in the adhesive, and (2) the energy input

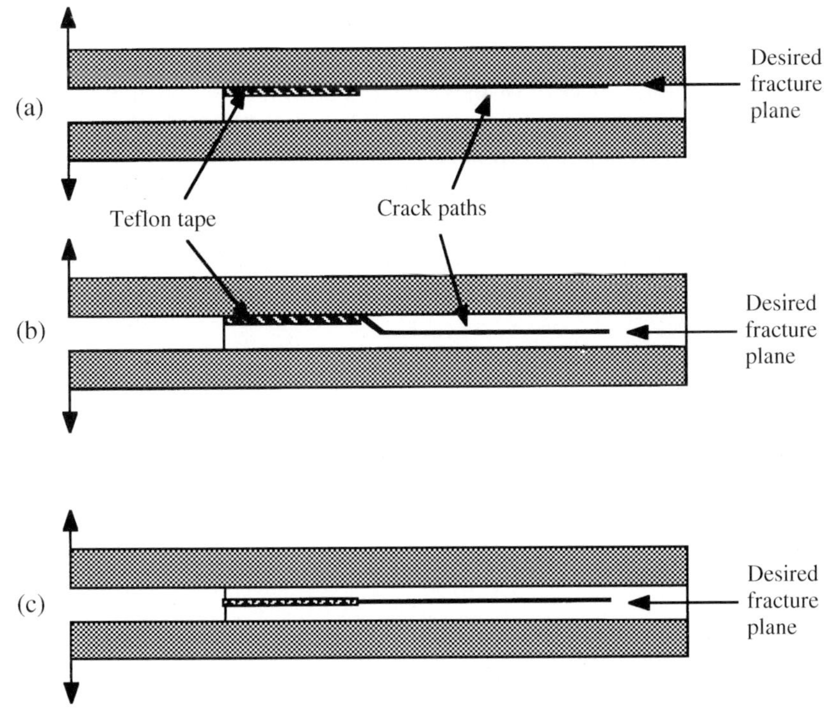

Figure 5. Schematics of the possible crack paths. (a) Shows an edge crack path assuming the initial crack was in the desired fracture plane. (b) Shows a center crack path assuming the initial crack was not in the desired fracture plane. (c) Shows the assumed location of the initial crack if the desired fracture plane was at the center.

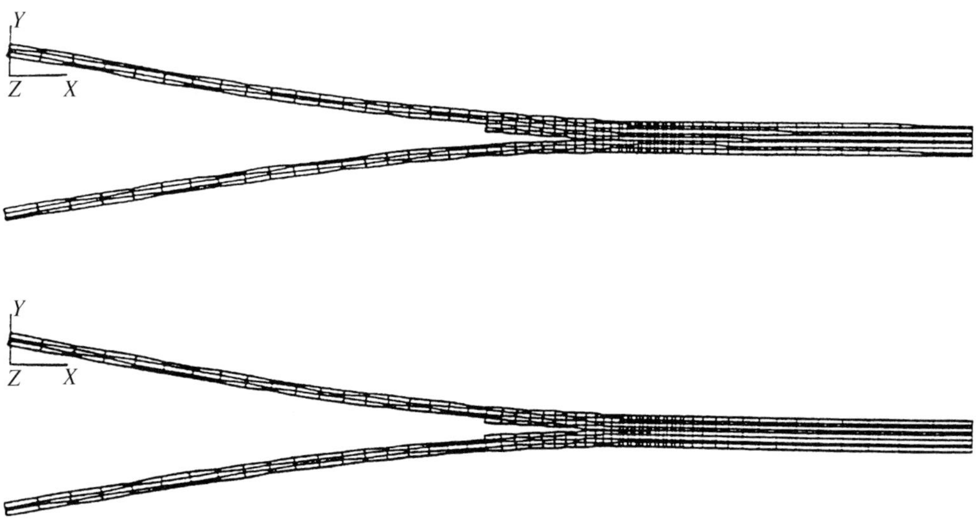

Figure 6. Deformed FEM grids used in the analysis of the Group 3A DCB geometry. The top grid was generated assuming the crack would advance down the adhesive–adherend interface. The bottom grid was generated assuming the crack would advance down the center of the adhesive.

associated with the non-rigid cantilever supports of the beams. The latter is due to rotation of the hypothetical supports and/or stresses in the beam beyond the 'support'. The relative contribution of both of these factors are functions of the slenderness ratios of the beams, the adhesive thickness and stiffness. The following mechanics of materials analysis illustrates these points.

The DCB specimen is two cantilever beams bonded together over much of their length (refer to Fig. 4). Assume each beam has a width b, a depth h, an unbonded region (free length from loading pin to the point of first adhesion) a, and end loads P. Applying the concepts of mechanics of materials, the elastic energy, U, stored in the two beams would be

$$U = 2\left[\int_0^a \frac{M^2}{2EI}\,\mathrm{d}x + \int_0^a \int_{\text{area}} \left(\frac{VQ}{Ib}\right)^2 \frac{\mathrm{d}v}{2G}\right] \tag{3}$$

$$= \left(\frac{P^2a^3}{3EI} + \frac{P^2h^2a}{10GI}\right), \tag{4}$$

where Q is the first moment of area (defined in most mechanics of materials texts), I is the moment of inertia, $\mathrm{d}v$ is a differential volume, E and G are the elastic and shear moduli, respectively, for the beam material, and the other terms are as defined above. The first term in these equations can be viewed as the energy due to the bending moment and the second that due to shear. The ERR would be the change in this energy due to an incremental increase in area, $\mathrm{d}A$, as the crack grows, $\mathrm{d}a$, ($\mathrm{d}A = b\,\mathrm{d}a$)

$$G_{\mathrm{c}} = \frac{1}{b}\frac{\partial U}{\partial a} = \frac{P^2}{EIb}\left[a^2 + \frac{E}{G}\frac{h^2}{10}\right]. \tag{5}$$

The second term above, which represents the shear energy, is only one form of this expression. Others can be found in the literature: see, for example, [40]. One of these was probably used by Mostovoy *et al.* in their pioneering efforts in this respect [18, 42, 43]. Irrespective of which of these are used, the basic form of equation (5) remains the same except for slight variations in constants.

Since

$$I = \frac{bh^3}{12} \quad \text{and} \quad G = \frac{E}{2(1+\nu)}, \tag{6}$$

equation (5) can be rewritten (assuming ν is constant and not material dependent) as

$$G_{\mathrm{c}} = \frac{4P^2}{E^{\mathrm{b}}h^3}\left[3a^2 + h^2\right], \tag{7}$$

which is the same as that found in ASTM D3433. Use of this equation is useful and informative for comparing similar adhesives if they have the same thickness and modulus or the adhesive thickness is very small. However, if one wants to obtain

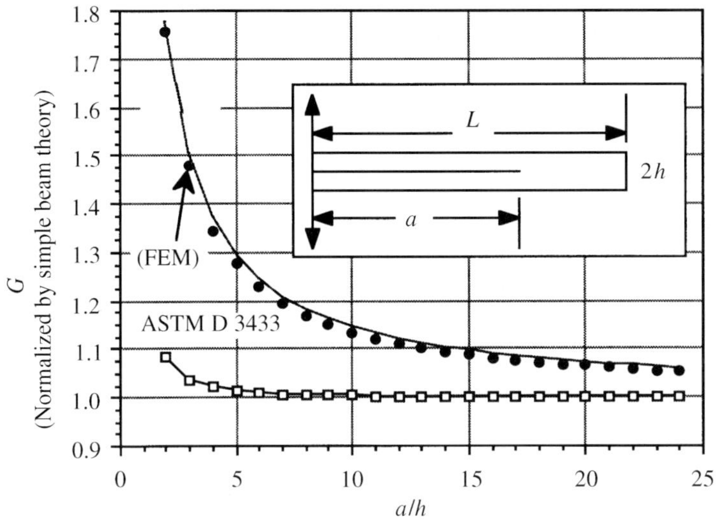

Figure 7. Normalized G values versus a/h ratios for the suggested geometry given in ASTM D3433. The values were calculated using the equation given in ASTM D3433 and FEA.

accurate absolute values of G_c for a given adhesive system or to explore such differences as thickness or other subtle features, the analysis is found lacking. Several energy effects are neglected in the above simple analysis. Careful consideration reveals that the shear and rotation can be large for typical crack lengths. Figure 7 shows a plot of normalized G values versus a/h ratios for a geometry with a constant beam cross-section using the equation given in ASTM D3433 and that determined by FEA. (The values were normalized by dividing through by the first term in equation (7), the bending term.) This emphasizes the contribution due to shear. The recommended geometry in D3433 has a total length of 330 mm and a beam height h of 12.7 mm. The smallest recommended crack length a is 25.4 mm. Note that at smaller crack lengths the ASTM D3433 value is about 50% less than the values determined using FEA. At $a/h = 14$ the difference is still 10% and the error does not fall below 5% until a/h exceeds 25.

The work by a force applied to an elastic cantilever beam can be equated to the energy stored in it. This is equivalent to the analysis just given. However, the usual boundary conditions at the ‘fixed’ end are zero deflection, zero rotation and no stresses beyond this hypothetical point. These conditions are extremely difficult to impose for even vise-type grips in laboratory bending tests such as are used in Izod impact tests. The mismatch between these assumed conditions and reality can be particularly severe for adhesive bonds.

As expected, Table 2 demonstrated that these solutions became more accurate the longer and more slender the beam, i.e. a/h increasing. The a/h conditions for the recommended ASTM D3433 specimens could be as small as 2 and over most of the test conditions would not exceed 20 before other boundary uncertainties would enter the picture as the crack approaches the other end of the sample. The methods of Classical Elasticity facilitate closed form solutions for cantilever beams of various lengths. However, matching boundary conditions on the 'fixed' end still poses serious problems (see, for example, Section 21 [40]). The effects of shear and, to a lesser extent, the rotation of the cantilevered end of such beams have been explored by a number of researchers [44–61] with some interesting conclusions. It also should be noted that the adhesive not only allows the beam to rotate away from the ideal cantilever end condition but can also be a 'sink' to absorb part of the work of the applied forces, P. This energy is ignored in most analyses, e.g. in ASTM D3433. It can, however, be significant and indeed has an absolutely essential contribution in the results to be subsequently presented.

The finite element analysis in this study included the energy stored in the adhesive as well as that in the adherends. The analysis also included the energy due to shear and that stored beyond the usual cantilevered fixed point and/or the effects of the support rotation. The analyses were conducted with different grid sizes to assure convergence of both the ERR value and the crack mode.

The basic hypothesis to be checked in this part of the study was that, other things being equal, fracture cracks would follow paths of greatest energy release rate. The premise being that if a variety of different crack fracture paths were assumed and the ERR calculated for each of these paths, that path with the largest ERR should be the preferred locus for failure.

The geometries of the samples were varied (varying adhesive thickness, varying adherend thicknesses, etc.) to produce specimens in which: (1) the calculated largest ERR was near the center of the adhesive between the adherends; (2) where the largest ERR was near the boundary with one or other of the adherends; and (3) where the ERR was essentially constant across the adhesive thickness.

Table 1 summarizes the geometries of the samples analyzed. Figure 8 shows a schematic of a typical DCB specimen to help understand the dimensions of Table 1. While the length dimensions of all the groups were essentially the same, the thicknesses varied as shown in Table 1. Starter cracks were at the locations listed in the table. For each case, samples were constructed to the dimensions assumed in the analysis and pulled to failure in an Instron machine with the following results:

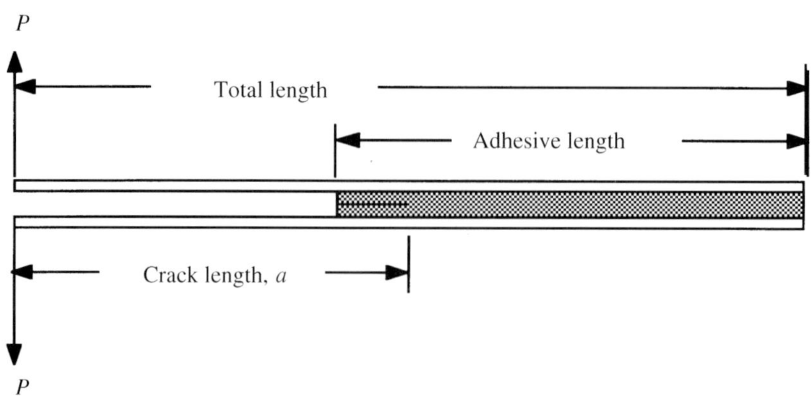

Figure 8. Schematic of typical DCB geometry adhesive joint fracture specimen with adherends of equal thickness analyzed and experimentally tested.

Table 1.
Dimensions and crack locations for the double cantilever beam samples

Sample group designation	Upper adherend thickness (mm)	Adhesive thickness (mm)	Lower adherend thickness (mm)	Initial distance to crack tip (a) (mm)	Initial a/h^*	Location of Teflon® tape (starter crack)
3A	1.98	1.25	1.98	101.6	51	[a]
3B	1.98	2.34	1.98	101.6	51	[a]
3C	1.98	1.25	1.98	50.8	26	[a]
3D	1.98	2.29	1.98	50.8	26	[a]
3E	1.98	1.52	1.98	75.4	38	[a]
3F	4.67	1.52	4.67	76.2	16	[a]
3G	1.98	5.08	1.98	76.2	38	[a]
3H	1.98	4.83	1.98	76.2	38	[a]
3I	1.98	3.23	1.98	76.2	38	[a]
3J	1.98	3.77	1.98	76.2	38	[a]
3K	1.98	3.96	1.98	76.2	38	[a]
3L	1.98	4.32	1.98	76.2	38	[a]
3M	1.98	5.08	1.98	Variable	Variable	[b]

[a] At one bondline between adherend and adhesive.

[b] No Teflon starter crack — rather a 0.36 mm saw was used to cut a short distance down the center line of adhesive and the starter crack was initiated with a razor blade.

As noted previously, starter cracks were introduced either with thin Teflon® tape (0.04 mm thick in this case) or by wedging open a short thin saw cut with a razor blade. The experimental results reported in the tables and figures are all for initiation at such starter cracks. In a number of cases, samples were pull to initiate a crack, unloaded and subsequently pulled to failure using the new crack length in the analysis. These results agreed quite well with those for the original starter crack but had more scatter, attributed in large part to difficulty in accurately determining the exact new crack length. It was felt that even though the tape-induced cracks might be somewhat blunt they were at least consistent in all the samples where they were used. The results were also consistent with the razor blade-induced cracks. Each group of samples contained six to ten specimens. The samples with 'edge debonds' were tested with the starter crack alternately on the top and bottom. The experimental results of sample lot 3A were used to calculate the reference energy release rate, G_c. It was calculated using finite elements, assuming plane strain conditions, with the geometries shown in Fig. 6. Since for this sample the center line crack path had the largest ERR, this path was used in the analysis. The G_c calculated from the results of this test was 0.265 kJ/m^2, with a SD of 0.038. Use of the equation from ASTM D3433 resulted in a $G_c = 0.274 \pm 0.039$ kJ/m^2. This differed by only 3% from the FEA calculated value. This is not surprising since an a/h of 51 is very large. As noted previously, this a/h is much larger than would ever be encountered using the ASTM suggested sample. The predictions of the load at failure for other samples, using this G_c, and the subsequent experimental results, are summarized in Table 2.

While fracture mechanics can be used to predict the failure load as shown in Table 2, this is not the usual purpose for ASTM D3433 specimens. Rather they are used to determine the G_{Ic} for a given adhesive system. It is informative to use the experimental results from the Instron tests to back-calculate G_{Ic}. If G_{Ic} is a material property, the values calculated for the various geometries should agree, barring changes in mode, amount of plasticity or other similar factors. These calculations were made and are summarized in Table 3. It may be noted that the finite element method gave very consistent results while calculations using equation (7) exhibited considerable variation. This would be expected since FEA includes energy from sources that are ignored in equation (7). It is also noted that the experimental samples used in this study had relatively large a/h values. The ignored factors should be even more important for more 'stubby' samples typical of those recommended in ASTM D3433.

For most of the samples, the maximum energy release rate as determined by the finite element analysis was along a line down the center of the adhesive. Experimentally it was observed that all these specimens failed along fracture paths that divided the adhesive into two nearly identical portions. Figure 9 is a photograph of a sample failure typical of this type of behavior. If the starter crack tape was placed at the boundary, the crack would rather quickly migrate to the path of higher ERR, i.e. the center. A few samples (e.g. 3G and 3H) exhibited larger calculated energy release rates along paths at the boundaries than at the center of the adhesive. These samples failed along paths close to the adhesive–adherend boundary. Typical behavior of this type is shown in Fig. 10. In most cases, Teflon® tape was attached to one of the adherends to form the starter crack near the interface. In other cases, a 0.36 mm saw

Table 2.
Comparison of predicted and experimentally observed failure loads and locus of failure

Sample group	Predicted failure load P_P (N)	Experimental failure load P_E (N)	% Difference between P_E and P_P	Locus of max. ERR	Experimentally obtained crack path	Predicted failure load using ASTM D3433, P_A	% Difference between P_E and P_A
3A	Reference	Reference	Reference	Down along center line	Down center of adhesive	Reference	Reference
3B	31.2	31.1 (±1.63)	0.3	Down along center line	Down center of adhesive	29.4	−5.5
3C	57.6	58.4 (±1.6)	−1	Down along center line	Down center of adhesive	58.8	−0.7
3D	60.8	61.1 (±2.25)	−0.5	Down along center line	Down center of adhesive	58.8	−3.8
3E	39.9	40.5 (±1.28)	−1.5	Down along center line	Down center of adhesive	39.6	−2.2
3F	133.9	137.5 (±8.9)	−2.6	Down along center line	Down center of adhesive	142	3.3
3G	48.2	48.6 (±2.8)	−0.8	Along bondline	In adhesive near metal	39.2	−19.3
3H	48.0	48.9 (±1.8)	−1.8	Along bondline	In adhesive near metal	39.2	−19.8
3I	44.1	45.8 (±0.8)	−3.7	Down along center line	Down center of adhesive	39.2	−14.4
3J	46.1	48.7 (±1.8)	−5.3	Center line and bondline were same	Wandered (see text)	39.2	−19.5
3K	46.6	47.8 (±1.5)	−2.5	Center 2% higher than bondline	Wandered	39.2	−18.0
3L	47.3	49.2 (±3.5)	−3.9	Long bondline	In adhesive near metal	39.2	−20.3

Table 3.
Comparison of G_c calculated from the various specimen geometries

Specimen	Location of largest ERR	G_c by FE (kJ/m^2)	% Difference from 3A	G_c by equation from ASTM D3433 (kJ/m^2)	% Difference from 3A
3A	Center	0.265 ± 0.038	Reference	0.274 ± 0.039	Reference
3B	Center	0.264 ± 0.029	−0.4	0.306 ± 0.033	11.7
3C	Center	0.273 ± 0.015	3.0	0.271 ± 0.015	−1.1
3D	Center	0.268 ± 0.020	1.5	0.296 ± 0.022	8.0
3E	Center	0.273 ± 0.017	3.0	0.286 ± 0.018	4.4
3F	Center	0.280 ± 0.036	5.7	0.258 ± 0.033	−5.8
3G	Near bondline	0.271 ± 0.018	2.3	0.422 ± 0.050	54.0
3H	Near bondline	0.275 ± 0.020	3.8	0.426 ± 0.031	55.5
3I	Center	0.286 ± 0.011	7.9	0.374 ± 0.014	36.5
3J	Center	0.296 ± 0.022	11.7	0.423 ± 0.031	54.4
3K	Near bondline	0.279 ± 0.018	5.3	0.407 ± 0.026	48.5
3L	Interface	0.288 ± 0.042	8.7	0.433 ± 0.063	58.0

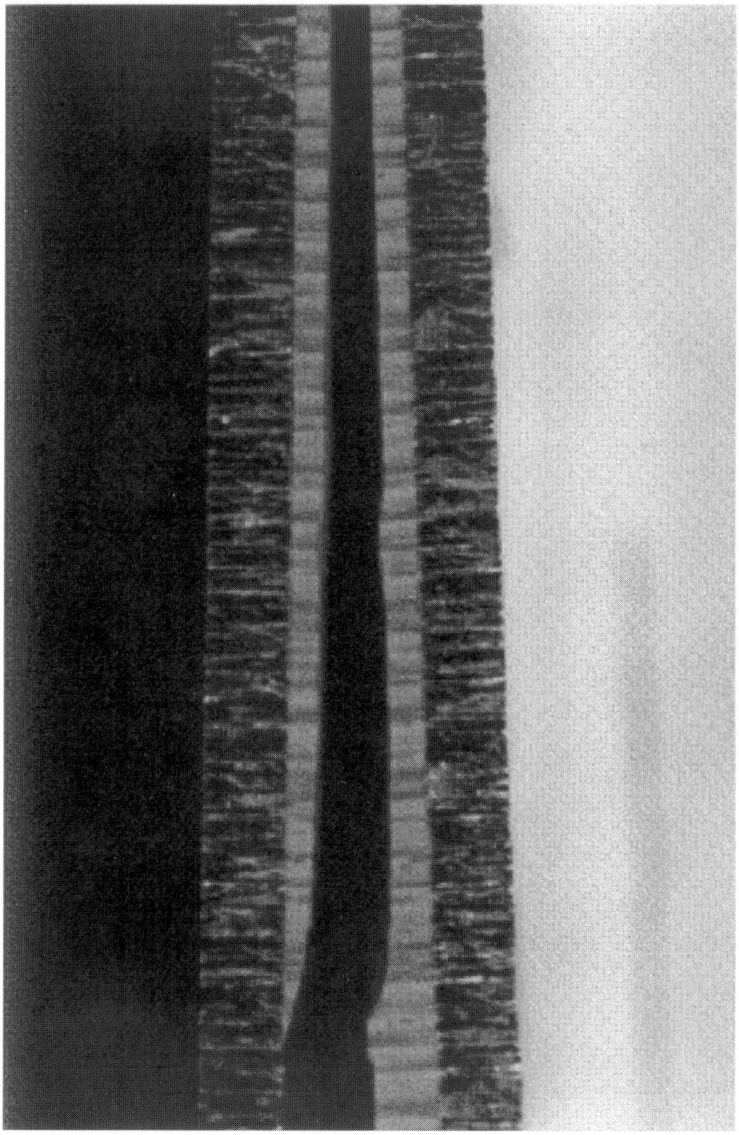

Figure 9. Photograph of a DCB specimen after testing. The calculated ERR was largest at the center and crack path was at center.

(a)

Figure 10. Photographs of two specimens for which the ERR was greatest at the edge and the crack advanced at or near the interface. The sample corresponding to the photograph (a) had ERRs that changed slowly as the interface was approached. The photo (b) is for a sample in which the ERR was significantly larger at the edge than at other planes in the adhesive. Note: for this sample one adherend was significantly thicker than the other. The results on this type of specimen are not summarized in the tables nor discussed in the text.

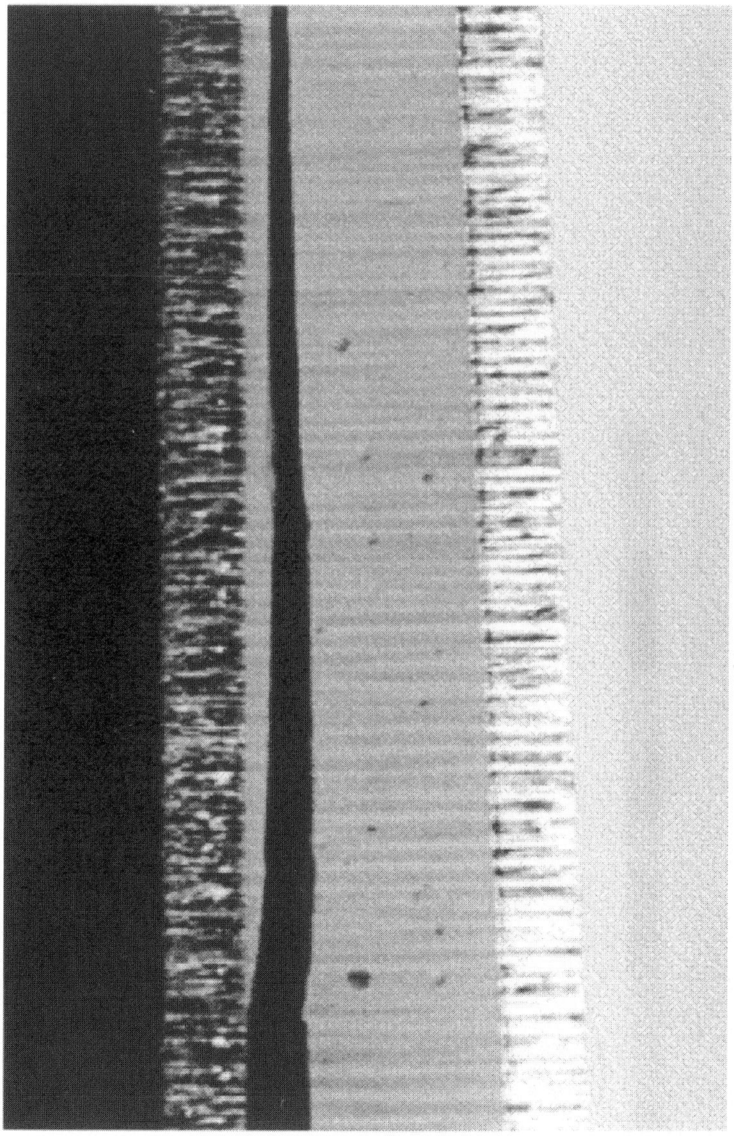

(b)

Figure 10. (Continued).

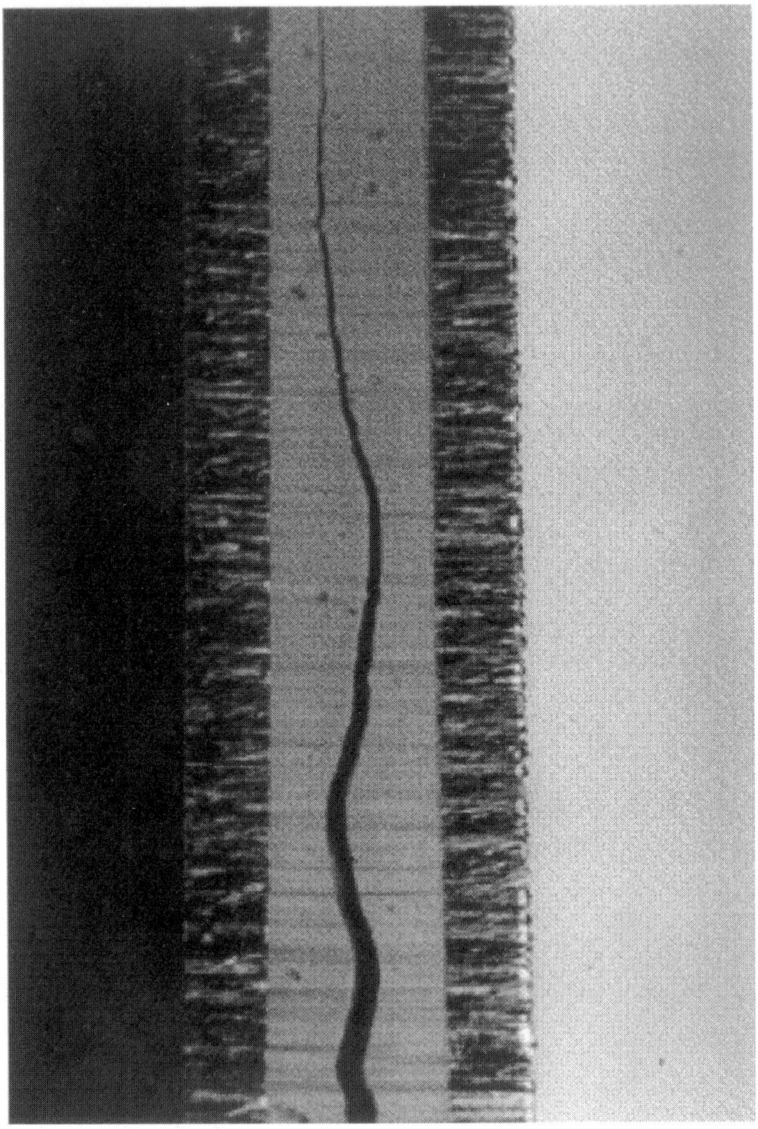

Figure 11. Photographs of a crack path for a specimen for which the ERR was constant across the adhesive thickness (from specimen group 3J).

was used to cut a short distance down the centerline of the adhesive and the starter crack was initiated with a razor blade. Neither of these locations was necessarily along the paths of highest ERR. In every case, when a crack was propagated, it quickly migrated to the path of higher ERR. This can be observed in the photograph of Fig. 9.

Some samples were designed such that the energy release rate calculated for paths along the adhesive center line was equal to that along the boundary (as well as for other parallel planes). In this case, the crack seemed to have difficulty in selecting a specific plane to follow and it proceeded in a wavy manner along the adhesive. Typical behavior of this type is shown in Fig. 11. The ERR for specimen group 3J was identical for central and boundary cracks and those for 3K differed by only one or two percent. All specimens in these two groups exhibited the wavy fracture path behavior.

The calculated energy release rate for all the 'center failing' samples converged to 100% Mode I. This was not necessarily true for the other samples. Some of these were still only approximately 96% Mode I. Quite clearly, the large differences between the moduli of the adherends and the adhesive would result in constraints that would induce shear stresses at the front of any crack proceeding close to and parallel with the boundary. If cracks have a strong preference for Mode I opening they would move some amount away from a boundary as was observed. These studies might then be viewed as supporting the Mode I preference hypothesis.

3. CONCLUSIONS

Fracture mechanics analysis provides insight into adhesive failure that has not been available through stress analysis alone. Such analyses making use of finite element methods were used:

(1) To explain why tapering of the adherends on lap shear specimens had a smaller effect on joint strength than might be anticipated from considerations of the reduction in the local shear stresses alone.

(2) To explain why reductions of the bonded area by 50% in lap joints does not result in decreases in strength anywhere near concomitant to the amount of bond area removed. This study also demonstrates the importance of knowing the fracture mode at the crack front. It lends credence to the hypothesis that cracks preferentially move along paths dominated by Mode I.

(3) In analyzing double cantilever beam specimens (and by inference other specimen geometries), where it is important to include all energy contributions. The energy in the adhesive, that associated with shear and rotation and/or stresses beyond the usual assumed ideal boundaries, can be significant. Use of standard equations that did not include these factors resulted in inconsistencies in determination of G_c from experimental values by as much as 50%. On the other hand, using finite element methods that encompassed these other factors resulted in very consistent results. For samples in which the energy release rate converged to Mode I,

the difference between the FEA calculated and experimentally determined G_c typically differed by only 5%. In one case studied, where it did not so converge, the errors were still only of the order of 10%.

(4) To demonstrate that cracks in double cantilever beam joints follow paths for which the energy release rate is a maximum. By changing sample geometry it is possible to change the paths along which the calculated energy release rate is a maximum. Specimens where this path runs along the center-line failed along a rather flat path leaving equivalent amounts of adhesive on both adherends. Those for which the maximum energy release rate was along the boundary consistently failed near the adherend. Samples where the energy release rate was independent of path exhibited cracks that followed a wavy locus through the adhesive.

Acknowledgement

The authors appreciate the support of the National Science Foundation (DMR-9014565 and CMS-9522743) for this research.

REFERENCES

1. P. R. Borgmeier and K. L. DeVries, *J. Adhesion Sci. Technol.* **7**, 967 (1993).
2. E. F. Rybicki and M. F. Kanninen, *Eng. Frac. Mech.* **9**, 931–938 (1977).
3. G. P. Anderson and K. L. DeVries, *Intl J. Frac.* **39**, 191–199 (1989).
4. L. J. Hart-Smith, *Adhesives Age* **30**, 28–32, April (1987).
5. A. J. Kinloch, *Adhesion and Adhesives: Science and Technology*. Chapman and Hall, London (1987).
6. F. Thamm, *J. Adhesion* **7**, 301 (1976).
7. R. D. Adams, R. W. Atkins, J. A. Harris and A. J. Kinloch, *J. Adhesion* **20**, 29 (1986).
8. M. Goland and E. Reissner, *J. Appl. Mech.* **2**, A17–A27 (1944).
9. G. P. Anderson, S. Chandapeta and K. L. DeVries, in: *Adhesively Bonded Joints: Testing, Analysis and Design*, W. S. Johnson (Ed.). STP-81, ASTM, Philadelphia, PA (1988).
10. T. R. Guess, R. E. Allred and F. P. Gerstle, *J. Testing Evaluation* **5**, 84 (1977).
11. G. R. Woolley and D. R. Carver, *J. Aircraft* **8**, 817 (1971).
12. R. D. Adams and N. A. Peppiatt, *J. Strain Anal.* **9**, 185 (1974).
13. S. Amijima, A. Yoshida and T. Fujii, in: *Proc. Second Intl Conf. on Composite Materials*, Toronto, Canada, p. 1185. Metallurgical Society of AIME, New York (1978).
14. N. Sharpe and T. J. Muha, in: *Proc. Army Symposium on Solid Mech.*, AMMRC-MS075-8, p. 23. Army Materials and Mech. Research Center Watertown, MA (1974).
15. A. J. Kinloch and G. K. A. Kodokian, *J. Adhesion* **37**, 39 (1992).
16. G. P. Anderson, K. L. DeVries and M. L. Williams, *J. Colloid Interface Sci.* **47**, 600 (1974).
17. W. D. Bascom, C. O. Timmons and R. L. Jones, *J. Mater. Sci.* **10**, 1937–1948 (1975).
18. S. Mostovoy and E. J. Ripling, in: *Adhesion Science and Technology, Part B*, L. H. Lee (Ed.). Plenum Press, New York (1975).
19. G. P. Anderson, S. J. Bennett and K. L. DeVries, *Analysis and Testing of Adhesive Bonds*. Academic Press, New York (1981).
20. W. S. Johnson and P. D. Manzalgire, NASA Technical Memorandum 87571 (1985).
21. K. M. Liechti and T. Freda, in: *Advances in Adhesively Bonded Joints*, pp. 65–77. ASME (1988).
22. J. W. Hutchinson and Z. Suo, *Adv. Appl. Mech.* **29**, 65 (1992).
23. G. Fernlund and J. K. Spelt, *Compos. Sci. Technol.* **50**, 441–449 (1994).
24. G. Fernlund, M. Papini, D. McCammond and J. K. Spelt, *Compos. Sci. Technol.* **51**, 587–600 (1994).
25. A. J. Kinloch, personal communication.
26. R. J. Good and L. A. Girifalco, *J. Phys. Chem.* **61**, 904 (1957).

27. R. J. Good and L. A. Girifalco, *J. Phys. Chem.* **64**, 561 (1960).
28. F. M. Fowkes, *J. Phys. Chem.* **67**, 2538 (1963).
29. F. M. Fowkes, *Ind. Eng. Chem.* **56** (12), 40 (1964).
30. D. P. Graham, *Ind. Eng. Chem.* **57**, 4387 (1965).
31. J. W. Whalen, *J. Colloid Interface Sci.* **28**, 443 (1968).
32. K. W. Allen, *J. Adhesion* **21**, 261–277 (1987).
33. J. J. Bikerman, *The Science of Adhesive Joints*. Academic Press, New York (1968).
34. A. R. Akisanya and N. A. Fleck, *Intl J. Frac.* **55**, 29–45 (1992).
35. N. A. Fleck, J. W. Hutchinson and Z. Suo, *Intl J. Solid Structures* **27** (13), 1683–1703 (1991).
36. P. D. Mangalgiri, W. S. Johnson and R. A. Everett, Jr., *J. Adhesion* **23**, 263–288 (1987).
37. S. Mall and W. S. Johnson, in: *Composite Materials: Testing and Design (Seventh Conference)*, pp. 322–334. ASTM STP **893** (1986).
38. W. Brockmann, *J. Adhesion* **37**, 173–179 (1992).
39. A. N. Gent and C. W. Lin, *J. Adhesion* **32**, 113–125 (1990).
40. S. P. Timoshenko and J. N. Goodier, *Theory of Elasticity*, 3rd edn. McGraw Hill, New York, NY (1987).
41. E. J. Ripling, S. Mostovoy and R. L. Patrick, *Mater. Research and Standards* **4** (3), 129–134 (1964).
42. E. J. Ripling, S. Mostovoy and H. H. Corten, *J. Adhesion* **3**, 107–123 (1971).
43. S. Mostovoy and E. J. Ripling, *J. Appl. Polym. Sci.* **15**, 661–673 (1971).
44. G. R. Cowper, *J. Appl. Mech., Series E* **33** (2), 335–340 (1966).
45. W. J. O'Donnell, *J. Appl. Mech., Series E* **27** (3), 461–464 (1966).
46. N. I. Muskhelishvili, in: *Some Problems of the Mathematical Theory of Elasticity*, pp. 467–471. P. Noordhoff Ltd., Groningen, Holland (1953).
47. J. Raasch, *Intl J. Frac.* **7**, 289–300 (1971).
48. M. F. Kanninen, *Intl J. Frac.* **9**, 83–92 (1973).
49. J. W. Obreimoff, *Proc. Royal Soc. London, Series A* **127**, 290–297 (1930).
50. J. J. Benbow and F. C. Roesler, *Proc. Physical Soc. London, Series B* **70**, 201–211 (1956).
51. J. J. Gilman, *J. Appl. Phys.* **31**, 2208–2218 (1960).
52. P. L. Gutshall and G. E. Gross, *J. Appl. Phys.* **36**, 2459–2460 (1965).
53. A. R. C. Westwood and T. T. Hitch, *J. Appl. Phys.* **34**, 3085–3089 (1963).
54. P. P. Gillis, *J. Appl. Phys.* **34**, 2149 (1968).
55. J. P. Berry, *J. Appl. Phys.* **34**, 62–68 (1963).
56. P. P. Gillis, *J. Appl. Phys.* **35**, 647–658 (1964).
57. S. M. Wiederhorn, A. M. Shorb and R. L. Moses, *J. Appl. Phys.* **39**, 156–172 (1968).
58. J. E. Srawley and B. Gross, *Mater. Research Standard* **7** (4), 155–162 (1967).
59. P. P. Gillis, *J. Appl. Phys.* **36**, 1374–1376 (1965).
60. R. G. Hoagland, *J. Basic Engr., Series D* **89** (3), 525–532 (1967).
61. J. G. Williams, *Intl J. Frac.* **36**, 101–119 (1988).

Mittal Festschrift, pp. 641–653
W. J. Van Ooij and H. R. Anderson, Jr. (Eds)

Durability of automotive adhesive bonds: Test methods and failure mechanisms

R. A. DICKIE and S. M. WARD

Ford Research Laboratory MD 3083 SRL, Ford Motor Company, P.O. Box 2053, Dearborn, MI 48121, USA

Abstract—Structural adhesives have found increasing acceptance in the automotive industry in recent years, providing an enabling technology for the use of light weight materials and aiding in the optimization of structural performance of conventional materials. Assessment of adhesive bond performance and prediction of durability in automotive applications remain contentious issues. Material characterization and engineering performance test methods play key roles in resolving these issues. Selected engineering test methods for adhesive flow properties, bond durability and bonded structure impact performance are described. While these tests do not necessarily provide a fundamental characterization of adhesive material or system properties, they do allow a ranking of material (or system) performance. Basic studies of failure mechanisms have been conducted to help validate the durability and impact test methods.

Keywords: Adhesive; automotive; bonding processes; durability; impact.

1. INTRODUCTION

Adhesives and sealers are widely used in the automotive industry for applications ranging from windshield installation to nameplate attachment. Among the most challenging applications are those in which an adhesive is used either alone or in conjunction with mechanical fastening to build the primary body structure. The drive to reduce vehicle weight (and thereby improve fuel economy) has resulted in intensive investigation of light weight materials including aluminum and polymeric composites — materials for which adhesive bonding is often the joining method of choice. The engineering effectiveness of structural adhesive bonding and the potential benefits of their use in conventional steel body construction have long been known [1, 2]. The combination of structural bonding and spot welding in steel structures has been demonstrated to provide substantial increases in torsion and bending stiffness [3]. Nevertheless, extensive implementation of structural bonding technology in high volume automotive production has been slow in coming. Even when utility and cost effectiveness can be demonstrated, there remain concerns regarding the reliability of design and engineering methodologies, the sensitivity of bond performance to variations in process conditions and parameters, and the long-term durability of bonded joints. The development and validation of material characterization and engineering performance test methods play key roles in resolving these concerns.

This paper begins with an overview of automotive adhesive materials and processes. Following an outline of material characterization and engineering test needs for structural applications, recent progress in the on-going development of selected test methods for adhesive flow properties, bond durability and bonded structure impact performance is summarized.

2. AUTOMOTIVE MATERIALS AND PROCESSES

Although the details of the materials, processes, and requirements differ widely from company to company and plant to plant, some generalizations can be made. A generalized production sequence and supply pattern is shown schematically in Fig. 1, and a summary of typical substrate and adhesive material classes is given for typical applications in Tables 1 and 2. In the normal automotive assembly sequence, the basic body structure is formed by welding together sheet metal stampings. Closure panels — doors, hoods, and deck lids — are constructed off-line, sometimes in the body construction area of the assembly plant, sometimes in stamping plants where the basic metal cutting and forming operations are carried out, and sometimes (especially in the case of parts made from polymeric composites) at separate facilities. Closure panels are attached to the body during body construction; the unit then goes through paint operations and into final assembly and trim where functional systems and decorative

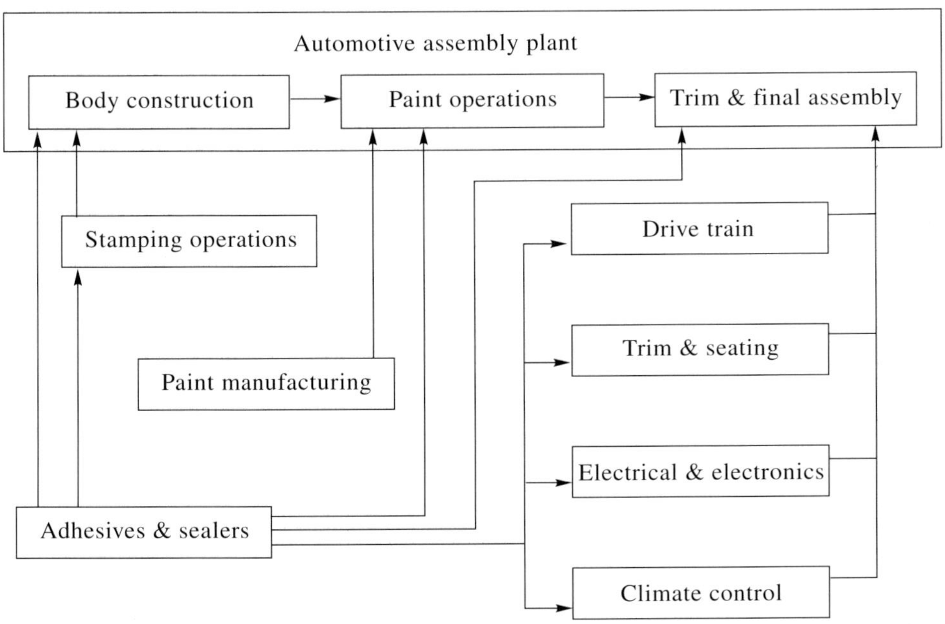

Figure 1. Schematic of adhesive and sealer supply to automotive component manufacturing and assembly operations.

Table 1.
Typical automotive adhesive applications in component and stamping plants

Component and subassembly plant adhesives			
Engine and drive train suppliers	Trim and plastic component suppliers	Climate control, electrical, and electronics suppliers	Stamping plants
Typical applications			
Thread lock Gaskets Oil seals Clutch plates Brakes	Lamp housings Bumpers SMC closure panels Engine mounts Window assemblies	Heat exchangers Circuit boards Ignition coils Electronic modules	Doors Hoods Deck lids
Typical substrates			
Steel Aluminum Cast iron Friction materials	Rubber Plastics Composites Glass	Aluminum Copper Plastics Composites	Steel Galvanized steel Aluminum
Typical adhesives			
Anaerobic acrylic Silicone Epoxy–novolac Epoxy–dicy	Urethane (2K) Urethane (H_2O cure) Silicone 2K epoxy PVB	Epoxy (1K) Epoxy (2K) Silicone	Epoxy (induction) Vinyl–epoxy Acrylic–epoxy (2K)

components are installed. There are adhesive bonding and sealing operations associated with each of the manufacturing stages, but the primary structural applications of interest are in closure panel fabrication and body construction.

Typical production rates in automotive assembly range from 40 to 100 jobs per hour; stamping operations (and associated closure panel fabrication) are typically several times faster (e.g. 200–400 jobs per hour). These productions rates mean that robust, high speed robotic dispensing is the preferred method of applying adhesives. In current practice, materials that require thermal cure must be formulated to react during paint stoving, although it should be noted that induction heating is commonly used to obtain an initial, partial cure of adhesives in closure panel fabrication [4]. Incorporation of separate stoving facilities for adhesive cure in body construction operations has been proposed [5], but carries significant capital and operating cost penalties.

The substrates of interest in current production are cold-rolled steel, galvanized steel (both hot-dipped and electro-galvanized), various alloy coated steels, aluminum, and, to a lesser extent, polymeric composites. For body structural and closure panel applications, typical adhesives are toughened epoxies, acrylic–epoxies, and, for sealing and lower performance structural applications, vinyl and vinyl–epoxy formulations. Most are applied as one- or two-part pastes at ambient or slightly elevated tempera-

Table 2.
Typical automotive adhesive applications in assembly plants

Assembly plant adhesives			
Body shop	Paint shop	Final assembly	Repair and service
Typical applications			
Weld through sealers Weld through adhesives Panel reinforcement	After-seal	Decorative trim Plastic fascia Plastic body panels Carpet and head liner Windshield	Windshield Trim Panel repair
Typical substrates			
Steel Galvanized steel Aluminum Composites	Painted metal	Painted metal Glass Plastics Composites Fabric	Glass Painted metal Composites Plastics
Typical adhesives			
Epoxy Vinyl–epoxy Plastisols	Vinyl plastisols Hot melt sealers	Acrylic p/s Elastomeric p/s Urethane	Various ambient cure

ture. Thermoset hot melt formulations, tapes, and pre-cut patches have been used for specific applications. Moisture-cure urethanes are used for installation of windshield glass, and two-component urethanes are widely used for bonding of composites and many plastic components.

3. TEST METHODS AND REQUIREMENTS

The primary goals of material characterization are to provide accurate material properties for structural analysis and process modeling and to determine key performance properties for material specification and process control. The standard material characterization methods of materials science, mechanics, and rheology can, in principle, provide the desired material properties. In some cases, however, the design of the required test specimens is not compatible with the materials and processes of interest and alternative test methods must be developed. There also is no shortage of engineering performance tests for strength, durability and impact performance of adhesive bonds, but the results of these tests are strongly dependent on details of specimen geometry and details of experimental procedures: what is obtained is a measure of performance of the particular structure tested rather than a measure of material response. In the absence of an understanding of failure modes and mechanisms in the test and in the intended application, the significance of engineering performance and durability tests is open to question and must be verified on a case-by-case basis.

There is a stark dichotomy between the demands of engineering performance specifications (which are typically concerned with system selection and qualification based on readily observable physical characteristics and *ad hoc* performance tests) and the material characterization needed for analytic modeling of structural response, simulation of dispensing and cure processes, and estimation of durability.

Typical engineering specifications for automotive adhesives describe requirements for basic physical characteristics (e.g. color, solids, ash, and density); place limits on flow characteristics (e.g. viscosity, sag resistance, slump resistance, bridging, and wash resistance); require demonstration of system compatibility with other materials and processes (e.g. lubricants, paints, welding operations); specify requirements for such other characteristics as flammability, odor, storage stability, and abrasive content, and, finally, prescribe cured material performance in terms of adhesion strength (typically in terms of lap shear failure load initially and after various environmental exposures), flexural strength, and peel strength. Fundamental measures of the mechanical properties of cured adhesives are conspicuous by their absence from this list and often are not included in data supplied by adhesive manufacturers.

For analytic modeling of structural response, measurements of tensile or shear modulus, the Poisson ratio, creep compliance, elastic/plastic stress–strain curves, and fracture energy as functions of temperature, time, frequency, and rate are needed and can generally be obtained using conventional measurement methods. Some of the appropriate tests and their use in finite element modeling of bonded structures are discussed in [6–8].

Most commonly used automotive adhesives are one- or two-component pastes. In order to model dispensing processes for pastes, flow characterization in terms of shear rate, shear history, and temperature is needed. Since the flow behavior of paste adhesives is generally non-Newtonian [9], the required rheological characterization can become rather involved.

Modeling of the cure process for thermoset materials requires knowledge of chemical kinetics of the cross-linking process and the heat transfer characteristics of the adhesive, substrates, and oven. Earlier work [10] on the cure of thermoset automotive paints suggests that the uniformity of cure cannot be assumed based solely on oven temperature and time: automotive paint stoving schedules do not result in uniform heat history over the painted surface of the body structure, and cannot be expected to do so in bonded joints throughout the body structure. Modeling of thermal response and chemical kinetics can be used to aid in defining appropriate process limits and operating ranges for adhesive cure during stoving.

4. ADHESIVE FLOW PROPERTIES AND DISPENSING

'Press flow' and other simple, single parameter measures of nominal viscosity do not capture the differences in flow behavior of non-Newtonian fluids such as paste adhesives. Flow characterization and its use in modeling of dispensing processes have

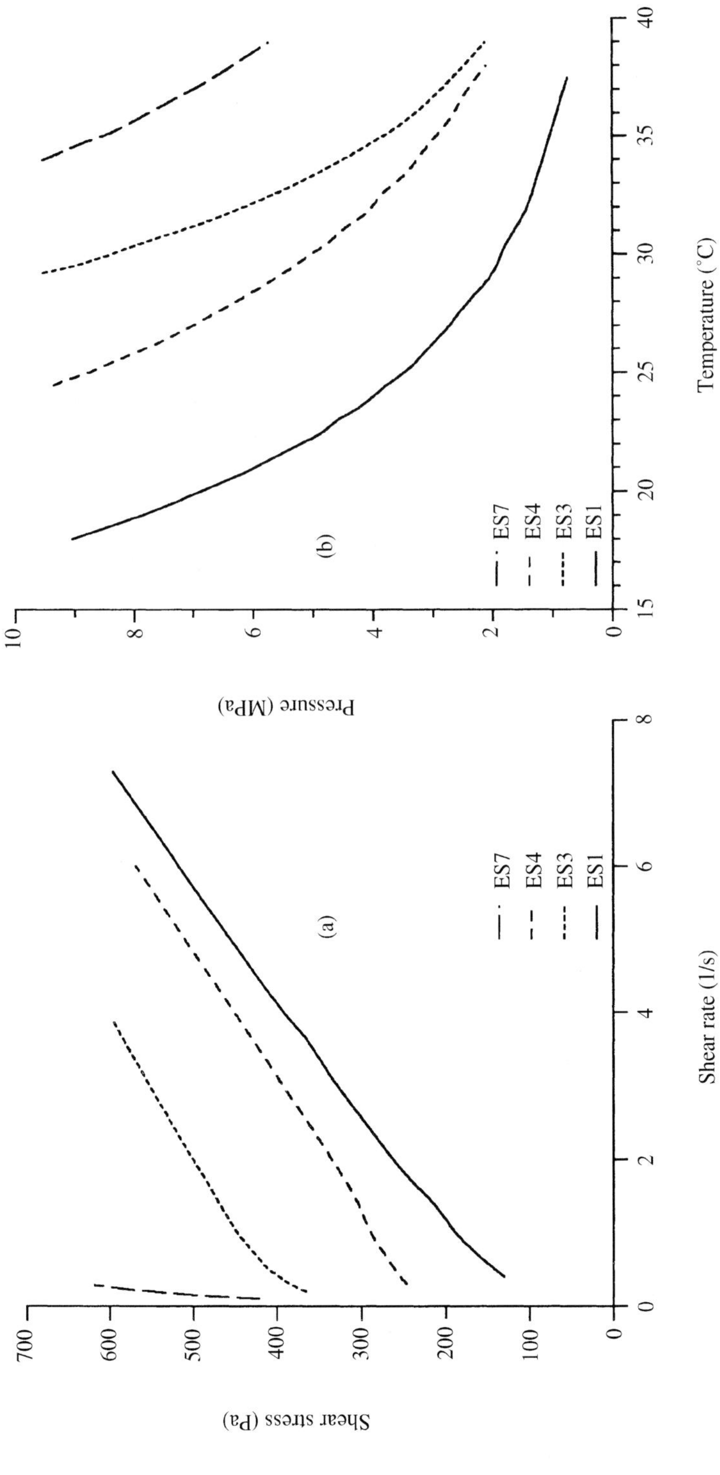

Figure 2. (a) Flow curves for a range of paste epoxy adhesives (adapted from [9]) and (b) constant volumetric output parameters (adapted from [11]).

been demonstrated [9, 11] for paste epoxy adhesives. The paste epoxy formulations discussed in [9] can be described, to a first approximation, as Bingham fluids: i.e. they exhibit apparent yield behavior at low shear rate, followed by a steady flow regime at higher rates (see Fig. 2a). The flow behavior of these materials can be described adequately in terms of two flow parameters (yield strength and steady shear viscosity) and two temperature factors (Arrhenius activation energies for yield and flow, respectively). Adhesive flow has also been modeled using a power law representation [12]. The strongly non-Newtonian character of the flow behavior affects the dispensing characteristics of these materials profoundly, and in ways that cannot be predicted from the single-point measurements of apparent viscosity often used for material specification.

For simple single-stage dispensing systems, analytic models of volumetric output have been developed that allow assessment of the processability of adhesives of differing flow and yield behavior in terms of dispensing pressure and temperature profiles for constant volume output. A set of such curves is shown in Fig. 2b for the adhesives of Fig. 2a. Effects of chemical reactivity and shear heating have also been incorporated into the dispensing model [11] to provide an estimate of process limits. The dispensing systems used in practice are often multiple stage, and can involve very high shear rates. For these systems, a detailed, closed form model may not be feasible. However, comparison and matching of carefully measured flow curves to those of a material known to dispense properly from a given dispensing system can be a useful basis for material selection and quality control.

5. DURABILITY OF ADHESIVE BONDS

Chemical and physical (mechanical) degradation must both be taken into account in evaluating the likely durability of adhesive bonds. Table 3 provides an overview of test method conditions and performance metrics used for evaluation of the chemical and physical durability of lap shear specimens.

'Strength' and 'retained strength' are often characterized in terms of the failure load under monotonic constant displacement loading. Tests exposing bonded specimens to high humidity or corrosive conditions have been used for many years under various names; 'time to failure' is the usual measure of durability, although retained strength as a function of exposure time and load may also be measured. 'Fatigue life' is

Table 3.
Performance metrics for chemical and physical durability of lap shear specimens

Load	Environmental exposure		
	Ambient lab	Humidity	Corrosion
None	Strength	Strength retention	Strength retention
Static	Creep	Time to failure	Time to failure
Cyclic	Fatigue life	Fatigue life	Fatigue life

ascertained as the number of reversals to failure under cyclic loading, typically using tension–tension loading and load control.

The nature of the observed failure processes, and the dominant material system characteristics governing performance, depend on the combination of mechanical loading and environmental exposure used. The performance in cyclic loading under ambient laboratory conditions is dominated by mechanical response, while strength retention under environmental exposure is dominated by the resistance of the structure to chemical attack.

In terms of automotive requirements, tests involving exposure of unstressed bonded test specimens to aggressive environments (such as high humidity or one of the many corrosion cycles commonly used in the industry) are appropriate and perhaps sufficient for evaluation of materials intended primarily as sealers (i.e. materials that are not expected to bear significant load during their useful life). Tests involving static and cyclic loading obviously become relevant for the evaluation of bonded structures used in load-bearing structural applications.

Strength retention measurements provide an early indication of potential problems with bond stability. In studies of bonded cold rolled steel and galvanized steel, for example, it has been found that the dominant mechanism of bond degradation is corrosion of the substrate, and specifically anodic attack on the substrate metal underlying the adhesive. The mechanism is similar to crevice corrosion [13–15]. Details of the interfacial chemistry during bond formation can strongly affect the rate of bond weakening. It has been suggested, for example, that degradation of poly(vinyl chloride), which is catalyzed by zinc, can lead to poor corrosion durability of adhesive formulations based on inadequately stabilized poly(vinyl chloride) [16, 17]. Presumably the degradation products, which include hydrogen chloride, contribute to interfacial corrosion. Conversely, the formation of stable interfacial reaction products has been postulated as a factor contributing to the relative stability of certain epoxy–galvanized steel bonds [18].

It has long been recognized in the aerospace industry that the deleterious effects of environmental exposure on bond integrity can be enhanced by simultaneous application of load. Typically, single lap shear bonds are subjected to tensile load at some fraction of the nominal failure load; the time to failure under specified exposure (typically a high humidity environment) is observed [19]. Essentially the same experimental techniques have been applied to evaluate the durability of bonds to galvanized steel. Use of an exposure cycle incorporating exposure to salt water greatly accelerates the failure of bonds to galvanized steel. The mechanism by which corrosion results in bond failure has been explored using an imaging X-ray photoelectron spectroscopy (XPS) technique to elucidate the interfacial chemistry [20, 21]. It has been found that there is an initial stage of crevice-like corrosion leading to a modest progressive loss of bond strength. At some critical stage of damage, the interfacial corrosion chemistry changes to include both localized anodic and localized cathodic sites; the rate of corrosion appears to increase dramatically, and spontaneous mechanical failure ensues.

The distribution of failure times at each load can be handled in conventional statistical terms, and the functional dependence of average failure times on load can

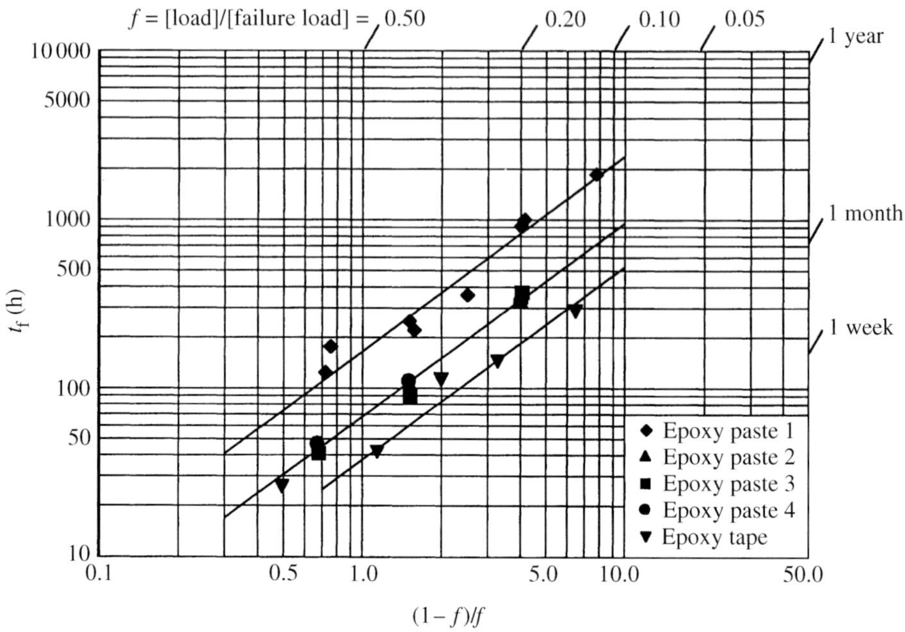

Figure 3. Time to failure, t_f, as a function of load parameter $(1 - f)/f$ for a range of commercial single-component epoxy paste and tape adhesives on galvanized steel under accelerated corrosion test conditions (adapted from [22]).

be analyzed [22]. Failure time, t_f, can be represented as a function of (fractional) load, f. For a power law dependence of the degradation process on f,

$$t_f = k(1 - f)/f^n.$$

Plotting failure time as a function of $(1 - f)/f$ allows an initial comparison of relative performance and degree of acceleration by load, as illustrated in Fig. 3 for a range of experimental and commercial epoxy adhesives tested under corrosive conditions. The experimental methods used to obtain the data of Fig. 3 have been described elsewhere [22]. Although the adhesives used are all similar in chemistry and mechanical properties, a wide variation in longevity and in the degree of acceleration of failure provided by the application of load is observed.

The results seem to suggest that the chemistry of the process is accelerated by the application of load. What is more likely is that the application of load changes details of how the corrosion process and ultimate failure occur [21]. For these bonded galvanized steel substrates, the effect of the applied load is most likely to prevent formation of an effective barrier of corrosion products, and thus to allow the corrosion process to proceed unimpeded. This suggests that the results of the experiment are likely to be critically specimen design dependent, and should also be strongly influenced by changes in substrate and corrosion inhibitor content of the adhesive. Partial confirmation is given by results obtained on bonded galvanized steel and aluminum substrates, which differ sharply in their functional dependence on load [22].

6. IMPACT PERFORMANCE OF BONDED STRUCTURES

Vehicle and component crash testing is expensive, time consuming, difficult to analyze, and does not provide material properties for modeling. Laboratory crush and impact tests on bonded test specimens (such as box beams) are also dominated by properties of structure, and, while they can be used to give qualitative material selection guidelines, do not provide material properties suitable for modeling [23, 24]. Measurements of fracture toughness, G_{Ic}, give unambiguous results for polymer properties, but pose serious challenges in specimen preparation for evaluation of materials of interest for automotive construction. Sheet metal substrates do not lend themselves to the preparation of compact tension or tapered double-cantilever beam specimens. Thin substrates of interest can be bonded to or inserted in backing beams to approximate the desired specimen geometry, but these methods are also not fully satisfactory [25].

There are many laboratory impact tests that have been used to obtain material performance rankings. One of some interest is the wedge peel test that is the subject of a recent ISO test method [26]. In the ISO method, a narrow wedge is dragged through a bonded specimen at a specified velocity (2 m/s). The force is recorded; total energy dissipated can be calculated. A simplified test configuration has been developed for use in a falling weight testing machine [27]; in this version of the test, the specimen is driven downwards over a stationary wedge. The performance rankings obtained are entirely comparable to those provided by the ISO method, but the specimen preparation and analysis of the deformation are substantially simplified. Direct comparisons of experimental results have shown only minor differences in measured values of total energy and average load in the two test configurations. Measurements of total energy dissipation have been used as a tool for comparison of material performance, and it has been suggested that the results are a reasonable predictor of the impact performance of bonded structures [28]. Impact peel energy

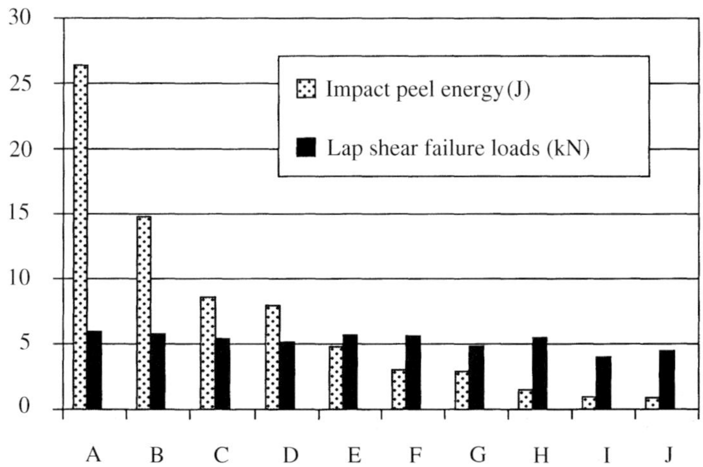

Figure 4. Impact peel energy (J) and lap shear failure loads (kN) for experimental high impact epoxy adhesives (A, B) and a range of commercial single-component epoxy paste adhesives on galvanized steel.

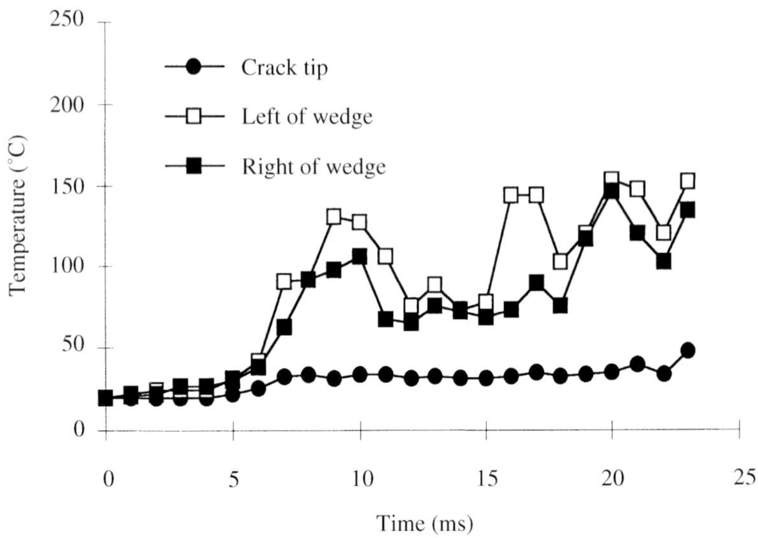

Figure 5. Temperature distribution in the impact peel experiment (data for adhesive A, Fig. 4).

absorption and lap shear strength data are shown in Fig. 4 for a range of commercial and experimental epoxy adhesives (the experimental details are described in [27]).

Processes contributing to energy dissipation in the test include fracture of the adhesive, plastic deformation of the substrate and friction of the arms of the test specimen against the wedge. High speed infrared images of the failure event indicate that temperature rises of at least 100 K can occur locally due to friction. The corresponding temperature rise at the crack tip in the adhesive is on the order of 10 K [27]. An example of the temperature distribution about the crack tip and wedge surface is shown in Fig. 5. Analysis of the deformation process suggests that about 10% of the total energy dissipation is directly attributable to adhesive fracture; about 30% to plastic deformation of the substrate; and about 60% to frictional heating [29].

The initial test results suggest that the impact peel test is a useful way of ranking materials, but that the energy dissipation numbers must be interpreted with caution due to the contributions of substrate plastic deformation and friction. While quantitative interpretation is therefore fraught with difficulty, the results do raise interesting questions about the role of processes other than adhesive fracture in the impact performance of bonded structures.

7. CONCLUDING REMARKS

Both fundamental tests providing material characterization and engineering tests for assessment of structural performance are necessary. Modeling and analysis require good fundamental material characterization. In principle, fundamental measures are preferable to engineering performance tests unless it can be shown that the engineering test has the same response as the actual process or structure to changes in material properties. In practice, however, the ease with which many of the engineering tests can be conducted earns them a place in specification and quality control provided their intrinsic limitations are recognized. Engineering tests are at least useful, and probably essential, in the assessment of system interactions and the durability of complex processes and structures. Understanding of the controlling parameters in fundamental terms does help to reduce errors in interpretation, and can lead to new insight into material and structure performance.

REFERENCES

1. J. R. Love, *Proc. 16th Annual Technical Convention*. American Society of Body Engineers, Detroit, MI (October 18–20, 1961).
2. S. S. Labana and F. R. Cotton, *Adhesives Age* **26** (6), 31 (1983).
3. D. A. Wagner, C. M. Cunningham and M. A. DeBolt, in: *Proc. International Body Engineering Conference 1993*, M. N. Uddin (Ed.), p. 123. International Body Engineering Conference, Ltd., Warren, MI (1993).
4. K. Drain, S. Chandrasekharan, in: *Engineered Materials Handbook, Vol. 3: Adhesives and Sealants*, H. F. Brinson (Technical Chairman), p. 551. ASM International (1990);
   H. Turner, in: *Engineered Materials Handbook, Vol. 3: Adhesives and Sealants*, H. F. Brinson (Technical Chairman), p. 716. ASM International (1990).
5. M. J. Wheeler, P. G. Sheasby and D. Kewley, SAE Paper No. 870146. Society of Automotive Engineers, Warrendale, PA (1987).
6. R. B. Krieger, Jr., in: *Adhesive Bonding*, L. H. Lee (Ed.), p. 291. Plenum Press, New York (1991).
7. S. M. Ward, M. A. Samus and G. L. Westerbeek, SAE Paper No. 960785. Society of Automotive Engineers, Warrendale, PA (1996).
8. J. N. Reddy and S. Roy, in: *Adhesive Bonding*, L.-H. Lee (Ed.), p. 359. Plenum Press, New York (1991).
9. C. Li, K. A. Mazich and R. A. Dickie, *J. Adhesion* **32**, 127 (1990).
10. D. R. Bauer and R. A. Dickie, *J. Coatings Technol.* **58** (738), 41 (1986).
11. C. Li and R. A. Dickie, *J. Adhesion* **35**, 115 (1991).
12. A. Razban and B. L. Davies, *J. Adhesion Sci. Technol.* **9**, 1435 (1995).
13. R. T. Foister, S. L. F. Niks and M. J. Barker, *J. Adhesion* **30**, 105 (1989).
14. J. W. Holubka, W. Chun and R. A. Dickie, *J. Adhesion* **30**, 173 (1989).
15. R. A. Dickie, J. W. Holubka and J. E. deVries, *J. Adhesion Sci. Technol.* **4**, 57 (1990).
16. J. E. deVries, J. W. Holubka and R. A. Dickie, *J. Adhesion Sci. Technol.* **3**, 189 (1989).
17. J. E. deVries, L. P. Haack, J. W. Holubka and R. A. Dickie, *J. Adhesion Sci. Technol.* **3**, 203 (1989).
18. R. O. Carter, III, R. A. Dickie and J. W. Holubka, *Ind. Eng. Chem. Res.* **28**, 48 (1989).
19. D. A. Wangsness, *J. Appl. Polym. Sci.: Appl. Polym. Symp.* **32**, 291 (1977).
20. L. P. Haack, M. A. DeBolt, S. L. Kaberline, J. E. deVries and R. A. Dickie, *Surf. Interface Anal.* **20**, 115 (1993).
21. R. A. Dickie, M. A. DeBolt, L. P. Haack and J. E. deVries, *J. Adhesion Sci. Technol.* **8**, 1413 (1994).
22. R. A. Dickie, M. A. DeBolt and S. M. Ward, SAE Paper 950128. Society of Automotive Engineers, Warrendale, PA (1995).
23. R. M. Rivett and S. T. Riches, Drop weight impact testing of spot welded, weldbonded and adhesively bonded box sections, Welding Institute Report 9431.01/85/478.2 (1986).
24. P. A. Fay and G. D. Suthurst, *Intl J. Adhesion Adhesives* **10**, 128 (1990).
25. J. K. Jethwa, A. J. Kinloch and G. Wallington, *J. Mater. Sci. Letters* **14**, 155 (1995).
26. Adhesives — Determination of dynamic resistance to cleavage of high strength adhesive bonds under impact conditions — Wedge impact method, International Standard ISO 11343: 1993(E). International Organization for Standardization, Geneva, Switzerland (1993).
27. M. A. DeBolt, S. M. Ward and R. A. Dickie, in preparation.

28. A. C. Taylor, B. R. K. Blackman, A. J. Kinloch and Y. Wang, *Proc. SAE IV*. Society of Automotive Engineers, Bristol, UK (July 1995).
29. M. Thouless, S. M. Ward and R. A. Dickie, *J. Materials Sci.* (submitted).

Part 5

Adhesion aspects of polymeric coatings and polymer-polymer interphase

Mittal Festschrift, pp. 657–671
W. J. Van Ooij and H. R. Anderson, Jr. (Eds)

Polyimide adhesion to the silica surface: Interfacial interactions

L. P. BUCHWALTER
IBM T. J. Watson Research Center, Yorktown Heights, NY 10598, USA

Abstract—Polyimide adhesion to a silica surface is initially good even without an adhesion promoter; however, its reliability is poor. The reliability of its adhesion to silica has been significantly improved by the use of an aminosilane adhesion promoter. The favorable chemical interactions at both polyimide–aminosilane and aminosilane–silica interfaces result in strong covalent bond formation. The peel locus of failure in samples with and without an adhesion promoter is always in the polyimide. Peel adhesion data analysis combined with locus of failure studies suggest that the improved peel adhesion is due to a change in polyimide mechanical properties in the interfacial region. This paper explores the interplay of chemical and physical factors in the adhesion of polyimide to the silica surface.

Keywords: Polyimide; adhesion.

1. INTRODUCTION

Polyimides (PIs) were introduced to microelectronics device applications in the early 1970s [1, 2]. The success of polyimides in these applications has been due primarily to their high thermal stability, toughness, high glass transition temperature, and excellent dielectric properties [3–5]. To achieve reliable structures using polyimides, adhesion of these interlayer dielectrics or passivation layers to a variety of inorganic surfaces must be achieved and maintained [3, 4]. The demands on the interface integrity change as the material properties making that interface change. In the case of 3,3′,4,4′-biphenyltetracarboxylic acid dianhydride–*p*-phenylenediamine (BPDA-PDA) PI–copper structures, the copper sidewalls are more prone to delamination than are the horizontal (parallel to chip surface) interfaces. The reason for the high driving forces for the vertical sidewall crack growth is, in large part, the high out-of-plane coefficient of thermal expansion (CTE) of the PI, while the in-plane driving forces are low [the in-plane CTE is close to that of silicon 5 ppm/°C and 2 ppm/°C at room temperature (RT), respectively] [6, 7]. The in-plane CTE is increased to 10 ppm/°C while the out-of-plane CTE increases from about 120 ppm/°C to about 400 ppm/°C at 400°C [7]. PMDA-ODA (pyromellitic dianhydride–oxydianiline) PI is more isotropic than BPDA-PDA. The RT in-plane CTE is 35 ppm/°C and the out-of-plane CTE is 120–130 ppm/°C increasing to about 300 ppm/°C at 400°C [8].

This paper will focus on one particularly important interface, i.e. between polyimide (PI) and silica, with the objective of elucidating a detailed understanding of the chemistry of interaction and a detailed mechanism of adhesion and its failure. The paper is organized as follows: (1) Introduction; (2) Acid–base concepts; (3) Polyimide chemistry; (4) Aminosilane chemistry; (5) Silica surface chemistry; (6) Interface chemistry; (7) Peel adhesion; (8) Summary.

2. ACID–BASE CONCEPTS

Polyimide adhesion to inorganic surfaces is affected by a number of factors. Fowkes *et al.* have discussed the importance of acid–base interactions in polymer adhesion to inorganic surfaces in a series of publications [9–11]. Adhesion of an acidic polymer to a basic sodium glass surface was stronger than that of a basic polymer to the same surface. This behavior was reversed by acid treatment of the glass surface. The intricate workings of interface formation can be rather complex in systems containing a polymer, a solvent and a substrate [9, 11]. *N*-Methyl-pyrrolidinone (NMP), which has been shown to complex with the PMDA-ODA polyamic acid (PAA) [12–14], can also compete with the PAA for the bonding sites on the silica substrate [9]. However, it will not be considered here as the thermal excursion to 400°C is well beyond the boiling point of NMP.

Acid–base concepts have been debated for over three centuries [15]. This paper, however, does not attempt to continue the debate. Arrhenius, or the classical acid–base concept, is shown in the following chemical reaction scheme [15]:

$$\text{acid} + \text{base} \longrightarrow \text{salt} + \text{water}.$$

The Arrhenius concept could not describe amphoterism, while the Brönsted–Lowry theory presented in 1923 is able to explain amphoterism [15]:

$$\text{acid 1} + \text{base 2} \rightleftarrows \text{base 1} + \text{acid 2}.$$

Brönsted called the above reaction 'protolysis' (proton transfer without requiring free proton formation). A Brönsted acid is a proton donor and a Brönsted base is a proton acceptor. Lewis [15, 16] proposed a more generalized acid–base concept:

$$\text{A} + \ :\text{B} \longrightarrow \text{A} : \text{B},$$

where A is acid, B is base, and (:) denotes an electron pair. A Lewis acid is an electron pair acceptor and a Lewis base is an electron pair donor. Much of the work published by Fowkes *et al.* [9–11] deals with the Lewis acid–base concept. The Brönsted–Lowry and Lewis acid–base theories do not agree in some cases. For example, Lewis acid–base theory does not allow ordering of acids based on their strength while the Brönsted–Lowry theory allows this. In this paper the reactivities at the different interfaces are considered using the Brönsted–Lowry acid–base concept. A reader desiring more information about the acid–base concepts, their advantages and disadvantages, are directed to the fine publications by Finston and Rychtman [15], Jensen [16], and Mittal and Anderson [17].

The reactivity of an organic acid or a base with an inorganic oxide surface may be addressed using Bolger's approach [18], which makes use of the Brönsted–Lowry acid–base concept:

for organic acid reactivity with a surface

$$\Delta_A = \text{IEPS} - \text{p}K_{A(A)};$$

for organic base reactivity with a surface

$$\Delta_B = \text{p}K_{A(B)} - \text{IEPS};$$

where IEPS is the isoelectric point of the surface, which is the pH of the aqueous oxide slurry at which the zeta potential is zero; $\text{p}K_{A(A)}$ and $\text{p}K_{A(B)}$ are the acid dissociation constants of an acid and a base, respectively.

Large positive Δ values may indicate too strong an interaction at the interface causing chemical attack or corrosion. For example, if PMDA-ODA PAA is brought in contact with copper, $\Delta_A = 9.5 - 3 = 6.5$, and copper particles are found in the polyimide film due to PAA reaction with copper and finally copper migration in the film caused by the NMP solvent [19–22]. Negative Δ values, on the other hand, indicate weak or no interaction between the organic material and the surface in question.

3. POLYIMIDE CHEMISTRY

Polyimides are generally applied from polyamic acid or ester solutions and are imidized thermally [23]. Pyromellitic dianhydride–oxydianiline (PMDA-ODA), the most commonly used and best characterized PI, is applied as a polyamic acid (PAA) from an *N*-methyl-pyrrolidinone (NMP, b.p. = 202°C) solution and cured thermally in a stepwise manner by exposing the film to temperatures up to 400°C [12, 23, 24]. The PAA has two types of reactive groups: the amide group and the carboxylic acid group. The acid dissociation constant ($\text{p}K_{A(A)}$) of the carboxylic acid moiety can be estimated as that of phthalic acid, and the acid dissociation constant ($\text{p}K_{A(B)}$) of the amide group as that of an acetamide [18]. The acid dissociation constant of the base is that of the conjugate acid:

$$\begin{array}{ll} \text{carboxylic acid (phthalic acid)} & \text{p}K_{A(A)} = 3, \\ \text{amide (acetamide)} & \text{p}K_{A(B)} = -1. \end{array}$$

The carboxylic acid is a relatively strong Brönsted acid, while the amide is a relatively weak Brönsted base.

Brekner and Feger [12–14] have shown that the NMP solvent complexes with the polyamic acid (NMP $\text{p}K_{A(A)}$ can be estimated as that of an acetamide ~ -1). The

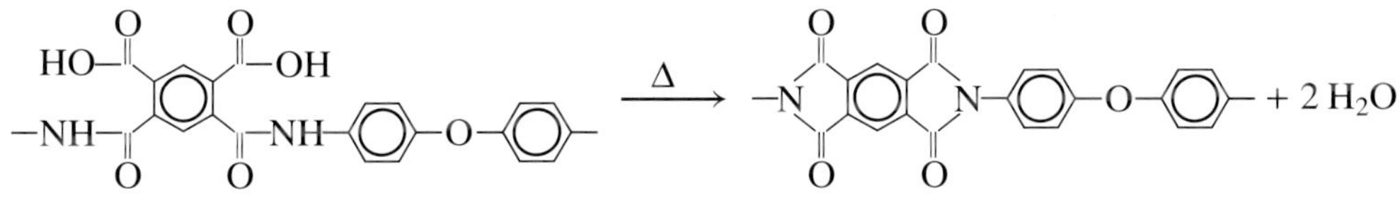

Figure 1. Simplified imidization process of PMDA-ODA polyamic acid (NMP decomplexation not shown).

NMP decomplexation temperatures are about 35 and 90°C. The onset of imidization is at about 150°C at which temperature the degree of imidization is about 50% [23]. Imidization is complete after exposure to a 200°C cure step, as indicated by infrared (IR) spectroscopy and X-ray photoelectron spectroscopy (XPS) analyses [23]. The imidization reaction is shown in Fig. 1.

4. AMINOSILANE CHEMISTRY

Even if the peel adhesion of PMDA-ODA/SiO_2 is initially high [24], the interface integrity is not guaranteed after exposure to elevated temperature and humidity (T&H) [25]. A spontaneous delamination of PMDA-ODA PI from the SiO_2 surface is detected when exposed to high humidity. Use of an aminosilane coupling agent at the PMDA-ODA/SiO_2 interface results in significantly improved interface reliability at T&H conditions [25]: γ-aminopropyltriethoxysilane (APS) is the most often used aminosilane adhesion promoter for polyimides derived from polyamic acids or polyamic esters [26, 27]. The APS, which is sold as a neat liquid, is hydrolyzed by exposure to water [26–29]:

$$H_2N(CH_2)_3Si(OEt)_3 + 3H_2O \longrightarrow H_2N(CH_2)_3Si(OH)_3 + 3EtOH$$

The APS is applied from a 0.1% deionized (DI) water solution by flood coating the silicon wafer and spinning it at 2000–3000 rpm for 30–60 s. The APS also has two reactive moieties in its structure: the free primary amine group and the silanol group [28, 29]. The acid dissociation constant of the basic amine group, $pK_{A(B)}$, is estimated as that of ethylamine, and of the silanol end of the molecule is assumed to be near that of silicic acid:

$$\begin{array}{ll} \text{free amine (ethylamine)} & pK_{A(B)} = 11, \\ \text{silanol group (silicic acid)} & pK_{A(A)} = 10. \end{array}$$

The free amine is a strong base readily accepting a proton, while the silanol end of the molecule exhibits weak acidic character. The pH of the 1% APS solution in DI water is 10.4 [29, 30].

5. SILICA SURFACE CHEMISTRY

A much used figure published by Bolger in 1983 [18] is adapted below. Figure 2 is drawn for a metal but works very well for the silicon surface also.

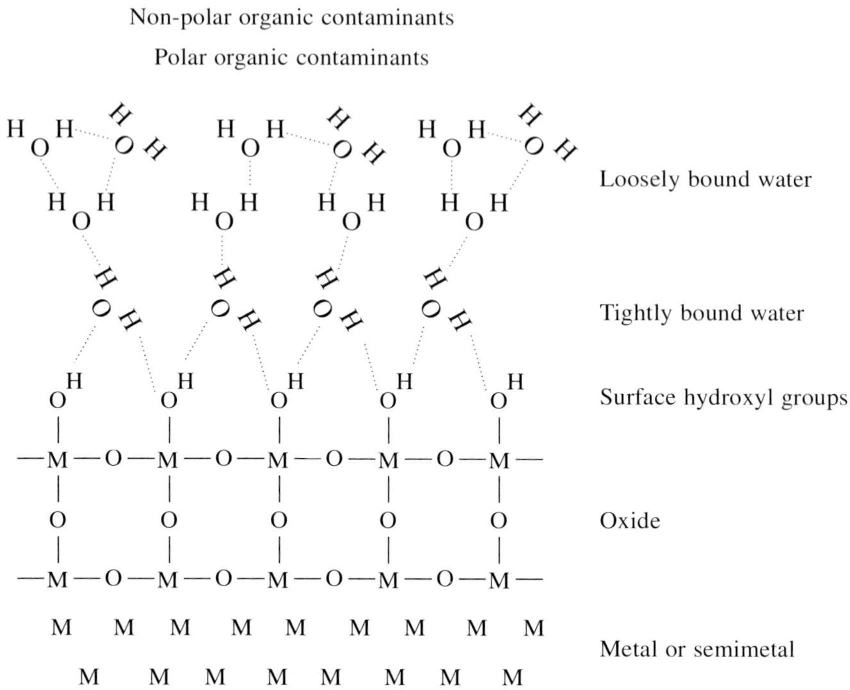

Figure 2. A schematic representation of a 'metal' surface.

Even a clean metal (or silicon) surface is covered with an oxide, the thickness of which is dependent on the nature of the metal. The silicon native oxide thickness is of the order of 1 nm [31]. The oxide surface is coated by hydroxyl groups, tightly bound water, loosely bound water, then polar organic contamination and at last non-polar organic contamination [31, 32]. Of course, the bonding of a coating should be with the actual surface and not with the surface contamination. Therefore, the importance of surface cleaning is clearly brought forth with this figure. Water will always be present, since the film applications are done in ambient conditions. Just for reference, a dry metal (Al, Fe, Cu) shows up to 20 molecular layers of water on the surface in ambient conditions [18]. The acid–base characteristics of the silica surface can be estimated for example by the isoelectric point of the surface (IEPS) [33–36]. For silica surfaces the IEPS is generally about 2.0 (pH), i.e. it is an acidic surface. There are two types of hydroxyl groups on a silica surface: hydrogen bonded hydroxyls (vicinal) and freely vibrating hydroxyl groups (isolated) [37]. SiO_2 surface chemistry is primarily due to the surface $-OH$ groups [37]. Hair [38] states that the reactivity of silica surface hydroxyl groups is similar to that of a phenolic or alcoholic hydroxyl group. The SiO_2 surface $-OH$ group $pK_A = 7.1$, which indicates only weakly acidic character [39, 40], and is in apparent contradiction with the IEPS of silica. However, Fowkes [9] reports differing degrees of acidity of the silica surface silanol-groups: the isolated SiOH-groups are much more acidic than the hydrogen-bonded vicinal SiOH-groups. Within the temperature range of interest to this paper, the following reactions occur at the silica surface as it is being exposed to higher temperatures (the studies referred to here were done using silica gel):

1. Temperatures up to 150°C — release of surface bound water [41].
2. Surface silanol groups start to condense at about 170°C [38].
3. The dehydration is completely reversible up to about 400°C [38].
4. At 400°C less than half of the surface hydroxyl groups have been eliminated [38].

Surface dehydration results in increased acidity according to Parks [35], which correlates with the statement made by Fowkes that isolated silica surface hydroxyls are more acidic than vicinal hydroxyls [9]. As an example, Parks [35] reports the drop in IEPS due to dehydration for Fe_2O_3, Al_2O_3, and TiO_2 by 1.9, 2.5, and 1.5, respectively.

Carboxylic acids, such as stearic acid, have long been known to undergo dehydration reactions with silica at temperatures above about 110°C [38, 42–45].

$$\mathrm{Si_s{-}OH + RCOOH \longrightarrow Si_s{-}OCOR + H_2O}$$

The maximum conversion may require temperatures in excess of 200°C [42]. The reason for the elevated temperature requirement may be the silica surface hydroxyl group removal initiation [38], which occurs at temperatures near 170°C. It has been suggested [40, 43] that polar groups would be formed immediately after thermal dehydration of silica:

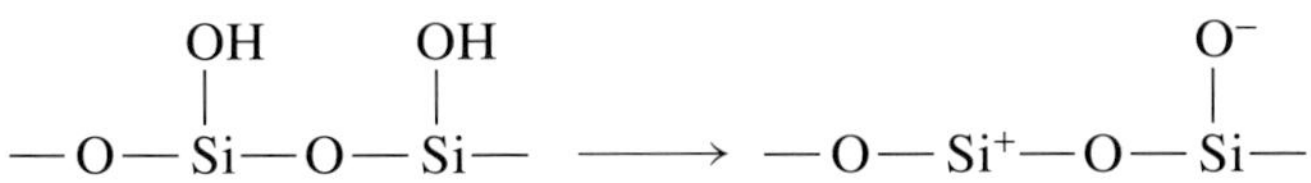

The silica surfaces discussed in this paper are exposed to oxygen-plasma immediately before coating with the polyamic acid solution and may therefore more closely relate to a silica surface with isolated SiOH-groups rather than the vicinal hydrogen bonded SiOH-groups. The maximum temperature to which the PI/SiO_2 samples are exposed is 400°C. It is clear that significant chemical rearrangement of the silica surface can occur during the high temperature excursions of this interface during PI processing.

6. INTERFACE CHEMISTRY

6.1. APS/SiO$_2$

Utilizing Bolger's concept [18], the acid–base reactivity of APS with SiO_2 surface can be estimated. Considering the APS silanol end reactivity with a silica surface:

$$\Delta_A = \mathrm{IEPS} - \mathrm{pK}_{A(A)} = 2 - 10 = -8.$$

It is apparent from the above that Brönsted–Lowry acid–base reactivity is not justified with the silanol end of the APS molecule and the silica surface silanol groups:

$$\mathrm{RSi{-}OH + HO{-}Si_s \rightleftarrows RSi{-}O^- + {}^+H_2O{-}Si_s}$$

However, a condensation reaction between the silica surface hydroxyl and the APS silanol is possible and will be discussed later.

Considering then the amine group Brönsted–Lowry acid–base reactivity with the silica surface silanol groups:

$$\Delta_B = pK_{A(B)} - IEPS = 11 - 2 = 9;$$

the amine end of the APS molecule shows strong acid-base reactivity with the silica surface silanol groups:

$$RNH_2 + HO{-}Si_s \rightleftarrows RNH_3^+ + {}^-O{-}Si_s$$

However, Eldridge *et al.* [46] have shown using time-of-flight static secondary ion mass spectrometry (TOF-SSIMS), Fourier transform infrared spectroscopy (FTIR) and XPS that the primary bonding scheme between APS and the silica surface hydroxyls is at the silanol end of the APS molecule. The N 1*s* XPS data collected with a Surface Science Laboratories (Fisons Instruments) small spot system with monochromatized Al K_α X-ray radiation and with take-off angle (TOA) of 35°, are shown in Table 1 below [46].

In this study, extreme care was taken to avoid any possibility of bicarbonate salt formation between moist CO_2 and the free amine of the APS. The water used for the hydrolysis of the neat APS was triply distilled and ambient exposure of the APS solution coated samples was less than ten minutes. Comparison of the XPS data given in Table 1 with that from Horner *et al.* [47] is shown in Table 2.

The differences between the two analyses are in the sample preparation: Eldridge *et al.* [46] used 0.1% APS solution in triply distilled water, while Horner *et al.* [47] used 1% APS solution in water. Silica substrate surface cleaning and APS application processes are also different in these two studies. However, the results of both studies show that free amine is found on the surface and that relatively high concentrations of protonated amines are also detected, which are primarily not located at the surface,

Table 1.
N 1*s* XPS data for APS on SiO_2

BE (eV)	Concentration (%)	Assignment
399.3	80	$-NH_2$
401.1	20	$-NH_3^+$

Table 2.
N 1*s* XPS data for APS on SiO_2 as a function of TOA

BE (eV)	TOA = 15° [47]	TOA = 35° [46]	TOA = 75° [47]	Assignment
399.3	73	80	57	$-NH_2$
401.1–401.3	27	20	43	$-NH_3^+$

Table 3.
PMDA-ODA/SiO_2 XPS locus of failure analysis. Elemental composition

PMDA-ODA side				SiO_2 side			
Sample	C (%)	O (%)	N (%)	C (%)	O (%)	N (%)	Si (%)
1	75	17	8	51	25	3	21
2	75	17	8	41	28	3	28

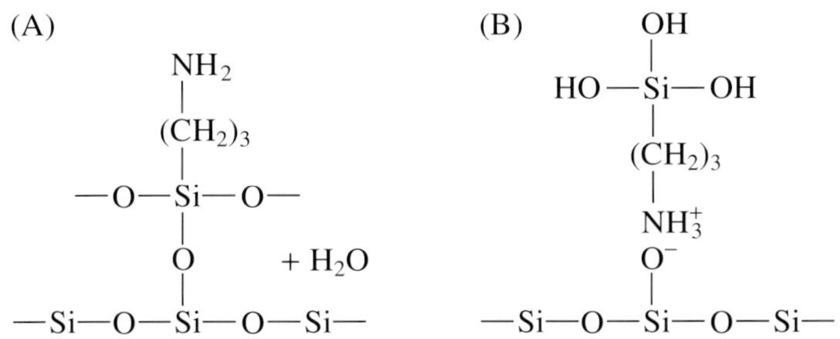

Figure 3. APS reaction with SiO_2 surface based on XPS N 1*s* data. (A) Condensation reaction and (B) Brönsted–Lowry acid–base reaction.

but are at the interface. Figure 3 shows bonding of APS to silica based on the above results as well as on earlier data from Plueddemann [28] and Ishida [29].

The result of the condensation reaction shown in Fig. 3A is an accepted form of APS bonding with this surface [26, 28, 29]. The Brönsted–Lowry acid–base reaction shown in Fig. 3B is reversible as long as water is present, while the condensation reaction product is much less reversible in particular when the three-dimensional network [26, 29] is formed. Solution condensation of the APS silanol groups occurs when the APS concentration is above about 0.2% by weight [29]. Since these samples were spin applied, the concentration of APS at the silica surface can quickly exceed the 0.2% by weight due to water evaporation during spinning as the starting solution is 0.1% by weight.

6.2. PMDA-ODA/APS

Linde and Gleason [26] have shown that PMDA-ODA PAA has a Brönsted–Lowry acid–base reaction with the free amine of the silica surface bound APS. The salt formation between the PMDA-ODA carboxylic acid group and the amine occurs at RT. Amide formation is realized at about 120°C and imide formation between the PMDA-ODA and APS is seen at about 200°C. This imidization process is associated with chain scission of the PMDA-ODA PI, which has an effect on the PI adhesion to the silica surface as shown by Greenblatt *et al.* [48].

6.3. PMDA-ODA/SiO$_2$

PMDA-ODA PAA interaction with silica surface hydroxyl groups may also be estimated using Bolger's [18] concept. The PAA has two reactive groups, as was discussed earlier: the carboxylic acid and the amide groups (Fig. 1). Considering first the carboxylic acid group acid–base reactions with the silica surface hydroxyl groups:

$$\Delta_A = \text{IEPS} - \text{pK}_{A(A)} = 2 - 3 = -1.$$

According to this concept the Brönsted–Lowry acid–base reactivity between the carboxylic acid and the silica surface hydroxyls is not expected to occur:

$$\text{RCOOH} + \text{HO–Si}_s \leftrightarrows \text{RCOO}^- + {}^+\text{H}_2\text{O–Si}_s.$$

Acid–base reactivity between the amide group of the PAA and the silica surface hydroxyl groups is not expected either, based on this concept [18]:

$$\Delta_B = \text{pK}_{A(B)} - \text{IEPS} = -1 - 2 = -3,$$

$$\text{RCONHR}_1 + \text{HO—Si}_s \leftrightarrows \text{RCO}\underset{\displaystyle |\atop \displaystyle \text{R}_1}{\text{N}}\text{H}_2^+ + {}^-\text{O—Si}_s$$

Based on the above estimations, the PMDA-ODA PAA should not react with silica surface hydroxyl groups, at least not in the Brönsted–Lowry acid–base manner. However, an analysis of the peel locus of failure of PMDA-ODA on native SiO_2 surface suggests reactivity. In the light of the apparent contradiction, a detailed look at the interfacial chemistry between the PMDA-ODA PAA and silica surface after PAA imidization completion is merited. The experiment was done in the following way: PMDA-ODA PAA was spin coated onto oxygen-plasma cleaned Si-wafer, cured in stepwise manner exposing the wafer with the coating to temperatures up to 400°C in N_2. The PMDA-ODA PI film was then cut to a 2 mm wide peel strips using a wafer dicer. After the 90° peel test using an Instron tester, the peel locus of failure was analyzed using XPS. The elemental composition on both the PI and the SiO_2 side are shown in Table 3.

Based on the presence of nitrogen on the SiO_2 side of the failure and the fact that the main N 1*s* peak binding energy (BE) is equivalent to that of imide nitrogens [23, 31, 49], it is clear that PI has been retained on the silica surface after the peel test. The C 1*s* XPS spectra show the three characteristic PI C 1*s* peaks on both sides of the failure locus as shown in Fig. 4.

The major differences between the two C 1*s* spectra are the relative ratio of the first and second peaks, and the relative concentration of the third peak (C=O, the imide carbonyl), are shown in Table 4.

The first observation that can be made from the above table is that the C=O concentration on the silica side has dropped by 33% as compared to that on the PI side. This suggests that C=O groups have been lost due to the interaction between the polymer and the substrate. Another supporting piece of evidence comes from the relative ratios of the peaks at 285.0 and 286.2 eV. The peak at 285.0 eV is associated with CH_x species, while the peak at 286.2 eV is due to three different C 1*s* environments: –C–O–, –C–N–, and the PMDA carbons. The 1.2 eV shift to higher binding energy (BE) of the PMDA carbon atoms is due to the four imide carbonyl groups

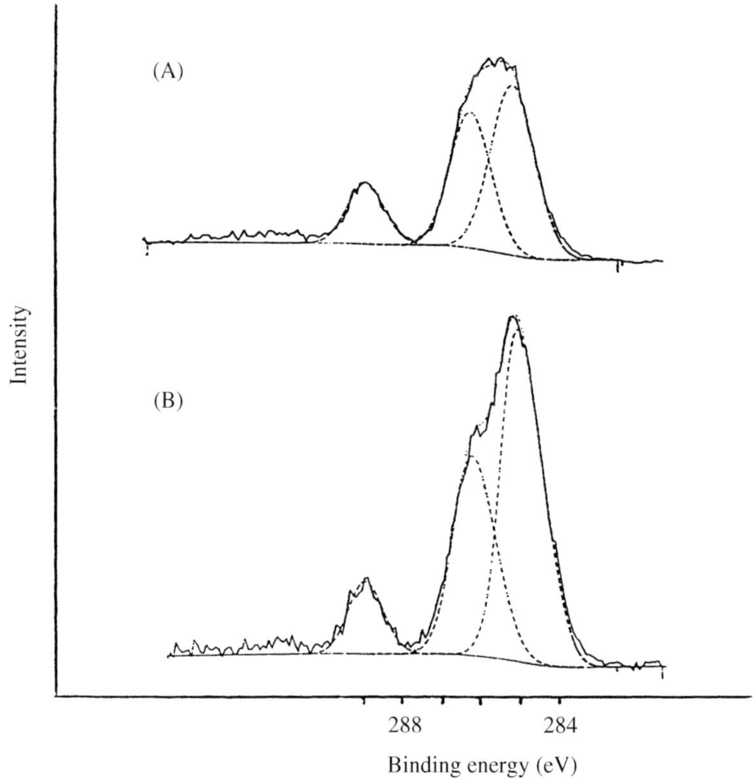

Figure 4. C 1*s* XPS of PMDA-ODA/SiO_2 locus of failure. (A) PMDA-ODA side and (B) SiO_2 side.

Table 4.
C 1*s* peak areas on PI and SiO_2 failure surfaces

C 1*s* peak BE (eV)	Peak area (% of total) PI side	Peak area (% of total) SiO_2 side
285.0	47	57
286.2	38	33
289.0	15	10

around the PMDA phenyl ring. The carbonyl groups reduce the electron density at the PMDA ring which results in increased C 1*s* BE for these atoms. If the number of C=O-groups bonded to the PMDA ring is reduced to two, this effect is not seen, while three is enough to cause the shift [49]. Table 5 presents the 286.2 eV peak ratio to the 285.0 eV peak, as well as the relative concentration of the imide C=O as a function of the analysis surface to underline the differences observed.

The Si 2*p* peak consists of four peaks: the elemental peaks at about 100 eV, a peak at 103.7 eV which is assigned to SiO_2, and a peak at 102.4 eV. The Si 2*p* peak at 102.4 eV is at the same BE as the Si 2*p* peak is for APS [46] and is associated with

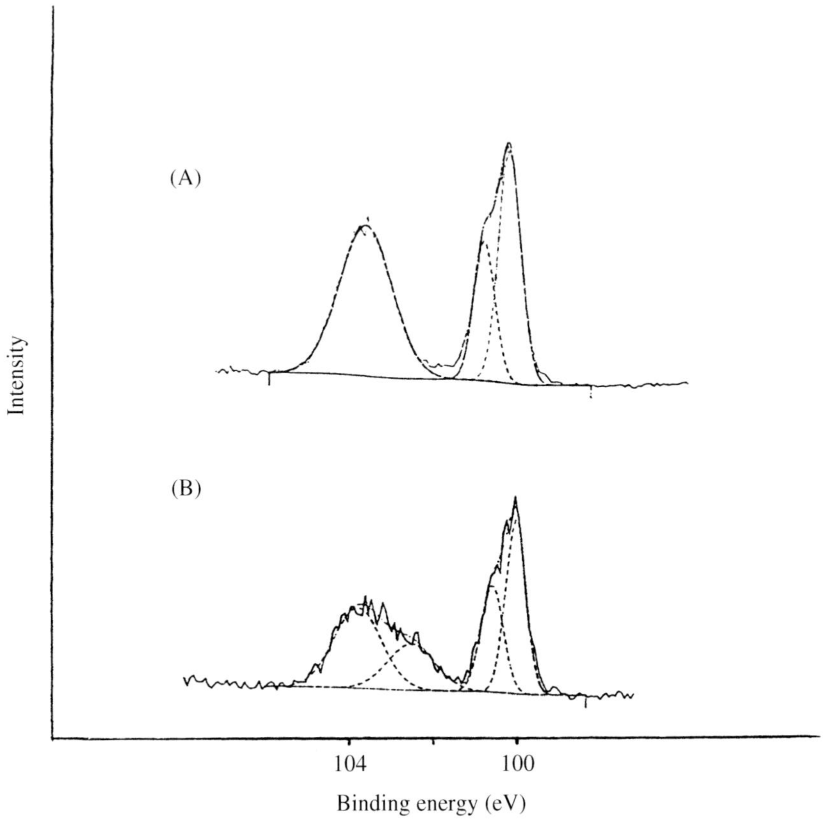

Figure 5. Si 2*p* XPS data for (A) oxygen plasma cleaned native SiO_2 surface and (B) substrate side after peeling PI off the silica surface.

Table 5.
C 1*s* peak ratios (286.2 eV to 285.0 eV) and at.% concentration of imide C=O groups

Analysis surface	$\frac{286.2 \text{ eV}}{285.0 \text{ eV}}$	[C=O] (at.%)
PI surface	1.18	15
PI interface	0.79	15
SiO_2 interface	0.59	10

a $-C-Si(-O)_3$ type Si environment. Figure 5 shows the Si 2*p* XPS peaks for the oxygen plasma cleaned silica surface and for the locus of failure on the SiO_2 side.

Based on the data presented, a hypothesis is proposed for a reaction scheme to form a Si bonded directly to the PMDA ring [50]. The interaction between the silica surface hydroxyl group and the carboxylic acid group of the PMDA-ODA PAA polymer results in a condensation reaction forming the corresponding ester [42]. This reaction is then followed by CO_2 elimination resulting in radical formation and recombination finally producing a Si species bonded directly to the PMDA aromatic ring.

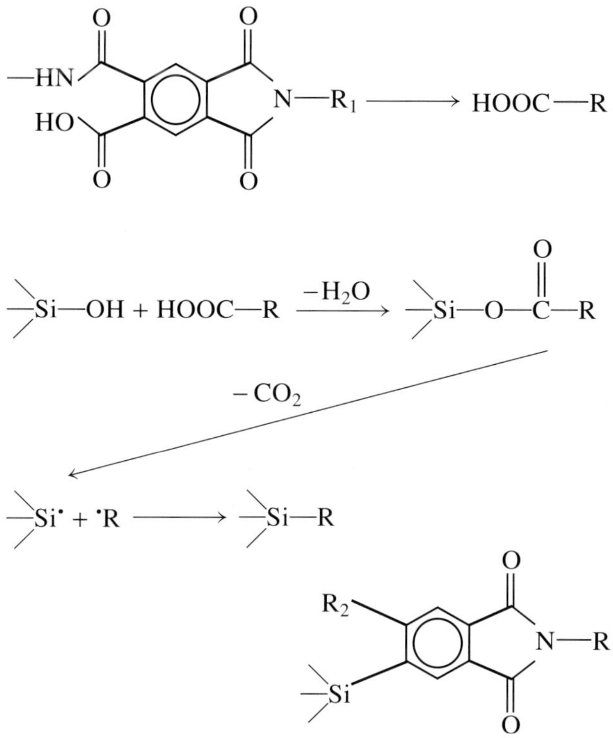

Figure 6. Proposed reaction route between partly imidized PMDA-ODA PAA and SiO_2 surface [50].

The first step of the proposed reaction scheme has been seen before with stearic acid [42, 43, 51]. The decarboxylation of esters does not generally occur for the all carbon analogs. No reference in the literature could be found to indicate whether such a reaction occurs in the silicon ester structure of Fig. 6. It is not clear, therefore, how one gets from the ester to the Si directly bonded to the phenyl ring, the bonding which is expected on the basis of the XPS Si $2p$ and C $1s$ data. The Si $2p$ data are unequivocal in that the peak at 102.4 eV must be a Si-species comparable to that found in APS, i.e. $-C-Si-(O)_3$. The decrease in the C $1s$ peak ratio (286.2 eV peak vs. 285.0 eV peak) is without doubt associated with increase in the electron density at the PMDA phenyl ring [49, 52] consistent with the reaction of Fig. 6. Combining these findings with the reduction of imide carbonyl group concentration at the silica surface suggests that the hypothesis presented in Fig. 6 may have some validity.

7. PEEL ADHESION

PMDA-ODA peel adhesion results are shown in Table 6. For the samples without APS four film thicknesses are shown, while the samples with APS show only one film thickness range taken from data by Anderson *et al.* [53].

In all cases reported above, the locus of failure after the 90° peel test is in the PI, some distance removed from the actual PI/SiO_2 interface [24, 31, 53, 54]. The

Table 6.
PMDA-ODA peel adhesion to SiO_2 with and without APS

APS	Film thickness (μm)	Peel strength (J/m^2)	N/Si on the silica surface	Reference
No	22	480	0.14	[24]
No	31	550		[31]
No	43	560		[31]
No	23	570	0.18	[54]
Yes	30–35	1440	1.51	[53]

presence of the PI at the SiO_2 surface is verified by the C 1*s* and N 1*s* XPS data. The N 1*s* in all cases shows the major peak (80–100% of the total N 1*s* envelope) at 400.8–401.1 eV, which is assigned to the imide nitrogen species [24, 31, 53, 55, 56]. The C 1*s* data show the three characteristic peaks due to $-CH_x$ (285.0 eV), $-C-O-$, $-C-N-$, and PMDA carbons (286.2 eV), and C=O (288.8 eV) [49]. The relative ratios of the first two peaks may differ due to the bonding discussed in the PMDA-ODA/SiO_2 case (the substrate can be seen through the residual PI), and due to the interfacial chemical characteristics in the PMDA-ODA/APS/SiO_2 case. On the PI side of the failure locus, the elemental composition is within experimental error of that of the PI, and the ratio of the two low BE C 1*s* peaks is comparable to that shown in Table 5. Therefore it is evident that the peel locus of failure is in the PI in all cases studied. However, the peel strength is over a factor of two higher in the case of APS than it is when the APS is not used. Since the failure is not interfacial it is not possible to attribute the difference in peel strength to differences in the chemical bond strengths at the interface. Moreover, in both cases there is evidence of strong covalent bond formation across the interface. Therefore, an explanation must be sought elsewhere. It has been proposed [56] that the reason for this improved peel strength was the difference in the polymer ordering in the interfacial region, which is affected by the interfacial chemistry. Therefore the timing of the interfacial bonding is of importance, i.e. RT chemical reaction as in the case of PMDA-ODA PAA/APS/SiO_2 vs. interfacial reaction requiring elevated temperature ($> 110°C$) as in the case of PMDA-ODA PAA/SiO_2. The interfacial ordering may entail higher degree of chain entanglements (physical crosslinking) with APS than without it. This physical crosslinking would then effectively change the interphasial fracture toughness of the polyimide near the actual interface resulting in higher peel strength.

8. SUMMARY

Chemical bond formation has been discussed at the PMDA-ODA/SiO_2 and PMDA-ODA/APS/SiO_2 interfaces. Acid–base chemistry is obvious between APS and PMDA-ODA PAA, while both acid–base and condensation reactions account for the bonding at the APS/SiO_2 interface. No obvious Brönsted–Lowry acid–base chemistry can be detected at the PMDA-ODA/SiO_2 interface. Based on the spectroscopic data of PMDA-ODA/SiO_2 peel locus of failure a hypothesis is presented suggesting condensation reaction at the interface followed by decarboxylation resulting in PMDA phenyl group carbon bonding directly with Si of the silica surface. The primary reason for the improved adhesion with APS is suggested to be due to interphasial fracture toughness increase caused by the initial chemical reaction between the APS treated silica surface and the PAA.

Acknowledgements

The author expresses her thanks to Dr. S. L. Buchwalter and Dr. A. Afzali for helpful discussions.

REFERENCES

1. D. S. Soane and Z. Martynenko, *Polymers in Microelectronics*. Elsevier, New York (1989).
2. R. R. Tummala, R. W. Keyes, W. D. Grobman and S. Kapur, in: *Microelectronics Packaging Handbook*, R. R. Tummala and E. J. Rymaszewski (Eds), p. 673. Van Nostrand Reinhold, New York (1989).
3. B. Luther, J. F. White, C. Uzoh, T. Cacouris, J. Hummel, W. Guthrie, N. Lustig, S. Greco, N. Greco, S. Zuhoski, E. Colgan, S. Mathad, L. Saraf, E. J. Weitzman, C.-K. Hu, F. Kaufman, M. Jaso, L. P. Buchwalter, S. Reynolds, C. Smart, D. Edelstein, E. Baran, S. Cohen, C. Knoedler, J. Malinowski, J. Horkans, L. Deligianni, J. Harper, P. C. Andricacos, J. Paraszczak, D. J. Pearson and M. Small, in: *Proceedings of the 10th International VLSI Multilevel Interconnection Conference*, p. 15 (1993).
4. J. M. E. Harper, E. G. Colgan, C.-K. Hu, J. P. Hummel, L. P. Buchwalter and C. E. Uzoh, *MRS Bulletin* **19**, 23 (August 1994).
5. J. Paraszczak, D. Edelstein, S. Cohen, E. Babich and J. Hummel, in: *Proceedings of the IEEE Electronic Devices Meeting*, p. 261 (1993).
6. R. H. Lacombe, L. P. Buchwalter and K. Holloway, *J. Adhesion Sci. Technol.* **7**, 1293 (1993).
7. S. T. Chen and H. H. Wagner, *J. Electr. Mater.* **22**, 797 (1992).
8. S. T. Chen, in: *Low-Dielectric Constant Materials — Synthesis and Applications in Microelectronics*, T. M. Lu, S. P. Murarka, T. S. Kuan and C. H. Ting (Eds), p. 141. Mater. Res. Soc., Pittsburgh, PA (1995).
9. F. M. Fowkes, in: *Microscopic Aspects of Adhesion and Lubrication*, G. M. Georges (Ed.), p. 119. Elsevier, Amsterdam, The Netherlands (1982).
10. F. M. Fowkes and M. A. Mostafa, *Ind. Eng. Chem. Prod. Res. Dev.* **17**, 3 (1978).
11. F. M. Fowkes, *J. Adhesion Sci. Technol.* **1**, 7 (1987).
12. C. Feger, *Polym. Eng. Sci.* **29**, 347 (1989).
13. M. Brekner and C. Feger, *J. Polym. Sci. Polym. Chem.* **25**, 2479 (1987).
14. M. Brekner and C. Feger, *J. Polym. Sci. Polym. Chem.* **25**, 2005 (1987).
15. H. L. Finston and A. C. Rychtman, *A New View of Current Acid–Base Theories*. John Wiley & Sons, New York (1982).
16. W. B. Jensen, *The Lewis Acid–Base Concepts. An Overview*. John Wiley & Sons, New York (1980).
17. K. L. Mittal and H. R. Anderson, Jr. (Eds), *Acid–Base Interactions: Relevance to Adhesion Science and Technology*. VSP, Utrecht, The Netherlands (1991).
18. J. C. Bolger, in: *Adhesion Aspects of Polymeric Coatings*, K. L. Mittal (Ed.), p. 3. Plenum Press, New York (1983).
19. J. Kim, S. P. Kowalczyk, Y. H. Kim, N. J. Chou and T. S. Oh, *Mater. Res. Soc. Symp. Proc.* **167**, 137 (1990).
20. Y. H. Kim, G. F. Walker, J. Kim and J. Park, *J. Adhesion Sci. Technol.* **1**, 331 (1987).

21. Y. H. Kim, J. Kim, G. F. Walker, C. Feger and S. P. Kowalczyk, *J. Adhesion Sci. Technol.* **2**, 95 (1988).
22. S. P. Kowalczyk, Y. H. Kim, G. F. Walker and J. Kim, *Appl. Phys. Lett.* **52**, 375 (1988).
23. L. P. Buchwalter, *J. Vac. Sci. Technol.* **A7**, 1772 (1989).
24. L. P. Buchwalter and J. Greenblatt, *J. Adhesion* **19**, 257 (1986).
25. L. P. Buchwalter, T. S. Oh and J. Kim, *J. Adhesion Sci. Technol.* **5**, 333 (1991).
26. H. Linde and T. Gleason, *J. Polym. Sci. Polym. Chem. Edn* **22**, 3043 (1984).
27. H. G. Linde, *J. Polym. Sci. Polym. Chem. Edn* **20**, 1031 (1982).
28. E. P. Plueddemann, *Silane Coupling Agents*. Plenum Press, New York (1982).
29. H. Ishida, in: *Adhesion Aspects of Polymeric Coatings*, K. L. Mittal (Ed.), p. 45. Plenum Press, New York (1983).
30. F. J. Boerio and C. A. Gosselin, in: *Polymer Characterization, Adv. Chem. Ser. No. 203*, C. D. Craver (Ed.), p. 541. American Chemical Society, Washington, DC (1983).
31. L. P. Buchwalter and R. H. Lacombe, *J. Adhesion Sci. Technol.* **2**, 463 (1988).
32. K. L. Mittal, in: *Surface Contamination: Genesis, Detection and Control*, K. L. Mittal (Ed.), p. 3. Plenum Press, New York (1979).
33. X. Demaret, G. Cristallo, P. Snauwaert, J. Riga and J. Verbist, *Synthetic Metals* **55–57**, 1051 (1993).
34. G. D. Parfitt, *Pure Appl. Chem.* **48**, 415 (1976).
35. G. A. Parks, *Chem. Rev.* **52**, 177 (1965).
36. G. A. Parks and P. L. Bruyn, *J. Phys. Chem.* **66**, 967 (1962).
37. M. L. Hair, in: *Silanes, Surfaces and Interfaces*, D. E. Leyden (Ed.), p. 25. Gordon and Breach Science Publishers, New York (1986).
38. M. L. Hair, *Infrared Spectroscopy in Surface Chemistry*. Marcel Dekker, New York (1967).
39. M. L. Hair and W. Hertl, *J. Phys. Chem.* **74**, 91 (1970).
40. H. P. Boehm, in: *Advances in Catalysis*, D. D. Eley, H. Pines and P. B. Weisz (Eds), p. 179. Academic Press, New York (1966).
41. M. Uhrova and O. Malicek, *Sci. Papers Prague Inst. Chem. Tech. Anal. Chem.* **H16**, 105 (1979).
42. F. P. Auteri, R. L. Belford, B. H. Robinson and R. B. Clarkson, *Colloids Surfaces* **A81**, 25 (1993).
43. R. K. Iler, *The Chemistry of Silica*. John Wiley and Sons, New York (1979).
44. K. Marshall and C. H. Rochester, *J. Chem. Soc. Faraday Trans.* **71**, 1754 (1975).
45. W. K. Lowen and E. C. Broge, *J. Phys. Chem.* **65**, 16 (1961).
46. B. N. Eldridge, L. P. Buchwalter, C. A. Chess, M. J. Goldberg, R. D. Goldblatt and F. P. Novak, in: *Silanes and Other Coupling Agents*, K. L. Mittal (Ed.). VSP, Utrecht, The Netherlands (1992).
47. M. R. Horner, F. J. Boerio and H. M. Clearfield, in: *Silanes and Other Coupling Agents*, K. L. Mittal (Ed.), p. 241. VSP, Utrecht, The Netherlands (1992).
48. J. Greenblatt, C. J. Araps and H. R. Anderson, Jr., in: *Polyimides: Synthesis, Characterization and Applications*, K. L. Mittal (Ed.), pp. 573–588. Plenum Press, New York (1984).
49. L. P. Buchwalter and A. I. Baise, in: *Polyimides: Synthesis, Characterization and Applications*, K. L. Mittal (Ed.), p. 537. Plenum Press, New York (1984).
50. S. L. Buchwalter, Private communication.
51. B. D. Ratner, J. J. Rosen, A. S. Hoffman and L. H. Scharpen, in: *Surface Contamination: Genesis, Detection and Control*, K. L. Mittal (Ed.), Vol. 2, p. 669. Plenum Press, New York (1979).
52. B. D. Silverman, J. W. Bartha, J. G. Clabes and P. S. Ho, *J. Polym. Sci. Part A: Polym. Chem.* **24**, 3325 (1986).
53. H. R. Anderson, Jr., M. M. Khojasteh, T. P. McAndrew and K. G. Sachdev, *IEEE Trans. CHMT* **9**, 364 (1986).
54. K. G. Sachdev, S. Chen, H. R. Anderson, Jr., M. M. Khojasteh and A. V. Patsis, in: *Proc. 36th Electronic Components Conf.*, p. 345 (1986).
55. C. D. Wagner, *NIST X-ray Photoelectron Spectroscopy Database*. US Department of Commerce, Gaithersburg, MD (1989).
56. L. P. Buchwalter, *J. Adhesion Sci. Technol.* **4**, 697 (1990).

Mittal Festschrift, pp. 673–685
W. J. Van Ooij and H. R. Anderson, Jr. (Eds)

Adhesion mechanisms of polyurethane to glass: Interphase formation and structure

RAJ K. AGRAWAL[1] and LAWRENCE T. DRZAL[2,*]
[1]*Donnelly Corporation, Holland, MI 49423-5368, USA*
[2]*Michigan State University, Department of Chemical Engineering, East Lansing, MI 48824-1326, USA*

Abstract—Polyurethanes are used in a variety of structural applications as adhesives to glass surfaces. A systematic experimental plan was conducted in which a segmented polyurethane system (caprolactone based polyols + toluene diisocyanate + 1,4-butanediol) with variations in hard and soft segment contents was bonded to the air side of float glass surfaces. Practical adhesion measurements showed that the structure of the interphase, in particular phase separation, played a predominant role in affecting the adhesion. Polyurethanes with higher modulus showed better practical adhesion than lower modulus formulations but those that had the greater degree of phase separation had superior adhesion to the glass. Analytical investigation of the failure surfaces indicated that covalent chemical bonding did not occur between the polyurethanes and the glass but hydrogen bonding through a butanediol-rich interphase layer was the dominant mechanism. When the practical adhesion values of all the polyurethane systems to glass were compared, a linear relationship between the polar component of the surface free energy of the polyurethanes and the adhesion level was found.

Keywords: Adhesion; interphase; glass; urethane; phase separation; hard segment; surface free energy.

1. INTRODUCTION

Polyurethanes are very versatile materials and are used widely in almost every major industry, including automotive and aerospace. Often, polyurethanes are reinforced with short and long glass fibers to improve their mechanical properties. Good adhesion between the glass reinforcements and the matrix is shown to be an important factor in determining overall mechanical properties of these composites [1–4]. In other applications, polyurethanes are used as adhesives or sealants where structural adhesion to glass surface is desired. For example, most automobile windshields are attached to the vehicles by direct bonding with polyurethane sealants. Yet another example is encapsulated modular windows for automobiles [5–7], in which polyurethane is directly molded onto glass windows and where structural adhesion between the polyurethane gasket and the glass panel is essential for successful performance.

---

*To whom correspondence should be addressed.

A substantial amount of work has been reported in the literature on the characterization and the morphology of polyurethanes [8–11], and also on the glass surface chemistry [12–16]. This study is an attempt to develop the relationship between polyurethane morphology and the glass surface chemistry in the interphase to develop an understanding of polyurethane/glass adhesion. Model cross-linked polyurethane matrix compositions with varying degrees of phase separation and the air side of soda-lime float glass plates are used as the substrates. Practical adhesion values determined in a shear mode are correlated with the moduli of the matrices and their phase separation. The failure surfaces are investigated analytically. This information, along with the surface free energy of the polyurethanes estimated by contact angle measurements are utilized to develop an understanding of the adhesion mechanism responsible for interphase formation and structure.

2. EXPERIMENTAL

2.1. Materials

The polyurethanes used in this study were based on caprolactone polyols available from Union Carbide under the trade name 'Tone.' The molecular weights of the Tone polyols 0301, 0305, and 0310 are 300, 540, and 900, respectively. Hard segments were made from a 80%–20% mixture of toluene 2,4-diisocyanate and toluene 2,6-diisocyanate (TDI, Aldrich Chemical Co.) and 1,4-butanediol (BDO, Aldrich Chemical Co.) as the chain extender. The various polyurethane formulations studied in this work are shown in Table 1. The formulations have sample designations based on the type of polyol used and the amount of hard segment content. The number portion of the sample designation represents the polyol type whereas the letter portion

Table 1.
Polyurethane formulations at isocyanate index of 1.0

Sample designation	Polyol type			BDO (mol%)	TDI (mol%)	Hard segment (wt%)
	Tone 0301	Tone 0305	Tone 0310			
10A	—	—	X	0	60	22
10B	—	—	X	22	56	37
10C	—	—	X	31	54	47
10D	—	—	X	34	53	60
10E	—	—	X	41	52	67
5B	—	X	—	7	58	37
5C	—	X	—	20	56	47
5D	—	X	—	31	54	59
5E	—	X	—	36	53	67
1C	X	—	—	0	60	47
1D	X	—	—	18	57	59
1E	X	—	—	26	55	67
HS	—	—	—	50	50	100

represents the hard segment content. Thus, if any two samples have the same letter in their designation, they have the same hard segment content. For adhesion testing, annealed soda-lime glass plaques were used. A typical soda-lime float glass plaque has two sides, an air side and a tin side. The air side of the glass plaques was used for the adhesion testing.

2.2. Adhesion sample preparation

The air sides of $50 \times 125 \times 6$ mm annealed soda-lime float glass plaques were cleaned with methyl ethyl ketone solvent and dried. Silicone molds with a $6 \times 6 \times 6$ mm cavity were placed on the air side of the glass plaques and degassed and homogenized mixtures of the polyol, the chain extender, and the isocyanate were poured into the $6 \times 6 \times 6$ mm cavities and then cured for 24 h at 90°C in an air circulating convection oven.

2.3. Adhesion testing

Adhesion of polyurethane matrix to glass surface was evaluated using soda-lime float glass plates as the substrate. Plate glass was chosen rather than fibers as the glass–matrix interface/interphase since the failure surface could be analyzed with relative ease using visual, microscopic, spectroscopic, and chemical means.

The conventional lap-shear configuration for adhesion testing with glass plates could not be used successfully due to the brittleness of the glass plates. To overcome this, ASTM test method D4501 for measuring shear strength of adhesive bonds between rigid substrates by a block-shear method was modified for this study. An Instron 4201, tabletop material testing machine, was utilized for the adhesion testing at 5 mm/min. The maximum load required for the detachment of the polyurethane block from the glass surface was recorded. At lease five samples were tested for each polyurethane formulation.

2.4. X-ray photoelectron spectroscopy (XPS)

Subsequent to adhesion testing, failed glass surfaces were analyzed using a Perkin–Elmer PHI 5400 X-ray photoelectron spectrometer equipped with a monochromotor. An area approximately 6×6 mm was sectioned from the failed glass surface and was placed inside the XPS chamber. Details of the procedure are provided in references [17–19].

2.5. Surface energy measurements

The various polyurethane formulations were cast on clean 75×50 mm glass slides. The final polyurethane coating thickness on the glass slides was about 1 to 2 mm. The glass slides were handled carefully throughout the sample preparation to avoid any surface contamination of the polyurethane surfaces.

The surface free energies of the samples were estimated from the experimentally determined contact angles of sessile drops using a Rame-Hart Model 100 goniometer.

Table 2.
Surface free energies of liquids used for contact angle measurements [20]

Liquid	Surface free energy (mJ/m^2)		
	γ^D	γ^P	γ
Water	21.8	51.0	72.8
Glycerol	37.0	26.4	63.4
Formamide	39.5	18.7	58.2
Methylene iodide	48.5	2.3	50.8
α-Bromonaphthalene	44.6	0.0	44.6

All experiments were conducted at room temperature ($\sim 23\,^{\circ}$C) and were completed as quickly as possible to minimize possible spurious effects due to swelling or surface rearrangement. The contact angle data were obtained with a series of liquids having varying surface free energies and polar/dispersive components as shown in Table 2. The liquids used covered a wide range of polarities from low (such as α-bromonaphthalene) to high (such as water) allowing for the determination of a comprehensive wettability profile for each polyurethane formulation. Liquids with very low surface free energies, such as n-alkanes, were not used in this study due to the difficulty and errors associated with measuring small ($< 10^{\circ}$) contact angles.

3. RESULTS AND DISCUSSION

3.1. Practical adhesion to glass

The practical adhesion of the various polyurethanes to the soda-lime glass plates was evaluated using a block-shear test method and the peak load values are shown in Table 3. The samples based on Tone 0310 and Tone 0305 polyols showed very good reproducibility, whereas Tone 0301 based samples had large variability, especially sample 1E. Thus, a large number of 1E samples were tested to obtain reliable mean and standard deviation values. It was observed that within each family of the polyurethane systems, adhesion values increased with increasing hard segment content. But for the same hard segment content (e.g. in samples 10C, 5C, and 1C or in samples 10D, 5D, and 1D), Tone 0305 based samples show the highest levels of shear strength followed by Tone 0310 and Tone 0301 based samples.

3.2. Relationship to polyurethane morphology

The polyurethane formulations used in this study have been characterized for phase separation in our previous studies [17, 18]. Differential scanning calorimetry, dynamic mechanical analysis, Fourier-transform infrared spectroscopy and near-infrared spectroscopy techniques were used to study phase separation in these samples [17]. A model was developed to predict phase separation in these polyurethane systems based on the miscibility of the hard and the soft segments. It was found that in these polyurethanes, higher molecular weight polyol phase separated more easily. Also,

Table 3.
Adhesion of various polyurethanes to glass surface

Sample designation	Shear strength (MPa)
10A	4.9
10B	11.0
10C	18.6
10D	32.0
10E	37.1
5B	28.7
5C	29.5
5D	36.2
5E	35.1
1C	5.7
1D	14.6
1E	20.8

Table 4.
Iosipescu shear modulus and phase separation characteristics of polyurethane systems

Polyol molecular weight	Sample designation	Iosipescu shear modulus (MPa) 2% secant	Phase separation
High	10A	—	No
	10B	—	Yes
	10C	460	Yes
	10D	1040	Yes
	10E	1240	Yes
Medium	5B	890	No
	5C	1040	Yes
	5D	1140	Yes
	5E	1190	Yes
Low	1C	1210	No
	1D	1280	No
	1E	1300	No

among the same polyol based polyurethane samples, phase separation increases with increasing hard segment content. Table 4 shows the phase separation characteristics of the various polyurethanes.

From the literature [21, 22], the adhesion of non-phase separated polymeric systems, such as epoxies to glass, is dependent on the shear modulus of the polymer matrix. The shear moduli of the various polyurethanes of this study were evaluated in our previous study [17] and are shown in Table 4. Figure 1 shows a plot of the adhesion to glass versus the shear modulus of the various polyurethanes. Also shown in the graph are the phase separation tendencies of the various polyurethane formulations. It can be seen that adhesion as measured by the shear strength increases with modulus

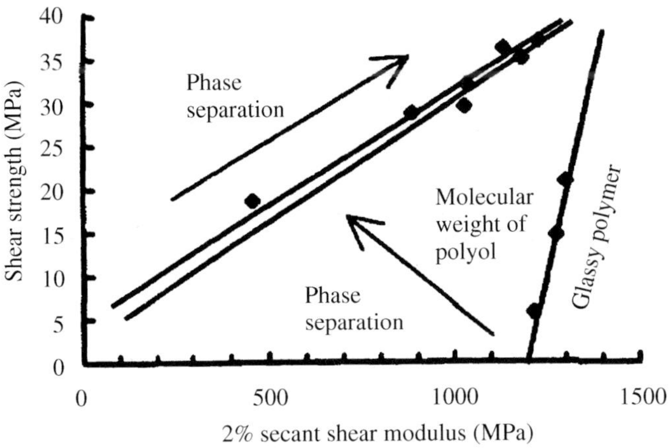

Figure 1. Adhesion (block shear strength) to glass as a function of the shear modulus of the various polyurethane formulations.

in all the polyurethane systems even though the different polyurethane systems show different dependencies. Higher modulus polyurethanes based on Tone 0301 show poor adhesion to glass. It is interesting to note that the same level of adhesion can be obtained with low modulus polyurethanes based on the results with the Tone 0310 polyol. Tone 0310 polyol has the highest molecular weight among all the polyols studied and a significant amount of phase separation. This suggests that along with modulus, phase separation in a segmented polyurethane system has a significant effect on practical adhesion to glass.

3.3. Failure surface analysis

After adhesion testing, glass surfaces of the samples were characterized as to the adhesion failure mode. Some of the samples had chunks of urethane left on the glass surface while other samples showed brittle failure and did not leave any visibly noticeable urethane. Scanning electron micrographs of Tone 0310 based urethane samples are shown in Fig. 2 and are typical of each series. These micrographs confirm that an interfacial failure mode predominates in samples 10A, 10B, and 10C while a cohesive failure mode characterizes the failure locus in samples 10D and 10E.

To further explore the effects of phase separation on adhesion, selected failed glass surfaces of the adhesion samples were analyzed by angular dependent XPS and electron microscopy. Samples 10D, 5D, and 1D, all of which have the same hard segment content, and 10B which has very low hard segment content and very low modulus were included in the XPS study. High resolution (pass energy 35.75 eV) angular dependent XPS scans of samples 10D, 5D, 1D, and 10B were obtained at three different photoelectron take-off angles 15°, 45°, and 90° to vary the sampling depth. The 15° angle scans provided the composition in the uppermost atomic layer ~2 nm whereas the sampling depth in the 90° angle scans could be ~10 nm and the 45° angle scans, somewhere in between. These spectra were charge corrected with the C 1*s* peak referenced to 285.0 eV.

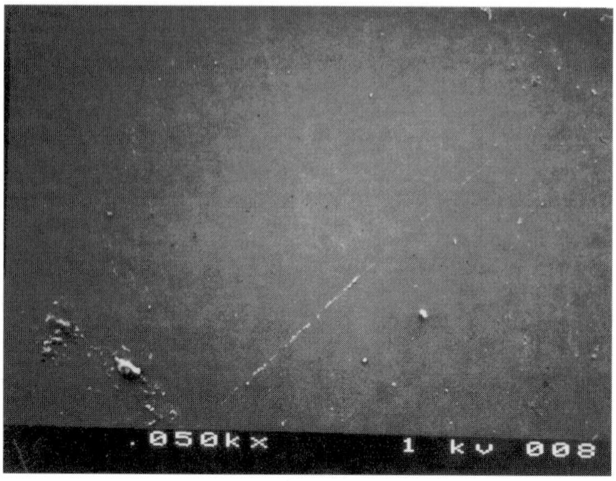

Sample 10A

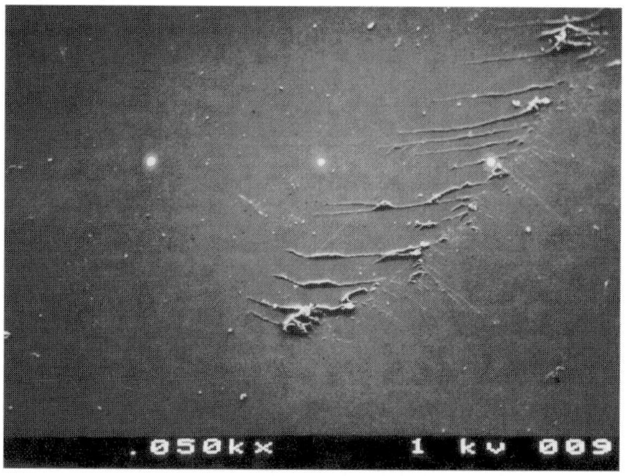

Sample 10B

Figure 2. Scanning electron micrograph of the glass side of the failure surfaces of the Tone 0310 series of polyurethanes.

The XPS spectra revealed that there were several binding states of carbon present on the samples. The higher binding energy component (~ 286.54 eV) could be due to the energy shift caused by C—O single bond environment. This might include C—O—C ether type linkage, unreacted free hydroxyls C—OH from polyols and chain extenders, and C—O—Si type linkages with the glass surface. Since the polyols used in these polyurethanes were the polyester type, the presence of ether linkages was unlikely. The even higher binding energy component at ~ 289.5 eV could be due to the polyurethane linkage [18] present on the samples. The lower binding energy component (~ 283.5 eV) is much smaller for the sample 10D than the other samples and was very high for the sample 10B. This might be due to the thickness variation of the residual film on the glass surfaces, with the film being thickest in the sample 10D and thinnest in the sample 10B.

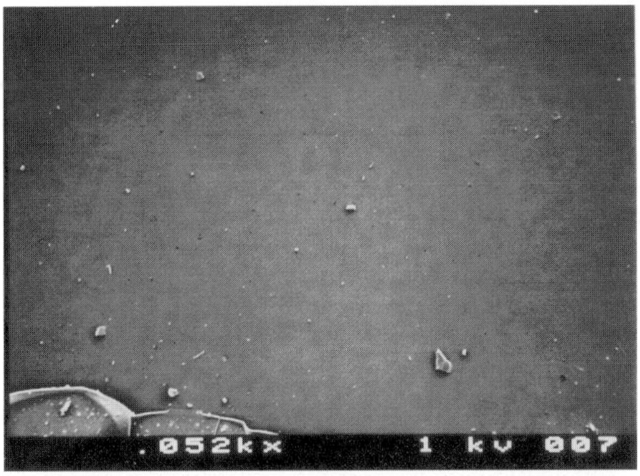

Sample 10C

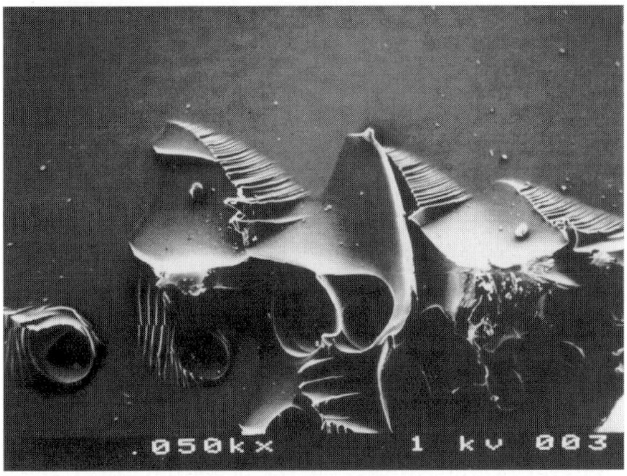

Sample 10D

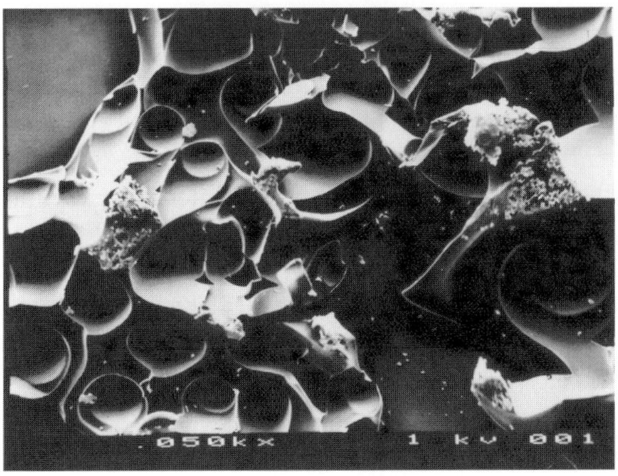

Sample 10E

Figure 2. (Continued).

Table 5.
O 1*s*, N 1*s*, C 1*s*, and Si 2*p* elemental compositions (%) at various photoelectron take-off angles

Sample designation	15°				45°				90°				Expected			
	O	N	C	Si	O	N	C	Si	O	N	C	Si	O	N	C	Si
10D	28	2	59	11	39	2	43	15	48	2	30	21	21	9	70	—
5D	30	2	55	14	41	3	40	17	53	2	25	21	21	9	70	—
1D	38	1	46	16	47	1	32	20	51	2	25	22	21	10	70	—
10B	48	1	36	15	55	1	23	21	46	2	29	23	21	9	70	—

The atomic percent concentrations of O, N, C, and Si obtained from the high resolution XPS spectra at different photoelectron take-off angles are shown in Table 5. It is clear from the data that as the sampling depth increases, the Si concentration increases, being highest at the 90° angle when it is close to that of the bare glass. The Si atomic concentration can also be used to compare the relative thicknesses of the residual films on the glass surfaces. The Si concentration in 10D is the lowest followed by 5D and 1D in the 15° angle scans. The Si concentrations in 1D and 10B are almost the same. The same trend can also be seen in the 45° and the 90° XPS scans. This suggests that the residual film is thicker in the sample 10D, followed by the samples 5D and 1D. The residual films in the samples 1D and 10B are almost of the same thickness. The C 1*s* concentrations in the various samples also indicate the same residual film thickness trend.

In Table 5, the stoichiometric atomic concentrations of the various polyurethane samples are also listed in the far right column. On comparing the XPS derived atomic concentrations with the stoichiometric values, it can be seen that the carbon and nitrogen atomic concentrations are much less than that of the bulk matrix.

Based on the elemental compositions of the residual films in all the samples and their variation with the sampling depths, the residual films on the glass substrates have a general thickness range 1–10 nm and have both inorganic and organic constituents. These residual films form an interphase region which has a composition intermediate to that of the glass substrate and the polyurethane matrices.

The curve fit C 1*s* spectra also revealed that the amount of the high energy component (~289.5 eV) carbon was in good agreement with the N 1*s* atomic concentration. This suggests that the 286.5 eV component in the interphase region is likely due to the presence of the butanediol rather than the polyols since the polyol should contribute to the ~289.5 eV peak.

The polyurethane matrix composition directly affects the proposed interphase region between the matrix and the glass substrate. The XPS data also suggest that the phase separation in the matrix also affects the composition and thickness of the interphase region. The high C 1*s* atomic concentration and the greater thickness of the interphase region in the sample 10D could be the result of phase separation, causing the preferential segregation of the polyol andor the low molecular weight chain extender butanediol to the surface. Whereas in the poorly phase separated sample 1D, preferential segregation of the polyol and/or the chain extender did not take place to any appreciable extent resulting in a thinner interphase region less rich in the C—O type linkages.

3.4. Modified interphase region

An adhesion experiment was designed to check the validity of the above observations. The air sides of clean soda-lime glass plaques were primed with a thin layer of a 2 wt% solution of 1,4-butanediol in acetone to facilitate phase separation in the interphase. Upon drying of the film, various polyurethane blocks were cast onto the glass surface. Upon curing, the samples were allowed to condition for a week prior to the adhesion testing. The adhesion of the polyurethanes to the glass substrate was evaluated using the block-shear method and the results are shown in Table 6. Also shown in Table 6 are the adhesion data for the polyurethanes without any primer on the glass substrates.

The data reveal that the practical adhesion values of the polyurethane samples 10D, 5D, and 10B are not significantly affected by the 2% BDO priming on the glass surface. On the other hand, the adhesion values of the samples 1D and 1C have significantly improved over the previously reported values obtained on bare glass. This experiment indicates that by priming the glass substrate with the 2% BDO solution, we have modified the interphase region in the samples 1D and 1C to be similar to the interphase regions in the other polyurethane samples. Interphase modification of samples 10D and 5D with the BDO primer showed little effect on the adhesion in these samples because the adhesion values obtained with or without the 2% BDO priming are all close to the cohesive strength of the polyurethane. The adhesion data for sample 10B remain unchanged because it has a low modulus characteristic of an elastomeric material as compared to the other polyurethane samples. This suggests that both the interphase region and the matrix are contributing factors in adhesion of polyurethanes to bare glass. The above experiment also provides insight regarding the adhesion mechanism. The BDO coating is not expected to covalently bond with the glass surface. Rather, it can easily hydrogen bond with the hydroxyl-rich glass surface. This indicates that a polar–polar interaction via hydrogen bonding is the dominant mechanism of adhesion through a butanediol-rich interphase region.

Table 6.
Adhesion of various polyurethanes to glass surface

Sample	2% secant shear modulus (MPa) [Iosipescu test]	Adhesion to bare glass substrate (MPa)	Adhesion to glass substrate primed with a 2% BDO solution (MPa)
10D	1040	31.7 ± 0.7	31.7 ± 0.6
5D	1140	35.9 ± 1.4	31.7 ± 1.9
1D	1280	14.5 ± 3.5	34.5 ± 5.9
1C	1210	5.7 ± 2.1	28.3 ± 0.9
10B	10*	11.0 ± 0.02	10.4 ± 0.4

*Tensile modulus, sample was too soft for Iosipescu testing.

3.5. Relationship to polyurethane surface free energy

The surface free energies of the various model polyurethanes have been estimated from the contact angle measurements of known surface tension liquids. A detailed procedure and calculation methodology have been described by Agrawal and Drzal [19]. The polar (γ^P) and dispersive (γ^D) free energy components estimated from contact angle measurements for the polyurethanes are shown in Table 7. As the hard segment content increases, the γ^D data do not change as significantly as the γ^P data. Also, the γ^P values for Tone 0310 based samples are higher than the Tone 0301 based samples for the same hard segment contents. This could be explained by the earlier observation regarding phase separation. The higher molecular weight polyol (Tone 0310) based polyurethanes phase separate which leads to preferential segregation of

Table 7.
Surface free energies of the polyurethanes based on contact angle measurements

Sample designation	Surface free energy (mJ/m^2)	
	γ^P	γ^D
10A	6.8	33.6
10B	6.5	34.2
10C	10.6	30.6
10D	13.6	30.5
10E	15.0	30.3
1C	7.7	34.2
1D	10.7	32.5
1E	11.1	32.5
HS	8.7	35.8
Glass	37.6	19.2

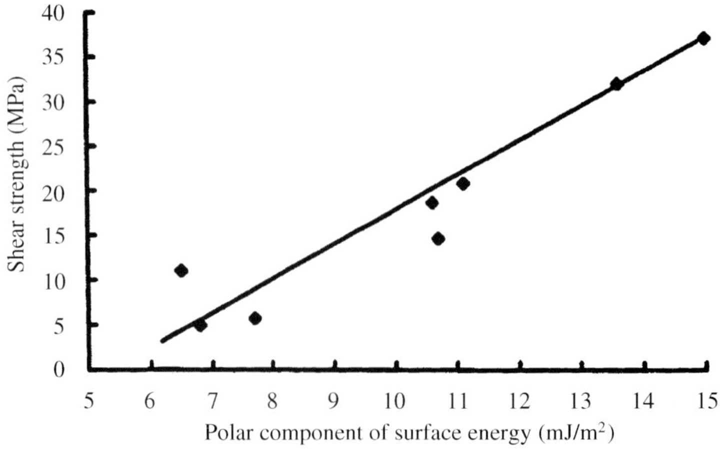

Figure 3. Adhesion (block shear strength) as a function of the polar component of the surface energy of Tone 0301 and Tone 0310 polyurethane formulations.

butanediol type species to the surface. Butanediol is a highly polar substance with $\gamma^{\mathrm{P}} = 14.6\ \mathrm{mJ/m^2}$ [23]. On the other hand, Tone 0301 based polyurethanes do not phase separate and thus their surfaces are not enriched with butanediol type species.

A plot of the adhesion values in shear as a function of γ^{P} is shown in Fig. 3. A linear relationship can be seen between shear adhesion values and the γ^{P}. This suggests that polyurethanes with higher polar components in the interphase region should exhibit better practical adhesion. This is consistent with the XPS analysis of the failure surfaces and also with the observations from the adhesion experiments with butanediol precoated glass plaques.

4. CONCLUSIONS

Adhesion of a segmented polyurethane system (caprolactone based polyol + toluene diisocyanate + 1,4-butanediol) to the air-side of soda-lime float glass was investigated emphasizing on the interphase chemistry and morphology. Practical adhesion measurements showed that the structure of the interphase, in particular phase separation, predominantly affected the practical adhesion to glass. Polyurethanes with higher shear modulus showed higher practical adhesion to glass than lower shear modulus formulations, but among the same shear modulus polyurethanes, phase separated formulations showed higher adhesion to glass than the phase mixed formulations.

SEM and XPS analyses of the failure surfaces revealed that the polyurethane matrix phase separation had a significant influence on the interphase chemistry. It was found that matrix phase separation caused preferential segregation of highly polar surface energy components such as butanediol type species in the interphase region.

The polar component of the surface free energy of the various polyurethanes correlated well with the XPS derived results regarding the butanediol enrichment of the interphase regions. A linear relationship was observed between the shear adhesion values and the polar surface free energies of the polyurethanes.

An adhesion experiment with BDO coated glass plaques demonstrated that the practical adhesion of the polyurethanes can be significantly improved by creating a 'beneficial' interphase rich in highly polar surface free energy components, i.e. phase separated hard segments. It was concluded that the most important mechanism responsible for adhesion between the polyurethanes and the glass surface of this study is the formation of an interphase region in which polar interactions such as hydrogen bonding between the butanediol-rich interphase region and the hydroxylated glass surface play a key role.

Acknowledgements

The authors wish to thank the Donnelly Corporation, the Office of Naval Research, and the Automotive Composites Consortium for supporting this work. The authors also acknowledge the assistance of D. Hook in conducting the XPS experiments in the study.

REFERENCES

1. J. R. Dawson and J. B. Shortall, *Cellular Polymers* **1**, 41–51 (1982).
2. C. Kau, A. Hiltner, E. Baer and L. Huber, *J. Reinf. Plastics Compos.* **8**, 18–39 (1989).
3. P. C. Yang and W. M. Lee, *J. Elastomers Plastics* **19**, 120–146 (1987).
4. E. G. Schwarz, F. E. Critchfield, L. P. Tackett and P. M. Tarin, *J. Elastomers Plastics* **11**, 280 (1979).
5. A. F. Reilly and J. L. Sanok, US Patent 4,755,339 (1988).
6. J. K. Fielder and R. Carswell, *SAE Intl Congress and Exposition*, Detroit, MI, Paper 900515 (1990).
7. R. K. Agrawal, A. Agrawal and J. Thomas, *SAE Intl Congress and Exposition*, Detroit, MI, Paper 910759 (1991).
8. R. J. Zdrahala, R. M. Gerkin, S. L. Hager and F. E. Critchfield, *J. Appl. Polym. Sci.* **24**, 2041–2050 (1979).
9. Z. S. Chen, W. P. Yang and C. W. Macosko, *Rubber Chem. Technol.* **61**, 86–99 (1987).
10. A. L. Change, R. M. Briber, E. L. Thomas, R. J. Zdrahala and F. E. Critchfield, *Polymer* **23**, 1060–1065 (1982).
11. R. E. Camargo, C. W. Macosko, M. Tirrell and S. T. Wellinghoff, *Polymer* **26**, 1145–1154 (1985).
12. C. G. Pantano, *Ceramic Bull.* **60** (11), 1154 (1981).
13. D. J. Vaughn and R. C. Peek, Jr., in: *The Society of Plastics Industry, Inc., 19th Annual Tech. Conf.*, Sec. 13-D, pp. 1–4. The Society of Plastics Industry (1974).
14. A. Lassas, L. E. Fagerholm, B. G. Stenlund and J. H. Nasman, *Polym. Compos.* **14** (9), 1 (1993).
15. S. Markus and Z. Priel, *J. Colloid Interface Sci.* **82**, 150 (1981).
16. M. Mohai, I. Bertoti and M. Revesz, *Surface Interface Anal.* **15**, 364–368 (1990).
17. R. K. Agrawal and L. T. Drzal, *J. Adhesion* **54**, 79–92 (1995).
18. R. K. Agrawal and L. T. Drzal, *J. Adhesion Sci. Technol.* **9**, 1381–1400 (1995).
19. R. K. Agrawal and L. T. Drzal, *J. Adhesion* **55**, 221–243 (1996).
20. D. W. Van Krevelen, *Properties of Polymers*, 3rd edn. Elsevier (1990).
21. V. Rao and L. T. Drzal, *Polym. Compos.* **12** (1), 48–56 (1991).
22. V. Rao and L. T. Drzal, *J. Adhesion* **37**, 83–95 (1992).
23. J. Brandrup and E. H. Immergut (Eds), *Polymer Handbook*. John Wiley & Sons (1989).

Mittal Festschrift, pp. 687–692
W. J. Van Ooij and H. R. Anderson, Jr. (Eds)

Kinetics of interphase formation between incompatible polymers

PHILIP Y. Z. CHU

Polaroid Corporation, 1265 Main Street, Building W4-4, Waltham, MA 02154, USA

Abstract—In this work, a theory for the kinetics of interphase formation between incompatible polymer pairs was developed. In this theoretical analysis, a local constitutive law for the diffusive flux is applied along the lines of the original Cahn–Hilliard equation. An approximate linear Cahn–Hilliard-type differential equation is derived which predicts interface adhesion energy G_a following two power law growth before reaching equilibrium: $G_a \propto t^{1/2}$ when $t < t_{rep}$ (t_{rep} is reptation time); and $G_a \propto t$ when $t_{rep} < t < t_{eq}$. This is in qualitative agreement with recent experimental data on the kinetics of interphase formation between a monodisperse, incompatible polymer couple (polyisoprene/poly(methyl methacrylate)) as a function of annealing time. The data show two distinct features in the plot of G_a versus $t^{1/2}$: at early times, a linearity is observed; at later times, there appears an abrupt upward turn followed by a downward curvature before reaching equilibrium.

Keywords: Incompatible polymers; adhesion energy; interphase; kinetics.

1. INTRODUCTION

The study of interdiffusion in polymeric materials is a vast subject that is directly connected to many fields such as processing of polymer blends, dynamics of phase separation in polymer mixtures, polymer adhesion and welding, kinetics of adhesion, and coatings. Understanding of diffusion processes in polymers is a key to successful production of polymers and applications of polymer products in industry. Due to relatively slow relaxation processes of polymer chains in comparison to systems of small molecules, polymers also offer the ideal system for fundamental studies of diffusion.

In the last few years, there has been intense and rather successful study of interdiffusion of compatible polymer couples, both experimentally and theoretically [1–7]. However, there has been somewhat less effort towards an understanding of interphase formation between incompatible polymer species [8–10]. This case is of particular interest since most polymer couples are immiscible over a wide range of temperature, including those typical of polymer processing conditions. Therefore, interdiffusion and adhesion between different species of polymers is an extremely important problem that has both great practical relevance and considerable academic significance.

In this investigation, a theoretical analysis of interdiffusion for incompatible polymer pairs follows the work of Harden and de Gennes [11, 12]. However, instead of applying a non-local constitutive law for the diffusive flux, we propose to use a simpler, local response, along the lines of the original Cahn–Hilliard equation [13]. As shown subsequently, this appears to provide a qualitatively correct description of interphase formation in incompatible polymer systems, based on currently available data [9, 10].

2. GENERAL THEORY

Consider interdiffusion between pure polymer A and pure polymer B. The local volume flux J of species A relative to the volume average velocity is related to the local gradient of a potential:

$$J = -\Lambda(\phi)\nabla\mu, \tag{1}$$

where $\Lambda(\phi)$ is the Onsager coefficient, ϕ is the volume fraction of species A, and μ is the local replacement potential given by the functional derivative of the total free energy F with respect to ϕ. F can be described by the Flory–Huggins free energy expression modified to account for slow spatial variations in polymer concentration normal to the interface [12]:

$$F = \frac{kT}{a^3}\int \mathrm{d}V\left[f(\phi) + \frac{\varkappa(\nabla\phi)^2}{2}\right], \tag{2}$$

where $f(\phi)$ is the free energy density for a homogeneous system given by the usual Flory–Huggins expression:

$$f(\phi) = \frac{\phi\ln\phi}{N_{\mathrm{A}}} + \frac{(1-\phi)\ln(1-\phi)}{N_{\mathrm{B}}} + \phi\chi(1-\phi), \tag{3}$$

where N_{A} and N_{B} are degrees of polymerization of A and B molecules, respectively. $\varkappa(\nabla\phi)^2/2$ is the Cahn–Hilliard 'gradient energy' associated with the spatial variation of the composition [13]. $\varkappa$ is the phenomenological parameter which plays an important role in the control and formation of interfaces and can be expressed as $a^2/[18\phi(1-\phi)]$, where a is the Kuhn segment size. Thus

$$\begin{aligned}\mu = \frac{\delta(F/kT)}{\delta\phi} \approx{}& \frac{\ln\phi+1}{N_{\mathrm{A}}} - \frac{\ln(1-\phi)+1}{N_{\mathrm{B}}} + \chi(1-2\phi)\\ &- \frac{a^2}{18\phi(1-\phi)}\nabla^2\phi + a^2\left[\frac{1}{\phi^2} - \frac{1}{(1-\phi)^2}\right]\frac{(\nabla\phi)^2}{36}\end{aligned} \tag{4}$$

and

$$\nabla\mu \approx \left[\frac{1}{\phi N_{\mathrm{A}}} + \frac{1}{(1-\phi)N_{\mathrm{B}}} - 2\chi\right]\nabla\phi - \varkappa\nabla^3\phi, \tag{5}$$

where higher order terms are neglected. Assume that the system follows the slow-mode theory [6, 10, 14]; then the Onsager coefficient in equation (1) can be expressed as

$$\Lambda = \frac{\Lambda_A \Lambda_B}{\Lambda_A + \Lambda_B}, \tag{6}$$

where Λ_A and Λ_B are the Onsager coefficients for A and B molecules, respectively. Following Brochard *et al.* [6], if both materials have $N > N_e$ (N_e is the critical N for entanglements), one may assume

$$\Lambda_A = \frac{\phi \Lambda_{0A} N_{eA}}{N_A}, \tag{7}$$

$$\Lambda_B = \frac{(1-\phi)\Lambda_{0B} N_{eB}}{N_B}, \tag{8}$$

where Λ_{0A} and Λ_{0B} are monomer mobilities for the pure A and B molecules. From equations (6)–(8), we have

$$\Lambda = \frac{\phi(1-\phi)\Lambda_{0B} N_{eB}}{N_B[\phi + (1-\phi)g]}, \tag{9}$$

where

$$g = \frac{\Lambda_{0B} N_{eB} N_A}{\Lambda_{0A} N_{eA} N_B}. \tag{10}$$

Combining equation (5) with equation (9) and using equation (1), we obtain an expression for the local diffusion current J of species A

$$J = -\Lambda(\phi)\nabla\mu = -\frac{\Lambda_{0B} N_{eB} N_A}{\Lambda_{0A} N_{eA} N_B}\left\{\left[\frac{1}{\phi N_A} + \frac{1}{(1-\phi)N_B} - 2\chi\right]\nabla\phi - \varkappa\nabla^3\phi\right\}. \tag{11}$$

Thus equation (11) together with the following continuity equation

$$\frac{D\phi}{Dt} + \nabla \cdot J = 0 \tag{12}$$

give a high order non-linear differential equation for ϕ, where D/Dt follows the volume average motion. Generally speaking, from the predicted profiles of ϕ, quantities characterizing the evolution of the interface can be determined.

3. APPROXIMATE LINEAR RESULTS

Consider the situation in which $N_A = N_B$ and $\Lambda_{0A} = \Lambda_{0B}$, that is, when the degrees of polymerization and monomer mobilities of the two species are equal. Assuming a mixture near the critical point and linearizing the chemical potential for small depth quenches in ϕ around the critical composition $\phi_c = 1/2$ allows one to approximate $\nabla\mu$ from equation (5) as [11]:

$$\nabla\mu \approx 2(\chi_c - \chi)\nabla\delta\phi + \frac{2a^2\nabla^3\delta\phi}{9}, \tag{13}$$

where $\delta\phi$ means the fluctuation of ϕ around $\phi_c = 1/2$ and $\chi_c = 2/N$. The Onsager coefficient estimated from that for a single chain transport in a melt of non-interacting Gaussian chains [15] at $\phi_c = 1/2$ is:

$$\Lambda = \Lambda_0\phi(1-\phi) = \frac{\Lambda_0}{4}, \tag{14}$$

where Λ_0 ($= \Lambda_{0A} = \Lambda_{0B}$) is independent of ϕ. Using equations (1), (12), (13) and (14), we obtain a linear Cahn–Hilliard-type differential equation for $\delta\phi$:

$$\frac{D\delta\phi}{Dt} = \frac{\Lambda_0}{2}\left[(\chi_c - \chi)\nabla^2\delta\phi + \frac{a^2\nabla^4\delta\phi}{9}\right]. \tag{15}$$

In a one-dimensional space z, the above equation can be solved by Fourier transform. The result in terms of ψ ($= \partial\delta\phi/\partial z$) is:

$$\psi(q,t) = \psi_0(q)\exp\left\{-\left(\frac{tq^2\Lambda_0}{2}\right)\left[(\chi_c - \chi) + \frac{(aq)^2}{9}\right]\right\}, \tag{16}$$

where $\psi(q,t)$ and $\psi_0(q)$ are the spatial Fourier transforms of $\psi(z,t)$ and $\psi(z,0)$, respectively. Thus, $\psi(z,t)$ is obtained by the inverse Fourier transformation of $\psi(q,t)$ as:

$$\psi(z,t) = \int dq\,\psi_0(q)\exp\left\{-\left(\frac{tq^2\Lambda_0}{2}\right)\left[(\chi_c - \chi) + \frac{(aq)^2}{9}\right]\right\}\exp(-iqz). \tag{17}$$

The interfacial thickness $d(t)$ can be estimated by $d(t) = \Delta\phi[\psi(0,t)]^{-1}$ [11] with equation (17):

$$d(t)^{-1} \approx (\Delta\phi)^{-1}\int dq\,\psi_0(q)\exp\left\{-\left(\frac{tq^2\Lambda_0}{2}\right)\left[(\chi_c - \chi) + \frac{(aq)^2}{9}\right]\right\}. \tag{18}$$

For a very sharp initial interface, $\psi_0(q)$ can be assumed to be approximately constant. At early times ($t < t_{rep}$, where t_{rep} is reptation time), short wavelength contributions, such that $qR_g \gg 1$ (where R_g is the radius of gyration of a polymer coil in melt conditions), dominate the integrand in equation (18) [11, 12]. Since

$$R_g = N^{1/2}a \tag{19}$$

one finds

$$\frac{(aq)^2}{9} \gg O\frac{1}{N} \approx \chi_c - \chi. \tag{20}$$

Therefore, $\chi_c - \chi$ can be neglected relative to $(aq)^2/9$ in the early stages of interphase formation. In this case, equation (18) has the Gaussian form $\exp(-x^4)$ with $x^4 \propto q^4$, indicating

$$d(t) \propto t^{1/4} \quad (t < t_{rep}). \tag{21}$$

On the other hand, at later times ($t < t_{rep}$), dominant contributions come from long wavelength, such that $qR_g \ll 1$ [11, 12]. This corresponds to

$$\frac{(aq)^2}{9} \ll O\frac{1}{N} \approx \chi_c - \chi. \tag{22}$$

As a result, $(aq)^2/9$ becomes negligible in the later stages of interphase formation. Thus equation (18) asymptotically approaches the form $\exp(-x^2)$ with $x^2 \propto q^2$. Consequently one has

$$d(t) \propto t^{1/2} \quad (t > t_{rep}) \tag{23}$$

at later times.

Now, according to de Gennes [8], if chain ends lie on the interface initially, one has

$$G_a \propto d^2, \tag{24}$$

where G_a is interface adhesion energy. Therefore, based on the above analysis of the kinetics of interphase growth, we have

$$G_a \propto t^{1/2} \quad (t < t_{rep}), \tag{25}$$

$$G_a \propto t \quad (t > t_{rep}). \tag{26}$$

The above results are summarized pictorially in Fig. 1.

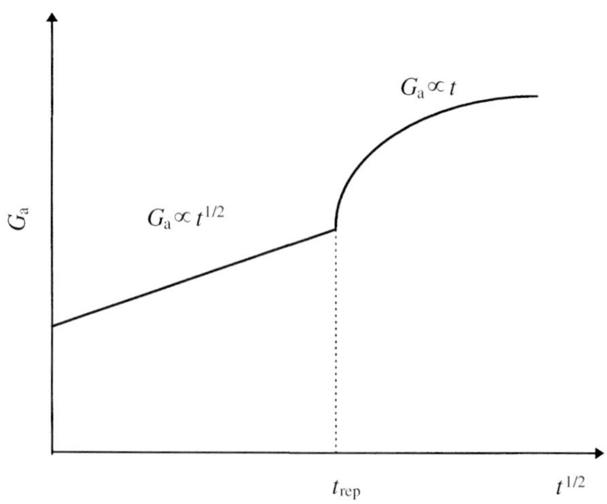

Figure 1. Sketch of G_a versus the square root of time ($t^{1/2}$) predicted by equation (15).

4. DISCUSSION

The results presented here are in qualitative agreement with recent experimental data on the kinetics of interphase formation between a monodisperse, incompatible polymer couple (polyisoprene/poly(methyl methacrylate)) as a function of annealing time [10]. Such data show two distinct features in the plot of G_a versus $t^{1/2}$: at early times, a linearity is observed; at later times, there appears an abrupt upward turn followed by a downward curvature before reaching equilibrium. This is certainly in qualitative agreement with equations (25) and (26), summarized in Fig. 1. Also, the results in equations (21) and (23) are consistent with recent experimental data on the kinetics of interphase formation between weakly incompatible protonated and deuterated polystyrene [9]. However, further refinement of the theoretical model is necessary before any detailed comparisons can be made.

Acknowledgements

I am grateful to Prof. C. J. Durning for very encouraging and informative discussion.

REFERENCES

1. P. G. de Gennes, *C. R. Acad. Sci., Ser. B* **291**, 219 (1980).
2. S. Prager and M. Tirrell, *J. Chem. Phys.* **75**, 5194 (1981).
3. K. Jud, H. H. Kausch and J. G. Williams, *J. Mater. Sci.* **16**, 204 (1981).
4. Y. H. Kim and R. P. Wool, *Macromolecules* **16**, 1115 (1983).
5. M. E. Fowler, J. W. Barlow and D. R. Paul, *Polymer* **28**, 1177 (1987).
6. F. Brochard, F. Jouffroy and P. Levinson, *Macromolecules* **16**, 1638 (1983).
7. E. J. Kramer, P. F. Green and C. Palmstrom, *Polymer* **25**, 473 (1984).
8. P. G. de Gennes, *J. Phys. (France)* **50**, 2551 (1989).
9. G. Reiter and U. Steiner, *J. Phys.* **II1**, 659 (1991).
10. P. Y. Z. Chu, PhD thesis, Columbia University (1993).
11. J. L. Harden, *J. Phys. (France)* **51**, 1777 (1990).
12. P. G. de Gennes, *J. Chem. Phys.* **72**, 4756 (1980).
13. J. W. Cahn and J. E. Hilliard, *J. Chem. Phys.* **28**, 258 (1958).
14. K. Binder, *Colloid Polym. Sci.* **265**, 273 (1987).
15. K. Binder, *J. Chem. Phys.* **79**, 6387 (1983).

Mittal Festschrift, pp. 693–702
W. J. Van Ooij and H. R. Anderson, Jr. (Eds)

Interfacial adhesion in incompatible polymer blends

E. OCCHIELLO, G. GIANNOTTA, M. PENCO and F. GARBASSI*
EniChem S.p.A.–Istituto Guido Donegani, Via G. Fauser 4, 28100 Novara, Italy

Abstract—Thermoplastic polymer blends are important in a number of applications and are required to provide optimal performance. Most polymer combinations are not miscible and frequently a high degree of incompatibility is actually desirable to best exploit the properties of all components. Interfacial interactions, including adhesion, are essential in determining both processing behavior and mechanical properties of blends. They are studied mainly by morphological observations and indirectly by end-use-related properties. However, some novel techniques are being developed to obtain more direct quantitative data about them.

The methods utilized for optimizing interfacial/interphasial behavior belong to two main categories: generating some chemical bonding between different phases, or utilizing compatibilizers (either as a separate ingredient or generating them *in situ*). Specific examples relative to poly(ethylene terephthalate) (PET) are considered here. The first case is the reactive mixing of poly(ethylene terephthalate) with polyolefins grafted with reactive monomers, which exploits the reactivity of PET hydroxyl and carboxyl end-groups for improving interfacial adhesion. The second case is the formation *in situ* of poly(ethylene terephthalate-co-2,6-naphthalene dicarboxylate) (PENT) in order to act as a compatibilizing agent for PET–PEN (poly(ethylene 2,6-naphthalene dicarboxylate)) blends.

Keywords: Polymer blends; interfacial adhesion; PET/PEN blends; toughened PET.

1. INTRODUCTION

Miscibility of thermoplastic polymer pairs (thermosets and interpenetrating networks are outside the scope of this work) is extremely rare. For many applications, it is not even desired, because often it is better to exploit the properties of dispersed phases. Typical examples are toughening of a polymer matrix by elastomeric phases [1] or self-reinforcing provided by liquid-crystalline polymer fibrils [2, 3]. Work on the interface/interphase of immiscible polymer blends has two main goals: improving the adhesion and the processing-related properties. Such goals are, of course, interconnected. A close analogy can be found in the optimization of formulations obtained by filling thermoplastic polymers with inorganic phases, in order to improve their mechanical properties. One may say that the two areas of adhesion and surfactancy, where K. L. Mittal provided important contributions [4–7], are both fundamental for this activity.

*To whom correspondence should be addressed. E-mail: garbassi@enichem.geis.com

A rather large set of literature references has been generated in academia on the above-mentioned topic [8–11], which includes studies aimed at modelling the interface behavior. Particularly noteworthy were the attempts at applying the interfacial tension criterion to blending [8, 12] and to study the role of molecular structure on polymer–polymer interactions [13, 14]. At the same time, there has been important activity in the industrial environment, which has resulted in an extensive patent literature.

Roughly, two principal approaches can be observed: (1) chemically generating connectivity between the two phases during processing, and (2) using a compatibilizer. In turn, the latter can be introduced as a separate component or generated *in situ* by reactive blending. Two case studies are described here relative to the aforementioned categories.

Analytical methods for systems with buried interfaces rely heavily on morphological studies, mostly carried out using SEM (see, for example, [15]) and on indirect observations of processing behavior (e.g. rheology, see [16]) and final properties, mainly mechanical [8–11, 17]. Recently, some novel techniques have been developed, some of which work on buried interfaces themselves, e.g. solid state NMR or scattering techniques [14, 18, 19], while others rely on etching methods for exposing the interface, such as ellipsometry or XPS [20, 21].

2. EXPERIMENTAL

2.1. Toughening experiments

Bottle grade PET (EniChem) was utilized as the matrix phase; it was vacuum dried (4 h at 120°C and 4 h at 180°C) before processing. The typical formulation included PET, a thermoplastic rubber and in some cases a combination of nucleating agents, 0.3% talcum powder and 0.2% sodium benzoate [22]. Blending was performed using a Werner ZSK 40 co-rotating twin screw extruder. Specimens for characterization were obtained using a Sandretto injection molding machine. Mechanical testing was performed using an Instron dynamometer and a CEAST pendulum, according to ASTM D638, D790 and D256. Micrographs were obtained using a Cambridge Stereoscan Scanning Electron Microscope (SEM). Rheological curves were obtained with a Goettfert Rheograph 2002 rheometer. The analysis of terminal carboxyl groups was performed as reported elsewhere [23].

2.2. Melt transesterification experiments

Bottle grade PET was a standard EniChem grade, while PEN was synthesised in-house as reported elsewhere [24]. Chemical structures of the two polymers are reported in Scheme 1. Before processing, the two polymers were dry blended and vacuum dried (4 h at 120°C, 4 h at 180°C). Blending experiments were performed using a Haake Rheomix 600 mixer, at 285°C, or a Berstorff ZE 25 co-rotating twin screw extruder, at 285°C and 300°C. NMR (Bruker AM 300) was used for the evaluation

Poly(ethylene terephthalate) PET

Poly(ethylene 2,6-naphthalene dicarboxylate) PEN

Scheme 1. Chemical structures of PET and PEN.

of block copolymer formation *via* transesterification [25], optical microscopy was used to determine cloud points, DSC (Perkin–Elmer DSC-7) provided information about phase transitions, and SEM (Cambridge Stereoscan 604) was used for specimen morphology observations.

3. RESULTS AND DISCUSSION

3.1. Toughening of PET with elastomers

Toughening of PET has not been studied extensively by academic researchers. However, quite a number of papers have appeared, dealing mostly with morphological studies and assessment of mechanical properties [26–28]. Recently some emphasis has been placed on blending of PET obtained from bottle scraps [29].

More extensive information is available from the patent literature, where combinations of PET with different elastomer types are reported. For instance, polyetheresters [30, 31], polyurethanes [32], acrylic rubbers [33], styrene rubbers [34], ethylene–acrylate–ionomer copolymers [35–37], ethylene–acrylate–glycidyl methacrylate copolymers [38, 39], grafted ethylene–propylene–diene rubbers [40–42], HDPE–acrylic copolymer mixtures [43–44] have been studied. In some of these cases [34–40], the importance of the chemical reaction between PET end-groups and functional groups grafted on the elastomer backbone is also discussed.

To assess the effect of forming chemical bonds between the rubbery phase and PET, blends of PET and different elastomers, both reactive and unreactive towards PET terminal hydroxyl and carboxyl groups, were prepared and injection molded. Results of mechanical testing on specimens are reported in Table 1.

While the modulus was always reduced by mixing with elastomers, as expected, and was independent of interfacial adhesion, impact strength, which critically depends on the ability to transfer the stress from the matrix to the rubbery particle, and hence on matrix–rubber adhesion, was dramatically increased for the rubber including epoxide groups. The latter are reactive with hydroxyl and carboxyl groups; the reaction can be catalyzed, for instance, by amines or alkaline carboxylates and indeed the impact resistance further increased upon the addition of *N*,*N*′-dimethylstearylamine.

Table 1.
Results obtained from 80 : 20 blends of PET with elastomers of different chemical composition (E = flexural modulus, $Izod$ = impact resistance)

Rubber	E (GPa)	$Izod$ (J/m)
Virgin PET	2.4	42
Kraton G 1652 (Shell) Styrene–butadiene block copolymer	2.3	37
Exxelor Va 1801 (Exxon) Ethylene–propylene rubber grafted with maleic anhydride	1.4	103
Lotril 3610 M (Orkem) Ethylene–ethyl acrylate copolymer	1.6	34
Lotader AX 8660 (Orkem) Ethylene–ethyl acrylate–glycidyl methacrylate copolymer	1.5	490
Same as above, with 0.3 wt% *N*,*N*-dimethylstearylamine	1.4	750

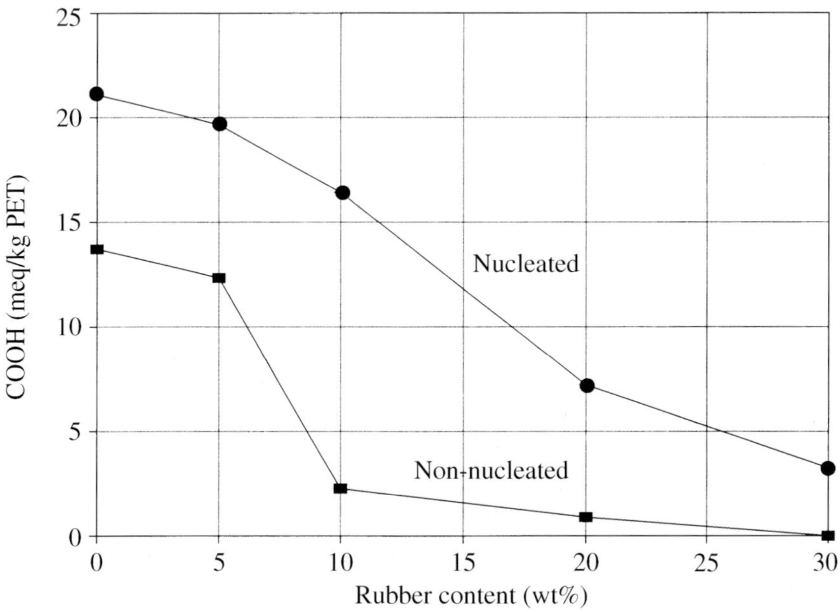

Figure 1. PET carboxyl end-group concentration for compositions with increasing rubber content.

Further to inferences from the improvement of impact resistance, direct evidence for the formation of chemical bonds was obtained from the determination of the concentration of carboxyl groups as a function of the glycidyl methacrylate containing copolymer, both in the presence and absence of nucleating additives (Fig. 1). Results

Table 2.
Calculation of interfacial tension (shear rate = 100 s^{-1}) using equation (1)

Rubber	Average particle radius (μm)	Viscosity ratio	Interfacial tension (mN/m)
Kraton G 1652	0.89	1.32	5.43
Lotril 3610 M	1.19	0.69	6.77
Lotader AX 8660 + *N*,*N*-dimethylstearylamine	0.15	1.77	0.74

clearly show that with increasing amount of rubber the number of carboxyl groups is reduced, thus suggesting the reaction of PET end-groups with the rubbery phase. In the case of nucleated formulations the reduction is less evident, due to concurrent end-group consumption by the elastomer and end-group formation by hydrolytic reactions induced by nucleants themselves.

As suggested in the Introduction, not only mechanical properties, but also processing-related characteristics, such as the average dimension of the rubbery phase particles, depend on compatibilization of matrix and rubbery phase. Actually (Table 2), in the presence of reactive, compatibilized rubbery particles much lower average particle sizes were observed. This effect is in agreement with an empirical equation derived by Wu [45], although not for situations where reactive compatibilization occurs:

$$\frac{G\eta_{\mathrm{m}}d}{\gamma} = 4\left(\frac{\eta_{\mathrm{r}}}{\eta_{\mathrm{m}}}\right)^{\beta} \tag{1}$$

where interfacial tension (γ), matrix (η_{m}) and rubber (η_{r}) melt viscosities, average particle diameter (d) and shear rate in the extruder (G) are related by a power expression; the β coefficient assumes the value of 0.84 for $\eta_{\mathrm{r}}/\eta_{\mathrm{m}} > 1$ and -0.84 for $\eta_{\mathrm{r}}/\eta_{\mathrm{m}} < 1$.

Another essential processing-related property connected to the formation of bonds between the two phases is the rheological behavior. As evident from the curves presented in Fig. 2, a progressive enhancement of melt viscosity is observed with increasing rubber content, even in a shear rate region where the elastomer is less viscous (Fig. 3), due to the formation of chemical bonding between PET and elastomer. A very good inverse correlation with the concentration of residual PET carboxyl end-groups (Fig. 1) has been demonstrated.

Again in the case of nucleated compositions, although the formation of strong interfacial bonds between phases is expected, based on the catalytic effect of the nucleant on the reaction between epoxide and carboxyl groups, a lower viscosity is observed, probably due to hydrolytic degradation of PET. Yet, when another important rheological characteristic, namely activation energy, is determined (Table 3), the striking effect of the increase in chemical bonding becomes evident, as suggested by its great increase.

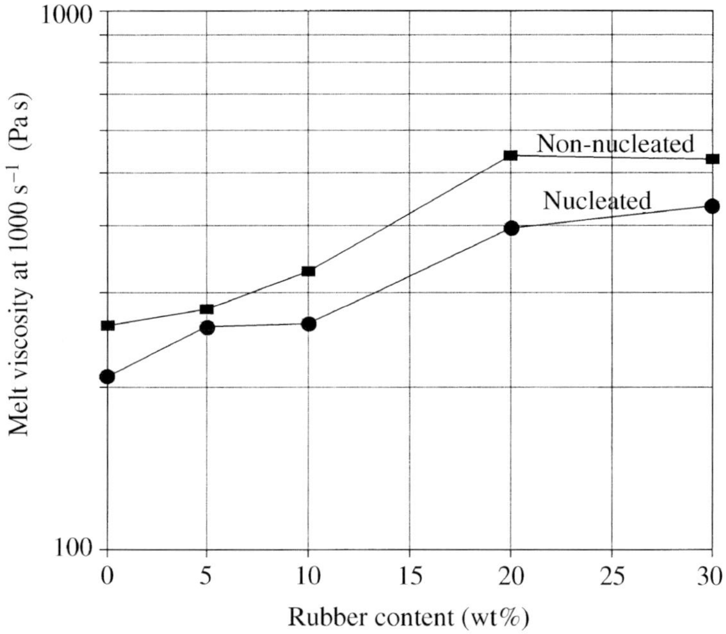

Figure 2. Melt viscosity at 1000 s^{-1} shear rate with increasing rubber content.

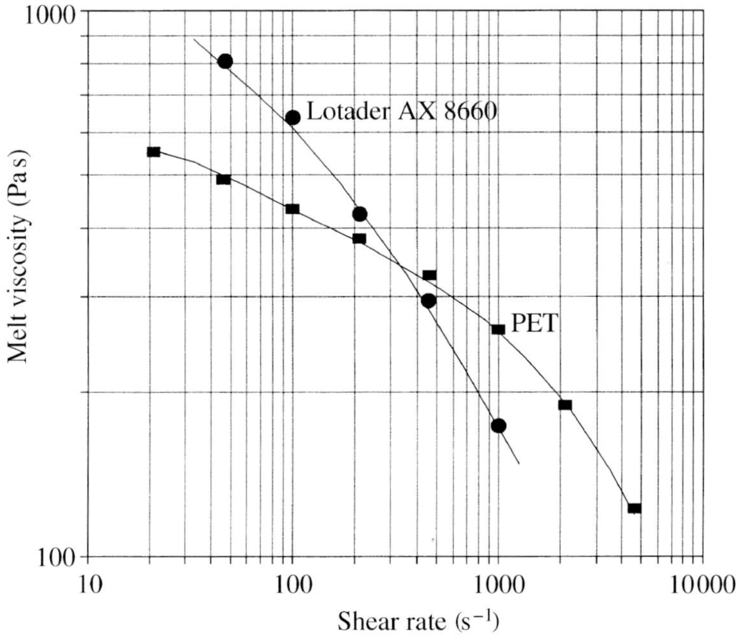

Figure 3. Melt flow curves (obtained with capillary rheometer at 280°C) for PET (squares) and Lotader AX 8660 (circles).

Table 3.
Activation energies of PET toughened with ethylene–ethyl acrylate–glycidyl methacrylate copolymers, from flow curves obtained between 270°C and 290°C

Material	Activation energy (kcal/mol)
PET	24
PET + 30% ethylene–ethyl acrylate–glycidyl methacrylate elastomer (no nucleant)	38
PET + 30% ethylene–ethyl acrylate–glycidyl methacrylate elastomer(with nucleant)	57

3.2. PET–PEN blends

PET–PEN blends are finding important application in a very different field, i.e. the production of heat-resistant plastic bottles. The mechanical mixture of PET and PEN is incompatible and leads to opaque specimens having inferior thermal and mechanical properties [46, 47]. Attempts at improving compatibility were first performed by utilizing PENT copolymers synthesized *ex situ* [48]. Further studies concerned the formation of PENT copolymers *in situ*, studying the effect of parameters such as temperature, catalyst residues, etc. [49] and claiming the effectiveness of compositions where randomization is limited using phosphorus transesterification inhibitors [50]. Finally a number of papers and patents were devoted to various aspects concerning PET–PEN properties [51, 52].

Compatibilization affects both processing-related properties, namely blowing behavior, and product-related properties, mainly transparency and thermomechanical behavior. Both are related to the dimensions of PEN domains in PET and to the characteristics of the interphase between the matrix and the inclusions.

In Fig. 4, the effect of contact time (in Haake Rheomix 600 mixer) upon the melt transesterification of PET–PEN blends is shown, using the disorder degree, as determined by NMR. Clearly, an increase of mixing time results in higher disorder and thus in better compatibility as suggested by a progressive improvement in transparency of molded objects obtained from the resulting products. In Table 4 the results of blending experiments in extruder at different temperatures are reported. In this case, the residence time is fixed and the reaction temperature is shown to increase the disorder degree and hence the amount of block copolymers able to act as compatibilizers. In Fig. 5 some DSC scans are reported, showing that for the blend extruded at lower temperature a melting endotherm typical of PEN is still visible, while it disappears when PEN is processed at higher temperature.

In terms of processing behavior, interesting information is provided by cloud points; in fact, injection molded pre-forms are blown at temperatures in the range of 110–125°C (for the compositions in Table 4). Thus, demixing effects, suggested by clouding, do lead to faulty bottles with poor thermomechanical properties [53]. Actually, SEM observations of bottles obtained from samples where mixing was not perfect do indicate inhomogeneities in different portions of the bottles, sometimes accompanied by delamination upon application of mechanical stress.

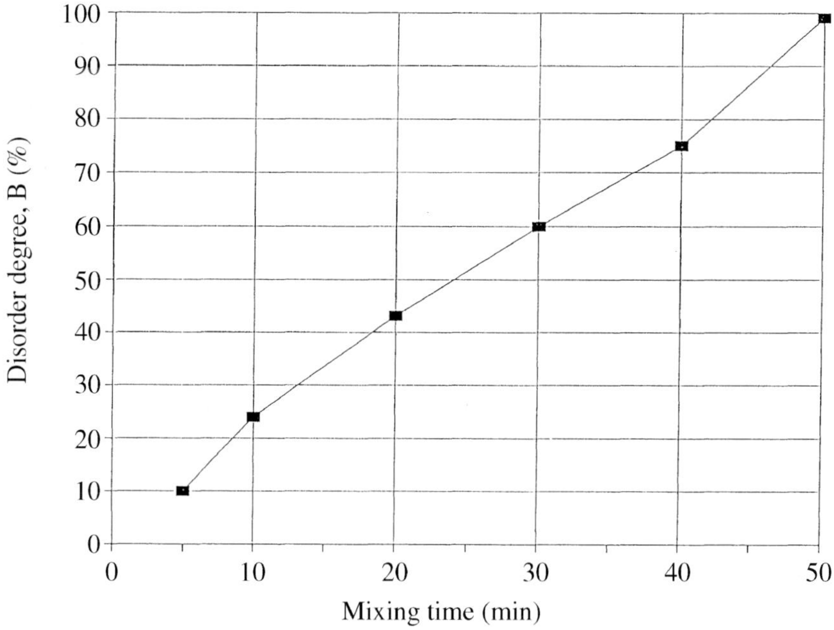

Figure 4. Disorder degree (B, from NMR) of PET–PEN mixtures as a function of mixing time.

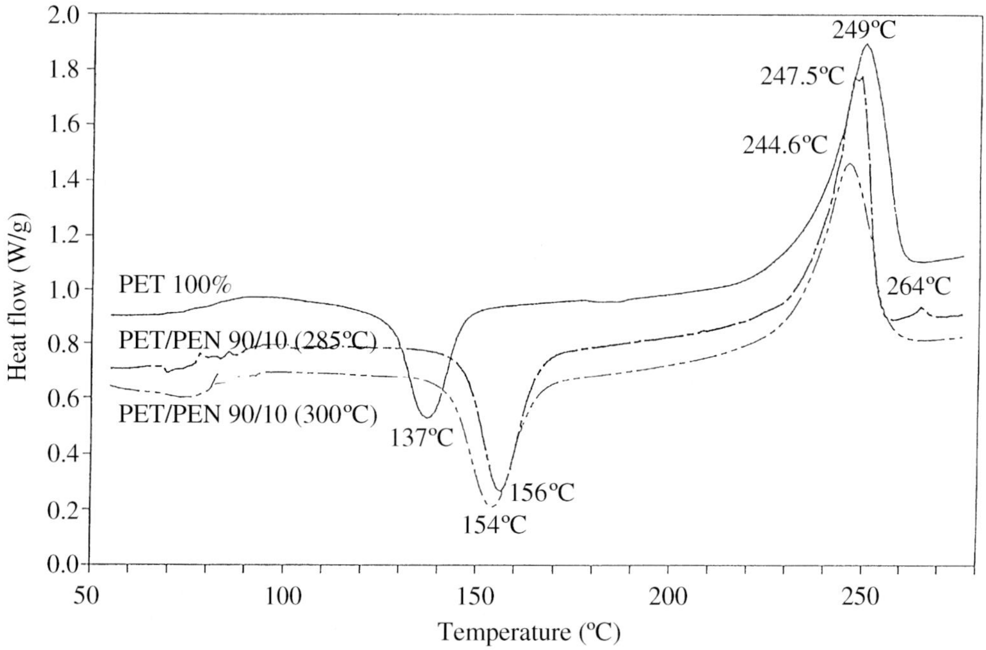

Figure 5. DSC thermograms for PET and for PET/PEN 90 : 10 blends extruded at two different temperatures.

Table 4.
Physico-chemical characteristics of PET–PEN blends processed at different temperatures

Sample	Extrusion temperature (°C)	Disorder degree from NMR (%)	Cloud point (°C)
PET–PEN 90 : 10	285	15	
PET–PEN 80 : 20	285	16	127
PET–PEN 90 : 10	300	24	
PET–PEN 80 : 20	300	26	131

A further comment is required: although transesterification and thus compatibilization is desirable, it should not be excessive; in fact the wanted properties (mainly better thermal resistance) induced by the presence of a PEN phase may be disrupted by an excessive randomization [54].

4. CONCLUSIONS

Results described in the examples provided confirm the relevance of interfacial adhesion for obtaining optimized blend-based materials starting from incompatible moieties. In fact, to improve the impact properties of PET, the use of a reactive elastomer able to react with PET end-groups was found very desirable, forming a true chemical bond. The increased formation of bonds due to the presence of a catalyst for such a reaction provides further support for this conclusion.

For PET/PEN blends, the formation *in situ* of PENT block copolymers acting as compatibilizers allows to obtain homogeneous and stable materials suitable to be transformed into transparent bottles.

REFERENCES

1. C. B. Bucknall, *Toughened Plastics*. Applied Science Publishers, London (1977).
2. W. F. Huang, D. R. Wiff, C. L. Benner, T. E. Helminiak and W. W. Adam, *Polym. Eng. Sci.* **23**, 784 (1983).
3. M. Takayanagi, *Pure Appl. Chem.* **55**, 819 (1983).
4. K. L. Mittal, *Pure Appl. Chem.* **52**, 1295 (1980).
5. K. L. Mittal, *Polym. Eng. Sci.* **17**, 467 (1977).
6. K. L. Mittal and D. O. Shah (Eds), *Surfactants in Solutions*, Vol. 11. Plenum Press, New York (1991).
7. K. L. Mittal (Ed.), *Surfactants in Solution*, Vol. 10. Plenum Press, New York (1989).
8. S. Wu, *Polymer Interface and Adhesion*. Marcel Dekker, New York (1982).
9. D. R. Paul and S. Newman (Eds), *Polymer Blends*, Vols I and II. Academic Press, New York (1978).
10. J. A. Manson and L. H. Sperling, *Polymer Blends and Composites*. Plenum Press, New York (1976).
11. E. Martuscelli, R. Palumbo and M. Kryszewki (Eds), *Polymer Blends: Processing, Morphology and Properties*. Plenum Press, New York (1979).
12. G. Schreyeck and P. Marie, *Comptes Rendus Acad. Sci., Ser. II* **320**, 653 (1995).
13. O. Olabisi, L. M. Robeson and M. T. Shaw, *Polymer–Polymer Miscibility*. Academic Press, New York (1979).
14. M. Stamm and D. W. Schubert, *Ann. Rev. Mater. Sci.* **25**, 325 (1995).

15. D. R. Paul and J. W. Barlow, *J. Macromol. Sci. Rev.* **18**, 109 (1980).
16. H. M. Lee and O. O. Park, *J. Rheol.* **38**, 1405 (1994).
17. C. C. Chen and J. L. White, *Polym. Eng. Sci.* **33**, 923 (1993).
18. M. Afeworki, R. A. McKay and J. Schaefer, *Mater. Sci. Eng.* **A162**, 221 (1993).
19. C. L. Wang, S. J. Wang, W. G. Zheng and Z. N. Qi, *Phys. Status Solidi A* **141**, 253 (1994).
20. S. Yukioka and T. Inoue, *Polymer* **35**, 1182 (1994).
21. S. Yukioka and T. Inoue, *Polymer* **35**, 1256 (1994).
22. M. Penco, M. A. Pastorino, E. Occhiello, F. Garbassi, R. Braglia and G. Giannotta, *J. Appl. Polym. Sci.* **57**, 329 (1995).
23. G. Giannotta, R. Pó, N. Cardi, E. Tampellini, E. Occhiello, F. Garbassi and L. Nicolais, *Polym. Eng. Sci.* **34**, 1219 (1994).
24. R. Pó, E. Occhiello, F. Garbassi and L. Pelosini, European Patent 528460 (1991).
25. R. Pó, P. Cioni, L. Abis, E. Occhiello and F. Garbassi, *Polym. Commun.* **32**, 208 (1991).
26. D. L. Wilfong, A. Hiltner and E. Baer, *J. Mater. Sci.* **21**, 2014 (1986).
27. T. L. Carte and A. Moet, *J. Appl. Polym. Sci.* **48**, 611 (1993).
28. L. P. Chen, C. L. Soles, H. Hristov, A. F. Yee, M. K. Akkapeddi and B. Van Buskirk, *Polym. Mater. Sci. Eng.* **70**, 198 (1995).
29. M. K. Akkapeddi, B. Van Buskirk, C. D. Mason, S. S. Chung and X. Swamikannu, *Polym. Eng. Sci.* **35**, 72 (1995).
30. R. W. William and T. E. Flora, Pat. Coop. Treaty Patent 89 07124 (1989).
31. R. Van der Meer, C. L. Verbraak, J. W. Heuseveldt, J. Van Helmond and H. H. T. M. Hendrikus, European Patent 609538 (1994).
32. H. Magerstedt, K.-J. Idel and W. Siebourg, European Patent 618264 (1994).
33. I. A. Abu-Isa, E. Eusebi and C. B. Jaynes, US Patent 4661546 (1987).
34. T. A. Bright and N. Niessner, European Patent 643104 (1995).
35. R. R. Light and J. A. L. McGill, European Patent 399922 (1990).
36. P. B. Blatz, Pat. Coop. Treaty Patent 92 03505 (1992).
37. F. M. Meyer, Jr., J. R. Dombroski, M. R. Tant and K. E. Breeding, Pat. Coop. Treaty Patent 94 06864 (1994).
38. V. Mehra and P. M. Subramanian, US Patent 4826897 (1989).
39. M. Penco, M. A. Pastorino and G. Giannotta, European Patent 529703 (1993).
40. F. C. Pratt, S. V. Phadke and E. J. Olivier, Pat. Coop. Treaty Patent 88 05452 (1988).
41. F. C. Pratt, Pat. Coop. Treaty Patent 88 07065 (1988).
42. T. J. Shea, A. J. Moffett, J. R. Campbell, M. E. J. Dekkers and F. F. Kouri, US Patent 5132361 (1992).
43. I. A. Abu-Isa and P. M. Graham, US Patent 5194468 (1993).
44. M. Penco, M. A. Pastorino and G. Giannotta, European Patent 535744 (1993).
45. S. Wu, *Polymer* **27**, 335 (1987).
46. R. R. Light and R. W. Seymour, *Polym. Eng. Sci.* **22**, 857 (1982).
47. E. Andresen and H. G. Zachmann, *Colloid Polym. Sci.* **272**, 1352 (1994).
48. F. A. Sheperd and R. R. Light, US Patent 5006613 (1991).
49. M. E. Stewart, A. J. Cox and D. M. Naylor, *Polymer* **34**, 4060 (1993).
50. A. J. Cox, M. E. Stewart, F. A. Sheperd and R. R. Light, Pat. Coop. Treaty Patent 92 02584 (1992).
51. N. Tindale, UK Patent 2228940 (1990).
52. H. G. Zachmann, D. Wiswe, R. Gehrke and C. Riekel, *Makromol. Chem. Suppl.* **12**, 175 (1985).
53. R. Pó, E. Occhiello, G. Giannotta, L. Pelosini and L. Abis, *Polym. Adv. Technol.* **7**, 365 (1996).
54. M. Guo and H. G. Zachmann, *Polymer* **34**, 2503 (1993).

Mittal Festschrift, pp. 703–714
W. J. Van Ooij and H. R. Anderson, Jr. (Eds)

Formation and durability of welded joints involving thermoplastics

H. H. KAUSCH, J. HILBORN and C. J. G. PLUMMER
Department of Materials Science and Engineering EPFL-MX-D, Laboratoire de Polymères MX-D-Ecublens, CH-1015 Lausanne, Switzerland

Abstract—Joining of different or identical materials has been a technical necessity ever since materials have been in use. For thermoplastics, three different principles in a large variety of techniques are employed: mechanical joining, physical joining (welding, adhesion) and chemical joining. In this presentation, the latter two principles are discussed in an attempt to identify the role of physical and chemical bonds in the stress transmission across an interface. The 'crack healing approach', employed, for example, by Kramer, Jud, Kausch and Brown, consisted in the determination of the interfacial fracture strength after chain interdiffusion. Results are reported on the kinetics of strength build-up, the critical depths of interpenetration and the number of main chain bonds necessary for full stress transmission across an interface. This information is of particular relevance with respect to the strength of welds or knit-lines or the effect of block copolymers as compatibilizers in blends.

An alternative approach to the use of block copolymers as compatibilizers is to form a block copolymer *in situ* when two polymer surfaces are brought into contact. In this presentation, results on the chemical bonding of maleic anhydride grafted polypropylene to polyamide 6 will be reported. In particular it is shown that in order to achieve a satisfactory bond energy (of 700 J/m^2) physical bonding through interdiffusion must accompany the formation of primary interfacial bonds through the chemical reaction.

Keywords: Strength of polymer interfaces; stress transmission by molecules; interdiffusion; compatibility.

1. INTRODUCTION

Joining of different or identical materials has been a technical necessity ever since materials have been in use. For thermoplastics, three different principles in a large variety of techniques are employed: mechanical joining, physical joining (welding, adhesion) and chemical joining. The physical origins of the adhesion forces have been described in many of the previous contributions in the literature [1–4]. In this presentation, joints are discussed where the transmission of stresses across an interface is achieved through the interdiffusion of chain molecules in the interfacial region. In such a microscopic situation the same basic principles apply as for the macroscopic joints since the chains in the interdiffusion region can interact mechanically (through newly formed entanglements), physically (by showing frictional resistance if displaced) or chemically (by forming chemical bonds to other chains). In this presentation an attempt has been made to identify the nature and the role of such bonds in three typical situations: establishment of strength through interfacial diffusion between identical (or compatible) polymers, through the use of block copolymers as compatibilizers in blends, and through chemical bonding at the interface (between maleic anhydride grafted polypropylene and polyamide 6).

2. ESTABLISHMENT OF STRENGTH THROUGH INTERFACIAL DIFFUSION

Diffusion phenomena at interfaces have been studied from very different viewpoints, such as the welding of plastic particles during processing, the mechanical effects of interdiffusion in elastomers or thermoplastics, or the physics of the diffusion of macromolecules. Reviews of the early studies have been given different in [5, 6]. More recent results concern the adhesion between polymers and different surfaces [7–11], the micromechanics of interfacial strength [12–18], surface analysis [19–21], and Monte Carlo simulations of polymer interfaces [22]. The short compilation in Table 1 gives a rapid overview on the aspects studied and their authors.

Table 1.
Diffusion at interfaces

1. Studies up to 1989 reviewed and referenced in [6].	
Welding of plastics	Merrill 1947
Tack between elastomers	Voyutski 1963; Gent 1969; Bister, Borchard 1977
Diffusion and relaxation of reptating chains	De Gennes 1971; Doi, Edwards 1978
Molecular weight dependence of diffusion velocity	Klein 1978
Mechanical effects (crack healing)	Kausch, Jud 1979 ff; Wool 1981 ff
Statistical analysis of distribution of chain ends and segments	De Gennes 1980; Prager, Tirrell 1981
Rheology and viscoelastic behavior	Moslé 1982; Baird 1983
Concentration profiles (compatibility, kinetics)	Brochard–Wyart 1983 ff; Green, Kramer 1985 ff
Diffusion in oriented polymers	Merrill, Tirrell 1989
2. More recent studies	
Adhesion between polymers and at surfaces	Brown [7]; de Gennes [8, 9]; Gent [11]; Kramer [10]
Induced interfacial strength through block copolymers	Brown [7]; Kramer, Creton, Hui, Washiyama *et al.* [12–18]
Surface and interface characerization, kinetics of interdiffusion	Russell [19]; Stamm [20]; Reiter [21]
Monte Carlo simulation of polymer interfaces	Binder [22]

A comprehensive review of this field has just been compiled by R. Wool in his book [27].

The 'crack healing approach', as employed, for example, by Jud, Kausch, Kramer, Brown and Wool, consisted in the determination of the build-up of interfacial strength through chain interdiffusion between two flat surfaces of identical (or compatible) polymers brought into contact at a temperature above the glass transition T_g. For such experiments appropriate fracture mechanics specimens are used and fracture toughness K_{Ic} or fracture energy G_{Ic} are taken as a measure of interfacial strength. It has been shown in all these experiments [7–18, 23–30] that the sequence of physical events in strength build-up is the following:

- establishment of (partial) mechanical contact;
- surface reorganization, full contact;
- interpenetration of segments according to compatibility and mechanical constraints;
- further interpenetration of whole chains and successive formation of entanglements;
- at rather long times the former interface will have disappeared completely.

In many of his publications and lectures de Gennes, e.g. [8, 9] has discussed the influences of chain length and chain end segregation on the kinetics of chain interdigitation and interfacial strength. Based on the results of crack healing experiments Kausch *et al.* [5, 23–26] and Wool [27] have pointed out the importance of the formation of new entanglements. Due to the statistical nature of the process of chain diffusion by reptation there will be a large local variation in the number of diffusing chain ends and in the curvilinear depth Δu of their interpenetration. This means that even at small values of $(\langle \Delta u^2 \rangle)^{1/2}$ there are some chains which have penetrated much further than the minimum distance $\Delta u_{\min}$ necessary to form an entanglement. During the crack healing process $n(t_h)$ new entanglements per unit surface area are formed in the interfacial region near the former crack surfaces. Unless the healing has been absolutely complete, $n(t_h)$ will be smaller than n_e, the corresponding concentration of entanglements in the virgin material.

The wriggling and translational motions of the chain segments within their confinement, the so-called *tube*, give rise to an average curvilinear displacement Δu of the chain ends which is described by the Einstein relation:

$$\langle \Delta u^2 \rangle = 2D_t t_h. \tag{1}$$

Following the argument of the above authors [23–27] that the fracture energy $G_c(t)$ of the healed interface is proportional to the number of entanglements formed, a relation between stress intensity factor K_c measured after healing and healing time t_h is obtained:

$$K_{c,rel} = \frac{K_c}{K_{Io}} = \left(\frac{G_c}{G_{co}}\right)^{1/2} = \left[\frac{n(t_h)}{n(\tau_o)}\right]^{1/2} = \left(\frac{t_h}{\tau_o}\right)^{1/4}, \tag{2}$$

where K_{Io} and G_{co} refer to the virgin material: τ_o is the time necessary for complete healing. It is this equation which has explained for the first time and in a straightforward manner the $t^{1/4}$ dependence of the increase of the stress-intensity factor. The

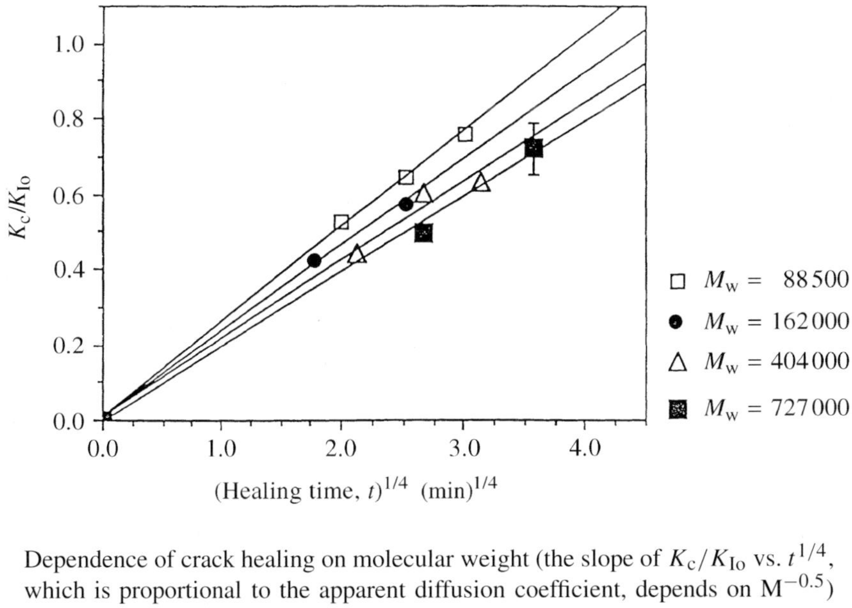

Figure 1. Relative stress intensity factor $K_{c,rel}$ of partially healed PMMA samples as a function of healing time t and molecular weight M_w (from [25]).

experimental data for PMMA of different molecular weights are represented in Fig. 1. These data show that K_c is in fact a linear function of $t^{1/4}$.

The second information derived from these experiments is the molecular weight dependence of the rate of healing. The data show that the time τ_o for complete healing increases in proportion to $M^{+0.5}$. This should be compared with $M^{+1.0}$ derived from reptation theory on the basis of the assumption that the average curvilinear depth of penetration at complete healing is independent of molecular weight. This assumption may not be true. The fraction of 'minor chains' too short to form an entanglement is certainly larger for low M_w material than for high M_w. It is quite conceivable, therefore, that the average curvilinear depth of penetration at complete healing is smaller for larger M_w. There are two other arguments which could explain the observation that long chains are apparently faster than predicted (or more efficient in crack healing). They are both related to the fact that in the crack healing method surfaces are brought into contact which were obtained by a first fracture event (used to determine K_o). Fracture in PMMA proceeds via the formation and rupture of craze fibrils. The latter occurs by disentanglement and/or chain scission. If chain scission plays the dominant role then there will be a reduction in molecular weight in the interfacial region. This was especially notable if the chains were broken more than once [8, 9]. The second argument is derived from the fact that fibril rupture leads to fracture surfaces having a certain roughness. Since it had been shown that the lower density and the stronger relaxation of the fibrillated material increased the rate of interdiffusion [25] it is easily understandable that the more highly fibrillated fracture surfaces of the higher M_w material also show a higher rate of interdiffusion.

Despite the quite satisfactory explanation of the time dependence of crack healing it should also be noted that the assumptions made in deriving the above model have led to the elimination of D_t from the final formula. Thus the above model does not account for a number of parameters which have been shown to influence the strength build-up in a healing interface such as:

- the absolute values of D_t or of the critical distance of interpenetration at complete healing;
- the nature of the diffusion mechanism (by reptation of the chain ends or by displacement of hairpin segments?);
- the absolute number of connector chains involved or the number of connections per chain.

A major breakthrough in the understanding of interfacial strength and an answer to some of these questions was obtained through the use of block copolymers as compatibilizing agents.

3. USE OF BLOCK COPOLYMERS AS COMPATIBILIZERS

The principle of using block copolymers as compatibilizing agents is straightforward. Two incompatible polymers A and B will generally form a clearly phase separated system. Figure 2 shows as an example a thin film of a blend of 10% polysulfone (PSF) particles dispersed in an (incompatible) polystyrene (PS) matrix. The interfaces between the two components are sharp and well defined. When such a film is deformed, large crazes open up in the matrix.

The interior of the crazes is characterised by a larger number of defects: local ruptures of craze fibrils which originate without exception at the surface of the PSF particles indicating the rather weak van der Waals attraction between the incompatible polymer pair PS–PSF. Such a PS–PSF blend can be compatibilized by adding a (PPO–PSF) tri-block copolymer to it. The blocks will interdiffuse into the phases with which they are compatible: the two PPO end-blocks into the PS matrix, and the central PSF block into the PSF particles. (From this interdiffusion result two strong covalent bonds per molecule having penetrated the interface.) The thus obtained interfacial strength is higher now than the crazing strength of the PSF particles. As evidenced by Fig. 3, the PSF particles break up when the craze is formed or extended.

The essential parameters controlling the interfacial strength of a compatibilized blend are evidently the number of covalent ties at the particle surface and the strength with which a block is anchored in its respective phase (which depends on block length). In this respect the most elegant early quantitative experiments on the build-up of interfacial strength through block copolymers were carried out by Brown *et al.* [7, 28, 29] in their elegant interdiffusion experiments using block-copolymers. They created a strong interfacial bond between two flat slabs of respectively PMMA and PPO by interdiffusing an interlayer of a PMMA–PS di-block copolymer. Each

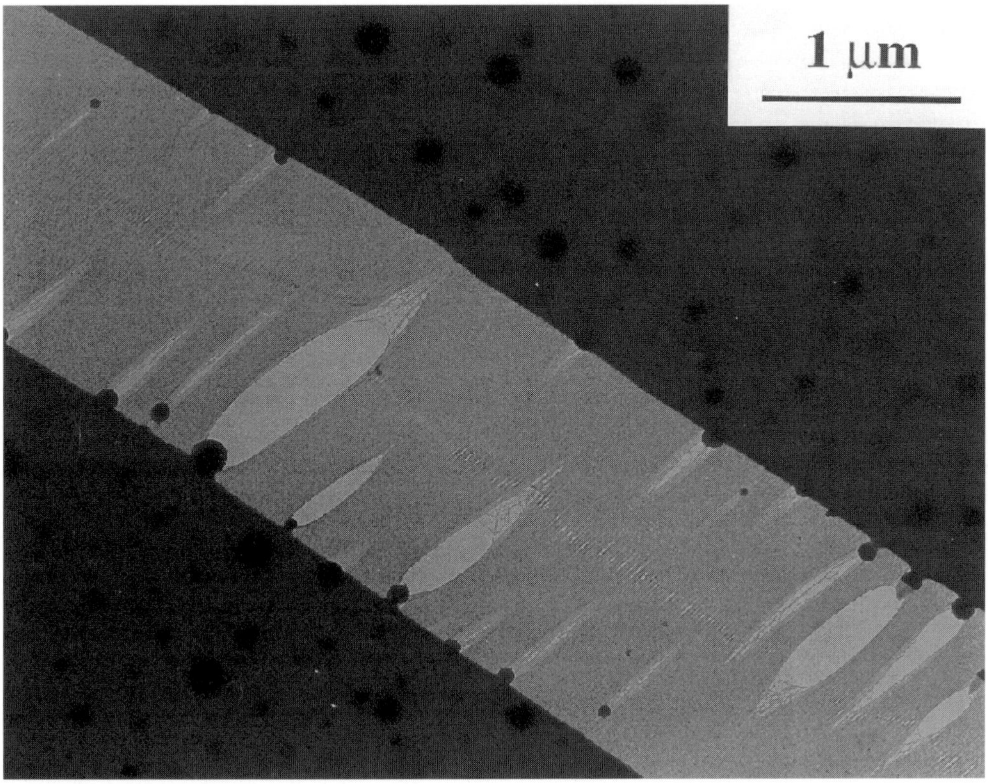

Figure 2. Craze in a thin film of an incompatible blend (10% PSF particles in a PS matrix) deformed at room temperature.

diblock thus provides a covalent bond between the slabs. These experiments provided for the first time the opportunity to virtually count the number N_{di} of primary bonds joining the two slabs. Using marked molecules, the authors [7, 28, 29] have also proven the fact that the PMMA section of the diblock interdiffuses only into the PMMA slab and the PS section only into the PPO. This way they are assured that the number of main chain crossings at the interface must be equal to the number N_{di} of diblock chains available at the interface. For a given thickness of the diblock interlayer — e.g. 25 nm — the number N_{di} obviously decreases with the M_w of the diblocks. As it turns out, the fracture energy of the interdiffused slabs reaches saturation values for the polymers having $M_w = 282\,000$ and $355\,000$ with N_{di} values of 22 and 17.5×10^{16} m^{-2}, respectively. The latter crossing densities are only 15 and 12%, respectively, of the maximum crossing density theoretically permitted by the chain cross section. With the largest molecular weight the crossing density even drops to 5%. This is evidently equivalent to saying that one needs only relatively few, well anchored chains crossing a given interface in order to achieve full mechanical interfacial strength. This result confirms the earlier observation of the beneficial effect of the presence of some 5–10% of high molecular weight material on the long term and fatigue strengths of thermoplastics (see [5] for references).

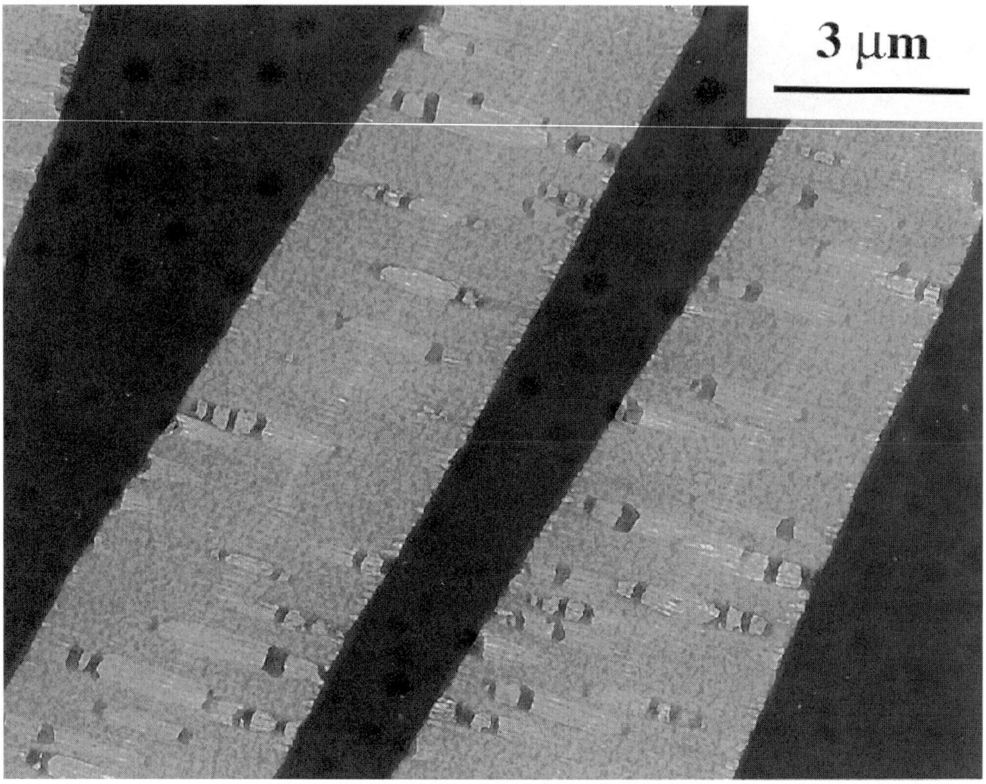

Figure 3. Craze in a compatibilized blend (11% free PSF particles in a PS matrix, PPO–PSF tri-block copolymer as compatibilizer).

4. MODEL OF INTERFACIAL FRACTURE INVOLVING AMORPHOUS POLYMERS

The experiments with block copolymers as strength-inducing links between incompatible polymers (such as PMMA/PPO, PS/PSF or PS/PVP) have been quite successful and have helped to establish a model of interfacial failure. All of the cited authors [5–30] consider the failure of such connected interfaces to be the result of local breakdown involving one (or more) of the three major micro-mechanisms for mechanical failure: (i) scission of connector chains, (ii) their slip with respect to the matrix, or (iii) matrix break-up. Depending on the nature of the dominating mechanism, interfacial cohesion will be limited by (i) the chain strength f_b, (ii) the frictional resistance of a chain to pull-out, or (iii) by the critical stress σ_{cr} for craze initiation. The use of a controlled number per unit interface of connector chains of known block length permits to vary individually the load per chain and its resistance to slippage as depicted in Fig. 4.

Creton *et al.* [12–15] have intensively investigated the system polystyrene/poly-(2-vinylpyridine) (PS/PVP). In the light of the discussed model for interfacial fracture they arrive at the following conclusions. For a degree of polymerization $N_{PVP} < N_{e\,PVP}$ (where N_e refers to the entanglement length), the interface is very weak and the

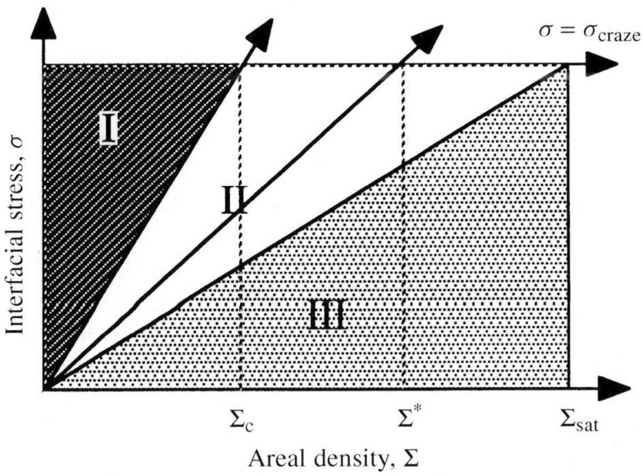

Figure 4. A micromechanical model describing the relative importance of chain scission, chain pull-out and crazing at polymer interfaces as a function of copolymer areal density Σ. If there are few copolymer connector chains then the force per chain, σ/Σ, is larger than the chain strength f_b which leads to chain scission (regime I). In regime II the chain forces are insufficient for chain scission and one observes chain pull-out or, if $\sigma > \sigma_{craze}$, crazing occurs. In regime III long chains are uncritically loaded, and short chains will be pulled out (after [18]).

chains are pulled out from the PVP side so that the only energy consuming mechanism is the friction energy from the viscous pull-out of the PVP block ($\sigma = \zeta_o \delta_o l_i \Sigma$, where ζ_o is the monomeric friction coefficient, δ_o is the rate of chain pull-out, and l_i is the length of block i). When N_{PVP} becomes larger than $N_{e\,PVP}$, fracture occurs by chain scission rather than chain disentanglement. At low copolymer areal density Σ, fracture occurs without a large amount of plastic deformation and the chains break close to their midpoint; while at high Σ, it occurs by formation of a wide craze, the fibrils of which subsequently fail by chain scission in the region of the brush between the PS and the PS block.

The authors give a scaling prediction for the chain pull-out case, namely $G_{Ic} \propto \Sigma l^2$, where l is the contour chain length of the block being pulled out. For all the 'short block' copolymers investigated $G_{Ic} \propto \Sigma$. The dependence on l although being qualitatively correct, does not follow exactly a square law however.

For the strong interfaces, Brown's model [7, 29] for craze breakdown predicts $G_{Ic} \propto \Sigma^2$. The data for the 'long block' case fall reasonably well within this model's prediction [14]. Very similar considerations do also apply to weak interfaces, such as those involving elastomers. De Gennes *et al.* [8, 9] as well as Brown [7, 30] have studied the role of viscous pull-out and of chain scission in such systems and have derived dynamical equations for interfacial separation.

In this section the load bearing capacity of chains at interfaces within a polymer solid has been investigated. It was shown that the notion of a chain as an entangled string of finite strength satisfies most model representations. We conclude that quite generally the strength of such interfaces depends on the amount of mechanical perturbation which the (highly) stressed connector chains provoke in the surrounding material before they disentangle or break. The fact that each chain easily perturbs an area some 10–20 times larger than its diameter also explains why the concentration of connector chains can be as small as 5–10%.

5. CHEMICAL BONDING: *IN SITU* BLOCK COPOLYMER FORMATION AT THE INTERFACE BETWEEN MALEIC ANHYDRIDE GRAFTED POLYPROPYLENE AND POLYAMIDE 6

As we have outlined above, for two essentially incompatible polymers that are brought in contact, the interpenetration of molecular segments over the interface into the other polymer is weak and insufficient to provide for the formation of entanglements and adhesion. The industrial needs to combine such polymers (in polymer blends and multilayer packaging films) without the addition of a third phase (a suitable block copolymer) has therefore resulted in the development of an alternative approach where adhesion is achieved at elevated temperatures by a chemical reaction between the two constituents through the accompanying formation of a di-block copolymer at the interface. This involves reactions at the contacting surfaces and has been used extensively for improving adhesion between polyamide and polyolefins [31–34]. Usually, modified polyolefins containing chemically linked anhydrides or acids are used to allow for reactions with the polyamide amine or carboxylic acid end groups.

In recent studies by Bideaux, Smith *et al.* [35–37] a series of experiments were performed where the fracture energy between two fusion bonded plates of maleic anhydride (MAH) grafted polypropylene (PP_g) and polyamide (PA) 6 at different temperatures was determined. The interfacial fracture energy G_c was shown to increase with increasing fusion temperature (Fig. 5). As concerns the resulting bond strength, three distinct temperature regions can be distinguished:

(i) a region below the melting temperature of PP_g where there is virtually no adhesion;

(ii) a region between the melting temperatures of PP_g and PA where the fracture energy increases with temperature; and

(iii) a region above the melting temperature of PA where the fracture energy is essentially constant.

In region (ii) there is a steady increase in G_c values. In this region the PP_g is completely molten and the polymer chains are now able to diffuse. Since there is a driving force for the enrichment of MAH segments at the interface leading to the reduction in interfacial surface energy it may be assumed that the polar MAH groups of the grafted PP have a tendency to concentrate on the higher energy PA surface. The MAH graft may then react at the common interface with the amine group in the amorphous regions of the PA explaining the observed increase in adhesion. In region (iii) the highest values for G_c are obtained. This is not surprising since both the PP_g and the PA are molten and thus interaction between the MAH and the PA becomes highly probable. The mobility of polyamide amine end-groups increases dramatically. Diffusion will occur over longer distances and anhydrides will react with the amine groups to saturate an interfacial region with block copolymer thus creating the number of covalent bonds necessary for strong adhesion.

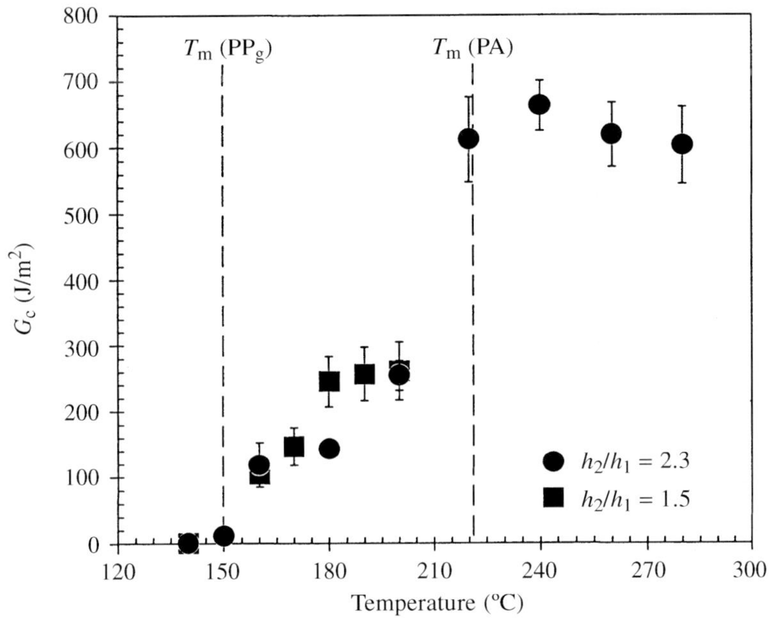

Figure 5. Strength G_c of the interfacial bond between flat PA- and PP_g-slabs as a function of mold temperature. Holding time: 9 min; h_1/h_2: ratio of the thicknesses of the PA and PP_g slabs (after [35]).

The absence of any significant adhesion at temperatures lower than 160°C indicates that the melting of at least one of the adherends, PP_g in this case, is a necessary condition to produce the intimate molecular contact and chain mobility needed for adhesion. This is consistent with a related study by Bothe and Rehage [38] on autohesion of crystalline elastomers. In highly crystalline materials a large number of chain ends are immobile because they are trapped in crystalline domains. Adhesion by mingling of chain segments can be effected, therefore, only by raising the temperature above the melting range.

When the bonding temperature is above the crystalline melting point of PA, not only the grafted PP chains, but also the PA chains are mobile. A larger number of reactions are possible which lead to the saturation of the interface with block copolymer. The dramatic increase in failure energy is accompanied by a change of failure mode and deformation mechanism. The fracture occurs cohesively in the PP_g, but is still relatively close to the interface. This is probably due to the existence, at the PA/PP_g interface, of a thin interphase with a high concentration of low molecular weight grafted chains. These low molecular weight chains can migrate easily and give rise to interface enrichment. An interphase formed of such low molecular weight chains has relatively low cohesion and therefore the crack tends to propagate through this zone [39, 40].

The larger interfacial fracture energies after bonding at high temperatures are also corroborated by microscopic analysis of the fracture surfaces. A large amount of ductile deformation with fibril formation was noted. It can be assumed that the dissipation of transmitted forces is largely improved by the incorporation of the formed copolymer blocks into the crystalline domains on either side of the interface — and not only into the (weaker) amorphous regions from which the connecting molecules may be pulled away without too strongly influencing the ambient matrix. The observed transition from very low adhesion to full interfacial strength is in some way comparable to the transition resulting from an increase in aerial density observed by Creton *et al.* [12] in particulate PS/PVP blends. In the case of relatively few, but well anchored chains, stress transfer to PVP particles was insufficient for their break-up.

Acknowledgements

The authors would like to acknowledge gratefully the frequent occasions for fruitful discussion with their colleagues C. Creton (Paris), E. J. Kramer (Ithaca), R. Wool (Newark) and J. A. Månson and G. D. Smith (Lausanne) who also have made available some of their manuscripts prior to publication.

REFERENCES

1. S. Wu, *Polymer Interface and Adhesion*. Marcel Dekker, New York (1982).
2. K. L. Mittal (Ed.), *Physicochemical Aspects of Polymer Surfaces*, Vols 1 and 2. Plenum Press, New York (1983).
3. F. Garbassi, M. Morra and E. Occhiello, *Polymer Surfaces — from Physics to Technology*. John Wiley, Chichester (1994).
4. J. Berg, 10th Lausanne Polymer Meeting on Polymer Thin Films, Surfaces and Interfaces, Lausanne, Sept. 1995.
5. H. H. Kausch, *Polymer Fracture*, 2nd edn.. Springer, Heidelberg (1987).
6. H. H. Kausch and M. Tirrell, *Ann. Rev. Mater. Sci.* **19**, 341–377 (1989).
7. H. R. Brown, *Ann. Rev. Mater. Sci.* **21**, 463 (1991).
8. P. G. de Gennes, Plenary Lecture, 1st Intern. Congress on Adhesion Science and Technology, held in honor of Dr. Kash Mittal on his 50th birthday, Amsterdam, 16–20 Oct. 1995.
9. Hong Ji and P. G. de Gennes, *Macromolecules* **26**, 520 (1993);
   P. G. de Gennes, *Comptes Rendus Acad. Sci., Série II* **320**, 193 (1995).
10. E. J. Kramer, Spring Meeting of the Am. Phys. Society, Seattle, 22–26 March 1993.
11. A. Gent, *Le Vide, les Couches Minces*, Supplément No. 258, 1 (1991).
12. C. Creton, E. J. Kramer and G. Hadziioannou, *Macromolecules* **24**, 1846 (1991).
13. D. B. Xu, C.-Y. Hui, E. J. Kramer and C. Creton, *Mechanics of Materials* **11**, 257 (1991).
14. C. Creton, E. J. Kramer, C.-Y. Hui and H. R. Brown, *Macromolecules* **25**, 3075 (1992).
15. C. Creton, E. J. Kramer and G. Hadziioannou, *Colloid Polym. Sci.* **270**, 399–404 (1992).
16. C.-A. Dai, B. J. Dair, K. H. Dai, C. K. Ober, E. J. Kramer, C.-Y. Hui and L. W. Jelinski, *Physical Rev. Lett.* **73**, 2472–2475 (1994).
17. J. Washiyama, C. Creton and E. J. Kramer, *Macromolecules* **25**, 4751–4758 (1992).
18. J. Washiyama, E. J. Kramer and C.-Y. Hui, *Macromolecules* **26**, 2928–2935 (1993).
19. T. P. Russell, *Ann. Rev. Mater. Sci.* **21**, 249 (1991).
20. M. Stamm, *Adv. Polym. Sci.* **100**, 357 (1992).
21. G. Reiter, S. Hüttenbach, M. Foster and M. Stamm, *Macromolecules* **24**, 1179 (1991).
22. K. Binder, *Physica Acta* **200**, 722 (1993).
23. K. Jud, H. H. Kausch and G. J. Williams, *J. Mater. Sci.* **16**, 204 (1981).
24. H. H. Kausch, D. Petrovska, R. F. Landel and L. Monnerie, *Polym. Eng. Sci.* **27**, 149–154 (1987).
25. D. Petrovska-Delacrétaz, Ph.D. thesis No. 866, EPFL, Lausanne (1990).
26. H. H. Kausch, T. Q. Nguyen and D. Petrovska-Delacrétaz, *Physica Scripta* **T35**, 57–60 (1991).

27. R. P. Wool, *Polymer Interfaces: Structure and Strength*, Hanser Publishers, Munich (1995); R. P. Wool and K. Connor, *J. Appl. Phys.* **52**, 5194–5201 (1981).
28. H. R. Brown, V. R. Deline and P. F. Green, *Nature* **341**, 221 (1989).
29. H. R. Brown, *Macromolecules* **24**, 2752 (1991).
30. H. R. Brown, Spring Meeting of the Am. Phys. Society, Seattle, 22–26 March 1993.
31. J. Duvall, C. Sellitti, C. Myers, A. Hiltner and E. Baer, *J. Appl. Polym. Sci.* **52**, 195–206 (1994).
32. J. Duvall, C. Sellitti, C. Myers, A. Hiltner and E. Baer, *J. Appl. Polym. Sci.* **52**, 207–216 (1994).
33. J. Duvall, C. Sellitti, V. Topolkaraev, A. Hiltner, E. Baer and C. Myers, *Polymer* **35**, 3948–3957 (1994).
34. T. Nishio, Y. Suzuki, K. Kojima and M. Kakugo, *J. Polym. Eng.* **19**, 123–149 (1991).
35. J. E. Bideaux, G. D. Smith, N. Bernet, J. Hilborn and J. A. Månson, 9th Lausanne Polymer Meeting 1994, accepted by *Polymer* (1994);
G. D. Smith, Thèse, Département des Matériaux, Ecole Polytechnique Fédérale de Lausanne, Lausanne (1996).
36. G. D. Smith, S. Stoll and J. A. Månson, *Flow Processes in Composite Materials 94*, Galway, Ireland, 423–433 (1994).
37. G. D. Smith, P.-E. Bourban and J. A. Månson, *Tenth International Conference on Composite Materials*, Whistler, Canada, 3, 357–364 (1995).
38. L. Bothe and G. Rehage, *Angew. Makrom. Chemie* **100**, 39 (1981).
39. J. Schulz, *J. Adhesion* **37**, 73–81 (1992).
40. J. Schulz, L. Lavielle, A. Carre and P. Comien, *J. Mater. Sci.* **24**, 4363–4369 (1989).

Part 6

Metal-polymer and metal-ceramic adhesion

Mittal Festschrift, pp. 717–746
W. J. Van Ooij and H. R. Anderson, Jr. (Eds)

Interactions and reactions at metal/self-assembled organic monolayer interfaces

D. R. JUNG* and A. W. CZANDERNA†

National Renewable Energy Laboratory, 1617 Cole Blvd., MS-3214, Golden, CO 80401, USA

Abstract—The purpose of research on a metal (M) deposited onto self-assembled monolayers (SAMs) is to understand the interactions between the metal and eventually metal oxide overlayers on well-ordered organic surfaces. Applications of M/SAM and inorganic/SAM research results to the understanding of real inorganic/organic interfaces in vacuum and under environmental conditions can potentially play a key role in the development of advanced devices with stable interfacial properties. The results of M/SAM studies to date are reviewed and M/SAM combinations ranked according to reactivity and penetration. As a set of specific examples, X-ray photoelectron spectroscopy (XPS) is used to identify the chemical shifts from species formed at M/SAM interfaces after evaporation of up to 1.0 nm of Cr overlayers on the CH_3, CN, COOH, and $COOCH_3$ end groups of appropriate SAMs on Au. The reactivity of Cr promotes growth of relatively smooth Cr layers without detectable penetration of the SAM. Chemical bonds are formed with carbon, nitrogen, and oxygen, which will enhance adhesion at Cr/polymer interfaces for polymers with the organic functional groups studied. The XPS spectra show intensities of carbide- and/or oxide-like species on the different end groups, all of which should enhance adhesion at the interface.

Keywords: Self-assembled monolayers (SAMs); metal–polymer interactions; metal–polymer interfaces; Cr interactions with CH_3, CN, COOH, and $COOCH_3$; XPS of metal/SAM interactions.

Acronyms used: ISS, ion scattering spectroscopy; M, metal; M/SAM, metal/self-assembled monolayer; OFG, organic functional group; SAM, self-assembled monolayer; UHV, ultra high vacuum; XPS, X-ray photoelectron spectroscopy.

1. INTRODUCTION

This review summarizes part of recent research [1–11] to identify metal/organic functional group interactions at interfaces of metal/self-assembled monolayers (SAMs), primarily by means of X-ray photoelectron spectroscopy (XPS) [1–5, 7–11], or to elucidate penetration rates of metal overlayers through the SAM using ion scattering spectroscopy (ISS) [5–7]. As illustrated in Fig. 1a, SAMs have fulfilled a need

*Present address: Science Applications International Corp., 1710 Goodridge Drive, McLean, VA 22101, USA.

†To whom correspondence should be addressed.

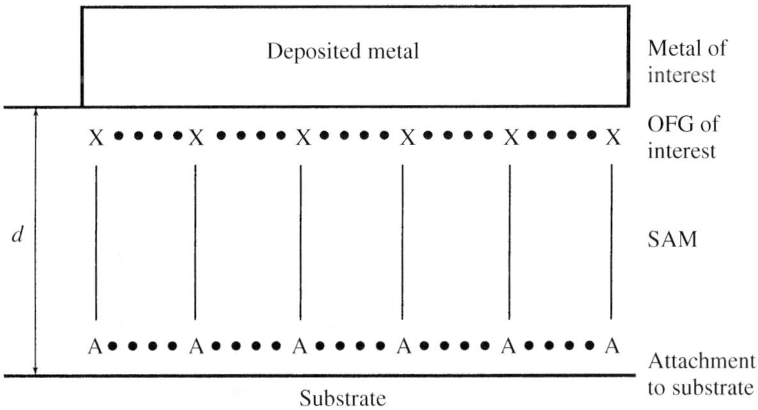

Figure 1a. Idealized scheme for using an organized molecular assembly to form an interface between a metal overlayer and an organic functional group. The thickness of the SAM is *d*, which is the length of the molecule at perpendicular attachment. For alkanethiol SAMs on gold, the attachment group is a thiolate and the methylene chain is 28° from the normal.

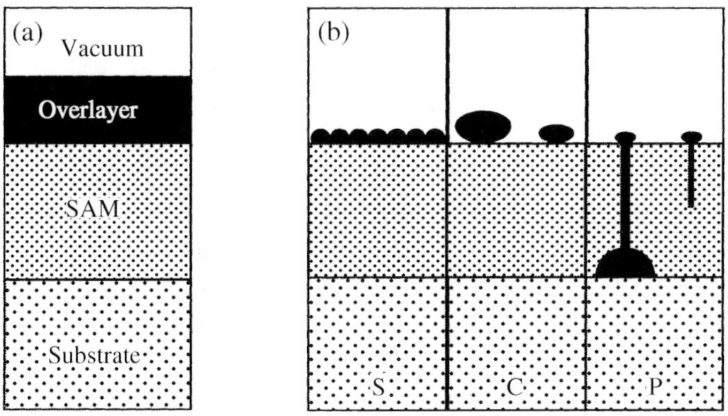

Figure 1b. Cross sections of the principally observed results to date in which there is (S) strong metal–OFG interaction, (C) weak metal–OFG interaction, and (P) very weak metal–OFG interaction. The designations S, C, and P stand for smooth, clustered, or penetrating metal, respectively.

for model systems with relatively well-ordered organic surfaces of uniform chemical composition for studying interactions at metal or metal oxide/organic interfaces [1–11]. Even though present state-of-the-art SAMs contain defects, such as rows, pits, and domains, atomically sharp interfaces can be formed, in principle, between a deposited metal overlayer and the SAM surface. Chemical bonding at metal/organic interfaces plays an important role in the reliability and durability of many technological devices [12]. Bonding at metal/organic or organic/metal interfaces has been typically studied either by surface and interface analysis of organic species adsorbed onto single crystal metal surfaces [13, 14], surface analysis of metallized polymers [15, 16] or by coating a metal oxide with a polymer. However, we point out that the chemical environment studied in the case of metal/organic interfaces first bears little resemblance to that present at polymer/metal (oxide) interfaces. In addition, the lack of structural and chemical order of polymer surfaces complicates the location and type of interactions with metal or oxide overlayers. Metal oxide or M/SAM studies have added a new approach in which the state of subdivision of the M or metal oxide overlayer is different and selection of highly ordered but different end-group organic surfaces is possible.

Reactive transition metals (e.g. Cr and Ti) are frequently used to promote adhesion at metal/polymer interfaces, and have been the subject of many studies [4, 8, 9, 11, 17–20]. Good metal adhesion on polymers with oxygen functionalities is well known and can be two to ten times better than on polymers without oxygen-bearing groups [15, 16]. Less reactive metals (e.g. Ag and Cu) are frequently used as metallization coatings on polymers for mirror applications or as grid lines in photovoltaic cells, which may contain polymer/metal interfaces. For studying the stability or reactivity of interfaces, SAMs are ideal because they form highly ordered, thermally stable arrays of close-packed aliphatic chains terminated by a surface of uniform chemical functionality as originally cited in the work of Nuzzo and Allara [21] and recently reviewed by others [22, 23]. SAMs with a variety of terminating functional groups are now available, making studies of the interaction between a known metal species and an organic surface of a known functionality possible. SAMs can be formed from alkyl sulfides, alkane acids, alkanethiols, and other more complex molecules on different surfaces, e.g. Ag, Au, SiO_2, Al_2O_3, and GaAs and present different tilt angles to the substrate, packing densities, and contact angles with water or hexadecane (Fig. 2). Progress in studying alkanethiol SAMs terminated with CH_3, OH, COOH, $COOCH_3$, and CN with Ti, Cr, Cu, Ni, Al, K, Na, Ag overlayers has been summarized and critically reviewed [10]. The specific metal/organic functional group (OFG) combinations studied are identified in Table 1. After briefly summarizing the objectives and critical questions in M/SAM research and the properties of SAMs, we will focus in this paper on our own studies of deposited Cr with five SAMs. More detail on aspects of our Cr/SAM studies is available [24].

Table 1.
Matrix of references of metal/organic functional group (OFG) systems ($X(CH_2)_nS/Au$, X = OFG) studied

	OFG				
Metal	CH_3	CH_2OH	$COOCH_3$	COOH	CN
Ag	6a, b, 7			6c	
Cu	33	3	8, 30, 33	1, 2, 22	3, 8
Ni				33	4
K			30		
Na	31, 32				
Al	5, 30		30		
Cr	11a		8, 11c, 30	11b	4, 8, 11a
Ti	9	9	9	9	

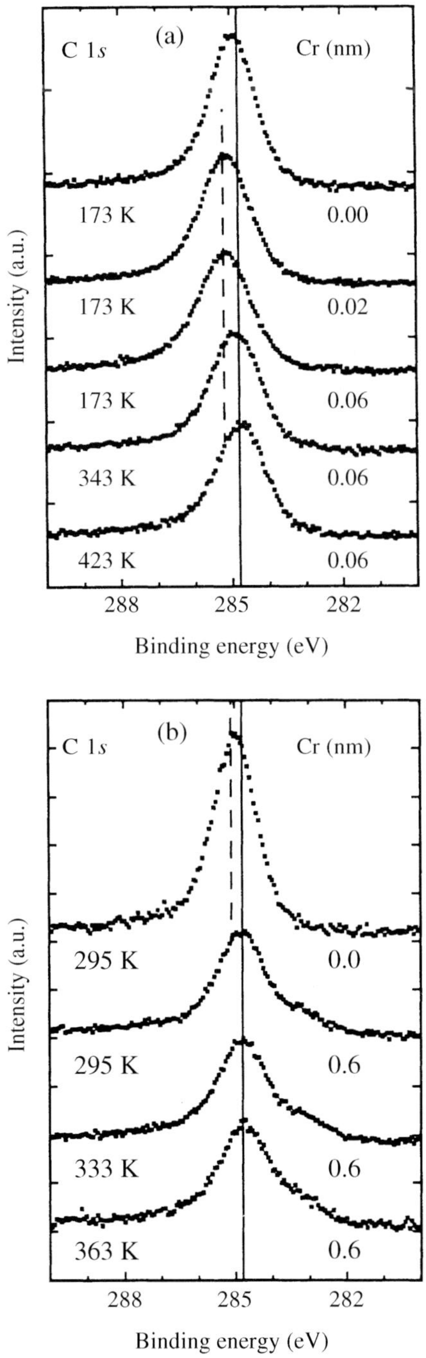

Figure 2. C 1*s* XPS line shapes for a $CH_3(CH_2)_{17}S$/Au SAM with (a) 0, 0.02, and 0.06 nm Cr coverages at 173 K and after heating to 343 and 423 K and (b) 0 and 0.6 nm Cr coverages at 295 K and after heating to 333 and 363 K in which the solid and dashed vertical lines are at 284.8 and 285.2 eV in both (a) and (b).

1.1. Objectives of recent research

In support of the main goal of recent M/SAM research (understanding the interactions between M overlayers on SAMs), the objectives of recent work have been to: (a) identify new chemical compounds or complexes, and intermediate stages in compound formation, if any; (b) establish which or what combination of the three main configurations of metal that may be present as a function of the ratio of deposited metal atoms to the total OFGs on the SAM, as illustrated in Fig. 1b: smooth metal (S) that is chemically bonded or complexed with the SAM and probably spread out on top of the SAM, clustered (C) nonbonded metal that is on top of the SAM and probably in the form of large clusters, and penetrating (P) nonbonded metal that significantly or totally penetrates the SAM and whose morphology could take many forms; (c) compare the quantities of smooth, clustered, or penetrated metal for deposition at different rates on surfaces at different temperatures, and thereby search for evidence of activated processes, metastable/nonequilibrium states, and results that depend on the degree of SAM ordering; (d) obtain images of clusters on the SAM (by scanning probe or electron microscopy), and use the distribution of cluster sizes, the rate of deposition, and the surface temperature as parameters to model the kinetics of cluster nucleation and growth; (e) characterize the degree of ordering and close packing of the SAM before and after metallization (are pinholes playing a role?); and (f) measure rates of penetration of metal through the SAM.

1.2. Critical questions in M/SAM research

The lowest free energy state of a M/SAM system is that for which all the deposited metal should be between the substrate (in this work, Au) and the SAM, because the surface free energies of metals are over two orders of magnitude higher than those of the organic end groups of the SAMs. Thus, two critical questions about M/SAM research are: are the configurations for the smooth or clustered metal of Fig. 1b stable at room temperature, and what factors determine that stability? The scheme depicted in Fig. 1b, and our corresponding identification of smooth metal overlayers with chemical bonding and (large) clustered or penetrated metals with nonbonding, are intended as idealizations (e.g. we neglect the role of metal carbide formation which could occur along the entire SAM chain).

Chemical interactions surely play a major role in the formation of a M/SAM interface. Studies that address compound formation or complexing in combination with objectives (b), (d), or (f) can yield information on the correlation of the chemistry of the M/SAM system with the way the metal is distributed on or in the SAM.

Objectives (c) and (e) are concerned with *how* a M/SAM system attains the observed chemical state and spatial distribution. The process of adlayer nucleation and growth, a competition that is kinetically dependent on surface temperature and deposition rate, may be considerably more complex than that found for two-dimensional systems if there is diffusion into (and not just over) the SAM. The mechanism of penetration may involve different possible rate-limiting steps, for example, lateral diffusion leading to penetration at defect sites vs. direct vertical diffusion that might be enabled by thermal motion (because of ambient thermal energy or transient thermal energy imparted by atom impact), chemical reaction with and saturation of the OFG, or disordering due to strong chemical effects.

The sorting of the various possibilities of chemical states and spatial distributions will require, in most cases, experimentation with both structural and spectroscopic probes under varying conditions of deposition rate and surface temperature. After this information is available, some general conclusions about processes at M/SAM interfaces may emerge that can be extended to predict processes at metal/polymer interfaces. Detailed process information may result only from quantitative comparison with theoretical calculations and simulations of these systems.

1.3. Experimental evidence for the order, structure, and stability of alkanethiols on gold

The properties of thiol monolayers on gold have been the subject of recent review articles [22, 23]. As implied, SAMs spontaneously self assemble but are also robust and stable. The characteristics of the ordered SAM include: (1) the presence of the desired OFG at the vacuum/SAM interface; (2) a low density of pinholes in the film; (3) uniform film thickness; (4) uniform average tilt angle of constituent molecules; (5) a low population of gauche bonds in the methylene chains; (6) lateral ordering (real-space Bravais lattice or translational symmetry vectors) and internal structure of the unit cell of the SAM monolayer; and (7) a domain size of coherent lateral order. Details about using experimental probes, e.g. contact angle measurements using sessile drops of liquids, scanning tunneling microscopy, atomic force microscopy, interfacial force microscopy, ISS, cyclic voltammetry, ellipsometry, Fourier-transform infrared reflection absorption spectroscopy, near-edge X-ray absorption fine structure, X-ray diffraction, helium atom scattering, and transmission electron microscopy are available [10]. As for stability, alkanethiol SAMs withstand extended atmospheric exposure, repeated washing, and modestly higher temperatures above 25°C without desorption: the latter depends on the thiol/substrate combination and the length of the alkanethiol molecules used. These and other film characteristics (e.g. their wettability by water and other liquids) are described further by Dubois and Nuzzo [22] and Ulman [23].

2. EXPERIMENTAL

A detailed list and critique of the experimental measurements used to characterize SAMs is available [10], including all those identified in the preceding section. For the typical results presented in this article, the use of XPS will be described. Brief mention will be made about ISS studies and detailed descriptions of our ISS experimental methods are published elsewhere [5, 6]. The sample preparation methods described below are similar to those reported by other groups.

2.1. SAM preparation

The SAMs were formed by immersing clean Au-coated Si ⟨100⟩ wafers in ethanolic solutions of octadecanethiol ($HS(CH_2)_{17}CH_3$), 12-mercaptododecanenitrile

($HS(CH_2)_{11}CN$), 16-mercaptohexadecanoate ($HS(CH_2)_{15}COOCH_3$), mercapto-undecanoic acid ($HS(CH_2)_{10}COOH$), and 16-mercaptohexadecanoic acid ($HS(CH_2)_{15}COOH$), using the well-known self-assembly process for alkyl thiols [21, 24, 25]. Some of the gold coatings were prepared by evaporating 80 nm Au directly onto Ar^+ ion bombardment-cleaned Si wafer coupons [26], while others were prepared by evaporating a 9 nm Cr adhesion layer followed by a 200 nm Au layer onto solvent-cleaned Si wafers. In most cases, the Au surfaces were placed in a ultra-violet ozone cleaner for 5 min to remove trace organic contaminants [27] prior to immersion in the thiol. The ozone treatment for 1 h produces a 1.7-nm thick Au_2O_3 oxide on the gold that is subsequently reduced during the self-assembly process [28]. Water contact angles of less than 10° are observed on the Au surface only if such a cleaning procedure is used. Following incubation times of 2–4 days at 295 K, a sample was removed, rinsed in ethanol, dried, and checked by water contact angle for the presence of the SAM. Contact angles of 10° to $20 \pm 2°$, $64 \pm 2°$, and $112 \pm 2°$ observed in this study for the $COOCH_3$-, COOH-, CN- and CH_3-terminated SAMs, respectively, are in good agreement with published results [25].

2.2. *Deposition of the Cr metal*

Deposition of the Cr metal onto the SAMs and subsequent X-ray photoelectron spectroscopy (XPS) and ISS analyses were carried out using a Leybold-Heraeus LHS-10 surface analysis system. The LHS-10 is equipped with an analysis chamber and a sample preparation chamber through which a sample introduction rod can be translated without breaking vacuum. In the sample preparation chamber, deposition is carried out at a sample-to-source distance of 38 cm and rates are measured with a quartz crystal microbalance. A complete description of this apparatus and its use for XPS and ISS have been presented previously [29].

With the sample mounted onto a hot/cold transfer rod (initially at room temperature) and inserted into the LHS-10 chamber that is maintained at a base pressure of $2–5 \times 10^{-8}$ Pa, deposition of the metal took place in an attached preparation chamber that permits *in situ* transfer of the sample to and from the XPS or ISS analysis chamber at constant temperature. The sample was allowed to outgas at room temperature for 1 h prior to heating or exposure to X rays. In our standard procedure, the sample was characterized by XPS at 295 K, heated or cooled as desired, recharacterized by XPS, exposed to the first incremental deposition of metal, and again characterized by XPS [1–6]. Repeated deposition/characterization cycles were performed in some cases, resulting in total coverages of up to 2 nm. Thus, the photoelectron escape depths permitted us to probe for M/OFG interface effects throughout the incremental deposition processes. For a frame of reference, we refer to the amount of metal deposited as a uniform coverage, e.g. 1.0 nm Cr, as though it is uniformly covering the surface of the SAM. In all experiments, less than 3 min elapsed between the end of metal deposition and the beginning of the XPS or ISS analysis.

The photoelectron spectra were recorded using a non-monochromatized Mg K_α X-ray source operated at 240 W. Narrow scans were taken using constant pass energies of 50 eV for C 1*s*, N 1*s*, O 1*s*, Au 4*f*, Cu 2*p*, Ag 3*d*, and Cr 2*p* (for coverages of 0.2 nm or more) and 200 eV for Cr $2p$ (for coverages below 0.2 nm) and N $1s$. Survey scans were taken in the constant transmission mode with a retardation ratio of 3. The XPS data are used to evaluate the interface stoichiometry from adjusted peak intensities, the 'chemical state' from binding energy shifts, and nonchemical binding energy shifts for partial to total penetration.

ISS compositional depth profiles were measured using 1 keV $^3He^+$ ions at a current density of 0.2 $\mu A/cm^2$ and temperatures from 113 to 295 K. The ion beam was rastered over a 2 × 2 mm area with the detected signal gated at a 70% aperture to minimize the signal from the etch-pit walls. During ISS analysis, He from the ion gun caused the chamber pressure to increase from a base pressure below 2×10^{-9} Torr to a carefully maintained pressure of 4×10^{-7} Torr. Each ISS spectrum was acquired over a period of approximately 40 s. Detection of C with $^3He^+$ is limited by the number of H atoms bonded to the C. Carbon in CH_3 or CH_2 is difficult to detect with ISS, but C or CH exposed to the ion beam are detectable. Consequently, we did not detect C in our compositional depth profiles. Sulfur from the thiol has been observed by ISS in previous work [26]. Accordingly, data were taken over an energy range E/E_0 of 0.850 to 1.0, to obtain depth profiles with good time resolution and good signal-to-noise ratios.

The Cr metal, which was outgassed extensively prior to use, was sublimed in very high vacuum from a Cr-coated tungsten rod (99.9%, R. D. Mathis, Long Beach, CA) onto the SAM. The quartz microbalance was used to establish and monitor the deposition rate for each increment; the average rate for all our experiments was 0.14 ± 0.05 nm/min. Deposited Cr on the walls of the preparation chamber gettered reactive gases, so that the ambient pressure was typically 2.0×10^{-6} Pa at the start of a series of experiments but gradually decreased to 2.0×10^{-7} Pa ($\sim 10^{-9}$ Torr) with the continued use of the source. The coverage of Cr on the SAM is expressed as the average thickness based on the bulk density and mass from the calibrated quartz crystal monitor, with no claim that the actual Cr overlayer is smooth, unclustered or homogeneous.

3. RESULTS AND DISCUSSION

We present and discuss in this section a summary of the M/OFG combinations studied to date and rank them according to their relative reactivity and degree of penetration into the SAM [10]. We then present and discuss typical results for the reactivity of Cr/CH_3, Cr/CN, Cr/COOH, and Cr/$COOCH_3$ interfaces. We include XPS spectra taken from 173 to 373 K for Cr/CH_3 and Cr/CN since little temperature dependent work has been done.

3.1. Ranking of M/OFG combinations according to reactivity and penetration

In Table 2, based on the XPS and Fourier transform infrared reflection absorption spectroscopy results, we have ranked the chemical reactivity of the various M/SAM systems according to the OFG of the SAM. The rankings are based on the degree

Table 2.
Ranking of the reactivity of M/SAM systems by metal and OFG*

Reactivity	Metal	Organic functional group	Type of bonding	Ref.
Very high	Ti	COOH, $COOCH_3$, CH_3OH	Oxide and carbide	9
Very high	Cr	$COOCH_3$	Oxide and carbide	8, 11c, 30
High	Ti, Cr	CH_3	Carbide	4, 8, 9
High	Cr	CN	Nitride and carbide	4, 8
Intermediate	Al	$COOCH_3$ (pre-heat treatment)	Oxide	30
Intermediate	Al	$COOCH_3$ (post-heat treatment)	Metal-carboxylate	30
Intermediate	Cu, Ni	COOH, CH_2OH	Unidentate metal oxide	1–3, 33
Intermediate	K, Cu	$COOCH_3$	Metal-carboxylate	8, 30, 33
Low	Ni, Cu	CN, $COOCH_3$	Unspecified	4, 8, 33
No reaction	Ag, Al, Na, Cu	CH_3, COOH (except Cu)	—	5–7, 31–33

*OFG is organic functional group. References are given for XPS and Fourier-transform infrared reflection absorption spectroscopy results, except for Na/CH_3. The degree of reactivity is based on the interpretation by DRJ and AWC of the results. The rankings are based on the degree of XPS or infrared evidence for binding energy shifts or attenuation/formation of absorbance from M/OFG interactions.

of XPS or infrared evidence for binding energy shifts or attenuation/formation of absorbance from M/OFG interactions. Likewise, in Table 3, we have ranked the penetration of the M/SAM systems according to the XPS results reported for the different OFGs and different chain lengths of the SAMs. A dependence of the degree of penetration on the temperature of the sample, that is, lower penetration for lower temperatures, has been found for the systems of Cr/CN, Ni/CN, and Cu/CN ([ref. 5], $T = 173$ K), Ag/CH_3 ([refs 6, 7], $T \geqslant 90$ K), and Ag/COOH [ref. 6c]. A dependence on the length of the alkanethiol molecule may also affect the XPS results shown in Table 3, that is, shorter molecules show evidence of permitting greater penetration.

We conclude from these rankings that the degree of penetration increases with decreasing reactivity. In order for the adsorbed metal atom or cluster to penetrate, it first must encounter the OFG. Penetration is unlikely if there is a reaction. If there is *no* reaction, then penetration can proceed in some cases. Rates of penetration have been measured only in the cases of Ag [6, 7] and Na [31, 32] on CH_3-terminated SAMs and for Ag on COOH-SAMs [33]. Table 2 shows that the reactivity is highest for the oxygen-containing SAM end groups (except for Ag/COOH), intermediate for the CN, and lowest for the CH_3. For the metals, Ti is the most reactive, followed by Cr, Al, K, Ni, Cu, Na, and Ag, although the data for Al, K, Ni, and Na are scant. The decreasing reactivity within the transition metal series, Ti, Ni, and Cu, and the low reactivity of the noble metals, Cu and Ag, is in the same order as the filling of the *d* orbitals. The relative reactivity of trivalent Al and the monovalent alkali metals (Na, K) is not so clear at this point because so few of these M/SAM systems have been explored. Evaporating oxygen-free Al films has not been successful, as yet [34]. The M/SAM studies involving Al evaporation may have been affected by rapid oxidation of the deposit [34], which would have had a severe effect on the chemical interactions at the interface.

Table 3.
XPS evidence for penetration in M/SAM systems (all M/OFG$(CH_2)_n$S/Au) at room temperature (RT) and at lower temperatures (LT)

n	Metal	OFG	Metal shift		C 1*s* attenuation		Penetration		Ref.
			RT	LT	RT	LT	RT	LT	
15	Ti	COOH	HBE	?	?	?	Very low	?	9
15	Ti	$COOCH_3$	HBE	?	?	?	Very low	?	9
15	Ti	CH_2OH	HBE	?	?	?	Very low	?	9
15	Cr	$COOCH_3$	HBE	?	?	?	Very low	?	8, 11c
15	Ti, Cr	CH_3	HBE	?	?	?	Low	?	4, 8, 9
15	Cr	CN	HBE	?	?	?	Low	?	4, 8
11	Cr	CN	HBE	HBE	Slight	OK	Low	Low	4
15	Cu	$COOCH_3$	HBE	?	?	?	Low/mod.	?	8, 33
10	Cu	COOH	Zero	?	?	?	Low	?	1, 2
10	Cu	CH_2OH	LBE	?	Low	?	Moderate	?	3
11	Ni	CN	Zero	HBE	Low	Slightly low	Moderate	Low	4
15	Cu	CN	HBE	?	?	?	Low	?	8
11	Cu	CN	LBE	HBE	Quite low	?	Moderate	Low	4
17	Ag, Cu	CH_3	LBE	HBE	Zero	?	High	Low	6, 7, 33
10	Ag	COOH	HBE	?	?	?	High	Low	6c
15	Ag	COOH	HBE	?	?	?	Mod./high	Low	6c

(i) The metal shift refers to whether a core level has been measured at low coverages to exhibit a shift to higher binding energy (HBE) or lower binding energy (LBE) with respect to the bulk metal core level.

(ii) C 1*s* attenuation refers to the attenuation of the C 1*s* level compared to that expected for a uniform metal overlayer.

(iii) References to the sources of these XPS results are given: question marks (?) indicate that the information has not been reported.

(iv) The degree of penetration is based on the interpretation by the authors (DRJ and AWC) of the available XPS results, in which 'very low' means the overlayer essentially resides on top (in the same plane) of the OFG and 'high' means the overlayer penetrates to the SAM/Au interface.

There is some evidence that the reactivity of the SAM with the incoming M is also dependent on the SAM temperature [5, 11]. Activation of M/SAM reactions may be because of electronic, steric, or structural factors and is not unexpected. Further temperature-dependent studies are warranted in order to learn more about activation energies in M/SAM reactions.

3.2. Reactive system: C 1s *XPS line shapes for Cr/$H_3C(CH_2)_{17}S$/Au*

The effects of post-deposition heating on Cr/$H_3C(CH_2)_{17}S$/Au interfaces formed at 173 and 295 K were studied for two purposes: (1) to investigate the reactivity of the CH_3 end group and (2) to determine if a carbide-like feature formed at 295 K could be increased by heating after metal deposition. The C 1*s* spectra are shown in Figs 2a and 2b. The corresponding peak positions and areas from curve-fitting have been tabulated [11]. Figure 2a shows C 1*s* spectra for a $H_3C(CH_2)_{17}S$/Au sample that was metallized with 0.02 nm and then 0.06 nm Cr at 173 K and then heated to 423 K. The C 1*s* line shifts from 284.9 eV at zero Cr coverage and 173 K to 285.2 eV at 0.06 nm coverage and 173 K, and then again to 284.8 eV at 0.06 nm coverage and 423 K. Figure 2b shows C 1*s* spectra for a $H_3C(CH_2)_{17}S$/Au sample that was metallized with 0.6 nm coverage at 295 K and then heated to 363 K. A carbide low binding energy shoulder, of intensity similar to that observed for Cr/$NC(CH_2)_{11}S$/Au (see Fig. 3) is found at 295 K, but after heating the carbide intensity is less than was found at similar temperatures when Cr was deposited onto the $NC(CH_2)_{11}S$/Au SAM held at 323 and 373 K, as also has been tabulated in detail [11]. The increase in the amount of carbide by post-deposition heating will be discussed in the section about collision activation.

3.3. Reactive system: C 1s *XPS line shapes for Cr/$NC(CH_2)_{11}S$/Au*

The C 1*s* line shapes are shown in Figs 3a and 3b for 0.06 and 0.6 nm Cr coverages, respectively, for deposition onto a series of different $NC(CH_2)_{11}S$/Au samples held at 173, 295, 323, and 373 K, along with the C 1*s* line shape for a bare SAM for comparison. The most striking difference in Fig. 3 is the steady increase from 295 to 373 K in the $\sim$283 eV low binding energy shoulder (Fig. 3b) that we assign to a Cr carbide species. A shift in the high binding energy peak also is observed.

The high binding energy C 1*s* peak of the CN-terminated SAMs is found at 286.85$\pm$ 0.1 eV for zero Cr coverage, and is attributed to the endmost, nitrile carbon and its nearest (α) neighbor along the chain. At 0.6 nm Cr coverage the high binding energy peak shifts to 286.45 $\pm$ 0.1 eV for lower temperatures (173–295 K) and to 286.25 $\pm$ 0.15 eV for higher temperatures (323–373 K). The larger shift of this high binding energy C 1*s* component (about 0.2 eV) shows there is a slight increase in Cr(CN) reactivity in the higher temperature range. Similar results are found at lower Cr coverages, as shown in Fig. 3a. The shift with Cr deposition probably results from a net charge transfer from Cr to CN molecular orbitals that increases the charge in the valence levels of the two endmost carbons. More detail about curve fitting of the C 1*s* line shapes, estimates of the chromium carbide stoichiometry (Cr_3C_2), and control experiments for assessing the influence of X-ray damage is available [11].

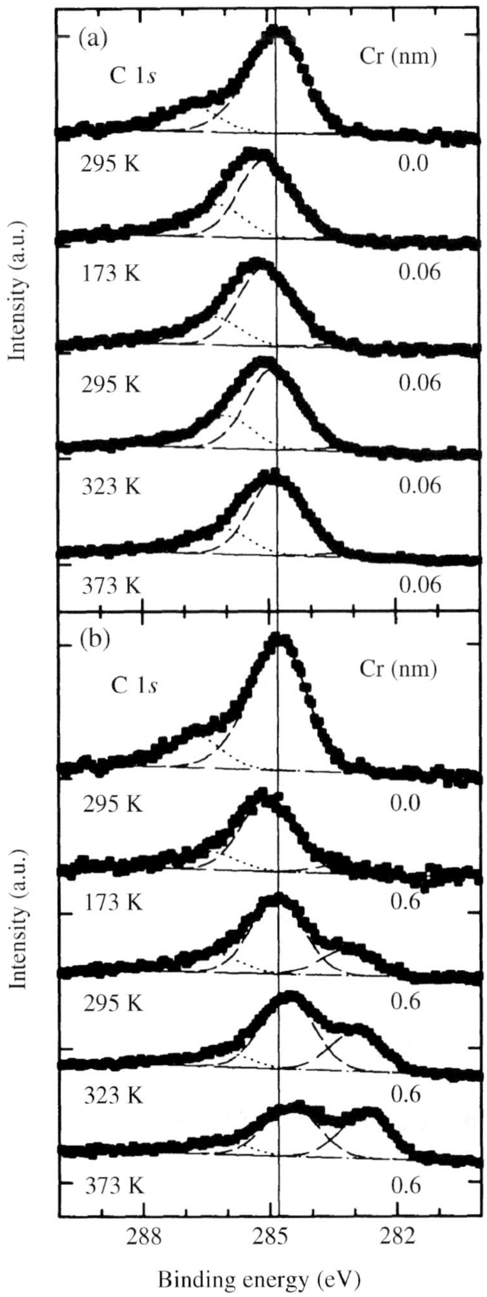

Figure 3. C 1*s* XPS line shapes for the bare $NC(CH_2)_{11}S/Au$ SAM at 295 K (top of (a) and (b)), and for (a) 0.06 nm Cr and (b) 0.6 nm Cr deposited on a series of $NC(CH_2)_{11}S/Au$ SAMs at 173, 295, 323, and 373 K, as labeled. The plots at 0.6 nm coverage have been increased in intensity by a factor of 1.5 for clarity. The position of the solid vertical line is 284.8 eV.

3.4. Reactive system: binding energies of the XPS Cr 2p *levels for Cr/NC(CH₂)₁₁S/Au*

In [5], we discussed the correlation of metal penetration into or possibly through the SAM with the position of the Cr 2*p* binding energies. The Cr 2*p* positions are shown in Fig. 4. They are shifted to higher binding energy for Cr deposited at 295 and 173 K onto the CN SAM (Fig. 2b), consistent with little or no penetration. They are shifted to lower binding energy for Cr deposited at 295 K onto Au (Fig. 3a), consistent with thin Cr deposits on a conducting substrate. They are not shifted, but are approximately at the bulk position, for Cr deposited at 323 and 373 K (Fig. 3b). These results show smaller, negligible high binding energy shifts for 323 and 373 K compared to 295 K. In terms of the core hole lifetime/final state effect reported by Wertheim for the shifts of metal clusters on insulating substrates [35], our result is consistent with some degree of penetration at 323 and 373 K.

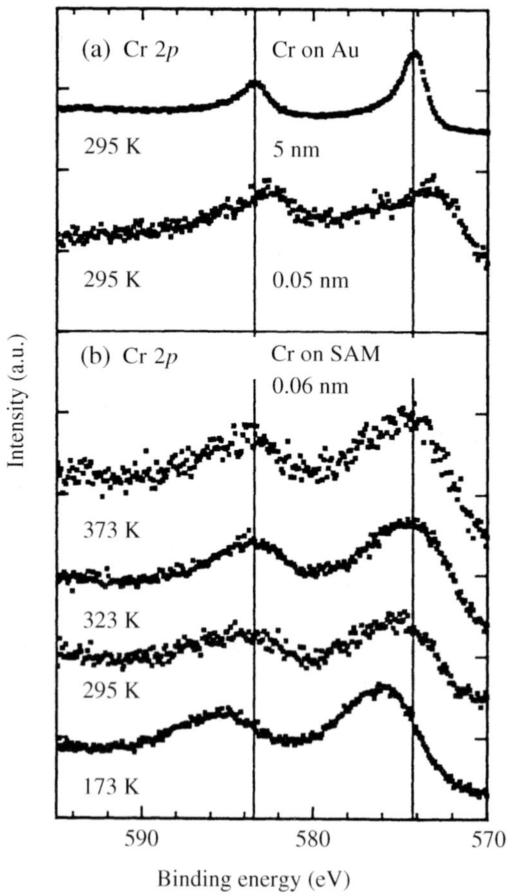

Figure 4. Cr 2*p* XPS line shapes (a) for 0.05 and 5.0 nm Cr on Au at 295 K, and (b) for 0.06 nm Cr deposited onto a series of $NC(CH_2)_{11}S/Au$ SAMs at 173, 295, 323, and 373 K, as labeled. The solid vertical lines at 583.4 and 574.2 eV mark the peak positions for bulk Cr 2*p*.

3.5. Reactive system: binding energies of the XPS N 1s *levels for $Cr/NC(CH_2)_{11}S/Au$*

Consistent with the shifts of the C 1*s* high binding energy peak, the N 1*s* peak shifts to an approximately 0.5 eV lower binding energy for deposition of 0.06 nm Cr at 373 K. Taken together, these data support an increased chemical reactivity, and a greater net transfer of electronic charge from Cr to the CN groups of the $NC(CH_2)_{11}S/Au$ SAM with increasing sample temperature. This is consistent with a thermally activated process, but our data are insufficient to determine an activation energy.

Furthermore, core level shifts alone do not provide sufficient information to specify the nature of the Cr–NC reaction. As has been found for CO adsorption on metal surfaces, the interplay of the molecular orbitals is not obvious, depends on the specific electronic states of the species involved, and requires additional experimental and theoretical work.

We also measured the effects of high temperature and X-rays on the bare SAMs and compared binding energies for SAMs with and without exposure to X-rays prior to Cr deposition to estimate contributions from sample degradation [11]. The results show that the measured N 1*s* binding energy at 295 K and 0.06 nm coverage is not affected by X-ray exposure prior to Cr deposition.

At this coverage, about 0.6 eV of the observed binding energy shift (1.2 eV) can be produced by holding the sample at 373 K for 12 h with a 3 h X-ray exposure. However, a 1 h exposure to 373 K and X-rays produced a much smaller shift (0.2 eV). Because our standard procedure involves about 1.5 h per incremental deposit of Cr, it is likely that the increased N 1*s* binding energy shift for $Cr/NC(CH_2)_{11}S/Au$ interfaces at 373 K is real, but does include a small contribution from the X-ray and high temperature exposures.

3.6. Reactive systems: collision-activation at $Cr/NC(CH_2)_{11}S/Au$ and $Cr/CH_3(CH_2)_{17}S/Au$ interfaces

Since more carbide is formed in both of the $Cr/CH_3(CH_2)_{17}S/Au$ and $Cr/NC(CH_2)_{11}S/Au$ systems during a deposition at a substrate temperature T than when the deposition occurs at a lower temperature followed by heating to temperature T, the collision of the Cr atoms with the surface must play a role in the reaction. Molecular beam studies in which organic molecules collide with metal surfaces have shown that deformation of bonds of the incident molecule is required for dissociative chemisorption to take place, and that sufficient bond deformation can be accomplished through impact, internal vibrational excitation, or both [36]. We believe that enhanced carbide formation during Cr deposition may result from deforming C–H bonds by collisional energy transfer from the incident Cr atoms. The effectiveness of Cr–SAM collisional energy transfer may be diminished at low temperature because the SAM is highly ordered and better able to distribute the collision energy among several neighboring chains.

The question of the role of the collisional bond deformation is related to how quickly the energy of the incident Cr atoms is accommodated by the SAM and to whether the Cr atoms are able to diffuse across the surface prior to bonding to the SAM or to other Cr atoms to form clusters; the amount of exothermic heat from the carbide forming reaction may also be a factor. If the Cr adatoms diffuse across the surface to a significant degree, then the size of Cr clusters should be smaller for deposition at lower substrate temperatures. Thus, greater Cr surface area would be available, and greater carbide peak heights would result from low temperature deposition followed by post-deposition heating, assuming the cluster sizes are unchanged by heating. Because we observe the opposite behavior implies that this scenario is incorrect, and suggests that a greater Cr surface area is available and smaller clusters are formed when the deposition is carried out at higher temperatures. Such an anomalous temperature dependence of cluster sizes is consistent with a collision-enhanced activation mechanism that becomes less effective as the sample temperature is lowered and the SAM packing is improved, allowing increased Cr adatom diffusion and larger clusters to form at lower temperatures.

For the Cr atoms emitted by our source at an estimated temperature $T_{source} = 1675 \pm 50$ K, the mean kinetic energy is $kT_{source} = 0.142 \pm 0.004$ eV/atom or 3.3 ± 0.1 kcal/mol. In a collision-enhanced process, most of the carbide would be formed from the 37% of the Cr atoms with energy greater than the mean. At 373 K on the CN SAM, N_{carb}/N_{Cr} is approximately 0.3. Assuming for simplicity 1 : 1 Cr–carbon stoichiometry, then 30% of Cr atoms bonded to carbide carbons is in fair agreement with 37% of Cr atoms having kinetic energy greater than kT_{source}. Thus, our data are consistent with a rather small activation barrier of about 3 kcal/mol.

The exothermic heat of reaction may also have an affect on the product distribution. In both the Cr(CN) and Cr–C (carbide) reactions, heat is transferred to the system as each Cr atom assumes a lower free energy state. The additional energy may be accommodated by the surface in the form of thermally induced motion of the outermost methylene end groups in the SAM. The additional disorder of this part of the SAM permits partial penetration of Cr deposits into the SAM. Because the greatest disorder of the SAMs is already occurring at the highest temperature (373 K), the additional disorder results from both the exothermic heat and the greater extent of reaction at the highest temperatures. When Cr deposition is at the lowest temperatures, the degree of penetration by Cr into the SAM is less than at higher temperatures because of the greater order in the SAMs. The heat of reaction will have a lesser effect because both the extent of reaction and the disorder of the SAMs are less. More carbide is formed at higher temperatures because a thermally-induced disorder allows depositing Cr atoms access to several of the outer methylene groups of the SAM; it is not clear from our data if the heat of reaction contributes significantly to increasing the disordering of the SAMs, even though it is plausible that it could.

To test the kinetic energy-dependence of carbide formation in Cr/SAM systems, the normal component of the Cr atom kinetic energy can be varied by means of the source temperature and the angle of incidence of the Cr flux. Studies by scanning tunneling microscopy, atomic force microscopy, or ISS of the degree of dispersion of Cr on SAMs as a function of deposition temperature can provide an additional test of the proposed carbide-forming mechanism.

The collision-enhanced activation of carbide formation has implications for the formation of well-bonded Cr/polymer interfaces. Chromium/polymer adhesion should be improved by depositing with high kinetic energies on to a warm ($\sim 100\,^\circ$C) polymer substrate, e.g. as found in sputter-deposition of metal films, when compared with lower energy deposition methods at lower substrate temperatures. The degree of improvement may, of course, be compromised if the higher polymer temperatures alter the properties of the interphase region and result in a cohesive failure mechanism.

3.7. Reactive system: Cr 2p *and C* 1s *XPS spectra for Cr/HOOC(CH$_2$)$_{15}$S/Au*

In general, the Cr $2p$ line shapes of Fig. 5 for Cr deposited onto Au, HOOC(CH_2)$_{15}$S/Au, or CH_3(CH_2)$_{17}$S/Au are asymmetric and may comprise more than one component. We are not able to satisfactorily decompose the spectra into metal, oxide, and carbide components by curve-fitting and therefore use the line shape maxima to identify approximate binding energies of the major components. A Cr $2p3/2-1/2$ splitting of

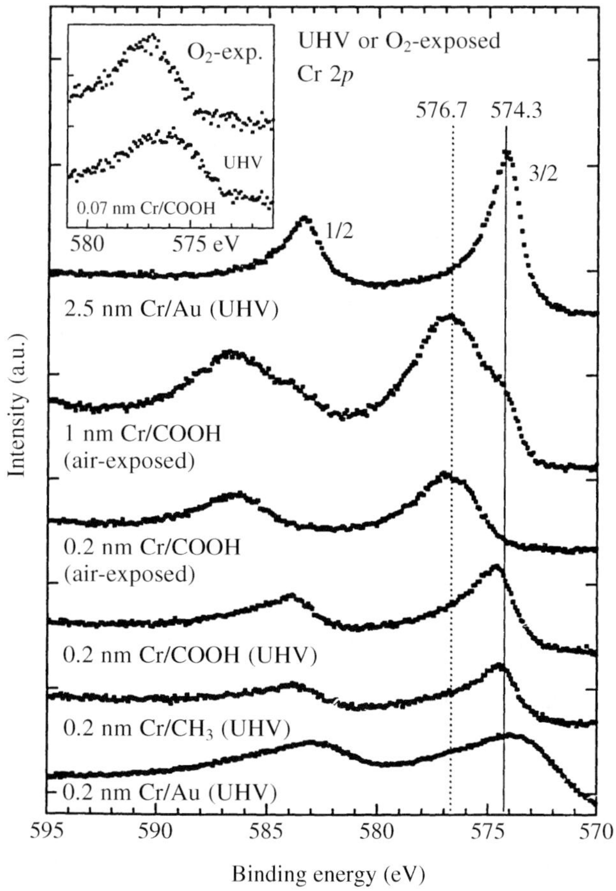

Figure 5. Cr $2p$ XPS data for the 0.2 nm Cr/Au, Cr/CH_3(CH_2)$_{17}$S/Au, and Cr/HOOC(CH_2)$_{15}$S/Au interfaces in UHV (lower), for air-exposed 0.2 nm and 1 nm Cr/HOOC(CH_2)$_{15}$S/Au interfaces (middle), and for the 2.5 nm Cr/Au interface in UHV (top). The pass energy was 200 eV for the bottom spectrum and 50 eV for all others. The inset shows spectra for the 0.07 nm Cr/HOOC(CH_2)$_{15}$S/Au interface in UHV and after oxygen-exposure. The pass energy was 50 eV. The data have not been smoothed.

9.2 eV for 2.5 nm Cr films on Au (top, Fig. 5) shows that *metallic Cr is produced* by our deposition procedure. In this spectrum, we observe Cr $2p3/2$ at 574.3 eV, in good agreement with the *Handbook of X-ray Photoelectron Spectroscopy* (574.4 eV) [37]. For 1 nm Cr/HOOC$(CH_2)_{15}$S/Au exposed to air, the Cr $2p3/2$ binding energy of 576.7 eV is consistent with Cr(III) but not with Cr(VI) [37]. For the case of 0.07 nm Cr/HOOC$(CH_2)_{15}$S/Au, the Cr $2p3/2$ position (inset of Fig. 5) and $2p3/2$–$1/2$ splitting (not shown) change upon oxygen-exposure: the Cr $2p3/2$ position shifts from 576.4 eV to a higher binding energy of 577.2 eV, while splitting increases from 9.6 to 9.7 eV, which differs slightly from the 9.8 eV splitting expected for Cr_2O_3. This is accompanied by a narrowing of the $2p3/2$ line shape from 4 to 3 eV. For this coverage, this indicates that Cr on the HOOC surface is not fully oxidized in UHV and may be composed of more than one Cr chemical state. There is not much difference between Cr $2p$ spectra for 0.2 nm Cr/HOOC$(CH_2)_{15}$S/Au and 0.2 nm Cr/$CH_3(CH_2)_{17}$S/Au samples in UHV.

For 0.2 nm of Cr/Au (bottom curve, Fig. 5), the Cr $2p$ peaks are slightly shifted to a lower binding energy compared to bulk metal. In Fig. 5 of [11], an approximately 1 eV Cr $2p3/2$ shift to a lower binding energy is shown for 0.05 nm Cr/Au compared to bulk Cr. This negative shift diminishes rapidly with increasing Cr film thickness, and may result, in part, from a surface core level shift observed for atoms at metal surfaces and for small metal clusters on conducting substrates [38]. The observed positive shifts (from the metallic binding energy) that decrease with increasing coverage indicate a lack of penetration of Cr through the HOOC$(CH_2)_{15}$S/Au SAM.

Alloying is another possible cause of the low binding energy Cr $2p$ shift for Cr/Au. We estimate a 5% atomic solubility for Cr in Au [39], but because of a 1.2 eV activation energy for Cr diffusion into Au [40], we expect alloying to be kinetically limited at 300 K. Opila *et al.* observed the Au $4f$ shoulder shifted by 1 eV to higher binding energy because of alloying between Ti overlayers and Au substrates [9]. We observe a negligible, 0.1 eV, increase in the Au $4f7/2$ core level for Cr/Au, which is consistent with little or no Cr–Au alloy formation.

In Fig. 6 we show C 1*s* spectra for Cr overlayer thicknesses up to 1 nm. The high binding energy component (assigned to O=C—O carbons) at 289.1 eV is weak and of gradually decreasing intensity but does not identify a specific interaction of Cr with the oxygens in COOH. The low binding energy shoulder at 283.3–283.0 eV is attributed to a Cr–carbide reaction as given above, as well as to the sample temperature dependence of this reaction. There is also a component at 286.4 eV that has been observed by other groups and is assigned to the β-carbon atom (adjacent to the carboxyl carbon) [25, 41].

3.8. Reactive system: O 1s *spectra of Cr/HOOC$(CH_2)_{15}$S/Au in UHV and after oxygen exposure*

In Fig. 7, we show O 1*s* results for Cr/HOOC$(CH_2)_{15}$S/Au interfaces in UHV; data for detailed binding energies, and peak heights, widths, and areas are available [11]. For 0.0 nm and 0.04 to 0.07 nm Cr/HOOC$(CH_2)_{15}$S/Au, the line shapes are fairly symmetrical, but grow increasingly asymmetrical for higher coverages. For the

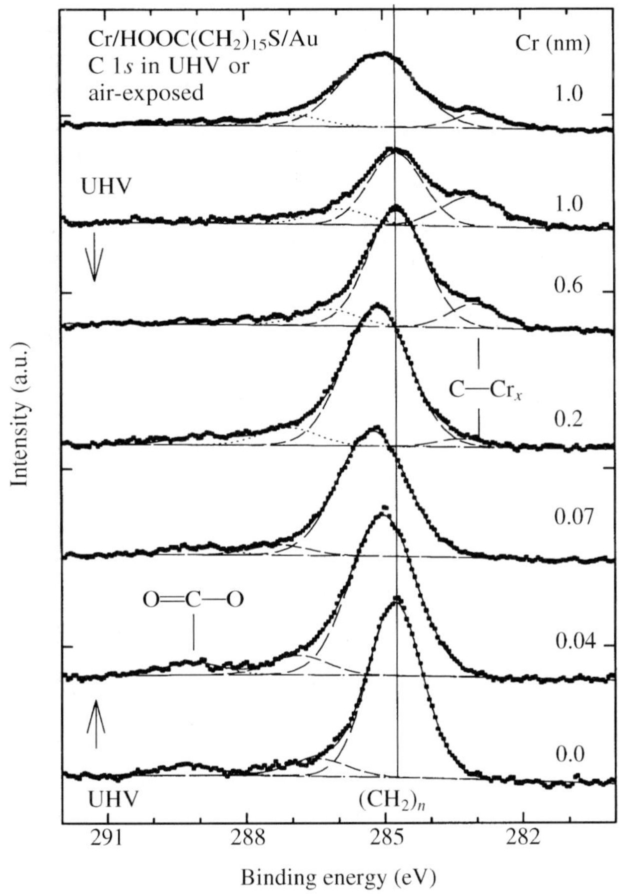

Figure 6. C 1*s* XPS results for Cr/HOOC$(CH_2)_{15}$S/Au interfaces in UHV with average Cr overlayer thicknesses of 0, 0.04, 0.07, 0.2, 0.6, and 1.0 nm. The upper spectrum shows reduced intensity in the *c*. 283 eV component because of exposure of the 1 nm Cr/HOOC$(CH_2)_{15}$S/Au sample to air. The data have been smoothed by a 10% polynomial interpolation program.

bare SAM, the O 1*s* spectrum was fitted using two components: P1 corresponds to oxygen in C−OH, and P2 represents the oxygen in C=O (Fig. 7, 0 nm). P1 and P2 are of approximately the same intensity and width and have binding energies that are separated by 1.4 eV, in agreement with [41].

For 0.04 nm Cr/HOOC$(CH_2)_{15}$S/Au, the overall line shape is narrowed and roughly between the original positions of P1 and P2, but nearer the original C=O feature (P2). We have fitted the data with two peaks, P3 and P4. Peak P3 is assigned to the oxygen in a Cr−COOH interaction. The predominance of P3 at 0.04 nm coverage indicates that a few unreacted COOH groups remain. We interpret the single 1.9 eV wide peak at 0.04 nm coverage to indicate that a few unreacted COOH groups remain. We interpret the single 1.9 eV width of P3, which is only about 0.15 eV wider than the original C−OH and C=O components, to indicate that two oxygens of

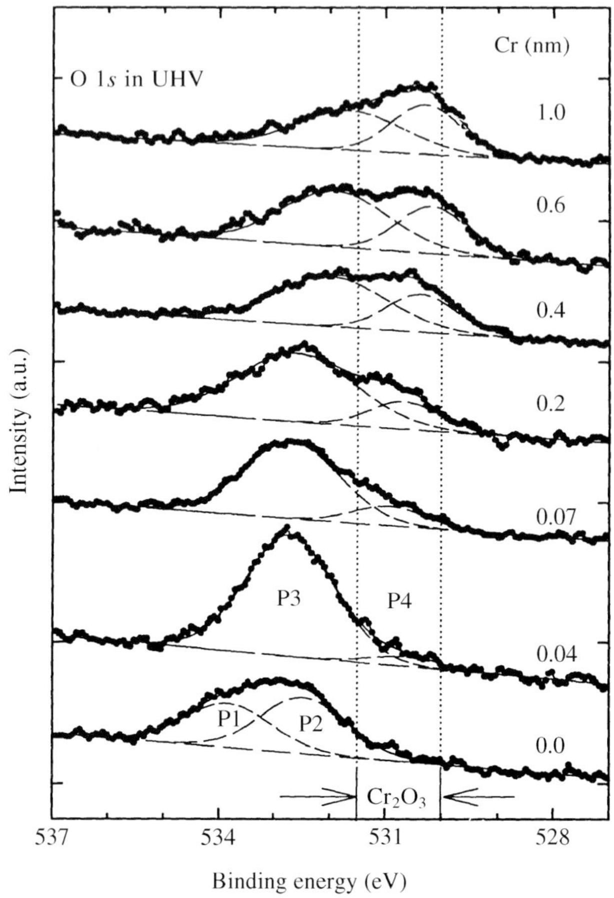

Figure 7. O 1*s* XPS results for Cr/HOOC$(CH_2)_{15}$S/Au interfaces in UHV with average Cr overlayer thicknesses of 0, 0.04, 0.07, 0.2, 0.6, and 1.0 nm. The parameters P3 and P4 of the fitted curves are given in Table 1 of [11c]. The data have been smoothed by a 10% polynomial interpolation program.

a single COOH endgroup are interacting with Cr and are indistinguishable by our instrument. We speculate that hydrogen evolution might result, as was concluded for Cu/HOOC$(CH_2)_{10}$S/Au [1], but no COOCr unidentate is formed.

To estimate an upper limit on the fraction of unreacted groups, we tried fitting with four peaks: two components with the original P1 and P2 peak positions and widths and of variable but equal heights, and two additional components P3 and P4 with free parameters as before. This yielded a reduction in the contribution of the 532.9 eV peak to only 81% of the total, and a 11% contribution from unreacted COOH. This four-peak fit may be somewhat more realistic, but we think that P3 represents the majority state of oxygen at this coverage. Contributions of the order of 10% are within the experimental error.

Based on the bulk density of a bcc, close-packed Cr film and a 4.7 chain/nm^2 packing density of the SAM, there is a Cr:O ratio of 0.35. With greater SAM surface area because of roughness, these ratios could be smaller. The onset of P4 in the O 1*s* spectra (Fig. 7) is roughly correlated with coverages of more than one Cr atom per COOH group, supporting the idea that, initially at least, P3 results from the interaction of Cr atoms with individual COOH groups.

At 0.6 nm and 1.0 nm coverages, P4 at 530.3 eV coincides well with literature values for Cr_2O_3 of 530 eV to 531.5 eV (as marked in Fig. 7) [37, 42–45]. We assign this feature to oxygen atoms that have been incorporated into a bulk-like oxide that is similar to Cr_2O_3 but perhaps not a fully-stoichiometric Cr_2O_3.

3.9. Reactive system: C 1s, *O* 1s, *and Cr* 2p *line shapes for Cr/$H_3COOC(CH_2)_{15}S$/Au*

The Cr 2*p* peak positions for Cr/$H_3COOC(CH_2)_{15}S$/Au in UHV approach those of metallic Cr for the higher coverages studied, as was also observed in the case of Cr/$HOOC(CH_2)_{15}S$/Au. For example, when 1 nm of Cr on $H_3COOC(CH_2)_{15}S$/Au is exposed to oxygen, the Cr 2*p*3/2 binding energy of 576.7 eV is consistent with Cr(III) but not with Cr(VI) [46].

For lower coverages in UHV, the lack of low binding energy shifts is consistent with a lack of penetration for Cr into $H_3COOC(CH_2)_{15}S$/Au [38]. The observed positive shifts with respect to metallic Cr are also the result of the chemical interaction of the Cr with the $COOCH_3$ oxygen atoms. However, the Cr 2*p*3/2 peak position and multiplet splitting change upon oxygen exposure for low coverages of Cr/$H_3COOC(CH_2)_{15}S$/Au, indicates the Cr is not fully oxidized in UHV. As for Cr/$HOOC(CH_2)_{15}S$/Au, information about the chemical state of Cr at low coverages is not easily obtained from the Cr 2*p* spectra.

The C 1*s* results for the bare $H_3COOC(CH_2)_{15}S$/Au SAM are shown in Fig. 8. A high binding energy component initially at 289.1 eV is assigned to O=C−O carbons. Initially, the area of this component is 6% of the total C 1*s* area, an amount that corresponds well to the simple stoichiometry (1 of 17 C atoms in $H_3COOC(CH_2)_{15}SH$ but slightly lower than that expected after considering the inelastic mean free path of the C 1*s* photoelectron escaping from each C in the $H_3COOC(CH_2)_{15}SH$ (8%). For the bare SAM, a component is also found at 286.8 eV that is assigned to carbons in OCH_3 and which provides 13% of the total area. It is not known why this component is so large, but it may include a contribution from the methylene part of the line shape (from vibrationally-induced asymmetry) or from the carbon atoms adjacent to the S headgroup of the molecule. With the first increment of Cr, a significant loss in intensity occurs for the 289.1 eV XPS peak, consistent with a reaction with the SAM endgroups. A low binding energy shoulder at *c*. 283.0 eV grows for average Cr thicknesses of 0.4 to 1.0 nm, which is attributed to a Cr−carbide reaction that is not the main focus of this work. For Cr/$H_3C(CH_2)_{17}S$/Au and Cr/$NC(CH_2)_{11}S$/Au interfaces, the dependence of this reaction on the sample temperature was discussed above [2]. The position of the main C 1*s* peak resulting from the 15 methylene carbon atoms shifts to a higher binding energy by 0.4 eV at about 0.07 nm Cr and then back to approximately its original position (284.9 eV) over the range of coverages up to 1.0 nm. These shifts have been observed before but are not understood [5–11, 37].

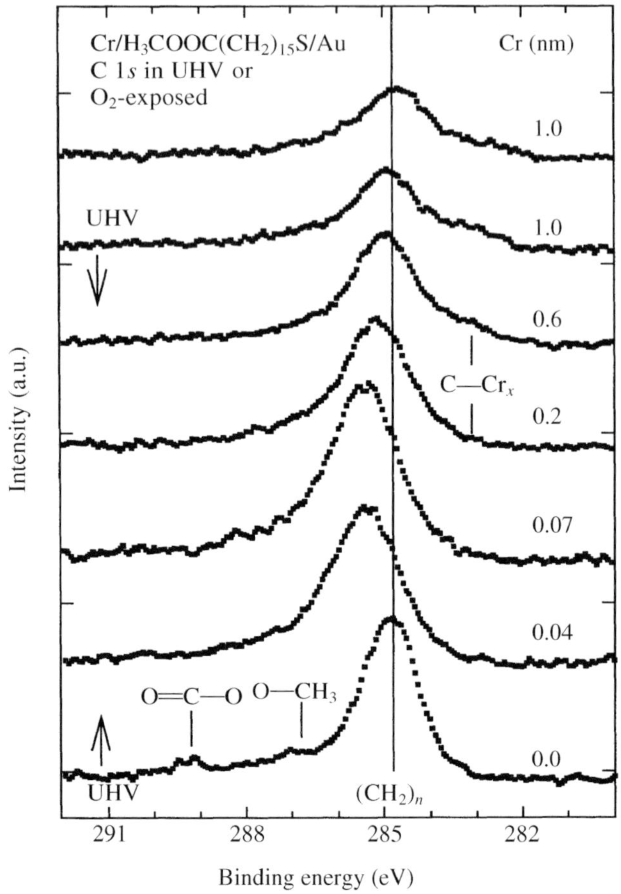

Figure 8. C 1*s* XPS results for Cr/H$_3$COOC(CH$_2$)$_{15}$S/Au interfaces in UHV with average Cr overlayer thicknesses of 0, 0.04, 0.07, 0.2, 0.6, and 1.0 nm. The upper spectrum shows reduced intensity in the *c.* 283 eV component because of exposure of the 1 nm Cr/H$_3$COOC(CH$_2$)$_{15}$S/Au sample to air. The data have been smoothed by a 10% polynomial interpolation program.

Curve-fitting the O 1*s* spectrum in Cr/H$_3$COOC(CH$_2$)$_{15}$SH/Au, which was done similarly to that for Cr/HOOC(CH$_2$)$_{15}$SH/Au, yields two nearly equal components at 532.4 eV (C=O) and 534.0 eV (C—O—CH$_3$) (P1 and P2, respectively) that are consistent with the presence of the corresponding O=C—OCH$_3$ endgroup at the surface. As the average thickness of the Cr deposit increases, the O 1*s* spectra of Fig. 9 show line shapes that are dominated by a central peak (P3) at 532.5–532.1 eV (see Table 1 in [11c]). In addition, one smaller peak has a lower binding energy of 531.0–530.5 eV (P4) and another has a higher binding energy of 534.8–534.1 eV (P5). With increasing Cr coverage, the low binding energy peak increases in area, while the high binding energy peak area decreases and disappears into the noise. The high binding energy peak (P5) is not included in the fits for 0.2 to 1.0 nm Cr coverage, because its inclusion in the 0.2 nm fit caused an abrupt and unlikely shift of P3 to a lower

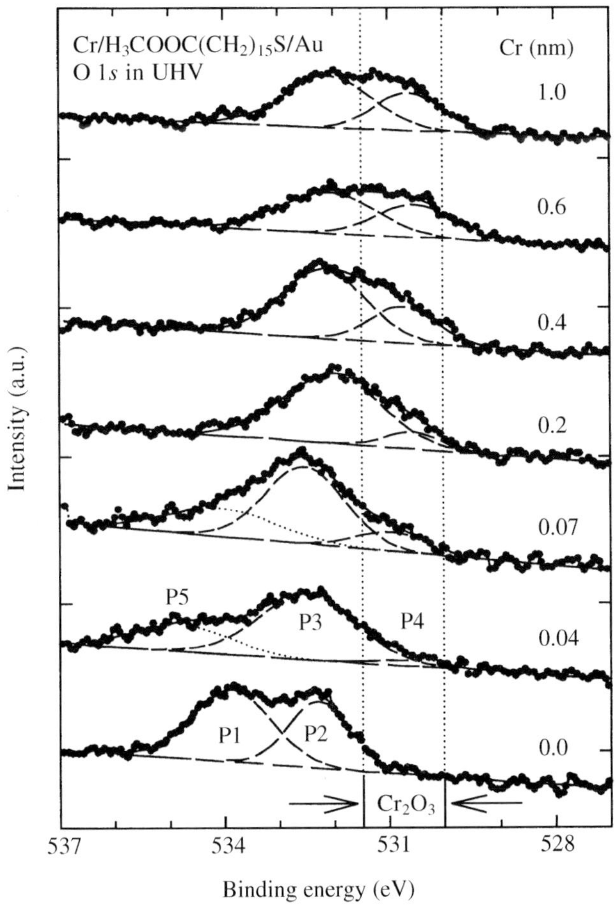

Figure 9. O 1*s* XPS results for $Cr/H_3COOC(CH_2)_{15}S/Au$ interfaces in UHV with average Cr overlayer thicknesses of 0, 0.04, 0.07, 0.2, 0.6, and 1.0 nm. The parameters of the fitted curves P3 and P4 are given in Table 1 of [11c]. The data have been smoothed by a 10% polynomial interpolation program.

binding energy followed by a high binding energy shift of P3 for 0.4 nm coverage. Although the line shape at 0.2 nm coverage has noticeable asymmetry that demands the higher binding energy component, including it in the fitting causes a change in line width of the central peak from 1.6 eV at 0.1 nm Cr coverage to 2.1 eV at 0.2 nm coverage. This abrupt change in line width is an artifact of fitting with only 2 or 3 components, when it is probable that a larger number of different chemical states exist. The components in Table 1 of [11c] should be taken as an approximation of the major types of chemical states, with each component comprising minor variations within its type.

Based on a simple electrostatic potential model for chemical shifts, the presence of a high binding energy peak with a binding energy (534.4 ± 0.4 eV) above that of the original $C-O-CH_3$ component (534.0 eV) indicates that a new oxygen species is formed with a reduced electronic charge compared to that of the $C-O-CH_3$ oxygen of the bare SAM. Based on the electronegativities of O and Cr, on the other hand, we expect electronic charge to be transferred to the endgroup oxygens rather than from them. Thus, this feature of the data is not understood and remains unassigned. (A similar three-component line shape containing a high binding energy component was observed in preliminary studies of 0.04 nm Cr on an HO-terminated SAM.) Our interpretation of the central peak and the low binding energy peak follows.

We interpret the three O 1*s* components in Fig. 9 by comparison with the results for Cr/HOOC$(CH_2)_{15}$SH/Au that showed only the central peak P3 and the low binding energy peak P4 (at approximately the same binding energies as found here) [11]. At the lowest Cr coverage (0.04 nm), the central component at 532.5 eV is assigned to a bonding interaction between Cr atoms and $COOCH_3$ groups, as will be discussed later. At 0.04 nm coverage, there are on average about 0.7 Cr atoms per endgroup, but there are 20 Cr atoms per endgroup at a coverage of 1.0 nm Cr. Over the range of coverages studied, it is likely that the initial interactions are between endgroup oxygens and isolated Cr atoms or with very small Cr clusters. At higher Cr coverages, the interaction is between endgroups in contact with (or incorporated into) relatively large islands of Cr. The low binding energy peak at 530.5 eV, which provides 35% to 40% of the total intensity at 0.6 nm and 1.0 nm coverages, is assigned to oxygens that have been incorporated into a bulk-like oxide that is similar to Cr_2O_3 but perhaps not a fully stoichiometric Cr_2O_3. As for Cr/HOOC$(CH_2)_{15}$SH/Au, this assignment is supported by literature values for the binding energy of the O 1*s* peak in Cr_2O_3 of 530–531.5 eV [42–47].

The central peak accounts for roughly 70% of the total O 1*s* peak intensity at 0.04 nm coverage, compared with 80% to 95% that was obtained for Cr/HOOC$(CH_2)_{15}$SH/Au (depending on the curve-fitting procedure) [11]. We conclude that less of this species is formed at this coverage on the $COOCH_3$ SAM compared with Cr on the HOOC$(CH_2)_{15}$SH/Au SAM. (The difference results from the additional peak P5 found when using the methyl ester. Note also that the presence of P5 precludes the modeling of the number of unreacted endgroups that was tried in the case of the HOOC$(CH_2)_{15}$SH/Au SAM, in which P1 for the bare SAM had little overlap with the line shape at 0.04 nm coverage [11].)

For higher average overlayer thickness (0.2 to 1 nm), a negative 0.4 eV shift of the central peak indicates a change in the average Cr–O bonding of this species. The P3–P4 separation is 2 eV at 0.04 nm Cr coverage, and diminishes to 1.5 eV at 1.0 nm coverage. This splitting is similar to that found in a study of Cr thin films that were briefly exposed to air prior to XPS analysis, in which O 1*s* peaks attributed to a Cr_2O_3-like oxide and to OH groups on the surface oxide were separated by 2 eV [14]. Another study of the chemisorption of carbon monoxide on Cr(111) identified O 1*s* binding energies of 531.9 and 532.6 eV, respectively, for side-on and terminal bonding of CO at 100 K [47]. Thus, the low binding energy shift of P3 with increasing coverage might also be interpreted as a conversion from an end-on Cr–O–C interaction (Cr interacts primarily with the O) to a side-on Cr–(OC) interaction in which Cr interacts jointly with CO. A bonding geometry of Cr atoms with two or more O atoms has not been identified in the literature, but is another possible assignment for P3.

3.10. Adsorption of oxygen for UHV- and oxygen-exposed Cr/ROOC interfaces ($R = H$ or H_3C)

The O 1*s* to C 1*s* peak height ratios from survey scan data (adjusted with sensitivity factors) have been given for Cr/HOOC$(CH_2)_{15}$SH/Au samples in UHV and after oxygen exposure [11]. A ratio of roughly 0.13 is obtained, as shown by Fig. 10, and it remains nearly constant with increasing overlayer thickness of Cr up to 0.6 nm. In contrast, the results for samples exposed to air or oxygen show a steady increase in the O 1*s* to C 1*s* ratio that reaches a value of 1 at 1 nm Cr coverage. These data were taken to estimate if oxygen adsorption from the residual oxygen-containing gases in our UHV system affected the XPS results for the Cr/HOOC$(CH_2)_{15}$SH/Au and Cr/H_3COOC$(CH_2)_{15}$SH/Au interfaces studied. The uptake in air is as expected, and is much greater than for the samples residing only in our UHV chamber. We calculated that the O 1*s* to C 1*s* XPS peak area ratio for these SAMs should be 0.16. Our measured value is about 0.13, which agrees well with our calculation and should be lower by about 15% because of the X-ray damage that can occur over the duration of the experiment. Thus, we found no statistically significant evidence that contamination of the Cr/SAM interfaces results from any residual oxygen-containing gases in our experimental system.

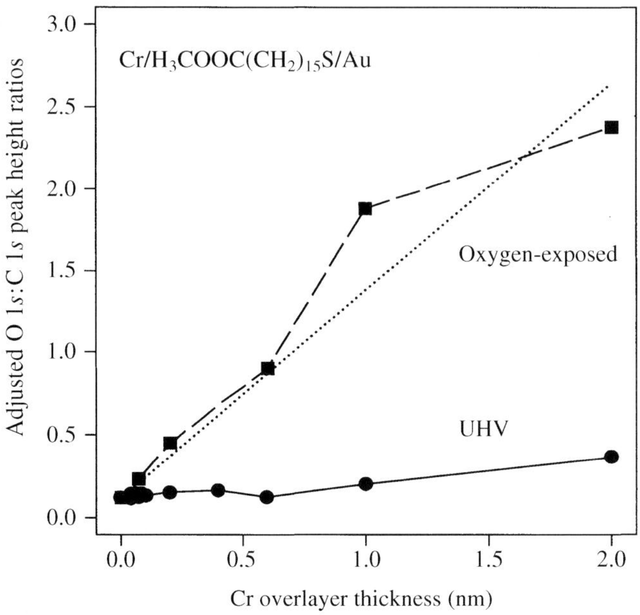

Figure 10. Ratios of O 1*s* to C 1*s* adjusted X-ray photoelectron peak heights from survey scan data taken in UHV (circles) and after oxygen exposure (squares). The solid and dashed lines are guides to the eye and the dotted line is a linear fit. The survey scan peak heights were measured relative to a linear background. They were then adjusted by using element- and core-level specific sensitivity factors determined by the manufacturer of the LHS-10 (in the constant pass energy mode) and the corresponding photoelectron kinetic energies to account for the energy dependence of the analyzer sensitivity in the constant transmission mode. An uncertainty of $\pm$ 14% is estimated for these ratios.

3.11. Modeling of the distribution of Cr at Cr/ROOC interfaces

Figures 11 and 12 show the Cr 2*p*, Au 4*f*, C 1*s*, and O 1*s* intensities as a function of Cr coverage that were determined by curve-fitting the survey scan data for $Cr/HOOC(CH_2)_{15}SH/Au$ and $Cr/H_3COOC(CH_2)_{15}SH/Au$, respectively. The experimental intensities are scaled by constant factors to permit direct comparison with the rates of attenuation we calculated for a model in which a uniform overlayer of Cr resides on top of the SAM and the known inelastic mean free paths for Au, Cr, C, and O are used [48]. In this model, the calculated intensities, I, are normalized to bulk values, I_B.

For the $Cr/HOOC(CH_2)_{15}SH/Au$ in Fig. 11, the O 1*s* attenuation is in fair agreement with the model for 0.04 to 0.4 nm Cr coverages. At the 0.6 and 1.0 nm Cr overlayer coverages, however, the O 1*s* intensities are roughly equal to that of the bare SAM. The higher oxygen intensities for the 0.6 and 1.0 nm coverages are consistent with adsorption of an oxygen-containing species, perhaps up to 0.2 monolayers as has been estimated [11]. This is to be expected because these coverages were obtained by adding a second increment to the 0.04 and 0.4 nm $Cr/HOOC(CH_2)_{15}SH/Au$ samples. For the lower coverages, however, the O 1*s* intensities do not indicate a significant effect of any adsorption and/or reaction with an oxygen-containing species.

At 1.0 nm Cr coverage, the attenuation of the Au 4*f* intensity is slightly greater than that for the model, while the attenuation of the C 1*s* intensity is too low compared

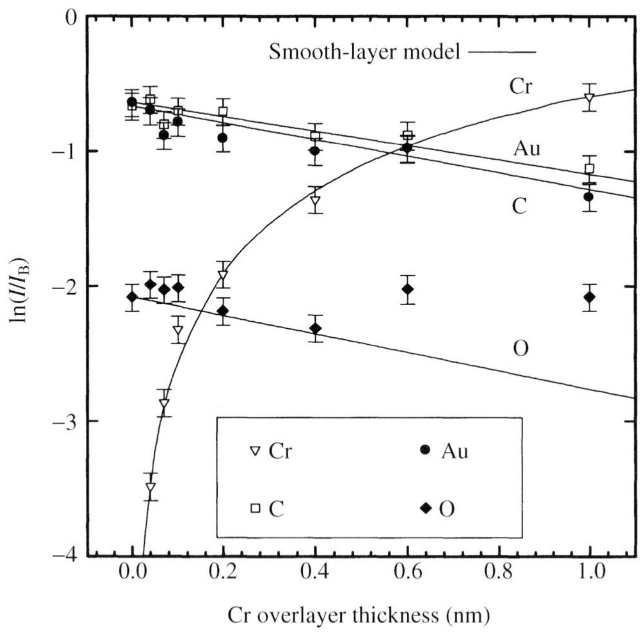

Figure 11. The logarithm of the ratio of the normalized calculated intensities, I, to the bulk values, I_B vs. Cr overlayer thickness. The peak areas, which were determined by curve fitting the XPS survey scan data, are shown for Cr (open triangle), C (open square), Au (filled circle), and O (filled diamond) for Cr overlayers of 0 to 1.0 nm average thickness on $HOOC(CH_2)_{15}S/Au$. The solid lines show the attenuation for the smooth-layer model. The peak areas have error bars of $\pm$ 10%.

with the model. The latter deviation may result from adsorption of a slight amount of carbon species, while the former may be caused by both oxygen and carbon adsorption.

The probable location and degree of clustering of the Cr deposit are addressed by the C and Cr data. Penetration and/or extreme clustering of the Cr would result in decreased C attenuation, which is not observed. Clustering would also affect the comparison of the Cr model curve with the data (that are scaled to agree with the model at 2 nm coverage). Extreme clustering would cause the data to fall below the model curve for the lower coverages, which is not observed. Taken together, the comparisons of each elemental data set with its corresponding model attenuation curve support little clustering or penetration of the Cr deposit. This is in contrast to the result of Opila *et al.* [9], who found that the highly reactive Ti/$H_3COOC(CH_2)_{15}SH$/Au system was extensively clustered over a range of 0 to 2 nm average Ti overlayer thickness.

In Fig. 12, the attenuation plots for Cr/$H_3COOC(CH_2)_{15}SH$/Au follow the same trends and these support our interpretation that the distribution of the Cr deposit on the two SAMs is similar. As for Cr/$HOOC(CH_2)_{15}SH$/Au, there are two data points for O 1*s* that lie significantly above the model attenuation curve. The data at 0.6 and 1.0 nm were obtained by first metallizing to 0.04 and 0.4 nm, respectively, and then adding a second increment to achieve the final coverage. The higher oxygen

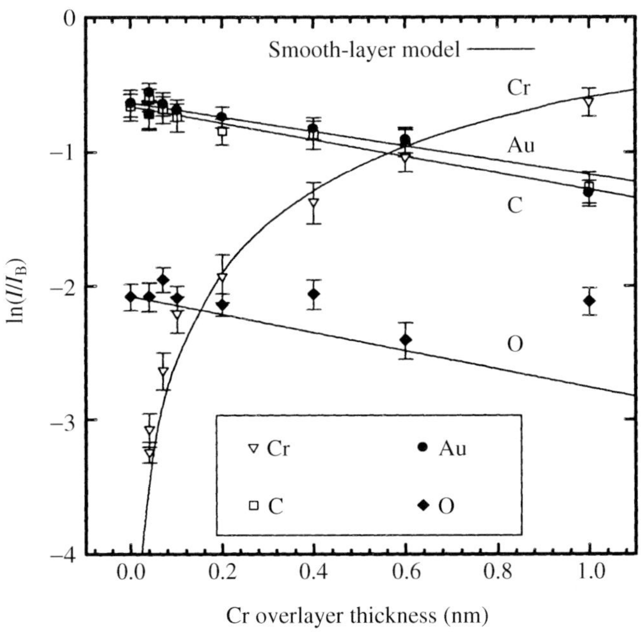

Figure 12. The logarithm of the ratio of the normalized calculated intensities, I, to the bulk values, I_B vs. Cr overlayer thickness. The peak areas, which were determined by curve fitting the XPS survey scan data, are shown for Cr (open triangle), C (open square), Au (filled circle), and O (filled diamond) for Cr overlayers of 0 to 1.0 nm average thickness on $H_3COOC(CH_2)_{15}S$/Au. The solid lines show the attenuation for the smooth-layer model. The peak areas have error bars of $\pm$ 10%.

intensities for the 0.4 and 1.0 nm coverages occurred on the same sample and may have been caused by some unknown difference in the experimental conditions, resulting in increased oxygen adsorption from the ambient vacuum [11].

3.12. Reactive system: concluding remarks about Cr/ROOC from XPS data

In this study of Cr interacting with relatively simple SAMs, although XPS has been used with great care within the limitations of our apparatus, we do not obtain a definitive description of the initial Cr—ROOC bond, for R = H or CH_3. Molecular orbital calculations of Cr interacting with ROOC groups are needed for comparison with the experimental data. Experiments with Cr/poly(acrylic acid) polymer films would also assist in clarifying the interaction.

We can make some comparisons with results from the Cr/polyimide literature. As an example, our O 1*s* spectra and those of [20] show common characteristics and dilemmas, even though the oxygen functionalities are different. The original O 1*s* components broaden and decrease in intensity, while a new peak appears at low binding energy that is Cr_2O_3-like and whose source may include oxygen from the substrate and oxygen from adventitious oxygen-containing species. Because of the broadening, it is not clear whether the evolution from a low coverage Cr—O species to the Cr_2O_3-like species is abrupt or involves multiple stages. Nevertheless, at intermediate coverages, incrementally-added Cr would appear to have two possibilities: to join an intermediate Cr—O species to form Cr_2O_3, or to react with carbon to form a carbide. From comparison between these two studies of the amounts of carbide, Cr_2O_3-like oxygen species, and unreacted oxygen or nitrogen species as a function of coverage, we suggest that carbide forms more rapidly for Cr/polyimide because of fewer available O or N groups, more available C atoms [20], or the absence of a Cr/C=O interaction [49]. For the $HOOC(CH_2)_{15}S$/Au and $H_3COOC(CH_2)_{15}S$/Au SAMs, the oxygen groups are more accessible than the carbons of the methylene chain, but for the polyimide films, reactivity and not structure would appear to be the primary factor. Here we have an opportunity, by using the layered structure of a SAM (possibly with buried rather than exposed oxygen groups), to control the accessibility of carbon and oxygen species, and observe the relative effects on the amounts of reactants formed.

3.13. Important concerns for future M/SAM studies

A discussion of several important and crucial questions for M/SAM studies is included in a recent review [10]. These include: (a) are contaminants present in or on the film; (b) what kind of defects are in the SAM and at what density; (c) does the degree of penetration (Table 5 in [10]) increase with increasing SAM defect density; (d) does the SAM disorder significantly even for low metal overlayer coverages; and (e) what problems are caused by X-ray and e-beam exposures? While none of the questions (a)–(e) seems unmanageable [10], addressing them should be a part of future M/SAM studies.

4. CONCLUDING REMARKS

In the research to date, the interactions of several M/SAM systems have been identified, along with qualitative relationships among the effects of chemical reactivity, temperature, and penetration of a metal/organic interface. The initial research on M/SAM systems has served as a test of their use as model systems, and of whether M/SAM research will have benefits in applied areas, for example, real metal/polymer interfaces. The results described above are clearly more readily interpreted and more detailed than most of their counterparts in the metal/polymer interface literature. Based on the presence of similar reactants at a metal/polymer interface, such as OFGs and metal atoms, this research approach has succeeded in identifying factors that affect the initial bond between metal and polymer as the interface is formed in vacuum.

For quantitative understanding, further more complete characterization is needed to develop (1) detailed, molecular orbital descriptions of new metal–organic bonds in a unique environment (because of the difference between bringing individual molecules to a metal surface compared with bringing individual metal atoms to an organic surface); (2) diffusion processes and rates for metals on organic surfaces; and (3) penetration processes and rates for metals into organic surfaces. Control and characterization of defects in the SAMs are necessary to understand the relative role of intrinsic vs. extrinsic effects in the chemical and physical processes in M/SAM systems.

Based on the initial results, a continued broad effort is warranted for studying up to 10 or 15 metals, metal oxides, and semiconductors with about 10 different OFGs on SAMs to evaluate the full potential of the approach and of this subfield of surface science. This should lead to the identification of submatrices targeted at specific classes of materials and applications.

Finally, the results of M/SAM studies may be applied to the understanding of metal/polymer interfaces, and for this reason, M/SAM interfaces are often described as 'model' metal/polymer interfaces. Further relevance to inorganic/organic interfaces actually used in practice and to interfacial reaction mechanisms under ambient conditions can be gained by exposure of inorganic/SAM interfaces to different atmospheres, temperature extremes, radiation, and chemical reagents. Inorganic/organic interfaces with well-defined and stable properties are needed in advanced devices [50–52]. Research to identify and modify reactions at inorganic/SAM interfaces will provide the fundamental understanding to fill that need.

Acknowledgements

The authors are pleased to acknowledge support by the US Department of Energy under Contract DE-AC36-83CH10093 through the Director's Development Fund at NREL. We are especially grateful for the helpful discussions with D. L. Allara, P. Zhang, R. W. Collins, and D. E. King and to D. L. Allara and G. Whitesides for providing most of the alkanethiols we used in our experimental work. We thank D. L. Allara, P. Zhang, R. L. Opila, K. Konstadinidis, M. D. Porter, M. Grunze, H. G. Rubahn, L. H. Dubois, R. G. Nuzzo, M. J. Tarlov and P. Zhang for permission to use figures from their work [10], and R. W. Collins, G. Scoles, A. J. Bard,

J. E. Houston, K. H. Gray, R. W. Linton, S. M. Lindsay, G. N. Robinson, A. N. Parikh, K. Edinger and F. Eisert for providing copies of their publications.

REFERENCES

1. A. W. Czanderna, D. E. King and D. Spaulding, *J. Vac. Sci. Technol.* **A9**, 2607 (1991).
2. D. E. King, A. W. Czanderna and D. Spaulding, in: *Metallized Plastics 3. Fundamental and Applied Aspects*, K. L. Mittal (Ed.), pp. 149–161. Plenum, New York (1992).
3. D. R. Jung, D. E. King and A. W. Czanderna, *Appl. Surf. Sci.* **70/71**, 127 (1993);
   D. R. Jung, D. E. King and A. W. Czanderna, *J. Vac. Sci. Technol.* **A11**, 2382 (1993).
4. D. R. Jung and A. W. Czanderna, *Mater. Res. Soc. Symp. Proc.* **304**, 131 (1993).
5. D. R. Jung, A. W. Czanderna and G. C. Herdt, *J. Vac. Sci. Technol.* **A14**, 1779 (1996).
6. (a) G. Herdt and A. W. Czanderna, *Surf. Sci. Lett.* **297**, L109 (1993);
   (b) G. Herdt and A. W. Czanderna, *J. Vac. Sci. Technol.* **A12**, 2410 (1994);
   (c) G. Herdt and A. W. Czanderna, *J. Vac. Sci. Technol.* **A13**, 1275 (1995).
7. M. J. Tarlov, *Langmuir* **8**, 80 (1992).
8. P. Zhang, Ph.D. thesis, Department of Materials Science and Engineering, Pennsylvania State University (1993).
9. R. L. Opila, K. Konstadinidis, D. L. Allara and P. Zhang, *Surface Sci.* **338**, 300 (1995).
10. D. R. Jung and A. W. Czanderna, *Crit. Rev. Solid State Mater. Sci.* **19**, 1 (1994).
11. (a) D. R. Jung and A. W. Czanderna, *J. Vac. Sci. Technol.* **A12**, 2402 (1994);
    (b) D. R. Jung and A. W. Czanderna, *J. Vac. Sci. Technol.* **A13**, 1337 (1995);
    (c) D. R. Jung and A. W. Czanderna, *Appl. Surface Sci.* **99**, 161 (1996).
12. E. Sacher, J. J. Pireaux and S. P. Kowalczyk (Eds), *Metallization of Polymers*, ACS Symposium Series, No. 440. American Chemical Society, Washington, DC (1990) (and refs therein).
13. F. P. Netzer and M. G. Ransey, *Crit. Rev. Solid State Mater. Sci.* **17**, 397 (1992).
14. M. R. Albert and J. T. Yates, *The Surface Scientist's Guide to Organometallic Chemistry*. American Chemical Society, Washington, DC (1987).
15. J. M. Burkstrand, *J. Vac. Sci. Technol.* **20**, 440 (1982);
    J. M. Burkstrand, *J. Appl. Phys.* **50**, 1152 (1978);
    J. M. Burkstrand, *J. Appl. Phys.* **52**, 4795 (1982).
16. K. L. Mittal and J. R. Susko (Eds), *Metallized Plastics 1: Fundamental and Applied Aspects*. Plenum Press, New York (1989);
    K. L. Mittal (Ed.), *Metallized Plastics 2: Fundamental and Applied Aspects*. Plenum Press, New York (1991);
    K. L. Mittal (Ed.), *Metallized Plastics 3: Fundamental and Applied Aspects*. Plenum Press, New York (1992).
17. K. Konstadinidis, R. L. Opila, J. A. Taylor and A. C. Miller, *Mater. Res. Soc. Symp. Proc.* **304**, 83–90 (1993).
18. S. G. Anderson, J. Leu, B. D. Silverman and P. S. Ho, *J. Vac. Sci. Technol.* **A11**, 368 (1993).
19. M. J. Goldberg, J. G. Clabes and C. A. Kovac, *J. Vac. Sci. Technol.* **A6**, 991 (1988).
20. J. L. Jordan, C. A. Kovac, J. F. Morar and R. A. Pollack, *Phys. Rev.* **B36**, 1369 (1987).
21. R. G. Nuzzo and D. L. Allara, *J. Am. Chem. Soc.* **105**, 4481 (1983);
    D. L. Allara and R. G. Nuzzo, *Langmuir* **1**, 45 (1985);
    D. L. Allara and R. G. Nuzzo, *Langmuir* **1**, 52 (1985).
22. L. H. Dubois and R. G. Nuzzo, *Annu. Rev. Phys. Chem.* **43**, 437 (1992).
23. A. Ulman, *An Introduction to Ultrathin Organic Films*. Academic Press, New York (1991);
    A. Ulman, *Chemtech* **25** (3), 22 (1995);
    A. Ulman, *MRS Bulletin* **20** (6), 46 (1995);
    A. Ulman, *Thin Solid Films* **273**, 48 (1996);
    A. Ulman, *Chem. Rev.* **96** (4), 1533 (1996).

24. D. R. Jung, A. W. Czanderna and G. C. Herdt, in: *Polymer Surfaces and Interfaces: Characterization, Modification, and Application*, K. L. Mittal and K.-W. Lee (Eds), pp. 189–221. VSP, Zeist, The Netherlands (1997).
25. C. D. Bain, E. B. Troughton, Y.-T. Tao, J. Evall, G. M. Whitesides and R. G. Nuzzo, *J. Am. Chem. Soc.* **111**, 321 (1989).
26. D. E. King and A. W. Czanderna, *Surf. Sci. Lett.* **235**, L329 (1990).
27. J. R. Vig, *J. Vac. Sci. Technol.* **A3**, 1027 (1985);
J. R. Vig, in: *Handbook of Semiconductor Wafer Cleaning Technology*, W. Kern (Ed.), pp. 233–273. Noyes Publications, Park Ridge, NJ (1993).
28. D. E. King, *J. Vac. Sci. Technol.* **A13**, 1247 (1995).
29. J. R. Pitts, Ph.D. dissertation, Department of Physics, University of Denver, CO (1985).
30. D. L. Allara, D. R. Jung and P. Zhang, Paper presented at 39th Natl Symp., American Vacuum Society, Chicago (1992) (D. L. Allara, private communication).
31. K. Bammel, J. Ellis and H.-G. Rubahn, *Chem. Phys. Lett.* **201**, 101 (1993).
32. F. Balzer, K. Bammel and H.-G. Rubahn, *J. Chem. Phys.* **98**, 7625 (1993).
33. G. Herdt, Ph.D. dissertation, Materials Science, University of Denver, CO (1996).
34. Independent observations by D. L. Allara *et al.* at Penn. State U., R. L. Opila *et al.* at AT&T-Bell Labs and A. W. Czanderna *et al.* at NREL.
35. G. K. Wertheim, *Z. Phys.* **B66**, 53 (1987).
36. S. T. Ceyer, *Science* **249**, 133 (1990) (and refs therein).
37. J. F. Moulder, W. F. Stickle, P. E. Sobol and K. D. Bomben, in: *Handbook of X-ray Photoelectron Spectroscopy*, J. Chastain (Ed.). Perkin–Elmer, Eden Prairie (1992).
38. G. K. Wertheim, S. B. DiCenzo, D. N. E. Buchanan and P. A. Bennett, *Solid State Comm.* **53**, 377 (1985) (and refs therein). The sign of the surface core level shift resulting from band narrowing is not easily predicted for Cr because the 3*d* band is half-filled.
39. H. Okamoto and T. B. Massalski, *Bull. Alloy Phase Diagrams* **6** (3) (June 1985).
40. R. E. Thomas and G. A. Haas, *J. Appl. Phys.* **43**, 4900 (1972).
41. G. Beamson and D. Briggs, *High Resolution XPS of Organic Polymers: The Scienta ESCA 300 Database*. Wiley, Chichester (1992). The separation between the two O 1*s* peaks is 1.34 eV for a poly(acrylic acid) thin film.
42. M. Gasgnier and L. Nevot, *Phys. Stat. Sol.* **A66**, 525 (1981).
43. K. Petkov, V. Krastev and Ts. Marinova, *Surf. Interface Anal.* **18**, 487 (1992).
44. E. Desimoni, C. Malitesta, P. G. Zambonin and J. C. Riviere, *Surf. Interface Anal.* **13**, 173 (1988).
45. I. Ikemoto, K. Ishii, S. Kinoshita, H. Kuroda, M. A. A. Franco and J. Thomas, *J. Solid State Chem.* **17**, 425 (1976).
46. Between 0 and 0.07 nm coverages, the methylene carbon peak binding energy shows a net shift of $+0.4$ eV after making slight corrections to energy scales based on the deviations of the Au $4f5/2$ binding energies from 84.0 eV ($+0.07$ and -0.04 eV, respectively). The mean Au $4f5/2$ binding energy for the entire set is 83.99 ± 0.05 eV. The variations appear to be random.
47. G. K. Wertheim, S. B. DiCenzom, D. N. E. Buchanan and P. A. Bennett, *Solid State Comm.* **53**, 377 (1985) (and refs therein). The sign of the surface core level shift resulting from band narrowing is not easily predicted for Cr because the 3*d* band is half-filled.
48. S. Tanuma, C. J. Powell and D. R. Penn, *Surf. Interf. Anal.* **17**, 911 (1991); and
P. E. Laibinis, C. D. Bain and G. M. Whitesides, *J. Phys. Chem.* **95**, 7017 (1991).
49. M. Grunze, private communication to AWC (February 1995).
50. A. W. Czanderna and R. J. Gottschall (Guest Eds), *Mater. Sci. Eng.* **53**, 1–168 (1982).
51. J. D. Swalen, D. L. Allara, J. D. Andrade, E. A. Chandross, S. Garoff, J. Israelachvili, T. J. McCarthy, R. Murray, R. F. Pease, J. F. Rabolt, K. J. Wynne and H. Yu, *Langmuir* **3**, 932 (1987).
52. A. W. Czanderna and A. R. Landgrebe (Guest Eds), *Crit. Rev. Surf. Chem.* **2** (1–4) (1993);
A. W. Czanderna and A. R. Landgrebe (Guest Eds), *Crit. Rev. Surf. Chem.* **3** (1) (1993).

Mittal Festschrift, pp. 747–761
W. J. Van Ooij and H. R. Anderson, Jr. (Eds)

Structure and formation of metal–polymer interfaces*

F. FAUPEL,[†] R. WILLECKE, A. THRAN, C. v. BECHTOLSHEIM, M. KIENE and T. STRUNSKUS

Lehrstuhl für Materialverbunde, Technische Fakultät, University of Kiel, Kaiserstr. 2, 24143 Kiel, Germany

Abstract—This paper reviews how structure and formation of metal–polymer interfaces are controlled by the preparation process and the nature of the materials involved. Emphasis is placed on noble-metal deposition onto fully cured polymers with a focus on results from radiotracer measurements of metal diffusion at the interface, structural investigations by means of electron microscopy, and computer simulations. Moreover, first results from X-ray photoemission spectroscopy and other techniques on the early stages of interface formation are presented. While reactive metals always seem to form relatively sharp interfaces with polymers, metals of lower reactivity diffuse into them at elevated temperatures and tend to agglomerate. The extent of diffusion appears to be determined by the initial stage of the deposition process and increases strongly at low deposition rates, where a large fraction of isolated metal atoms is able to diffuse into the polymer before being trapped by other atoms at or near the surface. No significant diffusion is expected from a continuous metal film as a consequence of the high cohesive energy of the metal.

Keywords: Diffusion; radiotracer measurements; computer simulations; cluster formation; electron microscopy; metal–polymer interfaces.

1. INTRODUCTION

Metallized polymers are nowadays of considerable technological importance and are widely used for products ranging from reflectors for car lights, compact discs and electrically shielded computer cases to foils for food packing. Particularly their application in microelectronics has stimulated intensive research during the last decade. Multilayers incorporating high-temperature polymers such as polyimides and copper wiring have substantially enhanced speed and density of packaging structures [1]. Polyimides provide a unique combination of properties encompassing low dielectric constant, ease of planarization and patterning by lithographic techniques, chemical resistance, and thermal stability. Polyimides are also increasingly employed as interlayer dielectrics in very large scale integrated circuits (VLSI) [2].

---

*Paper dedicated to Dr. Kash Mittal on the occasion of his 50th birthday.

†To whom correspondence should be addressed.

Both processing and operation of microelectronics components involves exposure to elevated temperatures. This may lead to crack formation and delamination induced by thermal–mechanical stress. In chip applications, diffusion from the polymer into silicon of even very small amounts of copper, acting as a deep-level impurity, is a concern. Therefore, much effort has been made to control the microstructure and thermal stability of metal–polymer interfaces, especially with the aim of improving adhesion of the metal coating and preventing its degradation [3, 4].

Irrespective of the application, investigation of how interfaces form between such dissimilar materials as metals and polymers and how they look is also very interesting from a fundamental point of view. While metals are densely packed crystalline materials with a high cohesive energy, polymers are made up of very long chains held together by weak van der Waals interactions in an open structure, which in amorphous polymers typically resembles a random-coil arrangement.

For comparison we first recall interface formation between two elements or compounds that are not too different and exhibit appreciable mutual solubility [5]. A corresponding simple phase diagram is sketched in Fig. 1. If A is deposited onto B at temperature T_0 allowing diffusion, a solid solution of B in A (α phase in Fig. 1) forms and the concentration c_A decreases to the value $c_{\alpha\beta}$, the solubility limit of B in A at temperature T_0, where it drops abruptly to $c_{\beta\alpha}$, the solubility limit of A in B. This drop comes about because the chemical potential is constant in the co-existence region $\alpha + \beta$ and hence the driving force for the formation of two-phase regions vanishes. As a consequence, two-phase regions cannot form during binary interdiffusion under isothermal conditions. The condition of constant temperature is very important here because precipitation may well occur if a supersaturated solution is cooled down.

We now assume that A is a metal and B a polymer. Since the cohesive energy of metals is typically two orders of magnitude higher than the cohesive energy of poly-

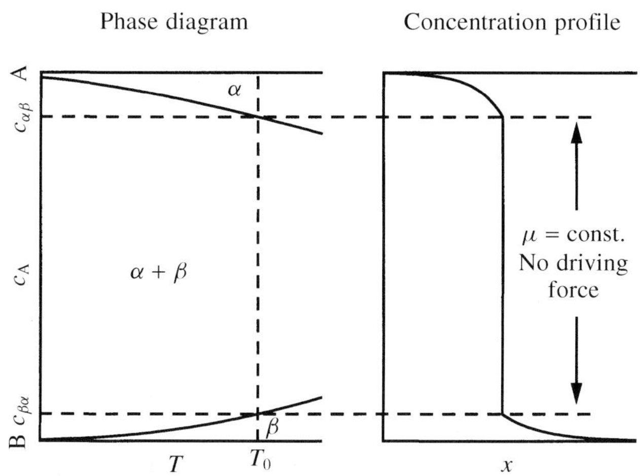

Figure 1. (Left) Scheme of a simple binary phase diagram for two elements or compounds A and B with mutual solid solubility and (right) concentration profile $c_A(x)$ expected after unidirectional isothermal interdiffusion at the temperature T_0. Note that the concentration changes abruptly from $c_{\alpha\beta}$ to $c_{\beta\alpha}$, and no two-phase region with co-existing phases α and β forms by diffusion.

mers and the interaction between metal and polymer is generally weak in comparison to the strong metal–metal binding forces, the solubility of metals in polymers should be very low and almost no intermixing should occur at metal–polymer interfaces.

In contrast to these expectations there is evidence now from various investigations, mainly in polyimides [6–10] and polycarbonates [11], that metals of low reactivity diffuse into polymers held at elevated temperatures during metal evaporation and even form metal clusters inside the polymer under isothermal conditions [12–14]. Only more reactive metals, particularly chromium and titanium, which interact strongly with polymers, appear to form sharp and thermally stable interfaces exhibiting no indications of intermixing [12, 13].

In this review we discuss how structure and formation of metal–polymer interfaces are controlled by the preparation process. Only interfaces prepared by metal deposition onto fully cured polymers are considered. For mechanisms taking place during deposition of polymer precursors onto metals the reader is referred to reference [15]. It is shown that intermixing between metal and polymer occurs under conditions being far from thermodynamic equilibrium and that no significant diffusion is to be expected when a continuous metal film is in contact with a polymer.

2. EFFECT OF METAL-DEPOSITION RATE

Strong evidence for the important role of the metal deposition rate in interface formation was obtained using the radiotracer technique in conjunction with high-resolution microsectioning. In this technique a certain fraction of the metal atoms are radioactive isotopes such as ^{67}Cu, ^{110m}Ag and ^{198}Au. The metal concentration profile in the polymer after metal deposition and heat treatment is obtained through 'cutting' sections from the sample perpendicular to the diffusion direction and counting the radioactivity of each section. A very high depth resolution of typically 3 to 4 nm can be achieved by use of ion-beam sputtering for sectioning. A scheme of the ion-beam sectioning apparatus used in our experiments is shown in Fig. 2. An intense beam of argon ions impinges onto the sample, which rotates to ensure uniform removal of material and to minimize surface roughening. In order to avoid artifacts from the sputtering process the ion energy should not exceed 250 eV in sectioning of polymers. An electron-emitting filament is placed in the beam near the ion gun to avoid surface charging. The material sputtered off is collected on a foil. After completion of a sputter section the foil is advanced like a film in a camera without interrupting the sputtering process. Usually a small portion of the sample is covered and a surface step is created in the sputter process. Its height, measured with a profilometer, allows one to determine the sputtering rates for depth calibration. Since the metal concentration in the polymer is very low, its effect on the sputtering rate can be neglected. After sputtering, the foil is cut into pieces corresponding to the individual sputter sections and the activity of each section is counted independently. The activity is proportional to the tracer concentration c, which can be measured absolutely if a calibration is carried out.

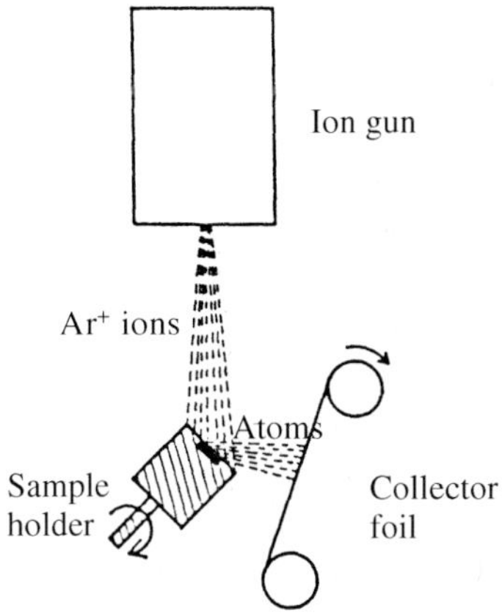

Figure 2. Sketch of a typical set-up used for ion-beam microsectioning. The whole arrangement is placed in a high-vacuum chamber.

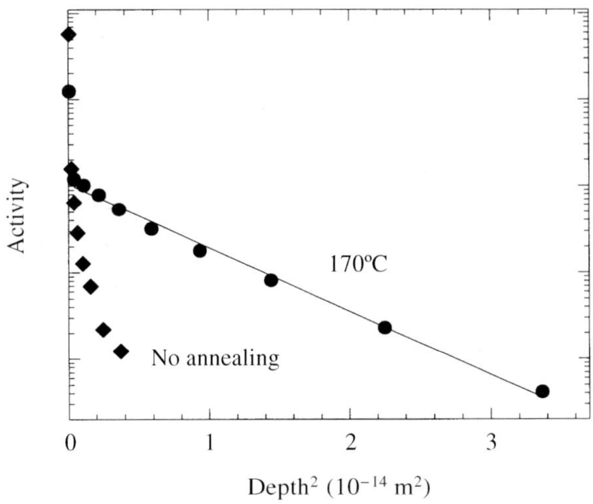

Figure 3. Typical penetration profile obtained after deposition of less than one monolayer (ML, nominal coverage) of gold onto BPA-PC at 170 °C at a rate of about 0.1 ML/min and subsequent annealing at 170 °C for 5800 s. A penetration profile of a sample after room-temperature deposition without annealing, which can be regarded as resolution function, is shown for comparison. The ^{198}Au activity in arbitrary units is plotted against the square of the penetration depth on a semilogarithmic scale.

A representative penetration profile is shown in Fig. 3. The logarithm of the activity of ^{198}Au is plotted versus the square of the penetration depth x. On this scale, ordinary diffusion of the tracer according to the thin-film solution of Fick's second law,

$$c(x) = const\ \exp\left(-\frac{x^2}{4Dt}\right), \tag{1}$$

leads to a straight line. In equation (1), D is the tracer diffusivity and t the diffusion time. With the penetration profile the resolution function of the present sectioning technique is displayed. It was obtained by sputtering through a very thin metal film deposited at room temperature. The real resolution function should be even narrower, because a perfectly flat surface and no intermixing for room-temperature deposition was assumed, but this is not necessarily the case.

It is obvious from Fig. 3 that gold diffuses deeply into bisphenol-A polycarbonate (BPA-PC) during deposition at elevated temperatures. However, most of the metal deposit is located at, or very close to, the surface, and the fraction of atoms that diffuse over large distances and contribute to the linear range in Fig. 3 is very small (note the logarithmic scale). Penetration profiles with a similar shape, i.e. a steep drop of the metal concentration at or near the polymer surface followed by a linear range, have also been found for noble-metal tracers in polyimides [7–9] and other polycarbonates [11, 16]. Only the ratio of metal in the near surface region and in the tail varied from case to case.

The fraction of metal atoms diffusing deeply into the polymer proved to be a function of the deposition rate. This is demonstrated in Fig. 4, where two profiles for silver diffusion into pyromellitic dianhydride–oxidianiline (PMDA-ODA) polyimide, recorded after deposition at extremely different rates, are compared. A low deposition rate results in nearly complete diffusion of silver into the polymer with little surface hold-up, while almost no diffusion occurs after flash evaporation. Here the metal concentration drops abruptly by more than three orders of magnitude at the polymer surface, and metal diffusion into the bulk is only detectable because of the high sensitivity of the radiotracer technique. (Nevertheless, diffusion of such small amount of copper may cause damage in VLSI chips if it is not effectively blocked by a diffusion barrier.)

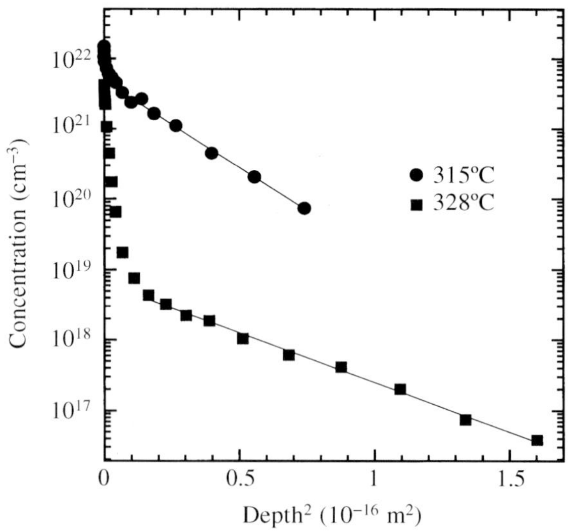

Figure 4. Effect of deposition rate on penetration profiles of ^{110m}Ag into PMDA-ODA. The tracers were deposited continuously at a rate of about 0.1 ML/min during annealing at 315°C and flash evaporated at the annealing temperature of 328°C, respectively.

This rate dependence, which has also been observed in transmission electron microscopy (TEM) studies (see below), is very unusual. Normally, the extent of diffusion into a material is not dependent on the deposition rate of the diffusant. Moreover, judging from the high cohesive energy of the metal, one should expect metal clustering only on top of the polymer but no appreciable diffusion into the bulk. One has to take into account, however, that during metal deposition isolated atoms hit the surface. (Some metals tend to form a certain fraction of aggregates of a few atoms already in the vapor phase.) Therefore the situation here is different from the case, where a continuous metal film is in contact with a polymer.

We are presently carrying out computer simulations to obtain further insight into interface formation during metal deposition onto polymers. In these simulations metal atoms deposited onto the surface are allowed to diffuse into the polymer by jumps of constant length l and random direction. Variations in the jump length caused by the amorphous structure are neglected, i.e. l is regarded as an average jump length. Atoms are deposited on a surface of $400\,l \times 400\,l$ of a half-space, which is periodically extended parallel to the surface. The size of the periodicities is large enough to not affect the results. Atoms encountering each other within a critical radius, taken as half of the jump length, are assumed to form stable immobile clusters. The assumption of stability was introduced in view of the high binding energy between metal atoms, e.g. 0.8 eV per atom for Ag_2 [17], which increases substantially for larger clusters. The mobility of the clusters was considered to be negligible in comparison with the mobility of single atoms because of the strong size dependence of the effective activation energy of diffusion [18]. While earlier computer simulations [19] aimed at explaining the shape of the metal clusters, the spherical shape of the clusters was taken as a well established input parameter from microstructural investigations (see below).

First results are depicted in Fig. 5, where the distribution of metal clusters in the polymer near the surface is shown after metal deposition at a moderate rate and subsequent 'annealing'. Marked metal aggregation has taken place and has effectively blocked metal diffusion. Isolated atoms dominate only at larger penetration depth and an appreciable fraction of clusters with more than three atoms is located in the near-surface region even at the very low nominal coverages simulated. Simulations for different rates show that the extent of metal clustering clearly increases at higher deposition rates. This explains why major diffusion into the bulk has been observed only at low deposition rates in the radiotracer studies. Surface diffusion, which is orders of magnitude faster than bulk diffusion [5], will enhance metal clustering at the surface and thus will lead to an even more pronounced retardation of diffusion and an increase in its dependence on deposition rate.

The pronounced clustering observed in the computer simulations already at very low metal concentrations strongly suggests that metal atoms break apart from clusters. Moreover, ongoing experiments point to appreciable mobility of even relatively large clusters in polymers above their glass transition temperature. Corresponding extensions of the present simulation are in progress.

Both our radiotracer experiments as well as the computer simulations were carried out under conditions that apply to the deposition of up to a few monolayers of metal.

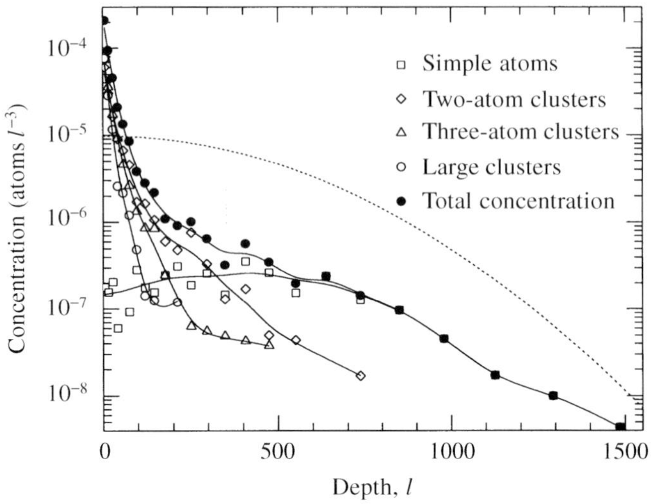

Figure 5. Computer simulations of depth profiles showing the distribution of isolated atoms and clusters after metal deposition and 'annealing'. The deposition time was 1/500 of the total 'annealing time' corresponding to 5×10^5 jumps per atom. The total 'metal' concentration and the ideal Gaussian profile obtained without metal–metal interaction are depicted, too.

Upon further deposition the observed metal clusters are expected to grow and become detectable in TEM. Metal clustering inside PMDA-ODA polyimide has indeed been observed in cross-sectional TEM studies for Cu by LeGoues *et al.* [12] and later for Ag by Foitzik and Faupel [14]. Marked cluster formation at considerable distances below the interface takes place only in the regime of elevated temperatures (though already well below T_g) and low deposition rates.

Our recent investigations of interface formation between gold and bisphenol–trimethyl cyclohexane polycarbonate (TMC-PC) demonstrate that the unusual phenomenon of metal aggregation inside polymers is not restricted to PMDA-ODA polyimide. In these investigations electron transparent samples were prepared by direct ultra-microtome cutting after metal deposition and annealing. No glue, which could lead to interface modifications through interaction with the polymer during the cutting procedure, was employed.

Figure 6a shows extensive metal aggregation resulting from Au deposition at 230°C at a very low rate of 0.05 nm/min and 6 h annealing. By contrast, flash evaporation at this temperature, being very close to the bulk glass transition temperature of the polymer, did not cause any metal clustering inside the polymer (Fig. 6b). Subsequent annealing of this sample for 6 h at 230°C did not produce any significant changes, judging from TEM observations. This demonstrates that metal clustering at the surface effectively impedes metal diffusion into the bulk and gives further evidence to the view that no significant metal diffusion into the polymer is expected for a continuous film or an arrangement of large clusters. In this connection we note that the metal films in Fig. 6 are still not continuous even at relatively large nominal metal coverages of about 30 monolayers. The impression of a continuous film is a consequence of the

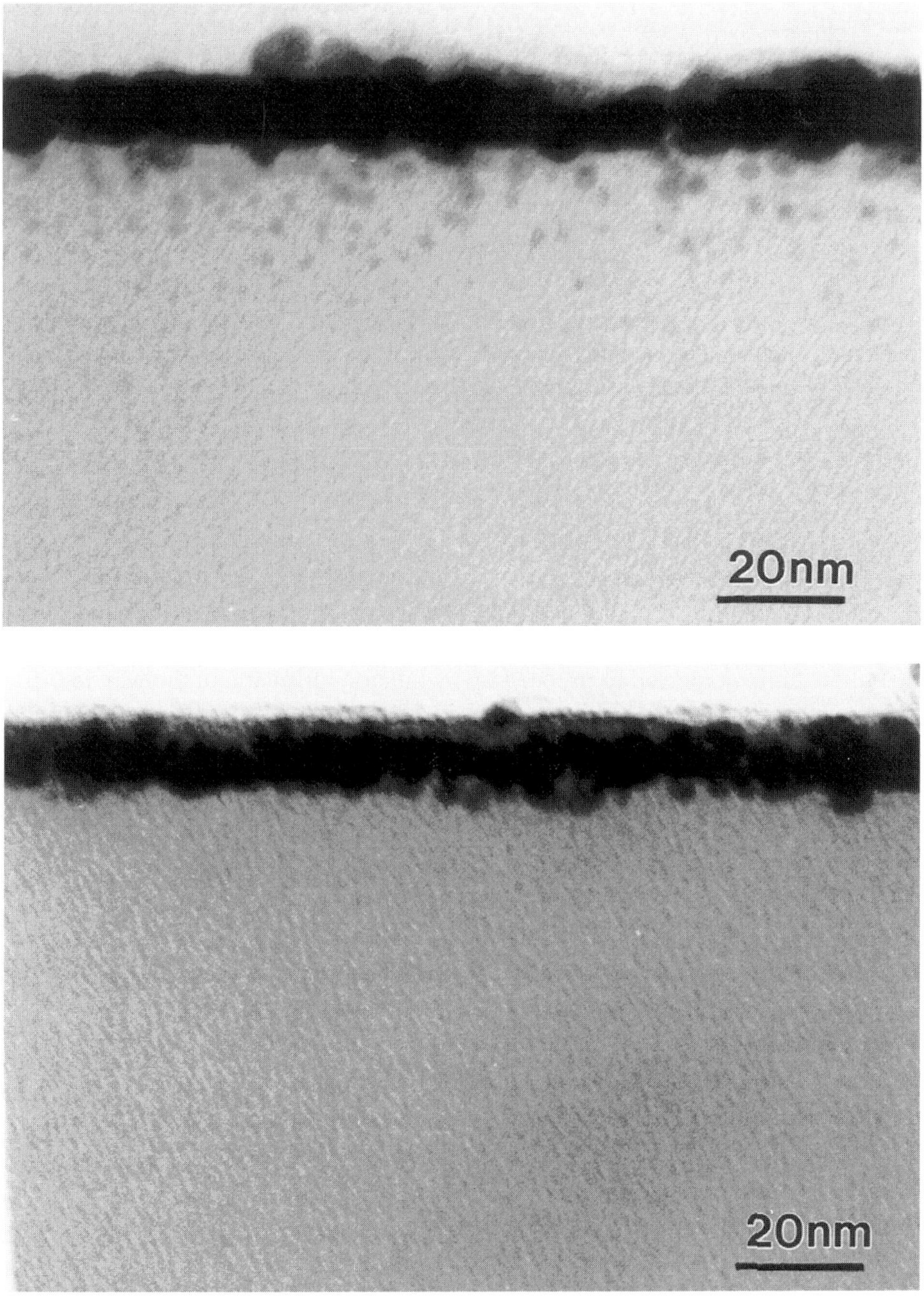

Figure 6. Cross-sectional transmission electron micrographs showing the effect of deposition rate on the formation of the interface between gold and TMC-PC. (a) A spread-out interface with extensive metal aggregation is obtained after Au evaporation at 0.05 nm/min at 230°C and subsequent annealing at this temperature for 6 h. (b) Flash evaporation at the same temperature prevents noticeable metal aggregation inside the polymer even after subsequent annealing for 6 h at 230°C very close to the bulk glass transition temperature of about 234°C.

final thickness of the samples of 40–100 nm, depending on the cutting procedure. Upon tilting, isolated and connected clusters are clearly visible.

The fact that noble metals may strongly diffuse into polymers and even form two-phase regions, despite their expected very low solubility, calls for further comments, particularly when contrasted with our discussion on binary interdiffusion (cf. Fig. 1). Apparently, there is a strong driving force for metal embedding, i.e. the Gibbs free energy of a metal particle inside the polymer is lower than that of the isolated particle at the surface. We have attributed this effect to the reduction of the high surface energy of the metal by embedding [10]. This model is clearly different from normal metal dissolution. Embedding of a metal particle is energetically favorable if the surface tension of the particle γ_M exceeds the sum of the interfacial tension γ_{MP} and the surface tension of the polymer γ_P [20]:

$$\gamma_M > \gamma_{MP} + \gamma_P. \tag{2}$$

Here γ_P is generally very small in comparison to γ_M (see above). Inequality (2) is obeyed by many metal/polymer systems, and sinking of metal particles, deposited onto polymers at low temperatures, has been observed after heating above the glass transition temperature [21]. In the glassy state the high viscosity certainly excludes sinking of large clusters from the surface into the polymer. Here cluster formation in the bulk is rather expected to proceed by metal atom diffusion in the way indicated by the computer simulations. This mechanism of cluster formation should also apply to the highly viscous rubbery state close to T_g, and specifically to the situation shown in Fig. 6. In the glassy state, formation of large clusters also requires effective relaxation of the stress field around the clusters by plastic deformation.

3. EFFECT OF TEMPERATURE

Considering the intermixing that takes place during deposition of noble metals onto polymers, a strong influence of temperature on interface formation is obvious. At temperatures well below the glass transition temperature, T_g, polymer chains as a whole are immobile, and only diffusion, not sinking of metal particles into the polymer is expected. Near and above T_g, polymer diffusion has to be taken into account, too. In this connection it is important to note that during interface formation the glass transition temperature at the surface comes into play. Here T_g can be depressed significantly. It has recently been suggested that a liquid-like layer exists at the surface of glassy polymers whose thickness diverges as the glass transition temperature is approached from below [22].

Meanwhile, metal diffusion has been studied by the radiotracer method, described above, for noble metals in polyimides [7–10, 23] and polycarbonates [11]. Furthermore, Rutherford backscattering has been utilized, despite its relatively poor depth resolution in polymers [24–26]. Tracer diffusivities were obtained from the linear tails of penetration profiles similar to those in Figs 3 and 4. (Deviations from the ideal thin-film solution, equation (1), are discussed in [11].) The extended linear tails in the

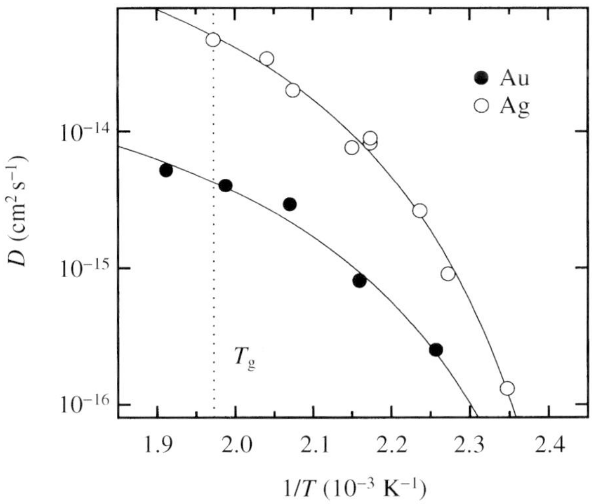

Figure 7. Arrhenius plot for diffusion of Ag and Au atoms in TMC-PC. Solid lines represent fits of the free-volume theory of Vrentas and Duda to the data (see Willecke and Faupel [28]). The glass transition temperature T_g (from DSC) is indicated.

low-concentration range of the profiles were attributed to diffusion of isolated metal atoms. While this seems to be reasonable in the cases cited, we recently measured profiles for Ag diffusion in tetramethyl bisphenol-A polycarbonate above T_g, where the linear tail seems to reflect the diffusivity of relatively large clusters, and isolated atoms are believed to be mostly gettered by clusters or to have penetrated to larger depths well above 100 nm [16].

An Arrhenius plot for diffusion of Ag and Au in TMC-PC is shown in Fig. 7. Like Arrhenius plots for diffusion of Cu and Ag in PMDA-ODA polyimide [23], it exhibits a distinct downward curvature. In both cases it was shown that the free-volume theory of Vrentas and Duda [27], which also applies to diffusion in the glassy state, can be fitted to the data with reasonable parameters (for details see [23, 28]). However, the free-volume theory (without incorporation of an additional term for diffusant–matrix interactions [27]) does not explain why the diffusivities of noble-metal atoms in polymers are many orders of magnitude smaller than those of simple gas molecules of comparable size [18]. This clearly points to important contributions from chemical interactions with the polymer even for noble metals. For more reactive metals, these chemical interactions are even more important and appear to block diffusion quite effectively, judging from the much lower mobility of Ni [25] and fruitless attempts in detecting any metal mobility by means of Rutherford backscattering for Mo [26], Cr, and Ti [25], as well as from the ability of Cr and Ti to form sharp interfaces without indications of intermixing [4].

Free-volume theory (without an additional term for thermally activated bond breaking [27]) also fails to account for the diffusion behavior of Au and Ag in bisphenol-A polycarbonate, a typical amorphous polymer with a large fractional free volume [11]. The glass transition of BPA-PC at about 150°C is much more pronounced than the

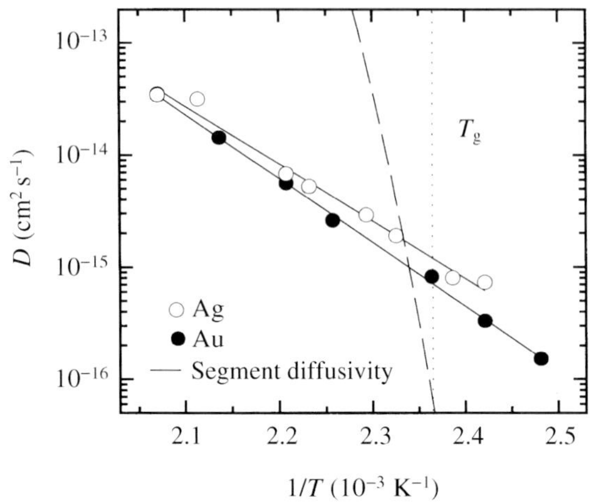

Figure 8. Arrhenius plot for diffusion of Ag and Au atoms in BPA-PC (from Willecke and Faupel [11]). The effective activation energies are 1.02 and 1.13 eV for Ag and Au, respectively. The dashed line represents the segment diffusivity of the polymer chains for comparison. The glass transition temperature T_g (from DSC) is indicated.

glass transition of PMDA-ODA which does not lead to a measurable effect in differential scanning calorimetry. As depicted in Fig. 8, a linear Arrhenius plot is observed over an extended temperature range with no change in slope in the glass transition region. Moreover, the temperature dependence is totally different from that of the diffusivity of the polymer chain segments (Fig. 8). Apparently, here metal diffusion is completely decoupled from polymer dynamics and, most remarkably, the metal-atom diffusivity is much lower than the segmental diffusivity at high temperatures above the glass transition. In view of current theories of diffusion in polymers one would expect that the diffusivity of the chain segments imposes a lower limit on the diffusion coefficient of small penetrants [18]. As a possible explanation of this unexpected behavior Willecke and Faupel [11] have suggested a substantial interaction of a metal atom with several polymer chains at the same time. This interaction should lead to a substantial depression of the local chain mobility through a kind of metal-atom-induced weak crosslinking. While this model certainly requires further validation, there are strong indications now that metal–polymer interaction is very important in controlling the diffusion behavior of metals, even of those of low reactivity. Particularly in polymers with a large fractional free volume the influence of the availability of free volume on diffusion appears to be less significant. Strunskus *et al.* [29] have pointed out that dipole/induced dipole interactions can lead to significant localization energies even for noble metals.

The above discussion on diffusion mechanisms and the diffusion coefficients displayed in Figs 7 and 8 refer to diffusion of isolated atoms. It has already been mentioned in Section 2 in the context of the rate dependence of diffusion that metal clustering may drastically reduce the amount of diffusion of metals into polymers. In order to get further insight into the depression of metal diffusion through aggregation

(a) (b)

Figure 9. TEM micrographs of a PMDA-ODA polyimide sample of about 0.1 μm thickness (a) after deposition of 1.6 nm of Cu at room temperature at a rate of 0.16 nm/min and (b) after subsequent annealing at 350 °C for 30 min. The field of view is 200×140 nm for each micrograph.

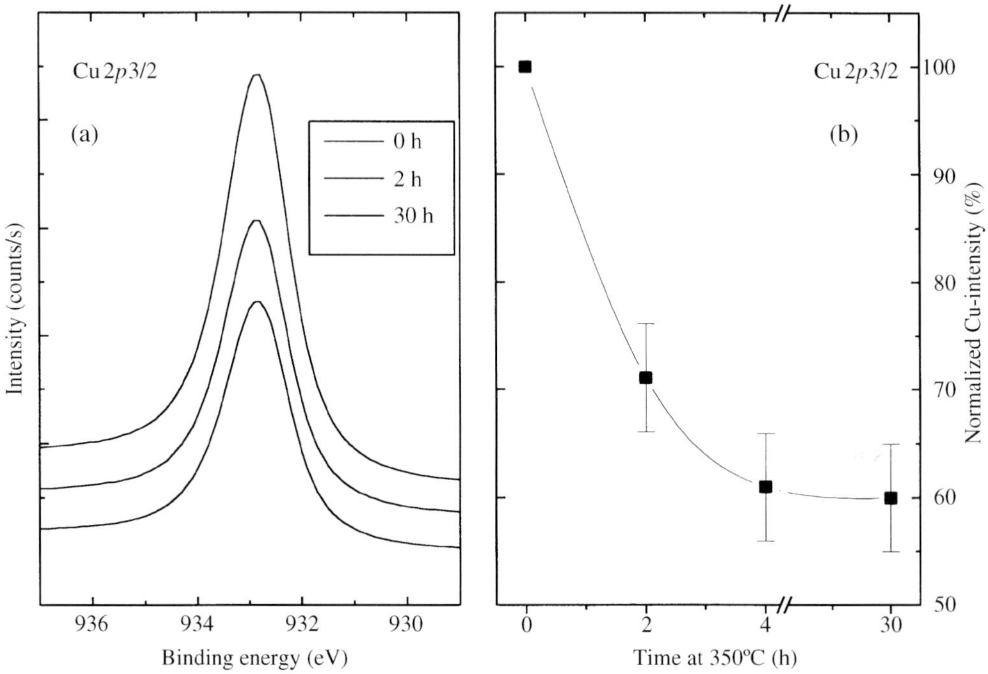

Figure 10. (a) Cu $2p3/2$ XPS spectra from PMDA-ODA polyimide samples of 20 nm thickness after room-temperature deposition of 1.6 nm Cu and different annealing times at 350°C. (b) Normalized Cu $2p3/2$ intensity versus annealing time. The leveling off of the Cu intensity is a clear indication of Cu immobilization.

and its role in the early stages of metal–polymer interface formation, we are presently performing experiments involving TEM, atomic force microscopy (AFM), and X-ray photoelectron spectroscopy (XPS). First results are shown in Figs 9 and 10.

Figure 9 depicts TEM micrographs of a PMDA-ODA polyimide sample (a) after deposition of 1.6 nm of Cu at room temperature at a rate of 0.16 nm/min and (b) after subsequent annealing at 350°C for 30 min. The polyimide was spin-coated onto a Si wafer and cured in the usual way with a final cure step of 60 min at 400°C in a nitrogen atmosphere. Cu evaporation and heat treatment were performed in ultrahigh vacuum (UHV). For the TEM investigations an electron transparent film of about 0.1 μm thickness was removed from the wafer prior to the metal deposition. After room temperature deposition small metal clusters are seen in the TEM micrograph. The size of the copper clusters is of the order of the TEM resolution limit of about 2 nm and thus they are not clearly resolved. Apparently, surface diffusion already leads to appreciable cluster formation even at temperatures near room temperature. This indicates a significant surface mobility of Cu atoms or small clusters. (The maximum heating of the polymer surface during metal deposition was estimated as less than 50°C.) Subsequent annealing at 350°C gives rise to substantial cluster growth to a relatively uniform size of about 6 nm. Moreover, smaller clusters of about 3 nm have formed. The larger clusters were also detected by means of AFM and are hence at the surface or only slightly embedded. The detectable clusters in the TEM micrograph do not add up to the nominal amount of deposited copper. It must be concluded that a fraction of the same order of magnitude is not visible in the TEM micrograph and therefore is in the form of atoms and/or smaller clusters (< 2 nm).

X-ray photoelectron spectroscopy particularly lends itself to studies of the growth mode of metals on polymers and of the diffusion behavior due to the short elastic mean free path of the photoelectrons. In order to avoid charging effects, polyimide films of only 20 nm thickness, spin-coated onto a silicon wafer, were used. These films are expected to exhibit a preferred molecular orientation parallel to the surface [30]. Figures 10a and 10b show Cu $2p3/2$ spectra and normalized intensities, respectively, for different annealing times. No indications of any chemical interaction were observed upon annealing. The copper intensity simply decreases by about 40% during annealing and finally approaches a plateau value. The occurrence of this plateau is in sharp contrast to ordinary diffusion of Cu into the polymer, which would lead to a complete drop of the Cu intensity down to zero. The plateau shows clearly that Cu does not diffuse significantly into the polymer upon heating but rather stays very close to the surface. The relatively small decrease of the Cu intensity can be almost entirely accounted for by the cluster formation process at the surface that was observed in TEM (Figs 9a and 9b). This does not rule out that the Cu particles are partly embedded into the polymer through polymer diffusion as discussed in connection with equation (2). However, the XPS data allow one to exclude that the metal clusters are embedded deeply inside the polymer, because this would cause a much stronger reduction of the Cu $2p3/2$ intensity. Apparently, motion of the Cu particles into the polymer bulk is simply blocked because of the high bulk viscosity at 350°C, still well below the bulk glass transition temperature.

It has to be stressed that the process just described, involving metal cluster formation on the polymer surface without major diffusion into the polymer, is quite distinct from cluster formation inside polymers during metal deposition at elevated temperatures and low rates. Under the latter conditions, which give rise to a rather spread-out interface, clusters appear to form by diffusion of metal atoms into the polymer. This illustrates how strongly interface formation may depend on the preparation conditions. Based on our present knowledge of how to control the interfacial structure on a microscopic level, we plan to carry out adhesion studies in order to obtain a deeper insight into the factors governing metal–polymer adhesion.

4. CONCLUSIONS

Noble metals strongly diffuse into polymers during their deposition at low rates and elevated temperatures. This process, which is accompanied by metal-cluster formation inside the polymer under isothermal conditions, is completely different from ordinary binary interdiffusion, where the formation of two-phase regions does not occur. It can be understood in terms of the very strong deviations from equilibrium in the early deposition stage. For high deposition rates, far above the monolayer/min range, very little metal diffusion takes place even at elevated temperatures. No significant diffusion is to be expected from a continuous metal film or large cluster. More reactive metals appear to form relatively sharp interfaces with the polymer with no indications of appreciable intermixing.

Acknowledgements

Financial support of our research into metal–polymer interfaces throughout the years by the Volkswagen-Stiftung is gratefully acknowledged. The ultra-microtome utilized for preparation of TEM samples was financed by the Technologie-Stiftung Schleswig-Holstein.

REFERENCES

1. R. R. Tummala and E. J. Rymaszewski (Eds), *Microelectronics Packaging Handbook*. Van Nostrand Reinhold, New York (1989).
2. P. Springer, *Semiconductor International* **52** (Nov. 1994).
3. S. P. Kowalczyk, in: *Metallization of Polymers*, E. Sacher, J. Pireaux and S. P. Kowalczyk (Eds), p. 10. ACS Symposium Series No. 440, American Chem. Soc., Washington, DC (1990).
4. P. S. Ho, R. Haight, R. C. White, B. D. Silverman and F. Faupel, in: *Fundamentals of Adhesion*, L. H. Lee (Ed.), p. 383. Plenum Press, New York (1991).
5. J. Philibert, *Atom Movements, Diffusion and Mass Transport in Solids*, p. 421. Les Editions de Physique, Les Ulis Cedex (1991).
6. R. M. Tromp, F. K. LeGoues and P. S. Ho, *J. Vac. Sci. Technol. A* **3**, 782–785 (1987).
7. F. Faupel, D. Gupta, B. D. Silverman and P. S. Ho, *Appl. Phys. Lett.* **55**, 357 (1989).
8. F. Faupel, D. Gupta, B. D. Silverman and P. S. Ho, in: *Advanced Materials and Processes*, H. E. Exner and V. Schumacher (Eds), p. 887. Deutsche Gesellschaft für Materialkunde, Oberursel (1990).
9. D. Gupta, F. Faupel and R. Willecke, in: *Diffusion in Amorphous Materials*, H. Jain and D. Gupta (Eds), pp. 189–200. The Minerals, Metals & Materials Society (1994).
10. F. Faupel, in: *Polymer–Solid Interfaces*, J. J. Pireaux, P. Bertrand and J. L. Brédas (Eds), pp. 171–187. IOP Publishing Ltd. (1992).
11. R. Willecke and F. Faupel, *Macromolecules* **30**, 567 (1997).
12. F. K. LeGoues, B. D. Silverman and P. S. Ho, *J. Vac. Sci. Technol. A* **6**, 2200–2204 (1988).
13. P. O. Hahn, G. W. Rubloff, J. W. Bartha, F. K. LeGoues and P. S. Ho, *Mater. Res. Soc. Symp. Proc.* **40**, 251 (1985).
14. A. Foitzik and F. Faupel, *Mater. Res. Soc. Symp. Proc.* **203**, 59 (1991).
15. S. P. Kowalczyk, Y. H. Kim, G. F. Walker and J. Kim, *Appl. Phys. Lett.* **52**, 375 (1988).
16. A. Thran, R. Willecke and F. Faupel, to be published.
17. V. Bonačić-Koutecký, L. Češpiva, P. Fantucci and J. Koutecký, *Z. Phys. D* **26**, 287–289 (1993).
18. D. R. Paul and Yu. P. Yampol'skii (Eds), *Polymeric Gas Separation Membranes*. CRC Press, Boca Raton, FL (1994).
19. B. D. Silverman, *Macromolecules* **24**, 2467–2474 (1991).
20. G. J. Kovacs and P. S. Vincett, *J. Colloid Interface Sci.* **90**, 335–351 (1982).
21. G. J. Kovacs and P. S. Vincett, C. Tremblay and A. L. Pundsack, *Thin Solid Films* **101**, 21–40 (1983).
22. J. L. Keddie, R. A. L. Jones and R. A. Cory, *Europhys. Lett.* **27** (1), 59–64 (1994).
23. F. Faupel, *Phys. Stat. Sol.* (*a*) **134**, 9–59 (1992).
24. K. Shanker and J. R. MacDonald, *J. Vac. Sci. Technol. A* **5**, 2894–2897 (1987).
25. J. H. Das and J. E. Morris, *IEEE Trans. CPMT B* **17**, 620–625 (1994).
26. J. H. Das and J. E. Morris, *J. Appl. Phys.* **66**, 5816–5820 (1989).
27. J. S. Vrentas and J. L. Duda, *J. Polym. Sci. Polym. Phys. Ed.* **15**, 417–439 (1977).
28. R. Willecke and F. Faupel, *J. Polym. Sci. Polym. Phys. Ed.* **35**, 1043 (1997).
29. T. Strunskus, M. Grunze, G. Kochendoerfer and Ch. Wöll, *Langmuir* **12**, 2712–2725 (1996).
30. T. P. Russell, *J. Polym. Sci. Polym. Phys. Ed.* **22**, 1105 (1984).

Mittal Festschrift, pp. 763–771
W. J. Van Ooij and H. R. Anderson, Jr. (Eds)

X-ray absorption spectroscopy at buried metal/polymer interfaces

R. L. OPILA,* K. KONSTADINIDIS, M. A. MARCUS and M. DU
Lucent Technologies, Bell Laboratories, 600 Mountain Avenue, Murray Hill, NJ 07974, USA

Abstract—X-ray absorption spectroscopy has been used to study the chemical evolution at metal (Cr or Ti)/polymer (polyimide or triazine) interfaces. The interfaces have been studied *in-situ*, by measuring the absorption of X-rays by the metal atom core levels. X-ray absorption near edge structure has been used to show that the oxidation state of the metal atoms in closest proximity to the polymer changes as a function of time, temperature, polymer, and polymer surface preparation. Extended X-ray absorption fine structure has been used to determine the structure of the thin metal overlayer by comparison with model systems. From previous XPS data, Ti is shown to form σ-bonds with the nitrogen, carbon, and oxygen in the polymer, and this distribution of compounds is shown to evolve to TiO_2. Cr is found to form σ-bonds and π-complexes with aromatic species in the polymer. These initial species convert to oxides with a local structure indistinguishable from the bulk oxide.

Keywords: X-ray absorption; triazine; polyimide; titanium; chromium; X-ray absorption near edge structure; extended X-ray absorption fine structure.

1. INTRODUCTION

As the electronics industry develops, metal/polymer interfaces become increasingly important. In electronic packaging, metal lines for carrying current are placed on printed wiring boards, and, in multichip modules, bare Si integrated circuits are interconnected by the deposition of metal on a polymer, often polyimide, followed by lithographic development of the circuitry. On future generations of integrated circuits, polymers may be used as the inner-layer dielectric, to replace SiO_2, because of the ease of planarization and low dielectric constant [1, 2]. Good adhesion at the metal/polymer interface is of prime importance. As a result, chemically-active metals like titanium and chromium are used to promote adhesion at such interfaces. Unfortunately, these metals do not have optimal conductivities, so metals like aluminum or copper are used as the actual conductor, and deposited on top of the adhesion promoting metal. Also, in structures like these, there exist two very different kinds of interfaces: metal/polymer where bare metal atoms are deposited directly on the

---

*To whom correspondence should be addressed.

polymer surface, and polymer/metal where polymer is deposited on a chemically passivated, usually oxidized, metal surface. In this work we will only be considering the deposition of metal atoms on polymer surfaces.

The chemistry that occurs at metal/polymer interfaces has been studied in several ways. First, X-ray photoelectron spectroscopy (XPS) has been used to study the chemical reactions that occur as metal atoms encounter the polymer surface [3–9]. Because this work is performed in an ultra-high vacuum system with extremely thin overlayers (< 10 nm), this technique suffers from the disadvantage that the effects of real subsequent processing cannot be studied. Yet practical experience suggests that the interface evolves after initial metal deposition. Metal atoms can diffuse into the polymer; oxygen or water can diffuse through the polymer and react at the interface; exposure to high temperatures or aggressive chemicals can accelerate either of these processes. Other studies of these interfaces can be performed using destructive techniques. For example, depth profiling using Ar^+ sputtering can be used to study an interface; however, the degradation of depth resolution and sputtering induced chemical changes make unambiguous interpretation of the results difficult. As Kash Mittal has pointed out, peeling of an interface provides some information about 'practical adhesion' [10], but oxidation at the peeled surface, and the usually inhomogeneous locus of failure make unambiguous determination of the chemistry difficult. In addition, all destructive measurements require that many samples be prepared if the chemical bonding or adhesion is to be studied as a function of another variable, such as time or temperature. For these reasons, a non-destructive technique to study the evolution of the chemistry at the interface is needed.

In this work, we have used X-rays as the probe for interfacial chemistry. By varying the wavelength of the incident X-ray beam we can study the absorption of X-rays by core electrons of the deposited metal atoms. Changes in the absorption spectrum reflect changes in the chemistry and structure of the metal atoms. If there are a sufficient number of metal atoms, the absorption spectrum can be determined directly. If not, the fluorescence of the atoms after emission of the core electron can be used. These X-rays are energetic enough to penetrate hundreds of nanometers of metal atoms, allowing study of the actual metal/polymer interfaces. Unfortunately, because the reaction products formed at metal/polymer interfaces are complex, comparison with model compounds is often necessary to determine the changes in the chemistry. Thus, we have studied a series of model compounds for the interfaces formed between Ti or Cr and polyimide or triazine. To understand the reaction between isolated metal atoms and the polymers, Ti or Cr ions were implanted into both polyimide and triazine. These reaction products will undoubtedly differ somewhat from isolated metal atoms deposited on a polymer, but will provide a mechanism of determining how isolated metal atoms might react with a polymer host. Finally, these model compounds have been compared to Cr/triazine and Cr/polyimide interfaces.

2. EXPERIMENTAL

Several sets of samples were prepared. In the first, thin films of polymer were coated on polycrystalline alumina substrates of the sort used in some multichip modules.

Either Ti or Cr (10 to 100 nm thick) was sputter deposited on the polymer at room temperature, followed by a layer of Pd (100 nm thick) in the same system without exposure to air [11]. The upper layer of Pd was deposited to ensure that the upper surface of the Cr or Ti was passivated and reaction with the atmosphere did not occur. Thicknesses were verified with Rutherford backscattering (RBS). These samples were used to study the effects of sample heating on the interfacial chemistry. The second set of samples was very similar to the first. For these samples, the alumina substrate was not included, but the metal was deposited directly onto the cured polymer. Only Cr was studied, and the Cr films deposited were 1 to 2 nm thick. Again, 100 nm of Pd was deposited on the Cr to passivate this outer surface. These samples allowed more detailed chemistry of the interface to be studied by minimizing scattering from the Al_2O_3 substrate. A third set of samples was formed by ion implanting Ti or Cr into the polymer substrate, as described in detail elsewhere [12]. This set of samples was to serve as model compounds to show how an isolated metal atom might react with a polymer matrix. Finally, fine powders of Ti and Cr model compounds were used as received and were evenly dispersed on Kapton® tape [13]. Air-sensitive model compounds were dispersed on filter paper which was then enclosed between two Be foils sealed with epoxy.

The preparation of the polymers has been described previously [3, 14]. Briefly, the triazine polymer was prepared by mixing the cyanate ester monomer with dibutyl tin dilaurate and curing under N_2 at 150°C for 3 h. The polyimide (PMDA-ODA) was cured under N_2 at 400°C for 2 h.

The X-ray absorption experiments were performed at beamlines X16C and X19A at the National Synchrotron Light Source at Brookhaven National Laboratory. X-ray absorption spectra from atomic 1*s* levels, or at the K-edge, and extended X-ray absorption fine structure (EXAFS) spectra of Ti and Cr were collected, mainly in fluorescence, although some of the model compounds were studied in transmission. The energy scale was calibrated periodically using 1 μm thick metal foils [12].

3. RESULTS AND DISCUSSION

K-edge X-ray absorption spectra for a series of Ti model compounds are shown in Fig. 1 and Cr model compounds in Fig. 2a. This series of compounds shows three basic trends. First, as the formal valence of the metal increases, the absorption edge moves to higher energy. For example, the edges of TiO, Ti_2O_3, and TiO_2 are shifted by 5, 9, and 13 eV, respectively, with respect to Ti metal. The X-ray absorption edges of Cr_2O_3 and CrO_3 are shifted by 2 and 7 eV, respectively, to higher energy relative to the absorption edge of Cr metal. When the degree of electron transfer is less than might be suggested by the formal valence, as in metal carbides and the Cr bis-benzene complex, this trend is less obvious. Second, the metal shows a metallic edge feature at low absorption energies, which disappears as the oxidation state of the metal increases. The apparent exceptions to this rule are the tetrahedral Ti compounds and the octahedral Cr compounds, where the symmetry allows mixing of the valence $3p$ and $3d$ energy levels, permitting previously forbidden transitions to the $3d$ levels.

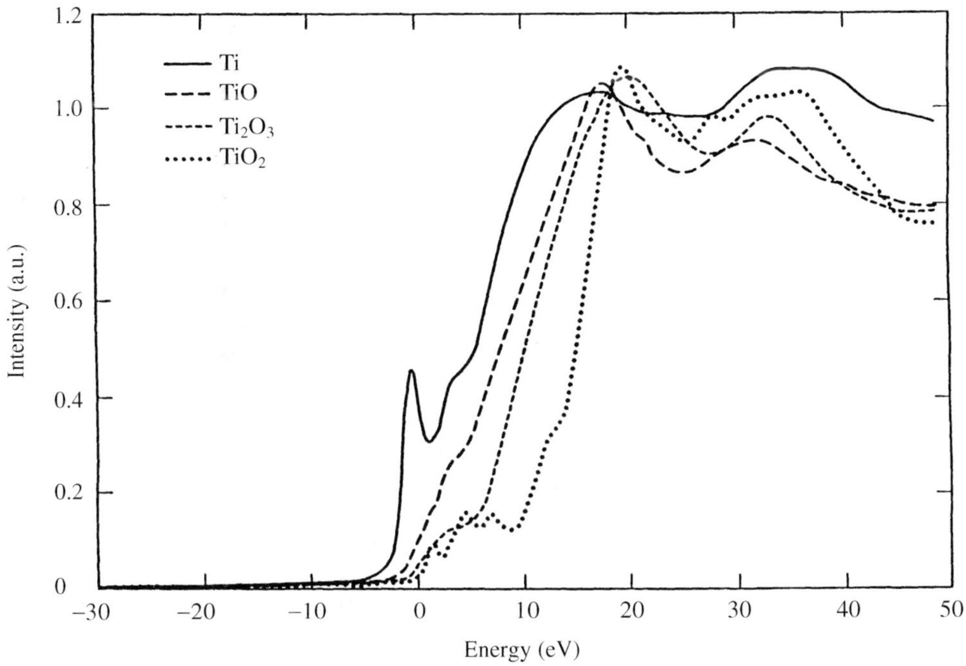

Figure 1. X-ray absorption spectra taken at the K-edge for Ti metal, TiO, Ti_2O_3, and TiO_2. Energy is taken relative to Ti metal K-edge at 4966.0 eV.

Such an exception is evident in the CrO_3 spectrum shown in Fig. 2a. The third trend is the increase in the variation of intensity of the features above the edge as the valence of the metal atom increases. This is clearly demonstrated by the Ti oxides in Fig. 1. In Fig. 2a, this trend is most evident in the absorption spectra of bis-benzene chromium and Cr_2O_3 compared to metallic Cr, over the energy range shown.

These trends have been used previously to study the reactions at buried metal/polymer interfaces [11, 12]. Reaction at the buried interface was studied by decreasing the thickness of the Cr or Ti metal layer (under 100 nm Pd) deposited on triazine or polyimide, which was coated over an alumina substrate (the first set of samples described in the experimental section). As a result of the decreasing thickness, metal reacted at the interface would have a larger percentage of overall signal. In other words, bulk metal would be a smaller fraction of the overall signal as thickness decreased. For the case of Ti/triazine, successive samples with decreasing Ti thickness showed that no metallic Ti was evident when 10 nm or less was deposited on a triazine surface [12]. In addition, temperature was used to drive reactions at the metal/polymer interface. Basically, as reaction proceeds, the metallic edge decreases in intensity, the edge moves to higher energies, and oscillations in edge intensity become more pronounced. Figure 3 shows the relative decrease in Cr metallic edge as a 30 nm thick layer on triazine polymer is heated. While the extent of reaction can be judged by looking at the decrease in metallic intensity, no determination of the reaction products can be made at this time.

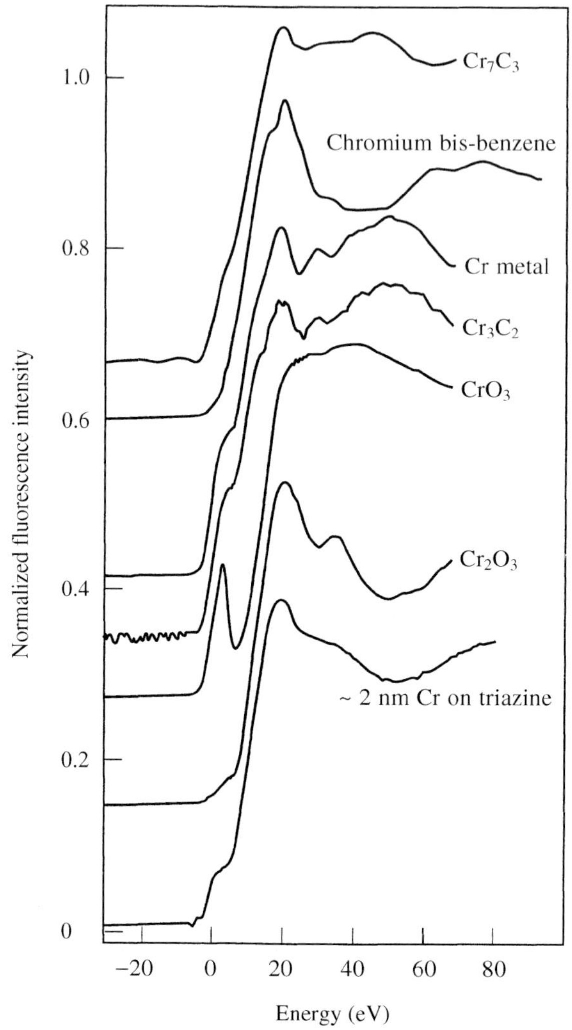

Figure 2a. X-ray absorption spectra taken at the K-edge for a series of Cr compounds. Energy is taken relative to Cr metal K-edge at 5989.0 eV.

The structure of the reaction products might be surmised by studying the extended X-ray absorption fine structure (EXAFS) of the reaction products. The fine structure is caused by interference of the final state of the emitted electron scattered from nearby atoms and interfering with itself. The variations in the intensity of the absorption edge up to 1000 eV above the edge can then yield the local structure around the emitting atom. This is best illustrated by some of our work on model compounds and the ion implanted samples described in the experimental section. When 1×10^{17} atoms of Cr are implanted into polyimide or triazine, EXAFS suggests that Cr has 6 nearest neighbors 0.203 nm away. In other words, Cr appears to be perched 0.162 nm above an aromatic ring. This structure is consistent with that proposed earlier by IBM workers for the structure of Cr adsorbed on polyimide [4, 5].

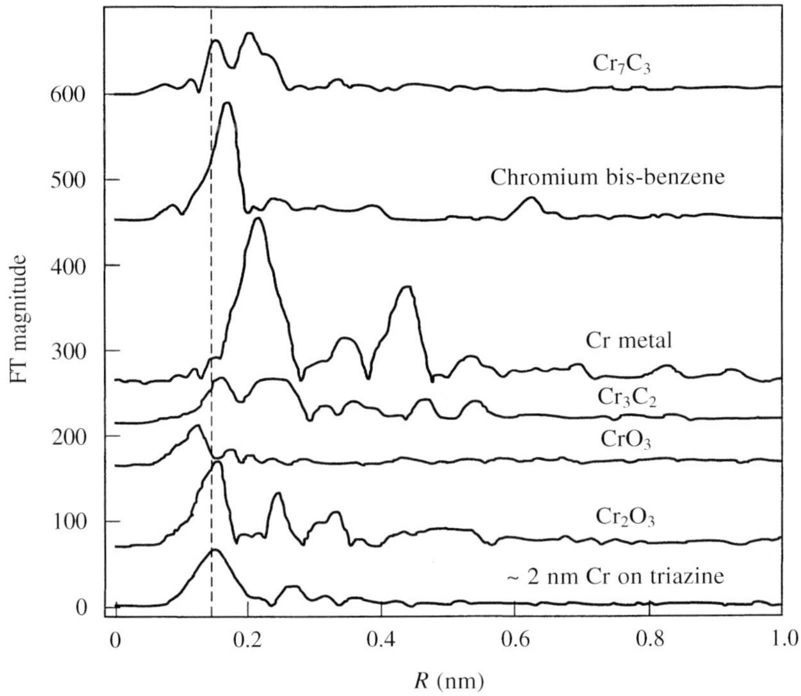

Figure 2b. EXAFS analyzed data for the same series of Cr compounds, where the Fourier Transform (FT) magnitude is plotted as a function of position (nm).

The EXAFS for Ti implanted into triazine or polyimide gave less satisfactory results. The near edge absorption structure did show that Ti formed a tetrahedral compound, probably a tetrahedral oxide upon implantation into the polymer. Our XPS results [3] suggest that Ti will form σ-bonds with any C, N, or O species in the polymer upon adsorption. However, peeling Ti/polymer interfaces in the vacuum system shows that Ti has evolved to form primarily TiO_2 at the interface. The implanted Ti appears to be an intermediate in this evolution from σ-bonded organometallic Ti, to tetrahedrally bonded Ti oxide species, and finally to bulk TiO_2.

Another study of a Cr model compound is illustrative. In this case, the EXAFS of Cr bis-benzene was studied, motivated by the Cr implant data above. The near edge absorption spectrum is shown in Fig. 4a, above, and it was acquired with the sample at 11 K, to minimize the effects of light atom vibrations. The oscillations in absorption intensity above 6 keV are obvious, but are also seen to consist of two components: a large low frequency component (occurring approximately every 0.1 keV starting at 6.05 keV) and a smaller high frequency component (occurring approximately every 0.02 keV). The low frequency component corresponds to scattering from nearest neighbors, and the high frequency component corresponds to scattering from next nearest neighbors. EXAFS analysis (Fig. 4b) shows the scattering from the nearest neighbor C atoms in the benzene rings attached immediately above and below the central Cr atom at a distance of 0.15 to 0.20 nm. In addition, the spectrum acquired at 11 K shows scattering from the nearby Cr atom, approximately 0.6 nm away. The

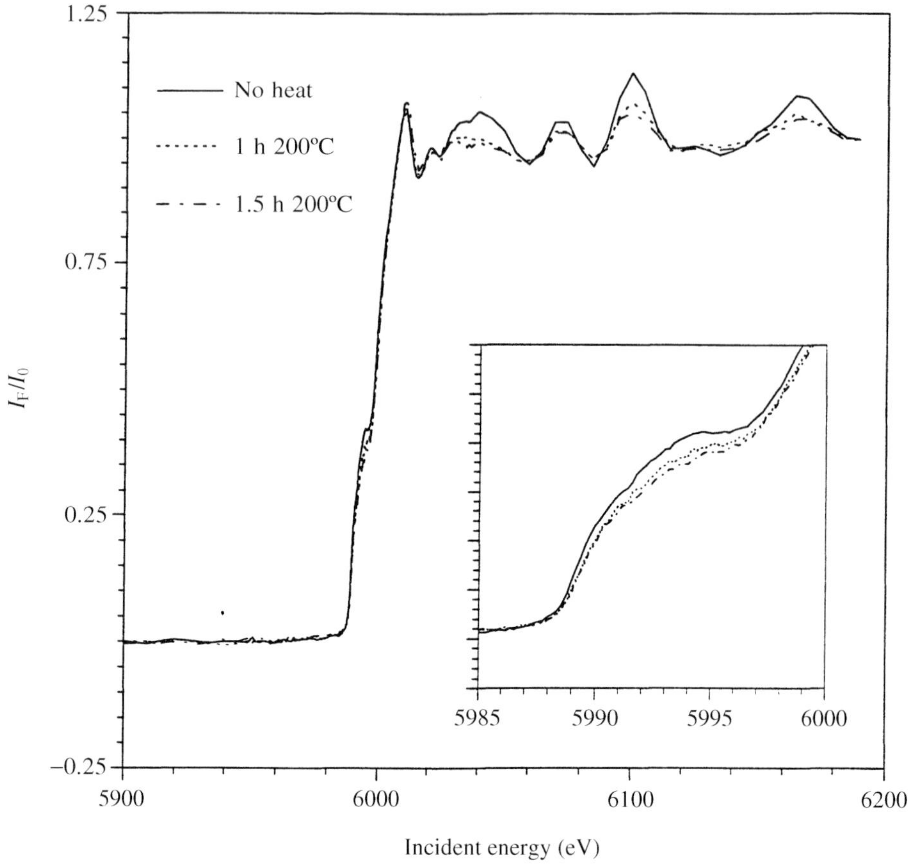

Figure 3. X-ray absorption spectra of 30 nm Cr on triazine: no heat (solid line), heat at 200°C for 1 h (dotted line), heat at 200°C for 1.5 h (dash-dotted line). Inset is the expanded edge region. The fluorescence intensity, I_F, is normalized to the incident X-ray intensity, I_0.

EXAFS structure is consistent with that found from X-ray diffraction [15]. In addition, the EXAFS for Cr bis-benzene is consistent with the EXAFS for Cr implanted into polyimide. However, it is clear that there is some more complicated structure in the implanted data.

In recent experiments we have been able to acquire the near edge spectra and EXAFS structure for extremely thin (2 nm) layers of Cr (under 100 nm Pd) on unsupported polyimide and triazine (the second set of samples described in the experimental section). Even at metal layers this thin, a substantial amount of Cr remains metallic, as seen by the metallic pre-edge feature in Fig. 2a. For a 2 nm film, approximately 30% remains metallic, yielding a stable reacted layer of approximately 1.2–1.5 nm. EXAFS of this thin layer most closely resembles Cr_2O_3 (Fig. 2b). Cooling the sample to 20 K shows no further structure, suggesting that Debye–Waller factors are not hiding further structure in this layer. Interestingly, the EXAFS and near edge spectra for Cr adsorbed on either polyimide or on triazine are identical, suggesting that for these high temperature polymers, Cr chemistry is dominating the structure of the interface.

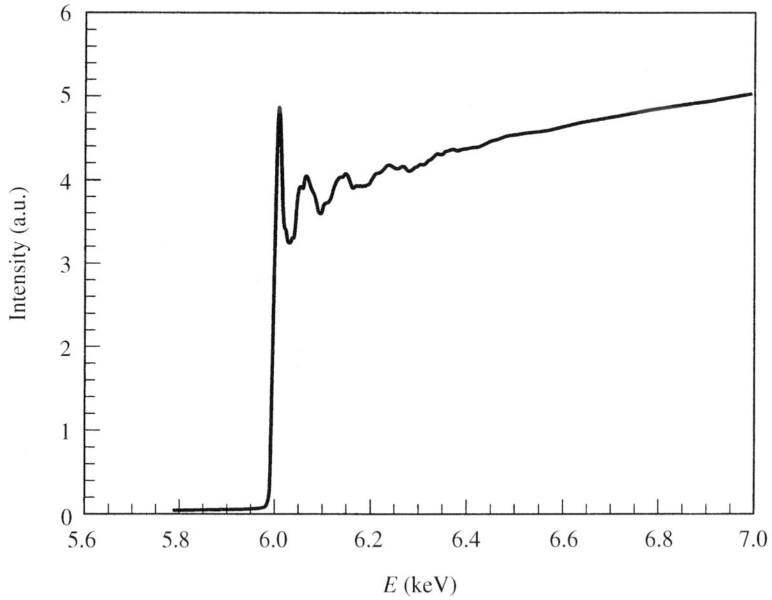

Figure 4a. X-ray absorption spectrum for Cr bis-benzene.

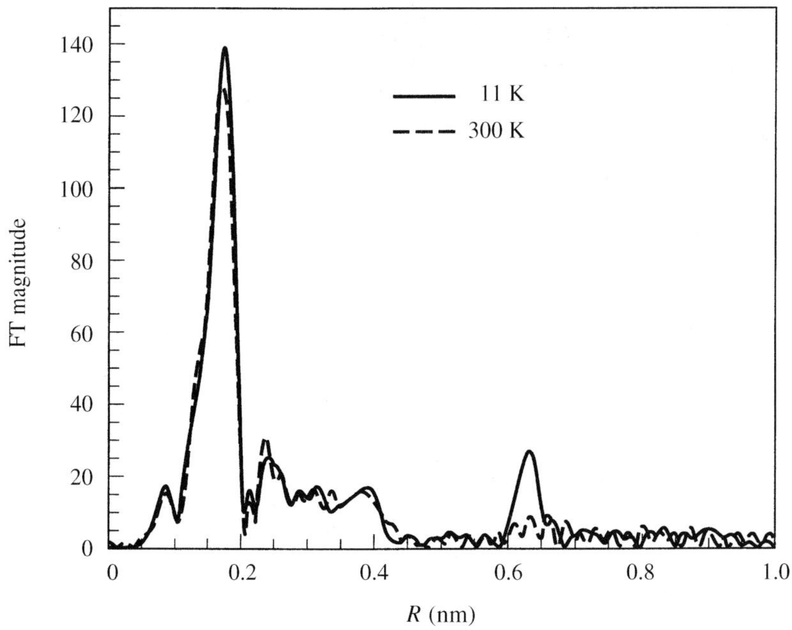

Figure 4b. EXAFS Fourier Transform (FT) magnitude is plotted as a function of position (nm) for Cr bis-benzene for spectra taken at 11 K (solid) and 300 K (dashed).

4. CONCLUSION

X-ray absorption spectroscopy results suggest that metal/polymer interfaces for Ti or Cr adsorbed on either triazine or polyimide evolve chemically. In other work, XPS shows that when these metals are deposited on polymers, Ti forms σ-bonds with C, N, and O, and that Cr also forms σ-bonds with C, N, and O, but in addition forms π-bonds with aromatic species. X-ray absorption shows that both metals eventually form stable oxides at the interface, due to diffusion of O_2 and H_2O to the interface. The final structure of the interface is dominated by this oxide formation, but also depends on the nature of the polymer, treatment of the polymer surface before deposition, and temperature history of the structure after metal deposition.

Acknowledgements

The authors acknowledge A. O. Ibidunni, A. J. Davenport, and H. S. Isaacs for their participation in early experiments. A. E. White and K. Short prepared the ion implanted samples. R. B. van Dover prepared the very thin Cr deposited on polyimide and triazine. D. Adler was of great help in acquiring the data from the very thin samples.

REFERENCES

1. S. P. Murarka, *Solid State Technol.* 83 (March 1996).
2. S. P. Jeng, R. H. Havemann and M. C. Chang, *Mater. Res. Soc. Symp. Proc.* **337**, 25 (1994).
3. K. Konstadinidis, R. L. Opila, A. C. Miller and J. A. Taylor, *J. Adhesion* **46**, 197 (1994).
4. P. S. Ho, P. O. Hahn, J. W. Bartha, G. W. Rubloff, F. K. Legoues and B. D. Silverman, *J. Vac. Sci. Technol.* **A3**, 739 (1985).
5. R. Haight, R. C. White, B. D. Silverman and P. S. Ho, *J. Vac. Sci. Technol.* **A6**, 2188 (1988).
6. J. W. Bartha, P. O. Hahn, F. Legoues and P. S. Ho, *J. Vac. Sci. Technol.* **A3**, 1390 (1985).
7. P. O. Hahn, G. W. Rubloff and P. S. Ho, *J. Vac. Sci. Technol.* **A2**, 756 (1984).
8. R. M. Tromp, F. Legoues and P. S. Ho, *J. Vac. Sci. Technol.* **A3**, 782 (1985).
9. B. D. Silverman, P. N. Sanda and P. S. Ho, *J. Polym. Sci. Polym. Chem. Ed.* **23**, 2857 (1985).
10. K. L. Mittal (Ed.), in: *Adhesion Measurement of Films and Coatings*, pp. 1–13. VSP, Utrecht, The Netherlands (1995).
11. R. L. Opila, *Mater. Res. Soc. Symp. Proc.* **264**, 225 (1992).
12. R. L. Opila, K. Konstadinidis, A. O. Ibidunni, A. J. Davenport and H. S. Isaacs, *Mater. Res. Soc. Symp. Proc.* **304**, 111 (1993).
13. M. A. Marcus and W. Flood, *Rev. Sci. Instrum.* **62**, 839 (1991).
14. R. L. Opila, K. Konstadinidis and S. O'Connor, in: *Polymer Surfaces and Interfaces: Characterization, Modification, and Application*, K. L. Mittal and K.-W. Lee (Eds), pp. 179–187. VSP, Utrecht, The Netherlands (1997).
15. E. Weiss and E. O. Fischer, *Z. Anorg. Allg. Chemie* **286**, 142 (1956).

Mittal Festschrift, pp. 773–790
W. J. Van Ooij and H. R. Anderson, Jr. (Eds)

Enhanced metal/ceramic adhesion by surface treatment with excimer UV radiation

HILMAR ESROM,[1,*] ANTHONY J. PEDRAZA,[2] RAJAGOPALAN A. KUMAR[2] and JUN-YING ZHANG[3]

[1]*Fachhochschule Mannheim, Institute of Technology Mannheim, 68163 Mannheim, Germany*
[2]*Department of Materials Science and Engineering, The University of Tennessee, Knoxville, TN 37996-2200, USA*
[3]*Department of Electronic and Electrical Engineering, University College London, Torrington Place, London, WCIE 7JE, UK*

Abstract—Controlled modification of surface properties by the interaction of photons is a field of considerable technical importance. It can be used to improve the adhesion properties of metal coatings by changing the morphology and chemical properties of the surface. Furthermore, new techniques for metal deposition on arbitrary materials at low temperature by UV stimulated processes are of growing interest. Due to the strong absorption of ultraviolet (UV) radiation by most ceramics, polymers and precursors for photo-induced metal deposition, excimer UV laser processing of these materials occurs efficiently. Recently, novel excimer UV lamps were shown to be very useful in surface processing. Both modifications of surfaces with high-intensity UV from excimer lasers and low-intensity UV from excimer lamps are appealing methods for interesting applications.

This paper describes the use of excimer lasers and excimer lamps for UV-induced modification and metallization of ceramics and discusses the characteristics of the processes concerning the adhesion of metal films on ceramics.

Keywords: Excimer UV laser processing; excimer lamp; surface treatment; modification of surfaces; metallization of ceramics; adhesion of metal films.

1. INTRODUCTION

UV laser light induced modifications of the near-surface region of insulators are caused by photo-chemical or thermal processes [1]. In the first case, the absorbed photon produces atomic or molecular excitations that eventually may lead to decomposition of the compound at the surface and removal of material by ablation [2]. It is known that under certain circumstances polymers and some ceramic materials such as SiO_2 can be photochemically etched by excimer UV laser and excimer UV lamp irradiation [3, 4]. There is also a large class of surface modifications that takes place by surface heating

*To whom correspondence should be addressed.

[5–7]. Melting, evaporation and/or decomposition of the near surface of insulators can take place as a result of the large heating produced by laser irradiation [8, 9]. There is a large body of experimental evidence that shows that laser irradiation, especially in the ultraviolet range, can produce permanent changes in dielectric materials [10–12]. Electronic states associated with damage sites pre-existent in the material (e.g. grain boundaries, dislocations and point defects) together with defects produced during irradiation, coupled with incoming radiation and sharply increased absorption of the laser radiation produces heat [13, 14]. The most important aspect of UV effects is that the damage and modifications produced at the surfaces of ceramics and polymers can alter their chemical and catalytic properties.

Metallorganic films may decompose during UV irradiation [15]. For instance, palladium acetate (Pd ac) films deposited on insulating substrates decompose when they are excimer laser and excimer lamp irradiated [16, 17]. In the case of excimer laser irradiation of Pd ac, the decomposition mechanism is thermally driven and a thin palladium film is left on the surface after irradiation [16, 17]. While pulsed excimer laser decomposition of Pd ac is mainly a pyrolytic process, the decomposition with incoherent excimer radiation of lamps is mainly photolytic [16, 17]. The Pd thin film produced by excimer lasers and lamps acts as catalyst for copper deposition when the irradiated substrate is immersed in an electroless solution [18]. In this two-stage process, namely, UV irradiation followed by electroless deposition, a conductive pattern can be made by either laser scanning or masking the substrate for the laser and lamp irradiation [16, 17]. One of the key problems for the application of this process is the adhesion strength between the metallic film and the substrate. In the first part of this paper we are going to analyze the irradiation conditions required to obtain maximum adhesion strength.

Ceramic materials can also be decomposed by excimer laser irradiation [19, 20]. The laser energy density required to decompose AlN is one order of magnitude larger than that required to decompose palladium acetate [19]. In this case metallic aluminum clusters are left on the surface [7]. Al_2O_3 also decomposes when it is irradiated in a reducing atmosphere, leaving metallic clusters on the surface [21]. Both AlN and Al_2O_3 become activated upon irradiation and, when immersed in a copper electroless bath, metal deposition takes place on the irradiated surface [21–23]. The effects of laser irradiation and irradiation atmosphere on the adhesion strength of electroless deposited copper on AlN and Al_2O_3 have been recently studied [24]. It was found that the adhesion strength of the electroless copper films deposited both on AlN and on Al_2O_3 could be significantly increased if a low temperature annealing was performed after deposition. Lower laser energy densities promote better adhesion upon subsequent annealing [24]. Another interesting effect is the strong adhesion enhancement in metallic films sputter-deposited on excimer laser irradiated substrates [25, 26]. In the second part of this paper we describe the effect of excimer laser irradiation of Al_2O_3 on the adhesion strength of thin gold films deposited on the irradiated Al_2O_3. We have measured the electrical resistance of the film after annealing. These studies, which allowed us to understand the relation between thermal stability of the films and adhesion strength, are presented at the end of this paper.

2. EXPERIMENTAL

Without surface pretreatment of the substrates, thin spin-on Pd ac films (50 nm in thickness) were deposited on Al_2O_3 substrates (Rubalit 708-Hoechst, roughness 0.6 μm maximum) and AlN substrates (Hoechst) using a solution of Pd ac dissolved in chloroform (concentration of about 0.1 mol/dm^3 [16]). The Pd ac films were irradiated in air with a 248 nm-wavelength excimer laser (EMG 103, Lambda Physik). The pulse energy was measured with a joule meter (Gen Tec PRJ-D, ED 200/500). In the photo-decomposition experiments with excimer lamps we used incoherent excimer radiation from a dielectric-barrier discharge (silent discharge) operating in a gas mixture of krypton/chlorine (pressure: about 0.06 MPa) providing intense narrow band radiation at the wavelength $\lambda = 222$ nm. The discharge was initiated in an annular gap between coaxial quartz tubes. The outer quartz tube had a diameter of 3 cm and an active length of 15 cm defined by the length of a grounded wire mesh acting as transparent outer electrode. The radial width of the annular discharge gap was about 0.4 cm. The excimer lamp was operated at 20 kHz and voltages up to 20 kV. More details about this special excimer UV lamp can be found in the literature [27, 28]. Irradiation of the samples was carried out in a small exposure reactor [29]. The distance between the samples and the excimer source was 1 cm. The UV intensity was measured using a vacuum phototube (R 765, Hamamatsu). Parts of the irradiation were performed with an adjustable rectangular mask to select the homogeneous part of the intensity distribution. Other specimens were excimer laser irradiated by scanning across the specimen a line-focused laser obtained using a cylindrical quartz lens ($f = 156$ mm). The focused line was 10 mm $\times$ 300 μm, the scan velocity was 1 mm/s and the scanned area was 2.5 cm^2.

A commercially available bath (Shipley CP-78) was used for the copper electroless deposition. For a qualitative estimation of the adhesion of Cu films to ceramic substrates a common Scotch tape test was used. The adhesion of metal films was also quantitatively measured using a Sebastian Adherence Pull Tester (Quad Group).

Al_2O_3 substrates (Coors Ceramic Co.) were irradiated with a 308 nm-wavelength excimer laser. After irradiation, gold films were ion-beam sputter-deposited on the irradiated substrates. The gold–Al_2O_3 couples were annealed for 1 h at 550°C under a pressure of 1×10^{-6} Torr. The adhesion strength of the gold–Al_2O_3 couples was measured by pull testing as described below using epoxy coated aluminum pins that have a strength rating of 70 MPa when metallic films other than gold films are used. An 80 nm-thick copper film was deposited on the gold–Al_2O_3 couples after the 550°C annealing because the epoxy does not adhere well to gold. The gold–copper interface did not pose any problem for the adhesion measurements because it is stronger than either the gold–Al_2O_3 or the copper–epoxy interface. The copper also acts as a convenient intermediate layer between the Al_2O_3 and the epoxy when the gold films clustered upon annealing. The bond strength of the copper film to Al_2O_3 is very small and we mostly measured the adhesion strength of the gold clusters to Al_2O_3. When gold clusters are present, they are the only anchoring points for the copper thin film to the substrate. The epoxy coated pins were placed on the copper film with an elastic clamp. The epoxy was cured by holding the assembly at 150°C for 90 min and slowly cooling it down to room temperature. During the pull test, perfect alignment of the pin in the pulling direction was ensured, as required to eliminate shear stresses. The pull tests were done at a speed of 3 μm/s with the help of a stepper motor and a computer-controlled system. The load was measured with a sensitivity of 5 g. After failure, pin and substrate were examined both visually and with a scanning electron microscope to establish whether the film had failed at the metal–ceramic interface or in the adhesive.

3. RESULTS

3.1. Adhesion of copper films

The laser induced thermal decomposition of Pd ac to produce metallic Pd depends on the excimer laser wavelength [16, 30]. We found in this study that no decomposition occurred if the laser energy density was equal to or less than 15 mJ/cm^2, in agreement with previous results [16, 30]. The Pd ac films deposited on Al_2O_3 substrates were irradiated at laser energy densities in the range of 15 to 250 mJ/cm^2. In contrast with excimer laser decomposition of Pd ac, no threshold was observed in the decomposition process of Pd ac initiated by incoherent excimer lamp radiation. Even at a low UV intensity of 30 mW/cm^2 complete decomposition of Pd ac was achieved after sufficiently long exposure times (about 5 min). The non-exposed portions of the Pd ac film were removed by immersing the substrates in chloroform. Following the removal of the Pd ac, the Pd–Al_2O_3 couples were immersed in a copper electroless solution where 0.1 μm-thick copper films were deposited on the laser or lamp-activated substrates.

Scotch tape tests performed on the electroless copper deposited on Pd-activated Al_2O_3 substrates showed that the adhesion strength increased as the laser energy density increased, reaching a maximum value at 25 mJ/cm^2 (ten pulses) (Fig. 1). After 25 mJ/cm^2 the bond between copper and Pd is strong enough so that 100% of the film remains on the substrate after the test. To overcome the limitation of this test and make a more quantitative evaluation, a pull tester was used on other irradiated specimens. Figure 2 shows the adhesion strength measured with a pull tester as a function of the laser energy density (ten pulses). The adhesion strength reaches a maximum at a laser energy density of 150 mJ/cm^2. Further increases in laser energy density produce a decrease in the adhesion strength. For comparison, the adhesion strength of electroless copper films deposited on Pd pre-nucleated Al_2O_3 substrates by Pd ac decomposition with the 222 nm-excimer lamp was low, i.e. 4.5 MPa. The adhesion strength of 0.1 μm-thick copper films deposited by conventional evaporation on the as-received Al_2O_3 substrates was determined to be 5 MPa by pull tests whilst those deposited by conventional sputtering had an adhesion strength of 7 MPa. The maximum value of 12 MPa obtained for electroless copper films deposited on the palladium-activated substrates compares favorably with that obtained when copper was deposited by evaporation, sputtering and excimer lamp processing.

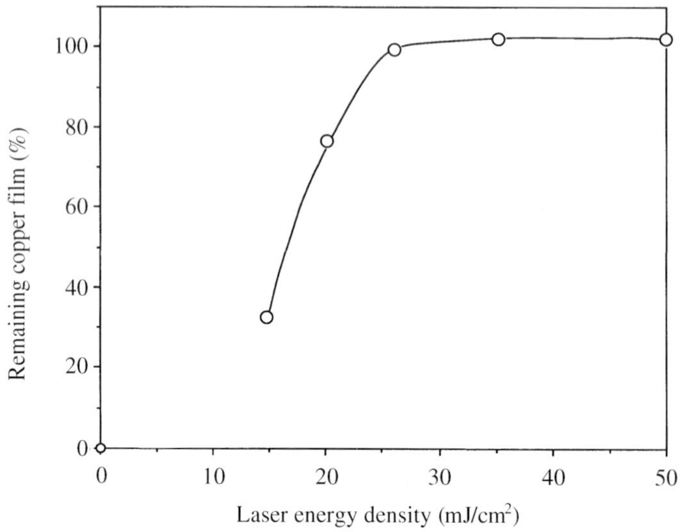

Figure 1. Percentage of copper film remaining on the substrate after tape testing, as a function of laser energy density.

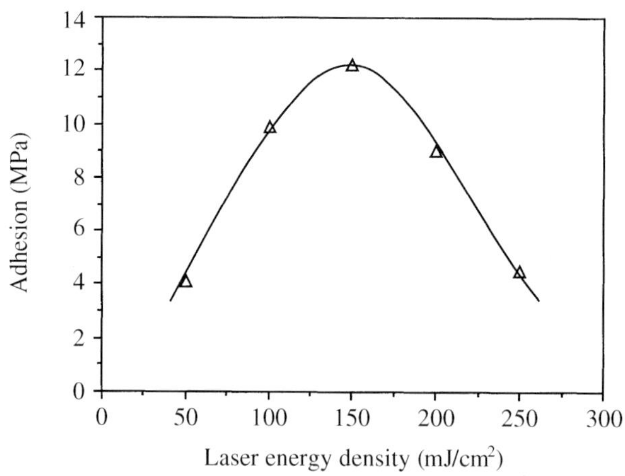

Figure 2. Adhesion strength of electroless copper deposited on Pd-activated Al_2O_3 as a function of laser energy density. The adhesion strength was measured with a pull tester.

Pd ac films were also deposited on AlN substrates using the spin-on technique. Electroless copper was deposited on these substrates using the same two-step procedure, as described above, with the excimer laser at optimum fluence of 150 mJ/cm^2. The adhesion strength using the pull tester gave very similar results to those shown in Fig. 2 again giving a maximum adhesion strength of 12 MPa. Similar excimer laser experiments were performed using a nickel instead of a copper electroless bath. The adhesion strength of Ni on both AlN and Al_2O_3 Pd-activated exhibit a similar behavior to that shown in Fig. 2.

Spin-on glass layers on the Al_2O_3 surface (Doduco) were used as samples in additional excimer laser experiments, in order to enhance the adhesion between the electroless copper and the substrate. The adhesion strength of the electroless copper to the glass/Al_2O_3 substrates was increased to 20 MPa. Further improvements in adhesion were obtained if the Al_2O_3 was chemically etched in H_3PO_4 for 30 min at a temperature of 240°C. This treatment, performed prior to spin-on deposition of Pd ac, increased the adhesion of the electroless copper film to 37 MPa.

The surface morphology characteristic of as-received Al_2O_3 substrates is shown in Fig. 3. On a larger scale, the topography of particles protruding on the surface of Al_2O_3 has disappeared after the deposition of a glass layer (Fig. 4). However smaller cavities, typically 0.1 μm in diameter, can be seen in the glass layer. Furthermore, at lower magnification it can be seen that the number of voids or cavities per unit area is high. Figure 5 is a SEM micrograph of the chemically etched substrate. The etching produced sharp edges and deep voids between the Al_2O_3 grains.

Figure 3. Surface of an as-received Al_2O_3 substrate. (SEM micrograph: scale shown = 1 μm.)

3.2. Adhesion of gold films

Excimer laser irradiation can produce chemical surface modifications of ceramic substrates if laser energy densities 4 to 10 times higher than those employed to decompose the Pd ac are used. A series of experiments were performed to measure the adhesion strength of Au films deposited on Al_2O_3, and at the same time to correlate the film integrity or stability with adhesion after annealing, Al_2O_3 substrates were laser treated in air, oxygen and Ar–4% H_2 atmospheres at different energy densities. Thin gold films 20 nm in thickness were ion-beam sputter deposited on the irradiated substrates. The gold–Al_2O_3 couples were then annealed at 550°C for 45 min. Six substrates identically treated were used to perform pull tests and the remainder were used to measure the electrical resistance and for SEM observations. Figures 6–8 show joint plots of the pull-test adhesion results and the two-probe resistance results for substrates irradiated in Ar–4% H_2, air and oxygen, respectively. In all cases, the highest adhesion strength corresponded to the lowest electrical resistance. The

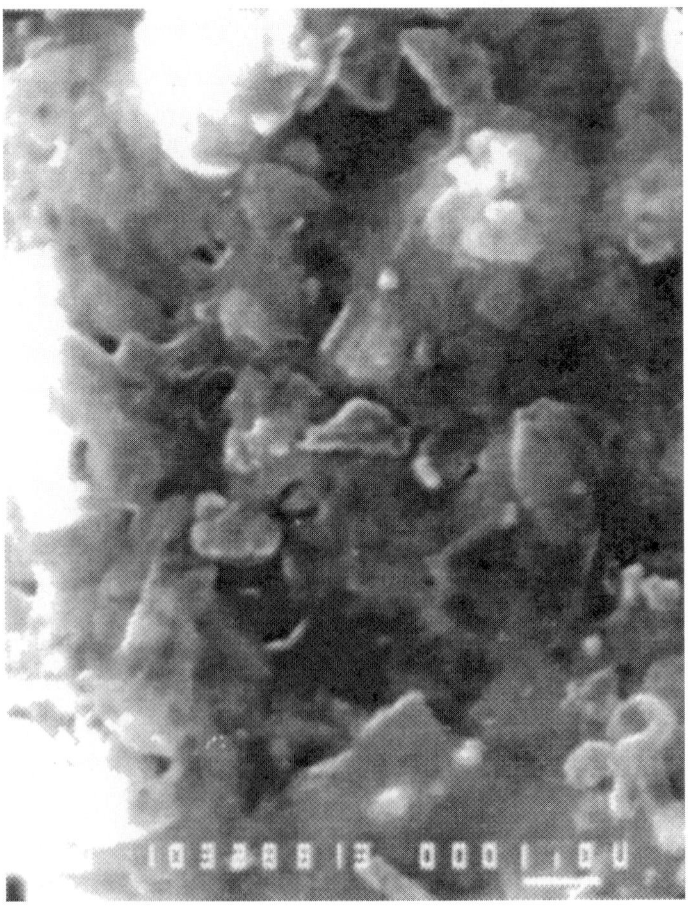

Figure 4. SEM micrograph of a spin-glass film deposited on the as-received Al_2O_3 substrate. (Scale shown = 1 μm.)

Figure 5. SEM micrograph of a chemically etched Al_2O_3 substrate. The etching was performed by immersing the Al_2O_3 substrate in H_3PO_4 heated at 240°C for 30 min.

adhesion strength has a maximum value at a laser energy density of ~ 1 J/cm^2 for all three laser irradiation atmospheres, while the resistance curves show a minimum value at the same energy density. The electrical resistance varied from 88 Ω for an adhesion strength of 50 MPa to 16 MΩ for an adhesion strength very close to zero. As found previously for 80 nm-thick gold films and lower annealing temperatures [25], the adhesion strength is strongly dependent on the irradiation atmosphere. When the irradiations were performed in a reducing atmosphere of Ar–4% H_2, the maximum value of the adhesion strength reached only 3 MPa, while in oxygen atmosphere the maximum adhesion strength was 50 MPa.

Figures 9 and 10 are SEM micrographs of gold films deposited on Al_2O_3 substrates irradiated in air at 1 and 2 J/cm^2, respectively, before deposition. The gold–Al_2O_3 couples were annealed at 550°C for 45 min. The film deposited on the substrate irradiated at 2 J/cm^2 completely broke down into small islands, locally agglomerating throughout on the substrate surface. This result is consistent with the very high

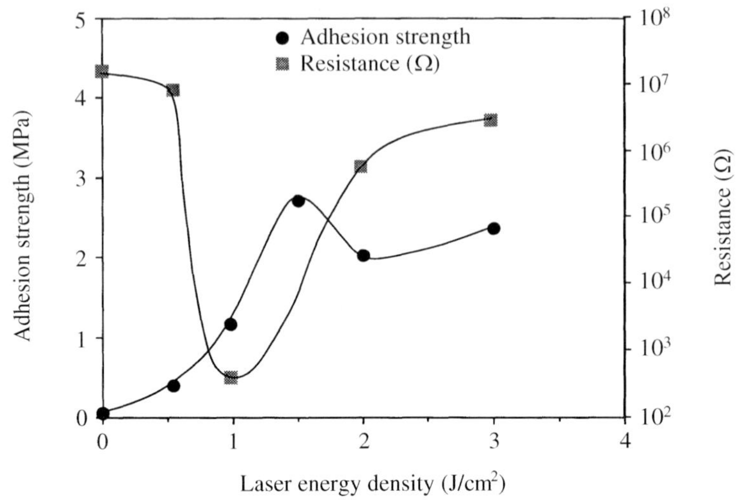

Figure 6. Adhesion strength and electrical resistance of 20 nm-thick gold films deposited on irradiated Al_2O_3 as a function of laser energy density. The substrates were irradiated in an Ar–4% H_2 atmosphere. The gold–Al_2O_3 couples were annealed at 550°C for 45 min before adhesion tests were performed with a customized pull tester.

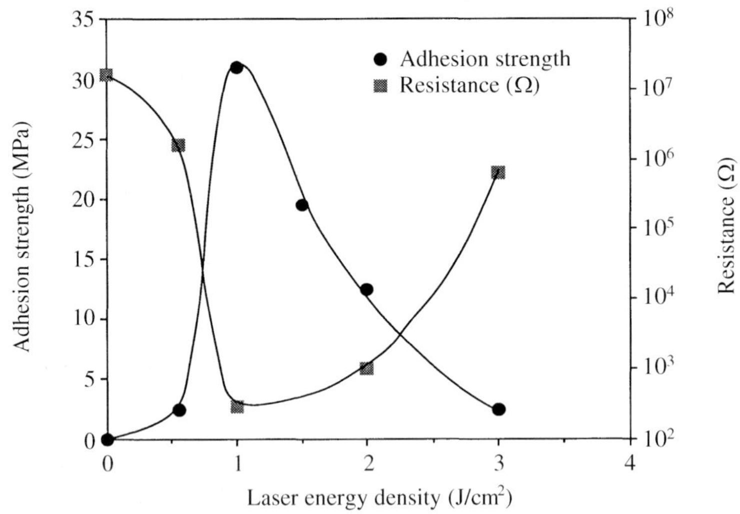

Figure 7. Adhesion strength and electrical resistance of 20 nm-thick gold films deposited on Al_2O_3 irradiated in air as a function of laser energy density. The gold–Al_2O_3 couples were annealed at 550°C for 45 min before adhesion tests were performed with a customized pull tester.

electrical resistance of the film. The film deposited on the substrate laser treated at 1 J/cm^2 showed much less breakdown and agglomeration. The film laser-treated in oxygen at 2 J/cm^2 is almost continuous after annealing with breakdowns only in the regions between Al_2O_3 particles (Fig. 11). The gold film is complete with no sign of breakdown in the Al_2O_3 substrate irradiated at 1 J/cm^2 (Fig. 12). These micrographs are completely consistent with the fact that the films that exhibit the higher adhesion

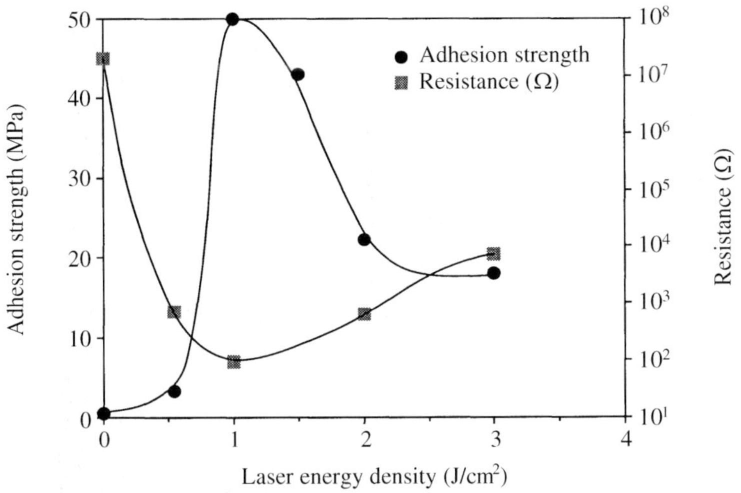

Figure 8. Adhesion strength and electrical resistance of 20 nm-thick gold films deposited on Al_2O_3 irradiated in an oxygen atmosphere, as a function of the laser energy density. The gold–Al_2O_3 couples were annealed at 550°C for 45 min before adhesion tests were performed with a customized pull tester.

strength and the lower electrical resistivity are those that have the least breakdown of the gold film into clusters after annealing.

4. DISCUSSION

4.1. Adhesion of copper films

Excimer laser and lamp irradiation decomposed Pd ac leaving a thin metallic film of Pd on the substrate surface. These prenucleated Al_2O_3 irradiated substrates are next immersed in a commercial bath and electroless copper deposition takes place on Pd because Pd acts as a catalyst for electroless deposition [18]. Two interfaces are then present between the Al_2O_3 substrates and the electroless copper, namely, Pd/Al_2O_3 and Cu/Pd. The dependence of the adhesion strength on the laser energy density seems to indicate that the interface controlling the adhesion is that between Pd and Al_2O_3, because the laser irradiation is performed on the Pd ac prior to metal deposition. This conclusion would be in agreement with experimental evidence and the results of theoretical analyses [31, 32] that show that bonding between noble metals and alumina should be weak.

During the 20 ns excimer laser pulse, the Pd ac film is heated and when the deposited energy is larger than 15 mJ/cm^2 thermal decomposition takes place [16, 30, 33]. At 20 mJ/cm^2 the Pd ac film decomposes and the thin Pd layer left on the surface clustered, leaving areas of the substrate uncovered (Fig. 13). In this case, the mean cluster diameter is smaller than 0.1 μm. When the laser energy density is increased, the decomposition takes place at earlier times in the irradiation process. The rest of the energy is used to heat the Pd film and the heat is transferred by conduction from

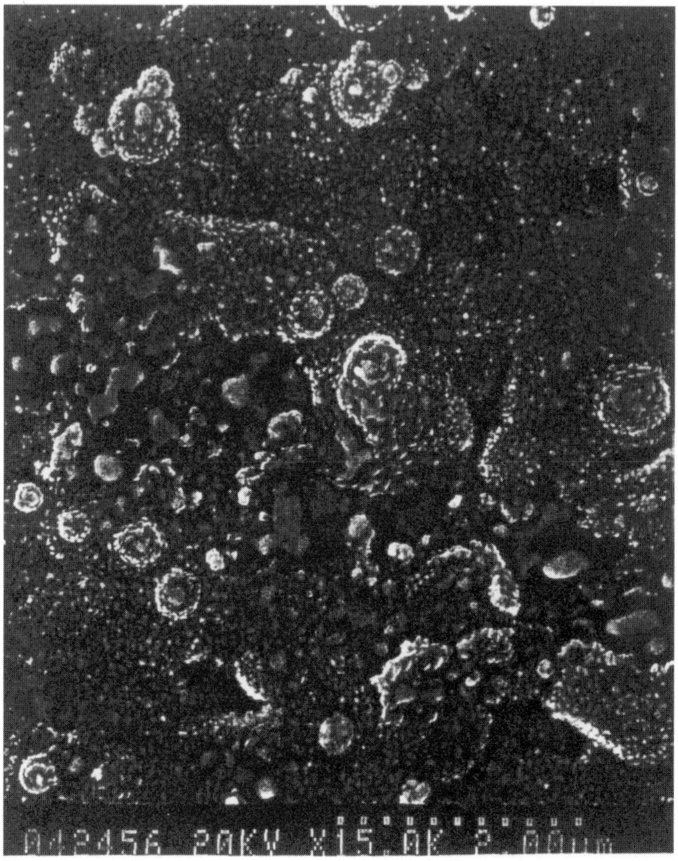

Figure 9. SEM micrograph of a gold film deposited on Al_2O_3 irradiated at 1 J/cm^2 in air. The couple was annealed at 550°C for 45 min at a pressure of 1×10^{-6} Torr.

the substrate into the bulk. The energy per unit area of the film needed to melt a 50 nm Pd film is ~32 mJ/cm^2 which, added to the 15 mJ/cm^2 required to decompose the Pd ac, gives a value of 47 mJ/cm^2. In this calculation no heat dissipation was allowed for, since Al_2O_3 is a very poor thermal conductor and most of the energy should be deposited initially in the film. From the results of this work, it is clear that melting of the Pd film takes place but it is very difficult to evaluate the overheating of the liquid Pd, because as soon as the Pd becomes liquid it tends to break into small clusters as observed by scanning electron microscopy (Fig. 14). The laser melting of Pd resulting in cluster films is necessary in order to improve the adhesion strength between Pd and Al_2O_3. In earlier SEM investigations we found that highly continuous Pd films without clusters were produced by the photolytic decomposition of Pd ac at low processing temperature (< 50°C), using excimer lamps [17]. The smoothness of Pd results in a low adhesion < 5 MPa of electroless deposited Cu on top of the continuous Pd films.

In previous experiments we found that copper melting during laser irradiation of a copper–alumina couple was a necessary condition to enhance the bonding of sputter-

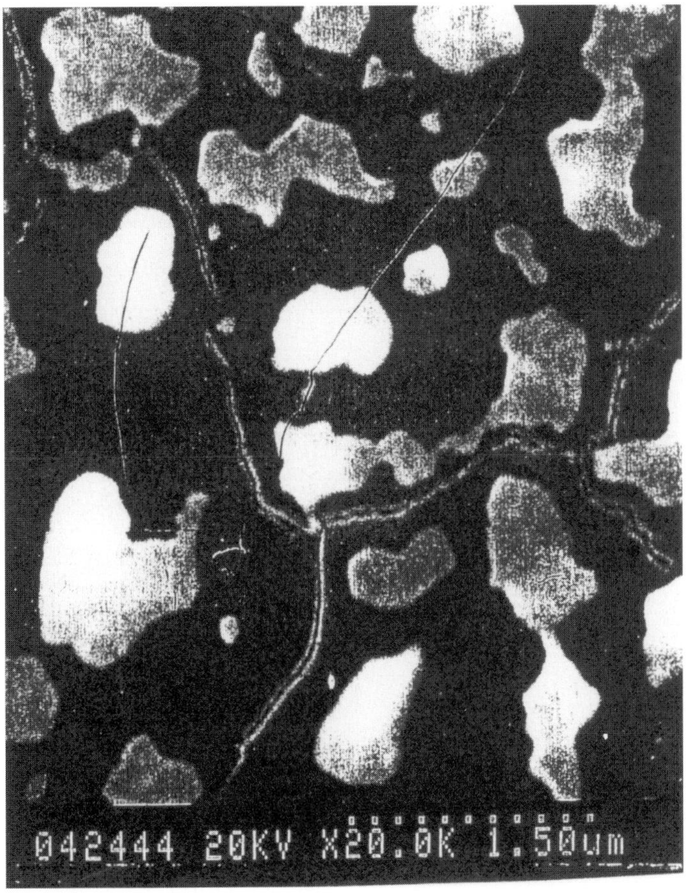

Figure 10. SEM micrograph of a gold film deposited on Al_2O_3 irradiated at 2 J/cm^2 in air. The couple was annealed at 550°C for 45 min at a pressure of 1×10^{-6} Torr.

deposited copper to alumina [9]. The calculations mentioned above suggest that at 150 mJ/cm^2 and above, the liquid is overheated. In the very short time of irradiation, the probability of a reaction and stronger bonding between metal and Al_2O_3 increases as the temperature increases. We should then expect an increase in bonding between metal and substrate as the laser energy increases. However the liquid that remains on the surface of the substrate breaks into clusters and the clusters become smaller as the laser energy increases [34]. At higher laser energy densities, melting and overheating of the liquid takes place at earlier stages in the irradiation process leaving long enough time for the liquid to break into small clusters. Also evaporation of the metal takes place, reducing the size and amount of metal clusters in contact with the substrate. The reduction of the surface area covered with palladium that remains on the surface makes the adhesion weaker at higher laser energy density. At fluences > 1 J/cm^2, much higher than the threshold fluence, complete removal of the Pd occurs by ablation and no electroless copper plating is possible in the laser beam treated region.

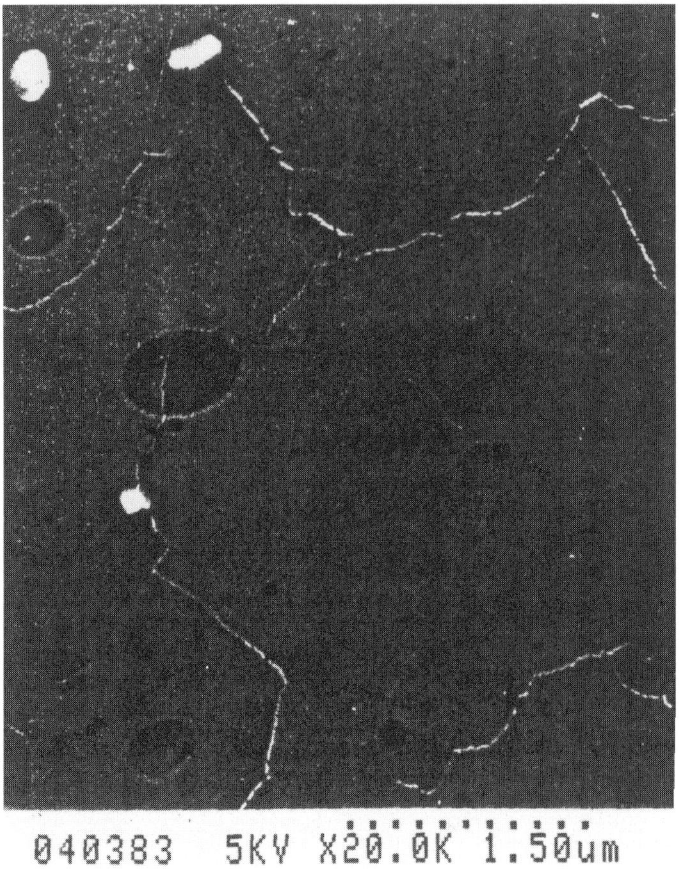

Figure 11. SEM micrograph of a gold film deposited on Al_2O_3 irradiated at 1 J/cm^2 in an oxygen atmosphere. The couple was annealed at 550°C for 45 min at a pressure of 1×10^{-6} Torr.

4.2. Adhesion of gold films

The surface of alumina itself was modified during laser irradiation at a laser energy density 4 to 10 times larger than that required for decomposition of the Pd ac. We have found using Auger electron spectroscopy and high resolution electron microscopy that laser irradiation in an Ar–4% H_2 atmosphere promotes the formation of nanoparticles of alumina and a layer of alumina that is oxygen deficient [20, 35]. In the case of irradiation in oxygen atmosphere both energy dispersive X-ray spectroscopy (EDS) [20] and Auger electron spectroscopy (AES) [35] detected an oxygen concentration in the irradiated area significantly larger than in untreated surface. A detailed AES analysis of the gold/Al_2O_3 interface showed that the Auger peaks from the film, and the oxygen and aluminum peaks from the Al_2O_3 substrate shifted by ~1.5, 1.6, and 1.4 eV, respectively, when the substrate was irradiated in an oxygen atmosphere [35]. On the other hand, no shift of the Auger peaks are observed at the interface when the irradiation is performed in argon–4% H_2 [35]. The AES results indicate that the

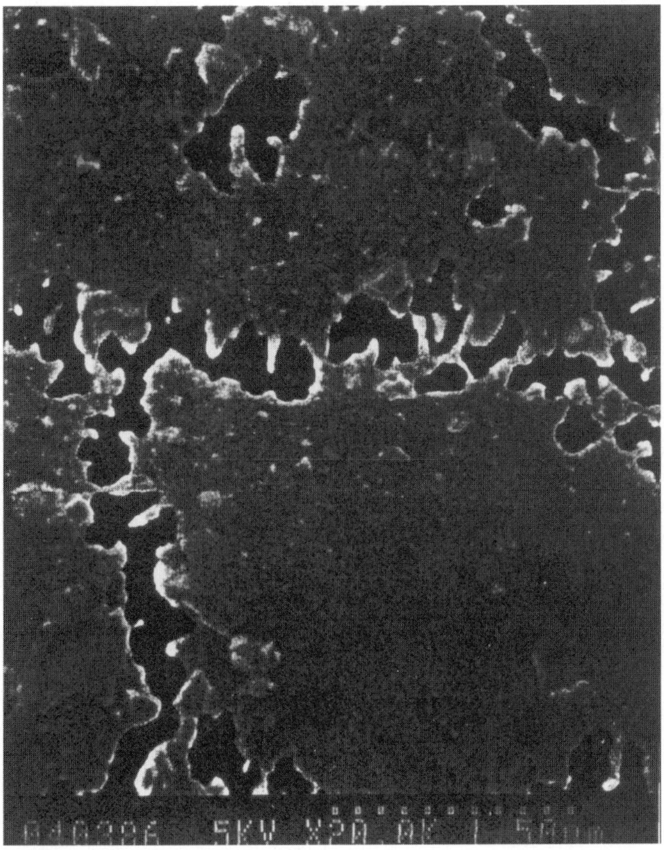

Figure 12. SEM micrograph of a gold film deposited on Al_2O_3 irradiated at 2 J/cm^2 in an oxygen atmosphere. The couple was annealed at 550°C for 45 min at a pressure of 1×10^{-6} Torr.

very strong adhesion detected here is due to a reaction that takes place between the oxygen-rich alumina and gold. Mattox [36] found that the presence of oxygen during sputtering deposition of gold promotes a strong bonding between the gold films and Al_2O_3. Moore and Thornton [37] also found that a strong bond between gold and fused silica developed only when the gold was melted in the presence of oxygen. Their results indicate that a chemical interaction takes place between molten gold and silica despite the apparent lack of reactivity of gold with oxygen. The AES results mentioned above are in agreement with those reported for silica and they are the first direct evidence that gold reacts with Al_2O_3 in the presence of oxygen. In our case, the interfacial gold–Al_2O_3 reaction is promoted by the highly metastable laser irradiated Al_2O_3 and not by the melting of gold.

The electrical resistance experiments show that there is a complete correlation between adhesion and thermal stability of the gold film deposited on irradiated Al_2O_3: the higher the adhesion strength, the more stable were the films after annealing. The films can maintain their integrity during thermal annealing for longer times because of the chemical reaction at the Au/Al_2O_3 interface during laser irradiation favored by

Figure 13. SEM micrograph of a Pd cluster film left by the laser-induced decomposition of a 50 nm-thick Pd ac film. The Pd ac film was deposited on rough Al_2O_3 and irradiated with a 248 nm-wavelength laser at an energy density of 20 mJ/cm^2. (Scale shown = 1 μm.)

the presence of oxygen. If this bonding were due to mechanical anchoring, only portions of the film would be attached to the substrate and the film would start breaking away in the weaker areas.

5. SUMMARY

The adhesion of the copper deposited on the irradiated surface when the Pd-activated Al_2O_3 substrate is immersed in an electroless bath is controlled by the Pd–Al_2O_3 interface. A laser energy threshold of 15 mJ/cm^2 is required to decompose a Pd ac film with a 248 nm-wavelength laser. The adhesion of electroless copper to the activated substrate is a strong function of the laser energy density used for activating the substrates. This adhesion exhibits maximum strength of 12 MPa at 150 mJ/cm^2. As the laser energy increases above 150 mJ/cm^2 the adhesion strength of copper films to

Figure 14. SEM micrograph of a Pd cluster film left by the laser-induced decomposition of a 50 nm-thick Pd ac film. The Pd ac film was deposited on rough Al_2O_3 and irradiated with a 248 nm-wavelength laser at an energy density of 150 mJ/cm^2. (Scale shown = 1 μm.)

Al_2O_3 decreases, probably because the area occupied by Pd continuously decreases as the Pd clusters become more rounded and occupy a smaller area. Also material ablated during decomposition and evaporated Pd can be re-deposited, impairing bonding. Chemical etching of the Al_2O_3 substrates prior to the optimized pre-nucleation process with excimer lasers leads to high adhesion (up to 37 MPa) of the electroless copper films.

After annealing for 1 h at 550°C, the bond strength of gold films to alumina is higher. The adhesion strength is a strong function of the laser energy density and the irradiation atmosphere. The maximum adhesion strength is obtained at a laser irradiation energy of 1 J/cm^2 for each of the irradiation atmospheres employed. Laser irradiation in an oxygen atmosphere strongly increases the adhesion strength. There is a direct correlation between adhesion strength and thermal stability of gold films deposited on irradiated substrates. SEM inspection shows that films irradiated at 1 J/cm^2 in oxygen are completely intact after annealing at 550°C for 45 min. By contrast, the films deposited on substrates irradiated in an Ar–4% H_2 atmosphere break down into small clusters during a similar annealing. A chemical reaction is produced between Al_2O_3 irradiated in an oxygen atmosphere and the deposited gold film. Part of the oxygen excess detected on the irradiated substrates by AES and EDS migrates to the gold–Al_2O_3 interface during the annealing, promoting the interfacial reaction.

Acknowledgement

This research was sponsored by the National Science Foundation Grant No. DMR-9116528.

REFERENCES

1. R. Srinivasan, in: *Laser Processing and Diagnostics*, D. Bauerle (Ed.), pp. 342–355. Springer, New York (1984).
2. R. Srinivasan and B. Braren, *J. Polym. Sci.* **22**, 2601–2609 (1984).
3. R. Srinivasan, B. Braren and R. W. Dreyfus, *J. Appl. Phys.* **61**, 372–380 (1987).
4. J.-Y. Zhang, H. Esrom, U. Kogelschatz and G. Emig, *J. Adhesion Sci. Technol.* **8**, 1179–1210 (1994).
5. R. F. Wood and G. E. Giles, *Phys. Rev. B* **23**, 2923–2942 (1981).
6. S. Cao, A. J. Pedraza, D. H. Lowndes and L. F. Allard, *Appl. Phys. Lett.* **65**, 2940–2942 (1994).
7. S. Cao, A. J. Pedraza and L. F. Allard, *J. Mater. Res.* **10**, 54–61 (1995).
8. A. J. Pedraza, in: *Laser Materials Processing III*, J. Mazumder and K. N. Mukherjee (Eds), pp. 183–195. TMS, Warrendale, PA (1989).
9. A. J. Pedraza and M. J. Godbole, *Metal. Trans. A* **23A**, 1095–1103 (1992).
10. T. E. Tsai, D. L. Griscom and E. J. Friebele, *Phys. Rev. Lett.* **61**, 444–446 (1988).
11. J. Ihlemann, B. Wolff and P. Simon, *Appl. Phys. A* 363–370 (1992).
12. C. H. Chen and M. P. McCann, *Opt. Comm.* **60**, 296–298 (1986).
13. R. L. Webb, C. Jensen, S. C. Langford and J. T. Dickinson, *J. Appl. Phys.* **74**, 2323–2337 (1994).
14. D. H. Lowndes, M. DeSilva, M. J. Godbole, A. J. Pedraza and D. B. Geohegan, in: *Laser Ablation in Materials Processing: Fundamentals and Applications*, B. Braren, J. J. Dubowski and D. P. Norton (Eds), pp. 191–196. Materials Research Society, Pittsburgh (1993).
15. R. C. Sausa, A. Gupta and J. R. White, *J. Electrochem. Soc.* **134**, 2707–2713 (1987).
16. H. Esrom and G. Wahl, *Chemtronics* **4**, 216–223 (1989).
17. H. Esrom and U. Kogelschatz, *Appl. Surf. Sci.* **49**, 158–162 (1990).
18. C. Y. Mak, *MRS Bulletin* **XIX** (8), 55–62 (1994).
19. H. Esrom, J.-Y. Zang and A. J. Pedraza, *Mater. Res. Soc. Symp. Proc.* **236**, 383–385 (1992).
20. C. Cao, A. J. Pedraza, D. H. Lowndes and L. Allard, unpublished results.
21. H. Esrom, *Mater. Res. Soc. Symp. Proc.* **204**, 457–462 (1991).
22. G. A. Shafeev, *Advanced Materials for Optics and Electronics* **2**, 183–189 (1993).
23. A. J. Pedraza, M. J. Godbole, M. DeSilva and D. H. Lowndes, in: *Laser Ablation in Materials Processing: Fundamentals and Applications*, B. Braren, J. J. Dubowski and D. P. Norton (Eds), pp. 203–208. Materials Research Society, Pittsburgh (1993).
24. M. J. DeSilva, A. J. Pedraza and D. H. Lowndes, *J. Mater. Res.* **9**, 1019–1027 (1994).
25. A. J. Pedraza, R. A. Kumar and D. J. Lowndes, *Appl. Phys. Lett.* **66**, 1065–1068 (1995).
26. A. J. Pedraza, J. DeSilva, R. A. Kumar and D. J. Lowndes, *J. Appl. Phys.* **77**, 5176–5179 (1995).
27. B. Eliasson and U. Kogelschatz, *Appl. Phys. B* **46**, 299–303 (1988).
28. U. Kogelschatz, *Appl. Surf. Sci.* **54**, 410–423 (1992).
29. H. Esrom, J. Demny and U. Kogelschatz, *Chemtronics* **4**, 202–208 (1989).
30. H. Esrom and U. Kogelschatz, *Thin Solid Films* **218**, 231–246 (1992).

31. W. A. Weyl, in: *Adhesion and Adhesives — Fundamentals and Practice*, J. E. Rutzler and R. L. Savage (Eds), pp. 36–47. John Wiley, New York (1953).
32. J. E. MacDonald and J. G. Eberhart, *Trans. Metal. Soc. AIME* **233**, 512–517 (1965).
33. H. Esrom and U. Kogelschatz, *Mater. Res. Soc. Symp. Proc.* **158**, 189–198 (1990).
34. M. J. Godbole, A. J. Pedraza, D. H. Lowndes and J. R. Thompson, Jr., *J. Mater. Res.* **7**, 1004–1010 (1992).
35. J. Park, A. J. Pedraza and W. R. Allen, *Mater. Res. Soc. Symp. Proc.* **357**, 59–64 (1995).
36. D. N. Mattox, *J. Appl. Phys.* **37**, 3613–3618 (1966).
37. D. C. Moore and H. R. Thornton, *J. Res. Natl Bur. Std.* **35** (17), 2345–2358 (1962).

Mittal Festschrift, pp. 791–801
W. J. Van Ooij and H. R. Anderson, Jr. (Eds)

Effect of titanium interlayers on adhesion enhancement of platinum films on titanium oxide films and on temperature coefficient of resistivity of platinum films

I. KONDO,[1] A. TAKAHASHI,[1] M. SHIRAI,[1] K. KONDO,[1] O. TAKENAKA,[1] E. KUSANO[2] and A. KINBARA[2,*]

[1]*Production Engineering R&D, Denso Co., 1-1, Showa-cho, Kariya, Aichi, 448 Japan*
[2]*Department of Applied Materials, Kanazawa Institute of Technology, Ohgigaoka, Nonoichi-machi, Kanazawa, Ishikawa, 921 Japan*

Abstract—The effect of insertion of a titanium (Ti) layer to enhance adhesion between platinum (Pt) and titanium oxide films has been investigated. Ti is known to be effective in enhancing adhesion, but it diffuses through the Pt films and oxidizes during heat-treatment in an oxygen atmosphere. Both diffusion and oxidation influence the temperature coefficient of resistivity (TCR) of the Pt films. High resolution electron microscopy revealed that the most of the Ti atoms in the titanium interlayer diffused through the grain boundaries of the Pt film and precipitated at the top surface grain boundaries of the film, while the rest of the Ti was oxidized inside the film. The optimum thickness of the adhesion-promoting Ti layer for enhancing adhesion and suppressing the decline of TCR was 10 nm.

Keywords: Adhesion; thin films; interlayer; Pt films; resistivity.

1. INTRODUCTION

Thin films are widely used in sensors and electrodes of many integrated circuit (IC) devices in the automotive industry, and the need for their reliability is steadily growing. Promotion of adhesion between the thin films and the substrates is necessary for reliability enhancement. In essence, adhesion controls the lifetime of thin film devices. Excluding mechanical treatments, three methods are well known [1] to promote adhesion of thin films: (1) cleaning of the substrate [2]; (2) high temperature deposition or heat-treatment after deposition [3, 4]; and (3) insertion of an adhesion-promoting layer [5].

The present work is focused on the insertion of an adhesion-promotion layer between a platinum (Pt) film and an oxide film on a silicon (Si) substrate. Pt films are used extensively as temperature sensing elements, because of their excellent corrosion resistance and high temperature coefficient of electrical resistivity (TCR). Since Pt films

---

*To whom correspondence should be addressed.

do not have good adhesion to oxide films, a thin titanium (Ti) layer is interposed to promote adhesion [5]. However, during heat treatment of the specimen, Ti atoms diffuse into the Pt films through the Pt grain boundaries [6] and affect their TCR.

In this paper, the behavior of Ti films deposited between Pt films and oxide films on Si (110) substrates is investigated in detail by high resolution scanning electron microscopy (HRSEM), Auger electron spectroscopy (AES), transmission electron microscopy (TEM) and other analytical methods to determine the optimum thickness that promotes adhesion.

2. EXPERIMENTAL PROCEDURE

2.1. Sample preparation

Si (110) wafers, passivated with thermally grown silicon dioxide layers, were cleaned using a chemical etching pre-treatment followed by a deionized water rinse. Titanium dioxide (50 nm), titanium (0–50 nm) and platinum (800 nm) films were deposited on the substrate by a vacuum evaporation technique using ULVAC EBX-12c system without breaking the vacuum. The deposition rates of TiO_2 (purity: greater than 99.9%), Ti (purity: greater than 99.9%), and Pt (purity: greater than 99.99%) were 0.5, 0.1, and 0.2 nm/s, respectively. The Si substrate temperature during deposition was about 573 K. The annealing was carried out in an oxygen atmosphere at 1373 K for 30 min after deposition.

2.2. Examination of film properties

There are many methods of measuring practical adhesion [7], and in this study the adhesion properties of the Pt films were evaluated by the thermosonic ball bonding (TBB) method [5], because this method is closely related to the actual bonding of lead wires in the production line. The TBB method is schematically illustrated in Fig. 1. A melted gold ball is formed at the tip of a gold wire by an electrical discharge. The sample is fixed on a heater, and the ball is welded to the Pt film surface, using a capillary tool and supersonic wave vibrations. The bonding spot size was about 40–80 μm in diameter. However, the spot size varied considerably depending on the generating power, heater temperature and the load applied to the ball. Adhesion for a given applied load was expressed in terms of the peel-off ratio, P, defined as:

$$P = \frac{\text{number of samples peeled off}}{\text{total number of samples examined}}.$$

We evaluated 20 samples under the same test conditions.

The TCR of the Pt films was calculated from the following equation:

$$\text{TCR} = \frac{(1/\rho_0)(\rho - \rho_0)}{T - T_0} = \frac{1}{\rho_0}\frac{\Delta\rho}{\Delta T},$$

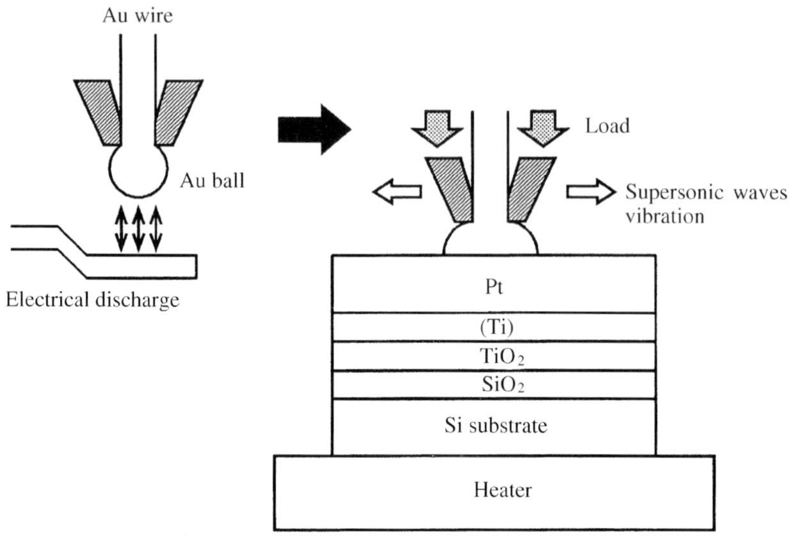

Figure 1. Schematic of thermosonic ball bonding method.

where ρ_0 is the resistivity at $T_0 = 273$ K and ρ, the resistivity at temperature T. We measured ρ in the temperature range from 273 K to 373 K, and in this range, $\Delta\rho/\Delta T$ was approximately constant.

HRSEM, AES, TEM, energy dispersive X-ray spectroscopy (EDS), and X-ray diffraction (XRD) were employed to study the behavior of Ti adhesion-promoting layers and the diffusion of Ti atoms into the Pt film. TEM samples were prepared by the focused ion beam (FIB) technique. By this method, a multilayered structure which has a variable sputtering rate was uniformly thinned. Time required for the preparation of a specimen was shorter than with conventional argon (Ar) ion milling [8].

3. RESULTS AND DISCUSSION

3.1. Interface structure and adhesion

Figure 2 shows the peeled-off surfaces observed by HRSEM in the absence of the adhesion-promoting layer. Small grains of a few hundred nm correspond to the TiO_2 grains. A rough horizontal line at the bottom half and a segment of line at the upper right corner are the traces of grain boundaries of the Pt film. The composition of each side was analyzed by AES. The AES spectra of the Pt film side and the substrate side are shown in Figs 3a and 3b, respectively. As the resolution of AES analysis in the depth direction is less than 1.5 nm, it can be concluded that the Pt film peels off the TiO_2 film indicating mechanical failure at the interface.

Figure 4 shows the results of the adhesion tests at two different applied loads on the wires as a function of Ti thickness. All the peeling occurred at the Pt/TiO_2 interfaces. From these results, it is confirmed that the Ti films are effective as adhesion-promoting

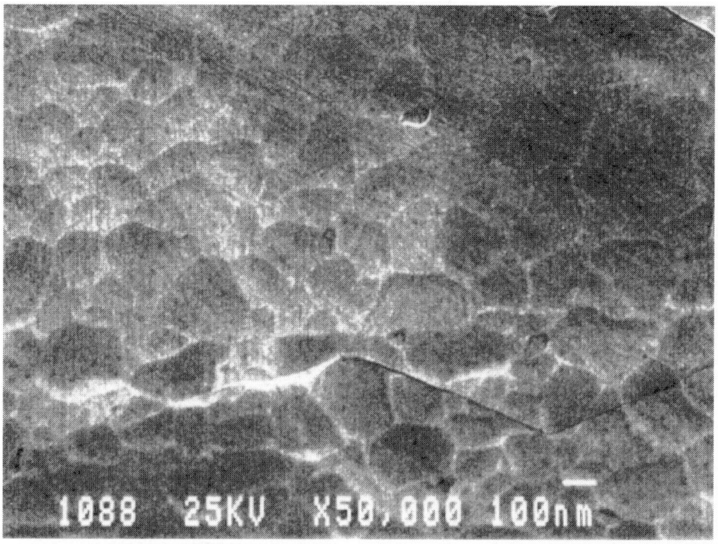

(a)

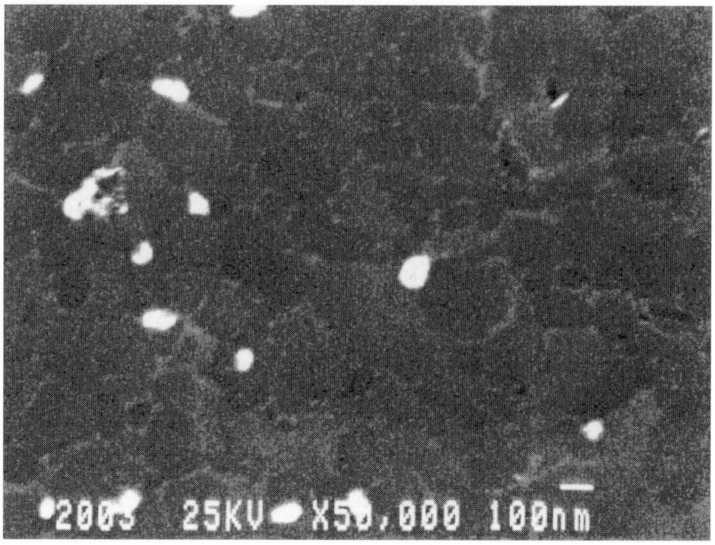

(b)

Figure 2. Scanning electron micrographs of the detached Pt film and the substrate after the TBB test in the absence of titanium interlayer. (a) Platinum side; (b) substrate side.

layers, even at a thickness of 2.5 nm. Of the two load conditions, a load of 0.6 N seems to yield better results than the load of 1.0 N. It may be assumed that for a 1.0 N load, the thermosonic power applied to the wire cannot be consumed at the interface effectively, because the load is too heavy to vibrate the film, but rather it propagates to the Si substrate and as a result the film peeling is hardly observed. In the case of a 0.6 N load, the power is effectively applied to the film/substrate interface and the thicker Ti film is better for adhesion promotion for a thickness less than 10 nm.

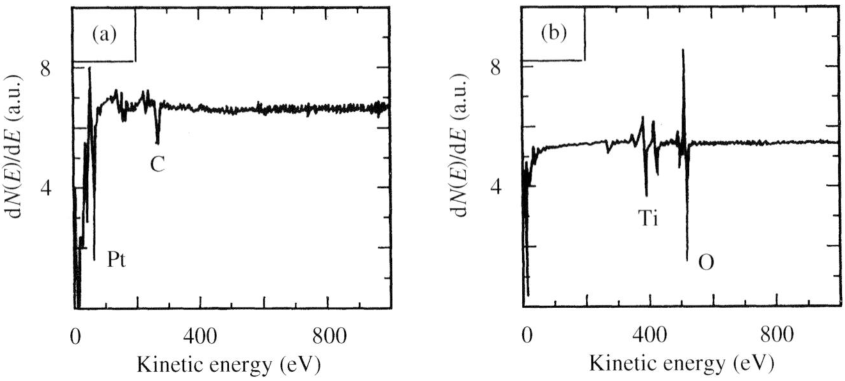

Figure 3. Auger electron spectra of the detached Pt film and the substrate after the TBB test. (a) Platinum side; (b) substrate side. Ti thickness is 5 nm.

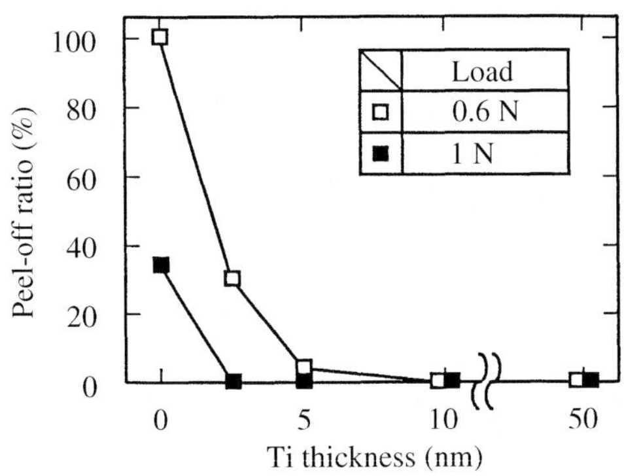

Figure 4. Effect of the Ti adhesion-promoting layer thickness on the peeling when using the TBB test. A thicker Ti layer is better for adhesion.

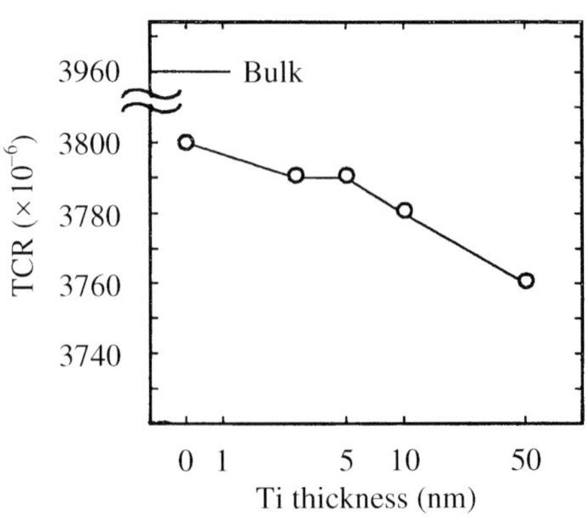

Figure 5. Measured TCR results as a function of the Ti thickness.

3.2. Temperature coefficient of resistivity

The effect of the Ti film on adhesion is clear, but its influence on the TCR of the Pt film cannot be ignored. Figure 5 shows measured results of TCR of Pt films as a function of Ti thickness. The TCR value of the Pt bulk is known to be 3960×10^{-6} [9], and in our experiment, the value of TCR for Pt films was 3800×10^{-6} in the absence of the Ti layer: it is known that there are many defects in Pt films such as dislocations which cause a reduction in the TCR [10]. In the presence of the Ti interlayer, TCR monotonously decreased with increasing Ti thickness.The decrease was about 1% for a thickness of 50 nm.

3.3. Behavior of Ti atoms

The behavior of Ti atoms was investigated to clarify their influence on both the adhesion and the TCR. Figure 6 shows the cross-section of Pt/Ti (5 nm)/TiO_2 structure observed by TEM. There are small islands and grains at the bottom, center and top of

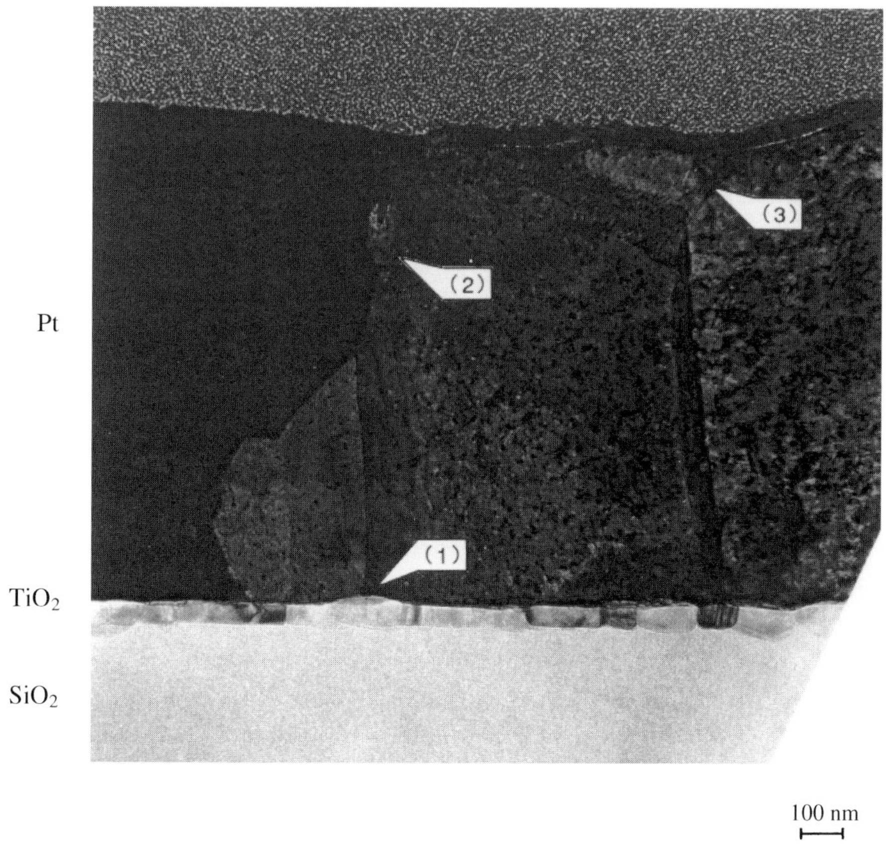

Figure 6. Cross-section morphology of the Pt/Ti (5 nm)/TiO_2/SiO_2 structure by TEM. There are small islands (1) and grains (2, 3) through the Pt grain boundaries.

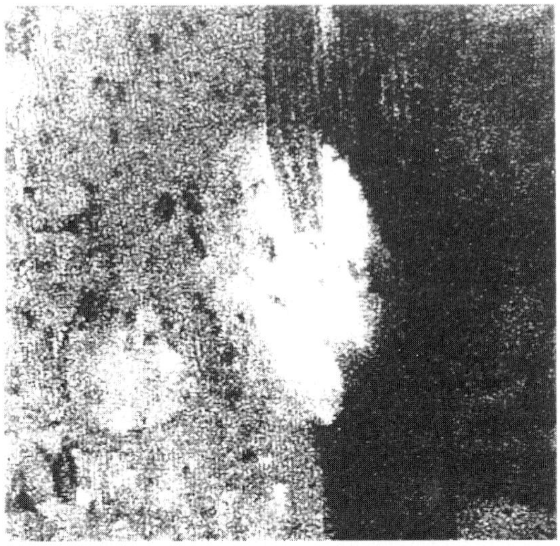

(a)

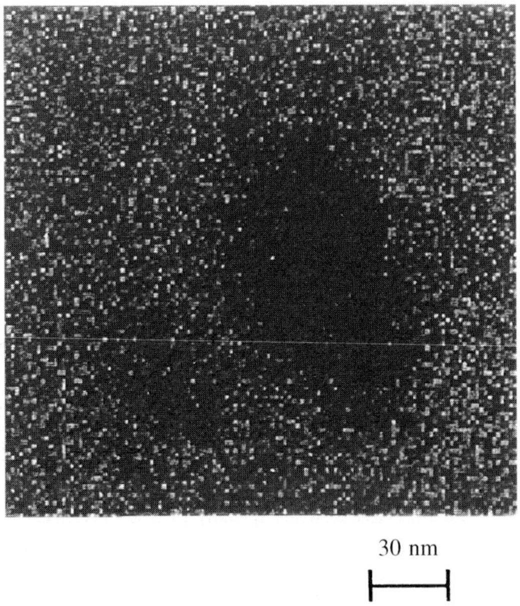

(b)

Figure 7. EDS mapping results of (2) in Fig. 6. (a) STEM image, (b) Pt, (c) Ti, (d) O. White points are the detected points.

the Pt grain boundaries as indicated by (1), (2) and (3) in the figure. Figure 7 shows the EDS mapping of the small grain at the center of the Pt grain boundary shown by (2) in Fig. 6. (a) is the STEM image, (b), (c) and (d) correspond to Pt, Ti and O, respectively. The existence of Ti and O, and the absence of Pt in the grain were confirmed. As a result, it may be concluded that the small grain is TiO_2. We also confirmed that the small island at the bottom (1), and the small grain at the top (3) also consist of TiO_2.

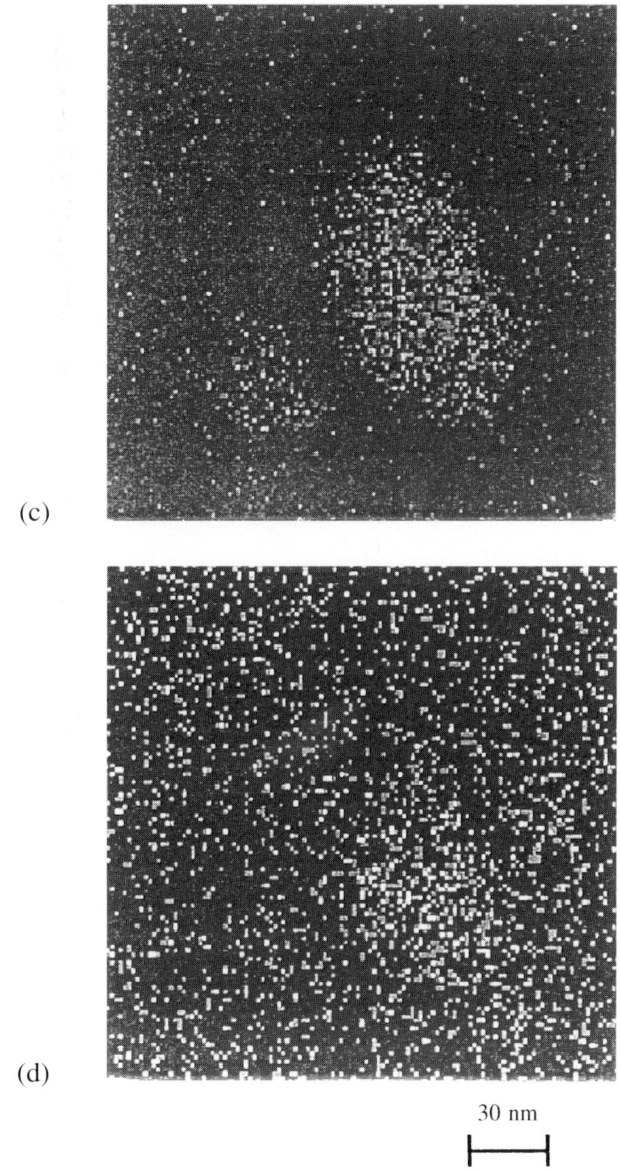

Figure 7. (Continued).

The results of XRD in Fig. 8 show that only the diffraction peak due to Pt is prominent. The TiO_2 layer appears to be mostly amorphous before the heat treatment in an oxygen atmosphere. As the binding energies of O–Ti, Pt–Pt, and Ti–Pt are 3.04, 0.77 and 0.98 eV, respectively [11, 12], it is concluded that some of the Ti atoms have already been oxidized on the TiO_2 layer during deposition, and have formed an oxide layer. From the observed results, it is clear that during heat treatment in the oxygen atmosphere, deposited Ti atoms reduced the existing TiO_2 layer and at the

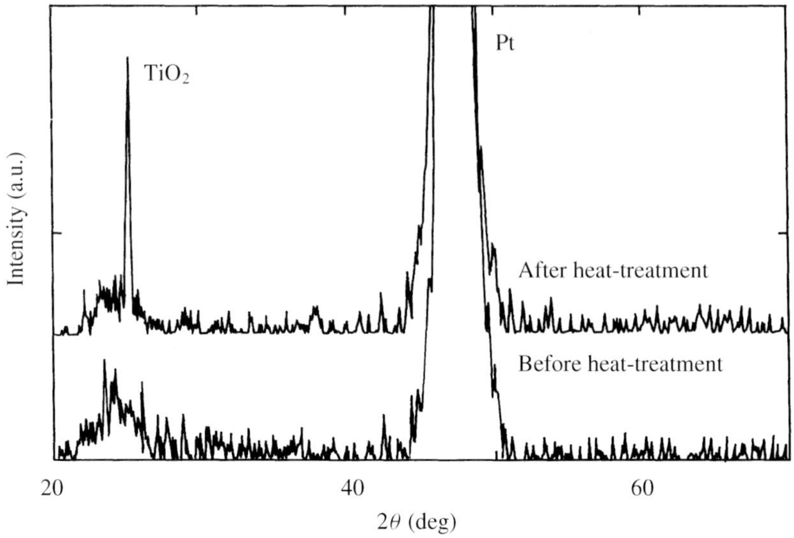

Figure 8. X-ray diffraction of TiO_2 and Pt. Ti thickness is 5 nm.

same time, they were also oxidized by the oxygen from the atmosphere that diffused through the Pt grain boundaries forming anatase. The HRTEM results also suggest that some Ti atoms move to Pt grain boundaries and some of these diffuse to the top of the Pt grain boundaries.

Prior to the heat treatment, the high adhesion of Pt film in the presence of Ti interlayer may be attributed to its bonding effect on TiO_2 and Pt. During the heat treatment, the oxygen diffusing along the Pt grain boundaries oxidizes most of the Ti atoms. However, the TBB indicates adhesion promotion (Fig. 4). Figure 9 shows the peeled-off surface after the heat treatment, (a) on the Pt film side and (b) on the substrate side. On the Pt film side, the Pt grain boundaries, the traces of the TiO_2 grains and a few depressions were observed. It is thought that the TiO_2 island shown by (1) in Fig. 6 may have generated the depressions in the Pt film, while remaining on the substrate side. On the other hand, on the TiO_2 side, only grain boundaries of TiO_2 are observed. The islands of TiO_2 shown by (1) in Fig. 6 could not be observed because of limited SEM resolution. From this observation, it can be concluded that the peeling occurred at the interface between the Pt film and the initially deposited TiO_2 film and, in addition, the TiO_2 islands resulted from oxidation of the Ti interlayer during heat treatment process. Figure 10 shows the AES results of peeled-off surfaces in the case of a 5 nm thick Ti insertion layer. Such results confirmed the existence of Ti on the Pt side. On the other hand, Ti on the substrate side seems to be oxidized. Hence, it appears that most of the Ti atoms diffused to and oxidized at the interface, as well as at Pt grain boundaries and Pt surface, whereas a small amount of Ti still remained at the interface even after the heat treatment to enhance the adhesion.

The TCR and resistivity values of Ti in its bulk were found to be 5460×10^{-6} and 5.5×10^{-7} Ω m [9], respectively. By comparison, the TCR and resistivity values

(a)

(b)

Figure 9. Scanning electron micrograph of the detached Pt film and the substrate after the TBB test. (a) Platinum side; (b) substrate side. Ti thickness is 5 nm.

of Pt in its bulk were found to be 3960×10^{-6} and 1.1×10^{-7} Ω m, respectively. By contrast, the resistivity of TiO_2 is more than 100 Ω m and it actually may be regarded as an insulator. In the case of the 5 nm thick Ti insertion layer, most of the TiO_2 appears at the bottom of the Pt film and only some appears at the top. The decrease of TCR caused by the presence of TiO_2 at the grain boundaries inside the Pt films is 10×10^{-6}. This decrease is acceptable for practical applications of the Pt films as temperature sensors.

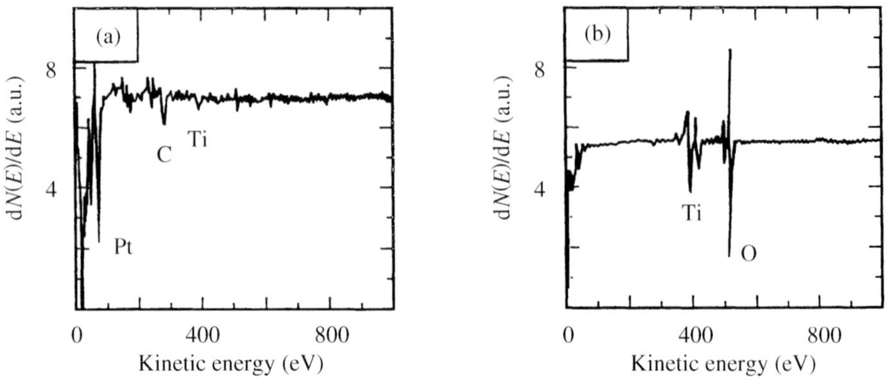

Figure 10. Auger electron spectra of the detached Pt film and the substrate after the TBB test. (a) Platinum side; (b) substrate side. Ti thickness is 5 nm.

4. SUMMARY

The effect of a Ti adhesion-promoting layer between Pt films and Ti oxide films has been investigated. After the heat treatment, most of the Ti adhesion-promoting layer oxidized, but small amounts of remaining Ti still enhanced adhesion, preventing the Pt film from peeling. The optimum Ti interlayer thickness was found to be 10 nm to both enhance adhesion and suppress the decline of TCR.

Acknowledgements

We wish to thank T. Yoneyama and M. Ohta of Nippondenso for helpful discussions. We would also like to thank H. Tanizawa and T. Hayasaki for focused ion beam treatment.

REFERENCES

1. A. Kinbara, *J. Surface Finishing Soc. Japan* **37**, 469 (1986).
2. I. Kondo, T. Yoneyama, K. Kondo, O. Takenaka and A. Kinbara, *J. Vac. Sci. Technol.* **A11**, 319 (1993).
3. A. Kinbara, S. Baba and N. Matuda, *Thin Solid Films* **141**, 229 (1986).
4. I. Kondo, T. Yoneyama, K. Kondo, O. Takenaka and A. Kinbara, *Thin Solid Films* **236**, 236 (1993).
5. I. Kondo, T. Yoneyama, O. Takenaka and A. Kinbara, *J. Vac. Sci. Technol.* **A10**, 3456 (1992).
6. T. C. Tisone and J. Drobek, *J. Vac. Sci. Technol.* **9**, 271 (1972).
7. K. L. Mittal (Ed.), in: *Adhesion Measurement of Films and Coatings*, pp. 1–13. VSP, Utrecht (1995).
8. A. Yamaguchi and T. Nishikawa, *J. Vac. Sci. Technol.* **B13**, 962 (1995).
9. N. F. Mott and H. Jones, *The Theory of the Properties of Metals and Alloys*, 2nd edn. Dover, New York (1958).
10. L. Maissel and G. Glang (Eds), in: *Handbook of Thin Film Technology*, p. 18. McGraw Hill, New York (1970).
11. M. S. Spencer, *Surface Sci.* **145**, 145 (1984).
12. M. S. Spencer, *Surface Sci.* **153**, 145 (1984).

Part 7

General papers

Mittal Festschrift, pp. 805–820
W. J. Van Ooij and H. R. Anderson, Jr. (Eds)

Electrochemical aspects of encapsulated bacteria adhesion to stainless steel surfaces

L. BOULANGÉ-PETERMANN,[1,2,*] M. N. BELLON-FONTAINE[2] and B. BAROUX[1,3]

[1] *Ugine-Savoie, Research Center, Avenue Paul Girod, F-73403 Ugine Cedex, France*
[2] *Institut National de la Recherche Agronomique, Laboratoire de Génie de l'Hygiène et des Procédés Alimentaires, 25 Avenue de la République, F-91300 Massy, France*
[3] *Centre National de la Recherche Scientifique, Laboratoire de Thermodynamique et Physico-Chimie Métallurgique, F-38402 Saint Martin d'Hères, France*

Abstract—The adhesion of *Leuconostoc mesenteroides* was studied on industrial stainless steel surfaces. Several surface conditions (BA, 2B mechanical polishing) were considered with a view to vary the hydrophobicity and roughness ($R_a \sim 0.02$ to $0.1\ \mu m$). In the absence of any electrochemical polarization, and in accordance with previous results obtained on polymers and glass, no correlation between the number of adhering *L. mesenteroides* and the water contact angle or the R_a value was found. On the other hand, with an electrochemical polarization, the number of adhering bacteria varies with the initial surface condition. Moreover, a cathodic pulse is observed when the bacteria are introduced to the solution. The same behavior is seen when dextran or other saccharides are introduced instead of *L. mesenteroides*, indicating a possible ionic and electronic exchange between the steel surfaces and the encapsulating polymer.

Keywords: *Leuconostoc mesenteroides*; stainless steel; polarization; electrochemistry; bioadhesion; dextran; saccharides.

1. INTRODUCTION

Microbial attachment to stainless steel surfaces is a well known source of potential contamination of food products and leads to significant economic losses and hygiene problems. Economic losses due to spoilage or the transmission of diseases may occur as a result of bacterial adhesion to equipment and food contact surfaces in the food industry [1–5].

Numerous studies dealing with the bacterial adhesion to metallic surfaces, and more particularly to stainless steels, have reported the effects of the bacterial cell hydrophobicity [6, 7], the solid surface wettability expressed in terms of surface free energy [8], and the suspending liquid [9]. The presence of a chloride-containing

---

*To whom correspondence should be addressed.

medium should modify the bacterial adhesion behavior, inducing variations in the bacterial cell surfaces [7, 9]. As far as the bacterial adhesion is concerned, the influence of the roughness and the surface topology have often been put forward as important parameters [2, 10–12].

The effect of some polysaccharides surrounding the bacterial cells and their involvement in bacterial adhesion have been studied by several authors, using scanning electron microscopy [2, 13, 14] or selected chemicals or enzymes [4].

Some authors [15] have shown two typical behaviors in bacterial attachment: non-encapsulated strains are sensitive to the substrate surface energy, whereas encapsulated strains are not sensitive. In the first case, it has been shown that variations of the energetic characteristics of the solid due to cleaning treatments can modify the bacterial adhesion [8]. Then, an elimination of biofouling onto metallic surfaces can be achieved by using the appropriate cleaning treatment. Moreover, appropriate anti-adhesive coatings [16] were shown to reduce bacterial attachment. In contrast, these methods cannot be applied to encapsulated strains since their adhesion is not sensitive to the solid surface free energy.

From another point of view, the stainless steel reactivity in various environments is controlled by the presence on the steel surface of a very thin (1–10 nm) oxide or hydroxide layer (the passive layer) [17]. The potential difference between the steel and the surrounding electrolyte controls the ionic and electronic exchanges throughout the passive film which can be explained in terms of surface acido-basicity.

This study was carried out to investigate the effect of this potential difference on the adhesion of an encapsulated bacterium (*Leuconostoc mesenteroides*) in some chloride-containing environments simulating suspending liquids which are encountered in the food industry. Different saccharides associated to the cell surface were used in order to simulate the interactions between the bacterial capsule and the stainless steel surfaces. The investigated subtrata were AISI 304 type stainless steels which are one of the most common materials used in equipment manufacturing and in domestic uses. AISI 434 grade was also used. The investigated bacterium was *L. mesenteroides*, which is employed in the production of dextran [18] but which can also have an adverse effect on the taste of dairy products [19].

2. EXPERIMENTAL

2.1. Bacterial strain

Leuconostoc mesenteroides NCDO 523 was originally isolated from a sugar refinery plant. Freeze-dried cells were suspended in 1 ml MRS broth [20] supplemented with 1% sucrose. This suspension was inoculated with fresh medium (9 ml), which was kept for 24 h at 22°C.

This culture was used to inoculate a second culture which was grown for 16 h and then harvested by centrifugation for 15 min at 8400 *g* and twice washed in a suspending liquid. The liquids used in this study were double distilled and sterile water, 0.01 M potassium phosphate buffer of pH 7 and a 0.02 M or 0.15 M NaCl solution of pH 6.6.

2.2. Saccharides

Some mono-, di- and polysaccharides supplied by Sigma were used in this study. The chemical formulas of the saccharides are given in Table 1, where some specific chemical functions of saccharides are also mentioned.

2.3. Substrates

Solids selected for this study were some AISI 304 and 434 type industrial stainless steel thin sheets (0.8 and 1 mm), the compositions of which are given in Table 2. The surface compositions were determined from ESCA analysis and are published in a previous paper [21].

These solids are covered by a 'passive film' which provides a high corrosion resistance to stainless steel. The composition and the thickness of the passive film depend markedly on the alloy composition and on the steelmaking finish process [17].

Concerning AISI 304, two surface conditions commonly used in the food industry were studied: the 2B condition corresponding to a pickled sheet and the BA referring to a bright annealed sheet. In addition to these two industrial surface conditions, a mechanically polished condition (SiC, grade 2000 under water) followed (PMP) or not (MP) by an electrochemical polarization (17 h at 200 mV/SCE in the NaCl-containing electrolyte defined in Section 2.4) was also considered in this work. After polishing, the samples were cleaned in ethanol, then aged 8 h in the laboratory ambience, since the procedure was found to stabilize the reactivity of the polished surfaces [22]. Before testing, all the samples were cleaned in an acetone/95% ethanol (50 : 50 ratio) solution and air-dried.

For AISI 434, two different bright annealed surfaces were prepared at Ugine Research Center. The preparation of these samples has been described in a previous paper [23]. Briefly, the samples were annealed for 15 min at 820°C in a moist hydrogen atmosphere and the atmospheric water content was carefully controlled. The oxidizing power of the annealing atmosphere was fixed by the pH_2O/pH_2 ratio. Two annealing atmospheres were used in this study which led to different passive film compositions on bright annealed surfaces: 1.3×10^{-5} and 2×10^{-6} atm corresponding, respectively, to dew points of −58°C (BA number 1) and −70°C (BA number 2). Further exposure to water completed the film formation [23].

The roughness of solid surfaces was studied with a surface tracing instrument, the Diavite DT 15 (Asmeto) equipped with a stylus of 5 μm radius. The surface roughness was quantified by the average roughness R_a expressed in μm.

Water contact angles on stainless steel surfaces were measured by the sessile drop technique. Immediately after cleaning of samples, droplets of 0.5 μl were released on the solid surface by means of a microsyringe. The contact angles were measured directly with a Krüss goniometer. The standard deviation in contact angles measurements did not exceed 4°.

2.4. Bioadhesion

- *Adhesion to steel surfaces.* 30 ml of the bacterial suspension was incubated with the steel surfaces for 3 h at 22°C. The suspending liquids were either sterile and

Table 1.
Chemical formulas and specific chemical functional groups of saccharides employed in this study

Saccharide	Chemical formula	Secondary alcohol functional group R-CHOH in C1	Primary alcohol functional group R-CH_2OH in C6	Comments
Monosaccharides				
Glucose	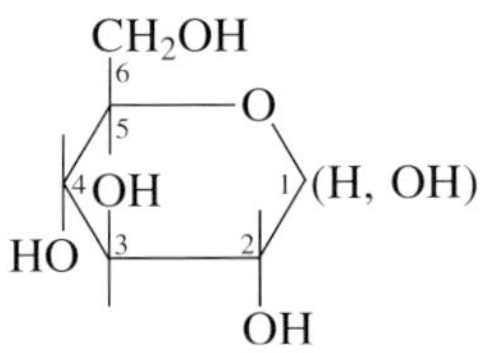	+	+	
Sorbitol	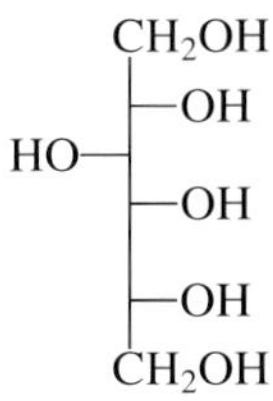	–	+	Primary alcohol group in C1
Glucosamine	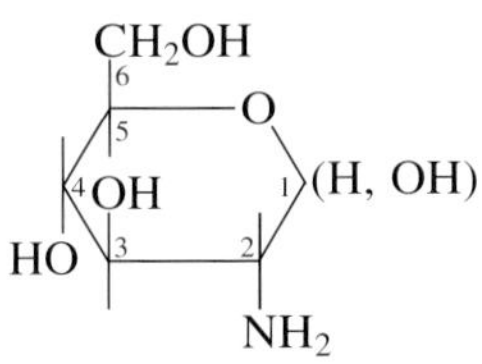	+	+	NH_2 group in C2
Glucuronic acid	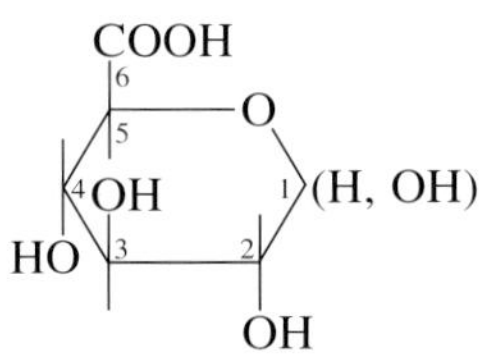	+	–	COOH group in C6

double distilled water or 0.01 M potassium phosphate buffer of pH 6.6. After the incubation, steel surfaces were washed for 15 min with a sterile water flow system (Filtron apparatus). The surface was air-dried in a laminar flow hood. In contrast with the polymeric surfaces [24], no bacterial clumping was observed on stainless steel surfaces. Adhering bacteria remaining on the surfaces after water rinsing were coloured with a 0.01% acridine solution and counted on a 100 epifluorescence microscope field. The total area corresponds to 1.5×10^{-2} mm^2. Results are expressed as the number of adhering bacteria per mm^2.

- *Adhesion to polarized steel surfaces.* The adhesion assays on polarized surfaces were realized in an electrochemical vessel shown in Fig. 1 and described in detail

Table 1.
(Continued)

Saccharide	Chemical formula	Secondary alcohol functional group R-CHOH in C1	Primary alcohol functional group R-CH_2OH in C6	Comments
Disaccharides				
Maltose	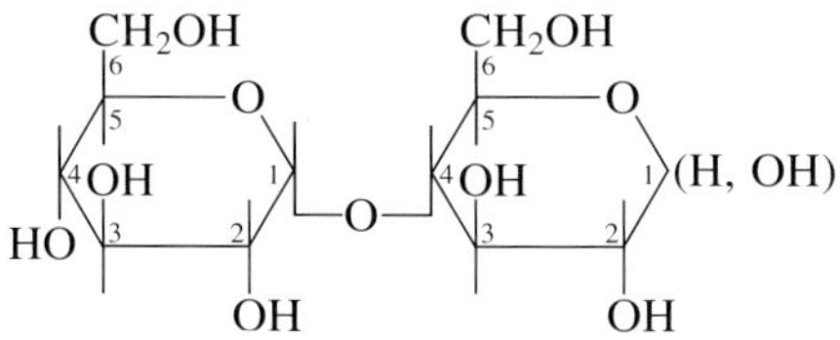	+	+	2 glucose units
Saccharose	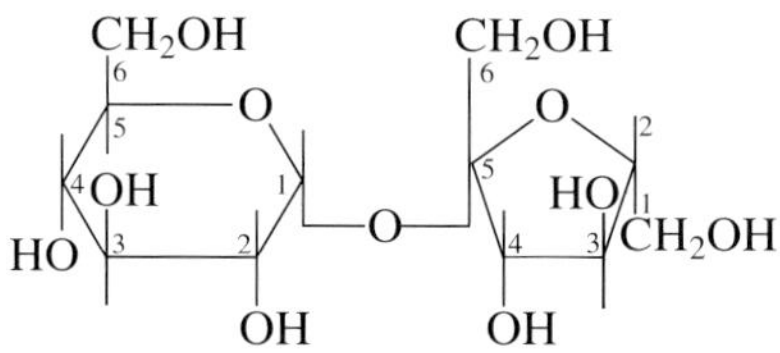	−	+	Primary alcohol group in C1
Polysaccharides				
Dextran	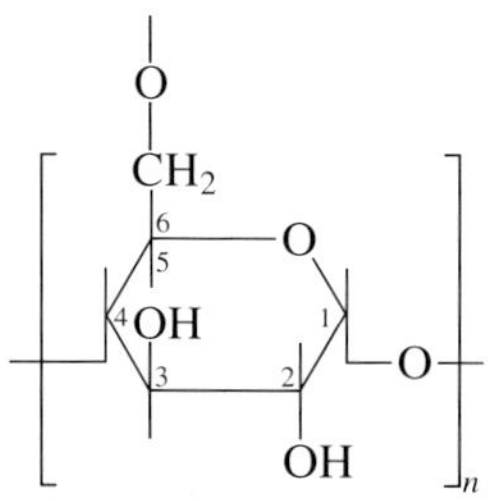	+	+	Glucose polymer

(+) and (−) correspond to the presence and the absence of the chemical functional group in the molecule, respectively.

Table 2.
Elemental composition of stainless steels used in this study (wt%)

Surface	Element							
	Cr	Ni	Mo	Si	Mn	P	S	C
AISI 304 (2B)	18.2	8.6	0.3	0.5	0.8	0.02	0.007	0.04
AISI 304 (BA)	17.2	8.3	0.09	0.5	1.3	0.03	0.02	0.05
AISI 434	16.6	—	1	0.3	0.4	0.02	0.002	0.05

previously [25]. The metallic sample was held horizontally in the vessel for the bacteria to deposit on the sample. The stainless steel surface was polarized using the potentiostatic method [26].

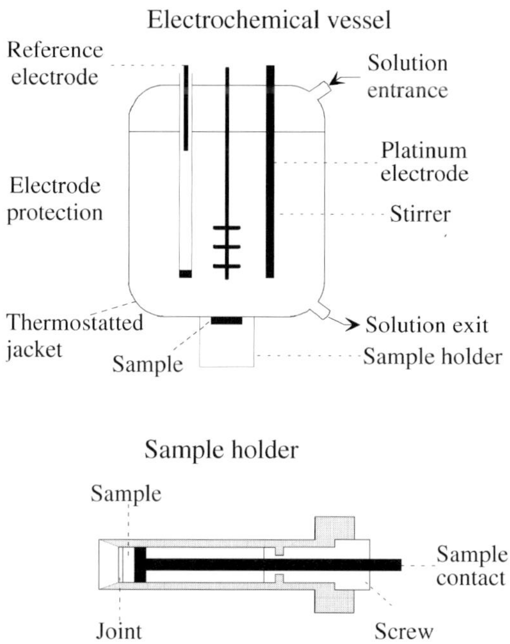

Figure 1. Schematic representation of the electrochemical vessel [18].

It is convenient first to recall some basic features of the surface reactivity of the stainless steels in aqueous media [17]. First, a metal immersed in water tends to convert to its oxidized form, the reaction involving the transfer of metallic cations from the metal into the solution. The intensity (i_A) of this anodic reaction depends on the difference in potential (V) established between the metal and the solution. Moreover, the electrons are consumed by the so called 'cathodic reaction', leading to a cathodic current (i_K), which also depends on the potential difference (V). In the steady state, the sum $I_G = i_A + i_K$ is obviously zero, and this occurs at the 'rest potential' V_{rest}. On departure from V_{rest}, for example in the presence of an imposed positive overpotential, $V - V_{rest}$, the anodic reaction becomes predominant and the current density markedly depends on the composition of the passive film.

The electrolytes were non-deaerated 0.15 M or 0.02 M NaCl solutions of pH 6.6 at 22°C. First, samples were left at rest potential for 30 min in order to stabilize the metal/solution interface, then they were anodically polarized for 3 h 15 min at 200 mV against a saturated calomel electrode. Bacteria were deposited on the steel surfaces after the first 15 min of sample polarization. The electrochemical current was recorded as a function of time.

Immediately after the electrochemical experiment, the steel surfaces were washed for 15 min with a sterile water flow system (Filtron Apparatus) and then dehydrated in an ethanol solution. Samples were coated with a layer of gold by vacuum evaporation. A Zeiss DSM 950 operated at 15 kV was used in this study at a 3000 magnification and 30 successive fields were examined each corresponding to 2.8×10^{-2} mm^2. Results are expressed as the number of adhering bacteria per mm^2.

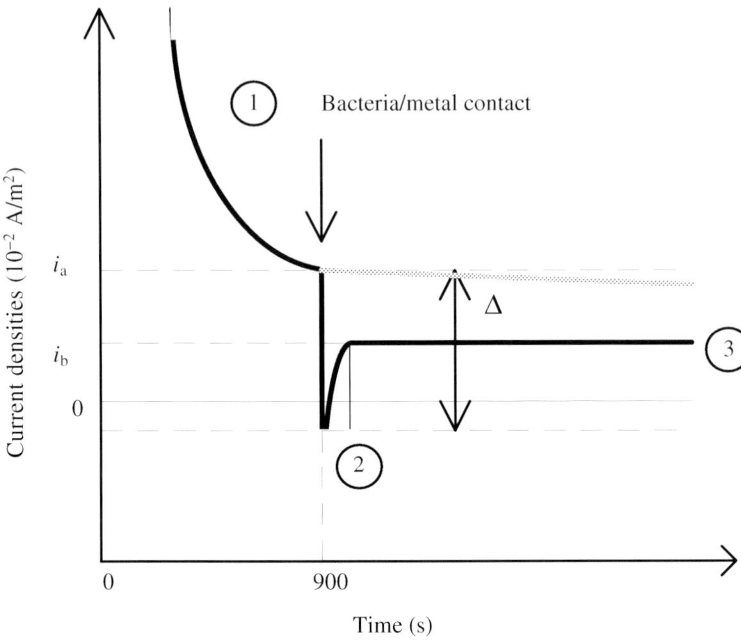

Figure 2. Schematic representation of current densities measured before the bacterial introduction (phase ①) and during the bacterial deposition (phase ②), followed by the irreversible adhesion (phase ③). Note that the gray line corresponds to the blank [18].

In a previous study [25], a typical electrochemical reponse was observed during the bacterial deposition. Briefly, as shown schematically in Fig. 2, as soon as the sample was polarized at a potential higher than the rest potential, the current density decreased. Once the bacteria were introduced in the vessel, a sharp decrease in anodic current was observed during some few seconds, followed by a relatively smoother increase up to a value i_b. Two electrochemical parameters were considered in this study: (i) the current density i_a measured just before the bacterial deposition which characterized the passive film electrochemistry and (ii) the decrease (Δ) in anodic current induced by the bacterial deposition (and defined as a cathodic pulse). Conventionally, an anodic pulse corresponding to a positive current is characteristic of an oxidation reaction whereas a cathodic pulse corresponds to a reduction reaction [17]. We have shown previously [25] that Δ depends on the biological material (bacterial strain, macromolecule), the bacterial concentration and on the polarization potential.

2.5. Electrochemical assays with saccharides

Some assays similar to ones mentioned above (with *Leuconostoc mesenteroides*) were realized with different saccharides. As previously, the sample was left at the rest potential for 30 min and then polarized for 1 h at 200 mV/SCE. The different saccharides were deposited with a dropper near the steel surfaces after the first 15 min of polarization. The substrate used in this part of the study was AISI 304 with BA finish.

The final saccharide concentrations were 1.2×10^{-2} M for monosaccharides, 6×10^{-3} M for disaccharides and 1.2×10^{-5} M for the polysaccharide.

The anodic or cathodic pulse induced by the addition of saccharide was measured and the corresponding charges (Q) were calculated as following:

$$Q = \int_0^t i \, dt,$$

where i is the current density and t the time; Q is expressed in nC/cm^2. Q is assumed to be the charge transfer associated with adsorption of the saccharides.

3. RESULTS

3.1. Surface characteristics

The R_a values and the water contact angles are presented in Table 3.

It is shown that the R_a value ranges from 35 to 150 nm, whereas contact angle varies from ~45° to ~80°.

The more hydrophilic surfaces were AISI 434 (BA) (contact angles lower than 50°). On the other hand, the AISI 304 industrial surfaces were more hydrophobic since contact angles were about 80°. Two types of surfaces can then be defined: hydrophilic surfaces with relatively high R_a value (AISI 434 and 304 MP) and hydrophobic surfaces AISI 304 (2B and BA) which are also smoother.

3.2. Bacterial adhesion

The number of adhering bacteria was not significantly different between hydrophilic surfaces ($\theta < 50°$), and hydrophobic surfaces ($\theta > 80°$) (Fig. 3). This result is in accordance with the previous data obtained on glass and polymer surfaces [15, 16] where *Leuconostoc mesenteroides* adhesion was shown to be non-substrata dependent.

The effect of roughness on the number of bacteria adhering to steel surfaces polarized at 200 mV/SCE was also analyzed (Fig. 4). In this case, *L. mesenteroides*

Table 3.
Surface roughness R_a (in μm) and water contact angles (in degrees) determined for stainless steel surfaces

Surface	R_a	Contact angle
AISI 304		
2B finish	0.080 ± 0.031	81
BA finish	0.035 ± 0.008	78
MP	0.120 ± 0.060	50
AISI 434		
BA no. 1	0.142 ± 0.020	44
BA no. 2	0.142 ± 0.020	45

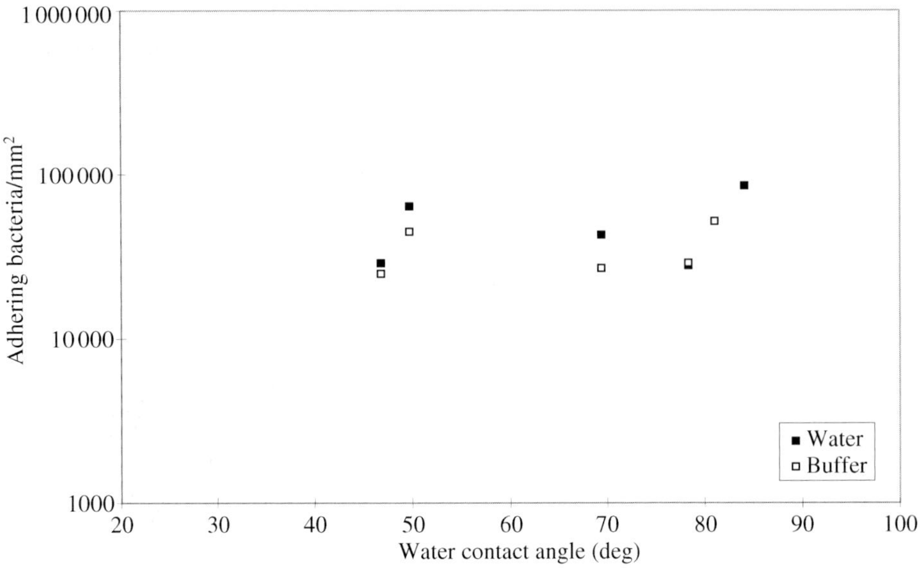

Figure 3. Number of adhering bacteria per mm^2 as a function of water contact angle (in degrees) measured on various stainless steel surfaces in contact with phosphate buffer (open square) or water (filled square).

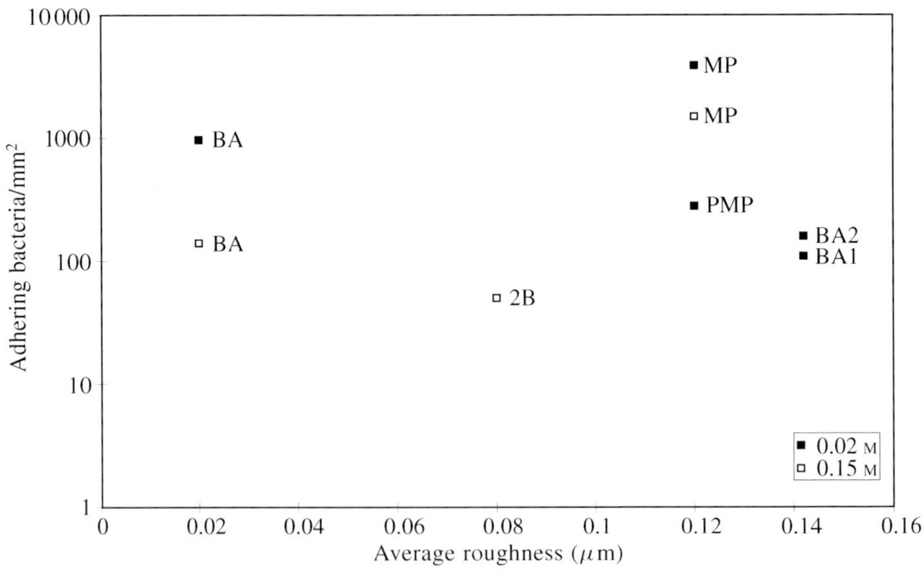

Figure 4. Number of adhering bacteria per mm^2 to stainless steel surfaces polarized in 0.02 M (filled square) and 0.15 M (open square) NaCl solution as a function of R_a value (μm), where BA1 and BA2 correspond, respectively, to AISI 434 stainless steel with a bright annealed surface obtained at −58 or −70°C dew point; and PMP denotes the AISI 304 stainless steel with a mechanical polish followed by a polarization at 200 mV/SCE for 17 h.

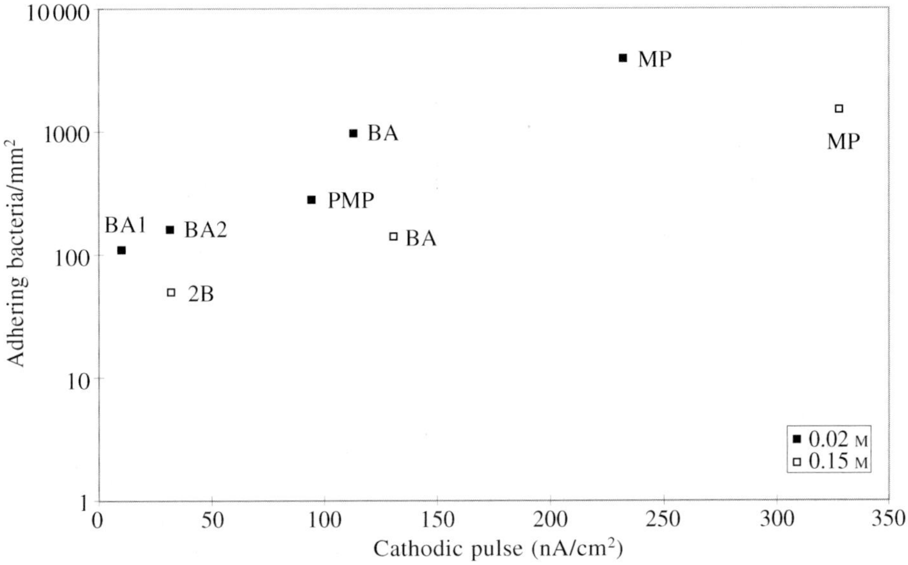

Figure 5. Number of adhering bacteria per mm^2 to stainless steel surfaces polarized at 200 mV/SCE in 0.02 M and 0.15 M NaCl solutions as a function of the cathodic pulse Δ_1 (10^{-2} A/m^2) where BA1 and BA2 correspond, respectively, to AISI 434 stainless steel with a bright annealed surface obtained at −58 or −70°C dew point; and PMP denotes the AISI 304 stainless steel with a mechanical polish followed by a polarization at 200 mV/SCE for 17 h.

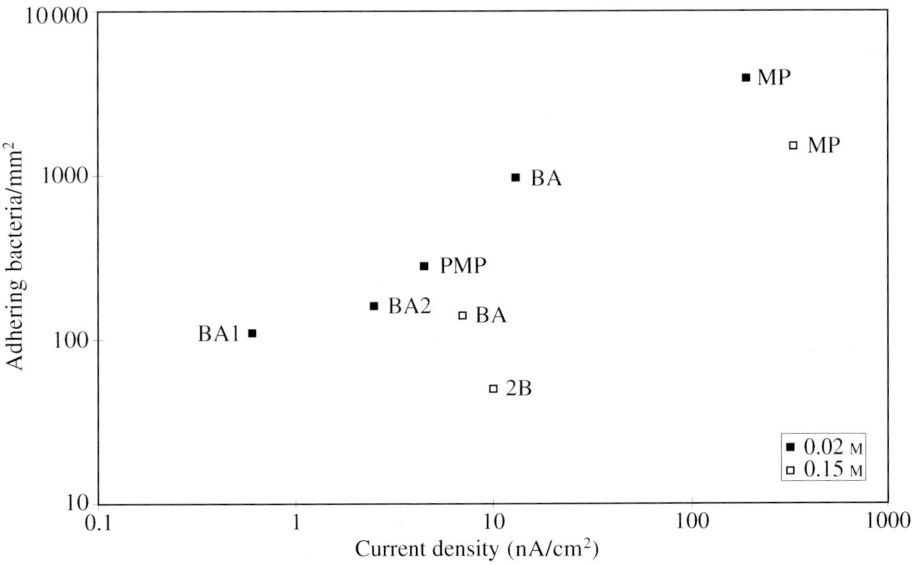

Figure 6. Number of adhering bacteria per mm^2 to stainless steels polarized at 200 mV/SCE in 0.02 M and 0.15 M NaCl solutions as a function of the anodic current density i_a (10^{-2} A/m^2) where BA1 and BA2 correspond, respectively, to AISI 434 stainless steel with a bright annealed surface obtained at −58 or −70°C dew point; and PMP denotes the AISI 304 stainless steel with a mechanical polish followed by a polarization at 200 mV/SCE for 17 h.

Table 4.
Charges (in nQ/cm^2) exchanged during the saccharide/stainless steel contact

Saccharide	Charge (nQ/cm^2)
Glucose	580
Sorbitol	300
Glucuronic acid	−9000
Glucosamine	−485
Maltose	740
Saccharose	400
Dextran	−290 + 40

showed a clear preference for the mechanically polished surfaces (MP), irrespective of the ionic strength of the chloride-containing medium. It is worth pointing out that (i) pre-polarized surfaces (PMP) retained approximately one log order less bacteria than the unpolarized surfaces (MP) and (ii) the number of adhering bacteria to AISI 434 (BA1 and BA2) was lower than that obtained on the AISI 304 (BA).

The following relations between electrochemical responses (Δ and i_a) and the number of adhering bacteria N were observed (Figs 5 and 6). First N increases with both Δ and i_a. Secondly, for given Δ or i_a value, N increases when the ionic strength decreases. Moreover, N was larger for MP than for other surface conditions. *L. mesenteroides* shows a slight preference for the AISI 304 BA surface than for the 2B one.

3.3. Electrochemical assays with saccharides

As for mono-, di- or polysaccharides adhesion, three kinds of behavior are observed when the metal/saccharide contact occurs: (i) an anodic pulse is produced with glucose, sorbitol, maltose or saccharose; (ii) a cathodic pulse is observed when glucosamine or glucuronic acid is introduced near the surface; (iii) a double pulse consisting of a large cathodic pulse followed by an anodic one is induced by the dextran molecule. Table 4 summarizes the corresponding electrical charges (Q). It is found that a larger charge exchange occurs for glucuronic acid than for other molecules. Moreover, glucose and maltose exchange twice more charge than sorbitol or saccharose.

4. DISCUSSION

4.1. Physico-chemical aspects of bacterial adhesion

4.1.1. Effect of the suspending liquid on bacterial adhesion. This study showed that according to previous results [15], the *Leuconostoc mesenteroides* adhesion to stainless steel (without any surface polarization) was not sensitive to solid surface hydrophobicity. The adhesion of *L. mesenteroides* can be explained by a possible adsorbing of dextran (the polymer produced by *L. mesenteroides*) on the solid substrata, which could mask the effect of the composition differences between the surfaces.

Concerning our results on bacterial adhesion to polarized surfaces, it was observed that adhesion of *L. mesenteroides* was dependent on the surface condition (2B, BA, MP, MPP) and the type of alloy (AISI 304, AISI 434). This is in accord with previous studies [6, 27] where the polarization was shown to modify the number of adhering bacteria. Moreover, it is worth noting that the number of *L. mesenteroides* adhering to stainless steel with BA finish increases with the potential imposed on the metal [25]. In this work, *L. mesenteroides* suspended in chloride solutions adhered more weakly to polarized surfaces than to non-polarized surfaces in water or phosphate buffer. This difference could be due either to the suspending liquid or to a polarization effect. It has been shown previously [7, 9] that bacterial adhesion to non-polarized AISI 304 stainless steel surfaces was weaker in chloride solutions than in water. It is worth pointing out that the microscopic techniques employed for the bacterial counting were different for unpolarized and polarized surfaces, since they were made by two different teams. However, equivalent surface areas were examined in the two cases; so, the bacterial numeration method could not explain the observed differences.

It is also worth noting that despite differences in liquid composition and ionic strengh, the same pH (around 7 for phosphate buffer, and chloride solution) was used in the experiments. Recent works [9, 11] have demonstrated for *Pseudomonas fragi* that these pH values correspond to the highest metabolic activity and to the greatest bacterial attachment.

For a given value of electrochemical response (Δ or i_a), the number of adhering bacteria depends on the ionic strength. In accordance with previous results [9], the adhesion of *L. mesenteroides* increases when the ionic strengh decreases. In the particular case of stainless steel/chloride solution interfaces, the theoretical model provided by the DLVO model clearly fails when predicting the number of adhering bacteria [28].

The bacterial surface hydrophobicity could explain this unexpected result [7, 9]. However, another idea may be put forward: the chloride ions are likely to adsorb on stainless steel surfaces leading to some significant changes at the passive film/solution interface, due to the formation of stable or transient complexes with metallic cations. The chloride ions can also modify the chemistry of the passive film/aqueous solution interface parameters such as the surface charge [21] or the wettability [29].

Surface polarization can act as an inhibitor of the dextran effect described above. A tentative explanation could be that a high electric field present at the film/solution interface modifies the conformation of the adsorbed dextran. In that case, the bacterial surface is no longer masked by dextran, and its surface energy should directly control the adhesion mechanism [25]. Such a conformational change was already proposed [30, 31] to explain the unexpected decrease of adsorption of some macromolecules on silicon surfaces in a NaCl aqueous solution, when the ionic strengh increased.

4.1.2. Influence of the surface properties. In addition to the influence of the suspending liquid, numerous studies have reported that the surface roughness and its water wettability can modify the adhesion phenomena [15, 32, 33].

In our study, according to the water contact angle and the R_a value measurements, two types of surfaces were characterized: hydrophobic (and smoother) surfaces and hydrophilic (and rougher) ones. Generally, the metallic surfaces are considered as hydrophilic [8, 33]. However, the industrial surfaces may be contaminated by hydrocarbons. Some monolayers (< 10 nm thickness) are not eliminated by a simple cleaning with solvent [34] due to very strong adhesion between the hydrocarbon layers and passive films. The samples obtained in our laboratory (434 BA 1 and 2) may have been less contaminated and thus were more hydrophilic than industrial ones. The similar numbers of adhering bacteria obtained on the two AISI 434 bright annealed surfaces, despite different passive film compositions [23], suggest that the outermost layer is probably more important that the passive film itself in adhesion phenomena.

It is worth noting that *Leuconostoc mesenteroides* adheres poorly to MPM or BA surface in comparison with MP surface. It is known that either BA or polarization produces some physico-chemical modifications in the passive film and also a passive film growth, whereas unaged MP surfaces exhibit poorly developed passive layers. The outermost layer is likely to be composed of hydrocarbon contamination. However, on non-polarized surfaces, it is shown that surface contamination can modify the bacterial adhesion [8], whereas on polarized ones, electrochemical interfacial reactions govern the adhesion phenomena. This work suggests that such passive film modification greatly influences the adhesion behavior. On the other hand, the surface roughness seems to have little effect, at least in the range of this study.

In the case of polarized steel surfaces, the surface wettability by water is not a reliable parameter to predict the bacterial adhesion ability since the steel surface wettability could change during the polarization. In contrast, a relation between the number of adhering bacteria and the electrochemical reponses (Δ and i_a) is clearly observed.

4.2. Interactions between saccharides/stainless steel

With saccharides such as glucose, sorbitol, maltose or saccharose, an anodic pulse is obtained and the charge Q varies between 300 and 740 nA/cm^2 which is characteristic of an oxidation reaction at the metal/saccharide interface [17].

Briefly, saccharides are organic compounds containing carbon, hydrogen and oxygen in the ratio 1 : 2 : 1. For example, the structural formula for glucose is $C_6H_{12}O_6$ where six carbon atoms are designated as C1, C2, C3, C4, C5 and C6 (see Table 1). Two structural formulas for hexoses (containing six carbon atoms) are possible: (i) a cycle containing one oxygen atom and five carbon atoms designed as a pyranose cycle (i.e. glucose) or, (ii) a cycle containing one oxygen and four carbon atoms designed as a furanose cycle (i.e. fructose in saccharose) [35].

The reaction could be explained by an oxidation of the primary alcohol species which are present in the pyranose cycle C6 position (monosaccharides) or in the furanose cycle C1 position and/or in the pyranose cycle C6 position in the saccharose formulation (see Table 1), the oxidation reaction being as follows:

$$R{-}CH_2OH + H_2O \Rightarrow R{-}COOH + 4e^- + 4H^+ \qquad (1)$$

A recent work [36] showed that saccharose was oxidized to 6-monocarboxylic acid of saccharose and to 1-monocarboxylic acid of saccharose at alkaline pH, for a potential of 400 mV/SCE on a gold electrode. It should be noted that the charges and the current densities are of the same order of magnitude as those obtained in the present work.

The cathodic pulse observed with the glucuronic acid or the glucosamine characterizes a reduction reaction. The reactive chemical functional group which is common for the two glucose derivatives is a secondary alcohol functionality in the pyranose cycle C1 position. The secondary alcohol group can be reduced as follows:

$$R{-}CHOH + 3e^{-} + 3H^{+} \Rightarrow R{-}CH_3 + H_2O \qquad (2)$$

In that case, some 1-desoxyoses can be obtained.

On the other hand, it should be noted that the charge Q is larger for the glucuronic acid ($Q = -9000\ nQ/cm^2$) than for glucosamine ($Q = -485\ nQ/cm^2$). It is assumed that the R−CHOH species in the C6 position is reduced; the R−COOH species can also be reactive and then reduced to $R{-}CH_2OH$ or $R{-}CH_3$.

The double pulse observed with the dextran can be explained by the two reactions (1) and (2) presented above: (i) the first cathodic pulse could be due to the reduction of the secondary alcohol functionality present at the chain extremities in the dextran structure according to reaction (2). This reaction could firstly occur because these functional groups are the more accessible sites in the dextran macromolecule; (ii) the second anodic pulse can be explained by the oxidation reaction of the primary alcohol species in C6 position on the pyranose cycle according to the reaction (1); it should be noted that these secondary alcohol functionalities are less accessible than the secondary alcohol species in the C1 position described previously.

In experiments with *Leuconostoc mesenteroides*, only a cathodic pulse was obtained which could be due to reaction (2). In that case, the dextran molecule could become more hydrophobic, because reaction (2) leads to a decrease of hydrogen bonding in the macromolecule structure. This is consistent with conformational modifications in the dextran macromolecule.

Moreover, in bacterial experiments, the absence of an anodic pulse can be explained by the stabilisation of the dextran molecule by some bacterial wall constituents such as peptidoglycan. In a few Gram positive species, it was shown that there was strong evidence for a covalent linkage between peptiglycane and polymers which form the bacterial surface [37].

5. CONCLUSIONS

(i) The hydrophobicity of stainless steel can be explained by a hydrocarbon layer contamination.

(ii) The adhesion of *Leuconostoc mesenteroides* was not influenced by the steel surface wettability by water, which was ascribed to its polymer excretion.

(iii) The adhesion of *L. mesenteroides* can be modified by electrochemical polarization. In this case, a relation between the electrochemical reponses and the number of adhering bacteria is shown.

(iv) The surface roughness is not a predominant parameter in the adhesion phenomena to polarized steel surfaces.

(v) It is suggested that some ionic and/or electronic exchanges may occur between the bacterial capsule polymers and the polarized steel surfaces. The polarization could induce some conformational changes in the dextran structure and, then, act as an inhibitor for dextran adsorption.

REFERENCES

1. F. M. Driessen, J. de Vries and F. Kingma, *J. Food Prot.* **47**, 848–852 (1984).
2. T. S. Schwach and E. A. Zottola, *J. Food Prot.* **47** (10), 756–759 (1984).
3. S. J. Lewis and A. Gilmour, *J. Appl. Bacteriol.* **62**, 327–333 (1987).
4. P. H. Herald and E. A. Zottola, *J. Food Sci.* **54** (2), 461–464 (1989).
5. A. A. Mafu, D. Roy, J. Goulet and P. Magny, *J. Food Prot.* **53**, 742–746 (1990).
6. B. J. Little, P. Wagner, J. S. Maki, M. Walch and R. Mitchell, *J. Adhesion* **20**, 187–210 (1986).
7. E. Vanhaecke, J. P. Remon, M. Moors, F. Raes, D. de Rudder and A. van Peteghem, *Appl. Environ. Microbiol.* **56** (3), 788–795 (1990).
8. L. Boulangé-Petermann, B. Baroux and M. N. Bellon-Fontaine, *J. Adhesion Sci. Technol.* **7** (3), 221–230 (1993).
9. P. M. Stanley, *Can. J. Microbiol.* **29**, 1493–1499 (1983).
10. L. M. P. Langeveld, A. C. Bolle and J. E. Vegter, *Neth. Milk Dairy J.* **26**, 149–154 (1972).
11. P. J. Herald and E. A. Zottola, *J. Food Prot.* **51**, 445–448 (1988).
12. J. T. Holah and R. H. Thorpe, *J. Appl. Bacteriol.* **69**, 599–608 (1990).
13. M. Fletcher and G. D. Floodgate, *J. Gen. Microbiol.* **74**, 325–334 (1973).
14. P. T. Zoltai, E. A. Zottola and L. L. McKay, *J. Food Prot.* **44** (3), 204–208 (1981).
15. M. N. Bellon-Fontaine, N. Mozes, H. C. van der Mei, J. Sjollema, O. Cerf, P. G. Rouxhet and H. J. Busscher, *Cell Biophysics* **17**, 93–106 (1990).
16. H. J. Busscher, M. N. Bellon-Fontaine, N. Mozes, H. C. van der Mei, O. Cerf and P. G. Rouxhet, *Biofouling* **67**, 55–63 (1990).
17. B. Baroux, G. Beranger and C. Lemaitre, in: *Stainless Steels*, P. Lacombe, B. Baroux and G. Beranger (Eds), pp. 161–182. Les Editions de Physique, Les Ulis, France (1993).
18. B. E. Brooker, *Arch. Microbiol.* **111**, 99–104 (1976).
19. J. J. Devoyod and F. Poullain, *Le Lait* **69**, 249–280 (1988).
20. J. C. De Mann, M. Rogosa and M. E. Sharpe, *J. Appl. Bacteriol.* **23**, 130–135 (1960).
21. L. Boulangé-Petermann, B. Baroux, A. Doren and M. N. Bellon-Fontaine, *J. Colloid Interface. Sci.* **171**, 179–186 (1995).
22. B. Baroux, *Corrosion Sci.* **27**, 969–986 (1988).
23. D. Gorse, J. C. Joud and B. Baroux, *Corrosion Sci.* **33**, 1455–1478 (1992).
24. A. W. J. van Pelt, A. H. Weerkamp, M. H. W. J. C. Uyen, H. J. Busscher, H. P. De Jong and J. Arends, *Appl. Environ. Microbiol.* **49**, 1270–1275.
25. L. Boulangé-Petermann, M. N. Bellon-Fontaine and B. Baroux, *J. Microbiol. Method.* **21**, 83–96 (1994).
26. R. Oltra and M. Keddam, in: *Stainless Steels*, P. Lacombe, B. Baroux and G. Beranger (Eds), pp. 183–212. Les Editions de Physique, Les Ulis, France (1993).
27. A. D. Maines, L. V. Evans and R. G. J. Edyvean, in: *Microbial Corrosion*, C. A. C. Sequeira and A. K. Tiller (Eds), pp. 213–220. The Institute of Materials, London (1992).

28. P. R. Rutter and B. Vincent, in: *Microbial Adhesion and Aggregation*, K. C. Marshall (Ed.), pp. 21–38. Springer-Verlag, Berlin (1984).
29. L. G. J. Fokkink and J. Ralston, *Colloids Surfaces* **36**, 69–76 (1989).
30. J. K. Luey, J. McGuire and R. D. Sproull, *J. Colloid Interface Sci.* **143**, 489–500 (1991).
31. S. J. Lewis, A. Gilmour and D. E. Johnston, *Intl J. Food Microbiol.* **8**, 155–164 (1989).
32. M. Fletcher and G. I. Loeb, *Appl. Environ. Microbiol.* **37**, 67–72 (1979).
33. M. Asther, M. N. Bellon-Fontaine, C. Capdevila and G. Corrieu, *Biotechnol. Bioeng.* **35**, 477–482 (1990).
34. M. Mantel and J. P. Whightman, in: *Proc. 17th Annual Meeting of The Adhesion Society*, held in Orlando, Feb. 20–23, 1994, p. 20.
35. T. D. Brock, M. T. Madigan, J. M. Martinko and J. Parker (Eds), *Biology of Microorganisms*, 7th edn, pp. 28–32. Prentice-Hall International Editions, Englewood Cliffs, NJ.
36. P. Parpot, B. Kokoh, B. Beden and C. Lamy, *Electrochimica Acta* **38** (12), 1679–1683 (1993).
37. I. C. Hancock, in: *Microbial Cell Surface Analysis*, N. Mozes, P. S. Handley, H. J. Busscher and P. G. Rouxhet (Eds), pp. 21–58. VCH Publishers, New York (1991).

Mittal Festschrift, pp. 821–836
W. J. Van Ooij and H. R. Anderson, Jr. (Eds)

The interaction of human endothelial cells with chemical gradient surfaces during exposure to flow

T. G. RUARDY,[1] H. E. MOORLAG,[1] J. M. SCHAKENRAAD,[2] J. VAN DER MEER,[3] H. C. VAN DER MEI[1] and H. J. BUSSCHER[1,*]

[1] *Laboratory for Materia Technica, University of Groningen, Bloemsingel 10, 9712 KZ Groningen, The Netherlands*
[2] *Department of Biomaterials, BMTC, University of Groningen, Bloemsingel 10, 9712 KZ Groningen, The Netherlands*
[3] *Division of Haemostasis, Thrombosis and Rheology, University of Groningen, Hanzeplein 1, 9713 GZ Groningen, The Netherlands*

Abstract—In this study, the position bound shape, spreading, detachment and migration of adhering HUVEC endothelial cells on dichlorodimethylsilane (DDS) chemical gradient surfaces was investigated during exposure to flow in a parallel plate flow chamber in the presence of serum proteins. Gradient surfaces were prepared by the diffusion method and characterized by the Wilhelmy plate technique for their wettability. In order to measure the possible influence of wettability steepness on the position bound cellular response, shallow and steep wettability gradients were prepared by applying two different diffusion times and DDS-concentrations. Quantitative analyses of the cellular response on the gradient surfaces showed that the position bound spread area and shape factor of adhering cells increased when going from the hydrophobic to the hydrophilic end, concurrent with the steepness of the gradient. After exposure to flow, cells detached immediately from the hydrophobic side and maintained their positions more easily on the hydrophilic side. Furthermore, it was shown that the position bound detachment and migration of cells depended on the steepness of the wettability gradient and the direction of the flow. When the flow was to the hydrophilic end, migration velocity did not vary along the length of the gradients, and was lower on the steep (approximately 2 μm/min) than on the shallow (approximately 3 μm/min) gradients. However, when flow was to the hydrophobic end, migration velocities appeared position bound and were higher on the hydrophobic ends of both gradients.

Keywords: Endothelial cells; adhesion; gradient surface; wettability; flow.

1. INTRODUCTION

Biomaterials are frequently used in biomedical devices as, for example, vascular grafts [1], artificial skin [2], contact lenses [3], bone prostheses [4], abdominal wall patches [5] or catheters [6]. Whereas biomaterials surface properties are usually engineered to discourage microbial adhesion in order to prevent infection of the device,

*To whom correspondence should be addressed. E-mail: H.J.Busscher@med.rug.nl

the materials surface properties are often also adjusted to stimulate or reduce cellular adhesion, spreading and growth through appropriate protein interactions [7, 8]. Many studies on cellular adhesion to biomaterials surfaces have been carried out under static conditions [9, 10], despite the fact that under clinical conditions varying shear forces may be exerted on adhering cells, as, for example, on endothelial cells seeded on a vascular graft [1]. Consequently, the strength of cellular adhesion may be a more important parameter to study than, say, the number of cells adhering under specific conditions. Following the development of flow chambers and image analysis procedures, the strength of cellular adhesion on substratum surfaces under flow has become an achievable parameter to determine [11–13].

The evaluation of different materials in cellular adhesion studies is usually time consuming and large biological variations over longer periods of time are common. Elwing *et al.* [14] circumvented this problem for a major part by using a so-called 'wettability gradient surface' for protein interaction studies. A gradient surface is a surface on which a continuously varying chemical composition exists over its length. The main advantage of gradient surfaces is that the effect of a continuum of physico-chemical properties can be measured in one experiment on a single sample surface. Adsorption of a variety of proteins on gradient surfaces, including albumin, fibrinogen and IgG, changes abruptly on the transition region of gradient surfaces [14, 15]. Unfortunately, it is not known at present whether this specific response of proteins on the transition region of a gradient is due to prevailing wettability, as a macroscopic reflection of the local chemical composition or due to microscopic chemical heterogeneity. Especially when gradient surfaces are used to study adsorption of molecular entities such as proteins, little is known about the influence of the microscopic chemical heterogeneities presumably existing on the transition parts of wettability gradient surfaces. This demonstrates how the advantage of gradient surfaces, as originally recognised, might be a simultaneous disadvantage.

Recently, gradient surfaces were also used to investigate cellular interaction phenomena [16–18]. We studied adhesion and spreading of human skin fibroblasts on a dimethyldichlorosilane (DDS)-gradient and observed preferential adhesion of fibroblasts on the steepest, transition part of the gradient. Williams *et al.* [17] evaluated the interfacial interactions between proteins and cells on DDS-gradient surfaces as well and concluded that competitive adsorption of proteins from mixed binary solutions led to conformational changes along the length of the gradient with a distinct influence on the adhesion and spreading of L929 fibroblasts.

Endothelial cells constitute the natural inner lining of blood vessels and as such possess antithrombogenic properties. These properties can be exploited in the application of vascular prostheses by seeding endothelial cells on the prosthesis. For this purpose, adhesion and spreading of endothelial cells to biomaterials must be studied under flow. Van Kooten *et al.* [19, 20] studied adhesion, spreading, migration and detachment of human skin fibroblasts and HUVEC endothelial cells to a number of substrata, like fluoroethylene–propylene, polystyrene and glass in a parallel plate flow chamber. Although the studies as done by Van Kooten *et al.* show an influence of substratum wettability on its interaction with endothelial cells, the experiments were too time consuming to yield a detailed relationship between cellular responses and biomaterials wettability. This can be obtained more conveniently using wettability gradient surfaces.

Therefore the aim of this work was to study the response, i.e. shape, spreading, detachment and migration of adhering human endothelial cells on chemical gradient surfaces during exposure to flow in a parallel plate flow chamber.

2. MATERIALS AND METHODS

2.1. Preparation of gradient surfaces

Wettability gradients were prepared according to the diffusion technique described by Elwing *et al.* [14] using two sizes of glass plates, namely, 7.6 × 5.0 × 0.2 cm (l × w × h) and 7.6 × 0.6 × 0.2 cm (l × w × h). The smaller plates were used for characterization by the Wilhelmy plate method, while the larger plates were employed for flow chamber experiments. The glass plates were rinsed thoroughly in trichloroethylene, methanol and Milli-Q filtered water. After drying in a vacuum desiccator for 30 min, each sample was glow discharge treated on both sides in an Edwards sputter coater S150B at 7.3 W and 0.1 mbar oxygen pressure for 2 min. The samples now showed complete wetting by water and contained little carbon contamination (about 10% C as analysed by X-ray photoelectron spectroscopy). After this cleaning procedure, the samples were immediately placed in a beaker filled with xylene. A solution of $Cl_2(CH_3)_2Si$ (dichlorodimethylsilane, DDS) in trichloroethylene was injected under the xylene phase through a drain in the beaker, allowing DDS to diffuse into the xylene phase and react with the glass surface (see Fig. 1). In order to obtain gradient surfaces with a steep gradient, a DDS concentration of 0.05% (w/w) was used in combination with a diffusion time of 45 min and a 0.025% (w/w) concentration with a diffusion time of 16.5 h for shallow gradients. After diffusion at room temperature, the solvents were removed through the drain of the beaker. The glass plates were then rinsed in trichloroethylene, methanol and Milli-Q water to remove unreacted DDS from the surface. Gradient surfaces prepared on the larger glass plates were stored in methanol prior to flow chamber experiments. The small plates were immediately characterized by the Wilhelmy plate method after drying in a vacuum desiccator for 10 min. Samples used for flow chamber experiments were never stored for longer than two weeks and were sonicated for 7 min in methanol and rinsed in Milli-Q water prior to the assemblage in the parallel plate flow chamber. All chemicals used (Merck, Darmstadt, Germany) were of P.A. grade.

2.2. Wettability characterization by the Wilhelmy plate method

Advancing and receding water contact angles were recorded along the length of the gradients using the Wilhelmy plate method [21]. Samples were fixed to a microbalance (Lauda tensiometer, Königshofen, Germany) and the force was continuously recorded while the plate was immersed into water at a constant speed of 27 mm/min. The

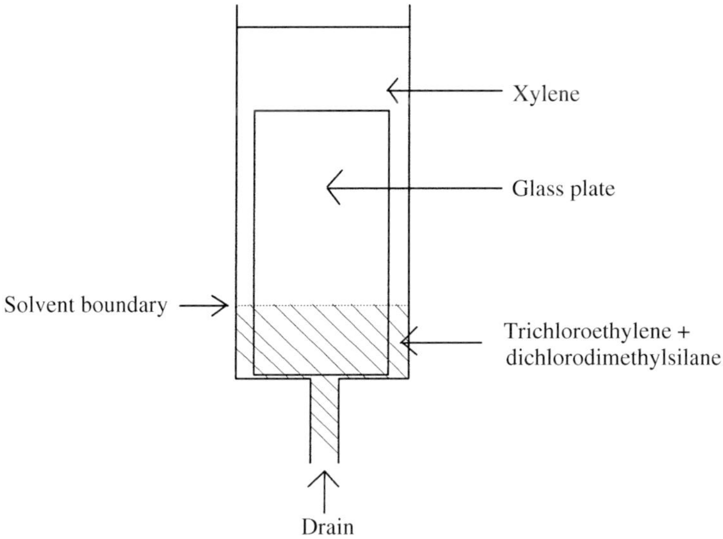

Figure 1. Schematic illustration of the method used to prepare chemical gradients on glass. Glass plates are placed in a beaker filled with xylene. A solution of $Cl_2(CH_3)_2Si$ in trichloroethylene is injected gently under the xylene solution. Dichlorodimethylsilane diffuses up into the xylene region and subsequently couples to the glass plate. The experiment is stopped by removal of the solutions through the drain.

contact angle θ was calculated as a function of the position along the length of the gradient from

$$F = p\gamma \cos\theta + (m - V\Delta\rho)g \tag{1}$$

in which F is the force on the plate, p the perimeter of the plate, γ the liquid surface tension, m the mass of the plate, $\Delta\rho$ is the liquid–vapour density difference, g the acceleration due to gravity and V the volume of water displaced by the immersed part of the plate [22].

2.3. Endothelial cell culturing and seeding

Human Umbilical Vein Endothelial Cells (HUVEC) were cultured as follows: cells were freshly harvested from the umbilical vein essentially as described by Mulder *et al.* [23]. The primary culture (passage P0) and subsequently passaged cultures were grown in RPMI 1640 medium, supplemented with 100 U/ml penicillin, 100 μg/ml streptomycin, 2 mM L-glutamine (all from Gibco Life Technologies, Breda, The Netherlands), 30 μg/ml Endothelial Cell Growth Factor (ECGF) and 10% pooled human serum (HPS) at 37°C in 5% CO_2. At confluency, cells were passaged using trypsin a splitting ratio at 1 : 3. Cells were routinely grown on tissue culture quality plastic flasks precoated with 0.2% gelatine in PBS (9.0 mM Na_2HPO_4, 1.3 mM KH_2PO_4, 140 mM NaCl). Cells between passage P1 and P5 were used for flow chamber experiments. Cells were harvested by trypsinization using trypsin–EDTA solution in modified Puck's saline A and seeded on the gradient surfaces positioned in the flow chamber to a density up to 200 cells/mm^2 and allowed to adhere and spread for 3 h at 37°C. Flow was established with RPMI 1640 medium without any supplements.

2.4. The flow chamber system

Figure 2 shows a schematic presentation of the flow chamber system. It contains three basic modules: the flow loop, the heating system, and the image analysis components [11]. RPMI 1640 medium is pumped through the flow loop by a peristaltic pump (Watson-Marlow 503U with 8 mm i.d. tubing) into a damping vessel. This vessel dampens the pulsations of the flow into a steady flow. Then the medium flows through a parallel plate flow chamber mounted on a phase contrast microscope stage. The medium is collected in a thermostatted double walled vessel. The flow chamber itself is heated directly with four power resistors (33 ohm each), connected in parallel to a power supply (10 V) with feedback from a Pt-100 thermocouple situated in the downstream compartment of the flow chamber.

The image supplied by the microscope system (Olympus BH-2 with 10 × A10PL phase contrast objective and NFK 3.3×LD photo ocular) is registered with a CCD camera (type MO, High Technology Holland BV, Eindhoven, The Netherlands) connected to the microscope with a MTV-3 (Olympus) connector and covers 0.28 mm^2 of the gradient surface. The signal from the CCD camera is digitized by a frame grabber (PCVision Plus, Imaging Technology Incorporated, USA) into an image consisting of 512 × 512 pixels, each pixel having 256 possible grey values. The series of images obtained over the length of the gradient surfaces can be stored on a hard disk for later analysis, requiring 256 kB of memory for each image. The frame grabber is

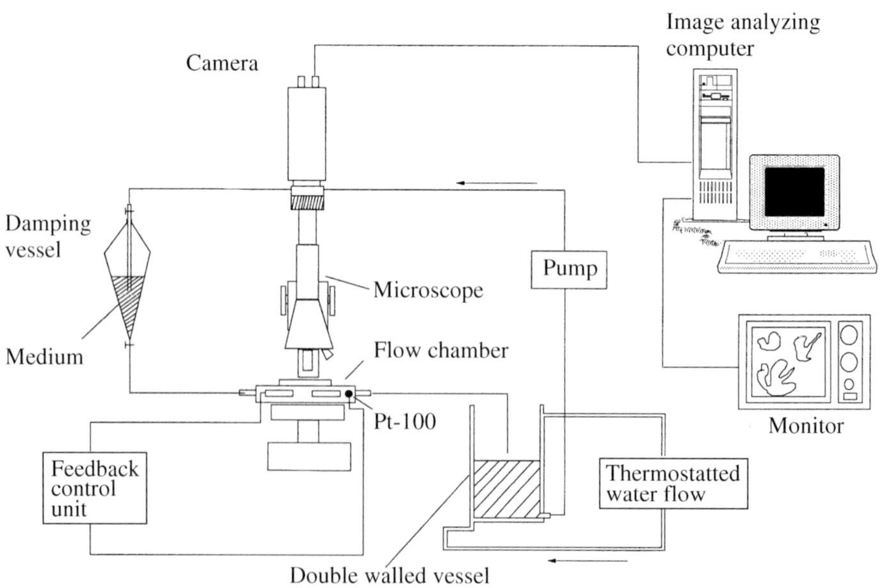

Figure 2. Overview of the complete flow system, including the flow chamber system, image analysis components and the heating system. The damping vessel serves to exclude pulsations from the flow.

controlled by image analysis software (TIM, Difa/TEA, The Netherlands) which also provides possibilities for image processing and programming.

2.5. Experimental setup

After allowing the cells to adhere and spread to a stationary state in 3 h [10], two essentially different types of experiments were carried out. In one type of experiment, flow was stopped at a 10 min time interval in order to grab images on 15 positions along the length of the gradient as a function of the cumulative flow time. Repositioning of the flow chamber was done employing the millimeter scale on the microscope stage as a reference. This repositioning was only approximate but fast, which allowed measurements on a high number of positions along the length of the gradient surface.

In the other type of experiment, flow was continuous and images were grabbed on only 4 positions along the length of the gradient for the analysis of flow-induced and position bound migration of the cells. This analysis required a much more accurate repositioning. To this end, a more time consuming but accurate (2 μm) repositioning was done by focusing onto a cell finder slide, attached to the bottom plate of the flow chamber.

All experiments were carried out at a fluid shear stress of 8.6 N/m^2. In order to observe a possible influence of the direction of flow on the cellular behaviour, flow was applied from the hydrophobic to the hydrophilic end of the gradient surface and *vice versa*.

2.6. Analysis of cellular response

In this study, the general expression 'cellular response' will be employed for four different, position bound parameters of the cellular interaction with the gradient, namely, the detachment of cells using a time dependent distribution function $g(x, t)$, the spread area of the adhering cells $A(x)$, the circularity or shape factor of the adhering cells $S(x)$ and the migration velocity $V(x)$ of the adhering cells during exposure to flow.

In order to be able to determine the detachment of adhering cells over the length of a gradient surface, the concept of linear distribution functions was introduced [18], by analogy with radial and angular pair distribution functions to analyze particle adhesion on collector surfaces [24]. The number of cells n in each image grabbed was counted and expressed as

$$g(x, t) = \frac{n(x, t)}{n_s(t)} \tag{2}$$

in which $g(x, t)$ is a time and position dependent linear distribution function, $n(x, t)$ is the number of cells in an image at distance x from the hydrophobic end of the glass plate at time t after starting the flow and $n_s(t)$ the mean number of cells over the length of the gradient surface within one series at time t after starting the flow. When adhering cells are uniformly adhering over the length of the gradient surface, $g(x, t)$ equals unity, but for regions where the number of adhering cells is more or less than average, $g(x, t)$ is larger or smaller than unity, respectively.

Prior to starting the flow, a number of randomly selected cells were outlined and the coordinates of the cells in each image were used to calculate a position bound spread area function $A_r(x)$ according to

$$A_r(x) = \frac{A(x)}{A_s} \tag{3}$$

in which $A(x)$ is the mean spread area of cells at distance x; A_s, the mean spread area of cells over the length of the gradient surface within one series; and $A_r(x)$ the relative spread area of the cells in an image at distance x.

Similarly, a position bound mean shape factor $S(x)$ was calculated by averaging the shape factors S_i of each individual adhering cell, which is defined as:

$$S_i = \frac{l_i}{2(\pi A_i)^{1/2}} \tag{4}$$

in which l_i and A_i are the perimeter and spread area of an individual cell, respectively. Since major detachment of adhering cells occurred on the hydrophobic side of the gradient surface almost immediately after the start of flow, position bound spread area functions and shape factors were calculated only for adhering cells prior to exposure to flow.

Finally, the position bound migration velocity of the adhering cells $V(x)$, as stimulated by the flow, was calculated by

$$V(x) = \frac{\Delta X}{\Delta t} \tag{5}$$

in which ΔX is the mean migration distance of the adhering cells. Cells that detached during the time period Δt were discarded.

3. RESULTS

3.1. Wettability characterization

Figure 3 shows the advancing and receding water contact angles as well as the contact angle hysteresis along the lengths of the steep and the shallow gradient surfaces used. On steep gradient surfaces, advancing water contact angles varied from 45 to 104 degrees over a length of approximately 8 mm, while for the shallow gradients advancing contact angles varied from 30 to 117 degrees over a length of approximately 33 mm. Receding water contact angles on the hydrophilic side of the gradient surfaces were similar and amounted to 20 degrees, but the receding water contact angle on the hydrophobic side of the shallow gradient (78 degrees) was higher than on the steep gradient surface (52 degrees). Consequently, a maximum contact angle hysteresis of around 65 degrees was observed in the transition region of both gradients.

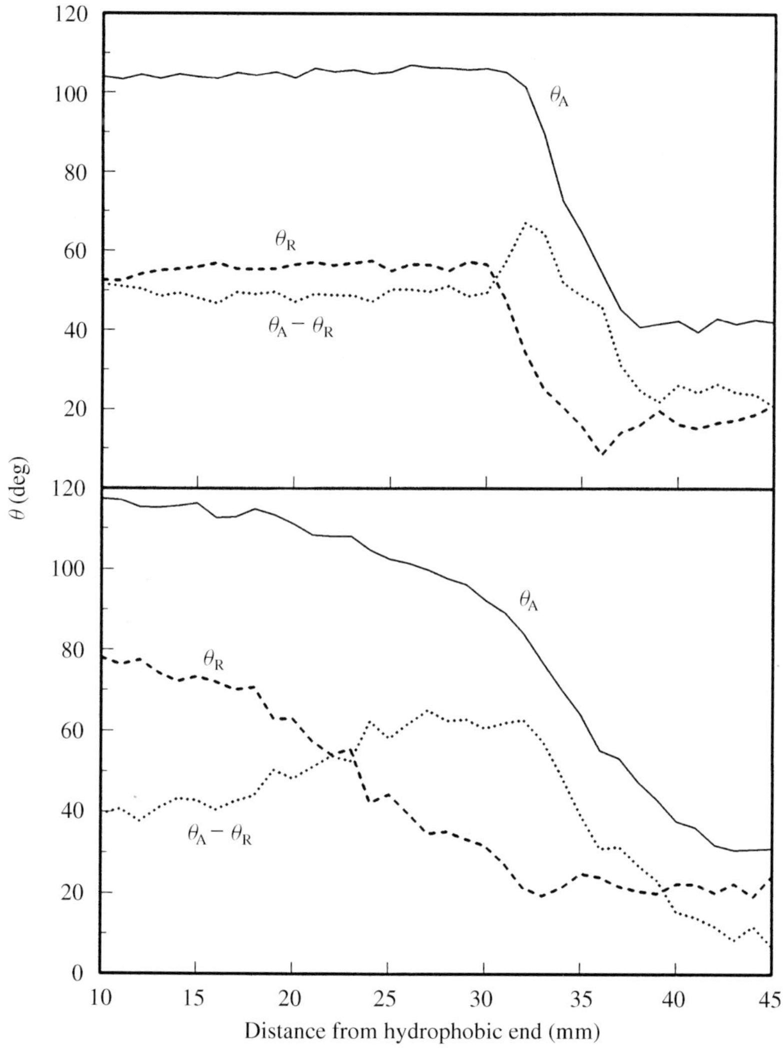

Figure 3. The advancing water contact angle θ_A, receding water contact angle θ_R and contact angle hysteresis $\theta_A - \theta_R$ as a function of the distance from the hydrophobic end of a DDS-chemical gradient surface on glass. Top: example of a steep gradient. Bottom: example of a shallow gradient.

3.2. Cellular spreading: spread area and shape factor

Figure 4 presents the relative spread areas of human endothelial cells along the lengths of the gradient surfaces. On the steep gradient, a sharp transition from non-spreading to complete-spreading (absolute spread area 2250 μm^2) is seen, when going from the hydrophobic to the hydrophilic end of the gradient. The advancing water contact angles at which spreading occurred varied between 74 and 90 degrees. A similar but less sharp transition in cell spreading was observed on the shallow gradient surface.

Figure 5 demonstrates the change in shape factor of spread endothelial cells occurring over the length of the gradient surfaces. The increase in shape factor observed

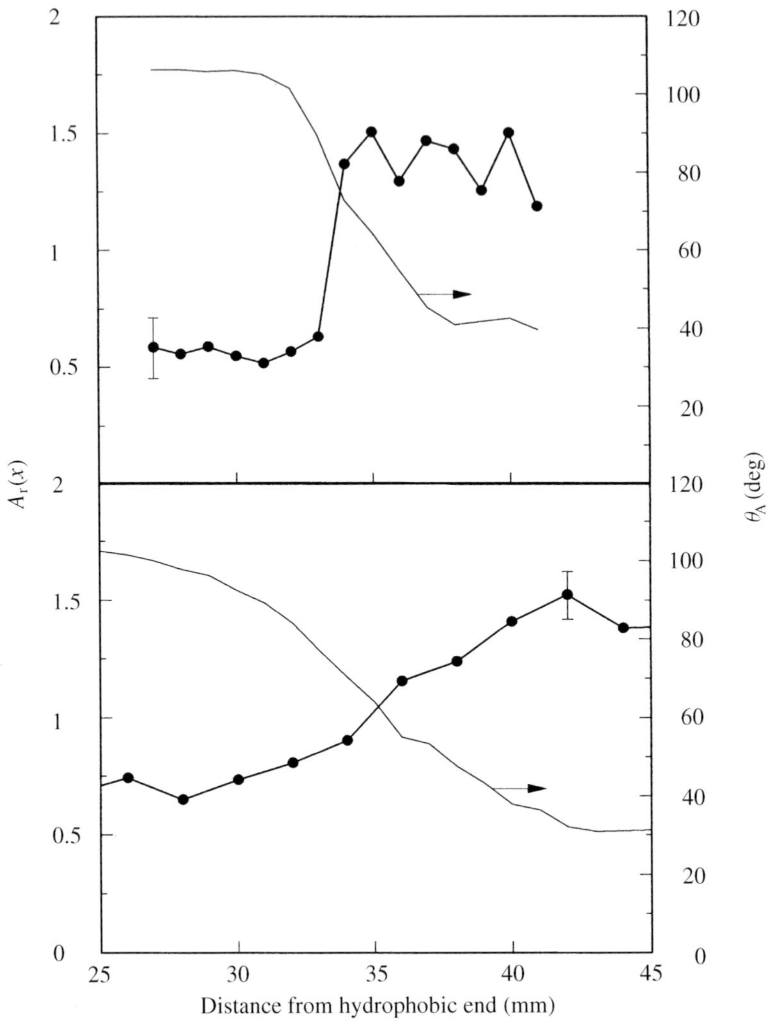

Figure 4. The relative spread area $A_r(x)$ of adhering human endothelial cells (individual data points, left axis) and the advancing water contact angle θ_A as a function of the distance from the hydrophobic end of a DDS-chemical gradient surface on glass. Top: data on the steep gradient. Bottom: data on the shallow gradient. Bars indicate mean values of the SD over four experiments, comprising a total of 80 cells per data point.

when going from the hydrophobic to the hydrophilic side of a gradient is fully concurrent with the increase in spread areas of the cells.

3.3. Cellular distribution during exposure to flow

Figure 6 presents the distribution of adhering cells along the length of the gradient surfaces after different periods of exposure to flow. The direction of the flow was from the hydrophobic to the hydrophilic side of the gradient or *vice versa*. Prior to

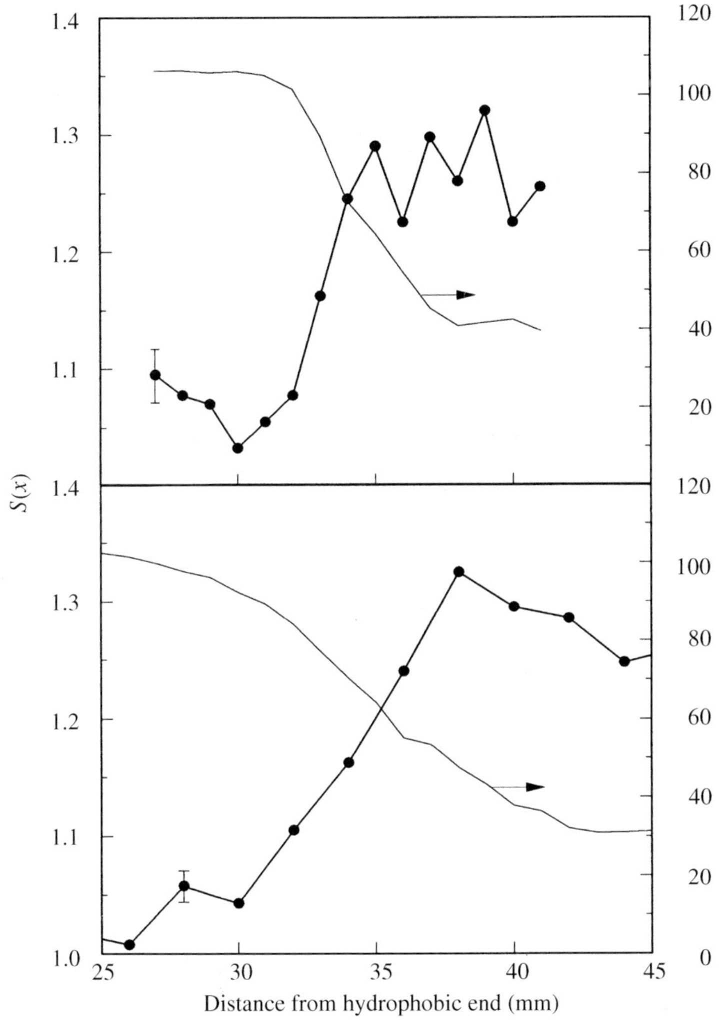

Figure 5. The shape factor $S(x)$ of adhering human endothelial cells (individual data points, left axis) and the advancing water contact angle θ_A as a function of the distance from the hydrophobic end of a DDS-chemical gradient surface on glass. Top: data on the steep gradient. Bottom: data on the shallow gradient. Bars indicate mean values of the SD over four experiments, comprising a total of 80 cells per data point.

exposure to flow, a uniform distribution of adhering cells is observed, i.e. $g(x, t) = 1$, both on the steep and on the shallow gradient surfaces. Immediately (10 s) after exposure to flow, cells detached from the hydrophobic side of the gradient surfaces, while they maintained their positions for at least 20 min on the hydrophilic sides of the gradients. After 20 min and 60 min exposure to flow, an influence of the flow direction on the distribution of adhering cells was observed. Adhering cells were better able to maintain their positions in the hydrophobic region of the gradients when flow was from the hydrophilic to the hydrophobic side of the gradients than *vice versa*.

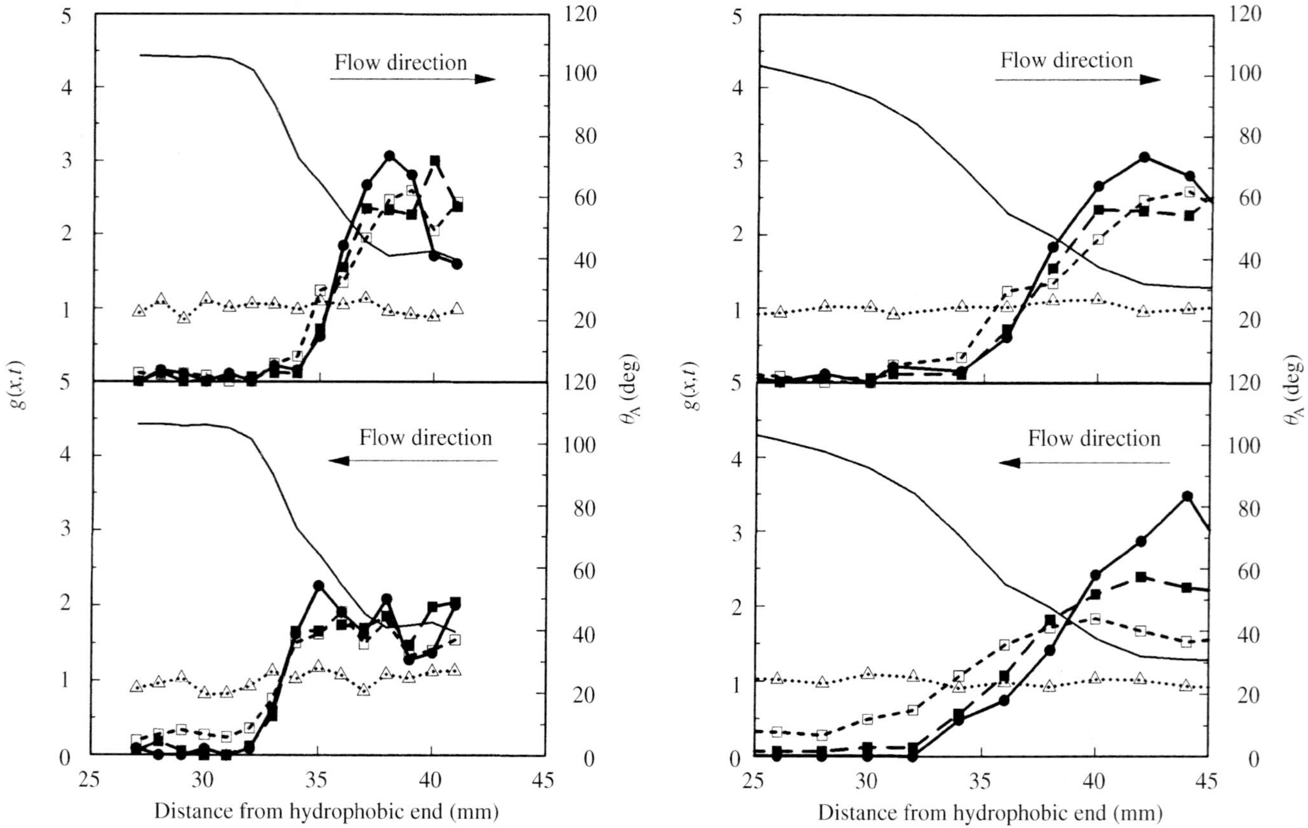

Figure 6. The distribution function $g(x, t)$ of adhering human endothelial cells (individual data points, left axis) and the advancing water contact angle θ_A (drawn line, right axis) as a function of the distance from the hydrophobic end of a DDS-chemical gradient surface on glass prior to (△) and after exposure to flow for 10 s (□), 20 min (■) and 60 min (●). Left: data on the steep gradient. Right: data on the shallow gradient. Data represent distribution functions obtained from two experiments, comprising a total of 50 cells per data point.

3.4. Cellular migration during exposure to flow

Figure 7 shows the influence of the direction of flow on the position bound migration velocity $V(x)$ along the length of the steep and shallow gradient surfaces after 60 min of exposure to flow. When flow is from the hydrophilic to the hydrophobic end, the migration velocity is approximately 2 μm/min for the steep gradient and 3 μm/min for the shallow gradient and no clear position bound effects are seen, but when flow was from the hydrophobic to the hydrophilic end, migration velocities decreased going to the hydrophilic side of both types of gradient surfaces.

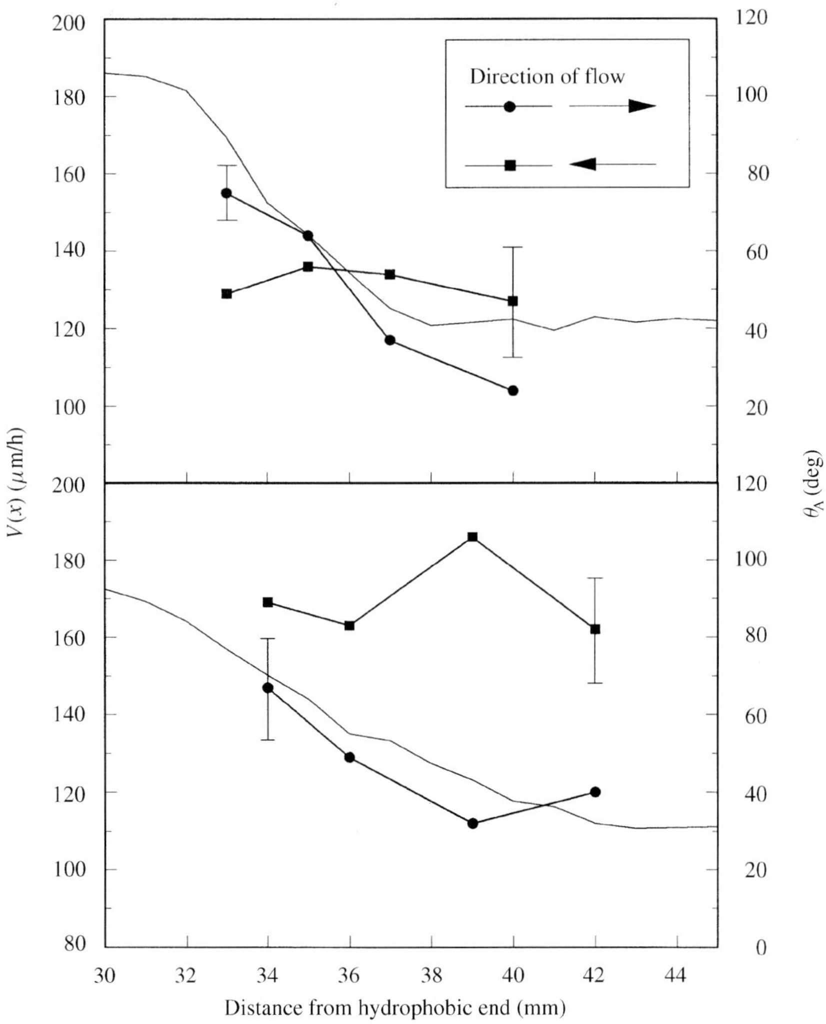

Figure 7. The migration velocity $V(x)$ of adhering human endothelial cells (individual data points, left axis) and the advancing water contact angle θ_A as a function of the distance from the hydrophobic end of a DDS-chemical gradient surface on glass. Top: data on the steep gradient. Bottom: data on the shallow gradient. Bars indicate mean value of the difference over two experiments, comprising a total of 30 cells per data point.

4. DISCUSSION

In this paper we studied the position bound shape, spreading, detachment and migration of adhering human endothelial cells (HUVEC) on chemical gradient surfaces during exposure of flow in a parallel plate flow chamber in the presence of serum proteins. The influences of the steepness of the gradient and the direction of flow along the gradient on the cellular behaviour were determined.

The chemical gradient surfaces were prepared in this study by the so-called diffusion method as originally introduced by Elwing *et al.* [14] and yield a continuous change in wettability along the length of a sample due to increased interaction of silane with the glass towards the hydrophobic end of the gradient. Consequently, it can be anticipated that a major, position bound, contact angle hysteresis will exist, as can be seen in Fig. 3. The increased water contact angle hysteresis on the steepest region of both steep and shallow gradient surfaces has been ascribed to chemical heterogeneities [15] and possibly to the presence of unreacted SiOH groups on the hydrophobic and transition regions of the gradient. Others have shown that clusters of hydrophobic patches are formed on DDS treated silicon wafers using Scanning Force Microscopy [25]. The size and density of these patches will be such that a maximum contact angle hysteresis results in the transition region of the gradients. The hydrophilic side of the gradient surfaces is easily contaminated by carbonaceous components during preparation of the gradients, leading to a higher contact angle hysteresis on the hydrophilic side of a gradient as compared to the hysteresis found on clean glass which is essentially zero. Studies on different, homogeneous surfaces have shown that cell–substratum interactions depend upon an interplay of several substratum properties, including surface roughness, charge and wettability reflecting the specific surface chemistry. Cells seldom adhere to bare substratum surfaces, but always to a so-called ‘conditioning film’ [26], consisting of adsorbed proteins from serum or plasma. In general, the properties of the bare substratum surface are reflected in the interfacial properties of the adsorbed protein layer [27]. Thus, the properties of the underlying substratum surface remain influential upon its interaction with cells, despite the presence of the conditioning film (some authors have termed this the ‘shining through’ effect [28]). Adsorption of the proteins involved in cellular adhesion to substratum surfaces, such as, for example, fibronectin and vitronectin [29, 30], on gradient surfaces has recently been studied, showing varying competition between adsorbing proteins along the length of a gradient surface [17]. In the present study, endothelial cells were allowed to adhere from medium supplemented with serum proteins and competitive adsorption of proteins will occur. The cellular responses along the length of the gradients occurred despite, or through, the presence of adsorbed proteins. The shape of the cells on the hydrophobic side of the gradients was almost circular which can be explained by the fact that the cells did not adhere very well to this side of the gradient. Schakenraad *et al.* [28] also observed a sigmoidal relationship between cell spreading and wettability while using several different materials with a widely varying chemistry. Recently we found similar results for human skin fibroblasts on DDS-gradient surfaces [18]. Although most studies hitherto have demonstrated good cell adhesion and spreading on hydrophilic substrata, Ueda-Yukoshi and Matsuda [31] have shown an inverse influence of wettability on cellular adhesion and spreading of Bovine Aortic Endothelial Cells (BAEC) on gradient surfaces with a continuous variation in carbonate and hydroxyl groups. Thus it can be concluded that the wettability and position bound cellular responses relate well within one type of gradient surface, but that comparison of results on gradients surfaces obtained using highly different chemical components and cell types may yield different results.

Adhering endothelial cells detached readily after exposure to flow from the hydrophobic side of the gradients, in line with the observation that the cells did not spread very well. However, the observation that the position bound detachment of cells not only depended upon wettability but also upon the direction of flow was surprising and leads to the conclusion that cells 'feel' the gradient and react with a biological response. The first report of an active response of cells on a steep chemical gradient under static conditions was given by Carter [32]. A gradient of palladium on cellulose acetate was prepared and a directional response of cells (haptotaxis) towards the more adhesive, palladium-rich region of the steep gradient was observed.

Flow induced detachment of adhering cells from a homogeneous surface has been shown to be preceded by migration over the surface, during which cells round up under the influence of fluid flow and their adhesion strength consequently decreases. Adhering cells actually detach when the shear induced lifting force overcomes the adhesion and gravitational force, which reportedly range from 10^{-8} to 10^{-10} N per cell [33]. In this study, detachment of adhering endothelial cells was stimulated by flow during a time span of one hour during which cells migrated over the surface, either in a direction more (towards the hydrophilic end of the gradient) or less (towards the hydrophobic end) suitable for their adhesion. Thus by sensing fluid flow, adhering cells tend to round up, thereby decreasing their adhesion strength [34]. When flow is to the hydrophilic side of the gradient, this is accompanied by the haptotactic effect [32] and the migration velocity decreases in the direction of the hydrophilic end (see Fig. 7). When flow is to the hydrophobic side of the gradient, cells tend to feel obstructed by this non-adhesive side and the migration pattern is more or less flat.

5. SUMMARY

The position bound behaviour of HUVEC endothelial cells adhering to DDS-chemical gradient surfaces in the presence of serum proteins varies along the length of the gradients. In addition, during exposure to flow, the position bound cellular migration and detachment depend on the direction of the flow, i.e. towards the hydrophobic or to the hydrophilic end of the gradients. These observations raise questions about the cellular behaviour generally reported for 'homogeneous' surfaces, because all 'real-life', homogeneous surfaces are in fact heterogeneous at the microscopic level.

Acknowledgement

This research has been supported by the ISP research grant IV-1 S91.003.

REFERENCES

1. J. M. Schakenraad, I. Stokroos, H. Bartels and H. J. Busscher, *Cells and Materials* **2**, 193–199 (1992).
2. S. Suzuki, K. Matsuda, T. Maruguchi, Y. Nishimura and Y. Ikada, *Br. J. Plast. Surg.* **48**, 222–229 (1995).
3. B. W. Morgan, V. A. Henry, E. S. Bennet and P. J. Caroline, *J. Am. Optom. Assoc.* **63**, 201–214 (1992).
4. S. Wachtl, G. Sennwald and M. Rodrigues, *Ann. Cher. Main. Memb. Super.* **13**, 36–41 (1994).
5. F. W. Grannis and L. D. Wagman, *Ann. Thorac. Surg.* **60**, 197–199 (1995).
6. G. Reid, H. J. Busscher, S. Sharma, M. W. Mittelman and S. McIntyre, *Surf. Sci. Rep.* **21**, 251–273 (1995).
7. J. H. Lee, J. W. Park and H. B. Lee, *Biomaterials* **12**, 443–448 (1991).
8. S. P. Massia and J. A. Hubbell, *J. Biomed. Mater. Res.* **25**, 223–242 (1991).
9. P. B. van Wachem, J. M. Schakenraad, J. Feyen, T. Beugeling, W. G. van Aken, E. H. Blaauw, P. Nieuwenhuis and I. Molenaar, *Biomaterials* **10**, 532–539 (1989).
10. J. M. Schakenraad, J. Arends, H. J. Busscher, F. Dijk, P. B. van Wachem and Ch. R. Wildevuur, *Biomaterials* **10**, 43–50 (1989).
11. T. G. van Kooten, J. M. Schakenraad, H. C. van der Mei and H. J. Busscher, *J. Biomed. Mater. Res.* **26**, 725–738 (1992).
12. G. A. Truskey and J. S. Pirone, *J. Biomed. Mater. Res.* **24**, 1333–1353 (1990).
13. P. Zilla, R. Eoval, P. Preiss, M. Kadletz, M. Deutsch, H. Schima, S. Tsangaris and P. Groscurth, *Surgery* **105**, 515–522 (1989).
14. H. Elwing, S. Welin, A. Askendal, U. Nilsson and I. Lundström, *J. Colloid Interface Sci.* **119**, 203–210 (1987).
15. C. G. Gölander, Y. S. Lin, V. Hlady and J. D. Andrade, *Colloids Surfaces* **49**, 289–302 (1990).
16. J. H. Lee and H. B. Lee, *J. Biomater. Sci. Polymer Edn.* **4**, 467–481 (1993).
17. R. L. Williams, J. A. Hunt and P. Tengvall, *J. Biomed. Mater. Res.* **29**, 1545–1555 (1995).
18. T. G. Ruardy, J. M. Schakenraad, H. C. van der Mei and H. J. Busscher, *J. Biomed. Mater. Res.* **29**, 1415–1423 (1995).
19. T. G. van Kooten, J. M. Schakenraad, H. C. van der Mei, A. Dekker, C. J. Kirkpatrick and H. J. Busscher, *Med. Eng. Phys.* **16**, 506–512 (1994).
20. T. G. van Kooten, J. M. Schakenraad, H. C. van der Mei, A. Dekker, C. J. Kirkpatrick, M. Walter, D. Korzec, J. Engemann and H. J. Busscher, *Colloids Surfaces B: Biointerfaces* **3**, 147–158 (1995).
21. R. E. Johnson and R. H. Dettre, in: *Surface and Colloid Science*, E. Matijevic (Ed.), Vol. 2, pp. 85–154. Wiley-Interscience, New York (1969).
22. C. G. Gölander, K. Caldwell and Y. S. Lin, *Colloids Surfaces* **42**, 165–172 (1989).
23. A. B. Mulder, N. R. Blom, J. W. Smit, M. H. J. Ruiters, J. van der Meer, M. R. Halie and V. J. J. Bom, *Thromb. Res.* **80**, 399–411 (1995).
24. J. Sjollema and H. J. Busscher, *Colloids Surfaces* **47**, 337–352 (1990).
25. P. Tengvall, L. Olsson, A. Askendal, I. Lundström and H. Elwing, in: *Biomaterials–Tissue Interfaces, Advances in Biomaterials*, P. J. Doherty (Ed.), Vol. 10, pp. 511–519. Elsevier Science Publishers, Amsterdam, The Netherlands (1992).
26. R. E. Baier, A. E. Meyer, J. R. Natiella, R. R. Natiella and J. M. Carter, *J. Biomed. Mater. Res.* **18**, 327–355 (1984).
27. H. J. Busscher, W. van der Vegt, J. Noordmans, J. M. Schakenraad and H. C. van der Mei, *Colloids Surfaces* **58**, 229–237 (1991).
28. J. M. Schakenraad, H. J. Busscher, Ch. R. Wildevuur and J. Arends, *J. Biomed. Mater. Res.* **20**, 773–784 (1986).
29. E. G. Hayman, M. D. Pierschbacher, S. Suzuki and E. Ruoslahti, *Exp. Cell. Res.* **160**, 245–258 (1985).
30. P. A. Underwood and F. A. Bennet, *J. Cell. Sci.* **93**, 641–649 (1989).
31. T. Ueda-Yukoshi and T. Matsuda, *Langmuir* **11**, 4135–4140 (1995).

32. S. B. Carter, *Nature* **213**, 256–261 (1965).
33. M. A. Hubbe, *Progr. Surf. Sci.* **11**, 65–138 (1981).
34. T. G. van Kooten, H. C. van der Mei, J. M. Schakenraad and H. J. Busscher, *Biofouling* **5**, 239–246 (1992).

Mittal Festschrift, pp. 837–855
W. J. Van Ooij and H. R. Anderson, Jr. (Eds)

The effect of radiofrequency electric field on the properties of dispersed systems: Relevance to adhesion

EMIL CHIBOWSKI* and AGNIESZKA SOŁTYS

Department of Physical Chemistry, Maria Curie-Skłodowska University, 20 031 Lublin, Poland

Abstract—The use of 'non-chemical' methods for modification of the properties of dispersed systems is of growing interest. Among other methods, irradiation with radiofrequency (RF) electric field may affect their properties. An example of a practical application would be to prevent hard scale formation in industrial pipes working at elevated temperatures. In the current study, a 44 MHz (60 V peak-to-peak, no-load amplitude) radiofrequency field was applied to aluminum oxide dispersion in water or NaCl solutions and then the zeta potential and the effective diameter were determined using a ZetaPlus instrument (Brookhaven Co.). Also, the oxide thin layer deposited on a glass plate was subjected to the field and then the free energy components were determined by the thin-layer wicking method. Moreover, the effect of RF field was also studied for the Al_2O_3 plates with preadsorbed Mg^{2+}, Al^{3+}, and OH^- ions. It was found that the RF field affected all studied parameters; especially large changes in the acid–base free energy components were observed for the oxide surface possessing the preadsorbed ions.

Keywords: Waste water; radiofrequency electric field; dispersed systems; aluminum oxide; acid–base.

1. INTRODUCTION

The application possibility of 'non-chemical methods' for treatment of waste waters or some other dispersed systems sounds very exciting from the point of view of environmental protection. There have been many examples of so-called magnetizers which, according to manufacturers, prevent hard scale formation inside the industrial pipes working at higher than room temperatures [1]. However, there is lack of systematic fundamental studies to clarify whether this phenomenon exists at all. On the other hand, we have found [2–5] that high frequency electric field (44 MHz, 34 or 60 V, peak-to-peak, no-load amplitude) altered some parameters characterizing the properties of dispersed systems. One may expect that an RF field applied properly can also affect adhesion of solid particles to solid substrata, which is the case of hard scale formation in pipes.

There are established theories on the behavior of dispersed systems in alternating high frequency electric field, i.e. dielectric dispersion (decrease of the observed dielectric constant). Recently, O'Brien [6] stated that 'it is an area in which the theory,

*To whom correspondence should be addressed. E-mail: emil@hermes.umcs.lublin.pl

which is likely to develop rapidly in the next couple of years, already outweighs experiment'. Lateral migration of ions inside the Stern layer of the electrical double layer is considered [7, 8] to be the effect of the field interaction. It is also reflected in the measured zeta potentials. In the region between the shear plane and the particle surface the ions migrate due to gradients in electrochemical potentials. The lateral movement of the ions is balanced against their normal flux toward the particle surface. However, there is an imbalance between the two directions; therefore, accumulation of the species can occur. This approach is termed 'the dynamic Stern layer model' [8]. The tangential flux across the shear plane is proportional to the surface conductivity, which depends on the equilibrium surface ion density. According to Saville [8] the treatment of the suspension 'dielectric behavior which takes account of ion migration behind the shear plane... would be useful'.

Although it may seem that there are no theoretical predictions for residual changes in the properties of suspension (zeta potential, conductivity, etc.) which has been exposed to RF field, the above cited approaches [6–8] point to the possibility of such changes, which in fact have been found [2–5]. The changes may persist even up to 90 min after the field cessation [3]. Among other things, an RF field was applied to study whether it could affect *in situ* calcium carbonate deposition onto internal walls of glass capillaries [9]. For this purpose, $CaCO_3$ was precipitated just at the inlet to the capillary. A method for deposition studies has been elaborated using a single capillary for each experiment. To standardize the experiments, relationships between the deposition and the parameters which could affect the deposited amount were established [10]. It was found that in a defined range of the flow (0.45–0.75 ml/s) the deposition was independent of it. A maximum in the deposition occurred around the temperature of 70°C, which might be due to the phase transition of the precipitating calcium carbonate from calcite to aragonite taking place about this temperature [10]. The adhesion of $CaCO_3$ particles was enhanced when it was precipitated from Ca^{2+} excess. The effect of impurity ions (Zn^{2+}, Fe^{2+}, Mg^{2+}, Fe^{3+}, and Al^{3+}) was also studied [9]. The presence of Zn^{2+} in the precipitating system caused the largest decrease in the $CaCO_3$ deposition. As to the effect of RF field on the zeta potential, it appeared [2–5] that the changes of zeta potential were the largest for the samples treated for 10–15 min with the field. The zeta potential oscillated as a function of time after the field cessation. On the basis of the above mentioned results it was concluded that an RF field could affect the solid particles' adhesion or the stability of dispersed systems.

According to the present approach [11–14], the total interaction energy $\Delta G_{132}^{\mathrm{TOT}}$ between particle (1)–particle (2) or particle (1)–planar substrate (2) in a liquid medium (3) can be considered as the sum of three contributions, i.e. electrostatic (EL), which results from the electrical double layers present around the particles, the Lifshitz–van der Waals (LW) contribution due to London, Debye, and Keesom's interactions, and solvating forces involving the Lewis acid–base (electron donor and electron acceptor) (AB) interactions:

$$\Delta G_{132}^{\mathrm{TOT}} = \Delta G_{132}^{\mathrm{EL}} + \Delta G_{132}^{\mathrm{LW}} + \Delta G_{132}^{\mathrm{AB}}. \tag{1}$$

In a polar liquid, like water, the AB interactions are about 100 times larger than LW ones, and even about 10 times greater than EL interactions. Therefore, any RF field effect on the particle adhesion may be studied by measurements of zeta potential and the surface free energy components, i.e. apolar Lifshitz–van der Waals, γ_s^{LW}, and polar, i.e. electron donor, γ_s^-, and electron acceptor, γ_s^+. In this paper these parameters were measured under and without RF field irradiation of the samples.

2. EXPERIMENTAL

2.1. Materials and methods

Most of the studies presented in this paper were carried out using aluminum oxide (150 neutral, Type T, Merck), as well as commercial glass plates covered with Al_2O_3 (60 F, Type E, Merck) for thin-layer chromatography. Some zeta potential measurements were done on aluminum oxide C, Degussa AG product. Zeta potentials and multimodal size distribution were determined using a ZetaPlus instrument from Brookhaven Instrument Co. (USA), which applies dynamic light scattering for determination of the parameters. The radiofrequency electric field was applied from a 44 MHz (60 V, peak-to-peak, no-load amplitude) generator which was constructed at the Department of Physics, Maria Curie-Skłodowska University, Lublin, Poland. An oscilloscope was used to control the maximum signal. The experiments were conducted at room temperature, $20 \pm 1\,^{\circ}$C.

2.2. Free energy components determination

For studies of RF field effect on the solid surface free energy, the commercial plates were chosen as a model system because their thin layer of Al_2O_3 had a uniform thickness and interparticle pores. This is important in the so called 'thin-layer wicking' method [14–19], which was applied in these studies. Previously, such RF effect was found for calcium carbonate [2]. To determine the surface free energy components of a solid by this technique, three liquids have to be used for wicking of the thin porous layer. The method is based on Washburn's equation [20], which originally related the rate of a liquid penetration with its surface tension. However, in such form, the equation is valid only if the liquid completely wets the solid surface (no wetting contact angle) and the surface has been pre-covered with an equilibrium film of the liquid. From our studies [16–19], it resulted that a more appropriate form of Washburn's equation reads:

$$x^2 = \frac{Rt}{2\eta}\Delta G, \qquad (2)$$

where x^2 is the squared distance penetrated by the liquid in time t, η is the liquid viscosity, R is the effective radius of the interparticle pores of the thin layer, and ΔG is the specific free energy accompanying the process. It should be mentioned that in this notation, a positive value of the energy means that the process is spontaneous.

The value of ΔG depends on the surface tension of the liquid used and the surface free energy of the solid [16–19]. To determine the three components γ_s^{LW}, γ_s^-, and γ_s^+, at least six plates covered with the tested solid must be used for wicking by three liquids: nonpolar n-alkane (nonane, decane), and polar water and formamide. Three of the six plates should be equilibrated with the liquid vapors and then the liquids allowed to penetrate. In the other three wicking experiments, bare plates must be used. From the results of the six penetrations [as a function of $x^2 = f(t)$] it is possible to determine the three components. The details of the method were published elsewhere [16–19].

To find whether RF field influences the free energy components, before wicking experiments, the plates were placed for 15 min between the copper plates to which RF field was connected. Besides these experiments, in other series the plates were first wetted with 1 ml of 0.001 M solution of $MgCl_2$, or $AlCl_3$, or with 1 ml of NaOH solution at pH = 11. After the plates had been dried at room temperature they were placed in an oven at 110°C for 1.5 h and then kept in a desiccator for further use in wicking experiments.

2.3. Zeta potential and effective diameter determinations

For zeta potential and effective diameter determinations stock suspensions 0.01% v/v of aluminium oxide were prepared by dispersing the sample in water (Millipore System) or in 0.01 M NaCl solution using polyethylene bottles. The suspensions were treated for 2 min in an ultrasonic bath and then mixed for one hour. The suspension in water was prepared 50 days before using it for measurements and that in NaCl solution 7 days before. Immediately before the measurements, the suspensions were shaken for 2 min and allowed to stand for 1 h to avoid any sedimentation effects during the measurements. The suspensions were exposed to the RF field for different times (up to 30 min) by placing them in a glass tube 20 cm long and 1.2 cm i.d. to the ends of which the field was applied using platinum electrodes [4, 5]. After the treatment, zeta potential was measured using the ZetaPlus apparatus. For comparison, similar experiments were conducted for TiO_2 (P25, Degussa). The zeta potentials were calculated automatically by the apparatus (software V.3.21) using the Smoluchowski equation. For the Al_2O_3, particle radius a is about 350 nm (see Results). Hence, in 0.01 M NaCl solution the product of Debye double layer parameter κ and a is $\kappa a = 115$, while in 0.001 M NaCl the $\kappa a = 36$. This means that the Smoluchowski equation could be applied for calculation of the zeta potentials with a good approximation, even in 0.001 M NaCl solution. The correction factor from Henry's equation is 0.93 for 0.001 M [21]. Because we were interested in relative changes of the zeta potential resulting from the RF field interaction, no corrections were applied.

With the same apparatus it was possible to determine changes in the effective diameter of aluminum oxide particles in the suspensions due to RF field interaction. The determination is also based on the dynamic light scattering (software V.6.03). The measurements were conducted at the wavelength $\lambda = 670$ nm and the angle $\theta = 90°$.

3. RESULTS AND DISCUSSION

As discussed above, zeta potential and surface free energy components are the parameters by determination of which the RF field effect can be studied. In these studies, such experiments were carried out using aluminum oxide suspensions and glass plates covered with a thin layer of this oxide. In Fig. 1, the relationship is shown between the RF field treatment time and zeta potential for Al_2O_3 suspensions in 0.001 M NaCl solution (natural pH = 6.3) [5]. The potentials were measured after 5 and 10 min of the field cessation. The dashed lines show standard deviation for the untreated sample. Here, the largest RF effect appears for the samples treated for 15 and 20 min and the changes oscillate on the time scale (Fig. 1) both as a function of the exposure time and time after the field has ceased. A very similar behavior has been found for calcium carbonate and other oxides [4, 5]. It seemed to us interesting to learn about the field effect on the pH scale. Such experiments were conducted for aluminum (Merck) suspension in 0.01 M NaCl. The pH value was regulated with HCl or NaOH and then one portion was directly used for zeta potential measurements and another was exposed to the RF field for 15 min and after that the zeta potential was determined. The measured values for untreated and RF treated samples are presented in Fig. 2. As can be seen, the isoelectric point (i.e.p.) for the untreated sample lies at $pH_{iep} = 7.0$, because the Al_2O_3 used was neutral. Irradiation of the suspension with the field causes the i.e.p. shift to $pH_{iep} = 6.85$. Although the shift was small, the trend was distinct, especially as the measurements (3–5 for each pH, Fig. 2) for the samples treated with the RF field followed those for the untreated ones. To learn

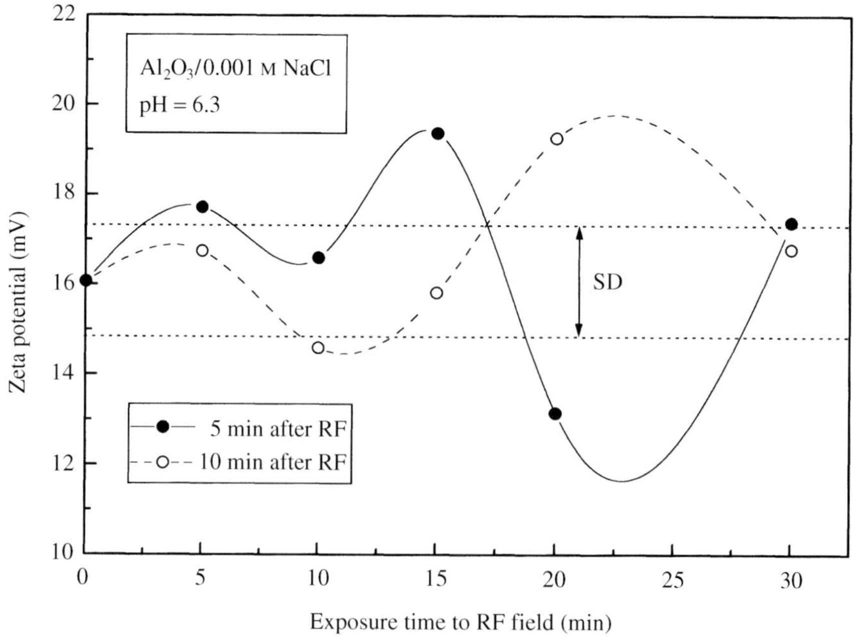

Figure 1. Zeta potentials of alumina particles as a function of the exposure time to RF field.

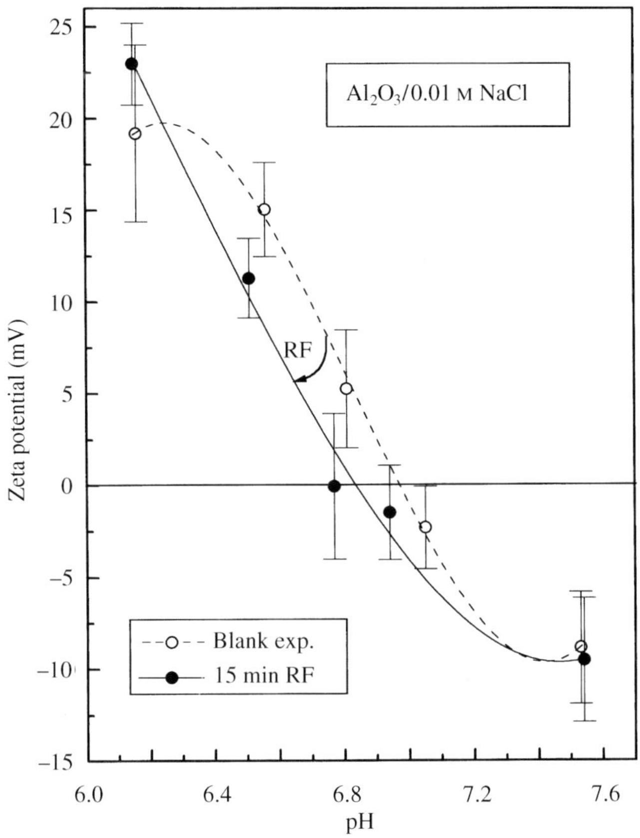

Figure 2. RF field effect on zeta potentials of alumina as a function of the suspension pH.

whether this effect would appear for other oxides, similar experiments were carried out for the titanium oxide suspension in 0.001 M NaCl. The results presented in Fig. 3 show that also for this oxide the effect is the same. The standard deviations were also similar. The pH of i.e.p. was shifted from *c.* 5.4 to 5.1 after the RF field interaction for 15 min. The values of zeta potentials (Figs 2 and 3) were measured 5 min after the field had ceased. The shift in pH_{iep} means that in the shear plane, the diffuse charge is less positive (or more negative for $pH > pH_{iep}$) after the field energy absorption. Because in the discussed systems no specific adsorption of the ions takes place, the shift may be interpreted as the H^+ ion's release from the oxide surface or the OH^- ion's adsorption from the solution onto the surface, as the i.e.p. occurs about neutral pH [21]. Moreover, in the previous studies on TiO_2 (Degussa), we have found that for the suspensions in 10^{-2}–10^{-4} M NaCl treated for 30 min with the field, the shift in pH of the point of zero charge was much more pronounced [22] than the shift in pH_{iep} observed in this paper. For that sample [22] untreated with RF field, the pH_{pzc} was 5.75, while it shifted to $pH_{pzc} = 6.5$ for the 30 min RF treated sample, if it was determined in the period of 2 to 9 min after the field had ceased, and to $pH_{pzc} = 5.0$, and then to $pH_{pzc} = 5.5$, if the potentiometric titrations were conducted

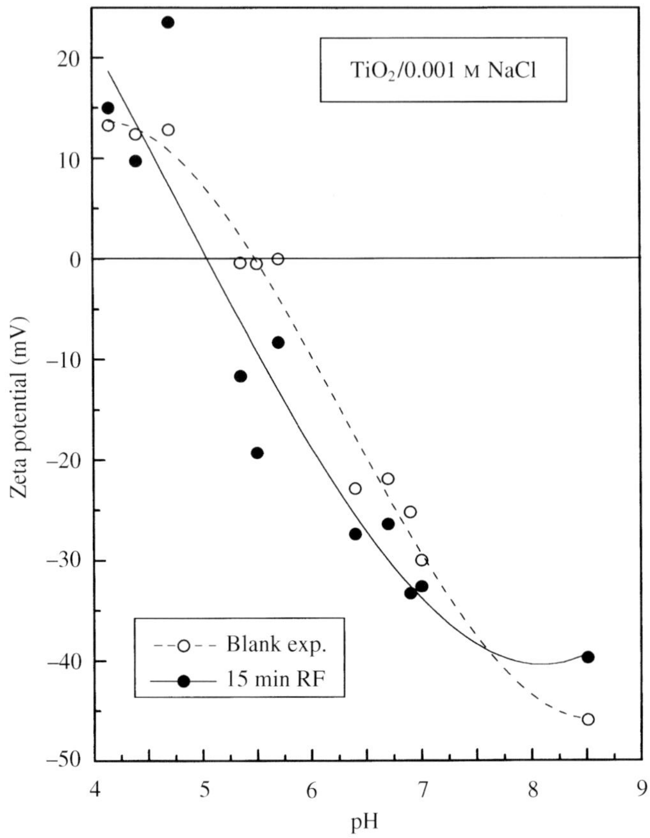

Figure 3. RF field effect on zeta potentials of titanium oxide as a function of pH.

between 15–24 min, and 30–40 min, respectively, [22]. It is worth mentioning that pH_{iep} determined for that sample [22] was 5.85, which was somewhat higher than that obtained for the TiO_2 sample used in this paper ($pH_{iep} = 5.4$, Fig. 3). The pH_{iep} and pH_{pzc} values determined clearly show that the absorbed energy of the RF field alters the adsorption equilibrium of potential determining ions (H^+ and OH^-), and then the system returns to the original adsorption equilibrium in an oscillatory manner. This 'relaxation' is probably a relatively slow process and takes up to the order of several tens of minutes [22].

As mentioned above, experiments were carried out to determine the RF field effect on the solid surface free energy components γ_s^{LW}, γ_s^-, and γ_s^+, by the 'thin-layer wicking' technique [16–19]. As an example, Fig. 4 shows a set of linear relationships between wicking time and squared distances for the three liquids (*n*-nonane, water, and formamide) penetrating aluminum plates both bare and pre-contacted with the vapors (as the time was measured for the particular distances, the x and t axes were interchanged for convenience). This set allows calculation of the three components of the surface free energy. It is worth noticing that for the same liquid the penetration times are different for bare and pre-contacted plates. The change in the rate of

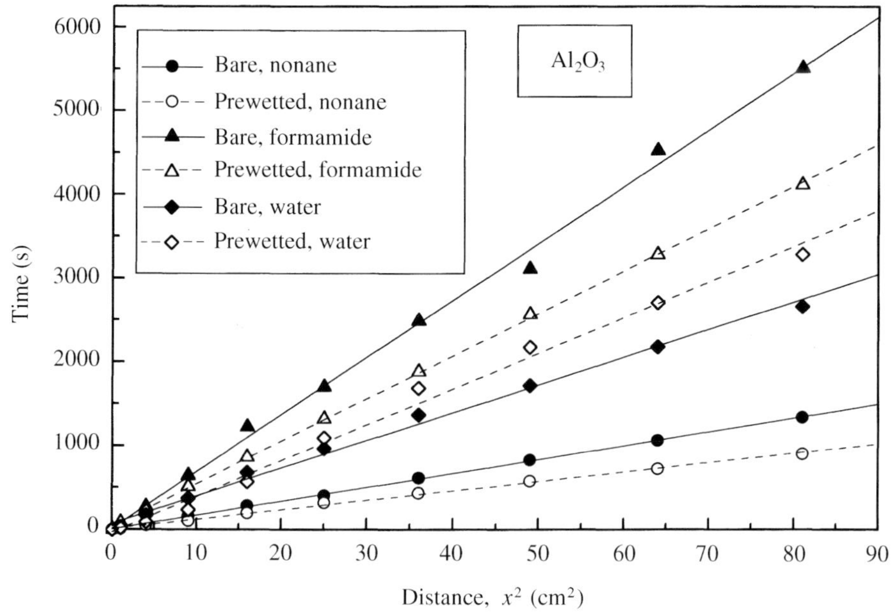

Figure 4. Wicking time of the probe liquids on the alumina plates versus squared distance.

wicking depends upon the difference between the liquid work of adhesion to the solid surface and the liquid work of cohesion. Such sets of experiments were carried out for all studied systems, i.e. with the adsorbed Mg^{2+}, Al^{3+}, and OH^- ions and for RF field treated and untreated plates. To show the RF field effect in Figs 5–7, wicking results are presented for bare aluminum plates and prewetted with the vapors. As can be seen, except for formamide on the prewetted plate (Fig. 6), RF field caused a slower penetration of the probe liquids. The changes in the wicking rate are most probably due to reorientation of the surface groups and/or the water dipoles hydrating the adsorbed ions. In general, the reproducibility of 2–3 wicking experiments was good. For the sake of clarity, the standard deviations are not presented in Figs 5–7. From these results, as well as analogous results for the rest of the tested systems, not presented here, the components of surface free energy, γ_s^{LW}, γ_s^-, and γ_s^+, were calculated for aluminum oxide samples pretreated in different ways as well as exposed to RF field for 15 min.

The calculated results are listed in Table 1. Analysis of the results shows that the unmodified surface of the tested aluminum sample has a relatively high electron donor, γ_s^-, interaction, equal to 72 mJ/m^2, and a very small electron acceptor interaction, γ_s^+, equal to 0.1 mJ/m^2. The apolar component, γ_s^{LW}, is found to be 41.0 mJ/m^2. Very similar values of the components were determined previously [16] for aluminum oxide, although the γ_s^- component was about 8 mJ/m^2 higher than that determined in this paper. The previous results [16] were obtained by the same technique (thin-layer wicking); however, the plates for wicking were prepared in this laboratory using Merck's α-aluminum reagent. With regard to the RF field effect on the aluminum surface free energy, from Table 1 it is seen that 15 min RF irradiation of the surface causes

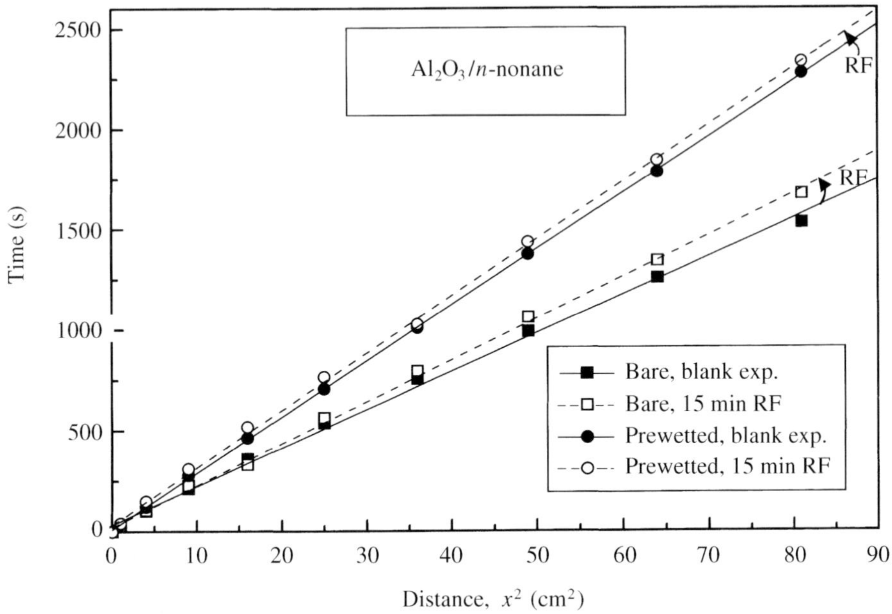

Figure 5. RF field effect on the wicking time of n-nonane versus squared distance on the alumina plates.

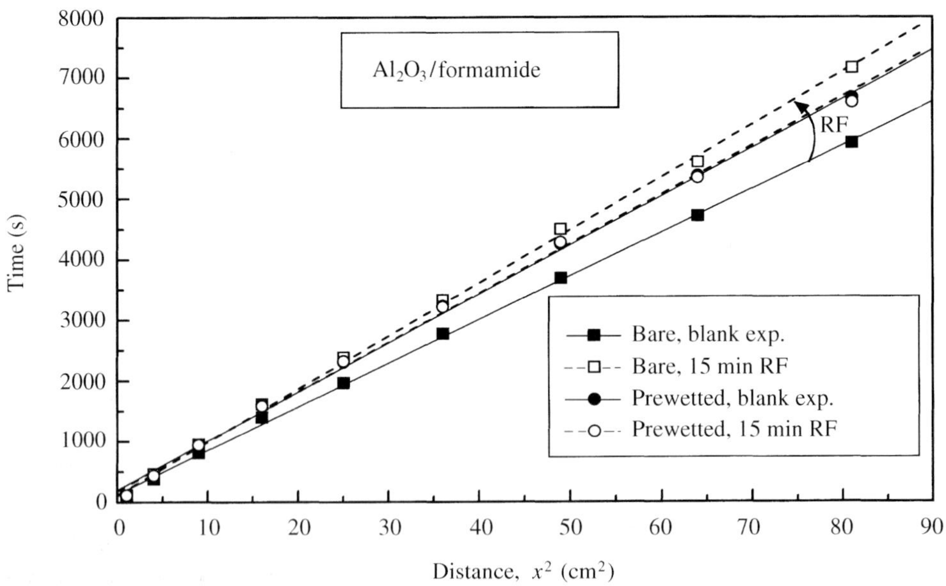

Figure 6. RF field effect on the wicking time of formamide versus the distance on alumina plates.

an essential decrease in the electron donor interaction, γ_s^-, from 71.8 to 52.6 mJ/m^2, and small increase in the apolar and electron acceptor components. A similar behavior, as to the acid–base components, was found for calcium carbonate sample treated

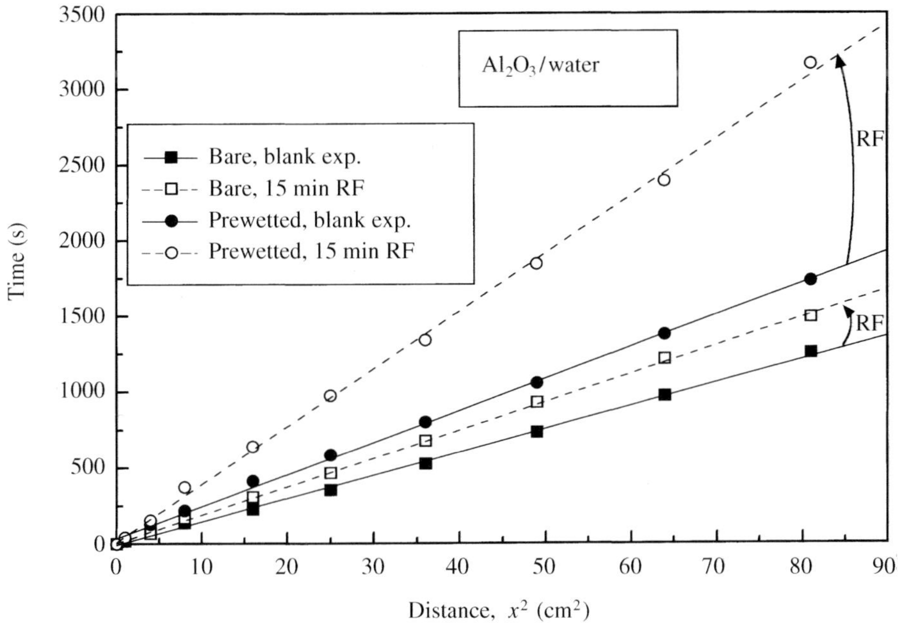

Figure 7. RF field effect on the wicking time of water versus the distance on alumina plates.

Table 1.
Surface free energy components of alumina samples pretreated in different ways

Alumina sample	γ_s^{LW} (mJ/m^2)	γ_s^- (mJ/m^2)	γ_s^+ (mJ/m^2)
Bare	41.0	71.8	0.1
Bare, 15 min RF	42.9	52.6	0.5
Mg^{2+}	41.1	174.8	0
Mg^{2+}, 15 min RF	45.4	59.5	0.9
Al^{3+}	46.2	201.0	0
Al^{3+}, 15 min RF	47.3	14.0	3.5
pH = 11 (NaOH)	44.6	80.6	0.7
pH = 11, 15 min RF	42.0	0	8.8

for 15 min with the field [2]. The electron donor component for 15 min treated calcium carbonate dropped from 79.0 (untreated sample) to 40.7 mJ/m^2, while the electron acceptor component increased from 0 to 1.34 mJ/m^2. Practically no change in the apolar component magnitude occurred for this solid.

The effect of RF field on the free energy components of aluminum surface with the adsorbed ions appeared to be very significant. Moreover, the effect of the tested ions themselves on the components was also very large. As can be seen from Table 1, wetting the aluminum surface with 1 ml of 10^{-3} M Mg^{2+} caused γ_s^- increase to 174.8 mJ/m^2, and the surface wetted with the same amount of Al^{3+} ions caused the γ_s^- component to be as high as 201.0 mJ/m^2; whereas the same concentration of OH^- ions increased the γ_s^- component only slightly to 80.6 mJ/m^2. The electron acceptor γ_s^+ component remained practically unchanged. Also, the apolar γ_s^{LW} component was influenced a little. The RF field effect on the acid–base components of the modified aluminum surface appeared to be appreciable. It caused a decrease of the electron donor component and an increase of the electron acceptor component. In the case of the surface prewetted with NaOH (pH = 11), the γ_s^- component was diminished up to zero while the γ_s^+ component increased from 0.7 to 8.8 mJ/m^2. It seems that the changes may be discussed in terms of water dipole rearrangement around the ions adsorbed on the surface. This may lead to change in the strength of the hydrogen bonding which is responsible for γ_s^- and γ_s^+ interactions. The hydrogen bonding strength lies between 10 and 40 kJ/mol [23]. For example, if one would like to ascribe the total $\gamma_s^- = 201$ mJ/m^2 (Al^{3+} modified surface, Table 1) to that resulting solely from the hydrogen bonding interaction, and assuming a bond strength of 20 kJ/mol, and taking 7.35 Å^2 for the —OH group effective surface, the surface fraction occupied by these groups would be about 0.45, which seems to be a very reasonable quantity for oxide surface [24]. From the results presented in Table 1 it is possible to speculate about some rearrangements of water molecules hydrating the adsorbed ions. Thus, the observed decrease in the γ_s^- magnitude after RF treatment may be considered as a partial 'mutual saturation' of the hydrogen bonding interactions inside the hydration shell of the adsorbed ions (Al^{3+}, Mg^{2+}, OH^-, as well as Cl^- and Na^+). Then, the zero value of γ_s^- for the aluminum surface prewetted with NaOH could result from 'mutual saturation' of the interactions between —OH groups and the adsorbed water molecules due to the RF field energy absorbed. The appearing electron acceptor γ_s^+ component (8.8 mJ/m^2) in such an approach should result from the interactions of 'stretching' hydrogen atoms of water molecules toward the bulk phase of the wicking water or formamide. However, to verify these speculations more systems have to be studied.

Finally, to check directly the RF field effect on the particles' stability, measurements have been carried out in which the effective diameter was determined using the ZetaPlus instrument. The results for aluminum oxide in water at pH = 6.5 and pH = 10.5 are presented in Figs 8 and 9, respectively, as a function of the exposure time to the RF field. The measurements were completed in about 3 min and then 8 min after the field had ceased. The effective diameter was determined as a function of the exposure time. The effective diameter should reflect the aggregation of particles, which, in turn, depends upon the balance of interaction forces, i.e. electrostatic and those resulting from the surface free energy. As can be seen from Fig. 8 at pH = 6.5, the prolonged exposure to RF field causes continuous decrease in the effective diameter of the particles when determined after 3 min, and increases when determined after 8 min, but it remains smaller than that for the untreated sample. Similar behavior was observed for the suspension at pH = 10.5 (Fig. 9), except for the sample treated for 5 min with the field, where the effective diameter increased after the treatment. Because at both pH values significant changes in the effective

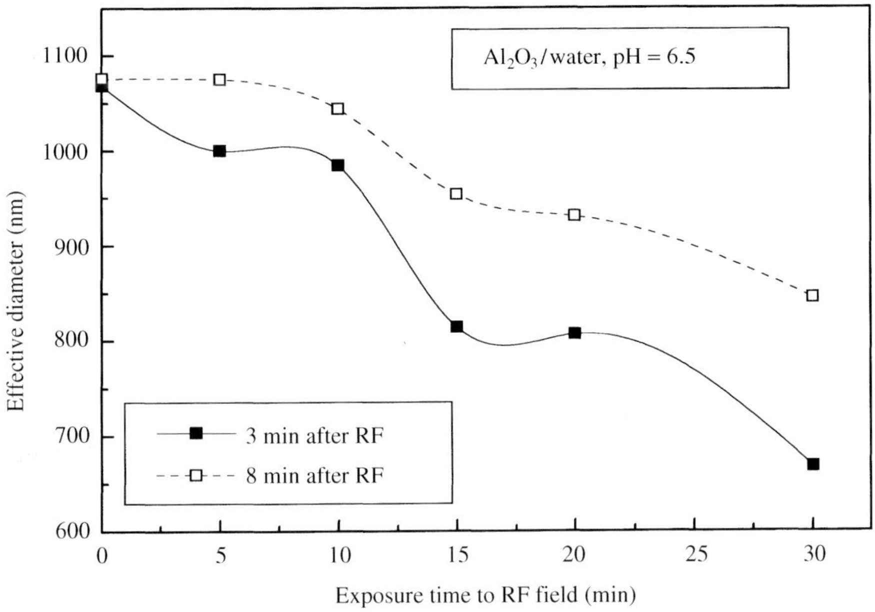

Figure 8. Effective diameter of alumina particles in water (pH = 6.5) as a function of the exposure time to RF field.

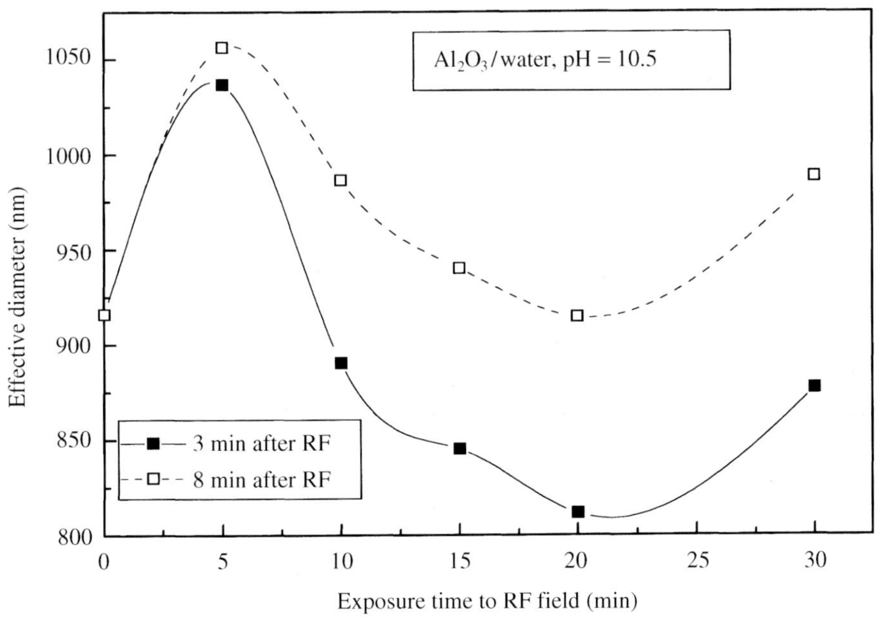

Figure 9. Effective diameter of alumina particles in water (pH = 10.5) as a function of the exposure time to RF field.

diameter of the particles versus the time after the field had ceased (Figs 8 and 9) were observed, experiments were carried out in which the suspensions were exposed for

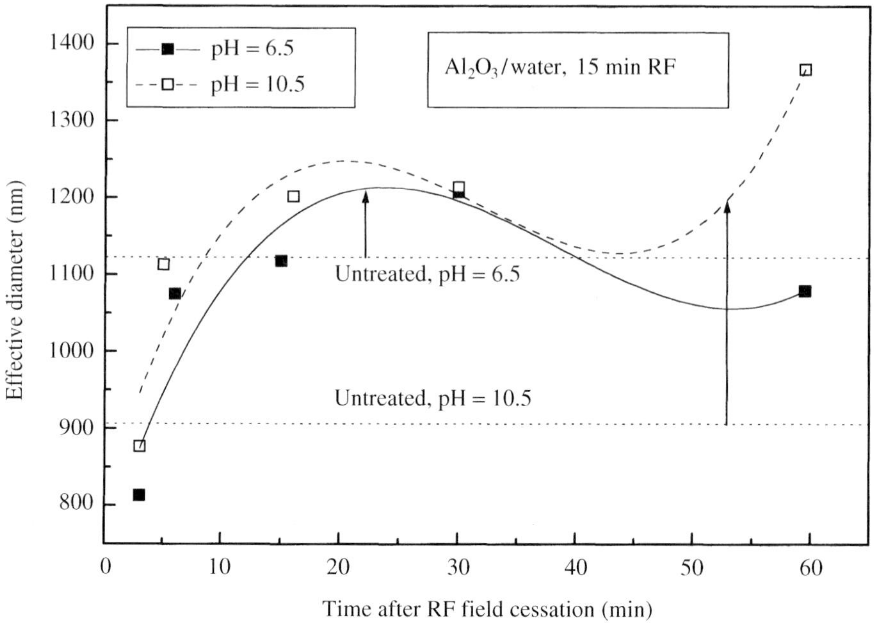

Figure 10. Effective diameter of alumina particles treated with RF field for 15 min a function of the time after the field has ceased.

15 min to the RF field and then the effective diameter was determined as a function of time after the field had ceased. These results are presented in Fig. 10. It was found that at both pHs (6.5 and 10.5), only during the first 3 min was the effective diameter of the particles smaller than in the untreated suspensions. Then it increased above it and was much higher in the case of pH = 10.5. A better insight into the RF field effect on the particle size can be obtained from the multimodal size distributions, which are collected in Fig. 11, for pH = 6.5 and in Fig. 12, for pH = 10.5, for different time of the exposure to RF field. As it is seen (Fig. 11) the original suspension of aluminum in water at pH = 6.5 was mostly monodisperse with the particles around 1 μm and some particles of about 4 μm. After 15 min RF treatment, two different populations of the particles are easily visible, the main one of 0.5 μm and a smaller one of about 1.8 μm, giving an effective diameter of 0.86 μm. The 30 min treated suspension again consists of two populations of particles, but of smaller average sizes. In the case of the aluminum suspension at pH = 10.5, the original suspension was very monodisperse (Fig. 12), and the particle size was actually about the same size as the main population of the particles at pH = 6.5 (Fig. 11). The 5 min treatment (Fig. 12) increased the diameter by about 120 nm, and 20 min exposure to RF field 'split' the particles into two populations, i.e. around 0.5 and 1 μm. The above discussed particle size distributions were measured within 3 min after the field had ceased. However, as seen in Figs 8–10, the measured effective diameter was a function of time after the field removal. To show multimodal size distribution corresponding to the results presented in Fig. 10, in Figs 13 and 14 are shown the distribution for the pH = 6.5 and 10.5, respectively, as a function of time

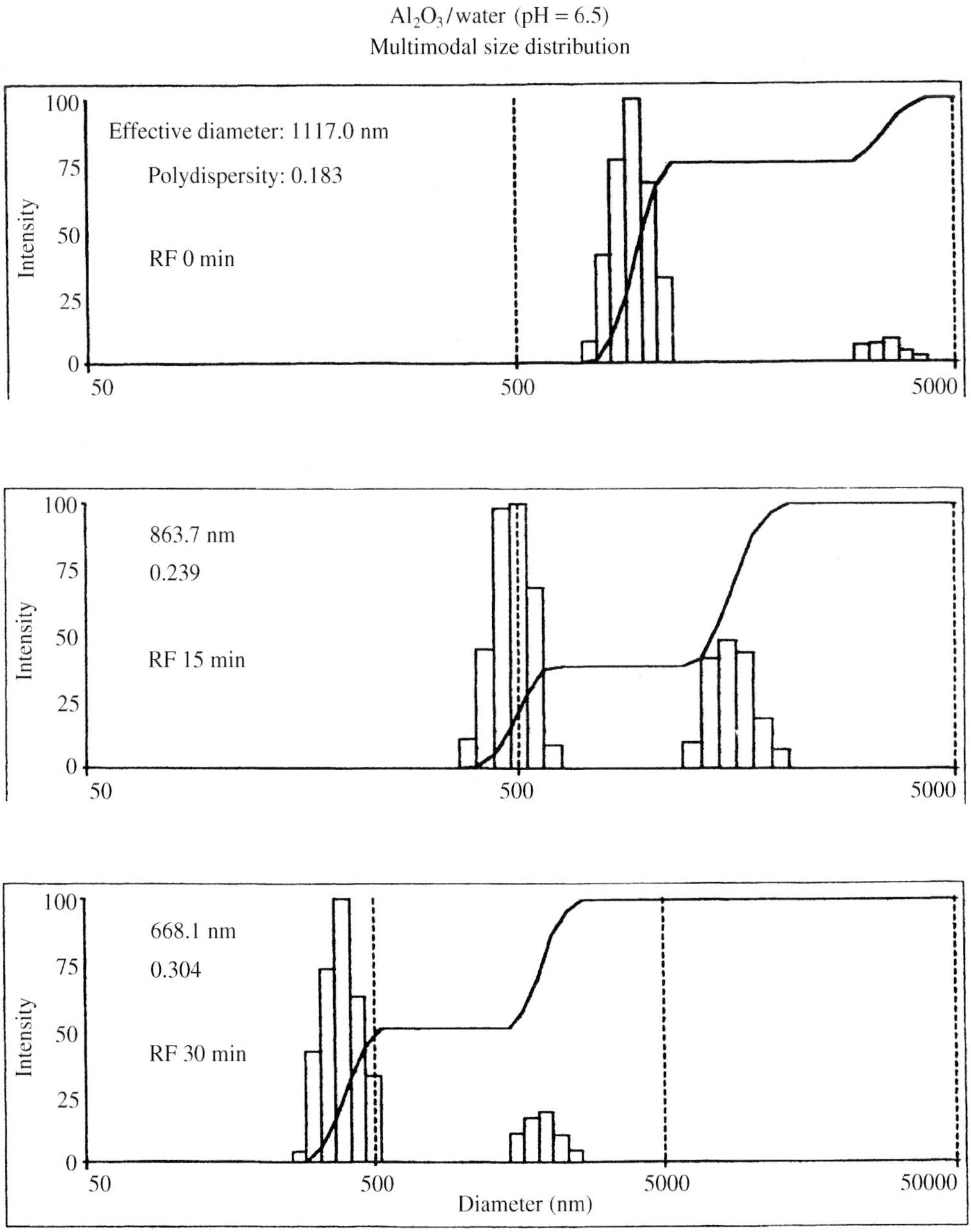

Figure 11. Multimodal size distribution of aqueous Al_2O_3 suspensions treated with RF field for different times (pH = 6.5).

after the field removal. The suspensions were treated with the RF for 15 min. At both pHs (Figs 13 and 14) a population of smaller particle size (aggregates) appears in the first 3 min after the field has ceased. For a longer time, the size distribution depends on the pH, but after 30 min it can be seen from Figs 13 and 14 that

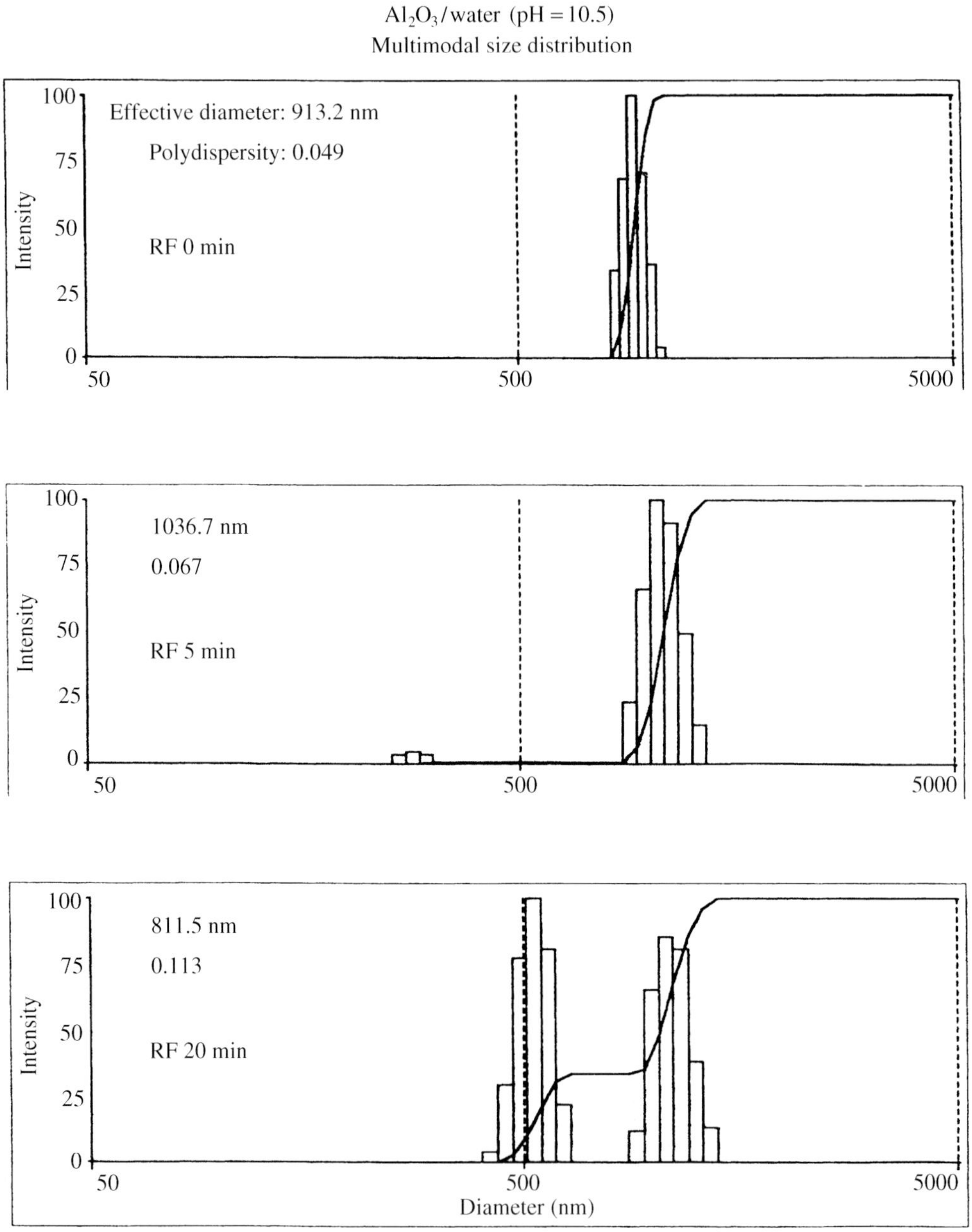

Figure 12. Multimodal size distribution of aqueous Al_2O_3 suspensions treated with RF field for different times (pH = 10.5).

in both suspensions, particle size distributions are more monodisperse than for the untreated suspensions. The determined particle size may be considered here as a result of aggregation of the individual particles. As mentioned in the Experimental section, any sedimentation of the larger particles during the measurements could

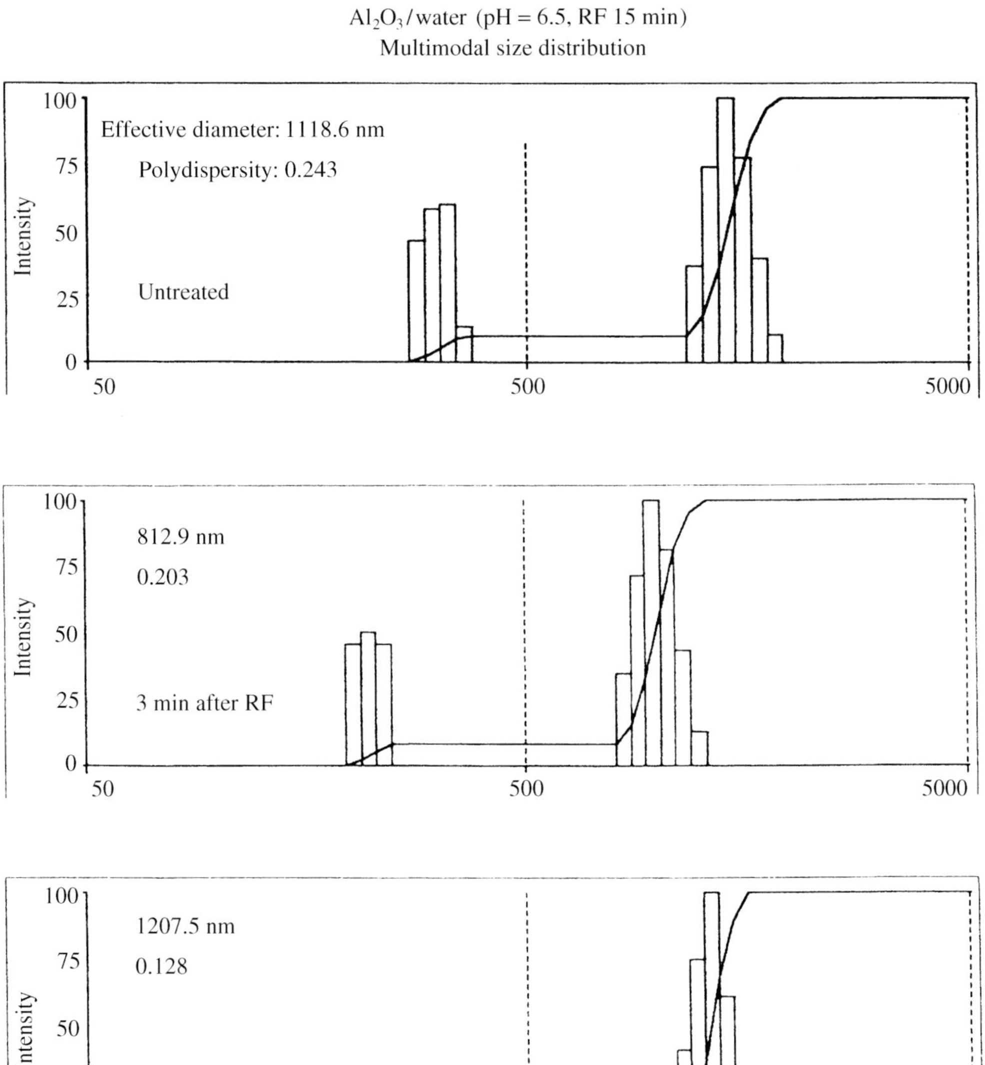

Figure 13. Multimodal size distribution of aqueous Al_2O_3 suspensions treated with RF field for 15 min at different times after the field has ceased (pH = 6.5).

be excluded. In general, the RF field effect is a very dynamic one with respect to the aggregation state of dispersed systems. It seems that not all the particles absorb the same quantity of the field energy. The experiments with a monodisperse system would be helpful to verify this hypothesis. Such experiments will be conducted next.

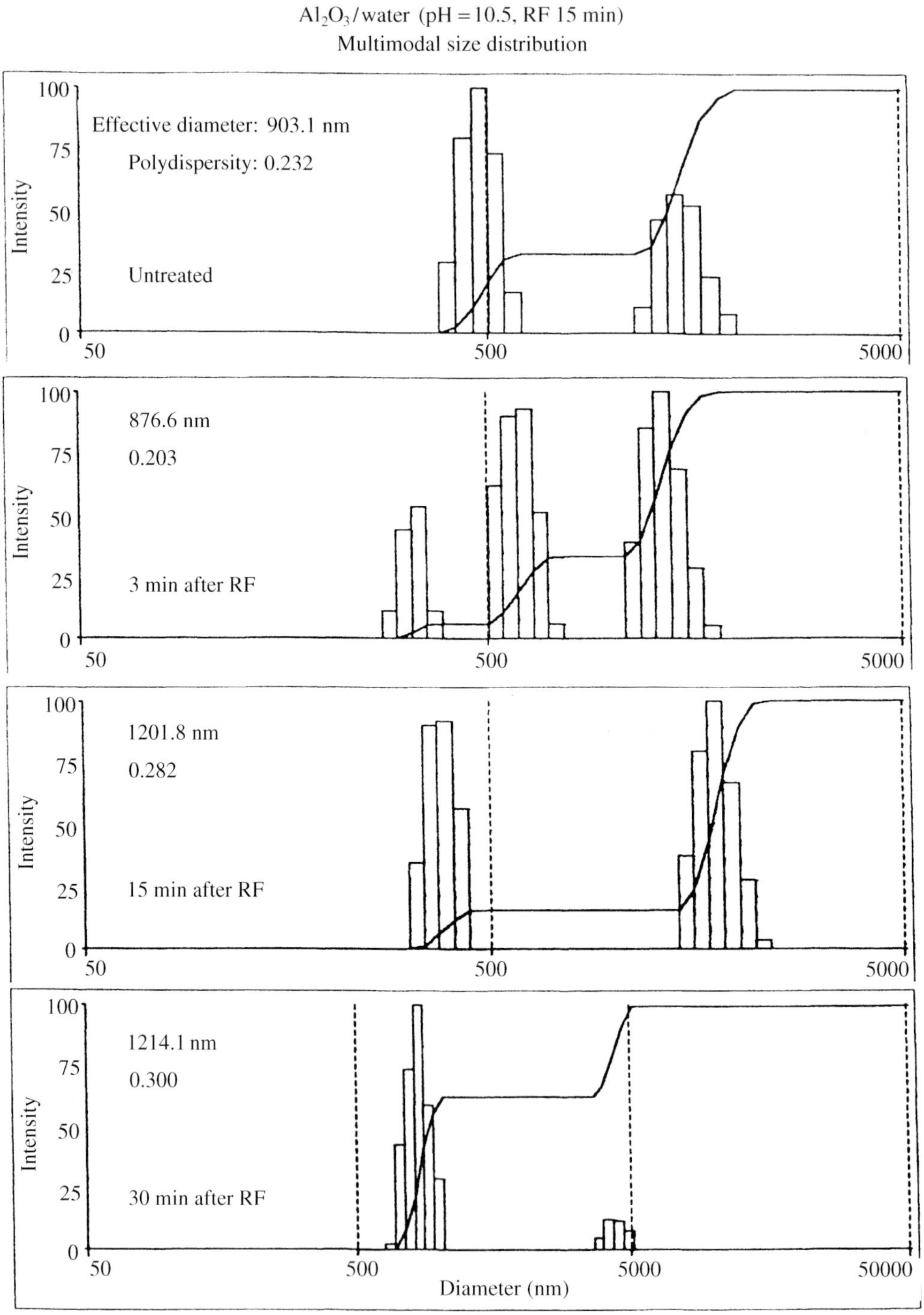

Figure 14. Multimodal size distribution of aqueous Al_2O_3 suspensions treated with RF field for 15 min at different time after the field has ceased (pH = 10.5).

4. SUMMARY

- Radiofrequency electric field affects the properties of dispersed systems.
- The field effect is not a linear function of time but it oscillates with treatment time as well as the time after the field has ceased.
- The RF field effect reflects changes in zeta potential, surface charge, surface free energy components, particle size and particle adhesion of the suspension.
- In the case of aluminium oxide surface, the field reduces the surface acidity, most probably by proton desorption.
- A very sharp effect of the field appears in the case of electron donor interaction. The field decreases it greatly, especially in the presence of the adsorbed ions on the aluminum surface.
- In many studied systems the biggest changes in the measured parameters occurred for 10–20 min RF field treated samples.
- Based on the above, it may be concluded that practical applications of the field effect are possible.

Acknowledgements

Acknowledgements are due to Mrs. B. Dzikowska for the zeta potential and particle size measurements and to Mr. R. Ogonowski for wicking experiments. We also very much appreciate the help of Brookhaven Instrument Co. in purchasing the ZetaPlus instrument.

REFERENCES

1. R. E. Herzog, Q. Shi, J. N. Patil and L. Katz, *Langmuir* **5**, 861 (1989).
2. L. Hołysz and E. Chibowski, *J. Colloid Interface Sci.* **164**, 245 (1994).
3. E. Chibowski, S. Gopalakrishnan, K. W. Busch and M. A. Busch, *J. Colloid Interface Sci.* **139**, 43 (1990).
4. E. Chibowski and L. Hołysz, *Colloids Surfaces, A: Physicochem. Eng. Aspects* **105**, 211 (1995).
5. E. Chibowski and L. Hołysz, *Colloids Surfaces, A: Physicochem. Eng. Aspects* **101**, 99 (1995).
6. R. W. O'Brien, *J. Colloid Interface Sci.* **113**, 81 (1986).
7. L. A. Rosen, J. C. Baygents and D. A. Saville, *J. Chem. Phys.* **98**, 4183 (1993).
8. D. A. Saville, *Colloids Surfaces, A: Pysicochem. Eng. Aspects* **92**, 29 (1994).
9. E. Chibowski, L. Hołysz B. Dzikowska and R. Ogonowski, *J. Adhesion Sci. Technol.* **10**, 421 (1996).
10. L. Hołysz, E. Chibowski, W. Wójcik, K. W. Busch and M. A. Busch, *J. Adhesion Sci. Technol.* **8**, 181 (1994).
11. C. J. van Oss, M. K. Chaudhury and R. J. Good, *Chem. Rev.* **88**, 927 (1988).
12. C. J. van Oss, R. F. Giese and P. M. Constanzo, *Clays Clay Minerals* **38**, 151 (1990).
13. C. J. van Oss, R. J. Good and M. K. Chaudhury, *Langmuir* **4**, 884 (1988).
14. C. J. van Oss, *Interfacial Forces in Aqueous Media*. Marcel Dekker, New York (1994).
15. R. F. Giese, P. M. Costanzo and C. J. van Oss, *J. Phys. Chem. Miner.* **17**, 611 (1991).
16. E. Chibowski and L. Hołysz, *Langmuir* **8**, 710 (1992).
17. L. Hołysz and E. Chibowski, *Langmuir* **8**, 717 (1992).
18. E. Chibowski, *J. Adhesion Sci. Technol.* **9**, 1069 (1992).
19. E. Chibowski and F. Gonzalez-Caballero, *Langmuir* **9**, 330 (1993).

20. W. Adamson, *Physical Chemistry of Surfaces*, 5th edn. John Wiley & Sons, New York (1990).
21. R. J. Hunter, *Zeta Potential in Colloid Science*. Academic Press (1981).
22. K. W. Busch, M. A. Busch, S. Gopalakrishnan and E. Chibowski, *Colloid Polym. Sci.* **273**, 1186 (1995).
23. J. N. Israelachvili, *Intermolecular and Surface Forces*. Academic Press, London (1985).
24. L. T. Zhuravlev, *Colloids Surfaces, A: Physicochem. Eng. Aspects* **74**, 71 (1993).

Mittal Festschrift, pp. 857–866
W. J. Van Ooij and H. R. Anderson, Jr. (Eds)

Production of high-density Fe–16 at.% Al alloy and Fe–16 at.% Al–15 vol% Al_2O_3 composite by mechanical smearing and hot isostatic pressing

S. BANDYOPADHYAY,[1,*] S. K. MUKHERJEE,[1] D. S. PERERA,[2] S. MORICCA,[2] M. V. SWAIN[3] and T. J. BELL[3]

[1]*School of Materials Science and Engineering, The University of New South Wales, Sydney, NSW 2052, Australia*
[2]*Australian Nuclear Science and Technology Organisation (ANSTO), Private Mail Bag 1, Menai, NSW 2234, Australia*
[3]*CSIRO, Division of Applied Physics, PO Box 218, Lindfield, NSW 2070, Australia*

Abstract—Fe–16 at.% Al alloy matrix and Fe–16 at.% Al–15 vol% Al_2O_3 composite have been made from iron, aluminium and alumina powder by mechanical smearing at room temperature followed by hot isostatic pressing (HIP) at 1200°C. The HIP compacts possess 97 and 99% theoretical densities respectively, resulting in very good hardness values. Ultra-micro-indentation techniques revealed that the composite has three distinct zones so far as hardness is concerned. These are Al_2O_3-rich grain boundaries, an interface region of functionally graded material, and a matrix of Fe–16 at.% Al. The incorporation of the Al_2O_3 phase significantly improves the hardness. From indentation fracture toughness tests, both materials showed an estimated minimum fracture toughness of 15 MPa m$^{1/2}$, which is good for these types of materials.

Keywords: Intermetallics; composites; mechanical smearing; sintering; hot isostatic pressing; compacts; Fe–Al system.

1. INTRODUCTION

The potential high-temperature applications for intermetallic compounds have stimulated considerable interest in these materials worldwide. Iron aluminide is one of the attractive candidates for medium- to high-temperature applications because of its low cost, high specific strength, and excellent corrosion resistance. Like many other intermetallic compounds, however, iron-based intermetallics are prone to embrittlement in moisture and hydrogen-containing atmospheres. The embrittlement involves the following chemical reaction at the metal surface:

$$2\,Al + 3\,H_2O = Al_2O_3 + 6\,H^+ + 6\,e^-$$

*To whom correspondence should be addressed. E-mail: s.bandyopadhyay@unsw.edu.au

The reaction of water vapour with aluminium atoms at the crack tip results in the formation of atomic hydrogen which diffuses into the metal and causes crack propagation. This is a serious drawback in the application of iron aluminides. Various researchers have attempted various means to overcome this limitation [1–7]. There have also been attempts to control the microstructure and alloy composition to improve environmental embrittlement [8, 9]. Although chromium is believed to be effective to some extent for nickel-based superalloys, its beneficial effect on iron aluminide is not well established [10]. Two of us recently developed a novel way to make intermetallic Fe_3Al and its composite with 4 vol% Al_2O_3 by mechanical smearing of the powders at room temperature followed by hot isostatic pressing (HIP) [11]. The compacts were found to have densities close to the theoretical densities of the respective materials and, as a consequence, very good mechanical properties. It was also observed that the bulk of the Al_2O_3 phase occupied the grain-boundary regions of the intermetallic compound [11, 12]. Success with intermetallic Fe_3Al and its composite led us to attempt to produce another constituent alloy of the Fe–Al phase system, namely Fe–16 at.% Al, which is attractive because iron-based alloys with 16 at.% (or less) aluminium are inherently not so sensitive to embrittlement [13]. But these alloys have lower strength and oxidation resistance than the intermetallic iron aluminide. However, it was considered possible to improve the strength and high-temperature properties by incorporating a ceramic second phase (in this case, Al_2O_3). There is no reported work in the literature on composites based on the Fe–16 at.% Al matrix. As such, in addition to making dense Fe–16 at.% Al alloy, the objective of the present work was also to develop a high-density composite of Fe–16 at.% Al with 15 vol% Al_2O_3.

2. EXPERIMENTAL

2.1. Premixing of powder

The matrix alloy composition selected was Fe–16 at.% Al. To make the control alloy, elemental iron powder (average size 65 μm) and aluminium powder (40 μm) were dry-mixed in a ball mill for 4 h using alumina balls of 19 mm and 25 mm diameter and a speed of 75 rpm. The charge-to-ball ratio was kept 1 : 1 by weight. For the composite (i.e. Fe–16 at.% Al–15 vol% Al_2O_3), iron, aluminium, and sub-micrometre size alumina powder were mixed under similar experimental conditions.

2.2. Consolidation

Sintering was considered as a first step in consolidation. For this purpose, the premixed powders were compacted at 250 MPa pressure in a cylindrical die and then sintered under argon atmosphere either at 600°C or at 625°C. The densification parameter, D, of the sintered compacts was calculated as follows:

$$D = \frac{\text{sintered density} - \text{green density}}{\text{theoretical density} - \text{green density}}$$

as an index of the densification process.

The dimensional change during sintering was also calculated. However, sintering did not give very high-density compacts. For this reason, the following method was adopted.

Hot isostatic pressing (HIP) was adopted as the main consolidation process. The premixed powders were encapsulated in a stainless steel tube after evacuation and sealing. HIP was carried out at 1200°C for 2 h under a pressure of 100 MPa, and again in an argon atmosphere.

2.3. *Characterization of the compacts*

The compact density was measured by a water displacement method. Scanning electron microscopy (SEM) and energy dispersive X-ray analysis (EDX) of the sintered and HIP compacts were also carried out. The conventional Vickers micro-hardness test method (200 g load) was employed to assess the hardness properties of the compacts. Further, the grain boundary to matrix hardness profile of the HIP compacts was examined using an Ultra-Micro-Indentation System (UMIS-2000) developed by CSIRO Division of Applied Physics–National Measurement Laboratory [14]. The equipment produces a plot of the load vs. the indenter penetration (nm–μm range) for every indentation made by the indenter onto the material — a Berkovich-type diamond indenter (triangular pyramid, face angle 65.3°) was used. The indenter was incrementally loaded to a pre-set value (a maximum of 300 mN was used in the present case) and then decrementally unloaded to zero load. UMIS plots were obtained for 25 different points for a single specimen and the average value was calculated using the standard deviation technique. From these plots, it was possible to obtain graphs of the hardness as a function of the depth of penetration, using the UMIS software [14]. Finally, the toughness of the compacts was examined by indenting the compacts in a Vickers macro-hardness tester with a diamond indenter under a load of 50 kg, to see if this caused cracking at the corners of the indentation.

3. RESULTS AND DISCUSSION

From the sintering studies, it was observed that 40% swelling of the compact resulted when it was sintered at 625°C. This behaviour is similar to that seen in the sintering of iron aluminide, where an exothermic reaction of elemental iron and aluminium powder mixture results in high porosity because of extensive swelling. Sintering at 600°C did not show any such problem. Table 1 lists the properties of the compacts at 600°C and it can be seen that at this sintering temperature, the Fe–16 at.% Al compacts show a very slight densification (i.e. a positive D value). In the case of the composite, sinterability is lower than for the matrix alloy, as $D = 0$. However, this also indicates that there is no swelling in the composite. The lower sinterability for the composite arises, plausibly, because of a number of metal–metal bonds being replaced by metal–ceramic and ceramic–ceramic bonds, whose diffusivity is generally lower than that of the original metal–metal bonds. Although the hardness values of the

Table 1.
Properties of Fe–16 at.% Al and its 15 vol% Al_2O_3 composite sintered at 600°C

	Green density (g/cm^3)	Sintered density (g/cm^3)	% Linear shrinkage	% Radial shrinkage	Densification parameter, D	Hardness, HV[a]
Fe–16 at.% Al	3.8	4.0	− 11.0	− 1.4	0.05	70
Fe–16 at.% Al + Al_2O_3	3.83	3.84	− 208	+ 0.5	0	85

[a]From micro-hardness tests.

Table 2.
HIP compact density and Vickers micro-hardness of HIP at 1200°C

	HIP density (g/cm^3)	% Theoretical density	Hardness, HV[a]
Fe–16 at.% Al	6.54	97	345
Fe–16 at.% Al + Al_2O_3	6.26	99	425

[a]From micro-hardness tests.

Figure 1. SEM micrograph of sintered Fe–16 at.% Al showing separate iron and aluminium grains.

sintered compacts are very low (Table 1), the composite registers some 20% higher hardness than the unreinforced matrix.

The SEM micrograph of sintered Fe–16 at.% Al (Fig. 1) shows separate iron and aluminium grains and the EDX result (Fig. 2) indicates that no significant reaction has taken place, between iron and aluminium, to form iron–aluminium alloy, as there are no significant amounts of aluminium or iron found in the respective iron or

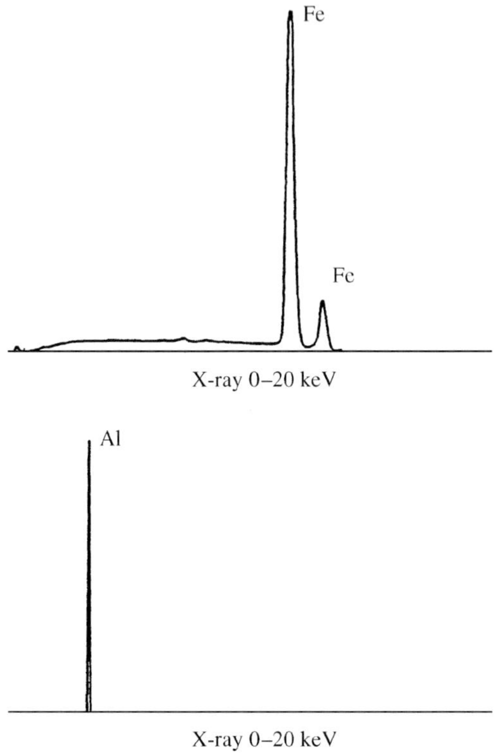

Figure 2. EDX result of sintered Fe–16 at.% Al showing only peaks of Fe and Al from the respective grains, indicating no significant reaction between Fe and Al atoms.

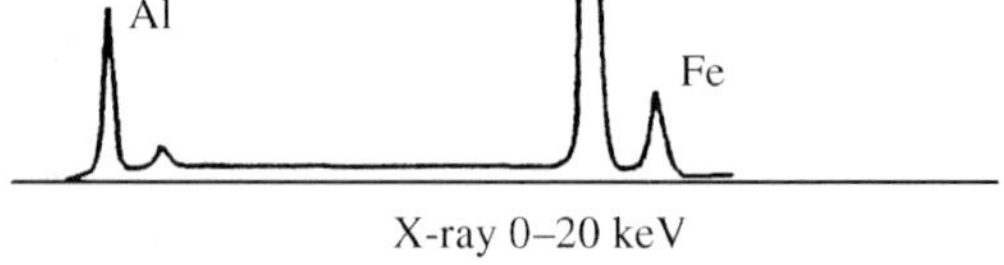

Figure 3. EDX result of HIP Fe–16 at.% Al showing peaks of Fe and Al from the same grain, indicating reaction has taken place.

aluminium grains as indicated by EDX analysis. Also, conventional sintering produces low compact hardness (Table 1), obviously, due to a large amount of porosity.

HIP, on the other hand, produced very significant results, particularly with respect to the microhardness value, for both the matrix and the composite (Table 2). The dense nature of the compacts (density close to theoretical density) is one of the reasons for

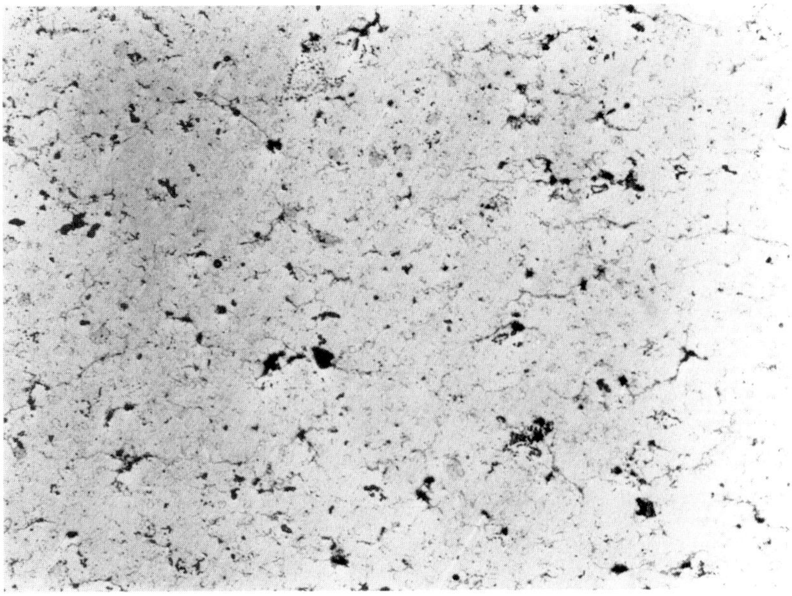

Figure 4. Unetched optical micrograph of HIP Fe–16 at.% Al showing some micro-porosity (× 100).

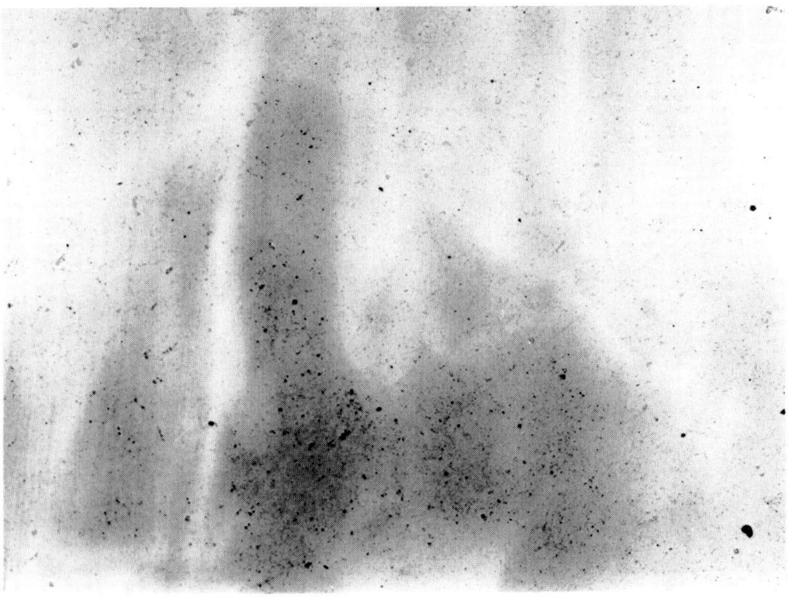

Figure 5. Unetched optical micrograph of Fe–16 at.% Al–15 vol% Al_2O_3 showing virtually no porosity (× 100).

the high hardness. Moreover, EDX analysis of the HIP compact (Fig. 3) shows the formation of iron–aluminium alloy, as strong peaks of both iron and aluminium are obtained from all grains. Unetched micrographs of the two compacts (Figs 4 and 5)

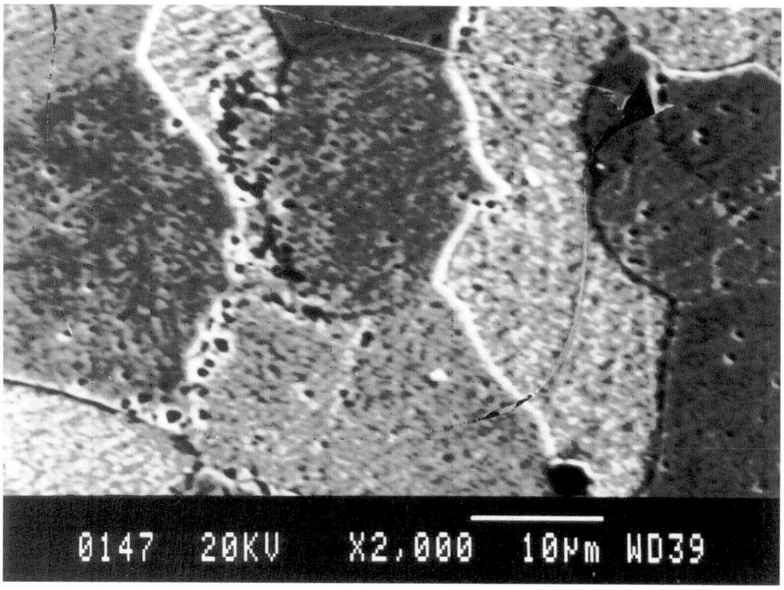

Figure 6. SEM micrograph of HIP Fe–16 at.% Al.

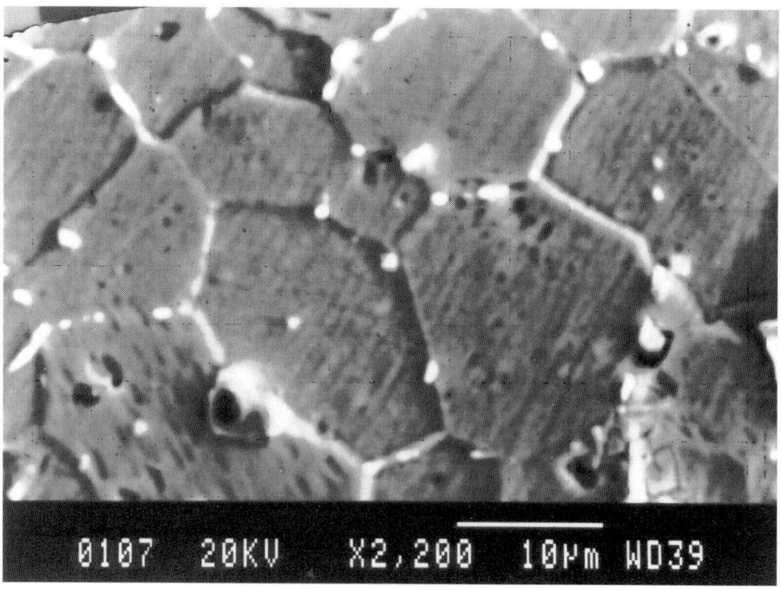

Figure 7. SEM micrograph of the HIP composite.

show the presence of some porosity in the matrix alloy (Fig. 4), whereas porosity is virtually non-existent in the composite.

Further, a 25% improvement in the hardness of the composite (Table 2) suggests that some type of strengthening mechanisms is present in the composite. Figures 6 and 7

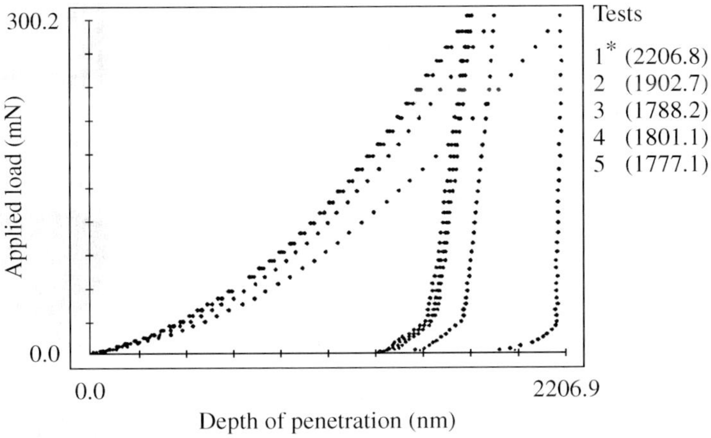

Figure 8. UMIS load (unload) vs. depth of penetration plot for HIP Fe–16 at.% Al.

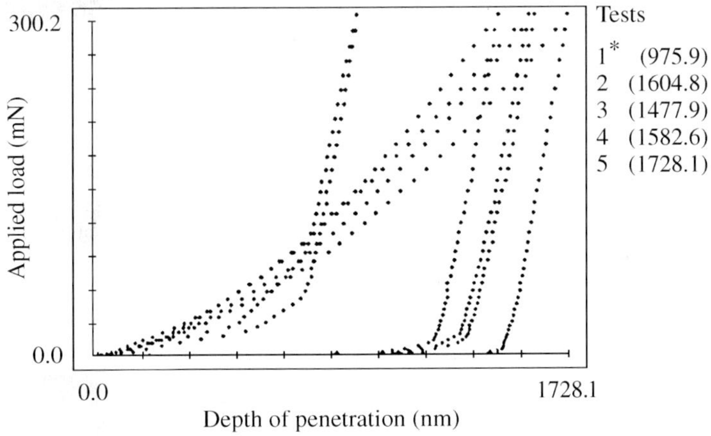

Figure 9. UMIS load (unload) vs. depth of penetration plot for the HIP composite.

show SEM micrographs of the HIP Fe–16 at.% Al and the composite, respectively. The presence of a large concentration of Al_2O_3 particles along the grain-boundary regions suggests grain-boundary strengthening as one of the mechanisms, while some evidence of particulate strengthening can also be seen in Fig. 7.

The results of the UMIS studies are given in Figs 8–11. Figures 8 and 9 are plots of the applied load vs. the depth of penetration during loading and unloading of the specimens, whereas the desired hardness vs. depth graphs for the matrix material and the composite as obtained using the UMIS software [14] are given in Figs 10 and 11, respectively. It is seen in Fig. 10 that for Fe–16 at.% Al, the hardness is virtually constant throughout the depth of penetration and the magnitude of the hardness is about 4.7 GPa. In contrast, the hardness pattern of the composite (Fig. 11) shows the presence of three distinct regions: (a) the grain boundary with a high concentration of Al_2O_3 showing a peak hardness of 15.2 GPa; (b) an intermediate hardness zone

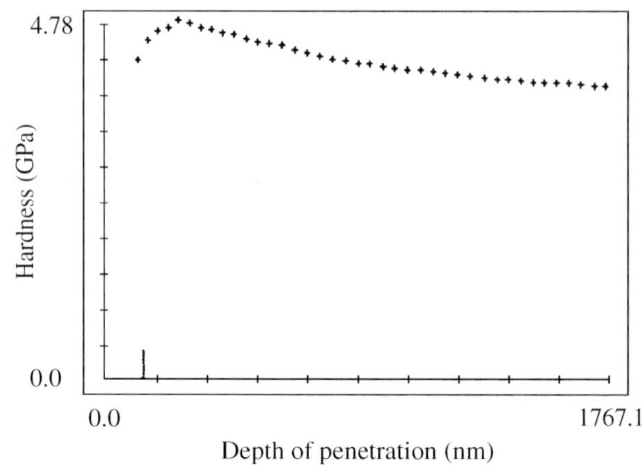

Figure 10. UMIS hardness vs. penetration depth for HIP Fe–16 at.% Al.

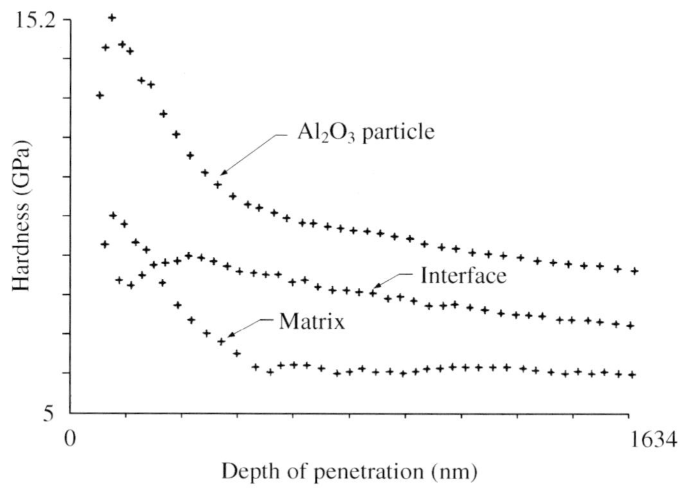

Figure 11. UMIS hardness vs. penetration depth for the HIP composite.

(7.6 GPa at peak) representing possibly the interface; and (c) a lower hardness region, reasonably constant at 6 GPa, corresponding to the composite alloy. Comparison of Figs 10 and 11 reveals that in the composite alloy, the 'matrix' hardness is some 25% higher than that in the unreinforced alloy, an observation parallel with the micro-hardness aspects of the two materials shown in Table 2. The hardness behaviour of the composite in Fig. 11 suggests the existence of a functionally graded material; however, more work needs to be done in this regard.

The Vickers macro-hardness test under 50 kg load revealed that for both Fe–16 at.% Al and Fe–16 at.% Al–15 vol% Al_2O_3, there were no cracks visible around the corners of the indentation, indicating that the materials are reasonably tough (ductile) with an estimated minimum fracture toughness of 15 MPa $m^{1/2}$.

4. CONCLUSIONS

(1) Sintering at 600°C does not produce the desired Fe–16 at.% Al alloy or the composite.

(2) By using hot isostatic pressing, compacts for Fe–16 at.% Al and Fe–16 at.% Al–15 vol% Al_2O_3 with a density close to the theoretical density were obtained. Consequently, high hardnesses resulted.

(3) The 'matrix' in the composite was about 25% stronger than the unreinforced alloy.

(4) The HIP compacts appeared to be reasonably tough, in that there was no cracking at the corner of the indentation under 50 kg load in a Vickers (macro-hardness) machine.

(5) The composite material apparently showed some characteristic feature of a functionally graded material.

Acknowledgement

Part of this work was presented by S. Bandyopadhyay during presentation of an invited paper, 'Interfaces in Composites', at the 1st International Congress on Adhesion Science and Technology (ICAST-1), Amsterdam, October 1995, on the occasion of the 50th birthday of Dr. Kash Mittal.

REFERENCES

1. N. S. Stoloff and R. G. Davis, *Prog. Mater. Sci.* **13**, 1 (1966).
2. C. T. Liu, E. H. Lee and C. G. McKamey, *Scr. Metall.* **23**, 875 (1989).
3. T. Takasugi and O. Izumi, *Acta Metall.* **34**, 607 (1986).
4. O. Izumi and T. Takasugi, *J. Mater. Res.* **3**, 442 (1986).
5. N. S. Stoloff, *J. Met.* **40**, 23 (1988).
6. C. T. Liu, C. G. McKamey and E. H. Lee, *Scr. Metall.* **24**, 385 (1990).
7. N. S. Stoloff and D. J. Duquette, *J. Met.* **45**, 30 (1993).
8. C. G. McKamey, P. J. Maziasz, G. M. Goodwin and T. Zacharia, *Mater. Sci. Eng.* **A174**, 59 (1994).
9. A. Castagna and N. S. Stoloff, *Mater. Des.* **13**, 385 (1990).
10. C. G. McKamey, J. A. Harton and C. T. Liu, *Scr. Metall.* **22**, 1679 (1988).
11. S. Mukherjee and S. Bandyopadhyay, *Mater. Sci. Eng.* **A202**, 123 (1995).
12. S. Mukherjee and S. Bandyopadhyay, *Composites Part B* **28B**, 45–48 (1997).
13. G. M. Campus, D. J. Duquette and N. S. Stoloff, *J. Met.* **5**, 950 (1990).
14. T. J. Bell, A. Bendeli, J. S. Field, M. V. Swain and E. Thawaite, *Metrologia* **28**, 463 (1991–1992).

Part 8

Titles of other papers presented

FUNDAMENTAL ASPECTS OF ADHESION, AND GENERAL TOPICS

Recent developments in molecular level adhesion monitored by rheo-photoacoustic FT-IR spectroscopy

MAREK W. URBAN

Department of Polymers and Coatings, North Dakota State University, Fargo, ND 58105, USA

Adhesion at metal-oxide interfaces: an electrostatics based model

J. W. MINTMIRE and F. H. STREITZ

Chemistry Division, Code 6179, Naval Research Laboratory, Washington, DC 20375, USA

Evolution of symmetric polymer interfaces: an insight into polymer dynamics

G. AGRAWAL[1] and R. P. WOOL[2]

Department of Materials Science, University of Illinois, Urbana, IL 61801, USA
Current addresses:
[1]*PPG Industries, 4325 Rosanna Drive, Allison Park, PA 15101, USA*
[2]*Center for Composite Materials, University of Delaware, Newark, DE 19716, USA*

Theory and measurement of contact angles

ABRAHAM MARMUR

Department of Chemical Engineering, Technion — Israel Institute of Technology, 32000 Haifa, Israel

Adhesion science and scientists in the Baltic states

M. ŠALKAUSKAS

Institute of Chemistry, Goštauto 9, 2600 Vilnius, Lithuania

Application of STM and AFM to adhesion science

K. S. BIRDI

Royal Danish School of Pharmacy, DK-2100 Copenhagen, Denmark

The role of interfacial adhesion in the stabilization of colloidal cobalt particles in polymer solutions

RINA TANNENBAUM

Department of Chemical Engineering and Materials Science, University of Minnesota, Minneapolis, MN 55455, USA

Evaluation of adhesion forces in liquids with an atomic force microscope

M. SEKOGUCHI,[1] T. SAKON,[1] K. SHIMANOE[2] and M. SAKASHITA[1]

[1]*Future & Frontier Field Laboratory, Advanced Technology Research Laboratories, Nippon Steel Corporation, Kawasaki 211, Japan*
[2]*Semiconductor Materials Laboratory, Electronics Research Laboratories, Nippon Steel Corporation, Hikari 743, Japan*

The dependence of glass melt adhesion with the lining on lining composition

B. S. BATALIN and N. PRAVINA

Department of Structural Materials and Special Technology, Perm State Technical University, 614000 Perm, Russia

Diffusion bonding in MC/Ni–B (M = W, V) composites

L. DIAZ,[1] J. LIRA OLIVARES,[1] M. VELEZ[1] and I. C. GRIGORESCU[2]

[1]*Universidad Simon Bolivar, Apdo. 89000, Caracas 1080-A, Venezuela*
[2]*INTEVEP, S. A., Apdo. 76343, Caracas 1070-A, Venezuela*

Adhesion phenomena in microobjects

J. EVDOKIMOV,[1] V. KESTELMAN[2] and E. KONDRATIEV[1]

[1]*Forest University, Moscow, Russia*
[2]*KVN Intl, 632 Jamie Circle, King of Prussia, PA 19406, USA*

Micro-Raman characterization of ceramic–metal interfaces

A. AMADEH, J. C. LABBE, T. MERLE-MEJEAN and P. QUINTARD

Laboratoire de Matériaux Céramiques et Traitements de Surfaces, L.M.C.T.S. URA-320 du CNRS, Faculté des Sciences, 123 Ave. Albert Thomas, F-87060 Limoges, France

Dispersive and acid–base properties of poly(dimethacrylate)s surfaces

A. VOELKEL, E. ANDRZEJEWSKA, R. MAGA and M. ANDRZEJEWSKI

Poznań University of Technology, Institute of Chemical Technology and Engineering, Pl. M. Skłodowskiej-Curie 2, 60-965 Poznań, Poland

New approach for the production of duplex composite engineering components

MOHAMED A. BAYOUMI[1] and FAWZIA H. KOUTA[2]

[1]*Department of Mechanical Engineering, Al-Azhar University, Cairo, Egypt*
[2]*The American University in Cairo, Cairo, Egypt*

CONTACT ANGLE, SURFACE ENERGETICS AND ADHESION

The use of the surface forces apparatus in the study of adhesion: polymer solid surface energies and the effect of surface treatment

V. S. MANGIPUDI,[1,*] M. TIRRELL[1] and A. V. POCIUS[2]

[1]*Department of Materials Science and Chemical Engineering, University of Minnesota, Minneapolis, MN 55455, USA*

[2]*3M Adhesive Technologies Center, St. Paul, MN 55144-1000, USA*

[*]*Current address: 3M Co./3M Center, St. Paul, MN 55144-1000, USA*

Contact angles on porous sheets

G. STRÖM,* A.-M. HÄRDIN and J. ELFTONSSON

Institute for Surface Chemistry, P.O. Box 5607, S-11486 Stockholm, Sweden

[*]*Current address: Stora Corporate Research AB, S-79180 Falun, Sweden*

The work of adhesion between water and aromatics interfaces: the role of H-bonding

H. YILDIRIM ERBIL

TUBITAK, Marmara Research Institute, Department of Chemistry, P.O. Box 21, 41470 Gebze, Kocaeli, Turkey

Relationship between the lifetime of thin films and the macroscopic parameters of wetting for apolar and polar liquids on non-wettable solids

A. T. JAMEEL[1] and A. SHARMA[2]

[1]*Department of Chemical Engineering, Aligarh Muslim University, Aligarh-202002, India*

[2]*Department of Chemical Engineering, Indian Institute of Technology at Kanpur, Kanpur-208016, India*

Wettability of quartz with Langmuir-Blodgett films of arachidic acid and/or cellulose didecanoate

K. GOTOH,[1] S. YOSHIMITSU[2] and M. TAGAWA[2]

[1]*Faculty of Education, Kyoto University of Education, Fushimi-ku, Kyoto 612, Japan*

[2]*Department of Textile & Apparel Science, Nara Women's University, Nara 630, Japan*

Effect of surface charge density on the kinetics of capillary wetting in a fibrous assembly

Y. OKAMURA,[1] K. GOTOH[2] and M. TAGAWA[3]

[1]*Suzugamine Women's College, 4-6-18 Inokuchi, Nishi-ku, Hiroshima 733, Japan*
[2]*Faculty of Education, Kyoto University of Education, Fushimi-ku, Kyoto 612, Japan*
[3]*Department of Textile & Apparel Science, Nara Women's University, Nara 630, Japan*

Assessment of the surface free energy of low-energy solids by means of contact angle measurements

N. P. WILLARD, A. R. BALKENENDE, H. J. A. P. VAN DE BOOGAARD and M. SCHOLTEN

Philips Research, Prof. Holstlaan 4, 5656 AA Eindhoven, The Netherlands

Adhesion between ceramic oxides and body liquids

S. AGATHOPOULOS and P. NIKOLOPOULOS

Chemical Engineering Department, University of Patras and Institute of Chemical Engineering and High Temperature Chemical Processes, GR-26500 Patras, Greece

Surface characterization of steel specimens: comparison of dynamic Wilhelmy balance (DWB) and X-ray photoelectron spectroscopy (XPS) results

L. TINUCCI,[1] L. DEL GAUDIO,[1] S. LUNETTO,[1] N. NISTICÒ,[1] S. CONTARINI,[1] R. SANTORELLI[2] and G. PONTI[3]

[1]*Eniricerche S.p.A., Via Maritano 26, 20097 S. Donato Milanese (Milano), Italy*
[2]*Euron S.p.A., Via Maritano 26, 20097 S. Donato Milanese (Milano), Italy*
[3]*Agip Petroli S.p.A., Via Laurentina 449, 00142 Roma, Italy*

Work of adhesion in zirconia–liquid metal system

D. SOTIROPOULOU, P. NIKOLOPOULOS and S. AGATHOPOULOS

Chemical Engineering Department, University of Patras and Institute of Chemical Engineering and High Temperature Chemical Processes, GR-26500 Patras, Greece

Microscopic hydrophilic and macroscopic hydrophobic behaviors of talc

F. VILLIÉRAS, L. J. MICHOT, M. FRANÇOIS, J. YVON and J. M. CASES

Laboratoire Environnement et Minéralurgie, URA 235 du CNRS, Ecole Nationale Supérieure de Géologie, BP 40, 54501 Vandoeuvre les Nancy cedex, France

Contact angle of a liquid drop on a fiber

TOSHIO OGAWA, TOMOHIKO SUGA and DAI KAKIWAKI

Faculty of Engineering, Kanazawa Institute of Technology, 7-1, Ohgigaoka Nonoichi, Ishikawa 921, Japan

Wetting on curved surfaces

T. BIEKER and S. DIETRICH

Institute for Theoretical Physics, University of Wuppertal, Gaussstr. 20, 42119 Wuppertal, Germany

Wettability of solid surfaces by melts of alkali metals

A. B. ALCHAGIROV, B. B. ALCHAGIROV, T. M. TAOVA, KH. B. KHOKONOV and B. S. KARAMURZOV

Department of Physics, Kabardino-Balkarian University, Chernyshevskogo Str. 173, Nalchik 360004, Russia

Wetting of polysulfone and modified polysulfone membranes

M. OLTEANU, G. NECHIFOR,[1] B. G. ALBU[1] and G. POPESCU[1]

Department of Physical Chemistry, University of Bucharest, Bd. Carol 13, 70246 Bucharest, Romania
[1]*Research Center for Macromolecular Materials and Membranes, Spl. Independentei 202, 79611 Bucharest, Romania*

Surface energy components of some zirconia powders and porous zirconia ceramics, studied by wetting techniques

A. B. A. PETTERSSON,[1] F. SHOJAI,[2] J. B. ROSENHOLM[1] and T. MÄNTYLÄ[2]

[1]*Department of Physical Chemistry, Åbo Akademi University, Porthansgatan 3-5, FIN-20500 Åbo, Finland*
[2]*Tampere University of Technology, Institute of Materials Science, PL 589, FIN-33101 Tampere, Finland*

Wetting of selected solid metals and oxides by liquid indium

J. JANDOVÁ

Department of Metals and Corrosion Engineering, University of Chemical Technology, 16628 Prague 6, Czech Republic

Wettability of reservoir rocks by surface science analytical techniques

P. G. TOLEDO, Y. C. ARAUJO, V. LEÓN and H. Y. GONZÁLEZ

INTEVEP, S.A., Apdo. 76343, Caracas 1070-A, Venezuela

Dynamic study of wetting: changes in surface properties of polymers in response to various pH's

A. DOREN, Y. ADRIAENSEN and P. G. ROUXHET

Unité de Chimie des Interfaces, Université Catholique de Louvain, Place Croix du Sud 2/18, B-1348 Louvain-la-Neuve, Belgium

Detection of a hydrogel layer at the surface of oxidized polyethylene by dynamics of wetting

E. TOMASETTI, A. DOREN and P. G. ROUXHET

Unité de Chimie des Interfaces, Université Catholique de Louvain, Place Croix du Sud 2/18, B-1348 Louvain-la-Neuve, Belgium

Interfacial energy of metallic alloy-submonolayer dielectrical coverage system

A. B. ALCHAGIROV, V. A. SOZAEV and KH. B. KHOKONOV

Kabardino-Balkarian University, Chernyshevskogo Str. 173, Nalchik 360004, Russia

Surface tension of sodium-potassium melts and wetting of surfaces of copper, nickel and steel by these melts

B. B. ALCHAGIROV, A. B. ALCHAGIROV, KH. B. KHOKONOV and B. S. KARAMURZOV

Kabardino-Balkarian University, Chernyshevskogo Str. 173, Nalchik 360004, Russia

Surface energy characteristics of metals investigated by inverse gas chromatography

E. PAPIRER,[1,*] E. BRENDLE,[1,*] H. BALARD[1,*] and J. LIGNIERES[2]

[1]*Centre de Recherche sur la Physico-Chimie des Surfaces Solides (CNRS), 24, Av. Président Kennedy, F-68200 Mulhouse, France*

[2]*L'AIR LIQUIDE, Centre de Recherche Claude Delorme, BP 126, F-78350 Jouy-en-Josas, France*

**New address: ICSI, 15 rue Jean Starcky, B.P. 2488, F-68057 Mulhouse Cedex, France*

Wettability of some nitride ceramics and alumina by Si-Ca treated steel

A. AMADEH, J. C. LABBE and P. QUINTARD

Laboratoire de Matériaux Céramiques et Traitements de Surface (URA 320 du CNRS), Faculté des Sciences, 123, Ave. Albert Thomas, F-87060 Limoges, France

SURFACE MODIFICATION: RELEVANCE TO ADHESION

Surface engineering of polymers for enhanced adhesion

D. Y. WU,[1] W. S. GUTOWSKI[1] and S. LI[1,2]

[1]*CSIRO Division of Building, Construction and Engineering, P.O. Box 56, Highett, Victoria 3190, Australia*

[2]*CSIRO Division of Chemicals and Polymers, Private Bag 10, Rosebank MDC, Clayton, Victoria 3169, Australia*

Surface engineering of rubbers for adhesive bonding with thermoset and contact adhesives

W. S. GUTOWSKI,[1] D. Y. WU[1] and S. LI[2]

[1]*CSIRO Division of Building, Construction and Engineering, P.O. Box 56, Highett, Victoria 3190, Australia*

[2]*CSIRO Division of Chemicals and Polymers, Private Bag 10, Rosebank MDC, Clayton, Victoria 3169, Australia*

Excimer laser treatment of thermoplastics for adhesive bonding

B. SADRAS, P. LAURENS and F. DECOBERT[1]

CREA/LALP, Unité Mixte ETCA-CNRS

[1]*CREA/PC, Etablissement Technique Central de l'Armement (ETCA), 16 bis av. Prieur de la Côte d'Or, 94114 Arcueil cedex, France*

TOF-SIMS and *in situ* XPS study of O_2-N_2 afterglow discharge plasma-modified PMMA, PE and hexatriacontane surfaces

D. LÉONARD,[1] P. BERTRAND,[1] A. SCHEUER,[2] R. PRAT[2] and J. P. DEVILLE[2]

[1]*Unité PCPM, Université Catholique de Louvain, 1 place Croix du Sud, B-1348 Louvain-la-Neuve, Belgium*

[2]*Institut de Physique des Matériaux de Strasbourg, 23 rue du Loess, F-67037 Strasbourg Cedex, France*

Plasma interactions with a polyethylene surface studied by AFM and XPS

S. O'KELL, S. D. PRINGLE and C. JONES

Department of Materials Science and Engineering, The University of Liverpool, P.O. Box 147, Liverpool, L69 3BX, UK

Surface dynamics and adhesive bonding of plasma-treated polyolefins and fluoropolymers

S. LI,[1] D. Y. WU,[2] W. S. GUTOWSKI[2] and H. J. GRIESSER[1]

[1]*Division of Chemicals and Polymers, CSIRO, Private Bag 10, Rosebank MDC, Clayton 3169, Australia*

[2]*Division of Building, Construction and Engineering, CSIRO, P.O. Box 56, Highett 3190, Australia*

Surface modifications of PEEK induced by a remote microwave plasma and its consequences on adhesion properties of carbon/PEEK composite with an epoxy amine system

P. RUMEAU,[1,2] J. P. SOULIER,[1] B. CHABERT[1] and P. ABJEAN[2]

[1]*Laboratoire d'Etudes des Matériaux Plastiques et des Biomatériaux, Université C. Bernard Lyon I, URA CNRS 507, 43 Bd du 11 Novembre 1918, 69622 Villeurbanne, France*

[2]*EUROCOPTER FRANCE, Etablissement de Marignane, 13725 Marignane, France*

Differences in surface oxidation of PP by corona, spark, and low-pressure oxygen discharge treatments and the relevance to adhesion

J. F. FRIEDRICH,[1] W. UNGER,[1] A. LIPPITZ,[1] L. WIGANT,[1] SH. GENG,[1] J. ERDMANN[2] and H.-V. GORSLER

[1]*Bundesanstalt für Materialforschung und -prüfung, Unter den Eichen 87, D-12200 Berlin, Germany*
[2]*Ahlbrandt System GmbH, D-36339 Lauterbach, Germany*

Modification of the physicochemical properties of oxygen plasma treated polypropylene

N. SHAHIDZADEH-AHMADI, F. AREFI-KHONSARI, M. M. CHEHIMI[1] and J. AMOUROUX

Laboratoire de Génie des Procédés Plasmas, ENSCP-Université Pierre et Marie Curie, 11 rue Pierre et Marie Curie, 75231 Paris Cedex 5, France
[1]*Institut de Topologie et Dynamique des Systèmes de l'Université Denis Diderot (Paris 7), associé au CNRS (URA 34), 1 rue Guy de la Brosse, 75005 Paris, France*

Surface modification of aramid fibers by excimer laser irradiation

H. WATANABE and T. TAKATA

Fiber & Textile Research Laboratories, Teijin Ltd., 3-4-1, Minohara, Ibaraki, Osaka 567, Japan

Adherence enhancement of polymers with low surface energy by excimer laser radiation

J. BREUER,[1] H. SCHÄFER,[2] V. SCHLETT,[2] S. METEV,[1] G. SEPOLD[1] and O.-D. HENNEMANN[2]

[1]*BIAS, Bremen Institut für Angewandte Strahltechnik, Klagenfurterstrasse 2, D-28359, Bremen, Germany*
[2]*IFAM, Fraunhofer Institut für Angewandte Materialforschung, Neuer Steindamm 2, D-28719, Bremen (Lesum), Germany*

Surface modification of carbon black by oxygen plasma processing and application to fuel cell electrodes

K. HORITA and Y. NAGAOSA

Faculty of Engineering, Fukui University, Fukui-shi 910, Japan

Adhesion improvement of ink to polymers by laser activation

PETER E. J. LEGIERSE

Philips Centre for Manufacturing Technology, P.O. Box 218, Blg. SAQ-p, 5600 MD Eindhoven, The Netherlands

Photochemical promotion of adhesion of light-sensitive layers to the surface of doped SiO_2

L. LAPČÍK, B. MIKULÁŠKOVÁ, E. PŘIBYLOVÁ and L. LAPČÍK, JR.

Faculty of Chemistry, Technical University of Brno, Veslařská 230, CZ-63700 Brno, Czech Republic

Strategy of functionalisation of poly(ethylene terephthalate) (PET) for the purpose of biological applications

M. ART, D. DELFORGE, B. GILLON, J. REMACLE and M. RAES*

Laboratoire de Biochimie Cellulaire, FUNDP, 5000 Namur, Belgium
**Research Associate of the FNRS, Belgium*

Surface modification of bioceramics and the effects of their surface characteristics on the adhesion of anchorage-dependent animal cells

K. NISHIZAWA, T. SUZUKI, M. TORIYAMA, Y. YOKOGAWA, Y. KAWAMOTO, F. NAGATA, M. R. MUCALO and T. KAMEYAMA

Bioceramics Laboratory, National Industrial Research Institute of Nagoya, 1 Hirate, Kita, Nagoya 462, Japan

FTIR/XPS analysis of treated glass fibres

P. ARORA, D. C. KEHOE, J. G. MATISONS, R. ST. C. SMART, A. E. E. JOKINEN[1] and J. B. ROSENHOLM[1]

Polymer Science Group, School of Chemical Technology, University of South Australia, The Levels, SA 5095, Australia
[1]*Department of Physical Chemistry, Åbo Akademi University, FIN-20500 Åbo, Finland*

Surface treated PET fibres — characterisation using AFM, HRSEM and XPS techniques

L. A. EMERSON,[1] E. CHONG,[2] N. L. ANDERSON[1,*] and M. R. PHILLIPS[2]

[1]*Cooperative Research Centre for Cardiac Technology: Telectronics Pacing Systems, 7 Sirius Road, Lane Cove, NSW 266, Australia*
[2]*University of Technology, Sydney, P.O. 123, Broadway, NSW 2007, Australia*
**Current address: Department of Materials Science, University of Technology, Sydney, P.O. Box 123, Broadway, NSW 2007, Australia*

Atomic–molecular construction of glass microspheres surfaces to improve adhesion to epoxy matrix

E. A. IVANOVA and M. N. TSVETKOVA

Department of Chemistry of Solids, St. Petersburg Technological Institute, Moscovsky Pr. 26, St. Petersburg 198013, Russia

SILANES AND OTHER ADHESION PROMOTORS

Plasma-deposited organosilane interfacial bonding layers for adhesive bonding of epoxy to aluminium

H. J. GRIESSER,[1] D. R. ARNOTT,[2] N. G. FARR[2] and A. N. RIDER[2]

[1]*Division of Chemicals and Polymers, CSIRO, Private Bag 10, Rosebank MDC, Clayton 3169, Australia*
[2]*DSTO Aeronautical and Maritime Research Laboratory, GPO Box 4331, Melbourne 3001, Australia*

Cold plasma activation and silane grafting on moving fiber glass bundle

B. DAS

PPG Industries Inc., Fiber Glass Research Center, P.O. Box 2844, Pittsburgh, PA 15230, USA

Silane and siloxane adsorption onto surfaces

D. C. KEHOE, J. G. MATISONS, A. E. E. JOKINEN[1] and J. B. ROSENHOLM[1]

Polymer Science Group, School of Chemical Technology, University of South Australia, The Levels 5095, SA, Australia
[1]*Department of Physical Chemistry, Åbo Akademi University, FIN-20500 Åbo, Finland*

Pretreatment of Al-43.4 Zn-1.6 Si-coated steel using organofunctional silanes

S. E. HÖRNSTRÖM,[1,*] J. KARLSSON,[1] W. J. VAN OOIJ[2] and NIE TANG[2]

[1]*Materials Centre, University College of Falun Borlange, Borlänge, Sweden*
[2]*Department of Materials Science and Engineering, University of Cincinnati, Cincinnati, OH 45221-0012, USA*
[*]*Current address: Center for Industrial Engineering and Development, Dalarna University College, S-78188 Borlange, Sweden*

Adhesion promotion by bifunctional organosilicon compounds: adhesion mechanisms at the molecular (functional group) level

L. W. JENNESKENS

Debye Institute, Department of Physical Organic Chemistry, Utrecht University, Padualaan 8, 3584 CH Utrecht, The Netherlands

Perfluoroalkyl end-capped oligomeric silane coupling agents: fluoroalkylation of polyester surface

T. KAWASE,[1] M. YAMANE,[1] T. FUJII,[1] M. MINAGAWA[1] and H. SAWADA[2]

[1]*Faculty of Human Life Science, Osaka City University, 3-3-138 Sugimoto, Sumiyoshiku, Osaka 558, Japan*

[2]*Department of Chemistry, Nara National College of Technology, Yata, Yamatokoriyama, Nara 639-11, Japan*

Chlorinated polyolefins — recent advances in adhesion theory and testing methodology

J. O. STOFFER,[1] W. L. DECHENT,[1] A. J. BLANKENSHIP,[2] R. L. EAGAN,[2] R. P. EVANS,[3] P. J. GREENE,[3] J. R. HANDLEY,[2] J. F. LAWNICZAK[3] and K. A. WILLIAMS[3]

[1]*Materials Research Center, University of Missouri at Rolla, Rolla, MO 65401, USA*

[2]*Arkansas Eastman Division, Eastman Chemical Co., P.O. Box 2357, Batesville, AR 72503-2357, USA*

[3]*Tennessee Eastman Division, Eastman Chemical Co., P.O. Box 431, Kingsport, TN 37662-5280, USA*

Reactivity of alkylsilanes to silica substrates: proposed structures of the monolayers

J. DUCHET,[1] J. P. CHAPEL,[1] B. CHABERT[1] and J. F. GÉRARD[2]

[1]*Laboratoire des Matériaux Plastiques et Biomatériaux, Université Lyon I, Bât 303, 69622 Villeurbanne Cedex, France*

[2]*Laboratoire des Matériaux Macromoléculaires, INSA Lyon, Bât 403, 69622 Villeurbanne Cedex, France*

Adhesion enhancement of polymer/aluminium interface by coupling agents

N. DUMONT and C. DEPECKER

Laboratoire de Structure et Propriétés de l'État Solide, Bât C6, URA CNRS 234, Université des Sciences et Téchnologie de Lille, 59650 Villeneuve d'Ascq., France

Study of wettability and structure of silane films on glass surfaces by XPS, SEM-BSE and AFM

Y. C. ARAUJO, P. G. TOLEDO, J. GUZMÁN, V. LEÓN and H. Y. GONZÁLEZ

INTEVEP, S.A., Apdo. 76343, Caracas 1070-A, Venezuela

Perfluoroalkyl end-capped oligomeric silane coupling agents: fluoroalkylation of glass surface

T. KAWASE,[1] T. FUJII,[1] M. MINAGAWA[1] and H. SAWADA[2]

[1]*Faculty of Human Life Science, Osaka City University, 3-3-138 Sugimoto, Sumiyoshiku, Osaka 558, Japan*

[2]*Department of Chemistry, Nara National College of Technology, Yata, Yamatokoriyama, Nara 639-11, Japan*

Adhesion mechanisms of organosilane/silica coatings to polydiethylene glycol bis allyl carbonate (CR39)

T. J. HORR,[1] R. HORN,[1] J. RALSTON,[1] G. REYNOLDS,[1] G. TOIKKA[1] and D. R. DIGGINS[2]
[1]*Ian Wark Research Institute, University of South Australia, The Levels, SA 5095, Australia*
[2]*SOLA International Holdings Limited, Sherriffs Road, Lonsdale, SA 5160, Australia*

Delamination of monomolecular adhesion promoters

M. STRATMANN and E. HORNUNG
Institut für Werkstoffwissenschaften, Lehrstuhl für Korrosion und Oberflächentechnik, Friedrich-Alexander-Universität Erlangen-Nürnberg, Martensstr. 7, 91058 Erlangen, Germany

ADHESIVE JOINTS: CHARACTERIZATION, TESTING AND DURABILITY

Characterization of the interphase, control element of the degree of interaction between continua, on continuum mechanics level

A. H. CARDON, J. DE VISSCHER and M. BRUGGEMAN
Composite Systems and Adhesion Research Group of the University Brussels (COSARGUB) VUB, (TW-KB) Pleinlaan 2, B-1050 Brussels, Belgium

Effect of interfacial slippage on viscoelastic adhesion

BI-MIN ZHANG NEWBY,[1] MANOJ K. CHAUDHURY[1] and HUGH R. BROWN[2,*]
[1]*Department of Chemical Engineering, Lehigh University, Bethlehem, PA 18015, USA*
[2]*IBM Research Division, Almaden Research Center, San Jose, CA 95120, USA*
[*]*Current address: BHP Institute for Steel Processing, University of Wollongong, Wollongong, NSW 2522, Australia*

Complex dynamic behavior in adhesive tape peeling

M. C. GANDUR,[1] M. U. KLEINKE[2] and F. GALEMBECK[3]
[1]*3M do Brasil Ltda., C.P. 123, 13001-970 Campinas-SP, Brazil*
[2]*Instituto de Fısica, UNICAMP, C.P. 6165, 13083-970 Campinas-SP, Brazil*
[3]*Instituto de Quımica, UNICAMP, C.P. 6154, 13083-970 Campinas-SP, Brazil*

On the observation of a fine structure associated with Schallamach's waves propagation produced by sliding friction of rubber-like materials

A. KOUDINE and M. BARQUINS

Laboratoire de Physique et de Mécanique des Milieux Hétérogènes, Ecole Supérieure de Physique et Chimie Industrielles, 10, rue Vauquelin, 75231 Paris Cedex 5, France

A non-linear fracture mechanics approach to strength analysis of adhesive joints

H. WERNERSSON[1] and P. J. GUSTAFSSON[2]

[1]*Perstorp Flooring AB, Box 1010, S-23125 Trelleborg, Sweden*
[2]*Lund University, Div. Structural Mechanics, Box 118, S-22100 Lund, Sweden*

Boundary and transition layers in polymeric adhesive/metal systems: new methods of characterization

A. SILBERMAN and O. FIGOVSKY

Polyadd Ltd., P.O. Box 678, Nazareth Illit 17106, Israel

Interfacial modifier to improve polyimide/silicone adhesive joint strength

P. SASIDHARAN ACHARY and R. RAMASWAMY

Polymers and Special Chemicals Division, Vikram Sarabhai Space Centre, Thiruvananthapuram-695022, India

Experimental and numerical analyses of the mechanical behavior of a new tensile-stripping adhesive-bonded ceramic system

D. AMARA,[1] F. LEVALLOIS,[2] Y. BAZIARD[1] and J. A. PETIT[1]

[1]*Laboratoire Génie de Production–Génie des Matériaux, Ecole Nationale d'Ingénieurs, B.P. 1629, 65016 Tarbes Cedex, France*
[2]*Matra Marconi Space, 31 rue des Cosmonautes, Z.I. du Palays, 31077 Toulouse Cedex, France*

Theoretical modelling of tension and bending of adherends in a four-step, single lap adhesively bonded joint

M. L. L. KLEIN

Groupe Analyse et Modélisation des Structures, Ecole Nationale Supérieure de Techniques Avancées/Laboratoire de Mécanique, 32, bd Victor, 75015 Paris, France

Investigation of the fatigue behaviour of adhesively-bonded joints using damage maps

C. BLANCHARD, A. CHATEAUMINOIS and L. VINCENT
Department M.M.P., Ecole Centrale de Lyon, 36 Avenue Guy de Collongues, 69131 Ecully Cedex, France

Mass-spectrometry investigation of surface layers in the vicinity of epoxyphenolic adhesive-metal interface

A. L. ZAITSEV and YU. M. PLESKACHEVSKII
Metal-Polymer Research Institute, Belarussian Academy of Sciences, 246652 Gomel, Belarus

The effect of rubber content on the bond strength of oil-accommodating adhesive

T. OGAWA and M. HONGO
Graduate School of Engineering, Kanazawa Institute of Technology, 7-1, Ohgigaoka Nonoichi, Ishikawa 921, Japan

A new test method for determining the decohesion energy of adhesively-bonded thin substrates

J.-Y. SENER,[1] F. VAN DOOREN[2] and F. DELANNAY[1]
[1]*Université Catholique de Louvain, Département des Sciences des Matériaux et des Procédés, PCIM, Place Sainte Barbe 2, B-1348 Louvain-la-Neuve, Belgium*
[2]*Cockerill Sambre, R&D, Boulevard de Colonster B57, B-4000 Liège, Belgium*

Adhesion of alkylresorcinol resins to wood

J.-A. STARKOPF and Ü. LUGA
Tallinn Technical University, Ehitajate tee 5, EE 00026 Tallinn, Estonia

Investigation of adhesive ability of textile materials

V. E. KUZMICHEV, E. V. TIMOFEEVA and S. G. EFANOV
State Textile Academy, Clothing Faculty, Ivanovo 153000, Russia

Use of acrylic adhesive for production of multilayer composite textile materials

S. G. EFANOV[1] and V. E. KUZMICHEV[2]
[1]*Melanz Corp., Sosnevo, Ivanovo 153006, Russia*
[2]*State Textile Academy, Ivanovo 153000, Russia*

Furan-amino resins as wood adhesives

F. F. ABD EL-MOHSEN, R. M. MOHSEN and Y. M. ABU AYANA
Department of Polymers and Pigments, National Research Centre, Cairo, Egypt

Adhesive bonding of microoptical components

R. EBERHARDT, G. TITTELBACH and V. GUYENOT
Fraunhofer Institute for Applied Optics and Precision Mechanics, Schillerstraße 1, 07745 Jena, Germany

ADHESION ASPECTS OF POLYMERIC COATINGS

Adhesion mechanics during low speed impact of 'rocks' onto a multi-layer painted automotive sheet steel substrate

A. C. RAMAMURTHY,[1] JOE HOLUBKA[1] and DENNIS J. MIHORA[2]
[1]*ACD & SRL Ford Motor Company, Dearborn, MI 48121-2053, USA*
[2]*FM Analysts, 5326 Plaza Aleman, Santa Barbara, CA 93111, USA*

Residual stresses and Young's modulus determination of thin and thick coatings applied onto substrates

M. BENABDI[1] and A. A. ROCHE[2]
[1]*Université Claude Bernard-Lyon I, Laboratoire d'Etude des Interphases (CNRS, ERS 069), 43 Boulevard du 11 Novembre 1918, F-69622 Villeurbanne Cédex, France*
[2]*Inst. National des Sciences Appliquées de Lyon, Lab. des Matériaux Macromoléc. (CNRS, URA 507), 20 Av. Albert Einstein, F-69621 Villeurbanne Cédex, France*

The peculiarities of adhesion of polymer blends and alloys and the mechanism of the formation of interphase layers

Y. S. LIPATOV
Institute of Macromolecular Chemistry, National Academy of Sciences of Ukraine, Kiev 253660, Ukraine

Atomic force microscopy as a tool for studying mechanical properties of polymer surfaces

W. N. UNERTL, X. JIN, A. LIN, W. KULNIS, A. B. EL BEDIWI, Y. LUO and D. WOODLAND
Laboratory for Surface Science and Technology, University of Maine, Orono, ME 04469, USA

Plasma polymerization of allyl alcohol: characterization and adhesion properties

A. NIHLSTRAND,[1] T. HJERTBERG[1] and K. S. JOHANSSON[2]

[1]*Department of Polymer Technology, Chalmers University of Technology, S-41296 Göteborg, Sweden*
[2]*Institute for Surface Chemistry (YKI), Box 5607, S-114 86 Stockholm, Sweden*

Development of a nondestructive method based on thermodepolarization and positron annihilation techniques for determination of adhesion in polymer–metal systems

A. G. LIPSON, D. M. SAKOV, I. I. BARDYSHEV and B. V. DERJAGUIN*

Institute of Physical Chemistry of the Russian Academy of Sciences, 31 Leninsky Prospect, 117915 Moscow GSP, Russia
*Deceased.

Surface ageing of aluminium 2024-T3 bare substrate after the CAA process for adhesion of polymeric coatings

E. S. J. MARGARITA,[1] B. VOGELESANG[2] and G. FRENS[3]

[1]*Adhesion Institute, Delft University of Technology, P.O. Box 5058, 2600 GB Delft, The Netherlands*
[2]*Laboratory of Production and Materials, Faculty of Aerospace Engineering, Delft University of Technology, P.O. Box 5058, 2600 GB Delft, The Netherlands*
[3]*Laboratory of Physical Chemistry, Delft University of Technology, Julianalaan 136, 2628 BL Delft, The Netherlands*

Raman scratch testing — a new method for paint adhesion evaluation

P. ALERS,[1] C. JULIA SCHMUTZ[1] and I. HAYWARD[2]

[1]*Swiss Centre for Electronics and Microtechnology (CSEM), P.O. Box 41, CH-2007 Neuchâtel, Switzerland*
[2]*Department of Physics, University of Leeds, Leeds, UK*

Factors affecting the adhesion at Cu/PA11 and Al/PA11 interfaces

M. GUNDJIAN and B. FISA

Centre de Recherche Appliquée sur les Polymères (CRASP), École Polytechnique de Montréal, P.O. Box 6079, Station Centre-ville, Montréal (Québec) H3C 3A7, Canada

Adhesion phenomena in (co)extrusion coating of paper and paperboard

J. T. KUUSIPALO and A. V. SAVOLAINEN

Tampere University of Technology, Institute of Paper Converting, P.O. Box 589, 33100 Tampere, Finland

Influence of ageing of bare anodised aluminium 2024 T3 on the adhesion of polyvinylchloride/vinylacetate

S. G. O. DE HAAS[1] and G. FRENS[2]

[1]*Adhesion Institute, Delft University of Technology, P.O. Box 5058, 2600 GB Delft, The Netherlands*
[2]*Laboratory of Physical Chemistry, Delft University of Technology, Julianalaan 136, 2628 BL Delft, The Netherlands*

Increase in wet adhesion stability of coatings by highly ordered polymers

J. GÄHDE,[1] R. MIX,[1] H. GOERING,[1] W. FUNKE[2] and U. HERMANN[2]

[1]*Institute of Applied Chemistry Berlin-Adlershof, Rudower Chaussee 5, D-12484 Berlin, Germany*
[2]*Institute of Technical Chemistry, University of Stuttgart, Pfaffenwaldring 55, D-70569 Stuttgart, Germany*

Analytical characterization of peel plies and the transfer of contaminations after peel ply removal and pretreatments of Hexel 655-2 BMI-laminates

A. BAALMANN,[1] V. SCHLETT,[1] M. KOLB[2] and K. SCHMIDTKE[2]

[1]*Fraunhofer-Institut für Angewandte Materialforschung, Bereich Klebtechnik und Polymere, Neuer Steindamm 2, D-28719 Bremen, Germany*
[2]*Daimler Benz AG, Forschung und Technik, Abt. F2K/ZTT43, Willi-Messerschmitt-Straße, D-85521 Ottobrunn, Germany*

Modified polyethylenes for improved adhesion to aluminium substrate

S. K. GHOSH

Parts and Materials Division, Systems Reliability Group, ISRO Satellite Centre, Airport Road, Bangalore 560017, India

Adhesion of UV-curable protective coatings for optical fibers

J. RAYSS

Laboratory of Optical Fibers Technology, Maria Curie-Skłodowska University, 20-031 Lublin, Poland

The properties of a chromium-free conversion coating for coil coated hot-dip galvanised steel

J. KARLSSON, S. E. HÖRNSTRÖM,* H. KLANG[1] and M. BACKLUND[1]

Materials Centre — HTM, University College of Falun/Borlänge, P.O. Box 764, S-78127 Borlänge, Sweden
[1]*SSAB Tunnplåt AB, S-78184 Borlänge, Sweden*
**New address: Center for Industrial Engineering and Development, Dalarna University College, S-78188 Borlänge, Sweden*

The influence of pretreatment process parameters on adhesion and corrosion properties of HDG steel strip

E. G. HEDLUND and S. E. HÖRNSTRÖM*

Materials Centre — HTM, University College of Falun/Borlänge, P.O. Box 764, S-78127 Borlänge, Sweden

*New address: *Center for Industrial Engineering and Development, Dalarna University College, S-78188 Borlänge, Sweden*

Stone impact damage and adhesion of thin organic films

M. RÖSLER, E. KLINKE and G. KUNZ

Research Institute of Pigments and Paints, Allmandring 37, D-70569 Stuttgart, Germany

The interactions of Kevlar* fibers with selected probes. Part I: Physico-chemical aspects

S. REBOUILLAT,[1] M. ESCOUBES[2] and R. GAUTHIER[2]

[1]*DuPont (UK) Limited, P.O. Box 15, Londonderry BT 47 1TU, Northern Ireland*

[2]*Laboratoire d'Etudes des Matériaux Plastiques, URA CNRS 507, Université Lyon 1, 43 Bd du 11 Novembre 1918, 69622 Villeurbanne, France*

DuPont's registered trademark for its high strength para-aramid fibers.

The interactions of Kevlar* fibers with selected probes. Part II: Water sorption and dynamic mechanical properties of fiber/resin samples

S. REBOUILLAT,[1] M. ESCOUBES[2] and R. GAUTHIER[2]

[1]*DuPont (UK) Limited, P.O. Box 15, Londonderry BT 47 1TU, Northern Ireland*

[2]*Laboratoire d'Etudes des Matériaux Plastiques, URA CNRS 507, Université Lyon 1, 43 Bd du 11 Novembre 1918, 69622 Villeurbanne, France*

DuPont's registered trademark for its high strength para-aramid fibers.

Effect of pretreatment on adhesion of electroless nickel-P-B alloy coating on aluminium

M. F. SHAFFEI,[1] S. THABET,[2] N. EL-EBIARI[1] and S. ABD-ELHALIM[1]

[1]*Chemical Engineering & Pilot Plant Department, National Research Centre, Dokki, Cairo, Egypt*

[2]*Faculty of Science, Ein-Chams University, Cairo, Egypt*

Adhesion and durability of polymeric coatings on steel structures in cold regions

V. YU. SETKOV and A. A. KOPYLOV

Norilsk Industrial Institute, Norilsk 663310, Russia

Study of adhesion of functionalized polyethylene to metals

A. I. KUZAVKOV,[1] O. V. KASPEROVICH,[2] V. V. KOZELSKAYA[1] and S. S. PESETSKII[1]
[1]*Metal–Polymer Research Institute of Belarussian Academy of Sciences, 32a Kirov Str., 246652 Gomel, Republic of Belarus*
[2]*Belarussian State Technological University, 13a Sverdlov Str., 220030 Minsk, Republic of Belarus*

Study of interphase adhesion and properties of polymer–polymer systems based on engineering thermoplastics

S. S. PESETSKII, V. D. FEDOROV and V. N. KOVAL
Metal–Polymer Research Institute of Belarussian Academy of Sciences, 32a Kirov Str., 246652 Gomel, Republic of Belarus

Friction induced paint damage to highly engineered painted plastic substrates: a preliminary contact mechanics perspective

A. C. RAMAMURTHY[1] and D. J. MIHORA[2]
[1]*ACD, Ford Motor Company, Dearborn, MI 48121-2053, USA*
[2]*FM Analysts, 5326 Plaza Aleman, Santa Barbara, CA 93111, USA*

Study on chlorine-free waterborne coatings for polyolefin substrates

K. FUJINO and H. NAMBA
Research Laboratory of Chemicals Development, Nippon Paper Industries Co., Ltd., 2-8-1 Iida-machi, Iwakuni, Yamaguchi 740, Japan

Electrical analysis of the interface between a polyester coating and anodized aluminium

P. H. MAASSEN VAN DEN BRINK[1] and J. VAN TURNHOUT[2]
[1]*Department of Polymer Technology, Delft University of Technology, Julianalaan 136, 2628 BL Delft, The Netherlands*
[2]*TNO Plastics and Rubber Institute, P.O. Box 6031, 2600 JA Delft, The Netherlands*

Adhesion/release alternative with self stratifying coatings

V. V. VERKHOLANTSEV and M. FLAVIAN
Tambour Ltd., P.O. Box 2238, Akko 24121, Israel

Examination of surfaces of polymers by inverse gas chromatography

A. VOELKEL, E. ANDRZEJEWSKA, R. MAGA and M. ANDRZEJEWSKI
Poznań University of Technology, Institute of Chemical Technology and Engineering, Pl. M. Skłodowskiej-Curie 2, 60-965 Poznań, Poland

ADHESION ASPECTS OF THIN FILMS

Novel approaches to improve adhesion of thin films on a variety of substrates

J. NARAYAN

Department of Materials Science and Engineering, North Carolina State University, Raleigh, NC 27695-7916, USA

Adhesion of diamond coatings on WC-Co based tools

AHARON INSPEKTOR

Kennametal Inc., P.O. Box 231, Latrobe, PA 15650, USA

Quantification of interface strength by a new evaluation of depth-sensing indentation measurements

B. ROTHER

Forschungsinstitut für Edelmetalle und Metallchemie, Katharinenstr. 17, D-73525 Schwäbisch Gmünd, Germany

Improved adhesion of plasma deposited films using a new variable duty cycle pulsed plasma gradient layering technique

J.-H. WANG,[1] C. L. RINSCH,[2] C. R. SAVAGE,[1] R. C. EBERHART[2] and R. B. TIMMONS[1]

[1]*Department of Chemistry and Biochemistry, The University of Texas at Arlington, Arlington, TX 76019-0065, USA*

[2]*Biomedical Engineering Program, University of Texas Southwestern Medical Center at Dallas, Dallas, TX 75235-9031, USA*

Measurement of adhesion strength of sputtered metal oxide and metal nitride thin films on glass by direct pull-off method

S. SUZUKI, Y. HAYASHI, K. SUZUKI and E. ANDO

Research Center, Asahi Glass Co. Ltd., 1150 Hazawa-cho, Kanagawa-ku, Yokohama 221, Japan

Measurement of interface toughness in multi-layer thin film structures using the four-point bending technique

QING MA,[1] HARRY FUJIMOTO[1] and REINHOLD H. DAUSKARDT[2]

[1]*Intel Corporation, Santa Clara, CA 95054, USA*

[2]*Department of Materials Science and Engineering, Stanford University, Stanford, CA 94305, USA*

Stress and scratch behaviors of (Ti, Al)N films deposited on tool steel by reactive sputtering

J. G. DUH and C. T. HUANG

Department of Materials Science and Engineering, National Tsing Hua University, Hsinchu, Taiwan

Laser-induced decohesion spectroscopy (LIDS): application to adhesion measurement in thin films systems

J. S. METH and S. BENNISON

CS&E, DuPont Co., Wilmington, DE 19880-0328, USA

Residual stress and adhesion in ion assisted hafnia coatings on glass

L. CANEVE,[1] S. SCAGLIONE,[1] D. FLORI,[1] M. C. CESILE[2] and S. MARTELLI[2]

[1] *Thin Film Optics Division, ENEA, CRE Casaccia, Via Anguillarese 301, 00060 Rome, Italy*
[2] *Division for Laser Spectroscopy and Laser Applications, ENEA Frascati, Via E. Fermi 27, 00044 Frascati (Rome), Italy*

Al_2O_3-ZrO_2 multilayer coatings for thermal barrier applications

PARVATI RAMASWAMY,[1] S. SEETHARAMU,[1] P. R. KRISHNAMOORTHY,[1] K. B. R. VARMA[2] and K. J. RAO[2]

[1] *Materials Technology Division, Central Power Research Institute, Bangalore-560094, India*
[2] *Materials Research Centre, Indian Institute of Science, Bangalore-560012, India*

Fabrication and mechanical properties of chemical vapour deposited boron-carbon-nitrogen films

S. NAKAMURA, S. YOSHIOKA and N. SAITO

Research Laboratory for Advanced Technology, Takamatsu National College of Technology, Takamatsu 761, Japan

Delamination of thin selenium layer from aluminium surface associated with the interruption of the Se/Al bond

L. LAPCIK, B. MIKULÁSKOVÁ, J. HRABOVSKÁ and L. LAPCIK, JR.

Faculty of Chemistry, Technical University of Brno, Veslarská 230, CZ-63700 Brno, Czech Republic

Measurement of adhesion of thin films deposited by the PVD method to ceramics and cemented carbides

M. KUPCZYK[1] and K. JÓŹWIAK[2]

[1] *Institute of Mechanical Technology, Poznań University of Technology, Piotrowo 3 Street, P.O. Box 5, 61-138 Poznań, Poland*
[2] *Chair of Metals Science, Poznań University of Technology, Pl. M. Skłodowskiej-Curie 5, P.O. Box 5, 60-965 Poznań, Poland*

Evaluation of coating adhesion to high speed steel by static and dynamic scratch tests

M. KUPCZYK and K. WIECZOROWSKI

Institute of Mechanical Technology, Poznań University of Technology, Piotrowo 3 Street, P.O. Box 5, 61-138 Poznań, Poland

Characterization of vapor phase deposited triazine derivates on silicon surfaces

S. DIECKHOFF,[1] V. SCHLETT,[1] W. POSSART,[1] O.-D. HENNEMANN,[1] J. GÜNSTER[2] and V. KEMPTER[2]

[1]*Fraunhofer-Institut für Angewandte Materialforschung, Bereich Klebtechnik und Polymere, Neuer Steindamm 2, D-28719 Bremen, Germany*

[2]*Physikalisches Institut der Technischen Universität Clausthal, Leibnizstraße 4, D-38678 Clausthal-Zellerfeld, Germany*

Formation of intermediate intermetallic layers at interfaces of electrodeposited tin, Ni-Fe, Ni-B coatings on copper substrate or of Cu, Ni-Fe sputtered onto ceramics

I. VITINA, I. PELECE, V. RUBENE, A. KORNE, V. BELMANE, M. LUBANE and A. KRUMINA

Latvian Academy of Sciences, Institute of Inorganic Chemistry, Salaspils, LV-2169, Latvia

Adhesion of thin metal (Au, Pd and Cu) deposits on α-alumina

B. EALET, A. ASSABAN and M. GILLET

SERMEC, Case 261, Faculté des Sciences et Techniques de St. Jérôme, Avenue Escadrille Normandie Niemen, 13397 Marseille Cedex 20, France

Factors affecting degradation of adhesion of metal films

S. STOLYAROVA,[1] A. SIMANOVSKIS[2] and Y. NEMIROVSKY[1]

[1]*Solid State Institute, Technion, Haifa 32000, Israel*

[2]*Solid State Physics Institute, Latvian University, Riga, Latvia*

ADHESION ASPECTS OF METALLIZED PLASTICS

Interfacial chemistry between evaporated silver and plasma-treated poly(ethylene terephthalate): relevance to adhesion

L. J. GERENSER

Analytical Technology Division, Eastman Kodak Company, Rochester, NY 14650-2132, USA

Physical and chemical interactions in polymer/metal interphases

M. GRUNZE

Angewandte Physikalische Chemie, INF 253, Universität Heidelberg, D-69210 Heidelberg, Germany

Mechanical characterization and failure analysis of diamond-like carbon and SiO_2 coatings on polycarbonate substrates

A. J. HSIEH

US Army Research Laboratory, Materials Directorate, AMSRL-MA-PB, Watertown, MA 02172-0001, USA
New address: US Army Research Laboratory AMSRL-WM-MA, Aberdeen Proving Ground, MD 21005-5069, USA

Structure, chemistry, and gas barrier at the AlO_x/PET interface

C. P. BARKER,[1] K.-H. KOCHEM,[2] K. M. REVELL,[3] R. S. A. KELLY[3] and J. P. S. BADYAL[1]

[1]*Department of Chemistry, Science Laboratories, Durham University, Durham DH1 3LE, UK*
[2]*HOECHST Aktiengesellschaft, Werk Kalle/Albert, Geschaftsbereich H, Rheingaustrasse 190–196, D-65174 Wiesbaden, Germany*
[3]*CAMVAC (Europe) Ltd., Burrell Way, Thetford, Norfolk IP24 3QY, UK*

Analysis and measurement of fracture toughness in metal/polyimide fine line structures

TZE WING POON,[1] JIHPERNG LEU[2] and PAUL S. HO[2]

[1]*Motorola Inc., Semiconductor Products Sector, 3501 Ed Bluestein Blvd., K-1, Austin, TX 78721, USA*
[2]*Center for Materials Science and Engineering, University of Texas at Austin, Austin, TX 78712, USA*

Interaction of chromium with polyphenylquinoxaline and amorphous carbon surfaces

L. BELLARD, C. FAUQUET-BEN AMMAR, J. M. THEMLIN and A. CROS

Faculté des Sciences de Luminy, URA CNRS 783, Département de Physique, Case 901, 13288 Marseille Cedex 9, France

N_2 and O_2 plasma treatments to enhance silicon oxide adhesion onto polypropylene

F. BODINO, L. KÖHLER, J.-J. PIREAUX and R. CAUDANO

Laboratoire Interdisciplinaire de Spectroscopie Electronique, Facultés Universitaires Notre Dame de la Paix, 61 rue de Bruxelles, 5000 Namur, Belgium

Aluminium/polymer interfaces: calculations on model systems and comparison to secondary ion mass spectrometry (SIMS) data

Y. TRAVALY, P. BERTRAND, X. GONZE, G.-M. RIGNANESE and J.-C. CHARLIER

Unité PCPM, Université Catholique de Louvain, 1 Place Croix du Sud, B-1348 Louvain-la-Neuve, Belgium

In situ XPS and TOF-SIMS studies of SF_6 plasma treated polycarbonate (PC) surfaces and adhesion of Al to PC

F. M. PETRAT,[1] D. WOLANY,[1] L. WIEDMANN,[1] A. BENNINGHOVEN[1] and K. REIHS[2]

[1] *Universität Münster, Physikalisches Institut, Wilhelm-Klemm-Str., D-48149 Münster, Germany*
[2] *Bayer AG, Zentrale Forschung und Entwicklung, D-51368 Leverkusen, Germany*

An adhesion study of SiO_x/PET films: a comparison between scratch and fragmentation tests

Y. PITTON,[1] S. D. HAMM,[1] F.-R. LANG,[1] Y. LETERRIER,[2] J.-A. E. MÅNSON[2] and H. J. MATHIEU[1]

[1] *Laboratoire de Métallurgie Chimique, DMX, Ecole Polytechnique Fédérale de Lausanne (EPFL), CH-1015 Lausanne, Switzerland*
[2] *Laboratoire de Technologie des Composites et Polymères, DMX, Ecole Polytechnique Fédérale de Lausanne (EPFL), CH-1015 Lausanne, Switzerland*

Reliability measurements of Cu-Cr alloy/polyimide films under 35°C and 85°C/85% R.H. environments

E. C. AHN, JIN YU and I. S. PARK

Department of Materials Science and Engineering, Korea Advanced Science and Technology (KAIST), Chongryang P.O. Box 201, Seoul, Korea

The PVD metallization of plastics for industrial applications: substrate surface morphology and film adhesion

H. KUPFER, G. HECHT and F. RICHTER

Institut of Physics, Technical University of Chemnitz-Zwickau, D-09107 Chemnitz, Germany

Peel strengths in Cu/Cr/polyimide films

I. S. PARK, JIN YU and E. C. AHN

Department of Materials Science and Engineering, Korea Advanced Science and Technology (KAIST), Chongryang P.O. Box 201, Seoul, Korea

Cu metallization of PET: influence of the deposition parameters

P. A. GOLLIER and P. BERTRAND

Unité PCPM, Université Catholique de Louvain, 1 place Croix du Sud, B-1348 Louvain-la-Neuve, Belgium

Adhesion aspects of metallized poly(ethylene terephthalate)

EK. DOBREVA, AL. LIRKOV and CHR. PETROV
Department of Chemistry, Technical University, Sofia 1000, Bulgaria

BIOADHESION

Adhesion of anchorage-dependent animal cells on bioceramics

T. SUZUKI, K. NISHIZAWA, M. TORIYAMA, Y. YOKOGAWA, Y. KAWAMOTO, F. NAGATA and M. R. MUCALO
Bioceramics Laboratory, National Industrial Research Institute of Nagoya, 1 Hirate, Kitaku, Nagoya 462, Japan

Characterization of conditioning films from natural waters

A. P. LEIS, R. S. PEMBREY and R. P. SCHNEIDER
School of Microbiology and Immunology, The University of New South Wales, Sydney, NSW 2052, Australia

Surface chemistry determines non-specific bacterial adhesion

R. S. PEMBREY and R. P. SCHNEIDER
School of Microbiology and Immunology, The University of New South Wales, Sydney, NSW 2052, Australia

Aggregation of proteins on polysulfone membrane surface

A. M. NECHIFOR, L. MOISE, G. NECHIFOR, N. LUCA and G. POPESCU
Research Center for Macromolecular Materials and Membranes, Spl. Independentei 202, 79611 Bucharest, Romania

Use of 'RGD'-cyclic peptides to modulate cell adhesion, proliferation and activation on poly(ethylene terephthalate) (PET)

B. GILLON, M. ART, D. DELFORGE, V. GEORGES, J. REMACLE and M. RAES
Laboratoire de Biochimie Cellulaire, FUNDP, B-5000 Namur, Belgium

Adhesion of bacteria to surface modified polymers

W. KOHNEN and B. JANSEN
Institute of Medical Microbiology and Hygiene, University of Cologne, Goldenfelsstr. 19-21, D-50935 Cologne, Germany

Investigation of the kinetics of adsorption (adhesion) of pharmaceutical steroid derived from plants on protein carrier using time-of-flight plasma desorption mass spectrometry

T. G. KALINICHENKO,[1] S. A. AKSYONOV,[1] V. D. CHIVANOV,[2] A. G. KALINICHENKO[1] and L. F. SUKHODUB[1]

[1]*Applied Physics Institute, Sumy 244030, Ukraine*
[2]*Sumy Agricultural Institute, Sumy 244030, Ukraine*

PARTICLE ADHESION

Time effects of the adhesion-induced deformation of submicrometer PSL spheres on silicon substrates

A. A. BUSNAINA

Microcontamination Research Laboratory, Clarkson University, Potsdam, NY 13699-5729, USA

Removal of fine particle and other contamination from the wafer surface in semiconductor industries: challenges and opportunities

RAJ. C. SHAH*

SEMATECH/IBM, 2706 Montopolis Drive, Austin, TX 78741, USA
**Current address: IBM Corp., Austin, TX 78741, USA*

Air-bubble detachment of adhering particles and microorganisms from solids

H. J. BUSSCHER, J. NOORDMANS, P. J. WIT and H. C. VAN DER MEI

Materia Technica, University of Groningen, Bloemsingel 10, 9712 KZ Groningen, The Netherlands

Oblique impact of microspheres with adhesion

P. F. DUNN and R. M. BRACH

Department of Aerospace and Mechanical Engineering, University of Notre Dame, Notre Dame, IN 46556, USA

The adhesion strength of salmeterol base and a series of salmeterol salts to compacted lactose monohydrate surfaces

F. PODCZECK,[1] J. M. NEWTON[1] and M. B. JAMES[2]

[1]*The School of Pharmacy, Department of Pharmaceutics, University of London, 29/39 Brunswick Square, London WC1N 1AX, UK*
[2]*Glaxo Research & Development, Ltd., Ware SG12 0DJ, UK*

The influence of physical properties of the materials in contact on the adhesion strength of particles of salmeterol base and salmeterol salts to various surface materials

F. PODCZECK,[1] J. M. NEWTON[1] and M. B. JAMES[2]

[1]*The School of Pharmacy, Department of Pharmaceutics, University of London, 29/39 Brunswick Square, London WC1N 1AX, UK*
[2]*Glaxo Research & Development, Ltd., Ware SG12 0DJ, UK*

Adhesion of latex particles to model inorganic surfaces

V. GRANIER and A. SARTRE

RHONE-POULENC RECHERCHES, 52 Rue de la Haie Coq, 93308 Aubervilliers Cédex, France

Adhesion of colloidal ZnS to silicon. Effect of pH, ionic strength and radiofrequency electric field

A. ONTIVEROS,[1] J. D. G. DURÁN,[1] E. CHIBOWSKI[2] and F. GONZÁLEZ-CABALLERO[1]

[1]*Departamento de Física Aplicada, Universidad de Granada, 18071 Granada, Spain*
[2]*Department of Physical Chemistry, Maria Curie-Skłodowska University, 20031 Lublin, Poland*

INTERFACES IN COMPOSITES

Adhesion at thermoplastic composite interfaces

RICHARD P. WOOL

Center for Composite Materials, University of Delaware, Newark, DE 19716, USA

Studies on mechanical properties of composites by X-ray diffraction: residual stress and mechanism of stress transmission

K. NAKAMAE and T. NISHINO

Department of Chemical Science and Engineering, Faculty of Engineering, Kobe University, Rokko, Nada, Kobe 657, Japan

Influence of the film former on the fibre-matrix adhesion and mechanical properties of glass-fibre reinforced thermoplastics

B. J. R. SCHOLTENS, J. C. BRACKMAN and R. A. C. DEBLIECK

DSM Research, P.O. Box 18, 6160 MD Geleen, The Netherlands

Effect of interfacial adhesion on the properties of short sisal fiber reinforced styrene butadiene rubber composites

R. PRASANTHA KUMAR and SABU THOMAS

School of Chemical Sciences, Mahatma Gandhi University, Priyadarshini Hills P.O., Kottayam-686560, Kerala, India

Influence of filler-rubber adhesion on the transport of aromatic hydrocarbons through natural rubber

G. UNNIKRISHNAN and SABU THOMAS

School of Chemical Sciences, Mahatma Gandhi University, Priyadarshini Hills P.O., Kottayam-686560, Kerala, India

The effect of fibre-matrix adhesion on the dielectric properties of sisal fibre reinforced low density polyethylene

S. THOMAS and AUGUSTINE PAUL

School of Chemical Sciences, Mahatma Gandhi University, Priyadarshini Hills P.O., Kottayam-686560, Kerala, India

Effect of chemical treatments on the mechanical properties of pineapple fiber-polyethylene composites

JAYAMOL GEORGE,[1] SABU THOMAS[1] and S. S. BHAGAWAN[2]

[1]*School of Chemical Sciences, Mahatma Gandhi University, Priyadarshini Hills P.O., Kottayam-686560, Kerala, India*
[2]*PED, VSSC, Thiruvananthapuram-695022, Kerala, India*

Polymer layered silicate nanocomposites

A. AKELAH

Chemistry Department, UAE University, P.O. Box 17551, Al-Ain, United Arab Emirates

Interfacial interaction between thermosetting matrices and aramid fibres (ARMOS)

G. S. SHUL[1] and YU. A. GORBATKINA[2]

[1]*State Research Centre of Russia, Obninsk Research and Production Enterprise 'Technologiya', Obninsk, Kaluga Region 249020, Russia*
[2]*Semenov Institute of Chemical Physics, Russian Academy of Sciences, 117977 Moscow, Russia*

Effect of the interface on the mechanical properties of glass fibre reinforced polypropylene

T. OGAWA and N. TAYA

Graduate School of Engineering, Kanazawa Institute of Technology, 7-1, Ohgigaoka Nonoichi, Ishikawa 921, Japan

Stability of filler reinforced composites based on poly(vinyl acetate) matrix

V. KOVAČEVIĆ, S. LUČIĆ and D. HACE

Faculty of Chemical Engineering and Technology, University of Zagreb, 10000 Zagreb, Croatia

ADHESIVE MATERIALS: CHARACTERISTICS AND PERFORMANCE

High solid asphalt adhesives: chemistry, technology, and bonding considerations

T. D. TACKETT and A. M. USMANI*

Firestone Building Products Company, Carmel, IN 46032, USA
**Current address: Usmani Development Company, 7318 Normandy Way, Indianapolis, IN 46278, USA*

Simulations of fracture of thin adhesive films

ARLETTE R. C. BALJON and MARK O. ROBBINS

Department of Physics and Astronomy, Johns Hopkins University, Baltimore, MD 21218, USA

In situ studies of the microstructure of polymeric adhesives/metal interfaces

GI XUE

Department of Chemistry, and the State Key Laboratory of Coordination Chemistry, Nanjing University, Nanjing, 210008, P.R. China

Some water-borne adhesives based on styrene-butyl acrylate copolymer latices

R. M. MOHSEN, H. E. NASR and A. S. BADRAN

Department of Polymers and Pigments, National Research Centre, Dokki, Cairo, Egypt

Effect of pre-vulcanization on the adhesive behavior of NR/SBR latex blends

JYOTHI T. VARKEY,[1] SABU THOMAS[1] and S. SOMESWARA RAO[2]

[1]*School of Chemical Sciences, Mahatma Gandhi University, Priyadarshini Hills P.O., Kottayam-686560, Kerala, India*
[2]*PED, VSSC, Thiruvananthapuram-695022, Kerala, India*

Cause and effect of void formation during vacuum bag cure of epoxy film adhesives

P. J. PEARCE, D. R. ARNOTT, G. I. MATHYS, A. CAMILLERI, M. R. KINDERMANN and A. R. WILSON

Department of Defence, Defence Science & Technology Organisation, Aeronautical & Maritime Research Laboratory, P.O. Box 4331, Melbourne, Victoria 3001, Australia

Novel air activated chemistry and its application to rapid curing one-component methacrylate adhesives

B. J. KNEAFSEY, J. GUTHRIE and D. MELODY

Research, Development & Engineering Group, Loctite (Ireland) Ltd., Tallaght Business Park, Whitestown, Dublin 24, Ireland

Study on fatigue crack growth properties of rubber modified epoxy adhesives under mode I and mode II loadings: effect of CNBR modification on fatigue crack growth

K. NAITO and T. FUJII

Department of Mechanical Engineering and Systems, Doshisha University, Tanabe, Kyoto, 610-03, Japan

Studies of phenol formaldehyde-methylol lignin resin systems in terms of pH, viscosity, wetting and gelation properties to find out satisfactory adhesive compositions for bonding wood-based products

A. SARKAR,[1] ASHIM DAS[2] and R. K. BHATTACHARYA[2]

[1]*Consultant, 10 Raja Raj Krishna Street, Calcutta-700006, India*
[2]*Allied Resine & Chemicals, Hyde Road, Calcutta-700088, India*